G-W
PUBLISHER

Industrial
Maintenance
and Mechatronics

Shawn A. Ballee | Gary R. Shearer

NIMS

Designed to help today's industrial maintenance
and mechatronics programs provide students with
the skills and knowledge required for entry-level
employment. Aligned with nine NIMS Industrial
Technology Maintenance Level 1 certifications.

Be Digital Ready on Day One with EduHub

EduHub provides a solid base of knowledge and instruction for digital and blended classrooms. This easy-to-use learning hub delivers the foundation and tools that improve student retention and facilitate instructor efficiency. For the student, EduHub offers an online collection of eBook content, interactive practice, and test preparation. Additionally, students have the ability to view and submit assessments, track personal performance, and view feedback via the Student Report option. For instructors, EduHub provides a turnkey, fully integrated solution with course management tools to deliver content, assessments, and feedback to students quickly and efficiently. The integrated approach results in improved student outcomes and instructor flexibility.

Michael Jung/Shutterstock.com

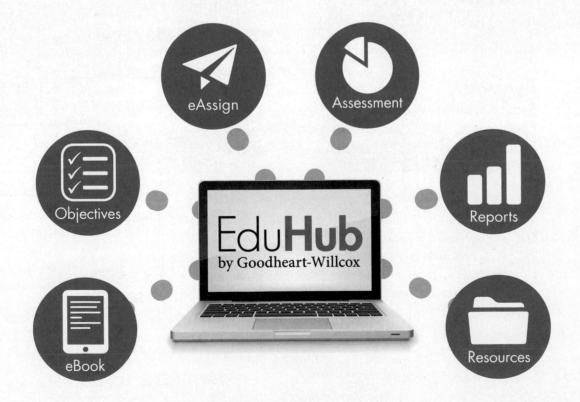

eBook

The EduHub eBook engages students by providing the ability to take notes, access the Text-to-Speech option to improve comprehension, and highlight key concepts to remember. In addition, the accessibility features enable students to customize font and color schemes for personal viewing, while links to videos and animations bring content to life.

Objectives

Course objectives at the beginning of each eBook chapter help students stay focused and provide benchmarks for instructors to evaluate student progress.

eAssign

eAssign makes it easy for instructors to assign, deliver, and assess student engagement. Coursework can be administered to individual students or the entire class.

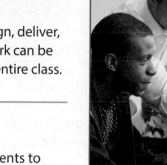

Monkey Business Images/Shutterstock.com

Assessment

Self-assessment opportunities enable students to gauge their understanding as they progress through the course. In addition, formative assessment tools for instructor use provide efficient evaluation of student mastery of content.

	🖶 Print ⬇ Export	
Score	Items	
100%	●	●
80%	●	●
100%	●	●
80%	●	●
100%	●	●
100%	●	●

Reports

Reports, for both students and instructors, provide performance results in an instant. Analytics reveal individual student and class achievements for easy monitoring of success.

Instructor Resources

Instructors will find all the support they need to make preparation and classroom instruction more efficient and easier than ever. Lesson plans, answer keys, and PowerPoint® presentations provide an organized, proven approach to classroom management.

Learn more about EduHub at www.g-w.com/eduhub

GUIDED TOUR

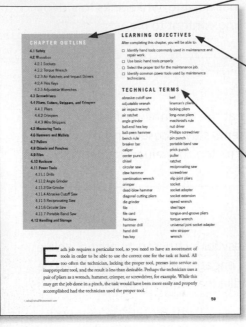

Chapter Outline
provides a preview of the chapter topics and serves as a review tool.

Learning Objectives
clearly identify the knowledge and skills to be obtained when the chapter is completed.

Technical Terms
list the key terms to be learned in the chapter.

Cautions
alert you to practices that could potentially damage equipment or instruments.

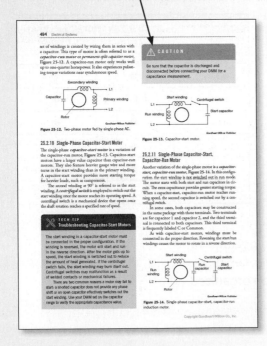

Illustrations
have been designed to clearly and simply show concepts, processes, and equipment in support of the text material.

Safety Notes
alert you to potentially dangerous practices and hazards.

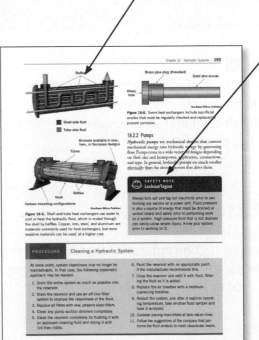

Procedures provide clear instructions for hands-on service activities.

Tech Tips provide advice and guidance with an "on-the-job" focus.

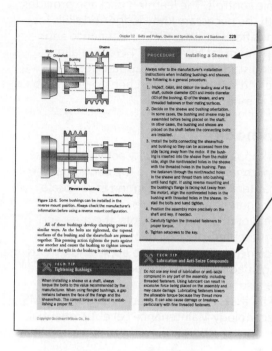

Thinking Green notes highlight items related to sustainability, energy efficiency, and environmental issues.

Summary feature provides an additional review tool for you and reinforces key chapter topics.

Review Questions allow you to demonstrate knowledge and comprehension of chapter material.

NIMS Credentialing Preparation Questions help prepare you to successfully achieve NIMS Industrial Technology Maintenance credentials.

EduHub

EduHub provides a solid base of knowledge and instruction for digital and blended classrooms. This easy-to-use learning hub provides the foundation and tools that improve student retention and facilitate instructor efficiency.

For the student, EduHub offers an online collection of eBook content, interactive practice, and test preparation. Additionally, students have the ability to view and submit assessments, track personal performance, and view feedback via the Student Report option. For instructors, EduHub provides a turnkey, fully integrated solution with course management tools to deliver content, assessments, and feedback to students quickly and efficiently. The integrated approach results in improved student outcomes and instructor flexibility. Be digital ready on day one with EduHub!

- **eBook content.** EduHub includes the textbook in an online, reflowable format. The interactive eBook includes highlighting, magnification, note-taking, and text-to-speech capabilities.

- **Lab Workbook content.** EduHub includes all of the Lab Workbook content in digital format, including chapter review questions (with auto-grading capability) and lab activities.

- **Interactive activities.** EduHub provides engaging activities to help students master the technical vocabulary, concepts, and procedures presented in the textbook.

Student Tools

Student Text

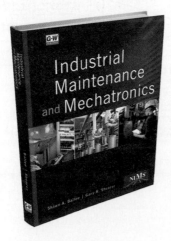

Industrial Maintenance and Mechatronics is a comprehensive text that provides curriculum support for a college-level Industrial Technology Maintenance (ITM) program. The text comprises sections that correspond to the principal industrial technology disciplines, with a special focus on electrical systems and electronic controls. The learning package provides students with the necessary knowledge and skills for entry-level positions in industrial maintenance and helps them prepare for NIMS Level 1 credentialing. Instructors and students alike will appreciate the convenience and value of a comprehensive text that can be used in multiple courses.

Lab Workbook

The Lab Workbook combines review activities and practical applications that relate to the content of the textbook chapters. Questions designed to reinforce the textbook content help students review their understanding of the terms, concepts, theories, and procedures presented in each chapter. Hands-on lab activities provide an opportunity to apply and extend knowledge gained from the textbook chapters. These lab activities also help to prepare students for the hands-on skills assessment required to achieve a NIMS Industrial Technology Maintenance Level 1 credential.

NIMS Industrial Technology Maintenance (ITM) Credentials

NIMS offers nine Industrial Technology Maintenance (ITM) Level 1 credentials. These credentials provide students and working technicians a recognized method to validate their knowledge and skills. ***Industrial Maintenance and Mechatronics*** is designed to work hand-in-glove with the duties and standards for each ITM credential. This standards-based learning package will help students pass the testing and performance requirements for NIMS credentialing. For more information about NIMS and their Industrial Technology Maintenance (ITM) credentials, visit www.nims-skills.org.

Instructor Tools

LMS Integration

Integrate Goodheart-Willcox content in your Learning Management System for a seamless user experience for both you and your students. Contact your G-W Educational Consultant for ordering information or visit www.g-w.com/lms-integration.

Instructor Resources

The Instructor Resources help to make preparation and classroom instruction easier than ever. Included are time-saving preparation tools, such as answer keys, editable lesson plans, and other teaching aids. In addition, presentations for PowerPoint® and assessment software with question banks are provided for your convenience. These resources can be accessed at school, at home, or on the go.

Instructor's Presentations for PowerPoint®

Instructor's Presentations for PowerPoint® provide a useful teaching tool when presenting concepts introduced in the text. These fully customizable and image-intensive slides help you teach and visually reinforce the key concepts from each chapter.

Assessment Software with Question Banks

Administer and manage assessments to meet your classroom needs. The following options are available through the Respondus Test Bank Network:

- A Respondus 4.0 license can be purchased directly from Respondus, which enables you to easily create tests that can be printed on paper or published directly to a variety of Learning Management Systems. Once the question files are published to an LMS, exams can be distributed to students with results reported directly to the LMS gradebook.

- Respondus LE is a limited version of Respondus 4.0. and is free with purchase of the Instructor Resources. It allows you to download test banks and create assessments that can be printed or saved as a paper test.

G-W Integrated Learning Solution

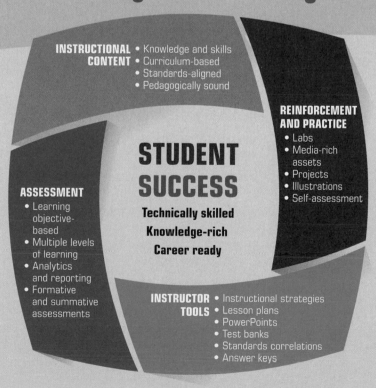

INSTRUCTIONAL CONTENT
- Knowledge and skills
- Curriculum-based
- Standards-aligned
- Pedagogically sound

REINFORCEMENT AND PRACTICE
- Labs
- Media-rich assets
- Projects
- Illustrations
- Self-assessment

STUDENT SUCCESS
Technically skilled
Knowledge-rich
Career ready

ASSESSMENT
- Learning objective-based
- Multiple levels of learning
- Analytics and reporting
- Formative and summative assessments

INSTRUCTOR TOOLS
- Instructional strategies
- Lesson plans
- PowerPoints
- Test banks
- Standards correlations
- Answer keys

The G-W Integrated Learning Solution offers easy-to-use resources that help students and instructors achieve success.

▶ **EXPERT AUTHORS**
▶ **TRUSTED REVIEWERS**
▶ **100 YEARS OF EXPERIENCE**

EMPLOYABILITY SKILLS · TECHNICAL SKILLS · ACADEMIC KNOWLEDGE · INDUSTRY RECOGNIZED STANDARDS

Industrial Maintenance and Mechatronics

Shawn A. Ballee
Assistant Professor of Industrial Systems Technology
Waubonsee Community College
Sugar Grove, Illinois

Gary R. Shearer
Instructor of Industrial Electrical Maintenance
Tennessee College of Applied Technology–Murfreesboro
Murfreesboro, Tennessee

Publisher
The Goodheart-Willcox Company, Inc.
Tinley Park, IL
www.g-w.com

The Goodheart-Willcox Company, Inc. Brand Disclaimer: Brand names, company names, and illustrations for products and services included in this text are provided for educational purposes only and do not represent or imply endorsement or recommendation by the author or the publisher.

The Goodheart-Willcox Company, Inc. Safety Notice: The reader is expressly advised to carefully read, understand, and apply all safety precautions and warnings described in this book or that might also be indicated in undertaking the activities and exercises described herein to minimize risk of personal injury or injury to others. Common sense and good judgment should also be exercised and applied to help avoid all potential hazards. The reader should always refer to the appropriate manufacturer's technical information, directions, and recommendations; then proceed with care to follow specific equipment operating instructions. The reader should understand these notices and cautions are not exhaustive.

The publisher makes no warranty or representation whatsoever, either expressed or implied, including but not limited to equipment, procedures, and applications described or referred to herein, their quality, performance, merchantability, or fitness for a particular purpose. The publisher assumes no responsibility for any changes, errors, or omissions in this book. The publisher specifically disclaims any liability whatsoever, including any direct, indirect, incidental, consequential, special, or exemplary damages resulting, in whole or in part, from the reader's use or reliance upon the information, instructions, procedures, warnings, cautions, applications, or other matter contained in this book. The publisher assumes no responsibility for the activities of the reader.

The Goodheart-Willcox Company, Inc. Internet Disclaimer: The Internet resources and listings in this Goodheart-Willcox Publisher product are provided solely as a convenience to you. These resources and listings were reviewed at the time of publication to provide you with accurate, safe, and appropriate information. Goodheart-Willcox Publisher has no control over the referenced websites and, due to the dynamic nature of the Internet, is not responsible or liable for the content, products, or performance of links to other websites or resources. Goodheart-Willcox Publisher makes no representation, either expressed or implied, regarding the content of these websites, and such references do not constitute an endorsement or recommendation of the information or content presented. It is your responsibility to take all protective measures to guard against inappropriate content, viruses, or other destructive elements.

Image Credits. Front cover: Andrey_Popov/Shutterstock.com (left); Mehmet Dilsiz/Shutterstock.com (center); Pressmaster/Shutterstock.com (right); ekapol sirachainan/Shutterstock.com (background). Chapter review background: T.Sumaetho/Shutterstock.com.

Library of Congress Cataloging-in-Publication Data

Names: Ballee, Shawn A., author. | Shearer, Gary R., author.
Title: Industrial Maintenance and Mechatronics / by Shawn A. Ballee, Gary R. Shearer.
Description: Tinley Park, IL : The Goodheart-Willcox Company, Inc., [2020] | Includes index.
Identifiers: LCCN 2018032043 | ISBN 9781635634273
Subjects: LCSH: Plant maintenance. | Industrial equipment--Maintenance and repair.
Classification: LCC TS192 .B356 2020 | DDC 658.2/02--dc23 LC record available at https://lccn.loc.gov/2018032043

Preface

Industrial Maintenance and Mechatronics covers the many aspects of industrial maintenance and prepares students for careers in a wide variety of industries and occupations. From theory to application, design to diagnosis, and installation to adjustment, you will gain a "big picture" perspective needed to be successful in a field that is constantly changing due to technological advancement. With straight-forward explanations and procedures, this text will serve as a valuable reference for you in both the classroom and the workplace.

Advancements in technology and automation have made mechatronics a critical component of industrial maintenance. Mechatronics combines the study of electronics, mechanics, control systems, robotics, and computer software. This mechatronics knowledge is used to design, operate, maintain, and repair "smart" devices and systems that incorporate sensors, actuators, instrumentation, process control, and automation. An understanding of mechatronics will prepare you for career opportunities in a wide range of industries that utilize industrial robots, automated systems, programmable logic controllers, and other mechanical systems.

Industrial Maintenance and Mechatronics is the first textbook specifically aligned with the NIMS Industrial Technology Maintenance (ITM) standards. These standards—developed with input from a nationwide team of industry and education leaders—provide a clear picture of the knowledge and skills needed by industrial maintenance technicians in today's workplace. Modeled specifically after these standards, *Industrial Maintenance and Mechatronics* applies not only to manufacturing, but to all fields that fall under the umbrella of mechanical, electrical, and electronic systems control and maintenance. Due to its broad coverage, *Industrial Maintenance and Mechatronics* can be used as a primary text for multiple courses in an Industrial Maintenance program, eliminating the need to purchase multiple costly textbooks.

Numerous in-text examples, procedures, review questions, diagrams, and artwork reinforce key concepts. Safety notes and "tech tips" highlight key bits of knowledge that are particularly applicable and critical in the workplace. Troubleshooting methodology is emphasized throughout *Industrial Maintenance and Mechatronics* (including several dedicated troubleshooting chapters) to ensure that you become a problem solver who can determine the true cause of a problem, rather than merely a "parts changer."

NIMS offers a set of nine national, industry-created, industry-recognized credentials based on Industrial Technology Maintenance Level 1 standards. In order to attain a NIMS credential, candidates must pass a written assessment on theory and also complete a hands-on skill assessment under a qualified instructor or supervisor. *Industrial Maintenance and Mechatronics* works hand-in-glove with the NIMS standards to help students successfully achieve NIMS credentials.

This area of study is both wide-ranging and challenging. While this text could have been literally thousands of pages, it has been condensed to highlight core knowledge in the key maintenance disciplines: maintenance operations, basic mechanical systems, basic hydraulic and pneumatic systems, electrical systems, electronic control systems, process control systems, maintenance welding, and maintenance piping. Technology is constantly advancing in many of these subjects, and it is expected that this book will undergo revisions and updates to keep pace with those advancements.

About the Authors

Shawn Ballee is Assistant Professor of Industrial Systems Technology at Waubonsee Community College in Sugar Grove, IL, where he has taught since 1995. His teaching responsibilities have included classes in motor controls, PLCs, hydraulics, pneumatics, mechanical power transmission, power distribution, and mathematics. He has extensive industry experience as an electrical and electronics technician and as a consulting maintenance technician for local manufacturers. Mr. Ballee holds a BS degree in Industrial Technology from Northern Illinois University and a MEd in Post-Secondary Mathematics from Concordia University. Mr. Ballee is a member of the NIMS Advisory Council and he has also served on MSSC's National Expert Panel.

    Gary Shearer is an Instructor of Industrial Electrical Maintenance at Tennessee College of Applied Technology (TCAT), Murfreesboro. Since joining the faculty of TCAT, Mr. Shearer has taught a wide range of technical classes, including electricity, electronics, mechanical systems, machine tool, welding, process control, motor control, PLCs, robotics, HVAC, fluid power, industrial/residential wiring, electrical test equipment, troubleshooting, safety, and print reading. Previously, he worked as a senior test engineer for ReMedPar, a medical equipment and parts organization, and as an R&D engineer at Vanderbilt University's W. M. Keck Foundation Free Electron Laser Center. Intellectual property created by Mr. Shearer includes a US patent and an extensive list of technical and scientific publications. He served as a member of the NIMS Technical Working Group that prepared the *Industrial Technology Maintenance Duties and Standards*.

Contributors

The author and publisher wish to thank the following teaching professionals for their valuable contributions to *Industrial Maintenance and Mechatronics*:

    James Mosman, Associate Professor of Welding Technology and Department Chair of Industrial Technology at Odessa College, Odessa, TX, for authoring the three chapters in the *Maintenance Welding* section.

    Dr. Paul Dettmann, Customized Training Representative for South Central College in North Mankato, MN, for contribution of technical edits, copy, and illustrations in parts of Chapter 3, *Maintenance Principles and Record Keeping*, and Chapter 6, *Print Reading*, and development of presentations and assessment question banks for the instructor resources.

Reviewers

The author and publisher wish to thank the following industry and teaching professionals for their valuable input into the development of *Industrial Maintenance and Mechatronics*:

Ramona Anand
Lorain County Community College
Elyria, OH

Dale Ballard
Prairie State College
Chicago Heights, IL

Bob Bender
Riverland Community College
Albert Lea, MN

Brent Childers
Amarillo College
Amarillo, TX

David Clark
Western Iowa Tech Community College
Sioux City, IA

Paul Dettmann
South Central College
North Mankato, MN

Charles Eckard
North Seattle College
Seattle, WA

Mike Gallimore
Tennessee College of Applied Technology
Knoxville, TN

Tom Groner
Muskegon Community College
Muskegon, MI

Gary E. Hall
Ivy Tech Community College
Fort Wayne, IN

William Hargrove
Florence-Darlington Technical College
Florence, SC

Frank R. Holcomb
Tennessee College of Applied Technology
Paris, TN

Pete Lomeli
Central Arizona College
Coolidge, AZ

Ron McGary
Industrial Systems Technical Trainer
Atlanta, GA

Thann Mughmaw
Ivy Tech Community College
Kokomo, IN

Gerald Pilliteri
Jefferson Technical and Community College
Shelbyville, KY

Mark Prosser
Ferris State University
Big Rapids, MI

Chris Sewalson
Western Iowa Tech Community College
Sioux City, IA

Richard Skelton
Jackson State Community College
Jackson, TN

Gregory D. Spence
Ivy Tech Community College
Logansport, IN

Tim Tewalt
Chippewa Valley Technical College
Eau Claire, WI

Ed VanAvery
Delgado Community College
New Orleans, LA

Acknowledgments

The author and publisher would like to thank the following companies, organizations, and individuals for their contribution of resource material, images, or other support in the development of *Industrial Maintenance and Mechatronics*.

Accuform
AdvancedHMI
AFC Cable Systems, Atkore International
Allen Manufacturing Company (Apex Tool Group)
American Cast Iron Pipe Company
Apex Tool Group
Armstrong (Apex Tool Group)
Atlas Copco
Automation Direct
Badger Meter, Inc.
Ballard, Shawn C.
Bimba Corporate
Bitzer
Bosch Rexroth Corporation
Boston Gear—Altra Industrial Motion
Brady Corp.
Campbell (Apex Tool Group)
Charlotte Pipe and Foundry Company
Columbus McKinnon Corporation
Crescent (Apex Tool Group)
Dettmann, Dr. Paul
Dotco (Apex Tool Group)
Eclipse Tools
Elliott Tool Technologies
EZAutomation
Field Controls
Fireye
Fluke Corporation
G.L. Huyett
GearWrench (Apex Tool Group)
General Tools & Instruments
Greyline Instruments
Hubbell, Inc.
Ideal Industries, Inc.
Johnson Controls
Kingsbury, Inc.
Klein Tools, Inc.

LUDECA Inc.
Lufkin (Apex Tool Group)
Macromatic Industrial Controls
Magid Glove and Safety
Martin Sprocket
Master Lock
McMaster-Carr
Migatron
Miller Electric Mfg Co.
Mosman, James
MSA Safety Inc.
National Lubricating Grease Institute
National Safety Apparel
Nicholson (Apex Tool Group)
NKK Switches
NOARK Electric North America
OSHA
Parker Hannifin
REXA, Inc.
Rheem Manufacturing Company
RIDGID, Inc.
Rockwell Automation, Inc.
Rotork
Schneider Electric
Siemens AG
SKF USA Inc.
SPIROL
Sporlan Division, Parker Hannifin Corporation
Starrett
TB Wood's—Altra Industrial Motion
The Lincoln Electric Co.
Thermadyne
Timken Belts
Uniweld Products Inc.
US Bureau of Labor Statistics
Werner Ladders

Brief Contents

Contents

ELECTRONIC CONTROL SYSTEMS

PROCESS CONTROL SYSTEMS

MAINTENANCE WELDING

MAINTENANCE PIPING

Feature Contents

Maintenance Operations

Congratulations! You are beginning your journey toward a rewarding career in industrial maintenance and mechatronics. The first section of this textbook, *Maintenance Operations*, begins by introducing you to some career opportunities in these fields. The remainder of this section addresses several basic topics—such as safety, tools, fasteners, print reading, troubleshooting, and rigging—that provide a foundation of knowledge for your full course of study in industrial maintenance and mechatronics.

The content in this section will help prepare you to earn the NIMS Industrial Technology Maintenance Level 1 Maintenance Operations credential. Credentials show employers proof of your knowledge and skills.

1 | Careers in Industrial Maintenance

LEARNING OBJECTIVES

After completing this chapter, you will be able to:

- ☐ Describe the different purposes of a technical school, community college, and university.
- ☐ Discuss the different continuing education options available to incumbent workers.
- ☐ Differentiate the four writing tasks essential for preemployment.
- ☐ List five steps you can take to prepare for an interview.
- ☐ Describe four personal behaviors that lead to success in the workplace.
- ☐ Discuss the certifications available for industrial maintenance personnel.

TECHNICAL TERMS

attitude

body language

conflict

conflict management

continuing education

cover letter

credentialing

critical-thinking skills

durable good

ethical behavior

initiative

integrity

leadership

National Institute for Metalworking Skills (NIMS)

negotiation

nondurable good

nonverbal communication

Occupational Safety and Health Administration (OSHA)

punctual

reference

résumé

self-motivation

team

verbal communication

"Believe you can and you're halfway there."

—Theodore Roosevelt

Hopefully you have a plan by now, and this course is part of your plan. Planning should include thinking about what you are doing, considering the outcomes, and envisioning where you want to be 5, 10, 20, or 30 years from now. Not planning and just hoping that coincidence will get you to your goal will lead you exactly nowhere.

Plan now. Expect to study. Expect to constantly learn new technologies and methods. Memorization is not studying; it does nothing to help you understand a concept. Learning and understanding a concept allows you to apply it over a wide variety of technologies. Experienced workers eventually see the same concepts applied over and over again in different areas. They see the big picture and can teach themselves. That is the point of education—to understand fundamental concepts and to be able to apply these concepts to solve problems.

1.1 MANUFACTURING IN THE UNITED STATES

Manufacturing is the process of converting materials into a product. In the United States, the manufacturing industry is responsible for greater than 10% of the economy. Manufacturing facilities range from large factories with thousands of workers to small shops with a handful of employees.

Manufactured products are often classified as either durable goods or nondurable goods, **Figure 1-1**. *Durable goods* are those that remain useful for at least three years, such as cars, appliances, and building materials. *Nondurable goods* are those that are consumed relatively quickly, such as food, shampoo, and gasoline.

In the United States there are more than 250,000 manufacturers and more than 12 million manufacturing workers. These 12 million workers make up about 10% of the US workforce. In addition, each job in manufacturing results in approximately three other jobs outside of the manufacturing sector. These positions could be involved in logistics, raw-material procurement, transportation, customer service, or any number of other supporting fields.

Manufacturing is not a low-paying field. The average manufacturing worker in the United States earns more than $25 an hour. Due to retirements, expansions, and lack of skilled talent, more than 2 million manufacturing jobs are expected to go unfilled between 2015 and 2025. Thus, many job openings will be available to skilled candidates. See **Figure 1-2**.

The increasing use of automation in manufacturing is driving the need for more highly skilled workers. Even entry-level maintenance technicians need to have both mechanical skills and basic automation skills. Even if robots and automation replace all factory floor workers, a large number of skilled technicians will be needed to maintain and repair the machines.

Durable Goods

Nondurable Goods

xieyuliang/Shutterstock.com; Semen Lixodeev/Shutterstock.com

Figure 1-1. Durable goods (such as cars) and nondurable goods (such as bread) are produced by machines and automated equipment in manufacturing facilities. Industrial maintenance technicians ensure that these machines and processes function properly.

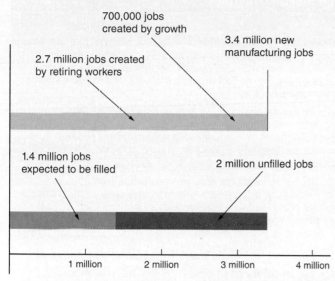

New Manufacturing Jobs, 2015–2025

700,000 jobs created by growth

2.7 million jobs created by retiring workers

3.4 million new manufacturing jobs

1.4 million jobs expected to be filled

2 million unfilled jobs

1 million   2 million   3 million   4 million

Deloitte analysis based on data from US Bureau of Labor Statistics and Gallup Survey

Figure 1-2. The retirement of manufacturing workers and economic growth is projected to create 3.4 million new jobs over a 10-year period. A shortage of skilled candidates is projected to leave 2 million of these jobs unfilled. By completing a certificate or degree program, these jobs and many others will be available to you.

1.2 EDUCATION

Education in industrial maintenance and mechatronics comes in many forms: on-the-job training, manufacturer seminars and online training, self-teaching, and classes at a local technical school or college. Technical schools tend to be more focused on a particular skill set for a specific occupation. While community colleges teach specific skills, those skills can be more widely applied (and this is how many employers view them).

TECH TIP
Educational Goals

Plan your education by setting your completion goal and determining the steps to achieve it. As you begin completing courses, allow for the possibility of discovering a specific area to excel in and alter goals accordingly.

1.2.1 Technical Schools

There are several private technical schools that are quite popular for automotive, welding, HVAC, and a few other areas. Relatively few technical schools offer programs in industrial maintenance and mechatronics.

When compared with community colleges, technical schools often offer a more focused curriculum and programs that can be completed in a shorter amount of time. Technical schools can be more expensive, and credits earned may not be transferable to other institutions.

TECH TIP
Do Your Research

Do some research before selecting a school or program. Always seek out opinions of graduated students from the program you are about to follow.

1.2.2 Community Colleges

Community colleges have four major educational functions:

- Act as transfer schools for students who plan to attend a four-year school. In this way, a student may take required general education classes closer to home and with less expense, usually for the first two years of school.

- Accomplish adult-education functions not available at a four-year school. This is a wide net that covers GED preparation, ESL (English as a second language), remedial mathematics, community education classes, and more.

- Provide the surrounding community with skilled workers with technical skills not typically taught at four-year institutions.

- Offer workforce education to employers in their districts for specific outcomes.

Figure 1-3 provides general information on typical community college programs and outcomes.

Typical Community College Program Offerings			
Program Type	Typical Time to Complete as Full-Time Student	Number of Credit Hours	Description
Certificate	1–2 semesters	10–30	Certificate programs generally require 3–10 classes. Classes normally focus on a single, specific technical area. Examples include a Fluid Power Technician certificate or a Process Controls Technician certificate.
Associate Degree	2 years	60	Associate degree programs include required technical classes, elective technical classes, and required general education classes. Examples of Associate degrees include Industrial Maintenance Technician, Mechatronics Technician, Mechanical Engineering Technician.
Bachelor Degree	4 years	120	Bachelor degree programs require transferring from the community college to a four-year university. Industrial maintenance programs could lead to bachelor degrees such as Mechanical Engineering, Industrial Technology.

Goodheart-Willcox Publisher

Figure 1-3. Three common community college program types. As you complete each level of programs, your range of employment opportunities widens as you become a candidate for higher-salary positions.

TECH TIP
Ask Your Instructors

Ask instructors about their contacts with local companies that need graduates. These connections are an important consideration when selecting a school because they can help fast-track postgraduation employment.

TECH TIP
Continuing Education

Always ask about tuition reimbursement during an interview (for seminars also). It shows the employer that you are interested in continuing your education.

1.2.3 Universities and Continuing Education

Many universities have programs specifically designed for students who transfer from a two-year technical program. Private engineering and technical schools may also transfer in many credits earned at a two-year institution. For example, Northern Illinois University has an excellent technology program that, when combined with an approved AAS, will allow a full-time student to complete his or her BS degree in one year. Many students have gone on to further their education in this way. Look for similar programs at your local schools.

Continuing education is education that occurs after formal education is complete. Continuing education may be related to your current position and is intended to improve your performance. This type of continuing education often involves learning about new technology or new processes. Continuing education is also used to learn new skills with the goal of advancing to a new position. This type of continuing education may involve learning technical skills or may focus on "soft skills" such as management skills or communication skills.

1.3 CAREERS

Training in industrial maintenance and mechatronics can prepare you for a wide range of careers. Due to the high demand for these skills in the United States, skilled technicians can go nearly anywhere and find a position.

1.3.1 Career Opportunities

The average person has had 10 different jobs by the time he or she is 40 years old. People typically spend two to three years in a position before they decide if the job is a good fit for them. This is perfectly normal. Just as very good employees are hard to find, a very good fit to an employer can be hard to find.

Many different types of companies employ workers with industrial maintenance and mechatronics skills. Many positions fall in the manufacturing sector, but private companies and public government facilities outside of manufacturing also need employees with these skills.

Working at a relatively small company, you may have a wide range of responsibilities, but less room for advancement. Working at a relatively large company, you may have more specialized responsibilities and more opportunity for advancement. Careers can span from entry level to supervision or from general technician to much more specific areas of expertise. See **Figure 1-4**. The path your career takes depends on your interests.

According to the *Occupational Outlook Handbook*, published by the US Bureau of Labor Statistics (BLS), median salaries for industrial maintenance technician occupations are in the range of $24–$27 per hour, or $50,000–$56,000 per year. In addition, the *Occupational Outlook Handbook* projects over 32,000 new industrial maintenance positions to be created in the

United States between 2016 and 2026. This increase will bring the total number of industrial maintenance workers to over 500,000. See **Figure 1-5**.

> ### TECH TIP
> **Median Salary**
>
> Median salary represents the middle of a range of salaries, with half earning less than the median salary and half earning more than the median salary.

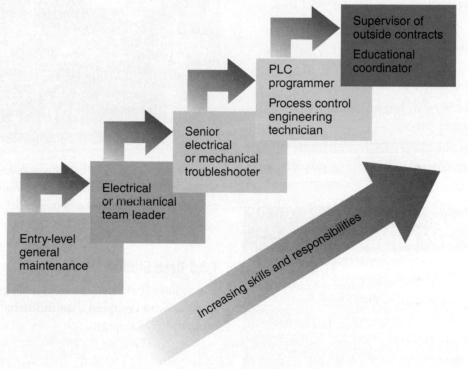

Goodheart-Willcox Publisher

Figure 1-4. Employment in the industrial maintenance field allows many pathways for advancements as you gain skills and experience. Your choices and actions will determine the path of your career.

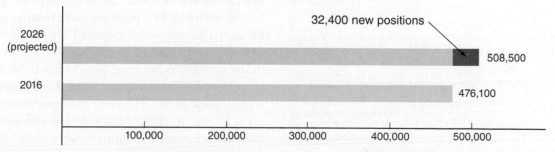

Goodheart-Willcox Publisher, data from US Bureau of Labor Statistics

Figure 1-5. The number of industrial maintenance technicians in the United States is projected to increase in coming years. In addition, retirement of older workers will create a large number of additional job openings.

Industrial maintenance and mechatronics encompasses a wide range of jobs. See **Figure 1-6**. Oftentimes, you can move from one position to another by completing additional formal training or through on-the-job training.

1.3.2 Work Conditions

A typical workday for an industrial maintenance technician could begin with a brief start-of-shift meeting to discuss work assignments and priorities. Depending on the scope of the assignment, work may be performed individually or by small teams. Common assignments include the following:

- Installing new machines, debugging machines, and bringing machines online.
- Troubleshooting and repairing malfunctioning equipment.
- Performing preventive maintenance.
- Evaluating processes to improve productivity and performance.
- Working with outside contractors or inside engineering on current or upcoming projects.

Positions Related to Industrial Maintenance and Mechatronics
Food service equipment technician
Machine installer
Lubrication technician
Industrial mechanic
City or municipal technician
Water treatment technician or operator
Forge technician
Plastic injection molding technician
Fluid power technician
Instrumentation technician
Instrumentation programmer/calibrator
Electrical technician
Electronic technician
PLC programmer
Communications technician
HVAC/refrigeration technician
Maintenance technician
Vibration technician

Goodheart-Willcox Publisher

Figure 1-6. The skills you gain in your industrial maintenance education make you a candidate for many different jobs.

Industrial maintenance work is often performed indoors in an industrial facility, **Figure 1-7**, but work may also be performed outdoors as well. Work schedules are typically 40 hours per week, but overtime work may be required at some facilities. Facilities may be operational 24 hours per day, so industrial maintenance work may need to be done during the day, at night, or during weekends.

Pressmaster/Shutterstock.com

Figure 1-7. Technicians have the skills to perform various job tasks in different work settings. This technician is performing mechanical maintenance with proper personal protective equipment and tools in a factory setting.

1.3.3 Case Studies

The following are two career case studies based on actual students who completed an industrial maintenance and mechatronics program.

Case Study #1: Jake

After graduating high school, Jake was employed in an entry-level maintenance position at a small plastic injection molding plant. The factory had an apprenticeship program that included completing specific industrial maintenance courses at a community college. Taking full advantage of this program, Jake went to school over the next two years and increased his hourly wage from $10 an hour to $20 an hour. He also earned a certificate in Industrial Maintenance, but did not finish his degree.

Over the next several years, Jake moved among several employers. He lost one position when the plant he was working at closed, and left another position when his responsibilities and required hours were increased without receiving an appropriate increase in salary. Jake finally returned to school and finished his two-year degree. With his degree in hand, he accepted a position at a larger international company specializing in plastic injection molding machines and tooling. Jake's territory is the central United States and includes 15 states. He travels to customers' locations and performs troubleshooting and repair on the company's machines. This work involves all aspects of industrial maintenance: automation, mechanical, fluid power, electrical, and electronic. Jake's workweek varies depending on callouts and can range from a minimal number of hours to as many as 60 hours a week. He earns more than $70,000 per year as a base salary, and his annual income can be near $100,000 if he works a good amount of overtime.

Case Study #2: Steve

Steve worked as a carpenter and cabinetmaker. While maintaining his full-time work schedule, he began taking courses in an HVACR program at a community college. During his courses, he found that he enjoyed his industrial maintenance courses as well. After a busy few years, Steve received certificates in both Industrial Maintenance and HVACR. Earning these two certificates made Steve an attractive candidate for positions. He accepted an offer to work as a maintenance mechanic at a large food manufacturing company. In three years, Steve was earning nearly $30 per hour ($62,400 per year). While overtime is not required, Steve usually works 60–70 hours a week, which results in him earning a six-figure income for the year.

Steve has also achieved certification for ammonia refrigeration and waste water operations. By completing this additional training, Steve gained skills that were valuable to the company, and his salary was increased. In a typical day, Steve checks on operations of the factory's steam and refrigeration plants, performs planned maintenance, and completes specific projects involving troubleshooting and repairing process control sensors. Initially intending to stay in a residential field involving carpentry or HVAC, Steve's $12,000 investment in his education has proved very valuable toward getting into the industrial field.

1.4 APPLYING FOR A POSITION

When you are ready to apply for employment, you will need to know the appropriate steps to take. Having a well-prepared résumé is an important first step. Knowing how to write an acceptable letter of application, also called a cover letter, is another goal. You will also need to put together a list of references—people you can trust and who know you to be a reliable person. Finally, you will want to practice your interviewing techniques.

1.4.1 Your Résumé

A *résumé* is a brief outline of your education, work experience, and other qualifications for work. A well-written résumé can help you get an interview. You will need to include several sections on your résumé. An example of a résumé appears in **Figure 1-8**. Make sure that your résumé is precise and without errors.

It is more likely that an employer will request that a résumé be sent as an attachment via e-mail (usually as a word-processing document or pdf file) or be uploaded during the application process on the employer's website. You can also post an electronic résumé to a number of online job-search sites. Employers may use the electronic file to search for key terms that match their descriptions of an ideal job candidate. Keep this in mind while compiling your résumé.

Along with the résumé, you need to develop a list of references. A *reference* is an individual who will provide important information about you to a prospective employer. A reference can be a teacher, school official, previous supervisor or coworker, or any other adult outside your family who knows you well.

You will need at least three references. Always get permission from each person to use his or her name as a reference before actually doing so. Your list of references, along with their titles, phone numbers, and addresses, should be kept private. Share this list only with an employer who has interviewed you and asks for your references.

1.4.2 Letter of Application / Cover Letter

The letter of application, or *cover letter*, is often the first contact you have with a potential employer. It can make a lasting impression. It should be neat and follow a standard form for business letters. You will likely be asked to upload or post the letter along with your résumé. As with your résumé, use a standard font to give the letter a professional look. You should use the same font for both documents. Be sure to check spelling and punctuation. Have several people read the letter and offer advice for improving it.

A sample letter of application appears in **Figure 1-9**. It is a good example to refer to when responding to a job ad. Your letter should be brief and to the point. It should include the following items:

- Title of the job you seek.
- Where you heard about the job.
- Your strengths, skills, and abilities that might apply to the job.
- Reasons you should be considered for the job.
- Request for an interview.

Michael J. Garcia

134 Lincoln Street   (212) 555-1234
Wilton, CA 93232   mjgarcia22@e-mail.com

Career Objective
To obtain an entry-level industrial maintenance or mechatronics position.

Professional Experience
Heavy Metal Ducts, Holloton, CA   August 2017–present
Sheet Metal Helper

   □ Perform general construction labor, including material loading and jobsite cleanup.
   □ Install ductwork systems.
   □ Help perform duct testing.

Simpson Supply Co., Wilton, CA   May 2016–August 2017
Parts Clerk

   □ Worked with customers at parts counter, checked inventory system,
     and obtained parts.
   □ Conducted daily and monthly inventory checks.
   □ General stocking and cleaning throughout store.
   □ Delivered and picked up parts and equipment.

Education
Associate Degree in Industrial Maintenance Technology   May 2019
Oceanside Community College

   □ GPA: 3.22/4.0
   □ Coursework included hydraulics, pneumatics, mechanical systems, industrial
     controls, and welding.
   □ Obtained three NIMS Industrial Technology Maintenance Level 1 Credentials:
     Basic Hydraulic Systems, Basic Pneumatic Systems, Electrical Systems, and
     Electronic Control Systems.

Community Service
Habitat for Humanity, volunteer, summers of 2017, 2018, 2019
Wilton Food Bank, volunteer, 2016–present

References
Available on request.

Figure 1-8. Your résumé highlights your work experience and educational accomplishments related to the desired position. Be sure to list applicable courses you have taken, activities you participated in, and your work experience.

Michael J. Garcia
134 Lincoln Street
Wilton, CA 93232
(212) 555-1234
mjgarcia22@e mail.com

April 23, 2019

Human Resource Director
Williamson Manufacturing
4392 East 134th Street
Wilton, CA 93232

Dear Ms. Wabarster:

The Industrial Maintenance Technician position you advertised on the Career Finder website is exactly the type of job I am seeking. After reviewing the job description and requirements, it was clear that my experience, skills, and interests are a perfect match for this opportunity.

While obtaining my associate's degree in Industrial Maintenance Technology from Oceanside Community College, I gained both the theoretical knowledge and the hands-on skills required for this position. While working as a parts clerk at Simpson Supply Company, I developed strong customer service skills and gained a better understanding of HVAC parts and systems. In my current position as a sheet metal helper, I've gained valuable experience working at a variety of job sites and with diverse teams. I am eager to apply the skills I have learned and to continue gaining new skills.

Please find my résumé enclosed with this letter. I would greatly appreciate an opportunity to interview for this position. Please contact me at your convenience by phone or e-mail to schedule an interview. I look forward to hearing from you.

Sincerely,

Michael J. Garcia

Michael J. Garcia

enclosure

Figure 1-9. When writing a letter of application (also known as a cover letter), include some information about the company you are applying to and position you are applying for.

1.4.3 Job Application Forms

Aside from the three main pieces of writing you will need—a résumé, cover letter, and references—a prospective employer may also ask you to complete a job application form before obtaining an interview. The job application form highlights the information the employer needs to know about you, your education, and your prior work experience. The appearance of an application form can give employers their first opinion about you. Fill out the form accurately, completely, and neatly. When asked about salary, you may write "open" or "negotiable." This means you are willing to consider offers. However, it is a good idea to research the typical salary range of the type of job you are applying for, so you have some idea of what to expect. Tips for completing the job application appear in **Figure 1-10**.

TECH TIP
Negotiating Salary

Many employers want to know your prior earnings. Realize that your prior salary may not take into account your newly acquired education. After obtaining a certificate or degree, your worth increases. Keep this in mind when reviewing any offer.

Most employers request electronic applications, either through their company website or independent job-search websites. When filling out an online application, it is extremely important to include key terms for which the employer may search. This will help your application stand out among the many other applications the employer will receive.

Tips for Completing a Job Application

- Follow the instructions for filling out the form. Many applications are completed online. If you are completing a printed application, be as neat as possible. The instructions may ask you to print or to use black ink. Be sure to follow these directions.

- Complete every question in the form. If some questions do not apply to you, draw a dash or write "NA" (for not applicable) so the employer knows you did not overlook it.

- If the application asks for a Social Security number, you may wish to ask if this can be provided after a job offer is made.

- You can write "open" or "negotiable" for any question regarding salary requirements.

- For each former job, there may be a question asking your reason for leaving the job. Avoid writing any negative comments about yourself or a former employer.

Goodheart-Willcox Publisher

Figure 1-10. Tips for completing a job application.

1.4.4 The Job Interview

The interview gives you the opportunity to learn more about a company and to convince the employer that you are the best person for the position. The employer wants to know if you have the skills needed for the job. Adequate preparation is essential for making a lasting, positive impression. Here are some ways to prepare for the interview:

- **Research the employer and the job.** Know the mission of the employer and specifics about the job. Also, try to learn what the company looks for when hiring new employees.

- **Be prepared to answer questions.** Go over the list in **Figure 1-11** and prepare answers for each question.

- **List the questions you want answered.** For example, do you want to know if there is on-the-job training? Are there opportunities for advancement?

- **List the materials you plan to take.** This seems simple enough. However, if you wait to grab items at the last minute, you will likely forget something important.

- **Decide what to wear.** Dress appropriately, usually one step above what is worn by your future coworkers. For instance, casual clothing is acceptable for individuals who will do manual labor or wear a company uniform. If the job involves greeting the public in an office environment, a suit is more appropriate. Always appear neat and clean.

- **Practice the interview.** Have a friend or family member interview you in front of a mirror until you are happy with your responses.

- **Know where to go for the interview.** Verify the address of the interview location by checking the site beforehand, if possible. Plan to arrive ready for the interview 10–15 minutes early.

Within 24 hours after the interview, send an e-mail or letter to each of the people with whom you interviewed, thanking them for meeting with you. If you get a job offer, respond to it quickly. If you do not receive an offer after several interviews, evaluate your interview techniques and seek ways to improve them.

1.5 SUCCEEDING IN THE WORKPLACE

After securing employment, adjusting to your new duties and responsibilities will occupy your first few weeks. Your supervisor and coworkers will help you learn the routine. It is common for new employees to receive an introduction to company policies and procedures, as well as the special safety rules that all employees must know.

Common Interview Questions and Responses	
Question	**Response**
What can you tell me about yourself?	Briefly summarize your abilities as they relate to the job qualifications or your career goals. Do not provide a general life history.
Why do you want to work for this company?	Tell what you know about the company. Explain how your abilities match the company's needs.
Why do you think you would like this kind of work?	Relate the job requirements to your successful past experiences.
What are some of the projects you worked on in school?	Briefly summarize a project or coursework relevant to the job qualifications.
What other jobs have you had?	Focus on jobs with skills that relate to the jobs you are seeking.
Why did you leave your last job?	Be honest. However, avoid saying anything negative about your previous employer.
Have you ever been fired from a job? If so, why?	Answer honestly. If you have been fired, share what you learned from the experience. Avoid trying to blame others.
What are your major strengths and weaknesses?	Select a strength that relates to the job qualifications. Be honest when selecting a weakness, but give an example of how you have worked to improve on it.
Have you ever had a conflict with a coworker? How did you handle it?	Briefly describe the situation and how you handled it. Avoid placing all the blame on the other person. Explain what you learned from the experience.
What do you expect to be paid?	If possible, determine the salary range before the interview. Say that you are willing to discuss the salary or state a range you feel comfortable with.
What are your future plans?	Describe how the need to learn and grow is important to you. Confine your answer to the company with which you are interviewing.

Goodheart-Willcox Publisher

Figure 1-11. Sample job interview questions. Prepare answers for these and other questions you might expect during your interview.

While your coworkers will be watching what you do, they will also pay attention to how you work. How to behave in the workplace is an important lesson all employees should learn. Making an effort to do your best will help you succeed.

1.5.1 Dress and Appearance

Some companies may provide work uniforms, laundry service, and a locker room. Other companies may provide nothing. Your job responsibilities dictate your work dress. If working in a lab setting, slacks and a collared shirt are appropriate (the company may provide lab coats). Working on a factory floor can include environments that range from dusty grinding and welding areas to a sterile clean room. Dress appropriately for your job requirements. If working near machinery, never wear anything that could get caught in a machine and result in injury—your workplace may have guidance on this. If your work requires safety shoes (such as steel-toed work boots), the company may either purchase them for you or reimburse you after your purchase. Clothing requirements are something to ask about during an interview.

While some employers may not have rules concerning hygiene, others will. This is dictated by the environment you work in and the responsibilities of the position. If you work in a food manufacturing plant, you may be required to be clean-shaven, have short hair, and use hairnets. You may also be required to be clean-shaven if you use a respirator or SCBA (self-contained breathing apparatus). In most companies, neatly trimmed facial hair and well-kept hair are the norm.

TECH TIP
Dress for Success

Even if wearing company-issued work uniforms, dress neatly. Showing up in a dirty or wrinkled uniform does not improve your image at the company.

1.5.2 Tools of the Trade

Companies will tell you whether you must supply your own tools or they will supply everything or, most frequently, a combination of the two. If the company requires you to have your own tools, ask them to be more specific. Normally, the company will require you to have hand tools in some or all of the following categories:

- Mechanical tools (sockets and wrenches) for 3/8″ and 1/2″ drives, up to 1″ fasteners.
- Basic electrical tools (screwdrivers, wire strippers and cutters, and basic electrical meters).
- Basic precision mechanical tools, such as a 0–6″ caliper.
- Your own toolbox or (eventually) stacked toolboxes.

Some companies may also reimburse you up to a certain yearly amount for tools and expenses. Consumable items (wire, tape, and other nondurable goods) are normally supplied by the company. If you are required to use your own tools, buy high-quality tools. You will use these tools for many years and do not want the hassle of having to repair or replace them because they are of poor quality.

1.5.3 Work Habits

Employers want employees who are punctual, dependable, and responsible. They want their employees to be capable of taking the initiative and working independently. Other desirable qualities of employees are that they be organized, accurate, and efficient.

A *punctual* employee is always prompt and on time. This includes not only when starting the workday, but also when returning from breaks and lunches. Being dependable means that people can rely on you to keep your word and meet your deadlines. If you are not well, be sure to call in and let the employer know right away. If there are reasons you cannot be at work, discuss this with your employer and work out an alternate arrangement. Many people have lost jobs by not checking with their supervisor about time off.

Taking the *initiative* means that you start activities on your own without being told. When you finish one task, you do not wait to hear what to do next. Individuals who take the initiative need much less supervision. They have *self-motivation*, or an inner urge to perform well. Generally, this motivation will drive you to set goals and accomplish them. All of these qualities together show that you are capable of working independently.

You are expected to be as accurate and error-free as possible in all that you do. This is why you were hired. Complete your work with precision and double-check it to assure accuracy. Your coworkers depend on the careful completion of your tasks.

1.5.4 Time Management

A good employee knows how to manage time wisely. This includes the ability to prioritize assignments and complete them in a timely fashion. It also involves not wasting time. Time-wasting behaviors include visiting with coworkers, making personal phone calls, texting, sending personal e-mails, or doing other nonwork activities during work hours.

While it is important to complete all of your work thoroughly, you must also be able to gauge which assignments are most important. Avoid putting excessive efforts into minor assignments when crucial matters require your attention. Even though you are still accomplishing work, this is another way of wasting time.

1.5.5 Attitude on the Job

Your attitude can often determine the amount of success you have in your job, **Figure 1-12**. Your *attitude* is your outlook on life. It is reflected in how you react to the events and people around you. A smile and courteous behavior can make customers and fellow employees feel good about themselves and you. Clients and customers prefer to do business in friendly environments. Being friendly may take some effort on your part, but it does pay off.

Steve Good/Shutterstock.com

Figure 1-12. Having a positive attitude on the job can help technicians accomplish tasks in an efficient manner and increase workplace morale.

1.5.6 Professional Behavior

You will be expected to behave professionally on the job. This includes showing respect for your boss and coworkers. Personal conversations and phone calls should be limited to break times or lunch. Act courteously; remember that others are focusing on their work. Interruptions can cause them to lose concentration.

Part of behaving professionally is responding appropriately to constructive criticism. Every employee, no matter how knowledgeable or experienced, can improve his or her performance. If you receive criticism from a supervisor or coworker, do not be offended. Instead, use the feedback to improve yourself. The more you improve, the more successful you will be in your work.

1.5.7 Decision-Making and Problem-Solving

Employers value workers who have the ability to make sound decisions. The decision-making process applies in the workplace as well as other aspects of life. The process involves identifying the issue, identifying possible solutions, making a decision, implementing the decision, and evaluating the results, **Figure 1-13**.

Having the ability to solve problems on the job shows an employer that you are able to handle more responsibility. Solving problems as a group can strengthen camaraderie and help employees feel more pride in their work.

The ability to make decisions and solve problems requires *critical-thinking skills*. These are higher-level skills that enable you to think beyond the obvious. You learn to interpret information and make judgments. Supervisors appreciate employees who can analyze problems and think of workable solutions.

A_stockphoto/Shutterstock.com

Figure 1-13. Basic problem-solving tasks on the job include reading and interpreting measurements. Understanding a problem is the first step in determining the best method for correcting it.

1.5.8 Communication Skills

Communication is the process of exchanging ideas, thoughts, or information. Communicating effectively with others is important for job success. Being a good communicator means that you can share information well with others. It also means you are a good listener.

The primary forms of communication are verbal and nonverbal. *Verbal communication* involves speaking, listening, and writing. *Nonverbal communication* is the sending and receiving of messages without the use of words. It involves *body language*, which includes your facial expressions and body posture.

Listening is an important part of communication. If you do not understand something or someone, be sure to ask questions. Also give feedback to let others know you understand them and are interested in what they have to say.

The message you convey in telephone communication involves your promptness, tone of voice, and attitude. Answering the phone quickly and with a pleasant voice and greeting conveys a positive image for the company. Learning to obtain accurate information from the caller without interrupting that person's message is important.

To be an effective employee, you need to know how to communicate well with the common tools of your workplace. When communicating by e-mail, carefully consider the message before sending it. Often messages are sent quickly without thought of how the recipient may interpret them. The same is true of voice mail.

1.5.9 Ethical Workplace Behavior

Ethical behavior on the job means conforming to accepted standards of fairness and good conduct. It is based on a person's sense of what is right to do. Individuals and society as a whole regard ethical behavior as highly important. Integrity, confidentiality, and honesty are crucial aspects of ethical workplace behavior. *Integrity* is firmly following your moral beliefs.

Unfortunately, employee theft is a major problem at some companies. Such theft can range from carrying office supplies home to stealing money or expensive equipment. Company policies are in place to address these concerns. In cases of criminal or serious misbehavior, people may lose their jobs. If proven, the charge of criminal behavior stays on the employee's record. With that record, an employee will have a difficult time finding another job.

1.5.10 Interpersonal Skills

Interpersonal skills involve interacting with others. Some workplace activities that involve these skills include teaching others, leading, negotiating, and working as a member of a team. Getting along well with others can require great effort on your part, but it is essential for accomplishing your employer's goals.

Teamwork

Employers seek employees who can effectively serve as good team members. Due to the nature of most work today, teamwork is necessary. A *team* is a small group of people working together for a common purpose. Often cooperation requires flexibility and a willingness to try new ways of doing things. If someone is uncooperative, it takes longer to accomplish tasks. When people do not get along, strained relationships may occur, which gets in the way of finishing tasks.

A big advantage of a team is its ability to develop plans and complete work faster than individuals working alone. In contrast, a team usually takes longer to reach a decision than an individual worker does. Team members need some time before they become comfortable with one another and function as a unit. You will be more desirable as an employee if you know how to be a team player.

Creative ideas often develop from building on another person's idea. Honesty and openness are essential. Also, trying to understand the ideas of others before trying to get others to understand your ideas is an effective skill to develop.

Leadership

All careers require leadership skills. *Leadership* is the ability to guide and motivate others to complete tasks or achieve goals, **Figure 1-14**. It involves communicating

ALPA PROD/Shutterstock.com

Figure 1-14. This technician exhibits strong leadership by directing her teammate to follow the proper steps in maintenance protocol.

well with others, accepting responsibility, and making decisions with confidence. Employees with leadership skills are most likely to be promoted to higher levels.

Leaders often seem to carry the most responsibility in a group. Other group members look to them for answers and direction. The most important role of leaders is to keep the team advancing toward its goal. Leaders do this by inspiring their groups and providing the motivation to keep everyone working together.

Good leaders encourage teamwork, because a team that is working together well is more likely to reach goals. They listen to the opinions of others and make sure all team members are included in projects. Leaders also want to set a good example by doing a fair share of the work. In these ways, leaders cultivate a sense of harmony in the group.

Conflict Management

When you work with others, disagreements are likely to occur. More serious disagreements can lead to conflict. *Conflict* is a hostile situation resulting from opposing views. It is important to know how to handle conflict to prevent it from becoming a destructive force in the workplace. This is called *conflict management*. A team leader has a special responsibility to prevent conflict among the team members. Several steps can be followed in managing conflict, **Figure 1-15**.

Sometimes the cause of a conflict is not so simple or easily understood. Use a positive approach and try to understand the problem from the other's point of view. Avoid jumping to conclusions and making snap judgments. Treat others with respect and in the same way you would like to be treated. Explore positive and negative aspects of each possible solution. If progress falls short of expectations, bring the parties back together and repeat the process. Many disagreements in the workplace can lead to productive change.

Steps in Managing Conflict
1. Know when to intervene.
2. Address the conflict.
3. Identify the source and the importance of the conflict.
4. Identify possible solutions.
5. Develop an acceptable solution.
6. Implement the solution and evaluate.

Goodheart-Willcox Publisher

Figure 1-15. Conflict can be healthy in the workplace, as long as the conflict remains professional and coworkers with differing opinions maintain an open mind. Resolving a conflict often leads to a better solution.

Negotiation

Sometimes employees and employers must negotiate on a task or work-related issue. *Negotiation* is the process of reaching an agreement that requires all parties to give and take. The goal is a "win-win" solution in which both parties get some or all of what they are seeking.

Negotiation begins with trying to understand the other party's interests. Possible solutions that meet the concerns of both sides can be developed. Often the best solution becomes clear when both parties have ample time to explain what they are trying to accomplish.

1.5.11 Staying Safety Conscious

Safety on the job is everyone's responsibility. Many workplace accidents occur because of careless behavior. Often poor attitudes can cause unsafe behavior, too. Common causes of accidents include the following:

- Taking chances.
- Showing off.
- Forgetting safety details.
- Disobeying company rules.
- Daydreaming.
- Losing your temper.
- Falling asleep.

Practicing good safety habits is essential for preventing accidents and injuries on the job. A healthy worker is more alert and less likely to make mistakes. Knowing how to use machines and tools properly is the responsibility of both the employer and employees. Wearing protective clothing and using safety equipment correctly helps keep workers safe. Your employer will emphasize the safety practices that employees must follow in your workplace.

The government agency that promotes safety in the workplace is the *Occupational Safety and Health Administration (OSHA)*. You will be required to follow the specific OSHA regulations that apply to your workplace.

SAFETY NOTE
Follow the CFR

Knowingly violating a safety code may not only get you fired from a job, but also personally fined by OSHA. Leaving a machine guard off after maintenance is completed is one instance you know is wrong and against the Code of Federal Regulations (CFR).

1.6 CERTIFICATIONS AND CREDENTIALING

Industry-recognized credentials and certifications are an important part of any technical field. *Credentialing* refers to establishing and documenting a specific set of qualifications, competencies, or skill standards. Certifications are one kind of credential. Having industry-recognized certifications shows potential employers that you are serious about your job performance, knowledge, and skills and that you meet accepted industry standards. All certifications require testing; most require both a written test and a performance or skills check. Successful completion of performance checks and written exams allows you to receive a certification and become credentialed. Possessing relevant credentials will improve your employability.

You can obtain some certifications through local educational institutions, and others you can pursue on your own. Some certifications are more universally accepted by employers than others. Certifications range from extremely expensive to very reasonable. The best certifications are those that are widely accepted and recognized, result in a tangible benefit (such as a higher wage), and are sought after. If you are not sure about which certification is best, ask your instructor or professor what most employers in the area value.

TECH TIP
The Right Certification

If an employer does not recognize or value a particular certification, then you may want to consider a different, more applicable certification. Do some research to determine the best certification for your needs.

1.6.1 NIMS

The *National Institute for Metalworking Skills (NIMS)* was started in 1995 to set industry skill standards, certify individual skills against those standards, and accredit training programs that meet NIMS quality requirements. Standards cover a wide range of advanced manufacturing occupations, from machining to maintenance, and include both theory and hands-on skills. Partnering with industry leaders, NIMS follows a rigorous process to develop and update standards and assessment tools.

NIMS offers more than 60 portable credentials in specific technical areas to certify individuals' skills against industry standards in three levels (Levels I, II, and III). Currently NIMS offers certification in Industrial Technology Maintenance (ITM) Level I. The ITM standards are developed by experts from industry and educational institutions. There are nine areas of NIMS standards in the ITM credentialing program:

- Maintenance Operations
- Basic Mechanical Systems
- Basic Hydraulic Systems
- Pneumatic Systems
- Electrical Systems
- Electronic Control Systems
- Process Control Systems
- Maintenance Welding
- Maintenance Piping

These nine areas compose a set of national, industry-created, industry-recognized credentials based on NIMS standards. NIMS requires candidates to not only take proctored assessments on theory, but also complete hands-on skill assessments with a qualified instructor or supervisor. This credentialing process is documented by registered evaluators who fill out competency achievement records (CARs), which are checklists of critical skills. You can learn more about NIMS at www.nims-skills.org. See **Figure 1-16**.

1.6.2 OSHA

OSHA (Occupational Safety and Health Administration) authorizes industry trainers to perform 10-hour and 30-hour training classes through the OSHA Outreach Training Program. The 10-hour training covers all aspects of safety that an incoming employee needs. The 30-hour training is intended for supervisors and those with responsibilities for safety at the workplace. Both in-person training and online training are available. Having your OSHA 10-hour card should be considered a requirement prior to employment in industry.

NIMS

About   Credentialing   Apprenticeship   Accreditation   Training   Resources   Log in / Register

Tools & Resources

Tools

Upload Affidavits

Send affidavits quickly and easily using the Affidavit Upload Portal. →

Upload Purchase Orders

Sending a PO? Use the PO Upload Portal for simple submission. →

Career Pathways

Explore some common positions in manufacturing and how you can get there. →

Resume Generator

Use our easy online tool to create a resume that gets you the job! (Login required) →

Resources

180 Results Found

Search

Filter by Topic ⌄

Filter by Type ⌄

Step-By-Step Guide: Candidate Registration

Type   TESTING GUIDE

A one-page instruction sheet on registering individuals as candidates

01/25/2018

Step-By-Step Guide: Online Testing

Type   TESTING GUIDE

A one-page instruction sheet on administering an online test

01/25/2018

Machining Application for Accreditation

Type   FORM

Application for first-time applicants and renewals (machining programs)

01/19/2018

Goodheart-Willcox Publisher

Figure 1-16. In addition to information about Industrial Technology Maintenance credentials, the NIMS website also provides information about manufacturing careers and a résumé generator.

1.6.3 Other Certifications

The Manufacturing Skill Standards Council (MSSC) offers certifications for production workers. Certified production technician (CPT) certification focuses on safety, quality practices and measurement, manufacturing processes and production, and maintenance awareness. Certified logistics technician (CLT) certification comprises two layers: certified logistics associate (CLA) and certified logistics technician (CLT). CLA subjects range from safety, logistics, and material handling to quality control, communication, and teamwork. CLT

training includes product receiving and storage, inventory management, safety, and transportation concepts.

The Society for Maintenance & Reliability Professionals (SMRP) is a nonprofit group that provides certification and education for maintenance professionals. SMRP offers a certification for a certified maintenance and reliability technician (CMRT).

Finally, certifications in other areas are sometimes very useful. For instance, most manufacturers view a certification in HVAC (heating, ventilation, and air conditioning) refrigerant handling (EPA Section 608) as a valuable asset. Certifications through your local Red Cross in first aid and CPR are also worthwhile.

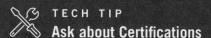

TECH TIP
Ask about Certifications

Ask your instructors which certifications they deem most valuable for local industries and employers. Also, take note of certifications mentioned in regard to jobs advertised locally and on industry websites. These sources will help you focus on the certifications that will add the most to your value as a potential employee.

CHAPTER WRAP-UP

You will most likely not spend your entire career at one location with one employer. With that in mind, you need to be flexible, have an overall plan, and keep your goal in sight. Some of the best technicians make more than $100,000 a year. This is only possible through hard work, study, discipline, and self-improvement. Employers are always looking for highly qualified and motivated technicians. Apply yourself so that you become that needed technician.

Chapter Review

SUMMARY

- Programs at private technical schools are usually faster paced and take less time to complete than a two-year degree, but they are also typically more expensive.

- If attending a two-year college, get to know your instructors and professors. They are the best sources of information about industry in your area.

- If you want to succeed, do not limit your job search to a specific area or a specific job title. All of the skills and knowledge in this text are widely applicable.

- Your résumé should appear neat and professional, with no errors, and should include keywords used in the industry.

- You should call your references ahead of time.

- A letter of application, or cover letter, should be short and to the point with no errors.

- A job application should be neat and professional looking. If your handwriting is poor, practice with neatness specifically in mind.

- Show up 10–15 minutes early for an interview.

- Dress appropriately and conservatively, as though you were about to give a presentation to a more mature crowd. If you want an employer to take you seriously, take yourself seriously.

- Even if you wear a company-supplied uniform, keep a neat and professional-looking appearance.

- Keep a professional, productive, and positive attitude on the job.

- When honing your troubleshooting skills, ask experienced team members around you for their input.

- Communicate in such a way that your intentions are clear and concise, and be prepared to explain your rationale if asked.

- If confronted with conflict at work, accept the possibility that the other person may have a valid point and ask for explanation or clarification.

- Safety should always be your top priority at work.

- Certifications show a future employer that you are serious about your education and work.

REVIEW QUESTIONS

Answer the following questions using the information provided in this chapter.

1. _____ goods are those that are consumed quickly, such as bread and fuel.

2. What is continuing education?

3. *True or False?* Since employers are expanding and more automation is taking place, highly skilled automation technicians are in demand.

4. List and describe the four pieces of writing you may have to produce for a job interview.

5. List five steps you can take to prepare for an interview.

6. What are four personal behaviors that lead to success in the workplace?

7. The ability to make decisions and solve problems requires _____ skills.

8. *True or False?* Body language is a part of nonverbal communication.

9. _____ is the process of reaching an agreement that requires all parties to give and take.

10. What are three common causes of accidents in the workplace?

11. How are NIMS standards verified?

NIMS CREDENTIALING PREPARATION QUESTIONS

The following questions will help you prepare for the NIMS Industrial Technology Maintenance Level I Maintenance Operations credentialing exam.

1. Manufacturing is responsible for approximately _____ of the US economy.
   A. 2%
   B. 10%
   C. 25%
   D. 50%

2. Education in the field of industrial maintenance and mechatronics can come from which of the following?

   A. Technical schools
   B. Community colleges
   C. Universities
   D. All of the above

3. A(n) _____ is a brief outline of your education, work experience, and other qualifications for work.

   A. résumé
   B. letter of application
   C. application for employment
   D. reference list

4. Which of the following is *not* an example of professional behavior?

   A. Dressing neatly and appropriately
   B. Showing respect for coworkers
   C. Making personal phone calls at work
   D. Showing up to work on time

5. What can happen if you knowingly violate a safety code?

   A. You will be required to take a 30-hour safety training course.
   B. OSHA can personally fine you.
   C. You will lose all of your certifications.
   D. None of the above.

2 | Industrial Safety and OSHA

CHAPTER OUTLINE

LEARNING OBJECTIVES

After completing this chapter, you will be able to:

- ☐ Explain OSHA and its reason for existence.
- ☐ Comprehend the reasons not to shortcut safety procedures.
- ☐ Demonstrate LOTO procedures and their importance.
- ☐ Understand the requirements of blocking and its necessity
- ☐ Comprehend pinch points and their dangers.
- ☐ Understand the various types of PPE and uses.
- ☐ Understand grounding and its importance.
- ☐ Outline safety precautions when storing and transferring flammable materials.

TECHNICAL TERMS

AED (automatic electrical defibrillator)
blocking
Environmental Protection Agency (EPA)
ground-fault circuit interrupter (GFCI)
grounding
job safety analysis (JSA)
"live-dead-live" (LDL) test
lockout/tagout (LOTO)
National Institute for Occupational Safety and Health (NIOSH)
Occupational Safety and Health Administration (OSHA)
personal protective equipment (PPE)
pinch point
safety data sheet (SDS)
6S program
zero energy state

◄ ndoeljindoel/Shutterstock.com

Safety is the most important aspect of your job. A choice to disregard safety practices may lead to serious injury for you or your coworkers. There is no job so urgent or important that you cannot follow proper safety practices. This chapter gives an overview of general safety procedures. Later chapters will provide topic-specific safety information.

As a maintenance technician, you may be exposed to potential hazards. If you are uncertain as to what safety procedures you should use, seek help rather than attempting to deal with a potentially unsafe situation on your own.

2.1 SAFETY STANDARDS AND REGULATIONS

Considering safety in everything you do reduces the chance for injury and damage to equipment and property, but it also ensures you and your workplace follow the law. Employers and employees must comply with federal safety standards and regulations.

2.1.1 OSHA

After Congress enacted the Occupational Health and Safety Act, the *Occupational Safety and Health Administration (OSHA)* was formed. OSHA has been tasked with ensuring that employers provide a safe and healthful workplace.

As a maintenance professional, you are responsible for your own safety and the safety of those around you. You must use proper personal protective equipment and remind others to do the same, **Figure 2-1**. You must communicate potential hazards to supervisors and coworkers.

MSA Safety Inc.

Figure 2-1. In training and on the job, use proper safety equipment and remind others to do so as well.

You must always take appropriate safety and hazard-containment steps to protect your health and safety, the health and safety of those around you, and the environment. As an employee, you must follow all safety procedures required by your employer. You must report any unsafe conditions that you are aware of to your supervisor without fear of reprisal. Any safety issues that have been reported by employees must be addressed.

Although there is some debate about the role of regulation in industry, OSHA is ultimately there to protect you. OSHA provides specific standards and training topics, **Figure 2-2**. Many organizations offer OSHA-approved 10-hour and 30-hour training courses and safety certifications. You can read the OSHA standards, download training materials, and learn about OSHA at the OSHA website.

OSHA works in conjunction with the *Environmental Protection Agency (EPA)* and the *National Institute for Occupational Safety and Health (NIOSH)*. The EPA is focused on the protection of public health and the environment. It works with OSHA in that capacity by assuring compliance with federal environmental regulations. NIOSH is the research arm of OSHA. It develops recommended occupational safety and health standards and conducts research and experimental programs to develop criteria for new and improved standards.

2.1.2 Hazards and Safety Data Sheets (SDSs)

Figure 2-3 summarizes common industrial hazards identified by OSHA. You should be able to recognize hazardous situations and apply proper procedures. This includes following guidelines concerning spill control and first aid, as well as the storage, handling, and protection of equipment.

Furthermore, OSHA requires additional steps when dealing with hazardous materials. *Safety data sheets (SDSs)* provide essential information about substances or mixtures used in workplace chemical management that may be classified as presenting physical, health, or environmental hazards. OSHA requires chemical manufacturers, distributors, and importers to provide an SDS for each hazardous chemical. SDSs contain information about hazards, including environmental hazards and safety procedures. SDSs are typically product specific, though they may contain workplace-specific protection measures.

OSHA's Hazard Communication Standard (HCS) requires that SDSs communicate the hazards of chemical products. SDSs should be in a uniform format and include the section numbers, headings, and information presented in **Figure 2-4**.

OSHA General Industry Training Program				
Safety and Health Topic	**10-Hour Training Program**		**30-Hour Training Program**	
	Mandatory	**Elective**	**Mandatory**	**Elective**
Introduction to OSHA	×		×	
Walking and working surfaces, including fall protection	×		×	
Exit routes, emergency action plans, fire prevention plans, and fire protection	×		×	
Electrical	×		×	
Personal protective equipment (PPE)	×		×	
Hazard communication	×		×	
Hazardous materials		×		×
Materials handling		×	×	
Machine guarding		×		×
Introduction to industrial hygiene		×		×
Bloodborne pathogens		×		×
Ergonomics		×		×
Safety and health programs		×		×
Fall protection		×		×
Managing safety and health			×	
Permit-required confined spaces				×
Lockout/tagout				×
Welding, cutting, and brazing				×
Powered industrial vehicles				×

Adapted from OSHA

Figure 2-2. OSHA offers a 10-hour and a 30-hour General Industry Training Program, with certain mandatory and elective topics for each program. You should be aware of OSHA safety requirements and understand that your job may require additional training specific to your workplace and duties.

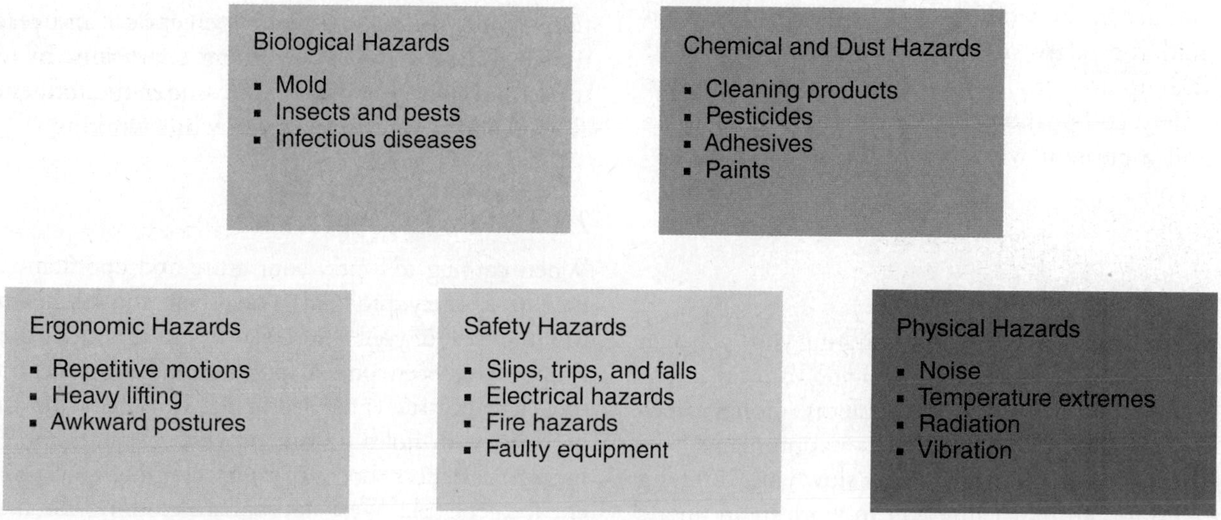

Biological Hazards
- Mold
- Insects and pests
- Infectious diseases

Chemical and Dust Hazards
- Cleaning products
- Pesticides
- Adhesives
- Paints

Ergonomic Hazards
- Repetitive motions
- Heavy lifting
- Awkward postures

Safety Hazards
- Slips, trips, and falls
- Electrical hazards
- Fire hazards
- Faulty equipment

Physical Hazards
- Noise
- Temperature extremes
- Radiation
- Vibration

Goodheart-Willcox Publisher

Figure 2-3. OSHA categorizes hazards into these broad classifications.

Safety Data Sheet (SDS) Sections	
Section	Description
Section 1. Identification	Includes a product identifier, contact information for the manufacturer or distributor, recommended use, and restrictions on use.
Section 2. Hazard(s) identification	Lists all hazards regarding the chemical and required label elements.
Section 3. Composition/information on ingredients	Lists information on chemical ingredients and trade secret claims.
Section 4. First-aid measures	Describes potential symptoms or effects and required treatment.
Section 5. Fire-fighting measures	Describes potential fire hazards and lists suitable extinguishing techniques and equipment.
Section 6. Accidental release measures	Lists emergency procedures, required protective equipment, and proper methods of containment and cleanup.
Section 7. Handling and storage	Lists precautions for safe handling and storage, including incompatibilities.
Section 8. Exposure controls/personal protection	Specifies exposure limits, including OSHA's Permissible Exposure Limits (PELs), ACGIH Threshold Limit Values (TLVs), and any other exposure limits recommended by the manufacturer, importer, or employer. Also includes recommended engineering controls and personal protective equipment (PPE).
Section 9. Physical and chemical properties	Specifies the chemical's characteristics.
Section 10. Stability and reactivity	Lists the stability of the chemical and the possibility for hazardous reactions.
Section 11. Toxicological information	Describes the toxicity of the chemical, including routes of exposure, symptoms of toxicity, acute and chronic effects of toxicity, and numerical measures of toxicity.
Section 12. Ecological information	Lists potential ecological impacts of the chemical.
Section 13. Disposal considerations	Explains how to dispose of the chemical safely.
Section 14. Transport information	Describes transportation requirements and safety considerations.
Section 15. Regulatory information	Lists regulations pertaining to the chemical.
Section 16. Other information	Lists the date of preparation or last revision of the SDS.

Adapted from OSHA

Figure 2-4. All safety data sheets (SDSs) contain these 16 sections of information.

2.2 PROFESSIONAL CONDUCT AND SAFETY

Your conduct in the workplace can directly affect your safety and the safety of your coworkers. All workers must live up to certain professional expectations to ensure safety, compliance with the law and an employer's rules, and a pleasant work environment for them and their coworkers.

2.2.1 Keep Your Mind on the Job

Everyone has personal problems. Bringing your personal problems to work keeps you from placing your full attention on your job. Relationship problems, money problems, legal problems, and other personal problems have no place at work. Lack of sleep may slow your thinking and reactions, and may cause you to work in an unsafe manner. If you work second or third shift, it may be difficult to acclimate your sleeping pattern to the hours you must work. Try to stick to a routine.

2.2.2 Drugs and Alcohol

Illegal and prescription drugs may impair your ability to focus your attention. Even over-the-counter medications could compromise your perception and reaction time, creating an unsafe working condition. In regard to alcohol, never arrive to work under the influence of alcohol and never drink alcohol while working.

2.2.3 Attire and Appearance

When coming to work, your attire and appearance may present a safety problem. Long hair should be tucked into the back of your shirt collar or put up out of the way. People have been pulled into running machinery when their hair became entangled in the workings. Short pants or pants with holes do not provide adequate protection in some instances. Loose or baggy clothing may also present a safety risk. Wear appropriate clothing for the job, **Figure 2-5.** Wearing jewelry in the workplace, especially around electrical equipment or rotating machinery, can create a safety hazard. Gold is an excellent conductor of

Phuangphech/Shutterstock.com

Figure 2-5. Choose appropriate work attire and consider the safety implications of your appearance.

electricity. Rings can get caught and present the risk of losing a finger. Chains and necklaces can get caught and pull you into machinery. Leave the jewelry in your car or at home.

TECH TIP
Work Clothes

Your grooming and dress should reflect appropriate safety practices for your job. Save the fashion statements for after work.

Accuform

Figure 2-6. Use appropriate signs and barricades to block work areas.

2.2.4 Shortcutting Safety

You cannot shortcut safety procedures. You may be tempted to take a shortcut around a safety procedure when performing a maintenance operation you have done many times before, or if you are facing pressure from a supervisor to quickly resolve a problem. However, it is only a matter of time before such safety shortcuts lead to far worse problems. Hasty work can lead to mistakes, which will cause delays and could harm workers and equipment. When you are feeling pressured to get something done quickly, take a deep breath and think. There is always time to follow safety procedures, and your supervisor does not want you to risk injury or damage to property.

2.3 SECURING THE WORK AREA

Before beginning work, make certain everyone in the area is aware of what is about to take place. Put up any necessary barricades, ropes, and placards before commencing work,

Figure 2-6. Perform lockout/tagout (LOTO) procedures to ensure your own safety and the safety of others during the repair process. Make sure the proper type of fire extinguisher is on hand when welding, grinding, or working with or near flammable materials.

2.3.1 Lockout/Tagout (LOTO) Procedures

Lockout/tagout (LOTO) procedures are used to isolate sources of energy from a piece of equipment while maintenance and repair operations are being performed. The lock prevents others from reenergizing equipment while you are working on it. The tag notifies others that the equipment has been taken out of service, who did it, and when it was done.

Before working on a machine, use LOTO procedures to lock out electrical energy, mechanical energy, and fluid energy. See **Figure 2-7.** You must ensure your system is at a *zero energy state*, which means it is safe from the possibility of becoming reenergized or experiencing a release

Master Lock

Figure 2-7. The equipment in this kit is used for performing LOTO procedures.

of contained internal energy from any source. This can be accomplished by isolating the system and checking that circuits are de-energized.

An example of electrical potential energy is a capacitor storing a charge. Even though its source of power has been locked out, it still has the potential to cause current flow given the correct circumstances. Another example of potential energy is a battery.

Pressurized fluids also exhibit mechanical potential energy. The source of pressure may be locked out, but pressure may remain. Examples of pressurized fluids include air, gases, steam, and hydraulic oil. You must exercise extreme caution to ensure the pressure has been safely released, even when the source of the pressure has been locked out.

The lock used should be of a color that is easily distinguished based on its purpose, **Figure 2-8**. It is best if the lock has only one key, which is retained when the lock is unlocked. Once the lock has been locked, the key should go in your pocket. Do not leave it near the lock and risk another person unlocking it and restoring power while you are working on the equipment.

Master Lock

Figure 2-8. LOTO locks should be easily distinguished by color.

There are several devices available to lock out cord-ended plugs. They usually involve a cover or prong lock, which prevents the device from being plugged into a power outlet. Various types of lockout devices are available for 208-volt, 3-phase plugs as well as conventional 120-volt, single-phase plugs. Refer to **Figure 2-9** for an example.

Figure 2-9. This LOTO device is used to lock out a cord-ended plug.

Valves may be locked in an open or closed position, **Figure 2-10**. Large handwheel valves may require a LOTO lock and chain in order to be secured. Some ball valves have a hole to which a LOTO lock may be applied.

If two or more people are working on the same system, a multilock hasp should be installed to enable each member of the maintenance team to install his or her own LOTO lock, **Figure 2-11**. This prevents any one person from removing his or her lock and restarting the system while others are still working on it. With a multilock hasp, all members of the maintenance team must remove their locks before the system may be restarted.

Shift changes pose a potential problem for LOTO. There must be an orderly transition from one shift to the next. You should not leave your lock on the system after your shift is over. During the handover, the new shift members apply their LOTO locks and the leaving shift members remove theirs. Do not remove your lock before the new shift members install theirs. Additionally, the leaving shift members should brief the new shift members as to the status of the repair.

2.3.2 Blocking

At times, LOTO is not sufficient to render a system safe to work on. Potential energy, such as the weight of a press ram or the weight of a robotic arm, presents a hazard even after the system has been locked out, requiring *blocking*. Blocking is the use of physical barriers to prevent a machine or part from moving unexpectedly.

Presses must be blocked and locked before working on the tooling. The motor, brake, clutch, and crank work together with the inertia of the flywheel to cycle the press. If the electrical system of the press has been shut off, any

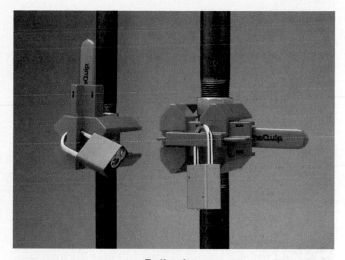

Ball valve

Gate valve

Figure 2-10. Fluid energy can be controlled with valve lockouts.

Figure 2-11. A multilock hasp allows several members of a maintenance team to install their own LOTO locks.

slipping or disengaging of the brake may cause the ram to fall. Steel blocks are located on the perimeter of the press, normally at each of the four corners of the ram. These blocks are attached with a chain so that they may not be carried away. When the blocks are placed between the ram and the bed of the press, they will prevent the ram from falling in the event of a brake release or slippage.

The same criteria hold true when working on an industrial robot. The arm components of an industrial robot are quite heavy. The servo motors driving the various joints of the robot have an integral brake. Normally, the brake is engaged when there is no power to the brake assembly. In order for the brake to be released, power to the brake must be applied. In instances where the motor is removed, the brake is also removed and the arm may fall. Failure to perform LOTO on a robot before starting maintenance work allows the brake to be disengaged from the teach pendant, which could cause injury.

TECH TIP
Machine Repair

When you are servicing a machine, you may need to remove a physical guard to effect the repair. Make sure you perform LOTO before removing such guards, and reinstall them before returning the machinery over to the operator after the repair has been completed.

2.3.3 Pinch Points

Any point where two mechanical parts come together is considered a *pinch point*. A *rotating pinch point*, also sometimes called an in-running nip point, is where two gears, a belt and sheave, or two rollers come together. A *pressing pinch point* is where two parts come near each other or together, such as the tooling in a press. Always keep your appendages and clothing well away from any pinch points.

Physical guards are often used to prevent an individual from coming too close to a pinch point. Other barriers, such as a light barrier, are used to stop machinery when the operator comes too close to a potential pinch-point hazard.

2.4 PERSONAL PROTECTIVE EQUIPMENT (PPE)

Depending on the job being performed, different *personal protective equipment (PPE)* is required. Your employer will advise you as to which type of PPE is required for various job functions.

2.4.1 Eye Protection

Generally, safety glasses are required PPE for most work situations. Eye protection is essential to prevent injury to your vision. In some cases a full face shield may be required, such as for grinding, operations involving chemical splatter, or electrical arc flash hazards. Welding operations require a welding hood to protect you from the high-intensity arc rays produced by the welding process. When using a plasma cutter or cutting torch, appropriate goggles or glasses should be worn to protect your eyes from the high-intensity light rays of the operation. Refer to **Figure 2-12** to see some examples of eye protection.

TECH TIP
Proper Use of PPE

Safety glasses do no good if worn on top of your head or baseball cap as a fashion accessory! Be sure to protect yourself and your vision by always using the proper safety equipment in the proper manner.

MSA Safety Inc. (top left and right); Reggie Lavoie/Shutterstock.com

Figure 2-12. Eye protection is required at all times when working. Different tasks require different types of protection.

2.4.2 Hard Hats

Hard hats help prevent injury from falling objects or if you accidentally bump your head. Hard hats designed for electrical work are constructed of high-impact plastic that provides insulation from electrical shock hazards. Some circumstances require wearing an insulated hard hat.

A suspension inside the hard hat allows it to adjust to the size of your head and holds the outer shell a safe distance away. Thus the hard hat, not your head, absorbs the shock of an impact.

Inspect a hard hat before use. Check for cracks or chips, and do not use it if it has been compromised in any way. Also check the suspension, and be sure it is properly adjusted to fit your head. In cases where the hard hat might fall off or get knocked off, a chin strap is provided.

The hard hat should be worn with the bill facing forward. This offers protection to your forehead as well. Refer to **Figure 2-13**.

2.4.3 Hearing Protection

Hearing protection, **Figure 2-14**, is essential in high-noise environments. On many factory floors, the noise level is so high that prolonged exposure will damage your hearing. Wearing earplugs or earmuffs will reduce the noise level to your ears and prevent hearing loss.

When using earplugs, roll the earplug between your fingers to compress it before inserting it in your ear. Make sure your fingers are clean when handling earplugs. If your hands are dirty, you will be putting dirt into your ear canals each time you insert them. Dirt in the ear canal may lead to an infection, which may also damage your hearing.

Earplugs

Earmuffs

MSA Safety Inc.

Figure 2-14. Hearing protection devices are required in high-noise environments.

MSA Safety Inc.

Figure 2-13. Always wear your hard hat level on your head with the bill facing forward. Never wear your hard hat backward or tilted back, forward, or to a side.

2.4.4 Safety Footwear

Safety shoes or steel-toed work boots are intended to protect your feet. For welding, high-top boots should be worn so your pants cover the tops of the boots. This will prevent hot sparks and molten metal from falling inside your boots and burning you.

Some companies require static dissipative footwear. Situations that require this might be work in an environment where a static discharge could cause an explosion, or work in electronic manufacturing where a static discharge could damage sensitive electronic components.

2.4.5 Protective Outerwear

Outerwear such as shop jackets, aprons, welding jackets, jumpsuits, and arc flash suits are used to protect from hazards you might encounter on the job. The appropriate type of outerwear for the job will put a barrier between you and the hazard. Some outerwear will protect you from contact with dangerous chemicals or other hazards. Other protective outerwear is useful when temperatures in the work environment are extremely high or extremely low.

In machining, the correct type of outerwear will protect you from flying metal chips that may be hot. When welding, a welding jacket or leather apron will protect you from hot sparks and bits of molten metal generated by the welding process. In both welding and machining, outerwear in the form of a jacket should not be tucked into your pants. On occasion, a hot chip or bit of molten metal may enter through the collar, and you certainly do not want it to become trapped inside and held against your clothing or skin.

Protective outerwear may be required when performing certain functions related to your job. Your employer will advise you as to what protective outerwear is required.

2.4.6 Gloves

Gloves protect your hands from hazards encountered in the workplace. Many varieties of gloves are available for specific uses. Be sure you have the right gloves for the materials you will be handling.

Latex or nitrile gloves serve to protect your hands from chemicals, grease, and oil. Constant contact with many oils, such as transformer oil, will cause dermatitis (inflammation of the skin). Trying to wash off oil will also wash off the natural oils in your skin, giving you extremely dry skin. Wear gloves when working with oils in order to prevent these problems.

Leather gloves can be used during welding operations or when handling steel or other materials with sharp edges. In the case of welding gloves, never pick up hot metal with your gloves. You can still be burned, and you will ruin a perfectly good pair of welding gloves. Use a pair of pliers to handle small pieces or wait until larger pieces cool before handling them. Welding gloves are made from various types of leather, so also make sure that your welding gloves can withstand the temperatures and type of welding you will be dealing with.

Rubber electrical insulating gloves with leather protector gloves worn over them provide some protection from electric shocks when you must work with the power on. Rubber electrical insulating gloves must be air tested before each use in order to check for holes. You can air test a rubber glove by rolling it up, beginning with the

cuff side, and trapping air inside. As you roll the glove, you compress the air. If the glove has a leak, you will easily be able to roll the glove completely as the air escapes through any holes.

Rubber insulating gloves come in ratings of 00, 0, 1, 2, 3, and 4, from lower (00 and 0 rating) to higher (4 rating) maximum voltage ratings. Be sure you have the proper amount of insulating protection before beginning any electrical work. Refer to **Figure 2-15** to see the class specifications and proper lengths for electrical insulating rubber gloves.

ASTM D120 Class Specifications for Insulating Rubber Gloves		
Glove Class	Proof Test Voltage	Label Color
Class 00	2500 AC/10,000 DC	Beige
Class 0	5000 AC/20,000 DC	Red
Class 1	10,000 AC/40,000 DC	White
Class 2	20,000 AC/50,000 DC	Yellow
Class 3	30,000 AC/60,000 DC	Green
Class 4	40,000 AC/70,000 DC	Orange

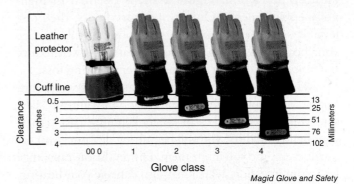

Magid Glove and Safety

Figure 2-15. These charts show the class specifications and proper length for electrical insulating rubber gloves. The insulating glove must be a certain length longer than the protector glove, based on the class of electrical protection.

2.5 SAFE LIFTING AND FALL PROTECTION

Many workplace injuries occur when employees fail to lift things in the proper manner. Proper planning and lifting technique can minimize this risk. Falling while on the job presents another workplace safety hazard. Using the proper safety equipment and taking necessary precautions can help prevent falls at the workplace.

2.5.1 Safe Lifting

When lifting heavy objects, use your leg muscles, not your back. Bend at the knees and keep your back straight while lifting. Bending over at the waist to lift puts undue strain on your back muscles and may result in a back injury. PPE is available for lifting if your job requires it. A lifting belt that supports the back and helps to keep it straight during a lift will help to prevent an injury. Follow the same procedure in reverse, using your leg muscles, when putting a heavy object back down.

If an item is too heavy to lift yourself, team lift the item with someone else. Forklifts, chain hoists, overhead cranes, and dollies are designed to lift and transport heavy objects.

TECH TIP
Heavy Lifting

Do not attempt heavy-lifting work better done by a machine just because you are in a hurry to get the job done—you are more likely to injure yourself and cause further delays in the end.

2.5.2 Using a Ladder

If used improperly, ladders can present a significant safety risk. The main types of ladder you are likely to encounter in the workplace are extension ladders and stepladders, **Figure 2-16**.

Before using any ladder, inspect it for damage. Do not use a damaged ladder. There should be a label on the ladder with manufacturer's instructions. Follow those instructions, and make sure the ladder can sustain the intended load weight before using it.

Ladders are usually constructed of wood, aluminum, or fiberglass. Wood ladders and aluminum ladders are about equally durable, but aluminum ladders are considerably lighter. Fiberglass ladders are a bit heavier than aluminum ladders, but they offer the insulating quality required for performing electrical work. If you are working on electrical systems, do not use an aluminum ladder.

Extension ladders are straight ladders with two sections, allowing for adjustable lengths. Check extension ladders for function of the latching mechanisms. Make certain the springs on each latching mechanism and the other parts are functional. Ensure that the extension ladder pull rope and pulley are in good operating condition. If the pull rope or latching mechanism is not in good order, the struggle in erecting the ladder might allow it to fall.

With an extension ladder, make certain the bottom of the ladder is far enough away from where the top of the ladder is supported. One-quarter of the ladder's length away from the wall is the rule of thumb. If the bottom

Werner Ladders

Figure 2-16. You are likely to use extension ladders and stepladders as a technician performing maintenance in an industrial setting.

is too close to where the top is supported, you have an increased risk of falling or of the ladder falling sideways when you reach the top. Make sure the base of the ladder is secure on a level surface and the side rails are evenly supported. Erect a barrier and block any nearby doors to ensure no coworkers accidentally bump the ladder.

When using a stepladder, make certain that it is fully opened and locked and that all legs are on a level surface. Any rocking may result in the ladder toppling. Use an appropriate method to level all legs if you must set up a ladder on unlevel ground. Never climb with your feet higher than the second step from the top.

Regardless of which type of ladder you use, be careful not to climb too close to the top. If you cannot reach whatever it is that you need to work on, use a taller ladder. Also, do not lean out to the side. If necessary, move the ladder so that you are always working over the centerline of the ladder. It is also a good idea—especially with an extension ladder—to have a coworker at the bottom of the ladder to help stabilize it while you are using it.

2.5.3 Fall Protection

When using ladders or working at even higher heights, such as on lifts or scaffolding, you must be mindful of protecting yourself against a fall. There are many options for fall protection that may be used, depending on your specific task:

- **Guardrail systems.** Guardrails can be used when walking and working at heights or to keep employees from falling into holes, **Figure 2-17**.
- **Safety net systems.** Make sure nets are checked for wear and are rated properly for safety.
- **Handrails or stair rail systems.** These will likely be used with raised platforms.
- **Personal fall protection systems.** These may include body belts, harnesses, and other components of personal fall arrest systems, work positioning systems, or travel restraint systems, **Figure 2-18**.
- **Lifelines and grab handles.** Lifelines are flexible lines that stop a hooked-in worker from falling too far. Grab handles provide a solid handle to hold onto when working at heights.

2.6 CONFINED SPACES

Confined spaces present a unique set of hazards for maintenance technicians. While inside such spaces, you can become trapped or injured by falling objects or structural collapses. Or, the atmosphere inside a confined space

MSA Safety Inc.

Figure 2-17. Guardrail systems can be used as a barrier between workers and a fall hazard.

MSA Safety Inc.

Figure 2-18. This worker is using a personal fall protection system, including a harness, lanyards, and connectors.

could be compromised by gas, hazardous chemical leaks, or fire. OSHA designates certain, more dangerous confined spaces as permit-required confined spaces, or permit spaces. Confined spaces OSHA designates as permit spaces may have any of the following hazards:

- A hazardous atmosphere or the potential for one.

- Material such as water or grain that could engulf a worker.

- Structural features that could trap a worker.

- Any other recognized hazard, including unguarded machines, the potential for open heat or flame, or other fall hazards.

Before working in a confined space, understand the potential risks and take proper safety precautions. Special PPE is available for confined spaces and could save your life, **Figure 2-19**.

2.7 ELECTRICAL SAFETY

Certain aspects of electrical safety have already been addressed in this chapter, such as the use of LOTO, flash suits, insulating gloves, and fiberglass ladders. Following chapters will address other specific safety practices. The most important part of electrical safety is keeping yourself from becoming a conductor in an electrical circuit. Keep one hand in your pocket while working on live circuits. This will lessen the chance that both hands can create a complete circuit through your heart, which could cause fibrillation and possible death.

It takes only a few milliamps of current to cause fibrillation of your heart. During fibrillation your heart loses its natural rhythm and its ability to pump blood through your body. Defibrillation is required to restart the heart in such cases. An ***AED (automatic electrical defibrillator)*** device, when used by a trained person,

MSA Safety Inc.

Figure 2-19. Various types of PPE are required for work in different kinds of confined spaces.

may be used to restart the heart and save an individual's life. The AED shocks the victim's heart back into a normal rhythm, allowing it to once again begin pumping blood. Some working conditions may require you to have AED certification and additional safety certifications, such as CPR, **Figure 2-20**.

Becoming a conductor through any body part can also result in fatal burns. Be sure to approach any energy source carefully, keeping the recommended distance based on the risk of an arc flash incident. In an arc flash, an electric current leaves its path and travels through the air from one conductor to another, often with violent results. Becoming the conductor of an arc flash incident can result in serious burns and death.

Whenever possible, turn the power off and check to be certain the circuit is de-energized before attempting repairs. Use a *"live-dead-live" (LDL) test* to verify the circuit is de-energized before proceeding with your work. First, use your voltmeter to measure a known live circuit. Next, use the same voltmeter to measure the circuit you wish to verify as dead. After that, once again measure the known live circuit. This method ensures that your meter is actually measuring correctly. Most equipment is under automatic control. Just because you measured a circuit and verified it is dead does not mean the control system will not restart it automatically. Always perform LOTO and a live-dead-live test to ensure that all circuits are dead before attempting repairs.

In some circumstances you will be required to work on a live circuit. You should not be scared of electricity, but you should have a healthy respect for any circuit that has even a remote possibility of being energized. If you must work on a live circuit, make certain that you have taken all safety precautions and use only insulated tools. Your company may require that you also wear arc flash PPE in certain circumstances, **Figure 2-21**.

National Safety Apparel

Figure 2-21. Arc flash PPE includes protective clothing, gloves, and hood.

Any practice that keeps you from accidentally becoming part of an electrical circuit would be a part of proper electrical safety procedures.

2.7.1 Grounding

Any discussion of electricity includes *grounding*. Grounding is the intentional electrical connection of items that could become a conductor, such as a metal frame, to earth (or ground). You will learn more about circuits and conductors in later chapters.

Grounding makes it impossible for there to be any current flow through an individual who touches both the frame or chassis and ground, as they are both at the same potential. In the case of an electric motor, for example, if you have the motor frame adequately grounded, electrical malfunctions may cause a fault, which blows fuses, but the motor will not present an electrocution hazard to anyone touching it, **Figure 2-22**.

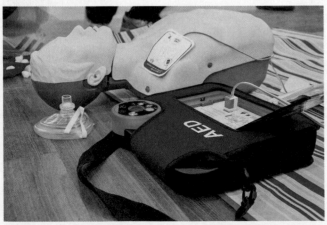

narin phapnam/Shutterstock.com

Figure 2-20. AEDs and test dummies are often used for CPR training, which your work environment may require.

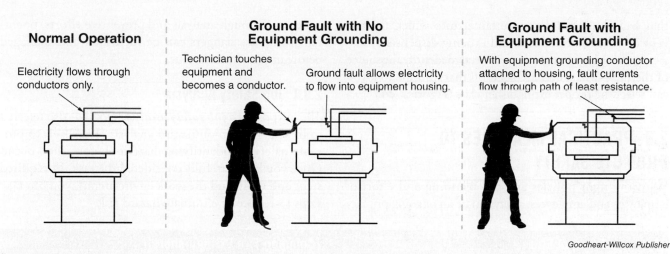

Normal Operation

Electricity flows through conductors only.

Ground Fault with No Equipment Grounding

Technician touches equipment and becomes a conductor.

Ground fault allows electricity to flow into equipment housing.

Ground Fault with Equipment Grounding

With equipment grounding conductor attached to housing, fault currents flow through path of least resistance.

Goodheart-Willcox Publisher

Figure 2-22. An electric motor is grounded to prevent an electrocution hazard.

2.7.2 Ground-Fault Circuit Interrupter (GFCI)

A device that protects individuals from the hazards of ground faults, as previously described, is a ***ground-fault circuit interrupter (GFCI)***. This device measures the current flowing in the hot (line) conductor of a circuit and compares it with the current flowing in the neutral (return) conductor. If there is a disparity where the neutral current is less than the current in the hot conductor, even by a few milliamps, the GFCI interrupts the circuit. GFCIs are designed to trip at a level that is below what is required to electrocute a human.

GFCIs are available in the form of a circuit breaker and also as an electrical outlet receptacle. You might also find one integrated into the plug of a blow-dryer to protect the user from electrocution hazards. Portable GFCIs may be used for temporary service. You will most likely find them on construction sites. The use of GFCIs

has saved countless lives from electrocution. Refer to **Figure 2-23** for an example of a GFCI circuit breaker and outlet receptacle.

2.8 FLAMMABLE MATERIALS

Flammable materials should be stored in an approved cabinet. Such a cabinet has a vent connection and, most importantly, a ground connection. The purpose of grounding in this case is to prevent the buildup of a static charge. A static discharge may cause a spark, which in turn could ignite the flammable material and cause a fire or explosion.

Containers used for dispensing flammable liquids, such as solvents or oils, should also be effectively grounded for the same reason. A good example of this type of container would be a 55-gallon solvent drum. Additionally, there should be a ground lead with a clamp

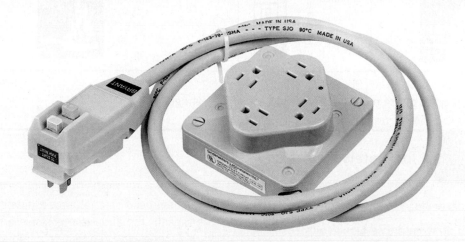

Hubbell Inc.

Figure 2-23. A GFCI circuit breaker and receptacle outlet can protect an individual from electrocution.

that will also ground the container into which the solvent or other flammable liquid is being dispensed.

In the event of a fire, you will need to identify the source of the fire—what fuels it? Based on that information, you can select the proper extinguisher class, **Figure 2-24**.

2.9 PRACTICAL MEASURES TO PROMOTE SAFETY

With the right policies and the planning and effort of employers and employees, safety can be an ongoing primary concern. Through analysis and preventive efforts, potential workplace dangers can be identified and managed before any accidents occur.

2.9.1 Job Safety Analysis

The purpose of a *job safety analysis (JSA)*—also referred to by OSHA as a job hazard analysis (JHA)—is to promote safety by identifying hazards before they occur. Once you have carefully considered the task, its required steps and tools, and the work environment, you can take steps to reduce or eliminate hazard risk.

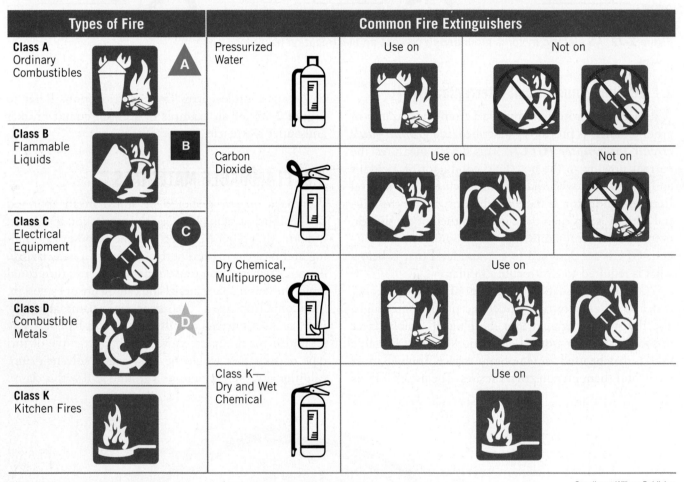

Goodheart-Willcox Publisher

Figure 2-24. Different types of fire require different classes of fire extinguisher.

PROCEDURE   Job Safety Analysis (JSA)

There are three basic steps to performing a JSA:

1. Break the job down into a sequence of steps.

2. Identify potential hazards associated with each step.

3. Recommend safe job procedures corresponding to each step and hazard. These procedures will be actions you can take to eliminate or reduce hazards and risks, **Figure 2-25**.

Job Safety Analysis Form

Job To Be Performed:

Tripping Pipe in Hole

Required PPE:

Hard hat, steel-toe boots, safety glasses

Step No.	Sequence of Basic Job Steps	Potential Hazards	Recommended Safe Job Procedures
1	Traveling block moving up derrick	Swinging blocks hitting sides of derrick. Tong counterweight line getting hooked on blocks or elevators.	Stabilize blocks and elevators. Do not put tongs on pipe too soon. Look up and live!
2	Put makeup tongs on and wrap spinning chain	Pinch points when latching tongs to pipe.	Keep hands and fingers on designated handles. Keep good tail on spinning chain. Keep control of chain.
3	Latching pipe into elevators	Pinch points of elevators and pipe. Dropping stand across derrick. Swinging pipe.	Derrickman should tail out pipe and stabilize stand after pickup. Floormen watch for snag or short stand.
4	Stabbing pipe	Slipping while tailing pipe. Pinch points of pipe and tongs. Missing box.	Get firm hold. Give driller clear view. Place hands and legs properly.
5	Throwing chain, torquing pipe, unlatching tongs	Chain breaking, stuck by chain, pinch points—getting hand or fingers in chain. Tongs slipping.	Make sure tongs are latched properly. Hold tongs out of way after unlatching. Stay clear of chain and out of swing of tongs.
6	Pulling slips	Strains	Proper lifting techniques. Lift together. Use moving pipe as leverage.
7	Lowering pipe	Hitting bridge, line parting brake, or hydromatic failure	Lower pipe at controlled speed. Watch weight indicator.
8	Set slips and unlatch elevators	Pinch points at slip handles, elevator links, and elevator latch.	Slow down pipe and set slips. All hands should work together. Proper lifting and hand placement.

Adapted from OSHA

Figure 2-25. Follow a job safety analysis sheet to eliminate or reduce the hazards and risks of a job.

2.9.2 6S Program

Many companies have implemented a 5S program to increase efficiency, reduce costs, and improve quality. A *6S program* adds a sixth S that stands for "Safety." The six S's in 6S stand for "Sort, Set to Order, Shine, Standardize, Sustain, and Safety." You might see other variations on the individual steps, but the overall message is the same—create and maintain a clean, orderly workplace to promote safety and efficiency. Any work environment can benefit from a 6S program. **Figure 2-26** shows a variation of a 6S program.

CHAPTER WRAP-UP

Safety is extremely important in your job, and it should always be the first priority. Shortcutting only leads to dangerous situations that cause injury, damage to machines and property, and even death. Always use the proper PPE for the task at hand. As a maintenance professional, you are responsible for your own safety and the safety of those around you. Make sure you are properly trained and certified, and always follow the safety guidelines as directed by OSHA and your employer.

6S Program	
Sort	Discard unnecessary items from the workplace and properly store items not in use.
Set to Order	Arrange tools, supplies, and equipment logically, in a way that best supports how they are used by workers. (Also sometimes called Straighten or Set in Order.)
Shine	Make sure the workplace is clean and uncluttered, and that items are working properly. (Also sometimes called Sweep.)
Standardize	Maintain a disciplined effort to continually pursue the first three S's.
Sustain	Be sure to follow these steps every day and create a workplace culture that supports them.
Safety	Eliminate all hazards and make safety a priority every day, in all operations, to ensure there are no accidents or injuries.

Goodheart-Willcox Publisher

Figure 2-26. A 6S program outline promotes safety and efficiency in a workplace.

Chapter Review

SUMMARY

- OSHA (Occupational Safety and Health Administration) is tasked with regulation and enforcement to promote a safe working environment for employees.

- OSHA provides safety standards and offers training and safety certifications.

- It is always important to keep your mind on the job. Personal problems and other factors interfere with your ability to concentrate on safe practices and the task at hand.

- Drugs and alcohol, and even some over-the-counter medications, may slow your reaction time and could result in a lapse of concentration, creating an unsafe situation.

- LOTO (lockout/tagout) does not just apply to electrical energy but to all sources of energy, such as mechanical energy and fluid energy.

- LOTO is required to prevent the injury or death of both the technician and other workers.

- LOTO locks are to be removed only by the individuals who installed them.

- Blocking is necessary to prevent a machine or part from moving unexpectedly and injuring technicians while they are working or have a body part in harm's way.

- Keep appendages and clothing well away from machine pinch points, which are points where two mechanical parts come together.

- You need to understand the various types of personal protective equipment (PPE), what jobs require PPE, and how each type of PPE is used.

- Hard hats should be worn when working under electrical hazards or items with the potential to fall.

- Hearing protection is required to prevent hearing loss when work must be performed in high-noise environments.

- Electrical insulating gloves prevent a technician from being electrocuted from accidentally touching an energized conductor.

- In order to prevent injury, lift with your legs and keep your back straight during the lift. Never bend over and lift with your back.

- Inspect all ladders before use and ensure they are properly set up to prevent falls and injury. Follow the manufacturer's instructions and never use a ladder that is not in good operating condition.

- Proper equipment grounding prevents the equipment chassis or frame from becoming energized and creating a ground-fault hazard.

- A GFCI (ground-fault circuit interrupter) will measure the difference in current between the line and neutral conductors. If the current in the neutral conductor is less than the current in the line conductor, the GFCI will interrupt the circuit.

- Flammable-materials safety cabinets must be effectively grounded in order to offer protection from a static discharge that could ignite the materials within.

- When transferring flammable liquids, both the supply container and the receiving container must be effectively grounded to prevent ignition by electrostatic discharge.

REVIEW QUESTIONS

Answer the following questions using the information provided in this chapter.

1. Explain the purpose of OSHA.

2. _____ provide essential information about substances or mixtures used in workplace chemical management.

3. *True or False?* Over-the-counter medications can impair your ability to safely and effectively complete tasks at work.

4. *True or False?* It is okay to shortcut safety procedures if your supervisor wants you to resolve a problem quickly.

5. Explain the functions of the lock and the tag of a LOTO procedure.

6. The three types of energy that may be locked out using LOTO procedures are electrical, mechanical, and _____.

7. When more than one person is working on the same system, a _____ hasp should be installed to enable each member of the team to install his or her own LOTO lock.

8. Explain what blocking is and when you need to use it.

9. A(n) _____ pinch point is where two parts come near each other or together, such as the tooling in a press.

10. *True or False?* It is important to keep your appendages and clothing away from any pinch points.

11. List five items of PPE and their uses.

12. *True or False?* Use welding gloves to pick up hot metal.

13. The main types of ladder you are likely to encounter in the workplace are the _____ ladder and the stepladder.

14. *True or False?* Permit-required confined spaces do *not* have the potential to cause death or serious harm.

15. When used by a trained person, a(n) _____ device may be used to restart the heart and save an individual's life.

16. A(n) _____ is designed to trip at a level that is below what is required to electrocute a human.

17. Define grounding and explain why it is important.

18. Explain why a flammable-materials safety cabinet should be grounded.

19. Containers used for dispensing flammable liquids should be _____.

20. List the three basic steps in performing a job safety analysis (JSA).

NIMS CREDENTIALING PREPARATION QUESTIONS

The following questions will help you prepare for the NIMS Industrial Technology Maintenance Level 1 Maintenance Operations credentialing exam.

1. What branch of OSHA conducts research and experimental programs to develop criteria for safety and health standards?

    A. Environmental Protection Agency (EPA)
    B. Job Safety Analysis (JSA)
    C. National Institute for Occupational Safety and Health (NIOSH)
    D. National Institute for Metalworking Skills (NIMS)

2. Which of the following must be provided by chemical manufacturers, distributors, and importers to anyone who will work with potentially dangerous materials?

    A. LOTO
    B. AED
    C. GFCI
    D. SDS

3. A LOTO procedure should be performed to lock out energy when working on which of the following systems?

    A. Electrical
    B. Mechanical
    C. Fluid power
    D. All of the above

4. Which of these personal protective equipment items prevents injury from falling objects?

    A. Leather gloves
    B. Tyvek jumpsuit
    C. Harness
    D. Hard hat

5. Which of the following is *not* an example of equipment used for fall protection?

    A. Arc flash suit
    B. Handrail
    C. Guardrail
    D. Safety net

6. What rating must a fire extinguisher have to be used for a fire involving live electrical equipment?

    A. Class A
    B. Class B
    C. Class C
    D. Class K

3 | Maintenance Principles and Record Keeping

CHAPTER OUTLINE

LEARNING OBJECTIVES

After completing this chapter, you will be able to:

- ☐ Check machine safety and operation.
- ☐ Monitor a machine and document performance.
- ☐ Define standard operating procedures.
- ☐ Identify examples of reactive maintenance and proactive maintenance.
- ☐ Demonstrate various forms of proactive maintenance.
- ☐ Explain the importance of maintenance record keeping.

TECHNICAL TERMS

autonomous maintenance
breakdown maintenance
computerized maintenance management system (CMMS)
condition monitoring
human-machine interface (HMI)
planned maintenance
predictive maintenance
preventive maintenance
proactive maintenance
reactive maintenance
reliability-centered maintenance (RCM)
root cause analysis (RCA)
scheduled maintenance
standard operating procedure (SOP)
total productive maintenance (TPM)
work order

In a perfect world, machines would never break down. In this world, machines do inevitably break down, and technicians must then fix them. Machine breakdowns and unscheduled shutdowns cost companies money, both for the repairs and from the loss in production. With an understanding of the normal operations of machines and their uses, you can identify signs of irregularity. By monitoring operations and performing regular maintenance, you can keep things running as smoothly as possible, avoid some breakdowns, and save time and money in the long run.

◄ ndoeljindoel/Shutterstock.com

3.1 MAINTENANCE METHODS

Maintenance methods may be divided into two major categories: proactive maintenance and reactive maintenance. **Proactive maintenance** is the type of maintenance activity you perform before something goes wrong. **Reactive maintenance** is the type of maintenance activity you perform after something goes wrong.

Automotive maintenance provides an example of these two methods. If you do not check the oil or have the oil and filter changed, eventually the engine oil will break down. It will no longer adequately lubricate the engine, and the car will start burning oil, **Figure 3-1**. Your engine will run out of oil and lock up, and the repair cost will include a new engine because the old one is no longer serviceable. The simple proactive maintenance of checking and replacing a car's oil and filter is the far better option.

On the other hand, you might get a flat tire after running over a nail or a pothole. You then have to change the punctured tire and have it repaired or replaced. This is an example of reactive maintenance.

Maintenance methods may be described with many different buzzwords, as each company applies different aspects of each method and gives it a title of its own. Whatever you call these approaches to maintenance, all companies have maintenance strategies defined by management. You must understand and abide by these maintenance strategies while working within the limitations of available resources (also allocated by management).

3.2 MAINTENANCE RESOURCES AND COSTS

Maintaining a company's machines and operations can be expensive, and there are several resources and variables to consider when developing a maintenance plan, including the following:

- **Maintenance personnel.** Maintenance personnel costs include the salaries and benefits for the maintenance technicians and managers.

- **Spare parts.** Spare parts include the number of spare parts either in inventory or on order, **Figure 3-2**. Many manufacturers have a recommended spare parts list, which is often divided into an A list and a B list. The A list contains all the parts that are most commonly replaced. The B list includes those parts that are critical but not replaced as often as the A list parts.

- **Outside services (as needed).** Outside services are defined as any services provided by a third party. Outside services are needed when the company does not have the resources or expertise to perform a particular operation. Outside services may include off-site repair services or on-site field service engineering.

- **Downtime required by maintenance activities.** Downtime is time when production is stopped and work cannot be accomplished. This happens, for example, if a part is on order and a machine cannot run without it. Downtime is one of the largest

Maintenance Methods

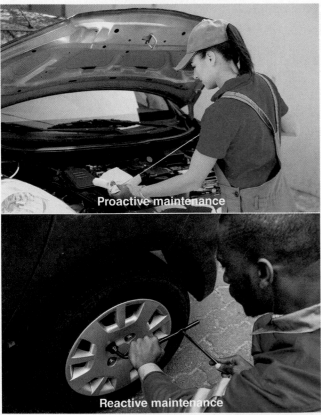

Africa Studio/Shutterstock.com; Andrey_Popov/Shutterstock.com

Figure 3-1. Most of us have experienced the proactive maintenance of checking and changing the oil in an automobile, as well as the reactive maintenance of changing a flat tire.

Syda Productions/Shutterstock.com

Figure 3-2. Spare parts inventories require careful organization to be effective.

maintenance costs. Machines are purchased to produce parts, and operators are hired to run the machines. When a breakdown happens, the machine operator is still being paid when not able to produce parts while waiting for the machine to be repaired.

- **Personnel training.** Personnel training is an upfront cost that yields increased productivity and

reduced downtime. With properly trained personnel, repair work may be completed correctly and in a shorter period of time.

- **Record keeping and analysis.** Record keeping and analysis is a system of recording machine maintenance and performance data to use for improving reliability, **Figure 3-3**. This record may be useful

Operator's Daily Checklist—Internal Combustion Engine Industrial Truck—Gas/LPG/Diesel Truck
Record of Fuel Added

Date	Jan. 19, 2019	Operator	J. McMasters
Truck#	14	Hour Meter	7:45 am
Department	Comm. delivery	Fuel	3/4 tank
Shift	1		

Safety and Operational Checks (Prior to Each Shift)
Have a **qualified** mechanic correct all problems.

Engine Off Checks	OK	Maintenance
Leaks—Fuel, Oil, or Coolant	✓	
Tires—Condition and Pressure		check front passenger tire psi
Forks—Check Condition	✓	
Load Backrest—Securely Attached	✓	
Hoses, Chains, Cables, and Stops—Check Visually	✓	
Overhead Guard—Attached	✓	
Finger Guards—Attached	✓	
Propane Tank (LP Gas Truck)—Rust Corrosion, Damage	NA	
Safety Warnings—Attached (Refer to Parts Manual for Location)	✓	
Battery—Check Level and Charge	✓	
All Engine Belts—Check Visually	✓	
Hydraulic Fluid Level—Check Level	NA	
Engine Oil Level—Dipstick	✓	
Transmission Fluid Level—Dipstick	✓	
Engine Air Cleaner—Check Dirt Trap	✓	
Radiator Coolant—Check Level	✓	
Operator's Manual—In Container	✓	
Nameplate—Attached	✓	
Seat Belt—Functioning Smoothly	✓	
Hood Latch—Adjusted and Securely Fastened	✓	
Brake Fluid—Check Level	✓	
Engine On Checks—Investigate Unusual Noises Immediately		
Accelerator—Functioning Smoothly	✓	
Service Brake—Functioning Smoothly	✓	
Parking Brake—Functioning Smoothly	✓	
Steering Operation—Functioning Smoothly	✓	
Drive Control—Forward/Reverse—Functioning Smoothly	✓	
Tilt Control—Forward and Back—Functioning Smoothly	✓	
Hoist and Lowering Control—Functioning Smoothly	NA	
Horn and Lights—Functioning	✓	
Cab—Heater, Defroster, Wipers—Functioning	✓	
Gauges—Functioning	✓	

Adapted from OSHA

Figure 3-3. Record keeping should include daily safety and operations checklists. Maintaining these records allows technicians to review data and determine the cause and duration of a malfunction.

when determining the root cause of a machine failure, needed repairs, and future maintenance plans.

- **Test equipment.** Test equipment can be considered any apparatus that quickly identifies or diagnoses equipment problems or is used to confirm that repairs have been successful, **Figure 3-4**.

- **Consumables.** Consumables are supplies that are consumed in the maintenance of equipment, such as lubricants, spare parts, safety equipment, and cleaning supplies, **Figure 3-5**.

Figure 3-4. Test equipment includes a variety of tools, including the multimeter shown here.

Figure 3-5. Consumables should be carefully cataloged and maintained. Consumables include a range of items, from cleaning supplies and lubricants to spare parts such as the filter elements shown here.

All of these variables should be considered when determining the company's maintenance strategy and defining an approach to maintenance operations.

3.3 OPERATING AND MONITORING A MACHINE

When a maintenance technician monitors a machine for proper operations it is essential to do so safely. This task starts with preplanning by reviewing the machine's safety operations procedures checklist, **Figure 3-6**, and the machine's start-up and shutdown procedures. Then, technicians should use their five human senses while the machine operator demonstrates the correct start-up, operation, and shutdown procedures. During this process, the technician is reading gauges and meters and looking and listening for loose connections or leaks. This step also enables the technician to observe that the operator knows how to properly start up, operate, and shut down the machine safely.

The technician can also observe the machine's operation in a process known as *condition monitoring*. In this process, the technician uses tools to examine such items as vibration, temperature, and voltage/current draw that help identify any symptoms of malfunctions. The technician can then take corrective actions as needed. This condition monitoring is part of a proactive maintenance program and is implemented on electric motors, pumps, presses, and other rotating equipment and machinery.

Five main techniques for condition monitoring are the following:

- Ultrasound analysis.
- Motor-current analysis.
- Thermal analysis.
- Vibrational analysis.
- Oil analysis.

Autonomous maintenance is part of a total productive maintenance or predictive approach in industrial maintenance. This activity involves the machine operator and the maintenance technician working as a team. The operator is trained to perform small routine maintenance tasks, such as general machine inspection, lubrication, and cleaning, as well as to detect abnormalities before they become failures in the machine's operation. This frees up the maintenance technician to spend more time on higher-level, value-added activities, such as technical troubleshooting and repair, thus keeping the machine in proper operating condition and ensuring the most efficient use of everyone's time.

Machine Safety Checklist

Description	Yes	No
Machine Guards		
Do the safeguards meet OSHA requirements?		
Do the safeguards prevent body parts from making contact with dangerous moving parts?		
Are the safeguards firmly secured?		
Do the safeguards prevent an object from falling into moving parts?		
Do the safeguards permit safe and comfortable operation of the machine?		
Can the machine be lubricated without removing the safeguards?		
Can the machine be shut down before safeguards are removed?		
Are starting and stopping controls within easy reach of the operator?		
Are safeguards provided for all hazardous moving parts?		
Electrical Hazards		
Is the machine properly grounded?		
Is the power supply correctly fused and protected?		
Are there any loose conduit fittings?		
Do workers ever receive shocks while operating the machine?		
Operation		
Do operators have the necessary training in how and why to use the safeguards?		
PPE		
Are operators safeguarded against noise hazards?		
Do operators have appropriate PPE for the job?		

Adapted from OSHA

Figure 3-6. Any machine maintenance task starts with preplanning by reviewing the machine's safety operations procedures checklist.

3.4 OPERATION AND MAINTENANCE DOCUMENTATION

The operation and maintenance documentation system comprises a thorough set of current maintenance records for each piece of equipment, including such items as performance specifications, safety requirements, standard operating procedures, and notes on past maintenance performed, **Figure 3-7.** A *standard operating procedure (SOP)* is an established and accepted method for performing a task. All documented maintenance records need to be accessible to the maintenance technician and operator. These documents can be in either hard copy or an electronic computer-based system such as a *computerized maintenance management system (CMMS)*. The hard copy of equipment documentation may be kept in various places, such as on the back of each corresponding machine, or in a central area, such as the supervisor's office. A digital-based system can be accessible from any computer or mobile device on the shop floor. Whether performing preventive or predictive maintenance, keeping track of equipment conditions and past and planned maintenance is critical to the success of a maintenance program.

Part of a successful maintenance program is the issuance of *work orders*. A work order is a basic document or form used for planning and controlling maintenance activities and tasks. Work orders provide information concerning the details of maintenance work for a piece of equipment, including such items as when and where the work takes place, how long the work takes, and who carries out the maintenance work. Work orders aid in the planning phase of the maintenance, **Figure 3-8.**

These records may be reviewed on a computer, on a mobile device, or physically in the central location where the records are kept. These records contain information regarding operation and safety procedures, any past maintenance performed on the machine, and general observations of the machine's operation. After examining the records, the technician plans the process of servicing the machine and then executes these plans. After completing the servicing of the machine, the technician closes out the work order by making notes as to what was done, what parts were replaced, and the amount of time spent repairing the machine. When technicians receive a work order from the CMMS for a piece of equipment, their first step in the maintenance process is to obtain the maintenance record and any other important information needed—or dictated by company policy—either electronically via computer or mobile device, or physically.

As an industrial maintenance technician in a modern manufacturing setting, you will most likely be using a *human-machine interface (HMI)* when dealing with machines. An HMI includes a device and software that allows you to interact with a machine. HMIs often incorporate digital touch displays. In companies using a CMMS, machine HMIs are usually integrated with that digital system.

Widgets Are Us
Equipment Maintenance Log Sheet

Name of the piece of equipment:	Raw Widget Maker
Label:	Widget #1
Serial number:	KXB397C855266
Manufacturer:	ACME Manufacturing Inc.
Manufacturer's contact person + contact details:	Jim Jones
Date of purchase:	1/18/2019
Date put into service:	6/1/2019
Person responsible for equipment:	John Hammerschmit
Location of equipment:	Production Floor Section C3
Physical condition:	Good
Service provider (for maintenance and calibration):	Plant Maintenance
Service provider contact person + contact details:	Plant Maintenance
Frequency of preventive maintenance:	Twice-a-year shutdowns

Date:	Description of maintenance:	Maintenance performed by:	Next maintenance planned on (date):	Remarks:
6/1/2019	Machine placed into service	Karl	12/20/2019	Machine placed into service
12/20/2019	Preventive maintenance, minor adjustments to chain drive	Karl	6/1/2020	Minor roller adjustments made to chain drive due to chain stretch
3/18/2020	Replace failed proximity sensor	David		Proximity sensor on opening handguard failed, check warranty
4/27/2020	Tighten feeder chain idler	David		Feeder chain idler was loose, causing slack in chain—tightened to torque specification
6/1/2020	Preventive maintenance per machine specifications	Karl/David	12/20/2020	Performed preventive maintenance per machine specifications

Dr. Paul Dettmann

Figure 3-7. The maintenance log includes descriptions of the equipment and completed maintenance, as well as additional maintenance notes.

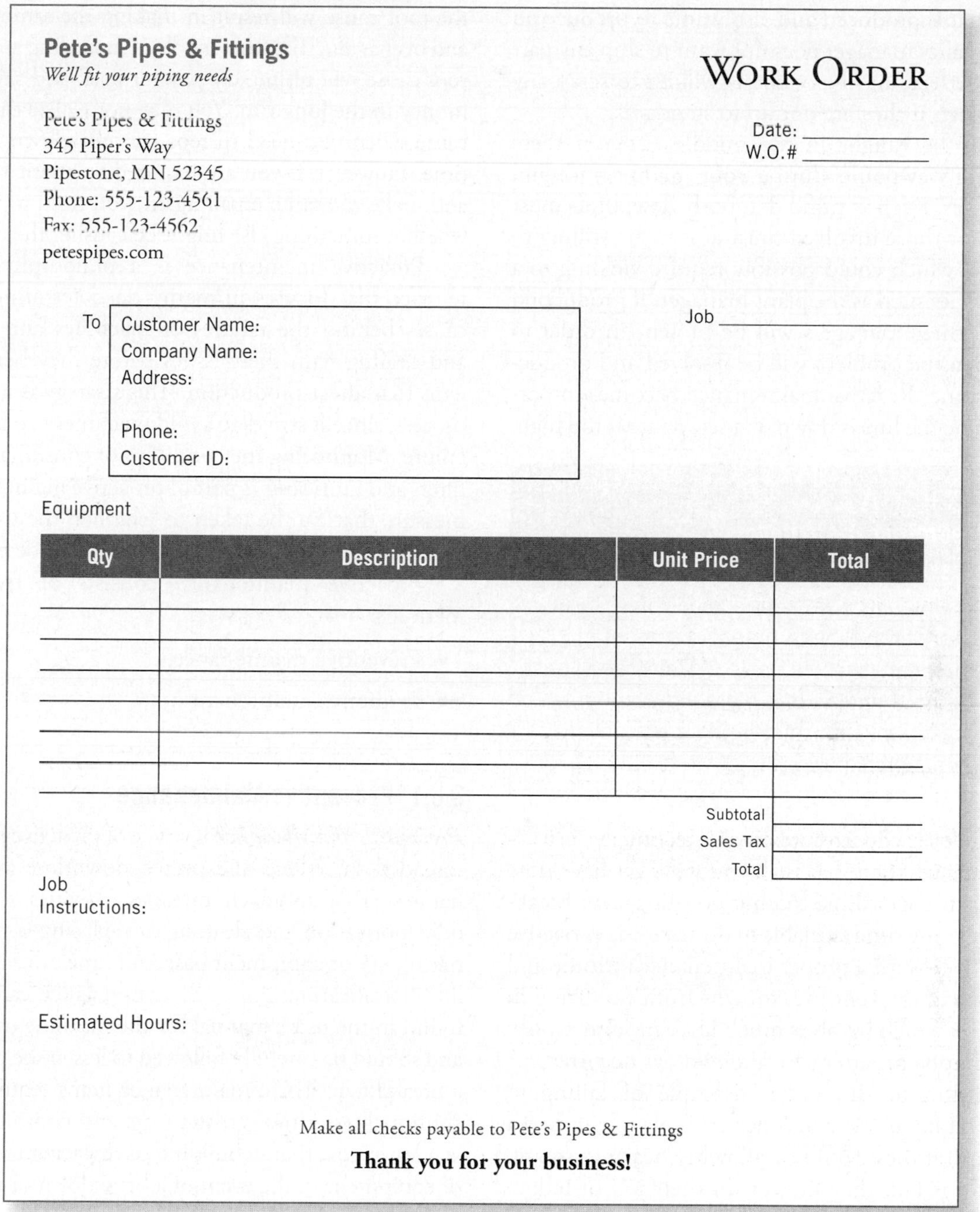

Pete's Pipes & Fittings
We'll fit your piping needs

Pete's Pipes & Fittings
345 Piper's Way
Pipestone, MN 52345
Phone: 555-123-4561
Fax: 555-123-4562
petespipes.com

WORK ORDER

Date: _____
W.O.# _____

To
Customer Name:
Company Name:
Address:

Phone:
Customer ID:

Job _____

Equipment

Qty	Description	Unit Price	Total
		Subtotal	
		Sales Tax	
		Total	

Job
Instructions:

Estimated Hours:

Make all checks payable to Pete's Pipes & Fittings
Thank you for your business!

Goodheart-Willcox Publisher

Figure 3-8. A basic work order includes the job information, customer details, a list of equipment required, and a cost estimate.

3.5 REACTIVE MAINTENANCE

Reactive maintenance is often referred to as ***breakdown maintenance***. With this type of maintenance, machinery components are repaired or replaced only when apparent problems happen. This can be due to unexpected events, machine failure, or simply a breakdown from wear.

Reactive maintenance may involve adjusting or replacing parts, which can cause a machine to run out of tolerance. If a machine is out of tolerance, it may produce defective parts, also called scrap. These low-quality parts could make it past quality control and be delivered to the customer. A situation like that can hurt a company's reputation and, eventually, its bottom line.

Not only does scrap production lead to a loss of saleable parts, but it is also a waste of raw materials. There are two managers in any production facility who have opposing viewpoints. The production manager wants to ship every part that is produced and also wants to produce no scrap. The quality manager does not want to ship any part unless it is perfect, and he or she is willing to scrap any number of parts if they are not up to standards.

You may be caught in the middle between these two opposing viewpoints during your maintenance and repair activities. Keep in mind that both viewpoints must be satisfied, or those involved must at least be willing to compromise, which could possibly require yielding to a higher authority such as the plant manager. If production is down, all three managers will be present in order to find out when the problem will be resolved and production can resume. Reactive maintenance becomes important every time the line is down or a scrap rate is too high.

TECH TIP
Predicting Breakdowns

Reactive maintenance is never a good thing, but it is necessary from time to time. No amount of proactive maintenance will prevent 100% of all breakdowns. It is impossible to predict exactly when a breakdown will occur, but it is possible to predict when a breakdown is likely to occur.

If all you ever concentrate on as a technician is reactive maintenance, then that is all you will ever have time to accomplish. You will be too busy dealing with breakdowns to have any time available to do anything proactive and preventive. With a proper maintenance attitude and strategy, you can prevent breakdowns from occurring in the first place, which involves much less time and money than fixing problems after a breakdown has occurred.

In addition to the earlier example of failing to change the oil in your car, another example of running a machine until it dies would be allowing bearings to fail. When bearings fail, they do not do their job of reducing friction, and they seize. If connected to an electric motor and the seized bearings cause the motor to stall, the windings of the motor will burn up.

Repairing a stalled motor caused by seized bearings is expensive. The better alternative is to simply replace the bearings before they seize. This takes a proactive approach in order to determine that the bearing failure is imminent.

3.6 PROACTIVE MAINTENANCE

After performing any repairs, conduct a ***root cause analysis (RCA)***, **Figure 3-9**, to determine the source, or root cause, of the problem. It is not enough to simply address the symptoms of a breakdown. Failure to repair

the root cause will result in making the same repair over and over again. The time involved in finding and fixing the root cause will ultimately pay for itself and save time and money in the long run. You may not realize there is a root cause when you need to repair a breakdown for the first time. However, if you are required to revisit the problem and make the same repair again, you need to ask yourself whether something else might be causing the failure.

Proactive maintenance is a philosophy for maintenance that focuses primarily on determining the root causes behind the failures of machines and equipment and dealing with these issues before they become problems that affect production. This strategy is a cost-saving process, since it strives to avoid machinery and equipment failure. Monitoring the condition of consumables such as filters and lubricants is part of proactive maintenance. Any measure that can be taken to lengthen the time before a failure might occur may be considered to be proactive.

Proactive maintenance consists of two primary methods:

- Preventive maintenance.
- Predictive maintenance.

3.6.1 Preventive Maintenance

Preventive maintenance is a type of proactive maintenance intended to reduce unexpected downtime and machine failures. This approach includes activities such as periodic lubrication and cleaning or replacing components of machinery or equipment based on time or usage to extend life. Specifications for such maintenance can usually be found in the user's manual for the machine or equipment and should be carefully followed to lessen the possibility of a breakdown. These maintenance items require time and consumables. However, the time and consumables come at a lower cost than rebuilding or replacing the machinery or equipment in the event of a breakdown or failure.

You are already familiar with the need for preventive maintenance in regard to changing the oil and filters in your automobile. Another example of preventive maintenance is making sure the operating system and software on your electronic devices are up-to-date. By making sure these items are up-to-date, you will hopefully avoid major problems and keep your computer running smoothly.

An industrial preventive maintenance plan requires you to replace machine parts or change the lubrication or filters at specified intervals whether or not a problem is perceived. You may also need to make sure any industrial automation software is up-to-date, so that machines run smoothly and are properly monitored. Such preventive

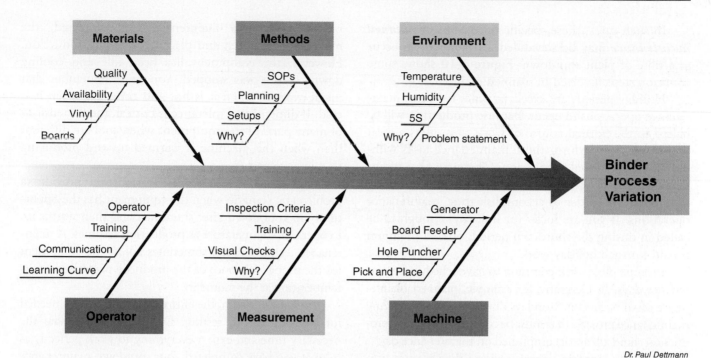

Figure 3-9. This fishbone diagram provides visual organization for root cause analysis.

Dr. Paul Dettmann

maintenance usually happens at certain specific intervals, based on the manufacturer's recommendations. Again, it is best to consult the user's manual to determine these details.

Scheduled Maintenance

Again take, for example, the user's manual for your automobile. It includes a section on preventive maintenance. Most often there is a table included that shows what regular maintenance must be done and the interval (or schedule) of when it should happen. For instance, aside from regular oil changes, other auto-related examples of scheduled maintenance might include checking the air pressure in the tires every 500 miles, changing the air filter every six months, checking the oil every fifth fill-up, checking the brake fluid and automatic transmission fluid every time you top off the oil, and checking the level of the coolant every 1000 miles.

These are examples of *scheduled maintenance*. With the scheduled maintenance approach, either time or the amount of use is the deciding factor as to what maintenance tasks should be performed.

An industrial example of a need for scheduled maintenance might be in relation to a standby generator. Most maintenance plans for standby generators require that you run the generator monthly to verify its ability to provide standby power. The maintenance plan could require that power from the utility company be interrupted. You then observe that the automatic transfer switch starts the generator and transfers the power supply from the utility

to the generator once the generator has started, and that the generator then provides stable power.

You will then need to verify that the generator's engine has oil pressure, there is an adequate fuel supply, temperatures are within specifications, and adequate voltage and current are being provided. The standby generator unit will have indicators that display the fuel level, voltage, current, temperature, and oil pressure. Additionally, it will have an indicator showing the total hours of operation. All readings should be recorded in the generator log and, if any remedial measures are taken, the test should be rerun and the readings once again recorded.

These scenarios should play out for every piece of equipment in the plant. Every user's manual or maintenance manual has a list of maintenance items and the frequency with which they should occur. Being vigilant and performing the recommended maintenance items for each machine in the plant at the proper intervals will go a long way toward preventing problems and breakdowns.

Planned Maintenance

Some maintenance operations are not as immediately critical as they might be in a breakdown situation. Nevertheless, these operations must occasionally be performed for the continued proper functioning of the equipment. As an important safety issue, many maintenance tasks require that production be halted to render moving machinery safe for performing maintenance.

In such cases, these specific operations of *planned maintenance* may be scheduled, or planned, to occur at a time of plant shutdown. **Figure 3-10** shows some common steps involved in planned maintenance.

Holiday periods are excellent times for these maintenance operations to occur. Because production will be halted in the normal course of work, this offers a great opportunity to perform those maintenance tasks without affecting normal production. Of course, that means that maintenance personnel must be brought in during the holidays in order to accomplish these maintenance operations. If you are lucky (or unlucky) enough to be called in during the shutdown period, often a premium is paid for such holiday work.

In some places, it is common to have scheduled maintenance days. In Germany, for example, planned maintenance often occurs on Mondays during the first shift. Any maintenance projects that must be completed during a production standstill are accomplished on maintenance day.

In either scenario, larger projects that require production to stop are planned. A list of projects is kept and prioritized, and the projects are then scheduled to occur during the allotted time.

Once planned maintenance is completed, the machines are started and placed back into production. However, the equipment has been idle and cooling down since it was stopped. You must remember that metal expands when it is hot and contracts when it is cold. If dimensional tolerance is critical, the dimensions of many parts of the equipment when cool are different than when the machine is warmed up and producing in-tolerance parts.

High-speed, close-tolerance production provides a problematic scenario when the equipment has the opportunity to cool down after shutdown and then warms up to operating temperature as production resumes. A maintenance technician can sometimes adjust the equipment for the proper tolerance of the produced part based on its temperature at the moment.

After a shutdown, the entire next shift is often needed for adjustment and testing. Do not underestimate the necessary time and expect everything to work perfectly as soon as you have completed your shutdown maintenance activities. Rarely does everything work as planned. Once the machines are warmed up, adjustments may be necessary to bring them back into acceptable tolerances.

Six Common Steps for Planned Maintenance

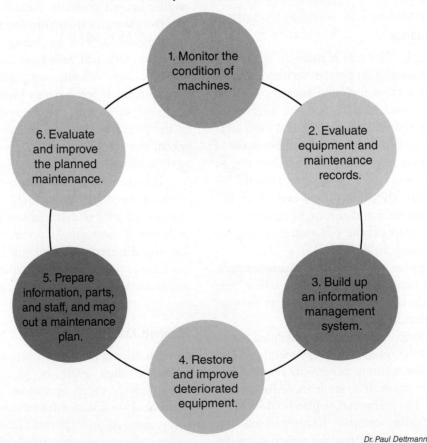

Dr. Paul Dettmann

Figure 3-10. A planned maintenance program involves these six common steps.

3.6.2 Predictive Maintenance

Preventive maintenance occurs on a regular schedule, regardless of a machine's performance. ***Predictive maintenance***, on the other hand, consists of testing and monitoring machines and equipment based on the amount of usage (hours, months, and cycles for example), mileage, operating conditions (humid, dry, and dusty for example), and other features of the machines or equipment in an effort to help predict breakdowns. This allows for the machine or equipment to be scheduled for repair and the necessary time to purchase needed parts and supplies, thus reducing the need for a high parts inventory and increased production time for the machines or equipment.

Record keeping in this case is essential, as is a method to analyze the performance data and determine when a failure is imminent. Many different types of data may be recorded and analyzed in order to determine when a failure has the highest probability of occurring. When technicians are not repairing machinery or equipment, they will be out on the floor collecting data. This data is analyzed to help predict failures.

There are five major steps to a predictive maintenance program (**Figure 3-11**):

1. Setup.
2. Testing.
3. Monitoring.
4. Repair.
5. Scheduling inspection.

Setup involves developing a list of critical processes and then determining how likely the related equipment is to fail. Maintenance records play an important part in the setup process.

Testing is the second step to a predictive maintenance plan. There are various tests that may be performed on your machinery. Test results must be recorded in the maintenance records. With this strategy, a single test will not tell you everything you need to know. The maintenance records and periodic testing will establish a baseline for the measurements.

The test data must be monitored over time to point out trends. In monitoring, you will be able to determine changes in operating conditions. Vibration, current, and temperature changes over time are all indicators that required maintenance action may be imminent.

Armed with the data, you need to determine if there are any warning signs to indicate that a repair may be necessary. The time left until failure may sometimes be calculated in order to estimate how soon the repair procedure must be performed.

Steps for a Predictive Maintenance Program

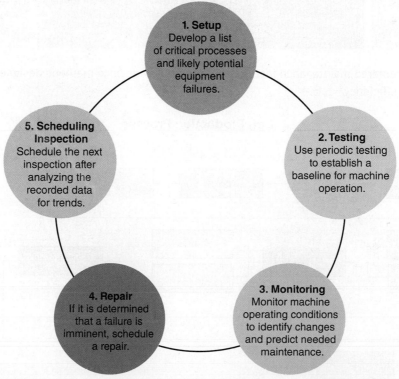

Figure 3-11. There are five steps in a predictive maintenance program.

If it is determined that a failure is imminent, a repair is scheduled. "Imminent" is the operative word for scheduling. You certainly do not want to make the repair if the machine would operate just fine for the next year or more, but you do not want to wait so long that the machine fails before the scheduled maintenance.

Measurement results concerning vibration, voltage, current, and temperature serve to indicate trends. Indicators showing trends will predict when a major failure is imminent. You cannot predict every failure, but you can greatly reduce major breakdowns by taking this proactive approach.

While measurements by themselves rarely tell the entire story, by recording and analyzing the data for trends, you can often determine where a failure is imminent on a piece of equipment, as well as which part of that equipment will soon fail. It is vital that an operation stress the importance of maintenance records.

3.7 RELIABILITY-CENTERED MAINTENANCE (RCM)

In the *reliability-centered maintenance (RCM)* concept, equipment is prioritized by its importance to the operation, **Figure 3-12**. Other factors to be considered are downtime cost, safety concerns, quality implications, and the cost of repair should a failure happen.

For an example of equipment prioritization, consider a metal beverage can production line. See **Figure 3-13**. A coil of metal is fed into a cupping press, which punches out disks. Then, a press forms the disks into cups, which have a diameter of approximately 1.75 times the diameter of the finished can. The height is about 30% of the finished can.

The "cups" are now distributed to any one of ten D&I (drawing and ironing) machines, which elongate the cups to slightly more than the height of a finished can. These machines are often referred to as "body makers" because they form the body of the can.

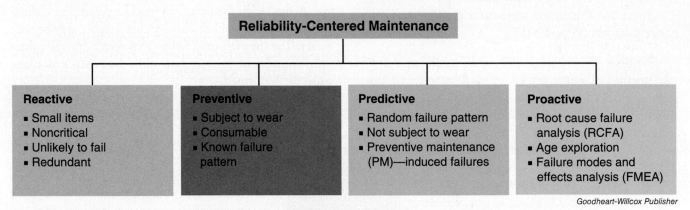

Goodheart-Willcox Publisher

Figure 3-12. Reliability-centered maintenance includes a range of maintenance methods designed to minimize downtime and maximize efficiency.

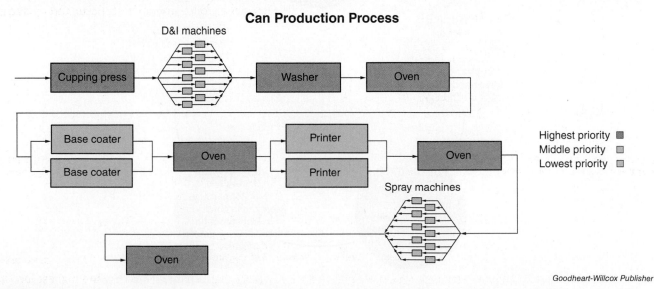

Goodheart-Willcox Publisher

Figure 3-13. This diagram shows the equipment prioritization in a can production process.

Directly after the D&I process, another machine cleanly trims the top edge of the can. The cans then file onto a mass conveyor and enter a washer that makes the metal bright and removes any lubricant left over from the D&I process.

After the washer, the cans travel on a mass conveyor into an oven that dries them. The cans now move onto one of two base coaters that print the white background onto the outside of the can. After the cans leave the base coater, they travel on a mass conveyor through an oven to dry the white lacquer.

When the cans emerge from the oven, they travel on to one of two printers, which print the outside color decoration onto the can. From the printers, the cans travel through an oven to dry the outside printing. Now the cans are sent to one of ten inside spray machines, which coat the inside of each can with a layer of protective lacquer. Once again, the cans travel by mass conveyor, to an oven that dries the inside lacquer.

The cupping press, should it fail, stops 100% of production. A D&I machine, should one go down, only stops 10% of production. The base coaters and printers, because there are two of each, would only inhibit 50% of production if either one were to fail.

A higher priority needs to be assigned to any entity of the process that would inhibit 100% of production in the event of a failure. On the other hand, a machine that would only inhibit 10% of production in the event of its failure would be ranked lower on the scale of critical equipment.

In the reliability-centered maintenance strategy, the machines and systems that are the most critical or expensive receive the most attention. Anything that is not critical receives a lower maintenance-effort priority.

3.8 TOTAL PRODUCTIVE MAINTENANCE

Total productive maintenance (TPM), **Figure 3-14**, is a maintenance method that involves a company-wide system for maintaining manufacturing equipment and facilities. It was developed from the concepts of preventive maintenance and productive maintenance. TPM is designed to increase production as well as employee job satisfaction and morale by involving all employees at all levels of the company in the process.

3.9 MAINTENANCE STRATEGIES

The maintenance strategy employers use may vary from one company to another. No one strategy is correct, and no one strategy is implemented the same in every company. The strategy used depends on management priorities and the size of the company. If your company is large, there are layers of management that have predetermined strategies as to what works best in their case.

If you are the only maintenance technician for a small manufacturer, you may be able to decide what strategy is employed. None of the strategies outlined in this chapter are right in every situation. Some are better than others for your specific situation.

CHAPTER WRAP-UP

Although breakdowns will occur, proactive maintenance is almost always better than reactive maintenance. It will save a company money in the long run. Keeping careful records based on manufacturers' recommendations and your ongoing observations will help you predict problems and keep things running smoothly. Preventive maintenance, predictive maintenance, and planned maintenance are the best strategies to follow for any technician and operation.

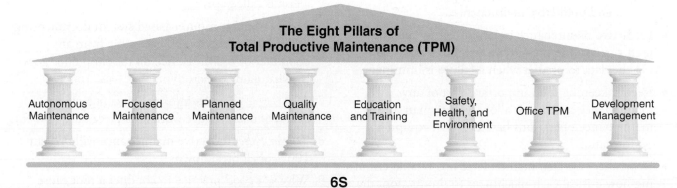

Figure 3-14. Total productive maintenance includes eight distinct elements that help a plant obtain the highest level of machine and work productivity.

Chapter Review

SUMMARY

- Maintenance strategies may be divided into two major categories: proactive maintenance and reactive maintenance.

- Condition monitoring is the process of observing a machine's operation using tools to help identify any symptoms of malfunctions.

- Autonomous maintenance is when the operator takes care of small maintenance tasks in a proactive way to eliminate or reduce breakdown maintenance.

- Maintenance records can be kept in digital form, accessible from any computer or mobile device.

- A maintenance work order is a document that is used for planning and controlling maintenance activities and tasks.

- Reactive maintenance is often referred to as breakdown maintenance. Reactive maintenance happens after a failure has occurred.

- Proactive maintenance activities are performed before a breakdown occurs.

- Reactive maintenance often costs much more than proactive maintenance, because a breakdown or failure requires more money to fix.

- It is important to determine the root cause of a failure so that measures may be taken to prevent future occurrences.

- Proactive maintenance includes preventive maintenance, scheduled maintenance, planned maintenance, and predictive maintenance.

- Predictive maintenance involves making measurements, recording measurement results, and analyzing the data to predict when a failure is imminent.

- Record keeping is an important part of any maintenance strategy. It allows you to plan future maintenance operations based on past equipment performance.

- Reliability-centered maintenance involves determining how critical each machine is to production, the cost of repair, the cost of downtime, and the likely time before failure, and then prioritizing and focusing maintenance efforts accordingly.

- The maintenance strategy employed by each company differs. A company may adopt one of the strategies mentioned in the text, or it may use a combination of various points from each strategy.

- Proactive maintenance is almost always better than reactive maintenance.

REVIEW QUESTIONS

Answer the following questions using the information provided in this chapter.

1. *True or False?* Proactive maintenance is the type of maintenance activity you perform after something goes wrong.

2. List four company resources or variables that should be considered when determining a maintenance strategy.

3. *True or False?* Condition monitoring is a process where a technician observes a machine's operation by examining such items as vibration, temperature, and voltage/current draw to help identify symptoms of malfunctions.

4. An approach to maintenance in which the operator is trained to perform routine maintenance tasks, such as general machine inspection, lubrication, and cleaning, as well as to detect abnormalities, describes the concept of _____ maintenance.

5. An established and accepted method for performing a task is called a(n) _____.

6. An electronic computer-based system documenting maintenance records and accessible from any computer or mobile device on the shop floor goes by what name or acronym?

7. A(n) _____ provides information concerning the details of maintenance work for a piece of equipment.

8. *True or False?* Proactive maintenance will prevent 100% of all breakdowns.

9. Why is it good practice to conduct a root cause analysis after any repairs?

10. *True or False?* A root cause analysis will ultimately cost more money than reactive maintenance.

11. The two primary methods of proactive maintenance are preventive maintenance and _____ maintenance.

12. Using your automobile as an example, provide two examples of scheduled maintenance.

13. *True or False?* Planned maintenance is a part of reactive maintenance.

14. What are the five major steps to a predictive maintenance program?

15. *True or False?* Reliability centered maintenance prioritizes equipment by its importance to the operation.

 NIMS CREDENTIALING PREPARATION QUESTIONS

The following questions will help you prepare for the NIMS Industrial Technology Maintenance Level 1 Maintenance Operations credentialing exam.

1. When should technicians use their five senses to determine symptoms of machine malfunctions?

   A. Start-up
   B. Operation
   C. Shutdown
   D. All of the above

2. What is an example of preventive maintenance?

   A. Testing for temperature changes
   B. Changing oil regularly
   C. Replacing a seized bearing
   D. All of the above

3. What is an example of predictive maintenance?

   A. Testing for temperature changes
   B. Changing oil regularly
   C. Replacing a seized bearing
   D. All of the above

4. What does HMI stand for?

   A. Human-machine incident
   B. Human manufacturing incident
   C. Human-machine interface
   D. Human manufacturing interface

5. Which of the following provides information concerning the details of maintenance work for a piece of equipment?

   A. Standard operating procedure
   B. Work order
   C. Root cause analysis
   D. Scheduled maintenance

6. In a company using a CMMS, where would a technician *not* be able to directly access maintenance records?

   A. An HMI
   B. A computer
   C. A machine operator
   D. A mobile device

4 | Maintenance Tools

LEARNING OBJECTIVES

After completing this chapter, you will be able to:

☐ Identify hand tools commonly used in maintenance and repair work.

☐ Use basic hand tools properly.

☐ Select the proper tool for the maintenance job.

☐ Identify common power tools used by maintenance technicians.

TECHNICAL TERMS

abrasive cutoff saw	kerf
adjustable wrench	lineman's pliers
air impact wrench	locking pliers
air ratchet	long-nose pliers
angle grinder	machinist's rule
ball-end hex key	nut driver
ball-peen hammer	Phillips screwdriver
bench rule	pin punch
breaker bar	portable band saw
caliper	prick punch
center punch	puller
chisel	ratchet
circular saw	reciprocating saw
claw hammer	screwdriver
combination wrench	slip-joint pliers
crimper	socket
dead-blow hammer	socket adapter
diagonal cutting pliers	socket extension
die grinder	speed wrench
file	steel tape
file card	tongue-and-groove pliers
hacksaw	torque wrench
hammer drill	universal joint socket adapter
hand drill	wire stripper
hex key	wrench

Each job requires a particular tool, so you need to have an assortment of tools in order to be able to use the correct one for the task at hand. All too often the technician, lacking the proper tool, presses into service an inappropriate tool, and the result is less than desirable. Perhaps the technician uses a pair of pliers as a wrench, hammer, crimper, or screwdriver, for example. While this may get the job done in a pinch, the task would have been more easily and properly accomplished had the technician used the proper tool.

Before the industrial revolution, hand tools were the only tools available. While machines are often used in everyday production and maintenance today, hand tools still play an important role in assembly, maintenance, and repair work. In this chapter, you will learn about a variety of tools and their proper or improper uses. This chapter will touch on power tools briefly. Further information about specialty tools is provided in later chapters.

It is not necessary to immediately spend thousands of dollars on a tool collection, but you will need to have certain tools in order to perform your basic duties. Your collection of tools will grow over time. As your responsibilities increase, so should the number of tools in your toolbox. Some companies supply technicians with all the tools appropriate to perform their required work. Other companies expect technicians to bring their own basic tools with them. Each company's requirements vary as to what tools you must have and what tools the company provides.

Professional tools are expensive. The 99-cent tools from the hardware store are not professional grade. Tools from a tool truck are often overpriced, even though such services offer easy credit to the purchaser. Unfortunately, those who resort to purchasing professional-grade tools from a tool truck on a payment plan often pay many times what they would to buy the tools outright from another seller, due to the interest paid and the higher initial cost. A good rule of thumb when purchasing professional tools is to buy the best tool you can afford. The best tool is not always the most expensive. Buy good tools when you need them, at a good price. Do a little research and shop around—it will pay off in the long run. Don't be fooled by an expensive lifetime guarantee. Most major brands of hand tools offer a lifetime guarantee, with the offer of a free replacement if the tool ever fails.

4.1 SAFETY

One of the most important points concerning tool use is safety, **Figure 4-1**. Review the safety instructions provided by the manufacturer, your employer, and OSHA. OSHA provides five basic safety rules:

- Maintain all tools to keep them in good working condition.
- Use the right tool for the job.
- Check your tools for damage, and do not use damaged tools.
- Use tools according to manufacturers' instructions.
- Use the right personal protective equipment (PPE). Safety glasses are always recommended, in addition to other job-specific PPE.

Hand Tool Safety

1. **Understand how to use a tool before attempting to use it for the first time.** Acquire the proper training to reduce personal injury and damage to tools.

2. **Use the correct and right size tool for the job.** Do not use a tool for something other than its intended purpose, as it can damage the tool and cause injury.

3. **Wear the appropriate personal protective equipment.** Always wear eye protection. When necessary, wear a face shield, steel-toed boots, or gloves.

4. **Keep floors clean and workspace clear of debris and volatile gases.** This helps prevent accidents and injury and reduces the chance of sparks causing a fire or explosion.

5. **Carry and store tools in a safe and secure manner and only transport them in toolboxes or tool belts.** Take extra care to safely store sharp and pointed tools and carry them in sheaths or holders. Never carry sharp or pointed tools in your pocket.

6. **Pass tools to others handle first.** Never throw tools to another person. These precautions prevent injury to yourself and the person to which you are handing the tool.

7. **Keep track of tools when working at heights.** A falling tool can kill a coworker. When using a ladder, transfer tools to the work area by rope or bucket.

8. **Cut away from your body.** Otherwise, if the cutting device slips, you can injure yourself.

9. **Make sure your grip and footing are secure when using hand tools.**

10. **Make sure tools are in good condition and do not use damaged or defective tools.** Keep tools clean and cutting tools sharp. Dull cutting tools or screwdrivers with worn tips can cause injury.

Goodheart-Willcox Publisher

Figure 4-1. Consider these safety guidelines before using hand tools. Be sure to follow manufacturer guidelines as well as OSHA regulations.

4.2 WRENCHES

Wrenches are tools used for tightening and loosening nuts and bolts. There are also specific wrenches designed for tightening and loosening pipes, fittings, filters, and other devices that are assembled with threaded couplings.

The most common wrench is the ***combination wrench***, also referred to as a box-end/open-end wrench, **Figure 4-2**. The box end of the wrench completely encompasses the bolt head or nut, providing many points of contact with the fastener and allowing it to have a better grip than the open end. Standard box ends have six points, and the better ones have 12 points. The 12-point variety is easier to use where there is limited space for wrench movement and a need for finer repositioning capabilities. Some box ends employ a ratcheting mechanism so you do not have to remove and replace the wrench when you need to reposition it.

Figure 4-2. Combination wrenches are available individually or in sets, as shown here, and have two ends—an open end (left) and box end (right)—for tightening or loosening nuts and bolts. The pictured combination wrenches are 12 point and vary in size from 3/4″ to 5/16″.

The open end of the wrench is slipped onto the bolt head or nut from the side. The flats of the open end grip onto the flats of the nut or bolt. Always select the correct size. If the wrench is too small, it will not fit on the nut or bolt. More importantly, if the wrench is too large, it may round off the points of the bolt head when you use it, making the fastener impossible to remove with a wrench.

Combination wrenches are sold in a large array of sizes, in both metric and SAE (Society of Automotive Engineers)—also known as standard—measurements. Be careful to determine if the fastener is metric or SAE before selecting a wrench. Mistaking SAE sizes for close metric sizes, and vice versa, could cause you to round off the fastener points. Combination wrenches are most often sold in sets.

4.2.1 Sockets

Sockets can be used to loosen and tighten bolts as an alternative to wrenches. A socket set contains sockets of various sizes and a ratchet handle to which the sockets attach. A ratchet wrench, or *ratchet*, **Figure 4-3**, consists of a handle and a wheel that allows motion in only one direction at a time. The sockets come in SAE and metric sizes and are sized according to the physical dimensions of the bolt head. For example, a 1/2″ socket

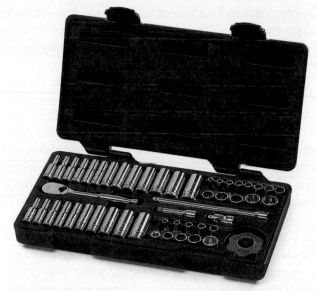

Figure 4-3. This socket set comes with standard and deep sockets, a ratchet wrench, an extender, and two adapters.

fits a bolt with a 1/2″ head. Sockets come in standard depth and deep-well varieties. The deep-well socket is especially useful in installing or removing nuts where the bolt protrudes a substantial amount past the nut. With sockets, there is a lesser likelihood of rounding off the points on a fastener, provided you use the correct size. Sockets are available in both metric and SAE sizes.

On the opposite side of where the socket meets the bolt is a recess for attaching the driving device, which may be a ratchet, air ratchet, pneumatic impact wrench, or breaker bar, **Figure 4-4**. Drive sizes are the physical dimensions of the square drive recess on the socket. Standard drive sizes are 1/4″, 3/8″, and 1/2″. *Socket adapters*

Figure 4-4. Ratchet wrenches work by applying turning force in one direction and freely turning in the other. This allows you to turn fasteners without having to remove and reposition the wrench after each turn. The lever on the head of the wrench reverses the ratcheting action.

allow for ratchets of different drive sizes to be used with sockets having a different drive recess. A small lever is provided to change the direction of the ratcheting action (loosening or tightening the bolt).

Often included is a *breaker bar*, which fits the socket and has a long bar to provide more torque when loosening especially difficult-to-remove fasteners. It is best to use a breaker bar for extremely stubborn bolts and nuts, because the ratchet may be damaged by excessive force. Some sets also include a *speed wrench*, which looks similar to a crank, allowing the user to rapidly install or remove a fastener with low torque.

Many socket sets also come with *socket extensions*, which can be installed between the socket and the ratchet wrench, allowing the ratchet to be positioned a distance away from the fastener. This is helpful when there is limited space around the fastener. *Universal joint socket adapters* allow for the ratchet to be used at an angle in situations where space is limited.

4.2.2 Torque Wrench

A *torque wrench* is a device that may be used with sockets. A torque wrench allows the technician to set the amount of tightening torque, **Figure 4-5**. It will click and stop tightening when the proper torque is achieved. A torque wrench is used when the technician must tighten a fastener to a specified torque. This is necessary especially on multiple-bolt flanges where all bolts must be tightened the same amount.

4.2.3 Air Ratchets and Impact Drivers

There are also pneumatic (air-powered) tools that can be used with sockets: an *air impact wrench* and an *air ratchet*, **Figure 4-6**. You may have seen and heard an auto technician use an air impact wrench to remove or install the lug nuts on the wheels when you were getting the tires changed on your automobile. An air impact wrench delivers short, concentrated bursts of torque and is capable of quickly tightening or loosening nuts or bolts that require extreme torque to install or remove.

SAFETY
Air Impact Wrenches

Only use sockets specifically designed for an air impact wrench. Sockets made for hand tools could break, causing injury. Also take care with smaller bolts, which will be more likely to break when using this power tool, which could also cause injury.

An air ratchet functions like a standard ratchet wrench, except that you only need to hold it stationary while the air mechanism takes care of the movement. Air ratchets have a small lever to change the direction of the ratcheting action and thus the turning direction. This is a great device to use when you have many nuts or bolts to install or remove or when there is little space for ratchet movement, so long as you don't mind the air hose trailing behind you.

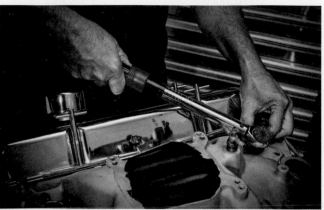

GearWrench (Apex Tool Group)

Figure 4-5. Torque wrenches allow the user to set the torque the tool will deliver, keeping the force even across multiple bolts.

GearWrench (Apex Tool Group)

Figure 4-6. Air impact wrenches are frequently used in the automotive industry, as shown here.

4.2.4 Hex Keys

Hex keys (often referred to as Allen wrenches) are commonly used with machine bolts and screws, **Figure 4-7**. They are also commonly used with setscrews. Screws and bolts that use hex keys have a hexagonal recess in the head. The hex key is inserted into this recess and may be turned to either tighten or loosen the bolt or screw.

Most hex keys are L shaped. Either end may be inserted into the recess of the screw. If you need more torque, insert the short end into the screw and pull on the long end. If your goal is to quickly run a screw in or out, insert the long end and pull on the short end.

Hex keys with a ball end on the long side, called ***ball-end hex keys***, are most useful because they allow the screw to be driven by an angularly positioned hex key. This comes in very handy in tight spaces. Hex keys are also available in a T-handle configuration, with either a hex end or ball end. All hex keys come in metric or SAE sizes.

Some hex key sets have a small plastic insert on the ball end that grips the screw. This is useful for installation in areas that are not finger accessible, such as gears, sheaves, knobs, or flywheels.

Every technician who works on machinery should have a complete set of hex-end and ball-end hex keys in both L-shaped and T-handle configurations. The loose-piece sets are less desirable, as it is easier to lose one or to not be able to rapidly locate the desired size.

For larger-size hex keys, there are socket hex keys available. A short section of hex key is mounted into a socket for use with a ratchet wrench. These hex key sockets are available in both SAE and metric sizes.

TECH TIP
Worn-Out Hex Keys

Hex keys should be replaced as soon as they become worn. Worn hex keys may easily round out the hexagonal recess of a screw, making it almost impossible to tighten or remove.

4.2.5 Adjustable Wrenches

An ***adjustable wrench*** (**Figure 4-8**)—often referred to as a crescent wrench or "knuckle buster"—has one fixed jaw and one movable jaw, with a screw-like adjustment located on the portion of the wrench between the jaws and the handle. The adjustable wrench offers the abilities of several open-end wrenches in a single tool. To adjust the wrench, locate the fixed jaw against the flat of the bolt head or nut. Turn the adjustment until the movable jaw comes against the opposite flat from the fixed jaw.

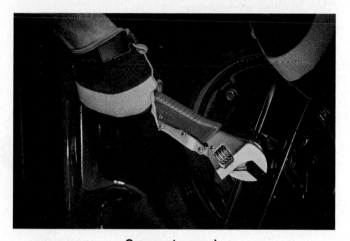

Crescent wrench

Pipe wrench

Crescent (Apex Tool Group); Armstrong (Apex Tool Group)

Figure 4-8. Adjustable or crescent wrenches can be adjusted with the dial on the head for use on fasteners of various sizes. The pipe wrench is a common adjustable wrench used for work on pipes or other round fittings, often in plumbing.

Allen Manufacturing Company (Apex Tool Group)

Figure 4-7. Hex key or Allen wrench sets, like other tool sets, include an array of sizes for use on different-sized fasteners.

The disadvantages of this tool are that the open end of this style of wrench is much larger than a conventional open-end wrench, and its adjustability may also allow the wrench to be misadjusted. If not properly adjusted, the wrench could slip, round off the points of the fastener, and cause injury to your knuckles and hand if they impact something in the process.

Always turn adjustable wrenches in the direction of the adjustable jaw. If this is not the desired direction, turn the wrench over and readjust the jaws. Failure to pay attention to this direction could allow the movable jaw to cock slightly, which in turn could possibly round off the points of the fastener.

4.3 SCREWDRIVERS

There are many different types of *screwdrivers* used to fasten the many different styles and sizes of screw. Screwdriver tips and lengths vary as well. The two most common styles of screwdriver are the slotted (or flat-tip) screwdriver and the Phillips screwdriver, **Figure 4-9**. The *Phillips screwdriver* fits the Phillips-head screw, which has a cross-like slot in the top. There are also other varieties of screw heads, and screwdriver tips to match, **Figure 4-10**.

> ### ⚠ CAUTION
>
> Dimensions of the recess or slot vary depending on the screw size. Be certain you select the proper style and size of screwdriver tip to appropriately fit the screw you are installing or removing. Selecting a screwdriver tip that is too small may result in stripping out the driving slot and rendering the screw impossible to remove. This point is especially important with hex and Phillips screws. With hex, it is important to determine whether the screw is metric or SAE. Many metric and SAE sizes are close enough to allow you to engage the screwdriver tip just enough to strip out the driving recess.

With Phillips head screws, it is easy to use a screwdriver tip that is too small. When the screwdriver tip is too small, it does not fully engage the screw. A good rule of thumb is to see which Phillips screwdriver head size appears to the eye as correct, and then try the next size larger first. If the screwdriver tip size is too large, it will not engage at all. If it is too small, it will most certainly strip out the recess.

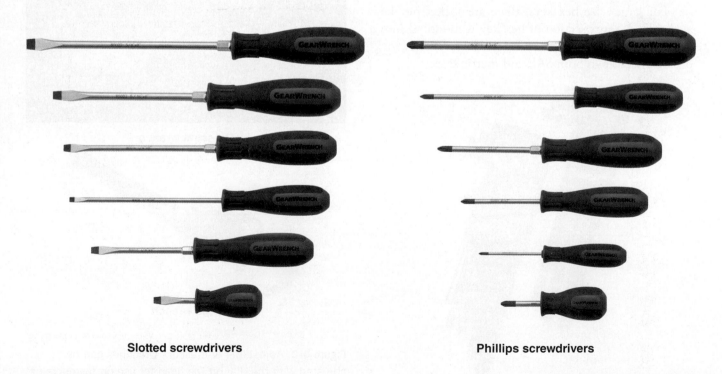

Slotted screwdrivers   **Phillips screwdrivers**

GearWrench (Apex Tool Group)

Figure 4-9. A variety of common and Phillips screwdrivers are pictured here.

Common Screw Heads	
Head	**Type**
	Square
	Hex
	Slotted
	Phillips
	Frearson
	Robertson/square recess
	Torx®
	Security Torx®
	Hex socket
	Security hex socket
	Tri-wing
	Pozidriv
	Spanner
	Torq-set
	Spline
	One-way
	Clutch

Goodheart-Willcox Publisher

Figure 4-10. This table shows some of the various styles of driving methods for screw heads.

TECH TIP
Small Screws

Remember that the smaller the screw is, the more difficult it is to use an alternative method of removal should the screw become damaged. Do not use an improperly sized screwdriver tip and risk damaging an extremely small screw head.

Various screwdrivers are available in different lengths, from long to stubby, depending on the intended use. The screwdriver shanks may be either round or square. Some round-shank screwdrivers have a hexagonal section near where the handle meets the shank. This hex section is for the use of a wrench when extra torque is required for particularly stubborn screws.

When the tip of a common screwdriver is worn, it often may be touched up with a file. Screwdrivers with worn tips should be replaced when the tip is either chipped or can no longer be dressed with a file. Screwdrivers are inexpensive enough to replace when they are worn, rather than risk using them and damaging screws.

Screwdrivers used for electrical work have a well-insulated handle and insulation along the shank, **Figure 4-11**. If you are working on live electrical connections, you should use an insulated screwdriver.

paveir/Shutterstock.com

Figure 4-11. Insulated screwdrivers are used to safely work around electricity.

Insulated screwdrivers are marked with the voltage rating they will provide protection for (commonly 1000 V). If you are working on electrical equipment, either energized or de-energized, you should use an appropriately rated insulated screwdriver.

Nut drivers or nut runners are like screwdrivers, but they have a hex socket at the tip, **Figure 4-12**. Nut drivers are used to loosen or tighten small nuts and bolts and are particularly suited for small spaces. Often, nut drivers have a hollow shaft, allowing the threaded part of the bolt to go up inside the handle in cases where the threaded portion of the bolt would protrude well beyond the nut. Insulated nut drivers are helpful when the termination of a wire is around a bolt being secured with a nut.

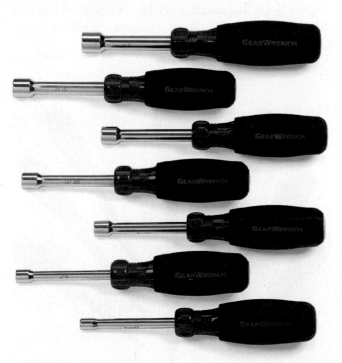

GearWrench (Apex Tool Group)

Figure 4-12. Nut drivers have a hex socket at the tip and are used to loosen or tighten small nuts and bolts in small spaces.

4.4 PLIERS, CUTTERS, STRIPPERS, AND CRIMPERS

Hand tools that fall under the category of pliers are many. Different types of pliers are available for holding, turning, twisting, bending, clamping, crimping, and cutting. See **Figure 4-13**. Some types of pliers combine several of the mentioned capabilities. Pliers are often misused in cases where a wrench would be a better option. Other misuses include crimping with a pair of pliers that were not designed for that purpose. Many types of pliers use a spring that opens the pliers when pressure on the handles is released. This reduces user fatigue for repetitive operations.

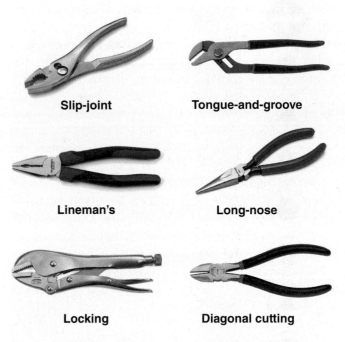

Slip-joint **Tongue-and-groove**

Lineman's **Long-nose**

Locking **Diagonal cutting**

Apex Tool Group

Figure 4-13. These are some of the common pliers used by technicians.

4.4.1 Pliers

When the term *pliers* is used, most people think of *slip-joint pliers*, which are often referred to as gas pliers. With slip-joint pliers, the joint where the two arms of the pliers are connected can be adjusted to change the gripping range of the jaws. Slip-joint pliers can be used for gripping and bending things like nails and bolts with the teeth of the pliers, and can also be used for quick loosening or tightening adjustments when a wrench is not handy.

Tongue-and-groove pliers are often referred to as Channellock™ pliers. The nose of the pliers is offset nearly 90° to the handles. The joint has multiple tongue-and-groove points, which are used to adjust the gripping range of the pliers. This tool is often used as a pipe wrench, but the deep serrations in the jaws will mar the surface of anything they are used on.

To use the tongue-and-groove pliers, orient them so that the jaws are facing away from you. Place the jaws over the item to be turned. While keeping slight pressure on the handle for the movable jaw, push the handle for the fixed jaw away from you. This action causes the movable jaw to bite more firmly. If this is not the direction you wish to turn the item being clamped in the jaws, remove the pliers and reorient them. Simply pulling instead of pushing on the pliers will only grind off the surface of whatever item you are trying to turn.

Lineman's pliers have wide, flat jaws and considerable weight. Many also have a wire-cutting portion on the jaws, which is useful for cutting heavy-gage wire. These pliers often have a crimp section for heavy crimp splices or a thread-holding section. These pliers are heavy-duty and quite versatile. However, do not misuse them as a hammer simply because they seem heavy and sturdy, as you risk damaging the pliers.

Long-nose pliers—also often called *needle-nose pliers*—have long, narrow, pointed jaws and are used for gripping and bending items such as wire. They come in many different sizes and shapes and are useful in tight places. Many long-nose pliers include a set of wire cutters located on the nose side of the hinge. Long-nose pliers for communication wiring often include a crimping or stripping notch in addition to the wire-cutting notch. Misuse occurs when the long-nose pliers are too large or too small for the job at hand. Unless they include a crimping notch, long-nose pliers should not be used to crimp terminals.

Locking pliers are often used for clamping and may be referred to as clamping pliers. Additionally, locking pliers can be used to remove damaged nuts and bolt heads that are rounded off and cannot be removed with wrenches. In these cases, the bolt or nut has been damaged and cannot be reused after removal. One of the major misuses of pliers is using them as a wrench on good nuts and bolts. Locking pliers will damage the surface of nuts and bolts, so they should only be used as a last resort for the removal of damaged fasteners.

Locking pliers are an excellent tool to clamp parts together for drilling or welding. Some locking pliers have wide, flat jaws for clamping and bending sheet metal. This variety of pliers has an adjusting knob at the end of one of the handles so you can adjust the width of the jaws and the clamping pressure. When adjusted properly, the pliers will lock when the handles are brought together. There is also a release lever between the two handles that may be operated to unlock the pliers for removal.

Diagonal cutting pliers have hardened jaws that are ground to a V shape for cutting copper wire and component leads. They also usually have insulated handles to protect against electric shock. Misuse occurs when a technician uses diagonal cutting pliers for cutting metal that is too hard, such as steel or stainless-steel wire. The pliers may be damaged when the jaws are nicked from cutting wire that is too hard.

The jaws of diagonal cutting pliers are set at an angle of 20°–30° from the handles. Some are ground for flush cutting of component leads after soldering the component into a circuit board.

4.4.2 Crimpers

Wire terminations come in a large variety of shapes and sizes, **Figure 4-14**. Some wire terminations are soldered onto the ends of wires, as they were most commonly in the past. However, currently most wire terminations are designed for crimp installation. A pliers-like tool called a *crimper*, **Figure 4-15**, is used to compress (crimp) a connector onto the stripped end of a wire to make a tight connection. That connection is then called a crimp, or crimped, connection.

Oil and Gas Photographer/Shutterstock.com

Figure 4-14. A variety of crimp terminals are available.

Courtesy of RIDGID®. RIDGID® is the registered trademark of RIDGID, Inc.

Figure 4-15. Wire crimpers are used to compress (crimp) a connector onto the stripped end of a wire to make a tight connection.

Crimping of terminations gives a positive contact, provided the appropriate crimpers are used. Such terminations include ring, fork, and spade terminals; pins and sockets; coaxial connectors; and many more. Some crimpers have interchangeable crimping dies, which allows for different terminations to be made with the same set of crimpers. Select the appropriate crimping dies for the termination to be crimped.

If done properly, crimped connectors are superior to soldered connectors. The time savings of crimping over soldering results in a connection that has a lower cost and higher reliability.

Never use conventional pliers to install crimp connections. When you complete a crimped connection, give it the pull test. If the terminal comes off during the pull test, it was not properly crimped.

4.4.3 Wire Strippers

Wire strippers remove the insulation from the ends of wires intended for termination. Many have different slots for various wire sizes, **Figure 4-16**. Always use the appropriate slot based on the wire size. If an inappropriate size slot is used, either the insulation will not come off cleanly (too large a slot), or the conductor will be nicked (too small a slot). If the wire is nicked, it may break at the point at which it was nicked. The stripper handles should be at 90° to the wire and pulled in the direction of the end of the wire. If the angle of pull is not correct, the wire will be nicked and soon break.

There are strippers made especially for coaxial cable and other nonstandard cable varieties. Always use the correct strippers for the job. While a knife may work, the risk of nicking the conductor is too great.

GearWrench (Apex Tool Group)

Figure 4-16. Wire strippers are used to strip the ends of wires before making a connection.

4.5 MEASURING TOOLS

You will do well to observe the old adage, "Measure twice and cut once." Maintenance technicians should have several different measuring tools in their toolboxes to measure for cutting or locating holes to be drilled, as well as for sizing replacement parts and verifying tolerance and wear.

The measurement tool to be used depends on the size of the object to be measured and the accuracy the measurement requires. For small parts that do not require much accuracy, the 6″ steel *machinist's rule* is convenient and useful, **Figure 4-17**. This rule has markings for 1/32 of an inch and 1/64 of an inch. The graduations take a good eye to read, especially the markings for 1/64 of an inch. If the job requires this level of accuracy, consider using more accurate measuring tools.

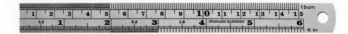

Manrit/Shutterstock.com

Figure 4-17. A machinist's 6″ pocket rule is a handy measuring device.

If more accuracy is required, a set of *calipers* would be the measuring tool to use. A set of calipers makes it easy to measure small parts, and the smaller the part, the more measurement accuracy is required. Digital calipers are accurate to 1/1000 (0.001) of an inch. They are capable of measuring external, internal, and depth dimensions, **Figure 4-18**. Every technician performing maintenance and repair work should have a set of calipers in his or her toolbox. Calipers are available in different sizes based on the maximum measurement capability, such as 6″, 8″, and 12″.

GearWrench (Apex Tool Group)

Figure 4-18. Digital calipers provide accurate measurement of external, internal, and depth dimensions.

For objects larger than 6″ that require a moderate amount of accuracy, a steel *bench rule* might be the measuring tool you select, **Figure 4-19**. Bench rules are available in 12″, 18″, 24″, and 36″ models. A 24″ or a 36″ bench rule probably will not fit into your toolbox. These rules have various marking graduations, and the proper one should be selected depending on the level of accuracy the measurement requires.

photo one/Shutterstock.com

Figure 4-19. Steel bench rules are useful tools for measuring objects larger than 6″ that require a moderate amount of accuracy.

A steel measuring tape, or **steel tape**, is used for many measuring jobs, **Figure 4-20**. Steel tapes are available in 10′, 12′, 25′, 30′, 50′, and many other lengths. The steel tape is not quite as accurate as a rule since the graduated markings are generally 1/16 or 1/32 of an inch on shorter tapes and coarser the longer the tape becomes. The steel tape is used more often than any other measuring tool for many maintenance measurements.

Lufkin (Apex Tool Group)

Figure 4-20. The convenient steel tape is used more often than any other measuring tool.

4.6 HAMMERS AND MALLETS

Hammers and mallets are available in many shapes and sizes for various purposes. See **Figure 4-21**. The **claw hammer** comes to mind when most people hear the word *hammer*. This hammer is useful for driving and removing nails. The claws on the hammer are split in the middle to encompass the head of a nail to be removed. The claws are also useful for prying pieces of wood apart when they have been nailed or stapled together. The claw hammer is used predominately when working with wood.

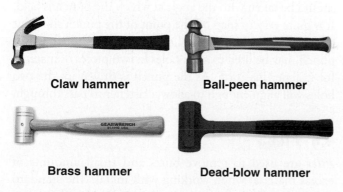

Claw hammer   **Ball-peen hammer**

Brass hammer   **Dead-blow hammer**

Apex Tool Group

Figure 4-21. Depending on the job at hand, there are many useful types of hammer.

The **ball-peen hammer** is one of the tools of choice when working with metal or heavy-duty mechanical assemblies. One side of the head resembles the striking surface of the claw hammer. The other side of the head is spherical in shape. Objects may be struck with either side of the head depending on the job at hand. Ball-peen hammers come in various weights and sizes. Be sure to select the proper-size hammer for your intended use.

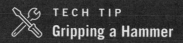

TECH TIP
Gripping a Hammer

All too often, technicians do not grip a hammer properly. A hammer should be held by the grip area of the handle. You will not gain full advantage of the hammer when you hold it too close to the head. Hammers are balanced to be held and used by the grip. If held properly, your striking accuracy will be greatly increased, as will the striking force.

Other styles of hammer include a brass or copper hammer, a dead-blow hammer, a 2-pound hammer, a rubber mallet, and a rawhide mallet. Brass or copper hammers are used where it is inadvisable to create sparks, such as when working around flammable or explosive vapors or when you do not wish to have the hammer mar the surface of the assembly you are working on.

Rubber mallets are sometimes too light or bounce too much when striking. Often, a better alternative is to use a **dead-blow hammer**. Neither hammer will mar the surface; however, the dead-blow hammer is filled with lead shot, giving it greater weight and preventing the annoying bounce when striking an object. A rawhide mallet does not bounce as much as a rubber mallet, and it is more solid and nonmarring. However, rawhide mallets are still rather light.

A 2-pound hammer has a head that weighs 2 pounds. It is like a miniature sledgehammer that can be used with only one hand. This hammer is used where some considerable striking force is required.

There are many varieties of hammer other than those mentioned in this chapter. Each has a purpose, and many come in larger and smaller sizes and weights. You need to select the appropriate hammer for the work you intend to do. Be certain to wear safety glasses when using hammers to prevent eye injuries.

4.7 PULLERS

Hammers are sometimes misused to remove bearings, gears, pulleys, or flywheels. These items are often stuck on a shaft and extremely difficult to remove. Many are held on with setscrews that dimple the shaft and create burrs, which makes removal difficult. The beating these items take by using a hammer for installation or removal will most certainly damage the assembly.

A *puller* is a better alternative to remove these items, **Figure 4-22**. Pullers are designed to remove gears, pulleys, bearings, and wheels without damaging machinery. Different types of puller can be used in different situations. As always, choose the right tool for the job and observe all relevant safety guidelines. Before installing a puller, use a file to remove any burrs from the shaft, which will allow the bearing or gear to slip on easily.

Apex Tool Group

Figure 4-22. Gear pullers are designed to remove gears, pulleys, bearings, and wheels without damaging machinery.

4.8 CHISELS AND PUNCHES

A *chisel*, often referred to as a cold chisel, is used for cutting or chipping, **Figure 4-23**. While not very elegant in its ability to cut metal, it will get the job done if you don't need a very precise or clean cut. There are various sizes and shapes of chisel points. Chisels commonly suffer from two different problems: The point often becomes dull and requires dressing with a file or grinder, and the struck end or head of the chisel can also become mushroomed from constant use. Mushroomed chisel heads need to be cleaned up using a file or grinder. A mushroomed head may result in the hammer glancing off the chisel head when striking

Armstrong (Apex Tool Group)

Figure 4-23. Cold chisels are used for cutting or chipping.

it and possibly injuring the hand you are holding the chisel with. A mushroomed chisel head may also chip, and one of the flying chips may cause injury.

Two of the most frequently used punches are the *prick punch* and the *center punch*, **Figure 4-24**. These two punches are used in combination to mark and dimple a metal object for drilling. The prick punch has a tip that is more pointed than the center punch. The prick punch is used to accurately locate the spot for drilling. The center punch is used to create a larger dimple at the location where the prick punch has marked. The dimple allows the tip of the twist drill to center itself better and allows for a more accurate hole location.

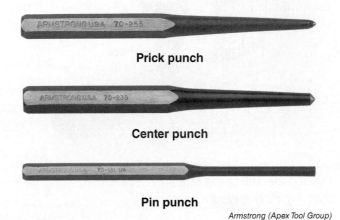

Prick punch

Center punch

Pin punch

Armstrong (Apex Tool Group)

Figure 4-24. The prick punch, center punch, and pin punch are commonly used punches.

Pin punches, or drift punches, are useful in removing roll pins or dowel pins. The tip of the pin punch is appropriately sized to the diameter of the pin. The shaft of the point is often tapered to provide greater strength, so that the tip does not break off. The point is placed on the pin that is to be removed, and a gentle tapping of a hammer on the striking end of the punch will drive the pin out. Care should be taken with the angle at which the punch is held. It is quite easy to snap off the point of the punch should it be misaligned when struck. Due to its tapered shaft, a drift punch may be used to align holes in two pieces of material for fastener insertion. As the punch is driven in, the two holes will align enough to allow a fastener to pass through.

4.9 FILES

Files are used to remove burrs and small amounts of excess material when working with metal. The standard parts of a file are the face, edge, cutting teeth, point, shoulder, and tang, **Figure 4-25**. The tang is designed so that a handle may be installed on the file. Files and handles are generally sold separately.

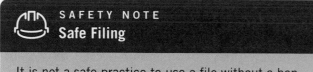

SAFETY NOTE
Safe Filing

It is not a safe practice to use a file without a handle. Without a handle, the tang may cause injury to your hand.

Files are often misused by applying pressure to the file and simply rubbing it back and forth. Files only cut in one direction; proper use of a file dictates that you allow the file to contact the surface of the material and push the file away from you. Pressure is not needed, as the file will do the work itself. If you find it takes too long

to remove an adequate amount of material with a file, consider using a different method of material removal.

Soft metals and other soft materials tend to fill up a file's grooves, creating a loaded file. This clogging of the file is also referred to as pinning, because the material shavings are called pins. Unfortunately, when a file is loaded, it ceases to cut. To remove the material that has loaded the file, a *file card* is used. A file card is a flat device with extremely short, straight bristles. It is pushed in the direction of the grooves in the file. After carding, the file will cut much better. If you have files, you will need a file card to keep your files in top cutting order.

Files vary in their cut (groove pattern) and shape. See **Figure 4-26**. Files range from very fine to coarse and thus remove different amounts of material. If a lot of material

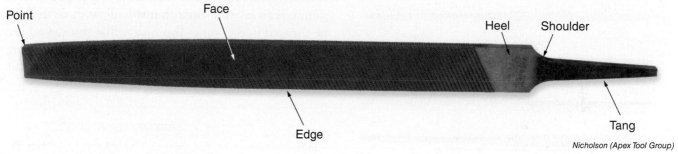

Nicholson (Apex Tool Group)

Figure 4-25. The standard parts of a file.

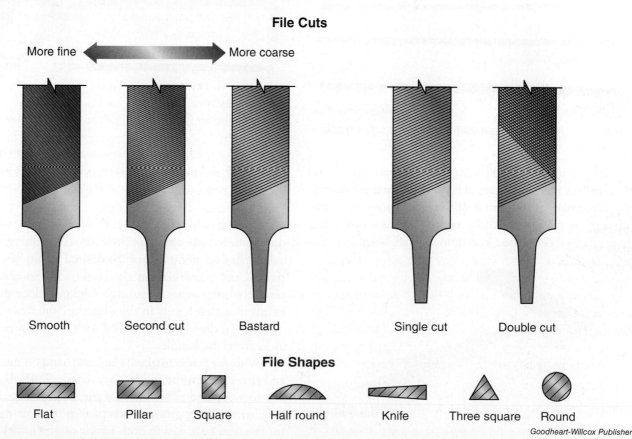

Goodheart-Willcox Publisher

Figure 4-26. These are some common file groove patterns and file shapes.

must be removed, start with a coarse file. Then, as you approach the desired dimensions and shape, use a fine or very fine file to produce a smooth finish. The various shapes make it easier to perform different filing jobs.

Needle files are extremely small files used to get into small spaces, holes, and corners, **Figure 4-27**. While not exactly needle sized, these files are necessary for some filing jobs. They are usually sold in a set.

Nicholson (Apex Tool Group)

Figure 4-27. Needle files are useful to get into small spaces.

A tip cleaner set, used for maintaining welding and cutting torches, is actually a set of round and extremely small, wire-sized files. See **Figure 4-28**. They are not only useful for cleaning the tips of cutting torches, but also when filing work must be performed in extremely small holes.

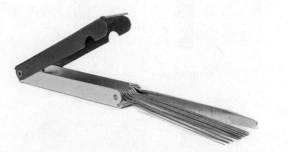

Photo Love/Shutterstock.com

Figure 4-28. A welding tip cleaner set is a set of round and extremely small, wire-sized files.

Diamond files are not actually files in the conventional sense. They are a file form that has been impregnated with industrial diamond grit. These files are extremely aggressive compared to conventional files. They are available in all shapes and sizes, and even as diamond sharpening stones. The grit ranges from coarse to extremely fine (much finer than a conventional file or stone).

4.10 HACKSAW

A *hacksaw* is a hand tool used to cut metal. It has two main parts: a blade and a handle, **Figure 4-29**. There are many different hacksaw blades available with different pitches. The pitch of a hacksaw blade is expressed by the teeth per inch, or TPI. The thickness of the metal to be cut determines the TPI of the blade. The general rule of thumb is to select a blade that will allow three teeth to be engaged in the work. This means that you should use a finer-pitch blade (a higher TPI) when you are cutting thinner metal. A hacksaw is not the tool of choice to cut thin sheet metal.

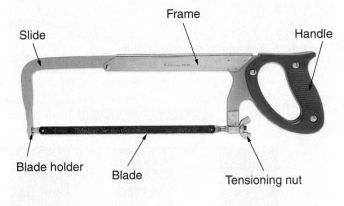

GearWrench (Apex Tool Group)

Figure 4-29. A hacksaw is used to cut metal, except for thin sheet metal.

As with a file, the hacksaw blade only cuts in one direction. Simply applying pressure and rubbing it back and forth does not produce the desired result. The direction of cut is marked on the blade with an arrow. Be certain when you are installing a hacksaw blade that you orient it so that it cuts in the direction you desire. Most commonly, the blade is installed with the arrow pointing away from the handle.

With the blade installed, place one hand on the handle and the other on the front of the frame. Push the hacksaw forward and allow it to do the work. Drag the saw back without any pressure to return to the starting point for the next pass. Using both hands on the hacksaw will allow you to guide it more skillfully.

4.11 POWER TOOLS

Saws, grinders, screwdrivers, wrenches, hammers, and drills operate using either electric or pneumatic power. Electric tools may be corded or cordless (battery-powered).

Electric power tools become warmer the more they are used. The opposite is true for pneumatic power tools; the more a pneumatic power tool is used, the colder it becomes. With plentiful air from an air compressor, pneumatic power tools are often the tool of choice for production work.

You must exercise extreme caution and control when working with power tools. While they work well for cutting metal, they are indiscriminate, and can just as easily cut you as cut the intended target. **Figure 4-30** lists some general safety rules to keep in mind when using power tools.

Power Tool Safety
1. **Understand the power tool.** It is important to read the owner's manual, including safety guidelines and warning labels, before operating any power tool for the first time.
2. **Choose the correct tool.** Do not use a tool that is not intended for the job at hand.
3. **Wear appropriate attire and use proper personal protective equipment.** Do not wear loose clothing or jewelry, as they can catch on moving parts. Tie back long hair. Do not wear gloves when operating certain power tools. Wear eye and hearing protection. When needed, wear a face shield, hard hat, gloves, or safety shoes.
4. **Be aware of your surroundings.** Identify hazards in the work area, such as power lines, electrical circuits, water pipes, and other mechanical hazards that may be hidden from view or below the work surface.
5. **Guard against electric shock.** Ground all tools unless they are double insulated.
6. **Do not abuse the cords on corded tools.** Never carry a tool by its power cord or yank the cord to disconnect it from a power source.
7. **Disconnect power tools** when not in use, before servicing, and when changing or installing accessories.
8. **Secure work.** Use clamps or vises to hold workpieces, as loose workpieces can cause injury.
9. **Avoid accidental starting.** Do not carry any tools with a finger on the switch. Be sure the switch is off before plugging a tool into a power outlet.
10. **Maintain all tools.** Inspect any tool for defects before using. Keep tools clean, sharp, and lubricated. Make sure handles are clean, dry, and free from any oil or grease. Change accessories according to the instructions, and have any repairs made by qualified service personnel.

Goodheart-Willcox Publisher

Figure 4-30. Here are some safety tips to keep in mind when using power tools.

4.11.1 Drills

Powered *hand drills* make drilling holes easier and are available in various sizes and styles. For production work, a pneumatic drill is often used. For repair work, or where a source of AC power is not readily available, a battery-operated drill is the tool of choice, **Figure 4-31**.

Battery-operated drill

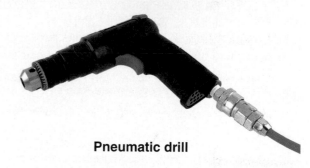

Pneumatic drill

Dinga/Shutterstock.com; Shutter Baby photo/Shutterstock.com

Figure 4-31. Depending on the job, both a battery-operated drill and pneumatic drill are useful power tools.

Hammer drills are almost always corded, **Figure 4-32**. They are used for drilling holes in concrete and masonry. In addition to the rotary action of other drills, hammer drills also incorporate a hammering action that helps break up the masonry material.

OlegSam/Shutterstock.com

Figure 4-32. Corded hammer drills incorporate a hammering action useful for drilling holes in concrete and masonry.

The part of the drill that holds the twist drill or bit is referred to as the chuck. Some chucks require a chuck key to tighten them, and others are keyless. Keyless chucks are tightened by hand and offer a much faster method of changing the twist drill bit.

Most drills provide variable speed, which allows the operator to select the best drilling speed for the type of material and size of hole being made. The trigger of the drill controls the rotary motion, and the further it is pulled, the faster the rotational speed.

Eimantas Buzas/Shutterstock.com

Figure 4-33. A variety of drill bits are pictured here, including a step bit (second from right).

SAFETY NOTE
Drill Safety

The following are some safety tips to keep in mind when using a power drill:

- Make sure the trigger works properly, turning the drill off when released.
- Make sure the chuck and bit are secured tightly and chuck key is removed.
- Unplug the drill before changing bits.
- Make sure there are no loose items that could catch in the rotating drill.
- Always wear safety goggles or a face shield.

⚠ CAUTION

For thicker steel or aluminum materials, it is advisable to use a small amount of cutting lubricant when drilling to help preserve the sharpness of the twist drill's cutting edge.

For deep holes, especially in aluminum, it is advisable to drill a small amount and then completely extract the twist drill to ensure that all the chips have been cleared. Repeat this process multiple times until the desired depth has been achieved. If chips are not cleared, excessive heat buildup due to clogging will present problems and further drilling will slow or stop. Excessive heat will also dull the twist drill and make it ineffective.

When a drill is referred to as a 1/4″, 3/8″, or 1/2″, this size makes reference to the maximum diameter of the twist drill bit it will accept. Normally, drills that accept larger twist drill bits are more powerful.

Drilling extremely thin material such as sheet metal does not produce good results unless the hole is rather small. A 1/2″ hole in sheet metal would cause the twist drill to hang and the sheet metal to tear. At best, the shape of the resulting hole would be quite ragged. It is best to punch sheet metal rather than drill it if the desired hole is at all large.

The same holds true for hard plastic materials such as polycarbonate or plexiglass. Once the twist drill penetrates a little more than halfway, it will most likely catch and crack the material unless the hole is relatively small. When drilling holes in such materials, first drill a very small hole with a twist drill bit, and then finish the job with a step drill bit, **Figure 4-33**. Step bits create increasingly larger holes in a step fashion. The cutting angle is such that it will not cause the plastic material to crack as the step bit goes through.

4.11.2 Angle Grinder

An *angle grinder*, **Figure 4-34**, is a handheld power tool that has an abrasive, motor-driven grinding wheel mounted at a 90° angle to the shaft. For welding and cutting, the

Dotco (Apex Tool Group)

Figure 4-34. The angle grinder is a valuable tool for cleaning up a cut, removing paint, and clearing rust.

angle grinder is a valuable tool. Cleaning up a cut, removing paint, and clearing rust are tasks easily accomplished with an angle grinder. An angle grinder is the tool to use when the object to be ground is too big, too heavy, or too long for the use of a pedestal or bench grinder.

Dotco (Apex Tool Group)

Figure 4-35. Die grinders are used for removing small amounts of material by hand with as much precision as possible.

SAFETY NOTE
Protect Your Hearing

Noise is a major factor with an angle grinder, so be sure to wear hearing protection when using one.

⚠ CAUTION

Take care to ensure the abrasive wheel of the angle grinder is free from chips and chunks. The speed at which the wheel turns can cause it to shatter if grinding operations are carried out with a chip or chunk out of the wheel.

4.11.3 Die Grinder

A *die grinder* is similar to an angle grinder, except the grinding wheel (or stone in some cases) is on the same axis as the drive mechanism, **Figure 4-35**. Die grinders are most often pneumatic and can range from pencil-sized to palm-sized. A die grinder is used for removing small amounts of material by hand with as much precision as possible. Palm-sized die grinders are also available in a right-angle configuration.

SAFETY NOTE
Die Grinders

- Be sure the switch is in the "off" position before you plug the grinder in.
- Always check that the grinding wheel is tight before use.
- Make sure the workpiece is securely clamped or held.
- Allow work surfaces to cool before touching them.
- Always point the grinder away from you and wear safety goggles.

4.11.4 Abrasive Cutoff Saw

An *abrasive cutoff saw* is similar to an angle grinder or die grinder, but it is fitted with a thin abrasive wheel used for metal cutting, **Figure 4-36**. The abrasive wheel is much thinner than a grinding wheel, so that it takes a narrower *kerf* (the portion of metal removed during the sawing process).

The abrasive cutoff saw cuts in a straight line. Any attempt to cut a curve or change angle with the cutting wheel engaged in the material may cause the wheel to shatter or chunk.

SAFETY NOTE
Chipped Cutting Wheel

Once a chunk or chip comes out of the cutting wheel of an abrasive cutoff saw, the wheel must be replaced, as it presents a danger to the user. Safety glasses or a face shield are a must when operating this tool.

Yupa Watchanakit/Shutterstock.com

Figure 4-36. The abrasive cutoff saw is similar to an angle grinder or die grinder, but it is fitted with a thin abrasive wheel used for cutting metal in a straight line.

4.11.5 Reciprocating Saw

In *reciprocating saws*, **Figure 4-37**, the blade of the saw is relatively short when compared to the size of the saw. The blade of a reciprocating saw moves in a reciprocating motion (in and out). Reciprocating saws are sometimes referred to by the trademarked name Sawzall™.

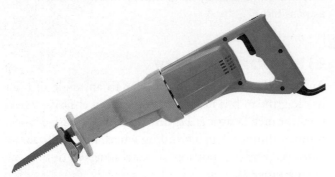

Charles Brutlag/Shutterstock.com

Figure 4-37. Select a blade for a reciprocating saw based on the type of material you intend to cut and required smoothness of the cut.

The better versions of these saws move the blade in an elliptical pattern (forward and backward as well as in and out), also called orbital action. The dual motion pushes the blade forward during the cutting portion of the stroke and pulls it back slightly while it returns to the starting position of the next cutting stroke. The amount of elliptical motion or orbital action is sometimes referred to as the "crowd," and on some models this motion is adjustable.

As with any cutting tool, excessive force is not required with a reciprocating saw. Allow the blade to do the cutting and advance the tool slightly to keep it engaged in the work. There is a wide array of blades available for cutting different materials. Select the appropriate blade depending on the smoothness of cut required and the type of material you intend to cut. It is important that you have a good selection of replacement blades. Safety glasses and hearing protection are required while using a reciprocating saw.

SAFETY NOTE
Reciprocating Saw

Here are some other safety tips for using a reciprocating saw:

- Use the shortest blade that will do the job, which will minimize the flexing of the blade.
- Before starting, be sure there are no hidden electrical wires or water pipes behind what you are cutting.
- Avoid cutting above shoulder height.

4.11.6 Circular Saw

A *circular saw* has a round (circular) blade. A motor rotates the blade, and the blade cuts the material, **Figure 4-38**. Circular saws are often used to cut wood, though they are capable of cutting other materials as well. A circular saw cuts the workpiece from the bottom up, so that the saw does not kick up during the cutting process. A table saw cuts from the top down, so that the workpiece does not kick up. Circular saws are placed on top of the material, while table saws are located underneath the material to be cut. This means that if you are cutting wood with a circular saw, you need to place the "good" side down. Conversely, if you are cutting wood using a table saw, you place the "good" side up. (When cutting wood, one side—the good side—will cut cleanly and one side will splinter slightly.)

phoMAKER/Shutterstock.com

Figure 4-38. The circular saw has a round (circular) blade that cuts the workpiece from the bottom up, so that the saw does not kick up during the cutting process.

SAFETY NOTE
Safe Use of a Circular Saw

The following are some safety tips to keep in mind when using a circular saw:

- Only use sharp blades. Dull blades can cause stalling and kickback.
- Do not use a saw too large or heavy for you to control.
- Check the guards before each use to make sure they work properly.
- Make sure the cord is long enough and not in the blade path before starting the saw.
- Make sure the workpiece is securely clamped.

4.11.7 Portable Band Saw

A band saw consists of a long toothed belt, or band, stretched between wheels and powered for continuous cutting. Stationary band saws are arranged with the blade in either a vertical or horizontal position. The ***portable band saw*** is designed for cutting metal, **Figure 4-39**. It is similarly configured to the horizontal band saw and is capable of making cuts that are impossible to make with a vertical band saw.

Susan Law Cain/Shutterstock.com

Figure 4-39. This worker is using a portable band saw to cut metal.

The throat of the portable band saw can cut material that is 4″ × 4″. This makes the portable band saw excellent for cutting small pipe, flat stock, extrusion, threaded rod, rod stock, and strut channel. The blade is pulled so that the cutting action is toward you. A material stop prevents the saw from jumping away from you when the cutting gets tough. Most portable band saws have a speed control trigger to adjust the cutting speed for optimal performance.

The portable band saw is a heavy device, and it requires practice to be able to cut along an intended line. The saw has a quick-change lever that slacks the band when pulled, so that the old blade may be easily removed and a new blade installed. The blade tension must be properly adjusted with the tension control. If the tension is too great, a blade break may occur. With too little tension, the blade may not provide accurate control of the cut. You also need to observe all safety precautions when using a portable band saw.

4.12 HANDLING AND STORAGE

Hand tools must be handled and stored properly if they are to last. Do not simply throw your tools in a drawer or toolbox. Instead, ensure that all tools are cleaned and properly stored according to manufacturers' recommendations. Proper maintenance will ensure your tools last as long as they are supposed to—often for a lifetime.

CHAPTER WRAP-UP

Each job you do will be best accomplished by a particular tool, and you should always use the proper tool for the particular job. Using the wrong tool can sometimes cause damage, which will make your job harder in the long run. Many tools will be provided by your employer, but you will also be expected to have some basic tools of your own. Purchase quality tools so that they will last. You do not have to buy hundreds of tools right away. As you grow in your career, so will your toolbox and the number of tools you own. When using any tools—especially power tools—always make safety the highest priority.

Chapter Review

SUMMARY

- Before using any tools, it is important to review the safety instructions provided by the manufacturer, your employer, and OSHA.

- You should understand the proper use of each tool in your toolbox and know and avoid the common misuses of maintenance tools.

- Wrenches are used for tightening and loosening nuts and bolts.

- Make certain you use the proper-size wrench for the fasteners you are working on.

- Many maintenance tools are available in both SAE and metric sizes. Make certain you are using the appropriately sized tool.

- If using sockets to remove stubborn bolts and nuts, the use of a breaker bar would be preferable to a ratchet wrench, as the ratchet may be damaged by excessive force.

- Replace hex keys as soon as they become worn to avoid rounding out the hexagonal recess of a screw.

- When using adjustable wrenches, make certain you are turning them in the appropriate direction to prevent fastener damage.

- You should select the proper point style and size of screwdriver based on the job at hand.

- When using tongue-and-groove pliers, make certain they are applied and turned in the proper direction to prevent damage.

- Always use the proper crimper to crimp terminals. Never use common pliers for crimping.

- Wire strippers are the appropriate tool to remove insulation from wires. The use of other tools, such as diagonal cutters or a knife, to remove insulation may nick the conductor and cause breakage.

- The measurement tool to be used depends on the size of the object to be measured and the accuracy the measurement requires.

- Hammers and mallets are available in many shapes and sizes for various purposes. Select the appropriate hammer for the work you intend to do, and wear safety glasses to prevent eye injuries.

- Make certain that the cutting edges of cold chisels are properly dressed and that the heads are not mushroomed.

- Files and hacksaws cut in one direction only; do not simply rub them back and forth. Minimal pressure should only be applied in the cutting direction (away from you).

- Air-operated (pneumatic) handheld power tools become colder with use, while electrically operated power tools become warmer.

- Powered hand drills are available in various sizes and styles and operate using power from a battery, a line cord, or compressed air.

- Hand tools must be handled correctly, cleaned, and stored properly to ensure they last.

REVIEW QUESTIONS

Answer the following questions using the information provided in this chapter.

1. If a bolt head is sized as metric and you only have SAE wrenches, what might happen if you attempt to use them on the metric bolt?

2. *True or False?* Sockets are sized according to the bolt thread size and not the physical dimensions of the bolt head.

3. List the various components you might find in a socket set and the use of each.

4. What is a torque wrench and when should you use it?

5. What are hex keys and how does a technician use them?

6. Selecting a screwdriver tip that is too _____ may result in stripping out the driving slot and rendering the screw impossible to remove.

7. What are two common misuses of pliers?

8. _____ pliers are used for cutting copper wire and component leads.

9. *True or False?* Since they are heavy and sturdy, it is acceptable to use lineman's pliers as a hammer.

10. Why is it important to use a crimper rather than a pair of pliers to crimp terminals?

11. A set of _____ makes it easy to accurately measure small parts.

12. *True or False?* Holding a hammer close to its head will greatly increase your striking accuracy.

13. A(n) _____ is designed to remove gears, pulleys, bearings, and wheels without damaging machinery.

14. What are the two issues chisels commonly experience?

15. *True or False?* The prick punch is used to accurately locate a spot for drilling.

16. List three standard parts of a file.

17. Explain how to use a hacksaw.

18. Why are pneumatic power tools often preferred for production work?

19. _____ drills are used for drilling holes in concrete and masonry.

20. *True or False?* Excessive force may be required to keep a reciprocating saw engaged in the work.

21. List the appropriate PPE you need when using a circular saw.

NIMS CREDENTIALING PREPARATION QUESTIONS

The following questions will help you prepare for the NIMS Industrial Technology Maintenance Level 1 Maintenance Operations credentialing exam.

1. Which of the following would you use to tighten the bolts on a multiple-bolt flange where all bolts must be tightened the exact same amount?

   A. Speed wrench
   B. Air impact wrench
   C. Torque wrench
   D. Combination wrench

2. What type of pliers can be used to remove damaged nuts and bolt heads that are rounded off?

   A. Locking pliers
   B. Long-nose pliers
   C. Lineman's pliers
   D. Diagonal cutting pliers

3. What measuring tool is best for objects larger than 6′ that require a moderate amount of accuracy?

   A. Steel tape
   B. Bench rule
   C. Calipers
   D. Machinist's pocket rule

4. Which of the following is designed to remove items such as gears and bearings without damaging machinery?

   A. Ball-peen hammer
   B. Puller
   C. Chisel
   D. File

5. What does it mean if a drill is pneumatic?

   A. It is battery-operated.
   B. It requires a line cord to operate.
   C. It requires a chuck key to operate.
   D. It uses compressed air to operate.

5 | Fasteners

LEARNING OBJECTIVES

After completing this chapter, you will be able to:

☐ Understand common terminology related to screw threads.

☐ Recognize designations used to specify inch based and metric threads.

☐ Identify different types of bolts and screws and describe how they are used.

☐ Identify common types of nuts and their uses.

☐ Explain typical methods of thread repair.

☐ Describe the purpose of washers.

☐ Recognize common types of non-threaded fasteners and their applications.

☐ Understand how to install and remove blind rivets.

TECHNICAL TERMS

bolt	nut
cable tie	pin
cap screw	pitch
clevis fastener	pitch diameter
clip	retaining ring
crest	rivet
die	rivet nut
fastener	root
key	tap
keyseat	thread class
keyway	thread form
lead	thread series
major diameter	washer
minor diameter	

n almost every aspect of your daily work, you will encounter fasteners in one form or another. A *fastener* is a device used to hold, or fasten, parts together. It is important to be aware of the different types of fasteners and their characteristics. Fasteners come in various forms and sizes. You must have the knowledge and tools to work with fasteners properly.

This chapter covers both threaded fasteners and non-threaded fasteners. To understand how threaded fasteners function, it is important to be aware of general terminology related to screw threads and characteristics of threads. The following sections introduce common terms and definitions for screw threads.

5.1 THREAD TERMINOLOGY

Screw threads consist of ridges that wrap around a cylindrical surface and follow a helical path. The basic shape of a screw thread is defined by the grooves formed on the machined part. Two basic types of screw thread are external thread and internal thread. Bolts and screws have external thread, where the grooves of the thread are on the outside of the fastener. Nuts and tapped holes have internal thread, where the grooves of the thread are on the inside of the fastener. External threads are considered to be "male" threads, while internal threads are considered to be "female" threads. The basic parts of a screw thread are shown in **Figure 5-1**. As you study the figure, make sure you understand the following terminology.

The **crest** is the top surface of a thread. The **root** is the bottom surface of the thread between two sides, or flanks, of the thread. The **major diameter** is the largest diameter of the thread, measured from crest to crest for external thread or from root to root for internal thread. The **minor diameter** is the smallest diameter of the thread, measured from root to root for external thread or from crest to crest for internal thread. The **pitch diameter** is the diameter corresponding to a theoretical cylinder passing through the points on the thread profile where the width of the thread ridge is equal to the width of the groove.

The **pitch** is the distance from one point on a thread to the corresponding point on the next thread. The pitch specification for inch-based screw threads is based on the number of threads per inch. The pitch is calculated by dividing 1″ by the number of threads per inch. For example, if the thread has 20 threads per inch, the pitch is equal to .05″ (1″ ÷ 20 = .05″). The pitch specification for metric screw threads is the actual pitch measurement.

The **lead** refers to the axial distance the thread advances in one complete revolution. On single-thread screws, the lead is equal to the pitch.

5.1.1 Thread Forms and Thread Series

A **thread form** is the standard profile of the thread. A number of thread forms based on approved American or metric standards are in use today. Standards for thread forms in the United States have been designated by the American National Standards Institute (ANSI). The most commonly used standard thread form for inch-based threads in the United States is the Unified system. A **thread series** is a classification of threads in a thread system. There are four main thread series in the Unified system: Unified Coarse (UNC), Unified Fine (UNF), Unified Extra Fine (UNEF), and Unified Constant Pitch (UN). Each series is used to designate combinations of diameter and pitch (pitch is specified in threads per inch).

UNC series threads are very common and are used for general applications requiring fast assembly and disassembly. Coarse threads have fewer threads per inch

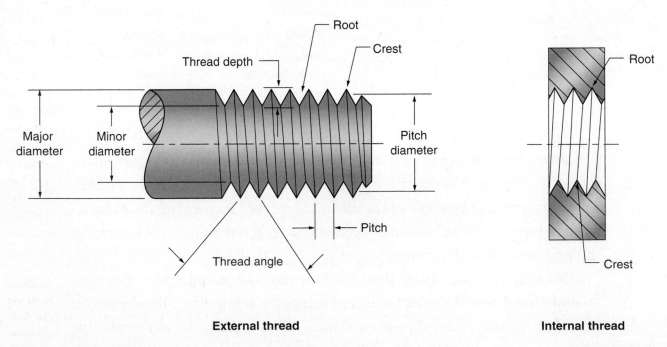

Goodheart-Willcox Publisher

Figure 5-1. The basic parts of external and internal screw threads.

than fine threads of the same diameter. UNF series threads have a smaller thread depth than coarse threads and are used where higher strength is required to bear greater loads. UNEF series threads are used in applications where finer thread is required. UN series threads are used for special purposes. Threads in the UN series are specified with a number preceding the letters *UN* to indicate the number of threads per inch for all diameters. For example, the designation *8UN* indicates the thread has eight threads per inch. UN series threads are used for applications where threads in the other series do not meet design requirements.

5.1.2 Thread Class

The ***thread class*** specifies the amount of tolerance permitted in engagement between mating threads. The tolerance refers to the total amount a part can vary from the design size. Each thread class includes the letter designation *A* or *B* to indicate external or internal thread. The Class 1A, 2A, and 3A designations are for external threads and the Class 1B, 2B, and 3B designations are for internal threads. Classes 1A and 1B have the greatest amount of tolerance, Classes 2A and 2B require a closer tolerance, and Classes 3A and 3B require the tightest tolerance.

5.2 THREAD SPECIFICATIONS

On drawings that include threaded features, such as threaded holes, thread notes are given to specify the required thread. See **Figure 5-2**. For inch-based threads, the thread note specifies, in order, the major diameter of the threads, number of threads per inch, thread form and series, and thread class. In **Figure 5-2**, the thread note 3/4–10UNC–2B indicates the threads are 3/4″ diameter with 10 threads per inch and are in the Unified Coarse series. The threads are internal threads with the Class 2B thread class designation. Note in the example shown that the thread depth is also specified, as indicated by the depth symbol and dimension. If the thread goes through the entire part, the depth is specified as THRU.

5.3 METRIC THREADS

You will frequently encounter metric fasteners on the job. Although metric thread is similar in appearance to inch-based thread, metric fasteners and inch-based fasteners are not interchangeable. On drawings, metric threads are identified using the metric thread symbol *M*. See **Figure 5-3**. In addition to the metric thread symbol,

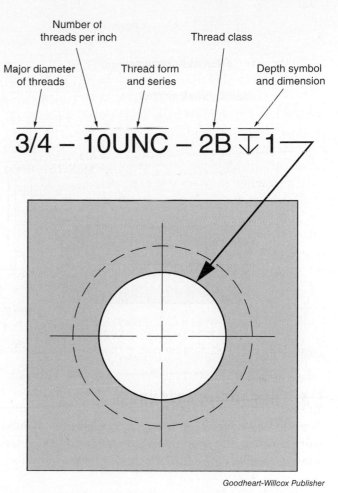

Goodheart-Willcox Publisher

Figure 5-2. Requirements for a threaded feature are specified with a thread note on the drawing. This example shows a thread note for inch-based internal threads.

a metric thread note specifies the nominal diameter, pitch, and tolerance class. The nominal diameter and pitch are specified in millimeters and separated by an × symbol. The pitch is the actual pitch measurement, as previously discussed. Standards for metric threads have been established by the International Organization for Standardization (ISO).

The tolerance class designation consists of separate specifications for the pitch diameter tolerance and the crest diameter tolerance. Each tolerance specification consists of a number followed by a letter. The number indicates the tolerance grade and the letter indicates the "position" of the tolerance. The tolerance position is indicated as "g" or "h" for external thread or as "G" or "H" for internal thread. The "h" or "H" position indicates zero allowance is permitted and the "g" or "G" position indicates a small allowance is permitted. In **Figure 5-3**, the specified tolerance class is 6h6g. The pitch diameter tolerance specification is 6h and the crest diameter tolerance specification is 6g.

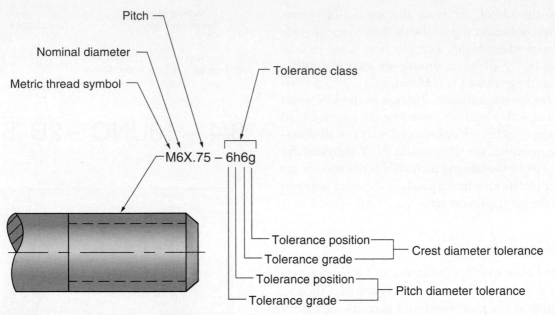

Figure 5-3. A thread note for metric threads. This specification is for external threads.

Goodheart-Willcox Publisher

5.4 THREADED FASTENERS

Threaded fasteners make use of screw threads to hold parts together. The most common types of threaded fasteners you will come across are bolts, screws, and nuts.

5.4.1 Bolts and Screws

Bolts and screws are threaded to screw into nuts or threaded holes. There are many types of bolts and screws available for various uses in industry. There is no official definition as to what constitutes a bolt or a screw. A *bolt* can be defined as a fastener with a head at one end and external threads at the other to accept a nut. The term *bolt* is generally used to describe a fastener that is inserted through non-threaded holes and affixed with a nut. The term *screw* is generally used to describe a fastener that screws into threads in an assembly. Typically, screws only fasten through one side of an assembly, whereas installing bolts requires access to both sides. For example, the term *cap screw* is commonly used to refer to a fastener that is similar to a bolt, but extends through a clearance hole in one part and screws into a second part without using a nut. However, some screws are installed with nuts in the same fashion as bolts when used to hold two parts together.

Common types of bolts and screws used for a variety of fastening purposes are shown in **Figure 5-4**. A description is given for each fastener. Note the different styles used for specific applications.

There are different head styles used for screws and bolts. Common head styles you will encounter are shown in **Figure 5-5**. Different head styles require different tools, such as an open-end or box-end wrench, socket wrench, hex key, hex bit socket, slotted screwdriver, or Phillips screwdriver. Common drive types used on screw and bolt heads are shown in **Figure 5-6**. You must use the appropriate tool when installing or removing fasteners based on the corresponding drive type. For more information about tools, refer to Chapter 4, *Maintenance Tools*.

A variety of screws are capable of cutting their own thread in the material they engage in. These are known as "self-tapping" screws. See **Figure 5-7**. Thread-cutting screws are self-tapping screws with a cutting edge at the end. They are installed by first drilling a pilot hole of the appropriate size. Then, the screw is driven into the hole. As the screw is driven in, it cuts its own threads. Another type of self-tapping screw is installed by first drilling an undersized hole. Then, the screw is driven into the hole. As the screw is driven, it displaces material. This type of screw is commonly called a *sheet metal screw*. Sheet metal screws are used to fasten together sheet metal and plastic parts. A third type of self-tapping screw, called a self-drilling, self-tapping screw, is used to drive into sheet metal without first cutting a pilot hole.

Lag screws and wood screws are most often used for fastening into wood. Prior to installation, it is necessary to drill a pilot hole to the correct size.

When working with bolts and screws, it is important to start with clean, undamaged fasteners. When fastening bolts, start by hand-threading to ensure that the threads are properly aligned. This helps prevent damage to the threads as the fastener is tightened further. You should also use lubricant, nuts, and washers as recommended by the manufacturer or to meet the requirements of a particular application.

When tightening several bolts on the same object (such as a flange), you should tighten them in a criss-cross pattern, rather than in an adjacent pattern. This is done to reduce preloading "crosstalk" between the bolts, **Figure 5-8**. If necessary, refer to an installation or service manual to determine the proper tightening sequence. Improper preloading can cause uneven compression and conditions such as leaks and premature failures, especially in critical applications.

Common Bolts and Screws		
Fastener Appearance	Name	Description
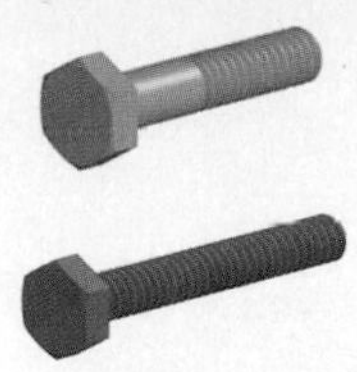	Hexagonal head bolts and screws	Commonly used with tapped holes or nuts. May be installed with an open-end wrench, box-end wrench, or socket wrench.
	Stud	A rod that usually has threads on both ends. May also be fully threaded or partially threaded on one end.
	Socket head screw	Similar to other bolts or screws, except it is installed and removed with a hex key. The head has a hexagonal recess to accept the hex key.
	Machine screws	Used for installation into a tapped hole or nut. The types shown are installed with a slotted screwdriver or a Phillips screwdriver.
	Shoulder bolt	Provides a shoulder in order to form a pivot point between two assemblies.
	Setscrew	Used to tighten collars, pulleys, gears, and other devices onto a shaft. Most common setscrews use a hex key or screwdriver for removal. Also used for other adjustments.
	Elevator bolt	Commonly used on conveyors or other assemblies as an adjustable height foot in order to level or change the height of a device.
	Carriage bolt	Has a rounded head with no driving method. It has a square collar beneath the head so that it may engage into an assembly.
	Eye bolt	Used as an attachment or lifting point. Often found on the top of equipment racks for safe lifting and placement.
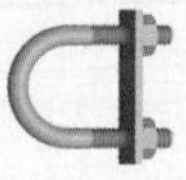	U-bolt	Used for clamping round objects, such as pipes, to flat surfaces. When used with a rounded mounting plate, can be used to clamp round objects together. A typical example is a muffler clamp.

Images: © McMaster-Carr 2018

Figure 5-4. Common types of bolts and screws used for fastening applications.

Fastener Head Types

Fastener Appearance	Name	Description
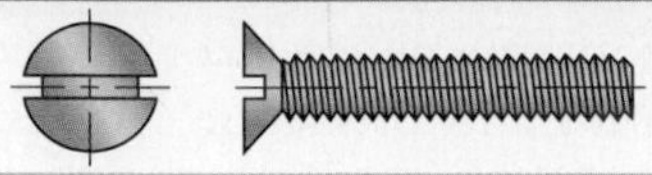	Flat head	A countersunk head with a flat top. When installed, the head of the screw is flush with the surface on which it is installed.
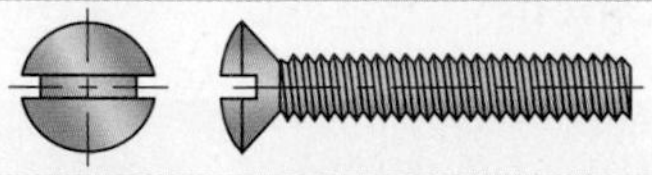	Oval head	A countersunk head with a rounded top.
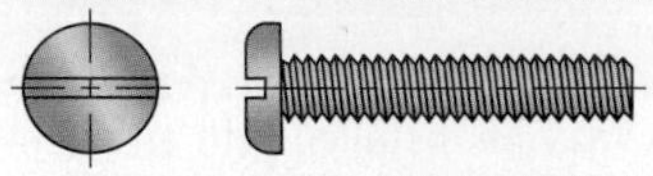	Pan head	A minimally rounded head with short vertical sides.
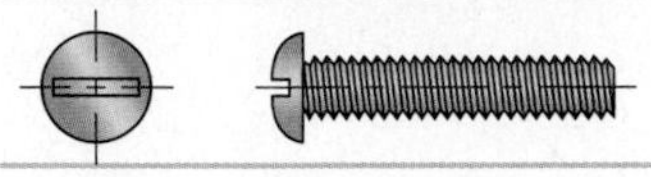	Round head	A round or domed top.
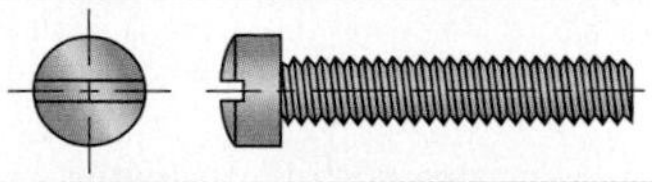	Fillister head	Has a smaller diameter than a round head.
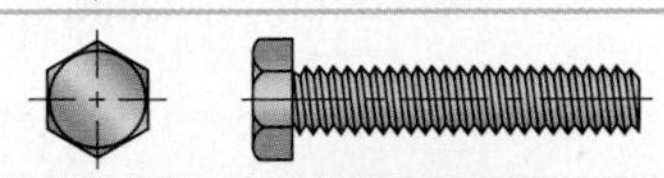	Hex head	A hexagonal-shaped head that allows a wrench or socket to be used for installation or removal.
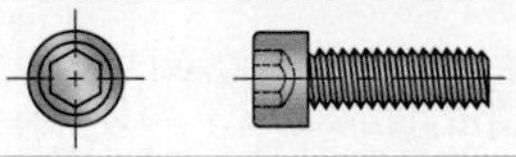	Hex socket cylindrical head	A small cylindrical head that accepts a hex key for installation or removal.
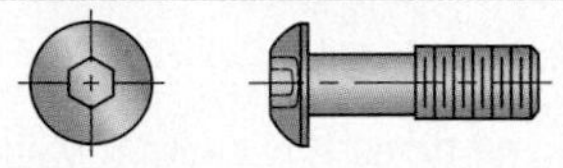	Hex socket button head	A low-profile head that accepts a hex key for installation or removal. The low profile is helpful in applications where space above the screw head is limited.

Goodheart-Willcox Publisher

Figure 5-5. Common styles of screw and bolt heads.

Fastener Drive Types

Fastener Appearance	Name	Description
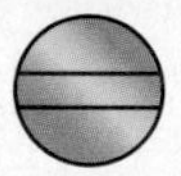	Slotted drive	Requires a slotted screwdriver to install or remove.
	Phillips drive	Phillips screwdrivers come in different sizes. You must use the correct size. Otherwise, you run the risk of stripping the screw head.
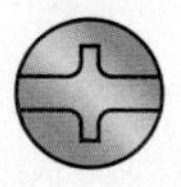	Combination drive	You can use either a Phillips or slotted screwdriver with this drive type.
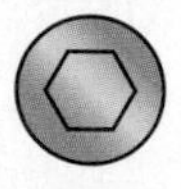	Hex drive	Installation or removal requires a hex key, also called a hex wrench or Allen wrench. Be careful to select the correct size tool to fit the fastener head. Hex keys are manufactured in both inch and metric sizes. Some hex key sets contain more keys in a greater range of sizes. This helps you select the correct size.
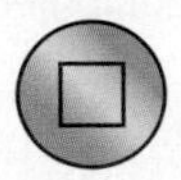	Square drive	Square drive fasteners are made in a variety of sizes. Make sure to use the correct size bit or screwdriver. This drive style is very popular in Canada.
	Torx drive	Torx drive screws are made in many different sizes. Make sure to select the proper size bit or screwdriver to prevent damage. This drive style is popular in the automotive industry and is also used in applications where screw removal is discouraged.

Goodheart-Willcox Publisher

Figure 5-6. Common drive types for fasteners.

Tapping and Wood Screws

Fastener Appearance	Name	Description
	Thread-cutting machine screws	Designed for self threading into blind holes that have been drilled to the appropriate size. They do not require that the hole be tapped, as they cut their own thread.
	Sheet metal screws	To install, a small hole is drilled to accept the screw point. As the screw is driven in, it cuts its own threads.
	Self-drilling, self-tapping screws	Designed to drill into sheet metal and cut thread during the installation process.
	Lag bolt	Used for fastening into wood. Be sure to drill the correct size pilot hole prior to installation.
	Wood screw	Most often used for fastening into wood.

Images: © McMaster-Carr 2018

Figure 5-7. Common types of metal and wood screws. The self-tapping types are designed to cut their own threads in metal as they are screwed in. These include thread-cutting machine screws, sheet metal screws, and self-drilling, self-tapping screws.

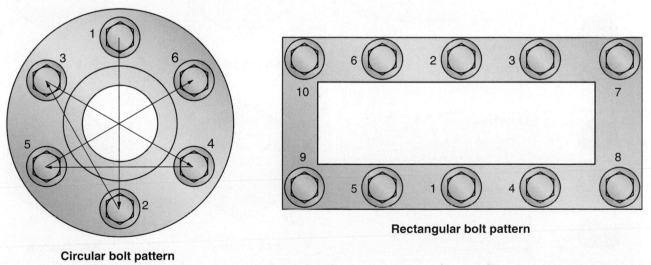

Circular bolt pattern

Rectangular bolt pattern

Goodheart-Willcox Publisher

Figure 5-8. A crisscross pattern is recommended when tightening multiple bolts on a part. The numbers indicate the sequence of tightening. The sequence for the rectangular pattern moves in a spiral pattern beginning at the center.

5.4.2 Nuts

Nuts are used in conjunction with bolts and screws in through-hole fastening applications. Generally defined, a **nut** is an internally threaded fastener screwed to a bolt to hold parts together. There are many types of nuts, **Figure 5-9**. Nuts are hexagonal or square in shape (hexagonal nuts are most common). Nuts are installed and removed with a wrench. The wrench size used corresponds to the measurement across the flats of the nut.

In addition to hexagonal and square nuts, there are other types of nuts with special characteristics for fastening applications. These include flange nuts, jam nuts, locknuts, tee nuts, and coupling nuts.

A flange nut has a flange that acts like a flat washer. This spreads the clamping force over a larger area. A jam nut is not as thick as a regular nut. It is tightened against another nut to lock it in place.

Locknuts are used to help prevent parts from loosening. A self-locking nut has deformed threads or a nylon insert to produce interference. After the nut is initially threaded, additional force must be used to tighten the assembly. Self-locking nuts are used for assemblies that must remain tight when subjected to high vibration. A lock washer nut has a free-spinning external lock washer to keep the nut from loosening once tightened. A slotted nut has slots where a cotter pin is installed to lock the bolt in place. After tightening, the installer inserts a cotter pin in one of the slots and through a hole drilled in the bolt.

Tee nuts are used when bolting a wooden part to another part. A tee nut has a collar with prongs that engage one side of the assembly and hold it in place. The collar is similar to a flat washer and allows the clamping force to be distributed to a larger area.

Coupling nuts are used to connect studs or bolts together. This lengthens the threaded assembly.

Assemblies that are fastened using nuts and bolts are generally stronger than those using screws. Always select the correct fasteners to meet the strength requirements. Bolt heads have markings identifying the grade, which indicates the tensile strength. Standard head markings have been designated by the Society of Automotive Engineers (SAE) and ASTM International. SAE head markings are shown in **Figure 5-10**. A higher grade indicates greater strength.

Types of Nuts	
Fastener Appearance	**Name**
	Hexagonal nut
	Flange nut
	Jam nut
	Self-locking nut with nylon insert
	Lock washer nut
	Slotted nut
	Tee nut
	Coupling nut

Images: © McMaster-Carr 2018

Figure 5-9. Common types of nuts.

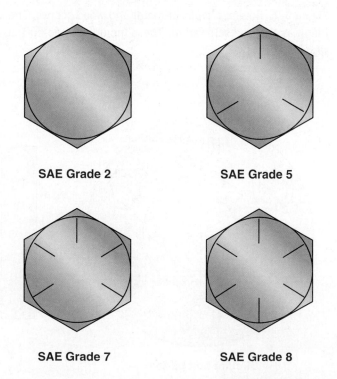

SAE Grade 2   SAE Grade 5

SAE Grade 7   SAE Grade 8

Goodheart-Willcox Publisher

Figure 5-10. Standard SAE head markings for bolts.

Bolts and nuts should be tightened to the specified torque using a torque wrench. Torque specifications for fasteners are given by the manufacturer. Applying the proper amount of torque produces a strong joint and prevents part failure. If too little torque is applied, the assembly can become loose. Too much torque can break the fastener or damage the thread. If manufacturer specifications are not available, refer to a torque value chart for the specific fastener being installed.

5.5 TAPS AND DIES

Taps and dies are used for thread cutting. A *tap* is used to cut internal threads. See **Figure 5-11**. A *die* is used to cut external threads. See **Figure 5-12**. These tools can be used when it is necessary to cut new threads on a part or repair a part with damaged threads. Often, damaged or stripped thread can be restored to a functional state through the use of a tap or die.

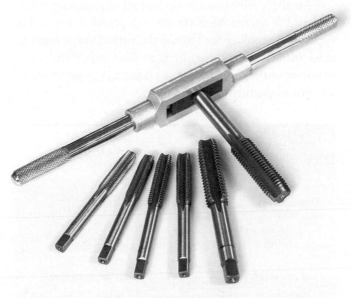

ampol sonthong/Shutterstock.com

Figure 5-11. Taps are used to cut internal threads. The tap is turned with a wrench attached to the head of the tool.

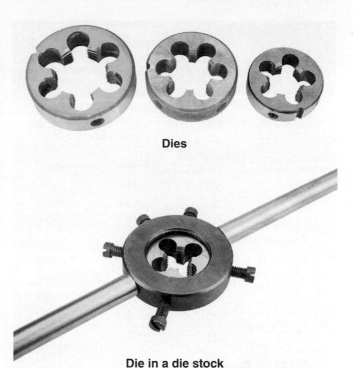

Dies

Die in a die stock

withgdd/Shutterstock.com
Michal Zduniak/Shutterstock.com

Figure 5-12. Dies are used to cut external threads. A die stock is used to hold the die and turn the tool.

Tapping a hole requires first drilling a hole with a tap drill. Select the appropriate tap drill for the thread size and form you wish to cut. A tap drill chart is used to determine the correct tap drill size, **Figure 5-13**. The chart shown is for inch-based coarse and fine threads. If you are working with metric threads, you will need to refer to a metric tap drill chart.

A tap is turned with a wrench attached to the head of the shank. There are two types of tap wrenches. A T-handle tap wrench is used for small tapping jobs. A hand tap wrench, shown in **Figure 5-11**, provides more leverage and is used for larger taps.

Internal threads may be cut in a through hole (a hole that passes entirely through the part) or a blind hole (a hole that does not pass entirely through the part). There are three types of taps used to cut internal threads. A taper tap has a tapered end of approximately eight to ten threads and is used to start the hole. A plug tap has a tapered end of approximately four threads and is used after using a taper tap. A bottom tap has no taper and is used to tap threads to the bottom of a blind hole. The three types of taps are used when tapping a blind hole. A taper tap may be sufficient when tapping a through hole.

When starting a hole with a tap, make sure the tap is perpendicular (90°) to the surface of the drilled hole. See **Figure 5-14**. Keep the tap square as it is turned.

Tap Drill Sizes Unified Coarse and Fine Threads		
Size	**Threads per Inch**	**Tap Drill 75% Thread**	**Decimal Equivalent**
2	56	50	.0700
	64	50	.0700
3	48	47	.0785
	56	45	.0820
4	40	43	.0890
	48	42	.0935
6	32	36	.1065
	40	33	.1130
8	32	29	.1360
	36	29	.1360
10	24	25	.1495
	32	21	.1590
12	24	16	.1770
	28	14	.1820
1/4	20	7	.2010
	28	3	.2130
5/16	18	F	.2570
	24	I	.2720
3/8	16	5/16	.3125
	24	Q	.3320
7/16	14	U	.3680
	20	25/64	.3906
1/2	13	27/64	.4219
	20	29/64	.4531
9/16	12	31/64	.4844
	18	33/64	.5156
5/8	11	17/32	.5312
	18	37/64	.5781
3/4	10	21/32	.6562
	16	11/16	.6875
7/8	9	49/64	.7656
	14	13/16	.8125
1	8	7/8	.8750
	14	15/16	.9375
1-1/8	7	63/64	.9844
	12	1- 3/64	1.0469
1-1/4	7	1- 7/64	1.1094
	12	1-11/64	1.1719
1-1/2	6	1-11/32	1.3437
	12	1-27/64	1.4219

Goodheart-Willcox Publisher

Figure 5-13. A tap drill chart for coarse and fine threads in the Unified system.

7th Son Studio/Shutterstock.com

Figure 5-14. The correct way to use a tap. Notice that the tool is held 90° to the work.

⚠ **CAUTION**

Taps are made of a very hard but brittle material. Off-axis torque can break the tap, rendering it unusable and causing damage to the part.

While tapping, it is necessary to keep the tool lubricated. Add the appropriate cutting fluid before starting and while cutting as needed. This cools the cutting surfaces of the tool and helps prevent it from binding or breaking.

An alternative to cutting external or internal threads by hand is to machine the threads using a lathe. The appropriate tooling for cutting the thread shape must be used. Internal thread cutting is possible only if the major diameter of the thread to be cut is large enough to accept the thread-cutting tooling.

PROCEDURE Tapping a Hole

Use the following procedure to tap threads for a through hole:

1. Mount the work solidly in a vise. Select the appropriate size tap drill and drill the tap hole. Wear appropriate eye protection. Make sure to drill through the work at a 90° angle. Remove chips with a brush or cloth. Use a file to remove burrs from the hole.

(continued)

2. Select the appropriate size tap and insert it into the tap wrench. Align the tap with the hole and start the hole by slowly turning the tap clockwise. After making a half turn into the material, turn the tap counterclockwise a quarter turn. Slowly turn the tool to allow metal chips to fall through the flutes of the tool. Then, resume turning clockwise.

3. Continue to work slowly and carefully, backing off the tool to clear chips every one or two turns. Do not try to force the tap through the work. Add cutting fluid as needed to lubricate the tap.

4. After threading the hole, remove the tap. Remove all chips with a brush and use a file to remove burrs from the hole. Test the fastener by threading it into the hole.

5.6 THREAD INSERTS

In some situations, threaded holes in assemblies become damaged. Rather than replace the assembly, it may be possible to repair the threaded portion with a thread insert. The repair involves drilling the threaded hole oversize with an appropriate drill, then using a tap to rethread the hole. Once the hole has been appropriately threaded, a thread insert is installed, **Figure 5-15**.

czoborraul/Shutterstock.com

Figure 5-15. Thread inserts are used to repair damaged threads.

Thread inserts are commonly available in stainless steel and can be used to provide added strength in parts made from softer metal.

The inside of the thread insert is the correct size for the original screw. The outside fits the threads cut by the tap. The thread insert is mounted to an installation tool and then threaded into the hole. The tang of the insert is broken off with another tool. In this manner, original-size threaded holes may be restored. This process is particularly useful when the assembly to be repaired is very large or expensive to replace.

Special thread inserts are available for use in plastic assemblies. These are available in brass or stainless steel and are designed to be installed during manufacturing. See **Figure 5-16**. These types of thread inserts are molded into plastic or installed after molding using hand tools, a manual press, or special installation equipment.

SPIROL

Figure 5-16. Thread inserts designed for use in plastic parts.

5.7 NON-THREADED FASTENERS

Non-threaded fasteners are used to fasten parts together without the use of screw threads. Common types of non-threaded fasteners are discussed in the following sections.

5.7.1 Washers

Washers are used in conjunction with bolts, nuts, and screws. In general, a **washer** is a device used with threaded fasteners to increase their contact area with the material being fastened. There are various types of washers used for different purposes, **Figure 5-17**. Washers distribute the applied force over a greater area and thereby strengthen the assembly. Washers may also be used to protect the surfaces being fastened from damage by the fasteners themselves.

Flat washers are primarily used to spread the clamping force over a larger area. The washer size is specified by the major diameter of the fastener it fits. Fender washers are often thinner than flat washers, but they have a larger outside diameter to spread the clamping force over an even greater area.

Lock washers are used in assemblies where vibration might loosen the fastener. A lock washer is often used in conjunction with a flat or fender washer. Three types of lock washers are common. A split lock washer is used to lock a nut or bolt in place and prevent it from turning due to stress or vibration. An external tooth lock washer has external teeth to prevent loosening. An internal tooth lock washer is similar to an external tooth lock washer, except the locking teeth are internal.

A finishing washer provides a more finished look to the fastener installation. This type of washer is primarily used for cosmetic purposes.

A sealing washer is composed of a rubber seal bonded to a metal ring, often made of stainless steel. This type of washer is used to seal around the fastener and prevent leakage.

> ⚠ **C A U T I O N**
>
> It is important to know the materials that fasteners are made from when assembling parts together. In environments where moisture is present, contact between dissimilar metals can cause galvanic corrosion. When required, use fasteners made from the same metal to prevent chemical reactions from occurring over time.

5.7.2 Pins

Pins are used in a variety of applications, usually in conjunction with other fastening devices, **Figure 5-18**. A *pin* is a cylindrical fastening device used to align and secure parts. Often, pins are used for fastening applications in which component parts are subjected to shear loads. Most pins are unthreaded. However, some pins have small threaded sections at the ends. The threaded sections allow the pins to be held in place using washers and nuts.

Types of Washers	
Fastener Appearance	**Name**
	Flat washer
	Fender washer
	Split lock washer
	External tooth lock washer
	Internal tooth lock washer
	Finishing washer
	Sealing washer

Images: © McMaster-Carr 2018

Figure 5-17. Common types of washers.

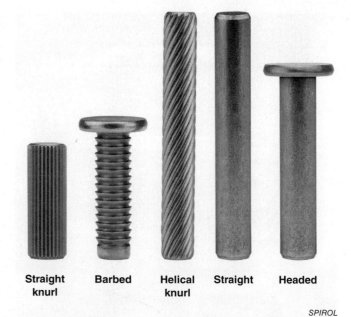

Straight knurl   Barbed   Helical knurl   Straight   Headed

SPIROL

Figure 5-18. Pins are used to align and secure parts. Shown are common types of pins with and without threading.

Clevis Pins

A clevis pin makes up one part of a clevis fastener. A ***clevis fastener*** is an assembly consisting of a clevis, a clevis pin, and a cotter pin. A clevis pin has a head at one end and a hole at the other. It is inserted through the ends of a clevis, or shackle, and anchored with a cotter pin. See **Figure 5-19**. In its most basic form, a clevis is C-shaped with a hole at each end. A clevis pin passes through the holes in the ends of the clevis, creating a *D* shape. Then, the clevis pin is secured with a cotter pin, sometimes called a split pin or cotter key. Clevis fasteners are often used in applications such as hitch assemblies or rigging, where the fastened parts need to rotate around the attachment point at the clevis pin. These assemblies are fixed and remain secure even under vibration, yet they are also easy to remove by simply removing the pins.

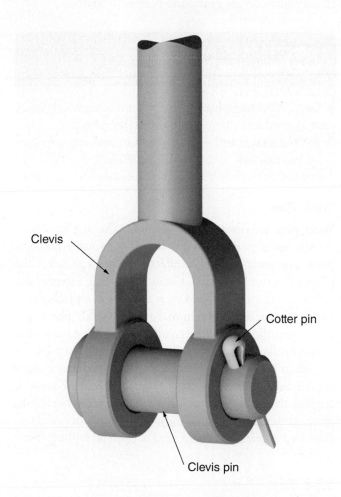

Clevis

Cotter pin

Clevis pin

Goodheart-Willcox Publisher

Figure 5-19. A clevis fastener consists of a clevis, a clevis pin, and a cotter pin.

Taper Pins

Taper pins taper slightly from one end to the other, **Figure 5-20**. One end of the pin is slightly smaller in diameter than the other end. Standard inch-based taper pins have a taper ratio of 1:48. This means that the diameter of the pin tapers 1″ for every 48″, or 1/4″ per foot. Metric taper pins have a taper ratio of 1:50. Taper pins are most often found in mechanical assemblies such as printing presses, where they are used to position parts and lock them together.

G.L. Huyett

Figure 5-20. Taper pins have a uniform taper from one end to the other.

Dowel Pins

Dowel pins are solid cylindrical pins, **Figure 5-21**. Dowel pins are precisely sized for use in machinery alignments. In these types of assemblies, holes are drilled in each of the objects to be fastened together. The appropriate size dowels are then inserted into both objects to attach them together. Dowel pins may also be used as guides to check tolerances in mechanical parts.

G.L. Huyett

Figure 5-21. Dowel pins are precisely sized to fasten parts together. They are available in a variety of end designs.

Spring Pins

Spring pins are hollow cylindrical pins. They are lighter and easier to install than solid pins. There are two major types of spring pins: slotted spring pins, **Figure 5-22**, and coiled spring pins, **Figure 5-23**. A slotted spring pin, also called a roll pin, is a roll of metal with a single slot

SPIROL

Figure 5-22. Slotted spring pins have a single slot extending the length of the pin.

SPIROL

Figure 5-23. Coiled spring pins are made up of several rolls of metal. They distribute stress more evenly than slotted spring pins.

extending the length of the pin. A coiled spring pin, also called a spiral pin, is made up of several rolls of metal with no slot. Spring pins are larger in diameter than the hole in which they are installed. When inserted, they compress to fit the smaller size of the hole. Spring pressure exerted against the surface of the hole holds the pin in place. Coiled spring pins have more uniform flexibility than slotted spring pins and distribute stress more evenly.

Spring pins offer several advantages. One is that they are self-locking. Spring pins are also very versatile and reusable, making them a cost-effective option for many applications. They have a variety of uses in different industries, including the automotive and electrical industries. Spring pins are used as hinge pins and pivots and are used to fasten gears to shafts.

Grooved Pins

Grooved pins have three parallel grooves extending longitudinally along the pin body, **Figure 5-24**. The grooves are uniformly spaced around the pin body. Each of the grooves

G.L. Huyett

Figure 5-24. Grooved pins have parallel grooves uniformly spaced around the pin body.

has a raised portion along the sides. When installed into a drilled hole, the raised portions are forced back into the pin grooves. This action exerts radial forces against the hole surface and locks the pin securely in place. Grooved pins are available in different design shapes with straight and tapered grooves of different lengths. Grooved pins are used in a variety of fastening applications. They are used as hinge, roller, and linkage pins and are commonly used to fasten collars, levers, gears, and pulleys to shafts.

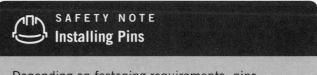

SAFETY NOTE
Installing Pins

Depending on fastening requirements, pins may need to be hammered or pressed in. Always wear eye protection when installing pins in this manner.

Shear Pins

Shear pins are designed to fail under the application of certain forces in order to protect more valuable machine parts. For shear pins to be effective, you must select the proper pin size and grade. Compared to the expense and difficulty of replacing a part such as a motor, replacing broken shear pins involves minimal cost and labor. Shear pins are frequently used as safeguards, analogous to electrical circuit breakers, in engines, augers, and towing setups.

Shear pins can also be used to prevent operation until certain forces are achieved. Once the required force is applied, the shear pin breaks, allowing the operation to take place. In such cases, the pin must be replaced after each use. Pins used as safeguards only need occasional replacement.

5.7.3 Keys

A *key* is a metal fastener used to prevent shaft rotation and transfer torque between parts. A common application is a key used to prevent rotation of a gear on a rotating shaft. A *keyseat* is the slot where the key fits on the shaft. A *keyway* is the slot where the key fits on the hub of the mating part. See **Figure 5-25**. A key and the corresponding keyseat and keyway must be carefully designed to fit properly and protect against key failures. Common types of keys include square, taper, and Woodruff keys. Keys are used in a variety of mechanical applications ranging from heavy equipment to automobile engines.

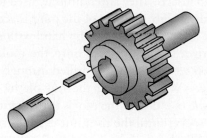

Figure 5-25. A keyed assembly of a shaft and gear using a square key.

5.7.4 Retaining Rings

Retaining rings are used to accurately position and hold mechanical parts together. Most retaining rings are installed into machined grooves in a hole or shaft. They provide a shoulder to hold parts in position securely. There are two major types of retaining rings: internal and external. See **Figure 5-26**. Internal retaining rings are installed into a bore (a hole), while external retaining rings are installed on the outside of a shaft. Retaining rings require special pliers to install. The pliers have tips designed to be inserted into the small holes at the ends, called lugs.

Retaining rings are cost-effective retainers used in place of more expensive shoulders or pin-based retainers. A variety of types are available, including wire rings, spiral rings, beveled rings, and self-locking rings. Retaining rings have numerous applications in industry. They are used in driveshafts, agricultural machinery, aircraft, electrical motors, hydraulic equipment, and computer equipment.

5.7.5 Clips

Clips are external ring fasteners installed into shaft grooves to hold mechanical parts in place. They are similar to retaining rings, but have larger side openings that allow

Internal retaining ring   **External retaining ring**

Figure 5-26. Retaining rings are used to accurately position assembly parts.

them to be seated directly into a groove. Clips are installed with a special applicator tool radially (vertically), along the side of the shaft. Clips are available in many varieties for different applications. Two common types are C-clips and E-clips. See **Figure 5-27**. E-clips have teeth that provide a larger shoulder and seat into the groove of the shaft.

C-clip   **E-clip**

Figure 5-27. Clips are external ring fasteners. They are installed radially along the side of a shaft.

5.7.6 Cable Ties

Cable ties, also known as tie wraps or zip ties, are inexpensive fasteners used in a variety of applications. See **Figure 5-28**. A *cable tie* is a thin, flexible nylon strap with teeth that engage with the open of the tie, preventing it from loosening. Cable ties are often used to organize and hold cables or electrical wires together. Specialized cable ties are made for applications in the medical, food, and security industries.

5.7.7 Rivets

Rivets are mechanical fasteners used in permanent installations. A *rivet* is a headed cylindrical shaft used to fasten parts together. When a rivet is installed, the end opposite the head, called the tail, is deformed to become larger. This anchors the rivet and produces a tight, secure joint. Rivets are classified by size, head type, and fastening method. They are commonly used in aircraft components, automobile parts, electronic devices, and sheet metal assemblies. Common types of rivets include solid rivets, semitubular and full tubular rivets, bifurcated (split) rivets, compression rivets, and blind rivets.

Figure 5-28. Cable ties, also known as tie wraps or zip ties, have teeth along their lengths to prevent them from being loosened once they are engaged.

Blind Rivets

Blind rivets are commonly used in applications where only one side of the joint is accessible. They are installed with a rivet tool, **Figure 5-29**. First, the blind rivet is inserted into a clearance hole drilled through the parts to be fastened. A blind rivet has a central rod inside the rivet body called a mandrel. When the mandrel is pulled by the rivet tool, the mandrel head expands the rivet body. The mandrel stem breaks off from the head when the required tension load is reached. See **Figure 5-30**. Blind rivets are most often used for permanent attachment of thin sheet metal pieces.

On occasion, it is necessary to remove a blind rivet in order to perform maintenance. To remove a blind rivet, first drill a hole approximately the size of the rivet shank. Drill the hole to nearly the depth of the rivet head. Next, use a chisel to remove the rivet head. Once the head is removed, use a hammer and center punch to punch the shank through the hole. Note that you generally do not want to use blind rivets on an assembly that is not permanent or semipermanent.

Rivet Nuts

Rivet nuts are used in applications where the material is too thin for conventional thread cutting to produce enough threads for a proper fastening job. A ***rivet nut*** is an internally threaded tubular rivet installed from one side of a joint, **Figure 5-31**. Rivet nuts are also known as blind thread inserts. They are commonly used for mounting parts where only one side of the joint is accessible.

A rivet nut can be installed by hand with a rivet nut tool. The rivet nut is threaded onto the tool, then the rivet nut is inserted into a hole drilled through the pieces to be joined. While holding the nut of the tool with a wrench, the mandrel of the tool is turned with another wrench or socket until the rivet nut expands sufficiently toward the blind end of the assembly. Rivet nuts can also be installed with special power tools.

At times, the thread of a rivet nut may become damaged. This requires removal of the rivet nut and then installation of a new one. The removal process is similar to removing a blind rivet, as previously discussed.

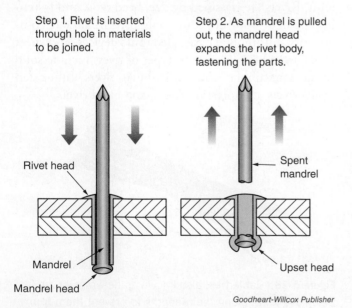

Maksym Sukhenko/Shutterstock.com

Figure 5-29. A rivet tool used to install blind rivets.

Step 1. Rivet is inserted through hole in materials to be joined.

Step 2. As mandrel is pulled out, the mandrel head expands the rivet body, fastening the parts.

Rivet head

Spent mandrel

Mandrel

Mandrel head

Upset head

Goodheart-Willcox Publisher

Figure 5-30. Installing a blind rivet.

G.L. Huyett

Figure 5-31. A rivet nut is used to provide internal threads in applications where only one side of the joint is accessible. Rivet nuts are also known as blind thread inserts.

CHAPTER WRAP-UP

This chapter provides an introduction to common types of fasteners. You will frequently encounter fasteners on the job. You must have the appropriate tools on hand to work with both threaded and non-threaded fasteners. Understanding the different types of fasteners and their characteristics will help you in assembling and disassembling mechanical parts.

On occasion, you will encounter damaged threads. Instead of replacing the related part or assembly, it is often possible to make a repair by cutting new threads. The ability to use taps, dies, and thread inserts to repair threaded parts is an essential skill for the technician.

Chapter Review

SUMMARY

- External threads and internal threads are the basic types of screw thread. Bolts and screws have external thread. Nuts and tapped holes have internal thread.

- The Unified system is the most commonly used standard thread form for inch-based threads in the United States.

- Unified Coarse (UNC) threads are used for general applications requiring fast assembly and disassembly. Unified Fine (UNF) threads have a smaller thread depth than coarse threads and are used where higher strength is required to bear greater loads.

- The term *bolt* generally refers to a fastener inserted through non-threaded holes and secured with a nut. The term *screw* generally refers to a fastener that screws into threads.

- Always use the proper driving tool for fastener installation and removal.

- Self-tapping screws are capable of cutting their own threads during installation.

- Damaged threads may be restored by cutting new threads with a tap or die. Damaged internal threads may be repaired with a thread insert.

- Washers are used in combination with bolts, nuts, and screws. Washers are generally used to spread the clamping force of a fastener over a larger area.

- Pins are used to align and secure parts. Pins are generally used in applications in which parts are subjected to shear loads.

- A key is a fastener used to prevent shaft rotation and transfer torque. A keyseat is the slot where the key fits on the shaft and a keyway is the slot where the key fits on the hub of the mating part.

- Rivets are used in permanent installations. When installed, the end opposite the head of the fastener is deformed to become larger.

- Blind rivets are often used to fasten pieces of sheet metal together, especially where access to only one side of the joint is available. For some repair work, blind rivets may be drilled out and removed.

REVIEW QUESTIONS

Answer the following questions using the information provided in this chapter.

1. Explain the difference between a thread form and a thread series.

2. The top surface of a screw thread is called the _____. The bottom surface of a screw thread between two sides of the thread is called the _____.

3. The _____ of a screw thread is the distance from one point on a thread to the corresponding point on the next thread.

4. Explain the difference between UNC and UNF series threads.

5. Explain the information indicated by the thread note 1/2–13UNC–2A.

6. Identify the following types of bolts and screws.

A.

B.

C.

D.

E.

7. Identify the following types of fastener drives.

A. ⊖   D. ⊛

B. ⬡   E. ▢

C. ✛

8. Identify the following types of nuts.

A.

B.

C.

D.

9. Internal threads are cut with a(n) _____. External threads are cut with a(n) _____.

10. Identify the following types of washers.

A.

B.

C.

D.

11. What is a clevis fastener?

12. What is the taper ratio of a standard inch-based taper pin?

13. A(n) _____ pin is larger in diameter than the hole in which it is installed and compresses to fit the smaller size of the hole.

14. The slot where a key fits on a shaft is called a(n) _____.

15. What is the difference between internal and external retaining rings?

16. Explain how a blind rivet is installed.

 NIMS CREDENTIALING PREPARATION QUESTIONS

The following questions will help you prepare for the NIMS Industrial Technology Maintenance Level 1 Maintenance Operations credentialing exam.

1. The _____ diameter of a screw thread is the diameter measured from root to root for external thread.
   A. major
   B. minor
   C. pitch
   D. outside

2. The thread _____ specifies the amount of tolerance permitted in engagement between mating threads.
   A. form
   B. series
   C. pitch
   D. class

3. Which of the following has a head at one end and external threads at the other to accept a nut?
   A. Self-tapping screw
   B. Washer
   C. Retaining ring
   D. Bolt

4. A _____ extends through a clearance hole in one part and threads into a second part without using a nut.
   A. washer
   B. cap screw
   C. setscrew
   D. blind rivet

5. Which of the following is installed with a slotted or Phillips screwdriver?
   A. Machine screw
   B. Washer
   C. Flange nut
   D. Key

6. Which of the following is an oval head fastener?

A. Choice A
B. Choice B
C. Choice C
D. Choice D

7. A _____ nut has deformed threads or a nylon insert to produce interference during tightening.

A. rivet
B. flange
C. self-locking
D. jam

8. Internal threads may be cut in a through hole with a _____.

A. rivet tool
B. hex key
C. die
D. tap

9. A lock washer is often used in conjunction with a _____.

A. C-clip
B. key
C. flat washer
D. taper pin

10. A _____ pin has a hollow center and compresses to fit the smaller diameter of a hole when installed.

A. coiled spring
B. dowel
C. taper
D. grooved

11. A clevis fastener is an assembly consisting of a clevis, a clevis pin, and a _____ pin.

A. shear
B. dowel
C. taper
D. cotter

12. The slot where a key fits on the hub of a mating part is called the _____.

A. keyseat
B. keyhole
C. keyway
D. key stop

13. Which of the following tools is used to install retaining rings?

A. Pliers
B. Socket wrench
C. Hex key
D. Slotted screwdriver

14. A(n) _____ is a thin, flexible nylon strap with teeth that prevent it from loosening when engaged.

A. E-clip
B. C-clip
C. cable tie
D. retaining ring

15. The central rod inside the body of a blind rivet is called the _____.

A. shank
B. mandrel
C. lug
D. head

6 | Print Reading

LEARNING OBJECTIVES

After completing this chapter, you will be able to:

☐ Explain the importance of industrial prints.

☐ Describe the purpose of a multiview drawing.

☐ Recognize dimensioning conventions used on mechanical drawings.

☐ Explain the purpose of geometric dimensioning and tolerancing (GD&T).

☐ Recognize and interpret electrical symbols.

☐ Understand how to read electrical diagrams.

☐ Recognize and interpret fluid power symbols.

☐ Understand how to read fluid power circuit diagrams.

☐ Recognize and interpret welding symbols.

TECHNICAL TERMS

alphabet of lines
assembly drawing
basic dimension
break line
centerline
clearance fit
computer-aided drafting (CAD)
cutting-plane line
datum
dimension
dimensioning
dimension line
extension line
feature control frame
fit
geometric dimensioning and tolerancing (GD&T)
hidden line
interference fit
ladder diagram
least material condition (LMC)
maximum material condition (MMC)
multiview drawing
object line
phantom line
scale
schematic diagram
section line
title block
tolerance
transition fit
weld symbol
welding symbol
wiring diagram

Print reading is a critical skill for a technician to learn. Consider it to be like learning how to read and write. If you don't have the ability to properly read a print, you are missing out on a wealth of information that will make your work easier. By being comfortable reading diagrams and prints, you will become a much more efficient and effective troubleshooter.

This chapter introduces you to the common types of prints you will encounter during your daily job functions. You could read an entire book on print reading and it would probably not cover every type of print or symbol you might see on the job. This chapter is intended to provide basic information you can apply to your everyday job tasks. Electrical, fluid power, and welding symbols and diagrams are introduced in this chapter and covered in much more detail in later sections.

6.1 PRINT READING BASICS

Imagine you were planning a trip from Chicago to Los Angeles and had to pick up your brother in Denver along the way. You would certainly need to be able to read a road map. Without a map, you could head west from Chicago, but you would most likely miss Denver and could end up in Walla Walla, Washington.

Just like map reading skills are important for someone driving a car on a long trip, print reading skills are important for anyone involved in industrial maintenance. If you are troubleshooting a problem on a malfunctioning machine, a print will give you directions on how to test the equipment at various stages of function and determine the location of the fault.

Long ago, prints were called "blueprints" because of their blue background. In the 19th century, blueprints were produced by exposing photosensitive copy paper under an original master drawing made on semitransparent paper. The resulting print had white lines on a blue background. In the 20th century, the diazo process emerged. This process produced a "blue line" print consisting of blue lines on a white background. Although the technologies used to make blueprints and diazo prints have been replaced by modern reproduction processes, the term "blueprint" is still sometimes used interchangeably with the term "print" when referring to a drawing made for manufacturing or construction.

Drawings and diagrams are normally prepared on a computer using a *computer-aided drafting (CAD)* program. They may be viewed on a computer or mobile device, may be attached to equipment, or may be printed on paper. See **Figure 6-1**.

Prints are usually printed to a standard size to make them easier to file and sort. See **Figure 6-2**. Standard sheet sizes are identified with a letter designation. Sizes A through F are used for preparing inch-based engineering and architectural drawings. The sizes used for architectural drawings are slightly larger than those for engineering drawings. Sizes A0 through A4 are metric sheet sizes established by the International Organization for Standardization (ISO).

Prints are necessary for installation of systems and also serve as a critical troubleshooting aid to help determine the location of a malfunction. When troubleshooting a problem, you can read the print, determine the flow of the system, and then test at various points to determine whether you have power, signal, or pressure.

Technicians must be able to interpret the information given on a print and use it to perform troubleshooting techniques. All too often, a technician guesses at the malfunctioning part and replaces it in the hope that it will correct the problem. This method is rarely effective.

Andrey Popov/Shutterstock.com

Figure 6-1. Technicians commonly refer to prints to help identify wiring connections when troubleshooting an electrical system.

Prints show how an electrical or electronic circuit is connected, how fluid power flows and is controlled, how a building is constructed, how piping is connected, and how a structure is to be fabricated and welded. There are assembly prints to illustrate how parts fit together in assemblies and subassemblies. For most anything that is constructed, wired, welded, or plumbed, there is a print that will illustrate how it is put together or connected.

6.2 TITLE BLOCKS

Most prints have a *title block*. The title block contains important information about the print, **Figure 6-3**. The title block typically includes the name of the manufacturer,

Standard Print Sizes

Letter Designation	Engineering Standard	Architectural Standard	Similar to ISO Standard
A	8.5×11	9×12	A4
B	11×17	12×18	A3
C	17×22	18×24	A2
D	22×34	24×36	A1
E	33×44	36×48	A0
E1		30×42	
F	28×40		

Goodheart-Willcox Publisher

Figure 6-2. Standard sheet sizes for prints.

integrator, or designer of the information contained on the print. The title block also typically includes the drawing number, title, and the sheet number. The sheet number indicates the number of the print in a set of multiple prints. In addition, the title block includes the sheet size and the scale of the drawing.

The **scale** is the proportional relationship between the actual, real-life size of an item and the size it is shown in the drawing. For example, a scale of 1:1 indicates that the part has been drawn at full size. A scale of 1:2 indicates that the part has been drawn at half the actual size of the part.

Usually, multiview drawings are drawn to scale. However, some types of drawings, such as electrical diagrams, do not have a drawing scale. Drawings that do not have a designated scale may have the abbreviation NTS noted somewhere on the drawing or the entry NONE in the scale area of the title block.

Some title blocks have areas to specify the part material, finish, and weight. When required, the weight of the part is specified as the actual or calculated weight.

An area for approvals is usually included on the title block. This area is used by individuals to indicate official approval of the drawing for release or manufacturing. The area provides spaces to enter approval signatures or initials and dates. In **Figure 6-3**, the title block has entry spaces to indicate the drawing has been completed by the drafter, checked, approved, and issued for use.

On prints used in manufacturing, the title block normally includes a tolerance block in the space on the left end. The tolerance block lists tolerances for the dimensions of the part. The **tolerance** is the total amount a dimension can vary. Tolerances define how much the final part can vary from the dimensions on the print. Dimensioning is discussed in more detail later in this chapter.

In **Figure 6-3**, notice that dimensions with one decimal place have a tolerance of ±.1″. This means that the actual size or location of a feature can vary from the design size by a total of one-tenth of an inch. This permits the part to be made smaller or larger than the design size. Dimensions with two decimal places have a tolerance of ±.03″ and dimensions with three decimal places have a tolerance of ±.005″. These tolerances are general tolerances for the entire print. If there is a different tolerance for a certain dimension, it is given next to the dimension. The tolerance block also specifies tolerances for dimensions expressed in angular and fractional units.

TECH TIP
Drawing Scale and Ruler Scale

The term *scale* also refers to a measuring device used to make measurements, such as a mechanical engineer's scale or an architect's scale. A scale (measuring instrument) can be used to make drawings at specific scales and to make measurements at different scales. However, the dimensions given on a print should normally be used to determine the measurements of a part. Measurements of a part should not be made by directly measuring distances or sizes on a print with a scale (measuring device).

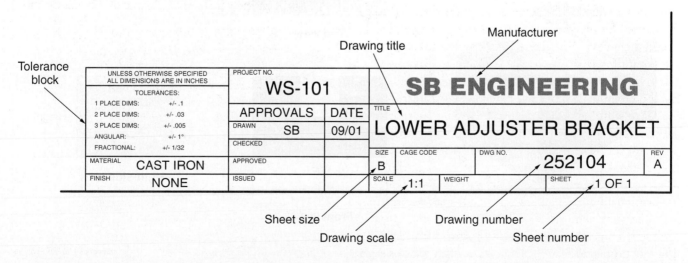

Goodheart-Willcox Publisher

Figure 6-3. A title block provides information about the print. Most title blocks have a similar format.

6.3 MULTIVIEW DRAWINGS

Multiview drawings are commonly prepared for mechanical parts produced in manufacturing. A *multiview drawing* is a representation that shows the different views of a part on one drawing.

The views in a multiview drawing are obtained using a method of projection called orthographic projection. The views show the different surfaces of the part. There are six principal views of a part (top, bottom, front, rear, right, and left). However, multiview drawings only include the number of views necessary to describe the part. Usually, two or three views are sufficient to provide a complete shape description.

Refer to **Figure 6-4**. The pictorial view is a two-dimensional representation that appears three-dimensional. The multiview drawing consists of three two-dimensional views showing the width, height, and depth of the object. A left-side view or bottom view can be added to show other important details of the part, if any. For this part, three views are sufficient. Notice that the views are placed in alignment with each other.

For some objects, such as cylindrical or prismatic-shaped objects, two views may be all that is needed to fully describe the object. See **Figure 6-5**. A single view and a note identifying thickness is sufficient for thin objects such as gaskets.

Standard line conventions used in drafting establish how lines appear in multiview drawings. Standard line conventions make up a system called the *alphabet of lines*. See **Figure 6-6**. The common types of lines that you will see on mechanical part drawings are described as follows:

- *Object lines.* Object lines represent the visible edges and contours of an object, **Figure 6-7**. Object lines are thick, continuous lines. They are also called *visible lines*.

- *Hidden lines.* Hidden lines represent object edges and contours that are located behind other features and not visible in a given view. Refer to **Figure 6-7**. Hidden lines are thin lines made up of short, closely spaced dashes.

- *Centerlines.* Centerlines represent axes of symmetrical objects, **Figure 6-8**. Centerlines are thin lines made up of alternating long lines and short dashes.

- *Cutting-plane lines.* A cutting-plane line indicates where an imaginary cut has been made through an object in order to show interior features. The cutting-plane line identifies the location of the cutting plane for a section view, **Figure 6-9**. Cutting-plane lines are thick, dashed lines. There are three

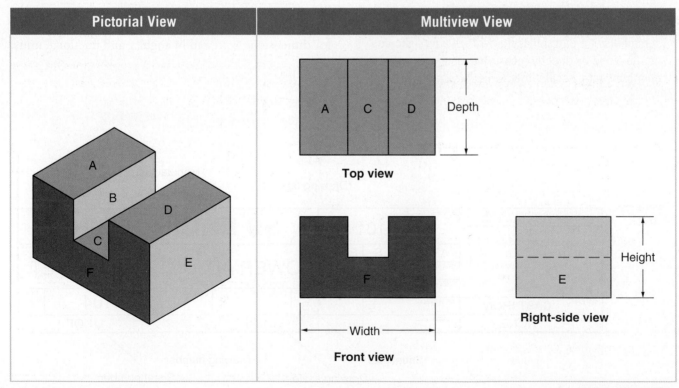

Figure 6-4. A pictorial view shows the part in three dimensions. A multiview drawing shows a three-dimensional object in two dimensions.

standard forms of lines used for cutting-plane lines. Refer to **Figure 6-6**. For each form, the ends are drawn at 90° and terminate with arrowheads to indicate the direction of sight.

- *Section lines.* Section lines are used in section views to show features of a part that have been cut by a cutting plane. Refer to **Figure 6-9**. Section views are used to clarify the internal details of a part. Section lines are thin, evenly spaced lines typically drawn at a 45° angle. Section lines are also used to represent the specific material of a part that has been sectioned. See **Figure 6-10**.

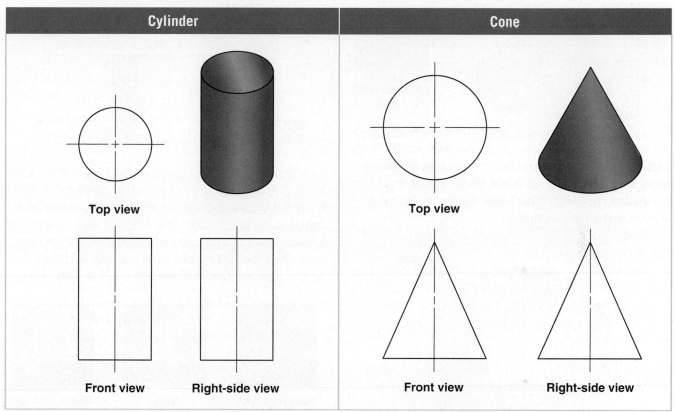

Goodheart-Willcox Publisher

Figure 6-5. Simple cylindrical objects can often be described with two views. For the cylinder and cone shown here, the right-side view is unnecessary. The objects are completely described with the top view and front view.

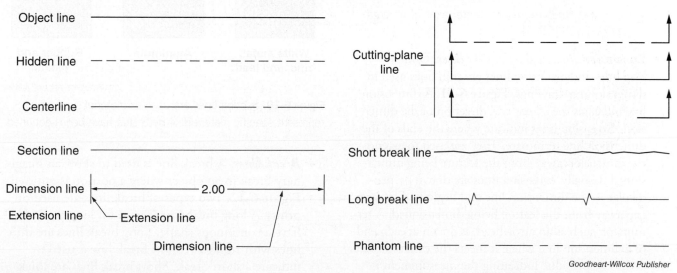

Goodheart-Willcox Publisher

Figure 6-6. The alphabet of lines.

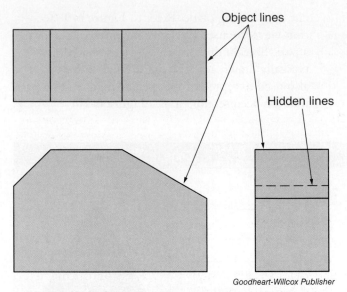

Object lines

Hidden lines

Goodheart-Willcox Publisher

Figure 6-7. Object lines are used to represent the visible edges of an object and are drawn thick. Hidden lines are used to represent object edges hidden behind other features and are drawn thin.

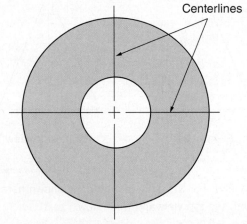

Centerlines

Goodheart-Willcox Publisher

Figure 6-8. Centerlines represent center axes of objects.

- ***Dimension lines*** and ***extension lines.*** Dimension lines and extension lines are thin lines used in dimensioning drawings, **Figure 6-11**. A dimension line indicates the extent and direction of the dimension. Extension lines indicate where the ends of the dimension line terminate. They are used to extend the dimension away from the feature being dimensioned. Usually, extension lines are drawn perpendicular to the dimension line and begin with a short gap away from the feature being dimensioned. A terminator, such as an arrowhead, is drawn at each end of a dimension line where it meets the extension line. A dimension value indicating the measurement is normally placed in a break along the dimension line.

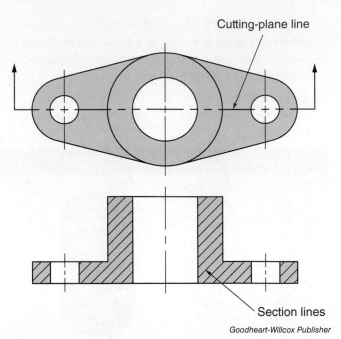

Cutting-plane line

Section lines

Goodheart-Willcox Publisher

Figure 6-9. A cutting-plane line indicates the location of a cutting plane for a section view. A section view clarifies the internal details of a part. In the section view, section lines show where the features of a part have been sectioned. Section lines are thin lines typically drawn at a 45° angle.

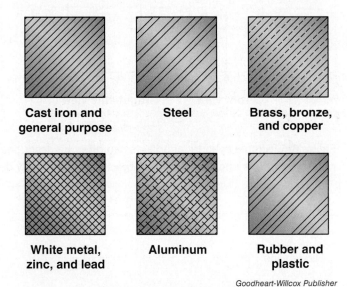

| Cast iron and general purpose | Steel | Brass, bronze, and copper |

| White metal, zinc, and lead | Aluminum | Rubber and plastic |

Goodheart-Willcox Publisher

Figure 6-10. Section lines are used as material symbols to represent specific materials of parts that have been sectioned.

- ***Break lines.*** A break line is used to show an imaginary break in an object where a portion is omitted, **Figure 6-12**. Two types of break lines are used on prints. A long break line is used for long parts that have a continuous shape. Long break lines are thin lines with zigzags. A short break line is used to indicate a short break. Short break lines are thick lines drawn freehand.

- *Phantom lines.* Phantom lines, **Figure 6-13**, are used to indicate alternate positions of moving parts, adjacent positions of related parts, or repeated detail. Phantom lines are thin lines made up of long dashes alternating with pairs of short dashes.

6.4 DIMENSIONS

Dimensioning is the process in which the designer adds dimensions to a part drawing to communicate the size and location of each feature. A *dimension* is the precise measurement of a feature. Dimensions appear on the print and supply the information necessary to make the part. The dimensions are generally used by the machinist in the production of a part or by the technician inspecting the part.

Linear dimensions are used to dimension straight distances and are usually expressed in decimal or fractional units. The example shown in **Figure 6-11** is dimensioned with linear dimensions. Angular dimensions are used to dimension angles and are usually expressed in degrees, minutes, and seconds. See **Figure 6-14**.

Leaders and notes are commonly used in dimensioning circular features, such as holes. When dimensioning the sizes of circular arcs and circles, standard practices and symbols are used. See **Figure 6-15**. Arcs are dimensioned by specifying the radius. The letter *R* precedes the dimension value to indicate a radius dimension. Circles and holes are dimensioned by specifying the diameter. The diameter symbol (∅) precedes the dimension value to indicate a diameter dimension. A technician would encounter circular features, for example, in parts that have holes drilled or bored into them.

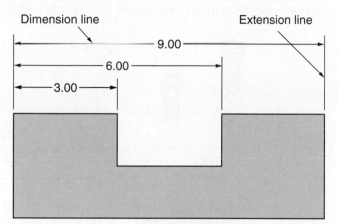

Goodheart-Willcox Publisher

Figure 6-11. Dimension lines and extension lines are used in dimensioning objects. A dimension line indicates the extents of a dimension. Extension lines indicate where the dimension terminates.

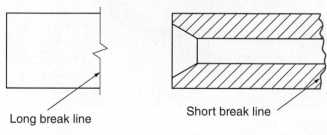

Goodheart-Willcox Publisher

Figure 6-12. Break lines are used to show where a portion of an object has been omitted.

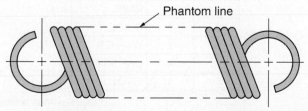

Goodheart-Willcox Publisher

Figure 6-13. Phantom lines are used to represent repeated detail in parts, such as the coils in a spring.

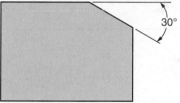

Goodheart-Willcox Publisher

Figure 6-14. The extension lines for an angular dimension extend from the sides of the angle.

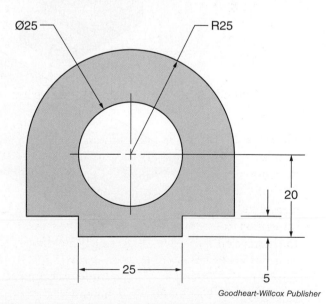

Goodheart-Willcox Publisher

Figure 6-15. Circular arcs and circles are dimensioned using standard practices and symbols. Arcs are dimensioned by specifying the radius. Circles and holes are dimensioned by specifying the diameter.

6.5 ASSEMBLIES, TOLERANCES, AND FITS

An *assembly drawing* shows the individual parts of an assembly and indicates how they are put together. See **Figure 6-16**. This type of drawing shows how one part connects to another part and indicates the functional relationship of the parts when assembled. Typically, each part making up the assembly is identified on the drawing. An assembly drawing in which the parts are shown "exploded" into individual components is called an exploded assembly view.

On engineering drawings made for manufacturing, tolerances are applied to dimensions to specify allowable variances from the design size. As previously discussed in

this chapter, a tolerance is the total amount a dimension can vary. Designers and engineers apply tolerances to part dimensions based on manufacturing requirements and functional requirements of mating parts. Tolerances are applied so that parts can be manufactured accurately within a specific tolerance range. This is essential in manufacturing because mass-produced parts must be interchangeable and must assemble properly with other parts. As a general practice, designers and engineers determine where tolerances should be tight and where they can be relaxed to achieve the desired design at a reasonable cost of production.

In addition to specifying tolerances, designers and engineers consider how fits should be specified for mating parts. *Fit* refers to the tightness or looseness between

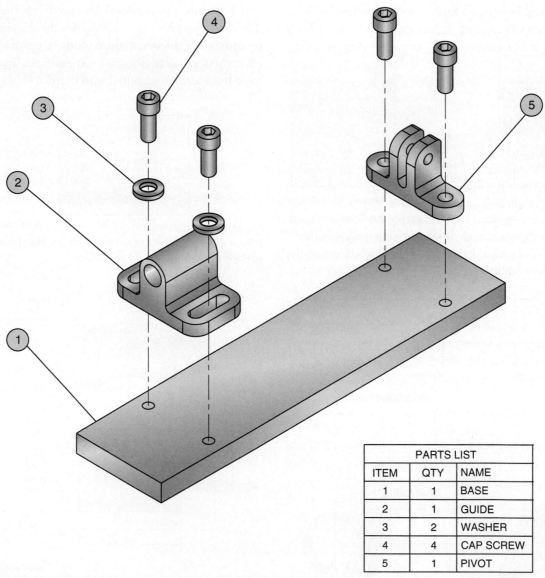

PARTS LIST		
ITEM	QTY	NAME
1	1	BASE
2	1	GUIDE
3	2	WASHER
4	4	CAP SCREW
5	1	PIVOT

Goodheart-Willcox Publisher

Figure 6-16. An assembly drawing shows how the individual parts making up an assembly are assembled. This type of assembly drawing is called an exploded assembly view because the parts are shown separated so they can be seen more clearly. In the assembly drawing, the circled numbers correspond to the item number in the parts list.

mating parts. There are three types of fits: clearance, interference, and transition.

A *clearance fit* is a fit in which clearance exists between two parts when assembled. A clearance fit occurs when the hole of a part is larger than a mating shaft that fits into the hole, thus enabling the two parts to slide or rotate when assembled. An example of this type of fit is an assembly of a wheel and an axle. The clearance fit allows the wheel to have 360° of rotation.

An *interference fit*, also called a *press fit*, is a fit in which the external dimension of one part is slightly larger than the internal dimension of the mating part. An example is the fitting of a shaft into bearings.

A *transition fit* is a fit in which a clearance or interference fit results when two parts are assembled. An example is the fitting of a shaft in a hole where location accuracy is important, but the size of the shaft results in a small amount of clearance or interference after assembly.

6.6 GEOMETRIC DIMENSIONING AND TOLERANCING (GD&T)

The application of tolerances to dimensions to establish allowable limits of size is traditionally referred to as conventional tolerancing. This dimensioning practice is suitable for certain applications. However, it does not allow for precise control of geometric relationships between features. In the manufacturing industry, functional requirements for mating parts are an important design consideration because parts are produced to be interchangeable. In addition, the most economic production of parts is an important design goal. To help meet these requirements, provisions are made in engineering to precisely define features with respect to feature relationships and function.

Geometric dimensioning and tolerancing (GD&T) is a dimensioning system used to control interpretation of tolerances defining geometric relationships of features. GD&T was developed to help organizations in the manufacturing industry communicate precise controls in a standardized language. Engineering drawings must provide the information needed for manufacturing a part and determining that the part meets specifications. The use of GD&T clarifies relationships between features and helps in verifying that parts meet functional requirements. The GD&T system does not replace conventional tolerancing. It is used in conjunction with standard dimensioning and tolerancing practices and places greater emphasis on the function and relationship of object features.

In the GD&T system, standard symbols are used to communicate information about geometric tolerancing requirements. These symbols include geometric characteristic symbols, datum feature symbols, feature control frames, and material condition modifiers. Geometric characteristic symbols used in the GD&T system are shown in **Figure 6-17**. These symbols define specific geometric conditions used to control feature geometry. For example, a form characteristic symbol indicates a form tolerance. Form tolerances are used to define specific characteristics of feature geometry on a part, such as the straightness of a line or flatness of a surface.

Datum feature symbols are used to identify datum features. A *datum* is an exact point, axis, or plane from which locations of other features and geometric controls are established. The datum feature symbol consists of a capital letter enclosed in a square frame with a leader attached to a triangle. See **Figure 6-18**.

Geometric Characteristic Symbols		
Type of Tolerance	**Geometric Characteristic**	**Symbol**
Form	Straightness	—
	Flatness	⬭
	Circularity	○
	Cylindricity	⌭
Orientation	Angularity	∠
	Perpendicularity	⊥
	Parallelism	∥
Location	Position	⊕
	Concentricity	◎
	Symmetry	⩵
Profile	Profile of a Line	⌒
	Profile of a Surface	⌓
Runout	Circular Runout	↗ or ↗
	Total Runout	↗↗ or ↗↗

Figure 6-17. Geometric characteristic symbols used in geometric dimensioning and tolerancing.

Figure 6-18. A datum feature symbol used to identify a datum surface.

A *feature control frame* is a rectangular box divided into compartments containing the geometric characteristic symbol, tolerance specification, and datum feature reference(s). See **Figure 6-19**. When applicable, a material condition modifier symbol follows the tolerance specification. Usage of material condition modifiers applies to features of size, such as a hole or shaft. The material condition modifier indicates whether the tolerance applies at maximum material condition, least material condition, or regardless of feature size. Material condition modifier symbols and other modifying symbols used in the GD&T system are shown in **Figure 6-20**. *Maximum material condition (MMC)* is the size condition of a feature containing the greatest amount of material within the tolerance limits. The symbol for maximum material

condition is the letter *M* enclosed in a circle. When the MMC modifier is given with the tolerance, it means that the tolerance applies at the maximum material condition of the feature. Examples of MMC values are the largest diameter of a pin and the smallest size of a hole. Note that the MMC value for a hole is the *smallest* size because the greatest amount of material exists in the part at this size.

Least material condition (LMC) is the size condition of a feature containing the least amount of material within the tolerance limits. The symbol for least material condition is the letter *L* enclosed in a circle. When the LMC modifier is given with the tolerance, it means that the tolerance applies at the least material condition of the feature. Examples of LMC values are the smallest diameter of a pin and the largest size of a hole. Note that the LMC value for a hole is the *largest* size because the least amount of material exists in the part at this size.

Regardless of feature size (RFS) is a condition in which the tolerance value applies at any size within the tolerance limits. When no material modifier symbol is given with the tolerance, regardless of feature size is assumed.

A material condition modifier that follows a datum feature reference is called a material boundary modifier. Maximum material boundary and least material boundary are noted in the same manner as maximum material condition and least material condition. When no symbol is given, regardless of material boundary is assumed.

On an engineering drawing, a feature control frame associated with a feature provides a complete specification of the geometric tolerancing requirement. Study the mechanical part drawing in **Figure 6-21**. Notice the linear dimensions that locate the hole. These are called basic dimensions. A *basic dimension* is a theoretically exact dimension. The appearance of a basic dimension indicates that a tolerance is associated with the dimensioned feature. Usually, the tolerance is given in a feature control frame. A basic dimension is indicated as a dimensional value enclosed in a rectangle.

Notice the dimension specifying the size of the hole. The dimension has a plus-and-minus tolerance allowing it to vary in the positive and negative directions. The plus-and-minus value is used to calculate the upper and lower limits of the dimension. The upper limit is 1.002 and the lower limit is .998. The tolerance is equal to the difference between the upper and lower limits and is calculated as $1.002 - .998 = .004$. The hole is at MMC when the produced size is $\varnothing.998$. The hole is at LMC when the produced size is $\varnothing1.002$.

The location of the hole is allowed to vary according to the geometric tolerance specified in the feature control frame. The feature control frame specifies a position tolerance for the hole relative to datum references A, B, and C.

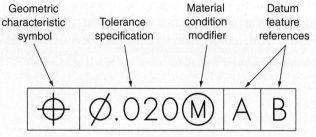

Geometric characteristic symbol   Tolerance specification   Material condition modifier   Datum feature references

Goodheart-Willcox Publisher

Figure 6-19. A feature control frame is used to specify a geometric tolerancing requirement.

Modifying Symbols	
At maximum material condition (applied to tolerance) or at maximum material boundary (applied to a datum reference)	Ⓜ
At least material condition (applied to tolerance) or at least material boundary (applied to a datum reference)	Ⓛ
Translation	▷
Projected tolerance zone	Ⓟ
Free state	Ⓕ
Tangent plane	Ⓣ
Unequally disposed profile	Ⓤ
Independency	Ⓘ
Statistical tolerance	⟨ST⟩
Continuous feature	⟨CF⟩
Between	↔
All around	⊶
All over	⊶

Goodheart-Willcox Publisher

Figure 6-20. Modifying symbols used in the GD&T system.

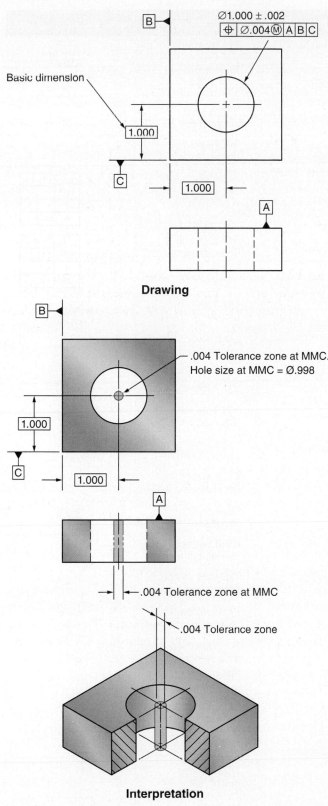

Drawing

Interpretation

.004 Tolerance zone at MMC.
Hole size at MMC = Ø.998

.004 Tolerance zone at MMC

.004 Tolerance zone

Goodheart-Willcox Publisher

Figure 6-21. A dimensioned mechanical part drawing with a specified geometric tolerance. The feature control frame specifies a position tolerance of Ø.004 at MMC. The MMC hole size is Ø.998. At this size, the hole axis must be located within a Ø.004 tolerance zone.

The tolerance zone is cylindrical, as indicated by the diameter symbol preceding the tolerance. The information in the feature control frame is read from left to right. The tolerance specification reads as follows: "The feature axis must be located within Ø.004 at maximum material condition relative to datums A, B, and C."

An illustrated interpretation of the geometric tolerance specification is shown at the bottom of **Figure 6-21**. Notice in the pictorial view that the tolerance zone is cylindrical in shape. The tolerance zone indicates the allowable position tolerance for the axis of the hole. When the part is manufactured, the axis must be within the Ø.004 tolerance zone when the hole is at maximum material condition. Note that the interpretation example is based on the tolerance zone at the MMC hole size. This is the smallest permitted hole size, equal to Ø.998. At other allowable produced sizes, the positional tolerance is allowed to increase by an amount equal to the departure in size from MMC. For example, when the produced size of the hole is at LMC (the largest permitted hole size, equal to Ø1.002), the positional tolerance is allowed to increase to Ø.008.

6.7 ELECTRICAL SYMBOLS AND DIAGRAMS

Electrical diagrams use graphic symbols to represent the various components in the system. Electrical symbols used on prints often follow a standard convention. Common electrical and electronic symbols used on drawings are shown in **Figure 6-22**. Standard symbols used to identify electrical devices have been established by the American National Standards Institute (ANSI), the National Electrical Manufacturers Association (NEMA), and the International Organization for Standardization (ISO).

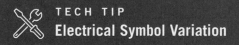

TECH TIP
Electrical Symbol Variation

Be aware that some electrical diagrams may use symbols different from those shown here. Some company standards may differ.

Electrical diagrams use symbols to identify devices in circuits and lines to represent the conductors connecting the devices. These diagrams identify the devices and equipment in the circuit and show how they are connected. Electrical circuits can be complicated, and an electrical diagram is a key tool for troubleshooting. Three common types of electrical diagrams are schematic diagrams, ladder diagrams, and wiring diagrams. These are discussed in the following sections.

Electrical and Electronic Symbols

Capacitors	Fixed	(symbol)
	Polarized	(symbol)
Contacts	Normally open	(symbol)
	Normally closed	(symbol)
	Timed open	(symbol) T.O.
	Timed closed	(symbol) T.C.
Diodes	Diode	(symbol)
	Zener diode	(symbol)
	Light-emitting diode	(symbol)
	Photodiode	(symbol)
Inductors	Air core	(symbol)
	Iron core	(symbol)
Miscellaneous Electrical Symbols	Bell	(symbol)
	Buzzer	(symbol)
	Lamp	(symbol)
	Pilot light	(symbol)
	Thermocouple	(symbol)
Miscellaneous Electronic Devices	Full-wave bridge rectifier	(symbol)
	Silicon controlled rectifier	(symbol)
	Triac	(symbol)
Sensors	Light dependent resistor	(symbol)
	Thermistor	(symbol)
	Electrostatic sensor	(symbol)

Category	Item	Symbol
Switches	Pushbutton	(symbol) Normally Open
	Pushbutton	(symbol) Normally Closed
	Mushroom head	(symbol)
	Single-pole single-throw (SPST)	(symbol)
	Single-pole, double-throw (SPDT)	(symbol)
	Selector switch Two-position	(symbol and table)
	Selector switch Three-position	(symbol and table)
	Limit switch (normally open)	(symbol)
	Limit switch (held closed)	(symbol)
	Limit switch (normally closed)	(symbol)
	Limit switch (held open)	(symbol)
	Float switch (normally open)	(symbol)
	Float switch (normally closed)	(symbol)
	Float switch (normally open)	(symbol)
	Float switch (normally closed)	(symbol)
	Pressure switch (normally open)	(symbol)
	Pressure switch (normally closed)	(symbol)
	Foot operated switch (normally open)	(symbol)
	Foot operated switch (normally closed)	(symbol)
	Solenoid	(symbol)
	Three-phase disconnect	(symbol)

Selector switch Two-position:

Letter Sym	Position 1	Position 2
A		X
B	X	

Selector switch Three-position:

Letter Sym	Position 1	Position 2	Position 3
A	X		
B			X

Goodheart-Willcox Publisher

Figure 6-22. Common symbols used on electrical diagrams. *(continued)*

Electrical and Electronic Symbols (continued)

Power Supply Symbols	DC supply		**Relays**	Relay coil	
	Constant current source			Motor starter	
	AC supply		**Resistors**	Fixed	
	Generator			Potentiometer	
	Battery		**Transistors**	Bipolar junction transistor (BJT) PNP type	
AC Motors	Single-phase			Bipolar junction transistor (BJT) NPN type	
	Three-phase			Junction field-effect transistor (JFET) N channel	
DC Motors	Armature			Junction field-effect transistor (JFET) P channel	
	Shunt field		**Transformers**	Air core	
	Series field			Iron core	
Wiring	Power			Current	
	Control		**Logic Gates**	AND gate	
	Ground			NAND gate	
	Terminal			OR gate	
Overload Protection Devices	Fuse			NOR gate	
	Circuit breaker			NOT gate (Inverter)	
	Thermal overload protection device			XOR gate	

Figure 6-22.

Goodheart-Willcox Publisher

6.7.1 Schematic Diagrams

A *schematic diagram* shows the electrical connections and operation of a circuit. It provides a simple representation of the sequence of functions in a circuit without showing the exact location of each component. The components in the circuit are represented with symbols and wiring is represented with lines. Component connections and interconnecting wiring are clearly indicated. A schematic diagram is also referred to as a *line diagram*.

You should know how to read a schematic diagram and visualize the flow of current through the circuit in order to understand the circuit's function. For example, study the basic "seal-in" motor starter schematic diagram shown in **Figure 6-23**. Note that different line thicknesses are used to represent the different parts of the circuit. Thick lines are used to represent the power circuit. Thin lines represent the control circuit.

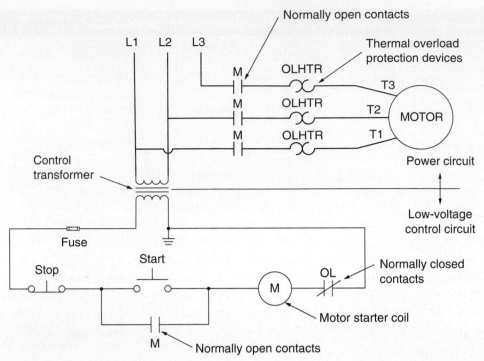

Figure 6-23. A motor starter schematic diagram.

The lines labeled *L1*, *L2*, and *L3* indicate the connections to the three-phase 480 V input connections to the circuit. The three sets of contacts labeled *M* are those of the motor starter relay. They all close in unison when the motor starter relay is energized, resulting in three-phase power being supplied to the motor. The three devices labeled *OLHTR* are the three thermal overload protection devices contained in the motor starter relay assembly. If too much current is drawn by the motor, they will open in unison and help protect the motor from damage.

The two primary windings of the control transformer are connected to L1 and L2 respectively. On one side of the secondary winding of the control transformer is a fuse. On the opposite side of the control transformer, the secondary winding is connected to ground.

The start pushbutton is normally open. When pressed, it completes the circuit, energizing the motor starter coil labeled *M* in the drawing. When coil M is energized, the set of contacts labeled *M* across the start pushbutton will close, maintaining the circuit in an energized state regardless of the position of the start pushbutton. The stop pushbutton is normally closed. When pressed, it interrupts the control voltage from the circuit.

When the start pushbutton is depressed, the circuit is sealed in and the motor will run. Once the stop pushbutton is depressed, the motor starter relay drops out and the circuit is reset to a de-energized state. At this point, the motor stops receiving power and coasts to a stop. If an overload

condition is detected by the thermal overloads, the normally closed set of contacts labeled *OL* will open, interrupting the circuit in a similar manner to the stop pushbutton.

You can state the entire description above if you are able to recognize all of the symbols, understand their function, and trace the path by which electricity flows through the circuit. Principles of electricity and motor starter principles will be discussed in later chapters. After reading those chapters, refer back to this chapter to gain a better understanding of print reading.

6.7.2 Ladder Diagrams

A *ladder diagram* is a type of schematic diagram that shows the function of a circuit and resembles a ladder. See **Figure 6-24.** The main power lines make up the vertical

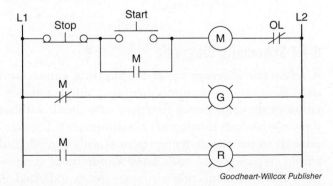

Figure 6-24. A ladder diagram.

rails of the ladder and the control devices in the circuit make up the horizontal rungs. A ladder diagram shows the basic operation of a circuit without showing the exact locations of components. Lines are organized to illustrate the flow of current and the drawing is designed to be read from top to bottom or left to right.

6.7.3 Wiring Diagrams

A *wiring diagram* shows the general arrangement and location of electrical components in a circuit. See **Figure 6-25**. This type of diagram is designed to show the actual location of components in the system and how the electrical connections are made. The wiring diagram shown in **Figure 6-25** is for a motor starter. Notice that this diagram is drawn differently in comparison to the motor starter schematic diagram shown in **Figure 6-23**.

Notice the dashed outline forming a box around the motor starter relay. The starter coil, all three sets of motor contacts, the overload devices, and the seal-in contacts are part of an assembly and are already wired. The start and stop pushbuttons are also in an assembled unit. However, connections between the assembled units are not wired for you, as indicated by the wires exiting each unit.

Based on this wiring diagram, you only need to make the connections that enter and exit the prewired units as well as the connections to the control transformer, fuse,

and circuit breakers. The layout in the wiring diagram indicates how the interconnecting wiring is run. Some diagrams may use a solid line rather than a dashed line for boxed outlines indicating prewired devices.

6.8 FLUID POWER SYMBOLS AND DIAGRAMS

Fluid power circuit diagrams use symbols to identify components making up a hydraulic or pneumatic circuit. Fluid power symbols are designed to describe characteristics and functions of a circuit, such as connections, flow direction, and method of operation. Fluid power symbols do not represent the actual physical appearance of a component. Rather, they use accepted conventions to convey information in graphic form. Standard symbols for use on fluid power diagrams have been established by the American National Standards Institute (ANSI) and the International Organization for Standardization (ISO).

Fluid power symbols are made up of basic geometric shapes, such as lines, squares, rectangles, triangles, and circles. Lines are used to represent flow paths. Triangles are used to show flow direction and the type of fluid used in a circuit. A filled triangle represents hydraulic flow and an open triangle represents pneumatic flow. Arrow symbols indicate characteristics, such as directional flow, rotational direction, and adjustment position.

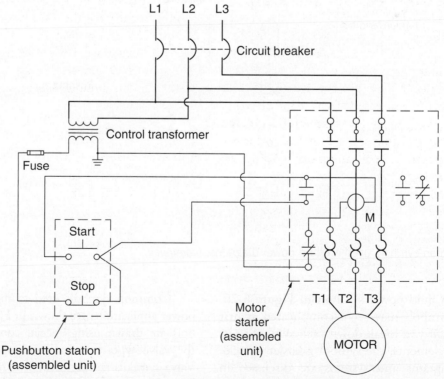

Goodheart-Willcox Publisher

Figure 6-25. A wiring diagram shows the actual location of electrical components in a circuit.

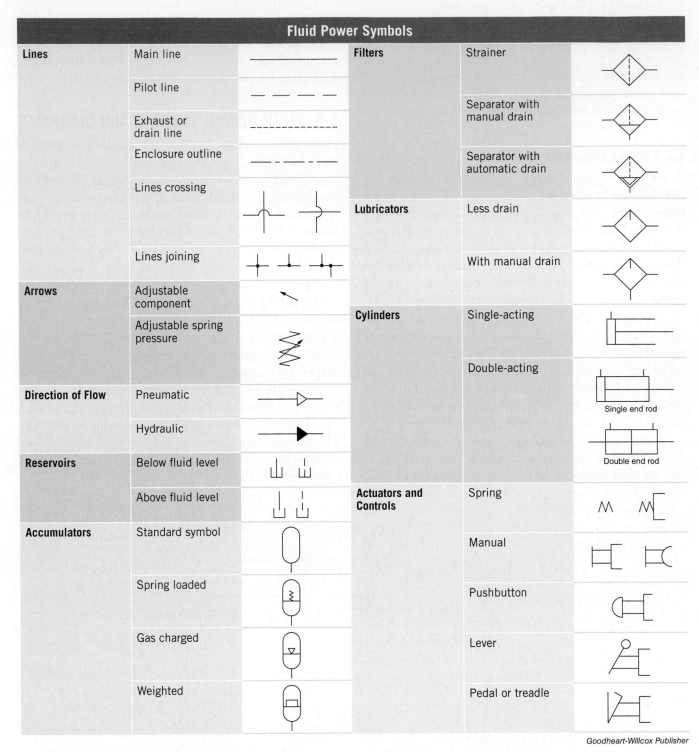

Figure 6-26. Common symbols used on fluid power diagrams. (*continued*)

See common fluid power symbols in **Figure 6-26**. Some of these symbols may seem complicated at first glance. However, if you break down each symbol to its basic individual elements, the symbol is easier to read. Fluid power diagrams and systems are discussed in greater detail in the *Basic Hydraulic and Pneumatic Systems* section, Chapters 15–18.

Common symbols used to represent valves in fluid power applications are shown in **Figure 6-27**. Valve symbols are drawn using special conventions that identify the valve type, function, and operating characteristics. Valve symbols consist of one or more boxes. The number of boxes indicates the number of operating positions. For example, two boxes indicate the valve has two

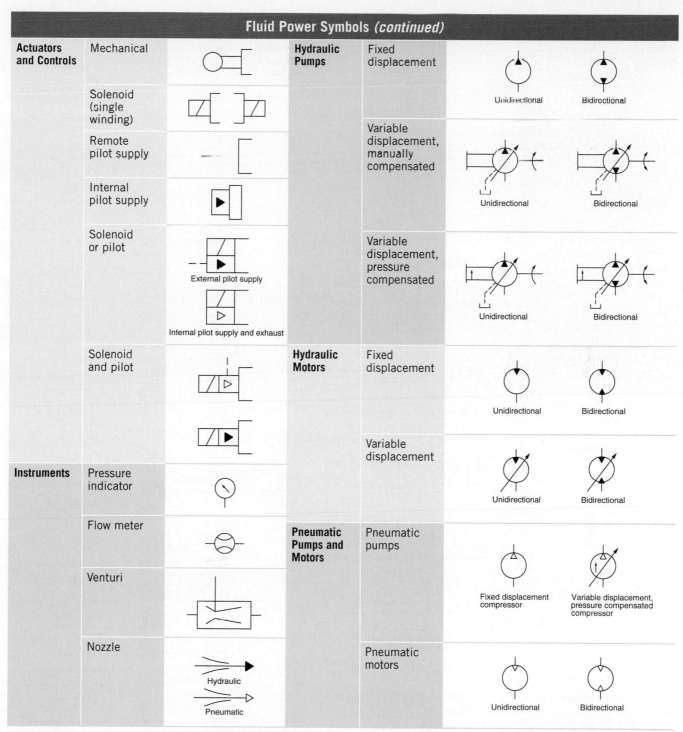

Figure 6-26.

Goodheart-Willcox Publisher

positions and three boxes indicate the valve has three positions. The center position of a three-position valve is the common default position.

Valve symbols also indicate flow paths, number of ports, and actuation method. Arrow symbols indicate the direction of flow. The number of ports typically indicates the number of "ways" in which fluid enters or exits the valve. For example, a two-way valve has two ports and a three-way valve has three ports. Ports are indicated

where flow lines intersect with the basic component symbol. A T-shaped symbol indicates a closed port at which flow is blocked.

The actuation method identifies the way in which a valve is operated. Common actuation devices include pushbuttons, levers, springs, mechanical rollers, solenoids, and pilot controls. The symbol representing the actuation or control device is indicated on the end of the component symbol.

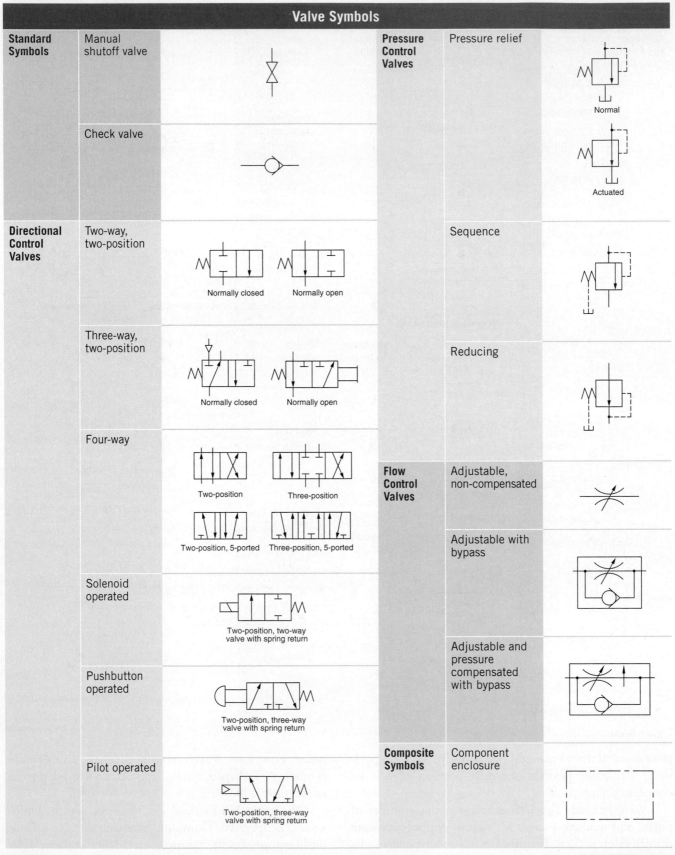

Figure 6-27. Common valve symbols used on fluid power diagrams. The symbol for a valve identifies the valve type, flow paths, number of ports, and actuation method.

Goodheart-Willcox Publisher

Fluid power circuit diagrams show the design and function of a hydraulic or pneumatic system. They indicate control components, connections, and flow paths and make it possible to analyze and troubleshoot a circuit.

A circuit diagram for a hydraulic system is shown in **Figure 6-28**. This is a basic circuit used to control the extension and retraction of a double-acting cylinder. The circuit has a hydraulic pump and a pressure relief valve to limit the maximum operating pressure in the system. A four-way, three-position directional control valve is used to direct pressurized fluid to the cap end (left end) and rod end (right end) of the cylinder. When the directional control valve is at the center position, as shown in the figure, the cylinder ports are blocked and fluid is returned to the reservoir. When the directional control valve is shifted to the first working position (A), fluid is directed to the cap end of the cylinder, forcing the cylinder to extend. Fluid at the rod end is directed to the reservoir. When the directional control valve is shifted to the second working position (B), fluid is directed to the rod end of the cylinder, forcing the cylinder to retract. Fluid at the cap end is directed to the reservoir.

A circuit diagram for a pneumatic system is shown in **Figure 6-29**. This circuit is used to control the operating speeds of a double-acting cylinder. A four-way, two-position directional control valve is used to control the extension and retraction of the cylinder. There are also two flow control valves in the circuit. Each valve is a needle valve with an integral check valve. The cylinder in the figure is shown in the retracted position, which

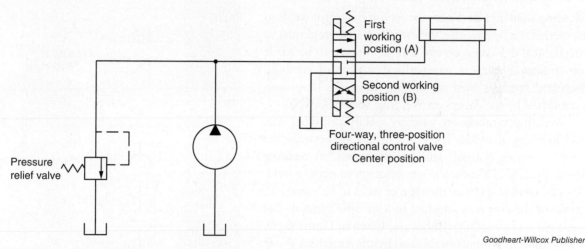

Goodheart-Willcox Publisher

Figure 6-28. A circuit diagram of a hydraulic power system. This system uses a four-way, three-position directional control valve to control the extension and retraction of the cylinder.

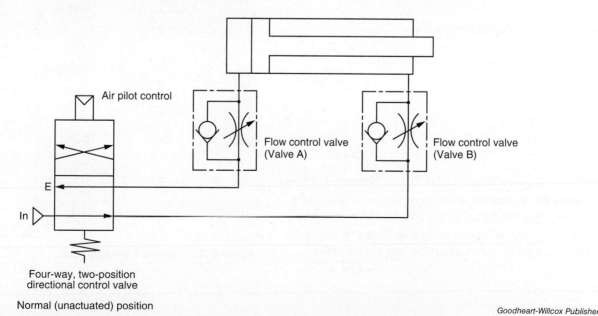

Goodheart-Willcox Publisher

Figure 6-29. A circuit diagram of a pneumatic power system. This system uses a four-way, two-position directional control valve and has flow control valves to control the extension and retraction speeds of the cylinder.

corresponds to the normal (unactuated) position of the directional control valve. When the directional control valve is actuated, the check valve in Valve A allows pressurized air to pass to the cap end (left end) of the cylinder, forcing the cylinder to extend. The exhaust air at the rod end of the cylinder is blocked by the check valve in Valve B. The exhaust air is forced through the needle valve, which meters the flow. This controls the extension speed of the cylinder. When the directional control valve is shifted back to the normal position, the check valve in Valve B allows air to pass, forcing the cylinder to retract. The needle valve in Valve A meters the flow of exhaust air and controls the retraction speed of the cylinder.

6.9 WELDING DRAWINGS

Welding drawings designate the requirements for welds to be carried out by a welder. Welding drawings are similar to mechanical drawings, except that weldments to be made are indicated using a standardized system of symbols. Standard symbols used on welding drawings have been established by the American Welding Society (AWS).

Welding symbols are used to specify the type, size, and location of welds. There is a difference between the terms "welding symbol" and "weld symbol." A **welding symbol** provides a complete specification to make a weld. A **weld symbol** specifies the type of weld to be made and is one of the elements attached to a welding symbol. The basic parts of a welding symbol are shown in **Figure 6-30**. Standard weld symbols established by the American Welding Society are shown in **Figure 6-31**.

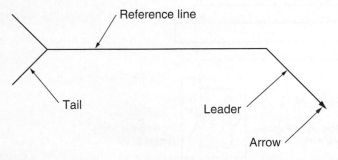

Goodheart-Willcox Publisher

Figure 6-30. A standard welding symbol is made up of a reference line connected to a leader line with an arrow. The tail is included when noting a welding process or additional specifications about the weld to be made.

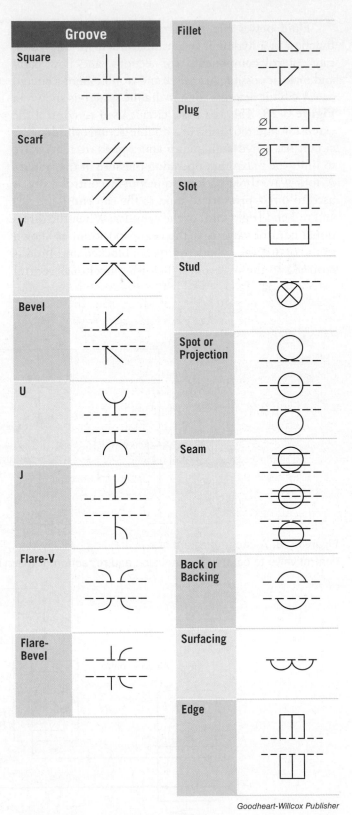

Goodheart-Willcox Publisher

Figure 6-31. Standard weld symbols.

The placement of the welding symbol indicates how the weld is to be made. The leader line and arrow of the welding symbol are connected to one side of the joint. This side is called the "arrow side" of the joint. The opposite side is called the "other side" of the joint. The placement of the weld symbol on the reference line of the welding symbol indicates the side of the joint where the weld is to be made. A weld symbol placed below the reference line indicates the weld is to be made on the arrow side. A weld symbol placed above the reference line indicates the weld is to be made on the other side. Weld symbols placed on both sides of the reference line indicate welds are to be made on both sides of the joint. See **Figure 6-32**.

Welding drawings are generally classified as assembly drawings because they show the assembly of multiple parts to be fastened by welding. You will normally see welding symbols used on metal fabrication drawings and drawings used in building construction.

CHAPTER WRAP-UP

Print reading is a critical skill for all aspects of industrial maintenance and mechatronics. By learning print reading, you also learn about system components and design. When you understand how systems are designed, troubleshooting becomes much easier. This chapter serves as a general introduction to print reading. Continue to develop your print reading skills throughout your coursework.

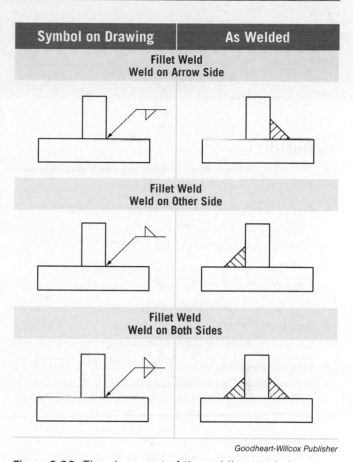

Symbol on Drawing	As Welded
Fillet Weld **Weld on Arrow Side**	
Fillet Weld **Weld on Other Side**	
Fillet Weld **Weld on Both Sides**	

Goodheart-Willcox Publisher

Figure 6-32. The placement of the welding symbol indicates whether the weld is to be made on the arrow side, the other side, or both sides of the joint.

Chapter Review

SUMMARY

- Technicians use prints to identify the functions of a system and troubleshoot problems.

- A title block supplies key information related to a print and typically includes the part name and number, drawing scale, and tolerance information.

- A multiview drawing provides a complete shape description of a mechanical part in a two-dimensional representation.

- An assembly drawing shows how the parts making up an assembly are assembled together.

- Geometric dimensioning and tolerancing (GD&T) is used in the manufacturing industry to specify tolerancing requirements and help verify that parts meet design specifications.

- Electrical diagrams use symbols to represent components and connections in an electrical circuit. A schematic diagram shows the sequence of functions in a circuit without showing the physical location of each component. A wiring diagram shows the location of components in a circuit and wiring connections.

- Fluid power circuit diagrams use graphic symbols to represent components, connections, and flow paths in a hydraulic or pneumatic system.

- Welding symbols are used on welding drawings to specify the type, size, and location of welds.

REVIEW QUESTIONS

Answer the following questions using the information provided in this chapter.

1. Briefly explain why prints serve as a key tool in industrial maintenance.

2. What does the scale of a drawing indicate?

3. What is a multiview drawing?

4. Lines that represent the visible edges and contours of an object are called _____ lines.

5. Lines used to indicate alternate positions of moving parts, adjacent positions of related parts, or repeated detail are called _____ lines.

6. On mechanical part drawings, arcs are dimensioned by specifying the _____. Circles and holes are dimensioned by specifying the _____.

7. What is an assembly drawing?

8. What is the difference between a clearance fit and an interference fit?

9. Briefly explain why geometric dimensioning and tolerancing (GD&T) is used in manufacturing.

10. A(n) _____ is an exact point, axis, or plane from which locations of other features and geometric controls are established.

11. Explain the difference between a schematic diagram and a wiring diagram.

12. Identify the following electrical symbols:

   A.

   B.

   C.

   D.
   CR

13. Identify the following fluid power symbols:

   A.   C.

   B.

   D.

14. Identify the three parts labeled A, B, and C in the following welding symbol.

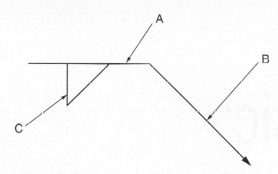

15. What is the difference between a welding symbol and a weld symbol?

NIMS CREDENTIALING PREPARATION QUESTIONS

The following questions will help you prepare for the NIMS Industrial Technology Maintenance Level 1 Maintenance Operations credentialing exam.

1. Which of the following is *not* true of a title block?

   A. It contains important information about a print.
   B. The format is the same from company to company.
   C. The size and placement can vary.
   D. An area for approvals is usually included.

2. What line type is used to indicate alternate positions of moving parts, adjacent positions of related parts, or repeated detail?

   A. Hidden
   B. Phantom
   C. Cutting-plane
   D. Object

3. What is the purpose of an extension line?

   A. Indicates where the end of a dimension line terminates
   B. Represents axes of symmetrical objects
   C. Shows an imaginary break in an object where a portion is omitted
   D. Represents object edges and contours located behind other features

4. What are linear dimensions usually expressed in?

   A. The diameter symbol (∅)
   B. Dimension of the radius
   C. Degrees, minutes, and seconds
   D. Decimal or fractional units

5. What does a thick line represent in a schematic diagram?

   A. The flow of current
   B. The motor
   C. The power circuit
   D. The control circuit

6. What does a filled triangle represent in a fluid power diagram?

   A. Hydraulic flow
   B. Pneumatic flow
   C. Electric flow
   D. Directional flow

7

Basic Troubleshooting Principles

LEARNING OBJECTIVES

After completing this chapter, you will be able to:

☐ Understand the importance of determining the root cause of a problem and correcting that cause as well as the symptoms.

☐ Develop communication methods and determine whom you should communicate with.

☐ Identify powers of observation that will pay big dividends and enhance your troubleshooting skills.

☐ Learn how to use a binary search to determine the source of a problem in the quickest possible manner.

☐ Understand the usefulness of troubleshooting aids, including prints and equipment used for monitoring and testing.

TECHNICAL TERMS

accelerometer	megohmmeter
belt tension gauge	micrometer
binary search	oscilloscope
clamp-on ammeter	pressure gauge
depth micrometer	root cause
digital multimeter (DMM)	thermal imager
flow meter	vernier scale
infrared radiation (IR)	vibration meter
infrared thermometer	

The most basic principle of troubleshooting involves determining the root cause of a problem, **Figure 7-1**. You may repair the obvious but, if you do not find and correct the *root cause*, or underlying source of the problem, the failure is destined to repeat. (You can review Chapter 3 to remind yourself how to use root cause analysis in conjunction with maintenance records.)

In many ways, failure to find the root cause and correct it is much like a doctor treating the symptoms and not the disease. Another example would be replacing tires on a car because they have an uneven wear pattern without checking and correcting the alignment problem that caused the uneven wear. The new set of tires will experience the same issues if the root cause of the problem is not remedied.

Root Cause Analysis

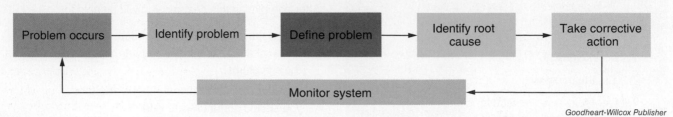

Figure 7-1. This diagram provides an overview of the process of root cause analysis.

A good technician troubleshoots a problem to determine the root cause before making any corrections, **Figure 7-2**. Once the root cause is identified, the technician corrects the problem along with any symptoms that it may present. A "parts changer," on the other hand, will keep replacing parts until the symptoms temporarily correct themselves, but may never find and correct the root cause. Your goal is not to be a parts changer, but rather to become a true technician!

Figure 7-2. These technicians are performing troubleshooting before taking any corrective action.

7.1 COMMUNICATION

It is important to communicate effectively with anyone who might have information that may prove helpful in allowing you to find the root cause of a problem. Communication may be either written or verbal. You can first consult the maintenance records that are kept on the piece of equipment presenting the problem. This might allow you to see what was done to correct similar problems in the past. You might also consult what is written on the service ticket requesting repair of a specific system.

Verbal communication is valuable for the same reason. Often, the person who reported the problem will have

some firsthand information that might assist you in finding a resolution. The maintenance supervisor who assigned the trouble call may also be able to provide a wealth of information. Remember, a supervisor has a vested interest in resolving the problem quickly in order to allow production to resume as soon as possible.

Has this problem occurred in the past? Who has worked on the problem before? What did those technicians do to resolve the problem? Does the same problem keep repeating? If so, maybe the root cause of the problem is yet to be found. You may need to look further than just repeating the same old fix. The answers to these types of questions may shed some light on how you may best resolve your problem in a more lasting or permanent way. Remember to review maintenance records to discover the answers to some of these questions, **Figure 7-3**.

For instance, maybe a belt was repaired but the tension was still excessive or unbalanced, causing the new belt to tear or separate at the splice. Repair of the belt and proper tensioning may be all that is needed to put things right permanently.

Communicating with the machine operator can provide valuable information to use toward problem resolution. The operator was most likely there and observed what

Figure 7-3. Good technicians will review maintenance records to determine the history of a problem.

happened leading up to the breakdown. Do not dismiss what operators have to say. They watch their machines run day in and day out, and will most certainly notice when the smallest abnormal thing occurs. The operators may not know how to repair the machine, but they will certainly know when things are going wrong, and they may even know exactly what went wrong, **Figure 7-4**.

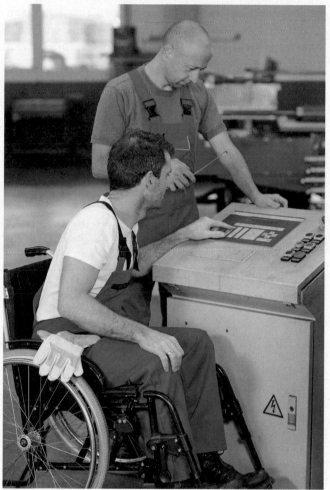

Firma V/Shutterstock.com

Figure 7-4. The machine operator has firsthand knowledge of the machine in need of maintenance and can describe what happened leading up to a breakdown.

7.2 OBSERVATION

Your powers of observation will pay dividends when attempting to determine the root cause of a problem. How will you be able to determine what is wrong if you do not know how things operate when they are right? Instead of sitting around the maintenance shop when things are slack, you should be out on the factory floor observing different pieces of equipment and taking notes. This will enable you to know what should be happening when everything is running correctly. Observation and notes, either written

or mental, will allow you to spend less time trying to find the root cause of a problem when one occurs.

In some instances, your supervisor may allow you to take digital photographs or videos. Company policy may prohibit the use of cell phones for any reason on the plant floor. However, some workplaces have company video cameras and digital cameras that may be used for documentation purposes. Videos, images, and notes can then be stored in the CMMS (computerized maintenance management system).

Cell phones and digital cameras are great for creating photographic and video documentation of wiring or how an assembly goes together. Once you have disconnected wiring or removed parts, it can be difficult to remember exactly how and where everything goes to put the item back together if it has not been properly documented. If digital cameras and devices are not permitted, you might have to settle for a self-made drawing as a documented reference to help you get everything back in the correct place when it is time for reassembly.

7.3 THE BINARY SEARCH

You should not randomly guess and measure to find a problem. Unless you are extremely lucky, such a method may require you to tear the entire system apart piece by piece before you arrive at the problem. Instead, you need to take a systematic approach to problem resolution.

One of the most useful troubleshooting techniques is called the ***binary search***. Some refer to it as the *divide-and-conquer approach* or *half-split method*. Regardless of what you call this technique, it is carried out by dividing a system in half and making a measurement to methodically determine where the problem is.

First, you must establish where the problem *does not* exist, and then you can focus on where the problem *does* exist. Take for example a loss of voltage. You must first establish where the voltage is good, possibly somewhere near the power source. Next, you need to take a measurement where the problem exists to verify the lack of voltage. Now, pick a point halfway in between where you have good voltage and where you do not. At this established point, take another measurement.

Depending on the result of your measurement, you can determine which direction to go looking for the missing voltage. If the midpoint shows good voltage, your problem is closer to the end point where you do not have good voltage. Your next point should then be halfway between that newly established good-voltage point and where you do not have good voltage. Continue on in this fashion, narrowing the potential area of where the problem might be.

Simply divide the area in half and take another measurement. The problem is always located between where you find the voltage to be good and where you find it to be bad. By successively dividing the area between good and bad in half, you will be on top of the problem area in no time at all.

To prove how effective the binary search technique is, you can play a number guessing game with your classmates. Have someone think of a number between 1 and 100 and write it down on a piece of paper. Tell that person you will guess the number within seven guesses.

Now, using the binary search technique you just learned, set about methodically guessing the number. The person who thought of the number has three possible responses to your guess: higher, lower, or correct.

Suppose, for example, the number is 37. Knowing the number could be between 1 and 100, you divide that range in half, and your first guess would be 50. The person who thought of the number would respond "lower." Now you divide the range between 1 and 50 in half, and your second guess would be 25. The person who thought of the number would respond "higher."

Your next guess would require dividing the range between 25 and 50 in half: $25 \div 2 = 12.5$, and $12.5 + 25 = 37.5$. Rounding up, you would then use 38, to which the person would respond "lower." You would then need to divide the range between 38 and 25 in half (arriving at 6.5), which would result in a guess of 32 (rounded up from 31.5). The person who thought of the number would then respond "higher." You then need to divide the range between 32 and 38 in half, arriving at 35, and the person who thought of the number would respond "higher." Finally, you divide the range between 35 and 38 in half, which yields 37 (rounded up from 36.5), and you have guessed correctly in five guesses!

The binary search method proves to be the best way to find a problem in a large network. It has been used for years by cable television companies. Their coaxial cable is spread out over an entire city, and it is difficult to get to each possible test point to even determine whether a problem exists. The binary search is often the best method to allow them to pinpoint the source of a problem and restore service in the quickest possible fashion.

The binary search method will often allow you to determine the source of a problem in record time.

7.4 ELECTRICAL TROUBLESHOOTING

The binary search method may not work for you in every case, especially with mechanical problems, but it will work every time with an electrical problem. Aside from missing voltage, you are likely to experience other occasions for electrical troubleshooting, such as blown fuses and open circuit breakers, for example.

Fuses and circuit breakers are used to interrupt the flow of current should it become excessive, thus avoiding major damage or harm. An open circuit breaker or a blown fuse may cause you to receive a trouble ticket or a call on the affected system. Again, you need to find the root cause of the blown fuse or open circuit breaker.

Fuses do not blow because they become defective. Something in the circuit drew too much current and caused the blown fuse. The same scenario holds true for circuit breakers, which give an obvious indication if they are tripped for any reason. Chapter 28, *Electrical Troubleshooting*, will give you more information on several methods of testing for blown fuses and open circuit breakers.

7.5 MECHANICAL TROUBLESHOOTING

You may be called upon to troubleshoot problems with mechanical drive systems, pneumatic systems, or hydraulic systems in the course of your duties. Mechanical problems may arise from the following features or areas:

- Bearings.
- Pumps.
- Piping.
- Compressed air and pneumatics.
- Hydraulics.

Be aware that problems may exist in the mechanical systems listed above or in their control systems. You should develop a sense of whether the problem lies in the mechanical system or is a result of a malfunction in the control system.

Bearing problems can be diagnosed with the aid of vibration analysis equipment or the use of a mechanic's stethoscope to determine if a bearing is running smoothly, **Figure 7-5**. The root cause of a bearing failure may be motor shaft currents, shaft misalignment, or simply lack of lubrication. You would do well to check for all three possible scenarios. Failure to find the root cause of bearing problems will put you in the position of simply being a parts changer, which will waste time and spare parts as the same problem inevitably reoccurs.

Pump problems can be bearing related, motor related, filter related, or control system related. Once again, you need to determine the area where the root cause exists. Repair both the root cause and any symptoms to fully complete the repair.

Vibration analysis equipment

Mechanic's stethoscope

Figure 7-5. Vibration analysis equipment and a mechanic's stethoscope are useful tools for diagnosing bearing issues and other mechanical problems.

Problems with a piping system may be the result of vibration, stresses, or improper installation. Make certain you know the ins and outs of proper installation methods before attempting to replace or repair piping. More piping system procedures will be covered in the section on *Maintenance Piping*.

SAFETY NOTE
Pneumatic Systems

Troubleshooting and repair work with pneumatic systems should be approached with care. Compressed air can be dangerous, and there is the potential for explosion. Always use the proper PPE and take appropriate safety measures before attempting any work.

Hydraulic hoses are susceptible to abrasion if not properly installed. Abrasion may lead to leaks or a hose rupture. Inspect hydraulic systems for any leakage around hoses and their connections. If leakage is detected, determine the cause and repair it before proceeding further.

7.6 TROUBLESHOOTING AIDS

There are many different aids available to assist you in the troubleshooting process. Take advantage of any such troubleshooting aids available, as they will help to shorten the time necessary to pinpoint the root cause of a malfunction. Depending on the equipment manufacturer and your company operations, some or all of the following troubleshooting aids may be available to you.

7.6.1 Prints

The most commonly used troubleshooting aids are prints. For example, schematic diagrams will assist you in determining the signal flow of electricity or the fluid flow of hydraulic or pneumatic systems.

Assembly drawings indicate how the various parts of a system are assembled or should be disassembled. These types of prints are extremely important when putting an assembly back together in order to ensure that you do not have any leftover pieces.

Multiview mechanical drawings are important for informing you of the various dimensions and tolerances of system parts. A quick part measurement and comparison against its specified tolerance will indicate if the part is too worn to be of further use.

7.6.2 Electrical Test Equipment

Knowing how to use electrical/electronic test equipment is one of the most important of the troubleshooting skills. Considering that you cannot see electricity flowing through conductors, test equipment will provide a picture as to how the current is flowing and the voltage behind it, as well as other circuit parameters.

The *digital multimeter (DMM)*, **Figure 7-6**, should be your first tool of choice when troubleshooting an electrical problem. It allows for voltage, current, resistance, diode, and capacitor testing. Some models are capable of measuring frequency as well. Some digital multimeters also have a peak hold memory, which will indicate the maximum, minimum, and average readings to let you know how much change the circuit is experiencing.

Voltage measurements are important. Low voltages are an indicator of poor connections and overload conditions (as an overload occurs, the voltage will sag). Voltage imbalances of greater than 2% on a three-phase circuit may reduce equipment performance and cause premature failure. Voltage drops across fuses or switches may also show up as a voltage imbalance.

Resistance measurements such as insulation resistance to ground may predict imminent failure of both motors and transformers. Wiring may also be checked for resistance to ground to determine if there is a problem ready to happen. A *megohmmeter* is a special ohmmeter device used to test for insulation resistance.

The *clamp-on ammeter* is an exceptionally important device, as it will measure higher currents than a DMM will. Most clamp-on ammeters also include a voltmeter, ohmmeter, diode tester, and capacitor tester. If you are contemplating the purchase of a clamp-on ammeter, invest in one that will measure DC as well as AC currents. The model shown in **Figure 7-7** also has a noncontact voltage detector, which detects the presence of voltage without touching any wires.

Goodheart-Willcox Publisher

Figure 7-6. A digital multimeter (DMM) is the tool of choice when troubleshooting an electrical problem.

Goodheart-Willcox Publisher

Figure 7-7. A clamp-on ammeter can measure currents and current imbalance and reveal valuable information regarding the health of equipment.

High currents and current imbalance on a three-phase circuit are indicators of problems. These measurements are easy to make with a clamp-on ammeter and often reveal valuable information about the health and well-being of the equipment.

Clamp-on ammeters do not require you to interrupt the circuit to take a current measurement. You simply clamp the meter around the conductor in which you wish to measure the current.

Current testing lets you determine if there is a problem involving either too much current or a total lack of current. If there is no current flow, it is a sign of an open circuit. If there is too much current flow, there might be a short or, in the case of three-phase circuits, a motor connected to the circuit could be single phasing, which is not a good condition. Single phasing may lead to motor burnout, which is an expensive repair proposition.

An *oscilloscope*, sometimes referred to simply as a scope, shows the change in electrical signal or voltage over time as a continuous graph, usually a wave. See **Figure 7-8**. An oscilloscope makes troubleshooting quicker and more definitive.

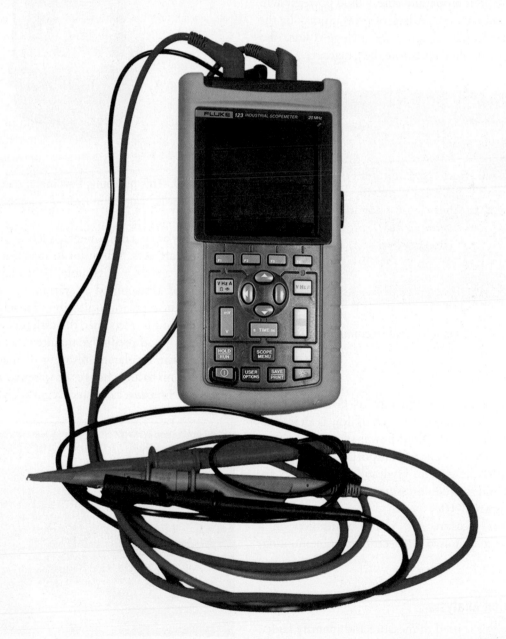

Figure 7-8. This ScopeMeter is a portable oscilloscope and DMM combination that makes troubleshooting quicker and more definitive.

In later chapters you will learn about many different types of electrical/electronic test equipment, what each piece of equipment is capable of testing for, and how to conduct the tests.

7.6.3 Fluid Power Testing

Pressure gauges and flow meters are essential for determining the health of a fluid power system, which is either a hydraulic (using hydraulic oil) or pneumatic (using compressed air) system. A *pressure gauge* is used to determine the pressure behind hydraulic fluid or compressed air, **Figure 7-9**. Without proper pressure, a fluid power system will not function correctly. A lack of pressure may be the result of an accidentally closed valve, a clogged line, or a clogged filter, if the filter sits before the gauge.

Figure 7-10. This hydraulic flow meter can withstand the high pressures involved in hydraulic systems.

Figure 7-9. These are some typical pneumatic pressure gauges.

Flow meters measure the rate of flow of a fluid through an area, such as a pipe. Lack of flow is an indication of a restriction in the fluid power system. The flow rate of hydraulic fluid is another important aspect in the health of a hydraulic system. Hydraulic flow meters are built to withstand the high pressures involved in hydraulic systems, **Figure 7-10**.

Technicians measure and record fluid flow in a method similar to that used for electrical troubleshooting, **Figure 7-11**.

7.6.4 Vibration Analysis

Vibration analysis is used to measure and identify faults, which aids in the prediction and prevention of failures in rotating machinery. In its simplest terms, vibration is simply the back-and-forth motion or oscillation of

a machine or its components, such as motors, pumps, fans, or gearboxes. Vibration in machinery can be both a sign and a source of trouble, or it may be just a normal part of a machined operation and not a cause for concern. The regular monitoring of machinery vibration will allow you to recognize the difference and should be part of any good predictive maintenance program.

With the regular monitoring of machine vibration and attention to any changes or irregular readings, maintenance technicians can detect items such as deteriorating

Figure 7-11. This technician is measuring fluid flow with a digital flow meter.

or defective bearings, looseness of the couplings or motor mounts, and worn or broken gears. Technicians may also detect misalignment and unbalanced shafts before they result in bearing or shaft deterioration. The vibrations can be measured using different sensors. The most commonly used sensor is the accelerometer. An ***accelerometer*** is a device that measures the changes in motion of an object, including vibrations. It is connected to an instrument called a vibration tester or ***vibration meter***, which measures and analyzes the vibrations emanating from the machine, **Figure 7-12**. The data collected from this instrument is then analyzed further by the technician or an engineer trained in the field of rotating machinery vibration to determine the possible cause or causes and the severity of the vibration.

Vibration can be a result of a number of conditions either acting in combination or alone, and can be caused by auxiliary equipment as well. The causes of vibration can be broken into four major categories:

- **Imbalance.** The unequal distribution of weight around the rotation point of a piece of machinery or equipment.

- **Misalignment, or shaft run out.** The shaft does not rotate exactly in line with the main axis.

- **Wear.** The removal of material on a surface as a result of mechanical action.

- **Looseness.** Structural looseness is generally caused by a piece of machinery not being rigidly attached to its foundation, for example a loose or broken motor mount, often referred to as "soft foot." Rotational looseness is excessive clearance between rotating and nonrotating components of a machine, such as bearings or couplings.

The effects of vibration may be severe enough to cause the accelerated wear of machinery bearings and damage to the machinery itself. The vibration may also create excess noise, thus causing an unsafe work environment or resulting in the machinery being taken out of production.

7.6.5 Thermal Imaging

Temperature monitoring and thermal imaging can reveal potential problems in machines. ***Infrared radiation (IR)*** is a part of the electromagnetic spectrum that is not visible to the human eye but is experienced as heat. Infrared radiation can be measured by an infrared thermometer or a thermal imager, which shows the temperature of a device. As the temperature of a device rises, the infrared radiation increases, which can be a sign of a problem at some point.

The ***infrared thermometer*** is a noncontact device that displays temperature. Scan the bearing mounts on motors, switches, circuit breakers, and wiring connections, holding your IR thermometer as close as safely possible to the target when making thermal measurements. Look for hot spots, and record and track your temperature readings. Make certain your readings indicate that the device you are measuring is within its temperature operating limits.

Reproduced with permission, Fluke Corporation

Figure 7-12. Vibration analysis can be used to find and measure faults in rotating machinery.

A *thermal imager*, **Figure 7-13**, differs from the IR thermometer in that it gives a picture of the various temperatures and their locations. The thermal imager shows a range of temperatures at the various locations within the captured image. This method allows for easier identification of hot spots.

Reproduced with permission, Fluke Corporation

Figure 7-13. Thermal imaging can be used to detect variations in temperature. For example, hot spots can indicate a problem that is causing equipment to operate out of its temperature operating limits.

While measurements by themselves rarely tell the entire story, by recording and analyzing the data for trends, you can often determine where a failure is imminent with a piece of equipment, and often which part of that equipment will soon fail. Again, the importance of maintenance records cannot be overstated.

7.6.6 Mechanical Test Equipment

Testing and measuring are essential to determine the health of a mechanical system. Belt tension is important, as well as shaft alignment. Bearing blocks and motor mounts should be properly shimmed to reduce bearing and gear wear.

A steel bench rule may be used for basic measurement, when extreme accuracy is not important, as well as to provide a straight edge. Steel bench rules are often used in combination with a *belt tension gauge*. A belt tension gauge will read the amount of deflection of a belt when used with a straight edge, and it will determine the tension of the belt.

A *caliper* can be used to measure an object when requiring an accuracy of up to 1/1000 of an inch. Dial calipers are traditional, but digital calipers, as shown in **Figure 7-14**, are common. An object is placed between the jaws of the caliper, which are then closed to hold the object, thus obtaining a size measurement. With a dial caliper, make sure the dial face reads as 0 when the jaws are completely closed, before measuring an object. Each graduation on the dial represents 0.001 inch. There are different sizes of caliper, including a span of up to 12″, 8″, or 6″, which is the one most often encountered. Use the smallest-span caliper to accomplish the measuring job at hand.

GearWrench (Apex Tool Group)

Figure 7-14. A digital caliper is a useful tool for quick, accurate measurements.

A *micrometer*, such as the digital micrometer shown in **Figure 7-15**, is more precise than a caliper. It has a calibrated screw used to close onto the sides of an object. A digital micrometer or a micrometer with a vernier scale is necessary when your required measurement accuracy is to within 1/10,000 of an inch. A *vernier scale* allows

Starrett

Figure 7-15. Use a digital micrometer when you need a very high level of accuracy.

a user to measure more precisely than can be done when reading a normal uniformly divided straight scale, such as you find on a bench rule. When reading a vernier micrometer, **Figure 7-16**, the line of the vernier scale that lines up closest to any of the lines on the thimble represents the finest part of the measurement (e.g., the "8" in a reading of 0.2448 inch). It is very difficult to visually measure to this level of accuracy. Make several measurements to be sure of consistent results. The range of a micrometer is normally 1″. There are various sizes of micrometer, such as a 0″–1″ span, 1″–2″ span, 2″–3″ span, and so on. Some micrometers come in a kit with spans up to 12″. You need to select a micrometer with the proper span for your measurement

application. A *depth micrometer*, **Figure 7-17**, is used when trying to measure the depth of a feature that would be impossible to measure with a conventional micrometer.

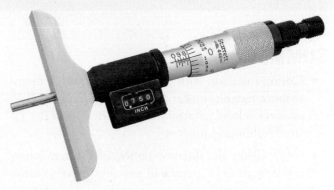

Starrett

Figure 7-17. Use a digital depth micrometer to measure the depth of a feature that would be impossible to measure with a conventional micrometer.

CHAPTER WRAP-UP

Maintaining proper maintenance records and constantly monitoring equipment will go a long way toward helping you troubleshoot problems when they occur. When you do encounter a problem, remember to narrow down the potential causes by using the binary search technique, and then be sure to address the root cause as well as the symptoms. And, whether troubleshooting electrical or mechanical problems, communicate with the operator who is familiar with the machine, and remember to always use the proper tool for the job. With these basic troubleshooting principles in mind, you will be a successful technician who keeps things running smoothly.

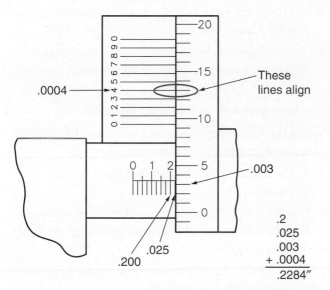

Goodheart-Willcox Publisher

Figure 7-16. This diagram shows the steps involved in reading a vernier micrometer.

Chapter Review

SUMMARY

- Communication with your supervisor, other maintenance technicians, and machine operators will provide helpful information to assist you in the troubleshooting process.

- Observation and documentation of a system when it is operating properly will give you valuable information that will assist you later when you need to troubleshoot any problems with it.

- A binary search will provide the fastest way to find the problem in a large network.

- A schematic diagram will assist you in troubleshooting problems with electrical, hydraulic, and pneumatic systems.

- A digital multimeter is the go-to tool for most electrical problems, as it will measure or test current, voltage, resistance, capacitance, diodes, and often other factors as well.

- A clamp-on ammeter is a good tool to determine if current is flowing through a conductor.

- Pressure gauges and flow meters help a technician determine the health of a fluid power system.

- Vibration analysis aids in the prediction and prevention of failures in machinery by measuring and identifying faults.

- Thermal imaging can reveal potential problems by measuring infrared radiation (IR).

- A bench rule may be used for coarse measurements as well as for a straight edge.

- Micrometers are the appropriate measuring tool where the measurement must be accurate to within 1/10,000 of an inch.

REVIEW QUESTIONS

Answer the following questions using the information provided in this chapter.

1. It is important to find and correct the _____ of a problem, or the failure will only continue to happen.

2. *True or False?* Communicating with the machine operator can provide valuable information to use toward solving a problem.

3. Explain how to troubleshoot a loss of voltage using the binary search technique.

4. *True or False?* Fuses blow because they become defective.

5. List three features or areas where mechanical problems may arise.

6. *True or False?* Multiview drawings are extremely important when putting an assembly back together.

7. A(n) _____ shows the change in electrical signal or voltage over time as a continuous graph.

8. List the four major categories the causes of vibration are broken into and explain what they are.

9. As the temperature of a device rises, the _____ increases.

10. If you needed to make a measurement to within 1/10,000 of an inch, would you use a digital caliper or a micrometer?

11. *True or False?* A steel bench rule is used when extreme accuracy is necessary.

12. A(n) _____ is used to determine the tension of a belt.

NIMS CREDENTIALING PREPARATION QUESTIONS

The following questions will help you prepare for the NIMS Industrial Technology Maintenance Level I Maintenance Operations credentialing exam.

1. Which of the following is *not* an observation technique you can use to help troubleshoot future problems?

   A. Digital photographs
   B. Self-made drawings
   C. Binary search
   D. Videos

2. Which of the following is used to determine the pressure behind hydraulic fluid or compressed air?

   A. Flow meter
   B. Pressure gauge
   C. Oscilloscope
   D. Micrometer

3. With a dial caliper, the dial face should read as 0 when the jaws are _____.

   A. completely closed
   B. set to 1/1000 inch
   C. set to 1 inch
   D. completely open

4. A digital or vernier micrometer is necessary when your required measurement accuracy is to within _____ of an inch.

   A. 1/10
   B. 1/100
   C. 1/1000
   D. 1/10,000

5. If a measurement taken using a vernier micrometer is 0.2864, which number lined up on the vernier scale?

   A. 2
   B. 8
   C. 6
   D. 4

6. What should be your first tool of choice when troubleshooting an electrical problem?

   A. Digital multimeter
   B. Oscilloscope
   C. Clamp-on ammeter
   D. Accelerometer

7. Which of the following can be performed with an accelerometer or mechanic's stethoscope and then used to identify faults, aiding in the prediction and prevention of failures in rotating machinery?

   A. Thermal imaging
   B. Fluid power testing
   C. Assembly drawing
   D. Vibration analysis

8 Mechanical Rigging and Installation

LEARNING OBJECTIVES

After completing this chapter, you will be able to:

☐ Discuss how to inspect rigging equipment for safety.

☐ Describe the most common types of rope used for lifting and rigging.

☐ Explain creep, elongation, and rope torque and how they relate to rigging.

☐ Discuss the relationship between safety factor, allowable working load, and minimum breaking strength.

☐ Identify the most common knots used with lifting.

☐ Describe how to hand splice wire rope.

☐ Identify the common grades of chain and chain slings.

☐ Describe the different types of synthetic slings and their use.

☐ Explain how sling hitch type and load angle affect load limit.

☐ Recognize and discuss the different types of rigging hardware such as eyebolts, hooks, and shackles, and their purpose.

☐ Describe hoists and how they provide a mechanical advantage when lifting loads.

☐ Demonstrate the basic hand signals used during a lift.

☐ Discuss how a foundation and grout can adversely affect a machine.

☐ Describe the purpose of machine anchor bolts and how anchoring is accomplished.

TECHNICAL TERMS

bowline knot	lang lay
bowline on a bight knot	lay
cat's-paw knot	lift capacity
clevis	lifting chain
come-along	minimum breaking strength
creep	plastic deformation
double bowline knot	reciprocating machine
ductility	resonance
elastic deformation	rope torque
elongation	rotating machine
endothermic reaction	safety factor
exothermic reaction	sine
eyebolt	sling angle
grout	splicing
hitch	thermal expansion
hitch knot	vector force
hoist	wire rope
laitance	working load limit (WLL)

Depending on your specific job, you may perform rigging operations to install or remove machines often or very rarely. Systems and equipment that you routinely work on may already be installed, but they may occasionally need to be removed for rebuild. When accidents occur in lifting and rigging, they tend to occur quickly and irreversibly, and injuries and equipment damage result. When equipment is installed incorrectly, this can prevent proper operation or cause long-term machinery problems that eventually need correction. Regardless of your exact type of work, knowing the fundamentals of rigging and installation will enable you to perform the job safely and effectively.

8.1 SAFETY

Three major categories of safety apply while lifting and rigging: wearing personal protective equipment (PPE), inspecting equipment used for the lift, and performing the lift properly. Personal protective equipment should include steel-toed boots, eye protection, a hard hat, work gloves, and proper clothing. The inspection of the equipment used in the lift will be discussed in each section in this chapter. Proper lifting techniques are also discussed in detail in a later section.

8.2 MATERIALS

Inspect your materials carefully when planning a lift. Each component of a lifting setup must be carefully inspected to ensure that it is not worn or damaged. Always verify the capacity of each component as well. Just one component that is weak will reduce the load capacity of the entire setup and may cause a catastrophic failure.

8.2.1 Pry Bars, Jacks, and Dollies

Some simple lifts of smaller items may not require complex rigging. A pry bar is a metal bar with flattened ends and a curve on one end. The curved end goes under the side of an item and the bar is then used for leverage to lift the item. With a firm, stable base, manual or hydraulic jacks can also be used to lift items, much as a jack is used to lift the corner of an automobile when you have to change a tire. Once the side of a smaller item is lifted, it can be placed onto a dolly and tied down for movement. A dolly is a wheeled platform used for moving heavy objects. These items are used only for smaller lifts and moves. Any larger industrial machinery will require more complex rigging and lifting, with the use of high-grade materials.

8.2.2 Rope

Rope comes in a wide variety of materials and weaves. Some materials are more common, depending on the environment in which the rope is to be used and the characteristics needed. Weave refers to how the fibers of the rope are formed together. **Figure 8-1** shows a comparison of four different weaves: twisted, solid-braided, double-braided, and hollow-braided rope. The twisted rope pictured is a three-strand rope, but other stranding is also available.

Twisted rope has a twist along its length, and no core is present. Since all braids are exposed to the outside, the rope fibers tend to fray under abrasive circumstances. Hollow-braided rope is made of an even number of strands—12-strand rope is the most common. This type of rope has no core and is therefore usually light enough to float. Individual strands are woven in two different directions to twist along the rope lengthwise. Solid-braided rope is similar to hollow-braided rope, but it has a core. The core and external shell of fibers alternate positions and are braided together lengthwise and in different directions. While this makes solid-braided rope very strong, it also makes it unable to be easily spliced. Double-braided rope is similar to solid-braided rope, but with two levels: both the inner core and outer layers of the rope are solid braided.

Figure 8-2 compares some of the most common rope materials used in lifting operations. Many types of rope use and combine high-end materials in different ways. Rope is also manufactured in various colors for specific purposes.

Safety Considerations

Creep, as used to describe manila rope, is the permanent separation of fibers over time. This occurs due to manila rope being made up of many smaller fibers intertwined together. Under a constant, heavy load, manila rope will fail (separate) in time. The time until failure depends on the static loading and can range from minutes to years.

A static load, or dead load, is a load that does not change (that is, a load that is not moving). Static loading produces a constant amount of stress in a rope. A dynamic load, or live load, is a load that is moving. A dynamic load causes the stress in the rope to vary.

The Code of Federal Regulations states that rope used for rigging must have a safety factor of 5:1 (meaning it must be able to support at least five times the intended load), and rotation-resistant rope must have a safety factor of 10:1. The *safety factor* (also called *design factor*) is a comparison of the minimum breaking strength to the working load limit (WLL). The *minimum breaking strength*, or breaking strength, is the amount of force required to break the rope. The *working load limit (WLL)* is the maximum

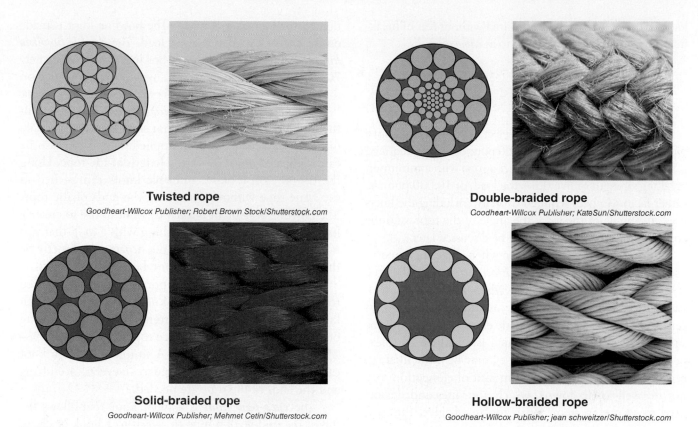

Twisted rope
Goodheart-Willcox Publisher; Robert Brown Stock/Shutterstock.com

Double-braided rope
Goodheart-Willcox Publisher; KateSun/Shutterstock.com

Solid-braided rope
Goodheart-Willcox Publisher; Mehmet Cetin/Shutterstock.com

Hollow-braided rope
Goodheart-Willcox Publisher; jean schweitzer/Shutterstock.com

Figure 8-1. These are four of the most common rope weaves.

Common Rope Materials				
Rope Material	**Common Sizes**	**Advantages**	**Disadvantages**	**Typical Strengths**
Manila	• 1/4"–1" in 1/8" increments • 1"–2" in 1/4" increments (larger available)	• Inexpensive • Commonly used and available	• Can creep and fail if left under load • Swells and shrinks with exposure to moisture • Easily degraded if not properly stored	• 500–27,000 lb breaking strength
Nylon	• 1/4"–2 1/2" (3 strand) • 1/4"–1" (double braid) • 1/2"–1" (8 strand) (others available)	• Strong • Flexible • Resistant to abrasion damage • Resists molds and mildew rot • Resists some chemicals	• Does not float • Loses some strength when wet • Absorbs moisture • Too much stretch in some cases for rigging	• 1500–100,000 lb • 2000–26,000 lb • 6000–16,000 lb
Polypropylene	• 1/4"–1 1/2" (3 strand) • 1"–1 1/2" (8 strand)	• Floats • Resistant to bases, solvents, and acids • Resists rot and mildew • Inexpensive	• Not as strong as other materials • Sensitive to UV light • May have too much stretch for rigging	• 1000–20,000 lb • Up to 36,000 lb
Polyester	• 1/4"–3/4" (double braid) • 1/4"–3/4" (3 strand) • 5/16"–3/4" (12 strand)	• Resistant to chemicals and UV • Low stretch • Stronger than nylon • Good for static loads • Abrasion resistant	• Does not float • Not good for shock loading • Some weaves (braids) are more difficult to work with	• 1000–16,000 lb • 1300–11,200 lb • 4000–18,000 lb

Goodheart-Willcox Publisher

Figure 8-2. This table lists common types of rope materials and sizes.

load, or mass, that can be safely lifted without fear of breaking. It is also sometimes referred to as *capacity*.

$$\text{safety factor} = \frac{\text{minimum breaking strength}}{\text{working load limit}}$$

(8-1)

As an example, if we were to select a nylon rope to hoist a piece of equipment weighing 2000 pounds and we required a 5:1 safety factor, we would select a rope with a minimum breaking strength of five times the load, or 10,000 pounds. Other factors impact required capacity, including the knots used, the angle of the load, the angle of the rope, and the condition of the rope. These will be discussed later.

Another factor to consider when selecting rope to use for rigging is elongation. *Elongation* describes the tendency of a rope to stretch when a load is applied. Elongation is specified by rope manufacturers as a percentage. Materials with very little elongation may not be suitable for shock loading, which is a sudden or unexpected load. With little give, they may be more likely to break. Materials with a great amount of elongation may not do well as tie-downs, which are ropes and devices used to secure an item. With too much give, the item may not stay held tightly enough in place.

SAFETY NOTE
Inspecting Rope

Always inspect rope before use for stretch, thin areas, abrasion, and chemical exposure. While newer rope is sometimes difficult to handle (because of stiffness), older rope is much easier to handle and tie. Rope can lose strength as it ages, so always inspect old rope carefully.

Rope torque or rope twist is the tendency of rope to turn when a load is applied. This turning action is dependent on rope material and manufacturing method. When lifting a load, rope torque can be offset by a spotter using a lighter-grade guide rope to steady the load.

Knots

Knots are a part of every rigging operation that uses rope. When a knot is tied in a rope, the strength of the rope is reduced—sometimes as much as 50%, depending on the knot.

One of the most common knots is the simple *bowline knot*. A properly tied bowline knot will not loosen or slip, which is to say that it will not slip down on the load and

tighten the loop that is made. The bowline knot is made on the end of a rope to secure a load. The *double bowline knot* is stronger than the simple bowline (by approximately 70%) and even less likely to slip. Again, this knot is tied on the end of a rope to secure a load, post, pipe, or other object.

The *bowline on a bight knot* creates a loop or eye in the middle of a rope that will not slip. This knot is easily untied after a load has been applied, and it can also be easily adjusted up or down the length of the rope. Using the bowline on a bight, multiple knots can be tied in the same rope without disturbing the ends of the rope. The bowline on a bight is a handy knot to use to create a handhold or to suspend something with a loop that will not tighten. **Figure 8-3** shows the process to properly tie the bowline, double bowline, and bowline on a bight.

A *hitch knot* forms a temporary noose that is used to secure a line. Using a hitch knot, such as a ring hitch knot, does not impact rope strength as much as using other knots. Hitch knots reduce rope strength by 20%–25%, depending on the type. A simple ring hitch knot can be used when attaching rope to an eyebolt for lifting and the ends of the rope are needed.

The *cat's-paw knot* is a hitch knot that allows the use of the middle of a rope to attach to a hook or clevis.

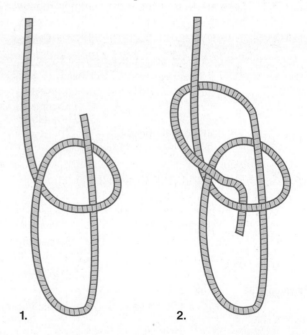

1.   2.

Bowline knot

(continued)

Goodheart-Willcox Publisher

Figure 8-3. The bowline knot is a commonly used, and misused, knot. The double bowline is stronger than the simple bowline by approximately 70%. The bowline on a bight is a handy knot to use to create a handhold or to suspend something with a loop that will not tighten.

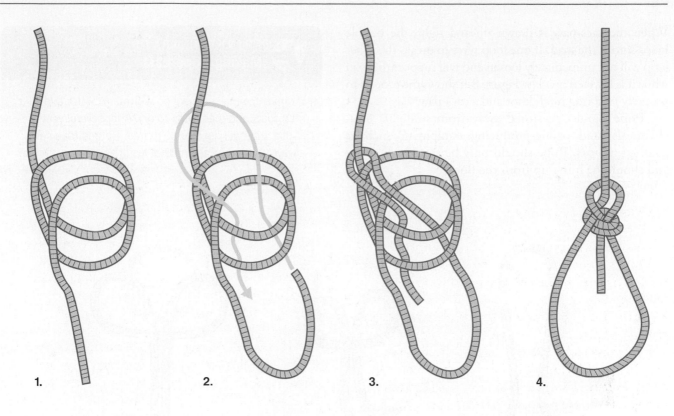

Double bowline knot

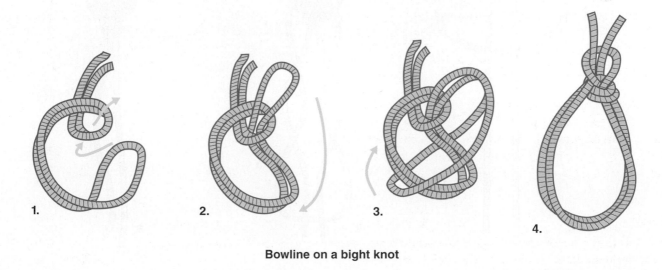

Bowline on a bight knot

Goodheart-Willcox Publisher

Figure 8-3.

When the cat's-paw is drawn up and tight, the double loops spread the load. If one loop were to break, the other loop will not immediately loosen and will support the load while it is lowered quickly. **Figure 8-4** shows the process to properly tie a ring hitch knot and a cat's-paw.

Rope should be stored away from sunlight, heat, chemicals, and ozone-producing equipment such as brushed motors. Rope should only be stored when dry and should be hung up from the floor.

THINKING GREEN
Making Rope Last

If dirty, rope may be cleaned with warm water and a mild soap, and then hung to dry. After several years of use, rope will need to be retired. Retired rope can always be used for noncritical lashings and for practice.

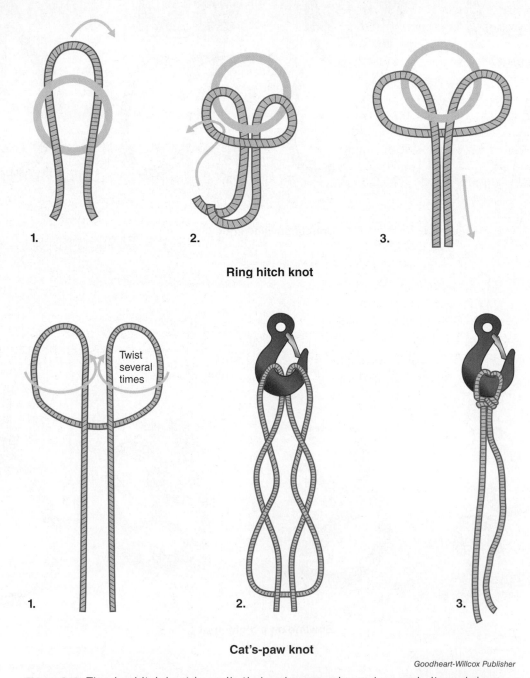

Ring hitch knot

Twist several times

Cat's-paw knot

Goodheart-Willcox Publisher

Figure 8-4. The ring hitch knot is easily tied and commonly used on eyebolts and rings. The cat's-paw is one of the quickest and easiest hitch knots to tie to secure a load to a hook or clevis.

8.2.3 Wire Rope

Wire rope is a common component of lifting and rigging. *Wire rope* is actually steel-braided rope with a core surrounded by wire strands made up of smaller steel wires. **Figure 8-5** is a cross section showing the basic components of a wire rope. The direction and method of how the strands are weaved around the core is called the *lay*. Wire ropes come in several types of lay: right-hand, left-hand, alternating, and nonrotating.

Each of the individual strands surrounding the core also has its own lay. When the lay of the wires in each strand is the same direction as the lay of the strands that form the wire, it is *lang lay*. When the lay of the wires in each strand is the opposite direction of the lay of the strands that form the wire, it is regular lay.

Wire rope with just one lay (a left-hand lay, for instance) tends to turn under a load. Using rotation-resistant wire rope can counteract this twisting. In rotation-resistant wire rope, several layers are used, with each layer twisted in the opposite direction of the one under it. With this design, as rotation-resistant wire rope is put under tension, the rotation of each lay cancels the other.

The core of a wire rope can be made in different ways:

- **Fiber core.** The core is made of either natural or synthetic fibers, which offers more flexibility but less strength.

- **Wire strand core (WSC).** This is a multiple wire strand core that offers more strength than a fiber core but less flexibility.

- **Independent wire rope core (IWRC).** This single wire-wound rope core is very strong but less flexible. It is generally the most durable and common core, used in many situations.

Wire rope is made from a specific type of steel called either improved plow steel (IPS) or extra-improved plow steel (EIPS). **Figure 8-6** shows the top eye end of a wire rope sling. Eyes are made by *splicing*, which connects the end of the wire rope back on itself to form the eye. We usually only see the metal sleeve, not the underlying splice of the wire rope. The wire rope is mechanically spliced, and then the sleeve is swaged over the splice. In swaging, a hammer or high pressure is used to create the shape. Mechanical splices are stronger than those done by hand. Hand splicing refers to using U-bolt clips to secure the rope end and form an eye. Normally with hand splicing, a thimble is used in the eye, and U-bolts are made so that the "U" portion clamps the end of the wire rope. The thimble holds the shape of the loop and prevents direct contact between the wire rope and the load. Do not use a hand splice with wire that is coated.

Artography/Shutterstock.com

Figure 8-6. Wire rope slings conform to ASME B30.9. Thimbles are used on the inside diameter of the eyes to help distribute the load evenly throughout the eye and prevent damage or distortion.

Mrs_ya/Shutterstock.com

Figure 8-5. This cross section of wire rope shows the core, strands, and wire.

PROCEDURE | Performing a Hand Splice

When performing a hand splice, use the proper number of clips for the rope size, and always use a thimble. Then, follow these procedures:

1. Match the clip size to the rope size and decide on the clip material.

2. Lay out the end of the wire rope and the turn-back properly to the first U-bolt. (Turnback is the amount of rope you double up to make the eye.)

3. Make the first U-bolt (on the dead end) to the full recommended torque.

4. Make additional needed U-bolts.

5. Test the strength of the splice on a noncritical lift prior to use.

Figure 8-7 shows clip sizes, number of clips required, proper amount of turnback, and recommended torque for common sizes of wire rope.

Wire Rope Specifications and Recommendations			
Clip and Wire Size	Number of Clips	Turnback	Recommended Torque
1/4″	2	4 3/4″	15 ft-lb
1/2″	3	12″	65 ft-lb
1″	5	26″	225 ft-lb
1 1/2″	8	54″	360 ft-lb

Goodheart-Willcox Publisher

Figure 8-7. This table lists wire rope specifications and recommendations.

8.2.4 Chain

Chain is manufactured by drawing wire through a die and then mechanically forming it into links. Some types of chain are left with links open. This type of chain is not meant for industrial lifting and rigging, but is instead used for residential purposes (suspending a ceiling light, for instance). In rigging and *lifting chain*, the links are welded closed, heat-treated, and tested. Never use open-link chain for lifting. Chain typically comes in surface treatments of bare, galvanized, lacquered, and powder coated.

The most common types of lifting chain are grade 80 and grade 100. Grade 80 chain is composed of alloy

steel, whereas grade 100 is heat-treated, hardened, and tempered. Grade 120 is also being manufactured and has even greater capacity than grade 100. Grade 70 chain is typically used for securing and tie-down purposes and not for lifting and rigging.

Figure 8-8 shows a comparison of the strengths of grade 80 and grade 100 chain at their typical sizes. It should be noted that grade 80 chain is an older grade and is generally being replaced by grade 100, which far outperforms grade 80 in capacity limits.

Capacity Comparison for Grade 80 and Grade 100 Chain		
Size	Grade	Capacity
3/8″	80	7100 lb
3/8″	100	8800 lb
1/2″	80	12,000 lb
1/2″	100	15,000 lb
3/4″	80	28,300 lb
3/4″	100	35,000 lb
1″	80	47,700 lb
1″	100	59,000 lb

Goodheart-Willcox Publisher

Figure 8-8. This table compares the capacities of grade 80 and grade 100 chain.

SAFETY NOTE
Matching Grades

Never use a lower-grade fitting (master link, hook, turnbuckle, or other fitting) with a higher-grade chain. The lower-grade fitting would render the greater capacity of the higher-grade chain useless.

Chain Slings

Lifting chain comes in a wide variety of lifting components. **Figure 8-9** shows a chain sling with four legs and sling hooks. Chain slings can also use grab hooks to secure yet another chain. The type of chain you will use greatly depends on your specific application. Lifting chain is very versatile, resistant to heat and environmental damage, and easily adapted to different lifts. If the chain is part of a sling, an identification tag is affixed that shows the manufacturer, serial number, working load limit, size, grade, and reach (length). Chain slings are specified by a three-letter designation system that is common to most manufacturers. **Figure 8-10** tabulates the letters in this system and their meanings.

Gearstd/Shutterstock.com

Figure 8-9. This chain sling has four legs with closing sling hooks.

Designations and Features of Chain Slings

Legs (First Letter)	Master Link (Second Letter)	Hook (Third Letter)
S = Single	O = Oblong	S = Sling hook
D = Double	P = Pear-shaped	G = Grab hook
T = Triple	S = Sling hook	F = Foundry hook
Q = Quadruple	G = Grab hook	L = Latchlock hook
C = Single leg with only chain link on the end (no hook)		O = Oblong link (no hook)
		H = Hammerlock link

Goodheart-Willcox Publisher

Figure 8-10. This table shows the letter designations and features for chain slings.

Just as with rope, wire, and synthetic slings, the angle of the load to the chain affects load limit. The effect is the same, and the same load calculation equation should be used. The horizontal lift angle should not be less than 30° between the load and chain. At that angle, a chain is capable of lifting only 50% of what it would be capable of lifting at a 90° angle (straight up and down). A load angle of less than 30° puts too much stress on the chain; the geometry of the lift should be changed to avoid this.

SAFETY NOTE
Before Lifting

Always inspect chain and lifting equipment regularly for signs of wear, corrosion, damage, and proper operation prior to use.

Chain Sling Baskets

Chain sling baskets come in a variety of assemblies, depending on their intended purpose. The adjustable basket sling is probably the most commonly used type. **Figure 8-11** shows a single adjustable chain sling with the grab hook attached to the master link with a chain, allowing for a customizable basket size. Grab hooks allow quick sling length adjustment, for a wide range of loads. Two chains and two grab hooks attached to a master link allow for a double basket.

Basket slings are also specified with a letter system that most manufacturers follow:

- Single basket (SB).
- Double basket (DB).
- Single adjustable loop (SAL).
- Double adjustable loop (DAL).

Columbus McKinnon Corporation

Figure 8-11. The grab hook attached by a chain to the oblong master link allows for more flexibility for a wider range of load geometries with a single basket hitch.

8.2.5 Synthetic Slings

Both nylon and polyester slings are commonly used, and each has its own advantages that should be considered when purchasing slings. While both materials are soft and will not mar or scratch the item being lifted, nylon web slings come in a wider arrangement of ply thicknesses. Plies are the number of strands or layers. With a larger number of plies, the lifting capacity is greater. Polyester slings tend to shrink only 30% as much as similar nylon slings. Polyester also tends to hold less moisture than nylon. Both materials are resistant to most chemicals, with the exception of acids.

With standard nylon or polyester webbing, stronger material is surrounded by the outer sheath. The stronger inner layer carries the load, while the outer layer protects against abrasion. Multiple plies are stitched together with multiple passes. Newer slings have an inner round or oval strength-bearing member, which is then surrounded by an outer protective sheath.

Synthetic slings are classified according to the following:

- Class (5–9).
- Dimensions (length and width).
- Ply thickness (1–4).
- Material.
- Type.

Figure 8-12 details the types of synthetic sling available and provides general information for each type. Each sling is labeled with the name of its manufacturer, the material it is made from, and its working load limits at various angles. Calculation of loading follows the same equation as other materials, with lift capacity being reduced as the angle to the load is reduced.

Figure 8-13 shows what some common types of synthetic sling look like and details a typical numbering system found on synthetic slings. Manufacturers may add additional numbers or suffixes depending on their customization. Note that the fittings on the ends of slings are made such that binding and pinching will not occur while lifting loads.

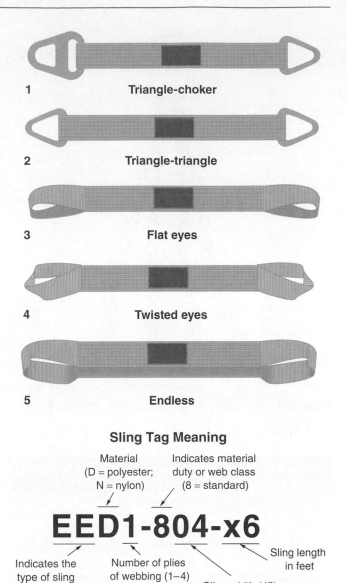

1. Triangle-choker
2. Triangle-triangle
3. Flat eyes
4. Twisted eyes
5. Endless

Sling Tag Meaning

Material
(D = polyester;
N = nylon)

Indicates material duty or web class
(8 = standard)

EED1-804-x6

Indicates the type of sling (eye and eye)

Number of plies of webbing (1–4)

Sling width (4″)

Sling length in feet

Goodheart-Willcox Publisher

Figure 8-13. Types 1–5 of synthetic slings are shown here. For its size, the endless loop is the strongest in a basket hitch. The tag on a sling provides useful information. While numbering systems are somewhat constant, manufacturers add prefixes or suffixes, depending on their specific manufacturing differences.

Overview of Synthetic Slings				
Type	Abbreviation	Description	Details	Lift Type
1	TC	Triangle and choker	A larger slotted triangular fitting at one end and a smaller triangular fitting at opposite end	Choker hitch
2	TT	Triangle and triangle	Similar triangle fittings at each end	Basket hitches
3	EE	Flat eyes	Eyes formed at each end of web material in the same plane	All hitch types
4	EE	Twisted eyes	Eyes formed at each end of web material and twisted at 90° to the main body	All hitch types
5	EN	Endless loop	Continuous loop of webbing material	All hitch types

Goodheart-Willcox Publisher

Figure 8-12. This table gives an overview of the five most common synthetic slings.

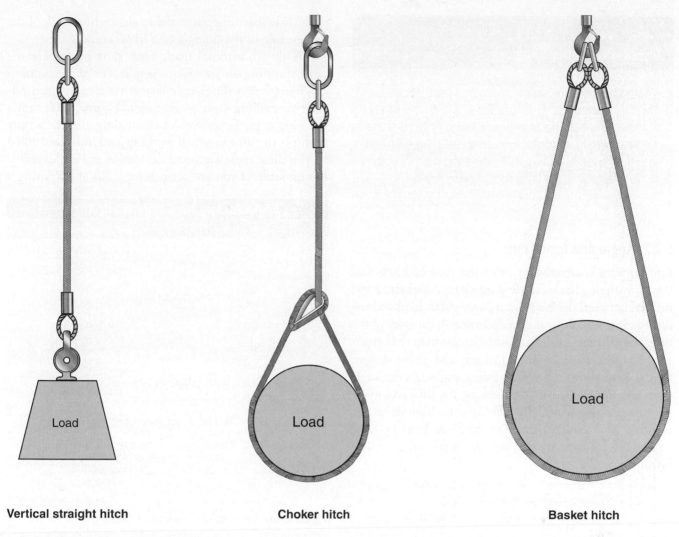

Vertical straight hitch   Choker hitch   Basket hitch

Goodheart-Willcox Publisher

Figure 8-14. The three common hitch types are vertical straight hitch, choker hitch, and basket hitch.

For further abrasion resistance, protective coverings can be added around the main webbing of the sling using Velcro® captures. When lifting or rigging, sharp corners and possible abrasive surfaces should be evaluated for any necessary covering prior to the lift. Sharp corners can be covered using wood, plastic, or sections of piping.

Standard eyebolts, hooks, and shackles that are designed for chain or rope should not be used with synthetic slings. Binding will result during the lift, which will reduce the lifting capacity of the sling. Specialty fittings made for sling rigging and lifting should always be used.

8.3   HITCH, ANGLE, AND LOAD LIMIT

When performing a lift, the configuration of the sling will affect its load limit. The angle between the sling and the load will also affect the load limit. To be safe, carefully consider both the sling hitch type and the sling angle before attempting a lift.

8.3.1   Hitch

A *hitch* is the manner in which a sling is configured for lifting. Three standard types of hitch are a vertical straight hitch, a choker hitch, and a basket hitch. **Figure 8-14** shows the different styles of hitch. *Lift capacity* is the size of the load that a sling or hitch can actually lift in its particular configuration. Lift capacity is also referred to as *load capacity*—the amount that can be lifted. The type of hitch affects the sling's lift capacity. With a vertical straight hitch, the sling can be used for its full working load limit. For a choker hitch, the lift capacity is reduced to 75% of the sling's working load limit. For a basket hitch, the lift capacity is double the normal working load limit, as long as the sling angle is 90°.

8.3.2 Angle and Load Limit

Lift capacity is affected by both the type of hitch and the sling angle. The term *sling angle* describes the angle formed between the horizontal plane of the load and the lifting sling. The sling angle affects the lift capacity of the sling for all types of lifting and rigging materials (such as rope, wire rope, synthetic slings, and chain slings). **Figure 8-15** shows examples of sling angles. As the angle from the sling to the load decreases, the lift capacity of the sling also decreases. The lift capacity depends on the working load limit of the sling or hitch and the sling angle. The lift capacity can be calculated using the following formula:

load at 100% × number of legs
× sine of sling angle = lift capacity

(8-2)

A sine is a trigonometric function of an angle of a triangle. The *sine* of an angle is the ratio of the length of the side that is opposite the angle to the length of the longest side of the triangle (the hypotenuse).

With a horizontal load, both sling angles are the same, forming an isosceles triangle. For that common configuration, a sling angle factor has already been calculated, making your work easier. **Figure 8-16** shows the sling angle factor for common sling angles. Simply multiply the sling's normal working load limit (at 100%) by the sling angle factor to determine the lift capacity. For example, if you are performing a lift at 45° using a

Angle Factors for Common Lift Angles	
Sling Angle	**Sling Angle Factor**
90°	1.000
85°	0.996
80°	0.985
75°	0.966
70°	0.940
65°	0.906
60°	0.866
55°	0.819
50°	0.766
45°	0.707
40°	0.643
35°	0.574
30°	0.500

Goodheart-Willcox Publisher

Figure 8-16. This table shows the sling angle factor for common sling angles. Simply multiply the sling's normal working load limit (at 100%) by the sling angle factor to determine the lift capacity.

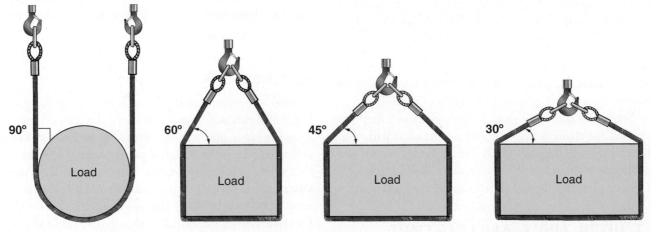

Goodheart-Willcox Publisher

Figure 8-15. Examples of sling angles are shown here. The minimum recommended sling angle is 30°.

sling with a normal working load limit of 1000 pounds, you could figure out the lift capacity as follows:

$$\text{load limit at } 100\% \times \text{sling angle factor} = \text{lift capacity}$$
$$1000 \text{ pounds} \times 0.707 = 707 \text{ pounds}$$

At a sling angle of 45°, this particular sling could only lift 707 pounds, as opposed to its rating of 1000 pounds at its 100% working load limit. Clearly, sling angle must be taken into account when performing any lift.

SAFETY NOTE
Safe Angle

OSHA regulations mandate that sling angles be 30° or greater. Sling angles less than 30° are unsafe because the lift capacity is greatly reduced at angles less than 30°, which can cause slings to become overloaded and fail.

8.4 EYEBOLTS, HOOKS, AND SHACKLES

Eyebolts, hooks, and shackles connect the main load-bearing materials to the load or lifting equipment. You should develop a general understanding about how each of these items is used, its intended purpose, and its limitations. These components should always be inspected prior to use. An *eyebolt* is a bolt with a loop at the end that is securely attached to an object so that ropes can be tied to it. In many instances, eyebolts are included on equipment from the factory (a pump or large motor, for instance). However, it should not be assumed that the included eyebolt is acceptable. Careful inspection must be performed to ensure no damage has taken place to included components over years of service.

8.4.1 Eyebolts

Eyebolts are manufactured in several categories:

- Regular eyebolt.
- Shoulder eyebolt, also called a machinery eyebolt.
- Eye nut.
- Open eyebolt, in which the circle is not welded closed.
- Eye tab.

Figure 8-17 shows the major categories of eyebolt. Open eyebolts are not used for lifting and rigging, only to secure a light load or for tie-downs. In all categories, eyebolts are made of a variety of materials, including steels and stainless steels. Manufacturing methods include forging, casting, and machining. Finishes can be plain, galvanized, and chrome.

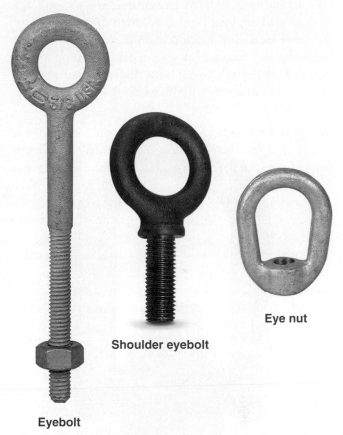

Eye nut

Shoulder eyebolt

Eyebolt

Campbell (Apex Tool Group); Columbus McKinnon Corporation; Campbell (Apex Tool Group)

Figure 8-17. A regular eyebolt, shoulder eyebolt, and eye nut are shown here.

SAFETY NOTE
Eyebolt Safety

Regular (nonshouldered) eyebolts are used only for vertical lifting at a 90° load angle. When using a regular eyebolt, the eyebolt threads should protrude through the material being lifted and the nut should be torqued to a proper level to ensure the eyebolt does not twist during a lift. All eyebolts must meet the ASME B30.26 safety standard for loading capacity.

Shoulder eyebolts are installed fully threaded into the equipment. If the threaded shank of the eyebolt is too long to fully thread into the load, washers are used to ensure that the shoulder fully transfers the load to the equipment. Because of the shoulder, these types of eyebolts can take a side load. As with other lifting equipment, as the load angle is reduced, so is the load capacity of the eyebolt. As an example, at a 45° pull, the eyebolt working load limit is adjusted down to 30% of the rated load. The specific manufacturer's limits need to be checked for each of your particular components. **Figure 8-18** shows a swivel hoist ring, which can be side loaded without a reduction in working load limit.

Columbus McKinnon Corporation

Figure 8-18. This swivel hoist ring allows for the alignment of the load, pick points, and lifting equipment.

SAFETY NOTE
Use a Shackle

Eyebolts should never be directly grabbed by lifting with a hook. The hook will not seat correctly and may fail. A shackle needs to be used between the eyebolt and hook.

8.4.2 Hooks

While specific lifting requirements determine what type of hook you will use, **Figure 8-19** shows the most common hooks. A *clevis* is a U-shaped hook that is attached at the end of a chain by use of a pin. Eye hooks are attached at the end of a rope or to a shackle. Grab hooks are used to hook a chain and prevent slippage. Sling hooks are used to hook slings or rope.

As with other components, hooks need to be checked prior to use. Manufacturer specifications must be checked to determine load limits and capacities. All lifting hooks should be forged steel hooks (preferably grade 100 or better).

As a hook is used over time, or if overloaded, it will have a tendency to stretch or twist. If any throat spreading or twisting is observed, the hook should be discarded and no longer used. The Department of Energy defines

Clevis grab hook   **Clevis sling hook**

Eye grab hook   **Eye sling hook**

Columbus McKinnon Corporation

Figure 8-19. These are common types of hooks used in lifting.

the limit before a hook is unusable as a 5% increase in throat opening or 1/4″, whichever is smaller. Wear is limited to 10% of the hook's original dimensions.

> **SAFETY NOTE**
> **Loading a Hook**
>
> A hook should be loaded in alignment with the load bearing member and at the bottom of the hook jaw. Hook loading beyond 45° from bottom center is not allowed and can cause failure of the hook.

8.4.3 Shackles

Shackles are used to connect rope to a hook, wire to a thimble in a hitch, chain to a hook, or a sling to a hook or load (eyebolt). Shackles can allow both connection points to move freely when lifting a load, or they can be used to secure one of the connection points. How a lift is performed depends on your specific situation and must be carefully considered.

Shackles should be die forged and receive heat treatment that improves their load-bearing characteristics, strength, and ductility. *Ductility* refers to a material's ability to bend and stretch rather than break. Improved ductility helps ensure that if a failure does happen with a shackle, it shows up as a deformation—the shackle stretching, for instance—and not a fracture, which would be more dangerous.

Figure 8-20 shows some of the most common types of shackle and their uses. Shackles are rated with a safety factor of at least 5 to 1, meaning that failure could occur at five times the working load limit. Shackles are rated in tons, with the working load limit shown on the bow of the shackle.

- Shackles will hinge (or move) in one plane only around the pin and should not be side loaded against the axis of the pin.

- Shackles should not be used to connect two load-bearing materials to extend the length in the bow (such as splicing two slings together). The pin is to be used as one load-bearing surface, the bow the other.

- Screw-pin shackles may be used in side-loaded applications, depending on the manufacturer.

- Shackles may include 45° marks to aid in loading checks.

- Side loading will require a reduction in load limit—check with the manufacturer.

Round pin shackles

Screw pin shackles

Bolt shackles

Columbus McKinnon Corporation

Figure 8-20. Round pin shackles use a cotter pin to secure the pin and still allow it to rotate. Screw pin shackles are typically used to connect to eyebolts or thimbles. Bolt-type shackles are used to further secure a load that may otherwise tend to rotate the pin. Only use shackles on which the working load limit is clearly visible.

Side loading refers to loading at an angle, rather than 100% vertical. If using synthetic slings, specially designed shackles and components need to be used to prevent damage to the slings by pinching and bunching when loaded. **Figure 8-21** shows some of the components available for synthetic slings.

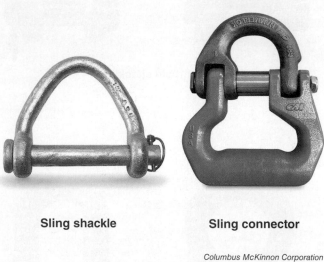

Sling shackle   **Sling connector**

Columbus McKinnon Corporation

Figure 8-21. This shackle and connector are designed for use with synthetic slings.

> ### TECH TIP
> ### Check Your Materials
>
> Planning for a lift requires time and is not an ad-hoc or last-minute venture. Check ahead of time to make sure you have the correct high-quality materials (slings, shackles, eyebolts) with the proper ratings to help ensure that a very expensive load will not be dropped and damaged. Even a 100-hp motor can cost between $5000 and $25,000 and is too expensive to risk using questionable lifting equipment.

8.5 HOISTS

Hoists are devices that provide a mechanical advantage when lifting heavy loads. For common lifts, hoists are attached to cranes that have either a cantilevered or gantry configuration. Jib cranes have a cantilevered configuration, which means they have a projecting overhead beam or girder, also called a boom or jib, that is anchored at one end to a vertical support structure. This allows the crane to turn as much as 180°, if mounted to a wall, or a full 360° if portable. Gantry cranes are similar but have vertical support structures at both ends. Unlike overhead bridge cranes, gantry cranes are portable. A hoist is attached to the overhead beam of either crane in order to make lifting the load easier. Hoists can be manually operated with a pull chain or mechanical lever or electrically or pneumatically driven.

8.5.1 Come-Alongs

Also known as a hand hoist or lever hoist, a *come-along* is a manually powered hoist that uses a lever and gear reducers to lift and suspend a heavy load from a chain. Some come-alongs can lift up to 9 tons. The gear reducers in a come-along reduce the amount of force that must be applied by the operator in order to develop the torque needed to rotate the chain about a sheave, which is a grooved wheel or pulley. **Figure 8-22** shows a come-along that features positive braking, allows free adjustment of the chain to take up slack, and includes a load-limit warning.

While many models have long chains, come-alongs are primarily intended for lifting over short distances. Lifting a load 10 feet or more would require substantial time due to the reduction in movement from the lever to the chain. However, longer-chain models are ideally suited for either making a short lift at a distance or applying force at a distance to move a load.

>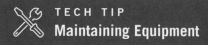
> ### TECH TIP
> ### Maintaining Equipment
>
> Chains, bearings, load pins, and latches need to be cleaned and lubricated at regular intervals. Always store come-alongs off the ground in a clean and protected place.

8.5.2 Chain Hoists

Chain hoists—also called hand chain hoists, chain falls, or chain blocks—develop their lifting capacity through multiple sheave and gear reductions, **Figure 8-23**. The load is lifted by pulling a separate hand chain that transfers torque to the main sheave and load chain. Some chain hoists have up to a 100-ton lifting capacity. With lower-capacity models, lifts can be accomplished quickly. As load capacity increases, sheave and gear reducers also reduce the speed of the lift chain in order to maintain the same speed for the hand chain. Chain hoists are typically limited to 12 feet of lifting, but they can be ordered with longer pull lengths.

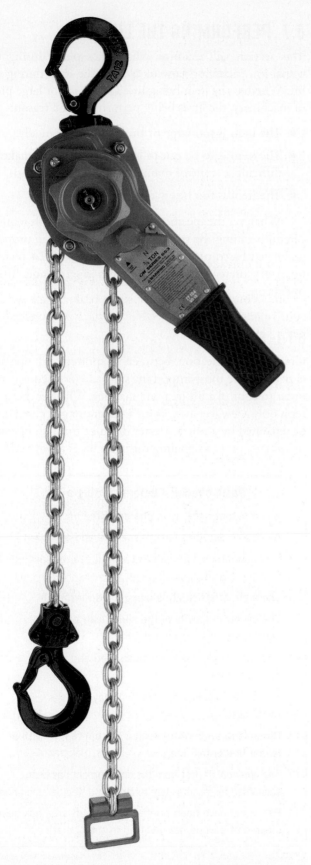

Figure 8-22. This come-along has top and bottom swivel hooks.

Figure 8-23. A chain hoist dramatically increases your lifting capability.

One of the first things to consider when using both come-alongs and chain hoists is the point of attachment for supporting the hoist. Along with all of the rigging needed to be able to handle the load, the point of attachment needs to handle the load plus the additional weight of the hoist. Do not assume that the point of attachment to support the hoist is a viable attachment point for the entire load.

SAFETY NOTE
Hoist Operation

All components of a hoist need to be regularly inspected—and some measured—to conform to safety standards set forth by the manufacturer. Reference your particular user's manual to ensure safe ongoing operation.

8.6 PACKING AND STORAGE

Parts being moved or stored must be packed correctly to protect them. Packing materials such as foam inserts can be used to prevent parts from moving, which helps prevent damage. Preventing parts from moving also ensures that the center of gravity for the box does not shift.

Storing a part in a box or other protective wrapping protects the part from dirt, moisture, and other contaminants. When removing a part from storage, always inspect the surface of the part carefully to ensure that it has not been contaminated. If the part has been contaminated in some way, clean the part to remove the contaminant if possible.

When moving items through environments containing contaminants, be sure the items are properly covered and protected. For example, when moving parts through an outdoor storage yard on a rainy day, be sure the parts are completely covered with plastic or a tarp so they remain dry. Similar protective measures are appropriate when moving parts through a dirty environment.

Transport items carefully to ensure no damage occurs to the container. When moving a hazardous liquid, review its safety data sheet (SDS) prior to the move so you are familiar with containment and cleanup procedures.

If items are stored outdoors, do not place the container directly in contact with the ground. Cover the container to protect it from dirt and moisture in the air.

After moving items, be sure to inspect and clean them, if necessary, before storing or installing them. Contaminants can cause a part to operate poorly or fail prematurely.

8.7 PERFORMING THE LIFT

This section will examine what takes place during the actual lift, including how to control the load during the lift. Whether the item being lifted is a small or large piece of machinery, the lift is being performed for a reason:

- The item is too large or bulky to lift manually.
- The item is to be put in a place that would make it difficult to control manually (such as overhead).
- The item is too heavy to lift manually.

Whatever the specific reason for a lift, each lift requires careful planning. Be sure the load limit has been properly calculated based on the sling angle. **Figure 8-24** reviews necessary prerequisites to starting a rigging and lifting operation.

8.7.1 Signaling

Good communication between a spotter and operator is perhaps the most important single item to consider in order to control a lift in a safe manner. Whether the spotter is only a few feet away from an operator of a chain hoist or separated by a much greater distance from an operator in a crane, proper signaling during a lift is always crucial.

Basic checklist before starting a lift:

☐ A competent rigger is in charge of the lift.

☐ All rigging gear has been inspected and approved.

☐ The working load limits (WLL) of all gear are acceptable for the lift.

☐ The appropriate hitch is being used for the lift.

☐ The center of gravity of the slung load has been considered and accounted for.

☐ The sling angle has been calculated and accounted for.

☐ There are no sharp corners that may damage the slings.

☐ Possible rotation of the load is accounted for with a leader line or tag line.

☐ Any potential impact from the environment has been accounted for, such as temperature and weather conditions.

☐ Personnel have been briefed and the lift area has been cleared of anyone not involved in the operation.

Goodheart-Willcox Publisher

Figure 8-24. Use this quick checklist prior to starting a lift. All things that could impact the lift must be considered.

Figure 8-25 shows the most common hand signals used during a lift. These signals may change depending on the type of machine performing the lift. Always talk with your heavy-equipment or crane operator prior to the lift and agree on possible alternate hand signals.

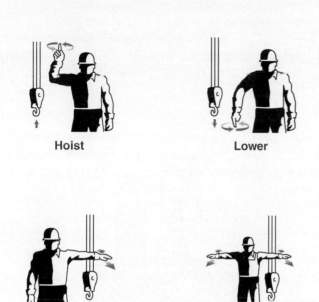

Hoist

Lower

Stop

Emergency stop

Columbus McKinnon Corporation

Figure 8-25. These standard hand signals apply to all lifts, regardless of lifting equipment.

8.7.2 Controlling the Load

The lift should be started slowly, by gradually putting weight on the hitch. The item being lifted may be stuck in place due to paint, gaskets, or seals. If the component does not immediately lift, it may need to be broken loose while under lift pressure. Gently tapping with a soft mallet or dead-blow hammer should be enough to break the item free. Excessive lift force should not be used, as this negates the point of performing the load calculation in the first place. Most cranes or hoists can easily overload lifting components.

Depending on the type of hitch and load being lifted, some loads may tend to rotate, tilt, or sway. Examining the center of gravity prior to the lift and designing the proper hitch to use will alleviate some if not all of this tendency. For example, a single-point lift using a standard wire rope would cause a load to rotate, but this can be compensated for by using either a rotation-resistant wire rope (and a tag line) or a two-point hitch. Forces that have not only magnitude but also direction, causing horizontal movement during a lift, are called ***vector forces***.

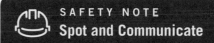

SAFETY NOTE
Spot and Communicate

Some lifts may require several spotters and several communicators speaking to the operator. Ensure that a briefing is held prior to the lift, so that everyone knows their role.

Once the load is lifted, no personnel should be closer than absolutely needed. The load should not be lifted higher than needed in order to put it at its final resting spot. Once over its intended resting place, the load should be lowered slowly and at a constant rate, with regular communication between any spotters and the operator. When the load is removed from the sling, do not leave the sling attached to the hook. If further lifts are to be performed, remove the sling from the hoist and manually move it to the new item to be lifted.

8.8 INSTALLATION

The installation of machinery may be necessary due to a company expansion, a consolidation of plants, the movement of an entire manufacturing plant, or simply when one piece of machinery is removed for replacement. Just as in building a house, the foundation affects everything it supports. Knowing the basics about machinery foundation may help you troubleshoot mechanical issues with your plant's machinery.

8.8.1 Foundations

The purpose of a foundation is to provide a stable base for the machinery being placed on it. A secondary purpose is to reduce the vibration that the machine induces on the foundation. Two basic types of machinery that are placed on foundations are centrifugal, or rotating, machines and reciprocating machines. A ***rotating machine*** is a machine that has one or more components that turn around an axis or point. A ***reciprocating machine*** is a machine that involves repetitive up-and-down or back-and-forth motion, such as through pistons. A large pump or turbine would be a good example of a rotating machine, while a diesel generator is a good example of a reciprocating machine. The foundation is not part of the surrounding structure of the building. Rather, the foundation has its own structure and is isolated from the building, so as not to transfer unwanted vibration.

Types and Methods

The type of foundation used depends on a range of factors, including the weight of the machine, the rpm of the machine, the materials available for the foundation, and the soil characteristics under the foundation. **Figure 8-26** shows some of the basic types of foundation.

In order to minimize vibration, a general rule of thumb for concrete foundations is for the weight of the foundation to be three times that of a centrifugal machine, or five times that of a reciprocating machine. This ensures that the natural ringing frequency of the foundation is low and will not reinforce the vibration of the machine. This natural ringing frequency is called *resonance*.

TECH TIP
Frequency Matters

If a machine, or machine and foundation combined, is operated near its natural resonant frequency, the amount of vibration will drastically increase and damage to the machine or foundation will eventually result.

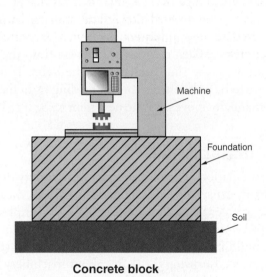

Concrete block

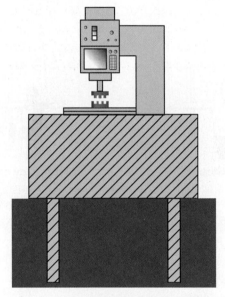

Concrete block supported by piles

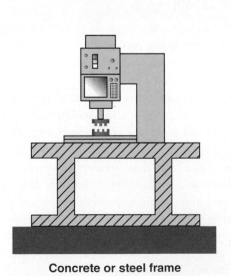

Concrete or steel frame

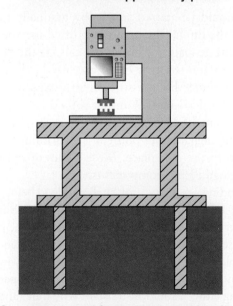

Concrete or steel frame supported by piles

Goodheart-Willcox Publisher

Figure 8-26. These are four common types of foundation for machinery.

Figure 8-27 outlines some of the most common methods of mounting machinery onto the foundation. The simpler baseplate method is used for small machines that are delivered in one complete package. Anchor bolts and leveling jacks are not shown, but they are discussed below.

Grouting

Since concrete cannot be poured to a finished height, the extra step of grouting is needed. **Grout** is a fluid form of concrete used to fill gaps as an adhesive element between materials. For machine installation, it serves the purpose of supporting the entire base of the machine and connecting the entire baseplate to the foundation. When concrete cures, the top layer tends to have a higher cement content. Also, the more the concrete is finished (troweled by hand), the more fine particles rise to the top of the pour. Because of these factors and the

need to achieve good adhesion of grout to the concrete foundation, the top layer of *laitance* must be removed prior to grouting. Laitance refers to the particles that accumulate on the surface. Laitance can easily be seen, for example, as sidewalks degrade and the top layer chips out. This top layer (2″–3″ thick) is removed using hand-held jackhammers on the top and corners of the foundation. Only then will the grout have a good adherence to the foundation.

While cementitious grout is widely used, epoxy grouts are becoming popular. Epoxy grout is made of several components that are mixed together and then cured through an *exothermic reaction*, which is a chemical reaction that releases heat. (An *endothermic reaction* is one in which heat is needed to cause the reaction to take place.) Epoxy grout has better resistance to oil, chemicals, and vibration, and it has better thermal properties than cementitious grout.

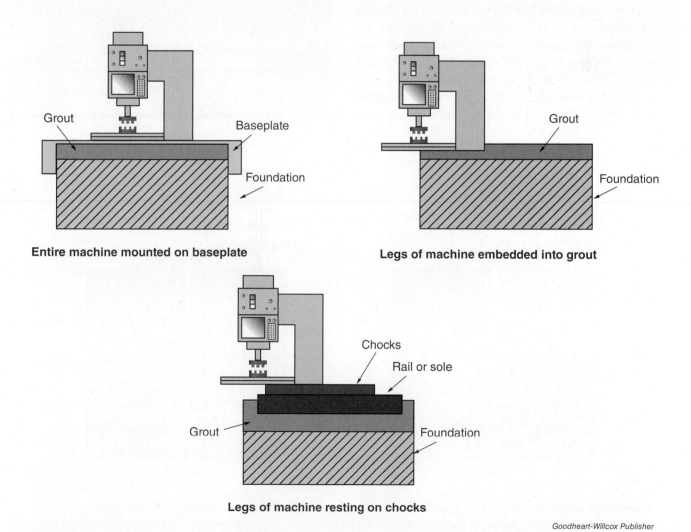

Entire machine mounted on baseplate

Legs of machine embedded into grout

Legs of machine resting on chocks

Figure 8-27. Low-horsepower machines often come from the supplier mounted on a baseplate. Medium-horsepower machines will have their legs embedded into grout when installed on site. Larger and more powerful machines may be mounted with their legs resting on chocks, which are wedges or blocks that prevent movement.

Anchoring Machinery

Machinery anchor bolts hold a machine in place, but they also allow for *thermal expansion*. Thermal expansion takes place as a machine and its foundation warm to operating temperature. On smaller machines or machines that do not generate much heat, this can be nearly disregarded. On larger machines that generate heat (such as compressors, engines, and large motors), the temperature difference between the surrounding environment and machine operating temperature can be significant, which causes an expansion or growth (depending on materials). Thermal expansion can affect shaft alignment, machinery vibration and noise, and machinery bearings due to loading. Properly installed machinery anchor bolts allow the stretching of the bolt. **Figure 8-28** shows two types of anchor bolt.

Proper torquing of the anchor bolts needs to be done initially and then checked periodically afterward. With improperly installed anchor bolts, such as bolts that are entirely cemented in, the longer length of the bolt will not be allowed to stretch. If this is the case, as the bolt is torqued, the stretch will only occur along a short length of the bolt. This stretch will be in the plastic deformation range instead of elastic deformation. *Plastic deformation* is the permanent deformation of a material due to its being stretched beyond its limit. *Elastic deformation* is only temporary, and the material will return to its original size. In this case, plastic deformation results in the bolt becoming loose, and then breaking, when retorqued.

8.8.2 Final Checks

After installation, a few final checks should be made before turning over a piece of machinery for operation.

PROCEDURE	Checklist before Operation

Before turning over a newly installed piece of machinery for operation, make these final checks:

1. Is there any piping strain that could affect alignment of the machine?
2. Is the baseplate of the machine properly grouted? Are there any hollow areas?
3. Is the machine properly lubricated?
4. Has the manufacturer's procedure been followed to place the machine into service?

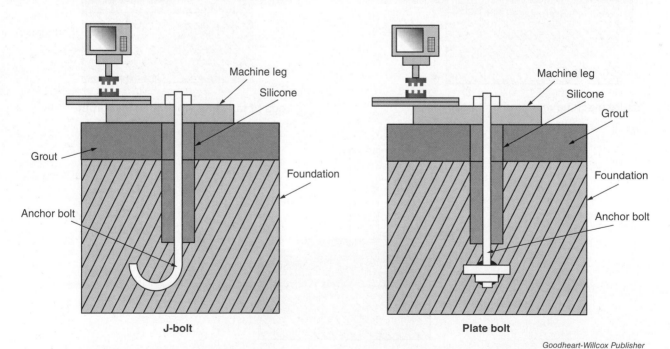

Goodheart-Willcox Publisher

Figure 8-28. J-type anchor bolts and anchor bolts using a round steel plate are common. Anchor bolts are only secured near the lower portion of the bolt to allow for stretch when properly torqued. The cavity around the bolt is filled with flexible silicone, which allows the bolt to stretch along most of its length.

Piping strain can and does affect shaft alignment. Piping strain includes any forces caused by unanchored, misaligned, or improperly installed piping. This should be checked for and corrected prior to turning over the machine (see Chapter 11, *Shafts, Couplings, and Alignment*).

The baseplate should be free of voids, which can easily be found by tapping with a hammer. If voids are found, they need to be filled properly prior to operation of the machine.

Machines are often shipped with no lubrication. Make sure that proper lubricant has been used and bearings have been checked. Running a machine without proper lubrication can cause a rapid failure.

Finally, the manufacturer's procedure should be checked. Many times the simple things are overlooked.

Pumps need to be vented, locks and tags must be removed, an appropriate valve lineup should be performed, and the conditions of the system need to be checked.

CHAPTER WRAP-UP

Rigging, lifting, and installation are not just activities that occur when a new manufacturing plant is being built. These activities take place in nearly every industrial plant on a regular basis, both when new equipment must be installed and when old equipment has to be rebuilt. Knowing the proper procedures, using the right equipment, planning ahead, and gaining practice will ensure that you possess the necessary skills to be successful in the field.

Chapter Review

SUMMARY

- Consider three main areas of safety with rigging: PPE, equipment inspection, and lifting the load properly.

- Any lifting equipment used must be tested, tagged, and inspected prior to use.

- The working load limit is calculated as the minimum breaking strength divided by the safety factor.

- Knots and hitches reduce the strength of a rope, but various types have characteristics that help with a successful rigging operation.

- Both rope and wire rope are very common components of lifting and rigging, with each type having advantages and disadvantages.

- Splicing connects the end of a wire rope back on itself to form an eye. Specific procedures should be followed if hand splicing a wire rope.

- Lifting chain uses a variety of components that are chosen depending on the specific application. Only properly graded lifting chain should be used for lifting.

- Synthetic slings are lighter and less expensive than chain equipment. The two main types of synthetic slings are nylon and polyester, with each having its own advantage.

- As the load angle of a sling is reduced, the working load limit is also reduced.

- Eyebolts, hooks, and shackles connect main load-bearing materials to the load or lifting equipment. Open eyebolts should not be used for lifting.

- Come-alongs are used for shorter lifting distances, while chain hoists, or chain falls, can be used for longer lifting.

- Communication during a lift is critical to performing the lift properly.

- Machinery installation is performed following procedures outlined by engineers with years of experience and training. Following the procedures ensures proper installation.

- The purposes of a foundation are to provide a stable base for machinery and reduce the vibrations that machines induce on the foundation. The two basic types of machinery that are placed on foundations are rotating machines and reciprocating machines.

- Grouting is necessary for machine installation and serves the purpose of supporting the entire base of the machine and connecting the entire baseplate to the foundation.

- Anchor bolts must be properly torqued at installation and then checked periodically for deformation.

REVIEW QUESTIONS

Answer the following questions using the information provided in this chapter.

1. Describe the difference between a solid-braided and hollow-braided rope.

2. Describe one problem with manila rope.

3. What safety factor must rotation-resistant rope have?

4. *True or False?* The safety factor for rope is calculated as allowable working load / minimum breaking strength.

5. What elements impact the safety factor?

6. What is elongation?

7. When a knot is tied in a rope, the strength of the rope is _____.

8. *True or False?* A hitch is used to permanently secure a line.

9. Which type of hitch knot allows the use of the middle of a rope to attach to a hook or clevis?

10. Discuss the difference between IWRC and WSC wire rope cores.

11. What is grade 70 chain used for? What is it *not* used for?

12. Which lifting chain has a greater capacity—grade 100 or grade 120 chain?

13. *True or False?* A chain is capable of lifting more than 50% at 30° compared to 90°.

14. What are the advantages and disadvantages of nylon and polyester slings?

15. How are synthetic slings classified?

16. The working load limit and sling angle determine _____.

17. What are the lifting limitations on a regular eyebolt?

18. What characteristics should a lifting hook have?

19. Shackles are used to connect rope to a _____.

20. What might overload deformation look like in hooks and shackles?

21. How do come-alongs and chain falls, or chain hoists, develop their lifting force?

22. Describe the four basic hand signals for lifting.

23. What are the two purposes of machinery foundation?

24. What is the purpose of grouting?

25. *True or False?* An exothermic reaction is a chemical reaction that releases heat.

26. How can thermal expansion affect machinery?

27. What is the difference between plastic deformation and elastic deformation?

NIMS CREDENTIALING PREPARATION QUESTIONS

The following questions will help you prepare for the NIMS Industrial Technology Maintenance Level 1 Maintenance Operations credentialing exam.

1. What does the safety factor of 5:1 indicate?

   A. A rigging rope must support at least five times the intended load.
   B. The lift capacity must be five times greater than the working load limit.
   C. For every five loads, all hoist components must be inspected at least once.
   D. A knot used for rigging must support five times the load capacity.

2. Which of the following wire rope core is the most durable?

   A. Fiber core
   B. Wire strand core
   C. Independent wire rope core
   D. Plow steel core

3. Which of the following lifting chains would be best for lifting and rigging?

   A. Grade 70
   B. Grade 80
   C. Grade 100
   D. Grade 120

4. Which angle between a load and chain would be capable of lifting the greatest load?

   A. 15°
   B. 30°
   C. 50°
   D. 90°

5. What two quantities determine the lift capacity?

   A. Allowable workload and minimum breaking strength
   B. Sling angle and minimum breaking strength
   C. Working load limit and sling angle
   D. Working load limit and lifting chain grade

6. As the sling angle is reduced, what happens to the load limit during the lift?

   A. It decreases.
   B. It increases.
   C. It stays the same.
   D. The load limit is not dependent on the sling angle.

7. Which of the following is *not* true about eyebolts?

   A. A shoulder eyebolt can handle a side load.
   B. Regular (nonshouldered) eyebolts are used only for vertical lifting at a 90° load angle.
   C. Open eyebolts are used for lifting and rigging only.
   D. The load capacity of an eyebolt is reduced as the load angle is reduced.

8. What is the purpose of a hoist?

   A. To provide mechanical advantage when lifting loads
   B. To connect the main load-bearing materials to the load or lifting equipment
   C. To prevent rope torque
   D. To eliminate the need for a hitch

9. What does a hoist attach to for a normal lift?

   A. Shoulder or open eyebolts
   B. Wire rope made from extra-improved plow steel only
   C. Cranes with either a cantilevered or gantry configuration
   D. The load itself

10. What is the purpose of grouting when installing machinery?

   A. It prevents resonance affecting the foundation.
   B. It connects the entire baseplate to the foundation.
   C. It removes the need to add a top layer of laitance to the foundation.
   D. It prevents an exothermic reaction from occurring near machinery.

Basic Mechanical Systems

Chapters in This Section

This section begins by covering the basic principles of mechanical transmission and the working principles of simple machines. Topics covered in this section include common types of bearings, seals, lubricants, industrial shafting, belt and chain drives, gear power transmission, and conveyor systems. This section also covers troubleshooting techniques used in evaluating mechanical systems.

The content in this section will help prepare you to earn the NIMS Industrial Technology Maintenance Level 1 Basic Mechanical Systems credential. By earning credentials, you are able to show employers and potential employers proof of your knowledge, skills, and abilities.

9 | Power Transmission Principles

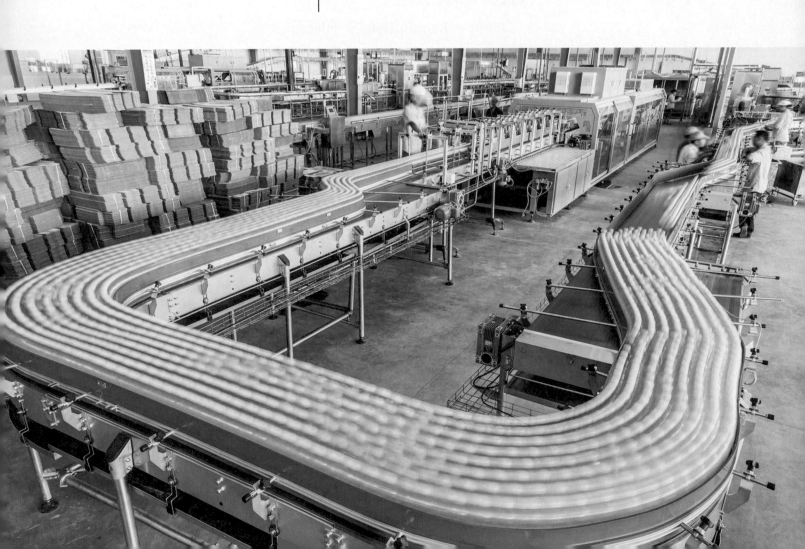

LEARNING OBJECTIVES

After completing this chapter, you will be able to:

☐ Understand the difference between kinetic energy and potential energy.

☐ Describe how mass and velocity affect kinetic energy.

☐ Explain how force is measured.

☐ Explain how torque is measured and understand the effect of lengthening the torque arm.

☐ Describe the relationship between work and power.

☐ Explain the working principles of the six simple machines and understand how they are used to develop mechanical advantage.

☐ Identify the three classes of levers.

☐ Describe Hooke's law.

☐ Explain how friction affects efficiency.

TECHNICAL TERMS

compression spring	pitch
elastic limit	potential energy
energy	power
extension spring	pressure
force	pulley
friction	screw
fulcrum	simple machine
Hooke's law	spring
horsepower	spring constant
inclined plane	static friction
kinetic energy	torque
kinetic friction	torque arm
lead	torsion spring
lever	wedge
mechanical advantage	wheel and axle
newton	work

From simple pulleys to advanced bearings designed for high thrust loads, mechanical devices and systems surround us in our daily life. The transmission of power by machines is an essential function in industry. Modern manufacturing methods and materials allow unprecedented machinery life and reliability—as long as regular maintenance is performed by the technician. **Figure 9-1** shows a tilting-pad thrust bearing. This bearing can support hundreds of thousands of pounds at high shaft speeds with lifetime expectancies of many decades—if lubricated properly.

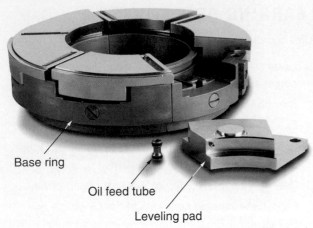

Base ring

Oil feed tube

Leveling pad

Photo courtesy of Kingsbury, Inc.

Figure 9-1. Tilting-pad thrust bearings are widely used in large industrial machinery.

All machines rely on proper maintenance to operate reliably and efficiently. Knowing how machines function helps the technician prevent failures and identify the proper steps to take when repairs are needed. In order to understand, troubleshoot, and maintain machinery, it is necessary to comprehend the fundamental operating principles of machines. This chapter introduces the basic types of energy and energy transfer, the principles of force, work, and power, and the mechanical principles of simple machines.

9.1 FUNDAMENTAL MECHANICS

Knowing how mechanical power is transmitted goes beyond knowing which way to turn a wrench. Understanding the basic types of energy and the core principles of force, work, and power will aid in your comprehension of how and why a machine performs its intended purpose.

9.1.1 Kinetic Energy and Potential Energy

Energy is the capacity to do work. Energy exists in many different forms, including mechanical energy, electrical energy, chemical energy, and heat. Each form of energy, however, can be classified into two broad categories—potential energy and kinetic energy. *Kinetic energy* is the energy of motion of an object. The amount of kinetic energy of an object is dependent on the mass and velocity of the object. If we were to compare two objects with the same mass traveling at different speeds (two cannonballs, for instance), common sense would tell us that the cannonball going faster has more energy. If we were to compare two objects with different mass traveling at the same speed, such as a cannonball and

a golf ball, common sense would tell us that the object with the greater mass (the cannonball) would have more energy. While it might hurt to be hit by the golf ball, one surely would not want to be hit by a cannonball. These simple examples illustrate the effects of mass and velocity on kinetic energy.

Kinetic energy is calculated using the following formula:

$$KE = \frac{1}{2} \times m \times v^2$$

(9-1)

where

KE = kinetic energy (measured in joules)
m = mass (measured in kilograms)
v = velocity (measured in meters per second)

Notice in this formula that kinetic energy is directly proportional to mass and velocity. If mass increases, kinetic energy increases. In addition, notice that an increase in velocity has more of an effect on the amount of kinetic energy. For example, if the velocity of the object is doubled, the kinetic energy increases by a factor of four.

Kinetic energy can be changed into other forms of energy. For example, kinetic energy from wind is converted to mechanical energy when wind spins the blades of a wind turbine. The mechanical energy from the turbine is used to turn a shaft connected to an electrical generator. The mechanical energy from the shaft is converted to electrical energy by the generator.

Potential energy is the stored energy of an object resulting from its position or internal stresses. A compressed spring is one example. In this condition, the energy of the spring is considered to be stored energy. Another example of potential energy is the pressure of fluid in a hydraulic system. Just like kinetic energy, stored energy can be changed into other forms of energy. For example, when a spring is released, the potential energy is converted to kinetic energy. The stored energy of a compressed spring is considered to be a form of elastic potential energy.

Another form of potential energy is gravitational potential energy. The higher an object is, the more gravitational potential energy it has. For example, assume there are two hammers resting on shelves at different heights, one near floor level and one above your head. The hammer on the lower shelf might barely be felt if it fell on your foot. However, the hammer on the higher shelf could break a bone if it were to fall on your foot from that height.

The amount of gravitational potential energy of an object is dependent on the mass, the acceleration due to gravity, and the height of the object. Gravitational potential energy is calculated using the following formula:

$$U = m \times g \times h$$

(9-2)

where

  U = gravitational potential energy (measured in joules)
  m – mass (measured in kilograms)
  g = acceleration due to gravity (9.8 meters per second squared)
  h = height (measured in meters)

Figure 9-2 shows two examples comparing the potential energy of objects with different mass and height. In **Figure 9-2A**, the ball at the higher height off the ground has twice as much potential energy as the ball at the lower height. **Figure 9-2B** shows two bricks with different mass falling from the same height. The brick with more mass has twice as much potential energy. If an object has more mass or is at a higher altitude, it will have more potential energy. The acceleration due to gravity does not change unless the object is at higher altitudes where the force of gravity becomes weaker (outer space, for example).

When an object is dropped, a change in energy occurs. The potential energy of the object is converted to kinetic energy as it falls to the ground. Consider dropping a penny from a very tall building. At the top of the building, the penny has a high amount of potential energy because of its height. When the penny is dropped, it gains in speed as it gets closer to the ground. The penny is losing potential energy and gaining kinetic energy. When the penny reaches ground level, it is traveling very fast and all of the potential energy has been changed to kinetic energy.

In machines, this type of energy transfer happens routinely. **Figure 9-3** shows the internal components of a motor-driven centrifugal pump. This type of centrifugal pump has a shaft turned by a motor, which adds energy to the system. The impeller of the pump rotates and spins fluid out into the volute casing, which directs the fluid to the discharge piping. The impeller is imparting kinetic energy to the fluid. The fluid is forced by the impeller outward to the volute casing, which expands in area as it winds to the discharge piping. This slows down the flow of fluid and converts the kinetic energy of the fluid to potential energy (pressure).

While the transfer of energy from one form to another is very common in machines, the amount of energy entering a system is equal to the amount of energy produced by the system. No energy is created or destroyed. This is the principle of the conservation of energy, which states that energy cannot be created or destroyed, only changed in form.

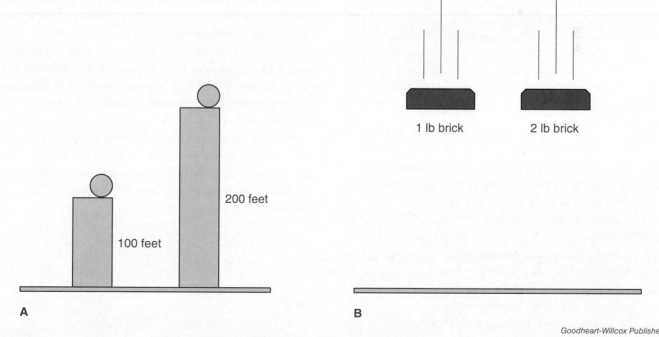

Figure 9-2. Object height, object mass, and acceleration due to gravity determine gravitational potential energy. A—These two balls have the same mass. The ball resting at 200 feet has twice as much potential energy as the ball resting at 100 feet. B—These two bricks are falling from the same height. As they fall, potential energy is converted into kinetic energy. The brick with the greater mass has twice as much potential energy. Both bricks fall at the same velocity.

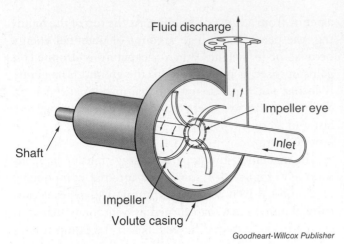

Fluid discharge

Impeller eye

Inlet

Shaft

Impeller

Volute casing

Goodheart-Willcox Publisher

Figure 9-3. A centrifugal pump converts rotational mechanical energy. It imparts velocity and pressure to fluid being pumped.

In everyday maintenance, technicians do not usually speak about energy in terms of potential energy and kinetic energy. Instead, the principles of force, work, and power are considered in stating mechanical relationships and measuring the application of energy.

TECH TIP
Temperature and Kinetic Energy

Temperature is a measure of the level of heat of a substance. In fluids and gases, temperature is a measure of kinetic energy. As the temperature of a fluid or gas increases, kinetic energy increases.

9.1.2 Force

Force is any pushing or pulling effort that changes or tries to change an object's motion. An example is the force exerted by pulling a rope. Force has direction and strength and may cause an object to move, slow down, stop, or change direction. For example, force is applied to accelerate or stop a vehicle. However, force can be applied without resulting in the movement of an object. For example, force can be applied by pushing against something that does not move (such as a concrete wall). Another example of force is mechanical stress caused internally in a machine. See **Figure 9-4**.

The following is a simplified formula used to calculate force:

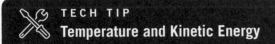

$$F = m \times a$$

(9-3)

Richard z/Shutterstock.com

Figure 9-4. Motor parts must be properly lubricated to reduce wear from vibration and other internal stresses. This technician is inspecting the shaft bearings in an induction motor for damage.

where

F = force (measured in newtons)

m = mass (measured in kilograms)

a = acceleration (measured in meters per second squared)

Notice in this formula that the mass of an object and its acceleration determine force. Another way to look at this relationship is that if force is increased while keeping the mass of an object constant, acceleration will increase.

The units used to measure force depend on the system of measurement you are using. When using metric units in the International System of Units (SI), force is measured in newtons. One *newton* is equal to the force needed to accelerate a mass of one kilogram at the rate of one meter per second squared. In the US Customary system, force is measured in pounds force (lbf) and poundals (ft-lb/s^2).

In fluid power systems, force is used to calculate pressure. *Pressure* is a measure of force applied to a unit area. Pressure in an enclosed system is exerted throughout the system in all directions equally. Pressure is calculated using the following formula:

$$p = \frac{F}{A}$$

(9-4)

where

p = pressure (measured in lb/in^2 or psi)

F = force (measured in lb)

A = area (measured in in^2)

In this instance, force is measured in pounds (lb). Pressure is measured in pounds per square inch (psi).

The formula for pressure can be used to calculate for force or area when the pressure and one of the other variables are known. Simply identify the variable you are trying to solve for and rearrange the formula. If you are solving for force, use the following formula:

$$F = p \times A$$

If you are solving for area, use the following formula:

$$A = \frac{F}{p}$$

9.1.3 Torque

Torque is a turning force applied to an object that is on a fixed axis. The turning force rotates the object around its fixed axis. Torque is applied when tightening a bolt or turning a nut around a bolt with a wrench, **Figure 9-5**. Another application is the rotation of the output shaft of a motor. Torque can result in movement, no movement, or nearly no movement, such as when an already tight fastener is being torqued to a higher specification. The distance from the axis of rotation to the point where force is applied and the amount of force applied are used to calculate torque. The distance from the axis of rotation to the point of force is called the *torque arm*.

Sarin Kunthong/Shutterstock.com

Figure 9-5. Torque is applied when tightening fasteners with a wrench. A torque wrench is used when fasteners must be installed to the specified torque.

Torque is calculated using the following formula:

$$\tau = F \times l$$

(9-5)

where

τ = torque
F = force (measured in lb)
l = length of torque arm (measured in ft)

Torque is usually measured in foot-pounds (ft-lb). When using metric units, torque is measured in newton-meters (N·m). Notice from the formula that increasing the force or increasing the length of the torque arm will increase the amount of torque.

> **TECH TIP**
> **Converting Torque Measurements**
>
> Some smaller fasteners have a torque limit expressed in inch-pounds. To convert inch-pounds to foot-pounds, divide by 12. To convert foot-pounds to inch-pounds, multiply by 12.

If a force of 20 lb is applied with a wrench that is 1′ long, the amount of torque applied is 20 ft-lb. By comparison, if a force of 10 lb is applied with a wrench that is 2′ long, the same amount of torque is applied. See **Figure 9-6**. It is easy to see that using a longer torque arm allows the application of more torque.

The previous calculations assume that the force applied is at a 90° angle to the torque arm. If this angle changes from 90°, then the sine of the angle between the torque arm and the applied force must be added to the formula for torque. Sine is a trigonometric function. It is

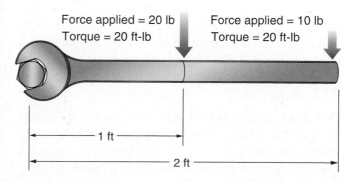

Goodheart-Willcox Publisher

Figure 9-6. Applying torque. When the torque arm is increased, less force is required to turn the wrench and produce the same amount of torque.

defined as the ratio of the side opposite the angle to the hypotenuse in a right triangle. Sine is noted as sin (θ) in the formula for torque:

$$\tau = F \times l \times \sin(\theta)$$

(9-6)

In machines with rotating parts, torque results in power output that is transferred through a shaft to turn a load. The machine bearings and the casing of the machine keep the shaft fixed in one plane.

TECH TIP
Torque and Electric Motors

In an electric motor, torque is developed through the interaction of magnetic fields. If the load on a motor increases, the magnetic fields are stressed, which draws more current. The increase in current makes the magnetic fields stronger, driving the higher load on the motor.

9.1.4 Work

Work is defined as the application of force through a distance to move an object. If weight is added to the object being moved or the distance is increased, the amount of work is increased. On the other hand, if something is not moved, then no work is done. Work is only accomplished while force is applied to produce motion. Work is calculated using the following formula:

$$\text{work} = \text{force} \times \text{distance}$$

(9-7)

The SI unit of work is the joule (one joule is equal to one newton-meter). In the US Customary system, work is measured in foot-pounds. For example, if a force of 200 lb is applied to move an object 10′, the amount of work produced is 2000 ft-lb.

9.1.5 Power

Power is a measure of work that takes place over a period of time. Calculating power provides a way to examine the relationship between work and time and determine mechanical power requirements. Power is calculated using the following formula:

$$\text{power} = \frac{\text{work}}{\text{time}}$$

(9-8)

When using US customary units, power is measured in foot-pounds per second (ft-lb/s). When using SI units, power is measured in watts. One watt is equal to one joule per second. As an example, suppose a 1000 lb object is moved a distance of 10′ in 20 seconds. The amount of power produced is 500 ft-lb/s. If the same object is moved 10′ in 10 seconds, the amount of power produced is 1000 ft-lb/s. **Figure 9-7** shows both calculations.

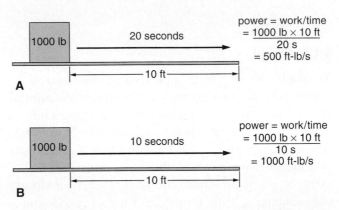

Goodheart-Willcox Publisher

Figure 9-7. Power is a measure of work produced over time. A—If a 1000 lb object is moved 10′ in 20 seconds, the amount of power produced is 500 ft-lb/s. B—More power is required to move the same object in less time.

Notice from these examples that when the amount of work is kept constant, time is inversely related to power. An inverse relationship is one in which one variable increases as the other decreases. Moving the same object the same distance in less time produces more power. In other words, the less time required to do work, the more power produced. Conversely, the more time required to do work, the less power produced.

A commonly used unit of measurement for power is horsepower (hp). One **horsepower** is equal to 550 ft-lb/s or 33,000 ft-lb/min. Referring to **Figure 9-7A**, it would require a power output of almost 1 hp to move the object. In order to move the same object in less time, more power is needed. Referring to **Figure 9-7B**, moving the object would require a power output of almost 2 hp.

Figure 9-8 shows a simplified analysis of how changes to variables affect power output. In each example, work is represented as the formula for work (force × distance). The up and down arrows indicate variables that increase or decrease. The horizontal arrows indicate variables that remain constant. In **Figure 9-8A**, force and distance are kept constant while time is reduced. This will require more power to move the object through the same distance. The opposite is also true. If the power input is increased, the amount of time required to move the object through the same distance decreases. In **Figure 9-8B**, both distance and time are kept constant. In this case, in order to move a heavier object through the same distance in the same amount of time, more power is required. In **Figure 9-8C**, both force and time are kept constant. Moving the same object through a shorter distance in the same amount of time will require less power. Study this analysis to gain a better understanding of the relationship between power, work, and time.

The table in **Figure 9-9** provides a summary of the key terms and formulas presented to this point in the chapter. Review the formulas to make sure you are familiar with the calculations required.

9.2 SIMPLE MACHINES

A **simple machine** is a mechanical device that takes an input force, changes it in one or more ways, and outputs a force to a load. Often, the principles of several simple machines are combined to make up more complex designs. There are six simple machines considered

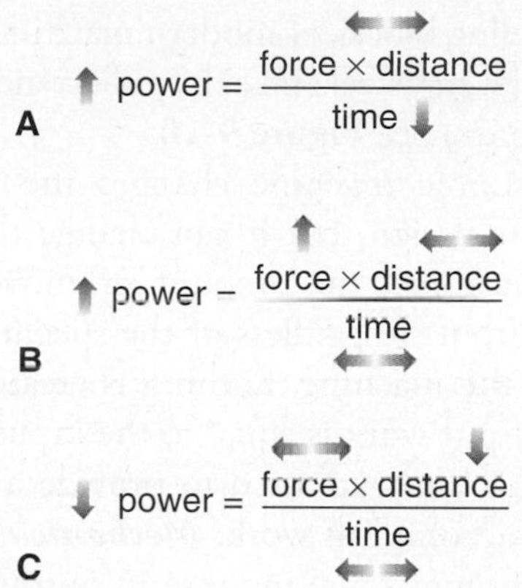

A $\uparrow$ power $= \dfrac{\text{force} \times \text{distance}}{\text{time} \downarrow}$

B $\uparrow$ power $= \dfrac{\text{force} \times \text{distance}}{\text{time}}$

C $\downarrow$ power $= \dfrac{\text{force} \times \text{distance} \downarrow}{\text{time}}$

Figure 9-8. Analyzing the relationship between power, work, and time. Work is represented as the formula for work (force × distance). A—More power is required to move an object in less time when force and distance are constant. B—More power is required to move a heavier object when distance and time are constant. C—Less power is required to move an object a shorter distance when force and time are constant.

Mechanical Formulas		
Key Term	**Formula**	**Definition**
Kinetic energy	$KE = \dfrac{1}{2} \times m \times v^2$ where KE = kinetic energy m = mass v = velocity	The energy of motion of an object.
Gravitational potential energy	$U = m \times g \times h$ where U = gravitational potential energy m = mass g = acceleration due to gravity h = height	The stored energy of an object resulting from its mass and height.
Force	$F = m \times a$ where F = force m = mass a = acceleration	Any pushing or pulling effort that changes or tries to change an object's motion.
Pressure	$p = \dfrac{F}{A}$ where p = pressure F = force A = area	A measure of force applied to a unit area.
Torque	$\tau = F \times l$ where τ = torque F = force l = length of torque arm	A turning force applied to an object that is on a fixed axis.
Work	work = force × distance	The application of force through a distance to move an object.
Power	power $= \dfrac{\text{work}}{\text{time}}$	A measure of work that takes place over a period of time.

Figure 9-9. Basic formulas relating to mechanical applications.

to be the building blocks of modern machines and tools. These are the lever, wheel and axle, pulley, inclined plane, wedge, and screw. See **Figure 9-10**.

How a simple machine changes the input force depends on its design, but it can change the direction of the output force, the amount of movement, and the torque output. Regardless of the specific change or advantage of the machine, nothing is created. In other words, the output work is equal to the input work.

Simple machines are used to provide a mechanical advantage to accomplish work. *Mechanical advantage* refers to the proportional increase in output force relative to the input force when moving a load. Mechanical advantage is expressed mathematically as the ratio of the output force to the input force. It is also expressed as the ratio of the input distance to the output distance. Gaining a mechanical advantage allows work to be accomplished more easily.

9.2.1 Lever

The *lever* is one of the most commonly used simple machines. It is composed of a rigid bar (such as a beam or plank) and a fulcrum. The *fulcrum* provides the axis about which the lever rotates to apply force and movement. Levers are classified according to the position of the fulcrum, input force, and load. See **Figure 9-11**. There are three classes of levers:

- First-class levers.
- Second-class levers.
- Third-class levers.

A first-class lever is one in which the fulcrum is between the input force and the load. The movement of the input force and the load are in different directions. In **Figure 9-11**, the input force and load are located the same distance from the fulcrum. When this is the

Simple Machines

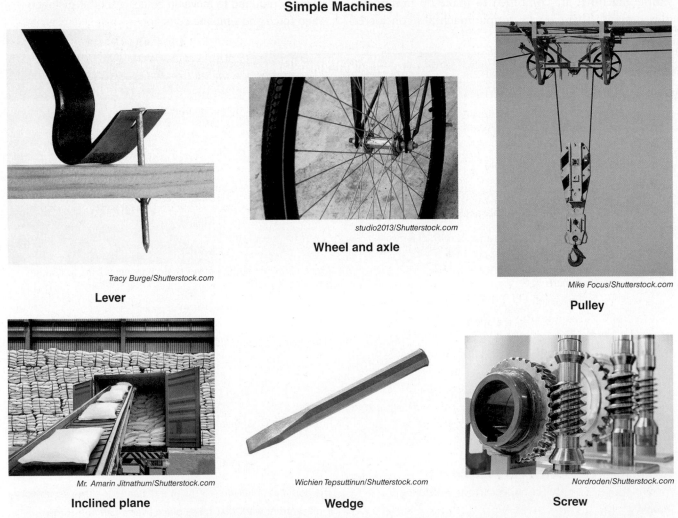

studio2013/Shutterstock.com
Wheel and axle

Tracy Burge/Shutterstock.com
Lever

Mike Focus/Shutterstock.com
Pulley

Mr. Amarin Jitnathum/Shutterstock.com
Inclined plane

Wichien Tepsuttinun/Shutterstock.com
Wedge

Nordroden/Shutterstock.com
Screw

Figure 9-10. The six simple machines include the lever, wheel and axle, pulley, inclined plane, wedge, and screw. Each machine provides a mechanical advantage to achieve work.

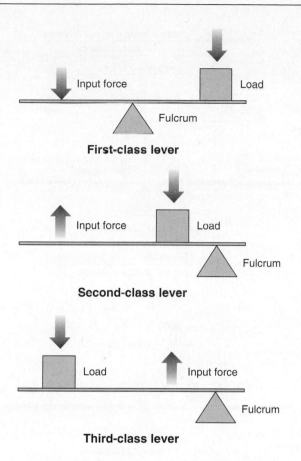

First-class lever

Second-class lever

Third-class lever

Goodheart-Willcox Publisher

Figure 9-11. There are three classes of levers. Levers are classified by the position of the fulcrum, the input force, and the load.

case, there is no mechanical advantage. If the fulcrum is moved closer to the load, the amount of force needed to move the load is reduced. If the fulcrum is moved closer to the input force, the load will travel a greater distance. An example of a first-class lever is a pry bar.

A second-class lever has a load at some point between the input force and the fulcrum. The movement of the input force and the load are in the same direction. The position of the input force and load determine the mechanical advantage. The longer the lever—and the greater the distance between the input force and the load—the greater the mechanical advantage. The length of travel by the input force is proportionally longer than that of the load. The output length of travel is sacrificed for a mechanical advantage. An example of a second-class lever is a wheelbarrow.

In a third-class lever, the input force is between the load and the fulcrum. The movement of the input force and the load are in the same direction. The position of the input force and load determine the mechanical advantage. This type of lever is used to produce a greater

length of travel by the load or to reduce the input force required. In this type of lever, the input force is always greater than the output force. An example of a third-class lever is a shovel.

As previously discussed, the mechanical advantage of a simple machine can be expressed as a mathematical ratio. For a lever, it is expressed as the ratio of the output force to the input force or the ratio of the input distance to the output distance. The output force refers to the load. The input distance is the distance between the fulcrum and the input force. The output distance is the distance between the fulcrum and the load. See **Figure 9-12**. Since the fulcrum is an axis about which the input force and load rotate, the input force and load travel in a curved path (not along a straight line). Therefore, the input and output distances are radial distances. For movement that occurs over a short distance, the rotational movement is not apparent. However, a greater amount of travel amplifies the effect. Note that if the input force and load are at an equal distance from the fulcrum, they will travel the same distance in opposite directions. If the input force and load are at different distances from the fulcrum, their length of travel will change proportionally.

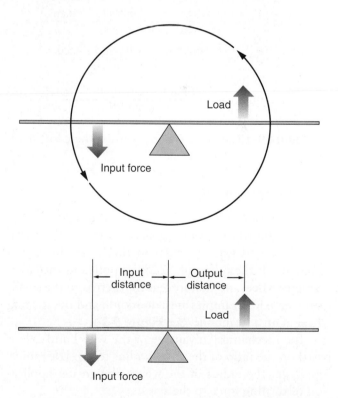

Goodheart-Willcox Publisher

Figure 9-12. The path of travel of the input force and the load is about an axis defined by the fulcrum. Thus, the input and output distances are radial distances.

Mathematical formulas can be used to calculate required forces and distances when determining the mechanical advantage of a lever. For example, the required input force and length of a lever to move a specific load can be calculated. These calculations are made using the following formula:

$$\text{input force} \times \text{input distance} = \text{output force} \times \text{output distance}$$

(9-9)

Forces may be measured in US Customary or SI units. This formula can be used in several ways:

- To determine the required input force when the output force and both distances are fixed.

- To determine the available output force when the input force and both distances are fixed.

- To determine one or both distances (or the ratio of the distances) when both forces are fixed.

Formulas used to calculate the input force, output force, and input distance are given below. Note that each variation is obtained by rearranging the original formula.

$$\text{input force} = \frac{\text{output force} \times \text{output distance}}{\text{input distance}}$$

$$\text{output force} = \frac{\text{input force} \times \text{input distance}}{\text{output distance}}$$

$$\text{input distance} = \frac{\text{output force} \times \text{output distance}}{\text{input force}}$$

Figure 9-13 shows calculations using these formulas.

9.2.2 Wheel and Axle

The **wheel and axle** is composed of a wheel attached to an axle. The wheel's center is fixed to the axle's center. The wheel and axle is similar to the lever (the shared center axis is the fulcrum). Force is applied to the large-diameter wheel and that force is transferred to the small-diameter axle. Examples are a doorknob and the steering wheel of an automobile. See **Figure 9-14**.

The mechanical advantage of the wheel and axle is based on the ratio of the wheel radius to the axle radius. The larger the radius of the wheel, the greater application of turning force to the axle.

In the case of an automobile wheel, the input force is reversed. The force is applied to the axle and transferred to the wheel. The wheel turns the same speed as the axle, but the wheel has a larger diameter. This allows the car to travel a greater distance.

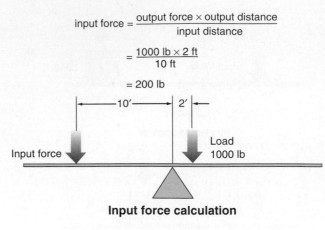

$$\text{input force} = \frac{\text{output force} \times \text{output distance}}{\text{input distance}}$$

$$= \frac{1000 \text{ lb} \times 2 \text{ ft}}{10 \text{ ft}}$$

$$= 200 \text{ lb}$$

Input force calculation

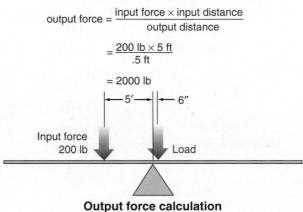

$$\text{output force} = \frac{\text{input force} \times \text{input distance}}{\text{output distance}}$$

$$= \frac{200 \text{ lb} \times 5 \text{ ft}}{.5 \text{ ft}}$$

$$= 2000 \text{ lb}$$

Output force calculation

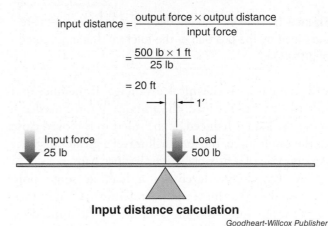

$$\text{input distance} = \frac{\text{output force} \times \text{output distance}}{\text{input force}}$$

$$= \frac{500 \text{ lb} \times 1 \text{ ft}}{25 \text{ lb}}$$

$$= 20 \text{ ft}$$

Input distance calculation

Goodheart-Willcox Publisher

Figure 9-13. Calculating required forces and distances for a first-class lever.

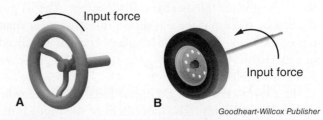

Goodheart-Willcox Publisher

Figure 9-14. Applications of the wheel and axle. A—The input force is applied to the wheel and transferred to the axle. B—The input force is applied to the axle and transferred to the wheel.

The mechanical advantage of the wheel and axle is equal to the ratio of the two radii. It is calculated using the following formula:

$$\text{mechanical advantage} = \frac{\text{radius of wheel}}{\text{radius of axle}}$$

(9-10)

Note that this formula is based on applying the input force to the wheel. Another way to examine the mechanical advantage of a wheel and axle is to consider the force exerted to move a load. The following formula can be used to calculate required loads based on different wheel and axle combinations:

$$\frac{\text{force on wheel}}{\text{force on axle}} = \frac{\text{radius of axle}}{\text{radius of wheel}}$$

This formula can also be rearranged as follows:

$$\text{force on wheel} \times \text{radius of wheel}$$
$$= \text{force on axle} \times \text{radius of axle}$$

In order to gain a large mechanical advantage, a very large wheel is used with a very small axle. Multiple reductions can also be made by using different wheel and axle combinations.

Just as with the lever, when applying the input force to the wheel, the length of travel of the wheel is increased as the mechanical advantage becomes greater. In addition, the force and torque applied to the axle are increased while length of travel is decreased. The opposite holds true when the input force is applied to the axle. In the case of a car, when the input torque is applied to the axle, the applied torque on the wheel is decreased while the length of travel is increased.

9.2.3 Pulley

The *pulley* is composed of a wheel that rotates on a shaft. The wheel may be grooved to hold the tensile member wound around the wheel. The tensile member is the part of the system to which force is applied and transmitted. It can be a rope, belt, or chain. Pulley systems are used to transfer force and change the direction of force. In construction, pulley systems are used by cranes to lift building materials. While belt drives also use pulleys, belt and pulley systems are different from pulleys classified as simple machines. Belt and pulley systems are covered in a later chapter.

Pulleys assembled into a system in order to move loads are called "blocks." A block can be a single pulley or a set of multiple pulleys on the same axle. Pulley systems develop mechanical advantage by combining multiple blocks or pulleys to share the load.

The most basic type of pulley system is a fixed pulley. This consists of a single pulley attached to a fixed point, such as a ceiling. See **Figure 9-15**. A fixed pulley is used to change the direction of force when lifting a load. However, there is no mechanical advantage. For example, if a 100 lb load is lifted, it is necessary to apply a force of 100 lb.

Another basic pulley system combines a fixed pulley and a movable pulley attached to the load. This is called a block and tackle assembly. See **Figure 9-16**. The block and tackle assembly shown is called a gun tackle assembly. In this assembly, the movable pulley is not fixed and is free to move up and down with the load. Note that there are two sections of rope supporting the load. The addition of the movable pulley divides the load. Each rope section supports half the load (50 lb). Thus, if a weight of 100 lb were to be lifted, a force of 50 lb would need to be applied to the end of the rope.

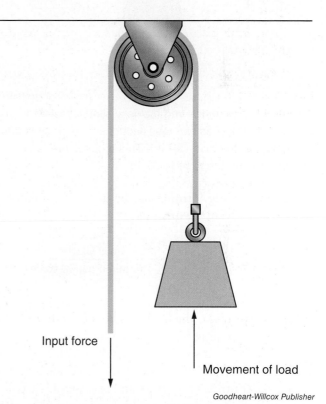

Input force

Movement of load

Goodheart-Willcox Publisher

Figure 9-15. A fixed pulley consists of a single pulley that changes the direction of force applied.

The mechanical advantage of a pulley system is equal to the number of rope sections that pass through a pulley. The gun tackle arrangement in **Figure 9-16** yields a mechanical advantage of 2. By extending the system to include 4 pulleys, the mechanical advantage is increased to 4. This reduces the load at the end of the rope to one-fourth of the weight. This type of system is called a double tackle assembly.

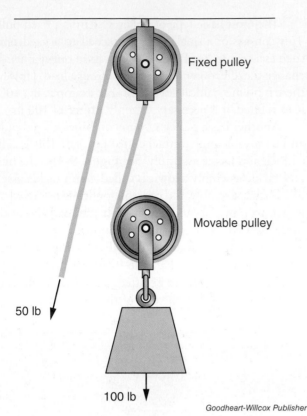

50 lb

100 lb

Goodheart-Willcox Publisher

Figure 9-16. This block and tackle assembly, called a gun tackle assembly, has two rope sections. Each rope section supports half the load (50 lb). This reduces the force required to pull the rope to 50 lb.

The following formula can be used to calculate the mechanical advantage of a pulley:

$$\text{mechanical advantage} = \frac{\text{load weight}}{\text{required pulling force}}$$

(9-11)

In this formula, the load weight is the output force. The required pulling force is the input force.

Using different size pulleys attached with a common shaft provides for further mechanical advantage. This arrangement utilizes the wheel and axle simple machine. The mechanical advantage is calculated in the same manner used when making a calculation for a wheel and axle.

9.2.4 Inclined Plane

The ***inclined plane*** is a sloped surface used to move an object in a vertical direction by applying force along a horizontal direction. The inclined plane in its simplest form is a ramp. The ramp makes it possible to move a heavy object to a certain height without lifting it. However, the work required to move the object by either means is the same. While the ramp does not reduce work, it does

reduce the force required to move the object. The force is reduced, but applied over a longer distance. The reduction in force required is the mechanical advantage.

The mechanical advantage of an inclined plane is calculated using the following formula:

$$\text{mechanical advantage} = \frac{\text{object weight}}{\text{force applied to move object}}$$

(9-12)

Another way to calculate the mechanical advantage is to divide the length of the inclined plane by the height. For a ramp, the following formula is used:

$$\text{mechanical advantage} = \frac{\text{ramp length}}{\text{ramp height}}$$

(9-13)

The preceding formulas can be combined as follows:

$$\frac{\text{object weight}}{\text{force applied to move object}} = \frac{\text{ramp length}}{\text{ramp height}}$$

Note that the variables in this formula represent the input force (force applied to move object), input distance (ramp length), output force (object weight), and output distance (ramp height). This formula can be used in several different ways:

- To determine the length of the ramp needed when the ramp height, object weight, and force to be used are known.
- To determine the maximum weight of an object that can be moved up a ramp when the ramp height, ramp length, and force to be used are known.
- To determine the force required to move an object when the dimensions of the ramp and the object weight are known.

Formulas used to calculate the ramp length, maximum object weight, and required moving force are given below. Each variation is obtained by rearranging the original formula.

$$\text{ramp length} = \frac{\text{object weight} \times \text{ramp height}}{\text{force applied to move object}}$$

$$\text{object weight} = \frac{\text{ramp length} \times \text{force applied to move object}}{\text{ramp height}}$$

$$\text{force applied to move object} = \frac{\text{object weight} \times \text{ramp height}}{\text{ramp length}}$$

Figure 9-17 shows several examples of these calculations.

These basic calculations do not account for friction, which must be considered when using ramps. When

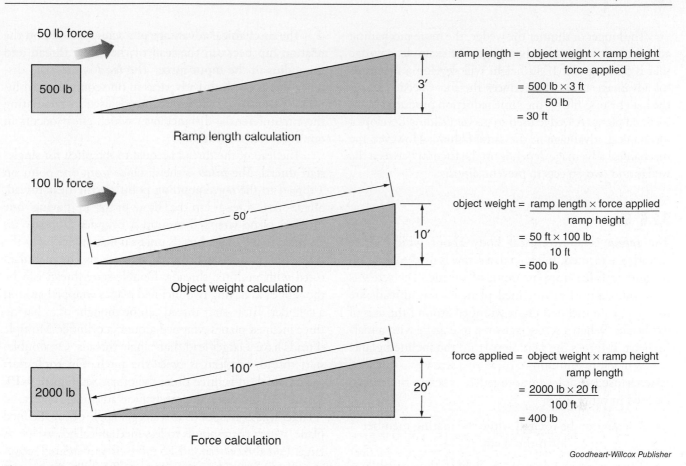

Figure 9-17. Sample calculations for using ramps to move objects.

Goodheart-Willcox Publisher

moving heavy loads by hand, rollers and bearings can be used to reduce friction of the load on the ramp surface. The friction between the load and the ramp works in two ways. First, friction adds to the load because it causes more effort to be exerted to move the load. Second, when the load is left on the ramp with no force applied, the friction will prevent the load from sliding back down the ramp. When there is less friction, the possibility of a load backsliding requires other tools to keep the load in place. In such cases, a ratcheting system or wedges can be used.

9.2.5 Wedge

The *wedge* is composed of two inclined planes joined back to back along a common base. The wedge is a smaller form of an inclined plane. As is the case with an inclined plane, a relatively small amount of force is applied over a long distance. However, the wedge moves into a load when force is applied. An example of a wedge is the cutting edge of an axe. When the axe is driven into a log, the input of force is transferred to the sharpened edge, which is very thin. The force therefore increases and allows an easier cut. Because the angle of the cutting blade is so small, the cutting blade develops a very large

mechanical advantage. When driven into the load, the wedge changes the direction of force applied. The force that is transferred is perpendicular to the sloped sides of the wedge. This force splits the object being cut. Other examples of wedges used to cut material are a chisel and a nail. Wedges can be used for the following tasks:

- Lifting an object a small distance in order to gain further advantage with a lever.
- Splitting or cutting apart objects in two.
- Preventing an object from moving or sliding back on a ramp by offsetting the ramp's angle.

Formulas used to calculate the mechanical advantage of a wedge are similar to those used for inclined planes. The following formulas are used for wedges designed to lift an object:

$$\text{mechanical advantage} = \frac{\text{weight of load being lifted}}{\text{force required to lift load}}$$

(9-14)

$$\text{mechanical advantage} = \frac{\text{length of wedge}}{\text{height of wedge}}$$

(9-15)

The longer or thinner the wedge, the more mechanical advantage developed by the wedge. For example, a wedge that is 5″ long and 1″ in height will develop a mechanical advantage of 5. This reduces the force needed to lift the load by one-fifth—the same reduction produced by an inclined plane. A wedge used to prevent sliding develops a mechanical advantage in the same fashion. However, the mechanical advantage depends on the friction between the wedge and two objects to prevent slippage.

9.2.6 Screw

The *screw* is a cylindrical body about which screw threads are formed. In general, a screw is used to develop motion or force from an input of torque. The screw is an application of an inclined plane. Screw threads are formed by an inclined plane wrapped around the axis of a cylinder. When a screw is turned to engage with a mating part, it forces the mating part up the inclined plane.

While it is common to think of a screw as a threaded fastener used to join parts together, a screw thread can be used in other ways:

- A screw can be rotated while the mating member moves up or down the shaft.

- The mating member can be held stationary while the screw moves in or out of the mating member.

- A screw can be used to move fluid in a positive-displacement pump.

A threaded screw assembly is made up of a screw and a mating part. The exact function a screw assembly serves depends on which piece is held stationary. **Figure 9-18** shows a ball screw assembly. This assembly consists of a screw and a movable nut. The assembly contains ball bearings that roll in the grooves between the screw and the movable nut. When the screw is rotated, the movable nut travels in a linear direction along the screw axis.

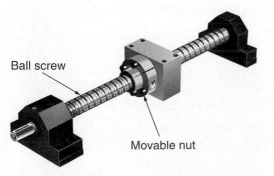

Ball screw

Movable nut

Image courtesy Bosch Rexroth Corporation. Used by Permission.

Figure 9-18. Ball screw assemblies are commonly used in motor-controlled positioning systems. The rotation of the ball screw produces linear movement of the movable nut.

The mechanical advantage of a screw is based on the relationship between the lead of the screw thread and the radius of the input force. The *lead* is the axial distance the screw thread advances in one complete revolution. The radius of the input force is used in calculating the circumferential distance over which effort occurs in one complete revolution.

The lead of the thread is equal to the pitch for single-start thread. The *pitch* is the distance from one point on a thread to the corresponding point on the next thread. Single-start thread can be thought of as having one inclined plane wrapped around a cylinder. Threads are assumed to be single threads unless noted otherwise in the thread specification. Threaded parts can also be manufactured with multiple threads. Double-start thread can be thought of as having two inclined planes wrapped around a cylinder. Triple-start thread can be thought of as having three inclined planes wrapped around a cylinder. Multiple threads have a larger lead than single threads. On double-start thread, the lead is twice the pitch. On triple-start thread, the lead is three times the pitch. See **Figure 9-19**. As the lead increases, the mechanical advantage decreases. A larger lead corresponds to a steeper angle of the inclined plane, which corresponds to less mechanical advantage. A larger lead also corresponds to more circumferential movement with each rotation of the part. Normally, most screw thread that you encounter will have single-start thread. When this is the case, the lead is equal to the pitch.

The following formulas are used to calculate the mechanical advantage of a screw. In the first formula, the radius of the input force is used in the calculation of circumferential movement in one complete revolution. The circumferential movement is found by multiplying pi (π) by twice the radius. Pi is the ratio of the circumference of a circle to its diameter.

$$\text{mechanical advantage} = \frac{2 \times \pi \times r}{\text{lead}}$$

(9-16)

$$\text{mechanical advantage} = \frac{\text{output force}}{\text{input force}}$$

(9-17)

These formulas do not account for friction, which is a significant factor with screw threads. Friction reduces the efficiency and mechanical advantage of a screw thread.

9.3 SPRINGS

A *spring* is a device that stores energy when compressed or extended by a force and exerts an equivalent amount of energy when released. While the spring is not considered a

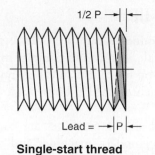
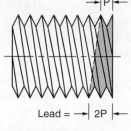
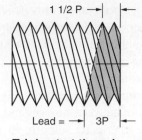

Single-start thread **Double-start thread** **Triple-start thread**

Goodheart-Willcox Publisher

Figure 9-19. Threads are assumed to be single-start threads unless the thread specification designates otherwise. For single-start threads, the lead is equal to the pitch. Notice that the slope of the thread on the near side of the cylinder represents one-half the pitch. The hidden line represents the slope of the thread on the opposite side of the cylinder.

simple machine, it is an important part of many machine assemblies. Springs are characterized by their shape and how a load is applied to them. Three common types of springs are compression, extension, and torsion springs:

- A *compression spring* resists axial shortening as a load is applied.

- An *extension spring* resists axial lengthening as a pulling load is applied.

- A *torsion spring* resists twisting as torque is applied about the central axis.

Each spring has a spring constant. The *spring constant* is a measure of the resisting force of the spring. It indicates the stiffness of a spring and is dependent on the spring's material and size. When using US Customary units, the spring constant is given in pounds of force per unit length (lbf/ft or lbf/in). In SI units, the spring constant is given in newton-meters (N·m) or newton-millimeters (N·mm).

Hooke's law is a principle that describes spring force. It states that the restoring force of a spring is proportional to the distance the spring is compressed or stretched. The restoring force is the force exerted by a spring to return to its rest position. It is opposite the applied force. Hooke's law is expressed mathematically as follows:

force = spring constant × change in length from rest

(9-18)

This relationship holds true as long as the spring is not stretched (or compressed) beyond its elastic limit. The *elastic limit* is the point beyond which an object will permanently deform when stretched or compressed. In evaluating Hooke's law, it can be seen that an increase in the spring constant or further tension (or compression) will result in a larger resisting force (more stored energy). Also, if a larger force is applied to a spring, the change in length of the spring can be easily calculated.

9.4 FRICTION AND EFFICIENCY

The examples given in this chapter did not include friction in any of the calculations. Instead, the examples were based on perfect or frictionless conditions. In actual practice, friction is an important factor and directly affects the efficiency of a process. *Friction* is a resistive force that acts counter to sliding motion. Friction is commonly thought of as a resistive force between two flat pieces moving against each other, but friction also occurs between fluids and the enclosed areas that contain them.

There are a number of cases in which friction is desirable and necessary. Examples include the following:

- An object resting on an inclined plane is only stationary because of the friction between the object and the inclined plane. If there was not enough friction, the object would slide down the inclined plane.

- A screw is self-locking because of the angle of the threads and because of friction. If there were no friction, the screw would be quickly loosened by the application of an opposing force.

- As is the case with an inclined plane, a wedge relies on friction. Without friction, the wedge would be pushed out from its resting point. For example, a door being held open would close.

- Friction is needed in order for the brakes of a car to slow it down. The brake pads press into the rotor, converting the kinetic energy into heat.

Machines are designed in almost all cases to reduce friction and thereby increase efficiency. When excessive friction occurs, energy that is being input into a machine component is converted into heat and wasted. This reduces efficiency.

There are two main types of friction: static friction and kinetic friction. *Static friction* is friction between two objects that are at rest. An example is a wedge holding a door open. *Kinetic friction* is friction between two objects that are moving in relation to each other. In most cases, the amount of kinetic friction between objects is less than static friction.

Friction can be reduced by minimizing or preventing contact between surfaces and making the surfaces that do come in contact smooth. In machine design, both of these principles are used when bearings and lubrication are included. Bearings reduce friction by using rolling elements, which are harder and smoother than the machine components that they support. Lubrication reduces friction by preventing metal-to-metal contact and by transferring heat away from the machine components.

CHAPTER WRAP-UP

This chapter has examined a number of core concepts related to mechanical energy and simple machines. Having a good grasp on how mechanical energy is used and transferred in machines will aid in troubleshooting mechanical systems.

The principles of simple machines are commonly used in the design of modern machines and tools. In this chapter, you have learned how simple machines are used to provide a mechanical advantage in order to accomplish work. Understanding how machines accomplish work will help you develop the skills to properly maintain mechanical systems.

Chapter Review

SUMMARY

- Kinetic energy is energy of motion. It is dependent on mass and velocity. Potential energy is stored energy due to the position or internal stress of an object.

- Force is an effort that changes or tries to change the motion of an object.

- Torque is a turning force that rotates an object around a fixed axis.

- Work is accomplished when force is applied to produce motion. Power is a measure of work that occurs over a period of time.

- Simple machines provide a mechanical advantage in order to accomplish work. The six simple machines are the lever, wheel and axle, pulley, inclined plane, wedge, and screw.

- Mechanical advantage is expressed as the ratio of the output force to the input force. It is also expressed as the ratio of the input distance to the output distance.

- Levers are classified by the position of the fulcrum, input force, and load. The lever develops a mechanical advantage by reducing the amount of force needed to move a load.

- The mechanical advantage of the wheel and axle is based on the ratio of the wheel radius to the axle radius.

- A single fixed pulley changes the direction of force with no mechanical advantage. Multiple pulleys provide a mechanical advantage by sharing the load.

- The inclined plane develops a mechanical advantage by allowing force to be applied over a longer distance.

- The wedge is a smaller form of an inclined plane. Wedges can be used to lift an object a small distance, cut an object in two, and prevent movement of an object on a ramp.

- The screw is used to produce motion or force from an application of torque. The mechanical advantage of the screw is based on the relationship between the lead and the radius of the input force.

- A spring stores energy. A spring's force is directly proportional to the distance the spring is compressed or stretched.

- Friction has a direct impact on efficiency. It is reduced by minimizing or preventing contact between parts.

REVIEW QUESTIONS

Answer the following questions using the information provided in this chapter.

1. Calculate the kinetic energy produced by a 35-kilogram object traveling at 4.85 meters per second.

2. How is kinetic energy affected if velocity of an object is doubled?

3. Give an example of an object that has potential energy.

4. What is the gravitational potential energy of a 92-kilogram object that is thrown off a 15-meter building?

5. A penny is dropped out of a plane. Describe how its energy changes as it is dropped.

6. *True or False?* A person still exerts force by pushing against a concrete wall that does not move.

7. What is the force required to accelerate a 25-kilogram object at 7.2 m/s²?

8. One _____ is equal to the force needed to accelerate a mass of one kilogram at the rate of one meter per second squared.

9. If 500 lb are distributed over a surface area of 12 in², what is the resulting pressure?

10. List two examples of torque being applied to an object.

11. Calculate the torque of a fastener that has a torque arm of 2.5 ft and force application of 12.25 lb.

12. If a fastener has a torque limit of 30 ft-lb, and you have a torque wrench calibrated in in-lb, what is the conversion?

13. A force of 75 lb is acting at an angle of 75° on a 36″ long lever. What is the resulting torque?

14. A 200-pound man claims he can drive a go-cart around a track faster than his 110-pound son. Explain why this is false.

15. How much power would be required for an engine to move an 1800-lb load 500′ in 1 hour?

16. Convert 12 hp to ft-lb/s.

17. If the power input is increased, the amount of time required to move an object through a distance _____. (increases, stays the same, decreases)

18. *True or False?* Mechanical advantage can be expressed as the ratio of the output distance to the input distance.

19. Give an example of each class of levers and explain their position of the fulcrum, input force, and load.

20. A first-class lever is used with an input force of 80 lb and length of 9′. The output lever length is 18″. What is the force produced?

21. *True or False?* In a wheel and axle machine, a greater wheel radius will produce less turning force to the axle.

22. Find the mechanical advantage of a wheel and axle with a wheel of 36″ and an axle of 6″.

23. If a 175 lb load is lifted on a fixed pulley, it is necessary to apply a force of _____ lb.

24. Describe how the mechanical advantage of a pulley system is calculated.

25. With an inclined plane, a maximum force of 100 lb is to be used to move a 510 lb stone to a height of 10′. How long must the inclined plane be?

26. List an example in which a wedge can be used to complete a particular task.

27. Calculate the mechanical advantage needed to lift an 85 lb load that requires 24 ft-lb/s² amount of force.

28. How are lead and pitch related in a single-start screw thread?

29. What is the mechanical advantage of a screw that has a lead of 3 and a 2″ radius?

30. Calculate the force of a spring that can be stretched 13″ and has a spring constant of 5.2 lbf/in.

31. How does doubling the stretch of a spring affect the spring force?

32. Give an example where friction may be needed to produce a desired function.

 NIMS CREDENTIALING PREPARATION QUESTIONS

The following questions will help you prepare for the NIMS Industrial Technology Maintenance Level 1 Basic Mechanical Systems credentialing exam.

1. Which object is storing gravitational potential energy?
   A. A bowling ball sitting on a 6-foot shelf.
   B. A compressed spring being released.
   C. A dropped penny gaining speed in midair.
   D. A load carried up a 15-meter hill.

2. What force is needed to accelerate a 2550-kilogram car at 18 m/s²?
   A. 22,950 N
   B. 45,900 N
   C. 413,100 N
   D. 449,820 N

3. What is the turning force applied to an object being rotated on a fixed axis?
   A. Energy
   B. Force
   C. Pressure
   D. Torque

4. Calculate the amount of work produced if a force of 125 lb is applied to move a 12-lb wagon across a 25′ distance.
   A. 5 ft-lb
   B. 1500 ft-lb
   C. 3125 ft-lb
   D. 37,500 ft-lb

5. Which action would produce a decrease in power output?
   A. Decreasing distance as force and time are held constant.
   B. Decreasing time as force and distance are held constant.
   C. Increasing force as distance and time are held constant.
   D. Increasing distance and force as time is held constant.

6. A lever is classified according to the position of what three components?

   A. Fulcrum, input force, and load.
   B. Fulcrum, output force, and load.
   C. Pulley, input force, and shaft.
   D. Pulley, output force, and shaft.

7. Calculate the mechanical advantage of a 2′-radius wheel and a 4″-radius axle.

   A. 0.5
   B. 1
   C. 2
   D. 6

8. In a single-start thread of a screw, what is the proportion between the lead and pitch?

   A. The lead of the thread is equal to the pitch.
   B. The lead of the thread is two times the pitch.
   C. The pitch of the thread is two times the lead.
   D. The pitch of the thread is three times the lead.

9. Which of the following is *not* an effective method to reduce friction and increase machine efficiency?

   A. Decreasing contact between machine surfaces.
   B. Heating up the machine to reduce output energy.
   C. Lubricating a machine to reduce metal-to-metal contact.
   D. Using rolling elements for bearings.

10

Bearings, Seals, and Lubrication

LEARNING OBJECTIVES

After completing this chapter, you will be able to:

☐ Describe the purpose of the rolling-element bearing.

☐ Recognize and label the common parts of a rolling-element bearing.

☐ Discuss the differences between the main types of rolling elements and their applications.

☐ Identify common designations in the standard numbering system for rolling-element bearings.

☐ Discuss proper methods of storage and handling of bearings.

☐ List preinstallation checks and discuss the importance of this process.

☐ Describe methods for bearing removal and installation.

☐ Explain the purpose of a run-in period after a rebuild or bearing replacement.

☐ Recognize and label the common parts of a radial seal.

☐ Discuss important considerations when installing a new seal.

☐ Explain specific limitations of seal design.

☐ List the purposes of lubrication.

☐ Describe the methods of lubrication.

☐ Discuss the purpose of various oil additives.

☐ Explain the characteristics and benefits of greases.

☐ Discuss the proper handling and storage of lubricants.

☐ Describe the procedures for various lubrication methods.

TECHNICAL TERMS

antioxidant	garter spring
anti-wear additive	grease
asperity	hydrodynamic lubrication
axial load	hydrostatic lubrication
axial movement	lip
back face	lubrication plan
ball bearing	mixed film lubrication
bearing	needle bearing
boundary film lubrication	non-contact seals
contact seals	outside diameter
corrosion inhibitor	plain bearing
cylindrical bearing	preloading
deactivator	radial load
demulsifier	radial movement
detergent	rolling-element
dispersant	bearing
elastohydrodynamic	sealing edge
lubrication	spherical bearing
emulsifier	tapered bearing
extreme pressure additive	thrust bearing
foam inhibitor	viscosity

The bearing, in its most basic form, has been in use for thousands of years. Current designs of bearings with rolling elements began being used in the 1700s. From a wristwatch to the propeller on a cargo ship, bearings are a part of nearly every machine that transfers mechanical power. Lubrication allows for higher machine efficiency and greater speeds with less wear. Seals isolate lubrication and bearings from their surroundings, and sometimes corrosive, environment. In this chapter, we will discuss industrial uses of rolling-element bearings, seals, and lubrication with regard to machine maintenance. This topic area is so vast that a book could easily be written on each subject.

10.1 ROLLING-ELEMENT BEARINGS

A ***bearing*** is a mechanical component that supports a moving part, generally a rotating shaft. The outer part of a bearing is secured to a nonmoving part of a machine, and the inside of the bearing supports the moving part.

Bearings perform two major functions: reduce friction and restrict motion. Without a bearing, friction would increase between machine elements to the point of reducing efficiency. This occurs when power is given off as heat. Bearings also restrict machine motion into one or more planes. For example, a bearing may allow a shaft to spin in order to transfer power, but not allow lateral movement. In the centrifugal pump shown in **Figure 10-1**, the bearings support the shaft in position and allow the shaft to rotate and transfer power from the motor to the impeller. Forces that act in directions against or other than the rotation of the shaft are transferred from the bearings into the casing of the pump, and then to the pump's anchor bolts and foundation.

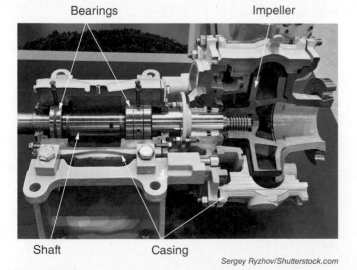
Sergey Ryzhov/Shutterstock.com

Figure 10-1. A pump's bearings provide support and allow motion in only one plane (rotation).

Two main types of bearings are rolling-element bearings and plain bearings. ***Rolling-element bearings*** are composed of rollers arranged between two races. ***Plain bearings***, or *sleeve bearings*, are bushings that support a shaft on a nonmoving, interior cylindrical surface. A plain bearing is shown in **Figure 10-2**. While plain bearings are used in larger machines, such as those used in power generation, it is much more likely that you will encounter and perform maintenance on rolling-element bearings. Because of this, we will limit our study to rolling-element bearings.

George_C/Shutterstock.com

Figure 10-2. A plain bearing contains no rolling elements.

Rolling-element bearings are classified by their load, rotating elements, construction materials, and other design features. The two basic categories of these bearings are ball bearings and roller bearings, as shown in **Figure 10-3**. The function and load, both amount and direction, largely determine the types of rolling elements used for the application.

Roller bearings

Ball bearings

Vladnik/Shutterstock.com

Figure 10-3. The two categories of rolling-element bearings are ball bearings and roller bearings.

A bearing typically includes at least three basic elements: inner and outer races, rolling elements, and a cage, as shown in **Figure 10-4**. The races transfer the load through the rolling elements to the other race, which then transfers the load to the machine casing. The cage keeps rolling elements equally spaced and receives no load.

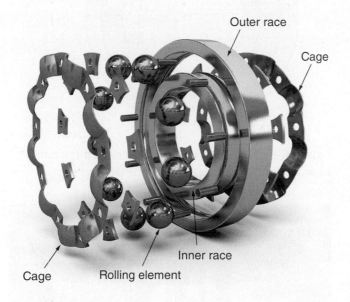

heromen30/Shutterstock.com

Figure 10-4. The basic components of rolling-element bearings.

Bearings are classified by the shape of their rolling elements. Common shapes of rolling elements, shown in **Figure 10-5**, include the following:

- ***Ball bearings.*** Sphere shape reduces friction and supports moderate loads.

- ***Cylindrical bearings.*** Cylinder shape increases contact area to distribute load.

- ***Spherical bearings.*** Barrel shaped and designed to carry heavy loads at low speeds.

- ***Tapered bearings.*** Conical shape, both races and rolling elements are tapered.

- ***Needle bearings.*** Long, narrow cylindrical shape, and many do not have inner races.

The load applied by the shaft determines the type of bearing required. Loading can be axial, radial, or a combination of the two, as shown in **Figure 10-6**. *Axial loads*, or *thrust*, exert force parallel to the shaft. ***Radial loads*** exert force at a right angle to the shaft. Shafts are also loaded by rotational torque, but this load is accepted by the bearing and friction is reduced.

Ball bearing
Ksander/Shutterstock.com

Cylindrical bearing
1989studio/Shutterstock.com

Tapered bearing
Aerodim/Shutterstock.com

Needle bearing
Stason4ik/Shutterstock.com

Spherical bearing

K. Kargona/Shutterstock.com

Figure 10-5. Common shapes of the rolling elements in bearings.

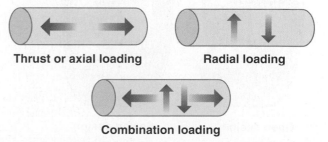

Thrust or axial loading   **Radial loading**

Combination loading

Goodheart-Willcox Publisher

Figure 10-6. Types of loading on a shaft supported by bearings.

10.1.1 Ball Bearings

The most common bearing you will work with is the ball bearing. **Figure 10-7** shows both an open design and a sealed design. A ball bearing is used when the shaft is loaded in a radial direction with little axial loading, such as in a machine with the shaft in a horizontal position. On a sealed bearing, the inner and outer races include a groove or shoulder for the bearing seal.

Ball bearings come in many styles and are classified by how they are assembled, the number of rows of rolling elements, and the loading capability. The following are some common construction designs of ball bearings:

- **Conrad bearing.** Construction allows the races to be offset for filling with the rolling elements. When rolling elements are distributed, races become concentric.

- **Slot-fill bearing.** A slot cut into the inner and outer races, **Figure 10-8**, allows more rolling elements to be loaded, and therefore more load can be applied. Slot-fill bearings are also called *maximum-capacity bearings*.

- **Angular-contact bearing.** Used in machines where both thrust and radial loads are present, **Figure 10-9**.

- **Self-aligning bearing.** Contains two rows of rolling elements to increase load capacity, as shown in **Figure 10-10**. Deep grooves in the race and a specially designed cage allows for slight misalignment.

- **Multiple rows of rotating elements.** Can offer smaller outside diameter and same load rating as a single row, or may provide a higher load rating due to more load-bearing members.

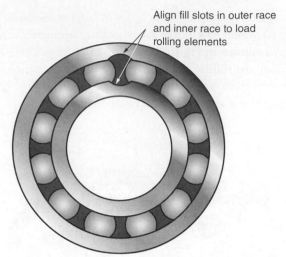

Align fill slots in outer race and inner race to load rolling elements

Goodheart-Willcox Publisher

Figure 10-8. A slot-fill bearing has grooves in the races to allow more rolling elements to be loaded.

Photo and Vector/Shutterstock.com

Figure 10-9. In this double row, angular-contact bearing, all the rolling elements are engaged and spread the force over a larger surface area to decrease friction and minimize wear.

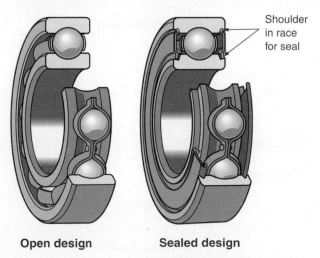

Shoulder in race for seal

Open design　　　**Sealed design**

Image published with the permission of SKF USA Inc.

Figure 10-7. Single-row ball bearings in open and sealed designs.

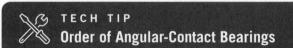

TECH TIP
Order of Angular-Contact Bearings

The direction of allowed thrust on angular-contact bearings may be indicated by a marking on the bearing. When replacing these bearings, be certain to place the new bearings in exactly the same order. Many times, multiple angular-contact bearings are installed back-to-back. This may be to provide for thrust in both directions or a higher thrust loading in one direction, depending on the arrangement. Multiple bearings are common in machine tools, such as lathes and computer numerical control (CNC) machines.

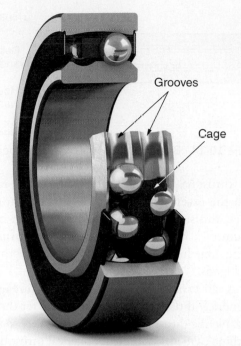

Grooves

Cage

Figure 10-10. Self-aligning bearings are available in various designs of rolling elements, seals, cages, and race mounting.

10.1.2 Cylindrical Bearings

Cylindrical bearings are designed for heavy radial loading. The allowable axial load for cylindrical bearings is generally 25% to 50% of the allowable radial load. **Figure 10-11** shows details of a caged cylindrical bearing. The cage keeps the rolling elements properly spaced so that the load is evenly distributed on the races. The integral flange on the inner edge of the outer race reduces stress on the edges of the rolling elements when axial thrust is applied.

Various designs of the inner and outer race flanges (edges) allow for axial loads from the housing and shaft, and possible displacements. If only one flange is present, thrust loading is allowed against that edge only and not away from it. Cylindrical bearings allow for only a small amount of shaft misalignment, generally less than one-tenth of 1°.

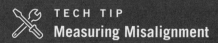

TECH TIP
Measuring Misalignment

In practice, the small amount of misalignment between a shaft and a cylindrical bearing is difficult to check even with alignment tools. Some mathematical calculations are required. If space permits, an alignment tool may be properly clamped to the shaft, which will allow a runout reading to be taken on the surface of the outer race.

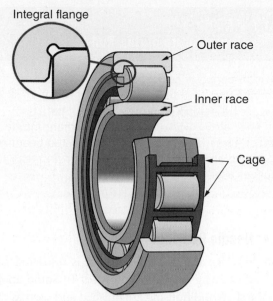

Integral flange

Outer race

Inner race

Cage

Figure 10-11. This caged cylindrical bearing includes integral flanges in the outer race.

10.1.3 Tapered Bearings

Tapered bearings are used for combination loads with both radial and axial loading. In these bearings, rolling elements and races are tapered to a similar angle, which determines how much loading the bearing is designed for. The diameter of the rolling element increases from one end to the other. See **Figure 10-12**.

Tapered bearings support axial loading in only one direction. If axial loads occur in both directions, a pair of tapered bearings (positioned face-to-face or back-to-back) is used.

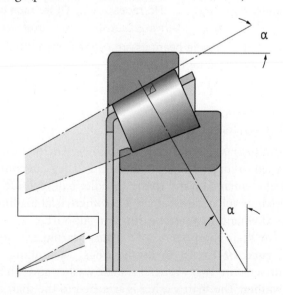

α

α

Figure 10-12. Tapered bearings have tapered rolling elements, as well as tapered inner and outer races.

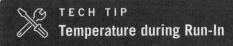

TECH TIP
Temperature during Run-In

A run-in period of several hours with tapered bearings results in higher levels of generated heat. Initial temperature during run-in periods can peak at 50–75% higher than normal running temperatures. The temperature of tapered bearings must be closely monitored during their first ten hours of use.

10.1.4 Needle Bearings

Needle bearings are used where space is restricted and a high load-carrying capacity is needed. In some applications, the shaft and housing are finished and used as races. These bearings come in a range of designs, as shown in **Figure 10-13**. Rolling elements have a small diameter and a high ratio of diameter to length. Self-aligning needle bearings can accommodate a small misalignment between the shaft and housing bore, typically less than one minute. (One minute is 1/60 of one angular degree.)

Maximum capacity   **No races**   **Outer race only**

Aerodim/Shutterstock.com   *Vasyl S/Shutterstock.com*   *Vasyl S/Shutterstock.com*

Figure 10-13. Needle bearings are needed when little space is available for the bearing.

10.1.5 Thrust Bearings

Thrust bearings are a type of rolling-element bearing designed to support high axial loading. They can contain rolling elements of any shape (needle, ball, cylindrical, tapered, or spherical) and are used when axial loading is beyond what a standard bearing can support.

The basic components of a thrust bearing are shown in **Figure 10-14**. In thrust bearings, the rolling elements separate two washers: a shaft washer and a housing washer. The shaft washer is attached to the shaft and transfers the axial load to the rolling elements. The housing washer is attached to the housing and receives the axial load from the rolling elements.

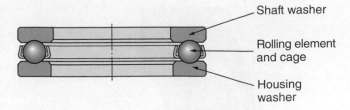

Shaft washer

Rolling element and cage

Housing washer

Image published with the permission of SKF USA Inc.

Figure 10-14. Basic components of a thrust bearing.

Thrust bearings cannot typically handle radial loads, with the exception of spherical thrust bearings. Many times, thrust bearings are used in tandem or multiple mountings, which combine radial and thrust bearings in the same housing in order to support both radial and axial loads.

A variety of thrust bearings are shown in **Figure 10-15**. Cylindrical thrust bearings are designed for thrust loading only. Needle thrust bearings are used where space between machine elements is limited. Spherical thrust bearings can support both radial and axial loading with heavy loads and high speeds. The specific type of thrust bearing used depends on a wide variety of factors that engineers take into account during machine design:

- Amount of loading.
- Types of loads.
- Space available for the bearing.
- Shaft speed.
- Whether or not the housing can be used as a race.
- Shaft shoulders and support for the bearing.
- Misalignment and possible shaft deflection under load.

Most thrust bearings are separable so their components can be installed one at a time. This allows some flexibility during installation, but during maintenance the components must be reassembled in the correct order and orientation. The following characteristics of the washers can help ensure correct reassembly:

- Shaft washers usually have a smaller inside diameter (ID) than housing washers.
- Shaft washers are ground (surface finish) on the inside diameter.
- Housing washers are ground on the outside diameter (OD).
- Surfaces of the shaft washer and housing washer that come in contact with the rolling elements are ground and polished to a higher degree.

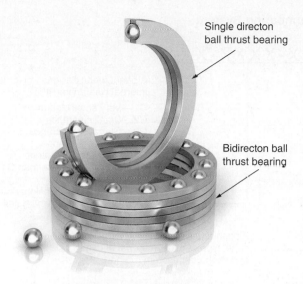

Ball thrust bearing

Single direcion ball thrust bearing

Bidirecton ball thrust bearing

studiovin/Shutterstock.com

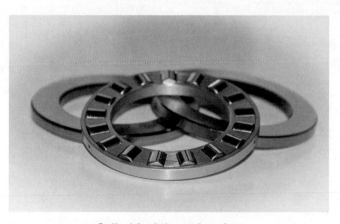

Cylindrical thrust bearing

K Kargona/Shutterstock.com

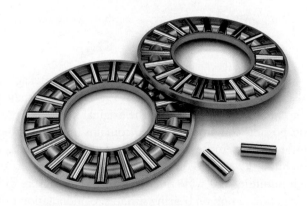

Needle thrust bearing

Ildarss/Shutterstock.com

Figure 10-15. Thrust bearings are manufactured with various rolling elements and are used for various loading scenarios.

10.2 NUMBERING SYSTEMS

The numbering system used to describe rolling-element bearings when comparing across different manufacturers has both common elements and a wide variety of add-ons. Because of this expansion of the numbering system and the wide variety of machines that use bearings from many different manufacturers, understanding the system can be difficult. As such, only common elements of the numbering system and certain unique aspects will be presented.

While some bearing manufacturers stamp or imprint the full bearing number on one side of the outside diameter, others may provide little or no information on the bearing itself. Having a proper manual for a specific machine before rebuilding ensures that you can research the proper bearings to be installed. **Figure 10-16** shows common numbering systems used in part or whole by many bearing manufacturers both inside and outside of the United States. Some bearing manufacturers outside of the United States may use a different numbering system.

The following are some aspects of bearing designations:

- Smaller bore, or inside diameter, sizes of 10, 12, 15, and 17 mm are identified with the bore number of 00, 01, 02, and 03 respectively.

- Some bore diameters are separated from other parts of the full number, but others may not be separated. Some nonstandard or custom manufactured bores are separated from the rest of the bearing designation and include decimal digits.

- The full bearing number is usually supplied on the packaging, but it may be omitted in part or whole on the bearing itself.

- If the bearing can be disassembled into component parts for installation, some manufacturers use markings on the components to aid in reassembling in the proper order and orientation.

TECH TIP

Bearing Library

As you become familiar with the common bearings used in your place of work, develop a library of bearing manufacturer's catalogs or design specifications. Many bearings can be interchanged between manufacturers. Your local supplier can help identify correct replacement bearings.

Common Bearing Numbering System

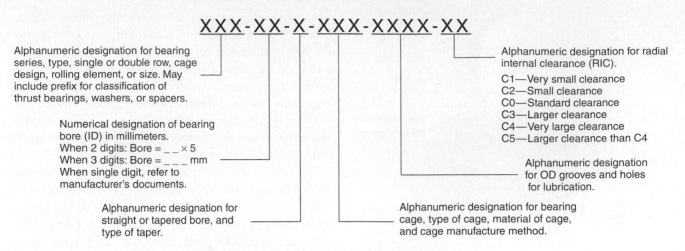

Alternative Bearing Numbering System

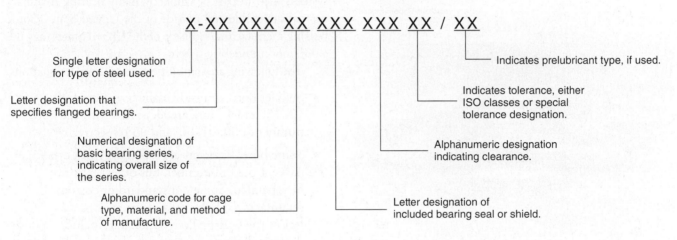

Goodheart-Willcox Publisher

Figure 10-16. The Common Bearing Numbering System detailed is typically used by bearing manufacturers in the United States. The Alternative Bearing Numbering System is used by some bearing manufacturers outside the United States. The entire bearing designation may or may not contain dashes, and the number of digits included for each item can vary.

10.3 MAINTENANCE AND REPAIR

Everything that happens to a bearing from the time it arrives at your site until it is properly installed can affect its life span. Bearings must be handled, stored, installed, removed, and maintained correctly.

10.3.1 Handling and Storage

While some bearings are relatively inexpensive, others are extremely costly. Any bearing, regardless of cost, can be ruined by improper storage. The most expensive bearing is the one that fails prior to ever having been installed.

A bearing should be stored in its original packaging and should not be removed until it is time to be installed or be checked for installation. The original packaging prevents possible contamination and contains the full bearing number for reference and identification. Store bearings in a humidity- and temperature-controlled atmosphere. High humidity and wide fluctuations in temperature can cause condensation on bearing surfaces. Condensation can lead to rust, which will make the bearing unfit for installation. Store bearings away from heat sources, such as welding and grinding equipment. Heat sources can damage the packaging and introduce contaminants.

When removing a bearing from its packaging for pre-installation checks, observe the following best practices:

- Do not clean the bearing unless specifically required by the manufacturer.

- Do not spin the bearing with pressurized air.

- Do not leave the bearing on a workbench where its cleanliness may be compromised.

- Do not lubricate the bearing prior to installation.

- Research appropriate locking compounds before starting installation.

Manufacturers often coat bearings in oil prior to shipping. This oil does not need to be removed unless specified by the manufacturer or required for the process (for example, food processing).

TECH TIP
Removing Cosmoline

Older bearings may have been coated in a hard grease called *cosmoline*, which appears brown and waxy. This type of preservative must be removed prior to use. Cosmoline can be removed by gently heating the bearing (150°F) and letting the hardened coating drip free of the bearing. A parts washer may also be used to remove cosmoline, if the fluid is heated sufficiently. Another method is to submerge the entire bearing in oil for several hours. The oil will penetrate the coating and soften it so it can be removed.

If a bearing needs to be cleaned prior to use, it also needs to be properly dried. Spinning an unloaded bearing to dry it can cause immediate bearing failure. Pressurized air may also contain excess moisture. This moisture will stay on the bearing races and could lead to early failure. Lubricating a bearing prior to heating is not recommended. Sealed bearings are lubricated and may be press-fitted without heat or heated to a low level. A light coating of oil on the shaft and the press-fit bearing race is appropriate.

10.3.2 Preinstallation Checks

Preinstallation checks begin with inspecting the shaft, which includes both a visual examination and measurement. Visually inspect the entire shaft for any wear, deformations, cracks, or corrosion. Visually inspect the

seating area and shaft shoulder of the bearing for wear, burrs, scratches, or general cleanliness issues. A shaft showing signs of any of these issues needs further work before a bearing can be mounted.

Accurate measurement of the shaft diameter where the bearing will seat is critical. Measurements should be taken at several points in this area to verify the diameter of the shaft. Machines that have had multiple rebuilds may have a worn shaft. This may be due to previous removal and installation of bearings multiple times with a press instead of heating. The outside diameter of the shaft must be compared to the bearing's inside diameter (ID) and the bearing's *radial internal clearance (RIC)* to find the final RIC. See **Figure 10-17**. As a bearing is stretched over a shaft through heating or pressing, the RIC of the bearing is reduced by the amount that the shaft diameter exceeds the ID of the bearing. If the bearing is too small or shaft too large, the RIC can be reduced to zero. This reduction will cause the bearing to fail.

If the RIC remains too great after a bearing is installed, the rolling elements may slide instead of roll. Sliding elements increase the amount of friction and heat produced by the bearing and decrease efficiency. One method of

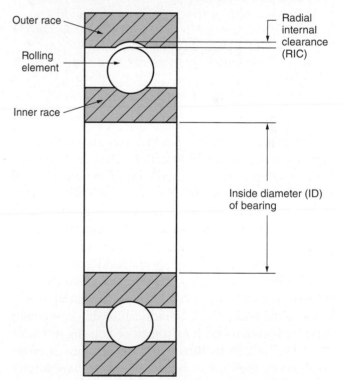

Outer race
Rolling element
Inner race

Radial internal clearance (RIC)

Inside diameter (ID) of bearing

Figure 10-17. When a bearing is press-fit on a shaft, the ID of the bearing increases and the RIC decreases. If the RIC becomes too small or reaches zero, the bearing will not operate properly, reducing the efficiency of the machine and potentially causing damage to the bearing.

reducing excessive RIC is preloading. ***Preloading*** is a practice in which the bearing has a load placed on it by outside means, such as the machine housing or an adjustment nut. As an example, consider the wheel bearings of a car. The wheel bearing (an inner and an outer on each wheel) mounts on the axle shaft and is held in place by a castellated nut. Load on the bearings depends on several factors:

- Fit of the bearings on the shaft.
- Load created by the mass of the vehicle.
- Load imparted by preloading the bearing by excessive tightening of the castellated nut.
- Load imparted by forces during driving.

Preloading, or tightening, the castellated nut decreases RIC.

TECH TIP
Exceeding Load Limits

Exceeding the load limits of a bearing, whether in preloading or operation, will cause short-term failure. Increasing load on a bearing reduces its life span drastically. Bearings operating at their limits of speed and load will have a shorter life than bearings that are oversized with greater load carrying capacity and higher speed ratings than required by the operation.

At this point in the process, you should have all tools on-hand, proper lubrication researched and on-hand, locking compounds on-hand (if to be used), and proper final position and direction of the bearing verified. If installing multiple bearings, verify position prior to heating. Remember that some bearings may take a thrust load in only one direction, and this should be clarified prior to installation. Some locking compounds are used for thread locking, while others are used to lock cylindrical pieces onto shafts. A press-fit bearing that is worn to the point that a cylindrical locking compound must be used to install it on a shaft or housing will most likely fail early in its lifetime. If the machine is excessively worn, the shaft or housing needs to have further corrective action taken prior to installing a bearing.

10.3.3 Installation and Removal

The most common methods of removal use a bearing puller (for smaller bearings), induction heater, or hydraulic press. Care needs to be taken to apply force to only the

race (and not to the rolling elements) if the bearing is to be used again. See **Figure 10-18**. Be certain that the shaft is not damaged in any way while removing a bearing.

Bearings that can be partially dismantled, leaving only the inner race on the shaft, can be heated for removal. An induction heater heats the inner race until it expands enough to be removed from the shaft without heating the shaft. Proper personal protective equipment, including heat resistant gloves, should be used to prevent possible injuries.

Using a hydraulic press to remove bearings must be done with caution. Special care must be taken to ensure the shaft is not damaged when such large forces are present. The bearing usually remains on the press plate while the shaft is pushed through the bearing. Netting or a catch mechanism must be used to prevent the shaft from free-falling and getting damaged. Pressing should be completed by squaring up the press on the shaft, applying a small force initially, and then pressing the shaft through the bearing bore in one consistent movement. If a cylindrical locking fluid was previously used, a large amount of force will be needed to break the bearing free.

TECH TIP
Never Use a Torch to Heat Bearings

Under no circumstances should a torch ever be used to heat a bearing for removal or installation. The heat from the flame can easily damage the shaft or bearing.

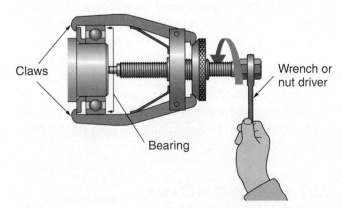

Claws   Wrench or nut driver

Bearing

Image published with the permission of SKF USA Inc.

Figure 10-18. A bearing puller is used to remove small to medium-sized press-fit bearings. In this example, the claws of the bearing puller are placed on the inner race of the bearing.

Removing a tapered bore bearing from a tapered sleeve or tapered shaft may be accomplished by first breaking the shaft sleeve free from the bearing using a puller or press, as previously described. However, be aware that the bearing and sleeve may rapidly loosen. Leaving the drive-up nut loosely on the shaft or shaft sleeve will help to prevent the complete removal and possible damage. Once the bearing or sleeve is broken free, the nut and bearing can be completely removed.

Larger bearings may be removed by oil injection if the shaft is fitted properly. **Figure 10-19** shows hydraulic removal of a tapered bore bearing from a tapered shaft using a hand pump. The shaft is fitted with oil holes specifically for the purpose of mounting and removing a bearing. By applying pressure between the inner race and shaft, the inner bore expands so the bearing can break free.

Installing a press-fit bearing can be accomplished using some of the same tools as removal. Bearings can be heated, pressed on using a hydraulic press, or installed with a handheld hydraulic pump, in the case of tapered bore bearings.

Heating can be used on bearings of all sizes, but it is required on larger bearings. Heating can be accomplished using an induction heater or oven-type heater. An oven-type heater may also use an oil bath, where the bearing is immersed into oil before heating. When heating a bearing, always use a temperature-controlled method. Do not use a torch or any kind of open flame. Controlling an open flame is unreliable, at best, and often results in uneven heating. Additionally, hot spots may develop that can damage the heat treatment of the races, rolling elements, or cages. A nonsealed bearing may be heated to a maximum of 250°F for mounting. An even temperature of 200°F is usually acceptable unless an extreme interference fit is noted. Sealed bearings should not be heated above 175°F.

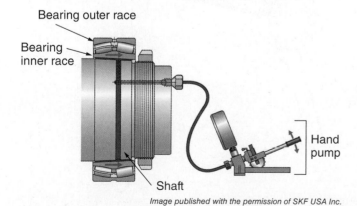

Figure 10-19. Using the oil injection method, pressurized oil is injected into holes in the shaft to expand the bore of the inner race.

Bearing outer race

Bearing inner race

Shaft

Hand pump

TECH TIP
Oven Heating

Just because the oven temperature reads 250°F does not mean that the entire bearing has evenly reached this temperature. Soak time is required to allow the entire bearing to reach the desired temperature.

If an induction heater is used, be sure to read the manufacturer's operating procedure. Some induction heaters can rapidly heat a bearing. Care must be taken not to exceed temperature limitations of the bearing.

When a bearing is properly heated, it should be quickly placed into position on the shaft. It must be forcefully and rapidly placed into position against the shaft shoulder. If too much time passes after heating, the bearing will cool too rapidly and need to be removed from the shaft. Be aware of the following potential complications with heat installation:

- The bearing will cool, contract, and tend to pull away from the shaft shoulder. This fitment must be checked with a feeler gage after the bearing is completely cooled.

- If the bearing is held with a drive-up nut, the nut should be tightened while the bearing cools.

- If an additional Allen screw holds the bearing's inner race to the shaft, do not tighten until the bearing has completely cooled.

If using a press to mount the bearing, cleanliness of the equipment is crucial. Force should never be felt through the rolling elements, as this can dent the race. The driver must be of sufficient diameter to clear the shaft and place the bearing into its proper position. The following are some tips for using a press to mount a bearing:

- Use a light oil to help reduce the force needed.

- Square up the bearing, supporting elements, and shaft or housing.

- Supply a slight pressure and check alignment of all elements.

- Press the bearing in one motion into its final place on the shaft shoulder or housing.

The installation of tapered bearings on a tapered sleeve or shaft requires driving a bearing up the sleeve or shaft until the proper reduction in internal clearance is obtained. With smaller bearings, this can be accomplished by hand. For larger bearings, hydraulics must be used.

PROCEDURE

Manually Mounting a Bearing

The following are general steps to manually mount a bearing. See **Figure 10-20** for some key elements of this procedure:

1. Check manufacturer's tables and recommendations for final clearance values.
2. Measure radial internal clearance using a feeler gauge. The typical reduction in clearance is about 50% of initial unmounted clearance.
3. Put a slight amount of light oil on the shaft and bearing.
4. With a fair amount of force, manually place the bearing on the shaft as far up the taper as possible.
5. Place the locknut and lock washer on the threads.
6. Use a spanner wrench to begin tightening the locknut, while checking the reduction in the radial internal clearance with a feeler gauge until the proper clearance is reached.
7. Bend over one flange of the lock washer to hold the nut in position.

After bearings are installed, proper lubrication can be performed. The standard amount of grease to use depends on the bearing size. A good rule of thumb is to fill the open space in the bearing 1/3 to 1/2 full of grease. Less than this will not provide proper lubrication. More than this amount may cause the bearing to run hot and can result in early failure.

10.3.4 Initial Operation and Run-In

After a rebuild or replacement, a run-in period should be performed where the bearing is lightly loaded and run at less than 100% speed. During this period, the bearings will initially increase in temperature and may peak, but will reduce until an equilibrium temperature is reached. *Equilibrium* is reached when the heat input (from the bearing's friction) and the heat output (transferred into the surrounding air and machine housing) are equal. If a bearing's temperature rises quickly and does not reduce to a lower level, secure the machine and check its lubrication. Oiled bearings tend to run at a lower temperature during both the break-in period and equilibrium period.

Any noise produced by the bearing should be a low, humming sound. Loud, sharp, or repeated noises indicate one of the following problems exists and needs to be solved before continuing the run-in:

- Improper installation (interference fit).
- Bearing not fully fitted to the shaft shoulder.

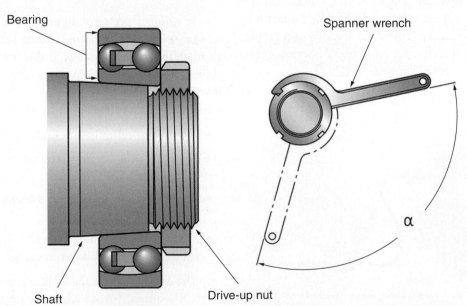

Image published with the permission of SKF USA Inc.

Figure 10-20. A dial indicator must be placed to determine the distance to drive up the shaft or shaft sleeve in order to obtain the proper reduction in internal clearance.

- Bearing and housing bore misalignment.
- Improper lubrication.
- Damaged bearing from improper installation method (dented races for instance).
- Machine shaft not in proper alignment with prime-mover (coupling alignment).

Having a proper run-in period and gradually loading the bearing will help to ensure a long service life of the rebuilt machine.

10.4 SEALS

Seals contain the lubricant within a system and protect lubricant from outside contaminants, **Figure 10-21**. The cleanliness of lubricant is an important factor in the longevity of system components. Two of the basic types of seals are *contact seals* and *non-contact seals*. **Contact seals** are those that come into contact with rolling or sliding surfaces. **Non-contact seals** create gaps or chambers between rotating and stationary components. Non-contact seals are used on large machinery, such as turbines, and not routinely dealt with as a maintenance technician.

While there is a wide range of seals used in industry, you will likely see *radial shaft seals* on a regular basis. Radial shaft seals are a type of contact seal. Factors including construction materials, design, the environment, and contacting materials all play a role in the engineering of industrial seals.

10.4.1 Seal Construction and Materials

The specific construction elements of a seal are determined by its main purpose. **Figure 10-22** shows the standard parts of a radial shaft seal. The main components of a seal include the following:

- **Outside diameter.** Seals against housing and provides rigidity to the seal.
- **Back face.** Provides transverse strength, normally faces out.
- **Lip.** Holds sealing edge against the shaft and is made of an elastomeric or other type of material.

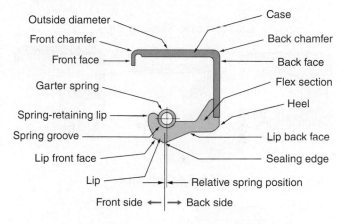

Image published with the permission of SKF USA Inc.

Figure 10-22. Cross-sectional view of a typical radial shaft seal.

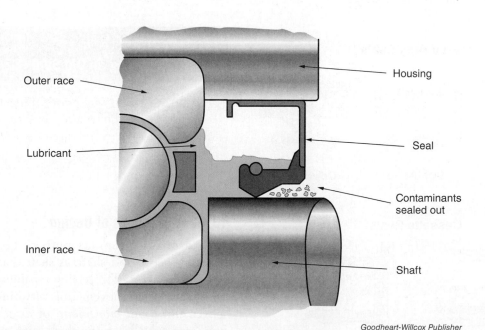

Goodheart-Willcox Publisher

Figure 10-21. Seals protect lubricant from outside contaminants, which also helps extend the service life of components.

- *Sealing edge.* Contacts shaft and provides seal.
- *Garter spring.* Provides additional pressure to keep seal against shaft.

All seals perform the same basic function. However, seals may differ in lip design, single- and double-seal lips, double-opposing or in-line seals, garter spring type and force applied, materials of construction, and type of shaft motion and sealing form (radial, axial, cassette). **Figure 10-23** shows several types of seals with various forms. Some forms and materials are more suited to specific applications, and this should be determined by an engineer with in-depth knowledge of the field.

Elastomeric Sealing Lips

PTFE Sealing Lips

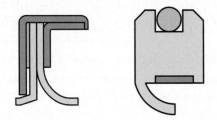

Heavy Duty Seals

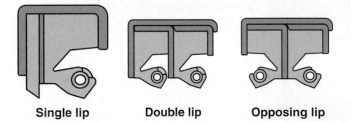

Single lip   Double lip   Opposing lip

Cassette Seals

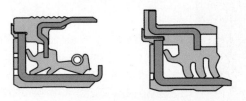

Images published with the permission of SKF USA Inc.

Figure 10-23. Commonly used seals.

Proper sealing requires a minimum amount of friction against the shaft, which depends on lubricant viscosity, shaft runout or deflection under load, shaft speed, lubricant temperature, and type of lubricant. Most sheet steel cases are coated on the outside diameter, at least, to help sealing between the inside housing bore of the machine and the seal. This coating can compensate for slight imperfections in the housing.

Seal lips are manufactured in a wide range of materials for various applications including nitrile rubber, acrylonitrile-butadiene rubber (ABR), silicone rubber, polyacrylate elastomer, and polytetrafluoroethylene (PTFE). Not all materials are resistant to all type of fluids or temperatures. Additionally, if process fluids or temperatures change significantly from the original design, sealing problems may occur. Consult a seal application engineer as well as the seal manufacturer's information for more detailed information.

When installing a new seal, the housing bore and shaft should be as clean as possible. If visible imperfections in the housing bore are apparent, a coating of another sealant should be applied. Wrap the shaft with a thin plastic sheet that has been lightly lubricated and slip the seal over the plastic and into position. This will prevent damaging the seal lip on a shaft key-way. Garter springs are typically made from steel wire. If corrosion is a concern, some seals are offered with stainless steel springs. The new seal can be gently pressed or tapped into place with a wooden block and dead-blow mallet, alternating taps back and forth across the back face.

TECH TIP
Removing an Old Seal

If removing an old seal proves difficult, drill a small hole or holes in the face of the seal with great care. Self-tapping screws will allow the seal to be levered out of the housing bore.

10.4.2 Limitations of Design

A shaft seal performs its purpose because of contact between the lip and shaft. A shaft that is too polished or too rough will affect sealing capability and the service life of a seal. Axial movement is very different from radial movement. *Axial movement*, or *axial runout*, refers to movement parallel to the shaft, such as a hydraulic rod moving in and out of a cylinder. *Radial movement*, or *radial runout*, refers to shaft deflection or runout that is perpendicular to the shaft and housing. All seals have their limitations and most cannot compensate for radical deflections in both planes.

Temperature limitations depend greatly on lip seal material. Most materials provide acceptable performance between −40°F and 210°F. Some materials have a wider range of acceptable temperature applications, but the manufacturer's recommendations should always be followed. In terms of wear resistance and long life, PTFE outperforms most other seal materials with proper application. PTFE also has a wider acceptable operating temperature range of −90°F to 390°F.

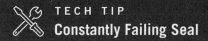

TECH TIP
Constantly Failing Seal

If a seal is constantly failing, it may not be an acceptable type for that application or is being operated beyond its design limitations. Contacting your local seal representative is the first step to finding a proper seal for the application.

Store seals in their original packaging in a temperature-controlled environment, away from direct heat and sunlight. Since some materials are natural, like rubber, seals should not be stored near ozone-producing equipment, such as brushed motors. Seals should be stored in a room with no fumes or vapors that could harm the sealing materials.

10.5 LUBRICATION

Lubrication using grease or oil serves the following purposes:

- Reduce friction and, therefore, the heat generated.
- Transfer heat.
- Prevent metal-to-metal contact.
- Remove particulates from lubricated area.
- Reduce or prevent corrosion.

Lubricants are made from a base oil, which can be organic or synthetic. Organic oils and greases begin with the refining of crude oil or plant-based oil, while synthetic oils are made at chemical plants. Each type has its unique advantages and disadvantages. Organic oils tend to be less expensive than synthetic oils, but they tend to have a shorter life and, depending on the additives, can be more easily affected by contaminants. Plant-based oils tend to be used in food-production facilities where product contamination is of greatest concern. A wide

array of additives are added to the base oil to provide more specific advantages, such as antifoaming, emulsibility, corrosion resistance, and other application-specific advantages.

10.5.1 Theory of Lubrication

Even when a surface appears smooth, it has minute asperities. *Asperities* are the small imperfections left on the surface of a material after machining. If the asperities on opposing surfaces contact each other, friction and heat develop. The most important purpose of lubricant is to inhibit metal-to-metal contact, which prevents friction and heat from developing. Some friction and heat development is expected during a run-in period. During this period, however, large asperities are smoothed and heat should lessen as the run-in continues.

Lubrication helps to reduce friction by forming a film that separates opposing surfaces. This separation can be created in different ways:

- ***Hydrostatic lubrication.*** Pressurized lubricant is injected between the two surfaces to prevent contact.

- ***Hydrodynamic lubrication.*** The lubricant film is maintained by motion of the surfaces themselves. The film develops as speed increases between surfaces, however no film is present at slow speeds or rest.

- ***Elastohydrodynamic lubrication.*** Opposing surfaces under very high loads can elastically deform temporarily, which causes increased lubricant pressure and viscosity in a localized area.

- ***Boundary film lubrication.*** Surfaces only separated by thin film of lubricant on asperities. Viscosity of film increases as asperities contact each other to prevent most metal-to-metal contact. Also called *thin film lubrication.*

- ***Mixed film lubrication.*** Combination of characteristics between elastohydrodynamic and boundary film lubrication. Distances between surfaces are smaller and some contact may occur.

10.5.2 Oils

The majority of oils are classified by their viscosity. *Viscosity* is a fluid's resistance to flow. It is related to the thickness of a fluid, although some thick fluids can flow freely. Viscosity is primarily measured in centipoise (cP) or centistokes (cSt) units, depending on whether dynamic or kinematic velocity is referenced. Saybolt Universal Seconds (SUS) is also used, as well as an ISO designation. In addition, the Society of Automotive Engineers (SAE) has a range of grades for both crankcase and gear oils. These viscosities are compared in **Figure 10-24**.

Each type of oil is intended and designed for a different purpose and with different additives. For example, oils with an ISO designation of 150, a SAE crankcase of 40, and SAE gear oil of 85W may appear with similar viscosities, but they are not interchangeable. The crankcase oil has a dual viscosity (10W-40) and contains a polymer to achieve this dual viscosity based on temperature. The gear oil may also have a dual classification, but polymer is not typically used. The ISO oil may be hydraulic oil with completely different additives and properties than the crankcase oil.

The final oil product that you see and use is a mixture of several components. These may include all or some of the following:

- A base oil.
- Wax or paraffin.
- Polymers to help control viscosity at high temperatures.
- Pour-point depressant to help control the solidification of wax.
- Viscosity-index improver to help maintain viscosity changes throughout temperature change.
- Additives to improve special properties.

Viscosity Designations			
ISO VG	**AGMA Grade**	**SAE Crankcase**	**SAE Gear**
1500	8A	60	250
1000	8	50	140
680	7	40	90
460	6	30	85W
320	5	20	80W
220	4	15W	75W
150	3	10W	
100	2	5W, 0W	
68	1		
46			
32			
22			
15			
10			

Goodheart-Willcox Publisher

Figure 10-24. A comparison of various viscosity numbering methods.

Oil additives make up a small percentage of the lubricant but play an important role in increasing oil life, improving lubrication properties, and reducing corrosion. The following are some of the most common additives used in industry:

- ***Antioxidants.*** Minimize thickening of lubricants when oxidation takes place.
- ***Dispersants.*** Prevent sludge from accumulating and clumping together.
- ***Detergents.*** Prevent contaminants from accumulating on surfaces and forming varnishes.
- ***Extreme pressure additives.*** Form protective thin boundary film for high-load equipment with low speeds.
- ***Anti-wear additives.*** Help reduce metal-to-metal contact with high load.
- ***Corrosion inhibitors.*** Prevent corrosion to metal surfaces when water or high humidity is present.
- ***Emulsifiers.*** Encourage the mixing and suspension of one liquid in another, such as water and oil.
- ***Demulsifiers.*** Separate emulsions; removed later by purification methods.
- ***Foam inhibitors.*** Reduce foaming in the lubricant sump, also helps trapped air escape from oil and settle out.
- ***Deactivators.*** Help protect non-ferrous metals from oxidation.

The specific additives used depend on the application. Some additives are used in both oil and greases.

THINKING GREEN
Re-Refining Used Oil

When lubricating oil is spent, typically the additives have broken down or the oil has become contaminated. The re-refining process removes the additives and contaminants from used oil and restores the chemical composition of the base oil. The base oil can then be used again to produce new lubricant products.

10.5.3 Greases

Greases contain the same base oils as oil lubricants and can be vegetable-based, mineral-based, or synthetic-based. ***Greases*** are soft or semisolid lubricants that have a larger amount of soap than oils have. The soap thickens the lubricant. Sodium, lithium, and calcium soaps are most commonly used. Grease initially has a high viscosity, but on application and pressure it tends to have a similar viscosity to the base oil from which it is made. The benefit of grease is that it tends to stay where it is placed. It is used in conditions where oil would not stay in position and, therefore, would not provide the necessary lubrication.

The National Lubricating Grease Institute (NLGI) developed standards and specifications for greases. The NLGI grade is the best-known standard. The consistency number is assigned based on the depth of penetration, measured in tenths of a millimeter, of a cone into a lubricating grease measured at 77°F (25°C). **Figure 10-25** summarizes the NLGI consistency numbers. Most greases fall into the range of 1–4 on the NLGI scale. These greases are typically used in rolling-element bearings. Use of grade 5 and grade 6 greases is rare, but may involve lubricating large, flat surfaces exposed to weather.

Purely synthetic greases are more expensive, but have some advantages over organic-based greases. Greases produced from a base oil of mineral oil may not be compatible with some seals, O-rings, and other chemicals, and the manufacturer should be consulted. Additives for extreme pressure include molybdenum disulfide (moly) and graphite.

10.5.4 Handling and Storage of Lubricants

Two primary concerns are considered when handling and storing lubricants. The first concern is safety. Most lubricants are hazardous and flammable materials, so preventing accidental leaks and avoiding ignition sources are critical. The second concern is ensuring that the lubricant is not degraded.

NLGI Grades for Lubricating Greases		
NGLI Consistency Number	Worked Penetration (mm/10)	Typical Appearance
000	445–475	Fluid
00	400–430	Semifluid
0	355–385	Very soft
1	310–340	Soft
2	265–295	Normal consistency
3	220–250	Firm
4	175–205	Very firm
5	130–160	Hard
6	85–115	Very hard

Adapted from National Lubricating Grease Institute

Figure 10-25. NLGI grade scale with typical grease appearance description.

Store lubricants in a clean, well-organized storage area. Arrange containers on shelves in the proper position. Be certain the area is free of ignition sources, and minimize the amount of flammable materials stored with the lubricants.

SAFETY NOTE
Lubricant Storage

Safety data sheets (SDSs) for lubricants and a fire extinguisher rated for combustible liquid (class B) fires should be placed in highly visible locations in the lubricant storage area.

If a lubricant container is stored outdoors, do not place the container directly in contact with the ground. Cover the container to protect it from the dirt and moisture in the air. Be sure that no welding occurs near lubricant containers.

Transport lubricants carefully to ensure no damage occurs to the container. If a lubricant spill occurs, first make sure there is no source of ignition in the area. Once the area is determined to be safe, the lubricant spill can be contained and cleaned up.

10.5.5 Lubrication Plan

A critical tool for industrial maintenance is a lubrication plan. A *lubrication plan* or *lubrication schedule* is a record of the lubrication requirements for all machines and equipment. A lubrication plan may include the following information for each item requiring lubrication:

- Frequency of lubrication.
- Type of lubricant.
- Method of lubrication.
- Person responsible for lubrication.
- Additional notes.

The lubrication plan may also include a historical record of completed lubrications and specific items of note, such as the presence of contaminants in a lubricant or an unexpected low lubricant level.

The user's manual for a machine or a piece of equipment is the best source of information regarding lubrication. Use the manual to find the frequency of lubrications, locations of lubrication, and the specification for the type of lubricant.

10.5.6 Lubrication Methods

The lubrication plan includes details of every item that requires lubrication. Several methods of lubrication are used, depending on the lubricant, required frequency of lubrication, access to the component, manufacturer's specifications, and other factors.

Some components are lubricated by automatic lubrication systems. An automatic lubrication system generally includes the following components:

- **Reservoir.** A reservoir contains the grease or oil to be used to lubricate parts.

- **Pump.** A pump moves the lubricant from the reservoir to the parts.

- **Feed lines.** The lubricant travels through this piping or tubing from the pump to the parts.

- **Metering device.** The metering device controls the amount of lubricant that is fed to a part. A system may include one metering device or multiple metering devices.

- **Injector.** The injector is located at the end of the feed line and dispenses the lubricant into the part.

When maintaining an automatic lubrication system, be sure the reservoir is loaded with a sufficient amount of lubricant. When adding lubricant, use care to ensure that no contaminants enter the reservoir. Inspect the system, watching for leaks. Monitor the parts being lubricated for any signs of overlubrication or underlubrication, such as unusual noise, heat, or vibration.

Manual lubrication methods are also common. Many parts include a grease fitting (also called a Zerk fitting). Grease fittings allow a component to be greased without removing the part from service. These fittings are normally threaded into the part and contain a port sealed by an internal force. When a grease gun is attached to the grease fitting, pressure from the grease gun overcomes the seal and injects grease into the part.

Some components are lubricated by a grease cup or oil cup. A grease cup is threaded to the component and continuously feeds lubricant into the grease chamber. Grease cups normally have a spring-pressured plate

PROCEDURE	Lubricating a Bearing

Before lubricating any item, make sure the grease gun contains the correct lubricant for the application. Always follow the manufacturer's procedure.

1. Clean the grease fitting and grease gun nozzle to ensure no contaminants are introduced to the bearing.
2. Attach the grease gun to the grease fitting.
3. If the bearing has a drain plug or vent plug, remove it and place a pan or container beneath it.
4. Add the proper amount of grease using the grease gun. Do not overfill the bearing.
5. Remove the grease gun from the grease fitting and clean both.
6. Clean drain plug or vent plug area and install the plug.

pushing the grease into the part. Oil cups may be gravity fed, and may include a wick to control the flow rate.

Grease cup systems and oil cup systems must be checked regularly to ensure the cup has a sufficient amount of lubricant. These systems must also be checked for leaks and monitored for proper operation.

CHAPTER WRAP-UP

The combination of bearings, seals, and lubricants all work to reduce friction in sliding and rotating machinery. By reducing friction, and therefore heat generated, machine efficiency is increased. Regardless of where a machine is located, contamination of the lubricant is always a factor in bearing life. Anything that comes into contact with a rolling-element bearing has an impact on its service life. Improper mounting, lubricating, and sealing can result in a very short bearing life. Always refer to manufacturer's information to determine the best fit, grease, and sealing for your application.

Chapter Review

SUMMARY

- Two main types of bearings are rolling-element bearings and plain bearings. Technicians are more likely to encounter and perform maintenance on rolling-element bearings.

- Bearings reduce both friction and restrict machine motion to one or more planes.

- The two basic categories of rolling-element bearings are ball bearings and roller bearings. The function and loading determine the best bearing type for an application.

- Bearings can be differentiated and classified by the shape of their rolling elements, such as ball, cylindrical, spherical, tapered, and needle.

- Loading of a bearing can be axial (also considered thrust), radial, or a combination of the two.

- Ball bearings come in many styles and are classified by how they are assembled, the number of rows of rolling elements, and loading capability.

- Cylindrical bearings are mainly designed for heavy radial loading, but can handle restricted axial loading.

- Tapered bearings are used for combination loads with both radial and axial loading.

- Needle bearings are used where space is restricted, but a high load-carrying capacity is needed.

- Thrust bearings can contain rolling elements of any shape and are used when axial loading is beyond what a standard bearing can support.

- A numbering system is used to describe rolling-element bearings and may include designations for outside diameter, radial internal clearance, bore dimensions, and bearing series. Bearing numbering systems may differ among manufacturers.

- Proper storage, prechecks, installation methods, and removal of bearings will help to ensure the life span and expected operation of bearings. Be sure to have proper documentation on hand and use the appropriate personal protective equipment.

- Seals contain the lubricant within a system and protect lubricant from outside contaminants. There is a wide range of industrial seals with various properties that all perform these basic functions. Two of the basic types of seals are contact seals and non-contact seals.

- All seals have their limitations and most cannot compensate for radical deflections in both planes.

- Lubrication using grease or oil reduces heat and friction, prevents metal-to-metal contact, and can reduce or prevent corrosion. Lubricants are made from either an organic or synthetic base oil.

- Opposing surfaces may be separated through hydrostatic lubrication, hydrodynamic lubrication, elastohydrodynamic lubrication, boundary film lubrication, or mixed film lubrication. The type of lubrication method the system develops depends on the viscosity of the lubricant, speed of the contacting surfaces, load, and coefficient of friction.

- Oils are classified by their viscosity, or resistance to flow. The Society of Automotive Engineers (SAE) has a range of grades for both crankcase and gear oils.

- Each type of oil is intended and designed for a different purpose and with different additives. Common additives include antioxidants, dispersants, detergents, corrosion inhibitors, emulsifiers, and foam inhibitors.

- Greases are soft or semisolid lubricants that have a larger amount of soap than oils. The benefit to grease is that it tends to stay where it is placed.

- The National Lubricating Grease Institute developed a grading scale for greases that assigns a consistency number from 000 to 6. Most greases fall into the range of 1–4 on the NLGI scale.

- The two primary concerns considered when handling and storing lubricants are safety and ensuring the lubricant does not degrade.

- A lubrication plan is a record of the lubrication requirements for all machines and equipment. It is a critical tool for industrial maintenance.

REVIEW QUESTIONS

Answer the following questions using the information provided in this chapter.

1. Identify the two major functions of a bearing.

2. Describe the two main types of bearings.

3. List the three basic elements of a bearing and describe the function of each.

4. What are the common shapes of rolling elements?

5. Describe the types of loading applied by the shaft.

6. The load applied by the shaft determines the type of _____ required.

7. *True or False?* A ball bearing is used to reduce friction and support light loads.

8. What are the five common construction designs of ball bearings?

9. What is the allowable axial load for cylindrical bearings?

10. What is the difference between shaft washers and housing washers for a thrust bearing?

11. _____ thrust bearings can handle radial loads.

12. What are the measurements that are considered small bore sizes? How are bearings with these sizes identified using a bearing designation system?

13. Describe the proper procedures for bearing storage.

14. *True or False?* A bearing should be stored in its original packaging and should *not* be removed until it is time to be installed or checked for installation.

15. Describe the two of the preinstallation checks to be performed on a shaft and bearing.

16. As a bearing is stretched over a shaft through heating or pressing, what is the result?

17. What is preloading?

18. List the most common methods used for bearing removal.

19. A nonsealed bearing may be heated to a maximum of _____ for mounting.

20. Identify some tips for using a press to mount a bearing.

21. Identify and compare the two basic types of seals.

22. List the main components of a seal.

23. What does proper sealing require?

24. *True or False?* Axial movement, or axial runout, refers to the shaft deflection or runout that is perpendicular to the shaft and housing.

25. Describe the purposes of lubrication.

26. _____ are small imperfections left on the surface of a material after machining.

27. What is the difference between hydrostatic lubrication and hydrodynamic lubrication?

28. *True or False?* Viscosity of film increases as asperities contact each other to prevent most metal-to-metal contact.

29. What is the purpose of an additive? List examples of common additives.

30. How do greases differ from oils?

31. What is the NGLI number?

32. What is a lubrication plan?

NIMS CREDENTIALING PREPARATION QUESTIONS

The following questions will help you prepare for the NIMS Industrial Technology Maintenance Level 1 Basic Mechanical Systems credentialing exam.

1. Which of the following is *not* a basic bearing element?

   A. Cage
   B. Inner and outer race
   C. Rolling element
   D. Sleeve

2. _____ bearings are a type of rolling-element bearing used when the shaft is loaded in a radial direction with little axial loading.

   A. Ball
   B. Needle
   C. Tapered
   D. Thrust

3. In the following thrust bearing illustration, which component is responsible for transferring the axial load to the rolling elements?

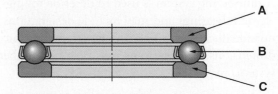

   A. Component A
   B. Component B
   C. Component C
   D. Component is not shown.

4. Why must the shaft diameter where a bearing is seated be measured accurately?

   A. Heating or pressing at this location can affect the bearings.
   B. Preloading may alter the diameter of the shaft.
   C. Radial internal clearance is affected if bearings are too small for the shaft.
   D. Shaft diameter will determine what lubricant is used.

5. Sealed bearings should not be heated above what temperature?

   A. 150°F
   B. 175°F
   C. 200°F
   D. 250°F

6. A seal operates based on its contact between what two components?

   A. Lip and garter spring
   B. Lip and shaft
   C. Shaft and back face
   D. Shaft and sealing edge

7. Which is *not* the purpose of an additive?

   A. To increase oil life.
   B. To improve lubrication properties.
   C. To reduce corrosion.
   D. To thicken lubricant by adding more soap.

8. The consistency number of grease, as depicted in the NLGI Grades for Lubricating Greases, is based on the _____ of a lubricating grease.

   A. additive
   B. depth of penetration
   C. flammability
   D. viscosity

9. Which of the following is a proper method for storing lubricants?

   A. Allow a lubricant to sit uncovered in a clean storage area prior to use.
   B. Place the container directly in contact with the ground when stored outdoors.
   C. Store the lubricant in a cool, dry area and away from welding equipment.
   D. Transfer lubricant into an uncontaminated, clean container prior to storage.

10. Which is a critical tool for maintenance that allows technicians to keep records of lubrication requirements for their machines and equipment?

   A. A lubrication plan
   B. A lubricant storage facility
   C. A NLGI grade chart
   D. A safety data sheet (SDS)

11

Shafts, Couplings, and Alignment

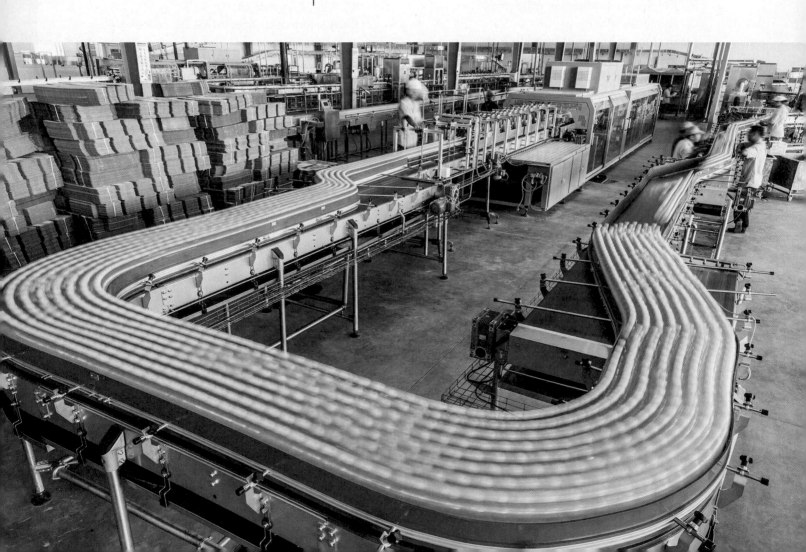

LEARNING OBJECTIVES

After completing this chapter, you will be able to:

☐ Describe how pillow blocks and flange bearings support shafts.

☐ Explain the role of universal joints in transmitting mechanical force.

☐ Discuss the function and operation of flexible couplings.

☐ Identify different types of flexible couplings.

☐ List the tasks involved in a prealignment check.

☐ Identify common causes of excessive shaft runout.

☐ Explain how bar sag impacts alignment measurements.

☐ Describe the use of common shaft alignment tools.

☐ Explain the use of shims in an initial rough alignment.

☐ Discuss the steps to perform a final alignment using the rim and face method.

TECHNICAL TERMS

balanced dial indicator	shaft runout
bar sag	soft foot
continuous dial indicator	spider
coupling	tapered adapter sleeve
dial indicator	total indicator reading (TIR)
eccentric locking collar	trunnion
end-float	universal joint
flange bearing	wiggler bar
pillow block	

The transmission of mechanical power has taken place since the first grain was ground into flour using a stone mill. The basic concepts of power transmission have been in use for thousands of years, but new design and engineering methods routinely refine these concepts. While the control of energy advances quickly, there will always be a need for a technician that has sound mechanical reasoning and hands-on skills. Mechanical assemblies may seem simple at first glance, but they require a high level of technical skill and knowledge to be properly maintained. This chapter examines some basic mechanical concepts and highly technical skills that every mechatronics technician should possess.

11.1 SHAFTS

Industrial shafts perform the function of transferring power. Power, torque, and movement are transferred from one machine to another, from human to machine, or from machine to its end use. From a simple valve stem that connects an operator to a valve from a distance, to the largest hollow driveshaft that moves a ship through the water, shafts transmit power.

11.1.1 Materials and Manufacturing

Shafts are available in a wide variety of materials, including plastics, ceramics, steels, and various alloys. In this context, we are most concerned with steel shafts used for mechanical power transmission and within machinery.

Steel shafts are made from various steels and finishes. Stainless steel is typically used where corrosion and resistance to chemical attack are needed, such as in a chemical pump. Chromium-vanadium steels are used where abrasion resistance is important, as in hydraulic cylinder rods.

While many internal machine shafts are solid, external power transmission shafts are often hollow. This reduces both the weight of the assembly and momentum.

Controlling the final diameter is perhaps the most important aspect of shaft manufacturing, which directly affects bearing life. Checking for shaft runout is the easiest way to find a bent or bowed shaft. ***Shaft runout*** is the amount a shaft deviates from centerline rotation, as shown in **Figure 11-1**. Runout is measured with a dial indicator placed on the outside diameter, while slowly rotating the shaft. If a seal is failing often, a bent shaft may be influencing its time-to-failure.

Shaft tolerance, or allowable variance, depends on the type of load, type of bearing, shaft diameter, and amount of axial load. Shaft tolerances, and the resulting interference or press fits, are usually tighter with increasing load or when shock loading is expected. Higher running accuracy with less shaft deviation and movement require a tighter fit.

Certain operations may require the shaft to be connected to another machine component, such as a coupling, pulley, or gear, for the purpose of transmitting torque or rotational force. This is accomplished through the use of a key, as shown in **Figure 11-2**. The length of the keyseat generally depends on the component being affixed to the shaft, but the width and depth dimensions of shaft keys are standardized. The most commonly seen types of industrial keys are square and rectangle.

The shoulder of a shaft is the section where the shaft changes diameter. The smaller the radius of the shoulder, the greater the concentration of stress at this point on the shaft. The radius of the shaft shoulder should be kept as large as possible, but must be small enough to allow proper fitment and support of the machine component that will be resting on the shoulder. **Figure 11-3** shows two examples of a shaft shoulder. With the proper shoulder radius, the bearing fully seats against the shaft and can properly transmit force. If the shoulder radius is too large or too small, the bearing or other machine component cannot fully seat on the shaft shoulder.

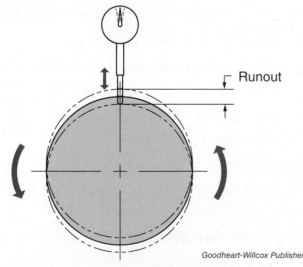
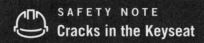

Figure 11-1. Runout is measured by placing a dial indicator on the outside diameter and then slowly rotating the shaft for one complete revolution. Shaft runout can impact the functionality and longevity of the bearings and seals.

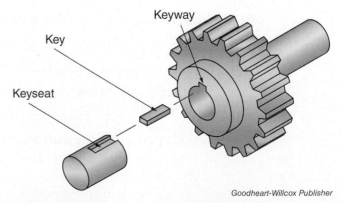

Goodheart-Willcox Publisher

Figure 11-2. A key can be used to fasten a shaft to a coupling or other mechanical component.

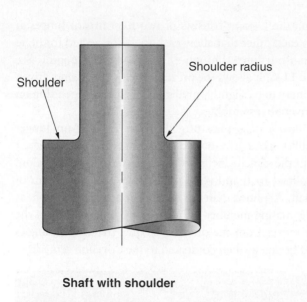

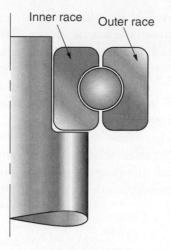

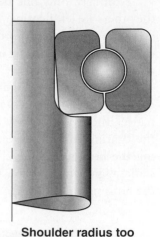

Shaft with shoulder

Bearing fits flush with small shoulder radius

Shoulder radius too large—bearing cannot sit flush on shaft

Goodheart-Willcox Publisher

Figure 11-3. If shoulder radius is too large, the bearing cannot seat properly.

11.1.2 Shaft Support

Shaft support within a machine is normally provided through the use of radial bearings. Power transmission shafts require support and connections to the driver, or prime mover, and load. This is accomplished using a range of pillow blocks, flange bearings, and universal joints. Pillow blocks and flange bearings provide support, and universal joints allow the transfer of power without exact alignment.

Pillow Blocks and Flange Bearings

A ***pillow block*** is a housing or bracket that contains a bearing and is used to support a rotating shaft, **Figure 11-4**. Pillow blocks mount to a surface parallel to the shaft supported and can be made of steel, iron, or aluminum. They are available in a one-piece design and a split-construction design. On a split-construction pillow block, the top portion of the housing is a separate, removable piece. This provides easy access to the bearing elements inside and allows for disassembly and maintenance.

The bearings used in a pillow block may be any general bearing type, but ball, cylindrical, and spherical are common. The bearing type used is determined by loading and the need for alignment. When spherical roller bearings are used, the bearings are self-aligning and provide *dynamic alignment*. When ball and roller bearings are used, *static alignment* is provided. In this arrangement, a curved inner housing or curved outer bearing race allows the bearing to slightly displace angularly within the housing.

The bearing-to-shaft fit is usually tight (slip-on) in the case of one-piece pillow blocks, and may be a tapered fit in the case of a split housing. Bearing-to-shaft fit can

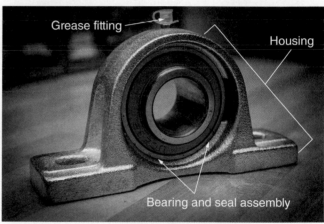

Bigdoug2005/Shutterstock.com

Figure 11-4. This one-piece pillow block includes a grease fitting at the top.

be accomplished through setscrews, an eccentric locking collar, or tapered adapter sleeve. An ***eccentric locking collar*** is used with bearings that have a machined, off-center shoulder to create a squeeze fit on the shaft. A ***tapered adapter sleeve*** is used in conjunction with a locknut and washer to lock a bearing with tapered bore onto the shaft.

Setscrews, or *grub screws*, fix the position of the inner race of a bearing against the shaft. In some instances, however, the outer race must be fixed against the housing. Longer shafts and shafts that undergo wide fluctuations in temperature will expand and contract. The outer race of one-piece housing units can be fixed with a retaining ring or snap ring fitted into the housing. Split housing units use a stabilizing ring that has an open design similar to a "C" ring to allow ease of fitting.

Normally, only one bearing on a shaft is fixed and the others are left floating. Many times, the bearing closest to the input force (motor) is fixed. The exact arrangement depends on bearing loads, possible expansion, and coupling arrangement.

Flange bearings, **Figure 11-5**, mount outside of the machine housing, which allows easy access for maintenance. These bearings use a cartridge system that allows the bearing to be removed and replaced while retaining the housing. Using a spherical roller bearing with static alignment allows a slight misalignment between the housing and bearing.

Two-bolt flanged bearing

rob3rt82/Shutterstock.com

Figure 11-5. The flange on the bearing housing helps with mounting and accurate positioning of the bearing. The bolts or fasteners must be properly torqued to prevent movement of the housing.

TECH TIP
Relubricating Bearings

Some bearings can be relubricated and have grease fittings and internal passages to disperse grease into the bearing. Always wipe the grease fitting clean before adding grease. If the grease fitting comes with a protective cover, be sure it is put back into place after relubricating.

Universal Joints

A *universal joint*, also called a *U-joint* or *Cardan joint*, is used to transmit mechanical force between two shafts, while allowing them to move out of alignment. The U-joint is commonly used in industrial power transmission. In general, the U-joint consists of two fulcrums or hinges at 90° to each other that allow each shaft connected to rotate and transmit force through a main body of the joint. See **Figure 11-6**. It allows the prime mover (motor) and driven machinery to be slightly misaligned or offset, depending on the drivetrain assembly.

Pillow blocks may provide shaft support on longer assemblies. On a driveshaft, the yokes are oriented with their arms in the same plane, which prevents excessive vibration. The splined shaft and yoke allow some axial movement of the shaft. A splined shaft has axial teeth machined into it, as does its mating member. When using U-joints to offset the prime mover from the driven machine, the U-joint angles should be equal when compared to the driveline centerline.

U-joint

Roman Korotkov/Shutterstock.com

Figure 11-6. A driveshaft includes U-joints at each end, and sometimes in the middle, to allow changes in the angle of the driveshaft when compensating for movement of connected components, such as a differential.

The U-joint itself, shown in **Figure 11-7**, comprises four needle bearings positioned on machined ends, called *trunnions*, of the journal cross. The bearing cap assembly is the outer race of the bearing and the inner race is the trunnion.

SAFETY NOTE
Drivetrain Repair

Follow lockout/tagout procedures prior to working on any rotating equipment. Support the driveshaft with proper slings or straps prior to unbolting yokes. Mark all driveline components to ensure they are reassembled in the same orientation to prevent vibration. If components are replaced, the entire drive assembly needs to be rebalanced.

Drivetrain Inspection

When inspecting a drivetrain, perform the following:

1. Check for any obvious damage and replace components, as needed.

2. Verify free movement of bearings when the shaft is turned by hand.

3. Pull on the shaft to check for bearing end play; there should be nearly no movement.

4. Identify any loose fasteners and tighten as needed.

5. Grease all components, including the yoke, U-joints, spline, and pillow block bearings.

6. Check for missing balancing weights.

Flexible couplings are available in a variety of configurations and materials. They may be manufactured using metal, plastic, rubber, neoprene, or polyurethane, among others. Any flexible coupling that employs a pliable material that can stretch or flex, such as rubber, is called an *elastomeric coupling*. See **Figure 11-8**. Elastomeric couplings allow misalignment like other flexible couplings, and also reduce vibration to the attached machinery. Common configuration types of flexible couplings include sleeve, gear, grid, and jaw couplings.

Photo courtesy TB Wood's—Altra industrial Motion

Figure 11-8. The elastomeric element in these couplings allow greater misalignment than other flexible couplings.

Needle bearings

Trunnion

Grease fitting

Bearing cap

Snap rings

Gunpreet/Shutterstock.com

Figure 11-7. A universal joint functions as a bearing mounted between two yokes.

11.2 FLEXIBLE COUPLINGS

Couplings are devices used to connect two shafts and allow the transfer power from one shaft to another. They are available in two basic forms: rigid couplings and flexible couplings. *Rigid couplings* form a sleeve around the connected shafts and provide precise alignment between the shafts. *Flexible couplings* create a connection between two shafts while allowing for some misalignment between the shafts.

TECH TIP
Flexible Couplings and Misalignment

Even though a certain amount of misalignment is tolerated, flexible couplings still transfer vibration due to misalignment. Parallel misalignment causes the driven machine shaft to bind and buck on each shaft revolution. Slight angular misalignment may not cause the coupling to fail, but will reduce the life span of the machine's bearing.

11.2.1 Sleeve Couplings

A sleeve coupling incorporates multiway flexing ability to compensate for various types of shock and misalignment. This type of coupling can be as simple as a flexible sleeve clamped to the ends of two connected shafts.

A common design of flexible sleeve couplings is comprised of two hubs and a flexible sleeve. In this arrangement, shown in **Figure 11-9**, torque is transferred from one hub into the sleeve, and then to the other hub. Each hub is anchored to the shaft using setscrews.

Hubs

Neoprene sleeve

Photo courtesy TB Wood's—Altra industrial Motion

Figure 11-9. On this flexible sleeve coupling, the neoprene sleeve is seated between two synthetic rubber hubs.

11.2.2 Gear Couplings

A gear coupling, **Figure 11-10**, transmits force through one or more gear meshes keyed to the attached shafts. One shaft is keyed to the inner element, which transfers torque to the adjacent flange or hub. Both flanges or hubs are fastened together, which allows the transfer of torque to the opposite shaft. Alignment affects how load is transferred through the gears. With very little misalignment, all gear teeth mesh and share loading. With greater misalignment, fewer gear teeth share loading and excessive wear can develop. Gear couplings require lubrication at regular intervals.

11.2.3 Grid Couplings

A grid coupling, shown in **Figure 11-11**, transfers power through a steel grid element that acts somewhat like a spring. The grid can accommodate various types of misalignment, including angular, parallel, and **end-float**, or axial displacement. The hubs are machined with slots into which the grid element fits. Grid couplings are lubricated and sealed, and are split horizontally or vertically depending on the manufacturer. The grease needs to be changed and the coupling should be relubricated on a regular basis to prevent excessive wear of the element and hubs. Hubs are typically provided in straight bore with spacers available for wider shaft separation.

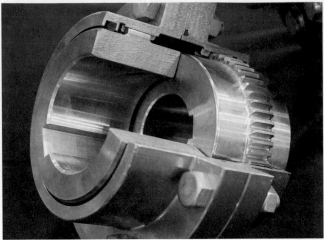

Mr. 1/Shutterstock.com

Figure 11-10. Gear couplings are keyed to the attached shafts.

Photo courtesy TB Wood's—Altra industrial Motion

Figure 11-11. Grid couplings are split to allow for maintenance and replacement of components when needed.

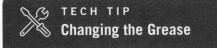

TECH TIP
Changing the Grease

When replacing grease, do not use solvents that could affect the seals. The seals are one piece. To replace them, either the hubs need to be removed or the seal is stretched over the hub and put in position.

11.2.4 Jaw Couplings

A jaw coupling is a three-piece unit that transmits power between two hubs that mesh together with the assistance of an intermediary element called a spider. The *spider*, made of metal or various polymers, allows some misalignment while reducing wear to the metallic hubs. See **Figure 11-12**. The material used for the spider can impact the coupling's ability to withstand vibration, changes in temperature, and chemicals in the environment. The type of spider used in a jaw coupling depends on the operation, environment, and manufacturer's specifications.

TECH TIP
Installation Instructions

Always read the manufacturer's installation instructions before installing couplings. These instructions contain important information, such as the proper torque for threaded fasteners, that will ensure correct and safe installation, as well as effective operation of the couplings.

Ildarss/Shutterstock.com

Figure 11-12. The spider element sits between the teeth or prongs of a jaw coupling and helps to reduce vibration and wear.

11.3 ALIGNMENT

While flexible couplings can withstand and compensate for some misalignment, they have a longer life when more precise alignment is maintained. Shaft misalignment can cause several machinery problems including the following:

- Excessive vibration.
- Early bearing failure.
- Excessive shaft wear at bearing mating surfaces and keyseats.

- Inefficiency due to frictional losses and higher motor amperage draw.
- Coupling wear and failure.

Shaft alignment is a technical skill that needs to be practiced to be maintained. Even with laser alignment systems, a technician still needs to know the basics of shaft alignment.

11.3.1 Prealignment Checks

Prealignment checks are a part of every alignment procedure. Checking certain basic conditions is crucial to successful machine alignment.

SAFETY NOTE
Lockout/Tagout

Prior to beginning any work, ensure the machine is locked out and tagged out. Let others know that you will be performing maintenance on the machine.

Piping Strain

Anything connected to the driven machinery may place excessive stress on that machinery. Piping that is not supported correctly is one cause of excessive stress. Piping should be designed and piped to the machine. Clamps, jacks, and other pulling tools should not be needed to make a connection. If excessive force is required to make connections to the machine, the piping needs to be reworked. To see if stresses are affecting alignment, set up for alignment and then loosen piping. If the dial indicator changes more than a few thousandths, there is too much strain on the machinery.

Baseplate

The larger and heavier the machine, the greater the forces transmitted through the anchor bolts to the foundation. Before performing a shaft alignment, check the baseplate to ensure certain conditions. The baseplate of both the motor and driven machinery should be all of the following:

- Free of voids, which can be tested with a hammer.
- Level in all directions (side-to-side, front-to-back, both diagonals).
- Free of cracks.
- Properly sealed from oil degradation.

- Properly anchored with appropriate hold-downs to the foundation.
- Clean of any debris, oil, old paint, and anything that could get under a machinery footing.

For example, if the baseplate that was initially installed was warped and forced into position, the baseplate is still warped and will affect alignment by placing stresses on the machinery.

Shaft Runout

Excessive shaft runout (greater than 0.003″) adversely affects final alignment. Check shaft runout first on the coupling and then on the shaft. If runout on the coupling is greater than 0.003″, check the shaft.

Couplings may have excessive runout due to any of the following problems:

- Machining off center.
- Machining the inside diameter (ID) at an angle other than true.
- Poor surface finish.
- Improper forming or machining, which resulted in eccentricity.
- Poor key fit-up in the key itself, keyseat, or keyway.

Poor surface finish may be corrected by machining the outer surface of the coupling a few thousandths or polishing on a lathe. Making a new key can address poor key fit-up. If the keyway or keyseat is unsatisfactory, components may need to be replaced. Otherwise, the coupling or a portion of the coupling must be replaced.

PROCEDURE | Runout Measurement

Follow these steps to perform a runout measurement:

1. Mount a magnetic dial indicator securely on the baseplate.
2. Verify that the dial indicator tip will not run into the keyseat. Start the measurement just past it.
3. Set the dial indicator to zero.
4. Slowly rotate the shaft or coupling through a full rotation while noting the maximum total indicator reading. The *total indicator reading (TIR)* is the difference between the minimum and maximum readings taken by an indicator.

5. Verify that the dial returns to zero at the end of the rotation, or close to it if the rotation is cut short due to a keyway or keyseat. If the dial does not return to zero, check the tightness of the dial indicator setup.
6. Repeat the procedure on the face of the coupling.

Alternatively, the face of the coupling can be checked by mounting the coupling in a lathe and affixing the dial indicator to the bed of the lathe. Check the alignment of the face to be measured. This reading may be internal or external, depending on the coupling.

Checking for runout on the rim of a coupling provides a runout reading for both the shaft and the coupling together. See **Figure 11-13**. Checking for runout on the face (or reverse face) of a coupling will reveal if the ID of the coupling is not square with the coupling rim.

Checking for shaft runout differentiates between coupling and shaft runout. If shaft runout is greater than 0.003″, the shaft may be bent or bowed and needs to be replaced. If not replaced, accurate shaft alignment to within specified limits may not be possible.

TECH TIP
Checking the Dial Indicator

Make sure that your setup to check runout is tight and that the dial returns to zero at the full revolution point. A simple check is to lightly tap the end of an extension rod and check that the dial indicator returns to its original position. If it does not, the setup is not tight.

Soft Foot

Soft foot is a condition where the feet of a machine do not make perfect and even contact with the baseplate or foundation. Soft foot can result from the following issues:

- One or more feet are short.
- One or more feet were improperly formed or machined during manufacturing.
- One or more feet are bent.
- The baseplate or foundation has an imperfection or is not level.

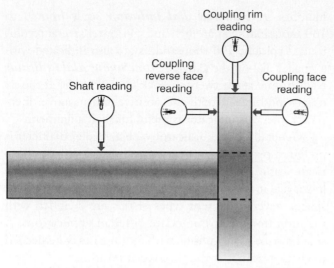

Figure 11-13. If runout on the coupling is excessive, the shaft must also be checked. The shaft may have been bent or bowed during shipping or installation.

Similar to a stool having a short leg, soft foot will be an annoying addition to shaft alignment. If not corrected, accurate alignment will be nearly impossible. **Figure 11-14** shows some of the possible soft foot conditions.

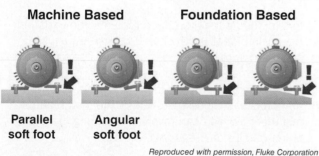

Figure 11-14. Soft foot can be caused by issues related to feet and legs of the machine or by conditions of the baseplate or foundation.

PROCEDURE Measuring and Correcting Soft Foot

To measure and correct soft foot, perform the following steps:

1. Torque down all feet to the baseplate.
2. Loosen one foot and use a dial indicator or feeler gauge to measure how far the foot lifts off the baseplate. If using a dial indicator, mount the dial indicator on the baseplate, zero the

indicator, and measure how much the foot rises when loosened.

3. If the measurement is greater than 0.002″, correct soft foot by inserting the corresponding amount of shim under the foot.
4. Retighten that foot and move to the next.
5. Repeat until all feet have been measured and corrected.
6. When complete, make a drawing that indicates how soft foot was corrected on each foot of the machine. Be sure to note the shims under each foot (graph paper works well for this). This drawing will be used for future alignments and should be kept for reference.

Bar Sag

Bar sag describes the amount that an indicator's mounting and supporting hardware bends as a result of gravity. This occurs to some degree with all alignment setups. See **Figure 11-15**. Bar sag is affected by the following:

- Weight of the measuring instrument, such as a dial indicator.
- Weight of the bars.
- Stiffness of the bars.
- Length of the bars.
- Possible loose friction clamps.
- Possible loose shaft clamp.
- Counteracting forces in the setup.

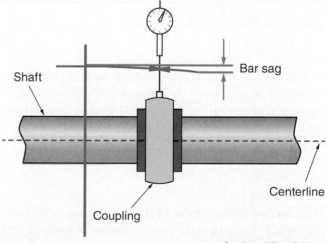

Goodheart-Willcox Publisher

Figure 11-15. Bar sag occurs due to gravity and affects some setups more than others.

When taking a measurement on the rim or face of a coupling, bar sag is corrected by adding it to the top dial reading on the rim by repositioning the dial face. Some setups have more bar sag than others. When a setup is changed, the amount of bar sag is affected. The only way to counteract the effects of bar sag on readings is by measuring and recording the amount of bar sag on each setup at several lengths.

11.3.2 Alignment Tools

If performing alignment as part of machine maintenance, you should have your own set of tools. Store tools properly in a vibration and shock-resistant case with interior padding. Do not expose tools to high humidity or environmental contaminants, such as dust, VOCs, or grinding sparks. These conditions can affect small moving parts in tools. Tools should be cleaned regularly, which includes applying a light coating of oil and wiping them clean.

TECH TIP
Alignment Tools Needed

When planning a shaft alignment, carefully examine the machine to determine what type of tool setup will work best. Make sure to have these tools on-hand before starting the work.

Shims

Shims are inserted between the foundation or baseplate and a machine's foot to achieve level. They usually come in sets that include a range of thicknesses. The material of the shim is an important consideration. Plastic and brass can deform, and plain steel is susceptible to corrosion. These weaknesses eventually lead to a change in alignment. Stainless steel is an appropriate shim material for machinery because it has good strength, resists corrosion, and can withstand fluctuations in temperature. These shims are available in a variety of sizes and thicknesses.

Dial Indicator

The *dial indicator*, **Figure 11-16**, is a gauge used to measure small linear distances for precision alignment and calibration. It is an important addition to a technician's precision toolbox. A digital indicator may seem appealing, but can be more difficult to read than a standard gauge

indicator. A ***balanced dial indicator*** reads from 0 to 100 thousandths of an inch and can travel several revolutions. Typically, 100 thousandths are also indicated with a smaller dial on the gauge. A ***continuous dial indicator*** reads from 0 at the twelve o'clock position to 50 at the six o'clock position in both the positive and negative direction. Either dial indicator is useful for shaft alignment.

Generally, a dial indicator with a smaller diameter is preferred over a larger, heavier indicator for shaft alignment work. Heavier indicators cause more bar sag and larger dial indicators with larger faces may not fit in tight spaces. Several different types of tips are included with the indicator to accommodate different applications. A small mirror or cell phone with a camera may be needed to check readings at the six o'clock position.

Store and handle dial indicators with care. When properly stored, a quality dial indicator will last through a lifetime of use. Be attentive when handling a dial indicator, as dropping it will destroy its smooth and accurate operation.

TECH TIP
Calibration of Instruments

Precision measuring instruments must be periodically calibrated to ensure their accuracy. The frequency of calibration may be based on several factors including the manufacturer's recommendation, the environment in which the instrument is used, and the nature of the measurements performed by the instrument. Calibration is typically performed by an expert instrument technician. The calibration process may be specified by the instrument's manufacturer and can involve measuring "masters" of known size, such as gage blocks.

Chase Clausen/Shutterstock.com

Figure 11-16. A dial indicator is a necessary tool for precision alignment.

Friction Clamps and Wiggler Bars

Friction clamps and wiggler bars are used to position measurement devices on the shaft. A ***wiggler bar*** is a tool support device used to take readings on the face of a coupling when there is limited space, as shown in **Figure 11-17**. While not always needed, it is an important item to have on hand when there is not enough space to take readings on the outside of the coupling face. Using multiple friction clamps on the shaft may be advisable. In some cases, taking readings off the shaft coupling may be difficult. It may be easier to install another friction clamp on the opposite shaft and take readings from it. **Figure 11-18** shows a setup for rim readings. In this example, hollow tubing is used for the extension rods to reduce weight, which reduces bar sag.

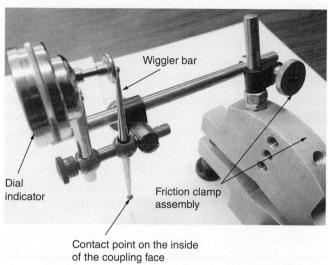

Figure 11-17. This setup is configured to take readings on the inside of the coupling face with a wiggler bar.

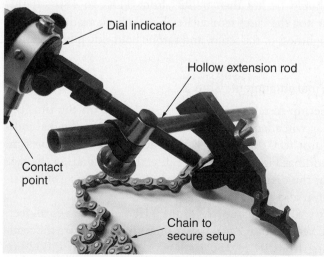

Figure 11-18. This setup is configured to take rim readings. The chain is used to clamp the setup to the shaft, and can be tightened using the wing nut.

Machinist's Rule and Level

A machinist's rule reads on several scales (16ths, 32nds, and 64ths) and is stiff enough so that it will not bend easily. It is required to perform an initial rough alignment. A basic level with a magnetic edge helps in checking a baseplate for level. Items to help you properly read a measurement, such as a flashlight, should also be in your toolbox.

> **TECH TIP**
> **Reading a Machinist's Rule**
>
> When reading a machinist's rule on the smaller scales, make sure your eyes are square to the rule and that enough light is present. It is also helpful to place a contrasting background, such as a piece of white paper, behind the rule. This reduces errors in reading the rule and, therefore, increases your accuracy.

11.3.3 Initial Rough Alignment

When performing an initial alignment, keep the following in mind:

- The motor is the movable item.
- The driven machinery is considered unmovable.
- The motor should start at a level 1/16″–1/8″ below the driven machinery.
- Initial alignment is performed with a straightedge and either shims or parallel bars.

The motor sits low so that if the driven machinery points down toward the motor, the alignment can still be accomplished. If the motor sat at the same height as the driven machine that was pointing down, the alignment could not be performed. In this case, if the front of the motor is brought up, the motor will be pointing up. If the back of the motor is brought up, the motor will be pointing down.

The most important aspect of an initial alignment is to get the motor and driven machine as parallel as possible, **Figure 11-19**. This is accomplished by measuring the distances between the faces of the coupling at 0°, or *top dead center (TDC)*, and 180°.

- If the faces of the coupling are farther apart at 0°, the back of the motor needs to be shimmed up.
- If the faces of the coupling are farther apart at 180°, the front of the motor needs to be shimmed up.

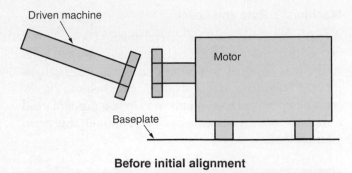

Before initial alignment

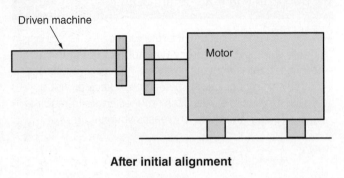

After initial alignment

Goodheart-Willcox Publisher

Figure 11-19. Once the shafts are in rough parallel alignment, half of the remaining vertical offset will bring the shafts into rough alignment.

Measure between the faces of the couplings after each shim is placed. When no noticeable difference is observed from 0° to 180°, the motor and driven machine are roughly parallel in the horizontal plane.

Next, measure across the rims of the coupling faces at TDC. This shows how far below the driven machinery the motor sits. Using two machinist's rules for this measurement will provide a good reading. Half of this reading represents how far the centerline of the motor shaft is positioned below the driven machine's shaft centerline. Carefully add shims in this amount under both the front and rear feet of the motor. Measure again to verify and refine, as necessary. Taking measurements from side to side may help, but this portion of the alignment will be affected when the final alignment is performed.

TECH TIP
Shim Stacks

When inserting shims, start with small shims. But, pay attention to the shim stack created under each foot. Remeasure and evaluate any foot with more than four total shims. Change the thickness of the shims

inserted to use the minimum number of shims possible. For example, a shim stack composed of 0.001″, 0.002″, 0.005″, and 0.020″ shims can be replaced with 0.025″ and 0.003″ shims. A stack that includes more than four shims can create a spring-like effect. This will change the alignment each time the hold-down bolts are torqued and loosened.

11.3.4 Final Alignment: Rim and Face Method

There are several methods to accomplish final shaft alignment, including graphical, dual indicators, and laser methods. However, the rim and face method that follows is easily understood and performed by beginning technicians.

Final Alignment: Step 1

Clamp the alignment setup to the motor coupling and take readings off the coupling face of the driven machine. Then, set the indicator to 0.000″ at 0° and take a reading at 180° on the opposite face of the coupling in order to make parallel, **Figure 11-20**. Bar sag will not affect this reading. Measure the center-to-center distance between front and back legs on the motor and the diameter of the path taken by the dial indicator.

Depending on your reading, either the front or back feet will be shimmed. Shim is equal to TIR multiplied by the center-to-center distance between front and rear motor legs, divided by the diameter of the indicator circle.

After placing shims, recheck the readings. Readings should be less than 0.002″ from TDC to 180°. If this is not the case, recalculate shims, check readings, check tightness of the setup, and torque hold-down bolts evenly.

Final Alignment: Step 2

Set up to take readings at 0° and 180° on the rim. Bar sag will affect this reading. So, after setting the dial indicator to 0.000″ at TDC on the opposite coupling rim, add bar sag in the positive direction by rotating the face of the dial indicator. Rotate the setup to the 180° position and take a reading.

Shim equal to 1/2 of the TIR. Add shims to all feet. Check readings again after shimming to ensure the reading from 0° to 180° is less than 0.002″. If 0.002″ or greater, reshim.

When finished shimming, sketch a picture and list the shims inserted on each foot. As an alternative, some facilities provide a standard alignment report.

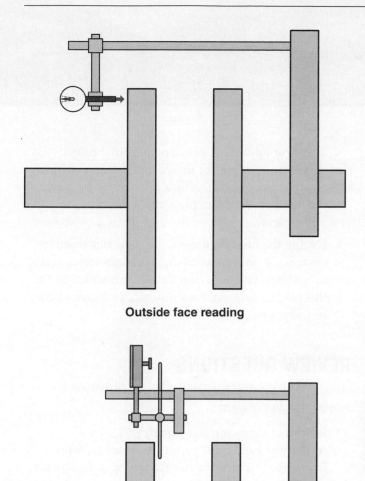

Outside face reading

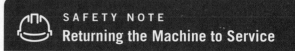

Inside face reading

Goodheart-Willcox Publisher

Figure 11-20. While performing the final alignment, take readings on the faces of the coupling halves. Readings are taken on either the inside or outside faces, depending on the coupling used and room available for tooling.

Final Alignment: Step 3

Set up to take face readings again on the opposite coupling face at 90° and 270°. Zero the dial indicator at 90° (clockwise of 0° TDC) and then rotate to 270°. This reading shows which way to swing the motor in order to align with the driven shaft.

Loosen three of the hold-down bolts completely, then loosen the last hold-down bolt until it is just less than snug. Pivot the motor around the last hold-down bolt until proper position is attained. Recheck readings and carefully reposition motor until readings result in less than 0.002″ difference. The shims are not changed at this point. The motor will be slightly offset from the driven machinery, but it is in angular alignment.

Final Alignment: Step 4

Set up to take rim readings at 90° and 270°. Set the dial to 0.000″ at 90° and take a reading at 270°. Half of this TIR is how far both the front and rear feet must be moved in the same direction. Some machines may have jacking bolts, which can be used to move the front and rear legs. Machines that do not have these bolts must be carefully moved. Move the machine and recheck until readings yield less than 0.002″. Recheck the 90° and 270° face readings at this point. Make only small movements to correct alignment.

> **SAFETY NOTE**
> ### Returning the Machine to Service
>
> Put all guards back in place before removing the lockout/tagout and returning the machine into operation. Personally observe machine startup to check vibration and sound levels.

CHAPTER WRAP-UP

Mechanical assemblies that transfer power can be very complex and require exact tolerances in order to function properly. Just as we would ensure that the foundation for a home is completed to specifications, everything about a shaft is controlled. Shaft speed, momentum, and vibration all impact components that come into contact with that shaft. Detailed and exacting measurement skills are needed to ensure maximum life from mechanical devices.

Chapter Review

SUMMARY

- Shafts transmit mechanical power and are available in a wide variety of materials. Steel shafts are manufactured through machining and forging operations. Controlling the final diameter is the most important aspect of shaft manufacturing.

- A key is used when the shaft must be connected to another machine component in order to transmit torque or rotational force. The most commonly seen types of industrial keys are square and rectangle.

- Shaft support within a machine is normally provided by radial bearings. External support of the shaft is accomplished using a range of pillow blocks, flange bearings, and universal joints. Pillow blocks and flange bearings mount outside the machine housing and provide shaft support. Universal joints allow the transfer of power without exact shaft alignment.

- Flexible couplings create a connection between two shafts to allow the transfer of power, while allowing for some misalignment. They can be made of metal, plastic, rubber, neoprene, or polyurethane, among others. Any flexible coupling that employs a pliable material that can stretch or flex is called an *elastomeric coupling*.

- Flexible couplings can withstand and compensate for some misalignment, but will have a longer life when more precise alignment is maintained. Every technician should know the basics of shaft alignment.

- Prealignment checks include checking for piping strain, checking the condition of the baseplate, measuring shaft runout, inspecting the finish of couplings, and inspecting for soft foot.

- Alignment tools should be properly stored and regularly cleaned to ensure accuracy and safe use.

- A dial indicator is used to measure small linear distances. For the purposes of shaft alignment, a dial indicator with a smaller diameter is preferred. A machinist's rule and magnetic level are additional measurement tools used for shaft alignment.

- Friction clamps and wiggler bars are used to position measurement devices on the shaft.

- The most important aspect of an initial rough alignment is to get the motor and driven machine as parallel as possible. Shims are inserted between the foundation or baseplate and a machine's feet to achieve level.

- During the final alignment, face and rim readings are taken several times as the machines are carefully moved into position. The shims are not changed at this point. Only small movements are made to correct alignment.

REVIEW QUESTIONS

Answer the following questions using the information provided in this chapter.

1. Industrial shafts transfer power, _____, and movement from one machine to another, from human to machine, or from machine to its end use.

2. Why are external mechanical power transmission shafts often hollow?

3. Explain the purpose and function of a key.

4. *True or False?* The radius of the shaft shoulder should be kept as small as possible.

5. Describe a split-construction pillow block.

6. *True or False?* A tapered adapter sleeve is used in conjunction with a locknut and washer to lock a bearing with tapered bore onto the shaft.

7. A(n) _____ joint is used to transmit mechanical force between two shafts, while allowing them to move out of alignment.

8. What is the difference between flexible couplings and rigid couplings?

9. Any flexible coupling that employs a pliable material that can stretch or flex is called a(n) _____ coupling.

10. How does alignment affect a gear coupling?

11. List three problems that can be caused by shaft misalignment.

12. *True or False?* Anything connected to the driven machinery may place excessive stress on that machinery.

222

13. What is the benefit of checking for runout on the rim of a coupling?

14. A condition where the feet of a machine do not make perfect and even contact with the baseplate or foundation is known as _____.

15. List three factors that can affect bar sag.

16. A(n) _____ dial indicator reads from 0 at the twelve o'clock position to 50 at the six o'clock position in both the positive and negative direction.

17. What factors affect how often precision measuring instruments need to be calibrated?

18. What is most important aspect of an initial shaft alignment?

19. What needs to be done if the faces of a coupling are farther apart at 0°?

20. *True or False?* When finished shimming, the technician should sketch a picture of the machinery and list the shims inserted on each foot.

NIMS CREDENTIALING PREPARATION QUESTIONS

The following questions will help you prepare for the NIMS Industrial Technology Maintenance Level 1 Basic Mechanical Systems credentialing exam.

1. Identify the shaft support device pictured below:

A. Sleeve
B. Flange bearing
C. Pillow block
D. Cardan joint

2. The _____ is the difference between the minimum and maximum readings taken by an indicator.

A. alignment
B. bar sag
C. total indicator reading
D. runout reading

3. Which of the following couplings uses a spider to assist in meshing and reduce wear?

A. Grid
B. Jaw
C. Gear
D. Sleeve

4. Which of the following is the best shim material for machinery?

A. Brass
B. Plain steel
C. Plastic
D. Stainless steel

5. When there is not enough space to take readings on the outside of a coupling face, use a _____.

A. wiggler bar
B. friction clamp
C. level
D. machinist's rule

6. Shim is equal to TIR multiplied by the center-to-center distance between front and rear motor legs, divided by the _____.

A. bar sag
B. distance between the faces of the couplings
C. circumference of the space the machine occupies
D. diameter of the indicator circle

12 | Belts and Pulleys, Chains and Sprockets, Gears and Gearboxes

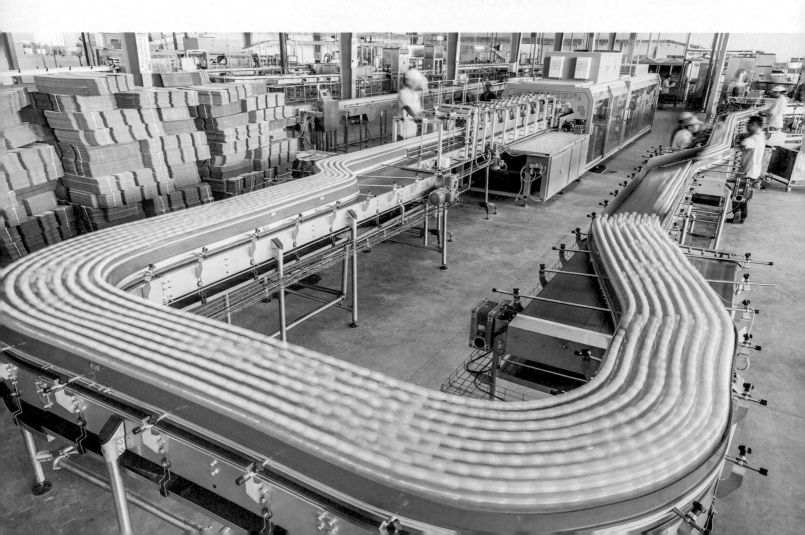

LEARNING OBJECTIVES

After completing this chapter, you will be able to:

☐ Describe how belt drives transfer power.

☐ Summarize the various types of belts.

☐ Describe attachment and removal methods for sheaves.

☐ Describe how to tension a belt.

☐ Describe how to align sheaves.

☐ Summarize the proper storage of belts.

☐ Identify common factors that lead to belt wear and failure.

☐ Summarize the operation of a chain drive.

☐ Identify the various types of chain.

☐ Explain the function of sprockets.

☐ Describe how to tension a chain drive.

☐ Discuss the common lubrication methods for chain drives.

☐ Summarize the function of gears.

☐ Identify the common elements and measurements of gears.

☐ Discuss the differences between various types of gears.

☐ Explain how a worm drive operates.

☐ Discuss considerations for gear lubrication.

☐ Explain the purpose of backlash and axial runout measurements as part of gear maintenance.

TECHNICAL TERMS

addendum	pitch
backlash	pitch circle
belt	positive drive system
chain pitch	pressure angle
chordal action	pulley
circular pitch	sheave
clearance	sprocket
dedendum	static balance
diametral pitch (DP)	undercutting
dynamic balance	working depth
gearbox	worm
lead	worm drive
overhung load	worm gear
pinch point	

Power transmission methods using belts, chains, and gears are highly engineered and have developed over many years with improvements along the way. While the methods are highly reliable, proper installation, maintenance, and inspection are key to the reliability. From the smallest miniature gear drive on an old clock to the largest gears used on power trains aboard ships, these types of drives surround us. In this chapter, we will discuss the use and components of belt drives, chain drives, and gear drives, as well as the proper maintenance of each.

12.1 BELT DRIVES

Any power transmission system relies on the transfer of power from a driver, through an intermediary element, and to the driven machine. The intermediary element can be a coupling, belt, chain, shaft, or even fluid medium. Due to the design of these systems, safety issues, such as pinch points, are an inherent danger.

The following are some specific advantages of belts compared to other methods of power transmission:

- Belts typically involve less regular maintenance.
- Belts are usually easier to replace.
- When properly aligned, belts have a very long life.
- Belts can be used over a wide range of speeds and horsepower.
- Using timing belts is effective when exact mechanical speed is needed.

However, belt drives inherently have a unique set of disadvantages:

- Over-tensioning a belt places excessive load on bearings.
- Exposure to environmental elements and other chemicals can significantly reduce belt life.
- If not properly applied and designed, a belt drive can induce vibration.
- Improperly storing belts can reduce their life.

Belt drives transfer power using the friction created between the belt and pulleys. The friction is developed through the contact of the sidewall of the belt to the inside area of the pulley. Power is transferred through the tensile cords of the belt to the other pulley through friction. Anything that damages the friction surface of the pulleys, the contacting belt surface, or the belt's tension members will affect belt life.

SAFETY NOTE
Pinch Points

Pinch points are places between moving parts or between a moving part and a stationary part within a machine where loose clothing and body parts can be caught. If not properly guarded, pinch points may cause serious bodily injury, in addition to damage to tools and clothing. Never operate a machine without the safety guards in place. Always reinstall the machine guards after maintenance and before starting any machine.

12.1.1 Belts

Belts are continuous bands used to transfer power from one pulley wheel to another. A wide range of belts are manufactured to match the variety of applications (**Figure 12-1**):

- **V-belts.** General-use belts made of rubber compounds, nylon, and various tensile members. The tapered sides of the belt contact the sheaves. Standard (classical) V-belts offer less contact area than narrow-groove (wedge) V-belts offer.
- **Double-sided belts.** Used for power transmission through multiple bidirectional loading sheaves. Can be octagonal, round, and double V-belt in design, and made of rubber compounds, nylon, and various tensile members.
- **Multiple belts.** Used in industry for higher horsepower applications. Made with multiple layers, stronger tensile members, and a backing that prevents the belt from rolling and turning over. Also called *banded belts*.
- **Flat belts.** Used with machinery that operates at lower speeds and can be a linked belt or truly flat. Belts may be made of leather, polyurethane, or other materials.
- **Timing belts.** Used specifically with timing sheaves and made of neoprene, nylon, and fiberglass.
- **Synthetic belts.** Lighter with a smaller cross section than other belts and can run at higher speeds. Can be made of a combination of materials including polyurethane, nylon, rubber, fiberglass, neoprene, and others. Also called *poly belts*.
- **Variable speed belts.** Used with variable speed sheaves and constructed of higher strength nylon and rubber compounds.

Most belts have embedded tension members. These tension members transmit most of the power from the drive sheave to the driven sheave. When measuring a belt's length, *pitch circle* represents the length of a belt at the location of the tension members. For a timing belt, *pitch* is the distance between the same location on adjacent teeth.

The numbering system used for belts depends on the belt type and the manufacturer. There are some standard designations for belts, but the numbering and identification system used for some belts is manufacturer-specific. See **Figure 12-2**. Always refer to the manufacturer for details about belt size, use, construction, and compatibility. Keep in mind that while some belts may have a similar cross section, they are likely not rated to transfer the same power.

Belt Cross Sections

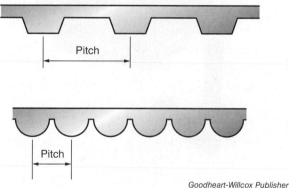

Tension members

Classical V-belt **Wedge V-belt** **Double-sided belt**

Braided V-belt

Plies

Flat belt

Timing Belts (Side View)

Pitch

Pitch

Goodheart-Willcox Publisher

Figure 12-1. The type of belt used in a belt drive system depends on criteria such as system design, the forces and speeds of the drive system, the length of the belt, the available space, and environmental conditions.

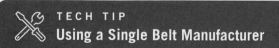

TECH TIP
Using a Single Belt Manufacturer

In industrial settings, there are benefits to sourcing all your belt needs to one company. Having numerous brands of belts operating various machines in a facility can increase the complexity of maintenance.

Belt Numbering and Identification	
Belt Type	**Designations**
V-belts	Fractional: FHP, 2L, 3L, 4L, 5L Standard or conventional: A, B, C, D Narrow V: 3V, 5V, 8V
Double sided	Numbering depends on the manufacturer
Multiple	Standard sizes: 3V, 5V, 8V, A, B, C, D
Synthetic	Imperial sizes: J, K, L, M Metric sizes: 3, 5, 7, 11
Variable speed	Identified by top width of belt, angle of belt, OD, and length
Timing	Specified by pitch, pitch length, and width

Goodheart-Willcox Publisher

Figure 12-2. There are some common designations for belt numbering and identification. However, always refer to the manufacturer for belt specifications and compatibility.

12.1.2 Pulleys and Sheaves

A *pulley* is a simple machine consisting of a wheel and axle assembly used to transfer power through a belt or cable that sits along the edge of the wheel. A *sheave* is the wheel component of a pulley that contains a groove or channel into which a belt or cable sits. Sheaves can be manufactured by casting, pressing, or machining, and can be made of steel, stainless steel, polyethylene, cast iron, aluminum, or other materials. Categories include single-belt sheaves, multiple-belt sheaves, poly V-belt sheaves, and variable speed sheaves. See **Figure 12-3**.

Larger sheaves and those meant for higher speeds are typically dynamically balanced from the factory. *Dynamic balance* is the ability of an object to stay in balance while rotating. Many smaller sheaves and those limited in speed capability are static balanced. *Static balance* is the ability of an object to maintain balance while stationary on its axis.

TECH TIP
Timing Belt Sprockets

Timing belt drive systems use sprockets rather than sheaves. Sprockets have teeth along their circumference that mesh with teeth on the timing belt. Sprockets are used in chain drive systems as well. When using a timing belt, the pitch of both sprockets must match that of the belt, just as two gears that mesh must have the same pitch. Timing belt drive systems are *positive drive systems*, meaning no slip occurs between the belt and the sprocket.

Attachment Methods

Attaching a sheave to a shaft is typically accomplished by one of the following methods:

- One or more setscrews with a key.
- Quick detachable (QD) bushing or split-taper bushing.
- Taper-lock bushing.

Setscrew methods are normally used with lighter loads, while applications that have heavier or shock loads use one of the bushing methods. QD bushings and split-taper bushings are both flanged bushings. QD bushings

have a single split through the bushing and the flange, while split-taper bushings have splits in the bushing but not the flange. Taper-lock bushings are flangeless bushings with a tapered outer surface and a split along their length. See **Figure 12-4**. Bushings have threaded and non-threaded holes around the tapered surface. These holes align with corresponding holes in the sheave or hub.

QD and split-taper bushings are normally installed using conventional mounting, but some bushings can be installed using reverse mounting. In conventional mounting, bolts are inserted through the sheave and threaded into the bushing's flange. The assembly is installed on the shaft with the bushing's flange toward the motor. In reverse mounting, the bolts are inserted through the bushing's flange and threaded into the sheave. The assembly is then installed with the bushing's flange facing away from the motor. See **Figure 12-5**.

Light duty, spoke-type single belt sheave

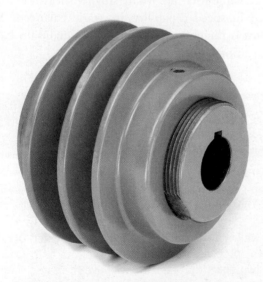

Variable pitch, light duty, double belt sheave

Photo courtesy TB Wood's–Altra Industrial Motion

Figure 12-3. Common sheave types found in industry.

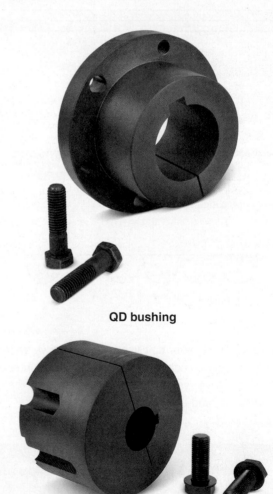

QD bushing

Taper-lock hub

Photo courtesy TB Wood's–Altra Industrial Motion

Figure 12-4. Both QD bushings and taper-lock hubs compress the hub onto the shaft to develop clamping force.

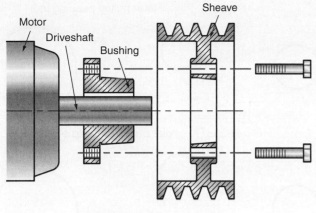

Conventional mounting

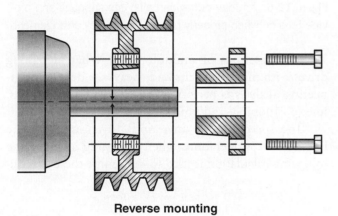

Reverse mounting

Goodheart-Willcox Publisher

Figure 12-5. Some bushings can be installed in the reverse mount position. Always check the manufacturer's information before using a reverse mount configuration.

All of these bushings develop clamping power in similar ways. As the bolts are tightened, the tapered surfaces of the bushing and the sheave/hub are pressed together. This pressing action tightens the parts against one another and causes the bushing to tighten around the shaft as the split in the bushing is compressed.

TECH TIP
Tightening Bushings

When installing a sheave on a shaft, always torque the bolts to the value recommended by the manufacturer. When using flanged bushings, a gap remains between the face of the flange and the sheave/hub. The correct torque is critical in establishing a proper fit.

PROCEDURE   Installing a Sheave

Always refer to the manufacturer's installation instructions when installing bushings and sheaves. The following is a general procedure:

1. Inspect, clean, and deburr the seating area of the shaft, outside diameter (OD) and inside diameter (ID) of the bushing, ID of the sheave, and any threaded fasteners or their mating surfaces.

2. Decide on the sheave and bushing orientation. In some cases, the bushing and sheave may be assembled before being placed on the shaft. In other cases, the bushing and sheave are placed on the shaft before the connecting bolts are installed.

3. Install the bolts connecting the sheave/hub and bushing so they can be accessed from the side facing away from the motor. If the bushing is inserted into the sheave from the motor side, align the non-threaded holes in the sheave with the threaded holes in the bushing. Pass the fasteners through the non-threaded holes in the sheave and thread them into bushing until hand tight. If using reverse mounting and the bushing's flange is facing out (away from the motor), align the non-threaded holes in the bushing with threaded holes in the sheave. Install the bolts and hand tighten.

4. Position the assembly more precisely on the shaft and key, if needed.

5. Carefully tighten the threaded fasteners to proper torque.

6. Tighten setscrews to the key.

TECH TIP
Lubrication and Anti-Seize Compounds

Do not use any kind of lubrication or anti-seize compound in any part of the assembly, including threaded fasteners. Using lubricant can result in excessive force being placed on the assembly and may cause damage. Lubricating fasteners lowers the allowable torque because they thread more easily. It can also cause damage or breakage, particularly with fine threaded fasteners.

Removing a Sheave

To remove a sheave from a shaft, complete the following steps:

1. Remove all threaded fasteners.

2. Loosen all setscrews.

3. Insert threaded fasteners into the appropriate jacking positions.

4. Slowly apply equal force to all jacking screws until the assembly separates. Be mindful that the assembly may come apart quickly. Do not allow the assembly to fall and impact anything.

5. Clean all mating surfaces, but do not apply lubricant.

6. If contamination is a concern, place components into a sealable plastic bag and seal a plastic bag over the shaft seating area with tape.

12.2 BELT DRIVE MAINTENANCE

The simplest type of drive consists of one drive pulley and one driven pulley, with no guides or tension members between them. As speed ratios increase, one pulley becomes substantially larger than the other. If the speed ratio becomes too high, the belt on the smaller pulley is more likely to slip, especially on start-up, due to decreased contact area.

An *idler pulley* guides or tightens a belt in a pulley arrangement. See **Figure 12-6**. When positioned to push in on a belt, the idler pulley can alleviate some slippage by increasing contact area. However, this positioning reduces belt life because the belt is forced out of its normal bend. An idler pulley can guide and provide tension when positioned to push out on a belt. This positioning does not reduce belt life, but does decrease the contact area on both the drive and driven pulleys.

When properly designing a drive, the following conditions are taken into account:

- The application, both the type of loading and service time.

- Speed of both the drive and driven machine.

- Amount of horsepower being transferred.

- Distance between the pulleys.

Operations that involve higher power, longer running times, and more severe duty typically require larger

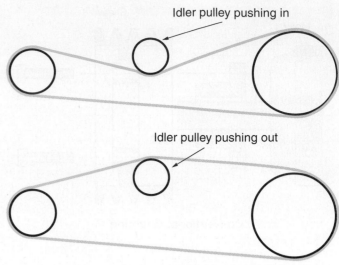

Idler pulley pushing in

Idler pulley pushing out

Figure 12-6. An idler pulley can alleviate slippage and provide tension when properly placed in a pulley arrangement.

belt cross sections or multiple belts. When designing drives with multiple shafts and sheaves, a detailed engineering analysis is performed to calculate belt lengths, idler positioning, and proper sheaves.

The speed of the drive and driven machine are important considerations when designing a drive. The following calculation is used to determine these speeds:

$$D_1 \times rpm_1 = D_2 \times rpm_2$$

(12-1)

where

D_1 = diameter of driver sheave
rpm_1 = driver speed
D_2 = diameter of driven machine sheave
rpm_2 = speed of driven machine.

Using this equation, knowing any three of the variables allows us to calculate the fourth variable. For example, if we know the driver speed (rpm_1), diameter of the driver sheave (D_1), and diameter of the driven machine sheave (D_2), we can calculate the speed of the driven machine (rpm_2). Note that the rpm of the driven machine does not affect the driver. The rpm of the driver is a fixed value, unless the motor is driven by a variable speed drive or other variable method.

12.2.1 Belt Tensioning

The most common method of belt tensioning is to allow 1/64th of an inch of belt deflection per inch of span, measured center to center, between the sheaves. This should include the proper amount of force placed at center span on the belt.

PROCEDURE   Belt Installation

For the initial installation of a belt, perform the following:

1. Move one of the machines to reduce the span distance to a point that the belt(s) may be easily placed on the sheave without stretching or prying.

2. Find the proper force deflection amount using the manufacturer's tables.

3. Slowly and evenly move the machines apart, while rotating the belt through the sheaves by hand. This will help seat the belts within the sheaves.

4. Align the sheaves before maximum tension is reached.

5. With sheaves aligned, increase tension to the proper amount to achieve belt deflection.

6. Recheck alignment.

7. After 24 hours of run time, recheck and readjust belt tension, if needed.

On larger drives, proper tension may be difficult to achieve without specialized measuring instruments. Belt suppliers have several belt tension instruments that can be used to very accurately measure deflection, force, and distance.

TECH TIP
Belt Tension

Timing belts and poly belts typically need less tension than conventional V-belts. Check with manufacturers to determine exact tension force and deflection.

Some belts may slip at start-up, especially with those that immediately accelerate to full speed (across-the-line starters) with high torque loads. Belt slippage at start-up may be inevitable, and excessive tension force should not be used to try to prevent it. Excessive tension force places further load on bearings because of the overhung load. An **overhung load** is any load perpendicular to the shaft that is applied beyond the outermost bearing of the shaft.

Belt tension may need to be adjusted as belts wear in and seat in the groove of the sheave. Also, the drive will develop a tight and slack side depending on rotational direction of the driver. It is perfectly normal for this to happen.

12.2.2 Alignment

Prior to alignment, check radial and axial runout of the sheave with a dial indicator. Excessive runout affects operation and could induce vibration. Radial runout should be limited to 0.010″ total indicator reading (TIR), and the axial runout limit is 0.005″ TIR. If runout checks of the sheave reveal excessive runout, check shaft runout to ensure the shaft is not bent.

Under optimal circumstances, both the drive and driven shafts are parallel to the baseplate. Misalignment is typically a combination of both angular and parallel discrepancies, **Figure 12-7**. Misalignment can result in sheave and belt wear and shortened belt life. The usual method is to align the face of sheaves with a straightedge, however the ultimate purpose of the alignment is to align the center of the sheave grooves.

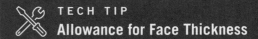

TECH TIP
Allowance for Face Thickness

If the faces of the sheave are of different thickness, make an allowance to offset the alignment to align the sheave grooves.

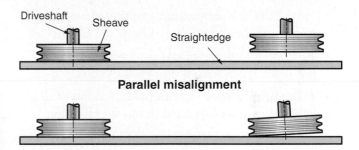

Driveshaft   Sheave   Straightedge

Parallel misalignment

Angular misalignment

Goodheart-Willcox Publisher

Figure 12-7. Sheaves should be aligned on both the parallel and angular planes. Parallel misalignment should be limited to less than 1/2°.

PROCEDURE   Sheave Alignment

The following is a general procedure for alignment:

1. Make both shafts parallel to the baseplate, if possible.

2. If shafts are vertical, check with a machinist's level.

3. Using the driven machinery as the stationary item, use a straightedge to project to the motor sheave.

(continued)

4. Move the motor until one edge of the sheave contacts the straightedge. Depending on the motor base, shims may be needed.

5. Pivot the motor until both edges of the sheave contact the straightedge. Shims may be needed again depending on orientation of the motor shaft.

6. Evaluate the need for offset if dealing with a variable sheave or sheaves with different face thicknesses.

SAFETY NOTE
Drive Guards

Always ensure that drive guards are back in place before starting the drive.

12.2.3 Troubleshooting

Properly storing belts will help maintain the integrity of the belts until they are needed for operation. Consider the following when storing belts:

- Store belts in a temperature-controlled area, away from sunlight and sources of heat.

- Keep belts in the original box if possible, and do not place excess weight on top of the box.

- Hang belts, if needed, with the full radius of the belt hanging unobstructed in its natural shape.

- Store belts away from ozone-producing machinery, such as brushed motors.

Belt wear and failure can be caused by some common factors. Broken belts or tensile members are the result of immediate loading or improper installation. Remember that prying a belt over or onto the sheave is not an acceptable practice. Worn belt edges point to alignment problems. Belt wear can also occur if the belt does not properly match the sheave. Top belt wear can result in idler bearing failure and the idler pulley sliding, instead of pushing, against the belt. Belt cracks and flaking can be caused by exposure to environmental factors and volatile compounds. The most common cause of drive failure is lack of or improper maintenance. Become fully familiar with the belt manufacturer's publications and recommendations to ensure a proper life span for the drives. Regularly scheduled drive inspections are an important step in increasing machinery reliability.

Check sheaves for wear each time a belt is replaced. Excessive sheave wear is often seen as polished sheave walls. Belt manufacturers may provide sheave gauges to check for groove wear, **Figure 12-8**. Sheave grooves should be clean, dry, and free of debris and lubricant. Variable speed sheaves may need lubricant on the threads that allow the sheave to expand and contract. Do not over lubricate the threads. If lubricant from the threads is deposited on the belt, it will reduce belt life.

TECH TIP
Hammering a Sheave

Never use a hammer on a sheave. Hammering a cast sheave or pulley can result in unseen cracks, which can fail when at speed and under load.

Sheave gauges   **Checking a sheave**

Timken Belts

Figure 12-8. Checking sheaves for wear is just as important as checking belts for wear.

12.3 CHAIN DRIVES

Chain drives are designed to lock into the sprocket of both the drive and driven machinery, which provides a definite speed ratio. Chains are used in various forms for applications such as conveyor systems, agricultural machines, escalators, bulk material handling, automotive, and industrial power transmission. Industrial power transmission chains are typically made from sheet steel by punching, stamping, and forming operations, followed by heat treatment.

THINKING GREEN
Recycling Steel

Steel is the most recycled resource in the world. The rate of recycling steel in the United States is around 85%. In addition to its high recycling rate, steel offers eco-friendly benefits throughout

(continued)

its lifecycle. For example, wastewater and waste material from steel manufacturing sites can be recovered and reused in other applications. A facility can reduce its environmental footprint by contributing to steel recycling programs. When tools and steel machine components fail or are at the end of their service life, collect and recycle them.

12.3.1 Chains

Using chain drives for industrial purposes provides several advantages. Chain drives are typically used for slower speed drives, but they can transfer more horsepower than a similarly sized belt drive. A chain drive is a positive drive that maintains an exact speed ratio. Since chains can be broken at any spot or a master link is included, maintenance is less time consuming. Once the drive is aligned, it rarely needs to be realigned. Additionally, chains are more resistant to environmental conditions than belts.

Types and Design

Standard roller chain is made from two different subassembly links. The first link contains the bushings, rollers, and inside side bars. The second link connects these together with pins and outside side bars. Pins may be hardened depending on the chain's intended use. **Figure 12-9** shows the components of standard roller chain.

Roller chain is identified using a common standard that generally consists of a two-digit number. The first digit in the chain identification number represents the chain pitch. ***Chain pitch*** is the distance between the

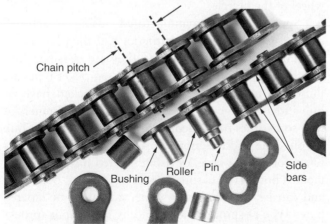

Chain pitch

Bushing   Roller   Pin   Side bars

Bayurov Alexander/Shutterstock.com

Figure 12-9. To assemble a link of standard roller chain, a pin is inserted through the link plate and both a bushing and roller slide onto the pin. A link plate is fastened on the end of the pin to complete the link assembly.

centers of the pins on each link. The first digit is the number of eighths of an inch (1/8″) in the chain pitch. For example, a chain with the number 80 would have a pitch of 1″ (eight one-eighths, or 8 × 1/8″).

The second digit indicates the type of chain: 0 for standard, 1 for light-duty, and 5 for rollerless. Multiple-strand chains are numbered in a similar manner, but include the number of strands in the designation. A 40-2 chain is composed of two side-to-side strands and has a pitch of 1/2″.

Single-strand chain strength, or ANSI average strength, is approximately the pitch squared multiplied by 2.25 lb. Drives are typically designed so that a safety ratio of strength to load is 5:1 or 6:1. Heavy chain is labeled with a standard number followed by the letter *H*. Heavy chain is approximately 120% stronger than standard chain, having heavier link plates and induction-hardened pins.

In addition to the standard types of chain, other types commonly used in industry include offset, leaf, and silent chain. *Offset chain* comes in a variety of standard pitch lengths, **Figure 12-10**. The side bars on each link of offset chain are offset to create a narrow end and a wide end when the side bars are paired. A bushing fits between the narrow side bars, which are then fit between the wide side bars and secured with a pin.

Pin   Bushing

Offset side bars

metwo/Shutterstock.com

Figure 12-10. The offset ends of chain fit together to create continuously connected links.

Leaf chain, **Figure 12-11**, is composed of heavier links with multiple side plates and strands to increase its strength rating. Leaf chain link plates are thicker than standard chain with the same pitch. The numbering system for leaf chain begins with the letters *BL* followed by three numbers that indicate pitch and the number of plates. For example, a BL 823 chain has a pitch of 1″ and follows a pattern of 2 × 3 plates.

Silent chain is widely used in conveying applications and power transmission where higher speeds are needed. These chains are similar in appearance to leaf chains, however the profile of the links includes two included angles of 60°, as shown in **Figure 12-12**. Driving links connect the flank links, and guide links ensure proper alignment and mesh with sprockets. Silent chain is available with several forms of guide links, including center, double, or outside guides. Other forms of guide

Figure 12-11. Leaf chain is often used in lifting applications, such as on forklifts.

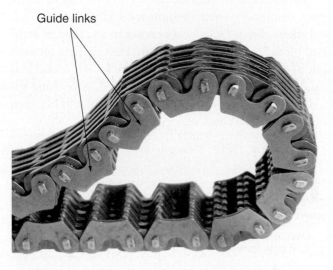

Guide links

Figure 12-12. The guide links on silent chain may be located on the outside edges of the chain or in the center of the row of links.

links are manufactured specifically for conveyor applications. These types can transfer power on both sides of the chain (duplex) and may not mesh with standard SC type of silent-chain sprockets. Always refer to and follow the manufacturer's recommendations.

Standard pitches of silent chain are available, but measuring pitch on silent chain is different from standard chain. Pitch of silent chain should be measured across three centers, or pins, and then divided by two. An example of how power transmission silent chain is labeled is shown in **Figure 12-13**.

Silent chain has less chordal action, and therefore less noise and vibration, than standard chains. *Chordal action* refers to the action that occurs when a straight

chain enters and revolves around a sprocket. As the chain enters the sprocket, it engages a tooth and then is displaced vertically, or tilted, as the sprocket turns through a revolution until the chain leaves the sprocket. This change in the link angle, as compared to the straight links, changes the velocity of the link and promotes noise and vibration.

Power Transmission Silent Chain Labeling

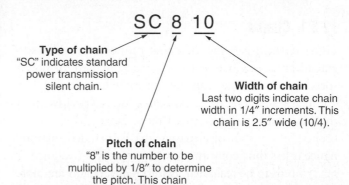

Type of chain
"SC" indicates standard power transmission silent chain.

Width of chain
Last two digits indicate chain width in 1/4″ increments. This chain is 2.5″ wide (10/4).

Pitch of chain
"8" is the number to be multiplied by 1/8″ to determine the pitch. This chain has a pitch of 1″.

Figure 12-13. The labeling of power transmission silent chain indicates the chain type, pitch, and width.

12.3.2 Sprockets

A *sprocket* is a gear or wheel with teeth that mesh with a chain in a chain drive system. In a chain drive system, the sprocket drives the chain to transmit rotary motion. A simple example of a chain drive system is a bicycle. The pedals drive a sprocket that moves the chain, which transmits rotary motion to sprockets that drive the rear wheel of the bicycle.

Sprockets are manufactured using machined steel, cast iron, or some synthetics and then hardened, depending on the application. They are classified into four different types: A, B, C, and D. Sprocket types are illustrated in **Figure 12-14**. Type A sprockets have no hub and are flange mounted using threaded fasteners or may sometimes be welded. Type B sprockets have a hub on one side and are mounted using a keyway and set-screw for a plain bore, or using tapered bushing with a tapered bore. Type C sprockets have a hub on both sides and can be mounted with either a straight or tapered bore. Type D sprockets are typically larger split sprockets, which allows ease of installation and maintenance.

The process of installing a sprocket on a shaft is identical to the process of installing a sheave described earlier in this chapter. Always follow the manufacturer's installation procedures.

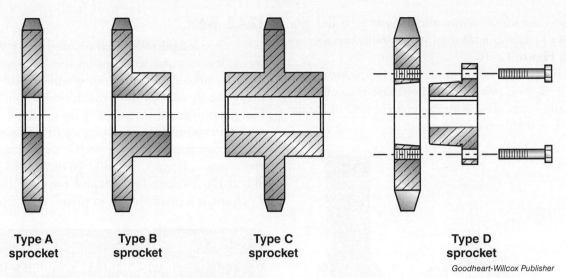

Goodheart-Willcox Publisher

Figure 12-14. Common types of sprockets may have no hub, a hub on one side, or hubs on both sides. Others may have a detachable hub/bushing.

> **TECH TIP**
> **Appropriate Sprockets**
>
> Silent chain meshes only with SC sprockets. Standard roller chain with multiple strands needs a specific sprocket with a corresponding number of rows of teeth.

12.4 CHAIN DRIVE MAINTENANCE

The two most common causes of chain drive failure are lack of lubrication and improper alignment. Both of these are regular tasks that should be familiar to all technicians. In addition, placing a chain drive in an environment for which it was not designed will impact chain and sprocket life.

> **SAFETY NOTE**
> **Lockout/Tagout**
>
> Never work on a chain drive unless it is locked out and tagged out. Always replace guards when maintenance is completed.

12.4.1 Tensioning and Initial Installation

On the initial installation of a chain, check the runout of the shaft and sprocket. Radial runout should be less than 0.004″ TIR. Face runout on machined sprockets

should be limited to 0.010″ on sprockets up to 10″ in root diameter, adding an additional 0.001″ of runout for every additional inch in root diameter. For example, a sprocket with an 18″ root diameter would have a limit of 0.018″ for face runout.

Both the drive and driven shafts of a machine must be relatively parallel. This can be performed before or after sprockets are installed using similar methods to aligning sheaves. The more accurate the alignment, the less it will affect chain and sprocket wear.

When placing the chain on the sprockets, it is usually easier if the connection is made on the top of the larger sprocket. The connection link depends on the manufacturer and size of the chain. Common methods of connecting links include using riveted pins, spring clips, retaining clips, and cotter pins. Spring clips and retaining clips, **Figure 12-15**, connect links using a flat clip that secures the pins to hold the link assembly of inside side bars, pins, and outside side bars together.

Spring clip   **Retaining clips**

Chromatic Studio/Shutterstock.com   Vasyl S/Shutterstock.com

Figure 12-15. Spring clips and retaining clips are common methods of connecting links of drive chain.

Cotter pins are inserted into the ends of each link pin to connect links and hold the link assembly together, as shown in **Figure 12-16**.

Initial chain tension should result in approximately 2 percent play on one side of the center-to-center distance of the shafts. This is approximately 1/4″ of sag for every foot of shaft separation.

Peter Sobolev/Shutterstock.com

Figure 12-16. Cotter pins can be used only with drive chain pins that have been machined with a slot or hole specifically for a cotter pin.

PROCEDURE	Checking Chain Tension

Tension should be checked using the following steps:

1. Place all slack on one side of the chain by slightly rotating sprockets in opposite directions.

2. Lay a straightedge across the top of the chain from one sprocket to another.

3. Apply pressure in the center of the span and measure the sag.

4. Adjust center distance to reduce sag while maintaining alignment.

5. Check chain tension after 100 hours of run time after initial installation. Subsequent periodic maintenance should include a tension check.

12.4.2 Wear

Wear on standard roller chain is not linear. As a chain wears, it wears and stretches on the links of the chain with the pins. Pitch length on the links with the bushings typically will not show much elongation, or stretch. Elongation can be measured using a longer section of the chain and an accurate tape measure. Some manufacturers provide plastic rulers with indentations to check chain wear. Elongation should be limited to 3 percent. When the increase in a chain's pitch is greater than 2 percent, it is practical to start planning its replacement.

TECH TIP
Measuring Chain Elongation

To measure a chain for elongation, pick an easy-to-use number of pitches. For instance, for a 1/2″ pitch chain, using 20 pitches should yield a measurement of 10″. The chain should be replaced if elongation reaches 3 percent, or just over a 1/4″ in this example. If 20 pitches measure over 10 1/4″, it is time to replace the chain.

To correct wear on a chain, replace the chain. Do not remove a link of the chain in order to eliminate elongation. Chain drives are designed for a certain chain pitch.

12.4.3 Lubrication

Proper lubrication involves the proper amount of lubrication, proper method, proper lubricant selection, and proper replacement schedule. There are four main methods of lubrication for a chain:

- Drip.
- Oil bath.
- Oil flinger.
- Forced lubrication.

Using the *drip method*, lubricant is dripped onto the lower chain before entering a sprocket. This allows some time for the lubricant to penetrate the chain into the bushings and pins. Drip rates can range from 5–20 or more drops per minute depending on the speed of the chain. The drip method is acceptable for slower speeds, as higher speeds require more lubrication.

In an *oil bath*, the lower span of chain is submersed into oil at a depth of at least 1/2″. Only a small part of the chain needs to be submersed, preferably prior to engaging a sprocket on the lower side. This method works well for low or medium speeds.

An *oil flinger* disc is attached to the lower shaft and rotates with it. As the disc turns, it spins through the lubricant and lubricant is flung onto the chain using centrifugal force. The flinger disc may be used if a drive cannot be designed to incorporate an oil bath.

Forced lubrication uses a small pump and motor to pressurize the lubricant and spray it onto the chain. Use of this method adds a mechanical maintenance item to the technician's routine checks to ensure proper operation of the pump and motor.

Follow recommendations from the manufacturer regarding the type and viscosity of lubricant to use. The operating environment, load, speed, and temperature all factor into the proper lubricant for the application.

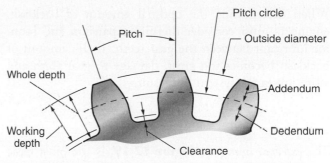

Figure 12-17. Technicians should become familiar with common gear nomenclature.

> ### TECH TIP
> ### Inspect Lubricant
>
> When changing the lubricant for a chain drive, inspect for any excessive metallic particles. Lubricant should be clean and bright. If the lubricant is contaminated, determine what changes need to be made to better protect the drive from environmental elements, such as dust, grinding, and salt.

12.5 GEARS AND GEARBOXES

Gears transmit power between machines while changing rotation, speed, and torque. Gears can also change rotational movement to linear movement through the use of a rack and pinion. Basic gears have been used for a long time. However, the advancement of engineering and manufacturing methods over time has allowed the use of internal gears, planetary gearing, and larger herringbone gears. In this section, we will examine the basic concepts of gear power transmission and maintenance concerns with gearboxes.

12.5.1 Gears

While gears come in a wide variety of sizes, there are common elements and measurements critical to their operation. Knowledge of these is important to understanding how a gear works and when troubleshooting system problems. Refer to **Figure 12-17** for common terms and dimensions related to gears.

On a gear, the *pitch circle* is a diameter measurement that includes only the portion of the gear that would make smooth contact with another gear. Typically, this includes the entire diameter of the gear below the ***addendum***, or the top half of the gear tooth. The ***dedendum*** is the length of

a gear tooth below the pitch circle. The ***working depth*** is the entire length of the tooth, addendum plus dedendum, minus the clearance. When gears mesh, there is a certain amount of space between the top of the tooth on one gear and the bottom of the mating space on the opposite gear. This space is called ***clearance***.

In order to mesh, gears must have the same diametral pitch and pressure angle. ***Diametral pitch (DP)*** is the number of teeth per inch of length on the pitch circle. A large diametral pitch measurement indicates that a gear has small teeth. ***Pressure angle*** is the angle at which gear teeth surfaces mesh together. It is typically 14.5° or 20°. Meshing teeth must also have the same circular pitch. ***Circular pitch*** is the distance between a point on the face of one tooth to the same point on the next tooth, measured along the pitch circle.

During operation, it is expected that gears will operate smoothly. To ensure that gear teeth do not bind when meshing, backlash is considered in the machine's design. ***Backlash*** is the side-to-side clearance between mating teeth, as shown in **Figure 12-18**. The smaller the amount of backlash, the more precisely gear teeth mesh.

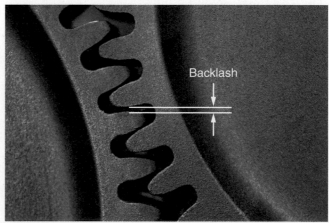

hayakato/Shutterstock.com

Figure 12-18. Having adequate backlash prevents gears from binding when meshing.

When determining the optimal amount of backlash, engineers must consider thermal expansion and room for lubricant between the gear teeth. If the amount of backlash becomes too great, damage to gear teeth and vibration in the system can result.

Gear Types

The *external spur gear*, **Figure 12-19**, is the most basic type of gear. These gears are used in offset parallel shafting. In order for gears to mesh, they must have the same diametral pitch and tooth pressure angle. An angle of 20° tends to be stronger and results in less undercutting. ***Undercutting*** is wear on the lower part of the dedendum. It can occur as a tooth enters the gap to mesh with another gear. When two spur gears are used, the shafts run in opposite directions. Using an intermediary idler gear between the input and output shafts causes the output shaft to rotate in the same direction as the input shaft.

Photo courtesy Boston Gear–Altra Industrial Motion

Figure 12-19. Notice the completely straight tooth profile on these external spur gears.

An *internal spur gear*, **Figure 12-20**, is used for parallel shafting. Either a pinion or a spur gear is used as the input. In this case, both the input and output turn in the same direction. This combination is used in planetary gear sets because of its close centers, parallel shafting, and high output torque.

The *rack and pinion*, **Figure 12-21**, is used to convert rotational movement to linear movement. In this case, the rack is movable and the pinion is held stationary. As the pinion rotates, the rack develops linear movement. An example of this can be seen in positioning systems, such as a manual lathe. Since spur gears are a straight cut gear, they can develop thrust if not properly aligned. Shafts must be close to parallel in order to not develop thrust.

Helical gears have teeth that are cut on a straight angle, or *helix angle*, as shown in **Figure 12-22**. The "hand" of a helical gear indicates the directional angle of the teeth. Helical gear may be right hand or left hand. The helix angle allows several teeth to engage at the same time, which reduces noise, produces smoother operation, and can handle higher speeds. Helical gears with opposite hands can

Internal spur gear

Jim Francis/Shutterstock.com

Figure 12-20. This internal spur gear is part of a planetary gear set.

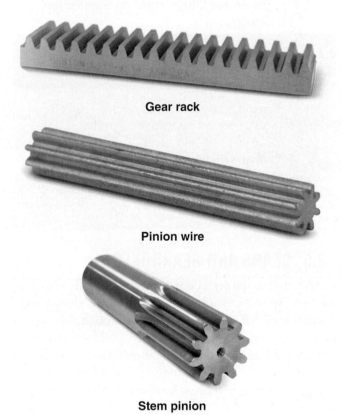

Gear rack

Pinion wire

Stem pinion

Photos courtesy Boston Gear–Altra Industrial Motion

Figure 12-21. Gear rack is used to convert rotary motion into linear motion. Pinion wire can be bored and fitted for specific applications. Stem pinions mate with external spur gears.

mesh on parallel shafting and produce opposite rotation. The helix angle results in axial thrust, so appropriate bearings must be used. If rotation of the input shaft is reversed, thrust will also be changed to the opposite direction.

Double-helical gears, **Figure 12-23**, are used in applications where axial thrust needs to be minimized. Since a double-helical gear has two opposite-handed helical gears on the same shaft, the thrust is canceled. This type of gear is used for high-torque applications such as those found in power generation and shipboard use.

Figure 12-22. The angle of helical gear teeth allows several teeth to engage at the same time.

Figure 12-23. A double helical gear is manufactured on the same hub with a space between the sets of teeth.

Figure 12-24. Larger herringbone gears are used in the power generation industry.

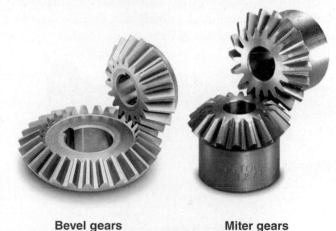

Bevel gears   **Miter gears**

Figure 12-25. Miter gears change the direction of rotation by 90°.

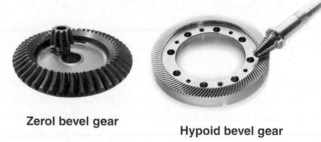

Zerol bevel gear   **Hypoid bevel gear**

Figure 12-26. Zerol and hypoid gears are engineered for specific applications.

The *herringbone gear*, **Figure 12-24**, is similar to the double helical, but it does not have space between the opposite hands. Herringbone gears are used in transmissions and high-torque applications. The lack of a center gap makes manufacturing more difficult and costly.

Bevel gears come in several forms and are used for shafts transferring power at 90° intersections. Straight bevel gearing made of sets of the same gear are called *miter gears*, **Figure 12-25**. These gears have a one-to-one speed ratio and only change the direction (90°) of rotation. Bevel gears are specified by pitch diameter, pitch angle, diametral pitch, and outside diameter.

While the most common bevel gear forms are the straight and spiral bevel gear, zerol and hypoid bevel gears are also available. See **Figure 12-26**. *Zerol bevel gears*

have teeth that are curved, not angled. These gears run smoother than straight bevel gearings with less noise, and do not produce thrust. The *hypoid bevel gear* and crown are offset. For this arrangement, the pinion may be larger and stronger than regular bevel gearing, which can allow high speed ratios and torque transfer.

12.5.2 Gearboxes

A *gearbox* is a set or system of gears that may or may not be contained within a casing. Gearboxes range from very small and lightweight, single-speed reduction to large high-horsepower, multispeed reduction units. A common type is a single-speed reduction unit in the 0.5–25 hp range made of worm gears.

A *worm drive* is used to transmit power between two shafts and change the direction of rotation by 90°. Worm drive gearboxes contain two types of gears: a worm and a worm gear, as shown in **Figure 12-27**. The *worm*, or *worm screw*, is a section of threaded rod that is the driving element of a worm drive. The *worm gear*, or *worm wheel*, is a wheel-shaped gear with meshing teeth that is the driven element of a worm drive. Worms are classified by diametral pitch, material of construction, and thread type (single start thread, double start thread, or triple start thread). Both the worm and worm gear can be right- or left- handed.

Worm

Worm gear

ra3m/Shutterstock.com

Figure 12-27. In a worm drive, the worm drives the worm gear to transmit power between shafts.

Pitch and lead are both important terms related to the operation of worm drives. On a worm drive, *pitch* refers to the center-to-center distance of threads on a worm. *Lead* is the number of starts on a worm multiplied by the pitch. A single-threaded worm, or single start thread, has a higher gear ratio than a triple start thread. The gear ratio of a worm drive is calculated using the following equation:

$$\text{gear ratio} = \frac{\text{number of teeth on the worm gear}}{\text{number of starts on the worm}}$$

(12-2)

Since the input is into the worm, calculating output speed and torque can be accomplished using the speed ratio. For example, if the number of teeth on a worm gear is 40 and the worm is a double start thread, the ratio is 20:1. The output speed is the input speed divided by 20.

12.6 GEAR MAINTENANCE

Ensuring proper lubrication is a key maintenance item with gears. At some point, you may have to disassemble a gear drive to investigate an issue. However, current drives do not have many adjustments that a technician can perform. Always refer to the manufacturer's manuals for specific maintenance.

12.6.1 Lubrication

Lubrication depends on speed of the gears, load, type of gearing, and temperature of both the lubricant and environment. Some slower gears may work well with grease, while faster gearing works better with mineral-based oil. Systems rely on a drip, splash, or pressurized lubrication. Follow manufacturer's recommendations for the type of oil or grease to be used. Both oil and grease should be regularly changed and should be included as a regularly scheduled maintenance item. Oils in the range of SAE 80–90 viscosity are recommended for splash systems.

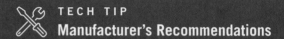

TECH TIP
Manufacturer's Recommendations

Review the manufacturer's recommendations on additives. EP greases and lubricants are not recommended with bronze gears.

Higher speed gear drives usually require that the lower gear be somewhat covered in lubricant. This can range from submerging part of the lower gear to completely covering the lower gear. When changing lubricant, check for abnormal wear. This is typically indicated by excessive metallic particles in the lubricant. If any larger pieces of metal are seen in the lubricant, immediately investigate to find the source.

TECH TIP
Replacing Gaskets

If a gearbox is to be opened, make sure you have appropriate gaskets prior to opening. Always replace the gasket and "soft" parts when performing maintenance.

12.6.2 Adjustments

Backlash is maintained by the center distance between the gears and typically cannot be adjusted in the field. However, taking a reading of backlash can provide a comparison to manufacturer specifications and may predict possible issues.

PROCEDURE — Backlash Measurement

To take a proper backlash measurement, perform the following steps:

1. Open the gearbox.
2. Hold one gear stationary.
3. Mount a dial indicator at a 90° angle to one of the opposing teeth.
4. Rock the movable gear back and forth to read total indicator displacement.

If backlash is beyond what the manufacturer specifies, it may be due to tooth wear or bearing wear. Some backlash is desired for lubricating purposes. Depending on the application, excessive backlash may not interfere with the operation of the drive.

Axial runout is another measurement that may indicate bearing concerns. Axial runout is the movement of the shaft and gear in direction with the shaft.

PROCEDURE — Axial Runout Measurement

Perform the following to take an axial runout measurement:

1. Hold other shafts stationary (one or more).
2. Set a dial indicator at one end of the shaft to be measured.
3. Jack the shaft in one direction, then set the indicator to zero.
4. Jack the same shaft back while keeping pressure on it.
5. Note the indicator reading.

Excessive axial runout can point to bearing wear, and may affect wear patterns and how gears mesh. Always check with the manufacturer for runout specifications on each drive.

Replacing worn lubricant seals can be a typical maintenance task on older drives. If the drive is not to be completely disassembled and only the seals are to be replaced, couplings need to be removed in order to remove oil seals. Lubricant may also need to be drained, depending on the level of the lubricant.

PROCEDURE — Removing and Installing Oil Seals

The following is a general procedure for removing and installing oil seals:

1. Cover the shaft with a thin plastic sheet that wraps around the shaft tightly and secure it with tape.
2. Carefully remove the oil seals by drilling two or three small holes in the face of the seal and insert screws. Use the screws to gently and evenly pry the seal free and off the shaft.
3. Completely clean the mating surface of the seal to the housing.
4. An additional light coating of sealant may need to be applied to the outside of the seal and inside the housing seating area, depending on the roughness of the housing.
5. Lightly lubricate the plastic sheeting.
6. Gently push the seal over the plastic sheeting until it starts to seat in the housing.
7. Use a wooden drift and hammer to evenly seat the seal into the housing.
8. Wipe off any excess sealant.
9. Remove the plastic sheeting.

CHAPTER WRAP-UP

Basic maintenance procedures, or the lack of, can greatly influence the life of a power transmission system. Consistent lubrication schedules, proper installation, and sound judgment go a long way to improve the reliability of machinery. When in any doubt, always research and verify proper methods and specifications with the manufacturer. Always have the proper information on-hand when performing any type of maintenance task.

Chapter Review

SUMMARY

- Power transmission systems rely on the transfer of power through an intermediary element. Due to the design of these systems, safety issues, such as pinch points, are an inherent danger. Never operate a machine without the safety guards in place.

- Belt drives transfer power using the friction created between the belt and pulleys. A variety of belts are available. Most belts have embedded tension members that transmit power from the drive sheave to the driven sheave.

- A pulley is a simple machine that transfers power through a wheel and a belt or cable. A sheave is the wheel component of a pulley where the belt or cable sits.

- Attaching a pulley to a shaft is accomplished using setscrews with a key, a tapered bore and bushing, or taper-lock sheaves and bushings.

- The most common method of belt tensioning is to allow 1/64th of an inch of belt deflection per inch of span, measured center to center, between the sheaves. Belt tension may need to be adjusted as belts wear in and seat in the groove of the sheave.

- Under optimal circumstances, both the drive and driven shafts are parallel to the baseplate. Misalignment can result in sheave and belt wear and shortened belt life.

- Properly storing belts will help maintain the integrity of the belts until they are needed for operation. Sheaves should be checked for wear each time a belt is replaced.

- Chain drives are designed to lock into the sprocket of both the drive and driven machinery. A chain drive is a positive drive that maintains an exact speed ratio.

- In addition to standard roller chain, other types commonly used in industry include offset, leaf, and silent chain.

- In a chain drive system, the sprocket drives the chain to transmit rotary motion.

- The two most common causes of chain drive failure are lack of lubrication and improper alignment. The more accurate the alignment between the drive and driven shafts, the less it will affect chain and sprocket wear.

- Initial chain tension should result in approximately 2 percent play on one side of the center-to-center distance of the shafts. Check chain tension after 100 hours of run time after initial installation.

- As a chain wears, it wears and stretches on the links of the chain with the pins. Chain should be replaced when elongation is 3 percent or greater.

- Four main methods of lubrication for a chain include drip, oil bath, flinger, and forced lubrication.

- Gears transmit power between machines while changing rotation, speed, and torque. Gears can also change rotational movement to linear movement. In order to mesh, gears must have the same diametral pitch and pressure angle.

- The most basic type of gear is the external spur gear. However, there are several types of gears that a technician may encounter including internal spur gears, rack and pinions, helical gears, herringbone gears, and bevel gears.

- A worm drive is used to transmit power between two shafts and change the direction of rotation by 90°. It is comprised of a worm that drives a worm gear.

- Ensuring proper lubrication is a key maintenance item with gears. Oil and grease should be changed regularly as part of scheduled maintenance. Follow manufacturer's recommendations for the type of oil or grease to be used.

- When backlash readings are compared to manufacturer specifications, they can reveal tooth wear or bearing wear. Excessive axial runout can point to bearing wear, and may affect wear patterns and how gears mesh.

REVIEW QUESTIONS

Answer the following questions using the information provided in this chapter.

1. Belt drives transfer power using _____ created between the belt and the pulleys.

2. Describe some of the advantages and disadvantages of belt drives.

3. What is a sheave?

4. Describe the most common method of belt tensioning.

5. Identify common factors that lead to belt wear and failure.

6. *True or False?* Chain drives can transfer more horsepower than a similarly sized belt drive.

7. How is silent chain different from standard chain?

8. Explain the function of a sprocket.

9. *True or False?* When placing chain on sprockets, it is generally easier if the connection is made on the top of the smaller sprocket.

10. Explain how to tension a chain drive.

11. Describe how the oil flinger method lubricates a chain.

12. Gears transmit _____ between machines while changing rotation, speed, and torque.

13. _____ is the distance between a point on the face of one gear tooth to the same point on the next tooth.

14. Identify some advantages that helical gears provide.

15. Name the four factors to consider when lubricating gears.

16. *True or False?* Depending on the application, excessive backlash may *not* interfere with the operation of the drive.

17. Why is it important to measure axial runout?

NIMS CREDENTIALING PREPARATION QUESTIONS

The following questions will help you prepare for the NIMS Industrial Technology Maintenance Level 1 Basic Mechanical Systems credentialing exam.

1. A _____ belt is used for power transmission through multiple bidirectional loading sheaves.
   A. flat
   B. double-sided
   C. timing
   D. V-belt

2. The _____ represents the length of a belt at the location of the tension members.
   A. pitch
   B. chordal action
   C. pitch circle
   D. elongation

3. The first digit in the two-digit chain identification number represents the _____.
   A. strength
   B. length
   C. type of chain
   D. chain pitch

4. Which type of chain is composed of heavier links with multiple side plates and strands to increase its strength rating?
   A. Roller chain
   B. Offset chain
   C. Leaf chain
   D. Silent chain

5. Type _____ sprockets have a hub on one side and are mounted using a keyway and setscrew for a plain bore or tapered bushing for a tapered bore.
   A. A
   B. B
   C. C
   D. D

6. The number of teeth per inch of length on the pitch circle of a gear is the _____.
   A. diametral pitch
   B. working depth
   C. circular pitch
   D. clearance

7. For what purpose is some amount of backlash desired in gear operation?
   A. Elongation
   B. Alignment
   C. Tensioning
   D. Lubrication

13 | Conveyor Systems

LEARNING OBJECTIVES

After completing this chapter, you will be able to:

☐ Discuss the benefits of using conveyor systems in industry.

☐ Explain how pulleys are used to operate a belt conveyor.

☐ Identify different types of belting.

☐ Describe the operation of roller conveyors.

☐ Describe the operation of chain conveyors.

☐ Explain the function of common subsystems used with conveyors.

☐ Discuss typical preventive maintenance tasks for conveyor systems.

☐ Discuss typical troubleshooting activities for conveyor systems.

TECHNICAL TERMS

accumulator	head pulley
backpressure	idler pulley
carcass	pick-and-pack operation
carryway	returnway rollers
catenary sag	snub pulley
conveyor	sorter
gravity conveyor	tail pulley
gravity flow rack	take-up pulley

Conveying systems serve many core functions in industry. These systems can be simple on/off controls or completely automated operations for distribution. With the continual improvement and rising complexity of conveying systems, both fundamental mechanical and electrical skills are needed to properly maintain and troubleshoot these systems. In this chapter, we examine many of the fundamental mechanical requirements and touch on some control aspects of these systems. Electrical controls are covered in later chapters.

13.1 CONVEYOR SYSTEMS IN INDUSTRY

Conveyors are mechanical systems that move materials. The uses for conveyor systems are as varied as the industries that employ them. Consider the use of conveyor systems in the following industries:

- Agriculture.
- Food processing.
- Order fulfillment.
- Material recycling.
- Automobile assembly.

The types of conveyor systems used in each of these areas may be different, but the controls and mechanics of all conveyor systems are similar.

The implementation of conveyor systems can benefit an operation in several ways. They minimize, or even eliminate, the manual handling of materials by employees, which can reduce injuries. These systems improve overall workplace efficiency by affording workers time to focus on more productive tasks. Conveying systems also increase the output rate and quality of production by limiting damage to the product. Common conveyor applications found in the manufacturing industry include pick-and-pack operations, manufacturing processes, and bulk material transportation.

13.1.1 Pick-and-Pack Operations

A *pick-and-pack operation* involves gathering and packaging various components for distribution, **Figure 13-1**. This is commonly found in a warehouse environment, often for e-commerce and retail order fulfillment. In this operation, employees pick the individual items of an order, assemble and package the order, and send it out for delivery.

Dmitry Kalinovsky/Shutterstock.com

Figure 13-1. Pick-and-pack operations can use feeder and main line conveyors to move items and orders through a warehouse.

13.1.2 Manufacturing Processes

The ways in which conveyor systems are used in manufacturing are as varied as the types of products that are manufactured, **Figure 13-2**. From food processing to product assembly, conveyor systems move products through various processes within a larger manufacturing system. These types of systems may include many smaller stand-alone systems that move product from machine to machine, or longer complete systems that travel through several processes.

SAFETY NOTE
Riding on Conveyor

Never walk on top of, sit on, or ride on a conveyor.

13.1.3 Bulk Transportation

Bulk transportation conveyors move large amounts of material and products from one area to another, **Figure 13-3**. These systems are characterized by large belts driven at one end by large drums or pulleys and support rollers equally spaced along the length of the belt. Bulk material can only be transported as fast as the belt will run. Because of the weight of the material, belt, and speed desired, these conveyors have larger motors with appropriate gear boxes.

13.2 TYPES OF CONVEYOR SYSTEMS

Conveyor systems can be identified and defined by the mechanism used to move material. Conveyor mechanisms include belts, buckets, rollers, chutes, chains, screws, and many others, as shown in **Figure 13-4**. The type of conveyor system implemented is dictated by the product and process. This section provides an overview of basic types of conveyor systems, including belt, roller, and chain conveyors.

SAFETY NOTE
Machine Guards

Regardless of the type of conveyor in use, never run a conveyor without all the guards in place.

Food processing

Budimir Jevtic/Shutterstock.com

Clothing manufacturing

panpote/Shutterstock.com

Publishing

zefart/Shutterstock.com

Automobile manufacturing

xieyuliang/Shutterstock.com

Figure 13-2. Almost every area of manufacturing makes use of conveyor systems in one or more aspects of operation.

Moving produce into processing

279photo Studio/Shutterstock.com

Airline luggage return

Nomad_Soul/Shutterstock.com

Figure 13-3. Large belts move bulk items in various industries, including agriculture and mass transit.

Slat conveyor

Christine Bird/Shutterstock.com

Chute conveyor

B Brown/Shutterstock.com

Vibrating conveyor

sspopov/Shutterstock.com

Roller conveyor

Baloncici/Shutterstock.com

Screw conveyor

WindVector/Shutterstock.com

Bucket conveyor

Tylinek/Shutterstock.com

Figure 13-4. A wide variety of conveyor mechanisms are used in industries of every size and type.

13.2.1 Belt Conveyors

A belt conveyor is a continuous length of belt that may be laid over rollers as guiding elements or over a low-friction surface. If laid over rollers, **Figure 13-5**, the rollers are driven by chain, O-ring, or belt. If laid over a low-friction surface, **Figure 13-6**, the conveying belt is driven by sprockets.

The belt is moved and controlled by a system of pulleys, as shown in **Figure 13-7**, including the head pulley, tail pulley, and snub pulley. The **head pulley**, or *drive pulley*, drives the conveyor belt. It is located at the discharge end of the conveyor and is typically larger in diameter than other pulleys in the system. The **tail pulley** directs the belt back toward the head pulley and is located at the input end of the conveyor. The **snub pulley** adjusts the amount that the belt wraps around the drive pulley. Positioning of the

snub pulley can increase or decrease contact area of the belt on the drive pulley, which impacts belt traction. Longer conveyor units have nondriving **idler pulleys** that redirect the belt and provide belt tension. The **take-up pulley** is a type of idler pulley that provides tension and can remove slack as the belt wears.

The belt that moves along a belt conveyor can be made of different materials depending on the application and environment. For example, the belt used on a conveyor in a food production application is very different from the belt used for bulk transportation of gravel. Rubber, PVC, coated fiberglass, metal, and plastic are some common materials used for conveyor belts.

Traditional rubber belting is manufactured in layers. Top and bottom layers of rubber surround the carcass of

Powered rollers   Conveyor belt

Johnny Habell/Shutterstock.com

Figure 13-5. Powered rollers may be used on belt conveyors as guiding elements.

Low-friction steel surface   Conveyor belt

Kokliang/Shutterstock.com

Figure 13-6. This conveyor belt is laid over a section of stainless steel, which provides a low-friction support surface.

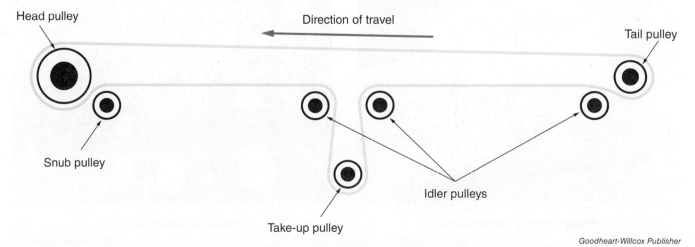

Goodheart-Willcox Publisher

Figure 13-7. A system of pulleys drives the belt of a conveyor and manages belt tension for optimal performance.

the belt. The *carcass* is the center layer of a belt that is composed of various reinforcing materials. This can be single or multiple layers of cotton, nylon, polyester, or even steel cord.

Metallic grid belts, **Figure 13-8**, are woven belts that are typically used for coating applications and in food processing operations. One particular advantage of this type of belt is the ability to withstand a wide range of temperatures, from cooking temperatures over 350° to below freezing. These belts are available in a variety of grid patterns and may be driven either by a roller or sprockets that engage on the edge of the metallic grid.

sspopov/Shutterstock.com

Figure 13-8. Metallic grid belts are commonly used in food processing, such as baking operations.

Plastic belting comes in a variety of sizes and designs and is used in applications ranging from food production to heavy manufacturing and distribution. See **Figure 13-9**. The belt is formed from individual segments that are interlocking

and hinged. Some common materials used to manufacture plastic belting include polyethylene, acetyl, polypropylene, and nylon.

The layout of a plastic belt conveyor is similar to other types of conveyors, with some changes in terminology and design. The belt rides on a low-friction top surface of wear strips. The top surface is called the *carryway*. *Returnway rollers* support the returning belt from the drive end of the conveyor, and ensure the belts wraps 180° around the sprockets. When older plastic belts are worn excessively, the sprockets will also be worn. Removing pitches of the belt does not correct the wear of the belt, even though the belt will be tightened. When replacing the belt, replace all the sprockets on that belt as well.

13.2.2 Roller Conveyors

Roller conveyors use either full length rollers or short wheels to move materials, **Figure 13-10**. The rollers may be powered or rely on gravity to move items. *Gravity conveyors*, also called *static conveyors*, are conveyor systems that have unpowered rollers and are pitched to use gravity to move material along the conveyor. Pitch is the angle at which the conveyor is situated and determines how fast the product travels.

SAFETY NOTE
Entanglement Hazard

Always keep hands, loose clothing, and hair away from the conveyor.

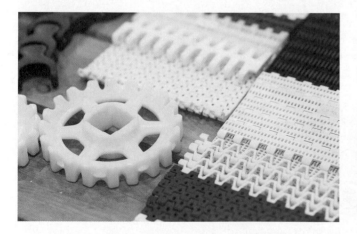

Plastic belting components

Aumm graphixphoto/Shutterstock.com

Plastic conveyor belt line

Marcin Balcerzak/Shutterstock.com

Figure 13-9. Interlocking plastic conveyor belting is available in many different patterns and typically offers longer belt life and less maintenance than other belt materials.

Figure 13-10. This section of the conveyor system includes both rollers and wheels to move material.

13.2.3 Chain Conveyors

Chain conveyors come in several designs and are used for various applications. The type and material of chain used is based on the following operation criteria:

- Strength, horsepower, and speed.
- Requirements for chain material.
- Chain movement for the application, such as roller carried or shoe supported.
- Movement of product for the application.

A chain conveyor is powered by a continuous chain and is typically used to move heavy loads, **Figure 13-11**. The product or material to be moved is placed on the same chain that powers the conveyor. In order to reduce friction and loading, the conveyor may be assembled with integral rollers. The conveyor chain can be quite long and needs a take-up method to adjust for stretch and wear. The take-up method may be spring-powered or pneumatic and places a preload on the chain to automatically adjust for slack as the chain wears.

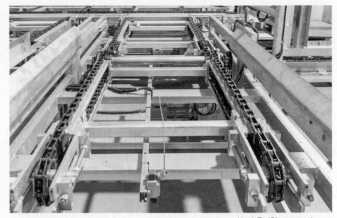

Figure 13-11. In a chain conveyor, product moves along the same chain used to drive the conveyor.

TECH TIP
Take-Up Pulley

Take-up pulleys are not normally used on chain conveyors, as they can excessively increase chain loading.

Not all products and applications are suited for chain conveyors. Items with soft or uneven bases may fall off a chain conveyor or become caught in the links of the chain. Applications that require products to be moved along an incline or around curves are not a good fit for chain conveyors because items can slip along and off the chain.

SAFETY NOTE
Chain Safety

Conveyor chain can become extremely heavy, especially with a load placed on it. Do not split a chain unless the driver is locked out and the chain and load are properly supported.

13.2.4 Conveyor Subsystems

In manufacturing and product distribution, conveyor systems can be simple manual conveyors or be a completely automated system. Local zones or subsystems of a larger conveyor system may have specific purposes, such as transferring product from one line to another, arranging product for palletizing, accumulating product for the next process, rotating the product, or spacing one product from the next. These may be stand-alone zones or they can flow from one conveyor to the next, connecting the entire factory.

Sorters

Sorters separate items on a conveyor line according to predetermined factors, such as destination, size, condition, and many others. Sorters are available in a variety of configurations. The type of sorter used depends on the type of material handled and the movement needed. Sorting subsystems may also include timing devices, sensors, and barcode readers for product identification and tracking.

Paddle sorters use a pivoting arm to redirect items to connecting conveyor lines. The arm moves into the item's line of travel, pushing it off the main conveyor to a side conveyor. Paddle sorters are available in a range of weight capacities to suit individual operations. Push sorters operate in a similar manner to paddle sorters. Instead of a pivoting arm, a bar pushes material perpendicularly off the main conveyor to a side conveyor, chute, or into a container.

Sliding shoe sorters use small, movable blocks positioned diagonally on a conveyor to push product to a take-away conveyor line. This type of sorter gently diverts products, which makes it suitable for a wide variety of applications.

Accumulators

Accumulators provide an overflow or staging area for products on a conveyor line. This type of subsystem can be as simple as a table on the side of a conveyor line or may be an extra wide section of the conveyor to allow items to accumulate. Accumulators are often used when one process feeds another and the output of the first process moves at a faster pace than the secondary process can handle, or when product is produced in batches. Accumulators may also be used to singulate products (place products in a single file line) in preparation for the next process.

With any conveying system, backpressure is an important consideration. *Backpressure*, or *line pressure*, is the amount of force applied between products on a line. If one product is pushed against another, pressure is placed on the product or its packaging. Some products can withstand a great deal of backpressure, such as a stack of bricks, while others are more easily damaged by backpressure, such as lightweight plastic bottles. Accumulators can be used within a conveyor system to reduce or eliminate backpressure and protect product on the conveyor lines.

Gravity Flow Racks

Gravity flow racks, also known as *first-in-first-out racks*, are an angled product storage system that allows product to roll or slide toward the front of the rack, **Figure 13-12**. This type of subsystem is commonly used in pick-and-pack operations where operators pick individual items from the racks to be packaged together. Gravity flow racks increase the efficiency of pick-and-pack operations by bringing a continuous flow of product to packing stations.

Baloncici/Shutterstock.com

Figure 13-12. Gravity flow racks are angled toward the work area to provide easy access to stocked items.

13.3 MAINTENANCE

Regular checks and preventive maintenance should be performed on all conveying equipment to prevent equipment failure and system downtime.

> **SAFETY NOTE**
> **Lockout and Tagout**
>
> Always lockout and tagout a conveyor before performing any work. Communicate with others in the area so they know what you are doing.

13.3.1 Preventive Maintenance

Preventive maintenance on conveyor systems starts the day the system is installed. The first step in the preventive maintenance process is to review the technical and engineering information supplied by the manufacturer. The second step is to regularly perform the recommended maintenance.

Alignment

On initial installation, all components of a conveyor system need to be level and aligned so that shafts are parallel and sprockets are in line. This can be accomplished using a rule and level for smaller systems. Larger systems have more distance between shafts, and a laser alignment tool can be valuable for accurate alignment, **Figure 13-13**. In addition, there are multiple sprockets that are aligned to specific points along the shaft. Both drive and driven sprockets must be aligned to each respective sprocket on the opposite shaft and timed with respect to each other. In any type of drive, misalignment will increase wear of the chain, sprockets, and guides. Alignment should be checked after an initial run-in period and at regular intervals thereafter.

Photo courtesy of LUDECA Inc.

Figure 13-13. This laser alignment tool mounts to the face of equipment and projects a laser line that allows the technician to evaluate any misalignment.

TECH TIP
Distance of Components

Altering distances outside of the manufacturer's specifications can result in sprocket disengagement, excessive chordal action, excessive wear, and belt slippage or binding.

Tensioning

Catenary sag is the amount of belt that hangs on the underside of the conveyor between the supporting rollers or pulleys. This excess length of belt helps with tension and the easy return of the conveyor belt. Manufacturer's recommendations for catenary sag vary, but it is common to allow sag equal to 3% of the distance between rollers. The amount of catenary sag changes with belt wear, loading on the belt, and temperature fluctuations of the belt.

Initial tensioning of the conveyor should be performed slowly and evenly on both sides of the take-up pulley until the proper amount of catenary sag is reached. During this time, the conveyor will need to be run to ensure tracking of the belt. If an automatic tensioning adjustment is used, review the manufacturer's literature on how to adjust it.

Inspection

Regularly inspecting equipment allows conditions to be addressed before problems arise, but also provides the technician with opportunities to become familiarized with the equipment and typical operating conditions.

SAFETY NOTE
Personal Protective Equipment

Always wear proper PPE while performing maintenance.

Depending on the drive type of the conveyor system, lubrication can be a key factor in preventing early wear. Some styles of rollers and drives are sealed, but those that are not should be checked for lubrication. Regularly inspecting and lubricating these elements will greatly extend the life of a chain.

Threaded fasteners are subjected to vibration and can loosen over time. This may affect chain tension, the machine's frame by causing twisting or becoming out of level, and even the mounting brackets of sensors. Periodically check all threaded fasteners for tightness. Consider using thread locker, where appropriate.

Heavy loading increases the wear rate of O-rings and belts that drive conveyor rollers. The O-rings and belts should be checked periodically to ensure power transfer. On plastic conveyor belts, wear strips support both the product and belt. These strips usually wear faster than the plastic conveyor belt and should be inspected periodically. Replacing worn wear strips extends the life of the conveyor belt.

The drive unit of the conveyor should also be regularly inspected. In general, follow the manufacturer's recommendations for the regular maintenance of drive units. Before any inspection or maintenance takes place, lock out the power source to the unit, **Figure 13-14**. A regular inspection includes checking the control cabinet for moisture, dirt, dust, rust, burned areas, or abnormal smells. Vacuum the cabinet clean and research the cause of any issues found. If applicable, check motor insulation bearing lubrication.

Check sprockets regularly for wear. Remember that a chain drive that always runs in the same direction will wear a sprocket on one side of each tooth. As the chain stretches from wear, it will wear the sprocket. As pitch elongates due to wear, it tends to ride up the forward tooth of the sprocket. Eventually, either the drive chain or conveyor chain will wear to the point of needed replacement.

SocoXbreed/Shutterstock.com

Figure 13-14. Lockout and tagout procedures should always be followed when inspecting or performing maintenance on any powered equipment.

SAFETY NOTE
Repair Safety

If performing a splice or repair, make sure the drive is locked out and tension is removed from the belt before removing any connecting pins.

13.3.2 Troubleshooting

Troubleshooting any process includes the following activities:

- Evaluating the problem.
- Correcting the problem.
- Identifying and correcting the cause of the problem.

Evaluating a problem with a conveyor system should include the following questions:

- Is the conveyor performing the way it was designed?
- If not, why not? How is the conveyor performing differently than designed?
- Is this a problem of performance, such as vibration, noise, or wear?
- Is this a problem of failure, such as drive failure, chain failure, or binding?
- What has changed since the conveyor was last working properly?

Under normal circumstances, equipment does not fail in a catastrophic way. It is more likely that wear has taken place over time. Eventually, this wear affects the whole system. It is important to be able to differentiate between normal wear and excessive wear. Manufacturers provide an expected lifetime for belts, drives, and other parts. However, failure due to excessive wear is caused by some abnormal condition that needs to be corrected.

Controls may be the cause of problems. Any type of control that is exposed to mechanical stress or impact will eventually fail. For example, a physically actuated limit switch used for sensing the presence of a package will eventually wear out, become loose, develop a high resistance across its electrical contacts, or become damaged. While nonimpact sensors may last longer, they may need more frequent adjustment.

Variable frequency drives that repeatedly trip must be investigated. Most drives store error codes that can be cross-referenced with the manufacturer's documentation. Overvoltage trips may indicate that the load is too heavy or the drive is trying to slow down the load too quickly. Under voltage trips are an indication of a problem with power supply to the drive. Consider what has

Common Mechanical Issues with Plastic Belt Conveyors		
Noticeable Issue	**Potential Causes**	**Solution**
Sprocket jumps or does not properly engage.	■ Too much catenary sag. ■ Improper roller spacing. ■ Sprockets not spaced evenly. ■ Sprockets not timed properly.	■ Level, align, and check all driven mechanical subassemblies. ■ Check manufacturer's documentation for proper spacing of sprockets and chain or belt sag.
Belts are wearing on sides (not tracking properly).	■ Shafts not level and square. ■ Sprockets not aligned to each other across shafts. ■ Debris on the belt. ■ Area under the belt is not clean.	■ Completely clean entire conveyor. ■ Inspect tightness of all fasteners. ■ Check alignment. ■ Check wear strips for wear and replace if needed.
Excessive sprocket wear.	■ Improper belt or chain tension. ■ Speed is too fast for design. ■ Load or weight of product has been increased. ■ Sprocket not properly lubricated and cleaned.	■ Increasing speed beyond design criteria results in much faster wear. ■ Increased load on belt or chain results in faster wear. ■ Check alignment.

Goodheart-Willcox Publisher

Figure 13-15. Knowing the common issues of equipment in the facility will help a technician become more effective and efficient.

changed since the drive was properly working. The most common drive problems are caused by the environment around the drive and not the drive itself.

Check the condition of metal oxide varistors (MOV) in the system. These components respond to voltage spikes and can short after a large spike, such as lightning or voltage surge. If they appear burnt or melted, this can prevent a drive from starting. Check with the manufacturer to determine a proper replacement.

Figure 13-15 summarizes some of the most common issues with plastic belt conveyors. It is always good practice to have the manufacturer's documentation readily available to check specs and recommendations for a unit.

CHAPTER WRAP-UP

When new systems are installed, the manufacturer typically offers training to employees. Always take advantage of this type of training. If representatives from the conveyor company are performing the initial installation, try to be involved. This is one of the best times to gain knowledge of your new system by asking the field representatives lots of questions. Broaden your knowledge base with further education, manufacturer's seminars, and training on new concepts and controls. These opportunities will keep you up-to-date on new technology in your area and make you an indispensable team member wherever you work.

> ### TECH TIP
> **Manufacturer's Manuals**
>
> Become familiar with the manufacturer of the plastic belts your company uses. Most companies have online technical manuals available for specific belt types. Manufacturer's procedures for installing, maintaining, splicing, and repairing belting should always be reviewed prior to starting a job.

Chapter Review

SUMMARY

- Conveyors are mechanical systems that move materials. They minimize, or even eliminate, the manual handling of materials, which can reduce injuries and improve workplace efficiency.

- Common conveyor applications found in the manufacturing industry include pick-and-pack operations, manufacturing processes, and bulk material transportation.

- Conveyor systems can be identified and defined by the mechanism used to move material, including belts, buckets, rollers, chutes, chains, screws, and many others.

- A belt conveyor is a continuous length of belt laid over rollers or a low-friction surface. The belt can be made of different materials and is moved and controlled by a system of pulleys.

- Roller conveyors may be powered or rely on gravity to move items.

- A chain conveyor is powered by a continuous chain. The product or material to be moved is placed on the same chain that powers the conveyor. Not all products and applications are suited for chain conveyors.

- Subsystems of larger conveyor systems may have specific purposes, such as transferring product from one line to another, arranging product for palletizing, accumulating product for the next process, rotating the product, or spacing one product from the next.

- Regular checks and preventive maintenance should be performed on all conveying equipment to prevent equipment failure and system downtime. The first step is to review the technical and engineering information supplied by the manufacturer.

- After installation, alignment should be checked after an initial run-in period and at regular intervals thereafter.

- Regularly inspecting equipment allows conditions to be addressed before problems arise. Inspection should include the drive unit, fasteners, lubrication points, and belts.

- Troubleshooting any process involves evaluating the problem, correcting the problem, and identifying and correcting the cause of the problem.

- It is important to be able to differentiate between normal wear and excessive wear. Excessive wear is caused by an abnormal condition that needs to be corrected.

REVIEW QUESTIONS

Answer the following questions using the information provided in this chapter.

1. Discuss the potential benefits of implementing a conveyor system.

2. Large amounts of material and products are moved from one area to another using _____ conveyors.

3. *True or False?* The tail pulley is usually larger in diameter than any other pulley in a conveyor system.

4. The _____ is the center layer of a belt that is composed of various reinforcing materials.

5. What is a gravity conveyor?

6. How is take-up applied to chain conveyors?

7. _____ sorters use a pivoting arm to redirect items to connecting conveyor lines.

8. What are the functions of accumulators in a conveyor line?

9. Why is backpressure an important consideration in conveyor systems?

10. What are the benefits of using gravity-flow racks?

11. *True or False?* The first step in the preventive maintenance process is to review the technical and engineering information supplied by the manufacturer.

12. What is catenary sag? What purpose does it serve?

13. What are the benefits of regularly inspecting equipment?

14. Threaded fasteners are subjected to _____ and can loosen over time.

15. What are the three activities included in troubleshooting any process?

16. *True or False?* Under voltage trips are an indication of a problem with power supply to the drive.

NIMS CREDENTIALING PREPARATION QUESTIONS

The following questions will help you prepare for the NIMS Industrial Technology Maintenance Level 1 Basic Mechanical Systems credentialing exam.

1. What type of pulley is item B in the following illustration?

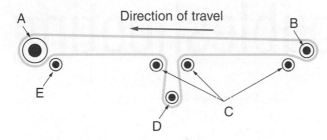

A. Head pulley
B. Tail pulley
C. Idler pulley
D. Snub pulley

2. The _____ pulley adjusts the amount that the belt wraps around the drive pulley.

A. snub
B. tail
C. take-up
D. idler

3. _____ conveyors are conveyor systems that have unpowered rollers.

A. Belt
B. Bulk transportation
C. Chain
D. Gravity

4. Items with soft or uneven bases that need to be moved along an incline are *not* suited for _____ conveyors.

A. roller
B. plastic
C. chain
D. belt

5. Which of the following provides an overflow or staging area for products on a conveyor line?

A. Sorters
B. Gravity flow racks
C. Palletizers
D. Accumulators

6. What is the amount of force applied between products on a line called?

A. Pitch
B. Catenary sag
C. Backpressure
D. Accumulator

7. It is common to allow catenary sag to equal _____% of the distance between the rollers of a conveyor.

A. 1
B. 2
C. 3
D. 4

14 | Mechanical Systems Troubleshooting

LEARNING OBJECTIVES

After completing this chapter, you will be able to:

☐ Explain how a bathtub curve illustrates the failure rate of products.

☐ Explain the meaning of points on a stress-strain curve.

☐ Discuss the differences between resolution, repeatability, and accuracy related to system performance.

☐ Describe how the various types of wear can cause mechanical failure.

☐ Explain how corrosion can cause mechanical failure.

☐ Identify the factors that cause vibration in machines.

☐ Discuss how the actions of human operators and technicians can contribute to mechanical failures.

☐ Explain the difference between preventive maintenance and planned maintenance.

☐ Discuss how the performance of technicians and the environment in which technicians work affect the quality of equipment maintenance.

☐ Discuss the importance of communication in maintenance and troubleshooting.

TECHNICAL TERMS

abrasive wear	galling
accuracy	galvanic corrosion
adhesive wear	hardness
bathtub curve	natural frequency
brittleness	pitting
corrosion	pitting corrosion
corrosive wear	repeatability
critical speed	resolution
damping	resonance
ductility	root cause analysis (RCA)
fishbone diagram	strength
fretting corrosion	stress-strain curve

When a system is not working correctly compared to how it previously operated, this issue in performance is referred to as a *failure*. Troubleshooting is the process of examining what has happened, why it happened, and what needs to be done to correct a failure. Further root-cause failure analysis can also help prevent the failure from recurring. ***Root cause analysis (RCA)*** is the process of examining a failure, following gathered evidence, and pinpointing a probable cause. Not all failures require RCA, as some mechanical systems are complex, while others are relatively simple and easy to diagnose.

14.1 MECHANICAL CONCEPTS

Just as a mechanical assembly is composed of subassemblies and components, a manufacturing operation includes many smaller individual processes. Each of the smaller manufacturing processes is slightly different from one to the next. In an ideal manufacturing operation, variations are kept to a minimum. Even if variations were nearly zero, failures would still occur due to outside influences. The highest quality component is still subject to application, installation, load, and environmental effects. Any of these influences can cause a mechanical component to fail early in its expected service life.

14.1.1 The Bathtub Curve

A *bathtub curve*, **Figure 14-1**, is a graphical representation of the failure rate of products, including mechanical and electronic equipment and components. The curve is divided into three regions:

- **Early-life failure.** Products that experience failure early in their expected service life. Also known as *infant mortality*.
- **Useful life.** Products that fail within the useful period of their service life.
- **Wearout.** Products that fail during the expected end of their service life.

The usable life span of manufactured components tend to fall along a normal bathtub curve. As a product ages, it wears, corrodes, and experiences repeated stress, which eventually lead to expected end-of-life failures. For most manufactured products, the normal life span can be calculated and is often tested as part of rigorous engineering practices and quality control. High-quality products that are

Bathtub Curve

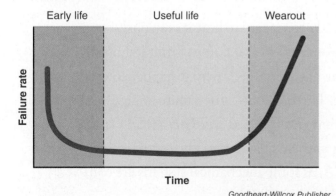

Goodheart-Willcox Publisher

Figure 14-1. A bathtub curve illustrates the failure rate of items early in their expected service life, within their useful life, and at the end of their service life.

manufactured to a high degree of robustness typically outlast inferior products. In general, this is the tradeoff between product life span and initial cost.

Early product failures can happen under the following conditions:

- At the far end of the normal bell curve of variability, the safety margin is small and the application stresses the product beyond what it can handle.
- The product is robust, but the application is incorrect.
- The product is robust, but the installation is performed incorrectly.
- The product is poorly manufactured with a low quality.

Early failures are typically covered by manufacturer guarantees of replacement. Failures throughout a product's service life can be influenced by installation practices, environmental effects, impact loading, improper maintenance, product quality, and many other factors. A technician's major areas of impact are proper installation and regular maintenance.

14.1.2 Materials and the Stress-Strain Curve

In troubleshooting a mechanical component, it is important to have a fundamental understanding of material properties and how the materials act under a load. There are a number of metallurgical properties that affect how materials act, including the following:

- *Ductility.* A measure of a material's ability to deform without breaking under a load.
- *Hardness.* A measure of a material's resistance to penetration, such as surface scratching, abrasion, or denting.
- *Strength.* A measure of how much stress can be applied to a material before it deforms.
- *Brittleness.* A material's tendency to fracture under load with little or no deformation before fracturing.

Depending on the application, one or more of these qualities may be desired. For instance, some materials may be very hard and have high strength, but are brittle. Finding a material with all of the best qualities is difficult.

The *stress-strain curve*, **Figure 14-2**, is a graphical representation of the relationship between the stress applied to a material and the strain produced in the material. All manufacturing materials are tested by applying stress and stretching the material until it breaks in order to determine the proper application. The following are important points along this curve:

- **Proportional limit.** A point on the curve where the values for stress and strain are in proportion to each other.

- **Elastic limit.** The maximum value of stress that can be applied that will still allow the material to return to its original dimensions when stress is removed.

- **Yield point.** The point at which plastic deformation begins to take place and some deformation becomes permanent.

- **Ultimate stress point.** The maximum amount of stress a material can bear before failure.

- **Breaking point.** The point at which the strength of a material fails and results in fracturing.

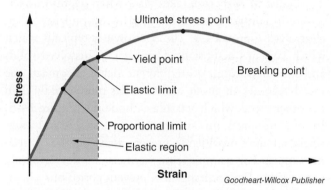

Goodheart-Willcox Publisher

Figure 14-2. The location of points along a stress-strain curve and the size of the elastic region will vary from material to material.

TECH TIP
Stress on Threaded Fasteners

Threaded fasteners have a maximum allowable torque because applying greater torque can result in placing the material within its yield point range where plastic deformation occurs. Threaded fasteners typically fail on the first thread inside the nut because this is where plastic deformation takes place.

Some materials, such as titanium alloys, have a steep elastic region with almost no elastic deformation. Other materials, like mild steel, have a definite elastic region. The application dictates the best-suited material. If we compared a threaded fastener to a metallic sealing gasket, for example, we would see that the threaded fastener needs high strength, while the gasket needs ductility in order to seal the mating parts.

In most mechanical applications, some components are engineered to wear. An example is the rolling-element bearing. The rolling elements in these bearings have a greater hardness than the bearing races. Under load, the races deform from the stress imposed by the rolling element. This results in the eventual wear of the races until the bearing is at its end of life. Another example is a worm drive, in which the worm is made of a harder material than the worm gear. The softer material of the worm gear, usually brass or bronze, wears throughout the lifetime of the worm drive.

14.1.3 Resolution, Repeatability, and Accuracy

In evaluating the performance of equipment, it is important to recognize and understand measures of resolution, repeatability, and accuracy. *Resolution* is the smallest increment a system can measure or recognize. Depending on the operation, resolution may be a measurement of temperature, distance, size, speed, thickness, or many other measurable variables that are critical in an operation. A high degree of resolution is important when determining tolerances, maintaining speed control, and performing surface finishing operations.

Repeatability expresses how well a system reproduces an established outcome under uniform conditions. This applies to any type of system that produces output, including manufacturing, measurement, and communication systems. To determine the repeatability of a manufacturing process, a large quantity of the product produced are measured. If the measurements remain consistent across all the products, the process is deemed to have good repeatability. Fluctuations in the product measurements indicate that a component in the manufacturing process likely needs maintenance or replacement.

Accuracy indicates how close a measurement is to a target or standard value. Consider an assembly operation that involves drilling holes for threaded fasteners and aligning components for final assembly. In this example, the location of the drilled holes would be compared to the programmed or design specifications to determine accuracy in their placement and size. The placement of individual components for product assembly may also be evaluated for accuracy. **Figure 14-3** illustrates the difference between repeatability and accuracy.

High repeatability but low accuracy   **Low repeatability and low accuracy**   **High repeatability and high accuracy**

Goodheart-Willcox Publisher

Figure 14-3. The placement of components in a manufacturing operation likely requires both high repeatability and high accuracy.

14.2 COMMON MECHANICAL FAILURES

Mechanical failures are typically a symptom of a deeper system issue. Technicians work to identify and address the root cause of mechanical failures, which can prevent repeated failures in the future. Common contributors to mechanical failure include wear, corrosion, machinery vibration, and human actions.

14.2.1 Wear

Wear is damage to mechanical components that occurs during normal equipment operation. Some wear is expected during the initial run-in of most machinery. This initial wear helps to polish mating surfaces and actually reduces surface roughness. Types of wear that can cause mechanical failure include adhesive wear, abrasive wear, and corrosive wear.

Adhesive wear occurs when sliding surfaces contact each other with enough force to remove material from one or both of the surfaces. This is often due to a breakdown in the lubrication layer between the sliding surfaces. If severe, galling can result. *Galling* is when two surfaces bond together due to extreme friction. When changing the lubricant, check for metallic particles to identify this type of wear.

TECH TIP
Cross Threading

Cross threading a screw and nut can result in galling. Hand-tighten the screw to make sure the threads are properly engaged before torquing.

Abrasive wear takes place when particles contained in lubricant scrape against the mating surfaces and remove material. These particles may form due to wear within the system itself or may be introduced by an external source. Abrasive wear takes place when the particles are larger than the thickness of the lubricant film, which allows the particles to contact both surfaces under force. Lubricant changes and filtering can prevent this from happening. When grease is used as a lubricant and abrasive wear is noticed, the components should be fully cleaned of old grease and have new grease applied.

Corrosive wear is caused by a chemical reaction between the material of a mechanical component and a corrosive agent. A corrosive agent can remove the protective coating from the surface of a material and cause pitting. *Pitting* is the formation of cavities or holes in the surface of a material. Corrosive wear often results when the material chosen for a system component is not appropriate for the operational environment.

14.2.2 Corrosion

Corrosion is a chemical reaction that causes the deterioration of metal components. It contributes to mechanical failure as corroded material weakens the structural integrity of components. Corrosion that occurs in lubricated systems is normally a combination of two chemical processes: reduction and oxidation. In the simplest terms, *oxidation* refers to the removal of electrons and *reduction* is when electrons are gained. Even with well lubricated components, rust can occur. Even very small quantities of water contained within the lubricant will accelerate corrosion. Water may be introduced through condensation. Moisture that is not visible in lubricant is a major factor in corrosion.

Galvanic corrosion takes place when dissimilar metals are in contact with each other in the presence of an electrolyte, **Figure 14-4**. The electrolyte is typically water, which contains a substantial amount of dissolved solids and other impurities. With different metals in contact, one metal becomes an anode and corrodes more rapidly than the other metal, which acts as a cathode. See **Figure 14-5**. In some instances, this arrangement is desirable and a sacrificial anode is provided to prevent attacking the component metal. For example, zinc sacrificial anodes have long been used in heat exchangers to prevent corrosion.

Pitting corrosion is a localized form of corrosion that produces cavities or holes on the surface of a metal. This

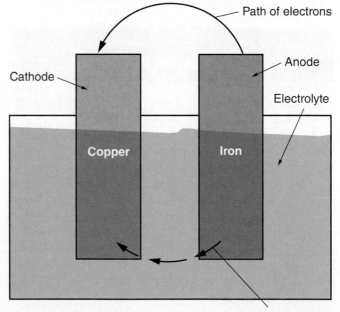
Goodheart-Willcox Publisher

Figure 14-4. In this example, the copper post acts as cathode and the iron post acts as the anode. These dissimilar metals are prone to galvanic corrosion as they sit in an electrolyte solution.

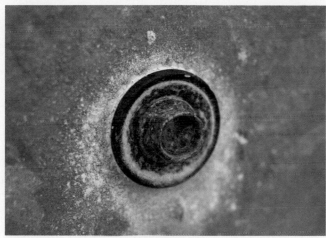

Figure 14-5. This illustrates the galvanic corrosion of an iron bolt in a galvanized steel structure.

type of corrosion starts when the protective surface coating is damaged or compromised. The untreated base metal is exposed and reacts with something in the environment. Pitting can progress deep into the base metal. Protective coatings, surface treatments, and removing chlorides to a very low level reduce the chance of pitting corrosion.

Fretting corrosion is a form of corrosion that takes place at the asperities of close-mating components. Fretting typically takes place under load conditions when the surfaces are in close contact with slight movement, typically vibration. The result of fretting corrosion is grooves in the surfaces and a fine, dry, red dust. This type of corrosion can eventually lead to failure on smaller components, such as threaded fasteners, if left unchecked. To prevent fretting corrosion, use lubricants or coatings on mating surfaces.

14.2.3 Vibration

Machinery vibration analysis can be used to monitor machine operation, check for component failures, and determine machine overhaul timelines. This is typically used on larger machines that are critical to the process and not easily replaced. Vibration analysis is used to evaluate changes in vibration and noise in machinery as compared to the initial installation. To develop this comparison, the following must be completed:

- An initial vibration reading when the machine is newly installed.
- Vibration measurements taken on a consistently spaced schedule.
- A running comparison between initial and current vibration readings.

The readings are compared to an established upper limit of vibration or noise. A qualified engineer evaluates the vibration readings, **Figure 14-6**, and determines the type of maintenance required.

All machines produce some amount of vibration related to the following factors:

- Balance of the rotating assembly.
- Rigidness of the machine mounting.
- Mass of the machine's foundation.
- Damping effects from mounting or attached elements.
- Natural frequency of the machine.

The ***natural frequency*** is a measure of how much an object vibrates without damping. ***Damping*** includes various methods used to reduce the transmission of vibration through machinery. Natural frequency can be observed by ringing a mechanical assembly with a hammer or blunt object and is affected by the weight and stiffness of an object. ***Resonance*** is vibration that occurs when an input force is equal to or close to the natural frequency of an object. At this frequency, vibration increases markedly.

Critical speed is the speed at which a rotating machine element reaches resonance. When operating a machine near critical speed, any vibration of the assembly is amplified and can result in rapid machine damage. Machine designers avoid operating near critical speeds by ensuring the stiffness of the assembly, employing damping methods, and specifying the mass of the foundation.

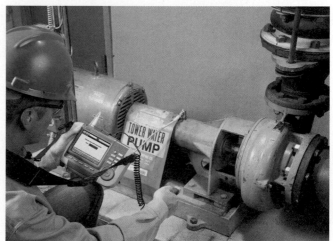

Figure 14-6. This tester produces vibration measurements and spectral diagrams, which allow technicians to quickly assess overall machine condition and maintenance requirements. Here, a technician is recording the shaft rotational speed using a laser tachometer.

PROCEDURE — Measuring Shaft Speed

A tachometer counts revolutions and can be used to measure shaft speed in determining critical speed. To use a noncontact photo tachometer to measure shaft speed, perform the following steps:

1. While the shaft is not rotating, mark a reflective spot on the shaft. Some tachometers may come with reflective tape for this purpose, otherwise a white paint marker can be used.

2. Aim the laser emitted by the photo tachometer at the reflective spot.

3. Initiate the operation that sets the shaft in motion.

4. The tachometer will provide a reading in revolutions per minute (rpm).

An imbalance is caused when the center of a rotating mass is not the same as the geometric center of the rotating assembly. Vibration that occurs due to an initial imbalance in mechanical assemblies can be caused by several factors, including the following:

- Issues within the materials themselves, such as cracks, poor castings, or nonhomogenous mixtures.

- Improper machining, such as an off-center bore, eccentric sheave, or a bore that is not square.

- Improper installation procedures, including mismatched keys and keyways, differing fasteners, and poor alignment.

While the initial balance of a machine may be acceptable, it can change over time in the course of regular operation. Identify the components of the system that are susceptible to wear and be sure to include them in regular maintenance checks. This will ensure that an imbalance condition does not become extreme.

14.3 HUMAN CONTRIBUTIONS TO FAILURES

While some amount of mechanical failure is caused by manufacturing defects, human actions or inaction also contribute to these failures. Operators, technicians, and engineers are involved in the design, installation, maintenance, and repair of machinery. If any of the critical tasks in these activities is neglected, mechanical failure can result.

14.3.1 Lubrication

Improper lubrication can reduce the service life of a component. One of the most common mistakes is applying too much lubrication. In a greased rolling-element bearing, for example, the internal space should never be completely filled with grease. If the bearing is entirely filled it will run hot, which may burn some of the grease and produce hard contamination particles. These particles can damage the contacting surfaces. However, a lack of grease can also lead to bearing noise, high operating temperatures, and early failures.

TECH TIP
Overgreasing

Never put so much grease into a rolling-element bearing that the seal is blown.

At some point in a bearing's life, old grease should be completely cleaned out and replaced, **Figure 14-7**. When greasing bearings on electric motors, examine the motor to see if there is a drain plug opposite the grease fitting. The fitting should be replaced if it is corroded, rusted, or otherwise damaged. Read and follow manufacturer's recommendations on how much grease to use and how often it should be applied.

Ployprapai Mongkolsamai/Shutterstock.com

Figure 14-7. In bearings, and other components where metal parts make contact, grease should be completely replaced at some point in their service life.

Depending on the lubricant and the component being lubricated, contamination can come from dirt, gas, contaminants within the system, water, or wear particles from the components themselves. Consistently scheduled lubricant changes and filter changes help reduce the effects of particulates in lubricant. Wear particles that are introduced into the lubricant can be filtered, if the system has a filter. Many small- to medium-sized gear drives and gearboxes do not have any type of filter. When changing lubricant in these gearboxes, look for the following:

- Obvious wear particles in the drained fluid. If present, use a paper filter to collect.

- The color of the lubricant should be clear and bright.

- Any visible moisture, which may appear as a small bubble at the bottom of the drained lubricant.

- Moisture that is emulsified in the lubricant makes the lubricant cloudy or even opaque, depending on the level of contamination.

14.3.2 Installation

Proper installation is an important factor that affects the service life and future maintenance requirements of a machine. Imperfect initial installation can lead to the following problems with a machine:

- Difficulty with shaft alignment.

- Excessive piping strain.

- Improper bearing lubrication.

- Long-term alignment problems due to a poor foundation.

- Failure of improper machinery hold-downs.

- Improperly seated bearings on shaft shoulders or in housing bores.

- Knowledge gaps for future technicians resulting from a lack of proper documentation.

Machine Foundation

While some machines may be self-supporting or movable, others are mounted on a cement foundation. A concrete foundation should have the following qualities:

- Provide enough mass to lower the natural frequency of the installation.

- Allow the machine to be secured in place with proper hold-down methods.

- Last the entire lifetime of the machine.

- Seal against oil degradation from possible leaks.

As a concrete foundation cures, fine particles rise to the top of the pour creating a layer called *laitance*. This layer does not have sufficient strength to support machinery. Therefore, anchor bolts should not be secured into this layer. Once the concrete is cured, the laitance is removed and replaced with grout to properly support the machine.

Machine anchor bolts are used to hold down the machinery to the foundation, **Figure 14-8**. Concrete screws, lead drop-in anchors, or similar fasteners are not acceptable methods of machinery anchoring. Anchor bolts should be cemented into place with enough free length to stretch in response to machine stress. Without enough free length to allow stretch, anchor bolts will eventually break.

Matee Nuserm/Shutterstock.com

Figure 14-8. This piece of machinery has been mounted to the foundation with anchor bolts. The jacking bolts are used to lift the machinery off the base, typically for alignment purposes.

Alignment

Proper alignment on initial installation is imperative for machinery reliability. Even if the machinery comes aligned on a baseplate from the manufacturer, the alignment still needs to be checked to ensure the following:

- The machine is level.

- Vertical shafts are plumb.

- Shafts that are coupled together are aligned.

- Sprockets, pulleys, and sheaves are aligned.

- Proper tension in any chains or belts.

- Alignment remains consistent after machine reaches an equilibrium temperature.

- No excessive noise or vibration is observed after reaching equilibrium temperature.

- Proper load on the motor (current draw) is observed after reaching equilibrium temperature.

An alignment report is prepared after the machine is installed. This report includes machine specifics, soft foot correction, alignment readings, runout readings, final shims, and shim locations under each machine foot. Coupling sizes, manufacturer, and part numbers are also noted on the report.

14.3.3 Preventive and Planned Maintenance

Preventive maintenance activities are intended to reduce unexpected downtime and machine failures, which improves machinery reliability. Preventive maintenance may cover a wide range of items that are normally performed as recommended by the manufacturer or are advisable based on past experience. Common preventive maintenance items may include the following:

- Lubricant changes.

- Filter replacements.

- Lubricant or fluid sampling.

- Checking operating specifics, such as temperatures and pressures, and making necessary adjustments.

- Tightening the packing on a centrifugal pump.

- Measuring runout on a shaft.

These types of maintenance tasks can be generated through a paper maintenance management system or computerized maintenance management system (CMMS). Regardless of how the maintenance task originated, feedback must be added to the maintenance management system. Feedback items technicians may want to track include these features:

- Consistent change in the color of lubrication.

- Sudden appearance of fine metal particles in recently changed lubricant.

- Abnormally clogged lubricant filter.

- Consistent reduction in output pressure from a pump over time.

- Increase in vibration or noise levels.

Each of these items may indicate a potential future problem. The ability to provide feedback into the system can reduce, if not prevent, future machinery failures. Even in the simplest paper recording system, feedback written into machinery history provides a reference for other technicians.

Planned maintenance items are typically scheduled to occur during regular downtime for the equipment, such as after regular work hours, but before an item shows signs of failure. The following are typical tasks performed during planned maintenance:

- Monitor the condition of machines.

- Evaluate equipment and maintenance records.

- Build up the data contained in an information management system.

- Fix and replace deteriorated equipment.

- Prepare information, parts, and staff and map out a maintenance plan.

- Evaluate and improve the planned maintenance.

THINKING GREEN
Proper Service Is Also Green Service

Service technicians that follow proper service procedures are already being green. Proper service procedures are designed to maximize system functionality and efficiency, while minimizing energy and power loss and untimely wear of components. By following proper service procedures, a service technician can make the system perform better and reduce its impact on the environment.

14.3.4 Attitude and Views

The quality of equipment maintenance is affected by the performance of the technician and the environment in which the technician works. A knowledgeable, competent technician is motivated to keep equipment data current, follow a regular maintenance schedule, and promptly address mechanical issues that arise. Technicians benefit from continually learning new methods, seeking new information, and learning from other experienced professionals.

The overall attitude toward maintenance in the work environment is also a factor in the quality of maintenance performed. Management that values quality, efficiency, and productivity generally encourages the regular use of maintenance resources. Some choose to view maintenance as just another expense. While maintenance is an expense, properly performed maintenance extends the life of machinery and prevents failures. The cost of maintaining a piece of machinery is an inherent cost throughout the lifetime of the machine. When maintenance is regularly performed, technicians can usually identify signs of wear or malfunction and work to avoid mechanical failure and work stoppage. There will always be true emergencies or failures, but these can be kept to a minimum by proper and thorough preventive maintenance and inspection.

14.3.5 Communication

Communication in the workplace is a powerful tool that should be used by maintenance technicians. Communicating ideas, problems, concepts, new learning, and failures are all part of an effective maintenance team.

One very useful team communication tool is the fishbone diagram, **Figure 14-9**. A *fishbone diagram*, or *Ishikawa diagram*, is a visual tool for cause-and-effect analysis to determine a root cause. When troubleshooting recurrent mechanical problems, many team members will have thoughts and experience about the problem to contribute. In this type of brainstorming session, a fishbone diagram can help capture and organize all the potential causes of the problem discussed. Seeing the possible relationships between causes and the problem can be helpful in identifying new approaches and solutions. Perhaps the most important part of this concept is to eventually settle on a target cause, or causes, and design an action plan to prevent the problem from happening again.

Collaborating to solve a problem also involves listening to and learning from more experienced, senior technicians. Technicians who have worked for many years maintaining the same equipment and processes have developed a broad base of knowledge and troubleshooting skills. Drawing on

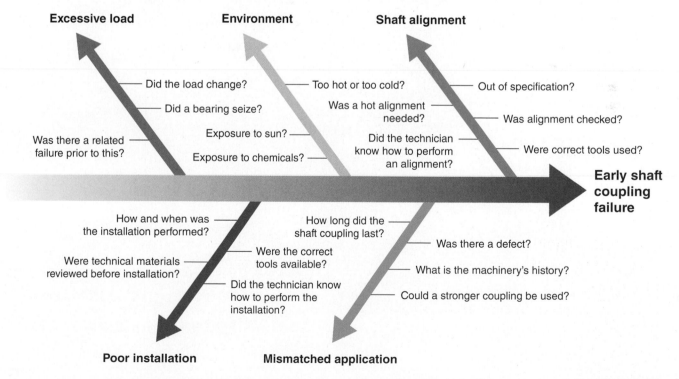

Figure 14-9. This fishbone diagram is an example of a brainstorming session about a shaft coupling failure. Possible general causes and related items to investigate branch off from the main problem.

that experience can be beneficial in bridging the knowledge gap of newer technicians.

CHAPTER WRAP-UP

In the previous chapters of this section, we have surveyed a wide range of mechanical components, assemblies, and applications found in nearly every industry. There are always new components and machines to study thanks to advances in engineering and design. The fundamentals of maintaining and troubleshooting are an important constant for mechanical systems. Review, investigate, measure, and stick to the FACTs: foundation, alignment, coupling, and transfer of power.

Chapter Review

SUMMARY

- Even the highest quality component is still subject to application, installation, load, and environmental effects. Any of these influences can cause a mechanical component to fail early in its expected service life.

- The bathtub curve represents failures of components throughout their expected lifetime. The curve contains three regions: early-life failure, useful life, and wearout.

- Certain metallurgical properties affect how materials act under a load. Depending on the mechanical application, one or more of these properties may be desired.

- The stress-strain curve is a graphical representation of the relationship between the stress applied to a material and the strain produced in the material. The important points along this curve include the proportional limit, elastic limit, yield point, ultimate stress point, and breaking point.

- Resolution, repeatability, and accuracy describe the regular performance of equipment.

- Common contributors to mechanical failure include wear, corrosion, machinery vibration, and human actions.

- Wear is damage to mechanical components that occurs during normal equipment operation. Types of wear that can cause mechanical failure include adhesive wear, abrasive wear, and corrosive wear.

- Corrosion contributes to mechanical failure as corroded material weakens the structural integrity of components. Even very small quantities of water contained within lubricant will accelerate corrosion.

- All machines produce some amount of vibration. Machinery vibration analysis can be used to monitor machine operation, check for component failures, and determine machine overhaul timelines.

- Human actions or inaction contribute to mechanical failures. This may include improper lubrication, imperfect installation and alignment.

- Both planned and preventive maintenance are important aspects of a maintenance program.

Preventive maintenance includes tasks normally performed as recommended by the manufacturer or that are advisable based on past experience. Planned maintenance tasks are scheduled to occur during regular downtime for equipment.

- The quality of equipment maintenance is affected by the performance of the technician and the environment in which the technician works.

- Communication and collaboration are powerful tools in effective maintenance and troubleshooting.

REVIEW QUESTIONS

Answer the following questions using the information provided in this chapter.

1. The process of examining a failure, following gathered evidence, and pinpointing a probable cause is called _____.

2. Briefly explain how a bathtub curve illustrates the failure rate of products.

3. How are the properties of ductility and brittleness different?

4. *True or False?* The elastic limit is the point on a stress-strain curve at which plastic deformation begins to take place and some deformation becomes permanent.

5. The point on a stress-strain curve where the values for stress and strain are in proportion to each other is the _____.

6. What is resolution? How is it measured?

7. Explain how the repeatability of a manufacturing process is evaluated.

8. List three types of wear that can cause mechanical failure.

9. The formation of cavities or holes in the surface of a material is called _____.

10. Corrosion that occurs in lubricated systems is normally a combination of which two chemical processes?

11. What are the signs of fretting corrosion?

12. List three factors that contribute to machinery vibration.

13. Briefly describe the relationship between resonance and critical speed.

14. *True or False?* Moisture that is emulsified in lubricant makes the lubricant cloudy.

15. Machine _____ is (are) used to hold down the machinery to the foundation.

16. How is preventive maintenance different from planned maintenance?

17. *True or False?* The cost of maintaining a piece of machinery is an inherent cost throughout the lifetime of the machine.

18. How is a fishbone diagram used as a communication tool?

 NIMS CREDENTIALING PREPARATION QUESTIONS

The following questions will help you prepare for the NIMS Industrial Technology Maintenance Level 1 Basic Mechanical Systems credentialing exam.

1. Which of the following is a measure of a material's ability to deform without breaking under a load?

   A. Ductility
   B. Hardness
   C. Strength
   D. Brittleness

2. The measure of how well a system reproduces an established outcome under uniform conditions is _____.

   A. resolution
   B. repeatability
   C. accuracy
   D. the bathtub curve

3. A technician notices localized cavities in the surface of a metal plate. This is evidence of which type of corrosion?

   A. Galvanic corrosion
   B. Fretting corrosion
   C. Pitting corrosion
   D. Abrasive corrosion

4. What is the natural frequency of a machine?

   A. A measure of how much an object vibrates without damping.
   B. Vibration that occurs when an input force is applied.
   C. An imbalance caused by vibration.
   D. Noise in machinery as compared to the initial installation.

5. Lubrication changes, checking and adjusting system pressures, and replacing filters are considered _____ tasks.

   A. troubleshooting
   B. planned maintenance
   C. installation
   D. preventive maintenance

Basic Hydraulic and Pneumatic Systems

This third section—*Basic Hydraulic and Pneumatic Systems*—introduces essential principles and components of hydraulic and pneumatic systems. You will learn about fluid power system diagrams and the important relationships between fluid flow and pressure in systems. Finally, some maintenance and troubleshooting tips will prepare you to work on these systems.

The content in this section will help prepare you to earn the NIMS Industrial Technology Maintenance Level 1 Basic Hydraulic Systems credential and the Pneumatic Systems credential. By earning credentials, you are able to show employers and potential employers proof of your knowledge, skills, and abilities.

15

Fluid Power Fundamentals

LEARNING OBJECTIVES

After completing this chapter, you will be able to:

☐ Describe the components of a basic hydraulic system and their purposes.

☐ Explain the application of Pascal's law in a fluid system.

☐ Discuss the basic components of a pneumatic system and their purposes.

☐ Describe the relationship between pressure, volume, and temperature in the combined gas law.

☐ List the purposes of a hydraulic fluid.

☐ Explain how fluid contamination occurs and how it is remedied.

☐ Describe the similarities and differences between hydraulic and pneumatic fittings.

☐ Interpret basic fluid power symbols.

TECHNICAL TERMS

actuator	Gay-Lussac's law
beta ratio	head loss
Boyle's law	kelvin
cavitation	Pascal's law
Charles's law	portable filtration cart
combined gas law	psia (pounds per square inch absolute)
creep	
C_v	psig (pounds per square inch gauge)
differential pressure (DP)	
drop length	solenoid
face velocity	stability
flow	swaging
foaming	viscosity
FRL unit	viscosity index (VI)

The next several chapters will examine fluid power systems, including how they transfer power and how this power is controlled, common preventive maintenance items, and tips for troubleshooting. Troubleshooting skills for both hydraulic and pneumatic systems are very important for a technician, as they are skills that are in very high demand.

15.1 PRINCIPLES OF FLUID POWER

Both liquids and gases are fluids. This means that they flow and will conform to whatever object contains them. It is through the pressurized containment and flow of fluid in a system that fluid power is generated.

Fluid power includes both hydraulics and pneumatics. Hydraulic systems use pressurized hydraulic fluid, which is an oily liquid, to generate power. Pneumatic systems use pressurized air to generate power. Both fluids are used to transfer power through conductors to an end use. An end use could be anything from powering heavy industrial machinery to powering simple hand tools.

While hydraulics and pneumatics can be used in similar ways, each method has its own advantages and disadvantages. Pneumatic systems are typically used for applications requiring higher speeds and lower pressures, such as for powering hand tools or clamping and stamping parts in an assembly line. Hydraulic systems are used where higher pressures are necessary to transfer more power, such as for lifting and powering heavy equipment.

15.1.1 Hydraulics

Hydraulic fluid's incompressibility makes it an ideal medium for power transmission. This characteristic makes it possible to accurately control the flow rate, and therefore accurately control the speed of actuators. *Actuators* are cylinders or motors that use the fluid power to create mechanical motion. They convert the power of the system into actual work.

Transfer of Power through a Fluid

The basic hydraulic system involves several essential parts:

- A reservoir for fluid storage.
- Piping or tubing to conduct the fluid.
- A pump to transfer energy into the fluid and move it through the system.
- Control valves to control the system.
- An actuator or actuators to convert the fluid energy into motion, or work.

Figure 15-1 shows a basic hydraulic system with these components, along with a diagram of the same system using the standard symbols for each component. In this simple system, the hydraulic pump takes suction from the reservoir and increases the fluid's pressure. This pressure is regulated on the discharge side of the pump using a pressure relief valve that limits the maximum pressure in the system. The control valve directs flow to either side of the actuator to either extend or retract the piston cylinder, performing work. Several specific features of the diagram should be noted:

- The symbol representing the hydraulic pump is shown with a solid triangle pointing out toward the piping. The hydraulic fluid flows out of the pump in that direction.

- The symbol for the relief valve shows that the relief valve "senses" pressure from the high-pressure inlet side, that it is normally closed, and that the pressure set point of the valve is set by spring pressure.

- The drawing of the control valve shows that it has three positions and four ports, it is spring centered, and it is actuated by solenoids. The center position blocks both ports of the actuator and lets the pump flow return to the reservoir. If extended, and the solenoids de-energized, the valve will return to its center position and hold the actuator in place.

A port is also called a way, which is why this valve is called a four-way valve. A *solenoid* is a cylindrical coil of wire that acts like a magnet when carrying a current. Solenoids can be used to turn something on or off or to open or close something.

With the pump running and the control valve in the center position, the pump flow is directed back to the reservoir through the valve, rather than on to the actuator. When fluid cannot flow through the whole system, pressure develops. The relief valve is set to a specific pressure for the system. Once the preset relief pressure level is obtained, the relief valve lifts, and the fluid is directed back to the tank, relieving the pressure from the system. In this way, the system's pressure is limited.

At some point, the control valve solenoid is energized and directs pressure (P) to the "blind end" of the cylinder (B), at the same time the rod end (A) is directed through the control valve to the tank (T). This is shown in the third position of the four-way control valve, represented in the diagram as the bottom position with crossing paths. The system pressure is reduced, and the relief valve shuts. At this point, all of the pump flow is being directed to the blind end of the actuator, and the cylinder reaches its fully extended position. When the cylinder reaches its full extension, system pressure rapidly increases, since the flow has nowhere to go, and the relief valve lifts. The control valve solenoid is then de-energized, and the valve returns to its original center position. Fluid is now being returned to the tank, and the relief valve may close slightly or completely, depending on whether any further

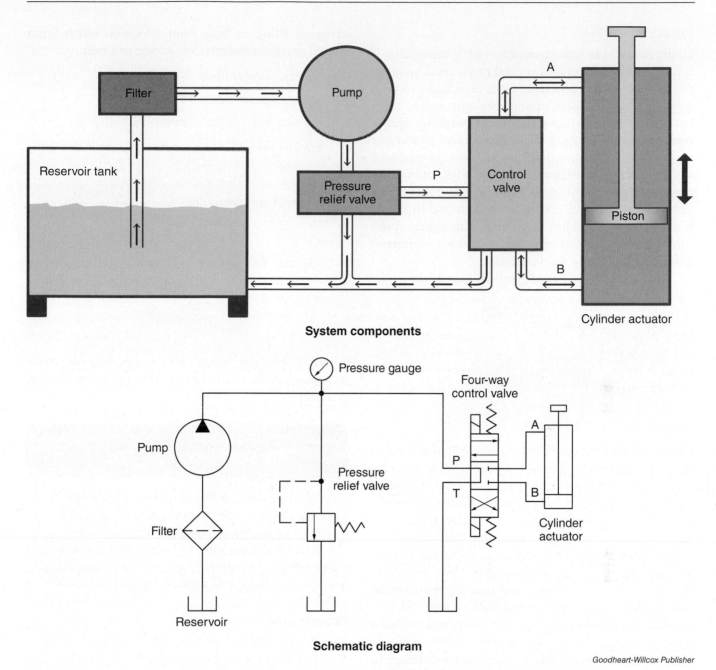

System components

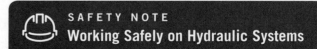

Schematic diagram

Figure 15-1. This drawing and diagram show a basic hydraulic system and its components.

pressure needs to be relieved from the system to maintain the preset pressure level.

If there is a load pushing back on the cylinder, such as when the actuator is being used to perform work, pressure is felt at point B. This pressure is being held back by the control valve. The amount of pressure depends on the amount of load pushing back and the area of the cylinder.

While this is a very basic system, it is important to study and consider the exact operation, flows, and pressures. Doing this with a simple system at first will lead to an understanding of—and therefore ability to troubleshoot—more complicated hydraulic systems.

SAFETY NOTE
Working Safely on Hydraulic Systems

Any work on a hydraulic system must be done only after the system has been locked out and the pressure relieved. Be careful when disassembling components, as they may still have unvented high pressure in them. Wear appropriate PPE when dealing with hydraulic fluids. Any drained fluid must be disposed of properly.

Pressure

Hydraulic fluid is incompressible, which means that if it is contained and force is applied to the fluid, pressure will increase. While pressure may be expressed in several units (pascal, kilopascal, pounds per square inch, torr), hydraulics normally uses psig. *Psig (pounds per square inch gauge)* is a measure of pressure relative to ambient air pressure, with ambient air pressure always measured as 0 psig. Absolute pressure, which includes atmospheric pressure, is abbreviated as *psia (pounds per square inch absolute)*. At sea level, atmospheric pressure is 14.7 psia. If you were to travel below sea level, atmospheric pressure would increase. If you were to climb a mountain, atmospheric pressure would decrease.

Most gauges are calibrated to include psia and show psig.

$$psia = psig + 14.7$$

(15-1)

$$psig = psia - 14.7$$

(15-2)

Flow

Flow refers to the amount of movement of fluid in a system in a certain time and is usually measured in gallons per minute (gpm) in a hydraulic circuit. Flow is caused by a difference in pressure within the system. Fluid flows toward the point of lower pressure. Flow through a valve is caused by the differential pressure across that valve. Some valves have smaller internal clearances and more restriction to flow, leading to higher pressure and lower flow. Other valves may have larger internal clearances and have less restriction to flow, leading to lower pressure and higher flow.

The loss of pressure through a component is called *head loss* or *differential pressure (DP)*. Valves are rated by a unitless number that reflects their restriction to flow. C_v, called "C sub v" or flow coefficient, gives an indication of how much flow would occur through a valve (gallons per minute) with a 1 psi pressure difference across the valve. By comparing C_v of valves, we can easily gauge their restriction to flow.

An increase in flow causes an increase in the fluid's velocity, or speed, through the conductor. As the velocity of the fluid increases, more head loss occurs (that is, the pressure decreases). In most hydraulic systems, we are not overly concerned with head loss because of the high pressures at which the systems operate. A small reduction in pressure will not affect the system's ability to provide power for the necessary operations. However, we are concerned with velocities at the suction and discharge sides of pumps. There are general rules that

designers follow to limit fluid velocities, which limits wasted energy in the form of friction and heat.

- Suction velocity limit: 2–4 ft/s
- Return line (to tank): 10–15 ft/s
- Pressure lines (outlet of pump): 10–25 ft/s

Velocity through a line can be easily calculated using the formula:

$$\text{velocity (in feet per second)} = \frac{\text{gpm} \times 0.3208}{\text{area (in square inches)}}$$

(15-3)

The area (A) is calculated as the area of a cross-section of the circular pipe, using the internal radius:

$$A = \pi \times r^2$$

(15-4)

As the pipe size increases for the same flow rate, velocity reduces. As flow rate increases through a pipe, velocity increases.

TECH TIP
Pipe Size

If replacing piping or fittings, ensure that the new piping and fittings do not cause more of a drop in pressure than the old piping and fittings. Never reduce the size of piping, tubing, or fittings, as this will increase velocity and pressure drop through them.

Pascal's Law

Pascal's law, defined by French mathematician Blaise Pascal in the mid-1600s, applies to a fluid that is confined. *Pascal's law* states that when there is an increase in pressure at one point, this pressure is transmitted through the fluid to all other points in the system.

As an example of how this applies to hydraulics, consider the basic jack in **Figure 15-2**. If a lever used in the jack provides a mechanical advantage of 10:1, then the operator could press down on the pilot cylinder with a force of 50 pounds and the lever would deliver 500 pounds of force onto the top of the pilot cylinder.

This pilot cylinder would develop pressure according to the following relationship:

$$\text{pressure} = \frac{\text{force}}{\text{area}}$$

(15-5)

Figure 15-2. A simple hydraulic jack increases lifting force.

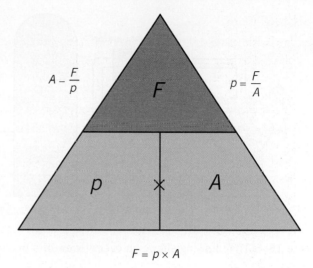

Goodheart-Willcox Publisher

Figure 15-3. This triangle shows the mathematical relationship between force, area, and pressure.

With an area of 0.1963 in² and 500 pounds of force, the resulting pressure would be 2547 psig.

This pressure transfers to the bottom of the main cylinder and presses the cylinder up, lifting the load. The resulting force the cylinder is able to move with the operator's initial force of 50 pounds is found by rearranging the equation:

$$\text{force} = \text{pressure} \times \text{area}$$

(15-6)

With the main cylinder having an area of 4.91 in², find the resulting force:

2547 psig × 4.91 in² = 12,506 pounds of lifting force

The lever supplies a mechanical advantage and demonstrates Pascal's law. **Figure 15-3** summarizes the relationship between force, pressure, and area.

TECH TIP
Small Increments

Note that while the jack will increase the force an operator can deliver because of the difference in areas of the cylinders, the volume to be filled differs. In order to move the main cylinder a small amount, the operator will have to pump the pilot cylinder many times.

15.1.2 Pneumatics

Compressed air is typically used for applications that require higher speed or reduced power compared to hydraulics, or in situations where hydraulic fluid is not compatible with the application. Plastic blow molding is one application where hydraulics may power the machine, but compressed air is used to fill the mold and shape the plastic. Since air is compressible, it is more difficult to exactly control actuator speeds with pneumatic systems. On the other hand, air surrounds us, so there is an unending supply. Where hydraulic fluid must be returned to the reservoir, compressed air that has done its job is simply vented at the actuator or control valve.

Power Transmission

A pneumatic system includes an air compressor. The air compressor introduces energy into the system by increasing the pressure. **Figure 15-4** summarizes the main components of a pneumatic system. For now, examine the system from the point of the **FRL unit**. FRL stands for filter, regulator, and lubricator. This is typically a point at which end use starts—the point where an actuator turns the power into work. Each end use may have its own FRL unit, which then delivers appropriate pressure to the control valve.

Figure 15-5 highlights the circuit after the FRL unit onward to the actuator, and uses graphical symbols similar to those found in a hydraulic diagram. Compressed air is controlled by the directional control valve (shown here as a two-position, five-port valve). This valve is actuated into one position by a solenoid and returned to its original position by a spring. Each line to the double-acting cylinder has a flow control valve, which throttles or limits flow in the reverse direction through a needle valve, while allowing unrestricted flow in the incoming direction by unseating a check valve. With a needle-shaped plunger and small orifice, needle valves allow

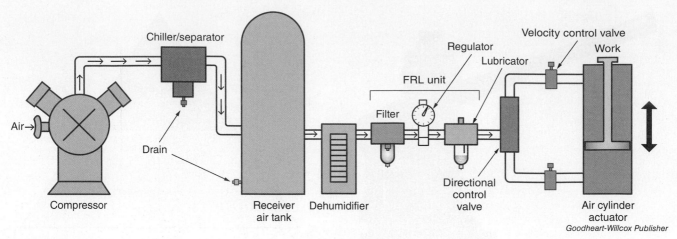

Figure 15-4. This drawing shows an overall view of a pneumatic system.

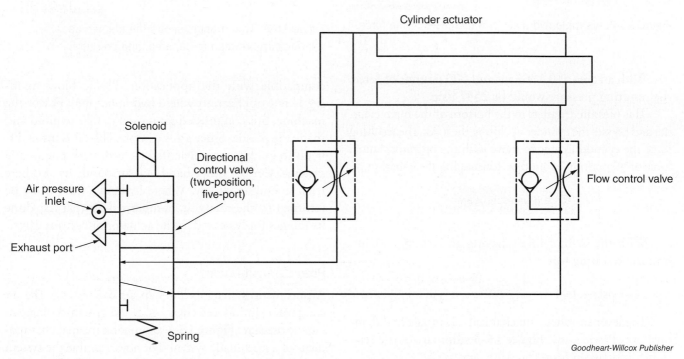

Figure 15-5. This is a graphical representation of the part of a pneumatic system after the FRL, showing a directional control valve and the cylinder.

precise control of the flow rate. Check valves work automatically and are unseated, or opened for flow, when the pressure reaches a certain level. When the solenoid of the directional control valve is actuated, compressed air flows through the flow control valve, unseats the check valve, and flows into the cap end of the cylinder. The rod end of the cylinder exhausts air through the flow control valve—seating the check valve portion and forcing the air to flow through the needle valve—limiting the velocity of the flow. The exhaust air is then directed through the directional control valve and vents through an exhaust port. It is a common practice to throttle or limit flow velocity on the discharge side of pneumatic

cylinders. This results in better speed control, as the cylinder extends to its full extension point.

When the solenoid of the directional control valve is de-energized, a spring return moves the valve back to its normal position. This normal position vents the air from the cap end of the cylinder and applies pressurized air to the rod end of the cylinder. Cap-end air velocity is controlled through the flow control valve, while air to the rod end unseats the check valve and passes unrestricted to the rod end, retracting the cylinder.

Other than understanding the basic flow paths within the system, you should also recognize the graphical symbols and understand their meanings, **Figure 15-6**.

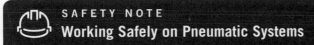
Pressure and Flow

Typical pressures in pneumatic systems range from 120 psig with single-stage compressors to much higher pressures with multiple-stage compressors. Manufacturing processes normally use a pressure of 120 psig. Just as in hydraulics, higher pressures result in higher flow rates through components. Higher flow rates also result in greater pressure losses. Basic calculations for area, force, and pressure for pneumatics are the same as with hydraulics.

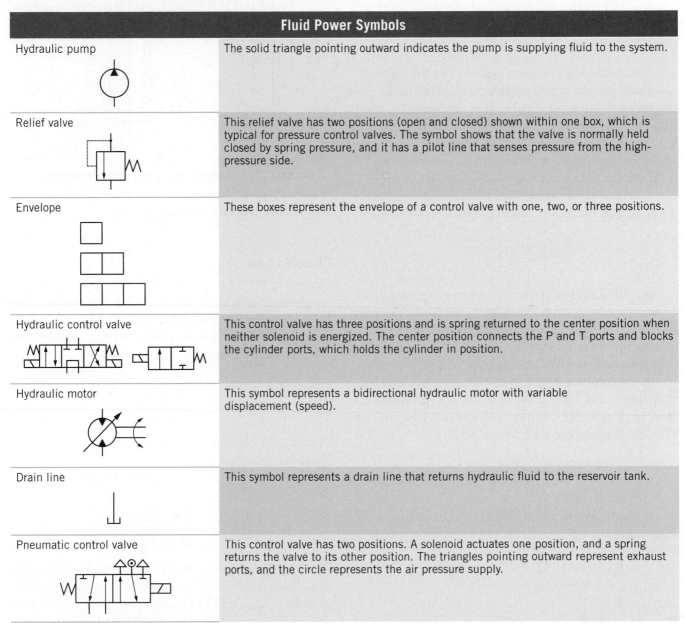

Fluid Power Symbols

Hydraulic pump	The solid triangle pointing outward indicates the pump is supplying fluid to the system.
Relief valve	This relief valve has two positions (open and closed) shown within one box, which is typical for pressure control valves. The symbol shows that the valve is normally held closed by spring pressure, and it has a pilot line that senses pressure from the high-pressure side.
Envelope	These boxes represent the envelope of a control valve with one, two, or three positions.
Hydraulic control valve	This control valve has three positions and is spring returned to the center position when neither solenoid is energized. The center position connects the P and T ports and blocks the cylinder ports, which holds the cylinder in position.
Hydraulic motor	This symbol represents a bidirectional hydraulic motor with variable displacement (speed).
Drain line	This symbol represents a drain line that returns hydraulic fluid to the reservoir tank.
Pneumatic control valve	This control valve has two positions. A solenoid actuates one position, and a spring returns the valve to its other position. The triangles pointing outward represent exhaust ports, and the circle represents the air pressure supply.

Goodheart-Willcox Publisher

Figure 15-6. Understanding graphical symbols used in diagrams is key to being able to troubleshoot a hydraulic or pneumatic system.

Boyle's Law

Boyle's law states that in a gas at a fixed temperature, pressure and volume are inversely proportional. Simply stated, if pressure on a gas is increased, its volume decreases (it is under higher pressure and occupies less space). Another way of explaining this would be to say that the product of the pressure and volume remain constant. As the volume of a fixed mass of gas at a fixed temperature goes up, the pressure goes down proportionally in a predictable fashion, and vice versa. The application of this law is visible today inside any engine or compressor, and even in our breathing. The equation for Boyle's law is as follows:

$$p \times V = k$$

(15-7)

This means pressure (p), measured in absolute pressure, multiplied by volume (V) equals a constant (k). In this case the constant, represented by k, is an unchanged product between two variable quantities: volume (V) and pressure (p). In this first example, with the product always being equal to a constant (k), as pressure (p) increases volume (V) must decrease in order to always arrive at the same result (k).

Not only can a gas be compared at an instantaneous point in time, but it can also be compared across an event:

$$p_1 \times V_1 = p_2 \times V_2$$

(15-8)

In this second equation, relating a system across an event, absolute pressure must again be used. Variables with a subscript of 1 indicate that this is before an event occurs, and variables with a subscript of 2 indicate that this is after the event occurs. **Figure 15-7** shows an example of how compression increases the pressure of air. Confirm Boyle's law by calculating p_2 for the given system. Start by finding the volume of the cylinder before and after the compression, which is area multiplied by height:

$$V = \pi \times r^2 \times h$$

(15-9)

In this example, the diameter of the cylinder is 3″, so the radius is 1.5″.

$V_1 = \pi \times (1.5″)^2 \times (24″)$
$V_1 = 169.6 \text{ in}^3$
$V_2 = \pi \times (1.5″)^2 \times (14″)$
$V_2 = 99.0 \text{ in}^3$

Now solve Boyle's law for p_2 and substitute the given values, including the value for p_1 given in **Figure 15-7**:

$$p_2 = \frac{p_1 \times V_1}{V_2}$$

$$p_2 = 14.7 \text{ psia} \times \frac{169.6 \text{ in}^3}{99.0 \text{ in}^3}$$

$$p_2 = 25.2 \text{ psia or } 10.5 \text{ psig}$$

Boyle's Law in Action

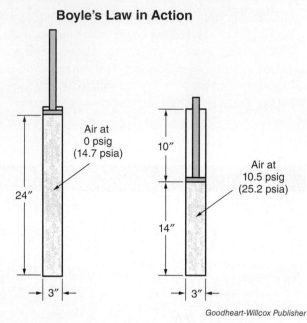

Goodheart-Willcox Publisher

Figure 15-7. The pressure increase due to compression is inversely proportional to the volume. As the air is compressed, pressure increases.

Charles's Law

In the late 1700s, it was found by experimentation that temperature also plays a part in the volume of a gas. Jacques Charles discovered that while pressure is constant, temperature and volume are directly related. **Charles's law** uses the following equation:

$$\frac{V}{T} = k$$

(15-10)

In other words, volume (V) divided by temperature (T) equals a constant (k).

Just as in Boyle's law, this can be rewritten to include an event:

$$\frac{V_1}{T_1} = \frac{V_2}{T_2}$$

(15-11)

Volume units must be the same (both expressed in cubic inches, for example), and temperature is expressed in **kelvin** (K), a unit of temperature measurement often used in the physical sciences. To convert between degrees Celsius (C) and kelvin (K), use the following equations:

$$K = C + 273$$

(15-12)

$$C = K - 273$$

(15-13)

To convert between degrees Fahrenheit (F) and kelvin (K), use these equations:

$$K = (F + 459.67) \times 0.556$$

(15-14)

$$F = [1.8 \times (K - 273)] + 32$$

(15-15)

Figure 15-8 summarizes Charles's law with an example. In this scenario, placing a gas-filled cylinder over heat while maintaining pressure causes the volume of the gas to increase. Use Charles's law to find the volume after heat is applied. Start by finding the initial volume for a cylinder with a diameter of 4″, radius of 2″, and height of 12″:

$$V_1 = \pi \times (2'')^2 \times (12'')$$
$$V_1 = 150.8 \text{ in}^3$$

Now solve Charles's law for V_2 using the values given in **Figure 15-8**:

$$V_2 = \frac{V_1 \times T_2}{T_1}$$
$$V_2 = \frac{150.8 \text{ in}^3 \times 316.7 \text{ K}}{255.6 \text{ K}}$$
$$V_2 = 186.8 \text{ in}^3$$

The volume has expanded nearly 40 in³, causing the piston to rise almost 3″.

Gay-Lussac's Law

In the early 1800s, the French chemist Joseph Louis Gay-Lussac noticed that when an enclosed gas is heated, it expands in a predictable way. If the gas is confined, the added energy to the system increases the pressure. Other scientists also recognized this, and so the law is called by several different names, but here it is referred to as **Gay-Lussac's law** for simplicity's sake. The basic law is that the pressure of an enclosed gas is directly proportional to its temperature, where k is a constant, and both pressure (p) and temperature (T) are in absolute units (psia and kelvin, respectively):

$$\frac{p}{T} = k$$

(15-16)

A change or event can also be examined and the results calculated using the following equation:

$$\frac{p_1}{T_1} = \frac{p_2}{T_2}$$

(15-17)

Figure 15-9 gives an example of Gay-Lussac's law. In this case, a steel cube cannot change shape or volume, even as temperature increases. Use Gay-Lussac's law to find the change in pressure that results from temperature increasing from 70°F (294.5 K) to 800°F (700.4 K).

$$p_2 = \frac{p_1 \times T_2}{T_1}$$
$$p_2 = \frac{114.7 \text{ psia} \times 700.4 \text{ K}}{294.5 \text{ K}}$$
$$p_2 = 272.8 \text{ psia or } 258.1 \text{ psig}$$

Charles's Law in Action

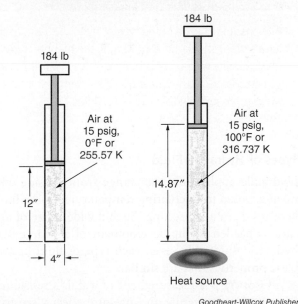

184 lb

184 lb

Air at 15 psig, 0°F or 255.57 K

Air at 15 psig, 100°F or 316.737 K

14.87″

12″

4″

Heat source

Goodheart-Willcox Publisher

Figure 15-8. As temperature is increased, volume increases proportionally.

Gay-Lussac's Law in Action

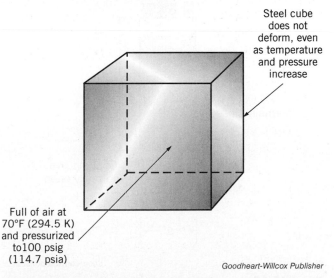

Steel cube does not deform, even as temperature and pressure increase

Full of air at 70°F (294.5 K) and pressurized to100 psig (114.7 psia)

Goodheart-Willcox Publisher

Figure 15-9. If volume is held constant, a temperature increase will cause a pressure increase.

Combined Gas Law

The three previous laws can be combined into one law that governs how a gas behaves with changes in pressure, volume, and temperature. The resulting law, referred to as the **combined gas law**, can be used to compare a system before and after an event.

$$\frac{p_1 \times V_1}{T_1} = \frac{p_2 \times V_2}{T_2}$$

(15-18)

Pressure and temperature should be in absolute scale (psia and kelvin), and the volumes must match each other in scale (cubic inches, for example). **Figure 15-10** shows one example of the use of the combined gas law. In this example, the air in the tank is pressurized to 120 psig (134.7 psia), and its volume is 2714 in³. Find the new volume of the air and the new air pressure after the water in the tank is drained to 4′ 6″. Start by finding V_2:

$V_2 = \pi \times (12'')^2 \times (18'')$
$V_2 = 8143$ in³

There is no temperature change, so the combined gas law can be simplified and used to solve for p_2:

$$p_2 = \frac{p_1 \times V_1}{V_2}$$

$$p_2 = \frac{134.7 \text{ psia} \times 2714 \text{ in}^3}{8143 \text{ in}^3}$$

$p_2 = 44.9$ psia or 30.2 psig

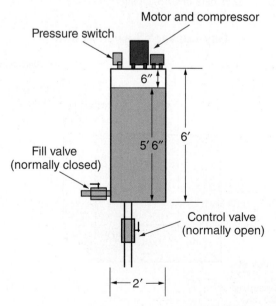

Combined Gas Law in Action

Motor and compressor

Pressure switch

6″

6′

5′ 6″

Fill valve
(normally closed)

Control valve
(normally open)

2′

Goodheart-Willcox Publisher

Figure 15-10. As water is drained from the tank, air pressure decreases.

15.2 FLUID PREPARATION

The preparation of the fluid through which power is transferred affects everything in the system. Contamination introduced from outside the system does not leave the system unless purposefully removed. While it is difficult to control the amount of internally generated particles produced from wear, the introduction of foreign matter can easily be prevented by following the proper procedures.

15.2.1 Hydraulics

In a hydraulic system, the hydraulic fluid performs several key functions:

- The transfer of power to actuators.
- The lubrication of all components internally.
- The transfer of heat generated from friction.
- The transfer of products from corrosion and particles due to wear to the reservoir and filter.
- The prevention of the corrosion of components.
- The sealing of components internally.

Hydraulic fluids range from simple petroleum-based fluids and mineral oils to completely synthetic fluids. Not all types of fluids are compatible with all types of materials used for O-rings, seals, and gaskets. Engineers and manufacturers give careful consideration to compatibility when designing a system.

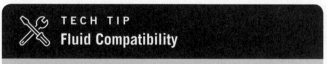

TECH TIP
Fluid Compatibility

Any possible change of the type of hydraulic fluid used for a system should be approved by the manufacturer. Compatibility with every O-ring, seal, and gasket must be considered. Some materials will soften and "melt" when exposed to certain hydraulic fluids.

Types of Hydraulic Fluid

Hydraulic applications can range from outside equipment exposed to fluctuating temperatures to equipment kept inside in a lab setting. Such a wide range of applications has led to the development of a wide range of hydraulic fluids. However, each type of fluid has some basic properties that are similar.

Viscosity is a measurement of a fluid's resistance to flow. Both temperature and pressure affect viscosity. As the fluid gets hotter, its viscosity lowers, which means it is less "sticky" and more likely to flow. As pressure becomes greater, viscosity increases slightly, meaning the fluid is more "sticky" and less likely to flow.

Viscosity index (VI) is a measure of the amount that a fluid's viscosity is affected by temperature change. A higher viscosity index means that the fluid's viscosity is not as affected by a temperature change. Viscosity index routinely ranges from 40 to 190, depending on the type of fluid.

The *stability* of a hydraulic fluid refers to its ability to keep performing its functions over time. A fluid's stability can be affected by temperature and environmental exposure. Additives in the fluid can react over time with metals in the system and create sludge, varnishing, and other deposits that affect system components. Oxidation inhibitors help prevent acids from forming in the fluid, which otherwise degrade the fluid's stability. Follow manufacturers' recommendations regarding the life span of their products.

As hydraulic fluids are compressed under high pressure, they tend to absorb air. This air is contained within the fluid as a solution under pressure, and it is released when the fluid returns to the reservoir. This gas coming out of solution can cause foaming in the reservoir and affect pump operation by causing cavitation. *Foaming* is the production of a mass of small bubbles in a froth. *Cavitation* is caused when the air bubbles implode and wear away the surface of components. Low suction pressure in a pump or the entrainment of gases into the pump may result in cavitation when the bubbles collapse under the pump's compression. This collapse of air bubbles sounds similar to rolling marbles in a tin can and is damaging to internal components. Antifoaming additives for hydraulic oil, proper reservoir design, and the avoidance of air entrainment into the system can help prevent foaming in the reservoir.

Figure 15-11 summarizes types of hydraulic fluid and their properties.

TECH TIP
Avoiding Gases

Gases in a hydraulic system can cause valves and cylinders to become spongy or to not operate at all. The removal of entrained gases may be as easy as ensuring that the system has a proper reservoir level, or as encompassing as checking every seal for leakage. If a seal can leak hydraulic oil out, it may also pull air into the system if placed under a vacuum.

Fluid Contamination

The contamination of hydraulic fluid is the major cause of hydraulic failures. Contamination can cause catastrophic failures, but it also causes erratic response, slower response and cycle times, and a drop in efficiency.

Contamination particles come from a variety of sources. External sources can include the following:

- Airborne dirt brought in through worn wiper seals or improper reservoir breathers.
- Paint chips and flakes.
- Sand from nearby sandblasting.
- Metallic particles from nearby metalwork.
- "Dirt" from the introduction of new hydraulic fluid that has not been filtered.
- "Dirt" from maintenance.
- Cloth and paper particles from the use of cloth and paper rags.

		Common Types of Hydraulic Fluid		
Type	**Subtype**	**Description**	**Common Uses**	
Fire resistant	Water based	■ Types: HFA, HFB, HFC, HWBF ■ Oil-in-water emulsions, water-in-oil emulsions, water-glycol mixtures ■ Contain 30%–95% water	Useful in applications near high-temperature processes.	
Fire resistant	Synthetic	■ Types: HFD, HFDR, HFDS ■ Phosphate esters or hydrocarbon based	Useful in applications near high-temperature processes. Special care must be taken to ensure compatibility with seals.	
Petroleum based	Similar to lubricant oils	■ Types: HL, HLP, HLPD ■ May be mineral oil based, paraffinic, or engine crankcase oil	Contain detergents and dispersants. Contaminants will not settle in reservoir and must be filtered. Tend to have higher water content.	
Biodegradable	Vegetable based	■ Canola oil, soybean oil, oleic vegetable oils, and biodegradable synthetic fluids	Useful in food manufacturing applications and applications where leaks may pose an environmental concern.	

Goodheart-Willcox Publisher

Figure 15-11. This table compares common types of hydraulic fluid.

Internal contamination can also come from many sources:

- Particles from the pump, valves, and cylinders due to wear.
- Corrosion particles, due to oxidation, that break free.
- Water suspended in the fluid, caused by improper reservoir breathers or changes in environmental temperature that lead to internal condensation.
- Internal parts of components becoming varnished.
- Particles generated due to wear on internal surfaces of fluid conductors.

Contamination particles are small—so small that you cannot see them—typically 5 microns in diameter. For comparison, a human hair is about 70 microns wide. However, even these small particles can have a large impact on a hydraulic system that has small internal clearances, especially in newer systems with cartridge valves and higher pressures.

Hydraulic fluid must be sampled to know the level of cleanliness your system is operating at. A lack of monitoring of hydraulic fluid will lead to failures and higher maintenance costs. Only by sampling, considering the results, and determining a corrective action will a cleanliness standard ever be met. The target cleanliness level should be fixed based on the most restrictive or sensitive component in the system. Brand-new hydraulic fluid needs to be properly filtered before adding it to the system.

Filters and Strainers

Fluid filters and strainers are the main method for removing system contamination. While strainers are normally used on the pump suction, filters can be used in a variety of positions in the system, **Figure 15-12**.

TECH TIP
Filters

Check filters every week to ensure they are not bypassing. Filters bypass due to clogging, and the hydraulic fluid then passes through unfiltered. Filters should be replaced before they start to bypass.

When adding new hydraulic fluid to the reservoir, it should be filtered from the container. A *portable filtration cart*, **Figure 15-13**, can serve this purpose. It can also be used as added filtration in a separate loop to clean systems that are not meeting cleanliness requirements. With a separate motor and series of filters, these devices can help remove unacceptable particle or moisture contamination.

TECH TIP
Cleaning Takes Time

Just because filters are changed often does not mean that the system is clean. Filters do not remove every particle in one pass. Particulates will diminish over time when proper procedures are followed.

Filters are rated using a *beta ratio*. The beta ratio compares the number of particles of a particular size upstream of the filter to the number downstream, thus indicating how many have been removed from the fluid that flows through the filter.

$$\text{beta ratio} = \frac{\text{particle count upstream of filter}}{\text{particle count downstream of filter}}$$

(15-19)

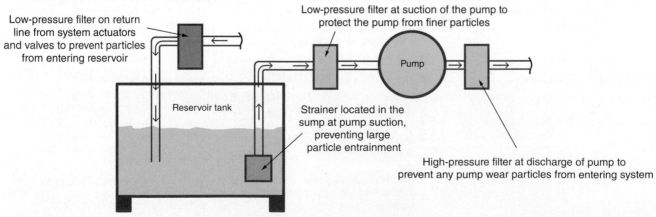

Figure 15-12. This drawing shows possible filter locations in a hydraulic system.

Low-pressure filter on return line from system actuators and valves to prevent particles from entering reservoir

Low-pressure filter at suction of the pump to protect the pump from finer particles

Pump

Reservoir tank

Strainer located in the sump at pump suction, preventing large particle entrainment

High-pressure filter at discharge of pump to prevent any pump wear particles from entering system

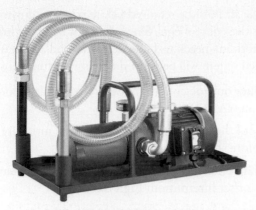

Filter unit

Filter cart

Images courtesy Bosch Rexroth Corporation. Used by Permission.

Figure 15-13. For a wide range of cleanliness applications with hydraulic systems, a portable filter unit or filter cart is a good investment.

A beta ratio of B12 = 100 would mean that 100 times the number of particles of 12 micrometers and larger were found upstream of the filter than downstream of the filter. The efficiency of a filter element is then determined by subtracting 1 from the beta ratio, dividing that result by the beta ratio, and then multiplying by 100 for a percentage. In this case:

$$\frac{100 - 1}{100} \times 100 = 99\%$$

This filter is 99% effective at removing particles 12 microns and larger.

Figure 15-14 shows some of the most common filter symbols used in hydraulic diagrams. Filters are made from a number of different materials and in several designs. The following should be taken into account when choosing the proper filter for your system:

- Flow.
- Pressure.
- Pressure drop.
- Holding capacity.
- Compatibility with the fluid.
- Beta ratio.
- System cleanliness target.

Hydraulic Filters

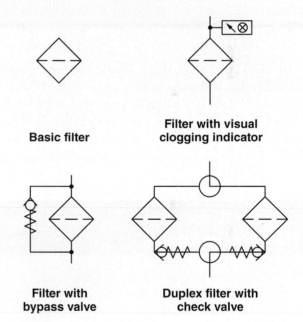

Basic filter

Filter with visual clogging indicator

Filter with bypass valve

Duplex filter with check valve

Goodheart-Willcox Publisher

Figure 15-14. Common symbols for hydraulic filters are shown here.

PROCEDURE	Changing a Filter

When changing a spin-off or housing filter, the following steps should be taken:

1. If a service indicator is included, inspect it and note the position. Is the filter being bypassed?
2. Turn off the system and lock it out.

(continued)

3. Release any remaining pressure in the system.

4. Carefully remove the housing or filter unit.

5. Remove the O-ring or gasket and replace it.

6. Clean all surfaces with lint-free cloths.

7. Replace the filter media and ensure the condition of the new filter.

8. If necessary, lubricate any seals or gaskets and thread to prevent galling.

9. Carefully tighten the filter or housing to make sure threads are not cross-threaded.

10. Slightly hand-tighten filters once seated.

11. Unlock, energize, and restart the system to vent air and check for leakage.

12. Carefully dispose of or recycle the filter element in accordance with local regulations.

In any operation, there should be one person responsible for hydraulic system cleanliness. If this is not the case where you work, be proactive and take this responsibility. Investment in a proactive hydraulic cleanliness program will pay dividends in the form of fewer failures, fewer necessary troubleshooting efforts, and less machine downtime.

TECH TIP
Lint-Free Cleaning

During any form of system maintenance, always use lint-free cloths. Clean your work area and lay down clean plastic sheets, if necessary, before disassembling a valve or cylinder. Before reassembly, use lint-free rags to clean all surfaces you can access. If you cannot complete the work in one shift, make sure exposed components are properly covered and protected from contamination. Wear clean, lint-free gloves.

15.2.2 Pneumatics

Since the medium that transfers power in a pneumatic system is compressed air, concern must be given to the cleanliness of the air. Factors affecting cleanliness include entrained particulates and moisture. Unlike the continuously filtered fluid in a hydraulic system, new air is constantly brought into a pneumatic system. Any contaminants brought in

with the air must be removed before traveling downstream to sensitive components. Since air is not naturally lubricating, lubricant needs to be added near end users. A well-designed system will perform the following functions:

- Filter outside contaminants from the air before compressing it.

- Cool the air and remove most of the moisture before storage.

- Have enough storage capacity in a receiver tank to offset intermittent high demands.

- Prevent excessive compressor run time by avoiding leaks.

- Distribute the air throughout the system with minimal pressure loss.

- Filter, regulate, and lubricate (FRL) compressed air prior to end use.

- Exhaust spent air in such a way as to reduce noise and not present a danger.

TECH TIP
Exhaust

Exhaust mufflers help exhaust spent air more quietly and prevent dirt from being sucked in during the cycling of valves. Valve ports should not be left open to the environment.

Intake Piping

Air intake may happen at the compressor, or air may be piped to the compressor from a remote location. Piping should not restrict the flow of air, which could cause the compressor suction to run at an excessive vacuum. Air intake piping should meet the following requirements:

- Be at least the size of the compressor inlet.

- Be increased in diameter if the length of the run or turns in the run will cause pressure drop.

- Be sloped properly or fitted with a drain leg to prevent large amounts of condensed moisture from entering the compressor.

- Be made of clean, proper material that is free of the possibility of rust, scaling, or the introduction of contaminants to the compressor.

- Take a suction above street level.

- Be positioned properly so as not to entrain moisture.

Figure 15-15 summarizes some of these requirements. Intake piping can be of any material that will remain clean. Galvanized piping is not recommended because of the possibility of flaking. Black iron pipe, stainless steel tubing, and even PVC are all good choices, as long as the internal surfaces are properly cleaned prior to use.

TECH TIP
Draining Moisture

If a drain leg is provided on the suction side of the compressor, do not drain the moisture while the compressor is running. If the valve is opened while the compressor is running, built-up liquid could be pulled into the compressor suction.

Intake Filters

One of the most important factors to consider in the selection of an air intake filter is the velocity of the air. As velocity increases, so does the pressure drop across the filter. The speed of the air passing across the filter is called *face velocity*. Manufacturers have specific recommendations on face velocity limits, which prevent excessive pressure drop across the filter. Of course, increasing the size of the filter will reduce face velocity.

Filters are specified based on micron rating and efficiency. A typical industrial pneumatic system should have at least a 5-micron filter with 99% efficiency. Filter sizes range from 0.1 micron to 100 microns and above with steel mesh. A brand-new intake filter will have some pressure drop across it. Intake piping should provide for test points, and this new pressure drop should be noted. As the filter becomes laden with contaminants, pressure drop across it will increase. If this pressure drop increases substantially, it will have an effect on compressor efficiency, which will increase operating costs. Follow filter manufacturer recommendations on when to clean or replace the filter. **Figure 15-16** shows a wide variety of intake filters.

Filter construction depends on the manufacturer, but most filters use a protective covering that surrounds a steel-reinforced paper or polyester element. The filter may be dry or oil impregnated. Some filter elements can be washed, while some cannot. Consult the manufacturer regarding recommendations for filter care and maintenance.

TECH TIP
Filter Maintenance

No filter can be washed over and over again indefinitely. Eventually filter elements must be replaced. Filter elements that become clogged quickly do so because they are performing their job. This does not mean they should be replaced with a larger-micron element, which could result in severe compressor problems and early failure. If a filter element is quickly clogged, it means the environment at the suction filter needs to be changed, air should be taken from a different location, or a prefilter should be used.

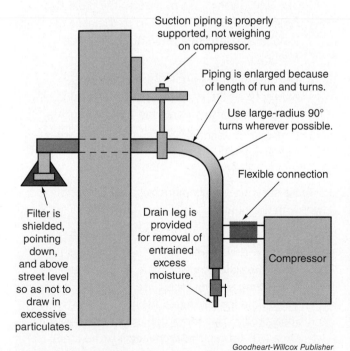

Goodheart-Willcox Publisher

Figure 15-15. Compressor intake piping recommendations depend on the type of compressor, and the manufacturer's recommendations should be followed.

Atlas Copco

Figure 15-16. Filters range from miniature to very large, with many types of media and different micron ratings.

15.3 CONDUCTORS AND CONNECTIONS

Piping, tubing, and fittings used for fluid power systems depend on the application, environment, pressure, flow, and acceptable cost. In a hydraulic system, pressure loss and flow become important factors to consider. In a pneumatic system, distribution and cleanliness may be the crucial focus.

15.3.1 Hydraulics

There is a wide array of fittings and piping used in hydraulic systems, and different types are often used within the same system. While the basics are examined here, you should get to know your local distributor and hydraulic shop if you have numerous hydraulic systems.

Piping

Piping includes both tubing and pipe. Black iron pipe (schedule 80), 304L stainless steel pipe, and 316 and 316L stainless steel pipe are common, but because of cleanliness concerns, black iron pipe is not normally used on newer systems. The majority of piping is threaded with either NPT or NPTF threads. Both thread types require the use of sealing compounds and rely on a metal-to-metal deformation fit as they are tightened. Since threads are deformed when tightened, they tend to leak when repeatedly assembled and disassembled. These threads can also be affected by vibrations and temperature changes that occur in a hydraulic system, and this can cause leakage. In older systems, copper, brass, or aluminum tubing was sometimes used. While they may meet pressure requirements, better materials are readily available and should be used.

Tubing materials are normally seamless or welded low-carbon steel or seamless stainless steel. Tubing is sized in both metric (mm) and standard (inches) measurements for outside diameter, while taking into account the wall thickness, working pressure, and burst pressure. The process for bending tubing is covered in the section on *Maintenance Piping*, along with pipe fitting tools.

TECH TIP
Replacing Tubing

If replacing tubing, always measure the exact outside diameter and wall thickness to ensure the tubing is replaced with the proper materials of the same pressure ratings.

Hoses

Hydraulic hoses come from a number of manufacturers, in a wide range of sizes, and with a wide range of pressure ratings. Both hoses and fittings follow a variety of standards set by recognized organizations:

- SAE (Society of Automotive Engineers).
- DIN (Deutsches Institut für Normung, or German Institute for Standardization).
- ISO (International Organization for Standardization).
- ENs (European Standards, or Norms), set by CEN (European Committee for Standardization).
- JIC (Joint Industry Council).

PROCEDURE | Determine the Proper Replacement Hydraulic Hose

Recognizing the proper type of hose for a replacement may be as simple as reading the original printed specification on the side of the hose to be replaced. If it has worn off, the following procedure should be followed:

1. Determine the dash size, which refers to the diameter of a hose or tubing in increments of 1/16″. In the case of a hose, the dash size refers to the inside diameter. In the case of tubing, the dash size refers to the outside diameter.

2. Be sure that the temperature rating of the replacement hose is higher than the temperature of the conveyed fluid.

3. Consider the application, including factors such as possible abrasion, required flexibility, and required fittings.

4. Be sure the replacement hose is compatible with the fluid.

5. Make sure the replacement hose has the proper pressure rating.

6. Determine the proper length, which is measured from sealing face to sealing face. If the hose includes an angled fitting, the length is measured to the centerline of the sealing face. On a 90° fitting, the *drop length* is the measurement from the centerline of the stem to the sealing face or seat.

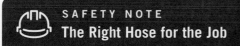

SAFETY NOTE
The Right Hose for the Job

Watch for DOT (Department of Transportation) and other safety ratings that are required for hoses used for specific applications, such as for brakes on vehicles. It is not safe to assume that something else will work or be good enough.

For pressurized hydraulic hose, pressure ratings range from 400 to 10,000 psig. Pressure rating is determined by the materials of construction, number of plies or braids, and type of reinforcement.

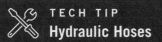

TECH TIP
Hydraulic Hoses

Blistering or leaking hydraulic hoses need to be replaced as soon as possible. Blisters appear due to the temperature limitations of a hose being exceeded. The improper swaging of fittings can cause a leak from the fitting back into the layers of the hose. *Swaging* is the act of joining a fitting to the hose through compression. Abrasion will affect pressure rating, and hoses should be protected from abrasion with hose covers.

Fittings

A large variety of fitting types, styles, and threads are used in hydraulic systems, **Figure 15-17**. In general, sealing types can be classified into tapered threads (interference fit), O-ring seal, metallic compression (flare or cone), and flange.

The best way to determine the type of fitting you are trying to replace is to measure the outside diameter of the midline of the threads using a caliper, and then use a thread gauge to determine the exact match in threads per inch.

15.3.2 Pneumatics

Pneumatic conductors and fittings are typically used with lower pressure than those for hydraulics. Because of this, a wider range of materials is used for pneumatic tubing and fittings.

Piping

Piping for large air distribution systems is normally black iron pipe, or flanged pipe if the system is very large. Smaller systems may use hard copper with sweated or brazed connections. In order to reduce pressure loss, pneumatic systems must be carefully engineered. For instance, fittings must be taken into account. Piping tees, which are T-shaped fittings, cause more pressure loss than Ys, which are Y-shaped connections. If a system

Common Fittings and Threads	
Thread or Fitting Type	**Description**
National pipe tapered (NPT) and national pipe tapered fuel (NPTF) Common in the United States	■ Same threads per inch for same size piping. ■ Must use some type of sealing compound. ■ Rely on interference fit as thread is tightened.
British standard pipe (BSP) and British standard pipe parallel (BSPP) Common on equipment from Europe	■ May be tapered or nontapered threads. ■ May have parallel threads. ■ May seat with a cone and male/female press fit, O-ring, gasket, or ring.
Society of Automotive Engineers (SAE) and Joint Industrial Commission (JIC) flare fittings Common in the United States for steel or copper tubing Sized by tubing size, which does not reflect actual fitting threads	■ SAE may be 37° or 45° flare; JIC is 37° flare. ■ JIC and SAE threads per inch often match, which makes it easy to mismatch components.
O-ring boss (ORB)	■ Same thread size as JIC fittings, but seal is made by O-ring on one fitting and machined seat on mating fitting.
Deutsches Institut für Normung (DIN) 24° and 60° cone fittings	■ Higher-pressure fittings use 24° cone and ferrule, while lower-pressure fittings use 60° cone and ferrule. ■ Threads are metric.

Goodheart-Willcox Publisher

Figure 15-17. This table details some of the more common fitting threads and sealing types. Most fluid tube and pipe sizes in the United States are identified by dash numbers. Dash numbers are the numerator in a fraction where 16 is the denominator. For example, a JIC connection with a dash number of -5 would have a tube size of 5/16″.

is expanded over time, the load on the system can go beyond what it was initially designed to handle. While some compressors are meant to have a 100% run time, others should not run more than 40%–50% of the time.

Connections on the discharge side of compressors should never be sweated copper, only brazed. Repeated vibration and heat can cause sweated copper joints to **creep**, meaning they gradually shift position or pull apart. Threaded pipe connections may also loosen, and an appropriate thread-locking compound should be used.

Pneumatic lines attached to the load may be metallic or plastic tubing. The appropriate materials depend on the application and the following:

- Vibration and movement of the load.
- Required flow and pressure loss.
- Cleanliness requirement of the air.
- Cleanliness requirement of the product the load is acting on.
- Any flexibility requirement.

Fittings

Pneumatic fittings may be metallic (steel, brass, aluminum) or plastic (acetal, polypropylene, polysulfone, PVC).

Fittings come in a variety of sizes and descriptions, including standard NPT threads, metric, quick disconnects, barbed, push-to-connect, compression, and flared sealing mechanisms. The types of fittings used in a system depend on the tubing material and hardness. Some fittings may be used on both plastic and metallic tubing, while others are strictly for metallic tubing. Tubing fittings are classified by the outside diameter of the tubing.

CHAPTER WRAP-UP

This chapter examined important basic concepts of hydraulics and pneumatics, including the parts and materials used and the importance of maintaining and replacing those parts in order to keep fluid power systems operating properly. The chapters that follow will expand on these ideas and apply them to the operation of hydraulic and pneumatic system components. While you explore these topics, compare the similarities and differences in how these systems operate with other systems you are familiar with. This act of reflection helps you reinforce ideas and develop a better understanding of the systems.

Chapter Review

SUMMARY

- Fluid power includes hydraulic systems, which use hydraulic fluid, and pneumatic systems, which use compressed air. Both transfer power through conductors.

- Hydraulic fluid is nearly incompressible, which makes it an excellent medium to use to transfer power.

- A pump transfers energy into the system by converting rotary motion into pressure, and the actuator converts this energy into linear or rotary motion to accomplish work.

- A relief valve limits the maximum pressure in a system.

- The flow rate into an actuator determines the actuator's speed.

- Psia is a measure of pressure that includes atmospheric pressure. Psig is a measure of pressure relative to ambient air pressure, with ambient air pressure always measured as 0 psig.

- Pressure is transferred through a fluid system equally and undiminished in all directions and at right angles to the container.

- Fluid velocities are limited in order to limit the amount of pressure loss across valves, fittings, and piping. This pressure loss is called head loss, or differential pressure.

- As the pipe size increases for the same flow rate, velocity is reduced.

- Pneumatic systems are typically used for applications that require higher speed or reduced power compared to hydraulics.

- To more accurately control the speed of actuation in a pneumatic system, the exhaust air is typically throttled or controlled.

- In order to accurately troubleshoot a hydraulic or pneumatic system, you must understand the diagrams and symbols used.

- Typical pressures in pneumatic systems range from 120 psig with single-stage compressors to much higher pressures with multiple-stage compressors. Manufacturing processes normally use a pressure of 120 psig.

- In an enclosed system, volume is inversely proportional to pressure for gas at a fixed temperature—as volume is reduced, pressure increases.

- Volume and temperature are directly related when pressure is held constant—as temperature of a gas is increased, volume will increase proportionally.

- Pressure and temperature are directly related in an enclosed system—as temperature is increased, pressure increases proportionally.

- With calculations concerning the gas laws, pressure and temperature must be in absolute units.

- Viscosity is a measurement of a fluid's resistance to flow. Both temperature and pressure affect viscosity.

- Viscosity index is a measure of the amount that a fluid's viscosity is affected by temperature change.

- The contamination of hydraulic fluid is the largest cause of failures.

- New hydraulic fluid should be filtered before adding it to the system.

- The beta ratio is a rating of the efficiency of a filter.

- A fluid cleanliness program will reduce machine downtime and failures.

- All air contains contaminants that have to be removed prior to use.

- Intake piping should not restrict airflow, which could cause excessive suction and add to the workload of the compressor.

- The speed of air passing across a filter is known as face velocity.

- Typical industrial pneumatic systems should have a 5-micron filter with 99% efficiency.

- As differential pressure increases across a filter because of contaminants, the efficiency of the air compressor drops.

- Hoses are measured by inside diameter. Tubing is measured by outside diameter.

- The systems you work on may have a wide variety of fittings. Research the types of fittings prior to performing a job.

- Pneumatic fittings are sized according to the outside diameter of the tubing they work with.

REVIEW QUESTIONS

Answer the following questions using the information provided in this chapter.

1. Describe the similarities and differences between hydraulics and pneumatics.

2. Describe how pressure is regulated on the discharge side of a pump in a hydraulic system.

3. What determines whether a cylinder extends or retracts in an actuator?

4. What two variables affect the pressure of a fluid?

5. How are psia and psig related?

6. *True or False?* Flow is caused by a difference in air compression.

7. The loss of pressure through components such as valves, fittings, and pipes is called _____, or differential pressure.

8. Calculate the velocity of fluid flowing at 30 gpm through a tube with an internal diameter of 1.5″.

9. According to Pascal's law, what occurs when there is an increase in pressure at one point in a system?

10. Calculate the resulting force if a pressure of 600 psig is applied to a cylinder with an area of 2.25 in^2.

11. FRL in a pneumatic unit stands for _____, _____, and _____.

12. In **Figure 15-5**, each line to the double-acting cylinder has a flow control valve. What is the purpose of the flow control valve?

13. What is the relationship between pressure and volume when gas temperature is held constant?

14. Calculate the resulting pressure if p_1 = 120 psig, V_1 = 175 cu in, and V_2 = 210 cu in.

15. Calculate the constant (k) if pressure of an enclosed gas is 475 psia and temperature is 22°C.

16. *True or False?* Pressure, volume, and temperature can be calculated at an instantaneous point but also across an event.

17. Assume an initial volume of 1075 in^3 and pressure of 120 psig at 315 K. If you drain the volume to 880 in^3 and the temperature drops to 280 K, what is the resulting air pressure?

18. What does a higher viscosity index mean?

19. List two examples of external and internal sources that can cause fluid contamination.

20. What should be taken into account when selecting a filter for hydraulics and pneumatics?

21. List two factors that can affect the cleanliness of air in a pneumatic system.

22. *True or False?* Lower velocities reduce the efficiency of the filter and affect face velocity.

23. How is the length of a hydraulic hose assembly measured?

24. What determines the piping size for a pneumatic system? For what reason?

NIMS CREDENTIALING PREPARATION QUESTIONS

The following questions will help you prepare for the NIMS Industrial Technology Maintenance Level 1 Basic Hydraulic Systems credentialing exam and Pneumatic Systems credentialing exam.

1. According to Pascal's law, if a lever used in a basic jack provides a mechanical advantage of 10:1 and the operator presses down on the pilot cylinder with a force of 10 pounds, how many pounds of force would the lever deliver?

   A. 10 pounds
   B. 100 pounds
   C. 1000 pounds
   D. It cannot be calculated without a temperature value.

2. In the following diagram, symbol 3 represents what hydraulic component?

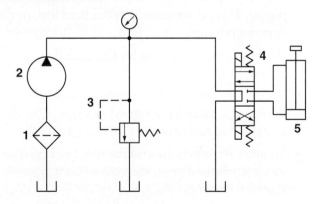

   A. Flow control valve
   B. Hydraulic pump
   C. Relief valve
   D. Solenoid

3. In the following diagram, when the control valve solenoid is energized and directs fluid flow to the "blind end" of the cylinder, which of the following describes the result?

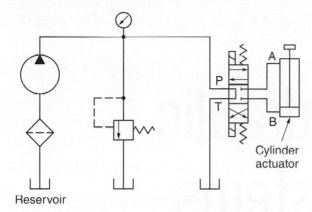

Reservoir

A. Fluid flows to side B of the actuator and the cylinder extends.
B. Fluid flows to side A of the actuator and the cylinder retracts.
C. Fluid flows directly from the pump to the reservoir.
D. Fluid flows back to the relief valve and to the tank.

4. What part of a pneumatic circuit is depicted in the following diagram?

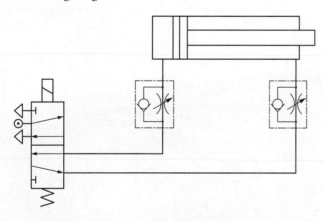

A. After the FRL unit onward to the actuator.
B. The relief valve onward to the single-acting cylinder.
C. The piston onward to the pressure regulator.
D. The compressor onward to the control valve.

5. In a pneumatic system, which valve throttles or limits flow in the reverse direction through a needle valve?

A. Directional control valve
B. Flow control valve
C. Relief valve
D. Solenoid

6. What is the purpose of a portable filtration cart in a hydraulic system?

A. To mechanically monitor the filter quality on a daily basis.
B. To test the level of system cleanliness by evaluating the percentage of particles of a particular size removed from a fluid.
C. To remove system contamination on the pump suction.
D. To replace hydraulic fluid and help clean systems not meeting cleanliness requirements.

7. What is the relationship between flow, velocity, and differential pressure?

A. An increase in flow will cause an increase in fluid velocity and differential pressure.
B. An increase in flow will cause a decrease in fluid velocity and increase in differential pressure.
C. A decrease in flow will cause an increase in fluid velocity and differential pressure.
D. A decrease in flow will cause a decrease in fluid velocity and increase in differential pressure.

8. Which of the following is true concerning viscosity?

A. Viscosity is a measurement of a fluid's flow.
B. Only temperature affects viscosity.
C. Viscosity index is a measurement of the amount that a fluid's viscosity is affected by temperature change.
D. Only pressure affects viscosity.

9. If a hydraulic hose includes a 90° fitting, how is the proper length taken?

A. From sealing face to sealing face.
B. From the centerline of the stem to the sealing face.
C. From the dash side of the hose to the sealing face.
D. From the fitting to the centerline of the stem.

10. What does a beta ratio of B8 = 100 indicate?

A. That 100 times the number of particles of 8 micrometers and larger were found upstream of a filter than downstream.
B. That 100 times the number of particles of 8 micrometers and larger were found downstream of a filter than upstream.
C. That 8 times the number of particles of 100 micrometers and larger were found upstream of a filter than downstream.
D. That 8 times the number of particles of 100 micrometers and larger were found downstream of a filter than upstream.

16 | Hydraulic Systems

LEARNING OBJECTIVES

After completing this chapter, you will be able to:

☐ Describe the purposes of a reservoir and how it performs these functions.

☐ Explain how a pump develops suction.

☐ Discuss the main valve types and how they perform their functions.

☐ Perform common calculations for both hydraulic pumps and motors.

☐ Draw common symbols for directional valves and their actuators.

☐ Describe why pressure and temperature compensation are needed.

☐ Discuss common inspection elements when rebuilding a cylinder.

☐ Describe a general method for rebuilding a hydraulic motor.

TECHNICAL TERMS

detent	positive-displacement pump
directional control valve	positive suction head
discharge pressure	pressure compensation
displacement	pressure control valve
envelope	pressure-reducing valve
flow control valve	relief valve
gear pump	revolutions per minute (rpm)
heat exchanger	sacrificial anode
heat sink	seal
horsepower (hp)	sequence valve
hydraulic cylinder	silting
hydraulic pump	spool
hydraulic reservoir	temperature compensation
lands	valve actuator
pilot valve	vane pump
piston	varnish
port	venturi

In this chapter, the major components of a hydraulic system are examined in detail. In order to properly troubleshoot a system, you must understand the purposes of each component, how the components perform their functions, possible ways the components can fail, and how each component fits into the system or subsystem. Troubleshooting a hydraulic system requires knowledge, logic, and thought based on what is happening to the system. Anyone can remove every valve in a manifold and check each for contamination, but a competent technician will be able to pinpoint the malfunctioning valve, and then remove and repair it.

16.1 BASIC HYDRAULIC SYSTEM

The basic hydraulic system, **Figure 16-1**, involves the following essential parts:

- A reservoir for fluid storage.
- Piping or tubing to conduct the fluid.
- A pump to transfer energy into the fluid and move it through the system.
- Control valves to control the system.
- An actuator or actuators to convert the fluid energy into motion, or work.

16.2 RESERVOIRS AND PUMPS

Hydraulic reservoirs do much more than just store fluid for the system. Proper design and maintenance ensure that contamination is not introduced. Hydraulic pumps are the most important component of the system. Without the pump, no energy is transferred into the system, and no work can be done.

16.2.1 Reservoirs

Hydraulic reservoirs store hydraulic fluid in a tank and perform several other functions in the system, including the removal of heat, settled particulates, air, and moisture. Typically, reservoirs are sized large enough to supply two to four times the pump flow rate. Reservoir

configuration is often based on physical constraints due to the machine or environment in which it is placed. **Figure 16-2** shows one of the most common configurations: a horizontal reservoir tank with the pump mounted on top.

Konstantynov_AA/Shutterstock.com

Figure 16-2. This hydraulic reservoir power unit is a common configuration, with a horizontal reservoir tank with the pump mounted on top.

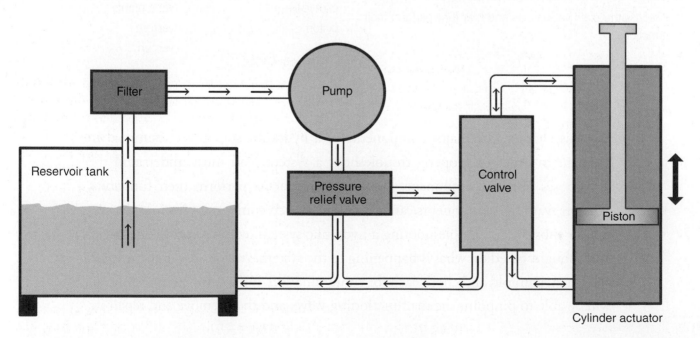

Goodheart-Willcox Publisher

Figure 16-1. This is a basic hydraulic system including essential components.

Purposes of the Reservoir

Figure 16-3 shows the internal details of a reservoir. Not only does the reservoir store fluid, but it also serves as a heat sink. A *heat sink* is a device that accepts heat and transfers it to the surrounding environment.

When starting a cold hydraulic system, the fluid temperature is the same as the surrounding environment. As the system runs, heat is transferred into the fluid. As fluid is returned to the reservoir, it mixes with cooler fluid. The average temperature of the hydraulic fluid in the reservoir will rise until a new equilibrium temperature is reached. Heat is transferred out of the fluid through the walls of the reservoir into the surrounding environment. The larger the surface area of the reservoir, the faster the heat will transfer out of the fluid.

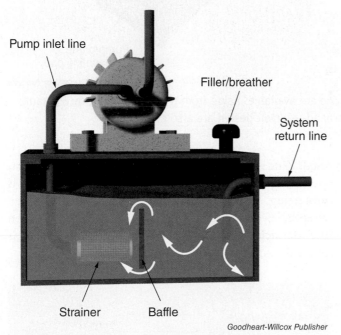

Pump inlet line

Filler/breather

System return line

Strainer

Baffle

Goodheart-Willcox Publisher

Figure 16-3. The reservoir is the first place to start a cleanliness program. Proper cleaning of the reservoir can greatly reduce the particles in the system.

⚠ CAUTION

The temperature of hydraulic oil should never reach 180°F (82°C). Temperatures higher than that will damage seals and negatively affect the system.

The reservoir also functions as a settling area for solids in the fluid. If designed properly, larger particles will settle out in the reservoir. Air can be entrained in the

fluid as it passes through the system, and the dwell time of the fluid in the reservoir helps to encourage the air to come back out of solution.

The reservoir tank can serve a number of additional purposes:

- Allow for the cleaning out of settled particulates and wear particles.
- Provide a positive suction head for the pump.
- Provide for a secondary cooling system and its connections.
- Provide for the addition of fluid to the system.
- Provide indicators for fluid level.
- Act as an air breather and moisture separator to prevent the addition of moisture due to condensation and to prevent a vacuum from forming due to changes in temperature.
- Provide extra connections for the addition of off-line and portable filtration.
- Act as a platform onto which other system elements can be mounted, allowing for a compact package.

A *positive suction head* is provided when the fluid height is above the suction height of the pump and there is minimal restriction to the pump. Without a positive suction head (or pressure), the pump must be able to draw fluid by creating a vacuum at its suction point. A pump performs more work when drawing a vacuum at its suction point, and some pumps are not capable of doing this.

Reservoir Design

Reservoir design is somewhat standardized for stand-alone power units, but it also depends on the application. Pumps may be internal or external to the reservoir, and pump suction, discharge, and return-line filters may all be present. Standard packages for the reservoir tank are in horizontal, vertical, L-shape, JIC (Joint Industry Conference), and overhead configurations (see **Figure 16-4**). The pump and motor can be mounted horizontally or vertically on or near the reservoir. While some designs allow access through the top of the reservoir, others have clean-out plates, also called clean-out covers, on the side. These are round plate covers that can be removed in order to clean out the tank. Heat exchangers may also be supplied as part of the integral package. *Heat exchangers* are devices that transfer heat from one medium to another. These heat exchangers may be fan-and-coil exchangers, or they may be shell-and-tube heat

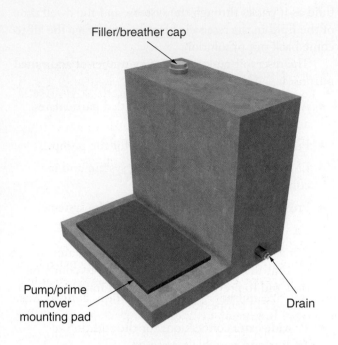

Filler/breather cap

Pump/prime mover mounting pad

Drain

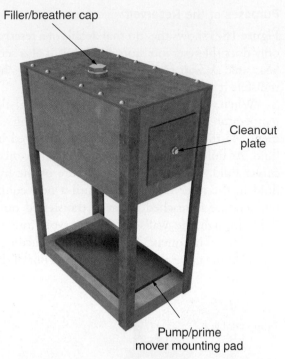

Filler/breather cap

Cleanout plate

Pump/prime mover mounting pad

Goodheart-Willcox Publisher

Figure 16-4. Several standard types of power unit configurations are available. Aside from the horizontal reservoir tank with the pump mounted on top, the L-shaped design and overhead-reservoir design are also common.

exchangers, as shown in **Figure 16-5.** Shell-and-tube heat exchangers may be single pass, double pass, or four pass, with water flowing through the tubes.

All heat exchangers require planned maintenance. Both fan-and-coil and shell-and-tube heat exchangers lose their heat-removal capacity as they become dirty, resulting in a higher fluid temperature. Use a spray-on coil cleaner to clean a fan-and-coil heat exchanger. Clean a shell-and-tube exchanger by removing the end plates and "punching" the water-side tubes with an appropriate-size scraper. Gaskets must be replaced, and any sacrificial anodes should be checked and replaced, as needed.

TECH TIP
Heat Exchangers

If heat exchangers are not cleaned as a part of regularly scheduled maintenance, the system will reach a higher equilibrium temperature. If left unchecked, this higher temperature can affect system operation, the fluid, and hoses with lower temperature ratings.

Corrosion is always a concern in a hydraulic pump system. To prevent the corrosion of more important and expensive materials, *sacrificial anodes* made of zinc are

used. Zinc is a highly active metal and will corrode first, sacrificing itself in place of the more valuable metal it is protecting. Some sacrificial anodes, **Figure 16-6,** have a weep hole. This hole will start to "weep," or leak, when the sacrificial anode has corroded sufficiently. At this point the zinc anode should be replaced.

⚠ **CAUTION**

Do not replace a leaking zinc sacrificial anode with a simple pipe plug. The leak is meant to tell you that it is time to replace the anode. If you use a pipe plug to stop the leak without the zinc sacrificial anode, this will lead to the corrosion of the metal of the component, which will result in the eventual failure of the component.

Reservoir Maintenance

A substantial investment is made for any hydraulic system or machine, and this investment needs to be maintained. The first step in this care is regular fluid sampling and evaluation. Most hydraulic failures are caused by a lack of fluid cleanliness.

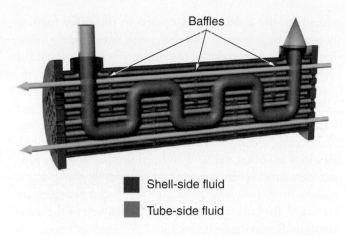

Baffles

■ Shell-side fluid

■ Tube-side fluid

Bonnets available in one-, two-, or four-pass designs

Tubes

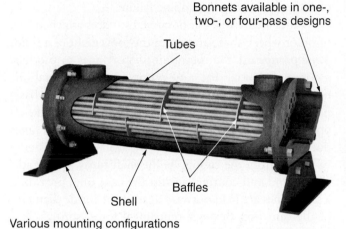

Baffles

Shell

Various mounting configurations

Goodheart-Willcox Publisher

Figure 16-5. Shell-and-tube heat exchangers use water to cool or heat the hydraulic fluid, which is routed through the shell by baffles. Copper, iron, steel, and aluminum are materials commonly used for heat exchangers, but more resistive materials can be used, at a higher cost.

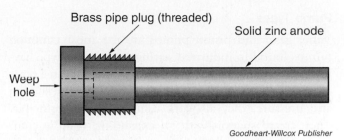

Brass pipe plug (threaded)

Solid zinc anode

Weep hole

Goodheart-Willcox Publisher

Figure 16-6. Some heat exchangers include sacrificial anodes that must be regularly checked and replaced to prevent corrosion.

16.2.2 Pumps

Hydraulic pumps are mechanical devices that convert mechanical energy into hydraulic energy by generating flow. Pumps come in a wide variety of designs depending on their size and horsepower, application, connections, and type. In general, hydraulic pumps are much smaller physically than the electric motors that drive them.

SAFETY NOTE
Lockout/Tagout

Always lock out and tag out electricity prior to performing any service on a power unit. Fluid pressure is also a source of energy that must be drained or vented slowly and safely prior to performing work on a system. High-pressure fluid that is not drained can easily cause severe injury. Know your system prior to working on it.

PROCEDURE   Cleaning a Hydraulic System

At some point, system cleanliness may no longer be maintainable. In that case, the following systematic approach may be needed:

1. Drain the entire system as much as possible into the reservoir.

2. Drain the reservoir and use an off-line filter system to improve the cleanliness of the fluid.

3. Replace all filters with new, properly sized filters.

4. Clean any pump suction strainers completely.

5. Clean the reservoir completely by flushing it with an approved cleaning fluid and drying it with lint-free cloths.

6. Paint the reservoir with an appropriate paint, if the manufacturer recommends this.

7. Close the reservoir and refill it with fluid, filtering the fluid as it is added.

8. Replace the air breather with a moisture-coalescing breather.

9. Restart the system, and after it reaches operating temperature, take another fluid sample and have it analyzed.

10. Consider placing more filters at tank return lines.

11. Follow the suggestions of the company that performs the fluid analysis to meet cleanliness levels.

Pump Types

Vane, gear, and piston pumps are the most common pumps found in industrial settings, with each type having advantages and disadvantages. While the three types of pumps have different methods for developing suction and pressure, they all work using the same principle. They create suction with an expanding volume coming in, and create discharge pressure with a reduction in that volume going out. ***Discharge pressure*** is the pressure of the fluid as it leaves the pump. These pumps are considered positive-displacement pumps. ***Displacement*** refers to the movement of something from one place to another. A ***positive-displacement pump*** moves a fluid by trapping a certain amount in a sealed area and then forcing that trapped volume into a discharge pipe. As fluid is forced out at the discharge side of the pump, pressure rises in the system as a resistance to that fluid flow. Controls limit this system pressure, and pumps can also be pressure compensated to limit their own flow.

Figure 16-7 shows how an external ***gear pump*** uses the rotary motion of meshing gears to directly move fluid through the unit. The pumping action of the mechanism results from the unmeshing and meshing of the gear teeth as the gears rotate. As the gears unmesh in the inlet chamber of the pump, the volume of the chamber increases, creating a low-pressure area. The low-pressure condition causes hydraulic fluid to be forced into the

chamber. The fluid is then carried in the space formed between the gear teeth and the closely fitted pump housing, and it is forced out of the pump when the volume of the discharge chamber is reduced as the gear teeth mesh.

The gear pump is typically the least expensive hydraulic pump and is normally easy to repair or rebuild, but wear usually takes place on both the gears and the casing. If the casing is badly worn, replacement is usually required. Bearings are often made of sintered bronze, with no rolling elements. When repairing or rebuilding gear pumps, gaskets must always be replaced. If threaded fasteners are damaged, they must be replaced with fasteners of the same strength. Replacing fasteners with lower-strength materials could lead to rapid catastrophic failure.

Vanes are narrow flat or curved surfaces attached to a rotor, or wheel, that turns from the movement of a fluid. ***Vane pumps*** use the rotary motion of vanes to produce a pumping action. The vane pump comes in several different configurations. **Figure 16-8** shows the most basic noncompensated vane pump. This type of pump develops the expanding and collapsing volume between vanes due to the eccentricity caused by the shaft and rotor being off-center from the casing. As the shaft turns, vanes radially expand and contract against the cam ring, providing a seal. Vanes are held outward against the inside diameter of the cam ring through centrifugal force, springs, and hydraulic pressure. In this most basic type of vane pump, a constant flow of fluid is pumped. That constant flow creates increasing pressure in the system, so further controls are needed to regulate system pressure.

Inlet

Power input shaft

Inlet chamber

Closely fitted pump housing

Driven gear

Drive gear

Discharge chamber

Outlet

■ Atmospheric pressure

■ Pump pressure

Goodheart-Willcox Publisher

Figure 16-7. Positive-displacement pumps output a constant flow at a constant rpm, regardless of discharge pressure.

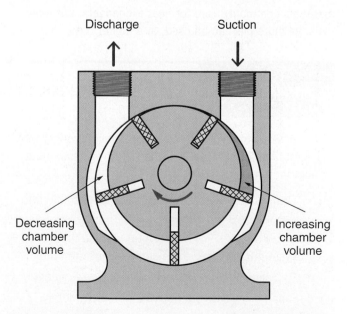

Discharge   Suction

Decreasing chamber volume

Increasing chamber volume

Goodheart-Willcox Publisher

Figure 16-8. Older noncompensated pumps, such as the vane pump pictured here, are being replaced by balanced and compensated pumps.

Figure 16-9 shows the simplified internal workings of a balanced vane pump. Because of the split suction and discharge ports, thrust forces on the rotor cancel each other out. A *port*—also called a way—is the open end of a passage that fluid passes through. Volume between adjacent vanes increases in the inlet quadrants and decreases in the outlet quadrants. Each vane has a guide rod and springs, which help keep the vane against the cam ring and guide it as it moves back and forth radially. Each vane also has ports on its contacting surface to help lubricate the vane against the cam ring. Vanes and internal casing surfaces eventually wear and need replacement.

> ### 🔧 TECH TIP
> ### Vane Pump Maintenance
>
> The cleaner your system is, the longer the time will be between necessary vane pump rebuilds. More contamination means more wear on the vanes and internal casing surface. If the system is badly contaminated, the surface of the internal casing will be badly scored and may need replacement or honing. Typical clearances can be in the range of 5–7 microns, so any dirt present can have a huge impact.

In a pressure-compensated vane pump, **Figure 16-10**, the compensator spring sets the initial adjustment for the maximum displacement of the cam ring and rotor. The maximum displacement, or the maximum amount the cam ring is off-center from the rotor, develops the maximum flow. As discharge pressure increases, the compensator piston overcomes the force of the spring and recenters the cam ring around the rotor to reduce the output flow of the pump. The outlet pressure feedback may be controlled directly from discharge pressure, or it can be controlled through a series of loading and unloading valves to achieve the desired system response and cost savings.

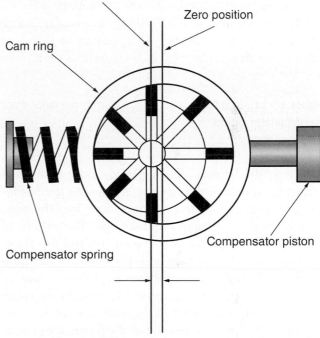

Goodheart-Willcox Publisher

Figure 16-10. In pressure-compensated vane pumps, compensator springs should be adjusted on initial start-up and checked periodically for response.

Axial piston pumps are the most expensive pumps for typical industrial use, but they typically have a longer life, produce less noise, and require less maintenance. **Figure 16-11** shows the internal workings of a piston pump, with some elements removed in order to understand the important basic components and how the pump develops pressure. A *piston* is made up of a cylindrical piece moved back and forth within a cylindrical chamber by means of fluid pressure. The term "axial" refers to the motion of the pistons. As the driveshaft is turned, the piston subassembly rotates with its back against the swash plate. This swash plate is at an angle, which is sometimes adjustable. The pistons are located

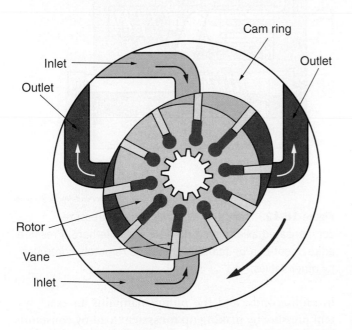

Goodheart-Willcox Publisher

Figure 16-9. Passages into and out of the pump, such as in the balanced vane pump pictured here, are carefully machined and may have multiple openings.

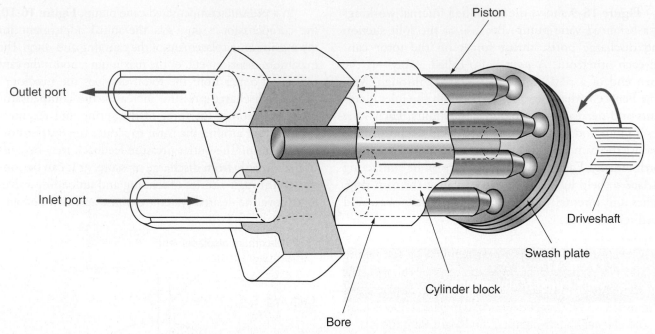

Goodheart-Willcox Publisher

Figure 16-11. In a basic axial piston pump, the axial displacement of the pistons provides for a positive-displacement pump.

within a cylinder bore. As the pistons move axially and retract from the bore, they draw in fluid. As the pistons rotate around the swash plate and move into the bore, they expel fluid through the discharge. The volume of fluid expelled is known as displacement, and this translates into flow. The amount of displacement or flow produced depends on the ***revolutions per minute (rpm)*** of the pump and the angle of the swash plate. As the swash plate angle approaches 90° to the shaft, less and less flow is produced. Rpm is a measure of the frequency of rotation, or the speed, of a rotating mechanism.

Figure 16-12 shows the internal workings of a pressure-compensated axial piston pump of inline design, where the motion is parallel to the input shaft. The swash plate is still at an angle in a pressure-compensated axial piston pump, but the angle is adjusted by a mechanical feedback mechanism involving a piston and a spring. A hydraulically moved piston is located on the side of the swash plate with the cylinder bore, and a return spring on the other side of the swash plate creates an opposing force. Pump output pressure is adjusted through a valve mounted directly on the pump. In this control valve, the spring is adjusted to oppose system pressure. When system pressure is increased enough, it overcomes the spring pressure that has been set and is directed to the underside of the swash plate piston. This piston then moves out against the swash plate, changing the swash plate angle until the pump produces no flow. This happens

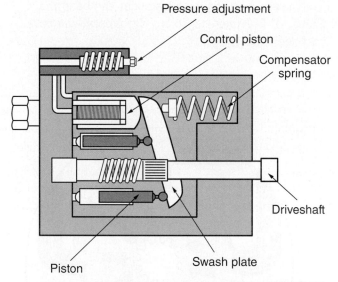

Goodheart-Willcox Publisher

Figure 16-12. The internal workings of a pressure-compensated axial piston pump are visible here. Pressure adjustment can be manual or may be remotely controlled by other valves.

in milliseconds, and the pump maintains an exact system pressure by making up for system load by constantly changing the angle of the swash plate. The mechanical feedback signal to determine the angle of the swash plate can also be remotely controlled. With a degraded system, feedback can become sluggish, and the pump may respond poorly to changing loads. Axial piston pumps also come in a bent-axis design, where the motion of the pistons occurs at a slight angle to the input shaft.

Pump Calculations

Output flow of a positive-displacement pump is determined by the pump's displacement per revolution and the speed at which the pump rotates. Flow is especially a factor in larger systems requiring a higher speed of actuation—work applications where quick movement is needed. When designing a system, make sure you have the right motor. Consider speed of actuation (rpm) and required pressure (psig). Then correlate those to the required pump flow (gpm), which in turn determines the input power required by the motor. This can be expressed in the equation that follows. The equation uses 231 as a conversion factor since displacement is measured in cubic inches per revolution and flow is measured in gallons per minute (1 US gallon = 231 in³).

$$\text{gpm} = \frac{\text{in}^3 \times \text{rpm}}{231}$$

(16-1)

If a pump's displacement is 2 cubic inches per revolution and the pump is driven at 1750 rpm:

$$\frac{2 \text{ in}^3 \times 1750 \text{ rpm}}{231} = 15.15 \text{ gpm}$$

Horsepower (hp) is a measure of the mechanical power of an engine or motor, or the rate at which work is accomplished. Input power from the motor to the pump is directly proportional to pump flow and pressure. It can be calculated in the following equation, using 1714 as a known approximate conversion factor required to yield pump input horsepower from flow (gpm) and pressure (psig):

$$\text{hp} = \frac{\text{gpm} \times \text{psig}}{1714}$$

(16-2)

Calculate hp using 15.15 gpm and a required pressure of 3000 psig:

$$\frac{15.15 \text{ gpm} \times 3000 \text{ psig}}{1714} = 26.5 \text{ hp}$$

Some efficiency is lost between the motor and pump, depending on how the power is transferred. With that in mind, and given the pump flow and pressure, a good selection for an electric motor to drive this pump would be a 30-hp motor.

Pump Maintenance

Before starting a power unit for the first time, review the manufacturer's start-up procedure, which should be followed.

PROCEDURE Power Unit Start-up

Start-up procedures generally include:

1. Fill the pump with filtered fluid.
2. Fill the pump casing with fluid, depending on the type of pump.
3. Ensure that the pressure setting is at its lowest.
4. Start and stop the system while venting any gases.
5. Adjust the pump output pressure to system requirements—but do not exceed the manufacturer's recommended pressures.
6. Run the pump for a few hours unloaded.
7. Check for any leaks and correct as necessary.
8. Observe the reservoir temperature, which should not exceed 150°F.

Once the system has been running for several hours, an equilibrium temperature will be established. If this temperature exceeds 150°F, you may need to install an after-market hydraulic oil cooler. As noted, a higher temperature can cause early failure of hoses, pumps, and other components.

Possible pump problems are usually preceded by the following signs:

- Excessive noise caused by constant cavitation.
- Excessive wear particles in a filter located just after the pump.
- A drop in output power or pump discharge pressure.
- A pressure-compensated pump not responding to load conditions (the amount of work taking place at actuators).

Cavitation will wear away metal at valve plates on the discharge side of the pump, where the pressure of the fluid is highest and trapped air bubbles collapse. Eventually this may affect discharge pressure. In this case, either the pump is entraining air in the reservoir or suction line, or it has an excessive restriction in the suction line. Wear particles on the filter on the discharge side indicate that the pump will soon fail, or has already.

TECH TIP
System Flush

If a pump mechanically fails and there is no discharge-side filter, wear particles and particles from failure are injected into the system. Just replacing the pump will not fix the problem at this point. The system should be flushed as much as possible, new filtered fluid should be used, and samples of the fluid should be taken and evaluated after putting the system back online.

If a pressure-compensated pump is not responding to load conditions, its pressure-adjustment piston, or swash plate piston, may be locked by foreign particles. Pumps that are not developing pressure or power in the system need further examination. Most manufacturers have pump rebuild kits that include the moving assembly and all replaceable soft parts. It is good to keep this type of rebuild kit on hand, especially if pump failure restricts production or results in severe monetary loss.

TECH TIP
Pump Rebuild

When performing a teardown and rebuild of a hydraulic pump, be sure you have a clean work area and any pertinent manuals, tools, and parts on hand. Any foreign material that touches the pump will end up in the hydraulic system and may cause a secondary failure, resulting in the loss of another component.

One pump element commonly needing replacement is a seal. A *seal* is a mechanical device that joins parts of a system together tightly to prevent leakage, contain pressure, and reduce contamination. **Figure 16-13** shows a close-up of a typical seal assembly. If a seal is leaking, it is leaking for a reason. The reason may be a misalignment with the prime mover, high radial run-out on the shaft (a bent shaft), the temperature exceeding the design of the seal, or foreign material wearing away the seal (environmental impact). A high temperature at the point where the lip of the seal contacts the shaft can

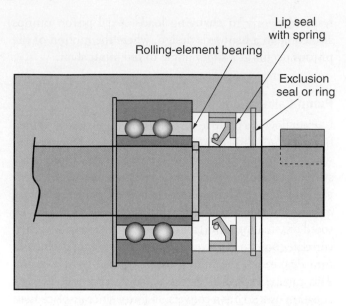

Goodheart-Willcox Publisher

Figure 16-13. This is a close-up of a typical seal assembly. A number of different issues can cause early seal failure.

be caused by friction due to high speed or pressure, or a high temperature of the pump and fluid. Excessive wear at the point of contact can be caused by a high temperature, excessive movement (radial or axial), or foreign contamination. The reason for the problem needs to be corrected in addition to simply replacing the seal—you want to address the root cause, not just the symptoms.

16.3 VALVES AND VALVE ACTUATORS

While most systems assembled today are made up of manifolds or stack valves, a large number of systems with older spool valves are still in use. An understanding of how these basic valves work will aid in understanding screw-in and slip-in cartridge valves used in manifolds. (Cartridge valves are covered in Chapter 18, *Advanced Fluid Power*.) While some valves automatically actuate based on internal system signals, external control of the system is provided through valve actuators. *Valve actuators* are manually operated or power-operated mechanisms used to control valves by opening or closing them.

16.3.1 Valves

The three main types of valves are pressure control valves, directional control valves, and flow control valves. Each of these types of valve has a specific purpose in the system. System and valve control may be accomplished using hydraulic pilot pressures or mechanical feedback mechanisms, through electrical controls using hardwired logic, and even through programming via a programmable logic controller (PLC). Older systems are typically hydraulically or mechanically controlled, while newer systems are usually controlled through programming logic. Very simple systems may still be completely controlled with electrical hardwired logic.

In system diagrams, relief valves, pressure reducing valves, and sequence valves are shown as one-position valves but in reality have two positions. Standard spool valves have two or three positions that are all shown. The *spool* in a spool valve is a cylindrical internal component with seals along its surface. The spool shifts within the valve, directing fluid flow to the various valve ports. The spool may direct one or multiple fluid paths at the same time, depending on the spool design and body of the valve. Most stand-alone valves come premounted to a baseplate with O-ring seals in between. They may have one or multiple flow-through ports for pressure (P), tank (T), output pressure A (A), and output pressure B (B).

Pressure Control Valves

Pressure control valves are used in hydraulic systems to keep pressure in the system below a set limit and to maintain a set pressure in a circuit. One of the most common types is the *relief valve*, shown in **Figure 16-14**. This valve serves the purpose of protecting the system in the case of overly high pressure. The relief valve allows excess flow from the pump to flow back to the reservoir. The pressure in a system is a reflection of the resistance to flow from the pump. Allowing for flow thus reduces system pressure. Common relief valves are manually adjustable but can also be remotely piloted, or controlled, with an additional poppet in the upper spring section. A poppet, or poppet valve, is a sliding shaft with a disk on the end that seals and unseals a port as it moves. It is often described as a mushroom-shaped valve due to its appearance. With a remote pilot signal, the signal can only raise the relief setting of the valve. The minimum pressure-relief setting would still be set locally

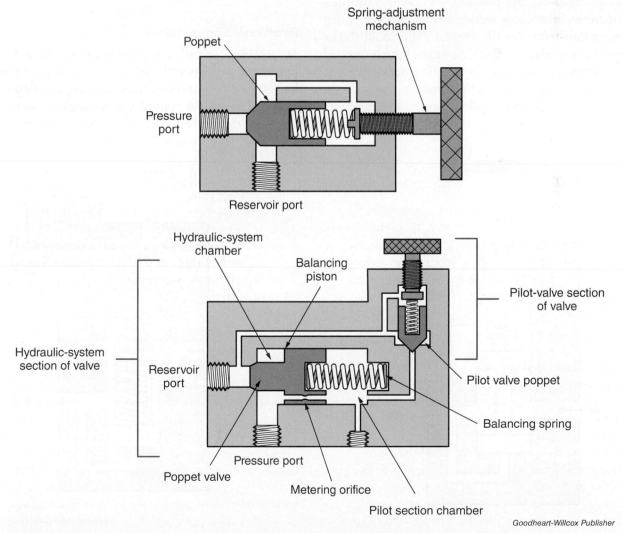

Goodheart-Willcox Publisher

Figure 16-14. The relief valve may be manually set, set by remote pilot pressure, or both.

by the manually adjusted spring tension. Depending on design, relief valves may be biased, or set as a default, to open slowly as pressure rises above the set point or open fully when the set point is reached, and to close slowly when the system pressure falls below the set point. The set point that triggers the relief valve is known as the cracking pressure.

The *pressure-reducing valve* is another type of pressure control valve. Pressure-reducing valves help to reduce upstream pressure to an acceptable level for downstream branches in the system. **Figure 16-15** shows a direct-operated and a pilot-operated pressure-reducing valve. When a valve is pilot operated, it means another valve, called a pilot valve, is used to then operate that main valve. *Pilot valves* are smaller, easily operated valves used to control operations that would require a great deal of force to otherwise operate directly. Notice that both pressure-reducing valves sense the reduced pressure downstream that acts against the main piston. In the direct-operated valve, downstream pressure offsets the spool against the spring pressure. As downstream pressure increases, the spool repositions against the spring and more flow returns to the reservoir. This could also be a flow-through valve, in which the excess fluid is directed back into the main system. In the pilot-operated valve, pilot pressure is set by pilot spring pressure. Downstream pressure is felt against the pilot poppet, which pushes against pilot spring pressure through the main spool.

If downstream pressure increases to a point higher than the spring setting, the pilot opens and drains the pilot area, allowing the main spool to reposition and decrease the flow to the valve outlet downstream. A parallel check valve is shown in the diagram. This check valve opens to allow flow back into the high-pressure line if the low-pressure side increases above the high-pressure side, which could happen with an actuator that is being forced back on itself, for instance.

A third type of pressure control valve is a sequence valve, shown in **Figure 16-16**. The *sequence valve* is similar in function to a pressure-reducing valve, but it senses upstream pressure to control a sequence of operations with more than one actuator. Once the first actuator has completed a cycle and the upstream pressure overcomes the spring force, the spool of the sequence valve allows flow to the second actuator in order for it to move. One or several sequence valves can be used to cause a sequence of cylinder or actuator extensions and retractions.

Directional Control Valves

Directional control valves are found in hydraulic and pneumatic systems and are used to control the flow of fluid from one or more sources into various paths in the system. Directional control valves include a spool that sits inside the cylinder body of the valve. The valve spool

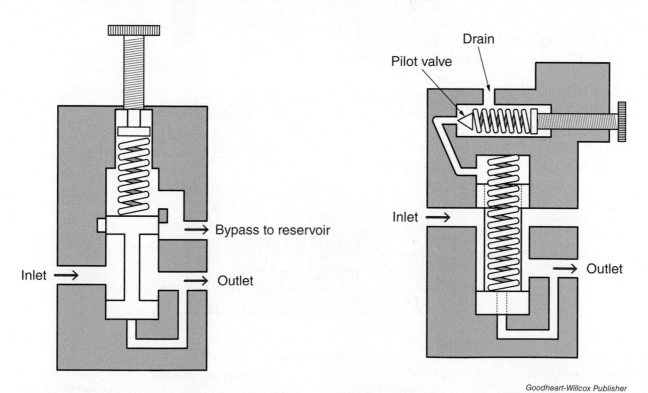

Goodheart-Willcox Publisher

Figure 16-15. The pressure-reducing valve may be direct acting (left) or have a pilot poppet (right).

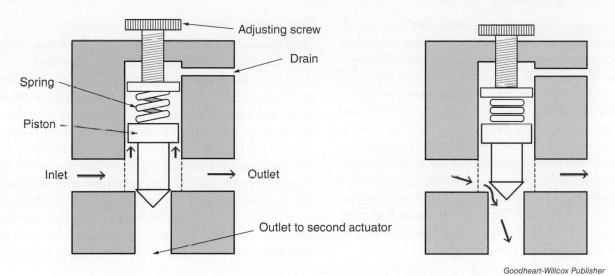

Figure 16-16. The sequence valve is similar in function to a pressure-reducing valve, but it senses upstream pressure. The sequence valve only opens once primary pressure exceeds the setting.

is a finely machined piece that barely clears the internal seals of the valve body. The spool fits into the body of the valve, and **lands**, which are the sealing areas, separate the different chambers of the valve body. The clearances between the lands and valve body are so close that very little leakage exists between chambers of the valve. Small grooves are cut into the lands to help reduce leakage between chambers, and also to help prevent hydraulic lock of the spool. One end is flat with a hole in order to attach a manual actuator to the spool. Older types of valves are mounted to a baseplate with appropriate labels and connections. Newer valves may be arranged in a stack configuration with other valves, with seals in between each valve.

Spool valves typically have two or three definite positions. (Some valves have "infinite" positions because the spool can be placed in a partial position.) Each position is represented in a diagram as one square part of the **envelope**, which is the larger rectangular box representing the component. The word "way" refers to the possible fluid paths through the valve. Some manufacturers and technicians refer to the number of ports on the valve, although this is typically in reference to pneumatic valves. **Figure 16-17** shows a simple manual directional control valve and its diagram symbol. In this instance, the center position is blocked, meaning all four ports are blocked when the valve is in the center position. If used for a cylinder, it will lock the cylinder in place when the valve is in the center position. When the spool is manually placed into either noncentered position, the pressure is sent to one of the output ports (A or B), while the opposite output port is vented back to the reservoir. Valves can have a blocked center position, open center position, or floating center

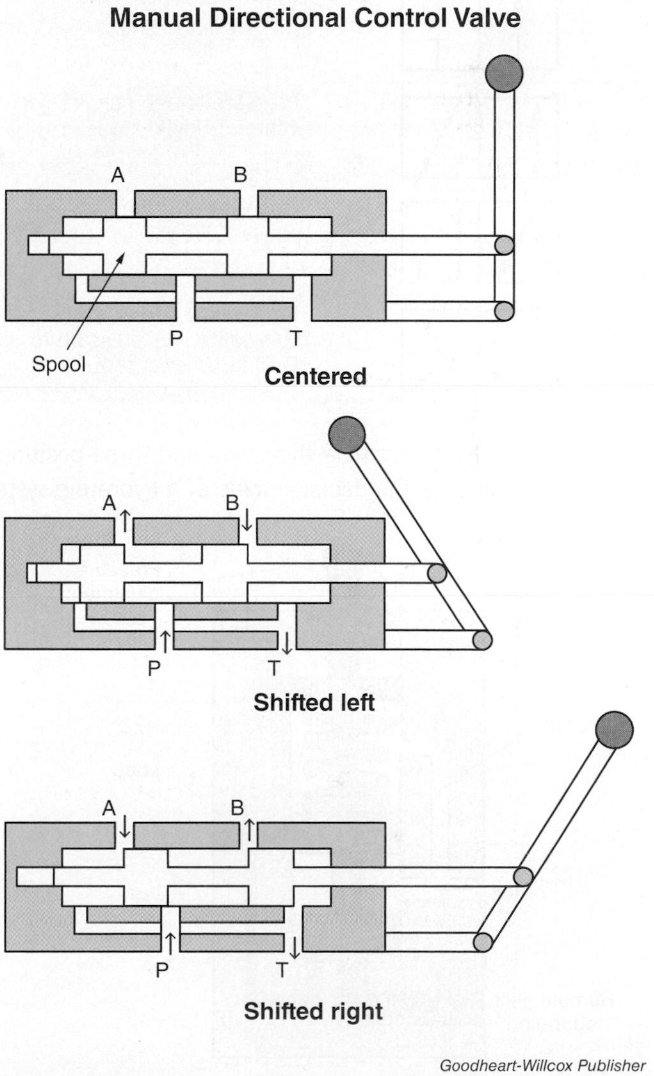

Manual Directional Control Valve

Centered

Shifted left

Shifted right

Figure 16-17. This three-position, four-way directional control valve has a blocked center position.

position. An open center position allows free flow back to the reservoir tank but blocks the actuator ports, so that the actuator cannot move. A valve with a floating center position allows flow between all four ports when in the center position, and the actuator is not held but is free to float.

Figure 16-18 describes several symbols for two- and three-position directional control valves without actuators. The type of valve used depends on system design and desired function. A circuit that requires a punching operation could use a two-position valve, while a circuit requiring a clamping action might use a three-position valve. Valves may be manually controlled, pilot controlled, or controlled with *solenoids*, which will be discussed later in this chapter.

Flow Control Valves

Flow control valves control the amount of flow of the fluid, which controls the speed of actuators. Such valves include simple check valves and needle valves, as well as more complicated counterbalance valves. **Figure 16-19** shows the internal workings of a counterbalance valve. This type of valve may have an integrated internal check valve or an external check valve. In this circuit, the valve is being used to suspend a platen, which is a heavy, flat plate. This heavy platen might normally drift downward due to its weight and any leak-by. The counterbalance valve creates an opposing hydraulic force that prevents this. The pressure developed on the rod side acts in opposition to the manually adjusted spring pressure

Directional Control Valves			
Two-Position Valves with Different Flow Paths		**Three-Position Valves with Different Center Positions**	
	On-off valve positions		Completely blocked center
	Pressure to one port, or port vented to tank		Pressure port and both A and B ports are connected to the tank (return) line
	Opposite positions as above valve		Both A and B ports are connected to the tank line
	Pressure to port A while B is vented, or pressure to B while A is vented		A and B ports are connected to each other

Figure 16-18. These symbols show two- and three-position directional control valves without actuators. The variety of flow paths makes up the decision logic of a hydraulic system.

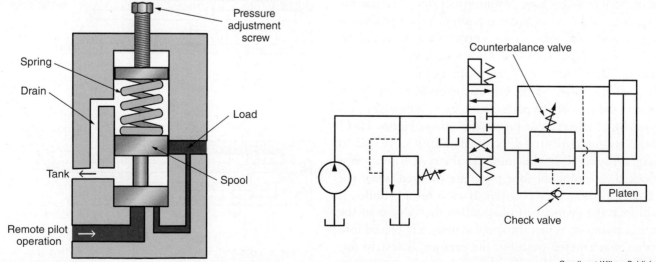

Figure 16-19. These are the internal workings of a counterbalance valve. While this diagram shows a manually adjusted spring, this may also be replaced with a secondary poppet and remotely controlled pilot pressure.

across the spool. When pressure is applied to the cap end of the cylinder to extend the cylinder, the pilot line pressure overcomes the spring resistance and the spool shifts, allowing flow from the rod end of the cylinder back to the tank. The counterbalance valve can also be used as a braking valve to prevent an overhauling load from running away, and it can be connected to receive a remotely controlled pilot signal. If a flow control valve cannot prevent drifting, leakage may be caused by leaking seals or another problem that needs to be fixed.

An increase in flow means a reduction in pressure, and vice versa. As the pressure drop across a valve increases, flow through the valve will increase. If a system needs constant flow to ensure actuator speed, such as for cylinder extension or motor rpm, pressure compensation must be used. *Pressure compensation* refers to the varying of an orifice, or opening, of a valve in order to maintain fluid flow regardless of differences in pressure between areas of the system. **Figure 16-20** shows a simplified pressure-compensated flow control valve, which is a valve that can be used to adjust the speed of an actuator. Flow across the venturi is kept constant by keeping the pressure differential across the venturi constant. A *venturi* is a short tube with a narrowing point of constriction in the center that creates an increase in fluid velocity and decrease in fluid pressure. In this case, the pressure drop across the venturi

is held fairly consistent due to feedback pilot pressures acting across a spool. The spring may be set as shown, have a manual adjustment, or have a pilot override.

In addition to pressure compensation, *temperature compensation* is needed. Temperature compensation refers to the use of a valve to compensate for changes in fluid temperature in order to manage flow rate. As a hydraulic fluid's temperature changes, the viscosity of the fluid also changes, which in turn affects flow rate. High temperatures bring lower viscosity and higher flow, and low temperatures bring higher viscosity and lower flow. **Figure 16-21** shows the use of a temperature-sensitive material to overcome this change in viscosity. The temperature-sensitive material will lengthen as temperature increases, slightly closing an orifice to compensate for the tendency of increased flow.

Valve Maintenance

One of the biggest concerns in systems with spool-type valves is silting. *Silting* is the accumulation of fine particles within the body of the valve. Silting may present itself in the form of valves that operate erratically, not at all, or only partially. Proper system cleanliness must be maintained in order to prevent these types of failure. Silting may also present itself as a solenoid failure. While the valve can be disassembled and cleaned, silting will continue and further failures will result unless system cleanliness is improved.

Needle-valve type flow-control device

Biasing spring

Compensator spool

Inlet port

Outlet port

Compensation chamber

Goodheart-Willcox Publisher

Figure 16-20. This simplified pressure-compensated flow control valve is used to maintain flow with a changing input pressure.

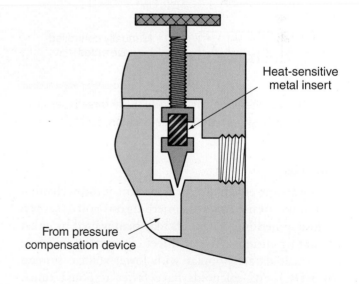

Heat-sensitive metal insert

From pressure compensation device

Goodheart-Willcox Publisher

Figure 16-21. Some flow control valves also incorporate temperature compensation by using a temperature-sensitive material that will lengthen as temperature increases, closing an orifice slightly to compensate for the tendency of increased flow.

Varnishing can also occur over extended time with the introduction of oxygen and moisture. *Varnish* is a hard coating in hydraulic components resulting from a breakdown of hydraulic fluid. Varnishing will increase diameters and reduce clearances, sometimes resulting in erratic operation. Varnishing is difficult to remove from spools and valve internals without scoring or damaging the spool.

16.3.2 Valve Actuators

The main methods for actuating a valve are by manual control, pilot control, and solenoid control. The symbols for these types of valve actuation are shown in **Figure 16-22**. Manually controlled valves may provide for a detent within the valve. A *detent* is a spring-loaded ball, which keeps the spool in position by providing pressure on one of several grooves. Some manual spool valves may not have a detent and may be spring centered or spring returned to the normal position. Valves may also be mechanically actuated by means of a roller or other component that is physically moved by something actuating in the process.

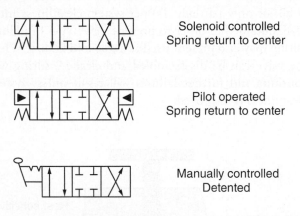

Solenoid controlled
Spring return to center

Pilot operated
Spring return to center

Manually controlled
Detented

Goodheart-Willcox Publisher

Figure 16-22. These symbols represent the three types of valve actuators, shown with a three-position, four-way valve.

Solenoids

Solenoids convert an electrical current into mechanical motion and are used to turn something on or off or to open or close something. Solenoids may be powered by either alternating current (AC) or direct current (DC). While DC solenoids are popular with lower-voltage systems (24 VDC), AC solenoids have faster response times. Both solenoids heat up over time, but AC solenoids have some specific operating characteristics. AC solenoids are more likely to fail when the valve does not fully operate. When the valve does fully operate, the plunger is drawn fully into the solenoid and reinforces the magnetic field.

This helps to reduce initial in-rush current. Likewise, if the plunger is not fully drawn into the solenoid, operating current remains high, causing excessive heat. This heat will eventually cause the solenoid to fail. Solenoid failure can also be caused by:

- Rapid cycling, above manufacturer recommendations.
- High or low voltage.
- A need for excessive mechanical force.

Most solenoids have a manual override, which the operator can use to reposition the spool with a small pin. Manual override of a solenoid must be done with care and an understanding of the operation and what will happen when the spool is repositioned. If excessive pressure is needed to manually override a solenoid, this may be an indication of a silted valve.

⚠ CAUTION

Manual override should only be done for purposes of diagnosing a problem. If the valve has a detent, return it to the inactive state after any testing.

Pilot Controls

There is a limit to the size of solenoids and the force they can develop. Because of this, larger valves may not be directly operated, but rather pilot operated with an additional smaller valve that is solenoid controlled. **Figure 16-23** shows the symbol for this type of valve. On some diagrams, the symbol may be simplified.

These pilot valves, also known as two-stage valves, may include:

- Drain lines from the pilot to prevent pressure back-feed from the main spool to the pilot.
- A pressure-regulating check valve within the main valve to provide lower pilot pressure to the pilot valve pressure port.
- Pilot chokes to slow down the actuation speed of the main valve and prevent hydraulic shock.
- Main-spool pushpins, which provide for faster main-spool actuation.
- Pilot manual overrides.

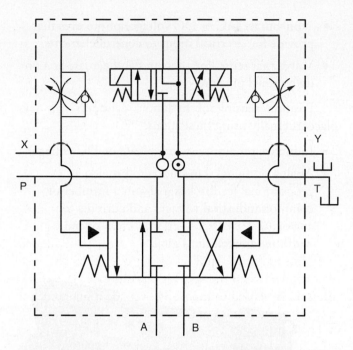

Figure 16-23. Larger valves with higher flows are controlled by smaller solenoid pilot valves. The top valve is a three-position, four-way solenoid pilot valve. The bottom valve is a three-position, four-way, pilot-operated valve with a spring return to center. The actuation speed of the main valve is controlled through a flow-control orifice.

16.4 CYLINDERS AND MOTORS

Hydraulic cylinders and motors convert fluid energy into motion. This section examines the construction of hydraulic motors and cylinders and discusses common maintenance practices for each component.

16.4.1 Cylinders

Hydraulic cylinders are a type of actuator that converts the energy from the fluid into linear motion and mechanical force. Hydraulic cylinders are often called linear actuators for this reason. The force output of a cylinder depends on the bore of the cylinder and is directly proportional to pressure and proportional to the radius of the cylinder squared. Because of the amount of force that hydraulic cylinders transfer, they must be properly mounted and momentum must be taken into consideration. The design elements of stop tubes, cushions, and deceleration techniques become more important as load and momentum increase.

Cylinder Construction

Depending on machine design, an array of mounting methods is available for cylinders. **Figure 16-24** shows some of the more common mounting methods. The mounting and load-bearing details have an effect on other components of the cylinder, such as the stop

NFPA Mounts

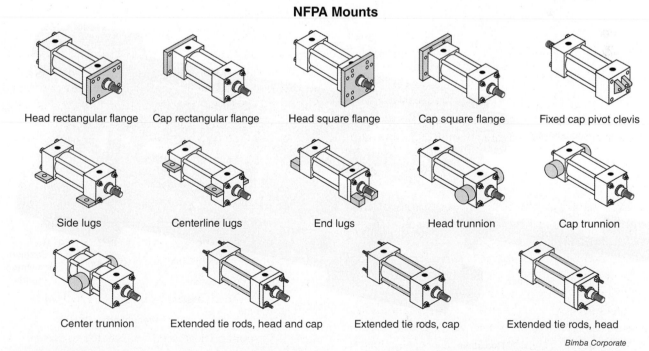

Figure 16-24. The method used for mounting a hydraulic cylinder is determined by the type of load, amount of load, and direction of travel for the load.

tube. **Figure 16-25** shows the internal construction of a typical cylinder using tie rods to seal end caps to the cylinder. Larger cylinders may also have welded seams to attach cap ends to the cylinder. Major components of the cylinder include the ports, cap ends, rod, piston, cushions, and seals. Seals may be made of different materials depending on the application. Nitrile or Viton™ seals are common in industrial systems.

Cylinder Position Sensing

The degree of position sensing needed for a cylinder greatly depends on the application and the required accuracy and repeatability. While a simple press might only need to sense line pressure to ensure the stroke has been completed, a servo system may need exact accuracy (in robotics, for instance). **Figure 16-26** shows a cross-section of a cylinder that monitors cylinder position through electronic feedback.

Position sensing of hydraulic cylinders can take place externally using the following:

- Mechanical contact-limit switches that electrically control the system.
- Mechanical contact actuators that directly control attached hydraulic valves.

- Noncontact (AC or DC voltage) photo eyes that provide for electrical input to control elements.
- Valves that sense line pressures and actuate at a predetermined pressure.

Position sensing of hydraulic cylinders can take place internally using these items:

- Inductive proximity switches (AC or DC voltage).
- Analog or digital transducers to provide a proportional feedback signal—in a number of common industrial voltages and currents for analog (0–10 VDC, 4–20 mA, etc.), or pulse-width modulated for digital.

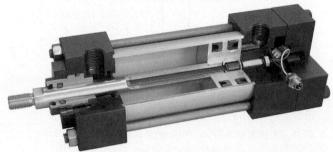

Bimba Corporate

Figure 16-26. This cylinder accurately monitors cylinder position through electronic feedback.

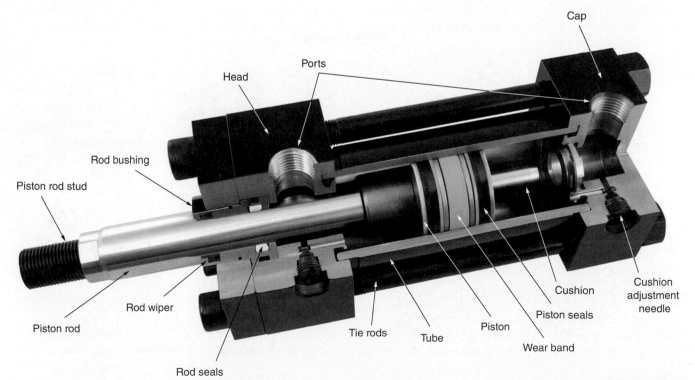

Bimba Corporate

Figure 16-25. This is the internal construction of a typical cylinder using tie rods to seal end caps to the cylinder. Seals and other soft components are what typically fail in a cylinder.

Cylinder Cushions and Stop Tubes

As a cylinder extends or retracts and approaches the end of stroke, a smaller-diameter cushion on either end enters a small cavity. This directs the fluid through a small needle valve. The needle valve restricts the flow of fluid and therefore reduces the velocity of the rod for the last part of the stroke. Cylinder cushions can help reduce noise and wear on the cylinder.

TECH TIP
Adjusting Needle Valves

Needle valves are usually adjusted with a hex key (Allen wrench) and may have a locking feature or cap that must first be removed in order to make an adjustment. Never fully close these needle valves during adjustment. Adjust the needle valve carefully, cycle the machine, and adjust again if needed. Make only small adjustments between cycles.

Stop tubes are used to restrict the extension of a cylinder on larger strokes. The stop tube allows for better distribution of force in order to minimize damage to seals and the possible buckling of the rod. **Figure 16-27** shows two common types of stop tube design, the single piston and double piston. The typical single-piston design may or may not have a cushion. The rod extends through the inside diameter (ID) of the stop tube, and the stop tube prevents overextension. The double-piston design typically has better operational characteristics and lasts longer. The length of the stop tube is based on several factors:

- Orientation of the application (horizontal versus vertical).
- Rod support and load support.
- Mass of the load.
- Velocity of the rod.
- Mounting and support of the cylinder.

Cylinder Maintenance

With loads that are not well supported and with longer rod lengths, rod bushings and cylinder seals can become damaged and eventually fail. As a rod extends, support comes from the seal between the piston and the cylinder wall, and from the rod bushing. If excessive side loading is present (the load is not properly guided), rod buckling could occur, which causes excessive force and wear on

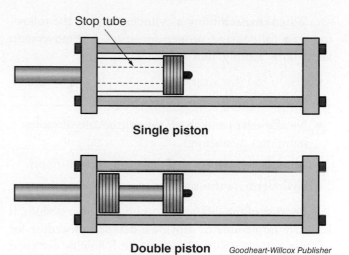

Stop tube

Single piston

Double piston   *Goodheart-Willcox Publisher*

Figure 16-27. Stop tubes come in several designs and help extend the life of the rod bushing and piston seals.

the rod bushing. In extreme cases, the piston can damage the inside surface of the cylinder.

Common maintenance corrections include replacing the following:

- Rod seals and bushings because of leakage.
- Piston seals due to leak-by or damage.
- Needle valves and/or ball check valves due to unreliable operation.
- The entire cylinder, if it cannot be repaired, due to damage or rod end blowout after overadjustment of needle valves.

With any situation, have the manufacturer's technical information on hand for reference. When disassembling a cylinder, the task should be performed in a clean and well-lit area. Cleanliness standards must be maintained so as not to introduce contamination into the system. All "soft" parts should be on hand before disassembly takes place.

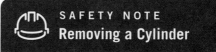

SAFETY NOTE
Removing a Cylinder

Larger cylinders produce a massive amount of force, and any maintenance must be performed carefully and safely. When removing a cylinder from its mounting, recognize that there may be extreme mechanical forces that are being supported by the cylinder, whether retracted or extended. Be prepared to mechanically block and support loads to prevent machine damage or personal injury.

When disassembling a cylinder, inspect the following areas for damage, proper operation, free movement, varnishing, scoring, and wear:

- Piston rod (for wear).
- Cushion (for wear and varnishing).
- Needle valves and check valves (for ease of movement and varnishing).
- All seals and wipers and their contacted surfaces.
- Cylinder bore (for scoring, wear, and varnishing).

Because of the differences among cylinder designs, it is nearly impossible to provide a detailed procedure for complete disassembly. However, the following are good practices that apply to nearly all designs:

- Be prepared with proper lubricants and locking fluids, based on the manufacturer's technical information.
- Many manufacturers use Loctite™, which may require heating up to 450°F to loosen fasteners, depending on the Loctite used.
- When replacing internal needle or check valves, follow the manufacturer's instructions exactly.
- Some components need to be lubricated prior to assembly. Check the instructions to make sure.
- When reassembling, use Loctite where recommended. Use a torque wrench to ensure fasteners receive the proper torque levels.

16.4.2 Motors

Motors convert fluid energy created by pressure and flow into rotary motion and torque. Torque is a measure of the turning force on an object rotating around an axis. In a hydraulic system, the rotary motion and torque created by the motor drives the pump, which converts the energy back into flow and pressure in the system, and the cycle repeats.

Types of Motor

Hydraulic motors are similar in function and assembly to hydraulic pumps but are concerned with controlling the load, as opposed to a pump, which reacts to system pressure. Overhauling loads require special precautions and are controlled hydraulically. One large difference between pumps and motors is that most motors are designed to rotate in both directions, depending on how the fluid flows through the motor, whereas pumps are not. Vane, radial piston, and gerotor/geroler motors are usually used as low-speed, high-torque motors. Bent-axis piston, axial piston, and gear motors are more suited for high-speed applications. **Figure 16-28** shows some of the different types of hydraulic motors.

Motor Calculations

The speed of the motor (rpm) is proportional to the flow into the motor (gpm), and is inversely proportional to the motor displacement (in³). Torque (usually measured in inch-pounds or foot-pounds) developed by the motor is directly proportional to the pressure (psig) supplied to the motor. It can be difficult to know when a hydraulic motor is not functioning properly and must be fixed or replaced. You may need to do some math to determine how efficiently the motor is running. Several motor-related calculations are common, and they can be rearranged to find an unknown value if the others are known. If the speed of the motor (rpm) and displacement (in³) are known, the required flow (gpm) into the motor can be calculated using the following formula:

$$\frac{\text{motor displacement} \times \text{rpm}}{231} = \text{gpm}$$

(16-3)

The conversion factor of 231 is again used since displacement is measured in cubic inches per revolution and flow is measured in gallons per minute (1 US gallon = 231 in³).

Vane motor   **Radial piston motor**   **Bent-axis piston motor**   **Axial piston motor**   **External gear motor**

Images courtesy Bosch Rexroth Corporation. Used by Permission.

Figure 16-28. While some motor designs are useful in low-speed, high-torque applications, other designs are used for high-speed applications.

If a motor's displacement is 2.75 cubic inches per revolution and needs to run at 1200 rpm, the required flow into the motor is:

$$\frac{2.75 \text{ in}^3 \times 1200 \text{ rpm}}{231} = 14.29 \text{ gpm}$$

Using the same equation rearranged, if the flow that the motor and system will provide to the pump is known, you can calculate the maximum rpm the motor will run at:

$$\frac{\text{gpm} \times 231}{\text{motor displacement}} = \text{rpm}$$

(16-4)

In this case, if you know the system can supply 10 gallons per minute to the motor and the motor displacement is 1.5 cubic inches, the motor's rpm can be calculated as follows:

$$\frac{10 \text{ gpm} \times 231}{1.5 \text{ in}^3} = 1540 \text{ rpm}$$

If you wanted to calculate the torque produced from a motor, or knew the torque required, you could use the following equation:

$$\frac{\text{psig} \times \text{motor displacement}}{2 \times \pi} = \text{torque produced}$$

(16-5)

There are two pi (π) radians in one revolution of a circle, and torque is a measurement of rotational force, which is why "$2 \times \pi$" is used in the equation. If supplying 1500 psig to a motor with a displacement of 1.75 cubic inches:

$$\frac{1500 \text{ psig} \times 1.75 \text{ in}^3}{2 \times \pi} = 417.8 \text{ inch-pounds}$$

If, in the last example, torque needs to be measured in foot-pounds rather than inch-pounds, use 24 instead of 2 in the denominator.

While most technicians will probably not be designing a hydraulic system, knowing the basic formulas will help in troubleshooting a power transmission problem. With a knowledge of the motor's characteristics at peak efficiency, its performance at any given time can be compared to that baseline to determine its current efficiency.

Motor Maintenance

Common maintenance issues with hydraulic motors include leaks and a nonresponsive motor. Many motors have integral relief valves, spooled control valves, and

check or poppet valves that can easily stop functioning if system cleanliness is not maintained. Before disassembly, the environment should be cleaned, and the outside surfaces of the motor should be cleaned and degreased to the greatest extent possible. Anything that comes in contact with the motor internals will contaminate the hydraulic system and add to possible future component failures. Only use lint free wipes. Have the motor manufacturer's technical information on hand before disassembly. Keep the work area clean and well organized because most motors have a large number of internal parts and seals.

Most manufacturers recommend the removal and inspection of any integral control valves. This may indicate the problem without having to further disassemble the motor. Inspect the removed components (spool, poppets, balls, shuttles, and internal surfaces) for burrs, scoring and wear, varnishing, and contaminants. Then get to work with any necessary repairs or replacement of the components. A motor's internal construction is complex, with many replaceable seals and soft parts. Ensure that you have all soft parts on hand prior to disassembly (O-rings, paper gaskets, metallic gaskets, seals, and springs). Review **Figure 16-29** to gain a better understanding of the intricacies involved in the sealing systems in a hydraulic motor.

Inspect the rotor and all moving parts for wear, scoring, varnishing, and foreign material. If the rotor or other mating parts are badly worn, they may need to be replaced. Most paper gaskets should not be lubricated, and the sealing surfaces should be carefully cleaned and dried. O-rings should be lightly coated with a lubricant specifically designed for O-rings, and their seating areas should be thoroughly cleaned. Most machined fit surfaces should be cleaned, and then a light coating of hydraulic fluid should be used to lubricate them before assembly.

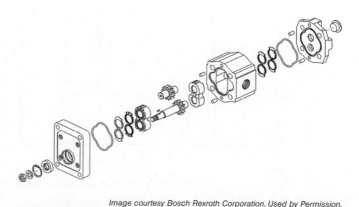

Image courtesy Bosch Rexroth Corporation. Used by Permission.

Figure 16-29. This exploded view is a good tool for understanding the assembly of a hydraulic motor.

16.5 BASIC CIRCUITS

Examining and understanding a few of the most basic systems is the first step in developing a useful routine for troubleshooting. Only by studying a number of systems can you build a pool of knowledge. This accumulated knowledge will help you troubleshoot more efficiently. This section will examine two common basic systems and discuss how to evaluate them for possible failures.

16.5.1 Hi-Lo Pump Control

In many hydraulic systems, the loading of the system changes and different flow rates are needed. This is because different amounts of work are being done at different actuators at different times, depending on the job at hand. In a plastic injection molding system, high flow may be needed to bring the two halves of a mold together, but very little flow is needed to keep pressure on the system until ready to remove the plastic form. Hi-lo pump circuits are ideal for these situations. **Figure 16-30** shows an older version of a circuit to control the flow from two pumps driven off the same shaft.

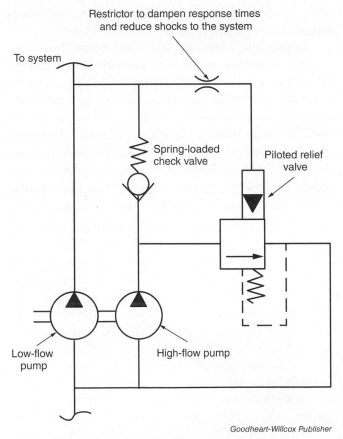

Restrictor to dampen response times and reduce shocks to the system

To system

Spring-loaded check valve

Piloted relief valve

Low-flow pump

High-flow pump

Goodheart-Willcox Publisher

Figure 16-30. This circuit controls the flow from two pumps driven off the same shaft. A hi-lo pump control system is common for larger hydraulic circuits.

The normally loaded pump is a low-flow pump, while the relieved pump is a high-flow pump. During system operation, a low loading condition has the pilot-operated relief valve unloading the high-flow pump back to the reservoir. The low-flow pump supplies needed pressure to the system. As system load increases and pressure drops, the relief valve closes, and the high-flow pump unseats the check valve and supplies flow into the system. As the system load decreases, pressure increases, the check valve shuts, and the relief valve opens again.

A hi-lo pump circuit allows for a lower amperage draw from the motor and increases system efficiency. Newer systems use cartridge valves and multiple settings to match system requirements.

16.5.2 Sequence Control

Sometimes a sequence of operations is desired, in which actions occur in order one after another. For those situations, sequence valves can be used with different pressure settings or the same setting. **Figure 16-31** shows a sequential circuit with sequence valves set so that the extension of each cylinder has enough force for that particular operation. Notice the numbering scheme on the diagram. Anything that will affect the cylinder is labeled in bold with a corresponding number. Valve 1.0 will supply pressure to and vent cylinder 1, and valve 2.0 will supply pressure to and vent cylinder 2.

All sequence valves have tags that show where the connections are made. For instance, the sequence valve that senses pressure from the cap end of cylinder 1 (A1) sends a pilot signal (2.0b) to reposition valve 2.0. By examining this schematic, and assuming the pump is supplying pressure and all flow is being directed over the relief valve back to the tank, you can build a sequence of events:

1. A solenoid is energized, and valve 3.0 changes position (3.0+).
2. Pressure is sent through valve 1.0 and extends the cylinder (cyl 1+).
3. Pressure builds on the cap end, and sequence valve 1.1 shifts (1.1+).
4. This sends a pilot signal to valve 2.0, which shifts (2.0+).
5. Pressure is sent to the cap end of cylinder 2, and the cylinder extends (cyl 2+).
6. Pressure builds on the cap end, and sequence valve 2.1 shifts (2.1+).
7. This sends a pilot signal to valve 1.0, which shifts (1.0+).

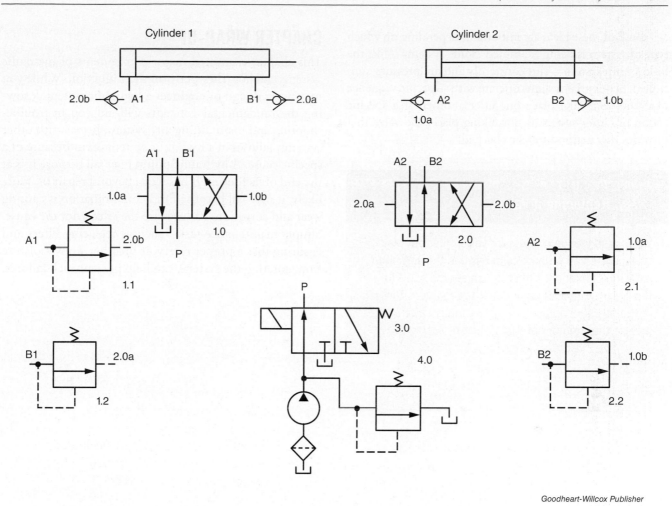

Figure 16-31. This diagram shows a sequential circuit for a repeating sequence of two cylinders.

8. This retracts the cylinder (cyl 1−).

9. Pressure builds on the rod end, and sequence valve 1.2 shifts (1.2+).

10. This sends a signal to valve 2.0, which shifts (2.0−).

11. This sends pressure and retracts cylinder 2 (cyl 2−).

12. Pressure builds on the rod end, and valve 2.2 shifts (2.2+).

13. This sends a signal to valve 1.0, and it shifts back (1.0−).

If the solenoid valve is still energized, the cycle will repeat. This sequence does not point out that pilot signals are vented once that sequence valve closes and the pressure of the pilot signal is vented to the tank. This type of drawing, without hydraulic lines shown, is becoming more popular. The ISO designations and numbering system are covered in Chapter 17, *Pneumatic Systems*. For now, the ideas are important. It should also be noted that when a valve spring-returns, it returns to the part of the envelope where the spring is attached.

16.5.3 Basic Troubleshooting

If the system in **Figure 16-30** was not unloading, regardless of system pressure or load, what component would you check? More than likely, you would find that the relief valve is stuck in the shut position. If this valve were stuck, the high-flow pump would not relieve. Inspecting this valve, through disassembly, might reveal contamination that is preventing the valve from operating properly.

If this same system was not developing any pressure with both pumps running, the check valve might be stuck in the open position in addition to the relief valve being open, thus preventing a buildup of pressure.

In **Figure 16-31**, a sequence of events takes place. If the system reaches a specific step and does not proceed, the valve that provides for the next step should be checked. For instance, if both cylinders extend and then nothing else happens, either sequence valve 2.1 is not responding and is stuck, or valve 1.0 is not responding to a pilot pressure signal and is stuck. By disassembling the sequence valve, you might find that contamination is preventing the valve from responding.

Both of these ideas are important, depending on which type of system is being examined. Some systems—like the hi-lo pump system—run constantly and are pressure controlled. Sequential systems operate with a specific sequence of events that take place one after another. If a specific action of either system is not taking place, the valve that provides that action must be checked.

TECH TIP
Replacing a Valve

When a sequence or pressure valve must be replaced, it can be hard to determine the setting in a given application. Either bench set the valve to the pressure specified, or count the rotations inward of the pressure adjustment of the faulty valve and reverse that number of turns when replacing it.

CHAPTER WRAP-UP

This chapter examined major components of hydraulic systems and how they perform their function. While you may never design or engineer a hydraulic system, knowing the fundamental concepts will aid you in troubleshooting and maintaining any system. Just as with other systems, failures in a hydraulic system occur because of a specific cause. A hydraulic pump may fail because it is at the end of its life and it has worn beyond repair or, more likely, it may fail because fluid contamination is causing wear and corrosion. Failures are the effect, not the cause. Simple maintenance tasks, such as replacing filters and ensuring that a proper reservoir breather stops moisture from entering the system, can help prevent such failures.

Chapter Review

SUMMARY

- Hydraulic reservoirs are typically sized large enough to supply two to four times the pump flow rate and are configured in various ways in relation to a mounted pump, depending on the physical constraints of the surroundings.

- The temperature of a hydraulic fluid rises during operation because of inefficiency, head loss, and friction.

- The primary functions of the reservoir include storing fluid, acting as a heat sink, and allowing contaminants and gas to settle out of solution.

- Heat exchangers increase the heat removal capacity of the reservoir.

- Suction is developed in a pump by expanding volume.

- A lack of cleanliness is the most common cause of component failure.

- A pressure-compensated pump will only pump enough to maintain a set pressure.

- Pump output flow is proportional to its displacement and speed.

- Input horsepower from the motor to the pump is proportional to the flow and pressure output of the pump.

- Pump failure is usually preceded by noise, higher rates of wear, a drop in output pressure, and the pump not responding to system load.

- Some valves automatically actuate, opening or closing, based on internal system signals. External control of the system is provided through manually operated or power-operated valve actuators.

- Pressure control valves respond to a combination of system pressure, springs, and pilot pressure to control system pressure. The three main types of pressure control valves are the relief valve, pressure-reducing valve, and sequence valve.

- Directional control valves change the flow path of fluid to accomplish actuator response or system control.

- Flow control valves either limit flow or maintain constant flow with changing system parameters.

- Varnishing and silting are two problems that can cause valves to fail. Both are caused by contamination.

- Valves may be controlled with solenoids, which convert an electrical current into mechanical motion to turn something on or off or to open or close something. Solenoids may be powered by either alternating current (AC) or direct current (DC).

- An AC solenoid that does not fully reposition the valve will continue to draw high current and will fail.

- Pilot controls operate larger valves with smaller signals, often in stages.

- Hydraulic cylinders convert fluid energy into linear motion and mechanical force.

- Stop tubes and cushions extend the life of rod bushings and seals. Needle valves restrict the flow of fluid and reduce the velocity of the rod for the last part of the stroke.

- Hydraulic motors convert fluid energy created by pressure and flow into rotary motion and torque.

- Numerous smaller integral and internal components can cause a motor to not respond to a load.

- A basic knowledge of circuits, including hi-lo pump circuits and sequence control, will ensure the proper troubleshooting and maintenance of hydraulic systems.

REVIEW QUESTIONS

Answer the following questions using the information provided in this chapter.

1. List and describe at least three purposes of a hydraulic reservoir.

2. *True or False?* Without a positive suction head, a pump must be able to draw fluid by creating a vacuum at its suction point.

3. Describe how a dirty heat exchanger can affect system operation.

4. How does a pump develop suction and pressure in order to work properly?

5. _____ is proportional to both the flow and pressure output of a pump.

6. *True or False?* A pressure-compensated vane pump uses the rotary motion of gears to directly move fluid through the unit.

7. In a vane pump, how is a seal provided between the vanes and casing?

8. A(n) _____ is the open end of a passage that fluid passes through.

9. What determines the amount of displacement produced in an axial piston pump?

10. Find the flow for a pump with a displacement of 3.25 cubic inches and speed of 1600 rpm.

11. Find the input horsepower required for a pump that is supplying 1200 psig in pressure and 18.75 gpm in flow.

12. What is the typical temperature limit on industrial hydraulic systems? If the temperature exceeds this limit, what may be needed?

13. List at least four possible causes of seal failure.

14. What is a spool?

15. Describe the three main types of valves and their purpose within a hydraulic system.

16. *True or False?* Relief valves allow excess pressure from the pump to return to the reservoir.

17. In a remotely piloted relief valve, how does the remote signal affect relief operation?

18. What are the differences and similarities between a direct-operated pressure-reducing valve and a pilot-operated pressure-reducing valve?

19. Describe how a floating spool center position and blocked spool center position are different.

20. Which of the following symbols represents a solenoid-operated, three-position, four-way, spring-centered valve with a floating center?

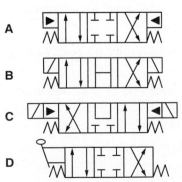

A

B

C

D

21. As pressure drop increases across a valve, flow through the valve will _____.

22. *True or False?* As a hydraulic fluid's temperature changes, the viscosity of the fluid also changes.

23. What is silting? How will silting of a valve affect operation?

24. What are solenoids? How are solenoids powered?

25. Hydraulic _____, also called linear actuators, transfer the energy from a fluid into linear motion and mechanical force.

26. Cylinder _____ can help reduce noise and wear on the cylinder.

27. When dealing with cylinder-related damage or wear, what are some common maintenance corrections?

28. *True or False?* Motors convert rotary motion and torque into fluid energy.

29. What is the difference between a hydraulic pump and motor?

30. Find the flow in gpm for a motor if its displacement is 1.75 cubic inches per revolution and it runs at 2300 rpm.

31. Calculate the rpm for a motor with a displacement of 3.25 cubic inches and that is supplied with a flow of 12 gpm.

32. With a pressure of 1250 psig and a displacement of 2.25 cubic inches, what torque, in inch-pounds, will a motor produce?

33. In a hydraulic system where the loading of the system changes and different flow rates are needed, a _____ pump circuit is ideal.

34. *True or False?* In a sequential system, all sequence valves have tags that show where pressure drops occur.

NIMS CREDENTIALING PREPARATION QUESTIONS

The following questions will help you prepare for the NIMS Industrial Technology Maintenance Level 1 Basic Hydraulic Systems credentialing exam.

1. Calculate the output flow for a pump with a displacement of 4.25 cubic inches that is rotating at a speed of 2025 rpm.

   A. 0.48 gpm   C. 37.26 gpm
   B. 5.02 gpm   D. 476.47 gpm

2. In a three-position, four-way directional control valve, what happens when the spool is shifted out of the center position?

   A. Pressure will be sent to output port A or B to the actuator while the opposite output port is vented back to the reservoir.

   B. The actuator ports (A and B) are blocked, but free flow is allowed back to the reservoir tank.

   C. Pressure will be sent to both output ports A and B to the actuator.

   D. All ports are blocked and the cylinder is locked in a position where pressure cannot enter.

3. Describe the valve indicated by the following symbol.

   A. Two-position valve in an on-off valve position

   B. Two-position valve where pressure is sent to port A while B is vented

   C. Three-position valve where both ports A and B are connected to the tank line in the center valve position

   D. Three-position valve with a floating center position

4. Which is responsible for protecting a system from overly high pressure and directing excess flow from the pump back to the reservoir?

   A. Pressure-reducing valve

   B. Sequence valve

   C. Relief valve

   D. Spool valve

5. When would pressure compensation be required for flow control?

   A. If a system needs constant flow to ensure actuator speed.

   B. If pump input power is equal to pump flow and pressure.

   C. If fluid temperature increases enough to affect viscosity.

   D. If a port is closed and will not allow fluid to pass through.

6. When a cylinder approaches the end of a stroke, what is the role of a needle valve?

   A. It increases the flow of the fluid and thus decreases the velocity of the rod.

   B. It increases the flow of the fluid and thus increases the velocity of the rod.

   C. It restricts the flow of fluid and thus increases the velocity of the rod.

   D. It restricts the flow of fluid and thus reduces the velocity of the rod.

7. How can actuator speed be adjusted in a hydraulic system?

   A. Using a directional control valve

   B. Using a detent

   C. Using a flow control valve

   D. Using a relief valve

8. Which does *not* describe a type of mount for a hydraulic cylinder?

   A. Flange

   B. Overhead

   C. Clevis

   D. Trunnion

9. Find the required flow for a motor that has a displacement of 1.25 cubic inches and is running at 2200 rpm.

   A. 0.13 gpm

   B. 11.9 gpm

   C. 119.4 gpm

   D. 406,560 gpm

10. What condition might indicate that a relief valve is stuck in the closed position?

    A. The system is not unloading, regardless of system pressure or load.

    B. The system does not develop any pressure.

    C. Certain events in a sequence do not take place.

    D. The hydraulic cylinder moves in only one direction.

17 | Pneumatic Systems

LEARNING OBJECTIVES

After completing this chapter, you will be able to:

☐ Describe common components of an air-treatment system.

☐ Describe types of receivers and calculate receiver size.

☐ Discuss types of pneumatic distribution systems.

☐ Explain the operation of a single-stage and two-stage reciprocating compressor.

☐ Explain the operation of pneumatic motors.

☐ Draw common symbols for valves and valve actuators.

☐ Label common elements of a pneumatic cylinder.

☐ Describe methods for cylinder-position sensing.

☐ Explain how meter-in and meter-out valves are used to control actuator speed.

☐ List a sequence of operations for a basic pneumatic circuit.

TECHNICAL TERMS

adsorption

airend

anaerobic adhesive

analog sensor

branched system

compressor

digital sensor

directional flow regulator

dynamic seal

FRL unit

intensifier

intercooler

lapped/shear

loop system

magnetic position sensor

magnetic reed sensor

O ring seal

parallel system

primary air treatment

quick exhaust valve

receiver

refrigerated air dryer

repeatability

run time (also called duty time)

secondary air treatment

shuttle valve

silencer

unloading

W hile industrial hydraulic systems may use pressures up to 5000 psi, industrial pneumatic systems typically use pressures of 80 to 100 psi. Since air is compressible, unlike hydraulic fluid, pneumatic systems are subjected to less shock loading than hydraulic systems. However, this same compressibility makes speed control more difficult with pneumatic systems than with hydraulic systems. Finally, while some specific pneumatic applications use very large cylinders, pneumatic systems are normally used for higher-speed applications requiring less force than is used in hydraulic systems. In a fashion similar to hydraulic systems, pneumatic systems use direction control valves and flow control valves to control the flow of fluid power, in this case compressed air, to actuators, which then perform work. **Figure 17-1** shows a basic pneumatic system, as a reminder of how the parts of the circuit fit together.

17.1 AIR TREATMENT

Before air can be used in a pneumatic system, it must be conditioned. Air treatment for pneumatic systems includes both initial treatment of the air before compression and treatment of the air after compression. The purposes for air treatment include:

- Removal of particulate contamination.
- Removal of moisture.
- Addition of lubricating properties for downstream machinery.
- Removal of heat.

Heat is generated as a result of the compression of the air. This heat is partially removed in multistage compressors, through the use of air or water coolers, between stages of compression. On larger systems, this removal of heat is performed by specific air dryers/coolers.

17.1.1 Primary Air Treatment

Pneumatic systems require clean and dry air. ***Primary air treatment*** is the first stage of air treatment or conditioning in a system, in which particles, moisture, and heat are removed from the system air. Primary air treatment starts at the outlet of the compressor with a water separator. Water separators cause the compressed air to travel in a high-velocity, rotational manner, using centrifugal force to remove condensed water from the airstream. **Figure 17-2** shows the internals of a moisture separator. Drains may be manual in smaller systems, or

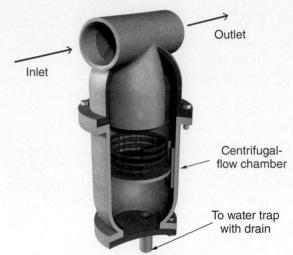

Inlet

Outlet

Centrifugal-flow chamber

To water trap with drain

Goodheart-Willcox Publisher

Figure 17-2. In a moisture separator, pressurized air is forced through vanes, which causes a vortex to form and moisture to fall outward onto the inside walls.

they may be automatically timed in larger systems that discharge to an oil/water separator. The water separator may include a filter for larger particles (40 microns and greater), and it may include differential pressure indicators to gauge periodic maintenance requirements.

On larger systems a wet receiver may be used. The receiver normally has a timed automatic drain that also drains excess moisture to the oil/water separator. Isolation valves and bypass valves should be present and should be of the full-port type. Full-port valves have openings the same size as the pipeline, allowing unrestricted flow and reducing possible pressure loss. Following the receiver may be one or multiple levels of particulate filters, ranging from 40 microns down to 5 microns.

The ***refrigerated air dryer*** is an important component of any pneumatic system. It removes moisture from the air by cooling the air and condensing moisture. Several

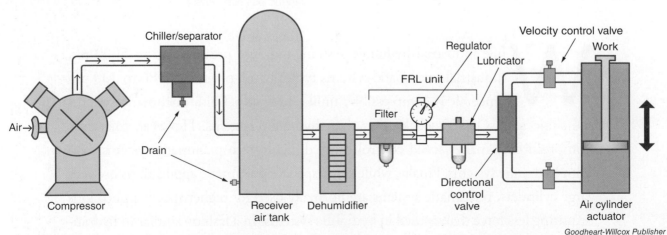

Chiller/separator

Velocity control valve

Regulator

Work

Lubricator

FRL unit

Air →

Filter

Drain

Compressor

Receiver air tank

Dehumidifier

Directional control valve

Air cylinder actuator

Goodheart-Willcox Publisher

Figure 17-1. This diagram shows a basic pneumatic system.

types of moisture removal systems are used depending on the size of the application and the needed purity of the compressed air. Using a refrigeration cycle, heat exchanger, and integrated automatic controls, the dryer maintains a nearly moisture-free supply of compressed air for system uses. **Figure 17-3** shows a variety of refrigeration dryers. Integrated controls feature alarm conditions for out-of-specification operation to allow for fast troubleshooting. Refrigeration units have automatic drains and are directed to the system's oil/water separator. Well-designed refrigeration units rarely require service, with the exception of occasional cleaning and checks of the drain valve.

In-line desiccant dryers are also popular for smaller systems. Desiccant is a substance that attracts water and is used as a drying agent. Shown in **Figure 17-4**, as air enters the dryer and descends through the silica gel desiccant, moisture is adsorbed. *Adsorption* is the process by which moisture is attached to the surface area of a media. Dry air then exits through a filter element and long tube. The silica turns from a clear orange color to a darker green color, at which time it can be replaced or regenerated by removing it and baking it to 275°F. The life of the silica depends greatly on the humidity of the air being supplied. In-line desiccant dryers usually have a higher pressure drop than refrigerator dryers and are used in smaller systems.

Figure 17-5 shows a regenerative heatless desiccant system. This system uses the same medium as the smaller desiccant units, but it alternates its use between the two columns. Air flows into the online column from the bottom, through the desiccant, and then out the top to the system. Meanwhile, the off-line column is being purged. In this process, a small amount of dry air from

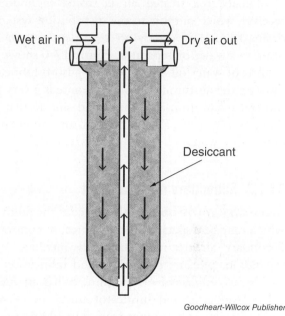

Goodheart-Willcox Publisher

Figure 17-4. The desiccant in this in-line dryer absorbs moisture as air moves through it.

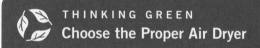

THINKING GREEN
Choose the Proper Air Dryer

Consider cost, efficiency, and the size and use of the pneumatic system before choosing what type of air dryer to use. Generating compressed air takes considerable energy, and choosing the right type of refrigerated dryer or desiccant dryer could yield big savings in energy and money over the life of the system.

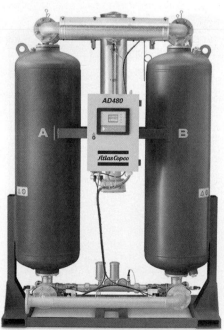

Atlas Copco

Figure 17-5. This heatless dryer regenerates desiccant by using the dry air from the opposite online column, switching between columns when necessary in order to continually regenerate the desiccant.

Atlas Copco

Figure 17-3. The refrigerated dryers pictured range in flow from 200 to 2400 standard cubic feet per minute (scfm) and use integrated controls to maximize efficiency.

the online column flows down through the desiccant in the off-line column, absorbs moisture from the wet desiccant, and exits out the exhaust at the bottom of the stack, thereby drying the off-line desiccant. Integrated controls sense moisture, control purge timing, and allow overall cycle control between the two columns. Desiccant rarely needs to be changed.

Finally, the treated air is stored in another dry receiver, which outputs to the distribution system. Any drains from the primary air treatment system should drain to a water/oil separator. **Figure 17-6** shows several models of water/oil separator that adsorb lubricants by filtering the air through zeolite. Zeolite is a very porous mineral made from aluminum and silicate that acts as a microsieve and is able to filter oil and other contaminants from water.

17.1.2 Secondary Air Treatment

Secondary air treatment takes place at the point of use, which may be a specific machine, area, or control panel. Secondary air treatment normally includes moisture separation, pressure regulation, and lubrication. These can be accomplished in one unit, called an ***FRL unit*** (filter, regulator, and lubricator unit), or in separate units. Standard filters range from 5 to 40 microns, while coalescing filters can be specified to 0.01 micron.

Sizing a filter properly depends on:

- System flow.
- Allowable pressure drop at that maximum flow.
- Piping size.
- Needed micron rating.

Filtration is the first step in an FRL unit. As the air enters the filter, it is forced into a swirling vortex, which uses centrifugal force to throw larger contaminants and moisture out and against the inner wall of the filter case. These contaminants fall down into the lower section of the filter. The air travels through the filter element, which removes smaller particulates before the air travels to the next part of the FRL. The filter drain may be manual or may be specified to a barbed or NPT (National Pipe Thread) fitting and be automatic, opening on its own once the liquid and contaminants reach a predetermined level.

The second stage in an FRL unit is the regulator. The regulator reduces the pressure from the distribution system to a pressure usable by the machine or local process. **Figure 17-7** shows the inner details of a manually set relieving regulator. Main spring pressure (and downstream pressure) is set by turning the adjustment knob. As this adjustment places increased spring pressure on the diaphragm and disc, the disc moves downward and opens the regulator. As downstream pressure increases, it opposes the main spring pressure across the diaphragm through the feedback orifice. Once set pressure is achieved, the disc closes and no airflow is allowed. As pressure downstream drops, the regulator opens to allow

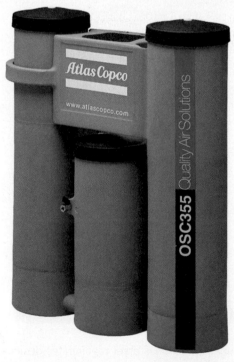

Atlas Copco

Figure 17-6. Using a high-quality water/oil separator prevents discharge of oil into the environment.

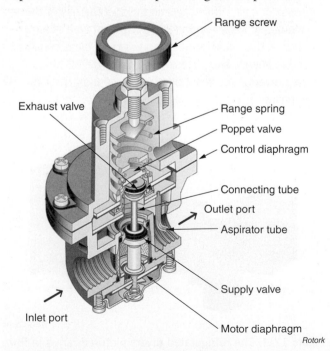

Rotork

Figure 17-7. The inner details of a manually set relieving pressure regulator are labeled in this diagram.

flow. If downstream pressure increases beyond the setting of the regulator, the diaphragm is forced off the top of the disc, and the pressure is relieved through the vent. Some FRL units combine filtering and regulating into one subunit.

The final stage in air preparation in an FRL unit is lubrication. In the lubricator, small oil droplets mix with the airstream as a mist, while larger oil particles fall down into the bowl at the bottom of the lubricator. The lubricated air then flows downstream through the workstation air supply line.

17.1.3 Maintenance

The majority of maintenance items with pneumatic systems are preventive, including regular checks to ensure that the best efficiency is maintained with filters, regulators, and lubricators. When excessive pressure drop is experienced at the load, increasing a regulator's pressure will not solve the problem. Start by cleaning or replacing filters, beginning with the filter closest to that load and working backward from there.

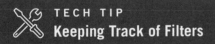

TECH TIP
Keeping Track of Filters

Often a company will use one model of an FRL but then use different filter elements for the various application requirements in the plant. Clearly label the filters so that you can perform proper maintenance. For example, you want to avoid causing airflow problems by using a fine filter in a coarse-filter application.

Take proper safety precautions when using FRL units with polycarbonate bowls or sight glasses. FRL units should not be exposed to excessive temperatures, fumes, direct sunlight, or the possibility of impact. Use regular, organic-based lubricants without additives. Some synthetic lubricants can affect nitrile seals and will cause them to fail over time.

SAFETY NOTE
Cleaning

Never use solvents or harsh chemicals to clean anything that will be placed in a pneumatic system. Polycarbonate bowls can rupture and cause harm if exposed to solvents. Use only a mild soap with water to clean.

17.2 COMPRESSORS AND MOTORS

Compressors produce compressed air for pneumatic systems by increasing the pressure of air by reducing its volume through one or several stages. Motors are the prime movers that drive the compressors by converting energy into torque and rotation. Compressors can range from very small fractional horsepower to hundreds of horsepower, depending on the application. Motors used in pneumatic systems are commonly small, except for some very specific applications. The type of compressor used is somewhat dictated by the power required. Reciprocating compressors use the linear movement of pistons to compress air and can be found in sizes up to around 30 hp. Rotary compressors use a rotating component to compress air and can be sized at several hundred horsepower. While there are several other types of compressors, reciprocating and rotary compressors are probably the most commonly used types in industry.

17.2.1 Compressors

The single-stage reciprocating compressor is the most common compressor and is often used in smaller industrial applications. The single-stage compressor draws in standard-pressure air and discharges compressed air with one or more pistons, with each piston driven by the same shaft but discharging independently. This type of compressor is efficient up to 130 psig. Compressing to a higher pressure with a single-stage compressor results in so much heat that the process becomes inefficient—air is heated to such a high temperature that it loses some of its pressure when it cools in the receiver. Two-stage reciprocating compressors are also common in industry and provide higher pressures. **Figure 17-8** shows a two-stage reciprocating piston air compressor and its common components.

For a two-stage compressor, the first stage discharges into the second stage, which performs the final compression. The air passes through an ***intercooler*** to cool it between compression stages. The intercooler may be air cooled or water cooled, and it helps to increase the efficiency of the second compression by removing some of the heat generated from the first stage. Lubrication may be provided by an attached oil pump (gear or gerotor type), which is directly mounted onto the crankshaft, or by splash means as the crankshaft dips into the oil sump. Suction and discharge valves are typically reed-type valves or disc-type valves that deform under suction or discharge pressure to allow flow through them in one direction.

Rotary screw compressors use two timed meshing screws that compress air as they mesh. These types of compressors are available in a wide range of horsepower

Atlas Copco

Figure 17-8. This is an example of a two-stage reciprocating compressor that contains two cylinders connected in series.

but are not always cost effective, except in larger systems. **Figure 17-9** shows details of the internals of a rotary screw compressor. Rotary compressors are designed for constant use and can provide pressures up to around 200 psig, depending on exact design. Pressure discharge may be constant, or it may cycle on and off by unloading means. Rotary compressors are oil flooded, providing sealing and lubrication directly within the casing. The oil cools the compressor and is then cooled by an air or water cooler. The screw element of the rotary compressor is also

called an *airend*. Oil-free rotary compressors use specially designed airends, and oil is not present in the compression chamber. The discharge volume of a rotary compressor can be adjusted by using a variable speed drive on the motor to adjust the speed (rpm) of the compressor, or by mechanical means with a valve that recirculates compressed air back to the suction for low-flow conditions.

Unloading is the act of removing the load from the mechanical side of the compressor when starting up or, in the case of a screw compressor, when reaching its maximum pressure. Smaller-capacity compressors may not unload, and are set to start and stop against a full-discharge head pressure using a pressure switch. Larger compressors typically need to reach a point close to full speed before being loaded against a discharge pressure. Unloading may be through means of mechanical or electrical control. For example, suppose a mechanical control is attached to the end of the crankshaft of the compressor in **Figure 17-8** and uses a centrifugal-driven switch for unloading. Initially the switch is open, which allows an external valve to be opened and the discharge side of the compressor to be unloaded. As the shaft spins faster, the weights move outward and close a port, allowing discharge pressure to build up. When the compressor is stopped, the crankshaft slows and the weights return to their initial position, venting the discharge of the compressor.

Larger compressors may have a pilot and main unloading valve. The pilot valve is a convenient, remotely located secondary control. It is connected to the pressure switch controlling the compressor and sends a pneumatic signal to the main unloading valve, which is connected between the compressor discharge line and receiver. **Figure 17-10** shows a small unloading valve used for smaller compressors, and **Figure 17-11** is a diagram of a pilot-operated unloading valve.

17.2.2 Motors

While vane motors are the type most commonly used in industrial pneumatic applications, other designs also mirror the types used in hydraulic systems: axial piston motors, radial piston motors, and gear motors. Piston motors are used for higher-torque applications, especially at lower speeds (rpm), while vane motors develop torque and power at higher speeds. Integral mechanical controls can be included to control speed, direction of rotation, and braking.

You will recall that torque is a measure of the turning force on an object rotating around an axis and is usually measured in inch pounds or foot pounds. In order to develop more torque at lower speeds, some pneumatic motors come as an integral unit with gear reducers. This configuration reduces the speed though the gears and outputs substantially more torque.

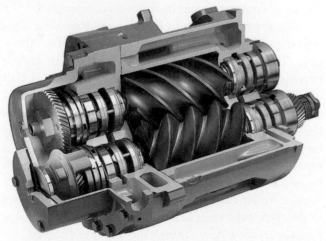

Atlas Copco

Figure 17-9. This internal view of a rotary screw compressor shows its main parts.

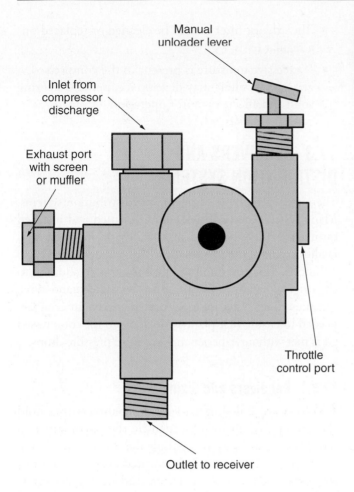

Manual unloader lever

Inlet from compressor discharge

Exhaust port with screen or muffler

Throttle control port

Outlet to receiver

Goodheart-Willcox Publisher

Figure 17-10. This small (1/2″ NPT) unloading valve is for smaller-horsepower compressors.

SAFETY NOTE
Lockout/Tagout

Just because something is mechanically driven (e.g., an air motor) does not mean that it can be worked on without being locked out and tagged out. Even a 2-horsepower air motor has more than enough power to do serious harm. With a motor of any size, make sure you properly perform the lockout and tagout, vent any residual pressure, and disassemble air piping to make sure nothing will happen.

17.2.3 Maintenance

With reciprocating compressors, excessive run times cause the early wear of intake and output reed valves and cause oil carryover. Staying aware of this and checking run times on a regular basis, and inspecting drained moisture for oil contamination on the wet receiver, will help you predict coming problems. Lubrication should be changed on a regular basis—follow the manufacturer's recommendations. The proper viscosity for the oil used depends on several factors, and you should follow the manufacturer's guidelines.

With rotary screw compressors, the airends can prematurely wear due to excessive heat, contaminated lubricant, and contaminated air. Again, proper lubrication and scheduled changes should follow the manufacturer's recommendations.

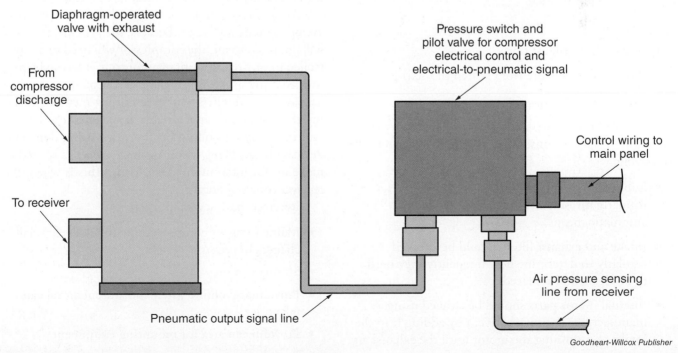

Diaphragm-operated valve with exhaust

From compressor discharge

To receiver

Pneumatic output signal line

Pressure switch and pilot valve for compressor electrical control and electrical-to-pneumatic signal

Control wiring to main panel

Air pressure sensing line from receiver

Goodheart-Willcox Publisher

Figure 17-11. Larger unloading valves may be pilot controlled.

Compressors often show signs of impending failure:

- Excessive heat generation.
- Excessive vibration.
- Excessive noise.
- Sharp, inconsistent noise.
- Trying to start against a full discharge head and tripping protective functions due to high current draw.

In all cases, the following items should be checked at least every 500 hours of run time as part of regular maintenance:

- Lubrication levels.
- Suction filter pressure drop.
- Alignment of the drivetrain.
- Belt condition (if used) and tension.

Unloading valves may fail in the open (vented) or the closed (nonvented) position. If an unloading valve fails in the open position, the compressor may not develop the full rated output pressure because some of the flow is being vented. If an unloading valve fails in the closed position, the compressor may trip over current protection on start-up. In either case, the unloading valve must be carefully disassembled. Depending on the type of valve, you should look for the following:

- Contamination preventing the valve from moving or causing it to stick.
- Ruptured diaphragms.
- Blockage of exhaust ports.
- Proper pilot pressure from switch (if used).
- Mechanical damage.

Many of the above items apply to air motors as well. In addition, special consideration should be given to the following:

- Lubricated air motors should be checked regularly for proper lubrication (manual or automatic means).
- Intake and exhaust filters should be checked regularly to determine how frequently they need to be cleaned or replaced.
- Internal motor parts should be cleaned using a manufacturer-approved solvent by adding it to the intake and running the motor until the exhaust air is void of solvent.

- The exhaust filter should be cleaned or replaced on a regular basis.
- If excessive moisture is present in the compressed air, exhaust filters may develop ice, perhaps freezing solid and affecting motor operation.

17.3 RECEIVERS AND DISTRIBUTION SYSTEMS

There are several types of pneumatic distribution systems. Many systems are plant-wide and designed and installed during the initial building of a facility. More often, as a facility expands, the pneumatic distribution system is also expanded. This expansion can take place as an addition to the original system or as an added separate, stand-alone compressed-air system. Proper design and engineering are needed to ensure that plant-wide demand for compressed air is met without experiencing excessive pressure drops.

17.3.1 Receivers and Sizing

Receivers are sealed, fixed-volume chambers that hold the compressed air until needed by the pneumatic system. They also assist in cooling the air and removing water vapor. Receivers may be located near the compressor outlet, before or after dryers, and near larger intermittent loads. Some compressors are mounted directly on the receiver. Receivers near the compressor are termed primary receivers. Receivers near a large intermittent load are considered secondary receivers.

A compressor must be larger than the maximum average system-wide load. Because the system-wide load will not be constant, the compressor will need to be controlled to vary the amount of air output. This can be performed using simple on-off control schemes, mechanical unloading methods, or variable output methods (driving the compressor at a variable rate). Load on the system may vary depending on the work being done and devices that are being used. A storage device is needed to make up for intermittent high loads. This is where the receiver comes in handy.

Receivers have several purposes:

- Maintaining air compression with relatively small changes in pressure.
- Storing compressed air.
- Providing a volume where moisture removal can take place.
- Providing an area for mounting equipment.
- Providing over-pressure relief for the system.

The proper size for a receiver depends on the variation in the demand on the system and the size of the compressor. Use the following equation to determine the necessary receiver volume:

$$\frac{\text{airflow} \times 14.7 \text{ psia} \times \text{time}}{\text{pressure band}} = \text{volume}$$

(17-1)

Where airflow is in cfm (cubic feet per minute), a standard air pressure of 14.7 psia (pounds per square inch absolute) is assumed, time is in minutes, the pressure band is the difference between the initial receiver air pressure and the final receiver air pressure (initial psig minus final psig), and volume is given in cubic feet. If we perform an arrow analysis—changing one variable while keeping others constant and examining the result—we can see that:

- A larger pressure band allows for a smaller receiver volume.

- With more airflow, a larger receiver is needed.

- The shorter the time to drain (with everything else constant), the smaller the volume needed.

The drain time represents the time it takes for the receiver to go from the upper to lower pressure limit, which represents the time the load is placed on the system. As an example to determine the needed volume size of a receiver, for a system with an average load of 750 cfm, pressure being held at 25 psig (175 initial–150 final), and a 0.5-minute drain time:

$$\frac{750 \text{ cfm} \times 14.7 \text{ psia} \times 0.5 \text{ minute}}{25 \text{ psig}} = 220.5 \text{ ft}^3$$

Manufacturers also follow their own guidelines when sizing a compressor and receiver as a single integrated unit, but you should still consult those guidelines regarding system demand to determine which unit best fits your needs.

Secondary receivers may be located very close to a high-demand and intermittent load. This tends to happen when the facility is expanded, or when more load than was originally designed for is placed on the system. A large load that is at the far end of the distribution system can draw down pressures in that area of the system. Careful planning and consideration has to be given to the compressed-air system before installing large loads. If pressures cannot be maintained in that part of the system due to pressure loss, there are few options. Main distribution lines leading to that part of the system can

be increased, a secondary air receiver can be installed if the load is intermittent, and additional compressors may also be installed near the new process, if needed.

17.3.2 Types of Distribution

Systems designed to distribute pneumatic power can have a straightforward layout for a simple shop or a more complex layout for a multibuilding facility, using one or several compressor stations. As expansions take place, additions and changes are often pieced together within a distribution system that was originally engineered with a smaller facility or specific purpose in mind.

Here are some general guidelines for a distribution system:

- Distribution headers and main air lines should be sized large enough so that pressure drop is kept to a minimum.

- Avoid the use of sharp turns. Long-radius 90° turns and 45° connections will reduce pressure drop in high-flow situations.

- Size the piping to include at least a 25% expansion.

- Use a loop distribution system if possible, to improve efficiency.

- Takeoff points to supply local loads should come from the top of the header so that the chance of entraining moisture is minimized.

- Piping should be sloped toward a drain point. The point of use should also have a drain port or collection point prior to entering sensitive equipment.

The three main types of distribution system are the loop system, parallel system, and branched system. **Figure 17-12** summarizes these types of piping systems. A ***loop system*** design has a main line forming a continuous loop, which provides maximum flow with a minimal pressure drop between workstations. With a loop system, isolation valves will allow larger sections of the loop to be isolated for maintenance while maintaining supply to the rest of the system. All individual drops should have isolation valves. Pressure loss is minimized by using oversized header piping with low velocity. Takeoff branches should come from the top of the main header and should include moisture separation. The main header should properly slope to points of moisture removal. Loop systems are most common for a medium to large single-building layout.

In a ***parallel system***, one large header supplies all loads in parallel down a single line. If the header is not sized properly, loads at the far end of the line will experience pressure drop. A good plant design should place

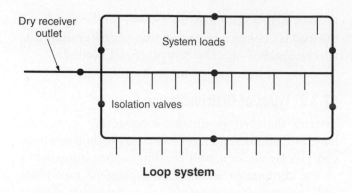

Loop system

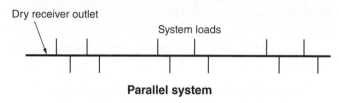

Parallel system

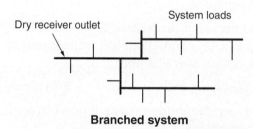

Branched system

Goodheart-Willcox Publisher

Figure 17-12. There are three common pneumatic distribution systems used in manufacturing: loop, parallel, and branched.

large loads closest to the receiver. Takeoff lines need isolation valves for proper maintenance. If any part of the header requires maintenance, the entire system is shut down and depressurized. Parallel systems are usually used for small to medium single-building applications.

Branched systems are small and are typically used for shop air supply. In a branched system, the main line branches off into other lines. Larger loads must be supplied from the main header, or unacceptable pressure drops can result. Each load must have an isolation valve, and any maintenance on main or subheaders requires complete shutdown and depressurization.

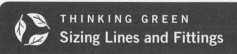

THINKING GREEN
Sizing Lines and Fittings

Pressure loss in a distribution system is a direct cost. Ensuring the proper sizing of lines and fittings for the system can substantially reduce your energy consumption and electrical costs.

17.3.3 Maintenance

Common maintenance items for distribution systems and receivers are preventive and include:

- Checking for consistent draining of all moisture.
- Routine checking for air leaks—especially at isolation valves.
- Checking for excessive pressure drops at system loads.
- Checking (manually lifting) receiver relief valves.
- Checking automatic (timed) drains for proper operation.
- Checking for the run time of the compressor.

Changes in the run time of the compressor will be a good indication of air leaks or excessive air consumption, and an indication that an upgrade may be warranted. *Run time*, or duty time, is the amount of time the compressor runs during the total cycle time, given as a percentage.

$$\frac{\text{run time of the compressor in minutes (or seconds)}}{\text{total cycle time in minutes (or seconds)}} \times 100 = \%\text{ run time}$$

(17-2)

As an example, suppose you have a reciprocating compressor that maintains 100–125 psig in the receiver and you time the cycle, keeping the units constant. The system's loads cause the receiver to draw down until the compressor turns on, taking 4 minutes. The compressor comes on at 100 psig and runs for 2.5 minutes, then turns off at 125 psig. Therefore, the total cycle time is 6.5 minutes. You can figure out the % run time:

$$\frac{2.5 \text{ minutes}}{6.5 \text{ minutes}} \times 100 = 38\%\text{ run time}$$

Run time for a reciprocating compressor is usually limited to about 50%. More run time than this will cause the early wear of discharge reed valves on the cylinder head and oil carryover. Higher run times do not allow the compressor and oil to cool. Low run times (less than 10%) may indicate that the compressor is oversized and may result in the compressor and oil not heating enough, allowing condensation to remain. With other types of compressors, the manufacturer's guidelines should be referenced.

17.4   VALVES AND SYMBOLS

Pneumatic valves have symbols and functions that are similar to those of hydraulic valves. One major difference between a pneumatic and a hydraulic system is the amount of power transferred. Since hydraulic systems use much higher pressures with noncompressible hydraulic fluid, the valves, fittings, and piping are normally much heavier and physically larger. A situation in a hydraulic system requiring a large block manifold may only take a small length of DIN rail in a pneumatic system. Pneumatic systems tend to be digital (using only on-off signals, with the exception of some environmental controls and large valve controls), whereas hydraulic systems regularly include pressure and flow controls and proportional and servo systems.

17.4.1 Types of Valve

There are several different types of pneumatic valve: poppet, diaphragm poppet, spoppet, and spool valves. The most common type of valve seen in industrial control is the spool valve, which can come with O-ring seals or as lapped/shear. *O-ring seals* are O-shaped or doughnut-shaped seals made of elastomer (rubber). A *lapped/shear* spool is a spool that has no seals. Spool valves with O-ring seals are the most commonly used valves in industrial systems. Static seals indicate there is no motion between the mated surfaces. *Dynamic seals* are seals that are used when there is motion between surfaces. They are usually made of more durable material and may require more lubrication to avoid tearing and wear from the movement they experience.

Common port sizes include a range of 1/8″ NPT–1/2″ NPT. Valve actuators are available in manual, external mechanical, air-pilot, and solenoid varieties. You will recall that a valve actuator is a mechanism used to control valves by opening or closing them. Valves may be mounted on a manifold or baseplate, or they may be stand-alone and cabinet mounted on standard DIN rail.

Valves are labeled by the number of "ways" and then number of positions. Ways refers to the valve's primary ports. A three-way, two-position valve would be referred to as a 3/2 valve. A four-way, two-position valve would be referred to as a 4/2 valve.

17.4.2 Common Valve Symbols

Recall that the International Organization for Standardization (ISO) develops and publishes international standards. While valve symbols come from standard ISO 1219-1, port numbering comes from ISO 11727, which most designers follow. Just as in hydraulic systems, pneumatic valves are represented by an envelope, with each position of the valve represented by one box of the envelope. **Figure 17-13** shows common valve symbols with valve actuators, while **Figure 17-14** details the port numbering system for valves. Main ports in a system are shown with single-digit numbers. Outputs are labeled with even, single-digit numbers, while inputs are labeled

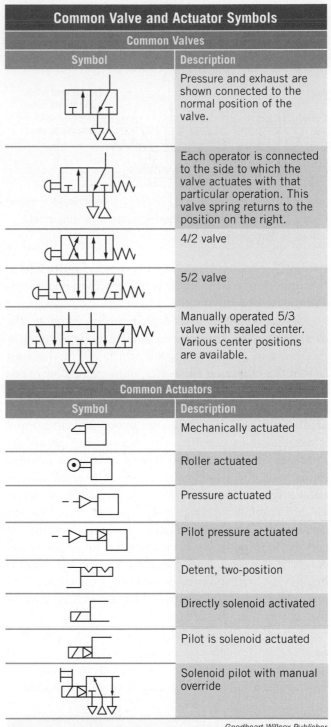

Common Valve and Actuator Symbols	
Common Valves	
Symbol	**Description**
	Pressure and exhaust are shown connected to the normal position of the valve.
	Each operator is connected to the side to which the valve actuates with that particular operation. This valve spring returns to the position on the right.
	4/2 valve
	5/2 valve
	Manually operated 5/3 valve with sealed center. Various center positions are available.
Common Actuators	
Symbol	**Description**
	Mechanically actuated
	Roller actuated
	Pressure actuated
	Pilot pressure actuated
	Detent, two-position
	Directly solenoid activated
	Pilot is solenoid actuated
	Solenoid pilot with manual override

Goodheart-Willcox Publisher

Figure 17-13. The symbols for pneumatic valves are similar to those for hydraulic systems, but air pressure is shown with open triangles or circles.

with odd, single-digit numbers. Actuators are shown with even, double-digit numbers, with the return position (normal position) being the lower number. In a two- or three-port valve the normal position is 10 and the actuator is 12. In four- or 5-port valves the normal position is 12 and the actuator is 14. Other common

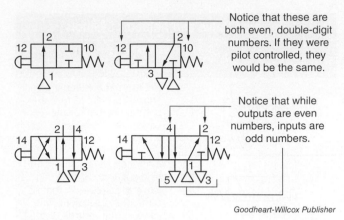

Notice that these are both even, double-digit numbers. If they were pilot controlled, they would be the same.

Notice that while outputs are even numbers, inputs are odd numbers.

Goodheart-Willcox Publisher

Figure 17-14. The port numbering system for pneumatic valves is demonstrated here.

symbols for control elements are shown in **Figure 17-15**. It is important to familiarize yourself with these symbols and the numbering system.

Several common valves deserve special attention. The ***directional flow regulator*** is a valve typically used in meter-in and meter-out circuits, which will be discussed later in this chapter. It allows free, unrestricted flow in one direction while regulating flow in the opposite direction with the use of a needle valve. The ***shuttle valve*** can be used as a logic element that allows an output if either input is received. (Logic elements are discussed in Chapter 18, *Advanced Fluid Power*.) This type of valve has a function similar to that of a quick exhaust valve. The ***quick exhaust valve*** is typically mounted close to or directly on an actuator (cylinder) for quick response in one direction. For instance, if a quick exhaust valve was mounted on the rod end and the cylinder was being extended, the rod air would pass into this valve, seating the check valve and exhausting out of a port through the silencer. If the cylinder was retracted, the pressurized air would seat the check valve on the exhaust port, preventing exhaust and pressurizing the rod end of the cylinder. ***Silencers*** reduce the noise from air escaping through exhaust valves and help protect the system from debris. Silencers can be made from a metallic mesh, sintered bronze, a fabric mesh, or other materials.

SAFETY NOTE
Valve Disassembly

Never disassemble a spring-loaded diaphragm valve without fully understanding the manufacturer's instructions. Some of the opposing springs are very large and store a large amount of mechanical energy, which must be slowly released in order to avoid injury.

17.5 CYLINDERS

Pneumatic cylinders can be small-diameter lightweight cylinders that are used to position parts on automated manufacturing lines, or large industrial-grade valve operators. Regardless of size, all pneumatic cylinders convert the energy in compressed air into linear motion. The system devices that convert energy into motion are also referred to as actuators. Other types of actuator include the rack and pinion and the slide. The rack and pinion converts the linear motion of a piston into rotational movement by use of a pinion shaft and gear. Pinions are smaller round gears, and pinion shafts are shafts with gear teeth machined onto them.

17.5.1 Construction

While the construction of a light-duty pneumatic cylinder is similar to that of a hydraulic cylinder, light-duty pneumatic cylinders are intended for light applications with little loading. Positioning lightweight objects in an automated manufacturing line does not require large cylinders with multiple seals. However, positioning smaller items does usually include a need for high cycle times, high rod speeds, and proper cushioning and deceleration. In larger applications (such as punch pressing), heavy-duty pneumatic cylinders are very similar to hydraulic cylinders and share many of the same components. Most heavy-duty pneumatic cylinders are made of steel. **Figure 17-16** shows a cutaway of a typical industrial pneumatic cylinder. In this case, stop tubes are not shown. Lighter-duty cylinders may be constructed of aluminum tubing, or square extruded aluminum. ISO 15552 establishes specific guidelines for metric series pneumatic cylinders for interchangeability purposes. With standard components, position sensing has also become easier. **Figure 17-17** shows a common type of ISO 15552 cylinder.

Sealing in a pneumatic cylinder is just as important as in a hydraulic cylinder. Multiple types of seal are used, from the rod end seal to the piston seal and end seal. Multiple elastomer materials can be used in the same cylinder.

Threaded components of a pneumatic cylinder are usually held with Loctite™ and may require large forces to unthread. Some types of Loctite require heating for removal, and some items should not be unthreaded. The piston-to-rod connection may use anaerobic Loctite. ***Anaerobic adhesives*** harden to form a tight seal under conditions without the presence of oxygen, and they take special care when unthreading.

Common Control Elements

Symbol	Description	Symbol	Description
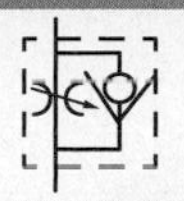	Flow regulator, unidirectional	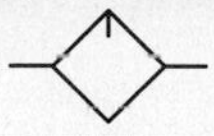	Lubricator
	Flow regulator, bidirectional, simplified	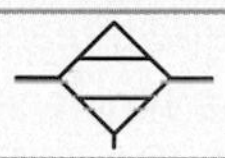	Dryer
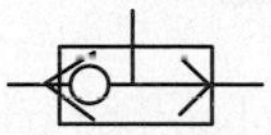	"OR" shuttle valve, simplified	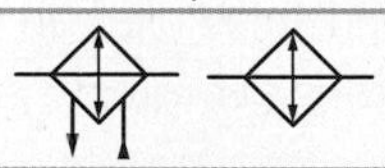	Cooler, with and without coolant flow lines
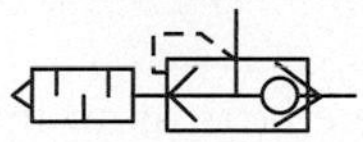	Quick-exhaust valve with silencer, simplified	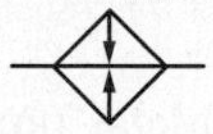	Heater
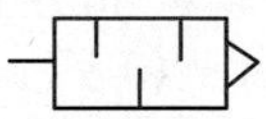	Silencer		Combined heater/cooler
	Pressure to electric switch, preset	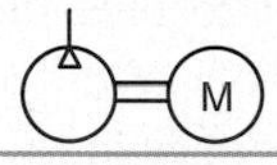	Compressor and electric motor
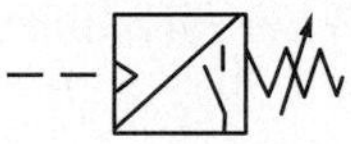	Pressure to electric switch, adjustable	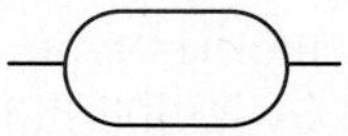	Air receiver
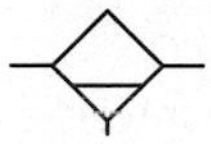	Water separator with manual drain		Isolating valve
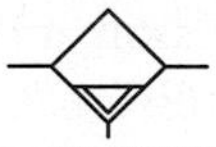	Water separator with automatic drain	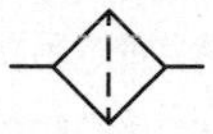	Air inlet filter
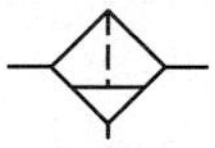	Filter with manual drain	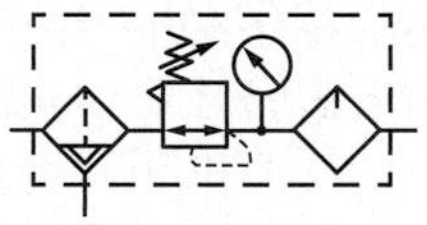	FRL combined unit
	Filter with automatic drain	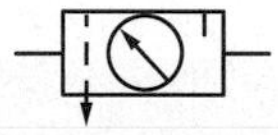	FRL simplified

Goodheart-Willcox Publisher

Figure 17-15. These are symbols for other common elements in a pneumatic system, as well as air preparation.

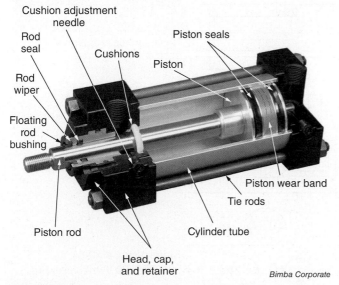

Bimba Corporate

Figure 17-16. This cutaway view shows the parts of a typical industrial pneumatic cylinder.

Bimba Corporate

Figure 17-17. ISO 15552 cylinders are typically shorter-stroke, light-duty cylinders.

> ⚠ **CAUTION**
>
> Avoid thread seal tape (also known as Teflon™ tape, PTFE tape, or plumber's tape) for sealing pipe threads in fluid power systems. It does not provide a secure seal and, more importantly, it easily tears, and small bits can enter and contaminate the system, causing damage.

Bimba Corporate

Figure 17-19. An air intensifier can quickly boost air pressure.

There are many types of pneumatic cylinder. Common cylinder symbols are shown in **Figure 17-18**. You should investigate and research the manufacturers of the cylinders that are common at your workplace prior to breakdown issues. Manufacturers typically offer a number of different types of rebuild kits, depending on the type of cylinder.

An *intensifier*, **Figure 17-19**, is another type of cylinder that may use compressed air, hydraulic fluid, or both in order to quickly boost pressure without the need for additional pumps or compressors, making the system more efficient. While the air pressure develops force on the rod through the large piston, the force is transferred into higher hydraulic pressure at the discharge end. Discharge of high-pressure hydraulic fluid from the high-pressure rod end is limited, and so must be carefully engineered to ensure that enough fluid flows to whatever load is being powered.

As an example demonstrating the effects of an intensifier, assume that the large piston bore is 4″ in diameter (2″ radius) and the rod is 1″ in diameter (0.5″ radius). If 150-psig air was allowed to act on the 4″-diameter piston, you could calculate the resulting pressure from the intensifier (rod) end. Recall the formulas for area, force, and pressure from Chapter 15, *Fluid Power Fundamentals*:

$$A = \pi \times r^2$$
$$F = p \times A$$
$$p = \frac{F}{A}$$

Substitute the values given for the piston and intensifier to calculate force, then resulting pressure.

$$F = p \times \pi \times r^2$$
$$F = 150 \text{ psig} \times \pi \times (2″)^2$$
$$F = 1885 \text{ lb}$$

	Common Cylinders			
Symbol	**Description**		**Symbol**	**Description**
	Spring returned			Nonmagnetic, double or through rod
	Spring extended			Nonmagnetic, adjustable cushion, through rod
	Magnet piston, spring returned			Analog position output
	Adjustable cushion, spring returned			Slide
	Nonmagnetic			Rotary actuator
	Nonmagnetic, adjustable cushion			Bidirectional and nonreversing motor
	Magnetic			

Goodheart-Willcox Publisher

Figure 17-18. These are the most common pneumatic cylinder symbols.

$$p = \frac{F}{A}$$

$$p = \frac{1885\ lb}{\pi \times (0.5'')^2}$$

$$p = 2400\ psig$$

This intensifier has an intensification ratio of 16:1. The resulting pressure (2400 psig) is 16 times the input pressure (150 psig) because of the ratio of the area of the large piston to the small rod. If a system needs only a small amount of linear travel but high pressure, this intensifier would be a good addition.

17.5.2 Cushions and Stop Tubes

Just as with a hydraulic cylinder, as a pneumatic cylinder extends fully, the distance between the rod bushing and cylinder is reduced, which increases the load felt on the bushing. Using a stop tube on long-stroke or horizontally mounted cylinders will increase the final distance, and therefore reduce bearing load. The sizes of cushions used are based on the momentum of the entire assembly (load, rod, etc.) and should be calculated by an experienced designer or engineer. The design and specification process for cushions typically incorporates the following considerations:

- Selecting bore size based on available pressure and load to be moved.

- Selecting piston rod diameter based on mounting method, load, and length of travel.

- Selecting the length of stop tube based on piston length and load.

- Selecting the cushion based on total load (including rod and piston) and the velocity of the load to be cushioned.

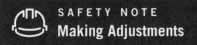

SAFETY NOTE
Making Adjustments

With any moving parts, there is always a safety risk. Just as for hydraulic cylinders, adjust end-of-travel limiters in pneumatic cylinders very slowly.

17.5.3 Position Sensing

While other processes may simply rely on the operator to verify a cylinder's position by eyesight (e.g., with earthmoving equipment), most industrial manufacturing processes require the ability to sense a cylinder's position.

For industrial processes, regardless of control type, a process is typically "told" to do something, and then feedback is needed to ensure that the action took place. In fluid-power systems this feedback may be a sensor, a pressure switch, a limit switch, or some other method of verifying that something took place. Depending on what is needed, both *digital sensors* and *analog sensors* are used. A digital sensor simply provides either an ON or OFF signal as feedback, while an analog sensor provides a signal that is proportional to what is being measured—in this case cylinder position. Which type of sensor is used depends on the receiver of the sensor signal.

Digital Sensors

One of the most popular types of digital sensor is a *magnetic position sensor* that uses a magnetic strip on the piston to sense its position. An electrical contact is closed when a switch senses the magnetic field as the piston moves. A *magnetic reed sensor* is a sensor with an electrical reed switch that operates through the use of a magnetic field. Sensors may come as an option from the manufacturer with integrated mounts, or they may be aftermarket additions clamped onto the cylinder.

Position sensing can also be accomplished by sensing pressure. Pressure will be lower while extending a cylinder, depending on the load, than once the piston reaches its end of travel. By sensing air pressure on the cap end, a sensor can be adjusted to "close" and send a signal once the cylinder is fully extended. This type of sensor uses a diaphragm and adjustable spring pressure against the line pressure to detect position, and it may send an output signal of air pressure or voltage. These pressure switches may be an integral part of the cylinder, or aftermarket pressure switches can be mounted to the cylinder externally.

Externally mounted sensors can be mechanically actuated such as limit switches, air-bleed sensors that are actuated when the air-bleed port is blocked by the presence of the rod, and nearly any type of industrial through-beam sensor or proximity sensor.

Analog Sensors

When the exact position of the cylinder is needed, an analog sensor may be used. Analog sensors may be integral to the cylinder and mounted on one end, as shown in **Figure 17-20**, or they may be external.

The sensor produces a proportional electrical signal based on the position of the piston. For example, a 0-10V sensor would produce 0 volts at full retraction

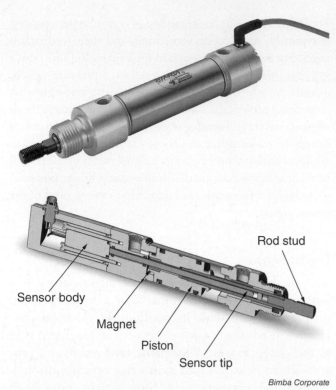

Rod stud

Sensor body

Magnet

Piston

Sensor tip

Bimba Corporate

Figure 17-20. This integral analog sensor varies output according to the magnetic field that is developed between the sensor body and magnet.

and 10 volts at full extension. Output of the sensor can be calibrated, and these types of sensor are normally very accurate (to within 0.001″), with high repeatability. *Repeatability* is the ability to produce the same output signal for the same position over many cycles.

Sensors input into a controller that then outputs to a valve to control the cylinder. While this type of control is not always needed, when a proportional output is needed for system control, this is an effective servo system.

Typical maintenance on analog sensors includes calibration by checking output at 0% position, 100% position, and at some intermediary position, and checking system response. Vibration or impact may change a sensor's position over time, and routine calibration may be needed to ensure proper system control. This type of control, called process control, is such a large topic that whole texts have been written on it. Chapter 34 examines process control in much more detail.

17.6 FLOW CONTROL AND BASIC CIRCUITS

There are several common basic circuits that are used in industry that are examined in the text that follows, but it is more important to understand the methods of control, the numbering systems, and the symbols presented here.

17.6.1 Meter-In and Meter-Out Control

Control of rod velocity, or speed, requires control of the airflow going into and out of a cylinder. This is accomplished by the use of a needle valve and parallel check valve. Since air is compressible, meter-out control is normally used. **Figure 17-21** shows a basic pneumatic circuit with meter-out control. In this circuit, when the directional valve is repositioned by the solenoid labeled as 14, pressure passes through the directional valve to the flow control valve. At the flow control valve, air pressure unseats the check valve and passes on to the cap end of the cylinder unrestricted, extending the cylinder. The air at the rod end is forced out to the flow control valve, seating the check valve, and is then throttled through the needle valve. Air then passes back through the directional valve to exhaust. When the solenoid is de-energized, the directional control valve repositions to the center position (spring returned), and the cylinder is held in place. The velocity of the extension is controlled by the throttling of the flow control valve connected to the rod end (the air flowing out of the cylinder). The more this needle valve is closed, the slower the extension speed.

When the solenoid labeled as 12 is energized, the directional valve repositions and pressurized air flows through the directional valve to the flow control valve at the rod end, unseating the check valve and freely flowing to the rod end, retracting the cylinder. Air at the cap end is forced out through the flow control valve, seating the check valve, and

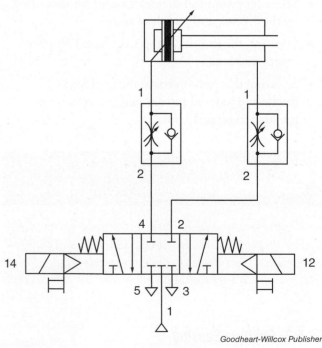

Goodheart-Willcox Publisher

Figure 17-21. This is a basic pneumatic circuit with meter-out control. Meter-out control of a cylinder controls both the extension and retraction speeds by controlling the exhaust air leaving the cylinder.

is restricted by the throttling of the needle valve, moving back through the directional valve to exhaust. The velocity of the retraction stroke is controlled by the flow control valve attached to the cap end of the cylinder.

With a meter-in circuit, the flow control valve restricts the flow of air going into, rather than out of, the cylinder.

17.6.2 Clamp and Stamp

Figure 17-22 illustrates a clamp-and-stamp circuit. In an operation using this type of circuit, one cylinder extends to perform a clamping action to hold an item (clamp), and the second cylinder activates a machining tool to perform a specific task (e.g., drilling, stamping, etc.) on the clamped item (stamp). When the manual valve (3.1) is repositioned, pilot pressure is sent to valve 3.0, which repositions. Pilot pressure is then sent to valve 1.0, which repositions, and cylinder 1 extends to clamp. When limit switch 2.1 is made, valve 2.1 repositions, sending pilot pressure to valve 2.0. Valve 2.0 repositions and extends cylinder 2.0 to stamp. The operator then releases valve 3.1, venting port 14 and sending pilot pressure to valve 3.0, port 12. This repositions and sends pilot pressure to both valve 1.0 and valve 2.0. Valve 2.0 repositions first and starts to retract cylinder 2.0. After a short delay, valve 1.0 repositions and cylinder 1 retracts. Both cylinders are held in the retract position when the manual valve is in its normal, spring-returned position. In a manufacturing operation, this is when the next work item would move into place for clamping and stamping, and the operation could be repeated.

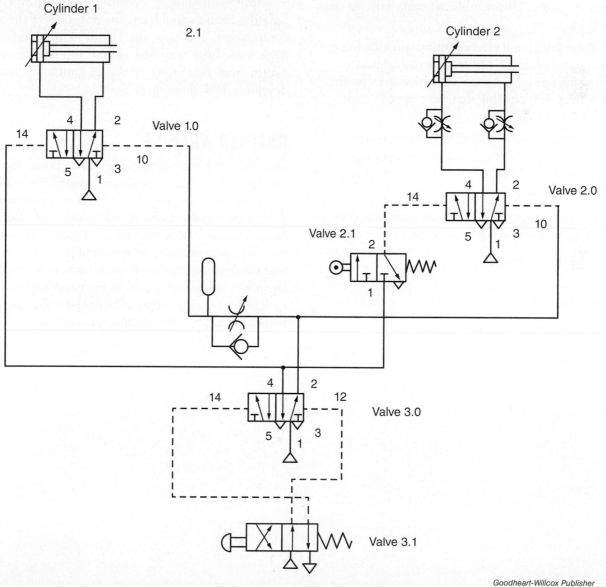

Figure 17-22. This is a simplified clamp-and-stamp pneumatic circuit.

17.6.3 Troubleshooting

Common problems with pneumatic circuits depend on the circuit logic. Troubleshooting a pneumatic sequencer is different than troubleshooting a clamp-and-stamp circuit. With the simple meter-out circuit, troubleshooting is straightforward:

- If the cylinder will not extend, check for proper solenoid operation.

- If a solenoid is energizing but the cylinder is not extending, use the manual bypass to check for proper valve operation. If the spool valve will operate manually, replace the solenoid.

- It is possible to energize a solenoid and for the solenoid to not be developing enough force.

- If the main valve is stuck in position, the cylinder will not move. Disassemble the main valve to clean and inspect it.

- If the cylinder will extend but not provide force at the end of stroke, ensure that the cylinder is receiving enough pressure and the system pressure is correct. It is possible that the cylinder's seals must be replaced.

- If rod speed is not controlled properly in either direction, check the appropriate flow control valve and check valve. The check valve may be stuck open or closed due to contamination.

For the clamp-and-stamp circuit, operations happen in a specific sequence because of the logic of the circuit.

1. Valve 3.1 + (manual)
2. Valve 3.0 +
3. Valve 1.0 +
4. Cylinder 1 +
5. Valve 2.1 +
6. Valve 2.0 +
7. Cylinder 2 +
8. Valve 3.1 − (manual)
9. Valve 3.0 −
10. Valve 2.0 −
11. Cylinder 2 −
12. Valve 1.0 −
13. Cylinder 1 −
14. Valve 2.1 −

If any of these events do not take place, look at the previous step and the signal sent to start the next step. For instance, if cylinder 1 extended but cylinder 2 did not, valve 2.1 should be examined.

Troubleshooting a circuit requires knowledge of how the circuit should act, the purposes of the circuit components, and how the circuit is not performing. Both knowledge acquired on the job and knowledge gained from examining circuit diagrams will help you develop troubleshooting skills.

CHAPTER WRAP-UP

This chapter examined many key concepts in relation to basic pneumatics and pneumatic systems, including the functions and symbols of key parts of pneumatic systems, as well as the makeup and processes of basic pneumatic circuits. These foundational concepts will help you to build, evaluate, and maintain pneumatic systems and troubleshoot common pneumatic failures you will encounter in the workplace. In the next chapter, we will look at advanced concepts in both hydraulics and pneumatics, including controls and automation.

Chapter Review

SUMMARY

- Primary air treatment consists of removing particulates, removing moisture, and possibly adding lubrication. Secondary air treatment takes place near the load, and includes filtering, pressure regulation, and lubrication.

- The removal of moisture can be accomplished by use of centrifugal moisture separators, refrigeration dryers, and desiccant dryers. Downstream pressure is increased by turning the adjustment knob of a regulator in the clockwise direction. Lubricators inject oil droplets so small that you cannot see them with the naked eye.

- Receivers are sealed chambers that hold compressed air until needed by the system. Sizing is based on flow, draw-down time, and pressure band.

- The main types of distribution system are loop, branched, and parallel.

- Compressor run time should be regularly checked and is figured as a percentage, calculated as the running time divided by the total cycle time.

- Single-stage reciprocating compressors draw in standard-pressure air and discharge compressed air. Two-stage reciprocating compressors provide higher pressures and are more efficient because of the intercooler that removes compression heat between stages.

- Unloading takes place at the beginning and end of a cycle to allow the compressor to start without pushing against a high pressure at low speed.

- Vane motors are the most common motors in pneumatic systems, while piston motors are used for applications requiring higher torque at lower speeds.

- The spool valve is the most common control valve. It can come with O-ring seals or can be lapped/shear without seals.

- Valves are referred to in terms of ways and positions.

- Pneumatic air cylinders are generally smaller and more lightweight compared to hydraulic cylinders. Intensifiers can use compressed air or hydraulic fluid to increase the pressure in a pneumatic system and add efficiency.

- When properly fitted by experienced personnel, cylinder cushions and stop tubes can increase the life of cylinders.

- Digital and analog sensors are available to monitor the position of cylinders.

- A proper examination of a circuit diagram starts with an understanding of the purposes of the components.

- Meter-in and meter-out controls are used to control rod velocity, or speed, by controlling the airflow going into and out of a cylinder.

- Clamp-and-stamp circuits are common in pneumatic systems, where one cylinder extends to perform a clamping action to hold an item (clamp), and the second cylinder activates a machining tool to perform a specific task (e.g., drilling, stamping, etc.) on the clamped item (stamp).

REVIEW QUESTIONS

Answer the following questions using the information provided in this chapter.

1. List three purposes for air treatment in a pneumatic system.

2. All moisture removed from a pneumatic system should be directed to what component? How is it removed?

3. Describe the process of adsorption.

4. How do in-line desiccant dryers differ from refrigerator dryers?

5. *True or False?* Secondary air treatment involves moisture separation, pressure regulation, and lubrication.

6. Describe how a regulator adjusts pressure from the distribution system to a pressure usable by a down-system machine.

7. *True or False?* A compressor must be smaller than the maximum average system-wide load.

8. Describe three purposes of receivers.

9. Calculate the volume of a receiver for a 500 cfm airflow, a 1-minute draw-down time, and a 30 psig pressure band.

10. Describe three relationships between the variables used to determine proper receiver size if one variable is changed as the others are held constant.

11. List the three types of distribution system layouts and describe in what setting each is used.

12. Calculate the run time percentage if a compressor runs for 2 minutes and then is off for 5 minutes.

13. What is the purpose of an intercooler in a two-stage compressor?

14. How is an oil-flooded rotary compressor different from an oil-free rotary compressor?

15. What is the purpose of unloading a compressor?

16. What type of motor is most commonly used in industrial pneumatic applications?

17. How will an unloading valve stuck in the open position affect compressor operation?

18. *True or False?* Valves, fittings, and piping in pneumatic systems are much heavier and physically larger due to the system using much higher pressures with noncompressible fluid.

19. What two types of O-ring seal can a spool come with, and how are they different?

20. When a valve is described as a 4/2 valve, what does that indicate?

21. Describe port numbering systems for valves.

22. A _____ allows free, unrestricted flow in one direction while regulating flow in the opposite direction with the use of a needle valve.

23. What is the purpose of a cylinder?

24. What does ISO 15552 enforce?

25. Find the final pressure output of an intensifier with a piston bore diameter of 1.5″ and a rod diameter of 0.4″ with input air having a pressure of 120 psig.

26. What is the result of using a stop tube on long-stroke or horizontally mounted cylinders?

27. Describe the difference between a digital and analog position sensor.

28. *True or False?* Repeatability is the ability to produce the same input signal for the same position over many cycles.

29. In a pneumatic circuit with meter-out control, what effect does the needle valve have on extension speed?

30. In a clamp-and-stamp circuit, a solenoid is energizing but the cylinder is not extending. What would you check when troubleshooting the problem?

NIMS CREDENTIALING PREPARATION QUESTIONS

The following questions will help you prepare for the NIMS Industrial Technology Maintenance Level 1 Pneumatic Systems credentialing exam.

1. When treating air at the point of use, where does moisture separation and pressure regulation occur?

   A. FRL unit
   B. Secondary air treatment storage
   C. Refrigerated air dryer
   D. In-line desiccant dryer

2. In a pneumatic relieving regulator, what occurs if downstream pressure increases beyond the setting of the regulator?

   A. Airflow through the venturi is reduced.
   B. Loads downstream increase, which forces the regulator to shut down airflow.
   C. Spring pressure increases on the diaphragm, and the disc restricts airflow.
   D. The diaphragm is forced off the top of the disc, and pressure is relieved through the vent.

3. What three quantities will determine the proper size for a receiver?

   A. Pressure band, airflow, and time to drain
   B. Pressure band, line size, and FRL location
   C. Airflow, FRL location, and compressor size
   D. Airflow, cylinder size, and temperature

4. In a distribution system, what can be done in order to minimize pressure drop?

   A. Do not use long-radius 90° turns and 45° connections
   B. Use undersized header piping with high velocity in a loop system
   C. Use one large header to supply all loads in a branched system
   D. Do not use isolation valves in any system

5. Calculate the run time percentage if a compressor runs for 3.25 minutes and then is off for 5 minutes.

   A. 4%
   B. 39%
   C. 61%
   D. 65%

6. What is responsible for controlling the compressor's pressure by sending a pneumatic signal to the main unloading valve?

   A. Rotary screw
   B. Crankshaft
   C. Pressure switch
   D. Relieving regulator

7. What can occur if an unloading valve fails in the open, or vented, position?

   A. The compressor may produce an unrestricted amount of output current.
   B. The compressor may trip overcurrent protection on start-up.
   C. The compressor may develop an unrestricted amount of output pressure.
   D. The compressor may not develop the full rated output pressure.

8. Which of the following can be used to verify the position of a cylinder?

   A. Stop tube
   B. Magnetic reed sensor
   C. Intensifier
   D. Primary receiver

9. In a pneumatic circuit with meter-out control, what two valves can alter the velocity of the extension stroke?

   A. Flow control valve and needle valve
   B. Flow control valve and directional valve
   C. Directional valve and needle valve
   D. Solenoid and needle valve

10. Which of the following is the correct trouble-shooting procedure when dealing with a cylinder that will extend but not provide force at the end of stroke?

   A. Disassemble the main valve to clean and inspect it.
   B. Test whether the cylinder is receiving enough pressure to determine whether the seals must be replaced.
   C. Check the cylinder's appropriate flow check valve to see if it is stuck open or closed due to contamination.
   D. Use the manual bypass to check if the spool operates manually. If it does, the solenoid must be replaced.

18

Advanced Fluid Power

LEARNING OBJECTIVES

After completing this chapter, you will be able to:

□ Describe the different types of manifolds commonly used.

□ Discuss the difference between a slip-in cartridge valve and a screw-in cartridge valve.

□ Explain the ports and connections of an industrial stacked valve group.

□ Explain major adjustments on a proportional valve.

□ Describe how a proportional and servo valve differ.

□ Describe indications used for troubleshooting a PLC-controlled pneumatic sequencer circuit.

□ Discuss how to troubleshoot a sequential operation controlled by a PAC.

□ Interpret and explain a Grafcet diagram.

TECHNICAL TERMS

cartridge valve

command signal

deadband

DIN rail

dither

error signal

feedback signal

gain

Grafcet

hunt

input/output (I/O) modules

linear variable differential transformer (LVDT)

logic function

manifold

manifold stack

monoblock

offset

P/E switch

programmable air controller (PAC)

programmable logic controller (PLC)

proportional valve

ramp

screw-in valve

sequential function chart (SFC)

slip-in valve

stack valve

zero point

J ust as electrical controls have advanced over the years, so has fluid power. In some instances, these advancements have standardized components and elements for ease of manufacturing. Other advancements in fluid power have made troubleshooting efforts more straightforward when problems occur. As manufacturing becomes increasingly automated, the control systems for fluid power become more complex. Many fluid power systems are now hooked up to a programmable logic controller (PLC)—basically a small computer that helps control the system. Manifolds with custom arrangements of valves have simplified systems and added to efficiency. Depending on the type of maintenance that you perform, you may see some or all of the components and ideas presented in this chapter.

18.1 MANIFOLDS AND VALVES

Nearly all new fluid power systems incorporate manifolds in some way. A *manifold*, or manifold block, is a component with numerous ports, or ways, to which valves and stacks of valves can be attached to form circuits and functions. Manifolds come in all sizes, from circuits that incorporate just a few valves to those with multiple stacks of valves. Servo and proportional valves are also increasingly popular, as they provide better control and feedback for systems that require finely tuned control.

18.1.1 Manifolds

A *logic function* is a simple processing function in a circuit that leads to the performance of a logical operation through the use of binary inputs and outputs. Basic logic functions provide system control and include AND (do this and this), OR (do this or this), and NOR (do neither this nor this) functions. Valves are logic elements that sense flow and pressure to provide the desired logic function. *Cartridge valves* are valves with a compact design that can be used in manifold systems. Manifolds provide for a common base on which cartridge valves can be mounted. A manifold can include as few as two logic elements (cartridge valves) or numerous elements. **Figure 18-1** shows three bar manifolds without any valves attached. Manifolds are usually made from steel or aluminum, and all internal passageways are machined.

Valves placed into the manifold can be screw-in type or slip-in type. Slip-in valves are backed up by a cover assembly, which is held to the manifold. Manifolds can be machined for either type of cartridge valve, sometimes on the same manifold. Programs help ensure that each manifold design places elements together in such a way as to minimize the overall size of the block, perform required system functions, and make sure that all cavities and passages can be machined. Manifolds contain pressure, tank return, and A and B ports for valves that require them, in addition to passageways that interconnect elements within the manifold. Passageways that are only used internally are still machined from the outside, and are then plugged after complete assembly.

Manifolds have some advantages inherent in their design:

- Manifolds include all logic and control in one place, reducing outside piping and possible leaks.

- The entire manifold, including elements, can be assembled and tested at the manufacturer prior to shipping.

- Valves attached to manifolds can quickly be removed, inspected, and replaced.

Figure 18-1. Manifolds can be machined for screw-in and slip-in cartridge valves, sometimes on the same manifold. While some common manifolds are mass-produced, others are custom designs.

SAFETY NOTE
Manifold Work

Never assume that a manifold is free of pressure just because the hydraulic pump has been locked out. Multiple actuators can be placing backpressure on valves, and some actuators have a large volume of fluid. Carefully review diagrams and operate under the assumption that there is pressure, venting manifolds as a safety precaution.

18.1.2 Manifold Stacks/Monoblocks

Manifold design depends on the application. A manifold block with several stacked valves attached is called a *manifold stack*, or simply *stack*. Manifolds include incoming pressure and tank ports, supplied to all attached valves, and output ports A and B from each valve. Stacks can be expanded to include multiple valves stacked on each position.

Figure 18-2 shows several types of manifold system. In smaller mobile equipment that only uses a limited number of valves, a *monoblock* is common. In a monoblock, a cast one-piece body houses all of the valve functions in one integrated unit. On larger mobile equipment, the monoblock is replaced by individual valves stacked together horizontally. Each valve has a separate machined housing that seals to the valves next to it by use of O-ring seals. The entire stack is held together by threaded fasteners, which pass through each segment of the block. Each valve is connected to a pressure port,

Bimba Corporate

Image courtesy Bosch Rexroth Corporation. Used by Permission.

Image courtesy Bosch Rexroth Corporation. Used by Permission.

Figure 18-2. Monoblocks and stacks reduce the footprint and allow more functions in a smaller space.

a tank (return) port, and two individual output ports. These blocks can be manually or solenoid operated, depending on the equipment.

Maintenance can be performed by removing plugs and withdrawing the spools, detaching each valve from the base, or disassembling the entire block. Since these blocks are used on mobile equipment, many times the valves are subjected to silting. Taking care of silting requires disassembly, cleaning, and the replacement of all sealing elements. A set of industrial control valves stacked vertically may be mounted to a manifold block as a complete set of functions for one actuator, such as a cylinder. Everything in a stack of valves and control elements is used to control that one actuator. The stack elements connect to each other, and common passages extend through each element. Each element may not act on each passageway. While one element may provide for flow control, another may provide for directional control.

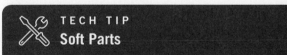

TECH TIP
Soft Parts

Always replace all soft parts (such as gaskets and O-rings) when disassembling valves and manifolds. If O-rings and other soft parts are not replaced, they may become pinched or damaged, and this could result in leaks.

18.1.3 Cartridge Valves

Cartridge valves may be either *slip-in valves* or *screw-in valves*, depending on how they attach to a manifold. Other specific valves are used for *stack valves*, which are valves stacked one on top of the other. **Figure 18-3** shows examples of various screw-in cartridge valves.

Each of these valves comes in many designs and can have a variety of purposes and characteristics, including:

- Overall function of the valve (flow control, directional control, etc.).
- Internal design (poppet or spool).
- Size (various standardized sizes).
- Ways (flow paths and connections).

Screw-In Valves

The cavities for screw-in valves are machined with standardized tooling for each type of cavity. Approximately 40 different cavities together accommodate all screw-in valve types, ways, and sizes. The number of ways may be two, three, or four, depending on the purpose of the valve. Nearly any type of function can be accomplished with one or multiple screw-in cartridge valves. Some common screw-in valves include: relief valves, sequence valves, pressure-reducing valves, check valves, flow regulators and needle valves, directional valves, unloading valves, and shuttle valves.

While the internals of screw-in valves look different than those of older, spool-type valves, the functionality is the same. You should understand the overall function of a valve by referencing its symbol. **Figure 18-4** shows

Image courtesy Bosch Rexroth Corporation. Used by Permission.

Figure 18-3. Screw-in cartridge valves come in standard sizes and perform a wide range of functions.

Common Cartridge Valves

Valve	Symbol	Description
Basic relief valve	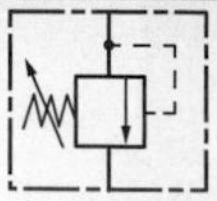	Direct-acting valve, used with lower pressures (<2000 psig).
Pilot relief or vented valve	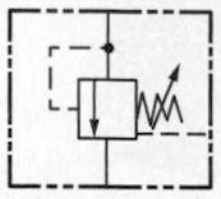	Pilot-operated spool valve, used to accurately control pressure up to 5000 psig.
Two-way relief, single cartridge valve	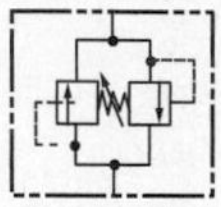	Can be used on bidirectional motors for overload protection.
Pilot-operated sequence valve	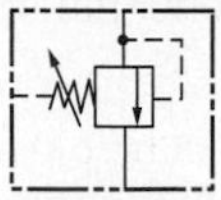	Typical sequence valve, but allows control over a small band.
Pilot-operated sequence valve with reverse flow	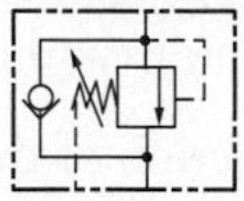	Allows reverse flow through bypass check valve.
Unloading sequence valve	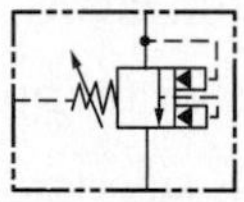	Opens and allows incoming pressure to fall to outlet pressure without closing. Closes once inlet pressure drops to 0.
Pressure-reducing valve	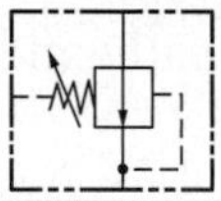	Standard pilot-operated reducing valve, with backflow versions available.
Over-center valve	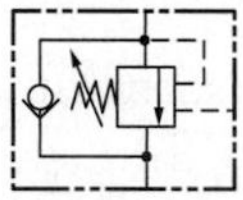	Used to prevent a load from running away or raised load from drifting down. Pilot opens valve and allows load to move.
Over-center balanced relief valve	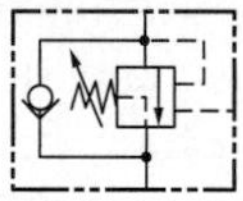	Includes relief function to prevent overpressure condition if load drastically increases. Will then let load fall by relieving pressure.
Dual over-center, fully balanced valve with pilot-assisted relief	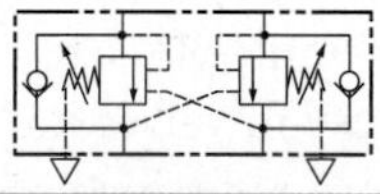	Applies load control to both directions of load movement with integral relief protection.
Dual pilot-operated check valve	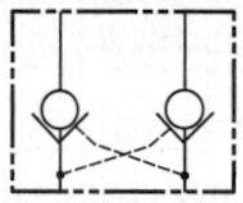	Must have pilot pressure to open check valve. Used to hold actuators in position.
Flow regulator/diverter valve	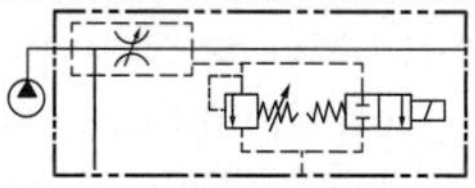	Regulates or diverts flow based on pilot pressure control. Pilot pressure is limited by relief valve, but also controlled by solenoid and setting.
Unloading valve	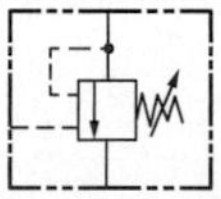	Opens and dumps flow to tank when set point is reached.
Pilot-operated unloading valve	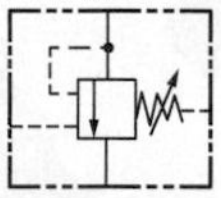	Unloads pump to tank when a secondary system pressure reaches set point.
Shuttle valve	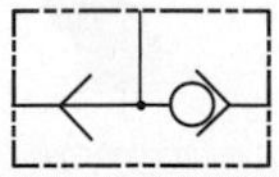	Depends on connections—can be used as an OR function or higher-pressure selection function.

Goodheart-Willcox Publisher

Figure 18-4. These are the symbols for common cartridge valves, along with a brief explanation of the function of each valve.

the symbols for common cartridge valves and provides a brief explanation of the function of each valve. Once you familiarize yourself with these valve symbols and functions, you will be better able to understand and interpret circuit diagrams for fluid power systems. **Figure 18-5** shows a simplified circuit for clamping a workpiece and drilling it using common cartridge valves. Examine this circuit and consider the following:

- The purposes of each valve in the circuit and how the valve performs its function.

- The order of operation when clamping and drilling.

- Possible failures that could occur if a component does not operate properly.

TECH TIP
Troubleshooting Valves

Only by recognizing the symbol and understanding the purpose of a valve can logical troubleshooting take place. When there is an issue, the valve in question can be removed and checked for obvious contamination that may be preventing its proper operation. Many cartridge valves contain smaller spools, poppets, or spring-loaded components. Care must be taken to reassemble the valve properly. Replacement of the element is often required. Each element has a specific manufacturer reference number, which should also be indicated on the print.

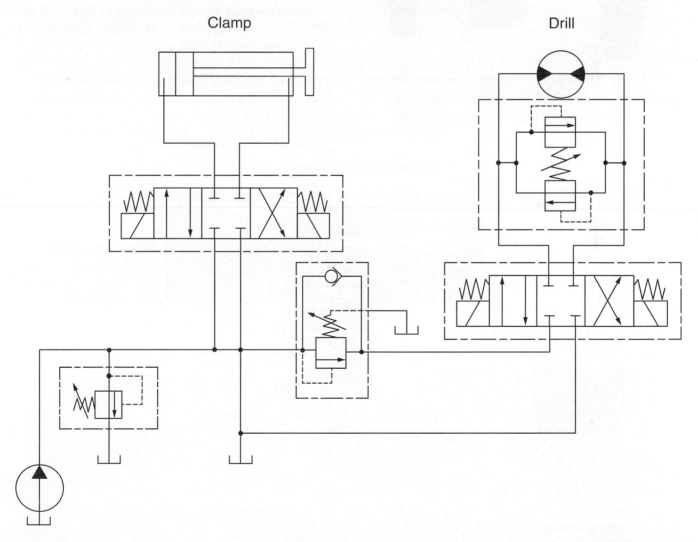

Figure 18-5. A simplified clamp-and-drill circuit (a circuit for clamping a workpiece and drilling it), using common cartridge valves.

Slip-In Valves

The elements of slip-in valves look similar to those in screw-in valves, but they are kept in place by a backing cover assembly. **Figure 18-6** shows examples of slip-in cartridge valves. Slip-in and screw-in cartridge valves perform many of the same functions, but slip-in valves are typically used for flow rates greater than 40 gpm. The backing cover of the slip-in valve is held in place on the manifold by threaded fasteners, and the manifold must be machined for this. Each valve on a manifold will perform a specific function for the system.

Slip-in valves have three ports and are two-position valves. Because of this, it takes several of these valves to perform the function of one spool-type valve. The cover assembly (detailed in **Figure 18-7**) can have other integrated elements within, allowing further control of the main valve. The cover assembly may have integral valves, solenoids, pilot controls, and flow controls. Orifice size and placement in the cover assembly dictate valve response time to system pressure changes. While **Figure 18-7** shows only one passage in the cover (labeled "X"), multiple passages can be used and may be connected through the manifold for other pilot or sensing lines. Depending on overall valve function, there may be several covers stacked.

Valves are sized for flow according to ISO 7368 and DIN 24342 standards and come in the following sizes: ISO 06 (16), ISO 08 (25), ISO 09 (32), ISO 10 (40), ISO 11 (50), ISO 12 (63), ISO 13 (80), and ISO 14 (100). Each size has different dimensions, from 5/8″ to 4″ in diameter, and covers a wide range of flows, from 60 gpm to 1450 gpm.

Image courtesy Bosch Rexroth Corporation. Used by Permission.

Figure 18-6. The function of a slip-in cartridge valve depends on both the cartridge and the backing cover.

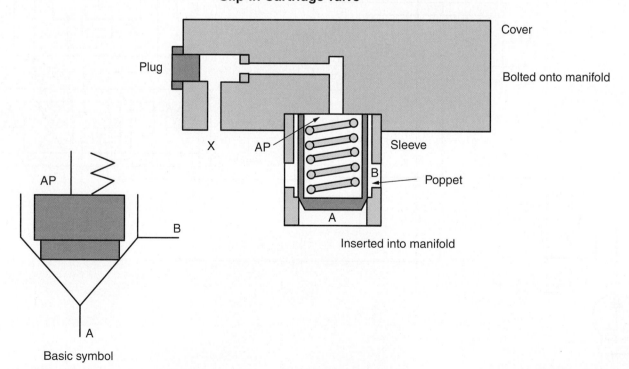

Slip-In Cartridge Valve

Goodheart-Willcox Publisher

Figure 18-7. The internals and the basic symbol for a slip-in valve are shown here. The cover assembly for a slip-in valve may have integral valves, solenoids, pilot controls, and flow controls.

Recall that a poppet, or poppet valve, is a sliding shaft with a disk on the end that seals and unseals a port as it moves. Several different poppets are used to achieve different responses. **Figure 18-8** summarizes the types of related symbols and area ratios. These four poppets vary the ratio of the surface area on which pressure can act. Areas on which pressure acts are A_A, A_B, and A_{AP}. Each of these areas corresponds to the area that pressure from that particular port can act on. Pressure from port A will act on A_A. Pressure from port B will act on A_B. Pressure from port AP will act on A_{AP}.

A simplified slip-in cartridge valve circuit is shown in **Figure 18-9**. In the lettering of ports, P represents a pressure port, T represents a tank port, and A and B represent user ports or actuators. When the main valve is in center position, both ports A and B are vented to the tank, which causes the AP areas of all cartridge valves to be vented. This allows nearly free floating of the actuator—only the spring force of one of the cartridge valves would need to be overcome by pressure. If solenoid A1 is energized, pressure is directed to port A. This pressurizes the first and third cartridge valves' AP areas through the flow controls that keep them closed. This also vents port B of the solenoid to the tank, and vents the second and fourth cartridge valves' AP areas.

System pressure, felt through the B port of cartridge valve 2, overcomes opposing spring pressure and pressurizes the cap end of the cylinder. Rod-end hydraulic

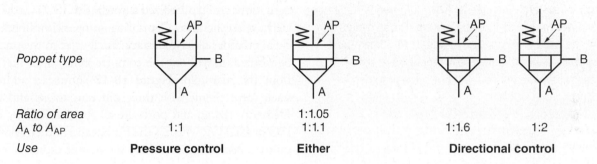

Poppet type				
Ratio of area A_A to A_{AP}	1:1	1:1.05 1:1.1	1:1.6	1:2
Use	**Pressure control**	**Either**	**Directional control**	

Goodheart-Willcox Publisher

Figure 18-8. These symbols represent types of poppet. The area ratio determines the pressure required to shift each valve.

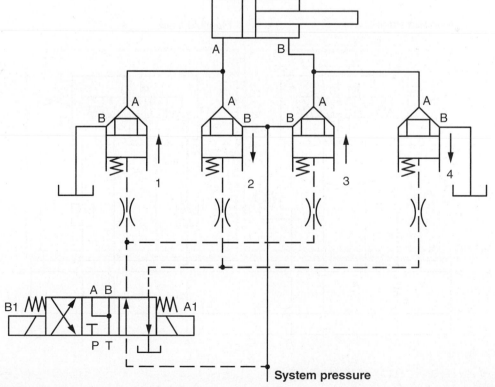

Goodheart-Willcox Publisher

Figure 18-9. This simplified circuit uses slip-in cartridge valves for directional control.

fluid acts against the A port of cartridge valve 4 and overcomes opposing spring pressure, allowing rod-end fluid to return to the reservoir. The cylinder extends.

If solenoid B1 is energized, both cartridge valves 2 and 4 are then held shut, while valves 1 and 3 are vented. Valve 3 opens, allowing system pressure to retract the rod, while cap-end fluid is vented back to the tank through valve 1.

TECH TIP
Keep It Clean

As mentioned, the most common cause of failure in a hydraulic system is contamination and silting due to a lack of system cleanliness. New fluid needs to be filtered prior to adding it to the system. Proper filters need to be used and changed regularly. Fluid samples need to be taken and analyzed. Without all of these actions, system cleanliness will not be maintained and components will fail.

Stack Valves

Industrial stack valves are common where a smaller manifold is needed, actuators can be controlled independently, and multiple control elements are needed for each actuator.

As with other cartridge valves, nearly any combination of elements can achieve proper design function. In stack valves, although all passages extend through each valve in the stack, not all passages are used by each valve, depending on the purpose of the valve. An examination of each of the stacks in **Figure 18-10** shows that in both stacks, pressure and tank lines run up through the stack to the directional valve, then output from the directional valve through the A and B ports to be controlled by flow control valves, and then exit the stack back into the manifold, where they will be connected to actuators. **Figure 18-11** shows an example of a stack valve symbol and port arrangements.

Since stack elements are connected to each other, O-rings provide a seal between each element and a seal to the manifold block at the bottom of the stack. Port configurations are standardized according to ISO standards, as are valve size and flow. Troubleshooting is simplified, since each actuator has its own stack. Each valve in the stack can be worked on without the complete removal of that stack from the manifold. **Figure 18-12** shows an additional stack valve circuit with three different stacks and loads. Elements, sizing, and ports are all standardized by either ISO or CETOP, which is the European Fluid Power Committee. Depending on where the manufacturer is located and where the elements are made, other standards may be involved: CHPSA (China), FPSI (India), JFPA (Japan), NFPA (USA), and TFPA (Taiwan).

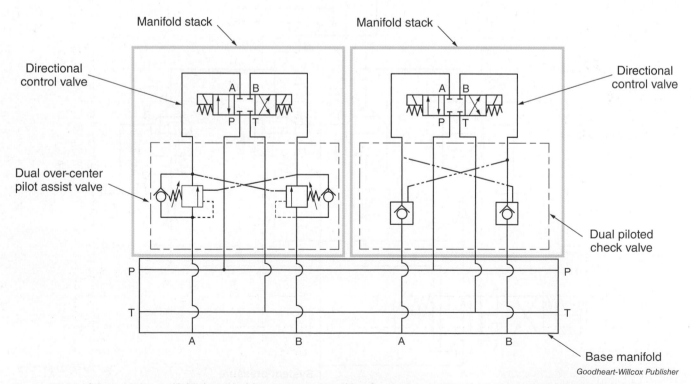

Goodheart-Willcox Publisher

Figure 18-10. A two-stack manifold that provides separate functions for each actuator. Each stack contains two elements. The base manifold has P, T, A, and B ports for each stack, with P and T connections common to all stacks in the block.

18.1.4 Proportional Valves

Proportional means that a response is directly related to an input. With proportional valves, the input may be a number of different types of signal, such as electrical or pressure, and the output is the position of the valve. *Proportional valves* allow for the control of the position of the spool, which allows for control of direction and speed using an electrical input signal. The two main types of proportional valve are classified by control: one classification is with feedback and the other is without feedback. Feedback means that output information returns to the valve input. Proportional valves can be mounted on baseplates

Dual Pilot Check Valve

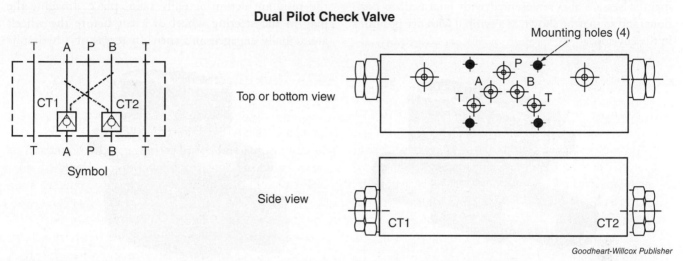

Goodheart-Willcox Publisher

Figure 18-11. An example of a stack valve symbol and port arrangements are shown here. Each check valve has its own element and can be serviced by removing the appropriate element from the side of the stack.

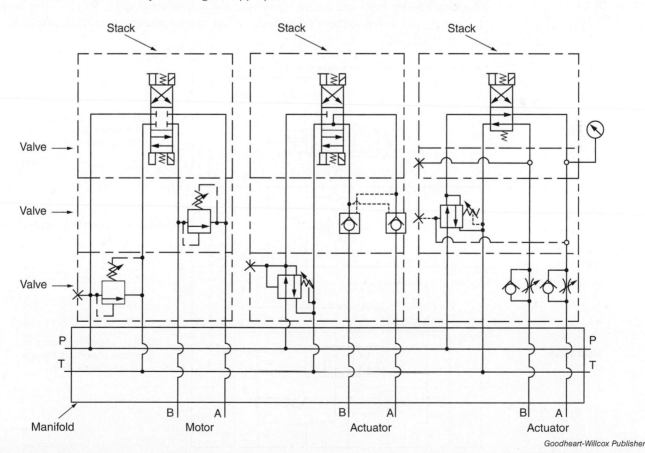

Goodheart-Willcox Publisher

Figure 18-12. This stack valve circuit has three different stacks and loads. Stack valve ports are standardized to ISO 4401 or CETOP standards.

or manifolds. **Figure 18-13** shows a direct-acting and a pilot-operated proportional valve. A main spool valve may be controlled by a pilot valve, which is positioned by two solenoids. The pilot spool is positioned depending on how much current each solenoid receives. **Figure 18-14** shows a stack for the control of a proportional valve. In this figure, the symbol shows a three-position valve. Proportional main valves are also represented with intermediate positions, and so may be shown as a symbol with five positions in the envelope.

A **_command signal_** is an input signal that determines, in this case, where the spool of the pilot valve will position. Controllers for the command signal usually have circuitry for deadband and gain. **_Deadband_** is the range of change of the input signal that does not show any change in output signal. In other words, it is the amount of give or movement that can occur before the resultant action actually takes place. Imagine the play in the steering wheel of a car before the wheels are actually engaged and turn. In terms of a hydraulic

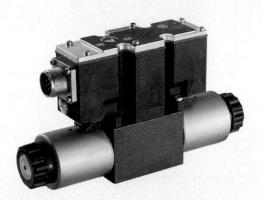

Images courtesy Bosch Rexroth Corporation. Used by Permission.

Figure 18-13. Pictured here are a direct-acting (left) and a pilot-operated (right) proportional valve. Proportional valves, with and without feedback, can be mounted on baseplates or manifolds. Valves can have multiple stacked elements, depending on the function and control needed.

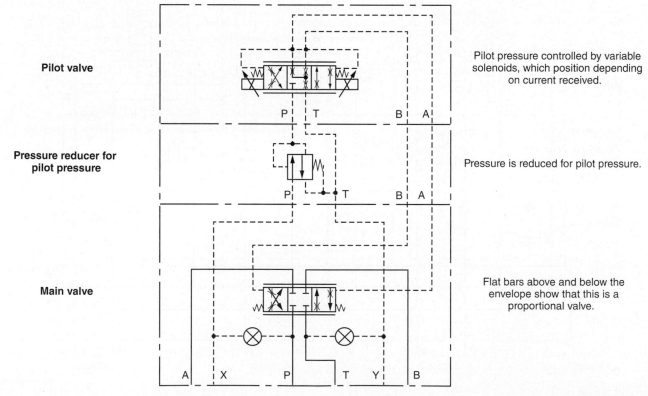

Goodheart-Willcox Publisher

Figure 18-14. This symbol shows a stack of three elements for the control of a three-position proportional valve.

valve, deadband is created by the amount of overlap of a section of the spool to a fluid cavity. Think of the deadband of a pressure switch that switches a compressor on and off. Deadband in a spool valve requires it to move slightly to effect a fluid flow change because of the overhang of the spool.

Gain is another name for sensitivity. The higher the gain, the more sensitive the system is to a changing input signal. However, too much gain can cause the system to *hunt*, or seesaw back and forth from fully open to fully closed, or fully on to fully off. Imagine a car with the cruise control set to 60 mph. With maximum gain, when the car slowed to 59 mph the engine would go full-bore, and when the car reached 61 mph the engine would idle. With gain set too low, the system will respond very slowly over a long time. Adjustments of gain (sensitivity) should only be performed by a process technician familiar with the specific valve in question. The difference between the command signal (60 mph in the example) and the point of response (59 mph in the example) is the deadband. Gain and deadband settings affect efficiency—you want power to run only when needed, and then only as strong as needed.

Other valve adjustments include ramp, zero point, dither, and offset, which are set at the factory. *Dither* is an AC signal placed on top of the DC control signal that causes the valve to constantly micromove, which helps to reduce friction between the spool and body of the valve. *Ramp* adjusts how quickly the signal is allowed to increase or decrease—also called acceleration or deceleration. *Zero point* adjustment aligns the zero position of the signal with valve position. *Offset* allows position change for the correction of a constant error.

TECH TIP
Troubleshooting Gain

If a system or valve response is very dramatic, the valve is either very contaminated or the gain has been adjusted too high. This can happen in any system with proportional controls.

Since flow through a valve depends on the pressure differential across the valve, the type of spool, and the position of the spool, the relationship of the flow characteristics to the command signal is not always exactly linear. In order to overcome this, valve spools may have notches or curves machined into them to help obtain a more linear flow characteristic through the chamber.

Valves with feedback provide that feedback through the position sensing of the main valve spool, not flow. One method of detecting spool position is by use of a *linear variable differential transformer (LVDT)*, which develops an output signal based on the position of the core that travels through the transformer as the main spool moves. This feedback signal is sent back to the controller. The controller compares the feedback signal with the command signal and develops an error signal that is proportional, which is then sent on to the pilot valve solenoids. The command signal is the set point of where you want to be—where you want the valve spool to be. The *feedback signal* is where you are—the actual position of the valve spool. The *error signal* is command minus feedback, and is the amount you need to change to get to where you want to be—the required change to get the valve spool to the desired position. This type of control is used in numerous electronic circuits, pneumatic circuits, and other control systems.

SAFETY NOTE
Proper PPE

High-pressure air from pneumatic systems can and will damage an eye. Always wear appropriate safety glasses when working with pneumatic systems. Prescription contacts and glasses do not offer enough protection. Air pressure greater than 30 psi can also propel sand, metal chips, and other debris likely to puncture skin and cause a dangerous infection. Always take care when working on pneumatic systems and use the proper PPE—air is not as benign as you might think!

18.1.5 Servo Valves

Like proportional valves, servo valves are also used to control flow and pressure within a system. Servo valves respond to a signal and are very similar in operation to proportional valves with feedback. Servo valves are highly accurate and are held to tighter tolerances than proportional valves. Servo valves are more sophisticated and expensive than proportional valves and are used in closed loop systems. The main difference between servo valves and proportional valves is that servo valves usually have less deadband than proportional valves, with less than 3% spool overlap. Spool overlap refers to how much overlap there is between a spool land and its complete closing or opening of a corresponding port. Servo valves are manufactured to mount to standard-sized manifolds or adapter plates. **Figure 18-15** shows several servo valves. The upper portion of the servo valve contains the torque motor assembly, while the lower portion contains the valve spool.

Figure 18-16 shows the internals of one type of servo valve. In this servo valve, hydraulic fluid is supplied under pressure to pilot areas of the main spool and to the nozzles. In most valves, this fluid has passed through an integral filter before entering the control portion of the valve. Fluid passes through each nozzle and is vented back to the reservoir. As a command signal dictates, the torque motor pulls the flapper and feedback wire toward one of the nozzles. As the flapper gets close to the nozzle a backpressure is felt, and this increases the pressure in the pilot area of the main spool.

This increase in pressure causes the main spool to shift. As the spool shifts, the feedback wire is pulled away from the nozzle and recenters. This recentering causes the two pilot pressures to equalize with a new spool position. The feedback wire is a stiff, solid, metallic machined wire that is pulled by energized electromagnets (the torque motors) at the top end and attached to the main spool in a slot.

If you need to perform maintenance on a servo valve, review the manufacturer's literature before taking the valve apart. Internal filters will eventually clog due to a lack of system cleanliness, and the valve will then respond poorly, if at all. Silting can also occur in the main spool in very contaminated systems. Disassembling a servo valve is not difficult, but knowledge about that specific valve is required for adjustments afterward.

18.2 PLCs, PACs, AND GRAFCET

A *programmable logic controller (PLC)* is a digital microprocessing device used for automating industrial control systems. Highly controlled PLC systems can incorporate simple logic functions (AND, OR, and NOR) into step sequencers. In fluid-power applications that must solely use pneumatics, logic elements can be built into an air-driven sequencer called a programmable air controller (PAC). While these pneumatic systems look complicated at first glance, they are easier to troubleshoot once the basics of operation are known. Grafcet is a useful graphical representation of a sequential process.

Sporlan Division, Parker Hannifin Corporation

Figure 18-15. Several different-sized servo valves are pictured here.

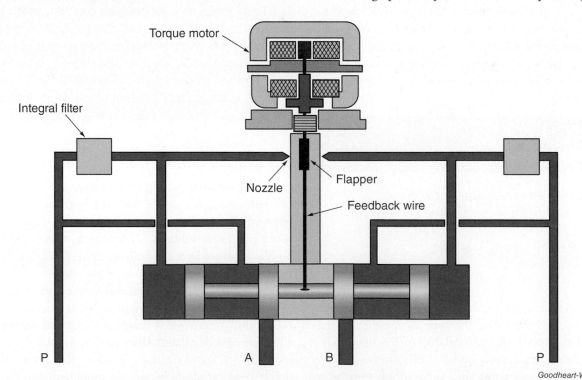

Goodheart-Willcox Publisher

Figure 18-16. The internals of one type of servo valve are pictured here. Servo valves contain internal filters that should be removed and replaced as a regularly scheduled maintenance item—especially in a contaminated system.

18.2.1 PLCs

Both hard-wired electrical control and PLC control are used to control fluid power circuits. The trend is to use PLCs, which make wiring, piping, and programming logic easier than with hard-wired control. PLCs can perform sequential functions and logical functions, and they have very fast response times. While there is a separate chapter covering PLCs, some very basic ideas are mentioned here for clarification. **Figure 18-17** highlights major portions of a PLC-controlled pneumatic system. Most of the logical sequences and decisions take place in the program of the PLC.

Layout, Piping, and Wiring

Input/output (I/O) modules are connected to PLCs to provide analog (also called discrete) or digital electronic connections between the PLC and external actions, such as the push of a button (analog) or an electronic signal. Pneumatic inputs, which are all wired in parallel, provide an output signal to the P/E (pneumatic-to-electric switches). The *P/E switches* convert the pneumatic signal to an electrical signal. The electrical signal is sent to the PLC, where the input signals are converted to a binary form (1s and 0s) in the PLC memory. The program evaluates the inputs and compares them with the programmed logic to determine which outputs should be energized. The output card receives the information (in binary) and turns on appropriate output

points, sending an electrical signal directly to outputs (indicators) or to solenoid-actuated valves. The solenoids actuate and perform the desired valve function. On larger systems, a human-machine interface (HMI) panel may be included just for indications or for operator control. If used for operator control, the HMI may have a touch screen or function keys and allow the operator to perform system commands. **Figure 18-18** shows typical components that might be involved in a PLC system.

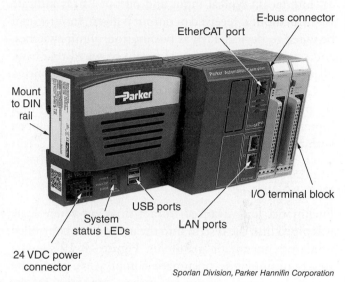

Figure 18-18. Parker's PAC (programmable automation controller) is an advanced PLC that allows programming in several languages and in graphical interfaces.

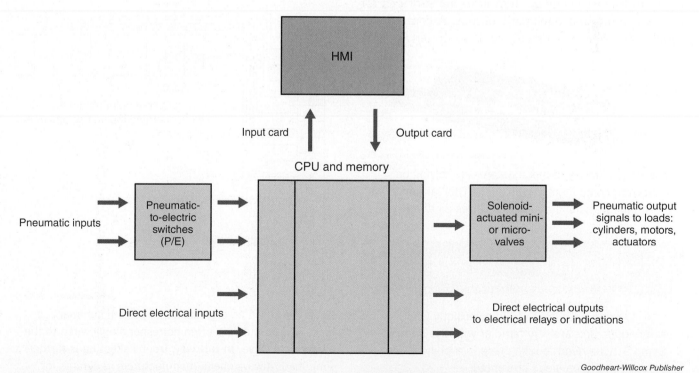

Figure 18-17. This basic overview highlights major portions of a PLC-controlled pneumatic system.

While some brands of PLC let you program in a graphical sequential function chart (SFC) for sequencer operations, others do not. Sequencers can still be programmed, but the interface is not as straightforward as with an SFC. A *sequential function chart (SFC)* is a graphical model that displays process flow as a diagram, allowing control of the sequential processes by describing the conditions and actions for each step. Some PLCs allow programming in several languages and in graphical interfaces. Typical input and output LEDs indicate when an input is received or output is given. Input would be wired to devices such as pushbuttons, limit switches, and sensors. The output card would be wired to a bank of miniature solenoids. It should be noted that once a PLC program is debugged and running, at least 90% of the problems are located in the real world (discrete) with inputs and outputs. Very rarely does a problem happen with the PLC itself.

Miniature Valves

Pneumatics has advanced from a situation where logic was piped into the system into the use of PLC-controlled miniature pneumatic manifolds. **Figure 18-19** shows a variety of miniature pneumatic components. Miniature solenoid-actuated pneumatic valves can take in the electrical signal output from a PLC and convert it to pneumatic action. Miniature valves can directly control smaller cylinders, and are used as pilot controls for valves to control larger cylinders. Indicators are provided for electrical input and pneumatic output. A common air supply and exhaust in a manifold setup simplify piping.

Sporlan Division, Parker Hannifin Corporation

Figure 18-19. More and more systems are relying on miniaturized components, such as the DIN rail-mounted set of miniature solenoid-actuated valves pictured here.

⚠ **C A U T I O N**

Only override outputs if you know exactly what will happen and are very familiar with the system. Some actions need to take place in a prescribed order so that damage is prevented.

Troubleshooting is simplified with this type of system. If an electrical input is shown by an indicator and that mini-valve is not actuating, the manual override can be used to actuate the mini-valve and check that the expected circuit operation then takes place. If the circuit operation takes place in this case, that particular miniature solenoid or valve would be replaced. Depending on the designer, PLC outputs are often correspondingly wired to the appropriate valve, **Figure 18-20**. In this way, if a technician knows the operation of the system, troubleshooting is further simplified. Further advances reduce the amount of wiring that needs to be done by substituting in a standard multiconductor cable and plug. Several manufacturers offer standardized PLC communication protocol, Ethernet/IP communication, both analog and digital inputs and outputs, and up to 32 solenoids per manifold. If your workplace uses these types of systems, it is highly suggested that you receive specific training on them.

18.2.2 PACs

A *programmable air controller (PAC)* is a pneumatic sequencer completely powered by air. All sensors, logic, and sequencing functions are mechanical. Because the pneumatic sequencer performs the logic functions, installation, troubleshooting, and piping are simplified when compared to traditional pneumatic logic circuits.

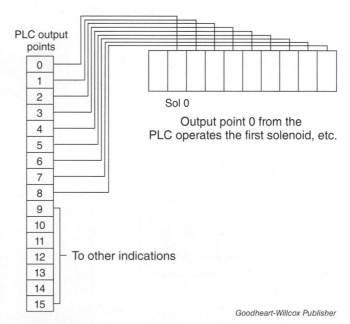

Goodheart-Willcox Publisher

Figure 18-20. In an organizational pattern summarized here, PLC outputs are often correspondingly wired to the appropriate valve. In this way, troubleshooting is further simplified. Always check manufacturers' diagrams for exact wiring before troubleshooting.

Figure 18-21 shows components of a DIN rail-mounted, four-step pneumatic sequencer. *DIN rail* is a metal rail that devices can be mounted to and that follows the standards set by the DIN (Deutsches Institut für Normung), which is the German Institute of Standardization.

18.2.3 Grafcet

Grafcet is a type of SFC that was developed in the mid-1970s by a group of industrial and process engineers who were researching machine control of sequential operations. This graphical representation of a sequential process includes each step of the process, conditions to prove a step has taken place, and logic that proves all previous steps have taken place in order for the next step to happen. By graphically designing a process with this method, applying a PAC to control the system directly relates, and future troubleshooting is straightforward. Grafcet is also used to program PLCs, and is excellent foundational knowledge for all technicians in the controls area.

Sporlan Division, Parker Hannifin Corporation

Figure 18-21. This DIN rail-mounted, four-step pneumatic sequencer performs four operations in order.

Figure 18-22 is a basic Grafcet outlining a two-step stamping process:

1. An operator presses a two-hand start button.
2. A sensor checks whether a part is in position.
3. The cylinder extends to stamp the part.
4. A sensor checks for cylinder extension.
5. The cylinder retracts.
6. A sensor checks for cylinder retraction.
7. The process starts over with step 1.

Each step has a transition that is proven by some input—a condition that must be met before the next step takes place. Notice in the accompanying diagram in **Figure 18-22** that the actual PAC has several elements to it. The following further explains elements of the Grafcet and the PAC diagram:

- A logic "&" symbol is placed in the diagram as a transition to start the cycle over. This "&" requires both inputs. The first input is sensor a0, which turns on when the cylinder is retracted. The second input is a series of normally open valves that, when all connections are made, will send an air signal to the "&." Only if all inputs are received will the sequence start over.

- While the Grafcet shows steps, it does not show exact logic, programming, wiring, or piping.

- The Grafcet is an organizational tool to get the sequence of the operation set, and it helps develop the PAC sequencer or PLC program.

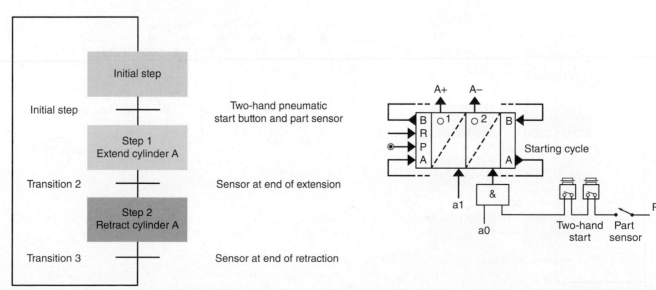

Goodheart-Willcox Publisher

Figure 18-22. This basic Grafcet and diagram outlines a two-step stamping process with a pneumatic sequencer. The example is for clarification only—usually sequencers must have at least three steps to work properly.

- The PAC sequencer diagram does not show exact pneumatic piping. For instance, in the first step, cylinder A is extended, and then proven to be extended with sensor a1. The air output signal from step 1 will probably be piped to a miniature DIN rail-mounted pneumatic valve pilot. When that valve repositions, cylinder A will extend.

A Grafcet (or SFC) can also include other functions, such as parallel paths, decision-making, or a skip forward in the sequence, as long as proper actions are proven. To go further with Grafcet, sequencers and logic components will be examined first.

Sequencers and Logic Components

A sequencer is composed of base modules containing input and output ports, step indications, and overrides. The base module makes all connections between modules with common porting. In order for a sequence to advance to the next step, the output must be given from the previous step, and the input must be received. This proves that the step actually took place, and the sequencer moves to the next action step. When the sequencer completes the last step, modules are reset, and the sequencer restarts, as long as the last

input is received. At each step, there is indication of input received, and step completion. This simplifies troubleshooting, as can be demonstrated in the Grafcet in **Figure 18-23**.

A sequencer needs to be designed to allow three cylinders to extend and then retract in the following order: Cylinder A, B, and then C should extend, and then the cylinders should retract in the order of C, B, and then A. The basic Grafcet is comprised of six sequential steps. Each step is proven with a sensor, and the final step is further verified by a manual pushbutton input to reset the sequence. Depending on the size of the cylinders, the outputs from the sequencer may go to pilot-controlled valves or directly to smaller cylinders.

Examining the sequencer closely, you could easily see the indication of which step the sequencer is performing. Steps are indicated by either small pop-up flags in windows or small red or green pop-up indicators. If the sequence stopped at any point, you would just examine that input needed from the previous step to make sure it was received. For instance, if cylinder A and cylinder B extended and the process stopped, you would check the b1 sensor that indicates if cylinder B is extended. If the signal is not reaching the sequencer, you would physically check the b1 sensor to find the problem, and then adjust or replace the sensor as necessary.

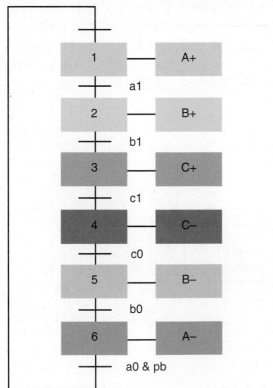

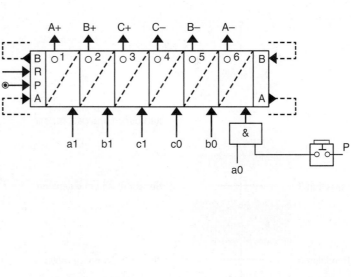

Figure 18-23. This Grafcet is for a sequencer designed to allow three cylinders to extend and then retract. The Grafcet diagram is the first step in designing and programming a process.

Basic logic elements make up the decision-making process of a sequencer circuit. Aside from the basic logic functions of AND, OR, and NOR, other common logic elements perform functions of YES, NOT, MEMORY, and TIME. **Figure 18-24** shows symbols for common logic elements available. There are also air-to-electric elements, elements that act as amplifiers, and more. Logic elements are typically mounted on DIN rail to make up circuit logic. Elements can be combined in series or parallel with a selector switch.

Possible inputs into the logic elements include pushbuttons, limit switches, position sensors (bleed sensors and pressure sensors), and other pneumatic components. All inputs are pneumatic and can act as normally open or normally closed. A normally open pushbutton would send a signal when pushed. A normally closed pushbutton would send a signal until it is pushed, and then would remove the output signal.

During the sequences, cylinder position sensing can be performed mechanically using limit switches or bleed sensors, or through adjustable pressure sensors at each end of the cylinder.

CHAPTER WRAP-UP

The range of fluid systems spans from the very simplistic, with only a few valves and actuators, to the more complex, with numerous valves, sequential logic, and multiple actuators. Since technology is always advancing, it is wise to regularly seek out more training from manufacturers on new control methods. If a new machine or system is being installed, ask for training, ask plenty of questions, and develop professional relationships with highly skilled and technical people. As a highly trained technician who is knowledgeable about the latest trends, devices, and methods, you will be very valuable to any company you work for.

Logic Elements		
Function	**Symbol**	**Description**
OR	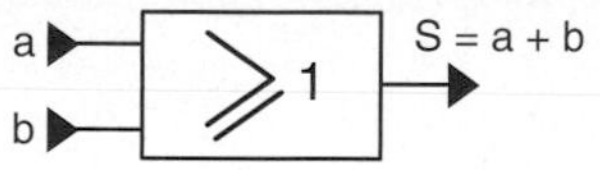 a, b → ≥1 → S = a + b	Similar to a shuttle valve, the output (S) will be on when either or both inputs (a and/or b) are on. If no input is on, the output is off.
AND	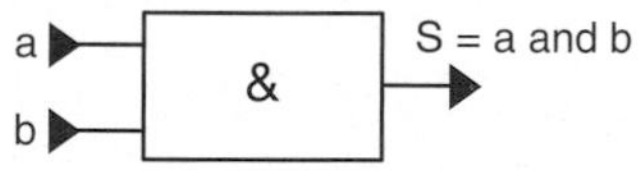 a, b → & → S = a and b	An output (S) is only given when both inputs (a and b) are received. Otherwise no output is given.
YES	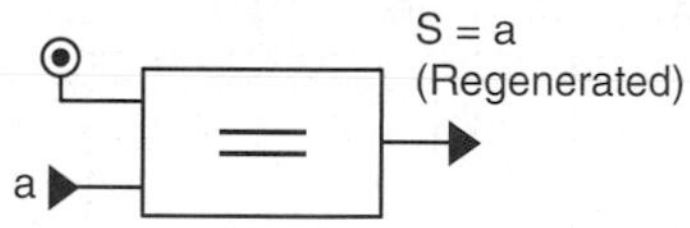 a → = → S = a (Regenerated)	When a low-pressure signal is received at the input (a), a higher-pressure output is sent (S).
NOT	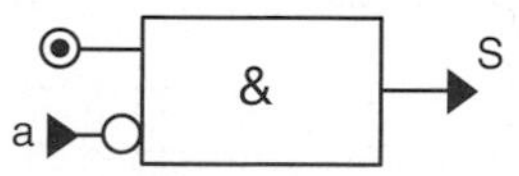 a → & → S	An output (S) is given when input pressure is sensed and the input (a) is not on.
MEMORY	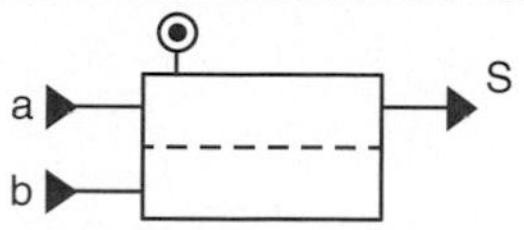 a, b → S	When an input is momentarily received at input a, the output (S) turns on. The output will stay on until an input is received at input b.
TIME	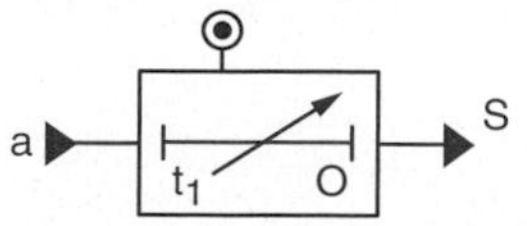 a → t_1 O → S	In the on-delay function, when an input is received at input a, after a set time period, an output (S) is given. When the input turns off, the output turns off.

Figure 18-24. These symbols represent common logic functions. Note the difference between the NOT function symbol and the symbol for the AND function.

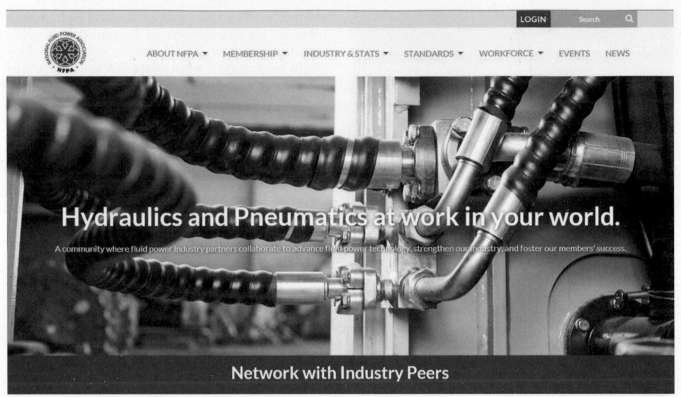

Goodheart-Willcox Publisher

The National Fluid Power Association (NFPA) supports and promotes the fluid power industry. Members include professionals, manufacturers, educators, and aspiring students. You will find links to scholarship and career opportunities on their website, as well as information about upcoming industry meetings, trade shows, and conferences.

Chapter Review

SUMMARY

- Manifolds allow custom arrangements of multiple valves and add to efficiency.

- Stack valves mount on top of manifolds.

- Monoblocks are commonly used on smaller mobile hydraulic systems. Smaller monoblocks are made from one casted piece. Larger monoblocks are pancaked valves held together to make a whole manifold of valves.

- Cartridge valves may be spool, ball, or poppet based and come as screw-in cartridge valves or slip-in cartridge valves, which are used for higher-flow systems.

- Generic manifolds share ports P and T for every valve, while ports A and B are separate for each valve.

- The function of a slip-in valve is determined by the valve and cover.

- In stacked industrial valve manifolds, the order of the stack is important for proper operation.

- A proportional valve positions its spool in direct response to current or voltage applied to solenoids.

- Deadband is caused by the overlap of a spool to flow areas. Gain is a measure of sensitivity to an input signal.

- Servo valves are used in a closed loop to provide better control and feedback for systems that require finely tuned control.

- In a PLC-controlled system, all logic takes place within the PLC program. Input/output (I/O) modules are connected to PLCs to provide analog or digital connections between the PLC and external actions. P/E switches convert pneumatic signals to electrical signals, which are converted to a binary form (1s and 0s) in the PLC memory.

- Pneumatics has advanced into the use of PLC-controlled miniature pneumatic manifolds for many systems.

- Grafcet is a graphical representation of a system's sequential process, including each step of the process, conditions to prove a step has taken place, and logic that proves all previous steps have taken place in order for the next step to happen.

- In order for a sequencer to advance to the next step, the previous step must be proved by an input.

- Common logic elements include AND, OR, NOR, YES, NOT, MEMORY, and TIME.

REVIEW QUESTIONS

Answer the following questions using the information provided in this chapter.

1. What is a manifold?

2. Valves are _____ elements that sense flow and pressure to provide the desired logic function.

3. List at least three advantages of a manifold control system.

4. Describe the ports and connections for a manifold used for stack valves.

5. What is the difference between a monoblock and horizontal stack of valves?

6. What is a cartridge valve?

7. This symbol represents which of the following valves?

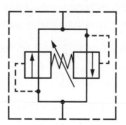

   A. Two-way relief

   B. Over-center balanced relief

   C. Shuttle

   D. Dual pilot-operated check

8. *True or False?* Screw-in valves are used for higher flow rates, typically greater than 40 gpm.

9. In slip-in valves, what is the purpose of a poppet?

10. What is the purpose of the orifice in the cover assembly of a slip-in valve?

11. In a simplified slip-in cartridge valve circuit, as shown in **Figure 18-7**, list the sequence of events if the main valve is in center position.

12. When are industrial stack valves used?

13. *True or False?* In stack valves, all passages extend through the stack and are used by all valves.

14. What occurs in the stacks of valves as depicted in **Figure 18-12**?

15. What is the role of a proportional valve?

16. List the two main types of proportional valves.

17. The _____ is an input signal that determines where the spool of the pilot valve will position.

18. In terms of a spool valve, what is deadband?

19. How is a linear variable differential transformer (LVDT) used to determine spool position?

20. How is a servo valve different from a proportional valve?

21. PLC stands for _____.

22. What are PLCs?

23. What are input/output (I/O) modules used for?

24. *True or False?* P/E switches convert an electrical signal to a pneumatic signal.

25. What are some indications that could be useful for troubleshooting a process involving miniature valves?

26. *True or False?* The Grafcet is an organizational tool to get the sequence of an operation set while helping develop the PAC sequencer or PLC program.

27. What must occur in order for a sequence to advance to the next step?

28. *True or False?* A sequencer can be designed to allow cylinders to extend and retract in a particular order.

29. Describe how an AND function is different from an OR function.

 NIMS CREDENTIALING PREPARATION QUESTIONS

The following questions will help you prepare for the NIMS Industrial Technology Maintenance Level 1 Basic Hydraulic Systems credentialing exam and Pneumatic Systems credentialing exam.

1. What does a manifold stack control?

   A. Smaller mobile equipment with limited number of valves
   B. All logic elements in a PLC
   C. The output ports A and B from each valve
   D. A complete set of functions for one actuator

2. Slip-in cartridge valves have how many ports and positions?

   A. 4 ports and 2 positions
   B. 3 ports and 2 positions
   C. 2 ports and 1 position
   D. 3 ports and 3 positions

3. In valves with feedback, what two things are compared that produce an error signal as a result?

   A. Feedback signal and command signal
   B. Zero point and controller
   C. Deadband and gain
   D. Pressure differential and flow

4. In a Grafcet sequence of several cylinders, if cylinder A and cylinder B are extended and the process stops, what should be checked?

   A. The sensor that indicates if cylinder C is extended
   B. The sensor that indicates if cylinder B is extended
   C. The sensor that indicates if cylinder A is extended
   D. The sensor that indicates if cylinder B is retracted

5. Which of the following is *not* a logic function that can be used for a PLC?

   A. AND
   B. OR
   C. NOR
   D. DIN

6. In a P/E switch, what does an input card convert the signal to that then goes to the PLC?

   A. Logic element
   B. Pneumatic signal
   C. Binary form
   D. Output LED

Electrical Systems

Most of the machines, equipment, and systems you will encounter as an industrial maintenance technician use electricity in some way. Electrical power provides the energy needed to run motors and illuminate lights. Electrical control systems use electrical signals to monitor systems and control the flow of power. Understanding electrical applications and the theory on which they operate is critical to your success.

The content in this section will help prepare you to earn the NIMS Industrial Technology Maintenance Level 1 Electrical Systems credential. Credentials show employers proof of your knowledge and skills.

19

Fundamentals of Electricity

LEARNING OBJECTIVES

After completing this chapter, you will be able to:

☐ Understand the building blocks of matter and how they fit together.

☐ Explain the law of charges and how different charges interact.

☐ Describe the differences between conductive and insulating (nonconductive) materials.

☐ Acquire an understanding of basic electrical circuit construction.

☐ Differentiate between series and parallel circuits.

☐ Comprehend the basics of magnetism and how electricity and magnetism are related.

TECHNICAL TERMS

alternating current (AC)	electrostatic field
ampere	inductor
anion	insulator
atom	magnet
battery	neutron
cation	parallel circuit
complete circuit	proton
complex circuit	resistance
conductor	series circuit
coulomb	short circuit
direct current (DC)	static electricity
electric current	switch
electromagnet	valence orbital
electromagnetic induction	volt
electron	voltage

This chapter outlines some of the fundamental principles of electricity. A strong knowledge of electrical systems starts with atomic theory, what electricity is and how it works, basic electrical circuits, and magnetism. As an industrial maintenance technician, this foundational material will help you understand and describe electricity and electrical systems.

19.1 ATOMS AND SUBATOMIC PARTICLES

All matter is made up of atoms. An *atom* is the smallest part of an element that still retains the characteristics of the element. Atoms are made of three main types of subatomic particles: protons, neutrons, and electrons, **Figure 19-1**. The nucleus (or center) of the atom contains *protons*, which are positively charged, and *neutrons*, which are electrically neutral (neither positive nor negative). Negatively charged *electrons* orbit around the nucleus. The number of protons, neutrons, and electrons define the atom.

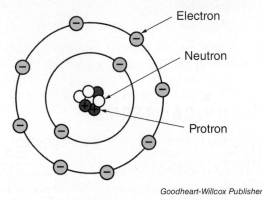

Goodheart-Willcox Publisher

Figure 19-1. Basic atomic arrangement. The atom consists of electrons, protons, and neutrons.

Electrons orbit in shells, commonly referred to as orbitals, around a nucleus. Each orbital has an energy level. The outermost orbital is called the *valence orbital*. Depending on the atom and the electron distribution between orbitals, the valence orbital may have only a few electrons, be completely full, or have almost enough to be full. The number of electrons in the valence orbital determines the electrical properties of an element or material.

Atoms normally have no net charge because the number of electrons (negative) and protons (positive) are equal. When an atom picks up extra electrons, negative charges outnumber the positive charges, and the atom becomes a negative ion, or *anion*. The reverse is true if an atom gives up some electrons. Its net charge becomes positive because the protons outnumber the electrons. Giving up electrons causes the atom to become a positive ion, or *cation*. Whether an atom picks up or gives up electrons depends on how tightly bound the electrons in the outer orbitals are. This in turn depends on how many electrons are normally in the valence orbital.

19.2 STATIC ELECTRICITY

Static electricity is a charge at rest. Static charges can be generated by simple friction, by the movement of materials against surfaces, or by surfaces separating. Nearly every time two materials are rubbed together, a static charge is generated. An object that picks up electrons from the surface it has been separated from is negatively charged. The object that gave up the electrons becomes positively charged.

All kinds of movement can create a static charge. Around the workshop, static is commonly generated from picking up items from the workbench, from having loose clothing rub together, or by scuffing the soles of your shoes. Many manufacturing plants that produce plastic or plastic-coated products have difficulties with static control. Static is often generated by the plastic rubbing against the conveyor belt as it is transported from one process to the other.

Refer to **Figure 19-2** for a list of some materials that generate static charges. Notice that materials at the top of the list more easily give up electrons and generate a net positive charge. The materials toward the bottom of the list more easily generate negative charges.

Static charges can accumulate on the surface of an object. When a carrier of static charge is brought near another charge carrier with an opposite charge, the electrons of the negatively charged object will "jump" to the positively charged object. This is an attempt for the two charges to equalize to a state of zero net charge.

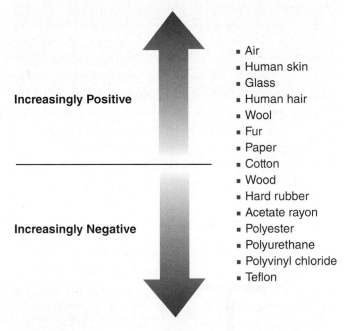

Goodheart-Willcox Publisher

Figure 19-2. A list of some materials that generate static charges. Note that the materials listed on the top half of the list tend to give up electrons and become positively charged. Those materials on the bottom of the list tend to collect electrons, therefore becoming more negatively charged. The closer to the top or bottom of the list a material is, the more easily it becomes charged.

19.3 COULOMB'S LAW OF CHARGES

Coulomb's law of charges states that there are two kinds of charges, positive and negative, and like charges repel while unlike charges attract. Coulomb's law also states that positively charged objects have more protons than electrons, and negatively charged objects have more electrons than protons. The **coulomb**, named after Charles-Augustin de Coulomb, is the fundamental unit of electrical charge. A coulomb is approximately equal to a charge of 6.241×10^{18} electrons and is represented by the unit symbol C.

To make sense of this abstract law about particles and charges, consider an everyday example. Have you ever noticed that when unpacking a box filled with foam peanuts, the peanuts stick to your hands and just about anything else they come in contact with? From the chart in **Figure 19-2**, you know your skin can develop a significant positive charge by giving up electrons. The packing peanuts are made of polystyrene, one of the polymer plastics near the bottom of the list, so they tend to develop a significant negative charge. This is a case of unlike charges attracting each other, causing the negatively charged peanuts to stick to your positively charged hands. Notice also that when the peanuts scatter, they move away from each other, making them more difficult to pick up. This is due to like charges repelling each other.

The field surrounding a charged object is referred to as an **electrostatic field**. The electrostatic field may be positive, due to losing electrons, or negative, due to gaining electrons. Similarly, when two or more charged particles or objects are near each other, an electrostatic field exists between them, **Figure 19-3**. Oppositely charged objects have an attractive force that increases the closer the objects are to each other. Two objects with the same charge have a repulsive force between them.

19.4 CONDUCTORS AND INSULATORS

Conductors are materials that allow electricity to flow through them easily. They have loosely held electrons in

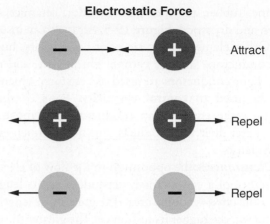

Electrostatic Force

Goodheart-Willcox Publisher

Figure 19-3. Opposite electrical charges attract each other, and like electrical charges repel each other. This natural occurrence is called electrostatic force.

their valence orbitals, so it is relatively easy for electrons to leave the outer orbitals and become free electrons. Free electrons are not attached to any atom. Conductors can also easily capture free electrons to fill their valence orbitals.

Insulators (sometimes referred to as **nonconductors**) have electrons in their valence orbitals that are tightly bound. They also have few free electrons. Thus, insulators do not permit the flow of electrons.

Electricians use both conductors and insulators to control where electricity flows. The appropriate use of conductors and insulators is the first step to harnessing electricity for useful purposes. Examples of good, commonly used conductors include copper, aluminum, gold, platinum, silver, nickel, zinc, iron, tin, and lead, **Figure 19-4**. Examples of insulators are compounds such as glass, oil,

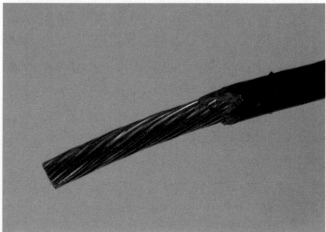

Goodheart-Willcox Publisher

Figure 19-4. Good conductors include metals such as copper, gold, and nickel. Copper conductors, such as the one shown here, are the industry standard for electrical wiring.

silicone, rubber, polymer plastics, mica, ceramics, sand, paper, and dry wood, **Figure 19-5**. Air is also an insulator.

Other elements that are poor conductors, but not nonconductors, include carbon, chromium, and tungsten. Poor conductors are used as resistors, which may also be used to control electricity. Poor conductors require more energy than conductors to remove electrons from their valence shells, but not as much energy as insulators.

Resistance is the opposition to the flow of electrons. Resistance of a conductor is dependent on its length as well as its cross-sectional area. The greater the length of a conductor, the higher its resistance. The greater the cross-sectional area of a conductor, the lower its resistance.

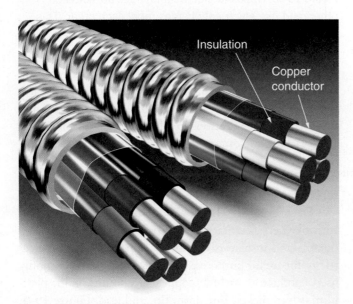

AFC Cable Systems, Atkore International

Figure 19-5. These electrical cables contain several conductors. The electric current flows through the solid copper wire conductors. Each conductor is surrounded by rubber or plastic insulation. The insulation prevents the electricity from leaving the conductor. Different colors of insulation are used to ensure that conductor ends are connected correctly. The metal outer casing provides some protection from damage.

19.5 BASIC ELECTRICAL CIRCUITS

Electrical circuits and systems can harness electricity and put it to some purposeful use. The most basic circuit consists of a source of electricity, a load to use the electricity, and conductors to deliver the electricity from the source to the load. Insulators must also be used to keep the electric current in the conductors, even if the only insulator used is air.

19.5.1 Electrical Source

The disparity in the number of electrons between the positive and negative poles (or terminals) of any source is the *potential difference*. The difference in electrical potential is what causes current to flow when a path between the two poles exists. The potential difference exerts *electromotive force (EMF)*, the force behind the flow of electrons. EMF is measured in **volts**, V, and is often referred to as the **voltage**.

A simple source of electricity is a **battery**, **Figure 19-6**. A battery has two terminals: a positive terminal (cathode) and a negative terminal (anode). The negative terminal has a surplus of electrons and the positive terminal has a deficit of electrons.

Other sources of electricity include generators and photovoltaic cells. Generators convert mechanical energy into electric energy. Combustion engines, water turbines, and steam engines are all examples of electric generators. Photovoltaic (PV) cells convert light into electricity. When light hits a PV cell, electrons in the cell absorb energy and begin to flow as current.

19.5.2 Current

In order for electrons to flow, there must be a complete path between the negative pole and the positive pole. This path is created using conductors. Unlike charges attract, so electrons flow from the negative pole to the positive pole.

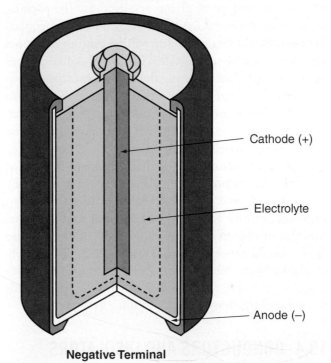

Positive Terminal

Cathode (+)

Electrolyte

Anode (–)

Negative Terminal

Goodheart-Willcox Publisher

Figure 19-6. The main parts of a battery.

When electrons are flowing through a circuit, the flow is referred to as *electric current*. Electric current is the rate of electron flow and is measured in *amperes*, often shortened to amps or indicated by the unit abbreviation, A. One ampere equals one coulomb per second.

In a circuit, electron flow occurs in one of two ways: direct current or alternating current. The electrical source determines which of the two occurs. *Direct current (DC)* flows in only one direction, from negative pole to positive pole. *Alternating current (AC)* changes direction periodically. The type of electricity supplied by a wall outlet is alternating current. Alternating current is discussed in greater depth in later chapters.

19.5.3 Circuit Loads

A load does not consume the electrons supplied by the source. Rather, the load runs on the energy of the electrons flowing through the circuit. For this reason, a complete path from the negative terminal to the load and then from the load to the positive terminal must exist for a circuit to work properly. This is referred to as a *complete circuit*, **Figure 19-7**.

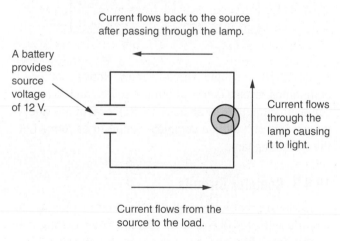

Current flows back to the source after passing through the lamp.

A battery provides source voltage of 12 V.

Current flows through the lamp causing it to light.

Current flows from the source to the load.

Goodheart-Willcox Publisher

Figure 19-7. Complete circuit containing a battery as the power source and a lamp as the load. Notice that the current flows from the battery, through the lamp, and back to the battery.

A load is used to transform electrical energy to another useful form. Energy derived from electricity may include magnetic force, heat, or light. Examples of devices that transform electrical energy include the following:

- A light bulb that converts electrical energy to light.

- A heating element that uses electrical energy to generate heat.

- An electric motor that converts electrical energy to mechanical energy through the force of a magnetic field.

Each load creates resistance in the circuit. The amount of resistance contributed by a load affects the amount of current flowing in the circuit.

19.5.4 Switches

It would be inconvenient to connect and disconnect a circuit from the source of power every time someone wished to turn the circuit on or off. A device called a *switch* can be added to the circuit to perform this task, **Figure 19-8**. The switch has two positions, open or closed. In the open position, the switch interrupts the circuit and no current flows. In the closed position, the switch completes the circuit and allows current to flow.

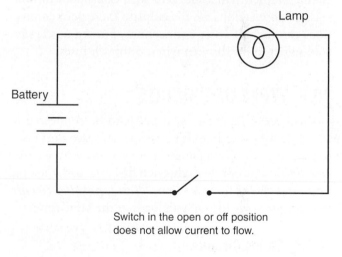

Lamp

Battery

Switch in the open or off position does not allow current to flow.

Goodheart-Willcox Publisher

Figure 19-8. Simple circuit with a switch. The switch enables you to connect or disconnect the lamp from the battery. When the switch is placed in the on position, a connection is made and the switch is considered closed. In the off or open position, the connection is broken and no current flows because the circuit is not complete.

19.5.5 Short Circuits and Open Circuits

A *short circuit* occurs when a path is created allowing electricity to return to the source without first going through a load. Electricity will always take "the path of least resistance," or the easiest path to lower electric potential.

All too often, people refer to any electrical problem as a "short circuit" or "short." In actuality, a short circuit only occurs when there is a path of least resistance around the load. A short may be due to something as simple as a wiring error or an insulator not doing its job, such as an insulation failure of a wire allowing two wires to "short out" against each other.

Often, an electrical system failure is due to an "open circuit," rather than a short. In such cases, poor or nonexistent connections are the culprit. These poor connections can include loose connections, burned out lamps, cold solder joints, broken wires, or blown fuses.

19.6 TYPES OF CIRCUITS

What if more than one load needs to be connected to the same power source? Electricians have two options: a series connection or a parallel connection. With a *series circuit*, the current flows through first one load, then the next, and the next, and so on. With a *parallel circuit*, the current flows through all loads at the same time.

19.6.1 Series Circuits

If two lamps are connected in a series circuit, the current flows through the first then the second, **Figure 19-9**. In the case where one of the lamps burns out, the other lamp would not illuminate because the burned out lamp interrupts the circuit. While the series connection offers more straightforward wiring, it has a major downside. If one of the lamps burns out, all the other lamps will also go out. This can be problematic when dealing with a long series of electrical loads.

19.6.2 Parallel Circuits

A parallel circuit allows multiple loads to be connected across the source. Each load forms a complete circuit, **Figure 19-10**. This scenario allows any single load to be connected or disconnected without affecting any of the other loads in the circuit. In this case, if one of the lamps were to burn out, the other would remain illuminated.

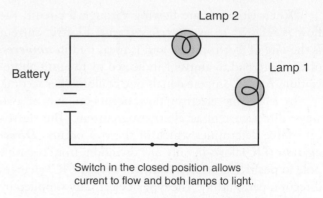

Switch in the closed position allows current to flow and both lamps to light.

Goodheart-Willcox Publisher

Figure 19-9. A simple series circuit. The current must flow from the battery, through the switch, through lamp 1, then through lamp 2, before flowing back to the battery. If any of the connections are open, both lamps will not light. Notice that the switch is also connected in series with the two lamps.

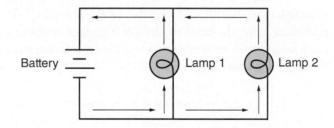

Goodheart-Willcox Publisher

Figure 19-10. A simple parallel circuit. Notice that both lamps receive current from the source and form a complete circuit at the same time. Removal of any one lamp would still allow a complete circuit to be formed by the remaining lamp.

19.6.3 Complex Circuits

Considering that series and parallel circuits operate differently with respect to how current flows through the circuit, it is often beneficial to use both types of connections in a single *complex circuit*. Complex circuits have some components connected in series and others connected in parallel within the same circuit, **Figure 19-11**. A switch, for example, should never be connected in parallel across the source. When a parallel-connected switch is closed, it presents a short circuit to the load and can cause unintended results. A switch should only be connected in series with the source in order to prevent a short circuit condition.

One of the advantages of a parallel circuit is that an open connection in any one leg does not affect what happens in the other legs. For example, a bad connection or a burned out lamp does not affect other loads in a parallel circuit, **Figure 19-12**.

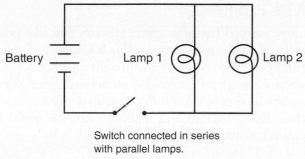

Switch connected in series
with parallel lamps.

Goodheart Willcox Publisher

Figure 19-11. A common complex circuit. Notice that the lamps are connected in parallel with each other. The switch is connected in series with the parallel lamp circuit. Both lamps illuminate when the switch is closed and extinguish when the switch is opened.

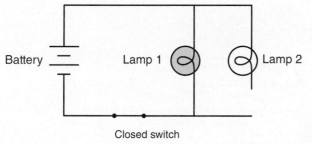

Closed switch

Goodheart-Willcox Publisher

Figure 19-12. In a parallel circuit, if one leg develops an open, the other leg remains normal. Look at the circuit for lamp 2, and notice it has an open. Also note that lamp 1 functions normally in this scenario. The open in the lamp 2 portion of the circuit does not affect the circuit for lamp 1.

The complex circuit in the two previous figures works well in cases where multiple loads need to be controlled with a single switch. However, that setup does not provide for individual control of each lamp. For a greater level of control, the circuit needs a switch for each lamp. Each switch must be carefully placed at a point in the circuit where it interrupts only the circuit related to a single lamp, **Figure 19-13**.

19.7 MAGNETISM

Electricity and magnetism are deeply interconnected. The flow of electricity through a conductor (for example, in any of the circuits described above) creates a magnetic field around the conductor. This induced magnetic field is the theoretical basis for how most motors operate. When current stops flowing in a conductor, the magnetic field collapses, and there is no longer a magnetic field around the conductor. Similarly, a changing magnetic field induces an electrical current in a wire. These phenomena are part of the area of science known as *electromagnetism*.

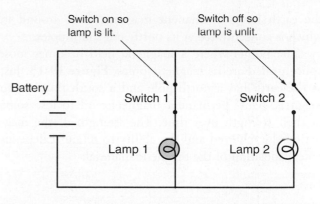

Goodheart-Willcox Publisher

Figure 19-13. A parallel lamp circuit with independent control of each lamp. Notice that opening the switch in the lamp 2 portion of the circuit does not extinguish lamp 1. Switch 1 controls only lamp 1, and switch 2 controls only lamp 2.

19.7.1 Magnetic Fields

A *magnet* is a material that has the power to attract iron, steel, and other ferromagnetic materials, **Figure 19-14**. A permanent magnet retains its magnetic field without the addition of any other outside influence. For example,

revers/Shutterstock.com

Figure 19-14. Magnets attract any ferromagnetic material.

the earth has a permanent magnetic field around it, which is concentrated at its north and south poles.

The points where a magnetic field becomes most concentrated are its magnetic poles, **Figure 19-15**. Just as the earth has a north pole and a south pole, so do magnets. Most permanent magnets tend to lose some of their strength over time. The strength of the magnetic field achieved and the ability to retain it depends on composition of the magnetic material.

19.7.2 Polarity

A fundamental law of magnetism states that like poles repel and unlike poles attract. This is similar to the law of charges. If two magnets are brought together, a north pole and a south pole physically attract each other, while two north poles or two south poles physically repel each other. This attraction and repulsion may be useful in applications such as electric motors.

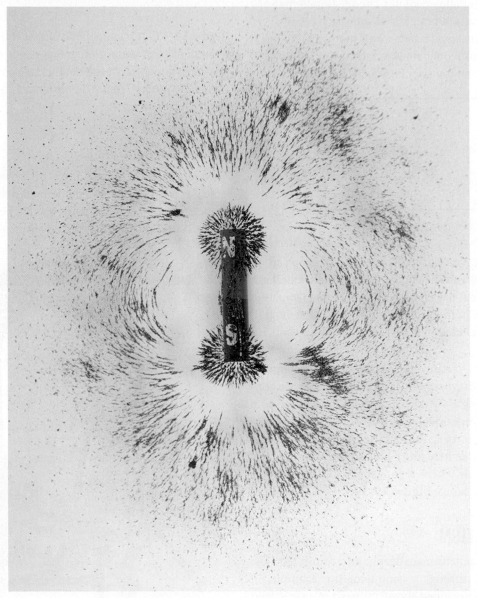

Figure 19-15. Magnetic lines of force surrounding a bar magnet. In this photograph, a sheet of paper is placed over a bar magnet and powdered iron is shaken over the surface of the paper. Notice how the powdered iron particles align themselves with the magnetic lines of force. Also observe that the lines of force are much closer together at the poles than at the center of the magnet.

19.7.3 Electromagnetic Induction

A moving magnetic field induces a voltage in a conductor and, if the conductor is part of a circuit, causes a current to flow, **Figure 19-16**. This phenomenon is known as ***electromagnetic induction***. *Induction* may also refer to the creation of a magnetic field by the presence of an electric current. Because a magnetic field induces a voltage, the conductor (often a coil of wire) is frequently referred to as an ***inductor***.

Electromagnetic induction principles are applied in generators, alternators, and transformers.

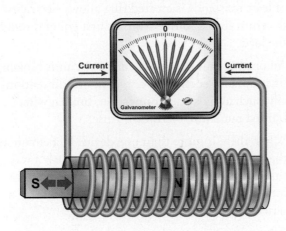

Fouad A. Saad/Shutterstock.com

Figure 19-16. Electromagnetic induction. A moving bar magnet induces current in a conducting coil.

19.7.4 Magnetic Fields Produced by Current Flow

As stated earlier, a current flowing through a conductor produces a magnetic field. If a conductor is wound in a coil form, the magnetic field is intensified. The greater the number of turns in the coil, the greater the magnetic field produced. To further concentrate the magnetic field, a coil may be wound around a core of ferromagnetic material, such as iron, to create an electromagnet. An ***electromagnet*** is a temporary magnet created by a current flowing in a conducting coil.

One of the most desirable traits of an electromagnet is its ability to be turned on and off. When current is flowing through the coil, it becomes a magnet. When current is not flowing through the coil, it is not magnetic. The strength of the magnet may also be controlled by controlling the amount of current allowed to flow through the electromagnet.

CHAPTER WRAP-UP

Understanding the fundamental theory behind electricity is an important first step in becoming an expert on electrical systems maintenance. In this chapter, you learned about how the flow of electrons generates electric current; how to direct the flow of electricity using conductors, insulators, and resistors; and how to create and describe basic electrical circuits. Keep these foundations in mind as you continue learning about electrical systems.

Chapter Review

SUMMARY

- Atoms are made up of protons, neutrons, and electrons. Protons have a positive charge, electrons have a negative charge, and neutrons have no charge.

- In the nucleus of an atom, there are protons and neutrons. Electrons orbit the nucleus in shells called orbitals.

- In normal atoms, the number of protons and the number of electrons are equal. When an atom picks up extra electrons, it becomes a negatively charged ion, or an anion. Conversely, if an atom gives up electrons, it becomes positively charged, or a cation.

- Static electricity can build up whenever two materials rub together.

- Like charges repel and unlike charges attract. This is called the law of charges.

- The coulomb is a unit of charge and is equal to 6.241×10^{18} electrons.

- Some materials easily conduct electricity and are known as conductors. Some materials tend not to conduct electricity and are referred to as non-conductors or insulators. Some materials conduct poorly and tend to oppose the flow of electricity. These materials offer resistance to the flow of electricity and may be used as resistors.

- Opposition to the flow of electrons is called resistance.

- Electromotive force (EMF), also referred to as voltage, is the force behind the flow of electrons and is measured in volts, V.

- Current is the flow of electrons and is measured in amperes, abbreviated as A or amps. One ampere equals one coulomb per second.

- Direct current (DC) flows in only one direction while alternating current (AC) periodically changes direction.

- In order for an electric current to flow, there must be a complete circuit. A complete circuit is formed when one side of the source is connected to one side of a load and the opposite side of the load is connected to the opposite side of the source. This forms a circle (circuit) for the electrons from the source to the load and back to the source.

- A short circuit occurs when an unintended path of least resistance is created that allows electricity to return to the source without first going through the load.

- An open circuit occurs when the circuit is broken intentionally, such as by a switch, or unintentionally, such as by a bad connection, broken wire, burned out lamp, or "blown" fuse.

- When there is more than one device in a circuit and the devices are connected together in such a way that current must flow through each load before returning to the source, the circuit is said to be a series circuit.

- When there is more than one device in a circuit and the devices are connected in such a way as to form distinct complete circuits, the circuit is said to be a parallel circuit.

- Circuits may have both series and parallel components. These circuits are referred to as complex or series-parallel circuits.

- A switch is a device used to control the flow of electricity. When it is turned off (not conducting), the switch is open. When a switch is turned on (conducting), it is closed.

- Magnets have a north pole and a south pole and retain a magnetic field without outside influence. Similar to the law of charges, like poles repel and unlike poles attract.

- A moving magnetic field induces an electric current in an induction coil, or inductor. This process is called electromagnetic induction and is used in generators, alternators, and transformers.

- Electric current flowing through a wire or a coil creates a magnetic field. The polarity of the magnetic field is determined by the direction of flow of the electrons.

REVIEW QUESTIONS

Answer the following questions using the information provided in this chapter.

1. Name the three subatomic particles that compose the atom.

2. *True or False?* Protons and neutrons are the two subatomic particles found in the nucleus.

3. *True or False?* An electron is electrically neutral.

4. If a material picks up electrons, it will have a _____ charge.

5. What kind of activity might generate static charge?

6. *True or False?* The coulomb is the fundamental unit of electrical charge.

7. If two charged particles, one with a positive charge and the other with a negative charge, were brought close to each other, what would happen?

8. What is the difference between a conductor and an insulator?

9. Name five materials that are good conductors.

10. Name five materials that are good insulators.

11. *True or False?* The opposition to current flow is measured in amperes.

12. In what units is electromotive force (EMF) measured?

13. _____ current flows in only one direction.

14. When a wire becomes unintentionally disconnected, does it represent a "short" or an "open"?

15. What is a complete circuit?

16. Give an example of a load used to transform electrical energy into another useful form.

17. Does current flow through a switch when it is open or closed?

18. In a circuit with two lamps in series, one lamp burns out. Will the other lamp remain illuminated? Explain the justification for your answer.

19. Like magnetic poles _____ and unlike magnetic poles _____.

20. *True or False?* All current-carrying conductors have a magnetic field around them.

21. What is the process of electromagnetic induction?

22. If a conductor is wound in a coil and the number of turns in the coil increases, how is the inductance and magnetic field affected?

 **NIMS CREDENTIALING PREPARATION QUESTIONS**

The following questions will help you prepare for the NIMS Industrial Technology Maintenance Level 1 Electrical Systems credentialing exam.

1. Resistance opposes the flow of _____.
   A. ohms
   B. electrons
   C. magnetic field
   D. protons

2. What is voltage?
   A. The difference in electric potential causing the flow of current.
   B. The gaining or losing of electrons in an atom.
   C. The rate of electron flow.
   D. The flow of electrons through a circuit.

3. Which of the following is *not* true about electric current?
   A. It is the flow of electrons through a circuit.
   B. It is measured in amperes, or amps.
   C. It can be induced by a magnetic field.
   D. It only moves from a positive terminal to a negative terminal.

4. In a series circuit, what occurs when a lamp in the series burns out?
   A. It depends on the direction of current.
   B. All lamps go out.
   C. Only the one lamp burns out.
   D. The circuit shorts out.

5. How does current flow differently in a parallel circuit than in a series circuit?
   A. Current flows through all loads at the same time.
   B. Current flows through only one load at a time.
   C. Current flows only if a switch is in the open position.
   D. Current flows only if there are multiple voltage sources.

6. What is the process in which a magnetic field activates voltage in a coil?
   A. Static electricity
   B. Resistance
   C. Electromagnetic induction
   D. Ionization

20 | Ohm's Law, Kirchhoff's Laws, and the Power Equation

LEARNING OBJECTIVES

After completing this chapter, you will be able to:

☐ Apply Ohm's law to calculate voltage, current, and resistance.

☐ Understand the basics of voltage, amperage, and resistance and how they are interrelated.

☐ Compute the total resistance of both series and parallel circuits.

☐ Use Kirchhoff's laws to find voltage drop, current, and resistance values of individual components in series and parallel circuits.

☐ Interpret a complex circuit and identify series and parallel groups in order to solve for voltage drop, current, and resistance in individual components.

☐ Convert between mechanical power and electrical power.

☐ Apply the power equation to compute power consumed by a circuit.

☐ Describe the relationships between voltage, current, resistance, and power.

☐ Use combinations of Ohm's law and the power equation to solve for unknown values.

TECHNICAL TERMS

complex circuit	power
joule	power equation
Kirchhoff's current law (KCL)	voltage divider
Kirchhoff's voltage law (KVL)	voltage drop
Ohm's law	watt

The basic features of electrical systems can be described mathematically with only a few fundamental laws. Ohm's law describes the relationship between voltage, current, and resistance. Kirchhoff's laws define the behavior of voltage and current in different kinds of circuits. Finally, the power equation builds on the other laws and introduces a method for finding the power consumed by a total circuit or individual components.

20.1 OHM'S LAW

Ohm's law is one of the most important laws of electricity and electronics. Ohm's law, named for German physicist Georg Ohm, describes the relationship between voltage (E), current (I), and resistance (R). Essentially, current passing through resistance equals the voltage applied to the circuit. Most technicians use Ohm's law in everyday electrical calculations, often in conjunction with the other formulas covered in this chapter: Kirchhoff's laws and the power equation. Ohm's law states:

$$E = I \times R$$

(20-1)

where

E = voltage measured in volts (V)
I = current measured in amperes (A)
R = resistance measured in ohms (Ω)

Through the application of algebra, other equations may be derived to find any one of the three values if the other two are known. The derived equations are as follows:

$$I = \frac{E}{R}$$

(20-2)

$$R = \frac{E}{I}$$

(20-3)

There is a simple method for remembering these three formulas. The Ohm's law triangle, **Figure 20-1**, shows the three variables contained in Ohm's law. Note the placement of each variable in relation to the others. To use the triangle, simply cover the variable you want to calculate, and the formula for that variable appears. For example, covering the E leaves $I \times R$, and covering the R leaves E / I.

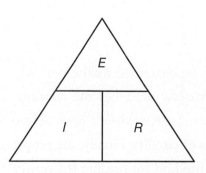

Goodheart-Willcox Publisher

Figure 20-1. The Ohm's law triangle is a useful tool for remembering the three formulas derived from the law.

Before any calculations are performed, all given values must be converted to base units. For example, convert 1 kΩ to 1000 Ω, 100 mA to 0.1 A, and 2.5 kV to 2500 V. Failure to convert the prefixed values to their base units is a common cause of incorrect results. This type of error can cause the result to be off by several orders of magnitude. Becoming proficient at conversion between SI prefixes will greatly improve your chances of obtaining correct results each time you make a calculation.

20.1.1 Finding Voltage with Ohm's Law

The following simple examples demonstrate the use of Ohm's law to find unknown values. Start with the standard version of Ohm's law, equation 20-1. This version of Ohm's law is set up to find the voltage when given the current and resistance of a circuit. Refer to the circuit in **Figure 20-2**. Both the resistance and current of the circuit are known, and the source voltage is unknown. To solve for voltage, use equation 20-1.

$E = I \times R$
$E = 1.5\,\text{A} \times 75\,\Omega$
$E = 112.5\,\text{V}$

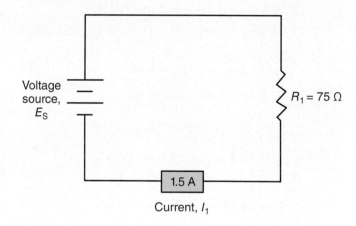

Goodheart-Willcox Publisher

Figure 20-2. Series circuit with unknown voltage source. The current, 1.5 A, is measured with a digital ammeter, and the single resistor, R_1, has a value of 75 Ω.

In this next example, refer to **Figure 20-3**, where once again the voltage is unknown while resistance and current are given. Apply equation 20-1 as in the previous example.

$E = I \times R$
$E = 0.25\,\text{A} \times 220\,\Omega$
$E = 55\,\text{V}$

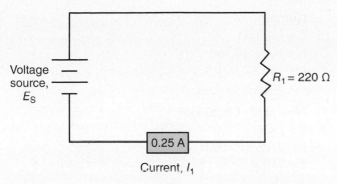

Goodheart-Willcox Publisher

Figure 20-3. Series circuit with unknown voltage source. The current, 0.25 A, is measured with a digital ammeter, and the single resistor, R_1, has a value of 220 Ω.

TECH TIP

Dimensional Analysis

Always label each of the quantities in any calculation with the proper units or dimensions, such as V, A, or Ω. Checking your units, also called dimensional analysis, is a good way to ensure calculations are correct. Make sure the value you calculate carries the proper units and is roughly the size you expect.

20.1.2 Finding Current with Ohm's Law

Refer to **Figure 20-4**. This example circuit has a different unknown than the circuits in the previous two figures. In this case, both the voltage and resistance are known, but the current is unknown. To find the current, use equation 20-2 as follows:

$$I = \frac{E}{R}$$

$$I = \frac{120 \text{ V}}{1000 \text{ Ω}}$$

$$I = 0.12 \text{ A or } 120 \text{ mA}$$

In **Figure 20-5**, the voltage and resistance are different than in **Figure 20-4**, but the solution method is the same because the current is the unknown value.

$$I = \frac{E}{R}$$

$$I = \frac{12 \text{ V}}{25 \text{ Ω}}$$

$$I = 0.48 \text{ A or } 480 \text{ mA}$$

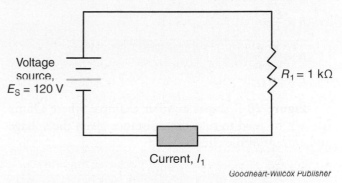

Goodheart-Willcox Publisher

Figure 20-4. Series circuit with 1 kΩ resistance, 120 V source voltage, and unknown current.

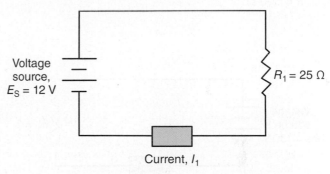

Goodheart-Willcox Publisher

Figure 20-5. Series circuit with 25 Ω resistance, 12 V source voltage, and unknown current.

In the last two examples, the current is expressed in both amperes (A) and milliamperes (mA). It is common practice to express currents of less than 1 A in mA or microamperes (μA), depending on the value.

20.1.3 Finding Resistance with Ohm's Law

Refer to the example circuit in **Figure 20-6**. The unknown value is the resistance, so the applicable version of Ohm's law is given by equation 20-3.

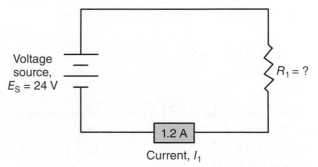

Goodheart-Willcox Publisher

Figure 20-6. Series circuit with unknown resistance. The current, 1.2 A, is measured with a digital ammeter, and the source voltage is 24 V.

$$R = \frac{E}{I}$$

$$R = \frac{24\ V}{1.2\ A}$$

$$R = 20\ \Omega$$

Figure 20-7 shows another example where Ohm's law can be used to find the resistance given the voltage and current of the circuit.

$$R = \frac{E}{I}$$

$$R = \frac{150\ V}{0.5\ A}$$

$$R = 300\ \Omega$$

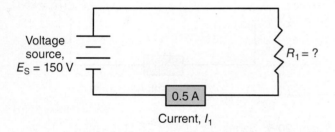

Goodheart-Willcox Publisher

Figure 20-7. Series circuit with unknown resistance. The current, 0.5 A, is measured with a digital ammeter, and the source voltage is 150 V.

20.1.4 Example Calculations with Ohm's Law

Now that you have seen how to apply Ohm's law to solve for voltage, current, and resistance, work through the following problems to practice using Ohm's law. Also remember to convert SI prefixes back to their base units.

Find Voltage Given Current and Resistance

Given:

$R = 2.2\ k\Omega$

$I = 100\ mA$

$E = ?$

First, convert the given resistance and current to base units. Resistance:

$R = 2.2\ k\Omega$

$R = 2.2 \times 10^{3}\ \Omega$

$R = 2.2 \times 1000\ \Omega$

$R = 2200\ \Omega$

Current:

$I = 100\ mA$

$I = 100 \times 10^{-3}\ A$

$I = 100 \times 0.001\ A$

$I = 0.1\ A$

Now apply Ohm's law:

$E = I \times R$

$E = 0.1\ A \times 2200\ \Omega$

$E = 220\ V$

Find Current Given Voltage and Resistance

Given:

$R = 1\ k\Omega$

$E = 120\ V$

$I = ?$

First, convert the given resistance to base units. Resistance:

$R = 1\ k\Omega$

$R = 1 \times 10^{3}\ \Omega$

$R = 1 \times 1000\ \Omega$

$R = 1000\ \Omega$

Voltage is already given in base units, V, so no conversion is necessary.

Now apply Ohm's law:

$$I = \frac{E}{R}$$

$$I = \frac{120\ V}{1000\ \Omega}$$

$I = 0.12\ A$

This value for current can be converted to milliamperes, mA.

$I = 0.12 \times 10^{3}\ mA$

$I = 0.12 \times 1000\ mA$

$I = 120\ mA$

Notice that 10^{3} is used to convert from base units to milliamps. This is the *inverse* of the process to convert from milliamps to base units. Check your results intuitively by remembering that milliamps are 1000 times smaller than amps.

Find Resistance Given Voltage and Current

Given:

$E = 1.5\ kV$

$I = 500\ mA$

$R = ?$

First, convert the given voltage and current to base units. Voltage:

$E = 1.5$ kV

$E - 1.5 \times 10^3$ V

$E = 1.5 \times 1000$ V

$E = 1500$ V

Current:

$I = 500$ mA

$I = 500 \times 10^{-3}$ A

$I = 500 \times 0.001$ A

$I = 0.5$ A

Now apply Ohm's law:

$R = \dfrac{E}{I}$

$R = \dfrac{1500 \text{ V}}{0.5 \text{ A}}$

$R = 3000 \ \Omega$

This value for resistance can be converted to kilohms, kΩ, as follows:

$R = 3000 \times 10^{-3}$ kΩ

$R = 3000 \times 0.001$ kΩ

$R = 3$ kΩ

Again, note that the *inverse* of the process to convert from kilohms to base units is required to move from base units to kilohms.

20.2 TOTAL RESISTANCE OF MULTIPLE RESISTOR NETWORKS

In circuits with more than one resistor, the total resistance must be calculated before other measurements can be taken. The calculations for finding total resistance vary depending on whether the resistors are in series or in parallel. Each type of circuit has distinct principles governing the combined behavior of circuit components. These principles may be selectively applied to complex circuits with a combination of series and parallel components.

20.2.1 Resistors in Series

Resistance in series is the simplest to calculate. Series resistance is the sum of the individual resistances.

$$R_T = R_1 + R_2 + \ldots + R_n$$

(20-4)

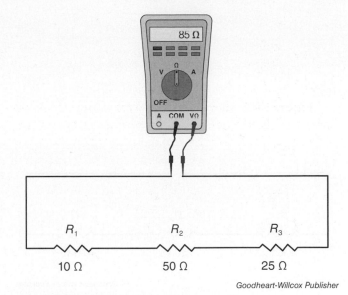

Goodheart-Willcox Publisher

Figure 20-8. Series circuit with three resistors valued at 25 Ω, 50 Ω, and 10 Ω. The total resistance is measured with a digital multimeter set to the ohmmeter mode.

where

R_T = total resistance

R_1, R_2, R_n = any number of individual resistances

Refer to the circuit in **Figure 20-8** with three resistors in series. In this illustration, a digital multimeter (a digital meter that has the capability to measure voltage, resistance, or current) has been connected and set to the ohms range to measure the total resistance of the circuit. To calculate the total resistance, use equation 20-4 for three resistors:

$R_T = R_1 + R_2 + R_3$

$R_T = 10 \ \Omega + 50 \ \Omega + 25 \ \Omega$

$R_T = 85 \ \Omega$

In the example shown in **Figure 20-9**, three resistors with values of 1 kΩ, 750 Ω, and 470 Ω are connected in series. Remember to convert 1 kΩ to 1000 Ω before attempting to solve the equation.

$R_T = R_1 + R_2 + R_3$

$R_T = 1$ kΩ + 750 Ω + 470 Ω

$R_T = 1000 \ \Omega + 750 \ \Omega + 470 \ \Omega$

$R_T = 2220 \ \Omega$ or 2.22 kΩ

20.2.2 Resistors in Parallel

Parallel resistance is a little trickier to calculate because there are three possible equations that might be applicable, depending on the individual resistors. One of the equations works every time. However, the other two only

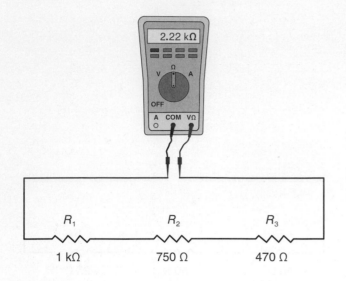

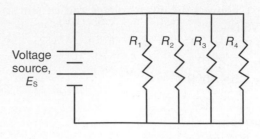

Goodheart-Willcox Publisher

Figure 20-10. Several resistors of equal value wired in parallel.

Goodheart-Willcox Publisher

Figure 20-9. Series circuit with three resistors valued at 470 Ω, 750 Ω, and 1 kΩ. The total resistance is measured with a digital multimeter set to the ohmmeter mode.

work in specific situations. Pay close attention to the setup of each circuit to determine which equation applies.

The parallel resistance equation that works in every situation is as follows:

$$R_T = \cfrac{1}{\cfrac{1}{R_1} + \cfrac{1}{R_2} + \cdots + \cfrac{1}{R_n}}$$

(20-5)

This equation is much more complicated than the formula for total resistance in series. In parallel, the total resistance (R_T) is the *reciprocal* (1/x) of the sum of the reciprocals of the resistances ($1/R_1$, $1/R_2$, and so on). This equation always works for calculating parallel resistance, no matter the particular resistances in the circuit.

Now imagine a circuit with any number of resistances in parallel and each resistor has the same value as all the others, **Figure 20-10**. In this case, the solution is less complicated. The simplified equation is as follows:

$$R_T = \frac{R}{n}$$

(20-6)

where

R = resistance of an individual resistor
n = number of resistors

Remember, this equation only works if all the resistors have the same resistance.

The third equation applies when only two resistors are in parallel, **Figure 20-11**. It does not matter if they have the same or different values. The equation for two resistors in parallel is as follows:

$$R_T = \frac{R_1 \times R_2}{R_1 + R_2}$$

(20-7)

Overall, the second equation is the simplest, and the third equation is not as complicated as the first. Equations 20-6 and 20-7 only work in specific instances, but equation 20-5 works every time.

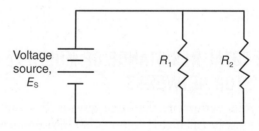

Goodheart-Willcox Publisher

Figure 20-11. Two resistors in parallel. These resistors, R_1 and R_2, may have any value. The resistance equation for two resistors in parallel works for any two resistances.

20.2.3 Example Calculations of Total Resistance in Parallel

Here are a few examples of parallel resistance calculations. Make note of the particular setup of each circuit. Practice deciding which of the three parallel resistance equations applies in a given situation.

Refer to **Figure 20-12**. Several features should immediately stand out. First, this is a parallel circuit with three resistors, each with a different value. Calculating the total resistance of three different resistors in parallel requires equation 20-5:

$$R_T = \cfrac{1}{\dfrac{1}{R_1} + \dfrac{1}{R_2} + \dfrac{1}{R_3} + \ldots + \dfrac{1}{R_n}}$$

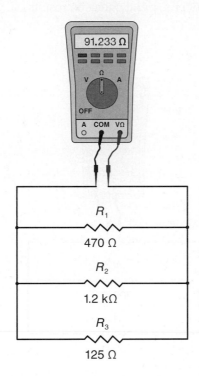

Goodheart-Willcox Publisher

Figure 20-12. Parallel circuit with three resistances of different values. The total resistance is measured with a digital multimeter set to the ohmmeter mode to measure resistance.

This equation looks complex at first, but it is easy to solve when broken down into several smaller pieces. The full equation with values for each resistor inserted in place of the resistor designations looks like this:

$$R_T = \cfrac{1}{\dfrac{1}{470\ \Omega} + \dfrac{1}{1.2\ \text{k}\Omega} + \dfrac{1}{125\ \Omega}}$$

Tackle the solution one step at a time. First, as in earlier calculations, convert SI prefixes to base units.

$$R_T = \cfrac{1}{\dfrac{1}{470\ \Omega} + \dfrac{1}{1200\ \Omega} + \dfrac{1}{125\ \Omega}}$$

Now work on the *denominator*, the part of the equation under the main fraction bar. Start by simplifying the term for each resistor.

$$R_1 = \frac{1}{470\ \Omega}$$
$$R_1 = 0.0021\ \Omega^{-1}$$
$$R_2 = \frac{1}{1200\ \Omega}$$
$$R_2 = 0.00083\ \Omega^{-1}$$
$$R_3 = \frac{1}{125\ \Omega}$$
$$R_3 = 0.008\ \Omega^{-1}$$

Observe that this part of the calculation generates reciprocals of both the resistance values and the units attached to them. In the next part of the calculation, another reciprocal gives the final result. Substituting the values calculated above, the equation now looks like this:

$$R_T = \frac{1}{0.0021\ \Omega^{-1} + 0.00083\ \Omega^{-1} + 0.008\ \Omega^{-1}}$$
$$R_T = \frac{1}{0.01093\ \Omega^{-1}}$$
$$R_T = 91\ \Omega$$

Solving this equation results in a total resistance of about 91 Ω. The results of the total resistance calculation may vary slightly depending on how many decimal places are used in the calculations for individual resistance terms. This discrepancy is referred to as "rounding error." The more decimal places retained in the calculations, the more precise they are. However, an overabundance of precision is unnecessary. A good rule of thumb is to use only as many significant digits (usually nonzero numbers) as appear in the values originally given in the problem. Accordingly, in this example, only two significant digits are included.

In most cases, a value ±1 Ω is of no consequence. If a 470 Ω resistor has 1% tolerance, the actual resistance measured may be ±4.7 Ω, or anywhere between 474.7 Ω and 465.3 Ω. The accuracy of the ohmmeter used to measure the resistance can also change the actual measurement even more. Measurement errors are additive.

The example circuit in **Figure 20-13** is much easier to solve. All the resistances in the circuit are the same, so it is simple to divide one of the individual resistances by the number of resistors to find the total resistance. Use equation 20-6:

$$R_T = \frac{R}{n}$$

$$R_T = \frac{300 \ \Omega}{3}$$

$$R_T = 100 \ \Omega$$

TECH TIP
Total Parallel Resistance

Notice that, when using any of the three equations, the total parallel resistance is less than the smallest resistor value in the circuit. This is the first check a good technician should make in order to ensure the calculation is correct. Computer programs are also available to make calculating complex equations simpler than using a calculator and can offer a useful check on results.

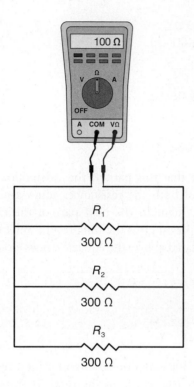

Goodheart-Willcox Publisher

Figure 20-13. Parallel circuit with three resistances of equal value. The total resistance is measured with a digital multimeter set to the ohmmeter mode to measure resistance.

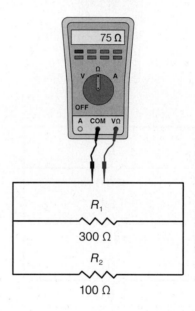

Goodheart-Willcox Publisher

Figure 20-14. Parallel circuit with two resistors of different values. The total resistance is measured with a digital multimeter set to the ohmmeter mode to measure resistance.

The final example of parallel resistance calculations involves two resistors in parallel, each with a different value. Refer to **Figure 20-14**.

In this scenario, use equation 20-7:

$$R_T = \frac{R_1 \times R_2}{R_1 + R_2}$$

$$R_T = \frac{300 \ \Omega \times 100 \ \Omega}{300 \ \Omega + 100 \ \Omega}$$

$$R_T = \frac{30000 \ \Omega^2}{400 \ \Omega}$$

$$R_T = 75 \ \Omega$$

20.3 KIRCHHOFF'S LAWS

Kirchhoff's laws are frequently stated as either a large group of complicated equations or two very general explanations, one pertaining to current and the other relevant to voltage. In their most general forms, Kirchhoff's laws are as follows:

- **Kirchhoff's current law (KCL).** The sum of the currents entering any point in a circuit must equal the sum of the currents leaving that same point. This is an expression of the conservation of charge.

- *Kirchhoff's voltage law (KVL).* The sum of the voltage sources and voltage drops around any closed circuit is zero.

For clarity, Kirchhoff's laws can be further boiled down to four rules:

- **Rule #1.** Current through any component in a series circuit is the same as the current through any other component in that circuit. That is, in a series circuit, the current is the same everywhere.

- **Rule #2.** The sum of all of the voltage drops in a series circuit equals the source voltage.

- **Rule #3.** The sum of the currents through each branch of a parallel circuit equals the total current of the parallel circuit.

- **Rule #4.** The voltage across each branch of a parallel circuit is the same as the source voltage.

20.3.1 Rule #1 Series Circuit Current

Figure 20-15 depicts the first rule related to Kirchhoff's current law. The current through every component in a series circuit is the same, $I_T = I_1 = I_2 = I_3 = I_4$. Find the total resistance of the series circuit by summing the individual resistances of the circuit using equation 20-4:

$R_T = R_1 + R_2 + R_3$
$R_T = 75\ \Omega + 100\ \Omega + 125\ \Omega$
$R_T = 300\ \Omega$

To calculate the total current, use Ohm's law with current isolated, equation 20-2:

$I_T = \dfrac{E}{R_T}$

$I_T = \dfrac{120\ V}{300\ \Omega}$

$I_T = 0.4\ A\ or\ 400\ mA$

As stated earlier, current is constant in a series circuit. Thus, the current passing through each resistor is the same as the total current, in this case 0.4 A.

20.3.2 Rule #2 Series Circuit Voltage

In **Figure 20-16**, the series circuit and resistor values are the same as in **Figure 20-15**. Only the metering has changed. The circuit still has an ammeter to measure the total current, which in a series circuit is the same at any point. Individual voltmeters have been connected to measure the *voltage drop* across each individual resistor. Think of voltage drop as the voltage that would be measured if a

voltmeter were connected across the component. Summing the voltage drops across all components in the series circuit should result in a total equal to the source voltage.

A series circuit is sometimes referred to as a *voltage divider* as it divides the source voltage into smaller increments. Use Ohm's law to calculate the voltage drop for each of the different resistor values.

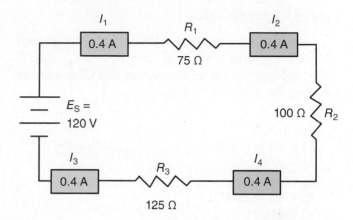

Goodheart-Willcox Publisher

Figure 20-15. Series circuit displaying currents measured at various points using several ammeters. This diagram illustrates Kirchhoff's current law as it relates to current in a series circuit.

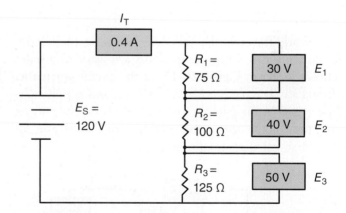

Goodheart-Willcox Publisher

Figure 20-16. The voltage drops across the resistances of a series circuit (also called a voltage divider). This diagram illustrates Kirchhoff's voltage law as it applies to voltage in a series circuit. Here the digital voltmeters read the change in voltage across each resistor. Note that the sum of all voltage drops equals the total source voltage.

First, calculate the total resistance of the circuit as in the previous example.

$R_T = 75\ \Omega + 100\ \Omega + 125\ \Omega$
$R_T = 300\ \Omega$

Again, calculate the total current.

$I_T = \dfrac{E}{R_T}$

$I_T = \dfrac{120\ V}{300\ \Omega}$

$I_T = 0.4\ A$ or $400\ mA$

Since this is a series circuit, the current through each resistor is the same as the total current, $I_T = I_1 = I_2 = I_3 = 0.4\ A$. Now, use the current values and Ohm's law to calculate the individual voltage drops.

$E = I \times R$
$E_1 = I_1 \times R_1$
$E_1 = 0.4\ A \times 75\ \Omega$
$E_1 = 30\ V$
$E_2 = I_2 \times R_2$
$E_2 = 0.4\ A \times 100\ \Omega$
$E_2 = 40\ V$
$E_3 = I_3 \times R_3$
$E_3 = 0.4\ A \times 125\ \Omega$
$E_3 = 50\ V$

As a check, add the individual voltage drops and see that the sum equals the source voltage.

$E_S = E_1 + E_2 + E_3$
$E_S = 30\ V + 40\ V + 50\ V$
$E_S = 120\ V$

Work through the following example to practice calculating the voltage drop across a resistor in a series circuit. Refer to **Figure 20-17** for the circuit schematic. $R_1 = 180\ \Omega$, $R_2 = 200\ \Omega$, $R_3 = 500\ \Omega$, and supply voltage = 220 V. Find total resistance, R_T; total current, I_T; and the voltage drop across each resistor, E_1, E_2, and E_3.

First, calculate total resistance.

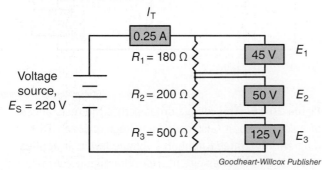

Goodheart-Willcox Publisher

Figure 20-17. Practice finding the total resistance, total current, and voltage drops across the resistances of a series circuit. The voltage drop across each resistance in the circuit is measured with a digital voltmeter.

$R_T = 180\ \Omega + 200\ \Omega + 500\ \Omega$
$R_T = 880\ \Omega$

Use the value for total resistance, the given source voltage, and Ohm's law to calculate total current.

$I_T = \dfrac{E}{R_T}$

$I_T = \dfrac{220\ V}{880\ \Omega}$

$I_T = 0.25\ A$ or $250\ mA$

Now, because $I_T = I_1 = I_2 = I_3 = 0.25\ A$, you can use Ohm's law and the value found for I_T to calculate the voltage drop across each resistor.

$E = I \times R$
$E_1 = I_1 \times R_1$
$E_1 = 0.25\ A \times 180\ \Omega$
$E_1 = 45\ V$
$E_2 = I_2 \times R_2$
$E_2 = 0.25\ A \times 200\ \Omega$
$E_2 = 50\ V$
$E_3 = I_3 \times R_3$
$E_3 = 0.25\ A \times 500\ \Omega$
$E_3 = 125\ V$

Finally, check that the individual voltage drops sum to the source voltage.

$E_S = E_1 + E_2 + E_3$
$E_S = 45\ V + 50\ V + 125\ V$
$E_S = 220\ V$

Here is another exercise for voltage divider calculations. This time there are four resistors, providing higher resistance, and there is a higher supply voltage of 480 V, **Figure 20-18**. Find total resistance, R_T; total current,

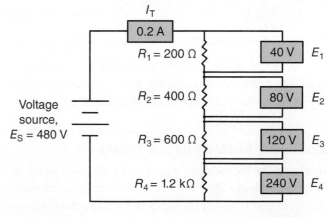

Goodheart-Willcox Publisher

Figure 20-18. This larger voltage divider has four resistors, valued at 200 Ω, 400 Ω, 600 Ω and 1.2 kΩ, and a voltage source providing 480 V. Find total resistance, R_T; total current, I_T; and the voltage drop across each resistor, E_1, E_2, E_3, and E_4.

I_T; and the voltage drop across each resistor, E_1, E_2, E_3, and E_4.

As in the previous example, start by calculating total resistance.

$R_T = 200\ \Omega + 400\ \Omega + 600\ \Omega + 1.2\ k\Omega$

$R_T = 200\ \Omega + 400\ \Omega + 600\ \Omega + 1200\ \Omega$

$R_T = 2400\ \Omega$

Use the value for total resistance, the given source voltage, and Ohm's law to calculate total current.

$I_T = \dfrac{E}{R_T}$

$I_T = \dfrac{480\ V}{2400\ \Omega}$

$I_T = 0.2$ A or 200 mA

Now, use Ohm's law and $I_T = I_1 = I_2 = I_3 = I_4 = 0.2$ A to calculate the voltage drop across each resistor.

$E = I \times R$

$E_1 = I_1 \times R_1$

$E_1 = 0.2\ A \times 200\ \Omega$

$E_1 = 40$ V

$E_2 = I_2 \times R_2$

$E_2 = 0.2\ A \times 400\ \Omega$

$E_2 = 80$ V

$E_3 = I_3 \times R_3$

$E_3 = 0.2\ A \times 600\ \Omega$

$E_3 = 120$ V

$E_4 = I_4 \times R_4$

$E_4 = 0.2\ A \times 1200\ \Omega$

$E_4 = 240$ V

Finally, check that the individual voltage drops sum to the source voltage.

$E_S = E_1 + E_2 + E_3 + E_4$

$E_S = 40\ V + 80\ V + 120\ V + 240\ V$

$E_S = 480$ V

Notice that as the resistance of a resistor is increased, the voltage drop across that resistor also increases.

20.3.3 Rule #3 Parallel Circuit Current

The sum of the currents through each branch of a parallel circuit equals the total current of the parallel circuit.

$$I_T = I_1 + I_2 + \ldots + I_n$$

(20-8)

where

I_T = total current

I_1, I_2, I_n = currents through any number of parallel branches

Parallel circuits, unlike series circuits, split the total current along the parallel branches. Visualize this by tracing the single path from one voltage source terminal to the other through a series circuit versus the several possible paths through a parallel circuit.

In order to calculate the current through each branch of a parallel circuit, use Ohm's law as follows for the circuit shown in **Figure 20-19**:

$I_1 = \dfrac{E_S}{R_1}$

$I_1 = \dfrac{120\ V}{50\ \Omega}$

$I_1 = 2.4$ A

$I_2 = \dfrac{E_S}{R_2}$

$I_2 = \dfrac{120\ V}{75\ \Omega}$

$I_2 = 1.6$ A

$I_3 = \dfrac{E_S}{R_3}$

$I_3 = \dfrac{120\ V}{100\ \Omega}$

$I_3 = 1.2$ A

$I_4 = \dfrac{E_S}{R_4}$

$I_4 = \dfrac{120\ V}{125\ \Omega}$

$I_4 = 0.96$ A

Use equation 20-8 to find the total current:

$I_T = I_1 + I_2 + I_3 + I_4$

$I_T = 2.4\ A + 1.6\ A + 1.2\ A + 0.96\ A$

$I_T = 6.16$ A

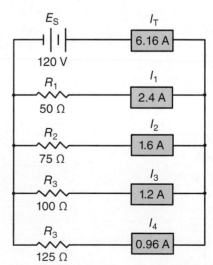

Goodheart-Willcox Publisher

Figure 20-19. This circuit illustrates how the currents in each leg of the parallel circuit sum to the total circuit current. Each current reading is accomplished with a digital ammeter.

As a check, find the total resistance of the circuit. In this case, use the formula for a parallel circuit with more than two resistors with different values.

$$R_T = \cfrac{1}{\dfrac{1}{R_1} + \dfrac{1}{R_2} + \dfrac{1}{R_3} + \dfrac{1}{R_4}}$$

$$R_T = \cfrac{1}{\dfrac{1}{50\ \Omega} + \dfrac{1}{75\ \Omega} + \dfrac{1}{100\ \Omega} + \dfrac{1}{125\ \Omega}}$$

$$R_T = 19.5\ \Omega$$

Now apply Ohm's law to find the total current as follows:

$$I_T = \frac{E}{R_T}$$

$$I_T = \frac{120\ V}{19.5\ \Omega}$$

$$I_T = 6.15\ A$$

The answer may vary slightly depending on the number of decimal places carried throughout the calculation. In this case, rounding to two or three significant digits is appropriate. Note that the value for total current calculated in the check is not exactly the value found in the previous calculation, but the two values are close enough to be attributed to rounding. The check confirms the initial calculation.

Work through the following exercise to practice calculating parallel resistance and currents. Follow the schematic diagram in **Figure 20-20**. Find total resistance, R_T; total current, I_T; and current through each resistor, I_1, I_2, I_3, and I_4.

$$I_1 = \frac{E_S}{R_1}$$

$$I_1 = \frac{240\ V}{175\ \Omega}$$

$$I_1 = 1.37\ A$$

$$I_2 = \frac{E_S}{R_2}$$

$$I_2 = \frac{240\ V}{200\ \Omega}$$

$$I_2 = 1.2\ A$$

$$I_3 = \frac{E_S}{R_3}$$

$$I_3 = \frac{240\ V}{500\ \Omega}$$

$$I_3 = 0.48\ A$$

$$I_4 = \frac{E_S}{R_4}$$

$$I_4 = \frac{240\ V}{750\ \Omega}$$

$$I_4 = 0.32\ A$$

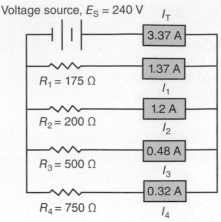

Figure 20-20. Practice calculating total resistance, total current, and individual currents in each leg of a parallel circuit.

Add all the individual currents in the parallel circuit to find total current using equation 20-8:

$$I_T = I_1 + I_2 + I_3 + I_4$$

$$I_T = 1.37\ A + 1.2\ A + 0.48\ A + 0.32\ A$$

$$I_T = 3.37\ A$$

As a check, find the total resistance of the circuit. In this case, as in the previous example, use the formula for a parallel circuit with more than two resistors with different values.

$$R_T = \cfrac{1}{\dfrac{1}{R_1} + \dfrac{1}{R_2} + \dfrac{1}{R_3} + \dfrac{1}{R_4}}$$

$$R_T = \cfrac{1}{\dfrac{1}{175\ \Omega} + \dfrac{1}{200\ \Omega} + \dfrac{1}{500\ \Omega} + \dfrac{1}{750\ \Omega}}$$

$$R_T = 71.2\ \Omega$$

Now apply Ohm's law to find the total current as follows:

$$I_T = \frac{E}{R_T}$$

$$I_T = \frac{240\ V}{71.2\ \Omega}$$

$$I_T = 3.37\ A$$

Again, two or three significant digits is enough precision to confirm the earlier calculation. Notice also that as the resistance decreases, the current will increase.

20.3.4 Rule #4 Parallel Circuit Voltage

Rule #3, the sum of the currents through each branch of a parallel circuit equals the total current of the parallel circuit, is closely connected with Rule #4. Rule #4 states that the voltage drop across each of the elements in a parallel circuit is the same.

In **Figure 20-21**, a voltmeter has been connected across each resistance in the circuit. No math is required to see that the voltage across any component of a parallel circuit is the same. Simply trace the circuit from the source, through each component, then back to the source again. Each component is connected directly across the source, and, in keeping with KVL, the sum of the source voltage and voltage drop across each individual closed circuit must be zero.

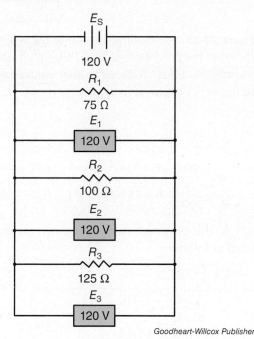

Figure 20-21. The voltage drop across each resistor in a parallel circuit is the same as the source voltage. Notice the voltage drop, regardless of the resistance value, is equal.

20.4 COMPLEX CIRCUITS

Not all circuits are completely parallel circuits or completely series circuits. Often, circuits encountered on the job are a combination of both series and parallel components. A circuit of this nature is a *complex circuit*, which may also be called a *combination circuit* or a *series-parallel circuit*.

In order to solve for unknown values in a complex circuit, first differentiate between the series and the parallel portions of the circuit. Then, appropriately apply the series and parallel resistance formulas and Kirchhoff's laws. Students often wonder, "Which should be solved first, series or parallel components?" Experienced technicians know the answer depends on the actual circuit. In some cases, a parallel portion must be solved first in order to solve the series portion. Other times, the series portion of the circuit has to be resolved before tackling the parallel portion.

Refer to the following examples of different approaches to complex circuits. It is often said that each concept in electronics is quite simple. Things only become complicated when many different concepts are used together. Understanding each concept in its simple form is an important first step toward dealing with more complex circuits.

20.4.1 Series Circuit with Parallel Components

The circuit in **Figure 20-22** is predominately a series circuit. However, R_2 and R_3 are in parallel with each other. The approach in this case is to start by solving the parallel combination. Simplify the circuit by solving for the resistance of the R_2, R_3 combination, then solve for the total resistance of the circuit by treating R_2 and R_3 as one resistor. Treating these two parallel resistors as a single resistor in series allows for a simpler calculation of R_T and I_T. These values can then be used to find the voltage drop across each resistor.

When combining the parallel resistors, remember to use the correct equation for parallel resistance based on the number of resistors in parallel and if their values are equal or not. Here the resistors are equal, so the parallel combination of R_2 and R_3 is calculated using a version of equation 20-6:

$$R_{parallel} = \frac{R_2}{2}$$

$$R_{parallel} = \frac{250 \ \Omega}{2}$$

$$R_{parallel} = 125 \ \Omega$$

Figure 20-22. A series circuit with parallel components. R_2 and R_3 are in parallel with each other.

The simplified circuit, **Figure 20-23**, is purely a series circuit. The total circuit resistance for a series circuit is calculated with equation 20-4:

$$R_T = R_1 + R_{parallel} + R_4$$
$$R_T = 125 \ \Omega + 125 \ \Omega + 125 \ \Omega$$
$$R_T = 375 \ \Omega$$

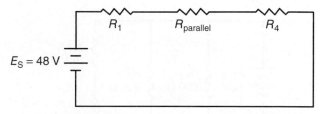

Figure 20-23. A simplified version of the original circuit shown in the previous figure. In this circuit, the parallel part has been reduced to an equivalent series resistance, so the entire circuit is in series.

Now use Ohm's law (equation 20-2) to calculate the total current using the value for total resistance and the source voltage.

$$I_T = \frac{E}{R_T}$$

$$I_T = \frac{48\ V}{375\ \Omega}$$

$$I_T = 0.128\ A\ or\ 128\ mA$$

The current in each component of a series circuit is the same. Use this property to calculate the voltage drop across each of the resistors. The voltage drops across R_1 and R_4 are easy to find using total current and individual resistance. When treated as a single unit, R_2 and R_3 have a combined resistance of 125 Ω, and, by KVL, the voltage drop across each resistor is the same because they are in parallel. In this case, the voltage drop across each resistor, regardless of its location in the circuit, is calculated using Ohm's law (equation 20-1) as follows:

$$E = I_T \times R$$
$$E = 0.128\ A \times 125\ \Omega$$
$$E = 16\ V$$

As a check, add the voltage drops ($16\ V + 16\ V + 16\ V$). The total equals the source voltage of 48 V as it should.

20.4.2 Parallel Circuit with Series Components

A predominantly parallel circuit with series components is depicted in **Figure 20-24**. In this example, first resolve the two series resistances, R_3 and R_4, into one equivalent resistance.

$$R_{series} = R_3 + R_4$$
$$R_{series} = 300\ \Omega + 600\ \Omega$$
$$R_{series} = 900\ \Omega$$

Next, calculate the total resistance of the circuit. Because the combined resistance of R_3 and R_4 is the same as the resistance of the other two resistors in parallel, **Figure 20-25**, the following equation may be used to solve for the total resistance of the circuit:

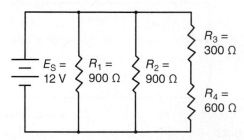

Goodheart-Willcox Publisher

Figure 20-24. A parallel circuit with series components. R_3 and R_4 are in series with each other.

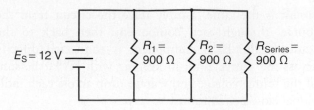

Goodheart-Willcox Publisher

Figure 20-25. A simplified version of the original circuit shown in the previous figure. In this circuit, the series resistors have been reduced to their equivalent resistance, so the entire circuit is in parallel.

$$R_T = \frac{R_1}{3}$$

$$R_T = \frac{900\ \Omega}{3}$$

$$R_T = 300\ \Omega$$

At this point the total circuit current may be calculated as follows:

$$I_T = \frac{E}{R_T}$$

$$I_T = \frac{12\ V}{300\ \Omega}$$

$$I_T = 0.04\ A\ or\ 40\ mA$$

The current for R_1 or R_2 may be calculated as follows:

$$I_1 = \frac{E}{R_1}$$

$$I_1 = \frac{12\ V}{900\ \Omega}$$

$$I_1 = 13.3\ mA$$

The current through the combination of R_3 and R_4 is the same as the current through either R_1 or R_2. KCL states that the current through all components of a series circuit is the same. So each resistor in this circuit experiences the same current calculated for R_1.

KVL states that the voltage drop across every parallel component in a circuit is the same. This means the voltage drop across the R_3, R_4 combination is also the same as the source voltage. The individual voltage drops of R_3 and R_4 may be calculated as follows:

$$E_3 = I \times R_3$$
$$E_3 = 0.0133\ A \times 300\ \Omega$$
$$E_3 = 4\ V$$
$$E_4 = I \times R_4$$
$$E_4 = 0.0133\ A \times 600\ \Omega$$
$$E_4 = 8\ V$$

Again, it is a simple matter to check that the total voltage for these two components in series adds up to the source voltage, 12 V.

20.5 THE POWER EQUATION

After calculating the unknown values for voltage drop, current, and resistance, the power equation may be used to calculate the power consumption of the entire circuit or of each individual component.

Mechanical power has already been presented in this text. Recall, work is the amount of force applied over a certain distance.

$$W = F \times d$$

(20-9)

where

    W = work in joules (J) or foot-pounds (ft-lb)

    F = force in newtons (N) or pounds (lb)

    d = distance in meters (m) or feet (ft)

For example, if a 10 lb object was lifted 2 ft, then the work done in accomplishing this task would be 20 ft-lb. In SI units, imagine a 50 N force is applied over a distance of 5 m. Then the resulting work would be 250 J. Note that the *joule* (J) is the SI unit for work as well as energy and is equivalent to the N·m.

Power is the amount of work done within a specific period of time. It can be expressed using the following equation:

$$P = \frac{W}{t}$$

(20-10)

Substitute the expression for work given in equation 20-9 into equation 20-10 to obtain the following expanded equation:

$$P = \frac{F \times d}{t}$$

(20-11)

where

    P = power in watts (W) or horsepower (hp)

    t = time in seconds (s)

For equations 20-10 and 20-11, W, F, and d are defined exactly as they are for equation 20-9.

Horsepower (hp) is the US customary unit for mechanical power, and one horsepower is equivalent to 550 foot-pounds per second (ft-lb/s). With electricity, the typical unit of power is the *watt* (W), named after James Watt, the inventor of the steam engine. The watt comes from the SI system of units and is defined as the amount of power required to move one volt of electrical potential one coulomb in one second. Remember that the movement of one coulomb (6.241×10^{18}) of electrons past a given point in a circuit is equal to an ampere of current. This means that electrical power depends on voltage and current.

The ***power equation*** used to calculate electrical power is defined as follows:

$$P = I \times E$$

(20-12)

where

    P = power in watts, W

    I = current in amperes, A

    E = voltage in volts, V

This equation is closely related to Ohm's law and, like Ohm's law, can be rearranged algebraically to solve for any of the three variables, **Figure 20-26**. For example, if a light bulb operates on 120 V and draws 0.83 A, find its rated wattage.

$P = I \times E$

$P = 0.83\ \text{A} \times 120\ \text{V}$

$P = 99.6\ \text{W}$

Both mechanical and electrical power are often involved in devices such as motors. For example, if a motor is rated at 3 hp of mechanical power, how many watts of electrical power are consumed? To change between electrical power and mechanical power, the following conversion applies:

    746 W = 1 hp

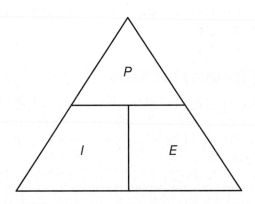

Goodheart-Willcox Publisher

Figure 20-26. The power triangle works just like the Ohm's law triangle. Cover the variable to be solved, and the equation for that variable appears.

20.5.1 Combining the Power Equation and Ohm's Law

In its given form, the power equation solves for power when current and voltage are known. However, it is easy to imagine cases with other unknowns. For example, perhaps only current and resistance are known and voltage is unknown. Ohm's law states that $E = I \times R$, so substituting $I \times R$ for E in the power equation results in $P = I \times I \times R$. Any number multiplied by itself is that number squared. Changing $I \times I$ to I^2 results in the following equation:

$$P = I^2 \times R$$

(20-13)

Use this equation to find the power required by a heating element drawing 2 A of current with a resistance of 25 Ω. How many watts of power does the heater consume?

$P = (2 \text{ A})^2 \times 25 \text{ Ω}$

$P = 4 \text{ A}^2 \times 25 \text{ Ω}$

$P = 100 \text{ W}$

The power equation may be further manipulated to derive the following equation where the value of I is unknown and the value of R is known:

$$P = \frac{E^2}{R}$$

(20-14)

If a toaster operates on 120 V and has a measured resistance of 30 Ω, how many watts of power does it consume? Apply equation 20-14.

$P = \dfrac{(120 \text{ V})^2}{30 \text{ Ω}}$

$P = \dfrac{14400 \text{ V}^2}{30 \text{ Ω}}$

$P = 480 \text{ W}$

20.5.2 The Ohm's Law Wheel

Algebraically manipulating Ohm's law and the power equation yields many possible combinations of known and unknown values. It would be a daunting task to memorize each of the possible equations. Similarly, without a strong command of algebra, it can be difficult to come up with the proper equation for every application using only the base equations. What really matters is understanding the application of each equation and becoming proficient with using each one.

In **Figure 20-27**, the Ohm's law wheel shows all possible combinations of Ohm's law and the power equation.

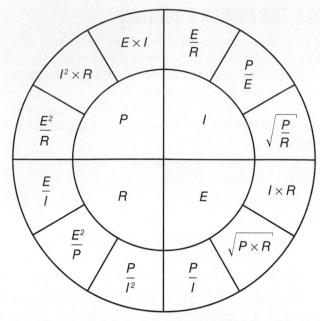

Goodheart-Willcox Publisher

Figure 20-27. The Ohm's law wheel presents all combinations of Ohm's law and the power equation. The inner wheel indicates the desired result. The outer wheel shows possible equations to obtain that result. Select the appropriate equation based on the known variables.

The inner wheel represents the desired value, either P (power in watts), I (current in amperes), R (resistance in ohms), or E (voltage in volts). In the outer wheel next to each of the four possible results are three different equations that may be used to obtain the result. Select the equation which contains the known variables and solve for the unknown variable.

CHAPTER WRAP-UP

Successful electrical maintenance technicians must understand how to work with primary circuit components and their measurements for a variety of circuit types. In this chapter, you learned about the fundamental relationships between voltage, current, resistance, and power in series, parallel, and complex circuits. You gained practice with Ohm's law, Kirchhoff's laws, and the power equation, which are essential to describing those relationships and calculating unknown values for any circuit or individual circuit component.

Chapter Review

SUMMARY

- Ohm's law describes the relationship between the voltage, current, and resistance of a circuit, $E = I \times R$. Voltage (E) is measured in volts (V); current (I) is measured in amperes (A); resistance (R) is measured in ohms (Ω).

- The method for calculating resistance depends on the type of circuit and the resistors themselves. There are three equations to solve for resistance in a parallel circuit, depending on the number of resistors and whether the resistance values are the same or different. There is only one equation for a series circuit where the sum of the resistances equals the total resistance.

- To solve for unknown values in a complex circuit, first break the circuit down into its smaller series and parallel components.

- Kirchhoff's current law (KCL) states that the sum of the currents entering any point in a circuit must equal the sum of the currents leaving that same point.

- KCL implies that the current at any point in a series circuit is the same as that of any other point in the circuit. KCL also implies that the current in each leg of a parallel circuit will sum to the total current through the circuit.

- Kirchhoff's voltage law (KVL) states that the sum of the voltage sources and voltage drops around any closed circuit is zero.

- KVL implies that the individual voltage drops in a series circuit sum to the source voltage of the circuit. KVL also implies that the voltage drop across each branch in a parallel circuit is equal to the source voltage.

- If the voltage of a circuit remains the same and the resistance increases, the current decreases.

- If the resistance of a component increases, the voltage drop across that component will increase.

- If all resistance values in a parallel circuit are the same, the resistance of one of the resistors divided by the total number of resistors equals the total resistance of the circuit.

- The unit of electrical power is the watt (W), represented in equations by the symbol P.

- The power equation states that power (P) equals the current (I) multiplied by the voltage (E), or $P = I \times E$.

- Ohm's law and the power equation may be combined to solve for unknown values.

REVIEW QUESTIONS

Answer the following questions using the information provided in this chapter.

1. *True or False?* Ohm's law can be used to calculate resistance as well as current and voltage.

2. What is the supply voltage of a circuit that draws 0.155 A and has a total resistance equal to 2 kΩ?

3. If the supply voltage of a circuit is 24 V and the total circuit resistance is equal to 750 Ω, what is the total current?

4. What is the total resistance of a circuit that draws 1.2 A at 100 V?

5. *True or False?* It is *not* necessary to convert the known quantities to their base units before solving for the unknown quantity.

6. In circuits with more than one resistor, _____ must be calculated before other calculations can be made.

7. Calculate the total resistance of a series circuit with 150 Ω, 1 kΩ, 2.2 kΩ, and 100 Ω resistors.

8. Calculate the total resistance of a parallel circuit with 175 Ω, 1 kΩ, 1.8 kΩ, and 500 Ω resistors.

9. Calculate the total resistance of a parallel circuit containing five resistors each with a value of 25 Ω.

10. Calculate the total resistance of a parallel circuit containing a 100 Ω resistor and a 250 Ω resistor.

11. The voltage drops in a _____ circuit sum to the source voltage of the circuit.

12. A series circuit consists of three resistors connected to a 60 V source. If $R_1 = 750$ Ω, $R_2 = 680$ Ω, and $R_3 = 1$ kΩ, calculate the voltage drop for each resistor.

13. A parallel circuit has 180 Ω, 1.2 kΩ, 2.7 kΩ, and 268 Ω resistors. The parallel circuit is connected to a 75 V power supply. Calculate the total resistance, current through each resistor, and the total circuit current.

14. *True or False?* The sum of the currents through each branch of a parallel circuit will be less than the total current of the parallel circuit.

15. Kirchhoff's voltage law (KVL) states that _____ across each branch in a parallel is equal to the source voltage.

16. What is the total resistance of the complex circuit shown below with parallel resistors in series?

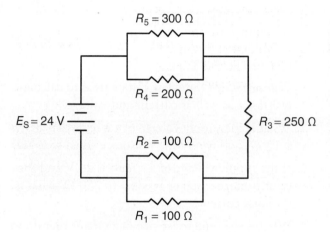

17. What is the total resistance of the complex circuit shown below with series resistors in parallel?

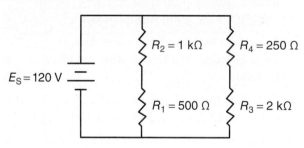

18. A watt measures the amount of _____ required to move one _____ of electrical potential one coulomb in one second.

19. 125 mA flows through a 270 Ω resistor. How many watts of heat does it dissipate?

20. What is the rated wattage for a heating element that draws 1.25 A when connected to a 120 V source?

21. *True or False?* Current is never included in the power equation.

22. How many watts does a toaster consume if it operates at 260 V with a resistance of 80 Ω?

23. What is the resistance for a 20 W light bulb with a 115 V source?

24. In Europe where the line voltage is 220 V, how much current does a 60 W lamp draw?

25. What would happen to the current from the previous question if the line voltage were lowered? Explain why.

NIMS NIMS CREDENTIALING PREPARATION QUESTIONS

The following questions will help you prepare for the NIMS Industrial Technology Maintenance Level 1 Electrical Systems credentialing exam.

1. Which of the following is *not* a correct equation of Ohm's law?
   A. $R = \frac{E}{I}$
   B. $I = \frac{E}{R}$
   C. $E = I \times R$
   D. $R = E \times I$

2. What is the current draw of a circuit operating from 48 V and having a total resistance of 350 Ω?
   A. 0.14 A
   B. 7.29 A
   C. 16.8 A
   D. 398 A

3. Which of the following affects the total resistance in a circuit?
   A. Number of resistors
   B. Voltage drop
   C. Temperature
   D. Direction of current

4. Where does current split to flow in different directions in the following parallel circuit?

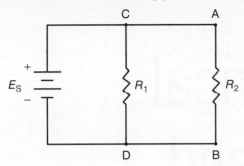

A. Point B
B. Point D
C. Points C and D
D. It is impossible to determine without more information.

5. When resistance in a circuit increases and voltage stays the same, what happens to current?

A. It decreases.
B. It increases.
C. It cannot be calculated unless voltage is known.
D. It stays the same.

6. As the resistance of a resistor increases, what happens to the voltage drop across that resistor?

A. It decreases.
B. It increases.
C. It cannot be calculated unless current is known.
D. It stays the same.

7. What is the total resistance of a parallel circuit with a 1 kΩ and a 2.7 kΩ resistor?

A. 0.73 Ω
B. 3.7 Ω
C. 729.7 Ω
D. 9,990 Ω

8. What is the conversion between mechanical power and electrical power?

A. 1 W = 746 V
B. 746 W = 1 A
C. 746 hp = 1 Ω
D. 1 hp = 746 W

9. Electrical power depends on what two quantities?

A. Temperature and resistance
B. Voltage and wattage
C. Current and voltage
D. Current and resistance

10. What is the wattage of a heating element that draws 5.25 A when connected to a 220 V source?

A. 0.02 W
B. 41.90 W
C. 225.25 W
D. 1155 W

21

Electrical Test and Measurement Equipment

LEARNING OBJECTIVES

After completing this chapter, you will be able to:

☐ Discuss test and measurement safety procedures.

☐ Identify assembly-level and component-level trouble-shooting methods.

☐ Explain the use of test lights, continuity testers, and receptacle testers.

☐ Understand the capabilities and uses of a digital multimeter.

☐ Demonstrate the use of oscilloscope controls and measurement methods.

☐ Use a clamp-on ammeter.

☐ Interpret oscilloscope calibration waveforms.

☐ Explain the use of power supplies and signal generators.

☐ Discuss the use of a phase sequence tester.

☐ Understand the need for calibration and the calibration process.

TECHNICAL TERMS

arbitrary function generator (AFG)

auto-ranging

clamp-on ammeter

continuity tester

cursor

digital multimeter (DMM)

infrared (IR) thermometer

International Electrotechnical Commission (IEC)

LCR meter

megohmmeter

noncontact voltage tester

oscilloscope

overvoltage

phantom voltage

phase sequence tester

receptacle tester

screwdriver voltage tester

test light

triggering

It is not possible to see the actual flow of electricity through a circuit—only the effects of it are visible. Electrical test and measurement equipment enables you to "look inside the circuit" and visualize what is happening. After taking electrical measurements, you can use basic electrical calculations to predict other values in that circuit. See **Figure 21-1**.

In this chapter, you will learn about different types of test equipment and the various measurements each is capable of taking. As you become proficient using measurement equipment and techniques, your troubleshooting abilities will continue to grow.

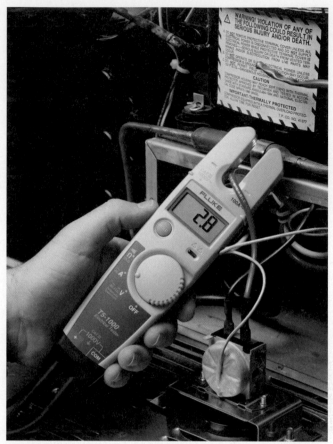

Figure 21-1. Electrical measurements provide information about how a system or equipment is operating. These measurements can be analyzed as part of the troubleshooting process.

Using testing and measurement, you can identify the faulty component in the circuit and replace it. This approach is far more efficient and affordable than guessing and swapping out part after part until the problem is resolved.

21.1 ELECTRICAL SAFETY

Safety is the most important part of testing and measurement. Be mindful of all safety procedures when making measurements. Anyone who makes their living working with electricity must develop a healthy respect for the potential dangers involved in dealing with live circuits.

21.1.1 Physical Condition

Your physical condition can lead to potentially unsafe and dangerous actions. Always be well rested when at work. If you are tired due to lack of sleep, your mind can be "foggy" and your reflexes slowed. This can lead to a poor decision or delayed action, resulting in a dangerous situation and potential injury.

Never work if your body is experiencing the effects of alcohol or drugs. Alcohol and drugs greatly reduce mental and physical abilities, which can lead to unsafe practices. In addition, arriving to work under the influence of alcohol or drugs is cause for immediate dismissal at many companies.

Illness can also lead to unsafe practices. Sickness may reduce mental and physical abilities, not unlike drugs or alcohol. If you do not feel you can do your work safely, call in sick.

Mental stress can also result in unsafe practices. The pressure of needing to get a vital piece of equipment back in operation or the time constraints of completing the job can cause stress. Even experienced technicians can make uncharacteristic mistakes when trying to work too quickly. These mistakes may result in fatal consequences. Never ignore safe practices in order to complete a job more quickly.

21.1.2 Electrical Testing Safety

When preparing to perform electrical tests and measurements, begin by wearing the proper personal protective equipment (PPE). See **Figure 21-2**. Your company or school will have policies regarding what PPE you should use based on the tests being performed, the equipment being tested, and the environment where the test is being conducted. OSHA also has guidelines about what PPE is appropriate. Refer to Chapter 2, *Industrial Safety and OSHA*, for additional information about PPE.

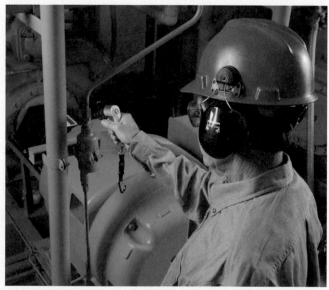

Figure 21-2. Always wear the appropriate personal protective equipment (PPE). This industrial maintenance technician is wearing hearing protection, safety glasses, and a hard hat while taking temperature measurements.

Whenever possible, de-energize the equipment prior to testing. Use proper lockout/tagout procedures, **Figure 21-3**. After disconnecting a device, always test to ensure that the equipment is truly de-energized. Working on "live" equipment that has been incorrectly de-energized creates an extremely dangerous situation. Refer to Chapter 2, *Industrial Safety and OSHA*, for additional information about lockout/tagout procedures.

Inspect test equipment before performing a measurement. Test leads are an important part of test equipment safety. Make certain that the CAT level (overvoltage category) of the test leads is appropriate for the job. **Figure 21-4** shows a set of test leads with the proper safety features. Look for test leads with double insulation, shrouded input connectors, finger guards, and a nonslip surface. Always inspect test leads and make certain they are not cracked, frayed, or worn. If the test leads are damaged or worn, replace them immediately.

Never exceed the maximum input values of the test equipment you are using. Test equipment is clearly marked with the maximum voltage and current adjacent to the input jacks, **Figure 21-5**. Always know the expected values before applying the test leads and set the appropriate range before testing.

Check the test equipment setting each time before applying the test leads. See **Figure 21-6**. If, for example,

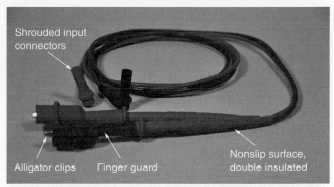

Goodheart-Willcox Publisher

Figure 21-4. Test leads with finger guards, nonslip surfaces, alligator clips, double insulation, and shrouded input connectors. By using alligator clips, you do not need to hold one or both of the test leads while testing.

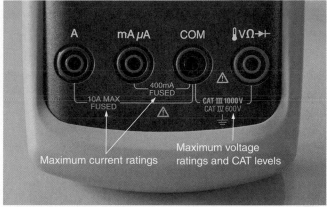

Goodheart-Willcox Publisher

Figure 21-5. Maximum voltage and current ratings are clearly marked next to the test equipment input connections. Always be sure that the CAT level of the test equipment is appropriate for the measurement.

Brady Corp.

Figure 21-3. Work on de-energized equipment whenever possible. Always follow proper lockout/tagout procedures.

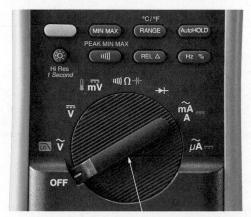

Verify correct setting before applying leads

Goodheart-Willcox Publisher

Figure 21-6. Always check the test equipment setting before applying the test leads. An incorrect setting can have disastrous results.

your meter is set to measure resistance and you apply the test leads to an energized circuit, catastrophic results may occur.

Use extra caution when using 1000:1 or 100:1 high-voltage probes. Make sure that you have the correct type of probe and that the voltage reaching the test equipment will be less than its rated voltage.

Follow the old electrician's rule: "Use only one hand when making a measurement and keep the other hand in your pocket." This lessens the possibility of a complete circuit being made from one hand to the other, with the current going through your heart.

Be extremely cautious of test lead placement. If the lead slips, it could come in contact with something that could cause a short circuit. Always move carefully and deliberately when performing a test.

PROCEDURE	Preparing to Perform an Electrical Test

The following are general steps to take before performing an electrical test or taking an electrical measurement:

1. If possible, de-energize the equipment before testing. Follow appropriate lockout/tagout procedures.

2. Wear appropriate PPE based on the test, equipment, and environment.

3. Select the appropriate test equipment based on the type of test and expected measurement values.

4. Inspect the test equipment for wear and damage.

5. Verify that the test equipment has an adequate voltage rating and CAT level.

6. Check that the leads are connected properly and that the test selector switch is set correctly.

21.1.3 Overvoltage Categories (CAT Level)

The ***International Electrotechnical Commission (IEC)*** is a nonprofit, nongovernmental international standards organization that prepares and publishes standards for electrical, electronic, and related technologies. The IEC standards for overvoltage protection categories are used by test equipment manufacturers. These manufacturers certify through testing that their test equipment adheres to IEC standards.

Overvoltages, or *transients*, are undesirable voltage spikes in excess of the intended voltage. If the test equipment you are using is not capable of dealing with the transients that may be present, the test equipment could explode, causing injury or death.

A table with descriptions and examples of the four CAT levels is shown in **Figure 21-7**. Test equipment must be labeled at the required CAT level or at a higher CAT level. For example, if the test requires a CAT II rating, equipment labeled for CAT II, CAT III, or CAT IV can be used safely.

21.1.4 Electrical Testing Safety Rules

Before testing:

- De-energize the circuit whenever possible. Live circuits present additional risks. When de-energizing a circuit, observe the proper lockout/tagout procedures.

- Use the proper PPE.

- Be certain you are using the correct test equipment for the job.

- Check your test equipment settings and connections before each measurement.

- Inspect your test leads for wear and tear before use and do not use any test leads that are damaged or in need of replacement.

During testing:

- Know the expected voltage and current before connecting test equipment to the device. Make sure test equipment is connected properly and set to the correct range.

- Avoid holding the test equipment while making measurements. This minimizes your exposure to an arc flash. Hang or rest the meter if possible. Use the "kickstand" if available.

- Use only one hand when performing the test. Keep one hand in your pocket while making measurements.

- Watch your test lead placement to avoid a short circuit.

- Stay alert and practice safe procedures when making measurements and working on electrical equipment.

Overvoltage Protection Categories		
Category	Description	Examples
CAT I	Connections to circuits in which measures are taken to limit transient overvoltages to an appropriately low level.	Electronic circuits with overvoltage protection.
CAT II	Energy-consuming equipment to be supplied from a fixed installation.	Appliances, portable tools, and other household and similar loads.
CAT III	In fixed installations and for cases where the reliability and the availability of the equipment is subject to special requirements.	Switchgear and polyphase motors, bus systems and industrial feeders; switches in fixed installation and equipment for industrial use with permanent connection to the fixed Installation.
CAT IV	Connections at the utility or origin of installation, and all outside connections.	Outside service entrance and drop from pole to building, wiring run from meter to panel. Electricity meters and primary overcurrent protection equipment.

Goodheart-Willcox Publisher

Figure 21-7. Standards for measurement categories are published by the IEC. Be sure the testing equipment you are using has the correct CAT level for the test being performed. If the equipment does not have an adequate CAT level, do not use the device—you could be seriously injured by a voltage spike in excess of what the device is capable of withstanding.

21.2 FIELD TESTING AND BENCH TESTING

There are two main types of troubleshooting. Assembly-level troubleshooting is usually accomplished on the plant floor. Component-level troubleshooting is often done at the test bench. A typical test bench has power supplies, signal generators, and test equipment to diagnose the component defect. Assemblies are often brought to the test bench for more in-depth diagnostics.

Many companies keep an inventory of critical assemblies. When an assembly fails, it can be removed from service and immediately replaced with another assembly from inventory. This allows the equipment to continue to operate while the assembly is repaired. Without a spare assembly, the equipment cannot operate until the assembly is repaired. The cost of having the equipment sit idle for an extended period of time can be very expensive.

After the assembly has been replaced and equipment is operating, the defective assembly is brought to the test bench, diagnosed to the component level, and repaired. Often, the repair is replacement of a defective component. The assembly is tested to confirm proper operation and then stored in inventory until needed for replacement.

The time needed to troubleshoot and repair an assembly is a significant factor in determining the cost of the repair. The reliability of the repaired assembly is also a factor. Relatively low-cost assemblies may be recycled rather than repaired.

Some companies outsource their repair work. A good technician who is capable of component-level repair can save a company significant money by completing repair work in-house.

21.3 ELECTRICAL MEASUREMENT TOOLS

Industrial maintenance technicians use many electrical measurement and testing tools. Some measurement equipment can perform multiple types of tests. Other, more specialized equipment performs a single test. Learning both the operation of and uses for each type of measuring equipment is critical.

21.3.1 Test Lights

The simplest form of test equipment is a test light, **Figure 21-8**. A ***test light*** is simply a lamp with two wires. It is used to test for the presence of electricity. The lamp lights when placed across a voltage source. A test light is a relatively limited electrical testing tool.

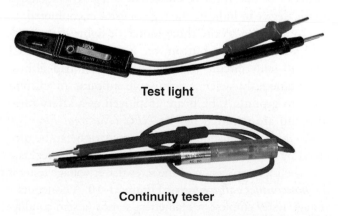

Test light

Continuity tester

Klein Tools, Inc.; Goodheart-Willcox Publisher

Figure 21-8. Test lights and continuity testers are simple electrical testing tools with limited application for an industrial maintenance technician.

21.3.2 Continuity Tester

A *continuity tester* is similar to a test light with a power source added to the lamp. See **Figure 21-8**. When the leads of the continuity tester are connected by an electrical path, the lamp lights to indicate a complete circuit. Like a test light, a continuity tester has limited application.

21.3.3 Receptacle Testers

Normal residential 120 VAC wiring consists of three wires. The "hot" wire (black), the "neutral" wire (white), and the "ground" conductor (uninsulated or green). The neutral wire is, by code requirements, bonded (connected) to ground at the point where the electrical service enters the building and again at the circuit breaker panel.

In order for a receptacle (electrical outlet) to be wired correctly, each conductor is connected to a specific terminal. However, in some cases, conductors are attached to incorrect terminals and the receptacle is not properly wired.

A common troubleshooting task for electricians is to check receptacle wiring. One device for testing a receptacle to determine the "hot" slot is a *screwdriver voltage tester*. This device consists of a pocket screwdriver that has a neon lamp and a current-limiting resistor connected between the screwdriver blade and the metal pocket clip. The electrician inserts the screwdriver into one slot of the receptacle while his hand is in contact with the metal pocket clip. If the screwdriver is inserted into the hot side of the receptacle, the neon lamp lights.

Receptacle testers, **Figure 21-9**, have replaced most screwdriver voltage testers. A receptacle tester has three indicator lamps. It is plugged into a receptacle and if the two yellow lamps illuminate, the outlet is wired correctly. Any other combination of illuminated lamps indicates a wiring error that must be corrected. A legend is included on the device to indicate the type of error encountered.

In addition to the three indicator lights, the receptacle tester has a test button that is used to trip a GFCI (ground-fault circuit interrupter). GFCIs trip almost instantaneously when a miniscule amount of current flows to ground. GFCIs are employed as a safety measure and are required by the *NEC (National Electrical Code)* in certain instances. GFCI protection can be provided by GFCI receptacles or by GFCI circuit breakers.

The modern version of the screwdriver voltage tester is the *noncontact voltage tester*, **Figure 21-10**. A noncontact voltage tester employs a Hall effect sensor (a semiconductor device that is sensitive to magnetic fields) and operates in a similar fashion to the electrician's screwdriver, except that no direct contact with a live conductor is required. This allows the user to verify the presence of voltage on a conductor even through the conductor insulation.

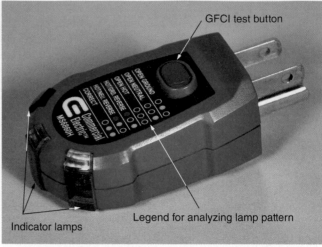

Indicator lamps · GFCI test button · Legend for analyzing lamp pattern

Goodheart-Willcox Publisher

Figure 21-9. A receptacle tester for analyzing the wiring connected to an electrical outlet.

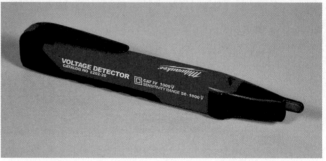

Goodheart-Willcox Publisher

Figure 21-10. A noncontact voltage tester lights in the presence of voltage.

Interestingly, if you drag the detector tip of a noncontact voltage tester along a power cord between an electronic device and a receptacle, you will notice that the detector senses the presence of voltage for a certain distance, then senses nothing. This occurs because the conductors in the cord are twisted. The twist places the hot conductor next to the sensor for a certain distance, and then the ground and neutral conductors next to the sensor for the remaining distance.

21.3.4 Digital Multimeter (DMM)

A *digital multimeter (DMM)* is an instrument used to measure a variety of electrical properties. A DMM can serve as a voltmeter for measuring electrical potential, an ohmmeter for measuring resistance, and an ammeter for measuring current. Due to its ease-of-use and versatility, a DMM is one of an industrial maintenance technician's most valuable tools.

Before taking a measurement with a DMM, select the appropriate type of measurement and range using the mode selector switch. On many DMMs, the mode selector switch either lacks or has few range values. See **Figure 21-11**. The DMM automatically determines the correct range without operator intervention. This automatic action is referred to as *auto-ranging*. As a result, you must be careful when reading the value in the display. At first glance, 100 mV might appear to be 100 V. Always check the suffix on the reading to verify that you are interpreting the displayed value correctly.

SAFETY NOTE
DMM Input Jacks

A common and dangerous mistake is to attempt a voltage measurement with the test leads plugged into the current input jacks. When configured for measuring current, the DMM creates a direct short to the voltage source. Always check to ensure that the test leads are connected correctly for the type of measurement being made, **Figure 21-12**.

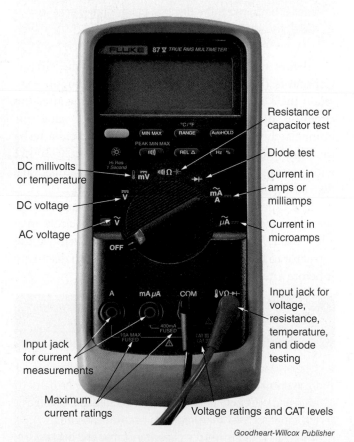

Figure 21-11. Most digital multimeters (DMMs) have features similar to those shown here.

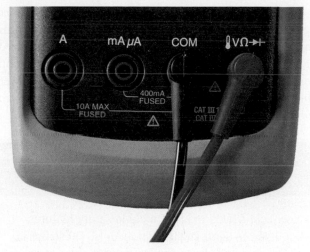

Figure 21-12. A digital multimeter with the test leads appropriately connected to measure voltage or resistance. Note the two current jacks to the left of the test leads.

When using a DMM, first set the mode selector switch to the correct setting for the type of measurement. Note that voltage and current normally have different mode settings for AC and DC. With the mode set, plug the leads into the corresponding input jacks. As mentioned earlier in this chapter, always check the voltage rating, current rating, and CAT level listed on the DMM to ensure that the device is appropriate for the expected measurement.

When measuring current, position the test leads so the DMM is connected in series with the load. See **Figure 21-13**.

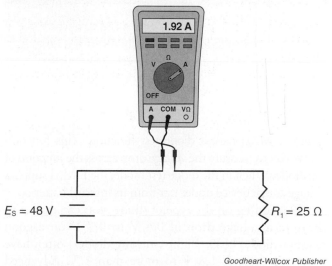

Figure 21-13. A DMM connected for measuring current. When measuring current, the DMM must be connected in series and set to the proper range. Never attempt to place the meter probes across the voltage source.

Many digital multimeters also have a min/max/average memory. With this feature, a DMM holds the minimum value measured, the maximum value measured, and the average value over a period of time. In addition, some advanced DMMs measure frequency and convert the current measurement of a 4-20 mA process control loop to a percentage reading.

TECH TIP
DMM Safety Features

Low-cost DMMs may not include safety features included in some higher-end models. These extra safety features may someday prevent damage to the device or (more importantly) injury to its user. When working in an industrial environment, the extra safety features are worth the extra cost.

Ohmmeter Mode

Before measuring the resistance of a device or component, disconnect electrical power from the device. The circuit must be de-energized before resistance can be measured. In the ohmmeter (resistance measurement) mode, the DMM provides voltage to the circuit under test. Most DMMs can measure resistance ranging from fractions of an ohm to several megohms.

SAFETY NOTE
Ohmmeter Mode

When measuring resistance with a DMM, never apply the test probes to an energized circuit. If you do, the DMM may be damaged and you could be injured.

Diode Test Mode

Often, DMMs have a diode test function. This function allows you to measure the voltage drop across the junction of the diode. When in the diode test mode, the DMM supplies voltage to the device under test from its internal batteries.

Depending on the type of diode, you can expect to measure a voltage drop of 0.4 V to 0.7 V on a good rectifier diode. LEDs (light-emitting diodes) often have a voltage drop of 1.8 V to more than 3 V. Advanced DMMs have sufficient voltage to cause LEDs to illuminate when tested in the forward direction. If a diode is good, it should show no conductivity (infinite voltage drop) in the reverse direction.

Capacitor Test Mode

Many DMMs also have a function for measuring capacitance. In the capacitor test mode, the DMM supplies voltage to the capacitor under test. The capacitance displayed on the DMM is typically measured in either microfarads (μF), nanofarads (nF), or picofarads (pF).

TECH TIP
SI Conversions

Capacitors are generally specified in either μF or pF, but the DMM may display the reading in a different unit. Therefore, you may occasionally need to convert from one unit to another. One microfarad is 1000 nanofarads, and one nanofarad is 1000 picofarads. To convert between these units, multiply or divide the number by 1000, which moves the decimal point three places. When converting from a larger unit to a smaller unit, the number increases. When converting from a smaller unit to larger unit, the number decreases. For example, to convert 4700 nF to microfarads (smaller unit to larger unit), move the decimal point three places to make the number smaller: 4.7 μF. To convert 6.8 nF to picofarads (larger unit to smaller unit), move the decimal point three places to make the number larger: 6800 pF.

Before testing a capacitor, de-energize the circuit. Capacitors that have been in an energized circuit may retain their charge for a certain period of time after the circuit is de-energized. Always discharge a capacitor immediately before testing by shorting it across a resistor. The *NEC* recommends using a resistor of 20-30 kΩ with a rating of 4 W, but your particular needs may vary depending on the capacitor being discharged.

If a capacitor remains in the circuit during testing, the influence of other components and capacitors may affect the DMM measurement. It is best to remove the capacitor to be tested from the circuit and then discharge it before attempting to test it.

SAFETY NOTE
Capacitor Test Mode

When measuring a capacitor, never apply the test probes to an energized circuit. If you do, the DMM may be damaged and you could be injured. Also, be sure to discharge the capacitor after the circuit has been de-energized. If you fail to discharge the capacitor, it may still hold a charge that can damage the DMM and cause you injury.

Continuity Test Mode

Some multimeters include a continuity test mode where an audible signal indicates the presence of a low resistance. This feature allows checking for continuity without having to look at the multimeter display. In the continuity test mode, the DMM provides voltage to the circuit under test.

> ⚠️ **CAUTION**
>
> Never connect a DMM in the continuity test mode to an energized circuit or the DMM may be irreparably damaged.

Advanced DMM Functions

Some multimeters are even capable of communicating with a computer, allowing you to plot periodic measurements and determine how the value measured changes over time.

To overcome the problem of slow variations in the value measured, many digital multimeters include a bar graph in the display. The bar graph simulates the movement of a meter needle so you can get a sense of the rate of change occurring while the display is continuously counting. See **Figure 21-14**.

Many multimeters are capable of measuring temperature with the aid of a thermocouple. An example of a DMM and a thermocouple is shown in **Figure 21-15**. Some multimeters even have a thermocouple provided in the accessory package that comes with the meter. The thermocouple is a contact-type temperature device. It measures the temperature of whatever it is contacting. It is capable of measuring the air temperature.

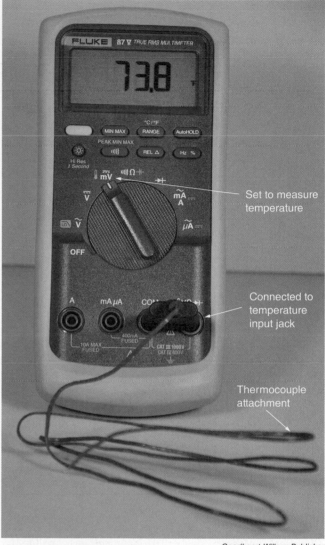

Set to measure temperature

Connected to temperature input jack

Thermocouple attachment

Goodheart-Willcox Publisher

Figure 21-15. With a thermocouple attached, this DMM can measure temperature.

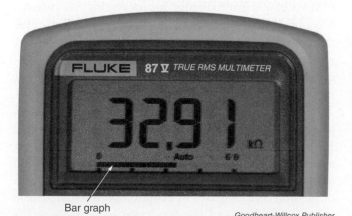

Bar graph

Goodheart-Willcox Publisher

Figure 21-14. This digital multimeter display includes a bar graph along the lower portion of the display area. The bar graph provides a visual representation of the fluctuations in the reading.

🔧 **TECH TIP**
Thermocouples

A thermocouple is a device that uses the junction of dissimilar metals to generate a voltage proportional to the temperature.

Most advanced multimeters are labeled *True RMS*. RMS refers to the term "root-mean-square," which is a type of averaging (integration) of a sinusoidal alternating current waveform. In simpler terms, it is the amount of AC voltage that would produce the same amount of power that a DC voltage would.

The RMS value is important because a VOM measures in RMS. A DMM can measure in peak voltage if it does not convert to RMS. Without any type of conversion, a DMM would read a voltage much higher than would a VOM.

It is not difficult to convert the peak value of a sine wave to RMS. You simply multiply the peak voltage value by 0.707 to obtain the RMS value. The problem is that with modern industrial electronics, the voltage may not be in a sine wave pattern.

Three-phase variable frequency motor drives do not produce a sine wave output. Switching power supplies and other electronic devices distort the normal sine wave of the AC line power. For this reason, true RMS algorithms (a part of a computer program) are used. With true RMS, the DMM is capable of measuring the RMS value of most any type of waveform.

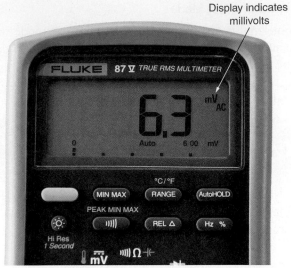

Display indicates millivolts

Goodheart-Willcox Publisher

Figure 21-16. A DMM displaying a phantom voltage. Be careful in interpreting the reading. Notice the mV indication in the display. The reading is 6.3 mV, not 6.3 V.

TECH TIP
True RMS DMMs

If you will be working with industrial electronics, be sure to have a DMM that reads true RMS in order to ensure your readings are correct.

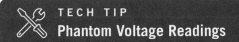

TECH TIP
Phantom Voltage Readings

Be careful when interpreting voltage readings and pay close attention to the range indication. Does the range show mV or does it indicate V? False interpretation might lead you to believe there is a real voltage present when it is only a result of a phantom voltage.

Phantom Voltages

Phantom voltages, also called *ghost voltages*, are DMM readings of electrical potential or voltage between conductors that have no actual voltage difference. These readings are a result of induction or IR drop (resistance) across conductors that are connected to the same point but have different resistances. In addition, the amplifiers inside the DMM are very sensitive, and the DMM will auto-range until it can obtain some sort of reading, even if that reading is so small as to be negligible.

Phantom voltages can range from a few millivolts to hundreds of volts depending on the situation. The DMM inputs have an extremely high impedance (opposition to the flow of alternating current). This high impedance means that, although the phantom voltage might be quite high, its current is extremely low, possibly in the microamp realm. The high impedance of the DMM does not present much of a load, so even these extremely small currents will cause the DMM to display a phantom voltage reading. An example of a phantom voltage reading is shown in **Figure 21-16**.

21.3.5 Clamp-On Ammeter

DMMs have a limited range for measuring AC current. Most DMMs are capable of safely measuring a maximum current of 10 A. Another drawback of using a DMM to measure current is that the DMM must be series-connected. This requires disconnecting a current-carrying conductor and placing the DMM into the circuit.

A **clamp-on ammeter** eliminates these shortcomings by providing a noncontact method of measuring current. See **Figure 21-17**. To accomplish this, some clamp-on ammeters use a current transformer that can be opened to insert the conductor to be measured. No disconnection or physical contact with a live connection is required. The magnetic field around the conductor to be measured is coupled into the clamp-on ammeter's current transformer.

Because older clamp-on ammeters use a current transformer, they are limited to measuring only AC currents.

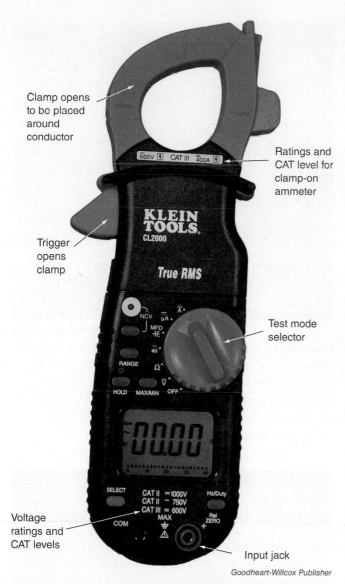

Figure 21-17. This clamp-on ammeter is capable of measuring AC or DC current. In addition, it has a DMM for measuring resistance, AC voltages, and DC voltages; a diode test function; a capacitor test function; and a noncontact voltage detector.

Modern clamp-on ammeters use a Hall effect sensor instead of a current transformer. A Hall effect sensor is a semiconductor device that is sensitive to magnetic fields. Hall effect sensors allow a clamp-on ammeter to measure both AC and DC current.

In general, low-cost models use a current transformer and more expensive models use a Hall effect sensor. When purchasing a clamp-on ammeter, consider whether the capability to measure DC current is needed.

Many models have a built-in DMM included to save the technician time and bother switching from one test instrument to another in the middle of troubleshooting a problem.

TECH TIP
Low Current Measurements

Clamp-on ammeters suffer from poor accuracy when measuring low current values. This problem can be minimized by placing a second turn of the conductor being measured through the clamp-on core. When this is done, the reading obtained will be twice the actual current in the conductor, so you must divide the reading by two. Higher currents do not present this accuracy problem.

21.3.6 Megohmmeter

A *megohmmeter* is an ohmmeter that uses high voltage to make resistance measurements. An example of a megohmmeter is shown in **Figure 21-18**. Megohmmeters are also called *meggers* or *insulation testers*.

Most DMMs can measure resistance in the megohm range. However, DMMs use a low voltage to accomplish this task. In some cases, shorts may not be readily apparent when tested at a low voltage.

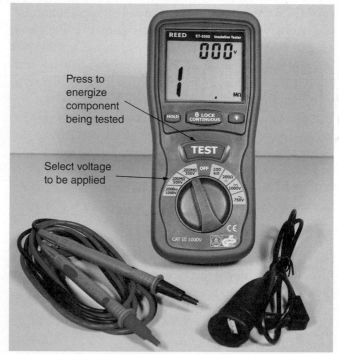

Goodheart-Willcox Publisher

Figure 21-18. Megohmmeters are also known as meggers or insulation testers. The range switch is set to the proper range, test leads are connected to the device under test, and then the test button is pressed to energize the high voltage and make the measurement.

The insulation in coil, transformer, and motor windings is relatively thin in order to conserve space. For this reason, the insulation rating is only marginally greater than the working voltage. In an overvoltage situation, the overvoltage present is greater than the insulation voltage rating of the wire and an arc-over may occur.

When an arc-over occurs, a carbon path is formed. This carbon path presents a relatively high resistance, which is unaffected by the low voltage of a DMM. However, the carbon path presents a significant problem at normal operating voltages.

Another scenario occurs when insulated wire comes in contact with the bare metal of the stator of a motor. Vibration eventually wears through the insulation of the winding and causes a high resistance circuit to ground. Again, this is relatively unaffected by the lower voltage output of the DMM in the ohmmeter mode.

In either scenario, when measured at operating voltage, substantial current flow may occur. By testing with the higher output voltage of a megohmmeter, you can identify high-resistance shorts such as those described.

21.3.7 LCR Meter

An *LCR meter* measures inductance (L), capacitance (C), and resistance (R). While a DMM is capable of making both resistance and capacitance measurements, an LCR meter is much more accurate. An LCR meter is also capable of making inductance measurements, which a DMM cannot. An example of an LCR meter is shown in **Figure 21-19**.

An LCR meter is more accurate than a DMM because an LCR meter has a higher measurement resolution and a variable frequency source. Taking capacitance and inductance measurements using a frequency at or near the operating frequency provides a more accurate measurement.

Inductance measurements are quite valuable when trying to diagnose a transformer, coil, or motor winding if a shorted turn is suspected. When a coil has a shorted turn, it will exhibit a substantially lower inductance than that of a coil without a shorted turn.

21.3.8 Oscilloscopes

An *oscilloscope* is a device that is capable of graphically representing a voltage waveform. This device is extremely valuable in troubleshooting problems that are otherwise invisible. A modern digital oscilloscope is shown in **Figure 21-20**.

Goodheart-Willcox Publisher

Figure 21-19. An LCR meter. Notice that there are various probes to more easily make component measurements. This unit comes with a probe that resembles a pair of tweezers for measuring tiny surface mount components. In place of a mode selector switch, this model uses buttons to navigate on-screen menus.

TECH TIP
Oscilloscopes

Some industrial maintenance technicians say, "I don't need an oscilloscope." This is often the case when the technician does not know how to use one. The more you learn about using an oscilloscope, the more indispensable this diagnostic tool becomes.

Divisions

The oscilloscope screen is divided into a grid. Refer to **Figure 21-21**. The Y (vertical) axis depicts voltage or amplitude of the incoming signal. If the trace deflects upward, it depicts a positive input voltage. If the trace deflects downward, it depicts a negative value. The X (horizontal) axis depicts time.

Each little "box" on the oscilloscope grid is referred to as a division. On the Y axis, the value of each division is in volts (volts/division). On the X axis, the value of each division is in time (time/division). Two controls on the

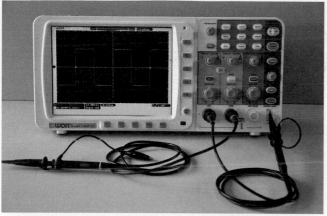

Goodheart-Willcox Publisher

Figure 21-20. A modern digital oscilloscope with probes. The probe for channel 1 is connected to the scope calibrator output. The scope screen displays the waveform output of the scope calibrator.

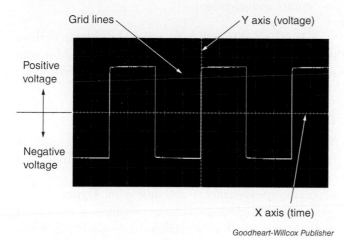

Goodheart-Willcox Publisher

Figure 21-21. A square waveform as shown on the oscilloscope screen.

oscilloscope set these values. The V/div (volts per division) control sets the value of a division on the Y axis, and the t/div (time per division) control sets the value of each division on the X axis. These controls are adjusted until the waveform can be conveniently measured.

For the waveform shown in **Figure 21-21**, the scales are set to 1 V/div and 200 μs/div. Count the number of divisions vertically from the positive peak of the waveform to the negative peak to arrive at the peak-to-peak value of 5 V. Count the number of divisions between any point on a cycle, such as where the wave begins to go positive starting from the center line, to the same point on the next cycle, to find 5 divisions. Multiply the number of divisions by the time/div (5 div × 200 μs/div) to calculate 1000 μs or 1 ms.

Vertical and Horizontal Adjustments

Many digital oscilloscopes have several knobs to adjust various parameters of the device. The most important are the V/div knobs, which set the voltage scale. Refer to **Figure 21-22**. Most oscilloscopes have two or four input channels. Each channel has a voltage scale (V/div) adjustment.

The math button allows you to perform math functions between the two channels. Unfortunately, unknowledgeable technicians press this button and receive a display that they are unable to interpret. This is one of the most common mistakes an operator can make. If you are unsure of the results of a setting or adjustment, do not use it.

Turning the vertical position or vertical offset knobs moves the trace higher or lower on the screen. For example, you may want to view channel 1 on the top half of the screen and channel 2 on the bottom half. On some oscilloscopes, pressing this knob centers the trace on the screen.

Horizontal controls include a time/division adjustment knob used to set the horizontal scale. A horizontal position adjustment knob allows you to shift the trace from side to side.

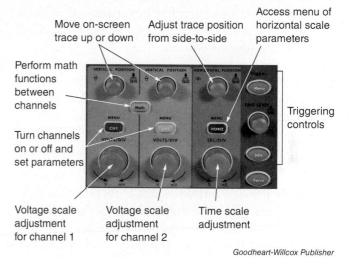

Goodheart-Willcox Publisher

Figure 21-22. Typical controls of a digital oscilloscope.

Triggering

Triggering allows the waveform shown on the screen to remain stationary without moving horizontally. Without triggering, the waveform drifts or rolls either right or left.

Triggering is the point in time at which the oscilloscope begins sweeping the displayed waveform from left to right. You can set the scope to trigger at a specific voltage. You can also specify if the trigger happens on the leading or trailing (rising or falling) edge of the waveform.

Most digital oscilloscopes have a plethora of triggering events to choose from. Read the user's manual for your oscilloscope and become familiar with what triggering options are available to you. Many oscilloscopes provide a button to set the trigger level to 50% of the value of the waveform. This method allows you to get into the "ball park" quickly. For most waveforms, the 50% setting works with no further adjustment of the trigger level control. More complicated waveforms may require more manipulation in order to lock the trace on the screen.

Additional Oscilloscope Functions

Oscilloscope controls typically include additional buttons. Button names may vary, but the functions are similar. Always refer to the user's manual of your specific oscilloscope for a complete explanation of available functions. The following are general descriptions of the functions of common buttons:

- **Run/Stop button.** Allows you to stop the sweep on the oscilloscope, "freezing" the waveform for closer examination. This feature is helpful when the waveform is changing too quickly to get a good look at what is happening. This button is not available on analog oscilloscopes.

- **Single (Sweep) button.** Takes a "snapshot" of the waveform. If the waveform frozen by the Run/Stop button is not useful, pressing the Single button takes another snapshot. This button is not available on analog oscilloscopes.

- **Autoscale button.** Allows the oscilloscope to automatically change the voltage scale (v/div). In some instances, it is beneficial to allow the scope to set this parameter.

- **Autoset (Auto Setup) button.** Measures the incoming signal, adjusts the vertical and horizontal scales, and sets the triggering to display a waveform. This option works most of the time on simple waveforms. However, more complex waveforms require additional adjustment. This button is not a substitute for the technician becoming proficient with oscilloscope setup and operation.

- **Save button.** Stores waveforms as a file that can be reloaded to the oscilloscope screen or as an image that can be printed. Most digital oscilloscopes have a USB port, so you can save waveforms to a flash drive.

- **Display button.** Allows you to set and control various aspects of the display.

- **Help button.** Provides general information on how to set up a function. The Help button is not a substitute for reading the user's manual.

- **Measure button.** Accesses a menu that allows you to set the scope to make voltage, time, frequency, pulse length, duty cycle, RMS value, and many other measurements. As many as two or three dozen measurement options may be available, depending on the model of the oscilloscope.

- **Multipurpose knob.** Allows menu-specific adjustments and selections.

Oscilloscopes may include buttons in addition to those described here. Always refer to the user's manual for your specific oscilloscope to learn about the available functions.

Cursors

With regard to oscilloscopes, *cursors* are a pair of lines, either horizontal or vertical, whose position may be changed in order to take measurements. Examples of cursor measurements are shown in **Figure 21-23**. Normally, a digital oscilloscope displays the absolute position of each cursor as well as the Δ value (difference) between the two cursors.

Portable Oscilloscopes

Many digital oscilloscopes are portable (handheld) models powered by batteries. See **Figure 21-24**. They either come with a rechargeable battery included or in some cases a battery may be added as an option.

The primary benefit of a portable oscilloscope is the ability to use the device on the plant floor without being tethered to a 120 V receptacle. This makes the scope more versatile and allows you to measure grounded circuits without the scope being connected to ground. This isolation is an important safety feature.

Oscilloscope Inputs and Probes

Oscilloscope inputs commonly have an impedance of 1 MΩ. For low frequency signals, the 1 MΩ input is

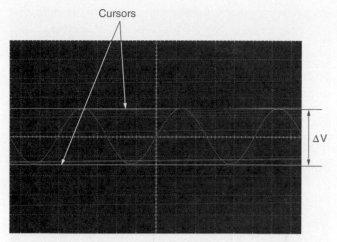

Voltage measurement

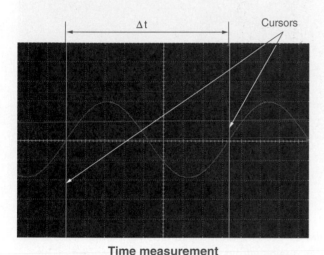

Time measurement

Goodheart-Willcox Publisher

Figure 21-23. Oscilloscope cursors serve as reference points for measuring voltage and time.

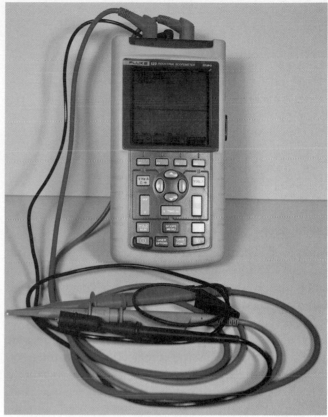

Goodheart-Willcox Publisher

Figure 21-24. A portable oscilloscope. Notice that this oscilloscope does not have any knobs. Instead, settings are accessed through the use of on-screen menus and navigation buttons. This portable oscilloscope also makes automatic measurements and functions as a multimeter.

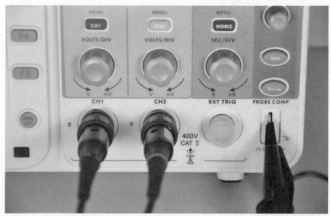

Goodheart-Willcox Publisher

Figure 21-25. Oscilloscope inputs and calibrator output.

often used with a 10:1 probe, resulting in circuit loading of 10 MΩ. Some oscilloscopes may also select an impedance of 50 Ω to match equipment with a coaxial cable output. With a direct coax cable connection between the scope and the equipment being measured, the impedance is matched and the scope should read accurately. **Figure 21-25** shows the input channel connections and the scope calibrator connection of a digital oscilloscope.

Often, oscilloscope probes have a switch labeled X1/X10 on the probe body. X1 is considered direct, and in this position the probe impedance is the same as the input impedance of the scope.

In order not to load a high impedance circuit under test, switch the probe to the X10 position. This position allows the probe to present a high impedance and not

load the circuit under test. When in the X10 position, multiply the reading on the oscilloscope screen by ten to obtain the actual voltage measured. Some oscilloscopes have a menu option that allows you to specify which probe is connected, such as X1, X10, X100, or X1000.

The X100 probe is a good accessory to have when measuring higher voltages. Always be careful to make certain the multiplier or switch position of the probe you are using is set appropriately. Check the menu setting for the probe multiplier as well.

Oscilloscope Calibrator

Most oscilloscopes have a built-in calibration source. This is a useful tool for checking the oscilloscope and probes to ensure that everything is in order. The calibrator output is commonly located next to the channel inputs on the scope.

The most common calibration signal is a 5 V peak-to-peak square wave at a frequency of 1 kHz. The probes are coaxial in design and have a compensation adjustment on them. In a perfect scenario, the coaxial cable inductance is equal to its capacitance and the two cancel each other out.

A probe compensation variable capacitor is built into the probe to allow you to zero out any inductance or capacitance. Connecting the probe to the calibrator and observing the square wave on the scope screen indicates if you need to make any adjustments to the compensation capacitor. On X1/X10 probes, the compensator is applied to only the X10 side of the probe, so be certain to check the switch.

21.3.9 Power Supplies

Assemblies require various AC and DC voltages at different levels. To bench test an assembly, you must provide it with the same type of power it receives in its usual operation.

There are many types of power supplies. Single-voltage DC power supplies provide one level of voltage. Variable DC power supplies can be adjusted among a range of voltage levels. AC power supplies include transformers and variable autotransformers.

21.3.10 Arbitrary Function Generators

An *arbitrary function generator (AFG)* is a digital electronic device capable of generating analog and digital signals. See **Figure 21-26**. An AFG, sometimes also called an "arb," can generate almost any imaginable analog or digital signal. AFGs can be physical devices or may be circuit cards controlled by software on a USB-connected computer, **Figure 21-27**.

Goodheart-Willcox Publisher

Figure 21-26. This arbitrary function generator (AFG) has an analog output and multiple digital outputs. This AFG is completely software controlled, so there are no controls on the unit. A USB port allows the AFG to connect to a computer.

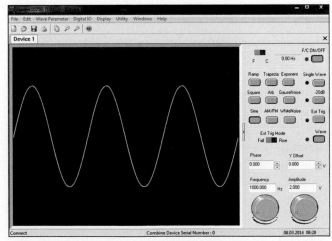

Goodheart-Willcox Publisher

Figure 21-27. This AFG control software provides all the controls to modify various waveform parameters of an AFG.

An AFG is capable of producing many different waveforms, including operator-specified waveforms. For example, you can save an oscilloscope waveform as a data file, load it into the AFG software, and output the waveform. An AFG is also capable of outputting digital signals and bit patterns.

21.3.11 Phase Sequence Tester

When connecting three-phase power to most equipment, especially electric motors, the three-phase conductors must be in the correct relationship to each other. If you were to label each phase A, B, and C, respectively, phase B would start 120° after phase A starts, phase C would start 120° after phase B starts, and phase A would start once again 120° after phase C starts. This phase relationship ensures all the phases are in the correct sequence. Note that each phase is 120° apart from the others.

If a three-phase motor is connected with all phases in the correct sequence, the motor rotates clockwise as viewed from the shaft end. If any one of the phases is out of sequence, the motor rotates counterclockwise. The phase sequence can be corrected by switching any two of the three phase conductors.

According to convention, phase conductors should be labeled L1, L2, and L3. The motor connections should be labeled T1, T2, and T3. Unfortunately, these conductors are often unlabeled or mislabeled. This situation can be corrected using the process of elimination by connecting the motor and seeing which direction it turns when energized. If the motor runs backward, reversing any two phases will correct the situation.

A *phase sequence tester* can be used to label or verify the labeling of the phase conductors. A typical phase sequence tester, **Figure 21-28**, has three inputs, three voltage-present

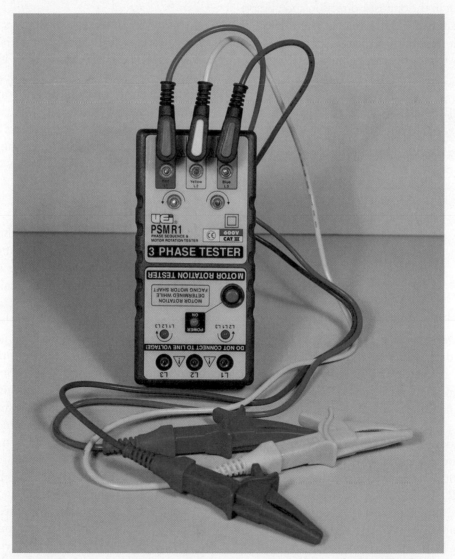

Goodheart-Willcox Publisher

Figure 21-28. A phase sequence and motor rotation tester is a relatively simple and valuable testing device.

indicators, two rotation indicators, and a test button. Simply connect the three input leads to the three phase conductors. The voltage-present indicators verify that voltage is present in each phase. The phase relationship is indicated by one of the two rotation indicators.

For phase sequence testing, some testers have a button that must be pressed to read the rotation. Some high-end units have a motor rotation test capability to identify the proper sequence of the motor leads.

For motor rotation testing, the tester has three input/output jacks, a power indicator, two rotation indicators, and a test button. To test, connect the three motor leads to the three input/output jacks. Press and hold the test button, and observe that the power indicator is illuminated. Looking at the shaft end of the motor, rotate the motor shaft and observe the rotation indicator. Verify that the rotation indicator shows the direction the motor is rotating. Rotate the motor in the opposite direction and observe the direction of rotation indicator that is illuminated. If the rotation indication is opposite the direction the shaft is rotating, reverse any two of the three conductors.

TECH TIP
Phase Sequence Tester Applications

A phase sequence tester is an extremely useful tool when working with conveying systems, pumps, compressors, and other rotating machinery using three motors.

21.3.12 Infrared Thermometer

An *infrared (IR) thermometer*, **Figure 21-29**, is a non-contact electronic device that measures infrared radiation emitted from a surface or object to determine its temperature. The infrared thermometer does not measure the temperature of the air, only the matter it is aimed at.

Most IR thermometers have an aiming laser to make it easy to determine the surface being measured. The area of measurement sensitivity is in the shape of a cone. The further the IR thermometer is from the surface being measured, the larger the area that is measured. The closer the IR thermometer is to the surface, the smaller the measured area. Most IR thermometers are marked with the distance vs area.

Reproduced with permission, Fluke Corporation

Figure 21-29. Infrared (IR) thermometers determine temperature by measuring the amount of infrared radiation emitted from an object.

This device proves useful in measuring the temperature of bearings to determine imminent failure. If the bearing on one end of a motor shaft is warmer than that on the other, watch out for a failure in the near future. High temperature of an electrical connection is a good indication that it is potentially loose and may soon fail. Transformer and motor temperatures may be checked to determine if they are within the manufacturer's specifications. Such devices may fail if they overheat.

21.4 CALIBRATION

Calibration is an important part of test and measurement equipment maintenance. Test equipment should be calibrated on a periodic basis to ensure its accuracy. Many companies are certified to various quality standards, such as ISO 9001. These quality system standards require that all test and measurement equipment be calibrated.

Calibration involves verifying that the test or measurement device measures within the manufacturer's specifications. Calibration does not necessarily mean that the test or measurement equipment must be corrected or recalibrated. Rather, the difference between the measurements the test equipment makes and the standard it is compared against is known and documented. Differences outside the manufacturer's specifications indicate that the equipment needs to be recalibrated and should not be used or relied upon.

Some large companies have their own calibration laboratory. Others employ calibration firms. Most calibration can be done on-site, but specialized test equipment may need to be shipped to a calibration lab.

Calibration labs are accredited by various organizations. They have calibration standards that must periodically be calibrated against standards maintained by NIST (National Institute of Standards and Technology). The calibration standards used by a calibration lab, when so calibrated, are considered to be traceable to NIST.

When a piece of test equipment has been calibrated, the company receives a certificate of calibration. Additionally, each device that has been calibrated receives a sticker indicating the date it was calibrated, who calibrated it, and when it is next due for calibration.

Always check the calibration date before using test equipment. Do not use equipment that is past due for calibration. The measurements provided by the equipment could be inaccurate.

CHAPTER WRAP-UP

Whether you are field testing or bench testing, safe practices are critically important when working with electricity. Safety should be a top priority for any electrical technician or responsible employer. Take time to familiarize yourself with the safety standards and practices required for a particular system before you start working on it.

Electrical technicians must be able to use many electrical measurement and testing tools. Learning how to use a variety of testing equipment in a careful and correct manner makes you invaluable for in-house troubleshooting. Keep your tools properly calibrated to ensure you are taking accurate measurements.

Chapter Review

SUMMARY

- Safety first! Always follow all safety procedures and use appropriate PPE before and while performing electrical testing.

- Check the settings on test equipment before connecting to ensure it can handle the voltages and transients present.

- Always de-energize a circuit before testing for resistance.

- A receptacle tester indicates if a receptacle has been properly wired and grounded.

- A digital multimeter (DMM) is one of the most versatile tools in the electrical technician's toolbox. Modes, inputs, safety features, and any advanced functions will vary based on your device.

- Always discharge a capacitor before testing.

- Connect a DMM in series with the source and the load when making current measurements.

- Clamp-on ammeters allow you to measure current without needing to disconnect conductors.

- A megohmmeter uses high voltage to make insulation resistance measurements.

- An oscilloscope shows voltage on the Y axis and time on the X axis.

- Triggering on an oscilloscope provides a way to lock the waveform in a stationary position on the scope screen so that waveforms may be observed and measured.

- The oscilloscope calibrator output allows you to verify proper operation of an oscilloscope and set the probe compensation of multiplier probes.

- An arbitrary function generator (AFG) can be used to produce various types of signals when bench testing an assembly.

- A phase sequence tester verifies that the phase conductors in a three-phase circuit are in the proper order. If the tester shows the phases to be reversed, switch the position of any two conductors. If a motor runs backward, switch the position of any two of the three motor leads.

- IR thermometers measure the infrared radiation being emitted from an object.

- Electrical testing equipment must be calibrated periodically to ensure accurate measurements.

REVIEW QUESTIONS

Answer the following questions using the information provided in this chapter.

1. *True or False?* Electrical equipment should be energized before testing.

2. Undesirable voltage spikes in excess of the intended voltage are known as _____ or _____.

3. What are the two main types of troubleshooting?

4. What does a receptacle tester tell a technician?

5. What is the modern version of the screwdriver voltage tester or receptacle tester and how does it operate?

6. If you want to measure the resistance of a circuit, what test instrument would you likely use? What would your first step be before taking the measurement?

7. List five measurements that a typical DMM can make.

8. What does the diode test mode of a test instrument measure?

9. *True or False?* Capacitors are generally specified in either µF or pF, but a DMM may display the reading in a different unit.

10. Convert 5800 nF to microfarads.

11. What test instrument replaces continuity testers with a continuity test mode? What does the continuity test measure, and how does it let the technician know the circuit has continuity?

12. Most advanced multimeters measure the amount of AC voltage that would produce the same amount of power that a DC voltage would. This is designated _____ on multimeters.

13. What is a phantom voltage?

14. What advantages does a clamp-on ammeter have compared to a DMM?

15. *True or False?* A megohmmeter uses low voltage to make resistance measurements.

16. If you need to measure the inductance of a motor, which test instrument could you use?

17. What does the oscilloscope depict on the X axis, and what does it depict on the Y axis?

18. Briefly describe which two controls on the oscilloscope you would use to set the X and Y values.

19. What is triggering?

20. The X10 position allows the oscilloscope probe to present a high impedance and not load the circuit under test. What would you need to do when interpreting the measurement?

21. If you need to simulate a signal while bench testing an assembly, which test instrument would you most likely use?

22. If you want to measure the voltage of a three-phase transformer, which test instrument would you use and how would you connect it?

23. An infrared (IR) thermometer is a noncontact electronic device that measures _____ emitted from a surface or object to determine its temperature.

24. *True or False?* Test or measurement equipment that is not identical to the calibration standard does not necessarily need to be recalibrated, but rather it must be documented and acknowledged.

NIMS CREDENTIALING PREPARATION QUESTIONS

The following questions will help you prepare for the NIMS Industrial Technology Maintenance Level 1 Electrical Systems credentialing exam.

1. Which of the following ratings identifies a test equipment's tolerance for voltage spikes and transients?

   A. Current rating
   B. CAT level
   C. RMS voltage
   D. PPE rating

2. A receptacle tester can determine:

   A. If a receptacle is grounded properly
   B. The amount of voltage available from a receptacle
   C. The current rating of a receptacle
   D. If a receptacle is weatherproof

3. A Hall effect sensor is likely to be found in:

   A. An oscilloscope
   B. Calibration certificates
   C. Noncontact test equipment
   D. Motor windings

4. The auto-ranging feature on a DMM may eliminate:

   A. The need to use test leads
   B. The need to recharge the DMM battery
   C. Voltage ratings
   D. Selection options on the mode selector switch

5. Convert 4600 pF to nanofarads.

   A. 0.046 nF
   B. 4.6 nF
   C. 4600 nF
   D. 4600000 nF

6. Which of the following is the correct abbreviation for microfarad?

   A. mF
   B. miF
   C. μF
   D. MF

7. A clamp-on ammeter provides a noncontact method for measuring what quantity?

   A. AC current
   B. Resistance
   C. Capacitance
   D. Voltage

8. Which of the following devices is also referred to as an insulation tester?

   A. Clamp-on ammeter
   B. DMM
   C. Megohmmeter
   D. VOM

22 | Alternating Current

LEARNING OBJECTIVES

After completing this chapter, you will be able to:

☐ Define basic principles of alternating current, including phase, cycle, and sine wave.

☐ Explain the relationship between time and frequency.

☐ Describe the relationship between peak, peak-to-peak, and RMS measurements.

☐ Identify reactive components and discuss how alternating current affects them.

☐ Compare the voltage and phase relationship changes caused by capacitance and inductance.

☐ Calculate the impedance of RL, RC, and RLC circuits.

☐ Differentiate between apparent, true, and reactive power.

☐ Describe the roles power, voltage, and current play in an AC circuit.

☐ Discuss how circuit components can be used to filter for certain frequencies or bandwidths.

☐ Discuss the phase relationships in three-phase circuits and their importance.

☐ Calculate the phase and line voltage and current for delta and wye connections.

TECHNICAL TERMS

alternating current (AC)	phase
apparent power	phase angle
band-pass filter	power factor
band-reject filter	quality factor
bandwidth	RC circuit
capacitance	reactance
capacitive reactance	reactive power
cycle	resonance
delta (Δ) connection	RLC circuit
farad	RL circuit
filter	root-mean-square (RMS)
frequency	series resonant circuit
henry	sine wave
hertz	three-phase alternating current
impedance	
inductance	true power
inductive reactance	tuned circuit
instantaneous voltage	wye (Y) connection
period	

Alternating current (AC) changes direction at a periodic rate, as opposed to direct current (DC), which always flows in the same direction. AC includes reactive components—reactance and impedance—that are similar to resistance. Resistance applies in both AC and DC circuits, but reactance and impedance are important considerations when dealing with AC circuits only. Building on what you have already learned, this chapter introduces additional principles related to AC.

22.1 PRINCIPLES OF ALTERNATING CURRENT

In *alternating current (AC)*, the current changes direction at regular time intervals, or periodically. AC is visually represented by a *sine wave*, **Figure 22-1**. Note that the ups and downs of the sine wave correspond to the current alternating directions. The periodic rate at which the current alternates is referred to as *frequency*. Frequency is measured in cycles per second or *hertz* in SI units, where 1 cycle per second equals 1 Hz. A full *cycle* of AC includes one complete rise and one complete fall of the wave, **Figure 22-2**. A cycle can start at any point and end at the next equivalent point along the waveform.

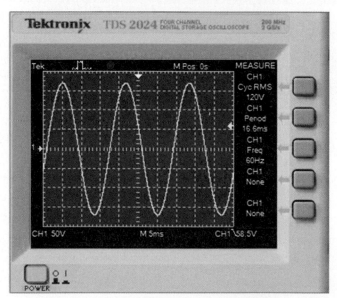

Goodheart-Willcox Publisher

Figure 22-1. Oscilloscope showing a single-phase alternating current waveform. The automatic measurement capabilities of the oscilloscope were set up to measure the frequency and the period of the waveform. The scope is set for 50 V/division vertical and 5 ms/division horizontal.

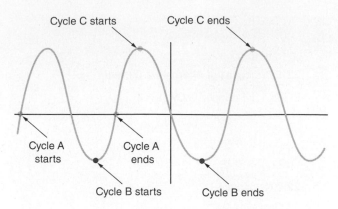

Goodheart-Willcox Publisher

Figure 22-2. Cycles can begin at any point on the waveform. A complete cycle is measured from the selected start point to the next equivalent point on the sine wave, as shown here for cycles A, B, and C.

22.1.1 Period of a Cycle

The *period* of a cycle is the amount of time it takes to complete the cycle. If the frequency is 1 Hz, the period is 1 second. The period can be found with the following equation:

$$T = \frac{1}{f}$$

(22-1)

where

T = period in seconds (s)
f = frequency in hertz (Hz)

For example, if the frequency is 60 Hz, the time it takes to complete one cycle is 0.016667 seconds. If the frequency is 50 Hz, the time required to complete one cycle is 0.02 seconds.

22.1.2 Phase

Figure 22-3 depicts one cycle of a sine wave. Notice the positive peak rises to +1 V, and the negative peak falls to –1 V. This results in the waveform measuring 2 V from peak to peak.

A sine wave may be divided into 360° with any specific point on the sine wave specified by the *phase*. Two waves are said to be "in phase" if they reach the same points at the same times. For example, both waves achieve their positive peak values at 90°, then drop to zero at 180°, and fall to their negative peaks at 270°.

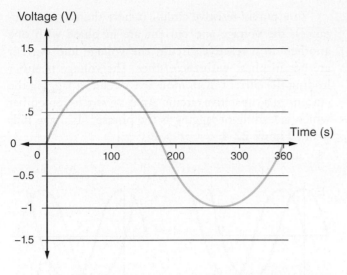

Voltage (V)

Time (s)

Goodheart-Willcox Publisher

Figure 22-3. Sine wave. Notice the X axis of the graph is divided into 360 units over the period of one cycle. Each unit along the X axis denotes one degree of phase angle.

22.1.3 Instantaneous Voltage

The voltage at any single time during the cycle is called the *instantaneous voltage*. In order to specify the exact point in time you wish to find the instantaneous voltage, reference the phase. With a sine wave, the instantaneous voltage at 90° is the positive peak. The instantaneous voltage at 270° is the negative peak. The instantaneous voltage at 0°, 180°, and 360° is 0 V.

To find the instantaneous voltage at any other phase, first determine the peak voltage of the waveform and the phase where you want to know the voltage. When measuring the peak voltage, it is often difficult to get exactly the correct reading due to symmetry and the position of the waveform on the oscilloscope screen. It is often easier to find the peak-to-peak value of the waveform. In order to find the peak voltage, divide the peak-to-peak value by 2.

$$E_{peak} = \frac{E_{peak\text{-}to\text{-}peak}}{2}$$

(22-2)

With the peak voltage and the phase, you can calculate the instantaneous voltage with the following equation:

$$E_{instantaneous} = \sin(\theta) \times E_{peak}$$

(22-3)

where

$E_{instantaneous}$ = instantaneous voltage (V)

$\sin(\theta)$ = sine of the phase angle

E_{peak} = peak voltage (V)

Consider the following example. To find the instantaneous voltage of a 150 V peak-to-peak sine wave at 120°, apply equation 22-2:

$$E_{peak} = \frac{E_{peak\text{-}to\text{-}peak}}{2}$$

$$E_{peak} = \frac{150 \text{ V}}{2}$$

$$E_{peak} = 75 \text{ V}$$

Then apply equation 22-3:

$$E_{instantaneous} = \sin(\theta) \times E_{peak}$$

$$E_{instantaneous} = \sin(120°) \times 75 \text{ V}$$

$$E_{instantaneous} = 65 \text{ V}$$

Practice calculating the instantaneous voltage at different phases. Make sure you know how to use your calculator to determine the sine of the various angles.

22.1.4 RMS

Without going into complex mathematics, *root-mean-square (RMS)* is the square root of the mean over time of the square of the peak value of a waveform. The RMS value is also called the *effective value*. For AC power, common practice is to specify RMS values of a sine wave. Because current alternates, the voltage provided varies from one instant to the next. The RMS value is a useful equivalence between AC and DC. RMS produces the same amount of power in a purely resistive element as a direct current of the same voltage would.

To calculate the RMS value of a sine wave given the peak value, use the following formula:

$$E_{RMS} = E_{peak} \times \sin(45°)$$

(22-4)

The sine of 45° is $1/\sqrt{2}$, approximately 0.707 when rounded. Substitute these values to obtain alternate versions of equation 22-4:

$$E_{RMS} = \frac{E_{peak}}{\sqrt{2}}$$

(22-5)

$$E_{RMS} = 0.707 \times E_{peak}$$

(22-6)

The previous three equations are equivalent. Understand how they relate, and know how to input the values for each into your calculator.

Depending on your available measurements, you may also need to convert RMS values into peak values. Algebraic manipulation of equation 22-4 yields the following:

$$E_{peak} = \frac{E_{RMS}}{\sin(45°)}$$

(22-7)

The peak value is equal to the RMS value divided by the sine of 45°. Observe that the inverse of $\sin(45°)$ equals $\sqrt{2}$, which is approximately 1.414, and substitute values as before to yield two more equations for peak value:

$$E_{peak} = E_{RMS} \times \sqrt{2}$$

(22-8)

$$E_{peak} = 1.414 \times E_{RMS}$$

(22-9)

Check your understanding by working through the following examples. If an oscilloscope measures a peak voltage of 170 V, what would the RMS voltage measured on a DMM show? Apply equation 22-6, which is relatively easy to input into a calculator.

$E_{RMS} = 0.707 \times E_{peak}$
$E_{RMS} = 0.707 \times 170$ V
$E_{RMS} = 120$ V

Conversely, if a DMM measures a voltage of 208 V, what would an oscilloscope show as peak voltage? Recall that almost all DMMs measure RMS voltage, and oscilloscopes show peak voltage. Convert from RMS to peak voltage using equation 22-9.

$E_{peak} = 1.414 \times E_{RMS}$
$E_{peak} = 1.414 \times 208$ V
$E_{peak} = 294$ V

22.2 REACTANCE

You already know resistance is the opposition to the flow of an electrical current. **Reactance** is the opposition to the change in an electrical current. Resistance affects both AC and DC similarly. However, reactance also affects AC circuits in which the current is changing constantly. Reactance is measured in ohms and designated by the variable X.

Reactance depends on the capacitance and the inductance of a circuit as well as the frequency. The reactance of an inductor is proportional to frequency while the reactance of a capacitor is inversely proportional to frequency.

In a purely resistive circuit (where there is no reactance), the voltage and current are in phase with one another. In a reactive circuit, the voltage and current are not in phase with each other. The voltage is either leading the current in an inductive circuit or lagging the current in a capacitive circuit. An easy way to remember which is leading or lagging is the phrase, "ELI the ICE man," **Figure 22-4**.

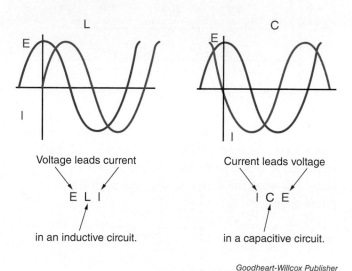

Voltage leads current
E L I
in an inductive circuit.

Current leads voltage
I C E
in a capacitive circuit.

Goodheart-Willcox Publisher

Figure 22-4. In this mnemonic device, E represents voltage, I represents current, L represents inductance, and C represents capacitance. Refer to the order of the letters in the phrase. In an inductive circuit (L in ELI), voltage leads current. E comes before I in ELI. In a capacitive circuit (C in ICE), the voltage lags the current.

22.2.1 Inductive Reactance

Inductance opposes any change in current and is designated by the variable L. The SI unit of inductance is the **henry**, H. **Inductive reactance**, X_L, describes the reactance of an inductor and is measured in ohms, Ω. Calculate inductive reactance with the following equation:

$$X_L = 2\pi \times f \times L$$

(22-10)

where

X_L = inductive reactance in ohms (Ω)
f = frequency in hertz (Hz)
L = inductance in henries (H)

Consider a circuit with a 75 mH inductor and a frequency of 60 Hz. What is the inductive reactance? Remember to convert mH to base units before applying equation 22-10.

$L = 75$ mH

$L = 0.075$ H

$X_L = 2\pi \times f \times L$

$X_L = 2\pi \times 60$ Hz $\times 0.075$ H

$X_L = 28.3\ \Omega$

Observe that as frequency increases, inductive reactance also increases.

22.2.2 Capacitive Reactance

Capacitance opposes any change in the voltage of a circuit and is designated by the variable C. The SI unit for capacitance is the *farad*, F, but capacitors are often valued in microfarads (μF) or picofarads (pF). *Capacitive reactance* measures the reactance of a capacitor. In capacitive reactance calculations, frequency must be in hertz (Hz), and capacitance must be in farads (F). Capacitive reactance is designated by the variable X_C and measured in ohms (Ω) like inductive reactance. Calculate capacitive reactance with the following equation:

$$X_C = \frac{1}{2\pi \times f \times C}$$

(22-11)

where

X_C = capacitive reactance in ohms (Ω)

f = frequency in hertz (Hz)

C = capacitance in farads (F)

Note that capacitive reactance is nearly the inverse of inductive reactance. Take care not to confuse these formulas.

Calculate the capacitive reactance for a circuit operating at 60 Hz with a capacitance of 250 μF. First, convert 250 μF to F using SI unit conversions.

$C = 250 \times 10^{-6}$ F

$C = 0.000250$ F

Now apply equation 22-11:

$X_C = \dfrac{1}{2\pi \times f \times C}$

$X_C = \dfrac{1}{2\pi \times 60\ \text{Hz} \times 0.000250\ \text{F}}$

$X_C = 10.6\ \Omega$

Now think about how reactance changes as capacitance changes. Test your intuition by calculating capacitive reactance for a circuit operating at 60 Hz with a capacitance of 25 μF.

$X_C = \dfrac{1}{2\pi \times f \times C}$

$X_C = \dfrac{1}{2\pi \times 60\ \text{Hz} \times 0.0000250\ \text{F}}$

$X_C = 106\ \Omega$

As capacitance decreases and frequency remains the same, capacitive reactance increases. Similar comparisons show that as frequency increases, capacitive reactance decreases.

22.3 RL CIRCUITS

An *RL circuit* has both resistance (R) and inductance (L). The simplest RL circuit contains a voltage source connected to an inductor and a resistor, **Figure 22-5**. The resistance in both components modifies how the circuit reacts to alternating current.

Most inductors are coils of conducting wire, and all conductors have some amount of resistance. The more turns of wire in an inductor coil, the higher its resistance. The greater the resistance, the more effect it has on the inductive circuit. Other factors affecting the resistance of a conducting wire are covered in later chapters.

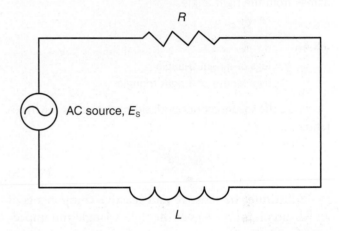

Goodheart-Willcox Publisher

Figure 22-5. The basic RL circuit consists of a voltage source, E_S, connected to an inductor, L, and a resistor, R.

22.3.1 Impedance in RL Circuits

Impedance, Z, is the total resistance to the flow of alternating current. Impedance is derived from the resistance and reactance of an AC circuit and measured in ohms. Calculating impedance is more complicated than simply adding the values for resistance and reactance. These two terms are out of phase with each other by 90°, so vector addition is required.

Vector addition is easiest to visualize using right triangles, **Figure 22-6**, and can be described mathematically with the Pythagorean theorem:

$$c^2 = a^2 + b^2$$

(22-12)

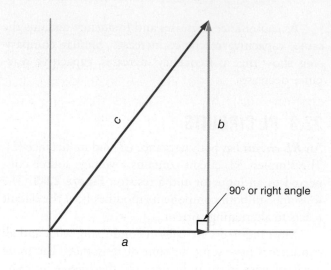

Goodheart-Willcox Publisher

Figure 22-6. A right triangle gets its name from the 90° or right angle it contains. Right triangles have two legs, *a* and *b*, adjacent to the right angle and a hypotenuse, *c*, across from the right angle.

where

a, b = legs of a right triangle

c = hypotenuse of a right triangle

Take the square root of both sides to solve for the hypotenuse, *c*.

$$c = \sqrt{a^2 + b^2}$$

(22-13)

Substitute the resistive and reactive components of an RL circuit for the legs of the right triangle and impedance for the hypotenuse to obtain the following equation for impedance:

$$Z = \sqrt{R^2 + X_L^2}$$

(22-14)

Practice finding impedance with the following example. Consider an RL circuit with 300 Ω resistance and inductive reactance of 400 Ω, **Figure 22-7.** Apply equation 22-14 to calculate impedance.

$Z = \sqrt{R^2 + X_L^2}$

$Z = \sqrt{(300 \ \Omega)^2 + (400 \ \Omega)^2}$

$Z = \sqrt{90000 \ \Omega^2 + 160000 \ \Omega^2}$

$Z = 500 \ \Omega$

22.3.2 Ohm's Law for AC Circuits

In an AC circuit, impedance, *Z*, combines the resistive and reactive components. Ohm's law for AC circuits takes this feature into account by substituting *Z* for *R*.

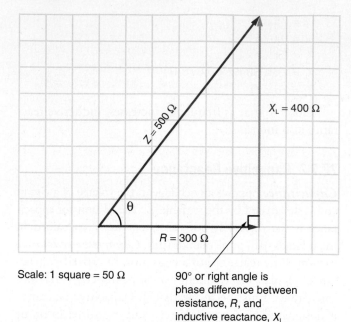

Scale: 1 square = 50 Ω

90° or right angle is phase difference between resistance, *R*, and inductive reactance, X_L

Goodheart-Willcox Publisher

Figure 22-7. The impedance triangle. Impedance may be calculated in an RL circuit using the Pythagorean theorem to find the hypotenuse of a right triangle. The hypotenuse represents the impedance value. The resistive and reactive components are represented by the legs of the triangle and are 90° out of phase with each other.

Ohm's law for AC circuits also assumes RMS or effective values for current and voltage.

$$E_{RMS} = I_{RMS} \times Z$$

(22-15)

$$I_{RMS} = \frac{E_{RMS}}{Z}$$

(22-16)

$$Z = \frac{E_{RMS}}{I_{RMS}}$$

(22-17)

where

E_{RMS} = effective voltage (V)

I_{RMS} = effective current (A)

Z = impedance (Ω)

Observe that Ohm's law is nearly identical for AC circuits as DC circuits, except that the reactive components are also considered. Accordingly, you must calculate impedance, *Z*, before applying Ohm's law to find effective voltage, E_{RMS}, or effective current, I_{RMS}.

22.3.3 Series RL Circuits

In a series DC circuit, the voltages across each component sum to the source voltage. This rule does not apply in a reactive circuit containing an inductor. Consider the circuit in **Figure 22-8**. This circuit has a 55 mH inductor and 125 Ω resistor connected in series with an AC source producing effective voltage of 120 V at a frequency of 60 Hz. To find the voltage drops across each component, calculate inductive reactance and impedance, then apply Ohm's law.

Inductive reactance is calculated using equation 22-10:

$X_L = 2\pi \times f \times L$
$X_L = 2\pi \times 60 \text{ Hz} \times 0.055 \text{ H}$
$X_L = 20.7 \text{ }\Omega$

Use the values for inductive reactance, 20.7 Ω, and resistance, 125 Ω, to calculate impedance using equation 22-14:

$Z = \sqrt{R^2 + X_L^2}$
$Z = \sqrt{(20.7 \text{ }\Omega)^2 + (125 \text{ }\Omega)^2}$
$Z = \sqrt{428.5 \text{ }\Omega^2 + 15625 \text{ }\Omega^2}$
$Z = 126.7 \text{ }\Omega$

Next, use Ohm's law for AC circuits to find the effective current, I_{RMS}.

$I_{RMS} = \dfrac{E_{RMS}}{Z}$

$I_{RMS} = \dfrac{120 \text{ V}}{126.7 \text{ }\Omega}$

$I_{RMS} = 0.947 \text{ A}$

As in series DC circuits, total current is the same through each component in series. Apply Ohm's law to each individual component to find the voltage drop across each one.

$E_R = I_{RMS} \times R$
$E_R = 0.947 \text{ A} \times 125 \text{ }\Omega$
$E_R = 118.4 \text{ V}$
$E_L = I_{RMS} \times X_L$
$E_L = 0.947 \text{ A} \times 20.7 \text{ }\Omega$
$E_L = 19.6 \text{ V}$

Notice that if you add the voltage drop across R and the voltage drop across L, the sum, 137.5 V, is greater than the source voltage, 120 V. The resistive and reactive voltages are out of phase with each other, so their combined voltage cannot be found with simple addition. The source voltage is a vector sum of the resistive and reactive voltage drops. The source voltage may be solved with the following equation:

$$E_S = \sqrt{E_R^2 + E_L^2}$$

(22-18)

where

E_S = source voltage (V)
E_R = resistive voltage drop (V)
E_L = reactive voltage drop (V)

As a check, solve for the source voltage in the circuit given in **Figure 22-8**.

$E_S = \sqrt{E_R^2 + E_L^2}$
$E_S = \sqrt{118.4^2 + 19.6^2}$
$E_S = 120 \text{ V}$

If the inductance increases, the voltage drop across the inductor increases. If the resistance increases, the voltage drop across the resistor increases. With source voltage constant, as the voltage drop across the inductor increases, the voltage drop across the resistor decreases.

22.3.4 Parallel RL Circuits

In a parallel RL circuit, impedance cannot be found through the vector method as previously described. In the resistive leg the current is in phase with the voltage, but in the inductive leg the current lags the voltage. Consider the circuit in **Figure 22-9**.

Start by finding inductive reactance, X_L, using equation 22-10:

$X_L = 2\pi \times f \times L$
$X_L = 2\pi \times 100 \text{ Hz} \times 5 \text{ H}$
$X_L = 3142 \text{ }\Omega$

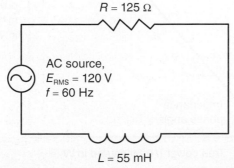

R = 125 Ω

AC source,
E_{RMS} = 120 V
f = 60 Hz

L = 55 mH

Figure 22-8. A series RL circuit.

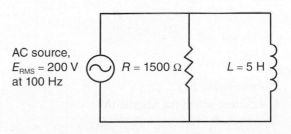

AC source,
E_{RMS} = 200 V
at 100 Hz

R = 1500 Ω

L = 5 H

Figure 22-9. A parallel RL circuit.

Now use Ohm's law to find the current in each leg of the parallel circuit.

$$I_R = \frac{E_{RMS}}{R}$$

$$I_R = \frac{200 \text{ V}}{1500 \text{ }\Omega}$$

$$I_R = 0.133 \text{ A}$$

$$I_L = \frac{E_{RMS}}{X_L}$$

$$I_L = \frac{200 \text{ V}}{3142 \text{ }\Omega}$$

$$I_L = 0.064 \text{ A}$$

To find the total current in a parallel RL circuit, first find the current in each leg of the circuit, then apply the following equation to take the phase difference into account:

$$I_{RMS} = \sqrt{I_R{}^2 + I_L{}^2}$$

(22-19)

where

I_R = resistive current (A)

I_L = inductive current (A)

$$I_{RMS} = \sqrt{I_R{}^2 + I_L{}^2}$$

$$I_{RMS} = \sqrt{(0.133 \text{ A})^2 + (0.064 \text{ A})^2}$$

$$I_{RMS} = 0.147 \text{ A}$$

Last, calculate the impedance using Ohm's law for AC circuits.

$$Z = \frac{E_{RMS}}{I_{RMS}}$$

$$Z = \frac{200 \text{ V}}{0.147 \text{ A}}$$

$$Z = 1356 \text{ }\Omega$$

22.3.5 Power in AC Circuits

Due to the reactive components, there are three different types of power in an AC circuit. **True power**, P, is measured in watts (W) and is equal to the product of voltage and current in a pure resistive circuit. True power is sometimes called *real power*.

$$P = I_R{}^2 \times R$$

(22-20)

where

P = true power (W)

I_R = current across the resistor (A)

R = resistance (Ω)

In a pure inductive circuit, voltage and current are out of phase with each other in such a way as to produce no true power. Power used by reactive components in an inductive circuit is **reactive power**, *VAR*, and is measured in volt-amperes-reactive (VAR). The formula for *VAR* is similar to the one for true power, except only reactive components are considered:

$$VAR = I_L{}^2 \times X_L$$

(22-21)

where

VAR = reactive power (VAR)

I_L = current through the inductor (A)

X_L = inductive reactance (Ω)

The third type of power is **apparent power**, S, which is the power consumed by the circuit. It is measured in volt-amperes (VA).

$$S = I_{RMS} \times E_{RMS}$$

(22-22)

where

S = apparent power (VA)

I_{RMS} = effective current through the circuit (A)

E_{RMS} = effective voltage through the circuit (V)

The three different types of power are depicted in the power triangle, **Figure 22-10**. When dealing with an AC circuit, apparent power is most relevant to circuit operation. If you were to calculate the total power consumed by an AC circuit by measuring the voltage and using the power equation, you would be calculating the apparent power, and your result would be in VA

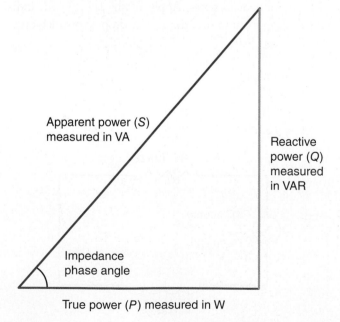

Goodheart-Willcox Publisher

Figure 22-10. The power triangle.

rather than W. Apparent power is the power the circuit consumes. True power is the power that does the work. Reactive power consumes power but does no work.

In addition to the equations for power given above, you can also use triangle solutions similar to those used to calculate the impedance to calculate the power.

$$S = \sqrt{P^2 + VAR^2}$$

(22-23)

Which equation or combination of equations you use to solve for power depends on the information available.

22.3.6 Power Factor and Phase Angle

Phase angle measures the difference in phase between two waves and is represented by the lowercase Greek letter theta, θ. The phase angle, θ, between the resistance and the impedance is the phase difference between the voltage and the current. Phase angle is equal to the angle between the hypotenuse (impedance) and the adjacent side (resistance) of the triangle representing the circuit, **Figure 22-11**.

The *power factor*, *PF*, is the ratio of true power to apparent power in an AC circuit and is equal to the cosine of the phase angle, $\cos(\theta)$. Power factor indicates the efficiency of the circuit. *PF* ranges from 0, where the circuit is purely reactive and least efficient, to 1, where the circuit is purely resistive and most efficient. Because *PF* is a ratio, it has no units. Recall from trigonometry that the cosine of an angle is equal to the adjacent side of a triangle divided by the hypotenuse. In the case of the reactive circuit depicted by **Figure 22-11**, $\cos(\theta)$, and therefore *PF*, is resistance divided by impedance.

$$PF = \frac{\text{true power}}{\text{apparent power}}$$

(22-24)

$$PF = \cos(\theta)$$

(22-25)

$$PF = \frac{R}{Z}$$

(22-26)

Find the power factor for a series RL circuit where the impedance is 50 Ω and the resistance is 30 Ω.

$$PF = \frac{R}{Z}$$

$$PF = \frac{30\ \Omega}{50\ \Omega}$$

$$PF = 0.6$$

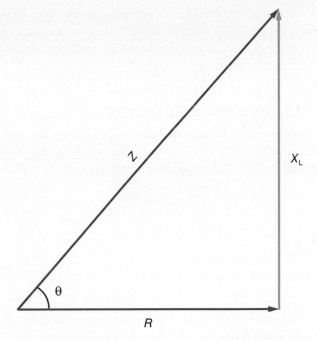

Figure 22-11. Triangle representation of a reactive circuit containing resistance and inductive reactance.

The power factor can be multiplied by 100 to yield a percentage. The power factor in the previous example indicates a circuit with 60% efficiency.

Once you know *PF*, you can take the inverse cosine to find the phase angle. Rearrange equation 22-24 to solve for θ.

$$\theta = \cos^{-1}(PF)$$
$$\theta = \cos^{-1}(0.6)$$
$$\theta = 53.1°$$

Remember "ELI the ICE man" from earlier in the chapter? This calculation tells you that the voltage leads the current by 53.1° in the inductive circuit in the previous example.

The out-of-phase relationship presents problems because of the higher consumption of power. This happens when inductive loads such as transformers and motors are present. The power provider charges a higher rate due to the higher demand. You will see later on how capacitance may be added to correct the power factor and lower the power consumption.

22.4 RC CIRCUITS

A simple *RC circuit* contains a resistor (R) and a capacitor (C) connected to a voltage source. Capacitance opposes any change in voltage. There are many different types of capacitors to choose from depending on their intended use. Capacitors are covered in greater depth in the next chapter.

Calculations for RC circuits are generally very similar to those for RL circuits. Simply substitute capacitive reactance for inductive reactance in formulas dealing with reactive elements.

22.4.1 Impedance in RC Circuits

The impedance calculations for an RC circuit are quite similar to those in an RL circuit. The following equation is used to calculate impedance in an RC circuit:

$$Z = \sqrt{R^2 + X_C^2}$$

(22-27)

where

R = resistance in Ω
X_C = capacitive reactance in Ω
Z = impedance in Ω

Consider the capacitive circuit represented by the triangle in **Figure 22-12**. Find the impedance.

$$Z = \sqrt{R^2 + X_C^2}$$
$$Z = \sqrt{(300\ \Omega)^2 + (400\ \Omega)^2}$$
$$Z = 500\ \Omega$$

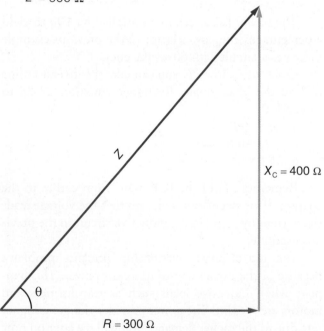

Figure 22-12. Triangle representation of a reactive circuit containing resistance and capacitive reactance.

22.4.2 Parallel RC Circuits

Ohm's law for AC circuits applies to RC circuits just as it does to RL circuits. As with parallel RL circuits, parallel RC circuits require you to account for phase differences in the resistive and reactive legs when calculating impedance.

Current and voltage are in phase in the resistive leg, but the current leads the voltage in the capacitive leg. Find the current through each leg and the impedance for the RC circuit in **Figure 22-13**.

Start by finding capacitive reactance, X_C, using equation 22-11:

$$X_C = \frac{1}{2\pi \times f \times C}$$
$$X_C = \frac{1}{2\pi \times 60\ \text{Hz} \times 0.000100\ \text{F}}$$
$$X_C = 26.5\ \Omega$$

Now use Ohm's law to find the current in each leg of the parallel circuit.

$$I_R = \frac{E_{RMS}}{R}$$
$$I_R = \frac{240\ \text{V}}{50\ \Omega}$$
$$I_R = 4.8\ \text{A}$$
$$I_C = \frac{E_{RMS}}{X_C}$$
$$I_C = \frac{240\ \text{V}}{26.5\ \Omega}$$
$$I_C = 9.1\ \text{A}$$

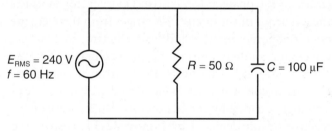

Figure 22-13. A parallel RC circuit.

To find the total current in a parallel RC circuit, first find the current in each leg of the circuit, then apply the following equation to take the phase difference into account:

$$I_{RMS} = \sqrt{I_R^2 + I_C^2}$$

(22-28)

where

I_R = resistive current (A)
I_C = capacitive current (A)

The total current, I_{RMS}, is a vector sum of the resistive and reactive currents using the Pythagorean theorem, **Figure 22-14**.

$$I_{RMS} = \sqrt{I_R^2 + I_C^2}$$
$$I_{RMS} = \sqrt{(4.8\ \text{A})^2 + (9.1\ \text{A})^2}$$
$$I_{RMS} = 10.3\ \text{A}$$

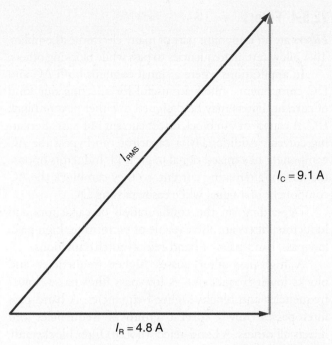

I_{RMS}

$I_C = 9.1$ A

$I_R = 4.8$ A

Figure 22-14. Triangle representation of the relationship between the currents through each leg of a parallel RC circuit and the total current.

At last, calculate the impedance using Ohm's law for AC circuits.

$$Z = \frac{E_{RMS}}{I_{RMS}}$$

$$Z = \frac{240 \text{ V}}{10.3 \text{ A}}$$

$$Z = 23.3 \ \Omega$$

Observe that this process is only slightly modified from the procedure used for RL circuits presented earlier.

22.4.3 Power in RC Circuits

Continue working with the RC circuit presented in **Figure 22-13**. Use the equations you already know to find power factor (*PF*) and phase angle (θ). Take care when sketching the trigonometric relationships between components in series and parallel circuits. Because of the phase relationships between circuits in parallel, it is a good idea to use equation 22-24 for *PF*:

$$PF = \frac{\text{true power}}{\text{apparent power}}$$

In this case, true power and apparent power are calculated using equations 22-20 and 22-22 as follows:

true power $= I_R^2 \times R$

true power $= 1152$ W

apparent power $= E_{RMS} \times I_{RMS}$

apparent power $= 2472$ VA

$$PF = \frac{1152 \text{ W}}{2472 \text{ VA}}$$

$$PF = 0.47$$

Once you know *PF*, you can take the inverse cosine to find the phase angle. Rearrange equation 22-24 to solve for θ.

θ = cos⁻¹(*PF*)

θ = cos⁻¹(0.47)

θ = 62°

This calculation tells you that the voltage lags the current by 62° in the capacitive circuit.

22.5 RLC CIRCUITS

Simple **RLC circuits** contain a resistor (R), an inductor (L), and a capacitor (C). One of the most important facts to understand about RLC circuits is that the reactive elements from the inductance and capacitance oppose each other. X_L counteracts X_C because X_L increases as the frequency increases, and X_C decreases as the frequency increases. The net reactance in an RLC circuit is either capacitive or inductive, depending on whether inductive or capacitive elements dominate.

22.5.1 Impedance in RLC Circuits

Since the inductive and capacitive reactances counteract each other, the impedance of an RLC circuit is calculated as follows:

$$Z = \sqrt{R^2 + (X_L - X_C)^2}$$

(22-29)

This equation is similar to the equations for impedance in RL and RC circuits, but it incorporates both X_L and X_C. In order to solve this equation, first calculate X_L and X_C. This equation is a simple one for basic RLC circuits. More complex RLC circuits may require more involved solutions. Remember that you can always apply Ohm's law to calculate the impedance of an RLC circuit, regardless of its complexity.

22.5.2 Resonance

If X_L and X_C are equal, the circuit is said to be at **resonance**. When a circuit is at resonance, there is no net reactance. A series or parallel RLC circuit at resonance is referred to as a **tuned circuit**. The resonant frequency of an RLC circuit may be calculated with the following equation:

$$f_0 = \frac{1}{2\pi \times \sqrt{L \times C}}$$

(22-30)

where

f_0 = resonant frequency in Hz

L = inductance in H

C = capacitance in F

Consider the following example: A series resonant circuit contains a 0.01 µF capacitor and a 15 mH inductor. Determine the resonant frequency using equation 22-29, and remember to convert to base units first.

$$f_0 = \frac{1}{2\pi \times \sqrt{L \times C}}$$

$$f_0 = \frac{1}{2\pi \times \sqrt{0.015 \text{ H} \times 0.00000001 \text{ F}}}$$

$$f_0 = 13,000 \text{ Hz}$$

$$f_0 = 13 \text{ kHz}$$

22.5.3 Bandwidth

The **bandwidth** of a circuit is the range of frequencies near resonant frequency, f_0, that produce resonance effects. Resistance of the RLC circuit causes the bandwidth or range of frequencies to be wider.

The **quality factor**, QF, is a unitless ratio of inductive reactance to resistance of the circuit. The equation for quality factor is as follows:

$$QF = \frac{X_L}{R}$$

(22-31)

where

QF = quality factor

X_L = inductive reactance in Ω

R = resistance in Ω

Because inductive and capacitive reactances are equal at resonance, this equation would be the same with X_C. However, resistance is most often associated with inductance, as the inductor coil has an inherent resistance from the conductor that makes up the coil itself.

The bandwidth of the resonant circuit depends on the QF of the circuit and can be calculated as follows:

$$BW = \frac{f_0}{QF}$$

(22-32)

where

BW = bandwidth in Hz

f_0 = resonant frequency in Hz

QF = quality factor

22.5.4 Filters

Filters are an important part of many electronic assemblies. They allow certain frequencies to pass while blocking others.

In applications where a signal contains both AC and DC components, filters are useful for selecting one kind of current. Filters may be designed to either pass or block DC. A capacitor can block direct current but pass alternating current. Additionally, a capacitor can bypass the AC component of a mixed signal to ground. Inductors oppose the flow of alternating current, so they can block the AC component of a signal while easily passing DC.

Depending on the configuration of capacitors and inductors, filters are also capable of performing high-pass, low-pass, band-pass, or band-reject (notch) functions.

A high-pass filter passes higher frequencies and blocks lower frequencies. A low-pass filter passes lower frequencies and blocks higher frequencies. A band-pass filter passes only a specific group of frequencies and rejects all others. A band-reject (notch) filter blocks only a specific frequency range.

Band-Pass Filter

A **series resonant circuit** is one where the resistance, inductance, and capacitance are in series, **Figure 22-15**. This type of circuit may be used as a **band-pass filter** or *acceptor circuit*. At resonance, the inductive and capacitive reactances are equal, resulting in zero net reactance and low impedance. This allows most of the current to flow through the resistance and produce output signal.

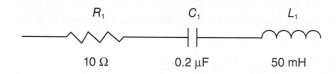

R_1 C_1 L_1

10 Ω   0.2 µF   50 mH

Goodheart-Willcox Publisher

Figure 22-15. A series resonant RLC circuit. This type of circuit contains a resistor, capacitor, and inductor in series and may be used as a band-pass filter.

As the frequency increases or decreases from the resonant frequency, the reactance of the circuit increases. As reactance increases, the amount of output decreases. By this mechanism, a band-pass filter passes frequencies at or near its resonant frequency and rejects those frequencies that are further from resonance.

Band-Reject Filter

A *band-reject filter* or *reject circuit* is a parallel resonant circuit, **Figure 22-16**. At resonance, this circuit rejects frequencies at or near the resonant frequency and passes frequencies further away from resonance.

The output of the circuit is across L_1, and, when the circuit is at resonance, the signal is shunted from the output, which causes the parallel resonant circuit to reject any signal at the resonant frequency.

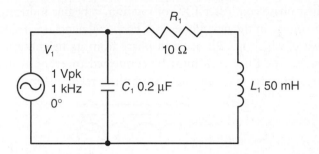

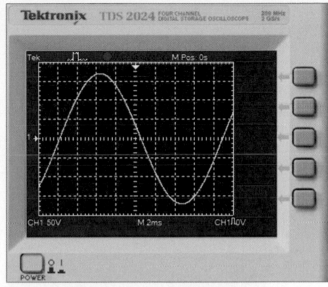

Goodheart-Willcox Publisher

Figure 22-17. A single-phase AC waveform.

Goodheart-Willcox Publisher

Figure 22-16. A parallel resonant RLC circuit that may be used as a band-reject filter.

22.6 THREE-PHASE ALTERNATING CURRENT

In this section so far, all discussion of alternating current has been about single-phase AC. A more efficient type of alternating current is *three-phase alternating current*. Three-phase alternating current combines three AC sources that are strategically out of phase with each other. There are three major reasons the efficiency of three-phase AC is superior to single-phase AC.

The power delivered by a single-phase system pulsates because it crosses zero three times during a cycle. Refer to **Figure 22-17**. Notice the single-phase power is at zero at the beginning of a cycle, once again in the middle, and also at the end.

Three-phase power also pulsates. However, it never falls to zero. Refer to **Figure 22-18**. Three distinct phases allow current to be flowing in at least one of the phases at any given time. This system delivers the same power to a load at any instant.

The kVA ratings, which measure the amount of apparent power that can safely flow, for three-phase motors and transformers are approximately 150% greater than that of similarly sized single-phase units. When a three-phase system is balanced, conductors may be sized at about only 75% of those required for a single-phase system.

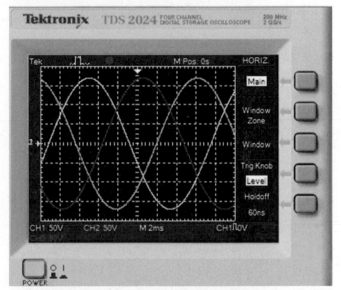

Goodheart-Willcox Publisher

Figure 22-18. A three-phase AC waveform.

22.6.1 Phasing

In a three-phase system, the phase conductors that supply power may be labeled A, B, and C. Each phase starts 120° after the preceding phase. This phase offset corresponds to dividing a circle (360°) into three equal parts.

It can be a little difficult to visualize the phase relationships when looking at all three phases at once. Breaking it down to a pair of phases makes the concept a bit easier to understand. **Figure 22-19** depicts A and B. Notice that B begins 120° after A starts. B and C are also 120° apart with C starting after B. Refer to **Figure 22-20** to see the B and C relationship. Phase A starts its second cycle 120° after C has started.

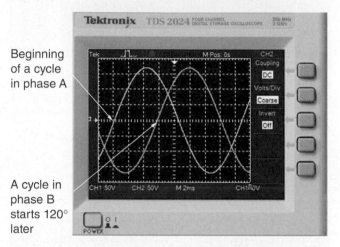

Beginning of a cycle in phase A

A cycle in phase B starts 120° later

Goodheart-Willcox Publisher

Figure 22-19. This waveform depicts the phase relationship between phases A and B of a three-phase circuit. The trace for A is shown in yellow, and the trace for B is blue.

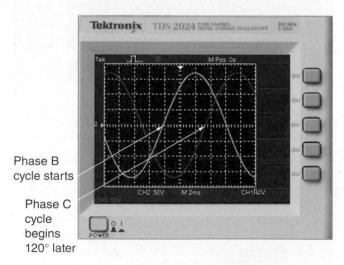

Phase B cycle starts

Phase C cycle begins 120° later

Goodheart-Willcox Publisher

Figure 22-20. This illustration depicts the phase relationship between phases B and C of a three-phase circuit. The waveform for B is depicted by the blue trace, and C is depicted as violet.

22.6.2 Generating Three-Phase Alternating Current

Single-phase alternating current can be generated when the magnetic field of a rotating magnet induces a voltage in a stationary coil, **Figure 22-21**. As one pole of the magnet rotates past the coil, the positive half of a sine wave is formed. As the opposite pole of the magnet rotates past the same coil, the negative half of the sine wave is formed.

In a three-phase generator, three coils are arranged in a triangular formation, **Figure 22-22**, 120° apart. When the magnet is rotated, it begins inducing a voltage in the first phase coil. After 120° of rotation, it begins inducing a voltage in the next phase coil. Each coil generates a sine wave, which is 120° apart in phase from its neighboring coils. The three coils must be connected together in the proper order to obtain the correct phase relationship.

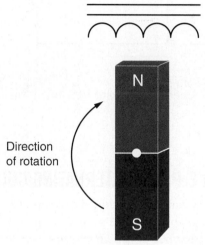

Direction of rotation

Goodheart-Willcox Publisher

Figure 22-21. Basic concept of a single-phase generator.

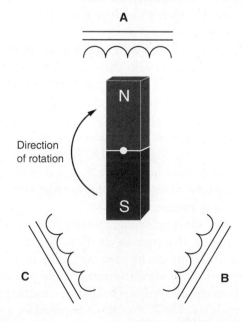

Direction of rotation

Goodheart-Willcox Publisher

Figure 22-22. Basic concept of a three-phase generator.

Chapter 22 Alternating Current **435**

22.6.3 Three-Phase Connections

There are two basic connection configurations for generators, motors, and transformers: either a delta (Δ) or wye (Y) arrangement. Each one has its own benefits.

Not only is it necessary to connect the three coils in the proper phase relationship, it is important to get the phasing correct on each of the individual coils as they are connected. Phasing refers to the start and end of each winding.

If the start and end of one of the windings are reversed during the connection process, it may produce catastrophic results. Reversing the start and end of a winding would cause the output of that coil to be 180° out of phase.

The Wye (Y) Connection

When the windings are connected in a Y shape, the connection is referred to as a ***wye (Y) connection***, **Figure 22-23**. The advantage of this connection method is that there are two separate voltages available.

In a wye connection, one end of each winding is connected to a central point referred to as the neutral. Normally with such a connection, four wires are provided, one for each of the phases and one for the neutral. In the case of a motor, only three wires are used and the neutral is not connected.

A, B, and C in **Figure 22-23** refer to the three phase conductors, which are typically labeled L1, L2, and L3 in the case of a transformer or a generator. In the case of a three-phase motor, they are labeled T1, T2, and T3.

If you were to measure 120 VAC across a phase winding (for example, from A to neutral), you would measure 208 VAC across any two phases (for example, from A to B). This is because the phase-to-phase voltage equals the winding voltage multiplied by $\sqrt{3}$.

$$E_{\text{phase-to-phase}} = E_{\text{winding}} \times \sqrt{3}$$

(22-33)

where

$E_{\text{phase-to-phase}}$ = phase-to-phase voltage (VAC)
E_{winding} = winding voltage (VAC)
$208 \text{ VAC} = 120 \text{ VAC} \times \sqrt{3}$

The wye connection is used in most commercial power systems where both 120 VAC and 208 VAC must be available in both single- and three-phase power. This is not the same method used for residential service. In a wye connected three-phase circuit, the line current and the phase winding current are the same even though the voltages are different.

The Delta (Δ) Connection

When the windings are connected in the shape of a delta (Δ), the start of one winding is connected to the end of the next winding. This is known as a ***delta (Δ) connection***, **Figure 22-24**. The delta connection method provides one voltage using three conductors.

In a delta connected three-phase circuit, the line voltage and phase winding voltages are the same unlike the wye connected three-phase circuit. However, the line and phase winding currents are different. The line current is equal to $\sqrt{3}$ times the phase winding current.

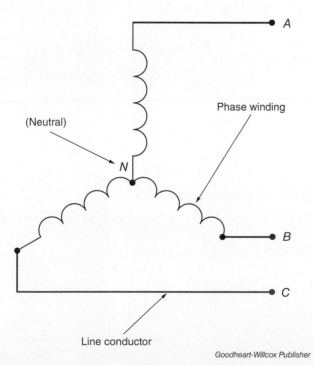

(Neutral)

N

Phase winding

A

B

C

Line conductor

Goodheart-Willcox Publisher

Figure 22-23. Windings connected in a wye (Y) arrangement.

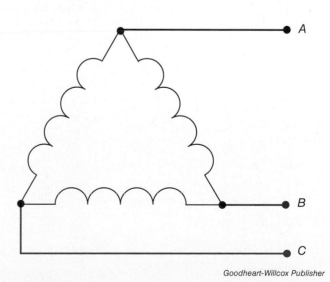

A

B

C

Goodheart-Willcox Publisher

Figure 22-24. Windings connected in a delta (Δ) arrangement.

Copyright Goodheart-Willcox Co., Inc.

The reason for the line current being higher is that current flows through different windings at different times. Each line conductor in a delta circuit is connected to two windings instead of one as in the wye connected three-phase circuit.

22.6.4 Power Calculations in Three-Phase Circuits

There are two different equations used to calculate power in a three-phase circuit. Both equations assume the three phases are balanced. This means all three phases have the same voltage, and all three phases have the same current. If you measure the voltage and current, the measured values are apparent rather than resistive or reactive.

To calculate line power, S, measured in volt-amps (VA), in a three-phase circuit, use one of the following equations:

$$S = \sqrt{3} \times E_{line} \times I_{line}$$

(22-34)

where

S = apparent power (VA)

E_{line} = line voltage (V)

I_{line} = line current (A)

$$S = 3 \times E_{phase} \times I_{phase}$$

(22-35)

where

S = apparent power (VA)

E_{phase} = phase voltage (V)

I_{phase} = phase current (A)

Notice the subtle difference between the two equations: on the line side $\sqrt{3}$ is used as the multiplier, while on the phase winding side 3 is used as the multiplier. Line measurements are most convenient with a DMM and clamp-on ammeter, so these are taken most often.

CHAPTER WRAP-UP

AC circuits build on the fundamental principles you already learned for DC circuits. Though AC circuits are a bit more complex, they offer major advantages over DC circuits in a variety of applications. Keep the fundamental principles of AC circuits in mind as you work through the following chapters on motors, transformers, and generators.

Chapter Review

SUMMARY

- Alternating current (AC) is represented by a sine wave. The rate at which current alternates is frequency, which is measured in cycles per second or hertz (Hz) in SI units.

- The period of a cycle is the amount of time it takes to complete the cycle. 1 cycle per second equals 1 Hz.

- A sine wave can be divided into 360° with any specific point on the sine wave specified by the phase, and peak values can be determined at any point in time.

- In an AC circuit containing only resistance, the voltage and current are in phase.

- In an AC circuit containing resistance and inductance, the voltage leads the current by less than 90°. In an AC circuit containing resistance and capacitance, the voltage lags the current by less than 90°.

- Inductance opposes change in the current of a circuit. Inductive reactance (X_L) describes the reactance of an inductor and is measured in ohms, Ω.

- Capacitance opposes change in the voltage of a circuit. Capacitive reactance (X_C) describes the reactance of a capacitor and is measured in ohms, Ω.

- Inductive and capacitive reactance counteract each other. A circuit combining inductive reactance and capacitive reactance is either inductively reactive or capacitively reactive, depending on the difference between the two reactances.

- The impedance (Z) of an AC circuit (inductive or capacitive) is the vector sum of the resistive and reactive components. Impedance may also be determined by Ohm's law for AC circuits.

- An RL circuit includes both resistance (R) and inductance (L). In series, the source voltage in an RL circuit is the vector sum of the resistive and reactive voltage drops.

- Apparent power (S) is the power the circuit consumes. True power (P) is the power that does the work. Reactive power (VAR) consumes power but does no work.

- The power factor (PF) indicates the efficiency of a circuit and is the ratio of true power to apparent power in an AC circuit.

- An RC circuit contains a resistor (R) and a capacitor (C) connected to a voltage source. Like parallel RL circuits, parallel RC circuits must account for the phase difference between resistive and reactive legs when calculating impedance.

- In an RLC circuit that includes both a capacitor and inductor, the reactive elements from the inductance and capacitance oppose each other.

- When X_L and X_C are equal, a circuit is at resonance and there is no net reactance.

- Filters allow certain frequencies to pass while blocking others. Two main types are band-pass filters and band-reject filters.

- In a three-phase alternating current system, each coil generates a sine wave that is 120° apart in phase and also induces voltages 120° apart.

- In a delta connected three-phase circuit, the phase voltages and line voltages are the same. Line current is equal to $\sqrt{3}$ times the phase current.

- In a wye connected three-phase circuit, the line current and the phase current are the same. Line voltage is equal to $\sqrt{3}$ times the phase voltage.

REVIEW QUESTIONS

Answer the following questions using the information provided in this chapter.

1. Alternating current (AC) is visually represented by a _____ that corresponds to the current changing directions.

2. What is the period of a cycle at a frequency of 120 Hz?

3. *True or False?* The instantaneous voltage at 270° is the positive peak.

4. Find the instantaneous voltage of a 120 V peak-to-peak sine wave at 110°.

5. If an oscilloscope measures a peak voltage of 210 V, what would the RMS voltage measured on a DMM show?

6. If a DMM measures a voltage and obtains a reading of 232 V, what would an oscilloscope show?

7. What is the difference between reactance and resistance?

8. Define inductance and inductive reactance, including their units.

9. What is the inductive reactance of an inductor with a value of 55 mH at a frequency of 50 Hz?

10. What do the variables X_L and X_C represent?

11. What is a farad?

12. What is the capacitive reactance of a capacitor when the frequency of the circuit is 105 Hz and the capacitance is 380 μF?

13. Calculate the impedance of an RL circuit with 430 Ω resistance and inductive reactance of 520 Ω.

14. The variable Z represents _____.

15. *True or False?* Ohm's law can be used for both AC and DC circuits.

16. In a reactive series circuit, how are the voltages across each resistor related to the source voltage?

17. What is the effective current of an RL circuit that has a 35 mH inductor and 115 Ω resistor connected in series with an AC source producing source voltage of 100 V at a frequency of 45 Hz?

18. What is the source voltage in a series RL circuit with a voltage drop across R of 338 V and a voltage drop across L of 25.5 V?

19. Explain how inductance, resistance, and source voltage each cause the voltage drop across a resistor to increase or decrease.

20. Find the impedance in an RL circuit with a source voltage of 85 V at a frequency of 68 Hz that has a 7 H inductor and 130 Ω resistor connected in parallel.

21. Explain the difference between true power, reactive power, and apparent power.

22. What is the true power of a circuit that has a 55 Ω resistor and produces 2 A across the resistor?

23. Calculate the apparent power of a circuit that has 0.85 A and 105 V through the circuit.

24. If the true power is 30 W and the reactive power is 40 VAR, what is the apparent power?

25. The phase angle between the resistance and the impedance is the phase difference between _____ and the current.

26. *True or False?* The power factor is equal to the cosine of the phase angle.

27. What is the power factor for a series RL circuit where the impedance is 80 Ω and the resistance is 20 Ω?

28. Find the phase angle for a circuit that has 47% efficiency.

29. Find the current through each leg, total current, and impedance for a capacitive circuit that has a 280 μF capacitor and 135 Ω resistor connected in parallel and source voltage of 177 V at a frequency of 52 Hz.

30. What is the phase angle of an RC circuit with true power of 1000 W and apparent power of 2400 VA?

31. How do inductance and capacitance oppose each other in an RLC circuit?

32. A series resonant circuit contains a 0.03 μF capacitor and an 18 mH inductor. What is the resonant frequency of the circuit?

33. What is the bandwidth of a circuit?

34. Calculate the quality factor of a circuit that has a resistance of 82 Ω and inductive reactance of 380 Ω.

35. What is the bandwidth of a resonant circuit that has a quality factor of 4.15 and 70 Hz?

36. Discuss the differences between a series resonant circuit and a parallel resonant circuit including which frequencies these circuits filter.

37. Why is three-phase alternating current more efficient than single-phase alternating current?

38. How many degrees apart are the phases of a three-phase circuit?

39. The two main types of three-phase connection are _____ and _____.

40. If the phase-to-neutral voltage of a wye-connected, three-phase circuit is 277 VAC, what is the phase-to phase-voltage?

41. If the phase current in a three-phase, delta-connected load is 2.5 A, and the phase voltage is 28 V, what is the apparent power consumed?

NIMS CREDENTIALING PREPARATION QUESTIONS

The following questions will help you prepare for the NIMS Industrial Technology Maintenance Level 1 Electrical Systems credentialing exam.

1. If frequency is 72 Hz, how much time does it take to complete one cycle?

   A. 0.0029 s
   B. 0.014 s
   C. 0.72 s
   D. 1.68 s

2. Ohm's law for AC circuits uses voltage and current to calculate what quantity?

   A. Impedance
   B. Capacitance
   C. Reactance
   D. Frequency

3. If an oscilloscope measures a peak voltage of 178 V, what would the RMS voltage measured on a DMM show?

   A. 39.7 V
   B. 87.58 V
   C. 125.85 V
   D. 251.69 V

4. What does capacitance measure?

   A. How much a circuit resists voltage change
   B. The amount of voltage increase possible in a circuit
   C. How much a circuit opposes current change
   D. The extent to which a capacitor allows current to flow

5. In parallel AC circuits with resistive and reactive legs, which of the following is true?

   A. Voltage and current are in phase in the resistive leg, but they are out of phase in the reactive leg.
   B. Voltage and current are out of phase in the resistive leg, but they are in phase in the reactive leg.
   C. Resistance and voltage are in phase in the resistive leg, but they are out of phase in the reactive leg.
   D. Resistance and voltage are out of phase in the resistive leg, but they are in phase in the reactive leg.

6. What is the efficiency of a series circuit where impedance is 45 Ω and resistance is 20 Ω?

   A. 2.25%
   B. 44.4%
   C. 63.9%
   D. 97.2%

7. What occurs when a circuit is at resonance?

   A. X_C counteracts X_L which decreases the frequency.
   B. X_L counteracts X_C which increases the frequency.
   C. X_L and X_C are equal and there is an inductive net reactance.
   D. X_L and X_C are equal and there is no net reactance.

8. What is the power consumed in a three-phase connection if line voltage is 17 V and line current is 0.95 A?

   A. 6.96 VA
   B. 16.15 VA
   C. 27.97 VA
   D. 48.45 VA

23

Electrical Components and Circuit Materials

LEARNING OBJECTIVES

After completing this chapter, you will be able to:

- Describe the resistor as a physical device and understand how it is rated.
- Explain the operation and use of variable resistors.
- Discuss capacitors, including physical construction, type, and working voltage.
- Perform series and parallel connection calculations for capacitors.
- Describe circuit boards and their characteristics.
- Compare fuse types, ratings, and troubleshooting methods.
- Explain the operation and use of circuit breakers.
- Describe the structure and function of relays.
- Discuss sizing, ampacity, and resistance of conductors.
- List common methods of termination and describe how to apply and test them.
- Demonstrate knowledge of soldering materials, equipment, and techniques.

TECHNICAL TERMS

American Wire Gage (AWG)	GFCI circuit breaker
ampacity	ground-fault circuit interrupter (GFCI)
arc-fault circuit interrupter (AFCI)	plates
cable	potentiometer
capacitor	power rating
ceramic capacitor	relay
circuit board	resistor
circuit breaker	rheostat
crimping	solder
dielectric	soldering
dielectric constant	surface-mount resistor
diode	termination
dual relay	through-hole resistor
electrolyte	tinning
electrolytic capacitor	tolerance
ferrule	wiper
flux	wire
fuse	wire connector
	working voltage

This chapter presents the essential components of electrical systems, including resistors, capacitors, fuses, circuit breakers, and relays. Methods for connecting these components with wires and cables and various methods of termination are also covered.

23.1 RESISTORS

Resistors are one of the most common components used in electronics. They are used to create voltage drops or limit current. Resistors are available in many different sizes and shapes for various applications, **Figure 23-1**. Higher wattage resistors may consist of a ceramic base with high-resistance wire wound around it. The winding is covered with a vitreous enamel coating, **Figure 23-2**.

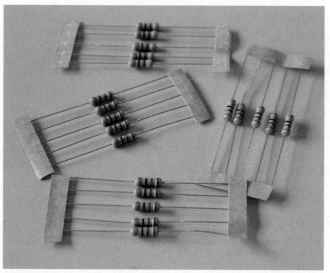

Goodheart-Willcox Publisher

Figure 23-1. Resistors with a thin carbon film. These resistors have a wattage rating of 1/4 W. The tape on the ends of the resistors is for feeding into an automatic insertion machine.

Goodheart-Willcox Publisher

Figure 23-2. High-power resistors formed by wrapping high-resistance wire around a ceramic core, then coating with enamel.

23.1.1 Tolerance

All resistors have a *tolerance*. This specification indicates how far from the published value an actual unit can be. A resistor with a resistance value of 100 Ω and a tolerance of ±10% could have an actual measurement that ranges from 90 Ω to 110 Ω.

Standard resistor tolerances are 1%, 5%, 10%, and 20%. It is unlikely when measuring a resistor that it exactly matches its published value. However, it should remain within the limits specified by its tolerance. Keep this in mind when measuring resistors to determine if they are defective or not.

> **TECH TIP**
> **Heating Can Change Resistor Value**
>
> Heat can cause a resistor to change value. The hotter a resistor becomes, the more the resistance increases. Depending on the resistor type, it may recover, or it may permanently remain at its higher value. Carbon resistors are extremely susceptible to value change. If you notice the color on a resistor has darkened, this may indicate the resistor has changed value due to overheating. Excessive current flowing through the resistor is the immediate cause of overheating. As a technician, you must identify and correct the root cause. Until you do, replacement units are also susceptible to overheating.

Smaller resistors may be of the through-hole or surface-mount variety. *Through-hole resistors* have wire leads that are inserted through the holes in a circuit board and soldered in place. *Surface-mount resistors* lay on the surface of a circuit board and are held in place by soldering. Surface-mount components are extremely small and require much greater care in placement. Soldering methods for these devices are different than for the through-hole variety. The smaller circuit "footprint" required by surface-mount devices makes it possible to place more circuitry on the same size circuit board.

23.1.2 Power Rating

All resistors have a *power rating*, sometimes called a *wattage rating*, to indicate the maximum amount of current the resistor can handle without overheating. Resistors have a voltage drop across them and a current passing through them, which generates heat. With too much heat, the unit may be degraded or destroyed.

23.1.3 Variable Resistors

A *potentiometer*, sometimes called a *pot* for short, is a three-terminal variable resistor used to control current by varying resistance, **Figure 23-3**. Two of the terminals connect to each end of the resistor while the third has a contact that can move across the surface of the resistor. The movable contact is referred to as the *wiper*. As the wiper is repositioned along the surface of the resistor, the resistance between the wiper and either end of the resistor changes. A *rheostat* is a type of variable resistor with only two terminals. Potentiometers may be used as rheostats by connecting only two of the three terminals.

Some potentiometers employ a carbon film resistive element while others are wire wound. The wire-wound devices are capable of withstanding higher currents without overheating. However, they are much coarser in their ability to make fine resistance adjustments. Potentiometers may have a knob or screwdriver slot to make the adjustment. Different models have the ability to be adjusted by either a single turn or multiple turns. A 10-turn potentiometer is commonly used where fine adjustments of resistance are required.

23.1.4 Tapped Resistors

Tapped resistors are often found on higher wattage units and operate similarly to several resistors connected in series with connections occurring at the junction of two resistors. There are also models available with adjustable taps, often referred to as adjustable resistors. In these

Goodheart-Willcox Publisher

Figure 23-3. Potentiometers. Notice that potentiometers have three terminals. The terminal in the center is connected to the wiper.

units, a screw is loosened and the tap slides to the new position. Once in the desired position, the screw is tightened to retain the value change.

23.2 CAPACITORS

A *capacitor* is formed when two conductors are in close proximity to each other and separated by an insulator, **Figure 23-4**. The conductors are referred to as the *plates* while the insulator is called the *dielectric*. Electrons cannot flow through the dielectric of a capacitor. Instead, electrons accumulate on the negative plate as electrons leave the positive plate. In this way, a capacitor can temporarily store electric charge and oppose changes in voltage. When a capacitor is charged, it blocks direct current.

Capacitance, C, is a measure of opposition to change in voltage and is measured in farads, F. For a specific capacitor, capacitance indicates the number of electrons the capacitor can store for each volt applied. Use the following equation to calculate capacitance:

$$C = \frac{Q}{E}$$

(23-1)

where

C = capacitance in farads (F)
Q = charge in coulombs (C)
E = voltage in volts (V)

This formula returns a value in farads, which is a large amount of capacitance. Though some modern capacitors are capable of holding a farad of charge, most capacitors are valued in microfarads, μF (10^{-6} F); nanofarads, nF (10^{-9} F); or picofarads, pF (10^{-12} F).

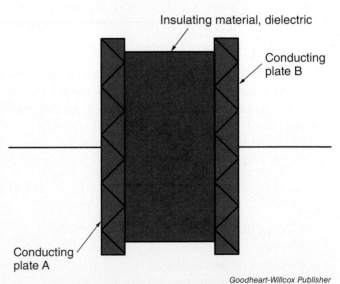

Insulating material, dielectric

Conducting plate B

Conducting plate A

Goodheart-Willcox Publisher

Figure 23-4. Basic structure of a parallel plate capacitor.

23.2.1 Dielectric Constant

The *dielectric constant*, *K*, is an important consideration when selecting the dielectric material for a capacitor. The dielectric constant is sometimes referred to as the *permittivity* of a material and describes an insulator's ability to support the electric field between the two oppositely charged conducting plates. Some dielectric constants for popular materials are tabulated in **Figure 23-5**. Note that the dielectric constant is unitless.

Constants for Common Dielectric Materials	
Material	**Dielectric Constant**
Vacuum	1
Dry air	1.00059
Polypropylene	1.5
Teflon (PTFE)	2.1
Transformer oil	2.2
Polyethylene	2.25
Polystyrene	2.4–2.7
Mylar	3.2
Polyimide	3.4
Paper	3.85
Pyrex (glass)	4.7
Porcelain	5.0
Mica	5.0–8.5
Rubber	7
Aluminum oxide	9.34
Tantalum pentoxide	26

Goodheart-Willcox Publisher

Figure 23-5. Dielectric constants for various materials.

23.2.2 Capacitance

The capacitance of a capacitor is determined by its physical features, such as the size of the plates and the type of dielectric. To calculate the capacitance of a parallel plate capacitor in picofarads, use the following equation:

$$C = \frac{K \times A}{d} \times 0.225 \text{ pF/in}$$

(23-2)

where

    C = capacitance in picofarads (pF)

    K = dielectric constant

    A = area of the plates (in²)

    d = distance between the plates or thickness of the dielectric (in)

The constant term in this equation, 0.225, accounts for the relationship between the dielectric constant and a physical property known as the permittivity of free space. This constant carries dimensions of picofarads per inch. Notice in the equation that greater plate area produces greater capacitance. Likewise, a higher dielectric constant or a thinner dielectric yields higher capacitance. Capacitance decreases as the distance between the plates increases. It is helpful to understand the relationship between these parameters when calculating capacitance and selecting capacitors.

23.2.3 Working Voltage

Working voltage is the maximum voltage that is safe to place across a capacitor without the dielectric material breaking down. The breakdown of the dielectric results in what is known as a "punch-through." When this happens, a short circuit occurs, creating one of the most common capacitor failures you are likely to encounter.

When replacing a capacitor, it is acceptable to replace the existing unit with one of a higher working voltage if the original value is not available. However, it is not appropriate to replace a capacitor with one of a lower working voltage. Often, capacitors with higher working voltages are also greater in physical size, so you should select a replacement unit as close to the original value as possible.

> **SAFETY NOTE**
> ### Discharge Capacitors before Work
>
> Capacitors may retain dangerous (even lethal) voltages long after the source of power has been removed and the unit is rendered "safe" by following lockout/tagout procedures. Before working on assemblies with capacitors, the capacitors must be discharged. Capacitors should be discharged using a high value "bleeder" resistor. Charged capacitors are not only dangerous to technicians, they may also damage test equipment.

23.2.4 Capacitors in Parallel

When wiring capacitors in parallel, the total capacitance equals the sum of the individual capacitances:

$$C_T = C_1 + C_2 + C_3 + \ldots + C_n$$

(23-3)

It is appropriate to use capacitors of the same working voltage rating when connecting them in parallel. Otherwise, the working voltage of the combined capacitors is the same as that of the least value in the combination.

23.2.5 Capacitors in Series

When connecting capacitors in series, the sum of the working voltages of the individual capacitors equals the total working voltage of the combination. It is good practice to have the same working voltage on all capacitors. High-voltage capacitors often contain a series combination of several capacitors with lower working voltages.

The equation for the total capacitance of a series of capacitors is similar to the calculation for total resistance of resistors in parallel:

$$C_T = \cfrac{1}{\cfrac{1}{C_1} + \cfrac{1}{C_2} + \ldots + \cfrac{1}{C_n}}$$

(23-4)

Refer to Chapter 20, *Ohm's Law, Kirchhoff's Laws, and the Power Equation*, for sample calculations involving reciprocals. Notice that when capacitors are connected in series, the total capacitance is less than that of the smallest individual capacitance in the combination.

23.2.6 Capacitor Classifications

Capacitors are available in many types and sizes. Capacitors are typically classified by the material used in their dielectric. Common capacitor types include ceramic capacitors and aluminum and tantalum electrolytic capacitors.

Ceramic Capacitors

For very low capacitance values, **ceramic capacitors** are common. They are often used when a high working voltage is required and are typically less expensive than other varieties. A ceramic capacitor is formed by plating a disk of ceramic material with silver on each side. Connections are made to the silver plates, and the entire unit is dipped in a phenolic or epoxy coating. Phenolic is a phenol formaldehyde resin and one of the earliest commercially available polymer resins. The ceramic capacitor is capable of a much higher working voltage for its size than many other capacitor styles.

Aluminum and Tantalum Electrolytic Capacitors

Electrolytic capacitors use **electrolytes**, nonmetallic conductors of electricity, to move charged ions. Electrolytic capacitors are further classified as either wet, containing a liquid electrolyte, or dry, containing a paste electrolyte. Aluminum and tantalum are two of the most common metals used in electrolytic capacitors.

Aluminum electrolytic capacitors have aluminum electrodes containing a layer of aluminum oxide, which is used as the dielectric, and contain a liquid electrolyte. Tantalum capacitors are similar to aluminum versions, except the electrodes are made of tantalum and the electrolyte is a dry manganese dioxide. Tantalum capacitors have the advantage of a dry electrolyte and are capable of greater capacitance in a smaller package than aluminum capacitors. The shelf life of tantalum units is also longer, and they have lower leakage currents.

Most electrolytic capacitors are polarized, which means they have a positive and negative terminal, **Figure 23-6**. Always observe the proper polarity when installing an electrolytic capacitor. A capacitor is likely to fail soon after initial operation if its polarity has been reversed.

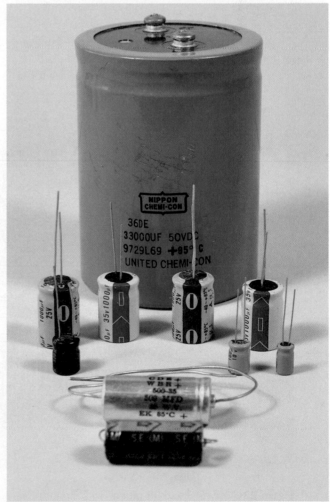

Figure 23-6. Aluminum electrolytic capacitors. On electrolytic capacitors, the polarity is clearly marked.

TECH TIP
Troubleshooting Electrolytic Capacitors

Heat is a consideration with aluminum electrolytic capacitors. A wet electrolyte may vaporize with heat. Once the pressure inside a capacitor builds to a certain point, the capacitor may rupture. For that reason, many wet electrolytic capacitors have a vent that opens when the pressure is too great inside. Repeated opening of the vent causes the capacitor to eventually dry out and fail. This common failure is often detectable by traces of dry electrolyte on the end of the capacitor or on the circuit board around the failed capacitor.

Some aluminum electrolytic capacitors have a cross-shaped vent stamped in the top to allow the capacitor top to bulge when the pressure becomes too great inside the can. Instead of being flat like a normal capacitor, the top has a slight pyramidal shape to it when failed.

Keen observation of electrolyte leakage or a bulging capacitor can indicate that the capacitor has failed.

23.3 CIRCUIT BOARDS

Circuit boards are used to connect electrical components, **Figure 23-7**. Commonly, boards are built up from layers of fiberglass mats and epoxy resin. There are various grades of glass epoxy circuit boards available, and the grade selected depends on voltage and mechanical requirements of the electrical system.

Circuit boards may be single sided, double sided, or multilayered. A single-sided board is clad with a copper layer on one side. The copper layer is covered with a photoresist layer. The photoresist is photographically exposed to light through a photo mask of the desired circuit board pattern. Once exposed, the circuit board is developed to remove the photoresist where no traces are required. The board is then placed in an etching solution, which removes the unneeded copper areas between the traces.

Once the circuit board has been etched, holes for components and pass-through connections are drilled. If the purpose of the hole is a pass-through conductor from one side of the board to another, carbon powder is blown through the holes, and a conductive metal is plated onto the carbon. Plated through-holes are used when a circuit board is multilayered or double sided. Component holes may also be plated through.

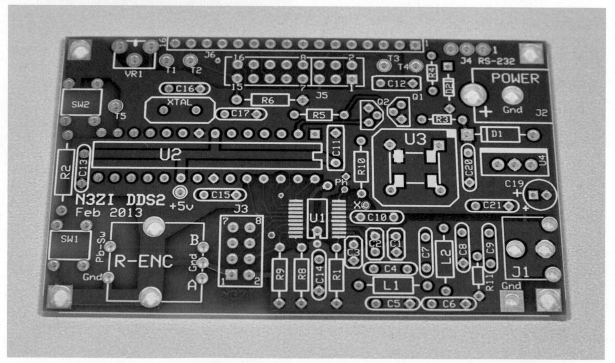

Goodheart-Willcox Publisher

Figure 23-7. Typical circuit board. This circuit board has provisions for both through-hole and surface-mount components.

Multilayered boards are single- and double-sided board combinations where the individual layers are sandwiched together and bonded. Surface-mount component boards only have holes drilled for pass through connections between layers or sides because holes are not required for component mounting.

23.4 FUSES AND CIRCUIT BREAKERS

Fuses and *circuit breakers* are overcurrent protection devices. Both devices interrupt the flow of current when it exceeds a predetermined amount. Typically, fuses are single-use devices, whereas circuit breakers can be reset. Overcurrent protection guards equipment from damage in the event of a fault or short circuit.

23.4.1 Fuses

Fuses are available in many different sizes, shapes, and ratings, **Figure 23-8**. Most are single-use devices. Once the fuse has interrupted the circuit, a new fuse must be installed. There are some varieties of larger fuses that are resettable.

The rated current of a fuse is the maximum current it can continuously conduct before interrupting the circuit. The voltage rating on a fuse is the maximum circuit voltage that would be encountered if the fuse were to open. Fuses are available in high-voltage (over 1000 V) and low-voltage (under 1000 V) types. The time-current characteristic of a fuse indicates the amount of current that can run for a given length of time before the fuse interrupts the circuit. A lag or time-delay fuse has a slower time-current characteristic than a fast-acting fuse. The time-current characteristic is important to prevent nuisance "blowing" or to protect equipment by interrupting at the least amount of overcurrent.

Replacing a Fuse

When replacing a fuse, make certain you replace it with a fuse of the same rating as the original. If in doubt, check the service manual of

the equipment to ensure you are replacing the fuse with a manufacturer-recommended device. Fuse ratings on equipment are generally written next to the fuse holder. If you are replacing a fuse in a disconnect box, make certain the replacement is appropriate for the size of wire being used and the current draw of the equipment connected to it.

Repeated blowing of a fuse indicates an underlying problem that must be corrected before replacement.

SAFETY NOTE
Tips for Replacing Fuses

- Ensure the power is disconnected before either installing or removing a fuse.
- Always use an approved fuse puller to remove fuses.
- Be certain the replacement fuse has the correct voltage rating, current rating, and time-current characteristics.

Goodheart-Willcox Publisher

Figure 23-8. Industrial cartridge fuses. Fuses are clearly marked with their amperage rating. Manufacturer's specifications include information about the parameters of the fuses. The three fuses in the lower-left corner have an indicator that enables you to visually determine if the fuse is good or open without needing to take an initial measurement.

Troubleshooting Fuses

Test a fuse by first removing it from the circuit and then measuring its resistance with an ohmmeter. If the fuse is good, it will show continuity. If not, it is open.

If it is safe to have the power on, check the fuse in circuit with a voltmeter. If there is voltage present on the line side of a fuse and not on the load side, the fuse is open. An alternate method is to parallel the fuse with a voltmeter, **Figure 23-9**. If voltage is present across the fuse, it is open.

You may also verify continuity of a fuse by reading the current on the load side with a clamp-on ammeter. If current is present, the fuse must be intact.

In a three-phase circuit, each of the phase conductors are fused. The loss of one fuse results in the loss of voltage on two phases. This happens because each of the three phases on the load side is connected between two phase conductors on the line side.

PROCEDURE | Three-Phase Fuse Test

To test the fuses in a three-phase circuit, test each combination of phases.

1. Check the voltage between phases 1 and 2, **Figure 23-10**.
2. Check the voltage between phases 2 and 3, **Figure 23-11**.
3. Check the voltage between phases 1 and 3, **Figure 23-12**.

This series of tests indicates that the fuse in phase 2 is open and must be replaced.

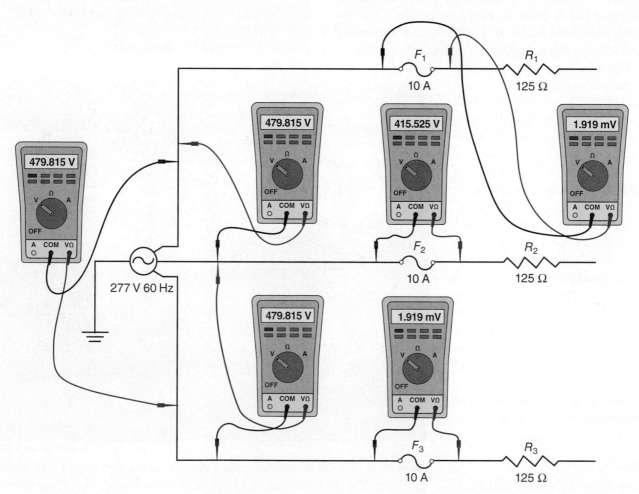

Figure 23-9. With this method, the voltage across each fuse is measured. An open fuse, F_2, reads close to the line voltage. In this case, R_1, R_2, and R_3 are the load. If the load is a motor, this method is unsafe because a motor can be damaged by delivering only a single phase. Notice that the voltmeters connected to the good fuses, F_1 and F_3, read a few millivolts. These are "ghost voltages" and do not indicate the fuse is defective.

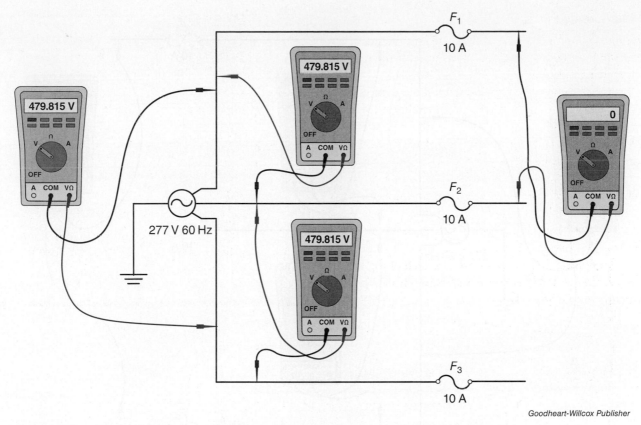

Goodheart-Willcox Publisher

Figure 23-10. Three-phase fuse testing. The voltmeter on the right indicates that F_1 or F_2 is open.

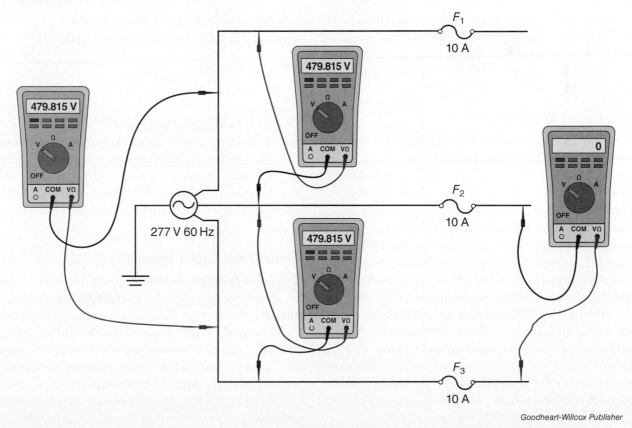

Goodheart-Willcox Publisher

Figure 23-11. Three-phase fuse testing. The voltmeter on the right indicates that either F_2 or F_3 is open.

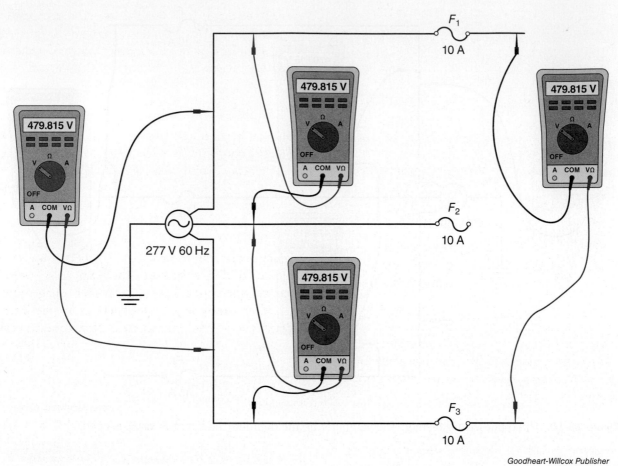

Figure 23-12. Three-phase fuse testing. The voltmeter on the right indicates that F_1 and F_3 are good because voltage is present. With the previous two tests indicating 0 V present, you can deduce that F_2 is the open fuse.

23.4.2 Circuit Breakers

A circuit breaker is a device to protect a circuit from overload, **Figure 23-13**. Unlike a fuse, a circuit breaker can be reset after tripping. If a circuit breaker "trips out," move the handle to the off position then back to the on position to reset it.

Normally, when a circuit breaker trips, the handle moves to a position halfway between off and on. Some circuit breakers also have an indicator signifying the breaker has tripped. There are two main varieties of tripping mechanisms employed in circuit breakers: bimetallic strips or a magnetic trip mechanism.

A bimetal is made from two different metals. Each metal has a different coefficient of expansion. As the temperature of the bimetallic strip increases due to a higher current through the circuit breaker, the strip bends, actuating the trip mechanism. The circuit breaker is set to trip out at a specific current. This type of circuit breaker must cool down before it can be reset.

The second type is magnetically operated. Inside the circuit breaker is a coil. As more current flows through the coil, the magnetic field around it increases. Once the magnetic field reaches a predetermined point, it actuates the trip mechanism. The magnetic circuit breaker is faster and more accurate than the bimetallic variety.

Ground-Fault Circuit Interrupter (GFCI)

A **ground-fault circuit interrupter (GFCI)** measures the difference between the current flowing into a circuit and the current flowing out through the neutral. If the neutral current is less than the hot current, the difference must have found a path through the ground conductor. The GFCI detects a difference between the hot and neutral current at currents as low as 4 mA. The GFCI reacts quickly, typically in less than 0.1 seconds, and trips out, disconnecting the source of current.

Goodheart-Willcox Publisher

Figure 23-13. Industrial circuit breakers are available in several sizes for various applications. Circuit breakers trip out when an overload current exists, but, unlike fuses, circuit breakers can be reset by switching them to the off position then back to the on position once the fault has been cleared.

A *GFCI circuit breaker* combines the protection of a standard circuit breaker and senses ground-fault current. GFCI circuit breakers trip out whenever current to ground is detected. A single-phase GFCI breaker requires a connection with the neutral conductor of the circuit being protected. GFCIs are also available as receptacles or portable devices. In a standard receptacle, there are three connections: hot, neutral, and ground.

GFCIs have a test button, which causes the device to trip when pressed. A GFCI circuit breaker is reset just like a conventional circuit breaker. GFCI receptacles have a reset pushbutton. If wired properly, a GFCI receptacle can protect all other devices after it. However, a conventional circuit breaker is required to protect the circuit from an overcurrent situation.

Arc-Fault Circuit Interrupter (AFCI)

An *arc-fault circuit interrupter (AFCI)* senses arcing and interrupts the circuit. Arcing is caused by loose or poor connections and is the root cause of many electrical fires.

23.5 RELAYS

Relays are switches controlled by electromagnets, often called by various names, including *motor starter*, *solenoid*, and *contactor*. Regardless of how they are named, relays have three main parts: the coil, contacts, and armature, **Figure 23-14**. A current flows through the coil, which creates a magnetic field. The magnetic field attracts the armature, which in turn actuates the contacts.

23.5.1 Relay Contacts

Contacts are defined as either normally open (NO) or normally closed (NC). NO and NC refer to the state of the contacts when the relay is de-energized, or in its "normal" state. When the relay coil is energized, a normally open contact closes and a normally closed contact opens.

Relay contacts have both a current and voltage rating. Typically, higher current requires larger contact surface area. This prevents the contacts from welding together during operation. The distance between the contacts when open is greater at higher voltage ratings.

Select a relay with contact ratings for the intended current and voltage encountered. Relay contacts may be rated differently for switching AC and for switching DC.

Goodheart-Willcox Publisher; Alexey Bukreev/Shutterstock.com

Figure 23-14. Relays. These two relays have different types of connectors but have the same basic features: coil, contacts, and armature.

23.5.2 Relay Coils

There is not much difference between the coils of an AC or DC relay. In either case, the relay coil functions as an inductor. However, the coil core and relay armature are quite different depending on whether the relay is intended for AC or DC operation.

DC relays often have a *diode*, a one-way "valve" for direct current, across the relay coil. The diode shorts out the inductive kick at the instant the current to the coil is switched off, and the magnetic field around the coil falls. The falling magnetic field induces a voltage in the coil in the opposite direction of the previously applied voltage. The diode is arranged to conduct only during the time when the inductive kick is present. When the coil is energized, the diode does not conduct. The diode in this application is referred to as a "snubber diode."

With a DC relay containing a snubber diode, you must observe the correct polarity. Polarity is clearly marked on the relay if a diode is involved. If you do not pay attention to the polarity markings and connect the relay in the reverse direction, the diode will present a direct short to the circuit that drives the relay.

Some relays contain a light-emitting diode (LED) and resistor combination connected across the relay coil. The LED illuminates when the relay should be energized. Once again, polarity is marked on the relay. Reversing polarity in this situation does not present a short circuit. However, the LED will not function.

Some relays, AC or DC, have an indicator flag connected to the armature to provide visual indication that the relay is energized. Some relays also have a pushbutton to mechanically force the relay into the energized state without any current flowing through the coil. This is handy in some cases for troubleshooting purposes but can be quite dangerous in others. Be sure you understand the ramifications of mechanically forcing a specific relay before you do it.

Common relay coil voltages are 120 VAC, 24 VAC, 24 VDC, 12 VDC, and 5 VDC. Other coil voltages are available for special applications.

TECH TIP
Troubleshooting Relays

Verify that a relay has voltage to the coil at the appropriate time before you declare it defective. An open coil cannot operate the relay when voltage is applied. Also check the contacts. Welded contacts cause the relay not to switch on or off. Burned contacts often do not conduct current.

If you suspect a relay is defective, check for voltage across the coil. If there is no voltage, check the control circuit. If voltage is present on the coil and the relay is not pulling in, remove it from the circuit to check coil continuity and contact continuity with your ohmmeter to determine the source of the defect.

23.5.3 Dual Relays

A *dual relay* is actually two relays in one with two coils and two sets of contacts. This type of relay is used to reverse the rotation of a three-phase motor by reversing the phase relationship of two of the three line-input phases. A mechanical interlock prohibits simultaneous operation of both relays. Operation of both relays at once would cause a catastrophic failure and arc flash. Often, dual relays have an extra set of contacts to serve as an electrical interlock as well.

23.6 WIRE

Wire is available in two basic types: stranded and solid. Stranded wire is more flexible than solid. The termination of stranded wire requires either tinning or terminals for a proper connection. Whether stranded or solid, wire is considered a single conductor, while multiple conductors within the same sheath or covering are referred to as cable.

The size of a conductor is determined by its cross-sectional area. The greater the cross-sectional area of a wire, the larger the amount of current it is capable of handling. In addition, greater cross-sectional area corresponds to lower resistance.

Circular mils are used to measure the cross-sectional area of a wire. One mil is equal to 1/1000 of an inch. The number of circular mils in a round conductor is equal to the diameter of the wire in mils squared, D^2. For example, if a wire has a diameter, D, of 10 mils, the cross-sectional area is 10^2 or 100 mils.

In North America, wire is typically classified by a system referred to as *American Wire Gage (AWG)*. AWG size corresponds to the diameter of the conductor in inches. This measurement only considers the wire diameter and does not include the diameter of the insulation. Notice that the larger the gage, the smaller the diameter of the wire, **Figure 23-15**.

Ampacity is defined as the amperage a wire is capable of safely carrying. The ampacity of a wire is affected by temperature. As the current in a conductor increases, so

AWG Sizing and Ampacity				
AWG Wire Size	Conductor Diameter in Inches	Ohms Per 1000 Feet	Max Amps Chassis Wiring	Max Amps Power Transmission
0000	0.46	0.049	380	302
000	0.4096	0.0618	328	239
00	0.3648	0.0779	283	190
0	0.3249	0.0983	245	150
1	0.2893	0.1239	211	119
2	0.2576	0.1563	181	94
3	0.2294	0.197	158	75
4	0.2043	0.2485	135	60
5	0.1819	0.3133	118	47
6	0.162	0.3951	101	37
7	0.1443	0.4982	89	30
8	0.1285	0.6282	73	24
9	0.1144	0.7921	64	19
10	0.1019	0.9989	55	15
11	0.0907	1.26	47	12
12	0.0808	1.588	41	9.3
13	0.072	2.003	35	7.4
14	0.0641	2.525	32	5.9
15	0.0571	3.184	28	4.7
16	0.0508	4.016	22	3.7
17	0.0453	5.064	19	2.9
18	0.0403	6.385	16	2.3

Goodheart-Willcox Publisher

Figure 23-15. Chart relating AWG size, wire diameter, resistance, and *NEC* specifications for ampacity.

does the heat that conductor dissipates. If the temperature is too high, it may damage the insulation of the wire. The *NEC* sets the ampacity for specific gages of wire based on raceway size, number of conductors, ambient temperature, and insulation type. Refer to the latest *NEC* handbook to determine the ampacity of a given wire.

23.7 CABLE

The term *cable* refers to multiple conductors within a sheath or outside covering. Industrial cable often contains twisted pairs of wire within a sheath, **Figure 23-16**. Each of the twisted pairs has a different number of twists per inch. The result is that any interference on one of the twisted wires is 180° out of phase with the other wire in the pair. This 180° phase shift effectively cancels out the majority of received interference.

Some better industrial cables have a foil shield surrounding the twisted pairs to minimize interference from other outside sources. Cables with foil shielding

Flegere/Shutterstock.com

Figure 23-16. Electrical cable. These cables contain copper wire with inner and outer insulation.

have a bare twisted conductor in contact with the foil to enable connections to the shield. This shielding wire is often referred to as the "drain wire."

Coaxial cable has a center conductor and a shield separated by a dielectric. Coaxial cable is used for the transmission of radio frequency AC signals or lower frequency signals where it is important to shield them from outside interference. This type of cable has constant impedance and is often found in 50 Ω and 75 Ω varieties.

There are many other custom or specialty cables for various purposes. Different wire gages and types can be bundled together in a common sheath to connect equipment with a range of electrical specifications.

23.8 METHODS OF TERMINATION

Termination refers to connecting a wire to a device or another wire. Properly terminating a wire results in a low-resistance and mechanically sound connection. Common termination methods include crimp terminals, wire connectors, and solder terminals. The methods for terminating stranded wire often differ from those employed with solid wire. For stranded wire, crimp terminals or soldering is necessary for high-reliability connections. Carefully strip wire using wire stripping tool, **Figure 23-17**, before splicing or terminating it.

> ### TECH TIP
> **Continuity Testing**
>
> Check connections made by any method of termination by performing a continuity test. Continuity testing requires simply using your ohmmeter to check any two points along the conductor for zero (or nearly zero) resistance.

Ideal Industries, Inc.

Figure 23-17. Wire strippers.

23.8.1 Crimping

Crimping refers to terminating wire by bending a metal crimp terminal securely around the wire. Crimping is often the best method of wire termination if done properly. A good technician needs a high-quality crimping tool specifically for the type of terminal to be crimped, **Figure 23-18**. All too often, technicians use pliers or a cheap crimper, which fail to make a sound connection. When improperly crimped, a terminal can easily pull off the end of the wire to which it was crimped.

After crimping a terminal onto a wire, perform the pull test by gently pulling on the terminal to ensure it is secure. If for any reason the terminal pulls off, reconsider the crimping tool you are using.

A **ferrule** is a variety of crimp terminal that, once crimped, resembles a pin and is an alternative for tinning stranded wire if it is to be used with a mechanical connection. Ring, fork, and spade terminals may be crimped on stranded wire and make solid connections. Bending a hook in stranded wire and placing it under a screw terminal is not an acceptable method of termination.

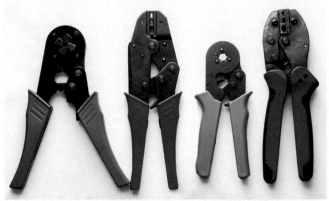

Romaset/Shutterstock.com

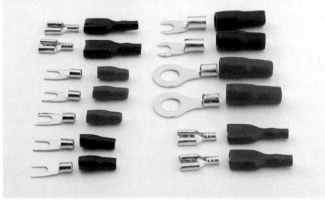

Mihancea Petru/Shutterstock.com

Figure 23-18. Crimpers and crimp terminals. Good crimping tools are well worth the investment.

23.8.2 Wire Connectors

Wire connectors are used to splice stranded wire to stranded wire. Wire connectors are also called *wire nuts, cone connectors*, or *twist-on wire connectors*. The stranded wire ends are twisted together, and the wire connector is screwed on to the twisted portion of the wire, making a tight and electrically sound connection. Wire connectors may also be employed with solid wire types.

23.8.3 Soldering

Soldering joins metals by a fusion of alloys known as solders, **Figure 23-19**. *Solders* are nonferrous metallic alloys with low melting points. When soldering, heat is quickly applied to the metals to be joined and solder is introduced to the heated base metal. *Flux* is used to remove oxides and prepare the joint for soldering. Clean joint surfaces allow the molten solder to create a strong bond. The mating surfaces of the joint and the solder form a metallurgical bond resulting in a good mechanical and electrical connection.

Solder

Common solder alloys include tin, lead, and silver. When various combinations of these metals are alloyed, the melting point may be set at the desired temperature for a particular application, **Figure 23-20**. Some tin/lead alloys have an even lower melting point than the individual metals themselves.

Pure tin solder is appealing for its low melting point, but tin alone can cause short circuits and arcing. The second

Melting Points for Tin/Lead Solder Alloys	
Tin/Lead, %	Melting Point, °F (°C)
40/60	460 (230)
50/50	418 (214)
60/40	374 (190)
63/37	364 (183)
95/5	434 (224)

Goodheart-Willcox Publisher

Figure 23-20. Melting points for tin/lead solder alloys by percent composition of each metal.

Goodheart-Willcox Publisher

Figure 23-19. Soldering supplies. From left to right, liquid flux dispenser with needle tip, three diameters of solder, and a bottle of liquid flux. In the foreground is a roll of solder wick used for desoldering purposes.

metal in the solder alloy, usually silver or lead, mitigates this problem. Following legislation prohibiting the use of lead solder in consumer electronics, the electronics industry has started using silver as the second alloy metal despite its higher melting temperature compared with tin/lead alloys.

The most common alloys used for electrical and electronic work are 60/40 and 63/37. The 63/37 alloy is referred to as eutectic or "fast freeze" solder. Fast freeze solder helps prevent any movement or vibration during solidification process. Movement can lead to a "cold solder joint," which has neither good mechanical stability nor good electrical conductivity. Movement, vibration, and temperature may cause an intermittent connection, which is good one minute and bad the next and particularly difficult to troubleshoot.

Flux

Flux is used to clean the surfaces to be soldered of oxidation. Flux does not clean the solder joint of dirt, grease, oil, or other contamination. Flux may be contained in the solder itself or applied separately as a liquid. Use the appropriate type of flux for the type of soldering you wish to accomplish.

Rosin flux is the preferred flux for electrical soldering. Unlike acid flux, which is not preferred for electrical systems, rosin flux does not attack the base metals being soldered. Some more aggressive fluxes are available for electronic work, but these must be removed immediately after soldering. Rosin fluxes are often referred to as "no clean" flux. However, good practice is to remove any flux after soldering is complete. Alcohols such as isopropanol or methanol do a good job of dissolving rosin flux.

Tinning

Component leads are often tinned prior to soldering. *Tinning* is the process of coating leads with solder to protect them and increase wetting during the soldering process. If the surface is plated, it still must be tinned. Tin stranded wire by using the tip of the soldering iron to apply a light coat of solder to the heated wire. Apply only enough solder to lightly coat the wire. Once tinned, the wire can be easily bent and remain in that shape without fraying. If too much solder is used, the wire may move during the soldering process.

Soldering Iron

The purpose of the soldering iron tip is to transfer heat from the heating element to the solder joint. The tip is made from a base metal with good thermal conductivity, such as copper. The mass and shape of the tip affect the way it performs. For best heat transfer, the tip should be

as large as possible but not too large for closely spaced connections. During the soldering process, the tip should be cleaned and tinned often, preferably before each solder joint is made and always after the soldering iron has been idle for a period of time.

TECH TIP
Cleaning Soldering Irons

Plated tips should never be treated with any abrasive method that might damage the plating. The best method of cleaning is wiping the tip on a sponge wet with water then immediately applying a light coating of rosin core solder to the tip. The clean and tinned tip provides the best thermal transfer to the joint to be soldered.

Soldering iron temperatures are normally set between 700°F and 800°F. Take care to apply the soldering tip quickly and accurately. When the tip of the soldering iron is applied to the joint, heat is transferred quickly from the tip to the work and the temperature of the tip drops. If not enough heat is available or if it is not adequately transferred to the joint, the solder will not melt. On the other hand, excess heating of adjacent conductors can cause wire insulation to burn or melt or traces to peel off the circuit board.

PROCEDURE Soldering

Consistent soldering requires attention to detail and a bit of practice. Select the correct solder, flux, and soldering iron for the job. A temperature controlled soldering station is recommended for soldering electronic components, **Figure 23-21**. With the exception of surface mount components, the following steps yield good results:

1. Set the desired temperature of the soldering iron, power up the soldering station, and allow it to come up to temperature.

2. Before making the solder joint, wipe the soldering iron tip on a wet sponge and immediately apply a light coat of solder to the tip. This is tinning the tip.

3. Immediately after tinning the tip, bring it in contact with the joint to be soldered.

4. Allow a few seconds for the joint to heat up.

(continued)

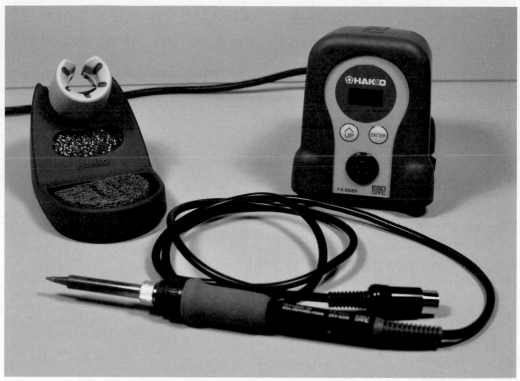

Goodheart-Willcox Publisher

Figure 23-21. Temperature-controlled soldering station. From left to right, soldering iron holder with sponge and tip cleaner, soldering iron, and power control unit with temperature control.

5. Apply solder to the heated joint and not to the soldering iron tip.

6. Continue applying heat. After the solder begins to flow, immediately remove the soldering iron from the work.

7. Do not move the soldered joint or any of its components until the solder has solidified or a "cold solder joint" will result.

8. After all soldering has been completed and everything has cooled down to room temperature, remove the remaining traces of flux with alcohol or flux remover.

9. Check the integrity of the solder joint by making sure the weld is secure and has an even appearance. Test the conductor for continuity.

Desoldering and Component Replacement

Desoldering defective components for replacement or just "lifting" one leg of a component for testing purposes is often necessary, **Figure 23-22**. Resoldering a cold solder joint might be required to correct a problem. When reheating a cold solder joint, a tiny drop of liquid flux may be applied to remove oxides and ensure maximum heat transfer is possible.

In circuit boards with plated through-holes, total solder removal may be problematic. All too often, maximum heat transfer is not achieved, and the solder is not completely removed. Without complete removal of the solder, component removal is impossible. Liquid flux or solder wick, which has been impregnated with flux, minimizes this type of problem.

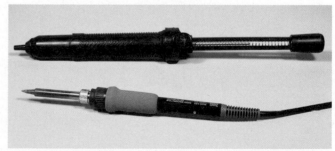

Goodheart-Willcox Publisher

Figure 23-22. A soldering iron (bottom) and a desoldering tool (top). The desoldering tool has the ability to "suck" molten solder from the connection, permitting component removal.

TECH TIP
Troubleshooting Solder Joints

Solder joints that are neither mechanically or electrically sound may result from a number of factors:

- The soldering iron temperature was too high, burning off the tin coating and allowing oxidation to occur.
- The soldering iron tip was improperly cleaned, allowing oxidation.
- Impure solder or solder with imperfections in the flux core was used.
- Insufficient tinning was applied when working with high temperatures.
- Highly corrosive fluxes caused rapid oxidation.
- Too mild flux insufficiently removed oxides from the tip.

CHAPTER WRAP-UP

Electrical systems are only as good as the basic components and connections that make them up. As an industrial maintenance technician, you must be able to evaluate these fundamental features of electrical systems and diagnose problems wherever they might occur.

Chapter Review

SUMMARY

- Resistors are used to create voltage drops or limit current. Every resistor has a tolerance and power rating.

- Potentiometers are variable resistors with three terminals. Potentiometers are used to control current by varying resistance.

- Capacitance (C) indicates the number of electrons a capacitor can store for each volt applied. It is measured in farads, F, although most capacitors are valued in microfarads, μF (10^{-6} F), or picofarads, pF (10^{-12} F).

- The dielectric constant (K), also called *permittivity*, of a material describes an insulator's ability to support an electric field between two oppositely charged conducting plates.

- Capacitors have a working voltage rating that indicates the maximum safe voltage the capacitor is capable of withstanding.

- The capacitance of capacitors in parallel is equal to the sum of the individual capacitances.

- For capacitors in series, the total working voltage equals the sum of the working voltages for the combination. The total capacitance is less than the least individual capacitance in the series.

- Electrolytic capacitors are polarity sensitive. Installing an electrolytic capacitor backward in a circuit may cause the capacitor to fail and rupture.

- Circuit boards are used to connect electrical components. Boards are built up from layers of fiberglass mats and epoxy resin and may be single sided, double sided, or multilayered.

- A fuse is a single-use device that opens a circuit when a current overload is present to prevent damage and possible fire.

- A fuse must be replaced with a fuse that has the same rating as the original. Fuses can be tested using a multimeter.

- A circuit breaker "trips out" when a current overload is present. A circuit breaker does not need to be replaced when it trips as it may be reset.

- Ground-fault circuit interrupters (GFCIs) can protect a circuit from a ground fault, and arc-fault circuit interrupters (AFCIs) prevent arcing that can cause electrical fires.

- Relays are switches controlled by electromagnets. Their three main parts include a coil, contacts, and armature.

- Relay contacts are either normally open (NO) or normally closed (NC) as they relate to the state of the contacts when a relay is de-energized, or in its "normal" state.

- Wire is available in two basic types: stranded and solid.

- The greater the cross-sectional area of a conductor, the greater its current-carrying capability and the lower its resistance.

- In the AWG (American Wire Gage), the smaller the size rating, the greater the cross-sectional area of the conductor.

- Termination refers to connecting a wire to a device or another wire. Crimping involves terminating a wire by bending a metal crimp terminal securely around it.

- Soldering joins metals with a fusion of alloys called solders. A solder is a nonferrous metallic alloy that joins other compatible surfaces together when melted.

- Flux is used to clean surfaces to be soldered of oxidation. Flux will not clean surfaces of grease, dirt, or other contamination.

- Common metals used in the solder alloy are tin, lead, and silver.

- Tinning is the process of coating leads with solder to protect them and increase wetting during the soldering process.

- Consistent soldering requires attention to detail and practice. The correct solder, flux, and soldering iron must be selected for the process.

REVIEW QUESTIONS

Answer the following questions using the information provided in this chapter.

1. Describe the two ratings assigned to resistors and what they indicate.

2. You find a resistor with a value higher than its tolerance allows. Explain one possible cause.

3. *True or False?* A potentiometer is a variable resistor with three terminals.

4. What is capacitance?

5. What is the capacitance of a circuit with a 2.9 mC charge and a produced voltage of 16 V?

6. What is permittivity of a material?

7. *True or False?* A higher dielectric constant or a thinner dielectric yields higher capacitance.

8. The _____ is the maximum voltage that is safe to place across a capacitor without the dielectric material breaking down.

9. When replacing a capacitor, what working voltage can be used if the original value is not available?

10. Compute the total capacitance of a 200 µF capacitor, a 300 µF capacitor, and a 10 µF capacitor in parallel.

11. Calculate the total capacitance of five capacitors in series with each having a capacitance of 100 µF.

12. What are the two most common materials used in constructing circuit boards? What are circuit boards used for?

13. What are fuses and circuit breakers?

14. You replace a fuse, and it blows immediately when the power is switched on. Can you use the old fuse? Should you replace it with a fuse with a higher rating?

15. Explain how to test a fuse with a voltmeter.

16. Describe the two main tripping mechanisms employed in circuit breakers.

17. A ground-fault circuit interrupter (GFCI) measures the difference between the current flowing into a circuit and the current flowing out through the _____ connection.

18. Why are arc-fault circuit interrupters (AFCIs) used in a circuit?

19. What are the three main parts of a relay?

20. What do normally open (NO) or normally closed (NC) contacts refer to?

21. *True or False?* DC relays often have a diode, a one-way valve for direct current, across the relay coil.

22. You suspect that a relay is defective. What steps should you take to confirm that the relay may need to be replaced?

23. Wire is available in two basic types: _____ and _____.

24. Would 750′ of #14 AWG wire have a higher or lower resistance and flow capability than 750′ of #8 AWG wire? Explain your answer.

25. *True or False?* Ampacity of a wire is affected by temperature.

26. Explain the three methods of termination discussed in this chapter.

27. How is flux used when soldering?

28. What is an acceptable temperature range for a soldering iron tip when soldering electronic components on a printed circuit board?

29. *True or False?* When soldering, the soldered joint and its components should be moved while the solder solidifies to prevent "cold solder joint."

NIMS CREDENTIALING PREPARATION QUESTIONS

The following questions will help you prepare for the NIMS Industrial Technology Maintenance Level 1 Electrical Systems credentialing exam.

1. What is the acceptable range of resistance values for a resistor that has a published resistance value of 150 Ω and tolerance of ±10%?

   A. 150 Ω to 165 Ω
   B. 135 Ω to 165 Ω
   C. 140 Ω to 160 Ω
   D. 135 Ω to 150 Ω

2. How does a potentiometer control current?

   A. It varies resistance using a three-terminal system.
   B. It varies resistance using a two-terminal system.
   C. It varies voltage using a three-terminal system.
   D. It varies voltage using a two-terminal system.

3. Calculate the total capacitance of three 225 μF capacitors in a series.

    A. 0.0015 μF

    B. 675 μF

    C. 225 μF

    D. 75 μF

4. The fuse in a circuit blows immediately after switching it on. What is the next step in troubleshooting?

    A. Check the bimetallic strip to determine if the circuit overheated.

    B. Install a new fuse with the same rating as the original.

    C. Install an arc-fault circuit interrupter and test the fuse again.

    D. Move the handle to the OFF position then back to the ON position to reset.

5. In a three-phase circuit, if voltage is lost in phase 1, what must be true about the voltage in phases 2 and 3?

    A. Voltage is lost in both phase 2 and phase 3.

    B. Voltage is lost in either phase 2 or phase 3.

    C. Voltage is lost in neither phase 2 nor phase 3.

    D. It depends on whether voltage was lost on the load side.

6. What does a ground-fault circuit interrupter measure to prevent a ground fault?

    A. Difference in current between ground and neutral conductors.

    B. Difference in current between hot and neutral conductors.

    C. Difference in temperature between neutral and hot conductors.

    D. Difference in temperature between neutral and ground conductors.

7. Which of the following is *not* true about a relay?

    A. Its contacts have both a current and voltage rating.

    B. Its normally open (NO) contacts remain open when the relay is energized.

    C. Its three main parts include a coil, contacts, and an armature.

    D. Its operation requires the conduction of a magnetic field.

8. Which of the following AWG wire sizes allows the highest ampacity?

    A. 00

    B. 4

    C. 13

    D. 22

9. Terminating a wire properly results in a _____ connection.

    A. high-amperage

    B. high-resistance

    C. low-amperage

    D. low-resistance

10. Which of the following steps must occur first when soldering?

    A. Apply solder to the heated joint.

    B. Bring tip in contact with the joint to be soldered.

    C. Remove traces of flux with alcohol or flux remover.

    D. Tin the tip of the metal to be soldered.

24 | Transformers

LEARNING OBJECTIVES

After completing this chapter, you will be able to:

☐ Explain the operating principles of a transformer.

☐ Understand various transformer constructions.

☐ Calculate turns ratio, voltage, current, and power for step up and step-down transformers.

☐ Describe the three types of transformer losses and calculate transformer efficiency.

☐ Understand transformer kVA ratings, sizing, and selection.

☐ Describe the structure and function of transformers with multiple windings and taps.

☐ Determine polarity and phasing of transformers and use this information to connect them.

☐ Understand various transformer types and their applications.

☐ Understand three-phase transformer applications and connection methods.

☐ Construct a three-phase transformer bank from single-phase transformers.

☐ Troubleshoot transformers.

TECHNICAL TERMS

autotransformer	phasing dots
copper losses	polarity
current ratio	saturation
current transformer	step-down transformer
eddy current losses	step-up transformer
ferroresonant transformer	tap
hysteresis losses	transformer
isolation transformer	turns ratio
mutual inductance	variable voltage transformer
phasing	voltage ratio

Transformers are used in power distribution to residential, commercial, and industrial customers. They may also be found inside commercial and industrial installations as well as in most of the electronic equipment that uses alternating current (AC) as a source of power. Transformers are used to change the voltage and current of AC power before it is finally delivered to the circuit where it is used. You will need to understand the various applications, connection methods, measurements, and troubleshooting procedures for transformers in many of your daily job assignments.

24.1 TRANSFORMER THEORY

A simple *transformer*, **Figure 24-1**, is a device consisting of two coils of wire wound on a common core. The transformer transfers energy from the primary winding to the secondary winding through mutual inductance. The primary winding may be considered as the input with the secondary winding as the output of the transformer. The primary winding of the transformer connects to the source, and the secondary winding connects to the load.

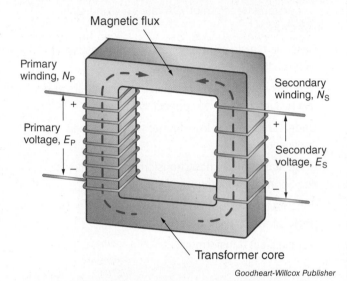

Figure 24-1. Simple transformer. Note the arrangement of primary and secondary coils around a common core.

Goodheart-Willcox Publisher

As a varying voltage is applied to the primary winding, a magnetic field rises and falls in the core. The lines of force of the magnetic field cut across the windings of the secondary coil inducing a voltage. This process, where two coils are brought close to each other and the varying magnetic field produced by one coil induces a voltage in the other, is called *mutual inductance*.

Generally, alternating current is used because direct current will not produce the rising and falling magnetic field required to transfer energy between the primary and secondary windings. In some transformer applications, pulsating DC is used on the primary. A transformer blocks DC and passes AC between the primary and secondary windings.

The core contains the magnetic field and couples it between the primary and secondary windings. Core materials are ferromagnetic and vary depending on the purpose of the transformer. Power transformers use a core of laminated iron, while pulse and radio frequency (RF) transformers use a core made of ferrite. Some RF transformers use an air core. The core material depends on the frequency of the current to be applied, **Figure 24-2**.

Ferrite

Fedorov Oleksiy/Shutterstock.com

Laminated iron

P A/Shutterstock.com

Silicon steel

Matee Nuserm/Shutterstock.com

Figure 24-2. Transformer cores may be made of any ferromagnetic material, including those shown here.

For this explanation, consider an industrial power transformer, operating at 60 Hz, with a laminated core. The laminations are made up of E- and I-shaped strips made from cold rolled silicon steel, **Figure 24-3**. Silicon steel, also known as electrical steel, generally contains 3.2% silicon. If the percentage of silicon were much higher, the steel would become too brittle. Each layer of lamination is insulated from its neighboring laminations by a thin coating of insulating material.

Laminated Core

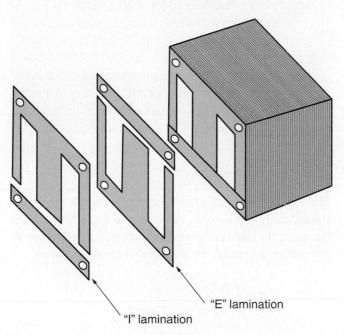

Goodheart-Willcox Publisher

Figure 24-3. E and I laminations.

The coils are wound, then the laminations are inserted to build up the core. Alternate lamination pairs are turned 180° from each other so as not to allow a gap through the entire core to form. The use of a laminated core rather than a solid one greatly reduces eddy current losses as the eddy currents are confined to a smaller area, resulting in fewer losses and making the transformer more efficient.

24.2 TRANSFORMER RATIOS

Three ratios determine how voltage and current change from a transformer's input (primary winding) to its output (secondary winding): turns ratio, voltage ratio, and current ratio. It is important to know how the three ratios relate to one another to understand the basic functions of a transformer.

24.2.1 Turns Ratio

Transformers can be step-up or step-down, **Figure 24-4**. In a ***step-up transformer***, the number of turns in the secondary is greater than the number of turns in the primary. This causes the output voltage in the secondary winding to be higher than the input voltage in the primary. Conversely, a ***step-down transformer*** has a greater number of turns in the primary than it does in the secondary and a lower output voltage in the secondary than the primary. The ratio of the number of turns in the primary to the number of turns in the secondary is the ***turns ratio***. This can be expressed as the number of turns in the primary divided by the number of turns in the secondary:

$$\text{turns ratio} = \frac{N_P}{N_S}$$

(24-1)

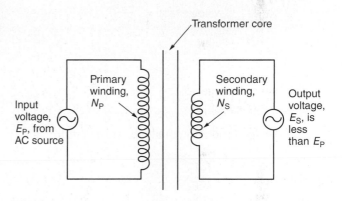

Step-down transformer

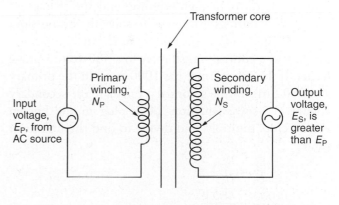

Step-up transformer

Goodheart-Willcox Publisher

Figure 24-4. Schematics for a step-down transformer and a step-up transformer.

where

N_P = number of turns in the primary winding
N_S = number of turns in the secondary winding

Consider the following example: A transformer has 1000 turns in the primary and 100 turns in the secondary. Calculate the turns ratio of the transformer.

$$\text{turns ratio} = \frac{N_P}{N_S}$$

$$\text{turns ratio} = \frac{1000}{100}$$

$$\text{turns ratio} = 10 \text{ or } 10:1$$

In the case where a transformer has a turns ratio of 1:1, it is referred to as an *isolation transformer*.

24.2.2 Voltage Ratio

The *voltage ratio* for a transformer is the ratio of primary voltage to secondary voltage:

$$\text{voltage ratio} = \frac{E_P}{E_S}$$

(24-2)

where

E_P = primary voltage
E_S = secondary voltage

The voltage ratio also equals the turns ratio:

$$\frac{E_P}{E_S} = \frac{N_P}{N_S}$$

(24-3)

The voltage in either winding may be solved if the number of turns in each winding and the voltage of one of the windings is known. To solve the proportion, cross multiply and divide. For example, you connect the primary of a control transformer to a 480-volt, 3-phase feeder. If the transformer has 1000 turns in the primary and 250 turns in the secondary, calculate the secondary voltage. To solve equation 24-3, plug in the known values, cross multiply, and divide to find the unknown value:

$$\frac{E_P}{E_S} = \frac{N_P}{N_S}$$

$$\frac{1000}{250} = \frac{480 \text{ V}}{E_S}$$

$$1000 \times E_S = 250 \times 480 \text{ V}$$

$$1000 \times E_S = 120{,}000 \text{ V}$$

$$E_S = \frac{120{,}000 \text{ V}}{1000}$$

$$E_S = 120 \text{ V}$$

TECH TIP
Connecting Transformers

For the transformer in the previous example, it is important to know which winding is the primary and which is the secondary. All too often, technicians connect the secondary of the transformer to the 480 VAC input voltage. With a 4:1 turns ratio, the reversed transformer now has 480 V × 4 on the output, which equals 1920 V and greatly exceeds the voltage rating of the winding insulation. The winding insulation is typically rated for approximately 600 V.

With this overvoltage condition, the windings arc over, creating a short circuit, and the transformer begins to burn as long as the voltage is applied. Once this happens, the transformer is permanently damaged. Always be certain you have the primary and secondary of the transformer properly connected before applying power.

24.2.3 Current Ratio

The *current ratio* of a transformer is the inverse of the voltage ratio or the turns ratio. As the secondary voltage increases, the secondary current decreases. As the secondary voltage decreases, the secondary current increases.

$$\frac{I_S}{I_P} = \frac{N_P}{N_S}$$

(24-4)

$$\frac{I_S}{I_P} = \frac{E_P}{E_S}$$

(24-5)

where

I_P = primary current
I_S = secondary current

Imagine you are installing a step-down transformer to provide AC voltage to a group of control relays. The number of turns on the primary is 1200, and the number of turns on the secondary is 120. You connect the primary of the transformer to 120 VAC. The current in the secondary is 5 A. Calculate the current drawn by the primary of the transformer.

$$\frac{I_S}{I_P} = \frac{N_P}{N_S}$$

$$\frac{5 \text{ A}}{I_P} = \frac{1200}{120}$$

$$1200 \times I_P = 5 \text{ A} \times 120$$

$$I_P = \frac{600 \text{ A}}{1200}$$

$$I_P = 0.5 \text{ A or } 500 \text{ mA}$$

Another example: A step-up transformer has 100 turns on the primary winding and 250 turns on the secondary winding. The primary current is 2 A. Calculate the secondary current.

$$\frac{I_S}{I_P} = \frac{N_P}{N_S}$$
$$\frac{I_S}{2\,A} = \frac{100}{250}$$
$$I_S \times 250 = 100 \times 2\,A$$
$$I_S = \frac{200\,A}{250}$$
$$I_S = 0.8\,A \text{ or } 800\,mA$$

Observe that current is stepped down as voltage is stepped up, and current is stepped up as voltage is stepped down.

24.3 TRANSFORMER POWER

The power on the primary (input) equals the power on the secondary (output) for both step-up and step-down transformers. This is often referred to as the conservation of power. Recall the power equation from Chapter 20, *Ohm's Law, Kirchhoff's Laws, and the Power Equation*:

$$P = E \times I$$

(24-6)

where

P = power in watts (W)
E = voltage in volts (V)
I = current in amperes (A)

Consider an ideal transformer with 400 turns in the primary and 100 turns in the secondary. The primary current is 1A and the secondary current is equal to 4 A. The primary voltage is 480 VAC while the secondary voltage is 120 VAC. The primary power and secondary power may be calculated as follows:

$P_P = E_P \times I_P$
$P_P = 480\,V \times 1\,A$
$P_P = 480\,W$
$P_S = E_S \times I_S$
$P_S = 120\,V \times 4\,A$
$P_S = 480\,W$

24.4 TRANSFORMER LOSSES

In real applications, transformers do not convert all input power into usable output power. That is, there is no such device as an ideal transformer. Transformers have losses in the form of eddy current losses, hysteresis losses, and copper losses. Transformer losses result in transformer heating. The greater the transformer losses, the lower the transformer efficiency.

Eddy current losses result from the magnetic field linking to other parts of the transformer, such as the core material, rather than the secondary winding. Because the core material is a conductor, it acts like a shorted turn, which consumes current. The eddy current losses manifest themselves in the form of heat. By using a laminated core as opposed to a solid core, **Figure 24-5**, the "shorted turns" are much smaller, consuming less current and reducing the total amount of eddy current losses.

Hysteresis losses result from the orientation and reorientation of the grains in the ferromagnetic material of the core. Each time the grains (also referred to as domains) must reorient themselves due to the change in the magnetic field, a small amount of power is consumed. The consumed power results in core heating. The type of core material used determines how much hysteresis loss is experienced. Electrical steel (cold-rolled, grain-oriented silicon steel) is used in modern transformers to lessen hysteresis losses.

Copper losses result from the resistance of the transformer windings. Conductors have a certain amount of resistance based on their cross-sectional diameter and length,

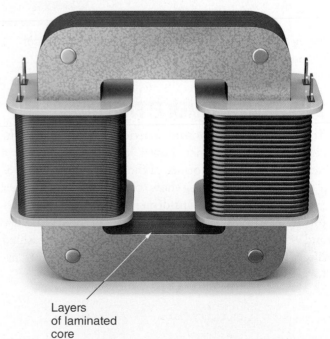

Layers of laminated core

Slavoljub Pantelic/Shutterstock.com

Figure 24-5. Transformer with a laminated core. The laminated layers help reduce eddy current losses.

Figure 24-6. As the current increases, copper losses also increase. The measured resistance of an unconnected transformer winding equals the copper loss of that winding.

Although there is not much you can do about transformer losses, you should be aware of their effects. The heat produced by transformer losses must be taken into consideration in selecting and installing transformers.

Copper Wire Resistance at 20°C (68°F)		
American Wire Gage	Diameter in Mils	Ohms per 1000' (305 m)
10	101.90	0.998 Ω
12	90.81	1.58 Ω
14	64.08	2.53 Ω
16	50.82	4.02 Ω
18	40.30	6.39 Ω
20	31.96	10.15 Ω
22	25.35	16.14 Ω
24	20.10	25.67 Ω
26	15.94	40.81 Ω
28	12.64	64.90 Ω
30	10.03	103.20 Ω
32	7.95	164.10 Ω
34	6.31	260.90 Ω
36	5.00	414.80 Ω
38	3.97	639.60 Ω
40	3.15	1049.00 Ω

Goodheart-Willcox Publisher

Figure 24-6. Resistances for various gages and sizes of copper wire.

24.5 TRANSFORMER EFFICIENCY

Transformer losses consume energy and reduce the efficiency of a transformer. Lower losses correspond to a more efficient transformer. Efficiency is expressed as a percentage indicating how closely the transformer approaches the ideal transformer model. Modern transformers typically have an efficiency of between 90% and 95%. While only copper losses may be measured directly, you can determine the efficiency of a real transformer with the following equation:

$$\text{efficiency (\%)} = \frac{P_{\text{out}}}{P_{\text{in}}} \times 100\%$$

(24-7)

where

P_{out} = power at the transformer output (secondary)

P_{in} = power at the transformer input (primary)

Only *true power* (watts) is used in transformer efficiency calculations and not *reactive power* or *apparent power* as losses are dissipative (heat producing) in nature.

Consider the following example: A control transformer has 480 V supplied to the primary winding with a current of 1.5 A. The voltage measured across the secondary winding is 120 V, and the measured current through the winding is 5.52 A. Calculate the efficiency of the transformer. First, calculate power in the primary:

$P_{\text{P}} = E_{\text{P}} \times I_{\text{P}}$

$P_{\text{P}} = 480 \text{ V} \times 1.5 \text{ A}$

$P_{\text{P}} = 720 \text{ W}$

Then, calculate power in the secondary:

$P_{\text{S}} = E_{\text{S}} \times I_{\text{S}}$

$P_{\text{S}} = 120 \text{ V} \times 5.52 \text{ A}$

$P_{\text{S}} = 662.4 \text{ W}$

Last, calculate efficiency using the values for primary and secondary power:

$$\text{efficiency} = \frac{P_{\text{out}}}{P_{\text{in}}} \times 100\%$$

$$\text{efficiency} = \frac{662.4}{720} \times 100\%$$

$$\text{efficiency} = 0.92 \times 100\%$$

$$\text{efficiency} = 92\%$$

At 92% efficiency, 57.6 W (720 W − 662.4 W = 57.6 W) of power are being lost by the transformer as heat. At higher currents with the same efficiency rating, the heat generated by the transformer may become quite significant. As a comparison, imagine a larger transformer with primary current of 15 A, secondary current of 55.2 A, primary power of 7200 W, and secondary power of 6624 W. That transformer would generate 576 W of power lost as heat.

24.6 TRANSFORMER RATINGS

If you read the specifications or look on the nameplate of a transformer, you will notice the transformer power is rated in VA (volt-amperes) or kVA (kilovolt-amperes). Transformer power is rated in VA rather than W to distinguish among three different types of power: *true power*, *reactive power*, and *apparent power*. Refer to Chapter 22, *Alternating Current*, for detailed definitions and methods for calculating these different types of power.

In Chapter 22, you learned that true power (W) is the power that does work. Because transformers have inductive elements, they have reactive power (VAR) used by reactive elements in an inductive component. Apparent power (VA) is the total power consumed by the circuit. It accounts for the power required to do work as well as the power consumed in internal reactive elements. Apparent power (VA) is used to rate transformers because it measures the actual power that must be supplied.

24.7 MULTIPLE WINDINGS AND TRANSFORMER TAPS

By adding multiple windings or taps to a transformer, you can adjust the turns ratio, apply different input voltages, or produce multiple output voltages using the same transformer. Multiple windings are placed on the same core and may be either primary or secondary depending on the transformer design. *Taps* are connections at one or more intermediate points within a winding, **Figure 24-7**.

One application for taps on a primary winding is to allow the adjustment of a transformer to accommodate for slight deviations in the input voltage from the nominal design value. For example, nominal input voltage might be 120 VAC, but actual line voltage might range from 110 VAC to 130 VAC. Primary taps allow you to select the input closest to the measured line voltage at the time of transformer installation. As a result, the actual secondary voltage is closer to the design value.

Consider the pole or pedestal transformer that the power company places near your house. The primary voltage feeding this transformer might be either 7.2 kV or 12.4 kV. The secondary has three connections to the triplex (three-wire) drop leading to your house. The secondary is wound to produce 240 VAC with a center tap for the neutral connection. The voltage between the start and finish of the winding is 240 VAC, while the connection between either the start or finish of the secondary winding and the neutral (center-tapped) connection is 120 VAC, **Figure 24-8**.

On a center-tapped winding, the start to center tap and the finish to center tap are 180° out of phase with each other.

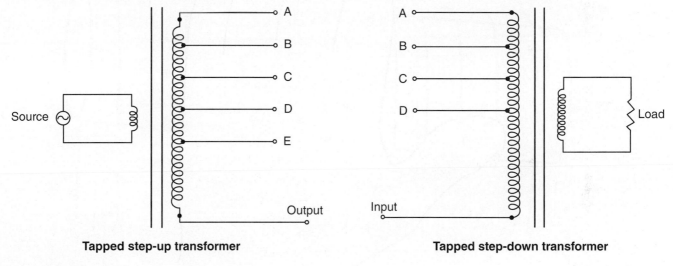

Tapped step-up transformer   **Tapped step-down transformer**

Goodheart-Willcox Publisher

Figure 24-7. Tapped step-up and step-down transformers. The voltage and turns ratios change depending on which tap the input and output leads connect to.

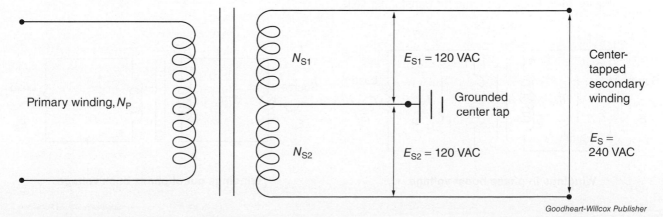

Goodheart-Willcox Publisher

Figure 24-8. Transformer with center-tapped secondary.

To visualize this, consider the voltage across the coils from the perspective of the center tap. The grounded center tap creates a 0 V point, so when the first half is nearing its low point, the second half approaches its peak voltage.

24.8 TRANSFORMER POLARITY AND PHASING

The direction of instantaneous current flow in a transformer winding is often referred to as its **polarity**. Polarity is also referred to as the *phase* of the winding. The direction you connect windings is referred to as **phasing**. The phasing of the two connected windings determines if they are in phase ("boosting" each other) or out of phase ("bucking" each other).

In a transformer, it is possible to wind the coils in the same or opposite directions. Apply the right-hand rule to determine the direction of the magnetic field at any instant, **Figure 24-9**. Using your right hand, wrap your fingers around the coil in the direction of the current. Your thumb points to the north pole of the magnetic field. Remember that with alternating current, the direction of current flow changes periodically, so the right-hand rule only applies instantaneously during the alternating current cycle. Knowing the direction of the magnetic field allows you to wire transformers together in such a way that their voltages either boost or buck each other, **Figure 24-10**.

Some manufacturers place a dot next to one of the leads on each winding. These **phasing dots** indicate the phase relationships between the windings. If currents are entering or leaving both dotted terminals at the same time, then

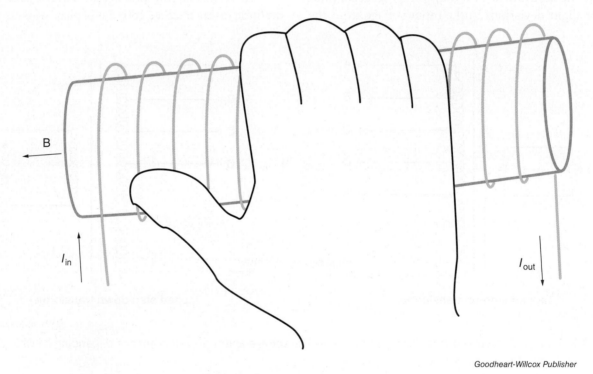

Goodheart-Willcox Publisher

Figure 24-9. Right-hand rule for transformer coils.

Windings in phase boost voltage

Windings out of phase buck voltage

Goodheart-Willcox Publisher

Figure 24-10. Boost and buck transformer connections. Note the use of phasing dots in this schematic.

the windings are in phase. Other manufacturers label the leads as H1, H2, H3, and H4 for the primary connections and X1, X2, X3, and X4 for the secondary connections. If the manufacturer places phasing dots next to the markings, all odd-numbered markings have a dot whereas even-numbered markings do not. If the connections are not marked to indicate phase relationships between the windings, you can test with a voltmeter.

PROCEDURE	Connect Transformers in Series

Use the following procedure to set up a series connection of secondary windings to achieve either a boost or buck voltage. Suppose you have a transformer with two secondary windings, one of which is 24 VAC and the other is 120 VAC, **Figure 24-11**.

1. Ensure the supply voltage is off with your voltmeter.
2. Connect the primary winding to the supply.
3. Ensure the output leads are not touching so as not to create a short circuit.
4. Switch on the power and measure the output voltage of each secondary winding.
5. Turn the power off and verify with your voltmeter.
6. Connect one lead of each winding together.

⚠ CAUTION

The remaining two leads should not be connected to the load at this time.

7. Switch the power on and measure the voltage between the remaining two unconnected leads.
8. If you measure 144 VAC, the phasing is correct for a voltage boost. If, however, you measure 96 VAC, the phasing is correct for a voltage buck.
9. If you have not achieved the desired voltage, perform the following steps:
   A. Turn the power off.
   B. Verify with your voltmeter.
   C. Exchange the connection of only one side of one of the transformer windings.
   D. Turn the power back on.
   E. Measure the two unconnected leads with your voltmeter to verify you have achieved the desired voltage as a result of correct phasing.
10. Turn the power off and verify with your voltmeter.
11. Connect the two remaining leads to your load.

⚒ TECH TIP
Transformer Phasing

If you achieve a voltage boost, the two windings are in phase with each other. If you achieve a voltage buck, the two windings are 180° out of phase with each other so that the voltage in one winding cancels the same amount of voltage in the other winding.

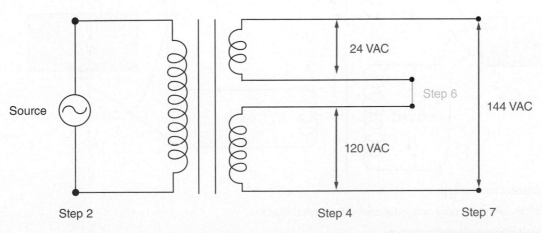

Figure 24-11. Transformer series connection.

A parallel connection of two secondary windings can be used to increase the current output ability of the transformer. The phasing of transformer windings connected in parallel is especially critical. In a parallel connection, if you reverse the phasing of either the primary or secondary, both windings dump all their power into the connecting secondary wiring, which may damage equipment.

PROCEDURE	Connect Transformers in Parallel

Two identical control transformers can be wired in parallel to provide the necessary current to power a load. Each transformer requires 480 VAC on the primary winding, and each is capable of 120 VAC on the secondary winding, **Figure 24-12**.

⚠ CAUTION

If you are connecting two transformers in parallel, it is imperative that they both have exactly the same specifications. Additionally, you must have the correct phase relationship between both secondary windings. This procedure will ensure that the windings of both transformers are properly phased. Follow the procedure exactly as outlined to prevent a circuit failure.

1. Turn off the supply power and verify this fact with your voltmeter.
2. Connect one primary lead of each transformer to one of the three input phases. In this example, call this phase A.
3. Connect the other primary lead of each transformer to another of the three input phases. In this example, call this phase B. Note that you are only using two of the three phases.
4. Ensure that all secondary transformer leads are disconnected and not touching anything, so as not to cause a short circuit.
5. Switch on the power and measure the two secondary windings to verify each has 120 VAC on the output.
6. Turn off the power and verify with your voltmeter.
7. Connect one lead of each secondary to the other.
8. Turn the power on and measure the voltage between the two remaining leads with your voltmeter.
9. If you measure 0 VAC between the two remaining leads, they are properly phased and in phase with each other.

(continued)

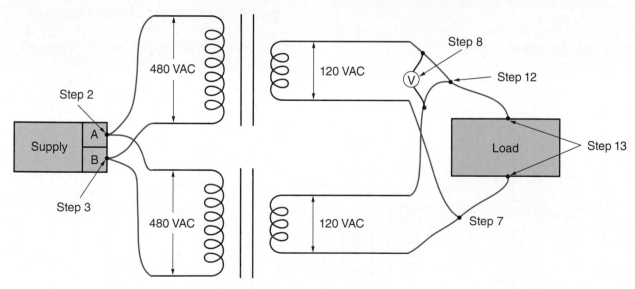

— Black = intial setup

— Blue = Primary connection for secondary initial check

— Red = Secondary connections

Goodheart-Willcox Publisher

Figure 24-12. Transformer parallel connection.

10. If you measure 240 VAC between the two remaining leads, they are 180° out of phase, and the connection must be corrected. If the windings are out of phase, perform the following steps:

   A. Turn the power off and verify with your voltmeter.

   B. Exchange the connected lead of only one of the secondary windings with the unconnected one. This exchange will reverse the phase of one of the windings and put both windings back in phase with each other.

   C. Turn the power back on and verify that you have a 0 VAC reading between the two unconnected leads.

11. Turn off the power, and verify with your voltmeter.

12. Connect the two remaining unconnected secondary leads together to form a parallel circuit between the two secondary windings.

13. Now connect the two junctions of the paralleled secondary windings to the load.

⚠ CAUTION

Ensure the windings are in phase with each other to connect them in parallel. If you do not read 0 VAC in step 9, do not connect the remaining leads until you have resolved this phasing issue.

24.9 SPECIALIZED TRANSFORMERS

In addition to the standard transformer types, there are many different transformers used for specialized applications. In electronics, transformers are used to match impedance between circuits, separate a mixed AC and DC signal and couple the AC component into the next circuit in line, couple pulses into other circuits, and other applications.

24.9.1 Autotransformers

An *autotransformer* differs from a standard transformer type in that it has only one winding and shares one connection between the primary and secondary. It may have one or more taps and may be used as a step-up or

step-down transformer, depending on how it is connected, **Figure 24-13**. In a simple autotransformer with only one tap, both the primary and secondary are connected to the start of the winding. The end of the winding and the tap serve as connections for the opposite sides of the primary and secondary. Primary and secondary connections are determined by the step-up or step-down function desired.

Autotransformers may have other taps in the primary or secondary side of the winding to adjust the turns ratio to achieve different voltages. Notice that in any case, the primary and secondary share a portion of the single winding. The turns ratio functions exactly as for multiple windings, as in a standard transformer. The calculations for turns ratio, voltage ratio, and current ratio are also the same as those for a standard transformer.

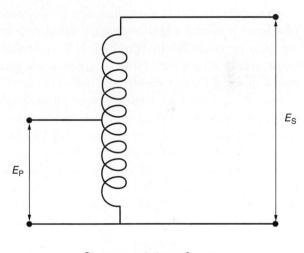

Step-up autotransformer

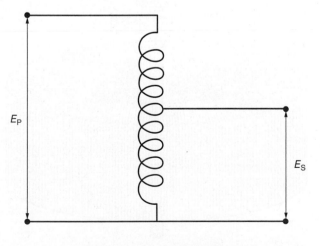

Step-down autotransformer

Goodheart-Willcox Publisher

Figure 24-13. Step-up and step-down autotransformer schematics.

The major benefit of an autotransformer is that it is much more economical to manufacture because it only has one winding. However, autotransformers provide no isolation between the primary and secondary. Standard transformers isolate the input and output through magnetic coupling, so there is no direct connection between the primary and secondary of the transformer. With an autotransformer, one side of both the primary and secondary is always connected to the line input. This may allow noise or transients on the line input to transfer to the output through the direct connection.

24.9.2 Variable Voltage Transformers

A **variable voltage transformer** is a variation of the autotransformer. It is wound on a toroidal (donut-shaped) core, **Figure 24-14**. The core material used by the variable voltage transformer is most often ferrite, a mixture of powdered iron and a binder material fused together under great pressure. Ferrite is quite brittle, so it easily chips or breaks if dropped. However, it makes a great magnetic material for transformer and inductor cores.

Once the coil has been tightly wound on the toroidal core, the winding is machined down to create a flat surface where the insulation of the wire has also been removed. A carbon "brush" is arranged so that it comes in contact with the machined surface forming a movable tap. Rotating the shaft of the variable voltage transformer allows repositioning of the tap, which creates a variable turns ratio. Moving the tap will change the output voltage proportional to its position.

Once again, the turns ratio, voltage ratio, and current ratio calculations for a standard transformer apply based on the position of the variable tap. Extra fixed taps may be employed to allow for variations in line input voltage or the range of the variable voltage. Three variable voltage transformers may be mechanically ganged in order to create a three-phase configuration.

24.9.3 Ferroresonant Transformers

A **ferroresonant transformer** is often referred to as a *constant voltage transformer* or *voltage regulating transformer*. It differs from a standard transformer in that it has an extra winding connected to a capacitor, **Figure 24-15**. The extra winding forms a resonant circuit when capacitive reactance from the capacitor and inductive reactance from the winding are equal. The value of the capacitor is selected so that its capacitive reactance matches the inductive reactance of the winding.

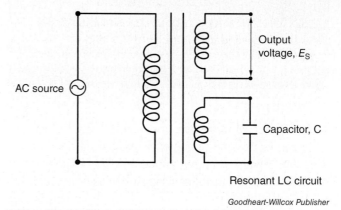

Goodheart-Willcox Publisher

Figure 24-15. Basic ferroresonant transformer.

The resonant circuit causes the transformer core to go into saturation at some voltage level. **Saturation** means that the magnetic field in the core can go no higher regardless of how high the input voltage to the primary becomes. Once the voltage level reaches the point where the core is in saturation, the peaks of the sinusoidal waveform begin to become clipped or flat topped, **Figure 24-16**. Increasing the voltage past the core saturation point clips the wave peaks more, and the flat top becomes wider. This clipping maintains peak-to-peak voltage at the same level regardless of the input voltage.

Ferroresonant transformers can also limit current. For example, if a dead short is placed across the output of a ferroresonant transformer, the current drops, making the entire unit short-circuit immune when the current demand becomes too great.

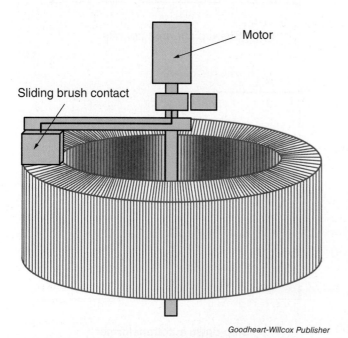

Goodheart-Willcox Publisher

Figure 24-14. Variable voltage transformer on a toroidal core. This variable transformer has a ferrite core and copper windings.

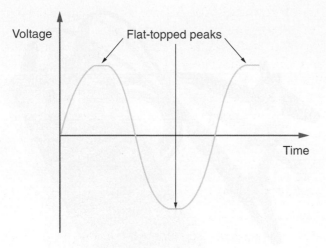

Figure 24-16. Basic ferroresonant transformer waveforms.

24.9.4 Current Transformers

A ***current transformer*** is not actually a transformer at all, as it is only a single winding on a toroidal form often made of ferrite, silicon steel, or nickel alloy. It becomes a transformer when a current-carrying conductor is passed through the "doughnut hole" of the toroidal coil, **Figure 24-17**. The conductor passing through the center of the coil becomes the primary and the coil itself becomes the secondary. The magnetic field surrounding the primary conductor induces a current, which flows through the secondary winding. A current transformer is often referred to as a *CT*. These devices are most commonly used in current metering or for making other measurements on an AC waveform. Some CTs have a split core that enables installation without disconnecting the conductor to be metered. The gap caused by splitting the core results in lower accuracy than a solid toroidal model. A common example of a split core CT is the one in a clamp-on ammeter.

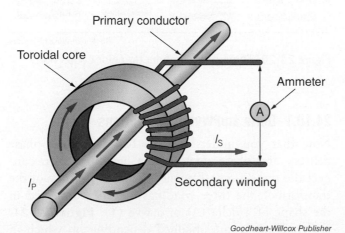

Figure 24-17. Current transformer.

Current transformers are rated by current ratio. Standard CTs range from 50:5 through 300:5 in various increments. For example, if you are using a 50:5 CT, 50 A of current flow in the primary conductor through the center of the CT, and 5 A flow through the secondary output of the CT. It is permissible to loop the primary conductor through the toroid a second time to create a new current ratio. For example, if a CT is originally 100:5, the second loop makes a new ratio of 50:5.

The output load is called the "burden" rather than the load to avoid confusion with the load connected to the current-carrying conductor that serves as the primary. The burden in most cases is a 5 A, full-scale ammeter or other current-measuring circuit.

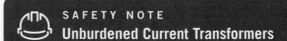

SAFETY NOTE
Unburdened Current Transformers

As the current in a transformer secondary decreases, the voltage increases proportionally to the turns ratio. This means a CT with 480 VAC and 100 A through the primary conductor creates extreme voltages across the burden connections. Never leave the current flowing through the primary of a CT while it is unconnected to a burden. The extremely high voltages produced may arc over, potentially damaging equipment and posing a serious safety risk. Current transformers are designed to handle the voltages and currents they are rated for only when an appropriate burden is connected.

24.10 THREE-PHASE TRANSFORMERS

Three-phase current is the standard for transmitting power and operating large loads. It is much more efficient than single-phase power for supplying large loads.

When the power company delivers electricity to commercial or industrial installations, they often use distribution transformers. Distribution transformers typically have a 3–500 kVA power rating with an input voltage of up to 15 kV. Transformers of 167 kVA or less are commonly pole mounted, and higher-rated transformers are most likely pad or platform mounted. Three-phase transformers may be a bank of three connected single-phase units or one three-phase unit with all windings on a common core.

Once power leaves the distribution transformer, the power company meters the electricity before it enters the building and terminates at the main disconnect.

At the entrance point, the power company supplies four conductors: three phase conductors and one ground conductor. The three phase conductors are commonly referred to as phase A, phase B, and phase C. The three phases are 120° apart with phase B following phase A, phase C following phase B, and phase A following phase C. See **Figure 24-18**.

All three phases enter the building via conductors with black insulation, and the power company does not identify which phase is which. The particular phase of each conductor does not matter, but the conductors must have the correct phase relationship.

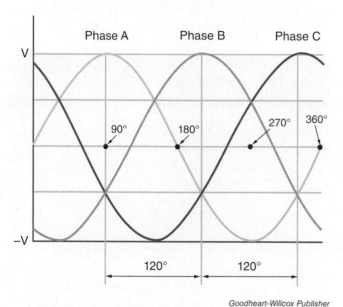

Goodheart-Willcox Publisher

Figure 24-18. Phase relationship in three-phase conductor.

Only two possibilities exist for the order of the three conductors: either forward (ABC) or backward (CBA). You will need to use a phase rotation tester, **Figure 24-19**, to determine if the three phases are in the forward order or the reverse order. Blindly hooking the phases up gives you a 50% chance of being correct. Verify the phase order with a phase rotation detector. If you have them in reverse order, switch the connections of any two conductors to correct the phase relationship.

Once you have the three phases in the correct order, identify the conductors with bands of the appropriate color of electrical tape. The *NEC* requires that all equipment grounding conductors be identified on each end with a green color, and any grounded (neutral) conductor must be identified with a white color. The identification colors of the phase conductors are not mandated by the *NEC*, other than they must not be green or white. Convention, however, provides for specific colors for phase and voltage. Please refer to the following color-coding chart, **Figure 24-20**.

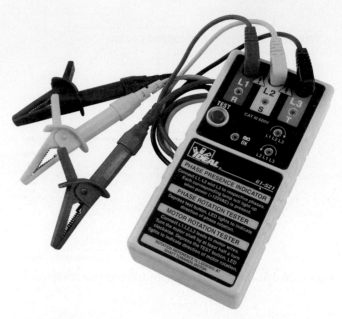

Ideal Industries, Inc.

Figure 24-19. Phase rotation detectors help maintenance technicians install and maintain transformers and motors.

Function	Color Code for 120/208/240 V	Color Code for 277/480 V
Three Phase Line A		
Three Phase Line B		
Three Phase Line C		
Grounded (Neutral)		
Equipment Grounding		

Goodheart-Willcox Publisher

Figure 24-20. Three-phase conductor color code chart.

24.10.1 Delta and Wye Combinations

Now that you understand the basics of three-phase delivery and phase rotation, the next step is to connect the transformer. Depending on the purpose of the transformer, the three windings may be connected in the shape of a delta (Δ) or a wye (Y), **Figure 24-21**. Different results are obtained depending on which of the forms you choose.

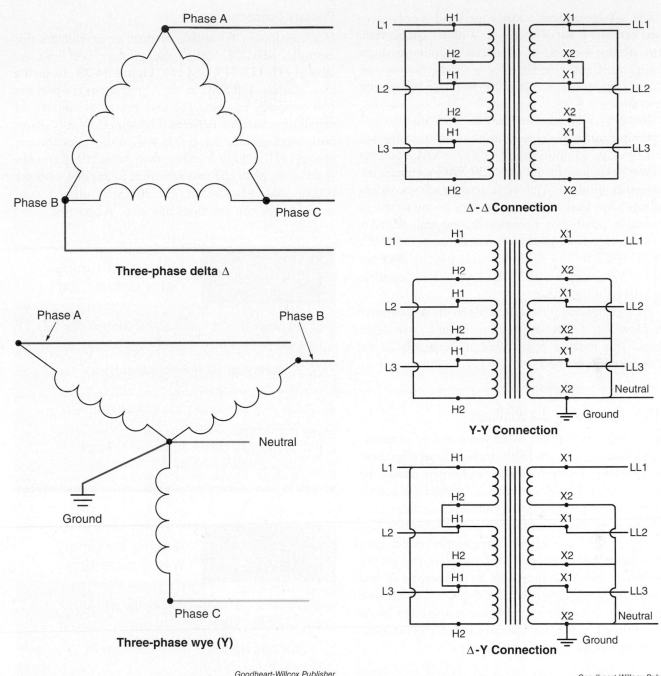

Three-phase delta Δ

Three-phase wye (Y)

Δ-Δ **Connection**

Y-Y **Connection**

Δ-Y **Connection**

Figure 24-21. Delta and wye configurations.

Figure 24-22. Delta and wye combinations.

Primary and secondary windings do not have to be wired in the same form. There are four possible form combinations for primary-secondary: Δ-Δ, Δ-Y, Y-Δ, and Y-Y, **Figure 24-22**. In connecting the windings in any of these combinations, proper phasing is important. As a general rule, the primary windings of each transformer or transformer section are labeled H1 and H2. The secondary windings are labeled X1 and X2. The 1 marking denotes the start of the winding, and the 2 denotes the end of the winding. This allows you to determine the phasing of each of the windings and properly connect them.

24.10.2 Three-Phase Transformer Connections and Phasing

The Δ-Δ connection is one of the most common. This configuration is advantageous because it can continue functioning even if one of the transformers in the bank fails or is removed. The remaining two transformers may operate in the open Δ or V configuration. In this situation, the transformer bank still develops voltages and currents in their correct phase configurations. However, the transformer bank capability is reduced.

The Y-Y connection is rarely used and is only mentioned here as a point of reference. With a Y connection, failure of any winding or connection results in single phasing, which is disastrous for a three-phase motor. In such a failure, only one phase-to-phase connection is operable.

The Δ-Y connection method is often used to create multiple output voltages from the same transformer. 208/120 VAC outputs and 480/277 VAC outputs are most common. In the 208/120 VAC scenario, the transformer produces 120 VAC across each secondary winding. One lead of each secondary is connected to a common point that becomes the neutral. Measuring the voltage from phase to phase returns 208 VAC, or 120 VAC × √3. A similar scenario is used for 480/277 VAC outputs. 277 VAC is commonly used for industrial lighting circuits.

The Y-Δ method is used similarly to the Δ-Y connection. However, it operates in reverse order to step down voltage. This method is not used as frequently as the Δ-Y connection method.

24.10.3 Connecting the Windings

When using transformers in an industrial or commercial installation, you are likely to be connecting them in a step-down configuration. Transformers should be selected to provide an appropriate VA capacity based on the demands of the loads to be connected to them.

Most three-phase transformers have a schematic diagram on the label to assist in making proper connections. Additionally, tables are provided to select the proper taps, if available, to compensate for differences in line voltage. Many single-phase transformers, which you might use to construct a three-phase transformer bank, also include a schematic of the transformer and information about its taps.

Connections for Dual Input Voltage Transformer

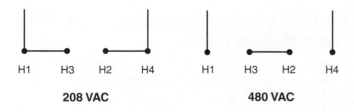

208 VAC   480 VAC

Goodheart-Willcox Publisher

Figure 24-23. Wiring schematic for transformer with dual input voltages, 208 VAC and 480 VAC.

Primary Connections

If a transformer is capable of dual input voltages (for example, 480/208 VAC), the primary windings are labeled H1, H2, H3, and H4, **Figure 24-23**. In such a case, if using 480 VAC as the input, series connect the two windings by using H1 and H4 as the input and installing a jumper between H2 and H3. This places both windings in series. If 208 VAC is the input voltage, connect H1 and H3 together, then connect H2 and H4 together to place the two windings in parallel with the proper phase relationship. The 208 VAC input is connected to the two junctions that were just created.

PROCEDURE	Making a Primary Delta Connection

If you chose Δ as the primary connection method, connect each primary between two phases:

1. Connect the first transformer between phase A and phase B.
2. Connect the second transformer between phase B and phase C.
3. Connect the third transformer between phase C and phase A.

PROCEDURE	Making a Primary Wye Connection

If you chose a Y connection for the primaries, complete the following steps:

1. Connect H1 of the first transformer to phase A.
2. Connect H1 of the second transformer to phase B.
3. Connect H1 of the third to phase C.
4. Connect all three H2 leads together.

It is not critical that the phasing of the primary windings is correct, as the phasing may be corrected when connecting the secondary windings. Good practice dictates that phasing should be preserved in the primary windings if at all possible.

TECH TIP
Connecting Transformers

Follow these guidelines when making transformer connections:

- Do not make any connections with the line power on.
- Correctly install a disconnecting means.
- Comply with all provisions of *NEC Article 450* regarding overcurrent protection and transformer installation.
- Practice necessary lockout/tagout (LOTO) and other safety procedures.

Secondary Connections

It is critical that the secondary windings be phased properly. Even if all leads are clearly identified and you have checked your wiring, mistakes are still possible. Any problem with phasing in the secondary windings can lead to catastrophic circuit failure.

PROCEDURE   **Preparing to Make a Secondary Connection**

The following steps outline proper phasing verification procedures. Check phasing before making secondary connections to keep both you and the equipment safe.

1. Once you have connected the primaries, ensure that the secondary windings are unconnected.

2. Next, switch on the power and measure each of the secondary windings to verify they have the proper output voltage.

3. After you perform these voltage checks, switch the power off, perform LOTO procedures, and proceed with secondary connections.

PROCEDURE   **Secondary Wye Configuration**

Consider a case where you want to connect transformers in a Δ-Y configuration. The transformers have turns ratio of 4:1 and 3-phase, 480 VAC input. The individual secondary windings should each read 120 VAC in your preparatory phasing check.

1. If the leads are marked, connect the X2 leads of two transformers together. If they are unmarked, make a guess.

2. Switch the power on and measure between the X1 leads of the transformers you just connected. If you read 208 VAC, your connections are correct. If you read 120 VAC, you will need to reverse the connections of one of the transformers.

3. Switch off the power, reverse the transformer leads of one transformer secondary if necessary, and connect the X2 lead of the remaining transformer to the junction of the other X2 leads.

4. Switch the power back on and measure phase-to-phase voltage on the X1 leads of each transformer. If you read 208 VAC on all three phase-to-phase measurements, you have connected all three transformers correctly. If, however, you have one reading of 208 VAC and two readings of 120 VAC, you must reverse the connections of the last transformer you connected.

When done properly, the common connection of all three transformers becomes the neutral. You will measure 208 VAC phase-to-phase and 120 VAC from any phase to neutral. Refer to the following diagram, **Figure 24-24**.

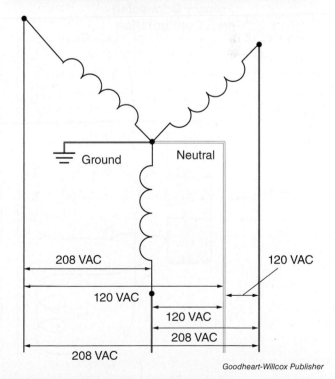

Goodheart-Willcox Publisher

Figure 24-24. Wye-configuration phasing.

The secondary delta configuration is a little more involved as it gives multiple opportunities to create an unwanted and unsafe outcome. It is important to double and triple check your connections and measurements.

PROCEDURE   Secondary Delta Configuration

For this example, you are making a Δ-Δ configuration with a turns ratio of 1:1 and 480 VAC input. You have already made your primary connections, and the unconnected secondary windings measure 480 VAC.

1. Begin by connecting the X1 lead of one transformer with the X2 lead of another.

2. Apply power and measure the voltage between the two unconnected leads of the transformers you just connected. If you measure 480 VAC, everything is connected properly, **Figure 24-25A**. If not, switch off the power and reverse the windings of one of the transformers.

3. When your voltage measurements are correct, connect only one lead of the remaining transformer to either of the unconnected leads of the connected transformers.

⚠ CAUTION

Do not make the final connection at this time!

4. Switch the power back on and use your voltmeter to make a voltage measurement between the remaining two unconnected leads. If you read 0 V, proceed to the next step, **Figure 24-25B**. If you measure any voltage other than 0 V, you need to reverse the leads of the last transformer you connected.

5. With the voltage readings correct and the power switched off, you may close the delta by connecting the two remaining unconnected leads.

6. Measure the phase-to-phase voltage from each corner of the delta connection, and verify that all are correct. Refer to **Figure 24-26**.

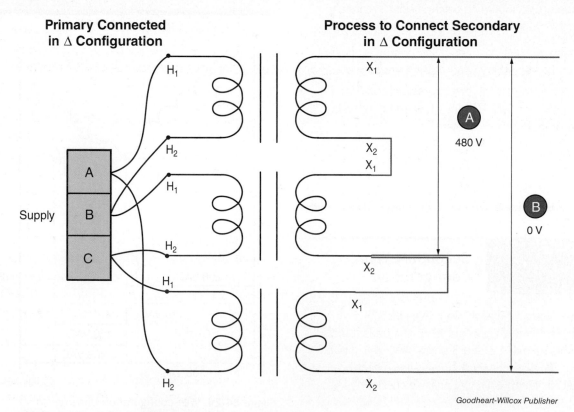

Primary Connected in Δ Configuration

Process to Connect Secondary in Δ Configuration

Goodheart-Willcox Publisher

Figure 24-25. Delta-configuration set-up procedure.

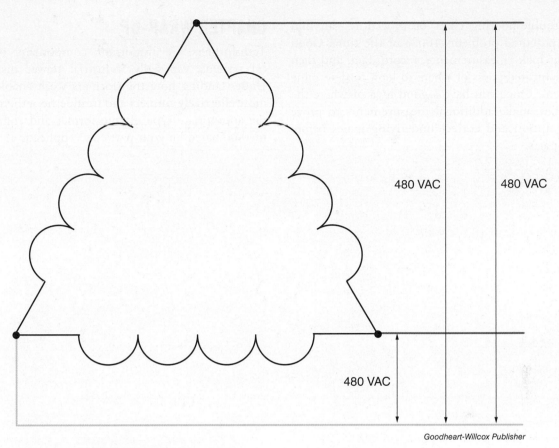

480 VAC

480 VAC

480 VAC

Goodheart-Willcox Publisher

Figure 24-26. Delta-configuration phasing.

24.11 TROUBLESHOOTING

Four basic measurement devices are commonly employed for troubleshooting power transformers: voltmeter, ohmmeter, clamp-on ammeter, and infrared thermometer. Many other measurement devices are helpful but used less often.

Checking the voltage in and voltage out is often the first step in troubleshooting. No voltage in results in no voltage out. If a transformer is delta-connected and one winding fails, you still measure the correct voltage across the open winding. If it is wye-connected and a single phase is open, you do not measure voltage across two of the three phases.

An ohmmeter can be used to look for an open winding or connection. Transformer windings should exhibit a low resistance. Shorts between the windings and ground are safety issues that require higher voltages to test than that provided by standard multimeters. Winding to ground shorts may be found with a megohmmeter. Remember, only use an ohmmeter on a circuit that has been switched off and is locked out and tagged out.

A clamp-on ammeter is a useful tool to check for current. In a three-phase transformer, the current on each phase should be approximately equal. Excessive current on the secondary also reflects excessive current on the primary. A shorted turn may show excessive current. Excess current can indicate that a motor connected as a load has a blown fuse and is operating in the single-phase condition. In this state, all the load current is shifted to one phase. Isolate the different loads connected to the transformer and observe which load isolation causes the currents to come back into balance. That load is the location of your problem.

Another useful tool is the infrared thermometer. This is a noncontact device that measures temperature. Poor connections tend to have a higher temperature than sound ones. Transformers have a temperature rating, which, if exceeded, can cause insulation failure in the windings. The temperature of the transformer depends on the load and cooling method. Inadequate airflow causes the temperature to increase. Also, if more circuits are added and consume greater current than the original design intended, this may also cause a temperature increase. Compare the measured temperature against other similar transformers and against the last time the measurements were made. Also, compare the temperature of each of the three-phase connections against the other two to determine if any of the connections are suspect.

No troubleshooting chart or procedure can find 100% of potential problems 100% of the time. Good technicians make measurements, record data, and then make an educated guess of where to look to determine the problem. Once you have a good idea of where the problem lies, make additional measurements to prove your assumption and correct underlying issues before replacing parts.

CHAPTER WRAP-UP

Transformers are important components of electrical systems, especially industrial power distribution. Understanding how transformers work enables you to more effectively connect and troubleshoot them. Choose the appropriate type of transformer and configuration method based on your particular application.

Chapter Review

SUMMARY

- Transformers are made by winding two coils of wire around a common core. The input or source winding of a transformer is called the primary, and the output or load winding of a transformer is called the secondary.

- Mutual induction is the principle responsible for the operation of a transformer. There is no physical connection between the primary and secondary of a transformer, only coupling of the magnetic field produced by the primary into the secondary.

- The turns ratio of a transformer is the proportion of the number of turns in the primary to the number of turns in the secondary.

- A transformer with more turns in the secondary than the primary is referred to as a step-up transformer, while a transformer with more turns in the primary than the secondary is referred to as a step-down transformer.

- An isolation transformer has a turns ratio of 1:1, which results in the primary and secondary voltages being the same. An isolation transformer is used to isolate direct connections between two circuits, which also blocks the DC component of the input circuit from coupling into the output circuit.

- The voltage ratio of a transformer is the ratio of primary voltage to secondary voltage and directly proportional to the turns ratio.

- The current ratio of a transformer is inversely proportional to both the turns ratio and the voltage ratio.

- An ideal transformer has the same amount of power on both the primary and secondary windings. That is, without transformer losses, a transformer consumes the same amount of power that it delivers to the load.

- Transformers are efficient, but they do exhibit losses in three main categories: eddy current losses, hysteresis losses, and copper losses. Eddy current and hysteresis losses are core losses, and copper losses result

from the resistance of the windings. Transformer losses cause transformer heating.

- Transformer efficiency is the output power divided by the input power. It is often converted to a percentage by multiplying the result by 100.

- Transformers have three different types of power: true power, reactive power, and apparent power.

- Transformers are rated in VA (volt-amperes) or kVA (kilovolt-amperes) rather than watts (W) because of the reactive components of the load. The ratings refer to apparent power only.

- Taps provide a method of creating multiple windings within a single winding. This allows for adjusting voltages on both the primary and secondary windings or creating two windings from a single winding.

- Additional windings on the secondary may be used to either boost or buck the voltage produced by the main secondary winding, depending on the phasing of the additional winding. A second transformer may also be used for this purpose.

- Phasing is extremely important when wiring transformer windings in series or parallel or when using separate transformers to create a three-phase transformer bank.

- Autotransformers have only one winding and share a common connection between the input and output portion of the single winding.

- Current transformers have a single winding on a toroidal core. A current-carrying conductor passes through the center of the toroid and forms the primary of the transformer.

- When wiring a three-phase transformer or three single-phase transformers, the primary windings and secondary windings may be connected in either a delta (Δ) or a wye (Y) configuration. The three phase voltages produced by these connection methods differ depending on the wiring configuration.

- Four of the most important troubleshooting tools for transformer problems are a voltmeter, ammeter, ohmmeter, and infrared thermometer.

REVIEW QUESTIONS

Answer the following questions using the information provided in this chapter.

1. Why is alternating current rather than direct current typically used in transformers?

2. *True or False?* In a step-up transformer, the number of turns in the primary is greater than the number of turns in the secondary.

3. If a transformer has 400 turns in the primary winding and 100 turns in the secondary winding, it is a _____ transformer.

4. Calculate the turns ratio of a transformer with 625 turns in the primary and 125 turns in the secondary.

5. *True or False?* The voltage in the primary or secondary winding can be solved if the number of turns in each winding and the voltage of one of the windings is known.

6. You connect the primary of a control transformer to a 240-volt, 3-phase feeder. If the transformer has 480 turns in the primary and 120 turns in the secondary, calculate the secondary voltage.

7. In current ratio, as the secondary voltage increases, the secondary current _____.

8. A step-up transformer has 300 turns on the primary winding and 750 turns on the secondary winding. If the primary current is 2 A, calculate the secondary current.

9. A step-down transformer has 1000 turns on the primary winding and 200 turns on the secondary winding. You connect the primary of the transformer to 200 VAC, and the current in the secondary is 3 A. What is the current drawn by the primary of the transformer? What is the secondary voltage?

10. If a transformer with a 4:1 turns ratio has 500 W on the input, what would the output wattage be?

11. Calculate both the primary and secondary power of a transformer with a primary current of 2 A, a secondary current of 4 A, a primary voltage of 480 VAC, and a secondary voltage of 240 VAC.

12. If a transformer has a turns ratio of 10:1 and you measure 120 VAC and 2.5 A on the input, what would the output wattage be if the load is purely resistive?

13. _____ losses occur when the magnetic field links to another part of the transformer instead of the secondary winding.

14. A control transformer has 120 VAC supplied to the primary winding with a current of 2 A and 48 VAC supplied to the secondary winding with a current of 4.31 A. Calculate the efficiency of the transformer.

15. In question 14, how many watts of power are being lost as heat?

16. _____ are connections at one or more intermediate points within a winding.

17. *True or False?* If two connected windings are "bucking" each other, they are out of phase.

18. Why is correct phasing critical for transformer windings connected in parallel?

19. A(n) _____ is different from a standard transformer type because it has one coil and shares one connection between the primary and secondary winding.

20. *True or False?* A ferroresonant transformer is also referred to as a variable voltage transformer.

21. If a current transformer is rated at 100:5, what does this indicate?

22. Three-phase transformers can be either a bank of three connected _____ units or one three-phase unit with all windings on a common core.

23. List three guidelines to follow when making transformer connections.

24. List four devices commonly used to troubleshoot power transformers.

 **NIMS CREDENTIALING PREPARATION QUESTIONS**

The following questions will help you prepare for the NIMS Industrial Technology Maintenance Level 1 Electrical Systems credentialing exam.

1. What is a transformer with a turns ratio of 1:1 called?

   A. Step-down transformer
   B. Variable voltage transformer
   C. Isolation transformer
   D. Autotransformer

2. If you measure 480 V on the primary of a transformer and 240 V on the secondary, what is the transformer's voltage ratio?

A. 2:1
B. 4:1
C. 6:1
D. 8:1

3. In current ratio, what does the variable I_P represent?

A. Primary current
B. Primary voltage
C. Secondary current
D. Secondary voltage

4. Transformer power ratings are given in what units?

A. Watts (W)
B. Amperes (A)
C. Volt-amperes (VA)
D. Hertz (Hz)

5. The *NEC* requires that all ground conductors be identified on each end with what color?

A. Blue
B. Green
C. Red
D. White

6. How many connection methods are there for connecting a three-phase transformer bank?

A. One
B. Two
C. Three
D. Four

7. Which of the following connections can continue to function even if one of the transformers in the bank fails or is removed?

A. Delta-delta (Δ-Δ)
B. Delta-wye (Δ-Y)
C. Wye-delta (Y-Δ)
D. Wye-wye (Y-Y)

8. Which of the following connections is rarely used because a failure of any winding or connection results in single phasing, which is detrimental for a three-phase motor?

A. Wye-delta (Y-Δ)
B. Wye-wye (Y-Y)
C. Delta-wye (Δ-Y)
D. Delta-delta (Δ-Δ)

25

Motors, Generators, and Alternators

LEARNING OBJECTIVES

After completing this chapter, you will be able to:

☐ Compare and contrast motors, generators, and alternators.

☐ Identify specifications listed on a DC and AC motor nameplate.

☐ Understand the operation of an AC motor.

☐ Explain the operation and connection methods of a three-phase induction motor.

☐ Describe how motor phases determine rotation in an induction motor.

☐ Determine motor speed and the relationship between motor speed and torque.

☐ Discuss how shaft currents are created in a motor and how to prevent them.

☐ Demonstrate the operation and connection methods of a single-phase induction motor.

☐ Identify the differences among a single-phase capacitor-run motor, capacitor-start motor, and capacitor-start, capacitor-run induction motor.

☐ Explain the operation of a permanent-magnet DC motor.

☐ Differentiate the operating principles of a series-wound and shunt-wound DC motor.

☐ Describe the operation of a stepper motor and its stepping sequence.

☐ Explain the operating principles of a self-excited generator.

☐ Describe the operating principles of an alternator.

☐ Identify the principles of general maintenance for AC and DC motors.

TECHNICAL TERMS

alternator	series-wound DC motor
capacitor-run motor	service factor
capacitor-start, capacitor-run motor	servo amplifier
	servo motor
capacitor-start motor	shaft current
centrifugal switch	shunt-wound DC motor
commutator	slip
generator	slip ring
ground ring	slip speed
induction motor	squirrel-cage rotor
microstepping	stator
motor nameplate	stepper motor
permanent-magnet motor	synchronous speed
	torque
rotor	universal motor
self-excited generator	wound rotor

Motors, generators, and alternators convert between rotational mechanical energy and electrical energy. A motor converts electrical energy to rotational mechanical energy. Generators and alternators operate like motors in reverse. They convert rotational mechanical energy to electrical energy. Different styles of motors are used for AC and DC sources. Generators and alternators differ in that a generator produces direct current, and an alternator produces alternating current. Careful installation and maintenance is required to keep motors, generators, and alternators in good repair.

25.1 MOTOR SPECIFICATIONS

Whether you are working on an AC or DC motor, you must be able to make sense of the specifications provided on a ***motor nameplate***, **Figure 25-1**. Motor nameplates are tags affixed to motors that contain information about the motor's construction and operation. These specifications are essential to effective maintenance.

Motor nameplates typically have the following sections:

- Manufacturer, type, model number, and serial number.
- Power rating given in horsepower (hp) or watts (W).
- Power factor (PF) or efficiency.

- Voltage (V); current or amperage (A), sometimes called full-load current (FLC) or full-load amperage (FLA); and frequency (Hz).
- Speed (rpm).
- ***Service factor*** (SF), a multiplier that indicates how much load beyond the rated load can safely be placed on the motor. SF may be left blank on a motor nameplate if it is 1.0.
- Phase, typically either single-phase or three-phase.
- Instructions for connecting the motor, including connection diagrams for delta and wye configurations.
- Details about bearing type and maintenance.
- Allowable operating temperature.
- Duty or time rating and usage specifications. These indicate how and for how long a motor should be operated.
- Enclosure type, usually *open drip-proof (ODP)*, *totally enclosed fan cooled (TEFC)*, or *totally enclosed non-vented (TENV)*.
- Frame type and size.
- Design code or National Electrical Manufacturers Association (NEMA) designation, **Figure 25-2**.
- Code to indicate starting or inrush current, **Figure 25-3**.

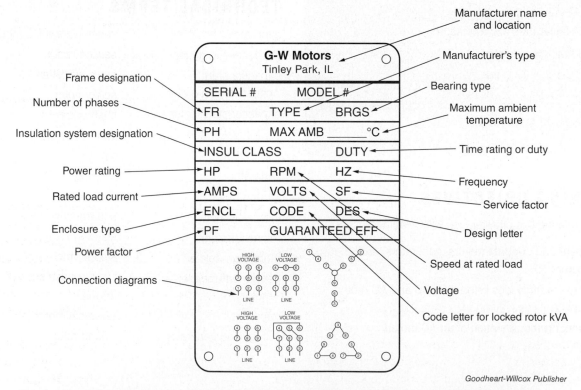

Goodheart-Willcox Publisher

Figure 25-1. The appearance of motor nameplates varies by manufacturer, but the information listed on them is essential for properly installing and maintaining electric motors.

Comparison of NEMA Motor Design Types						
NEMA Design Type	Starting Current	Starting Torque	Breakdown Torque	Slip	Efficiency	Applications
A	High	Medium	High	Low	High	Fans and pumps
B	Medium	Medium	Medium	Low	High	Fans, blowers, centrifugal pumps, and motor generator sets
C	Medium	High	High	Medium	Medium	Conveyors, crushers, and reciprocating compressors, such as for air-conditioning and refrigeration
D	Medium	Very High	Very High	High	Low	High peak loads, such as hoists, cranes, elevators, and extractors

Goodheart-Willcox Publisher

Figure 25-2. NEMA design codes are used to indicate basic features of motor operation, such as starting current, slip, torque, and efficiency.

Code Designations for Starting Current	
Code Letter	kVA per hp with Locked Rotor
A	0–3.14
B	3.15–3.54
C	3.55–3.99
D	4.00–4.49
E	4.50–4.99
F	5.00–5.59
G	5.60–6.29
H	6.30–7.09
J	7.10–7.99
K	8.00–8.99
L	9.00–9.99
M	10.00–11.19
N	11.20–12.49
P	12.50–13.99
R	14.00–15.99
S	16.00–17.99
T	18.00–19.99
U	20.00–22.39
V	22.40 and up

Goodheart-Willcox Publisher

Figure 25-3. Starting current codes for electric motors.

25.2 ALTERNATING CURRENT MOTORS

Alternating current motors are electric motors that run on AC power, **Figure 25-4**. AC motors consist of two main parts: a stator and rotor. The *stator* is made of stationary windings that produce electromagnetic fields when power is applied. Inside the stator, there is a rotating part called a *rotor*. The electromagnetic field from the stator induces a voltage in the rotor and causes a current to flow.

Oleksandr Kostiuchenko/Shutterstock.com

Figure 25-4. Electric motor.

25.2.1 Induction Motors

In an ***induction motor***, the rotor has no electrical connection to the power source. The power source is connected to the windings in the stator, and the current in the rotor is induced by the rotating magnetic field produced by the stator. Both the rotor and the stator are built up of laminations. The laminations in modern induction motors are made from silicon steel or electrical steel. However, laminations may also be made of iron.

Windings or coils are distributed around the stator in pole-phase groups. The windings may be connected in either a wye (Y) or delta (Δ) configuration in a three-phase induction motor. In a single-phase induction motor, the windings are connected in series.

There are two common kinds of rotors used in induction motors: *squirrel-cage rotors* and *wound rotors*. Both rotor varieties are made of laminated electrical steel. While wound rotors feature windings on a rotating cylinder, squirrel-cage rotors do not have windings. Rather, slots are provided in the rotor laminations for either aluminum or copper bars, **Figure 25-5**. Copper bars are pressed into the rotor lamination slots. Aluminum bars are cast into the laminations.

In the case of an aluminum squirrel cage, once the laminations are stacked, molten aluminum is forced into the voids or slots, and the casting mold is also used to form shorting rings on each end. The circumference of the rotor is then ground to provide a small gap between the rotor and the stator in the assembled motor. Shorting rings on each end of the rotor connect the aluminum or copper bars. The bars are skewed, rather than parallel, with the axis of the rotor. This skewing reduces mechanical noise while the motor is running and provides a more uniform starting torque.

25.2.2 Rotating Field

The induction motor depends on a rotating magnetic field for its operation. The rotating magnetic field of the stator crosses the conductors (aluminum or copper bars) of the rotor and induces a current in them. The electromotive force exerted by the rotating field on the conductors in the rotor causes it to turn.

The strength of the magnetic field in the stator is related to the current in its conductors and the shape of its laminations. The field of the stator has defined magnetic poles based on the instantaneous direction of the winding current. The magnetic field changes in step with the applied line frequency.

25.2.3 Three-Phase Stator

Consider a two-pole, three-phase stator. This form has three pairs of windings with each winding spaced 60° apart around the stator. Refer to **Figure 25-6**. A, B, and C represent the three phases and their order. Each phase has two windings directly across from each other, or 180° apart. The progression of the magnetic poles in the windings illustrates when each phase is at its positive peak.

The stator's magnetic field completes the second half of its rotation in **Figure 25-7**. These steps in the progression correspond to each phase at its maximum negative value. Take a look at all six steps and notice that the north pole jumps from one winding to the next in a clockwise direction. If any two phases were reversed, the magnetic field would rotate in a counterclockwise direction. The phasing determines the direction of rotation of the motor.

> **TECH TIP**
> ## Motor Phasing and Rotation
>
> Motor phasing matters. If a motor is connected to a drivetrain, do not correct rotational direction by experimenting. Instead, use a phase sequence tester to ensure the motor turns in the proper direction before power is applied, **Figure 25-8**. When a tester is connected to a three-phase motor and the motor shaft is turned by hand, the tester indicates which direction the motor is spinning. If the actual direction and the indicated direction do not agree, reverse any two of the three motor wires to correct the discrepancy. Industry standard is to swap phases 1 and 3 to reverse rotation of a three-phase motor.

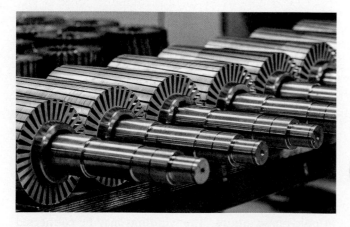

Smoczyslaw/Shutterstock.com; Oleksandr Kostiuchenko/Shutterstock.com

Figure 25-5. Most induction motors feature either a squirrel-cage rotor or wound rotor. While both types of rotors are made of laminated electrical steel, squirrel-cage rotors have bars of either copper or aluminum incorporated into the laminations.

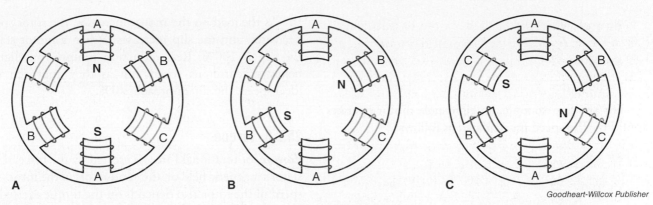

Goodheart-Willcox Publisher

Figure 25-6. A—Step 1 in field rotation sequence for a three-phase induction motor. B—Step 2 in field rotation sequence for a three-phase induction motor. C—Step 3 in field rotation sequence for a three-phase induction motor.

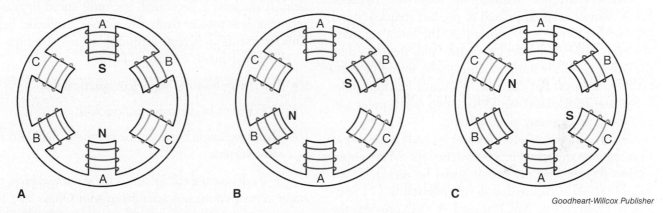

Goodheart-Willcox Publisher

Figure 25-7. A—Step 4 in field rotation sequence for a three-phase induction motor. B—Step 5 in field rotation sequence for a three-phase induction motor. C—Step 6 in field rotation sequence for a three-phase induction motor.

Reproduced with Permission, Fluke Corporation

Figure 25-8. Technician checking phase rotation on a compressor using a phase sequence tester.

The magnetic field rotates 360° in electrical phase as well as 360° mechanically. As the field rotates, the intensity of the field also rotates uniformly between one stator winding and the next. The moving field acts on the conducting bars in the squirrel cage, causing the rotor to rotate in the direction of the magnetic field rotation.

The relationship between the stator and rotor is similar to that of primary and secondary windings in a transformer. Each of the bars in the squirrel-cage rotor would act as a shorted turn in a transformer if the shaft did not rotate.

25.2.4 Motor Speed

In an AC induction motor, the rotational speed of the magnetic field (measured in revolutions per minute or rpm) depends on the number of poles and the frequency of input. The rotational speed of the magnetic field in the stator is also called the **synchronous speed** and can be calculated as follows:

$$s = 120 \times \frac{f}{P}$$

(25-1)

where

s = synchronous speed in rpm
f = frequency in Hz
P = number of poles

Consider a two-pole, three-phase motor operating from 60 Hz line frequency:

$$s = 120 \times \frac{f}{P}$$

$$s = 120 \times \frac{60}{2}$$

$$s = 3600 \text{ rpm}$$

For another example, an eight-pole motor operates at 60 Hz. The speed is calculated as follows:

$$s = 120 \times \frac{f}{P}$$

$$s = 120 \ \frac{60}{8}$$

$$s = 900 \text{ rpm}$$

These examples demonstrate that the magnetic field's rotational speed is fixed to the line frequency and the number of poles the motor has. The stator can have any even number of poles per phase. Poles are formed by the windings in the stator and are always found in pairs, north and south. For example, if a stator has four poles per phase, each phase would have two north poles and two south poles at any given instant.

A three-phase induction motor by itself is capable at running near the speed it was designed for. Many applications require that the motor speed be varied. When variable speed is required, a device referred to as a variable frequency drive (VFD) is used. A VFD provides the ability to control what speed the motor operates at by allowing you to vary the frequency of the three-phase alternating current feeding the motor. VFDs and their operation are discussed in a later chapter.

25.2.5 Slip

The rotational speed of the magnetic field or synchronous speed is not equal to the rotor speed in an induction motor. The difference between the rotor's actual speed and the synchronous speed is the *slip speed*. The slip speed divided by the synchronous speed may be expressed as a percentage, which is referred to as the *slip*.

For example, the synchronous speed of a motor is 3600 rpm. Under load, the rotor speed is measured at 3450 rpm. The slip of the motor is calculated as follows:

$$\text{slip} = \frac{\text{synchronous speed} - \text{rotor speed}}{\text{synchronous speed}} \times 100\%$$

(25-2)

$$\text{slip} = \frac{3600 \text{ rpm} - 3450 \text{ rpm}}{3600 \text{ rpm}} \times 100\%$$

$$\text{slip} = \frac{150 \text{ rpm}}{3600 \text{ rpm}} \times 100\%$$

$$\text{slip} = 0.0147 \times 100\%$$

$$\text{slip} = 1.47\%$$

As the load on the motor increases, the rotor speed decreases, and the slip increases. When a motor starts, the slip is 100%. Real induction motors run slightly below synchronous speed, but as they approach this ideal speed, the slip approaches 0%.

25.2.6 Torque

Torque, or rotational force, is created by the force of the stator's magnetic field on the conductors in the rotor. The speed of the rotor also depends on the torque exerted by the driven load. If the load is increased, the rotor speed decreases and receives more torque from the magnetic field. If the load is decreased, the rotor speed increases, receiving less torque from the stator's magnetic field. This represents an inverse torque/speed relationship.

Torque depends on several factors:

- The strength of the stator's magnetic field.

- The current in the rotor conductors.

- The phase angle between the magnetic field and the rotor current.

If you double the voltage across the stator windings, the stator current doubles in accordance with Ohm's law. The doubling of the stator current effectively doubles the rotor current. The doubling of the stator current and the rotor current increases the torque by a factor of four. That is, the torque varies as the square of the voltage applied to the stator.

25.2.7 Dual-Voltage Motor

Dual-voltage, three-phase motors are often used in industrial applications. The most common dual-voltage motor is the 440/208 VAC motor. The available line voltage determines how the motor is connected. Motors may be wired in a wye (Y) or delta (Δ) configuration. Each phase has two windings. In a three-phase motor, there are three pairs of windings. When operating from the higher of the two voltages, each pair of windings is connected in series. When operating from the lower of the two voltages, the windings are connected in parallel.

The National Electrical Manufacturers Association (NEMA) has established standard connections for dual-voltage, three-phase motors. These connections vary depending on whether the motor is wired in a wye, **Figure 25-9**, or delta, **Figure 25-10**, configuration. A simple ohmmeter check will reveal the difference. In a Y-connected motor, you will measure the same resistance between the wires numbered 7 and 8, 8 and 9, and 7 and 9. If the motor has been wired in a Δ connection, you will measure infinite resistance between these wires.

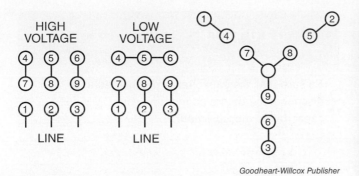

Figure 25-11. Bearing damaged by shaft currents.

Goodheart-Willcox Publisher

Figure 25-9. Y-connected, three-phase induction motor wiring diagram.

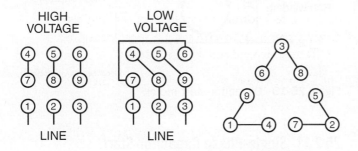

Goodheart-Willcox Publisher

Figure 25-10. Δ-connected, three-phase induction motor wiring diagram.

There are normally nine connections, which may be jumpered together before the three line input connections are made. Refer to the standard configuration diagrams to determine where to connect the line input phases and the jumpers. The manufacturer's data sheets and motor nameplate should also provide this information.

25.2.8 Shaft Currents

Some induction motors exhibit problems with *shaft currents*. When the motor shaft is in the center of a rotating magnetic field, voltages may be induced in the shaft. If each end of the shaft has a common electrical connection through the bearings to the motor frame, undesirable arcing and electrical shocks may result due to shaft current flow. These negative consequences can damage the motor and bearings.

Many variable frequency drives (VFDs) generate high-frequency transients (unwanted high-voltage bursts of energy caused by changes in state) via the switching transistors in the VFD. Because they are high frequency in nature, these transients easily travel through the lubricant in motor bearings. While seeking a path to ground, the shaft currents erode the rolling elements of the bearings, **Figure 25-11**.

Because bearings are the contact points completing the circuit through the motor frame, premature bearing failure is imminent. There are three possible solutions to this problem:

- Install bearings made from a nonconductive material. Such bearings are often made from ceramic material and are expensive.

- Insulate one or both ends of the shaft to interrupt the potential circuit for shaft currents through the motor frame.

- Install *ground rings* on the motor shaft. Ground rings are a method of conducting shaft currents to ground before they reach motor bearings. Unfortunately, ground rings are a sacrificial device and will fail after conducting shaft currents for a period of time. After a ground ring fails, the bearing on the failed side is susceptible to damage. If ground rings are employed, they must be closely monitored and replaced before they fail.

25.2.9 Single-Phase Induction Motor

A single-phase induction motor has a squirrel-cage rotor similar to the one used in a three-phase induction motor. A single winding of a single-phase induction motor does not produce a rotating magnetic field. Instead, it produces a pulsating magnetic field that peaks at $0°$ and $180°$. The motor develops little or no torque from a standstill. If the motor is rotated forward, the single-phase winding develops torque once the motor is started.

One method to overcome the problem of starting the motor is to create a two-phase motor. This is accomplished by placing a second set of windings $90°$ apart from the primary set. A phase shift of $90°$ for the second set of windings is created by wiring them in series with a capacitor. This type of motor is often referred to as a *capacitor-run motor* or *permanent-split capacitor motor*, **Figure 25-12**. A capacitor-run motor only works well up to one-quarter horsepower. It also experiences pulsating torque variations near synchronous speed.

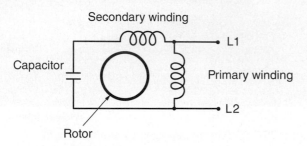

Goodheart-Willcox Publisher

Figure 25-12. Two-phase motor fed by single-phase AC.

25.2.10 Single-Phase Capacitor-Start Motor

The single-phase *capacitor-start motor* is a variation of the capacitor-run motor, **Figure 25-13**. Capacitor-start motors have a larger value capacitor than capacitor-run motors. They also feature heavier gauge wire and more turns in the start winding than in the primary winding. A capacitor-start motor provides more starting torque for heavier loads, such as compressors.

The second winding at 90° is referred to as the start winding. A *centrifugal switch* is employed to switch out the start winding once the motor reaches its operating speed. A centrifugal switch is a mechanical device that opens once the shaft rotation reaches a specified rate of speed.

TECH TIP
Troubleshooting Capacitor-Start Motors

The start winding in a capacitor-start motor must be connected in the proper configuration. If the winding is reversed, the motor will start and run in the reverse direction. After the motor gets up to speed, the start winding is switched out to reduce the amount of heat generated. If the centrifugal switch fails, the start winding may burn itself out. Centrifugal switches may malfunction as a result of welded contacts or mechanical failures.

There are two common reasons a motor may fail to start: a shorted capacitor does not provide any phase shift or an open capacitor effectively switches out the start winding. Use your DMM set on the capacitor range to verify the appropriate capacitance value.

⚠ CAUTION

Be sure that the capacitor is discharged and disconnected before connecting your DMM for a capacitance measurement.

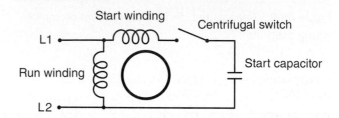

Goodheart-Willcox Publisher

Figure 25-13. Capacitor-start motor.

25.2.11 Single-Phase Capacitor-Start, Capacitor-Run Motor

Another variation of the single-phase motor is a *capacitor-start, capacitor-run motor*, **Figure 25-14**. In this configuration, the start winding is not switched out in run mode. The motor starts with both start and run capacitors in circuit. The extra capacitance provides greater starting torque. When a capacitor-start, capacitor-run motor reaches running speed, the second capacitor is switched out by a centrifugal switch.

In some cases, both capacitors may be constructed in the same package with three terminals. Two terminals are for capacitor 1 and capacitor 2, and the third terminal is connected to both capacitors. This third terminal is frequently labeled C or Common.

As with capacitor-start motors, windings must be connected in the proper direction. Reversing the start/run windings causes the motor to rotate in a reverse direction.

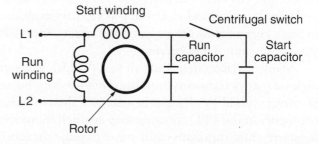

Goodheart-Willcox Publisher

Figure 25-14. Single-phase capacitor-start, capacitor-run induction motor.

25.3 DIRECT CURRENT MOTORS

Direct current (DC) motors are more complicated than AC induction motors because they do not have an inherent rotating magnetic field. With a DC motor, the field must be switched to cause rotation. In most cases, a stationary field is generated by the stator, which may be a permanent magnet or a winding fed by DC. The rotor is much different than that of an AC motor.

Rotors in DC motors have windings connected to a ***commutator***, **Figure 25-15**. Commutators control motor rotation, both speed and direction, by periodically switching the direction of current, thereby reversing the magnetic field in the rotor. A simple commutator reverses the polarity of the direct current flowing through the rotor winding once every 180° of mechanical rotation. Most commutators have multiple connections and one pair of brushes. This allows multiple windings to be energized throughout the rotation of the wound rotor.

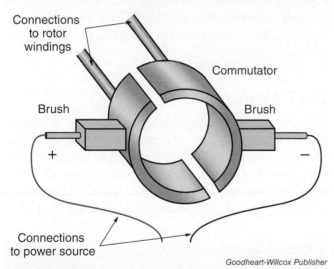

Figure 25-15. Simple commutator.

Goodheart-Willcox Publisher

25.3.1 Permanent-Magnet Motor

In a ***permanent-magnet motor***, there are no stator windings. Instead, the stator contains permanent magnets. The magnets are often ceramic or neodymium because they are capable of producing an extremely strong magnetic field.

The rotor has windings connected to the commutator. When the polarity of the stator matches the polarity of the winding in the rotor, the like poles repel each other and cause the rotor to rotate, **Figure 25-16**. Bearings, not shown in the illustration, hold the motor shaft and rotor in place. The only option is for the rotor to turn away from the center of the magnetic field created by the permanent magnets.

As soon as the rotor has traveled 180° mechanically, the commutator switches the rotor's magnetic polarity

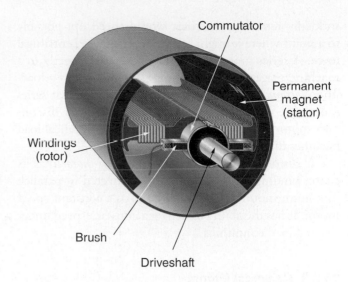

Fouad A. Saad/Shutterstock.com

Figure 25-16. Rotation begins with the north pole of the stator magnet repelling the north pole of the rotor. Likewise, the south pole of the stator repels the south pole of the rotor.

and it is once again repelled. This process repeats, causing the rotor to turn continuously. The speed depends on the voltage applied to the rotor and the resulting current. As the voltage is increased, the current in the rotor increases. Accordingly, the strength of the magnetic field increases causing the rotor to spin faster.

When a motor is switched off, it is often desirable to quickly stop its rotation. Dynamic braking may be employed by placing a load resistance across the motor and controlling it with a switch.

Because a permanent-magnet motor has a rotating coil within a magnetic field, it is possible for this type of motor to produce a voltage and subsequent current when the rotor is mechanically rotated. In this scenario, the motor becomes a generator, which converts mechanical motion into electricity. When an electrical load is placed on the motor and the rotor is mechanically rotated, mechanical resistance to the rotation is produced. This effect may be exploited for other purposes.

25.3.2 Series and Shunt DC Motors

Series-wound and shunt-wound DC motors are variations of the permanent-magnet DC motor. Instead of using permanent magnets as the stator, the stator has windings that form an electromagnet.

A ***series-wound DC motor*** (or *series motor*) has its rotor windings connected through the commutator in series with the stator windings. A series motor should never be operated without a load. If the load were to be suddenly removed, the rotor would speed up, possibly to a point where it might self-destruct due to centrifugal force. A series motor should be connected directly to a machine or gear-reduction unit so it always has a load. The *National Electrical Code* (*NEC*) requires that series motors be operated with a control system to disconnect motor power in the event that the mechanical load becomes disconnected from the motor.

A ***shunt-wound DC motor*** (or *shunt motor*) has its stator windings and commutator connected in parallel. The shunt motor is often referred to as a constant-speed motor. It has the ability to deliver a constant speed under varying load conditions.

25.3.3 Universal Motors

A ***universal motor*** can operate from either AC or DC. Universal motors have a wound rotor, commutator, and wound stator. Universal motors are most often used in single-phase AC power tools, vacuum cleaners, and other appliances. Universal motors also have carbon brushes connected to the commutator. The poles of the rotor and stator periodically reverse in lock step with the line frequency. Universal motors are not used in heavy-duty applications and most often have the rotor and stator or field windings connected in series.

25.3.4 Servo Motors

A ***servo motor*** is a DC motor used to control movement in a variety of industrial applications, such as manufacturing and robotics. Servo motors have a feedback device connected to the motor shaft to detect position, direction of travel, or distance of travel. Often a potentiometer or encoder is used as the feedback device.

The feedback signal is sent to a ***servo amplifier*** that reads the feedback device and controls the servo motor. The servo amplifier is given a set point (desired location) by the system controlling the servo motor, such as a computer or programmable logic controller (PLC). The servo amplifier reads the feedback device, reads the set point, and then drives the servo motor until it reaches the desired position.

25.3.5 Stepper Motors

Stepper motors are used to convert electrical impulses into discrete mechanical rotational movements. In a typical stepper motor, power is applied to two coils. The coils must only be energized long enough for the rotor to move to the next position. If the coil remains energized, it locks the shaft of the motor. This situation is commonly used for stopping or applying the brakes.

Error in a stepper motor is noncumulative. This means the positioning error is always the same whether the rotational movement is one step or 1000 steps. Since the step error is noncumulative, it averages to zero within a four-step sequence or 360 electrical degrees. The most accurate method of positioning is therefore obtained by stepping in multiples of 4.

Some stepper motor drivers are capable of generating microsteps. This is referred to as ***microstepping***. Microstepping allows many small steps to be taken in between the fixed mechanical steps a motor is normally able to take. Microstepping enables a stepper motor to perform fine positioning functions.

25.4 GENERATORS AND ALTERNATORS

Generators and ***alternators*** turn rotational mechanical energy into electricity. Often the term *generator* is used to apply to devices that produce DC and those that produce AC. More appropriately, a generator produces DC, and an alternator produces AC.

25.4.1 Generators

DC motors are also capable of producing DC power. For example, the stator of a permanent-magnet motor produces a magnetic field. If you turn the rotor, the magnetic lines of force from the stator cut across the rotor windings and induce a voltage in them. This scenario produces pulsating DC as depicted in **Figure 25-17**.

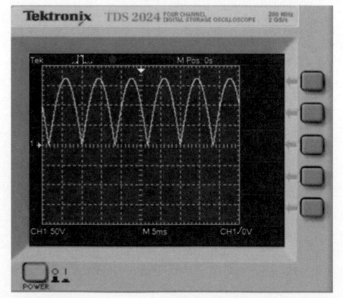

Goodheart-Willcox Publisher

Figure 25-17. Pulsating DC from the output of a generator.

Generators of this variety are often used when the required output power is low. Another option is a motor with a wound rotor and a wound stator. To provide the magnetic field required, provide a source of DC to the stator, causing it to become an electromagnet.

25.4.2 Self-Excited Generator

A shunt-wound DC motor can generate pulsating DC. In this type of generator, there is no voltage externally applied to the stator windings. The DC voltage needed to create the magnetic field in the stator comes from the rotor. For this reason, the generator is considered a *self-excited generator*. The self-excited generator relies on a small residual magnetic field in the rotor laminations. After being magnetized in operation, the rotor laminations retain enough of a magnetic field to start the process.

Notice that a self-excited generator outputs pulsating DC. This is fine for many purposes, such as battery charging. However, it is not suitable for running electronic equipment that requires smooth, ripple-free DC. A capacitor of a large enough value may be used to "smooth out" the pulsations and create DC that is acceptable for electronic circuits.

25.4.3 Alternators

An alternator functions much like a generator, except an alternator produces alternating current rather than direct current. The frequency of the AC depends on the number of windings and rotational speed.

A simple alternator is similar to a permanent-magnet motor with slip rings instead of commutators, **Figure 25-18**. *Slip rings* do not switch like a commutator. Rather, they allow constant contact with each end of the rotor winding. This creates an alternating current in the rotor as it passes each alternating north and south magnetic pole of the stator.

Common alternators have a wound rotor, wound stator, and slip rings. A small amount of DC is coupled to the winding of the rotor. As the rotor spins, its magnetic field induces a voltage in the windings of the stator, in turn producing electricity.

The stator may be wound with one, two, or three phases. In a two-phase unit, the voltages in each phase are spaced 90° apart, resulting in a higher output voltage than a single-phase winding by itself. Three-phase stators are most common. They may be wired in either a Y configuration, **Figure 25-19**, or a Δ configuration, **Figure 25-20**.

The Δ-connected alternator is more common than the Y variety. In a Δ-connected alternator, the DC excitation voltage is delivered to the rotor via the slip rings.

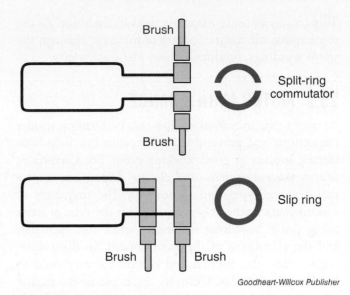

Goodheart-Willcox Publisher

Figure 25-18. Commutator and brush comparison. A comparison of commutators and slip rings.

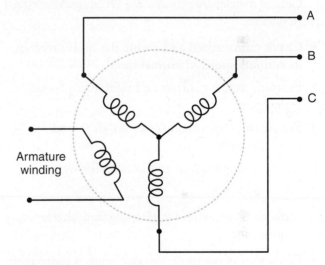

Goodheart-Willcox Publisher

Figure 25-19. Y-connected alternator schematic diagram.

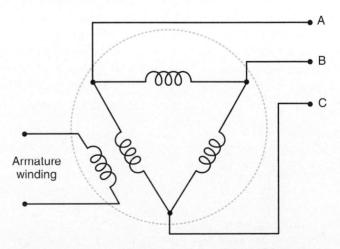

Goodheart-Willcox Publisher

Figure 25-20. Δ-connected, three-phase alternator.

This creates a steady magnetic field in the rotor. As the rotor spins, the magnetic lines of force cut through the stator windings, producing three-phase electricity.

25.5 MOTOR MAINTENANCE

As with other industrial equipment, performing regular inspections and preventive maintenance can help keep electric motors in good working order. See Chapter 3, *Maintenance Principles and Record Keeping*, to review maintenance types and techniques. The frequency of various maintenance tasks depends on the type of work being done, how long the motor must run each day, and the cleanliness of the environment. In dirty operating conditions, motors and equipment may need to be cleaned every day. Consider the needs of the facility where you work as well as motor manufacturers' specifications as you develop and carry out a maintenance plan.

General maintenance for AC and DC motors is required at least weekly:

- Check motor speed and ensure the motor reaches its running speed in normal time.
- Examine motor controls and tighten any loose connections.
- Ensure that the rotating parts are clean and rotate properly.
- Examine commutator and brushes for wear.
- Check oil levels.

Perform more extensive cleaning and checks every six months:

- Clean motor, commutator, and brushes thoroughly. Remove dirt from coils, vents, and other surfaces.

- Clean brush holders and ensure proper brush placement, pressure, and movement.
- Replace brushes that are more than halfway worn down.
- Check grease in ball or roller bearings. Clean out and replace oil in bearings as needed, according to manufacturer specifications.
- Compare input current with normal current.
- Check mounting bolts, gears, coupling mechanisms, setscrews, and keys.
- Examine the motor drive carefully for worn gears, chains, and belts.
- Confirm that the motor runs smoothly with no excess vibration.
- Check shaft end play.
- Inspect and tighten connections on the motor and controls.
- Ensure that all covers and guards are properly placed and securely fastened.

CHAPTER WRAP-UP

As an industrial maintenance technician, it is important for you to understand the basic function of various styles of motors, generators, and alternators. Your knowledge about how these machines work supports your implementation of an appropriate maintenance plan. Ensuring that the motors in your workplace are properly installed and maintained is vital to keeping your facility running smoothly.

Chapter Review

SUMMARY

- A motor converts electrical energy into rotational mechanical energy.

- Generators and alternators both convert mechanical rotational energy into electrical energy. They differ in that a generator produces direct current, and an alternator produces alternating current.

- Motor nameplates are tags affixed to motors that contain information about the motor's construction and operation. These specifications are essential to effective maintenance.

- Motors consist of a stator and rotor. A stator is made of stationary windings that produce an electromagnetic field when power is applied. The rotor is a rotating element located inside the stator.

- In an induction motor, the power source is connected to the windings in the stator, and the current in the rotor is induced by a rotating magnetic field produced by the stator.

- Windings or coils are distributed around the stator in pole-phase groups. Windings for three-phase induction motors may be wired in either a delta (Δ) or a wye (Y) configuration.

- Two common kinds of rotors used in induction motors are squirrel-cage rotors and wound rotors.

- Ensure that the phases of a three-phase supply are correctly oriented so motor rotation is in the intended direction. To change the direction of rotation of a three-phase induction motor, reverse any two line leads.

- The rotational speed of the magnetic field, or the synchronous speed, depends on the number of poles and the frequency of input. The slip speed is the difference between the rotor's actual speed and the synchronous speed.

- Torque, or rotational force, is created by the force of the stator's magnetic field on the conductors in the rotor. The speed of the rotor also depends on the torque exerted by the driven load.

- Some induction motors exhibit problems with shaft currents, which can cause arcing and electrical shocks. To prevent these problems, ground rings should be installed on the motor shaft.

- A single-phase induction motor produces a pulsating magnetic field. These motors have a start winding in a mechanical orientation of 90° to the run winding.

- A two-phase motor can be created from a single-phase by placing a second set of windings 90° apart from the primary set. A capacitor-run motor is created by wiring the windings in series with a capacitor.

- A centrifugal switch is used for a capacitor-start motor. The switch is closed when the rotor is stationary. Once the motor is at operating speed, it opens and switches out the starting capacitor.

- A capacitor-start, capacitor-run motor does not have its start winding switched out in run mode. The motor starts with both start and run capacitors in the circuit to provide extra capacitance and greater torque.

- A commutator switches the electrical polarity of rotor windings through its rotational movement, thus switching the magnetic polarity of the rotor poles.

- DC motors must be switched to cause rotation. Generally, a stationary field is generated by the stator, which may be a permanent magnet or winding fed by direct current.

- A permanent-magnet motor has no stator windings. Instead, the stator contains permanent magnets that produce an extremely strong magnetic field.

- A series-wound DC motor has its rotor windings connected through a commutator in series with the stator windings. A shunt-wound DC motor has its stator windings and commutator connected in parallel.

- A servo motor is used to control movement in a variety of industrial applications. A feedback signal is sent to a servo amplifier that reads the feedback device and controls the servo motor.

- Stepper motors convert electrical impulses into discrete mechanical rotational movements. This is commonly used for stopping or applying brakes.

- A self-excited generator relies on a small residual magnetic field in rotor laminations.

- A simple alternator is similar to a permanent-magnet motor, but it uses slip rings instead of commutators. Slip rings transfer the electrical connections between the mechanical rotation of the rotor and the source.

- General maintenance for AC and DC motors is required weekly, and extensive cleaning and checks are required every six months.

REVIEW QUESTIONS

Answer the following questions using the information provided in this chapter.

1. How do generators, alternators, and motors differ?

2. List five specifications that are included on a motor nameplate.

3. Describe the two main parts of an alternating current motor.

4. Explain the operation of an induction motor that causes the motor to turn.

5. Explain how a three-phase induction motor can be rotated in the opposite direction.

6. The rotational speed of the magnetic field in an AC induction motor depends on the number of _____ and the _____ of input.

7. Calculate the synchronous speed of a two-pole, three-phase motor operating from 80 Hz line frequency.

8. Calculate the slip of a motor if the synchronous speed is 2900 rpm and, under load, the rotor speed is measured at 2785 rpm.

9. *True or False?* As the load on a motor increases, the rotor speed decreases, and the slip increases.

10. What is torque? What is its relationship to rotor speed?

11. How many winding pairs are in a dual-voltage, three-phase motor? Explain when windings are connected in series rather than parallel.

12. Discuss the possible causes of shaft currents and potential problems that may result in a three-phase induction motor.

13. Explain how shaft currents that cause premature bearing failure can be prevented.

14. What are the similarities and differences between a single-phase induction motor and a three-phase induction motor?

15. *True or False?* Capacitor-start motors have a smaller value capacitor than capacitor-run motors.

16. What are the two common reasons a capacitor-start motor may fail to start?

17. Explain the operation of a capacitor-start, capacitor-run motor.

18. Rotors in DC motors have windings connected to a(n) _____.

19. Explain the principle behind what causes a permanent-magnet DC motor to turn.

20. *True or False?* A shunt-wound motor has its rotor windings connected through the commutator in series with the stator windings.

21. Servo motors have a(n) _____ connected to the motor shaft to detect position, direction of travel, or distance of travel.

22. Describe how a stepper motor is driven.

23. Generators and alternators turn _____ into electricity.

24. Explain the operation of a shunt-wound DC motor.

25. How do slip rings operate? How are they different from commutators?

26. List three general maintenance checks for AC and DC motors that should be completed weekly.

27. List three general maintenance checks for AC and DC motors that should be completed every six months.

NIMS CREDENTIALING PREPARATION QUESTIONS

The following questions will help you prepare for the NIMS Industrial Technology Maintenance Level 1 Electrical Systems credentialing exam.

1. Which of the following statements is true about motors, generators, and alternators?

    A. Alternators convert rotational mechanical energy to electrical energy.
    B. Alternators produce direct current.
    C. Generators produce alternating current.
    D. Motors and generators convert electrical energy to rotational mechanical energy.

2. What specification is *not* listed on a motor nameplate?

    A. Duty or time rating
    B. Rotor type
    C. Service factor
    D. Voltage and amperage

3. What two variables determine the motor speed of an AC induction motor?

    A. Number of phases and torque
    B. Number of poles and frequency of input
    C. Voltage and frequency of input
    D. Voltage and torque

4. What is a possible solution to prevent shaft current problems in induction motors?

   A. Increase the strength of the stator's magnetic field.
   B. Install ground rings on the motor shaft.
   C. Remove insulation from one or both ends of the shaft.
   D. Reduce the number of phases of the motor.

5. A single-phase induction motor that uses a centrifugal switch to switch out the start winding once it reaches its operating speed is called a _____.

   A. capacitor-start motor
   B. capacitor-start, capacitor-run motor
   C. capacitor-run motor
   D. single-voltage motor

6. Which is a difference between three-phase induction motors and single-phase induction motors?

   A. Single-phase motors produce a pulsating magnetic field, and three-phase motors produce a rotating magnetic field.
   B. Single-phase motors produce more torque than a three-phase motor.
   C. Three-phase windings are connected in series, and single-phase windings are connected in either wye or delta.
   D. A three-phase motor uses a squirrel-cage rotor, and a single-phase motor uses a wound rotor.

7. Series-wound DC motors and shunt-wound DC motors use _____ to form an electromagnet to begin operation.

   A. a commutator in parallel with the stator
   B. a commutator in series with the stator
   C. permanent magnets as the stator
   D. windings in the stator

8. What common application makes use of a stepper motor?

   A. Accelerating a diesel motor
   B. Applying brakes
   C. Starting an alternator
   D. Starting a generator

9. An alternator uses (a) _____ that create(s) an alternating current in the rotor as it passes each alternating north and south magnetic pole of the stator.

   A. commutator
   B. motor
   C. self-excited generator
   D. slip rings

26

Motor
Controls

LEARNING OBJECTIVES

After completing this chapter, you will be able to:

☐ Describe the basic functions of motor control circuits.

☐ Select and size fuses and circuit breakers to be used in a control circuit.

☐ Describe how to size conductors.

☐ Discuss input devices and types of contacts in a control circuit, including pushbuttons and indicators.

☐ Discuss types of switches used in a control circuit.

☐ Interpret control circuit symbols and internal wiring arrangement diagrams.

☐ Discuss motor starters and contactors used in a control circuit.

☐ Distinguish types of overloads and proper sizing.

☐ Understand reversing motor starters and reversing drum switches in DC, single-phase AC, and three-phase AC motors.

☐ Troubleshoot a motor starter to find possible failures in the motor and control circuit.

☐ Describe output device functions including relays and timers.

☐ Explain the operational difference between an on-delay and contact-controlled off-delay timer.

☐ Describe solenoids and how they may fail in a fluid system.

TECHNICAL TERMS

auxiliary contact	normally closed (NC)
bimetallic overload	normally closed, timed-open (NCTO) contact
contact-controlled timer	
contactor	normally open (NO)
deadband	normally open, timed-closed (NOTC) contact
dual-pole, dual-throw (DPDT)	
	off-delay timer
dual-voltage coil	on-delay timer
float switch	overload
hard contact	pressure switch
interlock	push-to-test
International Electrotechnical Commission (IEC)	reversing drum switch
	reversing motor starter
interrupt rating	seal-in current
limit switch	set point
magnetic overload	shading ring
magnetic reed contact	single-pole, dual-throw (SPDT)
maintained contact	
melting alloy overload	single-pole, single-throw (SPST)
molded-case circuit breaker	
momentary contact	solenoid
motor starter	temperature switch
National Electrical Manufacturers Association (NEMA)	thermistor
	thermowell

Just as in other areas of technology, motor controls have evolved from simple hardwired logic into programmed logic. Older or more basic systems typically rely on hardwired logic in which the decision-making of the circuit is physically wired into the system through switches and pushbuttons. Systems with a combination of hardwired logic and microprocessor controls use inputs to send signals to electronic boards for decision-making. The electronic board then provides outputs (usually through miniature relays) to discrete elements, such as solenoids, motor starters, or indicators. Newer systems nearly always use programmable logic controllers (PLCs) unless they are very rudimentary. In a PLC-controlled system, there is no external hardwired logic. All discrete elements input into the PLC. The processor and program determine the logic of the circuit and provide for output signals to energize discrete real-world outputs. This chapter examines various motor controls, including breakers, starters, switches, relays, and solenoids.

26.1 BASIC MOTOR CONTROLS

Motor control circuits are used to perform the following basic functions:

- Start and stop loads.
- Perform a sequence of events.
- Control an output relative to an input.
- Automatically control a system.

Most motor control circuits are digital, which means they obey binary logic and are either on or off. Digital controls may carry out timed events, sequences of events, and interlocks to make sure certain events happen or do not happen. Real-world digital devices, such as switches, are also referred to as discrete devices. Inputs into the circuit may be operator controlled (as with pushbuttons) or process controlled (as with limit switches, float switches, and sensors). Decision-making is performed by these types of inputs and depends on how they are wired into the system. Outputs from this decision-making may include timers, relays, motor starters, contactors, indicators, and solenoids.

26.2 FUSES IN CONTROL CIRCUITS

Recall from Chapter 23, *Electrical Components and Circuit Materials*, that fuses are one-time use devices categorized by voltage level, type (delay or fast acting), physical size, amperage rating, and interrupt rating. *Interrupt rating* is the maximum current at which a fuse is guaranteed

to clear a fault. Some of the different types of fuses are shown in **Figure 26-1**. For motor protection, low-voltage branch-circuit fuses are the most commonly used. Faulty fuses may be indicated by a marked increase in motor sound or other protective devices tripping.

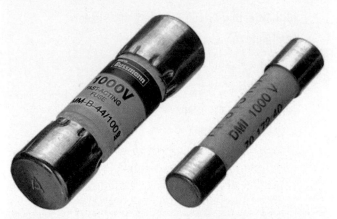

Ideal Industries, Inc.

Figure 26-1. Different fuses have different ratings, applications, and sizes. When replacing a fuse, be sure that the ratings of the replacement fuse are correct.

PROCEDURE	Removing and Inserting Fuses

Removing and inserting fuses should be performed according to the following procedures while wearing appropriate PPE.

For removal:

1. Use an appropriately sized fuse removal tool, **Figure 26-2**. Do not use a pry bar or screwdriver.
2. Ensure that power is off and has been locked out and tagged out.
3. Check with a multimeter to ensure power is off on both sides of the fuse.
4. Using a fuse puller tool, grab the fuse firmly at the center line of the fuse.
5. Firmly and quickly pull the fuse. Depending on the size of the fuse, this may take some effort.
6. Inspect the removed fuse.

For insertion:

1. Using the proper tool, make sure the markings and label on the fuse are in the correct position. Once the fuse is inserted into the holder, the fuse label should be clearly visible on the front.

(continued)

2. Firmly grab the fuse with the tool in the center of the fuse.

3. Make contact with the fuse holder on both sides of the fuse.

4. Inspect to make sure the ends of the fuse clear the ends of the fuse holder and the fuse is centered in the holder.

5. Firmly push the fuse into the fuse holder.

6. If only one end entered the fuse holder, use the tool to firmly grasp the other end and insert it into the holder.

7. Inspect the fuse to make sure it is centered and properly seated in the holder with the label clearly visible to the front of the fuse.

TECH TIP

Checking Fuses in Circuits with a Control Transformer

In motor control circuits with a control transformer, it is possible to have a blown fuse and read voltage on each side of the open fuse. The coil in the control transformer may slightly drop the voltage, and the open fuse should show a slight voltage drop across it. However, the best method for checking an open fuse in this situation is to remove the fuse and perform a resistance check.

In some older control circuits, the control transformer may have nonstandard wiring. Watch out for these common wiring scenarios:

- Normal wiring for a 120 VAC circuit means the "hot" is fused, and the "neutral" is grounded.
- For voltages higher than 120 VAC using two phases, both phases are fused on the downstream side of the control transformer. Be careful; one fuse could open and voltage still be present from the other phase.
- Older panels may have 120 VAC circuits in which both the "hot" and "neutral" are fused with the neutral, rather than grounded to the back panel. One fuse could open and still have voltage on the control circuit.

 CAUTION

Be careful when working inside a panel. Fully inspect the panel before taking readings so that you know what to expect.

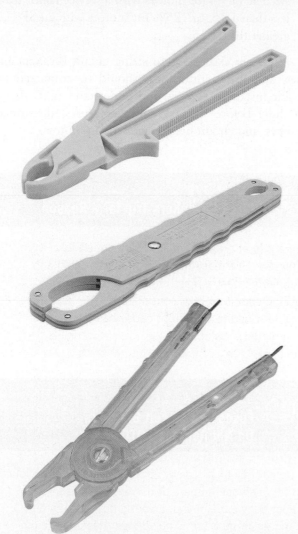

Ideal Industries, Inc.

Figure 26-2. Fuse pullers. Take care when selecting and installing fuses to choose the right tools for the job.

26.3 SELECTING CIRCUIT BREAKERS AND FUSES

The most common type of circuit breaker in a three-phase motor supply circuit is the ***molded-case circuit breaker*** (MCCB), **Figure 26-3**. These circuit breakers range in size up to 3000 A or more and are rated at various three-phase voltages. Circuit breakers not only protect from overloads and short circuits, but they also offer a wide variety of other tripping options and displays. Circuit breakers specifically used for motor protection

Figure 26-3. The molded-case circuit breaker (MCCB) is the most commonly used breaker in panels and motor control centers. A spring-loaded mechanism ensures quick and consistent opening and closing.

may sense overloads, detect phase imbalances, protect against phase loss and short circuits, and check current by thermal or magnetic means. While an overload trip can protect a motor if the circuit breaker has this option, a trip on a ground or short circuit also protects all components, conductors, and terminals downstream of the circuit breaker.

Sizing a breaker properly may also include sizing an integral overload for the breaker. Breakers, fuses, overloads, and other motor protection devices are sized according to *NEC Article 430*. Consider the following guidelines when sizing conductors and overload protection:

- The full-load current (FLC) rating of the motor must be known from *NEC* tables, not the motor nameplate. The FLC rating is then used to select conductor size, short circuit protection, and ground protection.

- The FLC on the motor nameplate is used to determine overload devices. Refer to the section later in this chapter on sizing motor starter overloads.

- Conductors are sized at no more than 125% of the full-load current rating. This means FLC × 1.25 must be less than the ampacity rating of the conductor.

- The breaker protecting the branch circuit is sized using the percent values listed in *Table 430.52* of the *NEC*. For example, imagine you are sizing a breaker for a motor with a current rating of 15.2 A and you want to use an inverse-time breaker (250% from *NEC 430.52*). You calculate 15.2 × 2.5 = 38 A. The next larger sized breaker would be used.

- Fuses are also sized depending on type. Non-delay fuses are sized at 300% of FLC, while dual-element time-delay fuses are sized at 175% of FLC for single-phase and three-phase motors with overload protection. Without overload protection, fuses are sized at 115% for motors with a service factor (SF) less than 1.15, or 125% for motors with an SF greater than 1.15.

There are exceptions to sizing circuit breakers and fuses, and the manufacturer should be contacted for specific instances. Review and understand *NEC Article 430* fully before sizing conductors, fuses, disconnect switches, and circuit breakers for use in motors.

TECH TIP
Circuit Breakers Trip for a Reason

If a circuit breaker trips, it has tripped for a reason. This reason should be investigated and corrected before resetting the circuit breaker. If the circuit breaker continues to trip, the underlying reason or root cause has not been corrected satisfactorily.

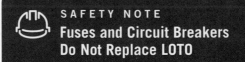

SAFETY NOTE
Fuses and Circuit Breakers Do Not Replace LOTO

Do not rely on a fuse or circuit breaker to save your life. Even one ampere of current passing through your body can kill you. Locking out, tagging out, and ensuring there is no voltage present on a circuit is the only sure method of protection.

26.4 INPUT DEVICES AND SYMBOLS

Input devices may be ***normally open (NO)*** or ***normally closed (NC)***. A normally open (NO) contact has maximum resistance when not actuated and nearly no resistance when actuated. A normally closed (NC) contact has no resistance when not actuated and maximum resistance when actuated. NO and NC describe the contacts in a de-energized or "normal" state. The representation on the diagram shows this state. Devices may also have multiple contacts controlled by a single mechanical means. For example, a normally closed STOP pushbutton allows current to flow through it when the button is not depressed. When the button is pressed, it opens the contact against spring force and stops current flow.

Most of the contacts discussed here should be considered ***hard contacts***, **Figure 26-4**. A hard contact is made of copper (sometimes with a silver coating) and has very little resistance when closed. When open, a hard contact should have infinite or nearly infinite resistance. Contacts including pushbuttons and switches may be ***momentary contacts*** (also called *snap-acting contacts*) or ***maintained contacts***, depending on the design of the element. In momentary contacts, when pressure is released, the contact returns quickly to its normal state. In maintained contacts, the contact remains active until it is deactivated.

Refer to **Figure 26-5** throughout this section to familiarize yourself with typical symbols for manual pushbuttons and switches.

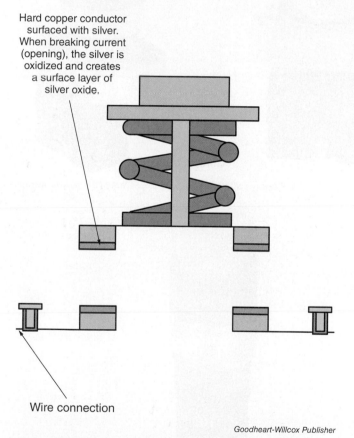

Hard copper conductor surfaced with silver. When breaking current (opening), the silver is oxidized and creates a surface layer of silver oxide.

Wire connection

Goodheart-Willcox Publisher

Figure 26-4. Common construction of a hard contact. This normally open pushbutton is a momentary hard contact that breaks current flow in two locations. When pressure is released, it spring returns to the open position. The silver oxide that develops on the contacts is normal and should not be removed. Cleaning or filing can damage the contacts in this and all modern silver-coated hard contacts.

Common Symbols for Pushbuttons and Switches	
Symbol	**Description**
	Normally open (NO) pushbutton
	Normally closed (NC) pushbutton
	Multiple contact pushbutton
	Multiple contact maintained switch
	Single-pole, single-throw (SPST) NO switch
	Single-pole, single-throw (SPST) NC switch
	Single-pole, dual-throw (SPDT) switch
	Dual-pole, single-throw (DPST) switch
	Dual-pole, dual-throw (DPDT) switch
	Normally open (NO) limit switch
	NO limit switch held closed
	Normally closed (NC) limit switch
	NC limit switch held open
	NO float switch
	NC float switch
	NO temperature switch
	NC temperature switch

Goodheart-Willcox Publisher

Figure 26-5. Pushbutton and switch symbols are similar and may show multiple contacts, alternate positions with dotted lines, and power supplies for contained indicators.

26.4.1 Pushbuttons and Indicators

The basic pushbutton comprises several parts, as shown in **Figure 26-6A**. Notice that the contact block is secured to the base with a threaded fastener. Contacts can be normally open or normally closed and stacked vertically and horizontally so multiple contact blocks may be mounted to one base.

Indicators or indicating lights use a power unit for the base but are otherwise similar to pushbuttons, **Figure 26-6B**. Indicators may be operators as well, where pushing the indicator activates connected contacts. Indicating lights are typically 24 VDC, but older systems may have 120 VAC or higher voltages. Indicating lights may be *push-to-test*, in which a contact is incorporated to allow a user to check that the indicating light is working.

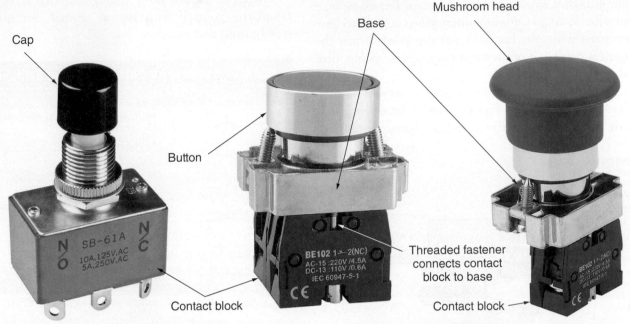

A

NKK Switches; NOARK Electric North America; NOARK Electric North America

Indicating light

Illuminating pushbutton switch

B

NOARK Electric North America; NKK Switches

Figure 26-6. A—The key parts of a basic pushbutton or switch. B—Indicators may also function as operators when part of an illuminated pushbutton switch.

26.4.2 Switches

Switches may be illuminated and include a power unit, have multiple positions (usually up to four), be maintained or spring-return momentary contacts, and have multiple contact blocks connected. **Figure 26-7** shows components and logic for a three-position switch with one contact block attached.

Switch contacts may be *single-pole, single-throw (SPST)*, *single-pole, dual-throw (SPDT)*, or have multiple poles. SPST switches have a single arm or pole that can make a single connection or throw. SPDT switches similarly have only a single arm, but they have two throw positions. Refer to **Figure 26-5** to review the symbols for various types of switches. For example, a light switch is a single-pole, single-throw switch that may be thrown to turn lights on. Some switches can be adjusted to open or close at a particular point called a *set point*. The *deadband* is the range of values between the make and break points for a switch or contact.

Limit Switches

Limit switches physically contact an element and convert mechanical movement into an electrical output. Rollers, whiskers, or plungers are used to actuate contacts. Since these are hard contacts, multiple contacts with multiple voltages may be used. Adjustment depends on model. Roller or whisker actuators can normally be adjusted by repositioning the actuator on the shaft or by adjusting the arc length. Some switches may also allow spring pressure adjustments. Switches may be the typical industrial size, **Figure 26-8**, or smaller miniature switches.

Chaowalit/Shutterstock.com

Figure 26-8. A basic limit switch. Limit switches are available with a variety of mounting styles, actuators, and ratings.

When adjusting limit switches, careful consideration must be given to placement and length of travel. Mechanical elements that actuate the switch must not over-range the switch actuator. Adjustment affects circuit operation and timing and must be expertly performed. Adjusting the mounting or overall position of the switch must also be carefully planned. Repositioning a switch that inputs into a system with a PLC changes how the control system receives feedback and may affect system operation. Each type of switch has different operating characteristics, and the manufacturer's manual should be consulted before adjustments.

Switches may come unwired with standard threads or prewired with a standard connector. **Figure 26-9** summarizes wiring for an SPDT switch using standard symbols.

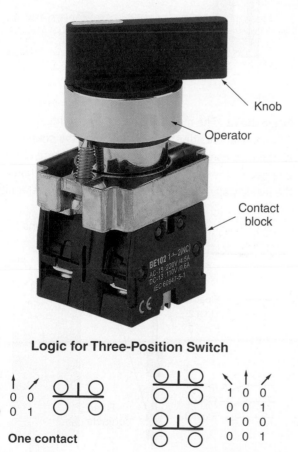

Logic for Three-Position Switch

One contact

Two contacts

NOARK Electric North America; Goodheart-Willcox Publisher

Figure 26-7. The operator determines the logic of the switch and the position in which it actuates contacts. The logic of a switch is indicated with logic tables.

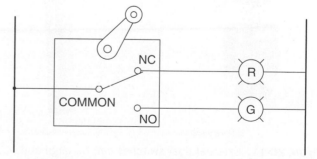

Goodheart-Willcox Publisher

Figure 26-9. Limit switches are not wired to switch on the neutral side. In the case of a single-pole, dual-throw (SPDT) switch, "common" refers to a common input to the two NO and NC outputs, rather than ground or neutral.

Some miniature switches may contain *magnetic reed contacts*, **Figure 26-10**. Reed contacts are sealed contacts actuated by an internal magnet. As the actuator arm is moved, the magnet approaches the sealed contacts and repositions them. These sealed contacts prevent contamination and have a longer life span, but they are limited in amperage rating.

Goodheart-Willcox Publisher

Figure 26-10. Magnetic reed contact.

Float, Temperature, and Pressure Switches

Float switches float in liquid and activate when the liquid reaches a certain level, **Figure 26-11**. Common types of float switches include external, internal, and magnetic reed. External float switches are mounted on a flange or bracket outside of the tank of liquid they are measuring. Some manufacturers allow field adjustment of set points, which may be a mechanical adjustment or an internal spring tension adjustment. Depending on the application, the contacts may directly operate a motor or a motor starter. Internal float switches are completely mounted inside the tank or liquid, and an internal microswitch changes state as the float is angled by a rising or falling level of liquid. **Figure 26-12** shows a common installation of an internal float switch.

Yuthtana artkla/Shutterstock.com; Winai Tepsuttinun/Shutterstock.com

Figure 26-11. External float switches can be large and flange-mounted or smaller and mounted with standard pipe fittings and threads. These switches may include multiple contacts that open or close on liquid level rise. Actuation level is adjustable over a small range. Extension rods are available to increase range.

Multiple floats can be used to control multiple loads (such as pumps) if spaced correctly to prevent interference. Magnetic reed level switches are also common on process systems, **Figure 26-13**. The floats have internally mounted magnets that actuate miniature reed contacts. If multiple reed contacts are used with multiple floats, the contacts may have a common input with separate outputs or individually wired contacts.

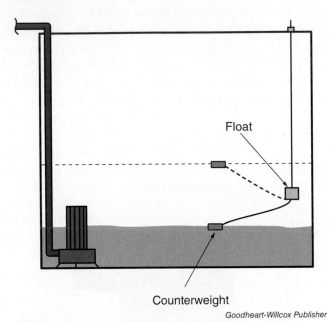

Goodheart-Willcox Publisher

Figure 26-12. The maximum on and off points of an internal float are adjusted by positioning the counterweight. As the float rises, it tilts, resulting in the closure of the internal microswitch. The float could also be anchored to the inside wall of the vessel, but this limits adjustability.

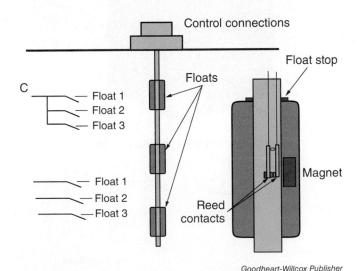

Goodheart-Willcox Publisher

Figure 26-13. A magnetic reed switch with an arrangement of floats on a single stalk. Magnetically actuated reed contacts are common on older systems and are more economical than some newer instruments.

Basic temperature sensing is accomplished using *temperature switches*, which open or close once a certain temperature is reached, **Figure 26-14**. Temperature switches are typically bimetallic or gas/vapor filled. Bimetallic elements contain two dissimilar metals that expand at different rates due to temperature change. Bimetallic contacts can be set by spring pressure to make

or break at the desired temperature. Gas/vapor-filled temperature sensors are more accurate. In this type of sensor, a gas fills the sensing element. As temperature changes, the gas expands or contracts, and this motion is transferred to a microswitch. The sensing element may be mounted into a *thermowell*, which isolates the element from the process liquid when required. Set point adjustment must be performed using special equipment according to the manufacturer's documentation.

Pressure switches most commonly use a bellows or diaphragm to sense pressure and convert it to mechanical movement against a spring. The movement is transferred to a miniature switch. For switches that operate on a pressure rise, this means that as pressure rises and reaches the set point, the contacts change state. Then, as pressure falls below the set point and deadband, contacts change back to normal position. Some pressure switches are field adjustable, and others are set from the factory and cannot be adjusted, **Figure 26-15**. Switches that operate at higher pressure tend to be less accurate and have a larger deadband.

Other types of switches and sensors are covered in Chapter 31, *Sensors and Variable Frequency Drives*, including capacitive sensors, inductive sensors, photoelectric sensors, Hall effect sensors, fiber-optic sensors, and light curtain switches.

Automation Direct

Figure 26-14. Temperature sensors are available with various features and adjustments to suit particular applications.

Sporlan Division, Parker Hannifin Corporation

Figure 26-15. Pressure switches are available in a range of pressure and current ratings. Manufacturer's documentation should be reviewed and understood prior to making any adjustments to set point or deadband. If exposed to environmental contamination, the switch should be de-energized, removed, and carefully cleaned.

26.5 MOTOR STARTERS AND CONTACTORS

Motor starters and *contactors* are very similar, **Figure 26-16**. A contactor provides resistance to a motor at startup and removes resistance when the motor reaches running speed. A motor starter operates like a contactor but also includes an overload to protect the motor from current spikes. Motor contactors may have multiple poles controlled by a single magnetic coil. The coil may be designed for the same voltage as the main contacts, or it may be energized by a lower voltage. Contactors and motor starters may have multiple auxiliary contacts mechanically connected to the main contacts but electrically separate.

26.5.1 Construction

Motor starters are available in different types, depending on the application, **Figure 26-17**. Motor starters are classified as manual starters or switches, magnetic starters, and definite purpose contactors. Toggle-switch motor starters are typically used for fractional horsepower motors, but they can be sized to handle several horsepower. Pushbutton manual starters are common on small single-phase, polyphase, or DC motors. Melting alloy overload units are standard. When tripped on overcurrent, the unit cannot re-engage without letting the overloads cool and then resetting them.

The standard motor starter contains contactor and overload sections, **Figure 26-18**. Both sections are sized according to their current interrupting capability. The ***National Electrical Manufacturers Association (NEMA)***, a US-based association of electrical equipment

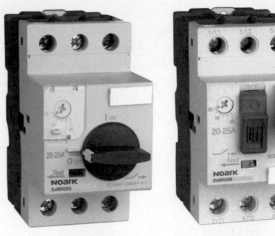

NOARK Electric North America

Figure 26-17. Manual starters and switches are popular on smaller fractional horsepower motors. They can include overload units, as many as three poles, reversing capability, and two-speed switching.

manufacturers, standardizes the sizes and ratings of starters, **Figure 26-19**. The ***International Electrotechnical Commission (IEC)*** provides international standards for electrical and electronic systems and related technologies, including motors and starters. IEC and NEMA starters have some physical differences as well as operating differences. The contactor shown in **Figure 26-16** is similar in appearance to an IEC starter, while **Figure 26-18** shows a NEMA starter. Typically an IEC starter is sized by power of the motor (kVA or kW), while the NEMA style is sized by horsepower.

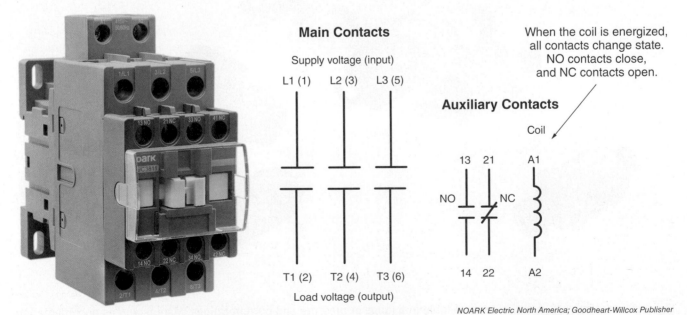

NOARK Electric North America; Goodheart-Willcox Publisher

Figure 26-16. A motor contactor and corresponding internal circuit diagram. Contactors and motor starters are rated by amperage, voltage, coil voltage, and maximum motor power.

Some magnetic motor starters are supplied with *dual-voltage coils*, **Figure 26-20**. The coil develops a magnetic field when energized, which shuts the main contacts against spring pressure and indirectly changes the state of any auxiliary contacts. When de-energizing, the spring pressure ensures contact opening and stopping of the motor.

26.5.2 Overloads and Sizing

The *overload* section of a motor starter provides protection to the motor by sensing motor current and opening a normally closed contact when current exceeds a predetermined value. While some motor starters provide for both normally open and normally closed overload contacts, the normally closed contact is wired in series with the coil of the motor starter. When the overload (OL) contact opens, the starter coil is de-energized and opens,

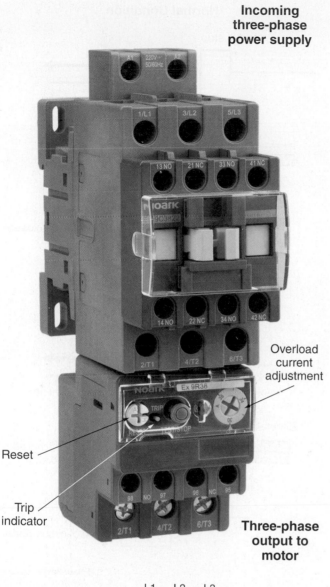

Incoming three-phase power supply

Overload current adjustment

Reset

Trip indicator

Three-phase output to motor

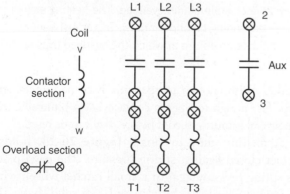

NOARK Electric North America; Goodheart-Willcox Publisher

Figure 26-18. Motor starters start and stop a motor and also provide protection on overcurrent conditions. Each brand of starter varies slightly but features the same major components. When encountering a new starter, carefully inspect and refer to the manufacturer's documentation.

NEMA Maximum Horsepower Ratings or Motor Starters				
NEMA Size	Single-Phase Motor at 115 V	Single-Phase Motor at 230 V	Three-Phase Motor at 230 V	Three-Phase Motor at 460 V
00	1/3	1	1.5	2
0	1	2	3	5
1	2	3	7.5	10
2	3	7.5	15	25
3	7.5	15	30	50
4			50	100
5			100	200
6			200	400
7			300	600

Goodheart-Willcox Publisher

Figure 26-19. NEMA starter sizes range from 00 to 7 and increase as horsepower and current increase.

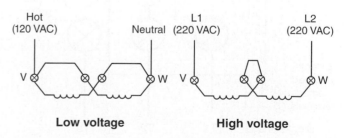

Low voltage   **High voltage**

Goodheart-Willcox Publisher

Figure 26-20. Dual-voltage coils are common on NEMA starters. For low voltage, the two internal coils are wired in parallel with each one receiving full voltage. For high voltage, coils are wired in series, so each coil has the same resistance and receives the same voltage drop and current.

shutting down the motor. Since there is mechanical and electrical feedback of this action and the auxiliary contact opens, the motor starter remains de-energized until the overloads are reset manually.

In a three-phase motor starter, the current of each phase is monitored. Each phase has its own normally closed contact. The three contacts are internally wired in series and appear as one contact externally. The overload section has two circuits, one to monitor current and the other to connect to the contacts. The high voltage passing through to the motor and the voltage to the motor starter coil are in separate circuits, **Figure 26-21**.

Some types of overloads sense the current magnetically, and others develop heat to trigger the overload contacts opening. Motor starting results in a very high initial current draw, so overloads have a time delay built into them. While overloads are sized to trip at a specific current draw, a slight overcurrent will take longer to trip the overloads than a large overcurrent.

Overloads that rely on current sensing through heat generation come in two types: the *bimetallic overload* and the *melting alloy overload*. A bimetallic overload, **Figure 26-22**, generates heat as current passes through the heating coil. A bimetallic strip located near the heating coil warps as temperature rises. This happens because two metals with different rates of expansion are sandwiched

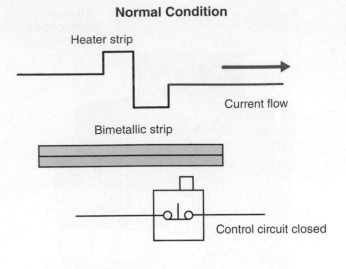

Normal Condition

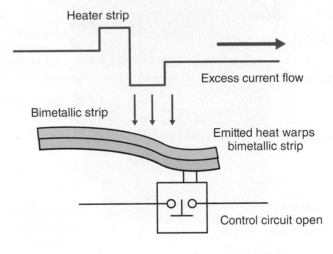

Overload Condition

Goodheart-Willcox Publisher

Figure 26-22. A bimetallic overload. The heating element is sized depending on current draw. Trip adjustment can change the temperature at which the overload trips slightly.

together. With enough temperature rise, the bimetal strip warps far enough to open a contact. Most bimetallic trips take several minutes to cool before they can be reset.

A melting alloy overload, **Figure 26-23**, holds a contact closed against spring pressure. The pawl places this spring pressure against a small ratchet wheel, while the ratchet wheel is held in place by its solid shaft. This shaft is made of an alloy that melts at a predetermined temperature. When enough current passes through the element and heats the shaft, the alloy melts, allowing the ratchet wheel to rotate, opening the contact. As the alloy cools, it solidifies, and the overload contact and pawl can be reset to the original position.

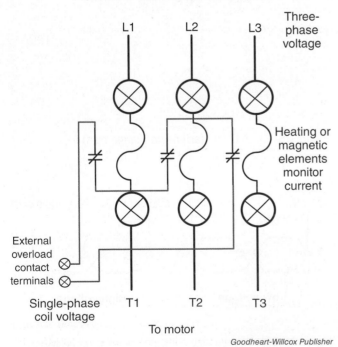

Goodheart-Willcox Publisher

Figure 26-21. This overload contact is shown as one normally closed contact on ladder logic, but it is actually three contacts in series.

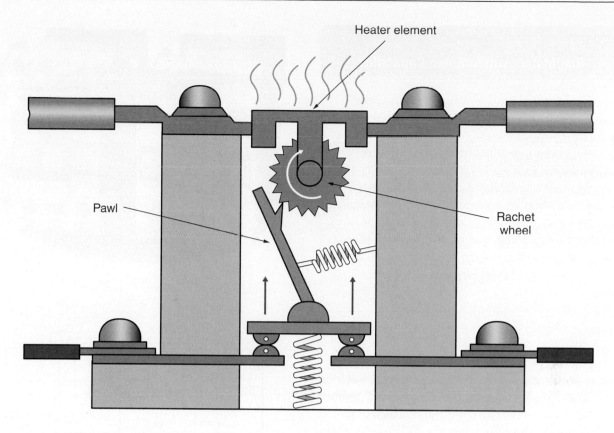

Goodheart-Willcox Publisher

Figure 26-23. Melting alloy overloads activate once a certain temperature is reached.

A *magnetic overload* works like a clamp-on amme-ter. Each phase to the motor passes through the over-load section in a transformer. The transformer generates a voltage proportional to the current to the motor. An electronic control senses this voltage and opens a contact when it exceeds the set point.

TECH TIP
Correct Underlying Problems

Increasing overload size because it routinely trips is ineffective and potentially dangerous. If an overload trips, the reason must be found and cor-rected. Correct the problem, not the symptom.

Properly sizing overloads to protect the motor is a crucial part of motor installation. Sizing depends on sev-eral factors:

- Service factor (SF) of the motor.
- Horsepower and resulting current draw.

- Environmental differences (including temperature) between the motor and motor starter.
- Manufacturer recommendations.

National Electrical Code (NEC) Article 430.32 states the maximum overload sizing as 125% of full-load cur-rent (FLC) for a motor with an SF of 1.15. With an SF of 1.0, the maximum is set at 115% of FLC. However, this is a maximum, and motor manufacturers are usually more stringent.

For an SF of 1.0, 90% of FLC is used to determine overload sizing. For a motor with an SF of 1.15–1.25, 100% of FLC can be used. Temperature differences between the starter and motor of more than 18°F must be taken into account. For example, if the starter is in an air-conditioned space at 68°F and the motor is in an unconditioned space at 90°F, the motor would be allowed to draw more current before the overloads trip. In this instance, the overloads should be downsized to the next lower size. In the opposite situation where the controller is hotter than the motor, overloads should be raised to the next larger size.

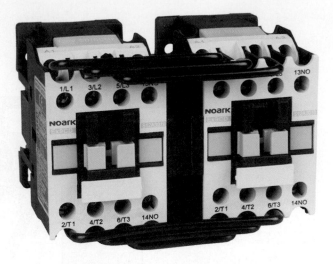

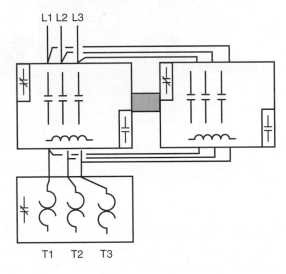

NOARK Electric North America; Goodheart-Willcox Publisher

Figure 26-24. Reversing starters may be oriented horizontally (as shown) or vertically. In a reversing starter, two of three phases (usually phases 1 and 3) are reversed either on top of the contactor section or the bottom, but not both. Reversing the applied phases reverses the rotation of a three-phase motor.

26.5.3 Reversing Starters

Reversing motor starters are used to reverse the rotation of DC, single-phase AC, and three-phase AC motors. **Figure 26-24** shows how two contactors are used to reverse the direction of a three-phase motor. When one starter energizes, it shuts main contacts to apply L1 to T1, L2 to T2, and L3 to T3. When the opposite starter energizes, it applies L3 to T1, L2 to T2, and L1 to T3 to accomplish motor reversal. If both contactors pulled in at the same time, L1 and L3 would be shorted together, resulting in equipment damage. To prevent this possible short, several *interlocks* are used.

An interlock can perform a number of functions:

- Prevent something from happening that may cause damage.

- Ensure that a sequence of events takes place in a specified safe order.

- If a failure occurs, ensure the system is shut down and placed in a safe mode.

- Prevent death or severe injury.

Because of these features, interlocks are never allowed to be defeated, changed, bypassed, removed, or altered. Doing so can result in damage or destruction to equipment or death to personnel. If an interlock's operation is in doubt, it should be tested with a multimeter or replaced.

Mechanical interlocks prevent two of the three phases on a reversing motor starter from shorting. This interlock mechanically prevents the opposite coil from shutting its contacts when a coil is energized and actuated. Even if the opposite coil is energized, the mechanical interlock prevents the closing of the contacts. Electrical interlocks

operate by several methods, including pushbuttons and normally closed contacts, **Figure 26-25**. Both interlock types are commonly used.

26.5.4 Reversing Drum Switches

Reversing drum switches are used to reverse the rotation of DC, single-phase AC, and three-phase AC motors. Reversing drum switches work by switching the line voltage to the motor and are typically mounted in the area where an operator can control or directly observe the motor, **Figure 26-26**. Since the motor starter is normally energized when using the drum switch, the reversing drum switch is directly switching a three-phase load. This arrangement is typical for older control circuits. Safer modern control circuits have the drum switch controlling the control circuit to energize the appropriate coil of a reversing starter or have the drum switch input to a PLC to control the reversing motor starter.

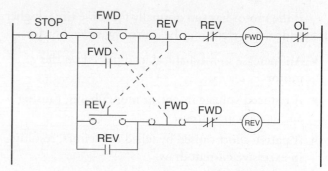

Goodheart-Willcox Publisher

Figure 26-25. Electrical methods may be used as interlocks. In this wiring diagram, both forward and reverse pushbuttons have normally open and normally closed contacts. Normally closed auxiliary contacts are provided on each contactor and wired in series with the opposite contactor coil. This prevents both coils from energizing simultaneously.

SAFETY NOTE
LOTO before Work on Drum Switches

Before working on a drum switch, ensure that the three-phase power is locked out and tagged out. If the drum switch is not working properly, high voltage could be present on the casing of the switch. Never open a drum switch with the power on.

26.5.5 Troubleshooting

The first step in troubleshooting a motor starter is understanding how it works. Before working on an unfamiliar motor starter, spend time reviewing manufacturer documentation. Before putting hands and DMM leads into a control cabinet, look at what you are doing. Carefully examine all visible components inside the cabinet without touching them. Use your sight and smell to discern any evidence of arcing, burning, and insulation breakdown. Examine the motor starter, recognize different parts, and mentally label all terminals. Review circuit diagrams to make sure you know how the circuit should operate, and verify how the circuit is not operating. Make sure power is locked out and tagged out. Some control cabinets may have multiple sources of power. This should be clearly marked on the front of the cabinet, but sometimes it is not.

Overloads

Tripped overloads are the most common issue with motor starters. While resetting overloads fixes the symptom, it does not fix the problem. Overloads trip for two basic reasons:

- Higher than normal current draw.
- Excessive heat in the environment.

Excessive heat in the environment can be seasonal, caused by a loss of cooling or air conditioning, or come from other components. Carefully consider causes of excessive heat and how to correct them. Placing a fan on the overloads to keep them cool is unacceptable and can result in eventual motor failure.

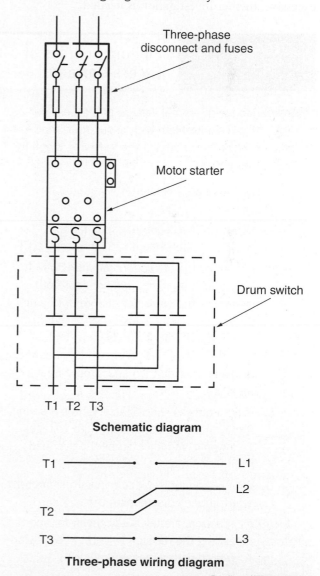

Goodheart-Willcox Publisher

Figure 26-26. Schematic diagram including drum switch and motor starter with internal wiring for drum switch.

If the motor has run normally for some time, higher than normal current may be caused by several factors:

- An increase in mechanical load placed on the motor.
- A reduced voltage in one or more phases, causing an increase in current.
- A partial short caused by failed capacitors, resulting in excessive current draw.
- A partial short in a winding caused by repeated overheating.

To check for excessive current, measure voltage and current for each phase of the motor. Compare these values with the motor's required voltage, previous voltage readings, and FLC. If voltage is correct and current is excessive, further investigation is needed.

PROCEDURE | Finding Sources of Excess Current

An overload with normal voltage and excess current may have a problem with capacitors, windings, or alignment. Perform the following steps to locate the problem:

1. If capacitors are installed, they should be checked first.

   A. Use a DMM to confirm capacitance readings match specifications within 10%.

   B. Discharge the capacitor, set your DMM for a resistance reading, and connect both leads. Let the capacitor charge until your DMM reads about 1 MΩ, then disconnect one lead, wait 30 seconds, and reattach it. If resistance has dropped, the capacitor is internally grounded and should be replaced.

   C. If both checks show the capacitor is sound, it may still be grounding at high voltages. This would be indicated by the overloads tripping on motor startup.

   D. When in doubt, replace capacitors carefully after properly discharging them.

2. If no capacitors are involved, check for shorts by measuring the resistance of the motor winding to ground with a megohmmeter.

3. If the windings are acceptable, the motor should be disconnected from the load and checked to see that it spins freely. The load should also be checked to see that it can spin freely.

4. Check the alignment of mechanical elements.

On some larger motors, other protections are wired in series with overloads or as part of a safety circuit. *Thermistors* sense winding temperatures and change resistance as temperature changes, **Figure 26-27**. Thermistors can have a positive or negative temperature coefficient. A positive coefficient results in the thermistor increasing resistance as temperature increases. These are installed within the motor and have separate connections. If a winding overheats, the safety circuit trips the motor starter and shuts down the motor. Occasionally, these protective devices may fail "open." This results in the motor starter not being able to engage, even though there is no over-temperature condition in the winding.

Coil

The motor starter coil has four basic issues. First, it can become noisy. Excessive electrical noise results from contacts not fully sealing in (or seating) when the coil is energized. The magnetic field generated by the coil rises and falls, just as the alternating current rises and falls. *Shading rings*, also called *shading coils*, offset the main coil and assist in maintaining the magnetic field of the coil. If the starter becomes noisy, check terminals for proper torque within manufacturer's recommendations. Check connections to ensure proper contact. Check main contacts and auxiliary contacts for excessive wear. Finally, if noise persists, replace the coil.

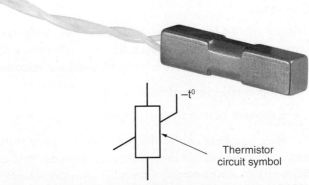

Thermistor

Thermistor circuit symbol

Sporlan Division, Parker Hannifin Corporation; Goodheart-Willcox Publisher

Figure 26-27. Thermistors are placed inside motor windings to sense temperature.

Second, open coils may show high resistance if the coil overheats and insulation breaks down. A resistance reading can indicate an open coil, which means the coil must be replaced. Typical resistances for size 00 and 0 starters range from 40–200 Ω. Larger starters have lower coil resistance readings.

Third, shorted coils can result in tripping a protective breaker or opening a protective fuse on the control circuit. Take a resistance reading with a DMM across the coil to test for this condition. A shorted coil may still show a resistance similar to a normal coil because the short may only show once energized with line voltage. Check the resistance versus a new coil of the same size and manufacturer. Replace when in doubt.

Finally, low voltage can cause coils to heat rapidly. Low voltage in a control system is unusual, but it could be caused by an improperly sized control transformer that cannot supply enough current to a system. Watch for starters or contactors intermittently pulling in, energizing and not pulling in, or severely chattering. In this case, the control transformer may need to be upgraded to a larger unit.

Main Contacts

If there is a voltage drop at the motor, all terminal connections in the three-phase circuit should be checked for proper torque. If the main three-phase contacts still show voltage drop, the contacts should be examined for excessive wear. Slight pitting and a black film is completely normal. Excessive pitting, weld spots, arcing, or resistance across the contact requires replacing all contacts.

Auxiliary Contacts

Auxiliary contacts are often used in motor control circuits with relay logic and may be directly operated by the motor coil or mechanically connected to the main action. This mechanical connection is typically made of plastic. With repeated operations in a high-heat environment, some plastics become weak. If the motor starter is operating but an attached auxiliary contact is not, a mechanical problem may exist. Replace all auxiliary contacts and check for proper operation. Contacts may also have a tendency to increase their resistance over time to the point of excessive voltage drop across them, which prevents the circuit from operating. Check contacts with a DMM. Typical hard contacts should have very low resistance when closed (less than a few ohms) and very high resistance when open (thousands or millions of ohms).

26.6 OUTPUT DEVICES AND SYMBOLS

Output devices perform functions such as extending circuit logic, timing sequences of events, converting electrical signal to physical movement, and indicating circuit operation.

26.6.1 Relays

Relays come in sizes from subminiature relays for circuit boards to high-amperage relays in power distribution systems. Relays convert input electrical signals to mechanical action of switches. Relay coils are specified by input voltage and come in all common control voltages. Relays may be SPST, SPDT, or contain multiple switches and contact positions as with a *dual-pole, dual-throw (DPDT)* relay. DPDT relays have two arms or poles that are each capable of making two connections or throws. One of the most common types of relays used is the general-purpose or cube relay, **Figure 26-28**. Standard relay models are two-pole (eight-pin) or three-pole (eleven-pin) relays. Relay diagrams, **Figure 26-29**, show switching arrangements and are important troubleshooting tools.

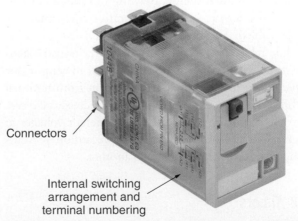

Connectors

Internal switching arrangement and terminal numbering

Automation Direct

Figure 26-28. Control relays are used for hardwired logic circuits.

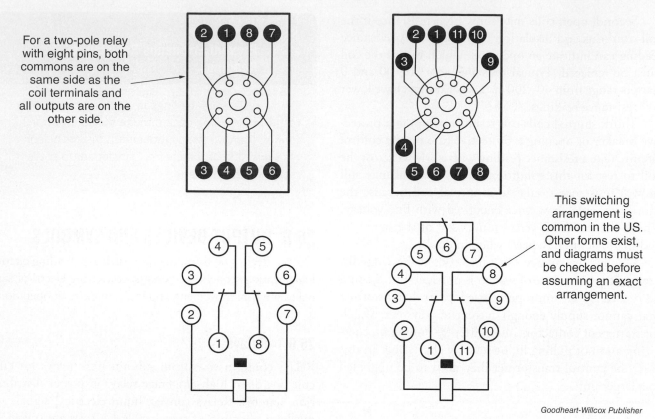

For a two-pole relay with eight pins, both commons are on the same side as the coil terminals and all outputs are on the other side.

This switching arrangement is common in the US. Other forms exist, and diagrams must be checked before assuming an exact arrangement.

Goodheart-Willcox Publisher

Figure 26-29. Standard relay bases with common relay terminal numbering and internal switching. Terminal 1 is a common input to outputs 3 and 4. Terminal 8 is a common to output 5 or 6. While coil polarity does not matter for AC, industry standard is terminal 2 hot and terminal 7 or 10 neutral. When troubleshooting, watch out for relays that may have inverted wiring with coil connections on the bottom.

Relays typically have a long life as long as ratings are adhered to, but drawing high current through contacts or supplying low voltage to the coil can cause early failure. Coil resistances should be checked against manufacturer's specifications. Resistances typically range from several hundred to several thousand ohms. An open coil prevents proper operation. High resistances across contacts may be caused by excessive arcing, drawing too much current, wearing of the contact surface, or vibration over time. Any of these problems can be checked with a DMM.

While relay contacts are not polarized, industry standard is to wire the contacts from common to output (for example, input into terminal 1 and output from terminal 3). While contacts are not electrically connected, take care to wire them properly when applying different voltages to the same contacts on one relay. This precaution prevents shorting from one voltage source to another.

26.6.2 Timers

The most common timers are electronic timers. While there are other older timers (such as dash-pot, motor-driven, or pneumatic timers), the majority seen in modern control cabinets are the electronic type. These timers mount into a standard two- or three-pole relay base. Some timers have selectable functions and selectable time ranges. Timers are offered in common control circuit voltages from 24 VDC to 120 VAC.

On-Delay Timer

An **on-delay timer** energizes an output after a preset time elapses. That is, it comes on after a delay. On-delay timers follow a specific series of events each time they are energized:

1. The "coil" is energized.
2. The timer counts to the specified time setting.
3. Contacts change state—closed contacts open and open contacts close.
4. At some predetermined later time, power is removed from the coil.
5. Contacts immediately change back to their normal de-energized state.

If power is removed from the coil before the time setting is reached, the timer resets to a zero time count and controlled contacts remain in their de-energized state.

Symbols for timed contacts are shown in **Figure 26-30**. Both *normally open, timed-closed (NOTC) contacts* and *normally closed, timed-open (NCTO) contacts* are shown with an up arrow indicating the direction of timing and contact movement. NOTC contacts are open when de-energized. NCTO contacts are closed when de-energized.

Just as for relays, timers have a maximum amperage limit for contacts. Timed contacts are not meant to control large inductive loads. They are meant to allow additional logic possibilities to control a circuit. Timer settings should not be disregarded and changed without consulting design specifications. If the operation of a timer is in doubt, observe operation while power is applied to determine if there is a malfunction. Timed contacts should have similar resistance readings as relay contacts when both open and closed.

1. Power is applied to the coil.
2. Contacts immediately change state.
3. At some later time, power is removed from the coil.
4. The timer counts to a specified time.
5. Timed contacts change back to original state.

For a contact-controlled timer the sequence is similar, but the timed contact's action is controlled by the opening of the control contact:

1. The timer coil is wired hot, and is always supplied voltage.
2. At some point the control contact closes.
3. Timed contacts immediately change state.
4. At some later time, the control contact opens.
5. The timer counts to a specified time.
6. Timed contacts change back to the original state.

On-Delay Contact Symbols

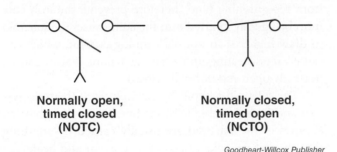

Normally open, timed closed (NOTC)

Normally closed, timed open (NCTO)

Goodheart-Willcox Publisher

Figure 26-30. Symbols for on-delay timer contacts.

Contact-Controlled Off-Delay Timer

Electronic *off-delay timers* de-energize after a set time has elapsed and come in two basic configurations. An older type is available in two- or three-pole design and activates when power is removed from the timer terminals. Some of these older timers rely on internal batteries for power during the timing function. The second type is a *contact-controlled timer*. This type may be controlled by an off-delay as well as more programmable functions and is characterized by several features:

- Multiple timer function settings.
- Multiple timer range settings.
- Three-pole terminal base.
- One pole reserved for a control contact, which determines timer actuation.

Figure 26-31 shows an example of an off-delay timer and a diagram for a three-pole contact-controlled timer. For a true off-delay timer, the following sequence of operational events occurs:

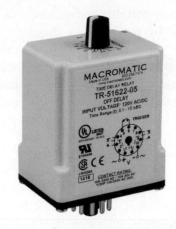

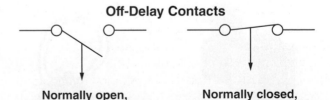

Control contact or trigger

Low voltage

120 VAC (hot)   Neutral

T1

Timer coil (wired hot)

Off-Delay Contacts

Normally open, timed open (NOTO)

Contact closes immediately with trigger, then times open when trigger opens

Normally closed, timed closed (NCTC)

Contact opens immediately with trigger, then times closed when trigger opens

Macromatic Industrial Controls; Goodheart-Willcox Publisher

Figure 26-31. A common two-pole off-delay timer. Off-delay timers may be controlled by the application and removal of power to the coil or by the closing and opening of the control contact. The control contact shown here can be any type of contact, including a limit switch, pushbutton, relay, or sensor.

Notice that the symbol for an off-delay contact uses a down arrow and indicates either a NOTO or NCTC contact. This is the only proper way to indicate an off-delay contact. Depending on the manufacturer, the front panel of the relay may indicate the power and condition of the trigger or control contact with LEDs. Refer to the manufacturer's documentation for more information.

In newer PLC-based control systems, the PLC programming performs the timing functions, but there are still many older systems that include relay timers. Other functions such as repeat, one-shot, interval, and others are sometimes used. Becoming familiar with your particular systems and timer functions is a part of becoming a successful troubleshooter.

26.6.3 Solenoids

A *solenoid* is any coil that converts current to mechanical movement by producing a magnetic field. While there are many types of solenoids used in electrical controls, this discussion is limited to solenoid-controlled valves, **Figure 26-32**. Hydraulic and pneumatic valves are covered in the *Basic Hydraulic and Pneumatic Systems* section. This section examines process control valves typically used for water or gas systems.

The solenoid itself is a conducting coil. Since a conductor carrying a current produces a magnetic field around it, the magnetic field can be concentrated by coiling the conductor. This magnetic field attracts ferromagnetic materials, such as iron, which is the basis of providing mechanical movement to a valve. Both AC and DC solenoids are common. Because of their design, solenoids tend to heat during prolonged operation. Current flowing through the solenoid coil heats the solenoid

Automation Direct; Field Controls

Figure 26-32. Common solenoid-controlled valves range in size up to 3/4″ NPT. Larger pilot-operated valves are also available.

based on the resistance of the coil. While some solenoids are designed for continuous operation, or rapid cycling, others are not. Solenoids eventually fail when subjected to excessive heat.

As a solenoid is initially energized, a magnetic field is developed to draw the soft iron core into the coil. Initial current draw can be several times the **seal-in current**, the normal running current. Initial current depends largely on the differential pressure across the valve and the size of the valve. Initial current can be as high as 2–5 times the seal-in current. Larger valves or larger differential pressures require larger solenoids with higher current draws. Larger valves require the use of internal pilot valves to reduce the differential pressure and therefore reduce the size of the solenoid. Initial current draw drops to a lower running current or seal-in current as the iron core is drawn into the coil and reinforces the magnetic field. The seal-in current drops further over time, as heating slightly increases the resistance of the coil. Anything that prevents the valve from repositioning (and therefore prevents the iron core from being fully inserted into the coil) causes the solenoid to draw higher than normal running current, which can cause excess heating and eventual failure. Valves can be normally open and normally closed.

The solenoid is typically isolated from the fluid system and is replaceable without breaking a fluid barrier. Failures of the solenoid are usually caused by anything that exposes the solenoid to more current and heat:

- Low supply voltage can lead to an increase in current and heat.

- Rapid cycling beyond what the valve was designed for.

- Failure of the valve to fully open allows initial start-up current to remain high.

Before replacing a solenoid, consider the above reasons for failure and correct the problem. Comparing the resistance of a suspected faulty coil to a new coil can easily determine if an open or short exists. If a valve is not fully opening due to some type of blockage, the reason for the blockage must be prevented if future failures are to be avoided.

CHAPTER WRAP-UP

Understanding how components perform their intended function and how they combine into a circuit to control a machine is the start of developing a sound troubleshooting method. Since technology is always improving and changing, this area of expertise always provides for new challenges and learning.

Chapter Review

SUMMARY

- Motor control circuits are used to start and stop loads, perform a sequence of events, control an output relative to an input, and automatically control a system.

- Fuses are single-use and can protect against overcurrent (overloads) or short circuits. They are categorized by several factors, including interrupt rating.

- Circuit breakers used for motor protection may sense overloads, detect phase imbalances, and protect against phase loss and short circuits.

- Conductors are sized at no greater than 125% of the full-load current (FLC) rating of a motor.

- Circuit breakers are sized according to *NEC Article 430*.

- Input devices may be normally open (NO) or normally closed (NC). These classifications describe the contacts in a de-energized or "normal" state.

- Most contacts in control circuits are considered hard contacts, which have little or no resistance when closed. Contacts may be momentary contacts or maintained contacts.

- Pushbuttons and indicators are both types of input devices that activate connected contacts.

- Switch contacts may be single-pole, single-throw (SPST), single-pole, dual-throw, or have multiple poles. Some switches can be adjusted to open or close at a specified point called a set point.

- Limit switches physically contact an element and convert mechanical movement into an electrical output.

- Float switches may be classified as external, internal, or magnetic reed. Other types of switches can be used to sense temperature and pressure.

- Motor starters and contactors are similar, but motor starters include motor overload protection.

- Overloads provide protection by sensing motor current and opening a normally closed contact when current exceeds a predetermined value. They may be bimetallic, melting alloy, or magnetic.

- Sizing motor overloads depends on motor service factor, current, temperature difference between motor and controller, and manufacturer's recommendations.

- Reversing motor starters change a motor's rotation by changing the applied phases. Interlocks are used to prevent possible shorts in the motor starter and must not be changed, disabled, or bypassed.

- Reversing drum switches typically switch on the main voltage lines supplying the motor.

- The most common cause of overloads tripping is excessive heat in the environment and higher than normal current draw.

- The motor starter coil has several issues that can occur and cause failure, including excessive noise, overheating, open coils, high resistance, shorted coils, and low voltage.

- Output devices can extend circuit logic, time sequences of events, convert electrical signal to physical movement, and indicate circuit operation.

- Relay and timer contacts are wired in the "forward" direction from a common input to a normally open or closed output.

- An on-delay timer energizes an output after a preset time elapses. An off-delay timer de-energizes after a set time has elapsed. Timed contacts can be normally open, timed-closed (NOTC) or normally closed, timed-open (NCTO).

- A contact-controlled timer has a low voltage circuit that controls the timing. Outside voltage should not be applied to this circuit.

- A solenoid is a coil that converts current to mechanical movement by producing a magnetic field. Failures of a solenoid are usually caused by anything that exposes the solenoid to excess current and heat.

REVIEW QUESTIONS

Answer the following questions using the information provided in this chapter.

1. Describe four basic functions of motor control circuits.

2. What is the purpose of digital controls?

3. What is an interrupt rating?

4. *True or False?* A fuse should be inserted with only one end entered into the fuse holder with a label clearly visible in front of the fuse.

5. What is the purpose of a circuit breaker specifically used for motor protection?

6. The _____ rating is used to select conductor size, short circuit protection, and ground protection.

7. If a motor has an SF of 1.25 and no overload protection, what size fuse should be selected?

8. *True or False?* A normally open (NO) contact has maximum resistance when not actuated and nearly no resistance when actuated.

9. Describe the two types of hard contacts.

10. _____ switches have a single arm or pole that can make a single contact or throw.

11. What is the difference between set point and deadband?

12. What is a limit switch?

13. How are magnetic reed contacts actuated?

14. Describe the difference between external float switches and internal float switches.

15. How does a bimetallic element work?

16. *True or False?* Switches that operate at higher pressure tend to be less accurate and have a larger deadband.

17. Discuss the similarities and differences between motor starters and contactors.

18. Describe the function of dual-voltage motor starter coils.

19. Name the two types of overloads that rely on current sensing through heat generation and describe how they function.

20. List four factors that determine the sizing of an overload.

21. *True or False?* For a motor with an SF of 1.15–1.25, 90% of full-load current (FLC) can be used.

22. When a reversing motor starter energizes, it shuts main contacts to apply L1 to _____, and applies L1 to _____ when the opposite starter energizes.

23. Describe the operation of an interlock.

24. Describe the operation of reversing drum switches.

25. What are the two reasons that overloads trip? Discuss how overload tripping can be prevented.

26. Describe three possible problems with motor starter coils.

27. What is the function of a relay? How is a relay specified?

28. Describe the sequence of events for an on-delay timer.

29. *True or False?* Off-delay timers energize after a set time has elapsed.

30. What is a solenoid?

31. What determines the amount of initial current draw for a solenoid?

NIMS CREDENTIALING PREPARATION QUESTIONS

The following questions will help you prepare for the NIMS Industrial Technology Maintenance Level 1 Electrical Systems credentialing exam.

1. Which of the following is an example of an input into a motor control circuit?

   A. Motor starter
   C. Relay
   B. Solenoid
   D. Switch

2. _____ switches physically contact an element and convert mechanical movement into an electrical output.

   A. Float
   B. Limit
   C. Magnetic reed
   D. Pressure

3. Which of the following is *not* a function of a motor starter?

   A. It includes an overload to protect the motor from current spikes.
   B. It provides resistance to the motor at startup.
   C. It removes resistance when the motor reaches running speed.
   D. It uses two magnetic coils to induce large amounts of voltage.

4. How does the overload of a motor provide protection when current exceeds a predetermined value?

   A. It closes a normally opened contact.
   B. It opens a normally closed contact.
   D. It de-energizes and closes the starter coil.
   C. It energizes and opens the starter coil.

5. An interlock prevents two of the three phases on a reversing motor from shorting by _____.

    A. de-energizing the coils on the two other phases

    B. preventing the opposite coil from shutting its contacts when a coil is energized

    C. shutting the contacts on the opposite coil and de-energizing the coil

    D. shutting the contacts on the opposite coil while keeping the coil energized

6. Which is a proper method for locating a problem with excess current in a motor starter?

    A. Check alignment of the starter elements.

    B. Check for shorts by measuring winding voltage with a megohmmeter.

    C. Observe whether a load can spin when connected to the motor.

    D. Use a DMM to confirm capacitance readings match specifications within 15%.

7. The following symbol shows what motor control component?

    A. Normally closed temperature switch

    B. Normally closed, timed-open contact

    C. Normally open float switch

    D. Normally open, timed-closed contact

8. The following schematic diagram shows a _____.

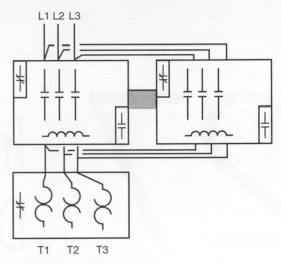

    A. reversing drum switch

    B. reversing motor starter

    C. three-phase motor starter

    D. three-pole relay

9. Which of the following is true about a solenoid?

    A. It is a conducting coil.

    B. It can produce a magnetic field.

    C. Both A and B.

    D. Neither A nor B.

10. What may cause a solenoid to fail?

    A. High supply voltage

    B. Low motor temperature

    C. Low start-up current

    D. Rapid cycling of the valve

27

Industrial Wiring Diagrams and Practices

LEARNING OBJECTIVES

After completing this chapter, you will be able to:

☐ Recognize the features of ladder diagrams and understand how to read them.

☐ Describe the four basic numbering systems used for ladder diagrams.

☐ Explain how various electrical diagrams are used to depict the operation of a control circuit as well as how they are applied in troubleshooting.

☐ Understand how the *National Electrical Code (NEC)* sets standards for safeguarding people and property against the hazards of electricity.

☐ Discuss *NEC* grounding specifications and why grounding is necessary in a system.

☐ List the differences between switchboards and panelboards.

☐ Describe system components that serve as disconnecting means.

☐ Understand the requirements for conduits, busways, and raceways.

TECHNICAL TERMS

ampacity	panelboard
busbar	panel layout diagram
busway	raceway
component terminal number	single-line diagram
cubicle	switchboard
disconnecting means	three-line diagram
highway diagram	Underwriters Laboratories (UL)
ladder diagram	
National Electrical Code (NEC)	wire run
	wiring diagram
National Fire Protection Association (NFPA)	

As you develop your skills as an industrial maintenance technician, you will learn to rely on various electrical diagrams as your map to effective troubleshooting. Diagrams contain essential information about a system, including how components are wired and installed. Electrical installations must also adhere to *NEC* guidelines and safety regulations. Familiarize yourself with this documentation as well as practical aspects of installing wiring to become an expert on any electrical system.

27.1 ELECTRICAL DIAGRAMS

There are several common electrical diagrams, each with their own use and purpose. While basic schematic diagrams are best used for understanding how a system works, other diagrams show exact terminal-to-terminal wiring. While some systems may only take 10–20 rungs on a ladder diagram, others may require hundreds of rungs. Once you understand the basic types of diagrams, you can properly use them to troubleshoot a modern control system. Expect to spend extensive time studying the control systems common to your line of work or area of employment. Without a clear understanding of electrical diagrams, a technician is simply guessing at troubleshooting.

27.1.1 Ladder Diagrams

A *ladder diagram*, **Figure 27-1**, is used to develop an understanding of a control circuit's operation and troubleshoot a circuit that is not operating properly. This type of diagram does not show exact terminal-to-terminal wiring. Instead, a ladder diagram shows wiring in a way that mimics how the actual circuit operates. Notice that the rungs of the ladder are made up of the control circuit, while the rails of the ladder are the hot and neutral of the system. Decisions are made on the left-hand part of the rungs, and actions or outputs occur on the right. When a complete path exists for current to flow, current will flow from the hot on the left, through the decisions, to the actions on the right.

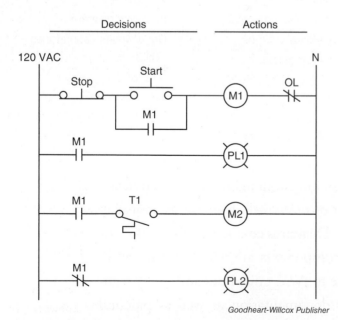

Goodheart-Willcox Publisher

Figure 27-1. Ladder logic diagrams show basic functions of a control circuit.

Be aware of the following features of ladder diagrams:

- Ladder diagrams are shown de-energized. Although a current path may exist, the output is not shown as energized. In **Figure 27-1**, the lowest rung contains a normally closed contact that turns on PL2 as soon as power is available to the system.

- A ladder diagram does not show exact wiring, only logic. On the first line there is a normally open contact, M1, bypassing the START pushbutton. The wires leading to M1 could be connected to the STOP pushbutton on one side and the M1 coil on the other, or both sides could connect to the terminals on the START pushbutton.

- Loads are not wired in series. Loads are wired in parallel so that each load receives full control circuit voltage. If two loads were required to energize at the same time, they would be wired closely in parallel.

- Because a ladder diagram shows the logic of the circuit, this is one of the first diagrams used for troubleshooting, though several other types are also used.

Rung Numbers and Contact Cross-Reference Numbers

Ladder diagrams use several numbering systems to indicate various features of a control circuit. **Figure 27-2** shows the application of rung numbers and contact cross-reference numbers. Rung numbers are used to label each rung of the ladder. Most diagrams place these numbers on the left-hand side of the diagram, but some list them on both sides. Contact cross-reference numbers are placed on the right-hand side of the rungs, appear in parentheses, and reference other rungs containing contacts that the current rung will activate. Rungs containing normally closed (NC) contacts are shown underlined, while rungs with normally open (NO) contacts are shown without underlining. By following the actions and rung numbers, technicians can develop a sequence of events. In **Figure 27-2**, the following events occur in order:

1. When power is applied to the circuit but before M1 is energized, PL2 lights because of the normally closed contact on rung 5.
2. When the START button is pressed, M1 energizes. This shuts NO contacts on rungs 2, 3, and 4, and opens the NC contact on rung 5.
3. The contact on rung 2 is the memory contact. It keeps M1 energized.
4. The contact on rung 3 causes PL1 to light.
5. The contact on rung 4 allows T1 to turn on M2 if a preset temperature is reached.
6. The contact on rung 5 opens, and PL2 turns off.

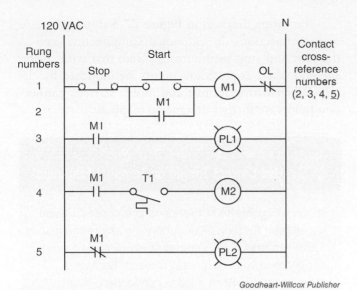

Figure 27-2. In ladder logic diagrams, rung numbers and contact cross-reference numbers help show the operational sequence of a circuit.

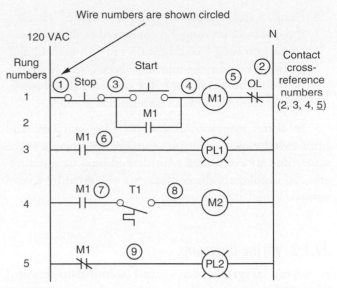

Goodheart-Willcox Publisher

Figure 27-3. Wire numbers are used by electrical technicians when wiring panels connecting to field devices.

Wire Numbers

Ladder diagrams also use a system of wire numbers. A wire number is given to each continuous wire. In other words, a different wire number is given any time a wire is broken by contacts or a load. In **Figure 27-3**, the hot wire is numbered 1, while the neutral is numbered as wire 2. Using 1 and 2 to number hot and neutral wires is common practice. Wire numbers are shown circled to differentiate them from other numbering systems. These numbers aid in troubleshooting, especially when dealing with wiring that leaves the control panel to a remote pushbutton or limit switch. For example, if there might be an issue with the temperature switch (T1), a technician can use the ladder diagram to know that when the switch is opened for inspection, wire 7 should be attached to the common, and wire 8 should be attached to the NO output terminals.

Terminal Numbers

Component terminal numbers can help clarify ladder diagrams when multiple terminals are used on a single component. This is common for pushbuttons with multiple contact blocks, relays, timers, or limit switches. This additional numbering system aids in troubleshooting that requires following a signal through a circuit, especially with larger and more complicated control systems. **Figure 27-4** shows an example of a ladder diagram with rung numbers, contact cross-reference numbers, and component terminal numbers. Component terminal numbers are imprinted directly on each component.

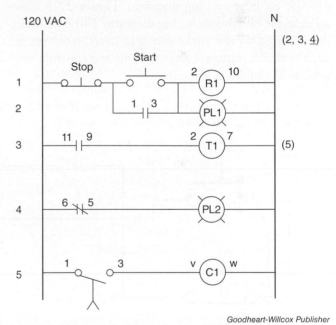

Goodheart-Willcox Publisher

Figure 27-4. Terminal numbers specify particular terminals on components with multiple terminals.

The numbers are shown as smaller numbers near each terminal on the diagram. Note the following features:

- Terminal 2 on relay coil R1 is hot. Terminal 10 is neutral. These hot and neutral terminal designations are industry standard.

- Contacts are wired going into the common and out of the output. Terminal 1 of R1 is connected to the wire between the two pushbuttons. Terminal 3 of R1 may be attached to terminal 2 of R1.

- Rungs 3 and 4 could be on the same "pole." That is, if the contact on rung 3 is connected as shown, the contact on rung 4 could connect to terminals 11 and 8. This is because the inputs of those contacts are both common to the hot. The normally closed contact on rung 4 could not be terminals 1 and 4.

In some instances, floats, pressure switches, and limit switches with multiple contacts that operate at the same time are connected with dashed lines or other referenced numbers if the contacts are separated by several rungs on the ladder diagram.

27.1.2 Wiring Diagrams

A *wiring diagram* shows exact terminal-to-terminal wiring. This type of diagram shows the placement of each wire termination and how many wires per terminal. While a wiring diagram could be used for troubleshooting smaller systems, it is difficult to show any semblance of logic on larger wiring diagrams. **Figure 27-5** shows an example of a simple wiring diagram. This type of diagram is used for the initial wiring of more basic systems, such as motor starters. Larger systems and control panels tend to use highway diagrams.

The wiring diagram in **Figure 27-5** shows where to make connections on individual components, such as the start and stop pushbuttons. Note that wires begin and end where connections points are indicated by circles. When wiring motors and motor starters, connection points are further designated by phase.

TECH TIP
Use Correct Torque on Terminal Fasteners

When wiring motor starters and components, never exceed the torque specifications on any terminals. Fasteners may be aluminum or aluminum-covered copper, and you could strip threads. Damaged threads can result in a loose connection, which may cause arc flash and result in burns or fires. Use an appropriate torque wrench.

27.1.3 Panel Layout Diagrams

A *panel layout diagram* shows the positions of all panel components without showing wiring, **Figure 27-6**. This type of diagram is used in the layout stage of designing

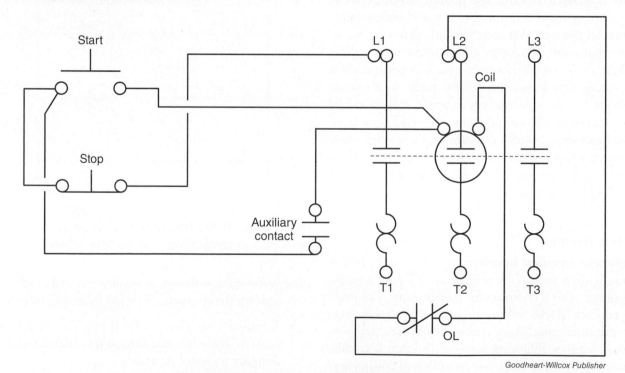

Goodheart-Willcox Publisher

Figure 27-5. A terminal-to-terminal representation of a start-stop circuit. This 240 VAC control circuit uses two of three phases as hot and neutral for the starter coil. This is more common in older systems and requires that the starter coil be wired for high voltage. Newer systems use the two phases to power a control transformer, which is then fused with the neutral and grounded.

and wiring a control panel. This type of diagram can also show the front of the panel and layout of all indicators, switches, and pushbuttons. While panel layout diagrams are somewhat useful for troubleshooting larger panels, especially when internal components are not labeled inside a panel, highway diagrams are typically used for troubleshooting purposes.

27.1.4 Three-Line and Single-Line Diagrams

A *three-line diagram* represents a three-phase system with all phases distinct. This system could be a large power distribution system, a distribution system for a complete site installation, a distribution system for a single building, or power to a three-phase motor as in **Figure 27-7**. A *single-line diagram* represents a three-line diagram with only one

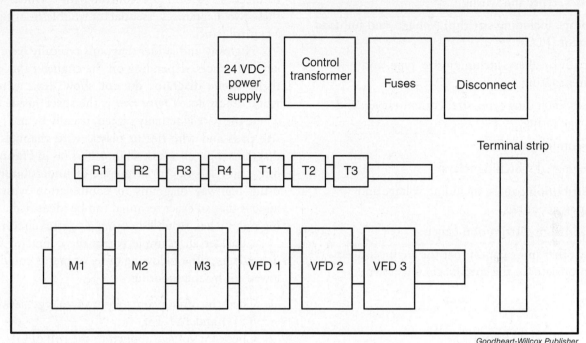

Goodheart-Willcox Publisher

Figure 27-6. A panel layout diagram is used for initial design and construction of a control panel.

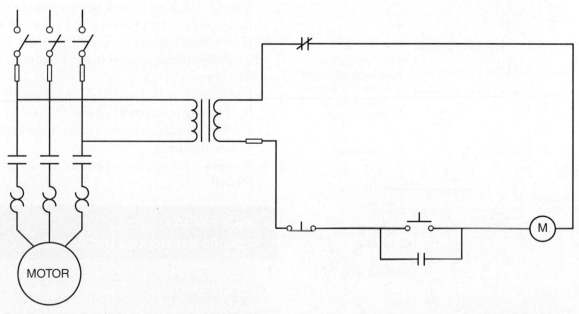

Goodheart-Willcox Publisher

Figure 27-7. Three-line schematic diagram of a motor power supply circuit typical in industrial applications. This three-line diagram is drawn using standard NEMA symbols. Familiarity with basic electrical symbols is essential for reading electrical prints.

line and can offer a simplified depiction of the same types of three-phase distribution systems, **Figure 27-8**. Single-line diagrams are more prevalent unless there are differences in specific phases.

Common components depicted in line diagrams include the following:

- Transformers, including size (kVA), connection phases, type, and wiring.
- Motors, including size (hp), voltage, and full-load current (FLC).
- Circuit breakers, including size, type, and current rating.
- Fuses, including type, size, current rating, and voltage rating.
- Disconnect switches.
- Automated switching relays.
- Distribution panels, including voltage and amperage ratings.
- Loads from distribution panels.
- Any other information that the engineer deems appropriate for the specific circuit shown.

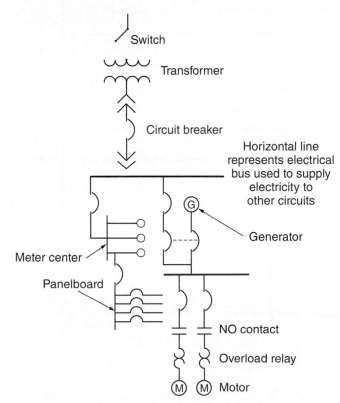

Goodheart-Willcox Publisher

Figure 27-8. This single-line diagram represents a three-phase system with connections to a backup emergency generator, meter center, and control panelboard.

27.1.5 Highway Diagrams

Highway diagrams are used by a panel builder to wire a control panel and make it ready for field connections, **Figure 27-9**. A highway diagram along with a ladder diagram, **Figure 27-10**, can be used for troubleshooting. While the ladder diagram shows the logic of a circuit, the highway diagram shows all locations, terminals, and connections within the control panel. Understanding these two diagrams is essential for troubleshooting inside a control panel.

Highway and ladder diagrams typically have several key differences, depending on the engineer that designs them. Some diagrams do not show exact wire runs, while others do. A **wire run** is the space inside a panel where the wire is actually placed, usually by use of adhesive pads and wire ties or plastic wire channels. Some diagrams show all wires, while others (as in **Figure 27-9**) only show labels. An advantage of understanding and using highway diagrams in combination with ladder logic is that an exact terminal can be identified as a testing point when troubleshooting a control circuit.

Consider an example where the contactor C1 will not energize. The following order of events could be followed for troubleshooting:

1. Check incoming control power voltage between TS1-1 and TS1-13.
2. Check for voltage applied to the coil of C1 between TS1-8 and TS1-12.
3. If no voltage is applied between TS1-8 and TS1-12, check for voltage being applied to the temperature switch between TS1-3 and TS1-13.
4. If no voltage appears between TS1-3 and TS1-13, check for voltage applied to the T1 contact at TS1-2 and TS1-13.
5. If voltage is applied to TS1-2, check the T1 coil by checking power at T1-2 and TS1-13.
6. If power is applied, but the timer is not timing, replace the timer.
7. If power is not applied, check for voltage at R1 coil.

> ⛑ **SAFETY NOTE**
> ## Do Not Work on Live Circuits
>
> Generally, no work is performed live. Only troubleshooting readings may be performed live with appropriate safety gear and supervision. Always lock out and tag out power sources before opening a control cabinet. Proper safety gear may include
>
> *(continued)*

all of the following: appropriate electrically insulated gloves, glove covers, insulated tools, faceshields, and safety mats.

Wearing safety gear and receiving arc-flash training does not enable you to perform live electrical work. The code of federal regulations (CFR) does not allow live work. Knowingly performing live work can end your life, get you fired, or get you personally fined by OSHA. Never disconnect wires, loosen connections or terminals, torque terminals, or make other adjustments under live conditions. Only take troubleshooting readings as needed.

27.2 NATIONAL ELECTRICAL CODE (NEC)

The *National Electrical Code (NEC)* is a standard for the safe installation of electrical wiring and equipment. The *NEC* is published by the *National Fire Protection Association (NFPA)*, a nonprofit organization that focuses on preventing fires due to electrical hazards, and is revised every three years. The *NEC* states: *"The purpose of this code is the practical safeguarding of persons and property from hazards arising from the use of electricity."* Further, *"This code is not intended as a design specification or an instruction manual for untrained persons."*

While *NEC* standards are not law, they may be adopted locally by an "authority having jurisdiction." Some authorities do not immediately adopt the latest

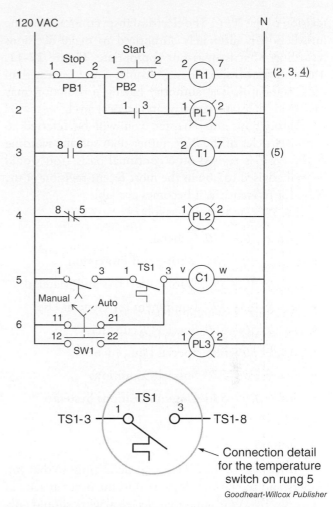

Figure 27-10. A ladder diagram supplements a highway diagram by showing the logic contained in the panel connections.

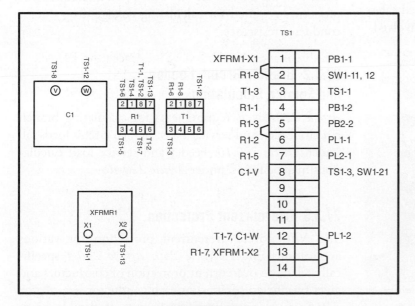

Back panel

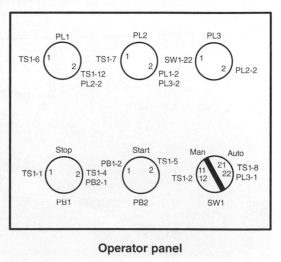

Operator panel

Goodheart-Willcox Publisher

Figure 27-9. The highway diagram shows internal panel components and all terminal connections.

revision of the *NEC*. The electrical inspector for a specific jurisdiction is ultimately authorized to make decisions regarding what is or is not appropriate, **Figure 27-11**. The electrical inspector may require more or less than *NEC* standards. The authority having jurisdiction may vary from the state, county, city, or local level.

Throughout this chapter, you will be referred to various articles of the *NEC* rather than specific wording as the *NEC* is revised on a continual basis. You would be well advised to obtain the most recent revision of the *NEC* for reference as it becomes available.

The *NEC* is arranged as follows:

- *Articles 100–110*, General
- *Articles 200–285*, Wiring and Protection
- *Articles 300–399*, Wiring Methods and Materials
- *Articles 400–490*, Equipment for General Use
- *Articles 500–590*, Special Occupancies
- *Articles 600–695*, Special Equipment
- *Articles 700–770*, Special Conditions
- *Articles 800–840*, Communications Systems
- Tables
- Annexes

Article 100 of the *NEC* covers definitions so that you understand the terminology used in the other articles in the code. The *NEC* states that electrical work should only be conducted by qualified personnel. *Article 100* defines a qualified person as "one who has skills and knowledge related to the construction and operation of the electrical equipment and installations and has received safety training to recognize and avoid the hazards involved."

Dmitry Kalinovsky/Shutterstock.com

Figure 27-11. Electrical inspectors are responsible for ensuring electrical installations adhere to *NEC* or other standards set with their local jurisdiction.

Article 110.12 states, "Electrical equipment shall be installed in a neat and workmanlike manner." The rest of the code follows these general principles.

Article 110 also addresses safety guidelines for installing and working on electrical equipment. For maintenance technicians, it is especially important to ensure there is sufficient space to safely approach and work on electrical systems. The *NEC* requires at least 3′–5′ of clearance, depending on the type of installation. Observe all required safety procedures for connecting, tightening, or disconnecting electrical conductors and for maintaining electrical systems in general.

27.3 WIRING STANDARDS

The *NEC* sets standards for wiring size and ***ampacity***, the ability of a certain size conductor to carry a certain amount of current. Also recall from Chapter 24, *Transformers*, that there are certain color codes used for conductors. Motors have similar color-coding guidelines as transformers, but motors have additional specifications depending on the number of phases and rated voltage.

27.3.1 Wire Sizing

Wire must minimally be large enough to deliver normal operating current under load conditions plus a factor for safety. *Article 310.15* gives ampacity standards for conductors rated 0–2000 V. The tables in *Section 310.15(B)* outline allowable ampacities based on temperature rating, ambient temperature, and installation method. Table 8 in the Tables section of the *NEC* lists conductor properties required for calculating voltage drop due to conductor resistance.

27.3.2 Branch Circuit, Feeder, and Service Calculations

Article 220 covers requirements for calculating branch circuit, feeder, and service loads. Branch circuit loads fall under *Article 220.10*. Feeder and service load calculations may be found under *Article 220.40*.

27.3.3 Overcurrent Protection

The *NEC* covers overcurrent protection for various applications under *Article 240*. *Article 240.4* specifically relates to overcurrent protection of conductors and gives specific ampacities for each conductor size. *Article 240.5* covers overcurrent protection of flexible cords and cables. *Article 240.24* covers overcurrent device accessibility.

27.3.4 Color Coding

The *NEC* has relatively few color-coding requirements. Certain conductors, such as neutrals and grounds, do have specific *NEC* color-coding requirements. Line or "hot" conductors may be any other color aside from the neutral or ground colors.

For line conductors, while there are no *NEC* requirements, conventional practice has established a color code. These color codes are generally based on the voltage and phase rotation of the power the conductor carries. To identify voltage and conductor purpose, conventional colors are used, **Figure 27-12**. In all cases, the *NEC* requires that the color code and meaning be permanently posted at each panelboard.

Notice that phases are indicated by X, Y, and Z with corresponding letters A, B, and C. Cord-ended connectors and receptacles are marked with the X, Y, and Z designations for the phase conductors. In addition, the screw of the connection for the ground connection is green and may also be designated by the letter *G*. The neutral conductor, if included, is connected to a silver screw and often designated with the letter *N*.

The color codes specified by the *NEC* for ground and neutral conductors may be found in *Article 200.4* and *Article 200.6*. Identification is based on the color of the insulation or tape, paint, or other permanent method allowed by the *NEC*.

27.3.5 Grounding

A proper grounding system provides required safeguards for the safety of workers and to prevent damage to equipment. Grounding requirements are specified in *NEC Article 250*. If the grounding system is not operating correctly, a hazardous and life-threatening condition may result.

Some of the components related to the grounding system include the following:

- **Grounded conductor.** A circuit conductor also called the *neutral*. Grounded conductors often have white insulation. A grounded conductor and an ungrounded (hot) conductor are often the two wires connected to an electrical device. Grounded conductors are connected to a *neutral busbar* in the service panel.

- **Equipment grounding conductor.** A conductor used to connect noncurrent-carrying metallic parts of equipment to the *grounding busbar* in the service panel. The equipment grounding conductor is part of the equipment grounding path. These conductors may be bare (uninsulated) copper wires or coated in green insulation.

- **Bonding.** Bonding is an intentional electrical connection between two or more items that are not intended to be part of an electrical circuit. Items such as metal machine enclosures, metal platforms and staircases, and steel columns may need to be bonded together. These items are then also connected to the equipment grounding conductor. Bonding and grounding of metal objects ensures that if a fault occurs and the metal becomes part of the energized circuit, the electrical current has a low-resistance path (to ground) to follow.

- **Main bonding jumper.** A conductor in the service panel that connects the neutral and grounding busbars. The main bonding jumper is the only connection between the grounded conductors and the equipment grounding conductors and ground path.

Wiring Color Codes for Various Motors and Voltages	
Three-Phase, Greater than 600 V	
Phase	**Color**
Phase X (A)	Black
Phase Y (B)	Red
Phase Z (C)	Blue
Three-Phase, 480/277 V	
Phase	**Color**
Phase X (A)	Brown
Phase Y (B)	Orange
Phase Z (C)	Yellow
Three-Phase, 208/120 V	
Phase	**Color**
Phase X (A)	Black
Phase Y (B)	Red
Phase Z (C)	Blue
Neutral	White
Ground	Green
Single-Phase, 240/120 V	
Phase	**Color**
Phase X (A)	Black
Phase Y (B)	Red
Neutral	White
Ground	Green

Goodheart-Willcox Publisher

Figure 27-12. Color codes for various circuit specifications.

- **Grounding electrode.** A device such as a ground rod or ground plate that establishes an electrical connection to the earth.

- **Grounding electrode conductor (GEC).** The conductor used to connect the grounding electrode(s) to the grounding busbar.

Industrial maintenance technicians must be sure that machines, equipment, and other metal items are bonded appropriately, and that bonded items are also connected to an equipment grounding conductor. The bonding and grounding connections and conductors are inspected regularly to ensure connections are solid and conductors are undamaged.

In some situations, the grounding system may need to be tested to ensure that its resistance is sufficiently low. If this task is performed by a maintenance technician, the company will provide a testing procedure developed by its electrical engineers.

27.4 SWITCHBOARDS AND PANELBOARDS

Switchboards and *panelboards* serve similar purposes. Both units are part of the electrical distribution system in an industrial environment and contain overcurrent protection devices and busbars. However, there are several key differences between switchboards and panelboards, which are defined in the *NEC*.

A switchboard, **Figure 27-13**, serves as a main disconnecting means for electrical service entering a building from the power company transformer. A switchboard is a free-standing, three-phase unit. Access is provided from both the front and rear of the unit. A switchboard may have fixed or draw-out power breakers, other overcurrent protection devices, and control switches. Switchboard units can handle much higher amperages compared to the capabilities of a panelboard.

Underwriters Laboratories (UL) is an independent testing organization that approves electrical devices related to safety. The following list shows overcurrent devices capable of being installed in a switchboard and their UL classifications:

- Molded-case circuit breakers, UL 489.

- Power circuit breakers, UL 1066.

- Fusible switches, UL 98.

- 5000 A maximum main.

- Group-mounted feeders up to 2000 A.

- Individually mounted feeders up to 4000 A.

Panelboards are wall-mounted or otherwise supported by mounting to some sort of structure, **Figure 27-14**. They have front access only. Panelboards have either single-phase or three-phase capabilities depending on the model.

Overcurrent devices used in panelboards and their UL classifications are as follows:

- Miniature and molded-case circuit breakers, UL 489.

- Fusible switches, UL 98.

- 1200 A maximum main.

- 1200 A maximum group-mounted feeders.

- Overcurrent devices, usually bolt-on style.

Control switches

Circuit breakers

SvedOliver/Shutterstock.com

Figure 27-13. An industrial switchboard with numerous circuit breakers and control switches.

Goodheart-Willcox Publisher

Figure 27-14. Panelboards like this one typically feature circuit breakers or other overcurrent protection devices.

27.5 DISCONNECTING MEANS

In some cases, a circuit breaker in a panelboard serves as a *disconnecting means*, a safety feature by which connected devices can be immediately disconnected from their source of power. With three-phase wiring, a fused disconnect box is often located near or at the machine being served, **Figure 27-15**. This disconnecting means allows easy access for lockout/tagout (LOTO) purposes. Additionally, equipment control cabinets require a disconnecting means on the outside to disconnect all electricity.

The *NEC* has various standards for disconnecting means. *NEC Article 430* details the standards for motors, motor circuits, and controllers.

27.6 BUSWAYS

A *busway*, often referred to as a *bus duct*, is typically a rectangular metal duct that encloses busbars. *Busbars* are often flat, heavy copper conductors capable of carrying high currents. Busways are most commonly run overhead and suspended from the ceiling by threaded rods, **Figure 27-16**. Many different methods of threaded attachment exist, such as concrete anchors or metal beam clamps.

Busways are manufactured in sections of standard sizes and either plug into or bolt onto each other. Various transitions are available, such as "T" sections or 90° corners, to enable routing busways where required. End-feed sections are available to connect busways to panelboards via conduits and cabling.

Along the length of a busway, often on both sides, there are removable access panels that allow the attachment of cubicles. A *cubicle* is a small box or cabinet that provides for the organized connection of drop cords or conduits containing conductors for connecting machinery. Cubicles also often contain overcurrent protection and disconnecting means.

> ### TECH TIP
> ### Disconnecting Means in Cubicles
>
> In good practice, a cubicle should not contain the sole disconnecting means for a machine. Because the cubicle is elevated out of normal reach, it may be inconvenient or impossible to use the overcurrent device in the cubicle to immediately disconnect power in an emergency. Another disconnecting means at ground level should be readily available.

Hubbell, Inc.

Figure 27-15. This industrial safety switch acts as a disconnecting means outside the main control cabinet. Circuit breakers and various types of motor starters may also serve as disconnecting means.

Attachments suspend the busway from the ceiling

Busway or bus duct

Jay Petersen/Shutterstock.com

Figure 27-16. Busways are often suspended overhead using various types of attachments.

Busways are an extremely flexible system to distribute power to various machines in the industrial environment. If it is necessary to move a specific machine location, reconfiguration of a busway system is relatively straightforward compared to reconfiguring separate conduits and conductors. The *NEC* sets standards for minimum installation height above the floor, distance between supporting members, and the presence of end closures on busways, in addition to other requirements. Refer to *NEC Article 368* for standards related to busways.

27.7 RACEWAYS

The *NEC* defines **raceway** as "an enclosed channel designed expressly for holding wires, cables, or busbars, with additional functions as permitted in this Code." Essentially, raceways help route and organize electrical infrastructure, **Figure 27-17**. Raceways can be constructed from various kinds of tubing or conduits and may be installed under floors, in walls, or overhead. Raceways can be used to install wiring between components within the same control cabinet or to connect components in different enclosures. Conduit or tubing fill indicates the maximum number of wires allowable in a raceway. These values are based on wire type, conduit type, and conduit size.

The designation of conduit size is based on *trade (nominal) size*. That is, a 1/2″ conduit is not actually 1/2″ in size. In fact, trade sizes can be noticeably different from actual measurements, **Figure 27-18**. The sizes of knockouts in metal boxes and panelboards also do not measure what their trade size indicates, **Figure 27-19**.

Raceway

zzMidnightzz/Shutterstock.com

Figure 27-17. Raceways like these are conduits for electrical wires or cables.

Actual Measurements for Various Types of Trade Size 1/2″ Raceways	
Type	Inside Diameter
Electrical metal tubing (EMT)	0.622″
Electrical nonmetallic tubing (ENT)	0.560″
Flexible metal conduit	0.635″
Rigid metal conduit	0.632″
Intermediate metal conduit	0.660″

Goodheart-Willcox Publisher

Figure 27-18. Actual measurements of 1/2″ trade size raceways of various types. Observe the disparity between the two values.

Trade Size versus Actual Size for Knockouts	
Trade Size Knockout	Inside Diameter of Knockout as Measured
1/2″	7/8″
3/4″	1 3/32″
1″	1 3/8″

Goodheart-Willcox Publisher

Figure 27-19. Trade size compared to actual measured size for various knockouts.

The *NEC* is quite specific about the maximum fill for conduits. The maximum fill takes into account the physical space of the conduit; ambient temperature of the air, earth, or building material around the conduit; temperature rise of the wire under load; and temperature rating of the insulation. Refer to the tables in *Article 310*.

Some *NEC* articles covering conduit types include *Article 344* for rigid metal conduits, *Article 348* for flexible metal conduits, *Article 358* for electrical metal tubing (EMT), and *Article 362* for electrical nonmetallic tubing (ENT). Specific standards to pay attention to are the standards for supporting each of the various types of conduit and the permissible spacing between supports.

27.8 CASE STUDY: INDUSTRIAL WIRING PRACTICES

A cylindrical grinder was installed in a machine shop. It was fed from an overhead busway approximately 15′ above the floor. The busway supplied 480 VAC, three-phase electricity through a disconnecting cubicle containing a circuit breaker as the overcurrent device. The busway was fed from a panelboard in a locked electrical vault some distance away from the machine shop.

The manufacturer of the cylindrical grinder originally designed the machine to operate using 208 VAC, three-phase current. To supply the grinder, the manufacturer installed a three-phase transformer ahead of the control cabinet disconnect. This left the closest disconnecting means on the busway cubicle 15' overhead and out of the immediate reach of the machine operator. The circuit breaker on the busway cubicle was rated at 30 A, which was appropriate for the wiring connected between the cubicle and the machine.

Some time after the original installation of the cylindrical grinder, a digital read out (DRO) was installed to assist the operator. The DRO operated from 120 VAC, single-phase current. The electrician who installed the DRO wired a duplex receptacle to the control transformer on the 120 V side. The primary of the transformer was directly wired to one of the phases of the three-phase transformer on the 208 VAC side.

The electrician installed a fuse between the duplex receptacle and the control transformer. The fuse was sized at 20 A, which was the appropriate size for the wire used and the receptacle selected, but not for the VA rating of the transformer, which should have required a 2 A fuse. With only about 100 mA current draw, the DRO was plugged into the receptacle and functioned well.

One warm day, a large portable fan was brought in to get a little air moving in the shop and make the workers more comfortable. The fan operated on 120 VAC and drew approximately 8 A when running. The worker who brought the fan in plugged it into the open side of the receptacle feeding the DRO.

After the fan was switched on, all went well for a few minutes. Then, workers began hearing a crackling and popping sound from the equipment cabinet before the fan quit running. Someone switched off the main disconnect on the equipment cabinet. The crackling and popping continued. The equipment cabinet door was opened, and a cloud of black smoke rolled out followed by fire. Workers scrambled to find a way to shut off the current since shutting off the main disconnect on the equipment cabinet seemed to have no effect.

Eventually, a crew brought in a man lift and shut off the circuit breaker on the busway cubicle. A fire extinguisher was used on the burning equipment cabinet contents. The fan was plugged into a different and more appropriate wall receptacle and used to extract the smoke from the shop.

A postmortem was conducted and the results were as follows:

- The fan consumed 8 A, which was considerably less than the rating of the receptacle, the wiring between the control transformer and the receptacle, and the 20 A fuse on the secondary of the control transformer.

- The transformer was incapable of providing 8 A on its output due to its low VA rating. The fuse should have been sized at 2 A to protect against excessive current draw from the transformer. The 8 A load on the secondary of the control transformer caused the secondary windings to overheat and eventually short after the insulation of the windings had burned off.

- The insulation on the wiring between the control transformer and the three-phase transformer was burned, as was the neighboring wiring in the same bundle. This was due to the excessive current draw after the control transformer shorted.

- One of the three secondary windings on the three-phase transformer was also burned and shorted because the shorted load of the control transformer was far beyond the VA rating of the three-phase transformer.

- There was not a blown fuse or tripped circuit breaker in the entire system due to incorrect sizing and lack of adequate overcurrent protection.

- It was inconvenient and time-consuming to disconnect the source of power to the machine via a disconnecting means 15' above the floor with no immediate means to reach it.

This litany of errors led to serious machine damage. A disconnect was installed at the machine within reach of the operator. Burned wiring was replaced. Both the three-phase transformer and control transformer were replaced. The duplex receptacle was replaced with a single version and moved inside the cabinet. Fuses were properly sized to protect equipment, wiring, and personnel.

CHAPTER WRAP-UP

Electrical diagrams are valuable troubleshooting tools involved in nearly every position in the maintenance area. Develop your skills in reading and using prints with study and practice. Understanding diagrams, *NEC* guidelines, and basic wiring practices will enable you to work on any electrical system you encounter.

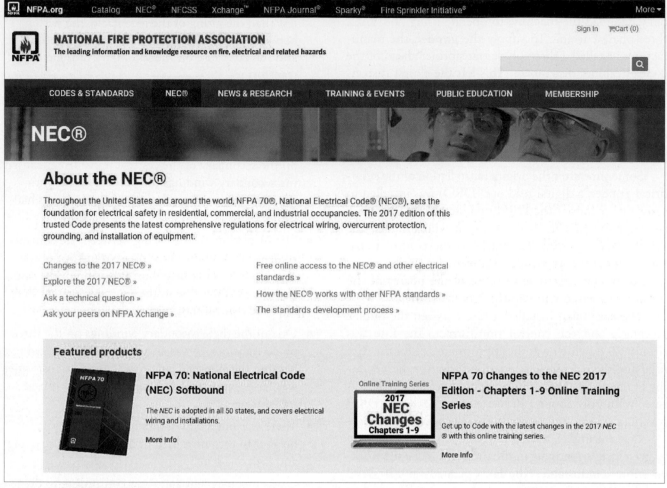

The National Fire Protection Association (NFPA) sets industry standards for electrical safety, specifically with the goal of avoiding possible fire hazards and preventing fires. The NFPA is a useful resource for electrical technicians who need to be familiar with the latest updates to the *National Electrical Code (NEC)*. The NFPA also offers training, professional development, and support through their website.

Chapter Review

SUMMARY

- A ladder diagram is used to develop an understanding of a circuit's control operation and troubleshoot a circuit that is not operating properly.

- A ladder diagram consists of rungs depicting the control circuit and rails representing the hot and neutral of a system.

- Rung numbers are used to label each rung of the ladder, and contact cross-reference numbers refer to other rungs containing contacts the current rung activates.

- Component terminal numbers help clarify ladder diagrams when multiple terminals are used on a single component.

- Wiring diagrams do not show circuit logic. Instead, they show the placement of each wire termination and how many wires exist per terminal.

- Panel layout diagrams are useful when a control panel is not fully labeled.

- Highway diagrams show exact terminal-to-terminal wiring inside a control panel and are used with ladder diagrams to troubleshoot.

- The *National Electrical Code (NEC)* is written by the National Fire Prevention Association (NFPA) and revised every three years. It states, "the purpose of this code is the practical safeguarding of persons and property from hazards arising from the use of electricity."

- The *NEC* has no specific requirements for conductor colors, other than those of ground and neutral. The *NEC* merely requires that phase conductors not use the same colors as ground and neutral wires and that the color code be permanently posted at each panelboard.

- Grounding requirements are specified in *NEC Article 250*. The bonding and grounding of metal objects ensures that if a fault occurs and the metal becomes part of the energized circuit, the electrical current has a low-resistance path to follow.

- A switchboard serves as a main disconnecting means for electrical service entering a building from the power company transformer. A switchboard is a free-standing, three-phase unit.

- The *NEC* has standards for disconnecting means. *NEC Article 430* provides standards for motors, motor circuits, and controllers.

- A busway, often referred to as a bus duct, is typically a rectangular metal duct that encloses busbars. Busbars are often flat, heavy copper conductors that are capable of carrying heavy currents.

- Raceways help route and organize electrical infrastructure. They can be used to install wiring between components within the same control cabinet or connect components in different enclosures.

REVIEW QUESTIONS

Answer the following questions using the information provided in this chapter.

1. What is a ladder diagram?

2. What are the two basic components that make up a ladder diagram?

3. *True or False?* Loads are wired in parallel so that each load receives full control circuit voltage.

4. Discuss the differences between rung numbers and contact cross-reference numbers.

5. When is a wire number given in a ladder diagram?

6. In the following diagram, describe the events that occur when PL1 lights.

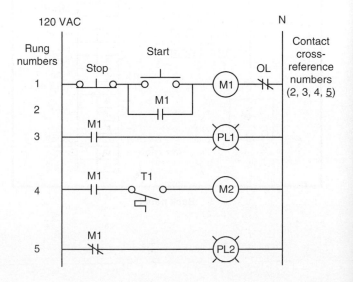

7. In the following diagram, M1 energizes, but PL2 stays lit. Describe a troubleshooting procedure.

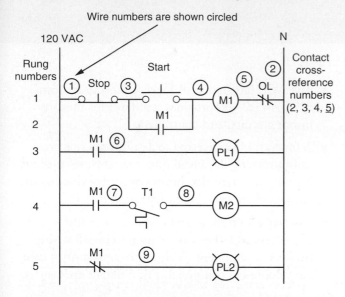

Wire numbers are shown circled

8. What is the purpose of component terminal numbers?

9. A(n) _____ shows exact terminal-to-terminal wiring, including the placement of each wire termination and how many wires per terminal.

10. *True or False?* Both three-line diagrams and single-line diagrams represent a three-phase system.

11. Compare and contrast highway diagrams and ladder diagrams.

Use the highway and ladder diagrams below for Questions 12–15. Indicate what terminal (on TS1) would be checked for the following:

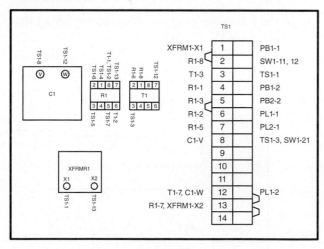

Back panel

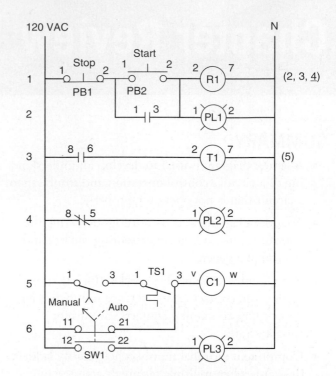

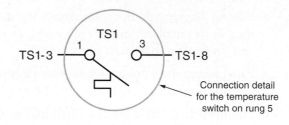

Connection detail for the temperature switch on rung 5

12. R1 coil voltage

13. C1 coil voltage

14. Voltage output from T1-3

15. Voltage to the control circuit

16. Explain the purpose of the *NEC* and who writes it.

17. List three wiring standards covered by the *NEC*.

18. What is the color code for a 480 V, three-phase circuit, and what does each color represent?

19. Compare and contrast switchboards and panelboards.

20. *True or False?* A busway, or bus duct, is a rectangular metal duct that encloses busbars.

21. What is a raceway and how is it constructed?

22. Explain how trade sizes of conduits and actual measurements differ.

NIMS CREDENTIALING PREPARATION QUESTIONS

The following questions will help you prepare for the NIMS Industrial Technology Maintenance Level 1 Electrical Systems credentialing exam.

1. Which of the following is *not* true about ladder diagrams?

   A. They are composed of rungs and rails.
   B. They are shown de-energized.
   C. They show current flow from hot on the right to action on the left.
   D. They show logic, not exact wiring.

2. In the given ladder diagram, which of the following events would *not* have occurred if PL1 is lit?

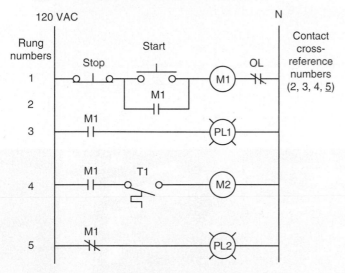

   A. M1 on rung 3 closes.
   B. M1 on rung 5 opens.
   C. PL2 lights.
   D. START button is depressed.

3. How do highway diagrams differ from ladder diagrams?

   A. Highway diagrams are used for troubleshooting, and ladder diagrams are not.
   B. Highway diagrams depict a control panel, and ladder diagrams do not.
   C. Highway diagrams do not show wire runs, and ladder diagrams do.
   D. Highway diagrams show exact wiring and connections, and ladder diagrams do not.

4. Which of the following is *not* a standard outlined by the *NEC*?

   A. Ampacity for conductors rated 0–2000 V
   B. Calculations for branch circuit, feeder, and service loads
   C. Color code for insulators in three-phase motors
   D. Overcurrent protection of flexible cords and standards

5. Based on *NEC* color-coding requirements, ground conductors must be designated by _____ and neutral conductors by _____.

   A. blue; green
   B. green; white
   C. red; white
   D. white; brown

6. In case a fault occurs and a metal enclosure becomes part of an energized circuit, what does bonding and grounding of metal objects ensure?

   A. The electrical current has a high-resistance path to follow.
   B. The electrical current has a low-resistance path to follow.
   C. The electrical voltage is decreased to complete its path.
   D. The electrical voltage is increased to complete its path.

7. Which of the following components does *not* serve as a disconnecting means?

   A. Busway
   B. Circuit breaker
   C. Panelboard
   D. Switchboard

8. Raceways are *not* used to _____.

   A. control the amount of voltage allowed in an electrical distribution system
   B. hold wires, cables, and busbars
   C. install wiring between components within the same control cabinet
   D. provide organized connection of conduits containing conductors

28 | Electrical Troubleshooting

LEARNING OBJECTIVES

After completing this chapter, you will be able to:

☐ Discuss the importance of observation as a troubleshooting tool.

☐ List the various troubleshooting aids available from a human-machine interface (HMI).

☐ Understand the basic functions of programmable logic controllers (PLCs) and how they are used for troubleshooting purposes.

☐ Discuss various sources of energy for industrial systems and their related sensors.

☐ Describe the types of documentation used as troubleshooting aids.

☐ Explain the uses of a digital multimeter (DMM) in the troubleshooting process.

☐ Explain the step-by-step troubleshooting methodology.

☐ Apply the binary search troubleshooting method.

☐ Discuss the importance of root cause analysis.

TECHNICAL TERMS

binary search

block diagram

dual-inline package (DIP) switch

five-why method

forcing

human-machine interface (HMI)

latency

light-emitting diode (LED)

line splitter

measurement by comparison (MBC)

observation

programmable logic controller (PLC)

root cause analysis (RCA)

schematic diagram

screwdriver-adjustable potentiometers

shotgun approach

step-by-step troubleshooting

G ood troubleshooting skills rely on technicians using the proper troubleshooting methodology. As a technician, you will need to develop an array of troubleshooting approaches to be able to determine quickly and effectively:

- What is the malfunction?

- What is needed to correct the problem?

- What is the root cause?

28.1 OBSERVATION

The most important tools in your troubleshooting toolbox are your powers of observation. Your eyes, ears, and sometimes nose can reveal where to start looking for a malfunction. *Observation* helps you discover obvious defects and gives you a better knowledge of how a machine operates. Principles of hands-on troubleshooting suggest that to understand a situation fully, you must go where work is done and see for yourself, **Figure 28-1**. This requires you to go to the shop floor, observe, and document what you have observed. Having a basic understanding of what "normal" is for a machine means you will also have an easier time determining what part of the machine is not functioning properly in a repair situation. If you do not understand how a system operates, it is almost impossible to troubleshoot.

Most systems have indicator lights to advise technicians of status conditions, good or bad. Newer systems may have a *human-machine interface (HMI)*, which often has a control panel and sensor status screen that tells technicians about the status of alarms, sensors, and other pertinent troubleshooting data, **Figure 28-2**. Proximity sensors and optical sensors typically have indicator lights to assist in the troubleshooting process. Such sensor indicators can show the presence of power and whether the device is sensing anything. It is important that you know where sensors on a system are located and what they sense.

28.2 CONTROLS, SETTINGS, AND ADJUSTMENTS

Many modern systems have numerous controls and adjustable settings. If you do not understand exactly what a control or adjustment does and how to properly set it, leave it alone until you can consult someone knowledgeable or find appropriate documentation on the adjustment.

28.2.1 Adjustable Components

Screwdriver-adjustable potentiometers, **Figure 28-3**, are particularly susceptible to improper adjustment. Recall from Chapter 23, *Electrical Components and Circuit Materials*, that potentiometers are variable resistors used to adjust circuit current by varying resistance. By design, screwdriver-adjustable potentiometers require special tools to adjust and so are difficult to change. Often, screwdriver adjustments require test equipment to determine the proper setting. In many cases, screwdriver-adjustable potentiometers have a drop of paint or glyptol (a substance similar to paint used to seal adjustments) to indicate that adjustment has been appropriately set and no further adjustment is necessary.

SeventyFour/Shutterstock.com

Figure 28-1. To perform hands-on troubleshooting, technicians must visit the factory floor, observe operations, and document their observations for later use.

Gumpanat/Shutterstock.com

Figure 28-2. A technician using an HMI sensor status screen to troubleshoot a machine.

Goodheart-Willcox Publisher

Figure 28-3. Screwdriver-adjustable potentiometers require tools to modify so they cannot be easily tampered with.

Dual-inline package (DIP) switches, **Figure 28-4**, may also be misadjusted by an inexperienced technician. DIP switches are manual electric switches packaged in groups and used to set various operating options or modes for equipment. Always consult documentation to determine the proper position of DIP switches. If you do not know the procedure for making an adjustment, do not attempt it. Many inexperienced technicians make adjustments in a random manner, resulting in additional problems that must be corrected before troubleshooting the original problem.

Many optical sensors and some proximity sensors have both DIP switches and screwdriver-adjustable potentiometers. DIP switches set the sense mode, and screwdriver potentiometers adjust sensitivity or threshold. More than 90% of all sensor problems are related to mechanical positioning adjustment, wiring damage, or lack of power. In these cases, no amount of electrical adjustment can cure the problem. If you must replace a sensor, have a data sheet at hand to make the necessary adjustments. If the data sheet is missing, search online and download the data sheet for the device.

Remember that making random adjustments only creates more problems requiring correction before the actual malfunction can be addressed. Unless someone else has tried to troubleshoot a piece of equipment and made random adjustments, you should probably not attempt adjustments as the first method of repair. Instead, exhaust other methods first and only use adjustments as a last resort. Except in cases where a component or part has been replaced or adjustment is a required step in normal operation, adjustment is likely not necessary.

28.2.2 Human-Machine Interface (HMI)

A human-machine interface (HMI) is a useful troubleshooting tool. Many HMIs provide operational controls for a machine via touch screen and include machine status indicators. Often there are several different screens providing different information, **Figure 28-5**.

The main screen typically includes virtual pushbutton controls for various machine operation modes and adjustments. It often has virtual gauges and indicator lights that indicate the status of critical machine parameters. The range of controls and indicators depends on the original programming of the HMI. From the main screen, there is generally a menu available to allow the operator to switch between screens.

Sensor status is important because it indicates what the machine perceives the state of each sensor to be. When the machine control perceives a sensor as malfunctioning, several explanations are possible. The sensor could be defective, the controls might have wiring or alignment problems, or the sensor could have been detached or damaged. The HMI display may not show exactly what the problem is, but it is an effective tool to point you in the right direction to look for problems.

Knowledge of HMI functions and sensor placement can also be useful in cases where a machine requires a manual reset of one or more functions. For example, a machine for pushing parts stops mid-cycle and cannot automatically reset. By your keen powers of observation, you know the conveyor will restart after the part fully advances and trips the part-present sensor. Using the manual control screen on the HMI, **Figure 28-6**, you press the virtual pushbutton to cycle the part pusher, advancing the part and tripping the appropriate sensor. The machine springs back to life and resumes operation. Your knowledge of sensor

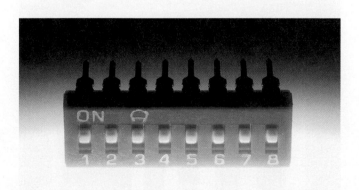

Figure 28-4. A dual-inline package (DIP) switch consists of a series of on/off switches.

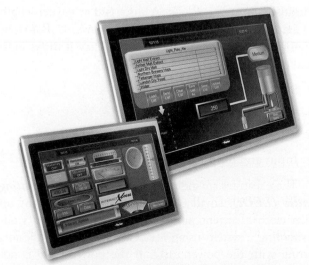

Figure 28-5. A typical HMI main screen often includes virtual pushbutton controls, gauges, and indicator lights.

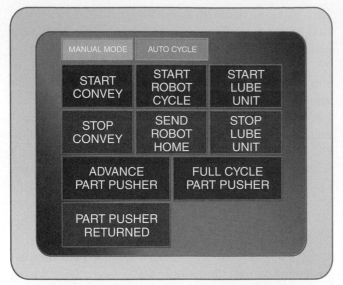

Figure 28-6. The HMI manual control screen can be used to manually test one or more functions on a machine.

locations, HMI controls, and the operating steps of the machine has allowed you to troubleshoot the problem and get the machine back online and into production quickly.

Notice at this point that you have not had to bring out any of your test equipment. All your troubleshooting has been done strictly by observation and using the troubleshooting aids built into the machine.

28.2.3 Programmable Logic Controllers (PLCs)

Programmable logic controllers (PLCs) are digital microprocessing devices used to control the inputs and outputs of industrial systems. PLCs are discussed at greater length in Chapter 32, *Programmable Logic Controllers*. For troubleshooting purposes, focus on several key features indicated on a PLC, **Figure 28-7**:

- System power.
- Fault condition.
- Run mode.
- Input and output status.

These features are indicated on PLCs by *light-emitting diodes (LEDs)*, small electronic devices that light up when electric current passes through them. When troubleshooting a system using a PLC, check LED indicators starting with the power LED. If the power LED is not illuminated, there is a loss of power at the PLC power input or in the PLC power supply. Determine why system power failed and correct the problem before doing further troubleshooting.

If the PLC fault light is illuminated, the PLC has a fault condition related to its firmware, processor, or memory. Many PLCs indicate a fault when communication has been interrupted during a download or an upload. When the PLC indicates a fault condition, it is said to be "bricked." Most often, this is due to software corruption, and a backup copy of the PLC program will need to be reloaded.

When illuminated, the run indicator LED on the PLC indicates that the PLC is executing a program. If it is not illuminated, program execution has been stopped either by the key switch on the front panel or by the remote programming computer.

The status LEDs on a PLC give detailed troubleshooting information. An illuminated LED in the PLC input section indicates an energized input. An illuminated LED in the PLC output section indicates the presence of an energized output. It is important to know what device is connected to each of the PLC inputs and outputs. If a normally open input pushbutton is not being pressed, and the corresponding input indicator is illuminated, then you should look for a short circuit in the pushbutton or associated wiring. Normally closed input pushbuttons, such as E-stop pushbuttons, may fail to illuminate due to an open circuit fault between the switch and PLC. Verify the pushbutton in either case by pressing it and observing if the LED lights up. Similarly, if an output LED is illuminated, there should be some sort of voltage present at the PLC output connection.

In general, if the PLC and its associated wiring worked once, you should not rewire or rearrange connections. Loose connections may be observable in the PLC

Light-emitting diodes (LEDs)

Figure 28-7. PLCs are used to control system inputs and outputs. They also include indicator LEDs that sense electric current.

indicator LEDs. Check wiring connections and correct any problems without disconnecting wires and attempting to reconnect them in a different configuration.

Likewise, if a program has run well in the past, it should always run well without programming changes. There may be cases when someone else made program changes they falsely believed would fix a problem. In such circumstances, make certain you have backed up any operational program so that you can recover a copy of the working program, download it, and get back to the job of trying to find the real problem.

Forcing a PLC program allows you to force the state of an input or an output. Forcing an input or output without regard to results may lead to catastrophe. For example, consider an overhead conveyor system moving product from one production line to another. Forcing an input or output on the PLC that controls product transfer can cause material to be released from the overhead conveyor prematurely and come crashing down on the floor below. The forced release not only destroys product, but it also presents a serious hazard for personnel working below.

Connecting a computer to a PLC may allow you to monitor the execution of the PLC program while it is running, **Figure 28-8**. Monitoring is a good way to determine why some part of the process is not functioning properly. PLC programs should be documented when they are written with labels indicating the function of inputs, outputs, timers, and counters. The monitoring computer should display the status of these features. Lack of labels within the program makes troubleshooting more difficult, as you may know which input is showing on the computer screen, but the program has no indication of what it is connected to or its function other than being an input.

When monitoring a PLC program, there is some amount of delay between the activation of an input or output and when the status change shows up on the computer screen. This delay is referred to as ***latency***. In some situations, latency is of no consequence. Problems with latency arise from how the PLC communicates with the monitoring computer using serial data communications. When a long and involved program is executing, it takes time for all the data to be communicated with the serial process. Normally, the state of all inputs, outputs, counters, and timers are communicated and the process completed before the next status update occurs. Latency is dependent on the size of the PLC program, serial communications speed, speed of change of input states, and other factors.

Ridvan Ozdemir/Shutterstock.com

Figure 28-8. A technician monitors a PLC program with a computer to evaluate a malfunction in the machine process.

28.3 SOURCES OF ENERGY

All systems and machinery require sources of energy. Electrical energy and fluid energy (such as air, hydraulics, compressed gases, cooling water, steam, and others) are often required to make a system function. One of your first checks should be to ensure that all sources of energy required for operation are functional. Be aware of required energy types and their related sensors.

Electrical energy can be lost if an emergency stop (or E-stop) is accidentally pressed. For safety, E-stop buttons are "in your face" by design, **Figure 28-9**. However, this design feature also makes it easy to walk by and unknowingly bump against an E-stop, resulting in a system shutdown. One of the first troubleshooting tasks is to check all E-stops. Resetting any depressed E-stops should restore energy and correct the malfunction. When there are several E-stop buttons in various locations, be sure to look for all E-stops in your observation and documentation phase.

Air pressure, oil pressure, hydraulic pressure, and cooling systems often have gauges enabling you to check their status, including whether pressure criteria are met, **Figure 28-10**. With a quick look at these gauges, you

should be able to determine if all required sources of power are present. Gauges may also point you toward problem resolution. Low or no pressure from a required source may trigger pressure switches to inhibit machine operation until correct pressure is attained.

The operator's control panel often has indicator lights indicating the presence or absence of all required sources of energy, **Figure 28-11**. Look at the control panel to determine if all sources of energy are present. In some situations, the indicators may show that certain pressures are too high, and that is the cause of the control system inhibiting system operation.

28.4 DOCUMENTATION

Depending on the situation, documentation may be a useful troubleshooting aid. The user's manual, technical manual, various electrical diagrams, and troubleshooting flowcharts can all assist you in the troubleshooting

videnko/Shutterstock.com

Figure 28-10. Sources of energy include electrical energy and fluid energy, such as air, water, steam, oil, and compressed gases. Check gauges to determine the status of these energy sources.

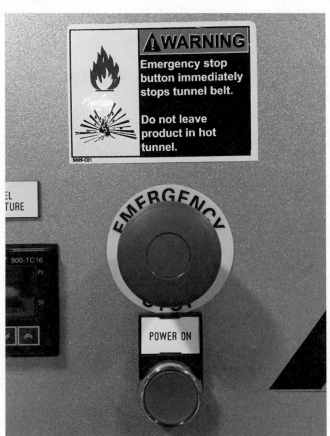

Shawn C. Ballard

Figure 28-9. E-stop buttons are easy to find and press by design. Make sure E-stops have not been accidentally activated before continuing to troubleshoot.

Nina Buchatska/Shutterstock.com

Figure 28-11. Operator's control panel with energy source and indicator lights.

process. However, all these troubleshooting aids are not always available to you. There are various reasons you may not have them in particular situations. For example, documentation may be missing for an old machine. Over the years, documentation tends to get lost or destroyed. Alternatively, if the equipment is proprietary, the manufacturer may not want to release too much information, lest it fall into the hands of a competitor. In other cases, manufacturers want to be the only entities with the ability to repair their equipment.

28.4.1 User's Manual

The user's manual is a valuable first resource. It outlines procedures for startup, shutdown, and basic adjustments for normal operation of the machine. You need to know these procedures before you begin troubleshooting. Speaking with the machine operator may reveal a wealth of information as to what is happening and where you might begin your troubleshooting process. The operator likely understands the normal functioning of the machine better than you do because they run the machine day in and day out. Pay attention to the operator, and find out what

they are experiencing. The operator may have generated the trouble call. Compare what the operator is telling you with the procedure in the user's manual.

28.4.2 Diagrams

In Chapter 27, *Industrial Wiring Diagrams and Practices*, you learned about various types of electrical diagrams and how they can be used to install and troubleshoot systems. Different diagrams may be more useful for particular systems and situations, but you can develop a general troubleshooting method using only a few essential diagrams.

A *block diagram* shows basic operations of a machine and steps related to machine function, **Figure 28-12**. It may describe which blocks of circuitry perform certain functions and the order of machine operation. This diagram will get you grounded as to the "pecking order" of the machine—what must happen first, second, and so on.

A ladder diagram, **Figure 28-13**, depicts an individual circuit on each rung with contacts, switches, sensors, and other inputs on the left-hand side and outputs on the right-hand side. If you examine one rung at a time, you can determine what must happen on the input side

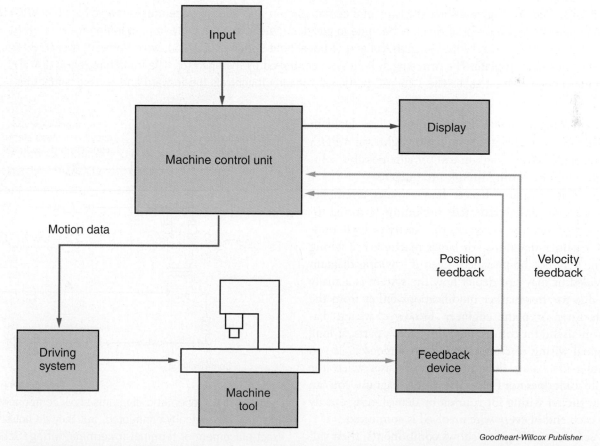

Goodheart-Willcox Publisher

Figure 28-12. Block diagram for a CNC machine. This diagram is a type of electrical diagram that can serve as a key troubleshooting tool.

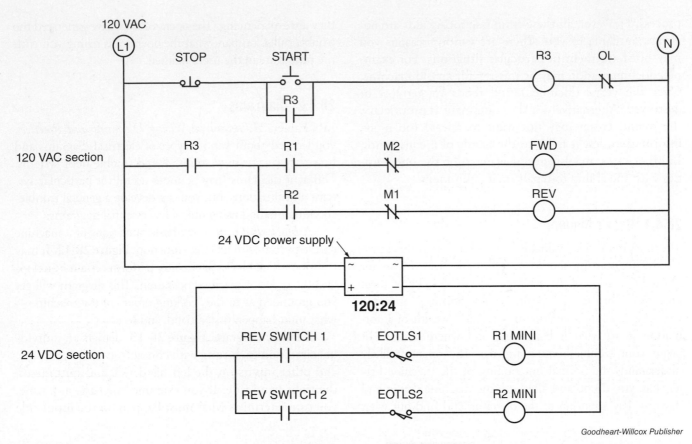

Figure 28-13. Ladder diagrams show the logic of a control circuit. This control system uses low voltage (24 VDC) to allow the operator to reverse or advance a machine to position plate steel on a table in preparation for cutting. In the 24 VDC section, real-world feedback occurs via end-of-travel limit switches (EOTLS), which prevent over-travel from the positioning of the steel plate. The reversing switches are controlled by the operator. The miniature relays (24 VDC) have one normally open contact each. The 120 VAC portion of the circuit controls the forward and reverse contactors.

before anything happens on the output side. Used in combination with the status indicator lights on a PLC, for example, a ladder diagram can provide possible solutions for malfunctioning inputs and outputs.

A wiring diagram depicts how various assemblies of a system are interconnected, including terminal-to-terminal connections, but lacks the clarity of logic present in a ladder diagram. For larger machinery, a wiring diagram may not be provided. Even if a wiring diagram is provided, it may not depict how the system is actually wired due to after-market modifications either from the manufacturer or plant engineer. However, wiring diagrams are useful for connecting replacement parts, as long as original wiring of components is documented. If you encounter inaccurate documentation or cases where the "as built" unit does not follow the wiring diagram, you can sort out proper wiring for yourself by determining exactly where each end of every wire involved is connected.

A *schematic diagram* shows components, their values, and how they are logically connected using symbols to depict individual components, **Figure 28-14.**

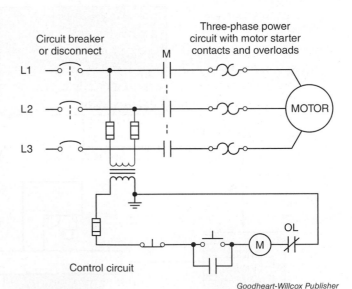

Figure 28-14. Schematic diagrams show components and how they are logically connected, but they do not show exact placement or terminal-to-terminal wiring. Schematic diagrams can help determine where a problem may exist within the system components.

Schematic diagrams do not show the physical arrangement of components, but they are helpful in pinpointing the correct component for further testing. With practice, you should be able to look at a schematic diagram, recognize various components by their standard symbols, and visualize the flow of current throughout the circuit.

28.4.3 Data Sheets

When you are unable to obtain schematic or wiring diagrams for equipment you must troubleshoot, there is another alternative. Components are almost always marked with either their value or part number. With the value of a passive component and an understanding of how such components function in a circuit, troubleshooting is possible without the aid of a schematic diagram.

However, semiconductors and integrated circuits pose a problem in that they are only marked with their part number. From the part number alone, it is often difficult to visually determine exactly what the part is and how it functions. If you search online by the part number, you can usually download a data sheet for that specific component. Armed with the data sheet, you can troubleshoot the component. The data sheet gives you specifications for a component and a connection diagram showing the function of each terminal. At least, this information allows you to test the component and verify its condition.

28.4.4 Troubleshooting Flowchart

A troubleshooting flowchart can guide you through checking common system errors using a step-by-step series of yes/no questions, **Figure 28-15**. Such troubleshooting charts cannot resolve every problem with every contingency, but they can help make you aware of major issues and point you to next steps, especially if you are unfamiliar with a certain machine or system.

Troubleshooting charts are often found in user's manuals and only cover the bare minimum of troublesome scenarios. Often, the ultimate result will be to call technical support.

28.5 CALLING TECHNICAL SUPPORT

Sometimes, a call to technical support can point you in the right direction to resolve problems. A tech support person may make you aware of common problems and how to resolve them. You also have the opportunity to ask questions that may not have been directly answered in the user's manual. Before calling tech support, read the manual and use whatever troubleshooting tips or instructions it provides.

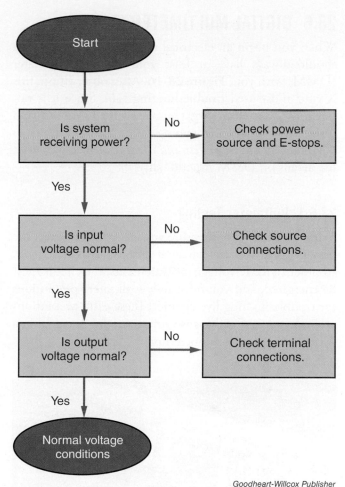

Goodheart-Willcox Publisher

Figure 28-15. An example of a troubleshooting flowchart. This chart provides the steps a technician should take to troubleshoot system power.

Be aware that not all companies offer tech support. Some want you to call or email for a return material authorization (RMA) and send the assembly back. This type of company often sends you a repair estimate after they receive the part. Other companies charge for tech support or require a paid contract for tech support. Others only provide tech support via email. Be sure you have exhausted all other avenues before you consider calling tech support.

You may find yourself speaking with a field service engineer when you call tech support. The service engineer will likely ask you a series of questions to check the condition of a part or setting on the machine. Many people call tech support out of frustration and have checked everything several times. Do not fail to check what the tech support person asks you to once more. Sometimes the order in which you check specific items and change settings can help you find the cause of the problem. Do not ignore what the tech support person asks because you think you know better.

28.6 DIGITAL MULTIMETER (DMM)

When you begin an electrical troubleshooting job, you should always have at least your digital multimeter (DMM) with you, **Figure 28-16**. After observation, this is your most used troubleshooting tool. Since it is not possible to see electricity, you need your DMM to gain insight into what is happening inside a circuit. Refer to Chapter 21, *Electrical Test and Measurement Equipment*, for details on DMM functionality.

28.6.1 Voltmeter Testing

With a DMM set to measure voltage, you can detect the presence or absence of voltage at various places in a circuit. In order to make voltage measurements, the circuit must be energized, and you must follow all safety precautions for troubleshooting live circuits. These extra precautions are worth the effort because voltage testing is useful in a variety of applications, including the following:

Reproduced with Permission, Fluke Corporation

Figure 28-16. A digital multimeter (DMM) is used for electrical troubleshooting and allows a technician to understand what is occurring within a circuit.

- Measure incoming voltage to a piece of equipment to determine if a source of energy is present.
- Measure voltage across a switch to determine if it is open or closed.
- Verify the output voltage of a DC power supply.
- Measure voltage across a fuse to give an indication if the fuse is open or good. If you measure voltage across the fuse, it is open. If no voltage is measured, it is probably good, as long as voltage is present to the input side of the fuse.
- Measure voltage across a relay coil. If the appropriate amount of voltage is present and the relay is not energized, you should look for a problem with the relay.

When performing these tests or any others, if you detect any problems indicated by your voltage measurements, you should power down the equipment and perform an ohmmeter check to verify your findings.

> ### ⚒ TECH TIP
> ### Measuring Ripple
>
> To check the amount of ripple current (AC riding on top of DC) of a DC power supply, set your DMM to AC and measure the supply output. This measurement indicates the amount of ripple on the DC supply without using an oscilloscope.
>
> Ripple should be a tiny percentage of the DC output voltage. For example, 500 mV of AC ripple on a 100 VDC power supply is acceptable. However, 500 mV of AC would not be acceptable on a 5 VDC supply. If in doubt, use an oscilloscope to measure the quality of power supply filtering.

Low voltages can indicate poor connections and overload conditions (as an overload occurs, voltage will sag). Voltage imbalances of greater than 2% on a three-phase circuit may reduce equipment performance and cause premature failure. Voltage drops across fuses or switches may also show up as a voltage imbalance. To find a voltage imbalance, measure the phase-to-phase voltage of each of the three phases. Once you have those three measurements, use simple math to calculate the imbalance:

$$\text{percent imbalance} = \frac{\text{largest imbalance}}{\text{average voltage}} \times 100$$

(28-1)

PROCEDURE Calculating Voltage Imbalance

The following is an example of how to perform the calculations to find voltage imbalance:

1. Measure the voltage between phases.

   Phase 1 = 218 V

   Phase 2 = 222 V

   Phase 3 = 225 V

2. Find the average of the three voltage readings.

   218 V + 222 V + 225 V = 665 V

   $\frac{665 \text{ V}}{3} = 221.67$ V

3. Find the imbalance for each phase, which is the difference between the measurement from step 1 and the average calculated in step 2.

   Phase 1: 221.67 − 218 = 3.67 V

   Phase 2: 222 − 221.67 = 0.33 V

   Phase 3: 225 − 221.67 = 3.33 V

4. Take the largest imbalance found in step 3, divide it by the average found in step 2, and then multiply by 100 to get a percentage.

   $\text{percent imbalance} = \frac{\text{largest imbalance}}{\text{average voltage}} \times 100$

   $\frac{3.67}{221.67} \times 100 = 1.66\%$ imbalance

   The imbalance is 1.66% in this three-phase circuit, which is less than 2% and therefore within acceptable limits.

28.6.2 Ohmmeter Testing

Ohmmeter testing should be conducted with the power off. Connecting a DMM set to measure resistance across an energized circuit may seriously damage the DMM and possibly create an arc flash event. It is often advisable to remove one or more connections to the component under test in order to get an accurate reading. Remember that other components may provide a parallel path that can influence your readings.

For example, two relays with their coils wired in parallel will cause problems for your measurement if one of them is not disconnected. In such a case, it is impossible to determine if one relay is open and the other is normal. The good relay provides a path for an ohmmeter measurement, which may mask the fact that the other relay is open. Disconnecting one lead to one of the coils will allow you to measure each component individually without one masking the other.

Resistors, capacitors, and diodes can also provide a parallel path and invalidate readings of your device under test (DUT). It is often best to isolate the DUT before making measurements with an ohmmeter. Also, be sure to discharge capacitors before measuring a circuit as capacitors may be holding a charge that can damage your ohmmeter. After isolating and discharging a capacitor, check resistance. A short may exist if the ohmmeter reading is low. When you encounter a low resistance reading across a capacitor, switch your DMM over to capacitor test mode and verify the capacitance value to determine if the capacitor is defective.

28.6.3 Capacitor Testing

Many DMMs can test capacitors. In capacitor test mode, the DMM reads the value of a capacitor in microfarads, nanofarads, or picofarads. Pay close attention to the DMM display when making capacitor measurements to determine exactly which range the DMM is measuring. A capacitor may not read exactly as marked. Most capacitors have a tolerance of ±20%. This means a typical capacitor marked as 100 μF may measure anywhere between 80 μF and 120 μF and still be within the 20% tolerance.

When measuring capacitors, make certain you disconnect the source of voltage from the circuit. Even after being disconnected from the power source, capacitors can store large amounts of charge for days, weeks, or even months, and are capable of producing substantial amounts of current when a discharge path, such as test equipment or your body, is provided. Be certain to properly discharge any capacitor before making measurements, disconnecting capacitor leads, or connecting DMM test leads in order to protect yourself and your test equipment.

SAFETY NOTE
Capacitor Storage

When storing large value capacitors, the leads should be shorted with a shorting wire so they do not build up a charge during storage. When capacitors are rapidly discharged and left to rest for a period of time, they can build up a charge and potentially injure the next person who handles them.

28.6.4 Diode Measurements

Most DMMs have a diode measurement mode. This mode is used to measure diodes and most semiconductors. In diode measurement mode, a DMM shows the voltage drop across the diode junction. When measuring a diode, you take two measurements. In one direction, the DMM should show an open circuit (OL or overload), and, in the other direction, it should indicate the voltage drop across the diode junction.

If you measure an open circuit in both directions, the diode is open. If you measure low or no voltage drop in both directions, the diode is shorted. Power semiconductors tend to fail in either open or shorted modes. Once you understand semiconductors a little better, you will find the diode test mode of the DMM invaluable in troubleshooting and testing semiconductors.

Diodes that must be capable of handling high voltages are actually stacks of diodes. Each diode in the stack must "hold off" a certain amount of voltage. For example, a diode stack capable of operation at 5 kV may be made up of 10 individual diodes stacked in series. Considering that each diode typically has a junction voltage drop between 0.4 V and 0.7 V, a stack of 10 diodes would have a voltage drop of 4–7 V. Many DMMs are incapable of measuring a voltage drop as low as 4 V and would indicate an open. Keep this in mind when measuring high-voltage diode stacks.

When high-voltage stacks begin to fail, one or more diodes fail shorted. As this progression continues, the diode stack eventually has enough failed diodes that the stack can no longer hold off the high voltage it was designed for. When this happens, the stack avalanches and the remaining diodes fail shorted. This condition is easily measured with a DMM in the diode test mode.

28.6.5 Current Measurements

Current measurements can quickly reveal an open. If you do not detect any current draw, you have an open circuit. If you detect excessive current draw, you have a problem. For example, in a three-phase feeder, you measure excessive current draw on one phase and no current on the other two. You can be almost certain that you have an open in one of the three phases. In a single-phase system, if you measure no current on a circuit, you can assume that somewhere in the circuit there is an open, such as an open fuse, poor connection, or open switch.

When using a DMM to take current measurements, disconnect a conductor and connect your DMM in series where the disconnected conductor was connected.

CAUTION

Be aware of the current limits of the DMM. One connection on the DMM is capable of only milliamps of current, while the other connection may be capable of up to 10 A. Be certain that the actual current you intend to measure falls within the maximum limits of your DMM. Otherwise, your DMM may be damaged.

An easier alternative method for taking current measurements is to use a clamp-on ammeter, **Figure 28-17**. This method, however, does have its limitations. All clamp-on ammeters can measure AC current. Better models are capable of measuring both AC and DC current because they are constructed with a Hall effect sensor, which measures the magnetic field around the conductor being measured.

Most clamp-on ammeters are not accurate when current is less than 1 A. Luckily, this problem has a simple workaround. A single conductor clamped in the current clamp displays the current reading times one. However, if you take an additional turn through the current clamp, the meter displays the current reading times two. If you take five turns through the current clamp, the device displays the current times five. With this trick, you can accurately measure current values less than 1 A as long as you remember to divide the reading by the number of turns taken through the current clamp.

TECH TIP
One Conductor per Current Clamp

When measuring current, you must have only one conductor at a time in the current clamp, though you may have multiple turns of the same conductor. For example, if you were to place a common electric cord with both a line (hot) and neutral conductor into the current clamp, you may not measure any current. The magnetic field of the hot conductor and the magnetic field of the neutral conductor are 180° out of phase with each other, and the two magnetic fields cancel out.

In such cases, you can use an adapter commonly referred to as a *line splitter*. The line splitter separates the hot and neutral conductors so that each may be individually placed into the current clamp.

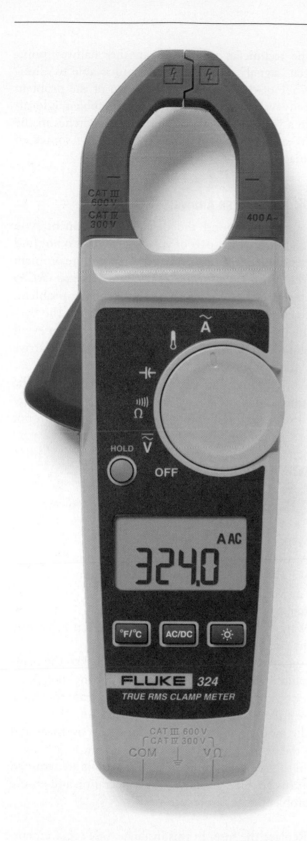

Figure 28-17. Clamp-on ammeters are an easier alternative to taking current measurements with a digital multimeter (DMM).

28.7 TROUBLESHOOTING METHODOLOGIES

As an industrial maintenance technician, you will need to develop a method for troubleshooting various electrical systems with an array of problems. Your method will likely encompass several approaches that may be useful depending on the particular situation. Basic troubleshooting skills, including observation and documentation, in combination with the following troubleshooting methods can help lead you to the solution.

28.7.1 Measurement by Comparison (MBC)

Often it is not clear exactly what a measurement should be. Some systems have a wide latitude as to what is a "go" or "no-go" condition. Other systems may require that a value be within a tight tolerance before a system is operable.

Even if you have a schematic of a circuit and especially if you do not, *measurement by comparison (MBC)* may help clarify whether a circuit is operating properly. MBC requires measuring a known good circuit then measuring a potentially faulty one. When using MBC, the two circuits must be exactly the same. Comparing the two measurements can reveal that the suspect circuit is operating properly if both match. If the two measurements do not match, you could conclude that the suspect circuit or the circuitry feeding it has a problem. On occasion, the load that the circuit drives may be at fault for the dissimilar measurements.

Dissimilar measurements indicate the presence of a problem but not exactly what the problem is. You still need to do further investigating, but MBC is useful because it can at least help you localize a problem.

28.7.2 Step-by-Step

Step-by-step troubleshooting is useful for determining exactly where a fault lies. This method requires you to start at the beginning of a circuit and follow it step-by-step when making measurements. As soon as you fail to find whatever you are measuring, the fault lies between that point and the previous correct measurement. The step-by-step method works especially well with small circuits and networks, but it may also be used in larger systems after other troubleshooting methods have localized a failure.

The step-by-step method is far preferable to making random measurements or using the *shotgun approach*. The shotgun method of troubleshooting is simply replacing random parts until the system functions again.

This random method requires considerably more time and money to find a problem, assuming you even find it at all. Save your resources and better invest your time in becoming an expert troubleshooter with systematic methodologies like the step-by-step method.

28.7.3 Binary Search

A method for large network troubleshooting is the ***binary search***, also known as the *divide-and-conquer approach* or *half-split method*. This method has its origins in computer programming and allows you to localize a problem in a comparatively short time. To conduct a binary search, start at the halfway point between the beginning and end of a system and make a measurement. If the measurement is good, divide the system in half again between your starting measurement and the end of the system. If the measurement is not good, move back to halfway between your starting measurement and the beginning of the system. Continue making measurements and splitting your potential fault zone in half until you have located the problem.

For example, a binary search would prove especially valuable for a technician troubleshooting a cable television outage in a large service network. A cable television system takes the form of a tree. The system begins with a trunk and branches multiple times into feeders before it reaches the outskirts of a service area. If an outage is reported on the outskirts of a network, the step-by-step method would prove impractical for locating the problem. Most cable is located high in the air on utility poles, and there is usually a connection at nearly every utility pole. It would be excessively time-consuming to use the step-by-step method, requiring a technician to climb every pole in order to make measurements.

Instead, a technician would employ a binary search to localize the problem in the shortest possible time. The technician knows the signal is good at the beginning of the system, but a customer on the outskirts has no signal. The technician consults a map book and picks a point in the system approximately halfway between where there is good signal and where there is no signal. The technician now makes a measurement at the halfway point to determine "signal" or "no signal."

If the technician finds signal at the halfway point, the fault lies between that point and the customer. Conversely, if there is no signal, the problem is located somewhere between the beginning and that halfway measurement point. With just one measurement, the technician has eliminated 50% of the cable system as the source of the problem.

The technician now picks another halfway point between "signal" and "no signal" and is able to eliminate 75% of the system as the source of the problem after only two measurements. Once the problem is localized to a very small area, the technician switches to the step-by-step method to finally identify the source of the outage.

28.7.4 Root Cause Analysis (RCA)

After you have found the defective component, you have completed only part of your job. If you do not find the root cause of the problem and correct it, the problem is likely to recur. Perform a ***root cause analysis (RCA)*** to determine the root cause or source of the problem. One method of RCA is the ***five-why method***. The five-why method involves asking why each successive event or failure happened. After five whys, you should arrive at the root cause.

Work through the following example of finding the root cause using the five-why method. Imagine you are tasked with troubleshooting a vertical mill that will not operate. Apply the five-why method as follows:

Statement: The mill will not operate.

Why #1: Why will the mill not operate?

Answer: The mill is not receiving power.

Why #2: Why is the mill not receiving power?

Answer: Because a fuse has blown.

Why #3: Why did the fuse blow?

Answer: Because the motor drew too much current.

Why #4: Why did the motor draw too much current?

Answer: Because a motor winding had burned and shorted.

Why #5: Why did a motor winding burn and create a short circuit?

Answer: An inexperienced operator locked the manual spindle brake in order to change a cutting tool. After the tool change, the operator forgot to release the brake before attempting to start the spindle motor. The stalled spindle motor caused a motor winding to overheat and burn the insulation, causing a short circuit.

In this scenario, three corrective actions are required to return the vertical mill to proper operation and ensure the problem does not recur:

1. Replace the fuse. In this instance, fuse replacement alone would result in repeated blowing of fuses.
2. Replace the motor or have the existing motor rewound. Rewinding takes longer than replacement, but rewinding costs are approximately

2/3 the cost of a new motor. Downtime costs and repair costs must be considered before either option is selected.

3. Retrain the operator or install an overload interrupter to prevent a stalled motor from self-destructing again.

If a technician does not determine the root cause of a problem, the scenario is destined to repeat itself. For this reason, a technician's job is not complete until the root cause has been addressed and corrective action has been taken.

CHAPTER WRAP-UP

Electrical troubleshooting skills are essential for any industrial maintenance technician. Troubleshooting skills include observation, documentation, and various testing methods. These skills are developed over time with careful study and practice. Familiarize yourself with a range of troubleshooting methodologies and where and when to best apply them. Your expertise will save your company time and money, making you an invaluable part of any operation.

Chapter Review

SUMMARY

- Observing a system in normal operation is one of the most important troubleshooting aids. Knowing how a system normally operates helps you localize problems when there is a malfunction.

- Screwdriver-adjustable potentiometers and dual-inline package (DIP) switches are susceptible to improper adjustment. Many inexperienced technicians make random adjustments, resulting in additional problems to correct before troubleshooting the original problem.

- HMIs offer significant troubleshooting resources, such as information on operational controls and machine status indicators, to help you pinpoint the source of a problem.

- PLCs are used to control the inputs and outputs of industrial systems. Indicator LEDs on PLCs offer valuable information regarding system power, fault condition, run mode, and input and output status.

- Forcing an input or output on a PLC without regard to results may lead to catastrophe. Latency of a PLC program, which depends on the size of the program and serial communications speed, should also be considered.

- Ensure that all sources of energy required for operation are functional. Always check all E-stop push-buttons before performing other troubleshooting procedures.

- Different diagrams may be useful for troubleshooting particular systems and situations. A block diagram can help describe which blocks of circuitry perform certain functions and the order of machine operation, while a schematic diagram may show components, values, and how they are connected.

- Learning to use all available documentation and how to obtain documentation when it is not readily available will speed the troubleshooting process.

- If you need to call technical support, make certain you have made all prerequisite checks and followed the instruction manual. After contacting tech support, follow their instructions even if you think you have already checked what they ask you to check.

- A digital multimeter (DMM) is used for electrical troubleshooting. DMMs can be used to test voltage, resistance, capacitors, and diodes in a circuit.

- Voltage imbalances can be measured with a DMM. An imbalance greater than 2% on a three-phase circuit may impair equipment performance and cause premature failure.

- Current measurements can quickly reveal an open. A DMM or clamp-on ammeter can be used to take current measurements.

- Basic troubleshooting skills, including observation and documentation, in combination with a troubleshooting method, can help find or localize a solution.

- Measurement by comparison (MBC) may help clarify whether a circuit is operating properly by measuring a known good circuit then measuring a potentially faulty one.

- The step-by-step troubleshooting method is best used for smaller systems or when the problem has been localized by other troubleshooting methods. The shotgun approach is not recommended when seeking an efficient solution.

- The binary search technique helps localize a problem on a large network. After the problem has been localized, use the step-by-step method to find the exact cause.

- Root cause analysis and correction is vital to ensure a failure does not recur. Use the five-why method to ask why each successive event or failure happened to find the root cause.

REVIEW QUESTIONS

Answer the following questions using the information provided in this chapter.

1. What are the principles of hands-on troubleshooting? What is required to do this troubleshooting?

2. List two components that are susceptible to improper adjustment. How are they sometimes misadjusted?

3. The majority of all sensor problems are related to _____, wire damage, or lack of power.

4. How can an HMI assist you in your troubleshooting process?

5. What key features should a technician focus on when troubleshooting a system using a PLC?

6. What are LEDs? How are LEDs used for troubleshooting?

7. *True or False?* Normally closed input pushbuttons, such as E-stop pushbuttons, may fail to illuminate due to an open circuit fault between the switch and PLC.

8. Give an example of forcing an input or output on a PLC and the repercussions that may follow.

9. _____ is the amount of delay between the activation of an input or output and when the status change shows up on the computer screen.

10. Explain three methods to check that all sources of energy are present before troubleshooting a system failure.

11. Describe how block diagrams and schematic diagrams can be used for general troubleshooting.

12. Explain how a DMM set to measure voltage may be used to test fuses and relay coils.

13. Calculate the voltage imbalance for each phase of a three-phase circuit and percent imbalance when phase 1 is 232 V, phase 2 is 238 V, and phase 3 is 228 V.

14. *True or False?* When ohmmeter testing, disconnecting a lead to a component wired in parallel prevents one good component from masking the other.

15. When measuring capacitors using a DMM, what should be disconnected first?

16. Explain what the diode measurement mode is set to measure. How is a measurement taken?

17. Besides a DMM, a(n) _____ can be used to take current measurements.

18. Discuss how a technician would perform a measurement by comparison (MBC).

19. How is step-by-step troubleshooting performed?

20. *True or False?* The shotgun approach is the most effective and time-efficient troubleshooting process.

21. What is the purpose of a binary search? How is it conducted?

22. Why is it important to determine the root cause of a system failure? Describe one method of RCA and how it is used.

NIMS CREDENTIALING PREPARATION QUESTIONS

The following questions will help you prepare for the NIMS Industrial Technology Maintenance Level 1 Electrical Systems credentialing exam.

1. Which of the following is *not* an appropriate troubleshooting practice?

   A. Adjust a dual-inline package (DIP) switch before it is needed to troubleshoot the original problem.
   B. Call technical support if a user's manual or troubleshooting flowchart is not sufficient.
   C. Refer to a block diagram to understand what blocks of circuitry perform certain functions and the order of machine operation.
   D. Use observation to evaluate a situation or machine fully.

2. Which of the following functions *cannot* be performed using an HMI?

   A. Checking sensor status
   B. Manually resetting a machine
   C. Reading voltage imbalances
   D. Virtually controlling a function

3. An illuminated LED in a PLC output section indicates _____.

   A. the presence of an energized input
   B. the presence of a de-energized input
   C. the presence of an energized output
   D. the presence of a de-energized output

4. In a PLC, what is first completed if a program that has run well in the past is no longer running properly?

   A. Download the faulty program status to a new system.
   B. Force the state of the input or output that pertains to the affected operation.
   C. Evaluate the latency of the program.
   D. Ensure an accidental change was not made to the program.

5. When using a DMM to measure voltage, which of the following scenarios indicates there is a problem in the circuit?

   A. Measuring appropriate voltage across a relay and the relay is not energized
   B. Measuring a voltage imbalance of 0.5% across a fuse or switch
   C. Measuring an output voltage of a DC power supply
   D. Measuring no voltage across a fuse

6. Which of the following is *not* a good practice when measuring current of a circuit?

   A. Checking the maximum limits of your DMM before connecting it to a circuit
   B. Connecting your DMM in parallel in place of the disconnected conductor
   C. Using a clamp-on ammeter to measure AC current
   D. Using a clamp-on ammeter to measure current that is more than 1 A

7. Step-by-step troubleshooting should be used to _____.

   A. localize a problem in a large network
   B. determine exactly where a fault lies
   C. determine whether a circuit is operating properly when compared to a known good circuit
   D. evaluate the root cause of an open fuse

8. What method would be best suited for troubleshooting a television cable outage?

   A. Binary search
   B. Measurement by comparison
   C. Root cause analysis
   D. Shotgun approach

Electronic Control Systems

This section introduces you to the use of DC power supplies, signal conditioning equipment, sensors, transistors, and variable frequency drives (VFDs). You will also learn about installing, programming, and troubleshooting programmable logic controllers (PLCs). Finally, you will learn about human-machine interfaces (HMIs)—devices and software that allow technicians to interact with industrial control systems.

The content in this section will help prepare you to earn the NIMS Industrial Technology Maintenance Level 1 Electronic Control Systems (ECS) credential. By earning credentials, you are able to show employers and potential employers proof of your knowledge, skills, and abilities.

29

Diodes, Rectifiers, and Power Supplies

LEARNING OBJECTIVES

After completing this chapter, you will be able to:

☐ Describe half-wave rectification methods using diodes.

☐ Describe full-wave rectification methods using diodes and diode bridges.

☐ Describe three-phase, half-wave rectification methods.

☐ Describe three-phase, full-wave rectification methods.

☐ Describe capacitor and inductor ripple filters.

☐ Perform diode testing and ripple measurements.

☐ Describe voltage regulation with zener diodes and transistors.

☐ Describe voltage regulation with three-terminal regulator integrated circuits.

☐ Describe switching power supply basics.

TECHNICAL TERMS

DC power supply

diode

filter capacitor

forward biased

full-wave bridge rectifier

full-wave rectification

half-wave rectification

pass transistor capacitance multiplier

peak inverse voltage (PIV)

pi (π) section filter

power factor correction (PFC)

pulse width modulation (PWM)

rectifier

semiconductor

switching power supplies

three-phase rectification

three-terminal voltage regulator integrated circuit (IC)

zener diode

M ost electronic circuits require some form of external power to operate. Usually this external power is in the form of direct current (DC). While an electronic circuit may process alternating current (AC) signals, it still requires some form of direct current to do its job. This direct current is normally provided by a DC power supply.

29.1 MAKING DC FROM THE AC WALL OUTLET

The wall receptacle, which provides power supplied from the electric company, supplies only AC. This power source often has a voltage of approximately 120 V, which is far more than most common semiconductors can utilize.

A transformer can be utilized to either reduce or increase an AC voltage, depending on the turns ratio employed. Transformers only work with AC or pulsating DC. With a transformer, you can reduce the wall receptacle voltage to a level that is more useful with today's electronic semiconductors. However, you still need to convert the AC to DC.

29.1.1 DC Power Supply

A **DC power supply** is a device that uses AC from a higher-voltage source—for example, the 120-volt, 60-Hz wall receptacle—and converts it to DC voltage appropriate to the circuitry requiring DC power. The power supply may contain a transformer to either raise or lower the AC voltage. Power supplies may also contain filtering circuits as well as voltage regulators. Some even include overcurrent shutdown capabilities.

Most power supplies use a circuit known as a **rectifier**. The rectifier circuit turns alternating current into pulsating DC. There are several types of rectifier circuits, and each has advantages, disadvantages, and applications.

Rectifiers may be simple or quite complex, depending on the circuit requirements. Most rectifier circuits include one or more solid-state devices known as diodes.

29.2 DIODES

A **diode** conducts when current flows in one direction and does not conduct when current attempts to flow in the opposite direction. You might think of a diode as a one-way gate for current. This characteristic of a diode allows for conduction during one-half of the AC cycle and not during the other half, when current wants to flow in the reverse direction. These properties make diodes useful for converting alternating current to direct current, and as protection against reverse current.

A simple diode is a two-terminal semiconductor device, **Figure 29-1**. One terminal is referred to as the

cathode, and the other terminal is the *anode*. Electrons flow from the cathode to the anode when a diode is conducting. When the diode is conducting it is said to be **forward biased**.

The term **semiconductor** indicates that under certain circumstances the device will conduct while in other circumstances the device is nonconducting. Electrons will flow from the cathode to the anode but not in the other direction.

The semiconductor material commonly used for rectifier diodes is silicon. Diodes used for small signal detection are most often made from the material germanium. There are many other types of diodes used for other purposes, such as zener diodes, varactor diodes, and PIN diodes.

Diodes have certain important parameters. One parameter is the **peak inverse voltage (PIV)** or *peak reverse voltage (PRV)*. This parameter indicates the highest voltage at which the diode can prevent current flow in the reverse direction. Once that peak inverse voltage is exceeded, the diode breaks down and conducts in the reverse direction. The diode must be replaced with one of equal or greater PIV rating.

When a higher PIV is required, the manufacturer stacks the appropriate number of diodes in series to obtain the desired breakdown, or standoff, voltage. While the manufacturer stacks several diodes in series inside the same package, you can do the same by stacking several diodes in series until you reach the necessary PIV.

> ### 🛠 TECH TIP
> ### Single Diode
>
> Using a single silicon diode to obtain DC from AC is cheap and simple, but it limits the efficiency of the power supply and leaves a large amount of AC ripple in the supplied power.

A diode's current rating is another parameter that must be considered. This is the maximum current that can flow through a diode while conducting in the forward direction. If a diode's maximum current rating is exceeded, it will be destroyed.

mosufoto/Shutterstock.com

Figure 29-1. A generic silicon diode is pictured here. The band on the right indicates the cathode.

29.2.1 Testing Diodes with a DMM

Another important parameter of a diode is its forward voltage drop, which is the voltage lost across the diode while conducting. Most silicon junction diodes are forward biased at 0.7 V, and some are designed for a lower voltage drop, such as 0.25–0.4 V for metal-silicon diodes. The diode test function of a digital multimeter (DMM) measures the voltage drop, or forward voltage. Unlike resistors, the voltage drop of a diode remains somewhat constant regardless of the voltage or current applied. When testing a diode with a DMM, connect it in one direction and take a reading, then reverse the test leads and take a second reading. You should measure a voltage drop in one direction and an open circuit in the opposite direction. The voltage drop for a good silicon diode is approximately 0.5 to 0.7 V.

High-voltage diodes may have a much greater voltage drop, as would a stack of silicon diodes. The voltage drop across a 5-kilovolt diode is 3–5 V. This often presents a problem when using a DMM to perform measurements, as any drop in excess of 3 V would be beyond the measurement capabilities of the test equipment.

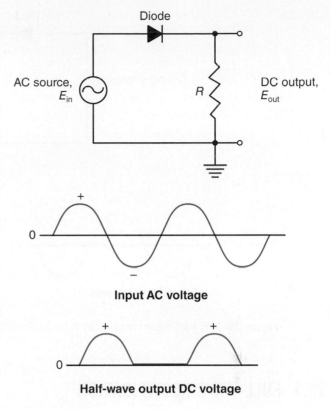

Input AC voltage

Half-wave output DC voltage

Goodheart-Willcox Publisher

Figure 29-2. This diagram shows a simple half-wave rectifier circuit using a silicon diode.

> ⚠ **CAUTION**
>
> The failure mode of most power semiconductors is to fail shorted. In some cases, a short will cause a diode to explode, causing an open failure mode. This open failure mode is quite easy to detect without test equipment, as only remnants of the diode—if that—are still present.

29.2.2 Rectification (Turning AC into Pulsating DC)

Half-wave rectification is the simplest of all configurations, in which a diode converts the positive half cycle of the input signal into a pulsating DC output signal. **Figure 29-2** depicts a simple half-wave rectifier circuit. An input sine wave enters the circuit, and the output signal includes only the positive half of the sine wave. The negative half of the sine wave is blocked by the diode.

To determine the no-load DC voltage (E_{DC}) of the half-wave rectifier, use the following equation:

$$E_{DC} = \frac{E_{peak}}{\pi}$$

(29-1)

For example, 15 volts peak / π = 4.8 volts DC. This equation is true in a no-load condition. With a load present, the DC voltage will be lower. While the half-wave

rectifier is appropriate for purposes of battery charging and other applications where pulsating DC is not a problem, it is not usable for powering electronic equipment without some form of filtering to smooth the signal.

29.2.3 Filtering Pulsating DC

With the addition of a capacitor across the load side of the diode rectifier, as shown in **Figure 29-3**, a certain amount of smoothing of the pulses may be obtained. The *filter capacitor* charges on the rising side of the DC pulse, then discharges on the falling side. This charging and discharging tends to provide some smoothing, bringing the pulsating DC closer to steady-state DC, in which there is little variation in the level of DC current.

The value of the capacitor is selected with consideration of how much of a load is present at the output and the amount of smoothing that is required. Single-phase, half-wave rectification is only used with loads that draw a little current, with a little ripple present.

The filtering action of the capacitor smooths out the DC pulses. Filtering results in less variation in the DC level. The trade-off of filtering is the resulting higher currents in the transformer secondary.

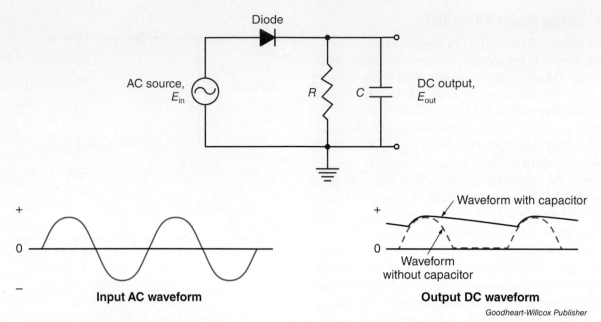

Figure 29-3. A half-wave rectifier with capacitor filtering (C) is diagrammed here.

29.3 FULL-WAVE RECTIFICATION

While half-wave rectification does an adequate job of converting AC to DC, it is not as efficient as other methods. *Full-wave rectification* uses both halves of the AC cycle to provide pulsating DC with more pulses and less time below peak.

In **Figure 29-4**, the full-wave rectifier utilizes a center-tapped transformer. This is a transformer with a contact point halfway along its secondary winding. The center tap is grounded. Either end of the secondary is positive during one-half of the AC cycle. If you use the conventional circuit to visualize the current flow, you will notice the upper end of the secondary is positive during the first half of the cycle, while the lower side of the secondary is positive during the second half of the AC cycle. With each diode conducting during its half cycle, the output waveform comprises two positive halves.

With full-wave rectification, the no-load DC voltage obtained is calculated as follows:

$$E_{DC} = \frac{2E_{peak}}{\pi}$$

(29-2)

So, with a peak voltage of 8 V, the DC voltage obtained with a full-wave rectifier would be 5.1 V.

The disadvantage of this full-wave rectification arrangement is that a center-tapped transformer is required, which adds expense.

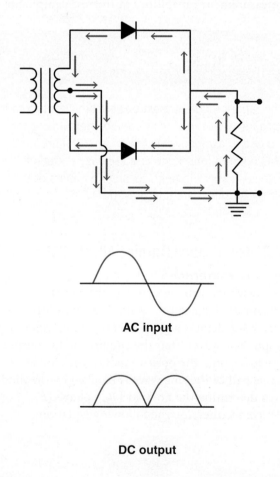

AC input

DC output

Goodheart-Willcox Publisher

Figure 29-4. This diagram shows a full-wave rectifier circuit with a center-tapped transformer.

29.3.1 Full-Wave Bridge Rectifier

Another way to accomplish full-wave rectification is with a full-wave bridge rectifier, **Figure 29-5**. The *full-wave bridge rectifier* uses four diodes to accomplish the same type of rectification that two diodes and a center-tapped transformer accomplish.

With the center-tapped transformer and two diodes depicted in **Figure 29-4**, each diode conducts during one-half of the AC cycle. With the full-wave bridge rectifier, the four diodes conduct in pairs during each half of the AC cycle. Without the smoothing capacitor, the waveform looks just like the output of the full-wave, center-tapped transformer setup.

The added capacitor allows for the smoothing of the pulsating DC, resulting in very little ripple. Full-wave rectification requires less filtering than half-wave rectification, as the resulting waveform is above 0 volts a much greater amount of time.

29.3.2 Three-Phase, Half-Wave Rectifier

Three-phase AC is the most common electrical power used in industry. It is often necessary to obtain DC from a three-phase AC source. Variable frequency drives often rectify three-phase AC to DC for later conversion back

to three-phase AC at the desired frequency. *Three-phase rectification*, in which the three-phase AC is converted to DC, is an important part of the variable frequency drive process.

Pulses occur at the top of each waveform in each cycle. Single-phase, half-wave rectification produces one pulse per cycle. Single-phase, full-wave rectification produces two pulses per cycle. Three-phase, half-wave rectification produces three pulses per cycle. This is because there are three overlapping waves that each pulse once in each cycle or period.

In **Figure 29-6**, notice that the resulting output from the three pulses per cycle never reaches 0 V, as the next pulse begins before the previous pulse drops to zero.

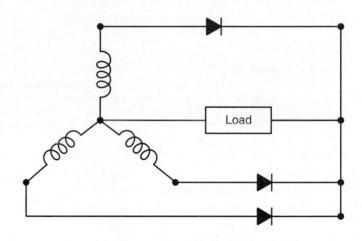

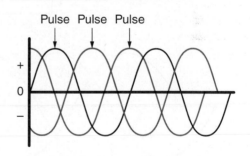

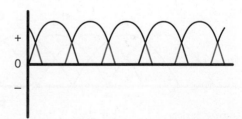

Figure 29-6. This diagram shows three-phase, half-wave rectification from a three-phase transformer with a wye-connected secondary. There are three overlapping waves in each cycle, which produces three pulses per cycle.

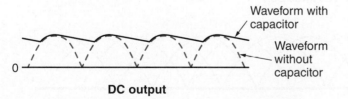

DC output

Figure 29-5. A full-wave bridge rectifier with filter capacitor (C) is pictured here, along with the resultant waveform. During the first half of the cycle, diodes 1 and 2 conduct. During the second half of the AC cycle, diodes 3 and 4 conduct.

29.3.3 Three-Phase, Full-Wave Rectifier

Notice that in **Figure 29-7**, three more diodes are added to create full-wave rectification. Instead of the three pulses per cycle obtained with the three-phase, half-wave rectifier, a three-phase, full-wave rectifier achieves six overlapping pulses per cycle. This rectifier circuit is also called a six-pulse unit. As is clearly visible by looking at the outputs, more diodes means more pulses and less ripple, so that less filtering is required.

The arrangement shown in **Figure 29-8** uses a transformer with two sets of secondary windings. One secondary winding is connected in a wye configuration, and the other is connected in a delta configuration. The connection of the primary windings is inconsequential, but in this case it is wired in a wye configuration. With 12 pulses per cycle, almost no filtering is required.

Electronic equipment requires smooth, ripple-free DC. While it is true that adding more and more capacitance will decrease the amount of ripple, it comes at the cost of higher current draw from the rectifier circuit. The current to fill in the gaps between the pulses must come from somewhere. Additionally, the higher the load current demand, the more capacitor filtering is required.

TECH TIP
Oscilloscope

An oscilloscope will display the shape of a voltage waveform and allow you to see electrical signals as they change over time. It is a valuable tool when troubleshooting electrical issues.

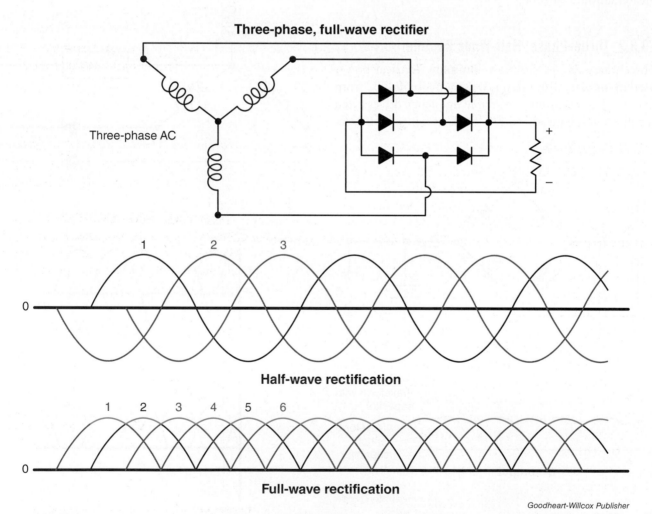

Three-phase, full-wave rectifier

Three-phase AC

Half-wave rectification

Full-wave rectification

Goodheart-Willcox Publisher

Figure 29-7. A three-phase, full-wave, six-pulse rectifier circuit is diagrammed here. Six diodes achieve full-wave rectification, with a smoother DC output of six overlapping pulses per cycle.

Primary   **Secondary**

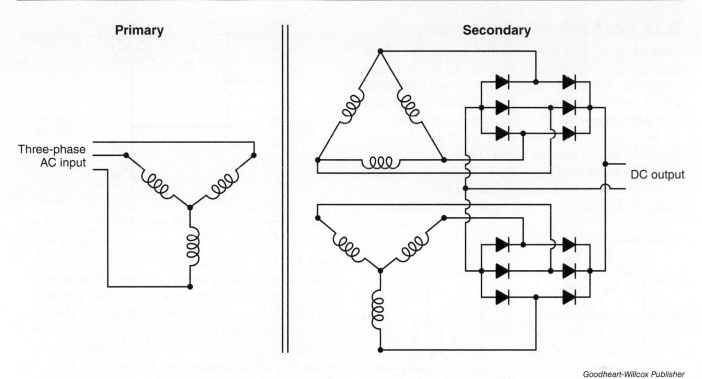

Three-phase
AC input

DC output

Goodheart-Willcox Publisher

Figure 29-8. This three-phase rectification circuit uses two full-wave rectifiers and two sets of transformer secondary windings, which provides a 12-pulse output.

29.3.4   Pi (π) Section Filter

Capacitors are not the only solution to filtering. With a capacitor in parallel with the rectifier output, the capacitor charges during the rising time of the pulse and stores potential energy in an electrostatic field. During the fall time of the pulse, the capacitor discharges by the electrostatic field collapsing and providing electrons as current.

An inductor may be placed in series with the rectifier output to provide filtering action. During the rising time of the pulse, the inductor begins to build a magnetic field as a magnetic flux develops. Flux is the measure of how much of a field—in this case, a magnetic field—is going through a certain area or surface. During the falling time of the pulse, the inductor's magnetic field begins to collapse and induce a voltage across the inductor's coil, which tends to fill in the gaps much like the capacitor does. The difference between the two is that a capacitor stores its charge in an electrostatic field, while the inductor performs nearly the same storage function using a magnetic field.

It is possible to build a filter called a ***pi (π) section filter*** by alternating capacitors in parallel with and inductors in series with the rectifier output. It is called a pi section filter because it is shaped like the Greek letter π. This pi section

filter has capacitors along the vertical sides of the π and inductors along the top. This pi network does an excellent job of filtering, but it occupies a considerable amount of space.

The pi section pictured in **Figure 29-9** is actually a low-pass filter. It passes lower frequencies and rejects higher ones. DC is lower frequency than the AC pulses, resulting in the filter passing the DC component while rejecting the AC component.

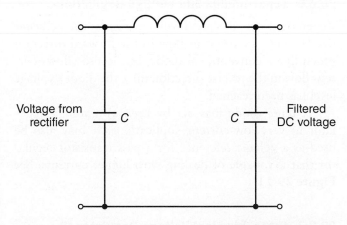

Voltage from
rectifier

C

C

Filtered
DC voltage

Goodheart-Willcox Publisher

Figure 29-9. A pi (π) section filter is diagrammed here.

29.3.5 Pass Transistor Capacitance Multiplier

There are times where there is insufficient space for a large-value capacitor, or possibly a capacitor value that is large enough for the circuit requirements is not readily available. A novel circuit called a *pass transistor capacitance multiplier*, **Figure 29-10**, could solve the problem.

In this circuit, the transistor acts as a simple emitter follower. The resistor provides bias for the base/emitter junction of the transistor. The capacitor connected to the base of the transistor smooths the ripple in the DC. This circuit multiplies the capacitance by the gain of the transistor.

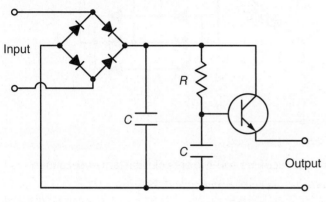

Goodheart-Willcox Publisher

Figure 29-10. This circuit diagram shows a pass transistor capacitance multiplier circuit.

Filtering is only one aspect that must be dealt with in a power supply. Most electronic circuits need a stable voltage source. Not only must input voltage variations be considered, but also those caused by variations in the power supply load.

29.3.6 Zener Diodes and Voltage Regulation

Zener diodes make good voltage regulators. A *zener diode* allows a current to flow in the forward direction much like a conventional diode, but it also allows current flow in the reverse direction once the diode's voltage level has been reached.

A zener diode may act by itself as a voltage regulator in very low-current applications. It may also be used as a voltage reference for a pass transistor regulator that is capable of dealing with higher currents. See **Figure 29-11**.

29.3.7 Three-Terminal Voltage Regulator IC

Another method of voltage regulation employs a *three-terminal voltage regulator integrated circuit (IC)*. The entire voltage regulator circuit, with the exception of a

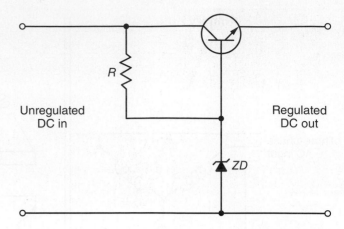

Goodheart-Willcox Publisher

Figure 29-11. In this pass transistor regulator circuit, ZD is the zener diode being used as a voltage reference in combination with R, which is the current limiting resistor. The transistor is capable of handling the current demand where the zener diode by itself would not. The voltage across the circuit would be able to vary, but the voltage across the zener diode would remain constant as long as the input voltage is above the zener diode's breakdown level.

couple of external capacitors, is included on one chip, **Figure 29-12**. These ICs are connected after the rectifier and filter circuitry. One external capacitor is used in the circuit to prevent the voltage regulator from breaking into oscillation. The other capacitor is used to improve the transient response of the circuit.

Refer to **Figure 29-13** to see the different package styles and the pin outs for standard three-terminal voltage regulators. Notice that each has an input, output, and ground pin. The manufacturer's data sheet will give important operating parameters, such as maximum output current, maximum input voltage, maximum operating temperature, and the output voltage range, to name a few. **Figure 29-14** shows the connection schematic for a three-terminal voltage regulator.

Adrian Stratulat/Shutterstock.com

Figure 29-12. A typical three-terminal voltage regulator is pictured here.

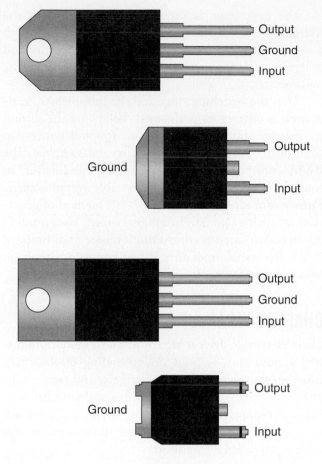

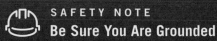

In some cases, the tab or back portion of a voltage regulator is connected to and will substitute for the ground terminal. Do not make dangerous assumptions. Check the data sheet for the specific type of regulator you have.

Goodheart-Willcox Publisher

Figure 29-13. Package outlines and pin outs for different styles of three-terminal voltage regulators are shown here.

Each three-terminal voltage regulator has a specific output voltage and a minimum input voltage to obtain the rated output voltage. Many applications of three-terminal regulators require a heat sink to remove heat and provide adequate cooling, as the difference in the output voltage and the input voltage is dissipated as heat during the regulating process.

Switching Power Supplies

Newer styles of DC power supplies employ switching technology. *Switching power supplies* are also referred to as *switching-mode power supplies*, *switch-mode power supplies*, *SMPSs*, *switched power supplies*, or simply *switchers*. Switching power supplies have several stages and a feedback path to regulate the output voltage. The thing that sets switching power supplies apart from linear power supplies is that they are much lighter due to the fact that they do not have a heavy transformer. Switching power supplies are also capable of much higher current ranges, and they have a much smaller footprint, **Figure 29-15**.

THINKING GREEN
Switching Power Supplies

Switching power supplies are used in a wide range of applications. Their advantages over other types of power supplies include a high level of efficiency and smaller size. Switching power supplies also run a bit cooler.

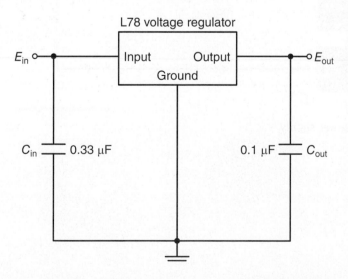

Goodheart-Willcox Publisher

Figure 29-14. This is a simple connection schematic for a three-terminal voltage regulator. L78 is a common series of voltage regulators.

Figure 29-16 is a block diagram of a switching power supply. Following the block diagram from left to right, the input power is listed first. In many cases, the input voltage may range from 100 VAC to 240 VAC. The second stage is transient filtering, to remove any spikes or transients (overvoltage) in the incoming power. The third stage is rectification, just like in a linear power

Automation Direct

Figure 29-15. These switching power supplies can be mounted on DIN rails and are used in industrial control applications.

supply. The third stage is where the difference between the two power supplies begins.

The active ***power factor correction (PFC)*** is the stage that adjusts the rectified pulsed DC from the rectifier to a standard voltage regardless of the input voltage. This stage allows for a wide input range and still maintains the same output voltage.

The next stage is the switcher, which turns the rectified DC voltage into AC. The AC then goes through a transformer. This transformer has a ferrite core and is lighter and more efficient than standard iron-core transformers.

After the switching stage and the transformer, rectification is once more performed. Following the second rectification, filtering is employed. You will notice two additional stages that feed back to the switcher. The PWM control stage uses pulse width modulation to control the switcher, resulting in a stable output voltage. ***Pulse width modulation (PWM)*** is a method of generating an analog signal from a digital source, thus producing an action (such as controlling a motor or dimming a light). The transformer after PWM control is simply an isolation transformer to separate the two stages.

CHAPTER WRAP-UP

Many electronic devices are common to modern industrial applications. A solid understanding of electronic components, such as diodes, rectifiers, and power supplies, will help you become a well-rounded technician. With this knowledge, you will be able to maintain and troubleshoot the electronic control systems where you work when called on to do so.

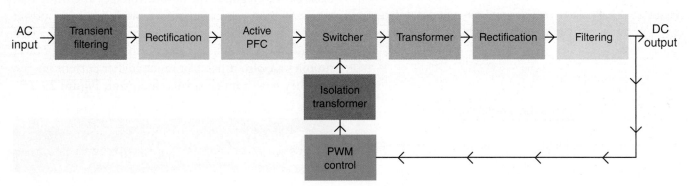

Goodheart-Willcox Publisher

Figure 29-16. This block diagram shows a typical switching power supply.

Chapter Review

SUMMARY

- Although electronic circuits can process alternating current signals, they still require some form of direct current to function. A DC power supply uses AC from a higher-voltage source and converts it to DC.

- Rectifier circuits are used in most power supplies and turn alternating current into pulsating direct current. Depending on the circuit requirements, they can be simple or complex.

- A diode is a solid-state device that conducts when current flows in one direction and does not conduct when current attempts to flow in the opposing direction. The simplest diode is a two-terminal semiconductor.

- A digital multimeter measures the voltage drop, or forward voltage, of a diode. However, any drop in excess of 3 V is beyond its measurement capabilities.

- Half-wave rectification is the simplest of all configurations. Only the positive half of the input sine wave is allowed through as output, while the diode blocks the negative half.

- A filter capacitor charges on the rising side of the DC pulse and discharges on the falling side. Filtering results in less variation in the DC level, but also results in higher currents in the transformer secondary.

- Full-wave rectification uses both halves of the AC cycle to provide DC with more pulses and less time below peak. It also utilizes a center-tapped transformer, where the center tap is grounded and either end of the secondary winding would be positive during one-half of the AC cycle.

- A full-wave bridge rectifier uses four diodes to accomplish the same rectification that two diodes and a center-tapped transformer does. The diodes conduct in pairs during each half of the AC cycle.

- Three-phase AC is the most common electrical power used in industry. With three-phase rectification, a greater number of diodes results in more pulses and less ripple, meaning less filtering is required.

- A pi (π) section filter alternates capacitors in parallel with and inductors in series with the rectifier output. It does a great job of filtering but occupies a large amount of space.

- Unlike a typical diode, a zener diode allows current to flow in both directions. In low-current applications, it can also act as a voltage regulator by itself.

- A standard three-terminal voltage regulator has input, output, and ground pins. Important operating parameters, such as maximum output current, maximum input voltage, and maximum operating temperature, are detailed on the manufacturer's data sheet.

- Switching power supplies have multiple stages and a feedback path to regulate the output voltage. Newer styles of DC power supplies use this kind of switching technology.

REVIEW QUESTIONS

Answer the following questions using the information provided in this chapter.

1. Describe the function of a diode.

2. *True or False?* Electrons will flow from the cathode to the anode in a semiconductor, but not the other way around.

3. What are two parameters you must consider when using a diode?

4. Explain how to test a diode with a DMM.

5. Calculate the no-load DC voltage of a half-wave rectifier with 20 V peak.

6. *True or False?* A filter capacitor charges on the falling side of the DC pulse.

7. Calculate the no-load DC voltage of a full-wave rectifier with 25 V peak.

8. Full-wave rectification requires less _____ than half-wave rectification.

9. *True or False?* Single-phase AC is the most common electrical power used in industry.

10. _____ is the measure of how much of a field is going through a certain area or surface.

11. What is the difference between a capacitor and an inductor?

12. How can you build a pi (π) section filter?

13. *True or False?* In very low-current applications, a zener diode can act as a voltage regulator by itself.

14. How are the two capacitors in a three-terminal voltage regulator integrated circuit used?

15. List three advantages switching power supplies have over linear power supplies.

16. In switching power supplies, the stage that adjusts the rectified pulsed DC from the rectifier to a standard voltage is called _____.

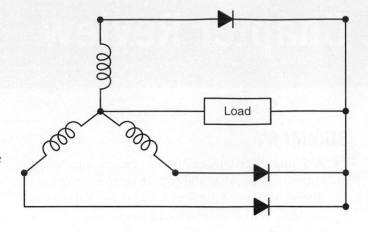

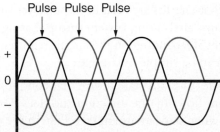

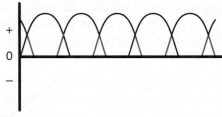

NIMS CREDENTIALING PREPARATION QUESTIONS

The following questions will help you prepare for the NIMS Industrial Technology Maintenance Level 1 Electronic Control Systems credentialing exam.

1. A _____ turns alternating current into pulsating direct current.
   A. diode
   B. filter capacitor
   C. rectifier circuit
   D. semiconductor

2. Which of the following is the most commonly used semiconductor material for rectifier diodes?
   A. Germanium
   B. Silicon
   C. Steel
   D. Plastic

3. How many pulses per cycle occur in single-phase, half-wave rectification?
   A. One
   B. Two
   C. Three
   D. Four

4. Which of the following rectifiers is shown in the image above?
   A. Single-phase, half-wave
   B. Single-phase, full-wave
   C. Three-phase, half-wave
   D. Three-phase, full-wave

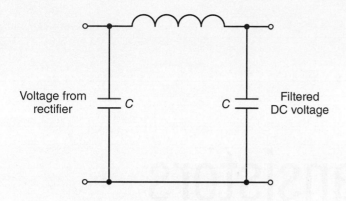

5. What is pictured in the image above?

   A. Zener diode
   B. Pass transistor capacitance multiplier
   C. Three-terminal voltage regulator
   D. Pi (π) section filter

6. In a switching power supply, what does transient filtering accomplish?

   A. Removes spikes or overvoltage in incoming power
   B. Rectifies AC to DC
   C. Changes the number of DC pulses
   D. Creates an analog signal

7. What stage in a switching power supply adjusts the rectified pulsed DC to a standard voltage regardless of the input voltage?

   A. Switcher
   B. Active PFC
   C. PWM Control
   D. Isolator

8. In a switching power supply, what stage controls the switcher, resulting in a stable output voltage?

   A. Rectification
   B. Filtering
   C. Switcher
   D. PWM Control

30

Transistors and Integrated Circuits

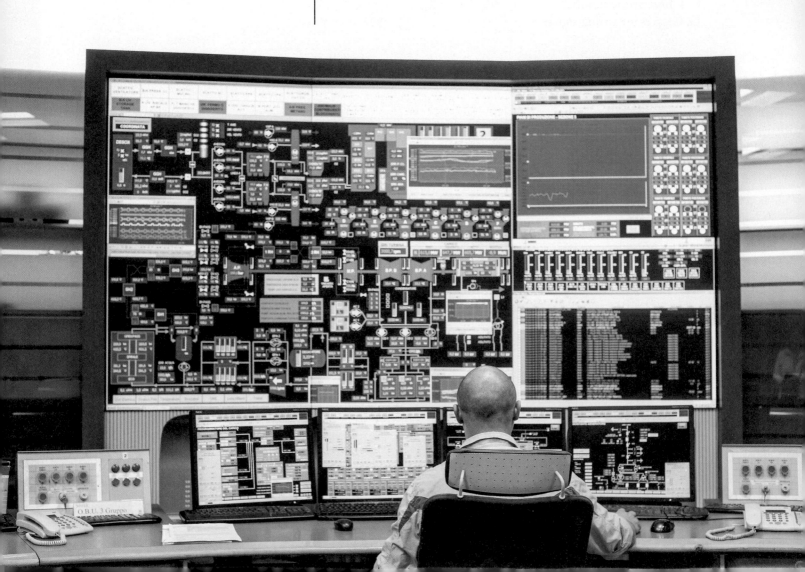

LEARNING OBJECTIVES

After completing this chapter, you will be able to:

☐ Understand the difference between an NPN and a PNP bipolar junction transistor.

☐ Understand linear and digital integrated circuits and describe their uses.

☐ Distinguish between common-emitter, common-collector, and common-base circuits.

☐ Identify a bipolar junction transistor (BJT) by its schematic symbol.

☐ Describe the operation of bipolar junction transistors (BJTs).

☐ Recognize a field-effect transistor (FET) by its schematic symbol.

☐ Describe the operation of field-effect transistors (FETs).

☐ Understand how a transistor can be used as an amplifier in an analog circuit or as a switch for digital operations.

☐ Recognize and understand the uses of various logic gates.

TECHNICAL TERMS

amplifier

bipolar junction transistor (BJT)

common-base circuit

common-collector circuit

common-emitter circuit

comparator

Darlington pair

differential amplifier

digital integrated circuits

doping

field-effect transistor (FET)

gain

insulated-gate bipolar transistor (IGBT)

inverter

junction field-effect transistor (JFET)

linear integrated circuits

logic gate

metal-oxide semiconductor field-effect transistor (MOSFET)

operational amplifier (op amp)

pull-down resistor

pull-up resistor

rail

transistor

I n Chapter 29, you were introduced to transistors that are used in voltage regulator and filter circuits. In addition, you were introduced to a very simple integrated circuit called an analog voltage regulator. Today, almost all electronic circuits contain one form or another of transistor. Integrated circuits often contain thousands of transistors in the same package.

There are two basic types of transistors, which are bipolar junction transistors (BJTs) and field-effect transistors (FETs). There are also further subdivisions of each of these two basic types of transistors.

Transistors used as voltage regulators, filters, amplifiers, or oscillators are said to be used as analog devices. Transistors can also be used as switches, in which case they are said to be used as digital devices. By combining the analog and digital modes of use, transistors may also be used to produce clocks and timers.

30.1 BIPOLAR JUNCTION TRANSISTORS

A *transistor* is a solid-state, three-terminal semiconductor device that is used as an amplifier or a switch. The most common type of transistor is the *bipolar junction transistor (BJT)*. A small current at two of the three terminals controls the current at another two of the three terminals. One of the three terminals is shared in common between the input and the output circuits.

The three terminals are called the *emitter*, the *collector*, and the *base*. A base current will control the emitter and collector current. In this configuration, the circuit is referred to as a *common-emitter circuit*.

There are two types of bipolar junction transistor, which are referred to as the NPN and the PNP type. The NPN bipolar junction transistor has P-type material sandwiched between two N-type regions, and the PNP has N-type material between two P-type regions. The schematic symbols for each are shown in **Figure 30-1**.

Notice that the base and the collector look the same on both transistors. The only difference between the NPN and PNP transistor symbols is the direction of the arrow on the emitter. One way to remember which symbol represents which transistor is to note that with the NPN, the arrow is Not Pointing iN, while the PNP transistor symbol has the arrow Pointing iN.

The current flows from the collector to the emitter in an NPN transistor, whereas it flows from the emitter to

the collector in a PNP transistor. Both types of BJTs are used in the same way, for signal amplification. The only difference is the biasing of the transistor and the polarity of the power supply. **Figure 30-2** illustrates the direction of current flow in the PNP and the NPN transistor.

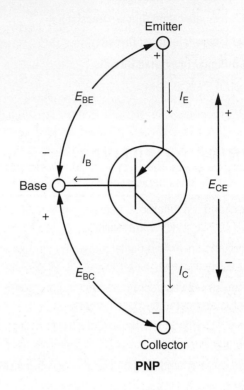

PNP

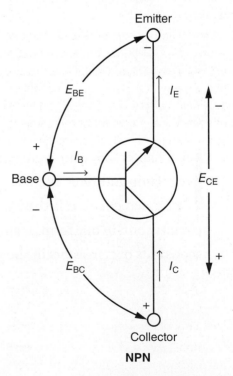

NPN

Figure 30-2. These diagrams show current flow and voltage direction in a PNP and an NPN bipolar junction transistor.

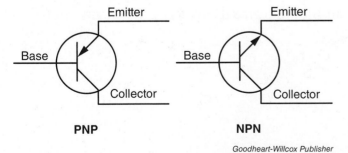

Figure 30-1. These are the schematic symbols for a PNP and NPN bipolar junction transistor. The transistors need not be drawn in the orientation shown in the illustration.

The P designates a positive-type material, while the N designates a negative-type material. These positive and negative electrical characteristics are created in a process called *doping*, in which materials called impurities are added to the silicon crystal. The illustration in **Figure 30-3** indicates the way the transistor is constructed. While not actually constructed of blocks of P and N material, the illustration serves to demonstrate how the layers are connected in both the PNP and NPN transistors. The P-N junctions are similar to those in a diode.

When testing a BJT, you can consider it to be represented as a pair of diodes connected as in **Figure 30-4**. There are not two diodes connected back to back in reality, nor can you construct a transistor with two diodes. But for testing purposes, the representation works just fine.

PROCEDURE — Testing a Transistor

Using the analogy illustrated in **Figure 30-4**, you may perform a simple diode test on each side of the transistor to determine if the transistor is good or bad. You need to disconnect the transistor from the circuit before performing the diode tests. While in circuit, the other circuit components will affect your readings.

1. Set the digital multimeter (DMM) to the diode test mode.

2. Connect the probes across the emitter-to-collector (E-to-C) terminals. Ensure that you have no conductivity in either direction. If you do have conductivity, the transistor is bad. The failure mode you will encounter most often is conductivity in both directions. Most bipolar transistors fail with the E-to-C terminals shorted together.

3. Connect the meter leads to the emitter-to-base (E-to-B) terminals. You should have conductivity in one direction only.

4. Connect the test leads to the B-to-C terminals. Again, you should have conductivity in only one direction.

5. If all tests pass, you may determine the type of transistor—NPN or PNP—by the direction of conductivity: collector to emitter in an NPN transistor, or emitter to collector in a PNP transistor.

This simple series of tests can identify most bad transistors. The tests may not be conclusive for marginally bad transistors, such as those suffering leakage or frequency issues, for example.

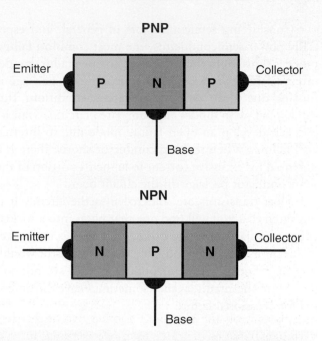

Goodheart-Willcox Publisher

Figure 30-3. The physical construction of a bipolar junction transistor is shown here.

SAFETY NOTE
Discharge Capacitors

Before performing DMM tests on transistors, make certain that the supply voltage to the circuit has been disconnected and all large-value capacitors are discharged. Capacitors that have not been discharged may store lethal voltages for a considerable time after the circuit has been disconnected from the power source.

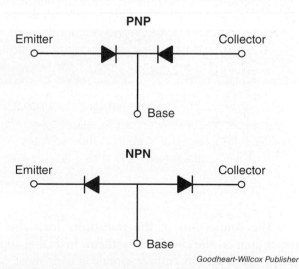

Goodheart-Willcox Publisher

Figure 30-4. Two-diode analogies of a PNP and NPN bipolar junction transistor are pictured here.

Considering semiconductors in general, and especially power semiconductors, the most common failure mode is shorted. In the rare case of an open failure mode, often a large portion of the semiconductor package is missing due to an earlier short circuit condition. This can happen with diodes and integrated circuits. Transistors can develop an open failure mode due to internal lead failure. When the semiconductor shorts, there is a potential for excessive current to launch a portion of the semiconductor package off the circuit board.

Most transistors are soldered into the circuit. It is not often that you will find one that plugs into a socket. Most manufacturers gave up that practice, as the socket connections would fail long before any transistor would. You may find power semiconductors that are not soldered into the circuit, as they sometimes have the potential for occasional failure.

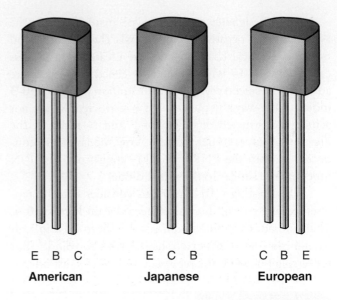

E B C
American

E C B
Japanese

C B E
European

Goodheart-Willcox Publisher

Figure 30-5. These are variations for TO-92 transistor packages based on the country of manufacture.

> ⚠ **CAUTION**
>
> It is important to know which terminal of the transistor belongs to the emitter, base, and collector. If a transistor is installed incorrectly into a circuit, it will most often be destroyed. Transistors are extremely polarity sensitive, as are diodes.

The illustration in **Figure 30-5** shows the same TO-92 bipolar transistor package for three different regions. The transistors look the same, but the pin location is different for each. You will need to look up the data sheet for your specific transistor to know for certain the arrangement of the emitter, base, and collector leads.

> 🛠 **TECH TIP**
> **Find the Data Sheet**
>
> You can search on the Internet for details concerning a specific transistor. For example, a search for "2N3906 data sheet" results in links to many different manufacturers' data sheets for that specific transistor.

The connection of the transistors may also vary depending on the circuit using them. In analog applications, the common-emitter circuit is used most often. The diagrams in **Figure 30-6** show common-emitter circuits for the PNP and NPN transistors.

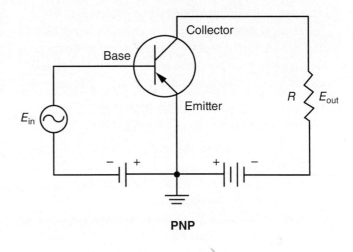

PNP

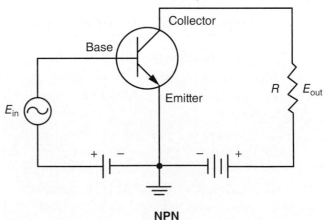

NPN

Goodheart-Willcox Publisher

Figure 30-6. In these common-emitter circuits, the emitter is common to both the input and output. Notice the polarity of the supply voltage in each case.

Another frequently used orientation is the ***common-collector circuit*** configuration, as diagramed in **Figure 30-7**. In this circuit, the input is applied to the base, the output is taken from the emitter, and the collector is shared (common) between the input and the output circuits. This circuit is also called an emitter follower.

One circuit that is used less often is the ***common-base circuit*** configuration, which is diagramed in **Figure 30-8**. In this configuration, the emitter input and collector output circuits share the base as a common terminal.

The discussion to this point has been of transistors as amplifiers. An ***amplifier***, or *amp*, is an electronic device that can increase the power of a signal. The amount of amplification provided by a transistor amplifier is referred to as ***gain***, which is the ratio of the signal output to the signal input. A circuit gain of 1 is referred to

as unity gain. This means that for every 1 unit of signal amplitude that appears on the input side of the circuit, 1 unit of signal appears on the transistor circuit output.

TECH TIP
Use of Amplifier Circuits

Each type of common connection arrangement of a transistor amplifier circuit has its own unique benefits for certain applications. The common-emitter configuration is useful for voltage and current amplification. The common-collector circuit can be useful as a current amplifier and voltage buffer. The common-base configuration works best as a voltage amplifier and current buffer.

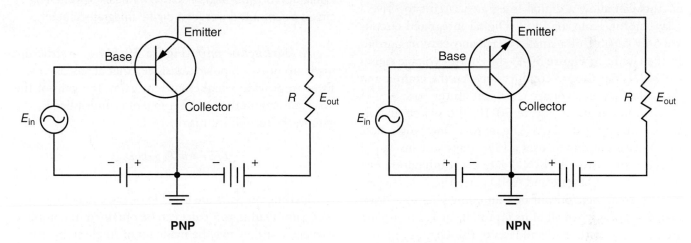

Figure 30-7. There are common-collector circuits for both PNP and NPN transistors. Observe the supply voltage polarities.

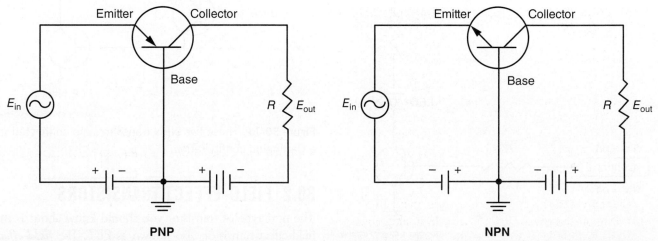

Figure 30-8. The common-base configuration for a PNP transistor is pictured here.

If a circuit has a gain of 10, for every 1 unit of signal appearing on the input, 10 units of signal appear on the output. Thus, the circuit amplifies the signal 10 times. Likewise, if a transistor has a gain of 20, for every 1 unit of signal that appears on the input, 20 units of signal appear on the output of the circuit.

Transistor data sheets include a parameter labeled *hFE*, Beta, or β (Greek letter beta). If you research the difference between the two indications, you will find a large range of opinions as to what exactly each represents and various nuances of explanation. Beta (β) is the ratio of collector current to base current. For the purposes of this discussion, consider each of these parameters to simply mean gain.

The application of transistors as amplifiers is analog. A digital application for a transistor is as a switch. Switching transistors use a small input current to control a much larger output current. Transistor switching circuits can allow a digital integrated circuit to drive a relay, motor, lamp, or LED. Digital integrated circuits are only capable of a small voltage and current output. In the circuit in **Figure 30-9**, the control voltage causes the NPN transistor to conduct between the emitter and the collector when voltage is present on the base.

In the circuit in **Figure 30-10**, the switch and R_1 are used to force the NAND gate to a "low" logic state (0). Notice this circuit uses a PNP transistor instead of an NPN transistor. A NAND gate is a combination of a digital logic NOT gate and AND gate. The NAND gate has a normal logic output of 1 and only goes to a "low" logic level of 0 when all of its inputs are at 1. Logic gates are discussed in more detail later in this chapter.

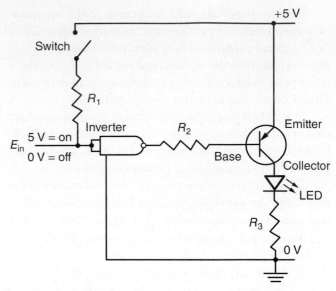

Goodheart-Willcox Publisher

Figure 30-10. This diagram shows a transistor switching circuit as an LED driver for a digital integrated circuit.

A ***Darlington pair*** (**Figure 30-11**) is a structure made up of two bipolar transistors that act as one transistor to provide a higher current gain. The gain of the first transistor in the Darlington pair is multiplied by the gain in the second transistor:

$$hFE_T = hFE_1 \times hFE_2$$

(30-1)

Darlington pairs are capable of extremely high current gain. Darlington pairs can be obtained in complete packages, or they may be made up of single transistors.

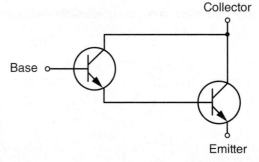

Goodheart-Willcox Publisher

Figure 30-11. These two NPN transistors are connected in a Darlington configuration.

30.2 FIELD-EFFECT TRANSISTORS

The next type of transistor you should know about is the field-effect transistor, also known as FET. The ***field-effect transistor (FET)*** is a three-terminal, voltage-controlled device that uses an input voltage applied to the gate terminal

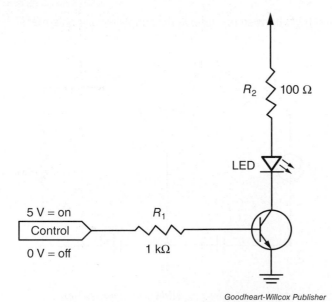

Goodheart-Willcox Publisher

Figure 30-9. An NPN transistor can be used as a switch, such as in this circuit, where it controls an LED.

to control output current in a source-to-drain circuit. There are various types of FETs available, each with its own schematic symbol. **Figure 30-12** compares a field-effect transistor with an NPN bipolar junction transistor. Notice on the FET that the gate is roughly the equivalent of the base on a BJT, the source is roughly equivalent to the BJT emitter, and the drain is roughly equivalent to the collector. The current path between the drain and the source in an FET is called the channel. The channel of an FET can be made of a P-type (positive) or N-type (negative) semiconductor material.

There are two major categories of FETs, which are *junction field-effect transistors (JFETs)* and *metal-oxide semiconductor field-effect transistors (MOSFETs)*.

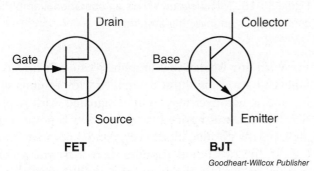

Goodheart-Willcox Publisher

Figure 30-12. These diagrams compare an FET and an NPN BJT.

MOSFETs are also sometimes called insulated-gate field-effect transistors (IGFETs). JFETs have three terminals and operate only in depletion mode. They are the simplest type of FET. MOSFETs have four terminals and operate in both depletion and enhancement mode. Generally, JFETs are used for small-signal applications, and MOSFETs are more expensive and used in VLSI circuits. VLSI stands for very large-scale integration, which describes the process of combining hundreds of thousands of transistors into a single chip, such as those found in computers.

Several different types of FETs and their depictions are shown in **Figure 30-13**.

TECH TIP
Which FET?

You cannot always depend on people sticking to the preferred representation of FETs when drawing schematic diagrams. However, you can often tell which type of FET is actually used by researching the FET identification number. Armed with this number, you are able to search for that part on the Internet and obtain a data sheet.

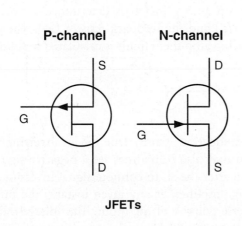

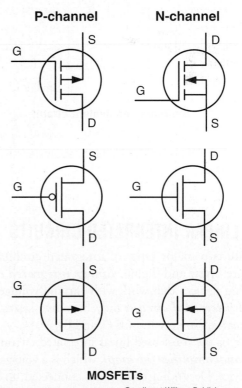

Goodheart-Willcox Publisher

Figure 30-13. Schematic symbols for various JFETs and MOSFETs are shown here. G represents the gate, S represents the source, and D represents the drain.

30.3 INSULATED-GATE BIPOLAR TRANSISTORS

The *insulated-gate bipolar transistor (IGBT)* is actually a cross between a BJT and an FET. The illustration in **Figure 30-14** depicts an IGBT and the BJT/FET equivalent circuit.

The IGBT is used primarily as an electronic switch. It is capable of high efficiency and fast switching. You will often find IGBTs in switching power supplies and variable frequency drives (VFDs) that control the speed of three-phase AC motors.

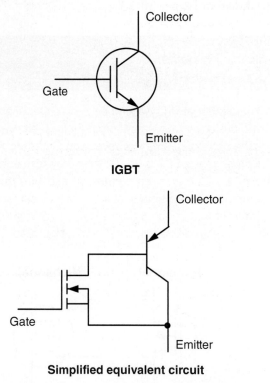

IGBT

Simplified equivalent circuit

Goodheart-Willcox Publisher

Figure 30-14. This is an IGBT and its equivalent BJT/FET representation.

30.4 LINEAR INTEGRATED CIRCUITS

There are two major types of integrated circuits (ICs), which are linear and digital. *Linear integrated circuits* are analog devices that work with varying voltage levels. *Digital integrated circuits* only have two states on the output, which are on and off (1 and 0).

One of the most used linear integrated circuits is the *operational amplifier (op amp)*, which is a voltage amplifying device. Depending on how it is connected, an op amp can perform many different functions, such as filtering, signal conditioning, and mathematical operations. The op amp circuit in **Figure 30-15** is the inverting amplifier.

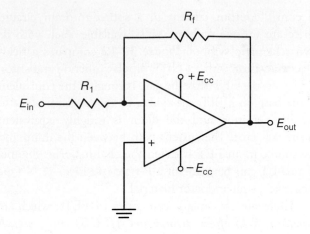

Goodheart-Willcox Publisher

Figure 30-15. This diagram shows an operational amplifier configured as an inverting amplifier.

With any operational amplifier, you will find two inputs: an inverting input designated with a minus sign (−) and a noninverting input designated with a plus sign (+). You will notice that there are two power supply inputs, a positive input ($+E_{cc}$) and a negative input ($−E_{cc}$). The two power supplies share a common connection, as well as the input and the output. This is considered to be a bipolar arrangement with the power supplies.

If you were to use 12 volts as the supply voltage, you would measure 24 volts between $+E_{cc}$ and $−E_{cc}$. The bipolar power supply arrangement allows the op amp to swing its output both positive and negative from the common terminal.

The op amp gain (A) is determined by the input resistor (R_1) and the feedback resistor (R_f). The gain of an inverting amplifier circuit is calculated as follows:

$$A = \frac{R_f}{R_1}$$

(30-2)

This equation is only true for an inverting amplifier. You may also sometimes see a negative sign in the equation after the A, to connote signal inversion. In an inverting amplifier, at any given instant, the output is the inverse polarity of the input. The voltage output of the inverting amplifier may be calculated as follows:

$$E_{out} = E_{in} \times A$$

(30-3)

The noninverting amplifier functions similarly to the inverting amplifier, except that its output is the same as its input at any given moment in time. The circuit for a noninverting amplifier is shown in **Figure 30-16**. Notice that the input resistor (R_1) is connected differently than it was with the inverting amplifier. The formula for calculating noninverting amplifier gain is as follows:

$$A = 1 + \frac{R_f}{R_1}$$

(30-4)

The voltage output of the noninverting amplifier may be calculated as follows:

$$E_{out} = E_{in} \times \left[1 + \frac{R_f}{R_1} \right]$$

(30-5)

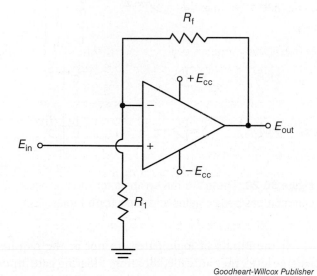

Figure 30-16. A noninverting amplifier circuit is diagrammed here.

The previous equation only works with a noninverting amplifier.

A ***differential amplifier*** is an operational amplifier that has been configured to amplify the difference between two input voltages while suppressing any voltage common to both inputs. **Figure 30-17** shows an op amp circuit for a differential amplifier. Notice that with the differential amplifier, you have two sets of input resistors (R_1) and two feedback resistors (R_2).

Another circuit using op amps is the comparator. The ***comparator*** is a decision-making circuit utilizing an operational amplifier. It is basically a one-bit, analog-to-digital converter. The comparator compares one analog voltage with another analog voltage. The op amp comparator compares two input signals and determines which one is larger. Often, one of the two voltages is a reference voltage, while the other is an actual input signal.

Two possible comparator circuits are shown in **Figure 30-18**. Notice that in each case, inverting and noninverting, R_1 and R_2 form a voltage divider circuit and create a reference voltage at the appropriate reference input of the circuit. In the noninverting comparator, when E_{in}

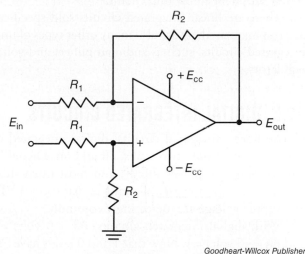

Figure 30-17. This diagram shows a differential amplifier circuit using an op amp.

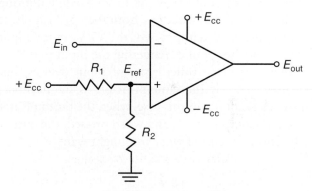

Inverting comparator

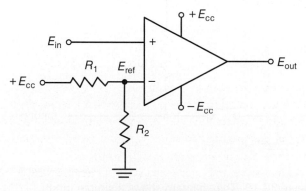

Noninverting comparator

Figure 30-18. Pictured here are an inverting comparator and a noninverting comparator, each using an op amp.

(the input voltage) is greater than E_{ref} (the reference voltage), the op amp output will saturate and swing to the positive supply rail. When E_{in} is less than E_{ref}, the output will saturate and change state, swinging to the negative supply rail. The term **rail** is another name used for the power supply in an op amp circuit.

There are linear integrated circuits sold specifically as comparators. There are also many other types of linear integrated circuits, such as audio amplifiers and voltage regulators.

30.5 DIGITAL INTEGRATED CIRCUITS

Unlike linear integrated circuits, digital integrated circuits work with only two states, off and on. Digital ICs require only one power supply, and most often this is +5 VDC. Today, however, digital ICs that use 3.3 VDC as a supply voltage are increasingly common.

With digital integrated circuits, off = 0 volts = 0, and on = +5 volts = 1. Now that 1 and 0 levels have been established, you may also want some form of input. The simplest input device is the pushbutton switch.

Referring to the circuit in **Figure 30-19**, with the pushbutton in the open position, the 10 kΩ resistor pulls the input pin of the IC up to +5 V while limiting the current the input can draw from the +5 V supply. If the pushbutton switch is closed, the +5 V on the input pin is shunted to ground, rendering the input to 0 V.

There are many other types of input devices, but each uses relatively the same principle as the pushbutton switch example. In this case, when the button is not pressed, the output is a 1, and when pressed, the output is a 0. This is inverse of what you might want.

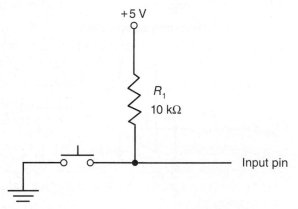

Goodheart-Willcox Publisher

Figure 30-19. This diagram shows a simple input device using a pushbutton switch and a 10 kΩ pull-up resistor. The circuit shows the pushbutton is normally open.

30.5.1 Logic Gates

Logic gates, **Figure 30-20**, are the building blocks of all digital logic ICs. Notice in the illustration that there is a truth table next to each specific logic gate. The truth table indicates the possible input states and the resulting output states. A **logic gate** is essentially a switching circuit that governs whether inputs pass to outputs, with input and output values being one of two states, on (1) or off (0).

AND	X	Y	Z
	0	0	0
	0	1	0
	1	0	0
	1	1	1

OR	X	Y	Z
	0	0	0
	0	1	1
	1	0	1
	1	1	1

NOT	X	Y
	0	1
	1	0

Buffer	X	Y
	0	0
	1	1

Goodheart-Willcox Publisher

Figure 30-20. These are the symbols for the four most common basic logic gates and their truth tables.

If the inputs of logic gates are not at the required voltage level, they are called floating. Floating gate inputs may create switching errors when voltage is incorrectly sensed as being a 0 (too low to be recognized as a 1) or 1 (too high to be recognized as 0). Highs must be recognized as highs, and lows as lows. To make sure this happens, pull-up and pull-down resistors are used. A **pull-up resistor** is used to connect input pins to the DC supply voltage to make sure the input is high enough to be recognized as a 1 by the logic gate. A **pull-down resistor** is used to connect input pins to ground to make sure the input is low and recognized as a 0 by the logic gate.

> ⚠ **CAUTION**
>
> Floating gate inputs can result in excessive current flow, increased power consumption, and damage to the circuit. This is why pull-up and pull-down resistors are used to ensure all inputs are held to a valid logic level.

The simplest logic gate is the inverter, which is often called a NOT gate. An *inverter* will output the opposite of the state of that which is on the input. Another simple logic gate is the buffer, which parrots what is on its input to its output: 0 in and 0 out, or 1 in and 1 out. Notice that the dot on the output of the NOT gate or inverter gate indicates an inverse output compared to that on the input. Notice also that the NOT gate and the buffer are the complement (opposite) of one another.

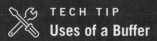

TECH TIP
Uses of a Buffer

Even though a buffer does not perform a logic function or change, it does have its uses. Buffer logic gates provide digital amplification of a signal. They can also be used to isolate other gates and parts of a circuit from each other, in case they might otherwise affect one another.

As you see in the truth table, for the AND gate, if both inputs are 0 or any input is 0, the output will be 0. The only way the AND gate will output a 1 is when both inputs are also 1. There is also the OR gate. Referring to its truth table, if either or both inputs are a 1, then the output is also 1. If both inputs are 0, the output is 0.

The next four logic gates, pictured in **Figure 30-21**, are variations of the AND and the OR gates. The first of these is the NAND or Negative AND gate. It is the same as the AND gate on the input side, with an inverter on the output side. This way, where the output would be a 1 in the case of an AND gate, the output would be a 0 in the case of the NAND gate.

Often, when designers need a NAND gate and have some spare AND and NOT gates, they wire together those spare gates to avoid having to create a space for a new IC. Small, basic gates often come several to an IC package. For example, a package often contains six inverters and is referred to as a HEX Inverter. AND and OR gates often come four to a package.

The NOR gate functions similarly to an OR gate with an inverter on its output. The XOR gate stands for eXclusive OR. Compare its truth table to that of an OR gate. You will notice the XOR gate will output a 0

NAND		X	Y	Z
		0	0	1
		0	1	1
		1	0	1
		1	1	0

NOR		X	Y	Z
		0	0	1
		0	1	0
		1	0	0
		1	1	0

XOR		X	Y	Z
		0	0	0
		0	1	1
		1	0	1
		1	1	0

XNOR		X	Y	Z
		0	0	1
		0	1	0
		1	0	0
		1	1	1

Goodheart-Willcox Publisher

Figure 30-21. Four other common basic logic gates are pictured here, along with their truth tables.

when both inputs are a 0 or when both are a 1. The XNOR gate functions like an XOR gate with an inverter attached to its output.

There are many other digital ICs, such as quad-input AND gates, various flip-flops, and counters. There are all sorts of encoders and decoders. There are digital-to-analog converters (DACs) and analog-to-digital converters (ADCs) to help interface with the outside, analog, real world. More discussion of interface digital ICs will be found in upcoming chapters.

CHAPTER WRAP-UP

The ability to control the flow of electrons is what makes electronics work for us in our daily lives. Transistors use semiconductors to control that flow. Integrated circuits make it possible to combine many transistors in a single component. Together, the transistor and integrated circuit have allowed modern electronic devices to be small, reliable, and affordable. Your knowledge of these important components will help you understand, build, and fix electrical systems and electronic control systems as an industry technician.

Chapter Review

SUMMARY

- Nearly all electronic circuits contain one form or another of transistor. When used as amplifiers and voltage regulators, they are used as analog devices. When used as switches, they are used as digital devices.

- The most common type of transistor is a bipolar junction transistor (BJT). The three terminals are called the base, collector, and emitter. A small current at two of the three terminals controls the current at another two of the three terminals, and one terminal is shared between the input and output circuits.

- There are two types of BJTs: NPN and PNP. The P designates a positive-type material, and the N designates a negative-type material.

- A field-effect transistor (FET) is a three-terminal, voltage-controlled device that uses an input voltage applied to the gate terminal to control output current in a source-to-drain circuit.

- There are two major categories of FETs: junction field-effect transistors (JFETs) and metal-oxide semiconductor field-effect transistors (MOSFETs). JFETs are used for small-signal applications. MOSFETs are more expensive and used in VLSI circuits.

- An insulated-gate bipolar transistor (IGBT) is a cross between a bipolar junction transistor and a field-effect transistor. It is capable of high efficiency and fast switching, and it is used primarily as an electronic switch.

- The two major types of integrated circuits (ICs) are linear and digital. Linear ICs are analog devices that work with varying voltage levels. Digital ICs only have two states on the output, which are on (1) and off (0).

- One of the most used linear ICs is an operational amplifier (op amp). It has two inputs, one being an inverting input designated with a minus sign (–), and the other a noninverting input designated with a plus sign (+).

- A logic gate is a switching circuit that governs whether inputs pass to outputs. The input and output values are one of two states, on (1) or off (0).

REVIEW QUESTIONS

Answer the following questions using the information provided in this chapter.

1. The arrow on the emitter in a PNP transistor symbol is pointing _____.

2. What does it mean if there is conductivity across the E-to-C terminals in a transistor when performing a diode test?

3. *True or False?* The most common failure mode for semiconductors is shorted.

4. Why are most transistors soldered into circuits rather than plugged into a socket?

5. The _____ circuit is used most often in analog applications.

6. *True or False?* A common-base configuration works best as a voltage buffer.

7. Explain how a transistor switching circuit works.

8. A _____ is a structure made up of two bipolar transistors that act as one transistor in order to provide a higher current gain.

9. *True or False?* The source on a field-effect transistor is roughly the equivalent of the collector on a bipolar junction transistor.

10. In what two applications will you most often find insulated-gate bipolar transistors?

11. What is the gain of an inverting amplifier circuit with an input resistance of 10 Ω and a feedback resistance of 5 Ω?

12. What is the difference between an inverting and a noninverting amplifier?

13. What is the gain of a noninverting amplifier circuit with an input resistance of 30 Ω and a feedback resistance of 15 Ω?

14. Calculate the voltage output of a noninverting amplifier circuit with a voltage input of 60 V, an input resistance of 20 Ω, and a feedback resistance of 12 Ω.

15. If the inputs of a logic gate are not at the required voltage level, they are called _____.

16. *True or False?* If an AND logic gate has one input of 1 and one input of 0, the output will be 1.

17. A _____ resistor is used to connect input pins to the DC supply voltage to ensure the input is high enough to be recognized as a 1 by a logic gate.

18. *True or False?* A NOT logic gate and a buffer are complements of each other.

NIMS CREDENTIALING PREPARATION QUESTIONS

The following questions will help you prepare for the NIMS Industrial Technology Maintenance Level 1 Electronic Control Systems credentialing exam.

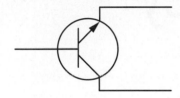

1. What type of transistor is pictured here?
   A. JFET
   B. MOSFET
   C. NPN BJT
   D. PNP BJT

2. Which of the following components of a bipolar junction transistor is roughly the equivalent of the drain on a field-effect transistor?
   A. Base
   B. Collector
   C. Gate
   D. Emitter

3. A(n) _____ amplifier is an op amp that has been configured to amplify the difference between two input voltages while suppressing any voltage common to both inputs.
   A. inverting
   B. comparator
   C. noninverting
   D. differential

4. Which of the following is a decision-making op amp circuit that converts analog signals into a binary digital output?
   A. Comparator
   B. Rail
   C. Gain
   D. Darlington pair

5. A logic gate with a truth table showing one input of 1, one input of 0, and an output of 1 is a(n) _____ logic gate.
   A. AND
   B. OR
   C. NOT
   D. buffer

6. A logic gate with a truth table showing two inputs of 0 and an output of 0 is a _____ logic gate.
   A. NAND
   B. NOR
   C. XOR
   D. XNOR

31

Sensors and Variable Frequency Drives

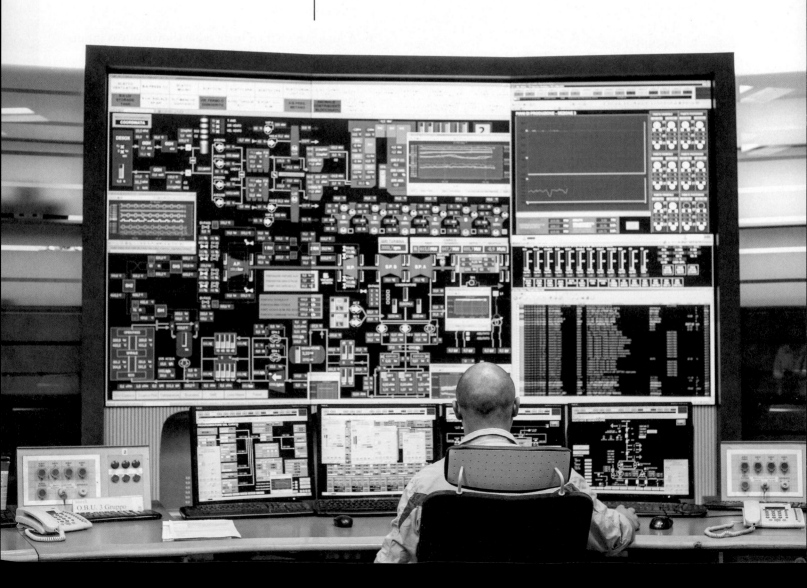

LEARNING OBJECTIVES

☐ Describe the different wiring used for AC and DC sensors.

☐ Discuss how the different categories of proximity sensors perform their purpose.

☐ Explain the different inputs and outputs on a laser curtain sensor.

☐ Describe the different categories of photo sensors.

☐ List the steps to adjust a photo sensor.

☐ Describe the operation of various temperature sensors.

☐ Describe the operation of pressure sensors and the piezoelectric effect.

☐ Explain the difference between open- and closed-loop control.

☐ Discuss how AC and DC drives differ.

☐ Describe the relationship between pulse width modulation and duty cycle.

☐ Discuss common drive parameters.

☐ Explain how the environment can impact a drive's performance.

TECHNICAL TERMS

blanking

bleed-off resistor

capacitive sensor

closed-loop system

current source inverter (CSI)

diffuse mode

duty cycle

electromagnetic interference (EMI)

Hall effect sensor

harmonic distortion

inductive sensor

laser curtain

line reactor

magnetic sensor

metal oxide varistor (MOV)

modulated light

negative slope

open-loop system

phase control

photo sensor

piezoelectric effect

polarized

positive slope

potentiometer

proximity sensor

retroreflective

resistance temperature detector (RTD)

rheostat

signal conditioning

slip

synchronous speed

teaching input

thermistor

thermocouple

thru-beam mode

total harmonic distortion (THD)

ultrasonic sensor

unmodulated light

variable voltage inverter (VVI)

Wheatstone bridge

T his chapter combines two subjects technicians deal with on a regular basis: sensors and variable frequency drives (VFDs). Tasks from installation, adjustment, and programming to troubleshooting and maintenance are critical components of modern motor and motion control.

Over time, both sensors and VFDs have increased in complexity and wiring, but also in reliability. While sensors used to rely on visible light, they now come in infrared, laser, or other LED-driven varieties. While VFDs used to be limited to large DC drives and AC-wound rotor motors, they can now control even the smallest motors using a variety of methods.

When sensors and VFDs perform reliably, we barely notice them. But when either of these elements fails, it can impact the entire automated control system.

31.1 SENSORS

Sensors are categorized by the method of sensing, range of material being sensed, and voltage used. The three-wire DC sensor is the most commonly used sensor. It comes in PNP and NPN configurations, and the appropriate choice will depend on the load.

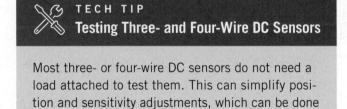

TECH TIP
Testing Three- and Four-Wire DC Sensors

Most three- or four-wire DC sensors do not need a load attached to test them. This can simplify position and sensitivity adjustments, which can be done without impacting the rest of the control system.

Sensors can be broken down into the broad categories of Hall effect sensors, proximity sensors, photo sensors, temperature sensors, and pressure sensors. A **Hall effect sensor** senses a voltage difference when in close proximity to a magnetic field. **Proximity sensors** sense a material based on its metallic or capacitive qualities or by using a generated sound wave and measuring the returned signal. **Photo sensors**—also known as *photoelectric sensors* or *photo eyes*—measure an

object's reflective ability, lack of reflection, presence of light, or lack of light. Various types of temperature sensors are also used during industrial operations, and pressure sensors are also common with industrial controls. Sensor wiring is summarized in **Figure 31-1** and illustrated in **Figure 31-2**.

Analog sensors often require signal conditioning. **Signal conditioning** is the manipulation of an analog signal so that it will meet the requirements of the next

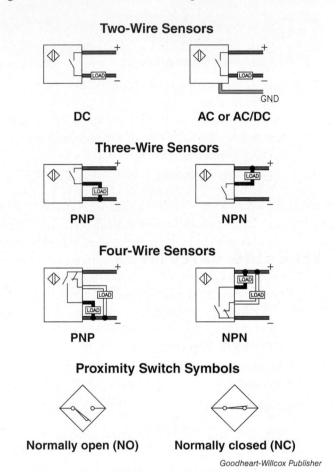

Figure 31-2. These diagrams depict wiring connections for sensors.

Sensor Wiring Types			
Wiring Type	Advantages	Disadvantages	Notes
Two-wire AC/DC	Simple wiring	Limited choice of models	Leakage through sensor to load, voltage drop through sensor when on
Two-wire DC	Simple wiring, sink or source depending on where load is placed	Limited models compared to three-wire DC	May have higher voltage drop when on and leakage current
Three-wire DC	Readily available and common with many models Load not required for testing	Proper connection (PNP or NPN)	Low leakage current
Four-wire DC	Available in many models with normally open or normally closed contacts Load not required for testing	Proper connection required with more complicated wiring	Low leakage current

Goodheart-Willcox Publisher

Figure 31-1. This table summarizes sensor wiring types and lists the advantages and disadvantages of each.

stage in the process. This can include amplifying, filtering, converting, isolating, or other conditioning to make the sensor output readable and usable in the system. Some sensors have built-in signal conditioning. When necessary, signal conditioners are added into a system. The zero point and range of signal conditioners can be adjusted to ensure the signal is properly interpreted.

31.1.1 Hall Effect Sensors

A Hall effect sensor registers a magnetic field and can be used in various ways. For example, a Hall effect sensor placed on a linear actuator can read position. When the sensor comes in close proximity to a magnetic field, it produces an output voltage. This output voltage relies on applied voltage and the level of the magnetic field.

Hall effect sensors are gaining in popularity, but their largest application is in electronic circuits and devices. Smaller electronic control systems such as servo-motor control, small positioning systems, robotics, and machine tool controls are common applications. The Hall effect sensor can also directly replace the contact limit switch in operations where the movement is accurate and repeatable. Packages range from board-mounted integrated circuits, **Figure 31-3**, to larger housings.

31.1.2 Proximity Sensors

Proximity sensors may be inductive, capacitive, ultrasonic, magnetic, or laser. Laser emitters and receivers may also be considered photo sensors, but they are covered in this section. Most proximity sensors follow standard wiring colors and are typically three-wire, DC-voltage sensors.

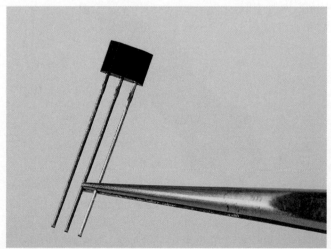

Kirill Volkov/Shutterstock.com

Figure 31-3. A wide range of Hall effect sensors are used for miniature and industrial applications.

Inductive Sensors

The *inductive sensor*, **Figure 31-4**, is comprised of a coil that develops a field around it when voltage is applied. As a ferrous metal (containing iron) is brought into the field, the inductance of the field changes, causing an output from the sensor. Nonferrous metals can be detected, but typically only at much closer ranges. Inductive sensors are best used to detect repeated movements of ferrous objects at close to very close distances, usually around 5–30 mm or less.

These types of sensors may be shielded or unshielded. Shielded sensors are used when flush mounting the front of the sensor and with very close detection ranges. The shielding helps eliminate the effects of surrounding metal. Unshielded sensors have longer detection ranges, but the head of the sensor must have clearance around it to help eliminate the effects of surrounding metal.

Capacitive Sensors

Capacitive sensors, **Figure 31-5**, work by detecting a change in capacitance. This change in capacitance can occur with nearly any object sensed, as long as it is close enough to the sensor. Sensor ranges depend largely on the diameter of the sensor. Capacitive sensors have an adjustable **potentiometer**, also called a *pot*, which is a resistor that measures voltage difference.

Automation Direct

Figure 31-4. Inductive proximity sensors come in a wide range of styles, depending on the application.

Automation Direct

Figure 31-5. Capacitive sensors can detect many objects but have a short range. Depending on the model, the adjustment potentiometer may have a screw cover that must be removed before adjustments can be made.

Capacitive sensors come in shielded and unshielded versions. Shielded sensors can sense nearly any material, while unshielded sensors are best for conductive materials, such as water and metals. Unshielded sensors can sense an object through another material, as long as the intervening material is not thicker than the sensing distance. Shielded sensors can be flush mounted. An unshielded sensor must not have material near its face that might falsely trigger the sensor output. If a sensor's LED indicators are not visible once the sensor is mounted, standard connectors and cables that have indicator LEDs are useful.

Capacitive sensors come with two or three wire connections, normally open or normally closed outputs, and voltages of 24–240 VAC and 10–48 VDC. Connections are offered in mini or micro connectors. Voltage drops for AC voltage sensors can reach 7.5 volts, and load specifications need to be considered. Sensors typically have two indicator LEDs: one indicates the sensor is receiving power, the other indicates detection of an object, or output. Here are some guidelines for installing a capacitive sensor:

1. Make sure the sensor is mounted rigidly. A thread-locking adhesive works well.

2. Apply power to the sensor. For a two-wire sensor, the load must also be wired.

3. The sensing distance must be within the sensor's range.

4. Place the object at the typical sensing distance, and check the LED for output.

5. If the output LED is not on, adjust the potentiometer slowly until the output LED is on. If the output is normally closed, adjust the potentiometer until the output LED turns off.

6. Check the sensor operation several times by removing and replacing the object and checking the output LED.

Ultrasonic Sensors

Ultrasonic sensors, **Figure 31-6**, emit a high-frequency sound and measure the time it takes for a return echo. These sensors may be used to detect solid objects or to sense liquids and liquid levels within a tank. Sensors may be stand-alone with digital outputs (on-off) or wired to a controller and have analog outputs (voltage or current).

When the object is within range, the sensor will give an output. The range is adjusted within a band by adjusting the time set point. Analog-output types will generate a signal proportional to the distance of the target, commonly 4–20 milliamps. Analog outputs may be configurable to a positive slope or negative slope. A ***positive slope*** causes the output to be lowest (4 mA) when the detected object is closest to the sensor and to increase (up to 20 mA) as the object moves away. A ***negative slope*** causes the output to be highest (20 mA) when the detected object is closest to the sensor and to fall as the object moves away. Controllers may have multiple settings for relay outputs and control functions.

Magnetic Sensors

Magnetic sensors appear very similar to inductive and capacitive sensors and are offered in several packages, **Figure 31-7**. They usually have longer sensing ranges than inductive sensors. Magnetic fields also penetrate nonmagnetic or nonmetallic materials, so these sensors can detect a magnet on the other side of materials. Magnetic sensors use a ***Wheatstone bridge***, which is an electrical bridge circuit used to measure an unknown electrical resistance on one of its sides. One side of the bridge changes resistance as a magnetic field is detected, to produce an output, which triggers the sensor to give an output. Sensing distances are usually less than 3″, and outputs can be normally open, normally closed, PNP, or NPN.

Laser Curtains

Laser curtains are presence-sensing devices that use lasers to detect and keep workers safe. Typically, laser curtains use a transmitter element and receiver unit,

Migatron

Figure 31-6. Ultrasonic sensors are commonly used to produce an analog output proportional to sensing distance. Sensors are made using a variety of materials and for different purposes and situations. Some sensors are used specifically to detect clear objects, some have analog and digital outputs, and some are washdown-proof.

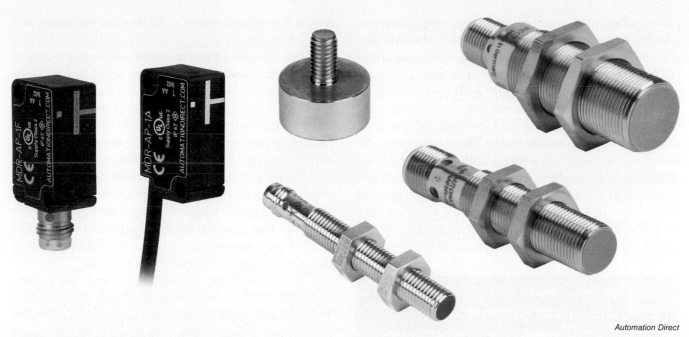

Automation Direct

Figure 31-7. With longer sensing distances than inductive sensors, magnetic sensors are commonly used in food-processing applications.

Figure 31-8. Laser sensors are most commonly used at machine points of operation, around sensitive or dangerous industrial robots, and at perimeter access points.

Point-of-operation applications include use with stamping or forming machines, where operators may be in close proximity to a potentially dangerous work point. The laser curtain ensures that the operator is outside of the pinch-point area before allowing machine operation.

Larger, perimeter-access curtains prevent workers from entering a work area while machine operations are occurring. These laser curtains may use mirrors to increase coverage and exclude entry into an entire area.

SAFETY NOTE
Proper Training Required

If intending to work on safety systems, laser curtains, or any related circuitry, you should receive specific training from the manufacturer.

Laser sensors typically have outputs for normally open (NO) and normally closed (NC) circuits, and some offer an analog output proportional to the size of the detected object. The emitter has typical 24 VDC

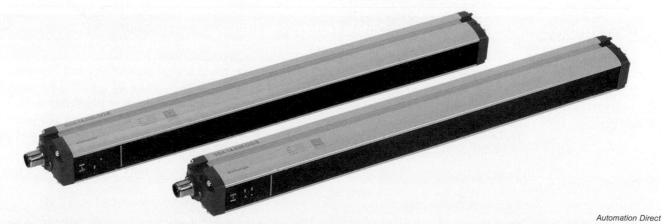

Automation Direct

Figure 31-8. With process control becoming more automated, fast and reliable safety measures must be used. The response time of point-of-operation laser curtains, such as the one pictured here, is 15–25 milliseconds. Safety contactors are used in machine control circuits that require positive feedback to isolate hazards to personnel. Safety controllers are similar to programmable logic controllers (PLCs), but are tested and proven to perform specifically for safety guarding systems.

power connections, may have additional wiring to synch with the receiver, and has a test function that will disable the emitter. The receiver also has power connections (24 VDC), normally open and normally closed outputs, a teaching input, analog output (0–10 V, 4–20 mA, or 0–5 V are common), and a feedback signal to the emitter. The *teaching input* is used to "teach" the receiver to recognize a new sensing range. The receiver may also have fine and coarse adjustments, and a blanking setting. *Blanking* is the disabling of a section of a laser curtain's sensing field. Common receiver connections are detailed in the table in **Figure 31-9**. All of these wires are connected to the receiver using a standard connector, but they must be wired at the PLC or load side.

> ### SAFETY NOTE
> ## Laser Curtain Maintenance
>
> Blanking is used to disable a section of a laser curtain, such as when material must move through it for a machine process. Muting refers to the complete disabling of a laser curtain, which may be done when maintenance work must be carried out on a machine. In either case, the technician must be sure the laser curtain is properly set to again protect workers after any maintenance is finished.

31.1.3 Photo Sensors

Photo sensors detect objects by use of visible or invisible light. The object may be detected in an on-off fashion, or the object may have specific attributes that are detected, such as the presence of a specific color. In general, all photo

sensors have a light source, a detector, internal logic, and output circuitry.

Both modulated and unmodulated light are used in the emitter. *Unmodulated light* is simply an unmodified light wave. *Modulated light* is a light wave that is varied in its amplitude or frequency in order to transmit information. Since most photo sensors use LEDs for the emitter, modulated light provides the advantage that background light is less likely to affect sensor operation.

The receiver portion of the sensor may be in a separate housing or mounted in the same housing as the emitter. The receiver senses the light by using a phototransistor that is paired to the emitter LED, while circuitry logic ensures that only the modulated light, at that specific frequency, results in an output.

The terms "light operate" (LO) and "dark operate" (DO) are both commonly used with photo sensors, **Figure 31-10**. Light operate refers to the sensor providing action when sufficient light is detected, at which point a normally open contact would close or a normally closed

Light Operate

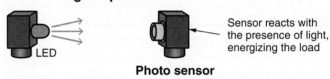

Sensor reacts with the presence of light, energizing the load

Dark Operate

Sensor reacts when no light is present, energizing the load

Goodheart-Willcox Publisher

Figure 31-10. Light and dark operation of photo sensors is dependent on whether or not they receive light.

Laser Curtain Features and Uses		
Receiver Connection	**Use**	**Notes**
Teaching fine	Small, transparent objects	Useful for clear plastics, clear glass bottles, and other glass.
Teaching coarse	Textured surfaces, large objects, opaque objects	The teaching settings are used to put the receiver in the correct mode depending on the material you are trying to detect.
Blanking	Used to "blank" part of the receiver that may be constantly blocked and should not be read	The emitter and receiver may be longer than the area, or may be partially blocked. This setting will nullify that portion of the receiver that is blocked, so you will not get a false detection signal.
Analog	Outputs an analog signal that is proportional to the size of the detected object, typically its height	This is also affected by the coarse and fine adjustment and blanking.
Normally open (NO) or normally closed (NC)	Standard outputs	Make sure the output load does not exceed the rating of the sensor.

Goodheart-Willcox Publisher

Figure 31-9. This table lists some features and uses of various laser curtains. Always refer to the manufacturer's documentation when wiring and setting up a laser curtain.

contact would open. Dark operate describes a sensor that provides an action when light is not present, with the output contacts changing state. The type of output—normally open (NO) or normally closed (NC)—has nothing to do with the light/dark operation of the sensor.

Several modes of sensing may be used, depending on the presence of background reflective interference, the reflective nature of sensed objects, a particular position or area of an object, or the use of a polarized beam. See **Figure 31-11**.

3. Rotate the transmitting element back in the opposite direction until it no longer senses the part or beam, and mark this spot.

4. Rotate the transmitting element to the center spot of both marks.

5. Perform this procedure in the vertical plane as well.

For sensitivity adjustment, the sensitivity should be great enough to register a sensor response accurately while not so great that background or other objects impact sensor performance.

PROCEDURE — Photo Sensor Adjustment

Use the following procedure when adjusting and aligning photo sensors:

1. Aim the transmitter and receiver at each other, or element at the reflector or object, until the sensor either sees the part or detects the beam.

2. Rotate the transmitting element until it no longer senses the part or beam, and mark this angle.

Fiber-Optic Sensors

Fiber-optic sensors, **Figure 31-12**, are used where the environment or electrical requirements do not allow a typical current-carrying sensor. Fiber-optic sensors have either thru-beam or diffuse modes of sensing. With *diffuse mode*, the light transmitted from the sensor shines on the object and is scattered, or diffused, in different directions, with a small portion reflected back to the sensor receiver.

Sensing Modes of Photo Sensors			
Mode of Sensing	Use	Notes	Operation
Thru-beam	• Activates receiver by sensing emitter light • Object breaks beam	• Emitter and receiver units • Longer distances • Alignment needed	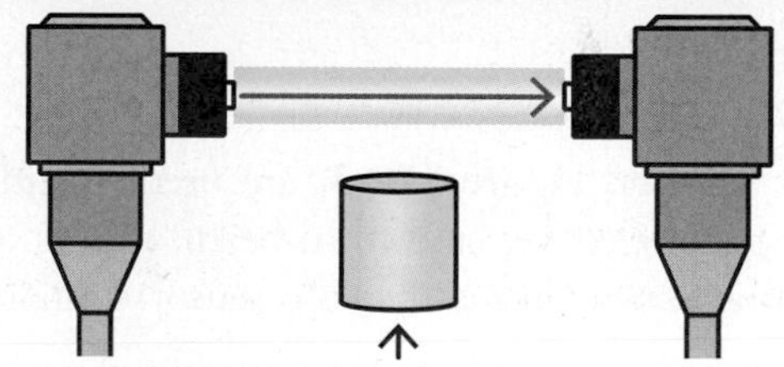
Retroreflective	• Uses a reflector • Activates from this beam being interrupted	• Shorter distances than thru-beam • Single transmitter/ receiver unit	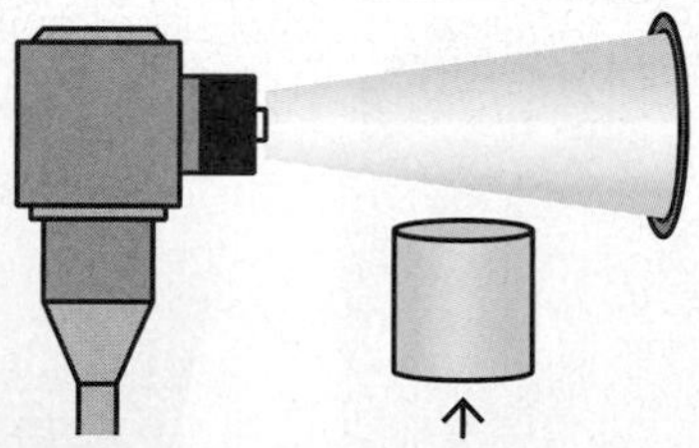
Diffuse	• Activates from reflected light from object	• Better for shorter ranges • Single transmitter/ receiver unit	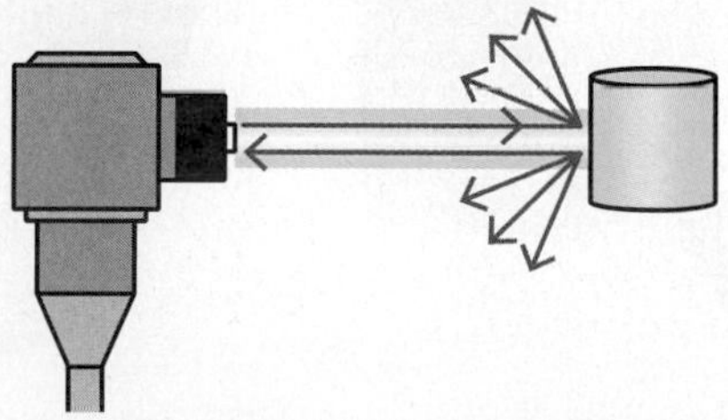

Goodheart-Willcox Publisher

Figure 31-11. This table details the three basic sensing modes of photo sensors.

Automation Direct

Figure 31-12. Fiber-optic sensors come as light on or dark on, with normally open (NO) or normally closed (NC) outputs, and with LED indications and sensitivity adjustments.

With ***thru-beam mode***, the transmitted light travels to a separate receiver and an object is detected when it interrupts that beam. The sensing light is transmitted from the sensor to the end point by fiber-optic cable. With fiber-optic sensors, the sensing body can be located remotely from the actual emission point.

Different sensing modes, sensing tips, and cables all need to be appropriately specified for a particular application. Closely follow manufacturers' specifications to ensure that these expensive sensors provide the longest life possible.

Laser Sensors

Laser sensors, **Figure 31-13**, are used in diffuse, polarized retroreflective, and thru-beam modes of sensing. ***Retroreflective*** means that the sensor contains the light

Automation Direct

Figure 31-13. Laser sensors come in nearly every sensing mode with digital or analog outputs.

source and receiver in one housing. ***Polarized*** means that polarized filters are used so that only the light reflected from the object is sensed. Outputs may be PNP or NPN, or analog, depending on the application. Analog outputs are usually used for range detection, while digital outputs are used for detecting object presence.

> **SAFETY NOTE**
> **Eye Safety with Lasers**
>
> ANSI Z136 standards provide guidance for the safe use of lasers, fiber optics, and required personal protective equipment (PPE). Lasers are classified by their energy into Class 1, Class 2, Class 2M, Class 3R, Class 3B, and Class 4. Laser sensors normally fall into Class 1 or Class 2, which are the safest, lowest-energy lasers. Lasers above Class 1 need to be properly identified and labeled. Eye protection is required for lasers in Class 3B or 4 only, but you should never let a direct beam enter your eye.

31.1.4 Temperature Sensors

Temperature sensors commonly used in industry today include thermistors, resistance temperature detectors (RTDs), and thermocouples. ***Thermistors*** are resistors that detect temperature changes through corresponding sharp but predictable changes in electrical resistance. Thermistors are accurate but have a limited temperature range. ***Resistance temperature detectors (RTDs)*** measure temperature by relying on predictable changes in the electrical resistance of metals that are proportional to changes in temperature. RTDs use a length of metal wire made from platinum (most common), nickel, or copper, and they are very precise. ***Thermocouples*** detect temperature based on the voltage produced when two metals are joined. One end of the thermocouple is the probe that measures the temperature change, and the other end is at a constant temperature and provides a reference point.

31.1.5 Pressure Sensors

Most pressure sensors work in a similar fashion. They have a diaphragm that is affected by an acting force, and the amount of deflection of the diaphragm is converted into a proportional electrical signal to measure pressure. In a resistive pressure sensor, the change in the diaphragm causes a change in a built-in resistor, and the change in electrical resistance is measured and converted into an electrical signal. In a capacitive pressure sensor, pressure is measured through changes in capacitance brought about by force applied to the diaphragm or membrane. Piezoelectric sensors use the piezoelectric effect to create an electrical signal. The ***piezoelectric effect*** is the ability of a material to generate electricity in response to applied pressure. In a piezoelectric pressure sensor, this often takes the form of a quartz crystal squeezed between two pressure plates to produce an electrical signal.

TECH TIP
Sensor Replacement

Replace any sensor with the exact same model. If a different model is used to improve some aspect of the process (speed or reliability, for example), it should be carefully considered and tested by engineering.

PROCEDURE | Troubleshooting Sensors

Pursue the following steps when troubleshooting a new sensor problem or failure:

1. Check that the sensor is receiving proper voltage, shown by the LED indicator. If sensors are wired in series, voltage drop could be an issue if voltage supply has dropped.

2. Check that the sensor is detecting the object, using the output LED. Also check the output voltage at the load the sensor is activating (the relay or PLC input).

3. Check that the sensor properly activates by placing and removing an object in the field of detection. Properly align and adjust the sensor if necessary.

4. Check the physical integrity of the sensor. If mechanical impact has taken place, replace the sensor.

5. If any of these items cannot be corrected, replace the sensor with an exact duplicate and align and adjust the new sensor.

31.2 VARIABLE FREQUENCY DRIVES

Variable speed drive (VSD) refers to either AC or DC drives with variable speed. Variable frequency drive (VFD) refers to AC drives only. A VFD is a motor controller that varies the speed of an AC motor by varying the input frequency to the motor. Not all VSDs are VFDs, but all VFDs are VSDs. Nonetheless, VFDs and VSDs are often referred to interchangeably.

Older methods of achieving a variable speed motor used DC motors, AC wound-rotor motors, or AC induction motors. In older AC induction motors meant for multiple speeds, numerous stator coil motor leads were connected and disconnected to change the number of poles by using a bank of contactors, thereby changing the synchronous speed of the motor.

For an AC motor, synchronous speed is calculated as follows, with frequency being the power supply frequency:

$$\text{synchronous speed} = \frac{\text{frequency} \times 120}{\text{number of poles}}$$

(31-1)

Output shaft rpm is determined by the synchronous speed minus the slip. ***Synchronous speed*** is the speed at which the magnetic field rotates about the stator, which is the stationary part of the rotary system. ***Slip*** is the amount that the rotor falls behind the rotation of the stator field, calculated as follows and expressed as a percent:

$$\text{slip} = \frac{\text{synchronous speed} - \text{rotor speed}}{\text{synchronous speed}} \times 100\%$$

(31-2)

AC wound-rotor motor speed could be controlled by controlling the voltage applied through brushes to the stator by external resistor banks or variable resistors. By changing the resistance of the rotor circuit, the current and magnetic field of the rotor was changed, which then changed the motor speed. DC motors—namely shunt motors—were controlled by controlling the field current through the use of variable resistors, thereby controlling the speed of the motor.

In modern applications, a combination of voltage and current control through the use of diodes, MOSFETs, IGBTs, and SCRs allows control of the motor speed. The major theoretical concepts, controls, programming, and maintenance of VFDs and other variable speed drives will be examined in this section.

31.2.1 Theory of Operation

All variable speed drives operate and maintain their speed in two general types of systems: open loop and closed loop, **Figure 31-14**. The *open-loop system* of control does not provide feedback to the drive from the system. A setting, or desired speed, is established through a manual input from a rheostat or programmed speed. A *rheostat* is an adjustable resistor used to control current. The drive outputs to the motor, following any other settings, such as ramp. Ramp refers to how quickly the speed is increased (ramp-up) or decreased (ramp-down). No feedback regarding motor rpm is given to the drive from an external source. A *closed-loop system* of control provides for direct feedback to the drive concerning the motor speed. In this case, a device such as a tachometer/generator, high-speed counter, or resolver, for example, generates a digital or proportional analog signal that is fed back to the drive. This feedback signal allows the drive to compare where it is with where it wants to be (the

setting or set point), and to develop an error signal proportional to the difference between the two points. This error signal tells the drive how to alter current or voltage to change output to the motor.

External feedback is a signal from an externally controlled system that is reacting to or changed by the drive. **Figure 31-15** shows one example of this type of system. Most newer, smaller-application drives operate without direct motor feedback but have internal feedback through the use of voltage and current sensors within the drive itself. **Figure 31-16** shows an example of internal feedback where the VFD is sensing and controlling the speed of the motor.

TECH TIP
VSD Control

Various names are given to the aforementioned types of VSD control schemes, such as open-loop control (also called *volts per hertz control* or *scalar control*), closed-loop vector control (with either feedback from the motor or internal feedback from voltage and current sensors), open sensorless control, and sensorless vector control.

DC Drives

DC drives control the speed of a DC motor by controlling the voltage to the field winding of the motor (on the rotor) by two main methods. The first method of voltage control uses silicon-controlled rectifiers (SCRs) to control the applied voltage—sometimes referred to as *phase control*, **Figure 31-17**. By controlling when the

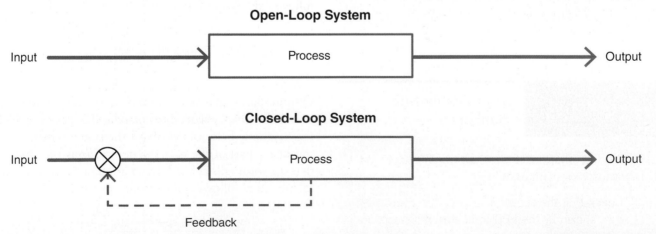

Figure 31-14. The open-loop control system has no built-in feedback. The operator can make manual adjustments. The closed-loop control system provides feedback, so the output changes based on the input set point and actual feedback.

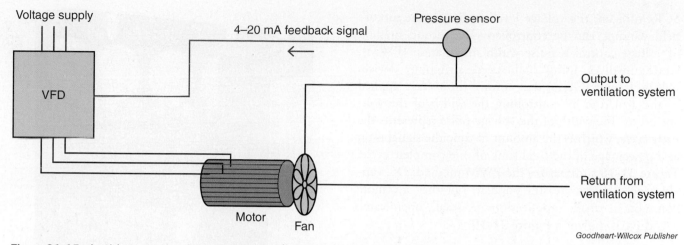

Goodheart-Willcox Publisher

Figure 31-15. In this example of external feedback, a pressure sensor senses duct pressure and sends a proportional signal back to the VFD. The VFD then changes the speed of the motor to maintain duct pressure. The motor speed is limited by the VFD's internal settings, and internal programming determines how the VFD reacts to the input signal.

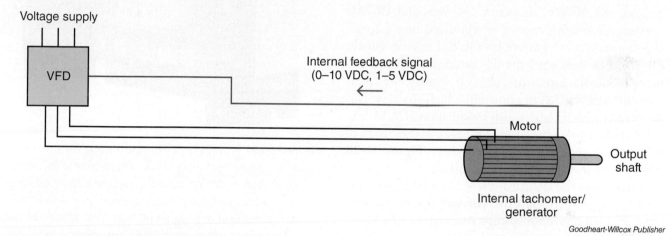

Goodheart-Willcox Publisher

Figure 31-16. In this example of internal feedback, motor rpm is controlled by the VFD by directly sensing motor speed (rpm).

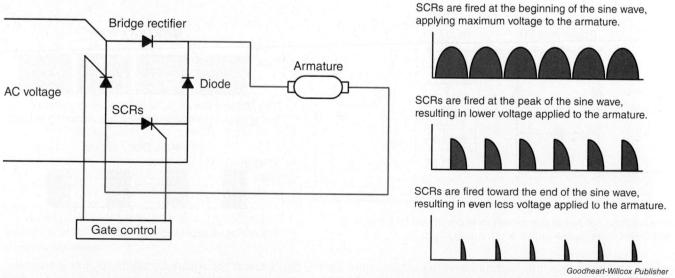

Goodheart-Willcox Publisher

Figure 31-17. This circuit demonstrates phase control of a DC motor. The firing angle of the SCRs is controlled by either a manual (via rheostat) or programmed input. The speed of the motor is changed by changing the applied voltage. DC drives may also contain capacitors to smooth the DC output from the bridge rectifier.

SCRs turn on, the voltage level applied to the motor's field winding can be controlled. The second method of voltage control is pulse width modulation (PWM), which involves using high-speed switching devices (MOSFETs or IGBTs) that control the voltage applied to the armature by controlling the width of the voltage pulse. The width of the voltage pulse represents the *duty cycle*, which is the amount of time the signal is on as a percentage of the total time of one complete cycle. **Figure 31-18** summarizes the PWM method. DC variable speed drives are still popular, but they are more common in smaller applications. A small, open-frame DC drive is shown in **Figure 31-19**.

AC Drives

AC variable frequency drives are composed of three basic sections: the AC/DC converter, DC bus, and DC/AC inverter. Alternating current is converted into a somewhat constant direct current by a bridge rectifier circuit. The DC bus then smooths the waveform through the use of inductors, capacitors, or both. Finally, the DC is inverted back into AC by controlling high-speed switching devices with a pulse width modulation (PWM) signal. The exact components used for each of these parts of the AC drive determine the type of drive. **Figure 31-20** shows the general layout of an AC VFD.

Aside from the standard PWM drive, the two other main types of drives are the *current source inverter (CSI)* and the *variable voltage inverter (VVI)*. While the current source inverter controls the current output,

Automation Direct

Figure 31-19. This inexpensive open-frame DC drive is suited for permanent-magnet DC, shunt-wound DC, and universal motors. Speed control is accomplished using a potentiometer input via a knob mounted on the front of the control cabinet or a signal of 4–20 mA. Minimum and maximum speed and current limits are adjustable.

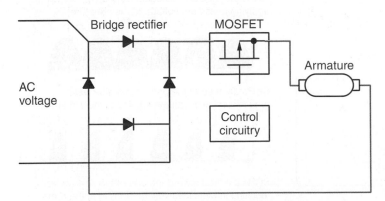

The voltage applied to the armature is controlled by controlling the width of the pulse (pulse width modulation).

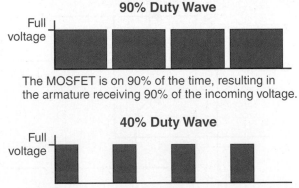

90% Duty Wave

The MOSFET is on 90% of the time, resulting in the armature receiving 90% of the incoming voltage.

40% Duty Wave

The MOSFET is on 40% of the time, resulting in the armature receiving 40% of the incoming voltage.

Goodheart-Willcox Publisher

Figure 31-18. Pulse width modulation is the act of controlling the duty cycle of the square waveform. By using another MOSFET in the circuit, the drive could be reversed. By using three more MOSFETs, the drive could run forward, in reverse, or with an overhauling load to act as an alternator. The control circuitry controls the gate of the MOSFET, and therefore the on time.

Layout of AC Variable Frequency Drive

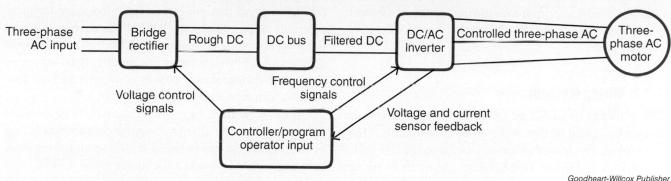

Goodheart-Willcox Publisher

Figure 31-20. This diagram shows the three main sections of an AC variable frequency drive.

the variable voltage inverter controls the voltage and frequency. The largest differences between these two types of drives are in the rectifier, also called the converter, and the DC bus, also called the DC link. Current source inverters (shown in **Figure 31-21**) control the rectifier section by using SCRs or other high-speed switching devices. Inductors in the DC bus circuit smooth the current. Variable voltage inverters (shown in **Figure 31-22**) typically use diodes as a bridge rectifier, and then both inductors and capacitors in the DC bus circuit to smooth the DC voltage applied to the inverter section. Smaller drives tend to use transistors in the inverter section.

Insulated gate bipolar transistors (IGBTs) are also used in the inverter section of drives. IGBTs tend to have a lower voltage drop and faster response time than the typical transistor used in VFDs. The IGBT is controlled by a PWM signal, just as is used in DC drives. A typical industrial IGBT module is pictured in **Figure 31-23**.

While older drives tend to be at a constant torque and maintain a linear volts/hertz ratio (except above

Variable Voltage Inverter

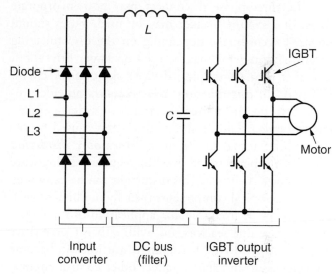

Goodheart-Willcox Publisher

Figure 31-22. In variable voltage inverters, diodes are used in the converter, while IGBTs are used in the inverter section.

Current Source Inverter

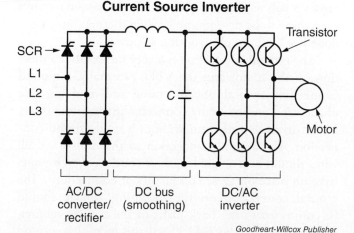

Goodheart-Willcox Publisher

Figure 31-21. This drive uses an SCR bridge rectifier to convert AC to DC.

Andrii Zhezhera/Shutterstock.com

Figure 31-23. This is a typical industrial IGBT module that might be seen in a VFD.

100% speed), newer drives with microprocessors and voltage and current output monitoring are able to control current and torque. This is called *vector control*, or *sensorless vector control*.

31.2.2 Wiring Methods

Wiring a VFD to an AC motor and relay control circuit involves line components and control components. Line components are those that are connected to the line voltage, such as in **Figure 31-24**, which shows three-phase voltage. Control components consist of any logic that starts and stops or controls the speed of the VFD. The control circuit may consist of only low-voltage components (24 VDC) or a combination of 120 VAC, 24 VDC, and proportional signals (0–10 VDC, 4–20 mA).

The three-phase disconnect may also incorporate fuses, or a circuit breaker may be used. Fuses should be sized accordingly, depending on the disconnecting means. A contactor may be used on larger drives, but not as a regular stopping device—the control circuit will perform this function. *Line reactors* provide protection from intermittent or transient voltage spikes, current spikes, and harmonic distortion caused by the drive and external to the drive. *Harmonic distortion* is a signal superimposed over the regular AC sine wave that distorts the wave. This distortion can be caused by external electrical components (usually nonlinear loads) or by the drive itself. Smaller applications do not typically require line reactors, but with either a large drive or many small drives, harmonic distortion can become a serious issue and start to affect other control circuits. Drives larger than 20 horsepower should use line reactors to reduce *total harmonic distortion (THD)*, which is the total, cumulative amount of distortion caused by harmonics in the electrical current.

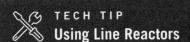

TECH TIP
Tackling Repeat Failures

If you are having repeated failures on the incoming power side in your drive, such as failed diodes, then you need to consider installing line reactors. Voltage line spikes can cause these failures, in addition to various trips by the VFD, such as overvoltage trips, for instance.

Filters such as RF filters and EMI filters are good investments when a system is installed in, around, or near instrumentation and possible interference. *Electromagnetic*

interference (EMI), also called *radio frequency interference (RFI)*, is a signal that can impact low-voltage controls and communication equipment. EMI affects circuits either by inducing a voltage/current on a nonattached conductor (induction), or by superimposing another signal on top of the voltage/current already present on the attached conductor (conduction).

Because of their fast switching, IGBTs cause voltage spikes that can damage regular noninverter-rated motors. Line reactors between the drive and motor can help to prevent damage to the motor from IGBTs.

TECH TIP
Using Line Reactors

Not all motors are inverter-duty rated. Inverter-rated motors have better insulation than noninverter motors, and some have better internal cooling capability. Placing an IGBT drive on a large older motor without load-side line reactors can lead to serious issues with the motor insulation, and eventual short circuits to ground. This is especially true when the wire run is greater than 75 feet. Line reactors should be used, and a voltage drop calculation should ensure the wire size is correct.

High Voltage

Line voltage wiring on VFDs is important, and all terminals should be properly torqued and checked on a regular basis. Thermal imaging can identify hot spots on line voltage terminals in order to correct a problem before it causes a failure. Physically checking the torque requires everything to be de-energized, while thermal imaging is done live, with no interruption of power.

The ground connection is a very important connection on the motor and the VFD. Normally, a ground connection on both the incoming and outgoing side of the drive are internally connected through the drive. Many drives will not run without a solid ground connection. The ground connection to the motor and all wires should be a single length of conductor. Wire nuts have no place on these conductors or connections. The ground connection on the incoming voltage should be connected to the back plane of the control cabinet, which is then connected to local and incoming ground conductors. Connections at the motor itself should match one of the following styles:

Line Components, Three-Phase Voltage, VFD Circuit

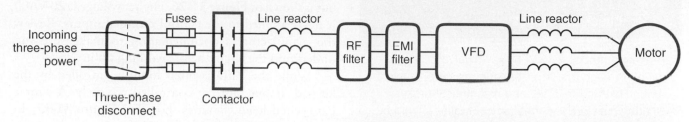

Figure 31-24. This diagram shows the important line components connected to the line voltage of a VFD circuit.

- Connectors on the supply side should be properly soldered and crimped, and they should be secured on a terminal connection on the motor side.

- Connections on both sides should be properly soldered and crimped, then bolted together.

- Connections should be properly bolted using split-bolt connectors.

- Newer connection methods that meet applicable standards can be used.

Any of the above methods should be followed by a proper application of self-fusing splicing tape (two coats), and then two coats of electrical tape.

VFDs generate high-frequency shaft currents that can cause damage to motor bearings. A grounding ring, either on the drive side or both sides of the motor, allows for the discharge of these VFD-induced currents, thus protecting the motor bearings and preventing the shaft current from passing down the line to other components.

Metal oxide varistors (MOVs) are attached on the incoming power side of most VFDs. Three MOVs are wired phase to phase across the incoming power. These small, usually circular components (**Figure 31-25**) protect against voltage surges. As applied voltage increases above a clamping voltage, current is allowed to flow. In this way, the drive is protected by eventually sacrificing the MOV.

MOVs can fail either open or shorted, depending on the mode of failure. A common cause of failure can occur when a MOV is repeatedly stressed over time by multiple voltage surges and spikes, eventually failing in a shorted condition. The outward appearance may seem normal, but this shorted condition may cause fuse failures, circuit breaker trips, or drive errors on start-up. The second common cause of failure is a massive overload when a single spike occurs, such as with a lightning strike. A surge that is beyond the capacity of the MOV may burn or completely destroy the MOV. Failed MOVs should be replaced.

> **SAFETY NOTE**
> **Proper Clearance**
>
> Mounting drives with the proper amount of clearance around them is important for heat dissipation. Ensure that drives mounted inside panels and multiple drives mounted close together meet manufacturer recommendations for clearance. Without proper clearance, heat will build up and may cause early drive failure and damage.

Drives commonly provide several protective features with regard to line voltage and motor current. Typical features involve motor overload protection on high current, similar to how motor starter overloads protect the motor. The time to trip depends on the amount of overcurrent. Lower overload current will increase trip time. Overvoltage and undervoltage protection are also common features. Motor protection current should be set by referencing the manufacturer's documentation.

Figure 31-25. One common style of metal oxide varistor (MOV) is pictured here.

Drive line connections are commonly labeled as *L1*, *L2*, and *L3* for incoming power (also sometimes labeled as *R*, *S*, and *T*), and *T1*, *T2*, and *T3* for motor connections (also sometimes labeled as *U*, *V*, and *W*).

Low Voltage

The low-voltage wiring of a VFD is comprised of the control circuit. This circuit may be made of hard-wired logic, PLC-programmed logic, proportional signals, or a combination of all three. A typical control terminal layout is shown in **Figure 31-26**. Higher voltage (120 VAC) is never applied to the control terminal inputs. Review the manufacturer's documentation to make sure you understand the circuit before working on it.

While the VFD is often locally controlled by the keypad, it may also be controlled remotely. A simple hard-wired logic circuit is shown in **Figure 31-27**. In this circuit, two-pole relays are used. One pole of the relay is reserved for the low-voltage VFD circuit. This circuit energizes relays to provide for input to the VFD. The VFD has internal logic that will not allow shorting to take place between phases.

Analog input control signals may come from a controller, instrumentation, or a PLC analog output. If from another controller or instrumentation, the shield of the shielded twisted pair wire should be grounded at the VFD terminal. If the signal is from a PLC analog output, the shield should be grounded at the PLC. Analog output signals should be grounded at the VFD. Other guidelines include:

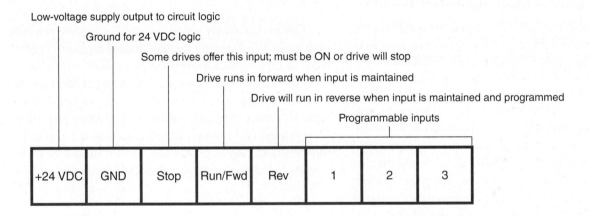

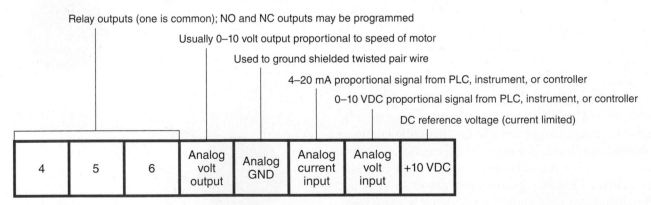

Figure 31-26. Even though control terminals are low-voltage DC, never work on them "live" or "hot," because you may damage internal circuitry or logic.

- Do not run analog signal wires in the same conduit as the power wiring—run a separate conduit.

- Use shielded twisted pair wire, or shielded and armored wire.

- Ground the shield at one end only—preferably the drive, controller, or PLC, and not the instrument that is transmitting. This will prevent a loop effect and will also make it easy to check grounds, rather than having to check every instrument location.

- If analog and power lines are running parallel, keep a separation distance as far as possible—at least 16″.

- If power lines and analog signal lines must cross, cross them at 90° and with as much separation as possible.

- Within control cabinets, try to separate signal lines from power lines. Maintain at least 6″ clearance if possible, and run lines in different trays.

- Always refer to the manufacturer's recommendations, and perform regular running tests to check for proper VFD response to a 0%, 50%, and 100% signal. Interference can induce current/voltage on top of the signal. The first symptom of this may be a drive not properly responding to an analog signal.

Drives normally include several programmable inputs and outputs. These programmable I/O (input/output) operations may be used for alarm condition reporting, overload trip alarms, drive reset functions, multiple speed functions, jogging, and other functions. Refer to your specific drive manual and program.

Alarms can be programmed to alert an operator or technician, usually via LED, that a condition needs to be addressed to avoid a VFD fault. The drive will function in the meantime, and the alarm will be cleared once the cause is addressed. Other built-in alarm conditions will prevent a VFD from operating until the cause is corrected.

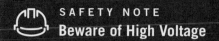

SAFETY NOTE
Beware of High Voltage

When powered, high voltage is applied to the drive even if it is not running. Internal capacitors can maintain a charge for several minutes after the drive has been de-energized and the power supply locked out. Refer to your drive's manual for safety hazards.

TECH TIP
Static Charge

Even if working on a de-energized drive, you may damage components if you are carrying a static charge and discharge it through the drive or onto new circuit boards. This can be avoided by wearing a wrist grounding strap (shown in **Figure 31-28**) while handling drive components.

Goodheart-Willcox Publisher

Figure 31-27. Drives can be easily controlled with standard hard-wired logic. Note that the 120 VAC is only used as an example, and this 120 VAC is never applied to the control terminals at the VFD. Only the internally generated low-voltage DC is applied back to the VFD control terminals. The 120-VAC circuit could easily be replaced with a 24-VDC circuit.

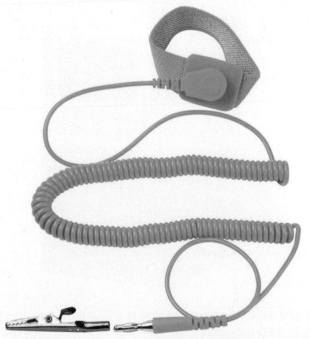

Eclipse Tools, ESD Wrist Strap #900-012

Figure 31-28. A grounding strap will ensure that you do not damage circuit boards with static discharges.

31.2.3 Programming and Settings

While a VFD can be programmed using the interface panel of a human-machine interface (HMI), many drives can be programmed using the manufacturer's software and the program downloaded to the drive. If you have multiple drives from the same manufacturer, the drive programming software is a good investment. Replacing an older drive with a new and different drive is always possible, but careful study of the old drive, its parameters, and its controls should first be done as an engineering function.

PROCEDURE	Programming VFD Parameters

Many parameters can be adjusted or changed while a VFD is operating, but many others can only be changed while the drive is stopped. Typical onboard HMI, **Figure 31-29**, allows the programmer to:

1. Change the drive to programming mode.
2. Page through the multitude of parameters.
3. Select a parameter to change.
4. Change the value of the parameter.
5. Enter the value.
6. Return the drive to operational mode.

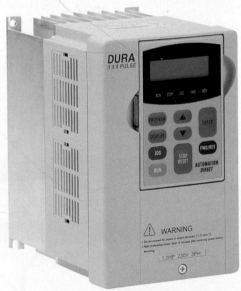

Automation Direct

Figure 31-29. The onboard HMI of a GS3 drive is pictured here. LED indicators show when the drive is receiving the signal to perform the indicated function. The local keypad can be detached as a remote control, and the buttons are used for viewing and changing parameters.

Typical adjustable parameters for a VFD are shown in **Figure 31-30**, although more may be available, depending on the specific drive. Exact parameter settings depend on the application. Multispeed inputs are also common, which allow multiple speeds to be preset if the drive is not running on an analog input signal. These digital inputs (on or off) determine which speed setting is used. A common terminal arrangement is to have three terminals for this function, **Figure 31-31**.

Analog inputs are also commonly used to control the speed of the drive. Analog inputs are typically 0–10 V or 4–20 mA. The input signal—along with parameter settings—determines the output signal. The relationship between the input (command) and speed output is a proportional relationship, as shown in **Figure 31-32**. The beginning and end points of the graph can be controlled through parameter settings. Proportional-integral-derivative (PID) settings are also common, which control the output response to the measured system variable. PID settings are discussed in Chapter 34, *Industrial Process Control*.

31.2.4 Preventive Maintenance

VFD maintenance items most manufacturers recommend pursuing on a consistent schedule include the following:

- Overall inspection, including cleanliness, air filters, and the surrounding environment.
- Inspection of incoming power for proper voltage and phase balance.
- Inspection of all terminations for proper torque and to be sure there is no apparent insulation breakdown, loosening due to vibration, or corrosion because of the environment.
- Inspection of drive internals to be sure there is no apparent damage, burning, smells, corrosion, dust, or overheating, and that proper grounding is maintained and cooling fans are operating properly.
- Inspection of displays to be sure the onboard HMI has no damage and is legible, indicating LEDs are working, and drive internal LEDs are operating properly.

Excessive ambient temperature is one of the major conditions that causes drive failures. The drive itself creates heat and disperses it into the surrounding environment using heat sinks and fans. Anything affecting this transfer of heat causes the drive to run at a higher temperature and will eventually result in failure. An increase in ambient temperature, a blockage of airflow, accumulated dust on the heat sink, or a fan not properly working can all cause early failure. In addition, capacitors will have a shortened life if operating in hotter ambient conditions.

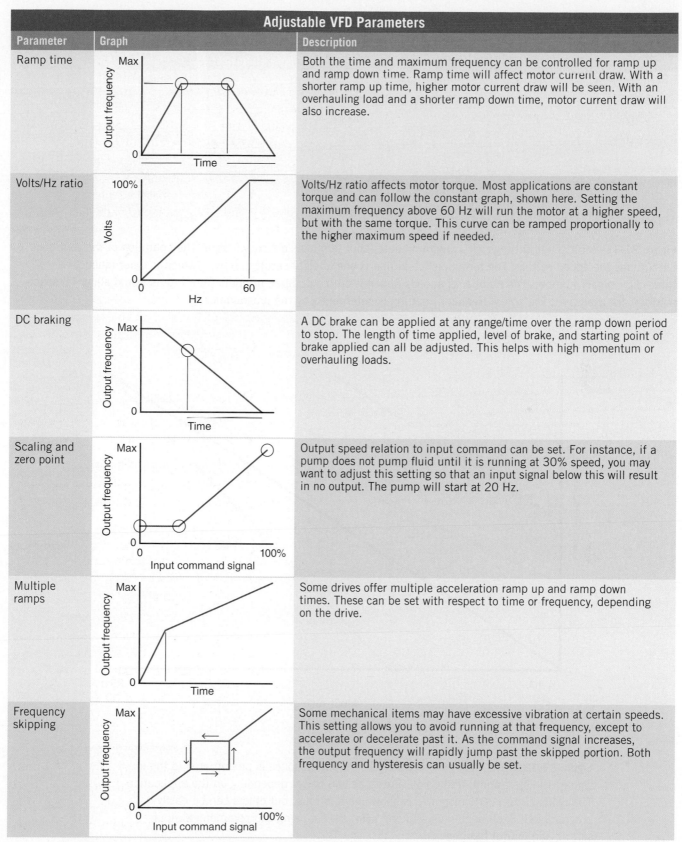

Adjustable VFD Parameters

Parameter	Graph	Description
Ramp time		Both the time and maximum frequency can be controlled for ramp up and ramp down time. Ramp time will affect motor current draw. With a shorter ramp up time, higher motor current draw will be seen. With an overhauling load and a shorter ramp down time, motor current draw will also increase.
Volts/Hz ratio		Volts/Hz ratio affects motor torque. Most applications are constant torque and can follow the constant graph, shown here. Setting the maximum frequency above 60 Hz will run the motor at a higher speed, but with the same torque. This curve can be ramped proportionally to the higher maximum speed if needed.
DC braking		A DC brake can be applied at any range/time over the ramp down period to stop. The length of time applied, level of brake, and starting point of brake applied can all be adjusted. This helps with high momentum or overhauling loads.
Scaling and zero point		Output speed relation to input command can be set. For instance, if a pump does not pump fluid until it is running at 30% speed, you may want to adjust this setting so that an input signal below this will result in no output. The pump will start at 20 Hz.
Multiple ramps		Some drives offer multiple acceleration ramp up and ramp down times. These can be set with respect to time or frequency, depending on the drive.
Frequency skipping		Some mechanical items may have excessive vibration at certain speeds. This setting allows you to avoid running at that frequency, except to accelerate or decelerate past it. As the command signal increases, the output frequency will rapidly jump past the skipped portion. Both frequency and hysteresis can usually be set.

Goodheart-Willcox Publisher

Figure 31-30. These are six common control functions of a VFD, along with a description of each.

Conditions with Three Terminal Inputs (0 = off, 1 = on)			Three-Terminal Binary and Base 10 Results		Three-Terminal Speed Settings	
Terminal 1	Terminal 2	Terminal 3	Binary Number	Base 10 Number	Speed	Setting (Hz)
0	0	0	000	0	Speed 0	0
0	0	1	001	1	Speed 1	10
0	1	0	010	2	Speed 2	20
0	1	1	011	3	Speed 3	30
1	0	0	100	4	Speed 4	40
1	0	1	101	5	Speed 5	50
1	1	0	110	6	Speed 6	60
1	1	1	111	7	Speed 7	90

Goodheart-Willcox Publisher

Figure 31-31. Multispeed settings with programmable input terminals are shown here. The condition of the terminals determines the speed setting. The terminals' condition is one of eight settings, 0 to 7, which is determined by a three-digit binary number. The number of speeds depends on the number of inputs (four inputs would allow 16 speed settings, for example). The actual frequency setting is determined by the programmer.

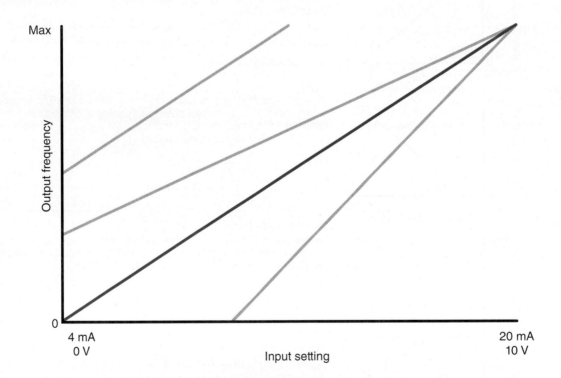

Goodheart-Willcox Publisher

Figure 31-32. With analog commands, the output response is proportional to the input signal. The start point, end point, and slope can be set depending on the application. The typical 0%–100% response is shown here in blue, but others can be easily programmed. Each beginning and end point are set using two parameters: input setting and output frequency.

Variable Frequency Drives (VFDs)

VFDs provide continuous motor control and allow for the fine-tuning of processes. This adjustability helps reduce the stress on the system and its parts, thereby extending their life and reducing repair costs. Reductions in motor speed, where applicable, can reduce energy usage and costs.

31.2.5 Troubleshooting

The four things that cause the most malfunctions with VFDs are:

- Issues in the environment.
- The driven load changing.
- Settings too restrictive for the load.
- A problem with the control circuit.

In the event of a failure, if the drive has been operating properly for weeks or months, something recently changed and caused the drive to fail. This root cause must be corrected along with the drive failure. The following are some troubleshooting scenarios you might encounter with VFDs.

Scenario 1: A plant expansion has taken place and has resulted in a drop in the voltage on L1, unbalancing the phase voltages. The VFD fails, and a technician determines that diodes have failed on the converter. In addition, the voltage supply board, which supplies all of the internal control voltages, has failed. A technician replaces all of the diodes and the voltage supply board, and then re-energizes the drive. Just a few days later, the drive fails again. More diodes have failed.

Correction: An imbalance in the phase voltages is causing the failures. Rebalance the phase voltages to address the true cause of failure.

Scenario 2: A contractor is boring a hole through the concrete foundation to provide for a drain line and accidentally cuts into a conduit. This grounds out one of the conductors on the control circuit for a VFD. The VFD refuses to show a "run" signal when the remote start button is pressed. In addition, a technician finds that the voltage supply board output fuse is blown. The fuse is replaced, but opens again. The technician carelessly replaces the output fuse with a larger fuse, which also opens. The technician then dangerously places a 12-gage wire across the fuse holder and re-energizes the drive. This unsafe practice results in the voltage supply board catching on fire.

Correction: Disconnect and use a megohmmeter to check the control wiring. Re-run the control conduit and wiring to the remote start-stop station. Replace the voltage supply board, and educate the technician in order to prevent future errors and possible electrical fires.

Scenario 3: A drive keeps tripping. The error code says that the cooling fan is not running. The cooling fan is replaced, and the drive starts. A month later, the same drive trips, and it is found that capacitors have short-circuited. The capacitors are replaced, and the drive runs. A month later, the drive trips again, and it is found that several IGBTs have failed. The IGBT package is replaced, and the drive runs. The supervisor finally investigates the problem and notices that the motor control center (MCC) room is very hot. While the air-conditioning vent is putting out cool air, the airflow is much lower than it used to be. Further investigation reveals that the filters on the air-handling unit for the building have not been changed in a decade and are completely laden with particulate and mold. The HVAC technician on staff did not replace the filters regularly because, at a company meeting, the head of maintenance had complained about preventive maintenance costs being too high.

Correction: Retrain the maintenance manager to more thoroughly and accurately consider the costs of preventive maintenance versus those of corrective maintenance in order to avoid future problems and higher corrective costs.

PROCEDURE Troubleshooting a VFD

If a drive trips (fails) and an error code is retrieved, look up the error code in the drive manual. Beyond that, the following step approach to troubleshooting a VFD is recommended:

1. Examine the environment around the drive for heat, moisture, and other potential issues.
2. Examine the input voltage source to the drive for proper voltage.
3. Check the drive trip history, and write down any displayed codes.
4. De-energize the drive, and lock out and tag out the power.
5. Open the drive and visually inspect the internals for any signs of damage.

Other Failures

Common failures of drive components usually take place in the power section (diodes, SCRs, IGBTs, and capacitors). These components fail due to excessive heat and current. While they can be replaced, the cause of the failure needs to be eliminated. See **Figure 31-33**.

SAFETY NOTE
VFD Capacitor Safety

Capacitors used in VFDs are much larger than run-start capacitors used on smaller motors. Use extreme care when working around these capacitors. On newer drives, a bleed-off resistor is normally installed across the capacitor leads for safety purposes. A *bleed-off resistor* drains the capacitor within a few minutes of it being de-energized. On older drives, a capacitor may have been replaced and the bleed-off resistor not put back on, or the bleed-off resistor may have been removed because a technician thought it was the source of a short. Be very careful and visually inspect large capacitors for this resistor.

Failures on the CPU board, voltage supply board, or I/O board can occur from the following:

- Lightning strikes without proper grounding.
- Voltage spikes.
- EMI or RFI.
- Grounds due to conductive dust on the board.
- Grounds on control circuit.
- Jumping an improper voltage to a control terminal.
- Static discharge.

VFD Trips and Possible Causes	
Type of Trip	Possible Causes
Overcurrent at any time (acceleration, deceleration)	Mechanical load on the motor has increased. Check driven load.
Overload trip	Load on motor has increased. Check driven load.
Undervoltage trip	Check incoming line voltage and converter section.
Overvoltage trip	Usually on shut down. Motor generates too much voltage because of an overhauling load. Check ramp settings and increase ramp down times.
Memory or CPU trip	Due to heat, electrical interference, noise. Check ambient temperature conditions.
Safety trip	External safety circuits trip or internal high temperature trip, caused by lack of cooling within the drive, lack of airflow, open door switches (larger drives).

Goodheart-Willcox Publisher

Figure 31-33. This table shows types of trips and possible causes. You should review your drive manual and become familiar with error codes and trips prior to having a failure.

Larger drives have multiple smaller fuses. These fuses may be located on the boards themselves, on control transformers, or on the terminal strips. The drive should be carefully inspected when installed, so that you know where all the fuses are located and can keep spares in inventory.

CHAPTER WRAP-UP

A serious study of either sensors or VFDs could be a course in itself. At this point, a technician should understand the basic theories and applications of sensors and VFDs, as well as applicable maintenance and troubleshooting ideas and techniques. Correcting the root cause of a problem will limit the occurrence of the same problem in the future. Carefully consider and study everything impacting a system to ensure its maximum reliability.

MOLPIX/Shutterstock.com

Assembly lines rely on sensors for many precise automated operations.

Chapter Review

SUMMARY

- Major categories of sensors include Hall effect, proximity, photo, temperature, and pressure.
- Analog sensors often require signal conditioning.
- The most commonly used sensor is the three-wire DC sensor.
- Three-wire DC sensors come in PNP and NPN. The appropriate sensor must be used depending on the load.
- Hall effect sensors are typically used in smaller, high-speed applications.
- The types of proximity sensors are inductive, capacitive, ultrasonic, magnetic, and laser curtain.
- Inductive sensors detect metallic objects.
- Capacitive sensors work by detecting a change in capacitance because of an object in the field of detection.
- Ultrasonic sensors use a high-frequency sound to bounce a signal off an object and measure the time it takes for the echo to return.
- Magnetic sensors use a magnetic field to sense objects at a distance of usually less than 3″. The magnetic field can penetrate nonmagnetic or non-metallic materials, so the sensor can detect a magnet on the other side.
- Laser curtains are typically used with safety systems and exclusion areas. Work on these systems requires extra training through the manufacturer.
- Photo sensors use modulated light so that visible light does not affect sensor reliability.
- Fiber-optic sensors are photo sensors that use a fiber-optic conductor to reach a remote area.
- Temperature sensors commonly used in industry today include thermistors, resistance temperature detectors (RTDs), and thermocouples.
- Piezoelectric pressure sensors use the piezoelectric effect, which is the ability of a material to generate electricity in response to applied pressure.
- While any motor can be driven by a VFD, not all motors are rated for it.

- A closed-loop feedback system adjusts output according to a comparison between a proportional input signal that reflects current state and a command or set point.
- Pulse width modulation (PWM) is the action of varying the duty cycle of a square waveform.
- The main parts of an AC drive are the rectifier or converter, the DC bus or DC link, and the inverter.
- Harmonic distortion, electromagnetic interference, and radio frequency interference are all problems that need to be considered for drive reliability.
- Become familiar with the drives installed at your facility prior to a drive failure.
- Preventive maintenance is important to ensure the proper reliability of a VFD.
- Troubleshooting a VFD failure includes finding the root cause and correcting it.

REVIEW QUESTIONS

Answer the following questions using the information provided in this chapter.

1. Name the broad categories of sensors and explain how they function.
2. *True or False?* Proximity sensors are usually three-wire, DC-voltage sensors.
3. _____ sensors are best used to detect repeated movements of ferrous objects at close to very close distances.
4. What is the purpose of unshielded capacitor sensors?
5. *True or False?* A negative slope causes the output to be lowest when the detected object is closest to the sensor.
6. At or around what three places are laser curtains used?
7. What is the difference between blanking and muting?
8. Magnetic sensors use a _____ to measure unknown electrical resistance on one of its sides.

9. *True or False?* Photo sensors use modulated light, since most of them use LEDs for the emitter.

10. Describe how to properly adjust a photo sensor.

11. In _____ mode, transmitted light from a fiber-optic sensor travels to a separate receiver and an object is detected when it interrupts that beam.

12. *True or False?* Nonsafety laser sensor outputs can only be PNP, not NPN.

13. Lasers are classified by their _____ into Class 1, Class 2, Class 2M, Class 3R, Class 3B, and Class 4.

14. Calculate the synchronous speed of an AC motor with four poles and a frequency of 50 Hz.

15. Explain the difference between an open-loop and closed-loop system of control.

16. _____ is a method of voltage control that uses SCRs to control the applied voltage of DC drives.

17. Describe how an alternating current VFD works.

18. *True or False?* Insulated gate bipolar transistors (IGBTs) tend to have a lower voltage drop than the typical transistor used in variable frequency drives.

19. _____ provide protection from intermittent voltage spikes, current spikes, and harmonic distortion caused by the drive and external to the drive.

20. Depending on the mode of failure, an MOV can fail either _____ or shorted.

21. What are two environmental causes of drive failure?

22. List the four factors that produce the most malfunctions with VFDs.

![NIMS] **NIMS CREDENTIALING PREPARATION QUESTIONS**

The following questions will help you prepare for the NIMS Industrial Technology Maintenance Level 1 Electronic Control Systems credentialing exam.

1. A(n) _____ sensor is most widely used in electronic circuits and devices.
   A. Hall effect
   B. inductive
   C. photo
   D. ultrasonic

2. The lasers in sensors that utilize lasers typically fall into Class _____ or _____.
   A. 1; 4
   B. 3R; 3B
   C. 2; 4
   D. 1; 2

3. _____ refers to a photo sensor that provides action when sufficient light is detected.
   A. Unmodulated light
   B. Light operate
   C. Modulated light
   D. Dark operate

4. Output shaft rpm is determined by synchronous speed and what other variable?
   A. Frequency
   B. Number of stator poles
   C. Slip
   D. Current

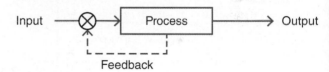

5. What is pictured in the image above?
   A. Current source inverter
   B. Open-loop system
   C. Closed-loop system
   D. Variable voltage inverter

6. When mounting drives, heat will build up and potentially cause early drive failure if there is not proper _____.
   A. clamping voltage
   B. shielding
   C. current flow
   D. clearance

32 | Programmable Logic Controllers

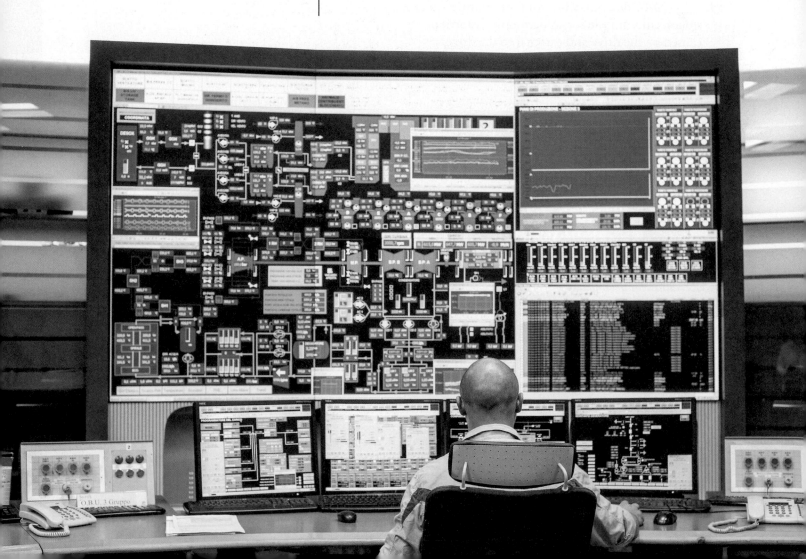

LEARNING OBJECTIVES

☐ Describe the main elements of a PLC.

☐ Discuss safety concerns with forced conditions on a PLC-controlled system.

☐ Explain the difference between a sinking and sourcing input module.

☐ Describe how grouped commons are wired on relay output modules.

☐ Discuss the purpose of a safety relay/contactor.

☐ Describe troubleshooting a suspected faulty input module.

☐ Describe troubleshooting a suspected faulty output module.

☐ List possible failures that would prevent an output from activating with interposing relays.

TECHNICAL TERMS

bit	off-delay timer (TOF)
discrete device	on-delay timer (TON)
EEPROM (electrically erasable programmable read-only memory)	optocoupler
	processor
	programmable logic controller (PLC)
force instruction	
grouped commons	relay output module
harmonics	resolution
input image table	retentive timer (RTO)
interference	shielded twisted pair wire
interposing relay	sinking module
isolation transformer	sourcing module
master control relay (MCR)	

U p to this point, you have studied hard-wired logic created by the order of wiring decision trees. This method of creating all but the simplest of logic is expensive and difficult to troubleshoot. The **_programmable logic controller (PLC)_** is a digital microprocessing device used to control the inputs and outputs of industrial systems. PLCs replace hard-wired logic by providing logic through programming. To this day, programming still resembles ladder logic.

Today's PLCs are capable of many inputs and outputs, advanced mathematics, high-speed manufacturing, motion control, and process control. The ability to understand basic programming and troubleshoot a PLC-controlled system is a requirement for all technicians starting in this field. In this chapter, we will examine basic programming, wiring, and troubleshooting of PLC-controlled systems, with an emphasis on the wiring methods and troubleshooting.

32.1 PLC COMPONENTS AND WIRING

The processor, I/O (input/output), and power supply are the major components of a PLC-controlled system. While an entire book could be written on the subject of PLCs, we will limit our examination to common components, with a focus on maintenance and troubleshooting. While there are several different types of PLCs, those that are most commonly used follow a predictable cycle:

1. Examine input conditions.
2. Check the input conditions against the programmed logic.
3. Change the output conditions as required by the programmed logic.
4. Perform more advanced functions (timers, counters, math).
5. Communication.

This cycle takes milliseconds in most programs and repeats over and over again.

The main components of a PLC-controlled system are shown in **Figure 32-1**. External devices—also called **discrete devices** or *field devices*—provide an input signal (voltage) to the input module. The input module converts these input signals into a binary representation (0 or 1) that reflects all of the input conditions. This record of the input conditions is called an **input image table** and is held in a specific section of the processor's memory.

The **processor** runs the operating system, performs programmed functions, and communicates with the human-machine interface (HMI) and remote input/output (I/O). The HMI is the software and graphic interface that allows interaction between the user and the machine. The processor also communicates with other peripherals, such as PCs and printers. The processor is programmed via built-in HMI or by using an appropriate cable to plug in a device such as a laptop, which can then be used to program the PLC through the corresponding PLC software. When running the user program, the processor compares the input conditions to the programmed logic to determine what output conditions should take place. The output modules then convert the low-level binary signal from the processor to a usable output signal.

The power supply converts incoming line voltage to a low-level DC voltage (usually 24 VDC) to run the PLC. On some models, the DC power supply is large enough to drive inputs and outputs. On other models, an external power supply is used. **Figure 32-2** shows a typical expandable PLC that incorporates all of the common elements of a smaller PLC. Other layouts may include remote inputs/outputs, multiple racks, or smaller, limited I/O.

In some instances, a relay breakout board (shown in **Figure 32-3**) may be used for outputs. This output strategy inserts small relays, called interposing relays, between the PLC output module and discrete output devices. An **interposing relay** is used to separate two different devices or circuits, such as in cases where each connected device requires operation at a different voltage. This setup reduces the output current that is imposed on the PLC output module—it only requires the current needed to energize a small relay. The interposing relay energizes the larger discrete device, such as a solenoid or motor starter. The output wiring may also include fuses for protection.

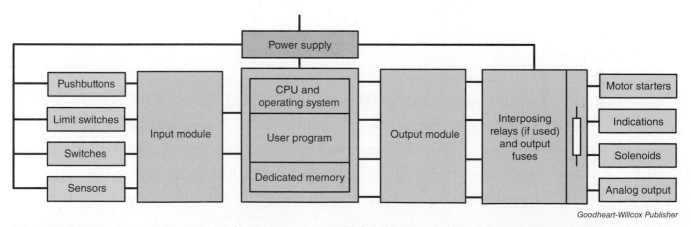

Goodheart-Willcox Publisher

Figure 32-1. This diagram shows the main components in a PLC-controlled system. External logic is limited in a PLC-controlled system.

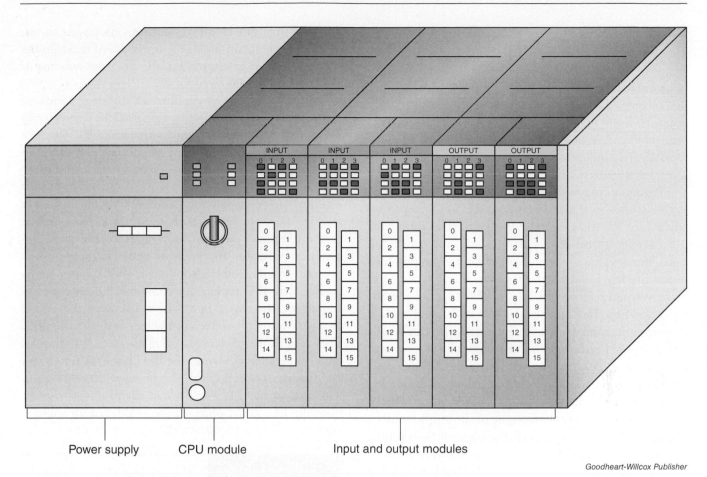

Goodheart-Willcox Publisher

Figure 32-2. The common components and module and terminal numbering of a PLC are shown here. The power supply section has a fuse and terminal connections for incoming AC. The CPU module is keyed for program/run mode and has communication ports for programming, uploading/downloading, and data transfer. The I/O modules have terminal connections and indication lights.

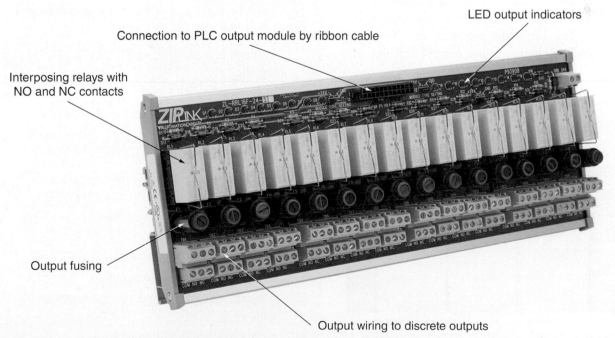

Automation Direct

Figure 32-3. This relay breakout board is set up for mounting on a DIN rail.

32.1.1 Processor

The processor of the PLC is also known as the CPU, which stands for central processing unit. There are a wide array of processors, from much older 8-bit processors with limited memory and I/O expansion, to much newer and significantly more advanced processors. **Figure 32-4** shows a processor module used in an expandable PLC. Processors typically have indications for power, battery, communications, force enabled, force, and a general fault.

Processors may also be key controlled, which means a key switch places the processor into RUN, PROGRAMMING, or REMOTE mode. While in RUN mode, the processor performs its regular duties with no limitations, but it cannot be remotely or locally programmed unless the key is switched. However, the PLC may still be monitored remotely. While in REMOTE mode, the processor function can be remotely controlled though a communications network. This allows the PLC mode to be changed remotely, and therefore allows programming remotely. Placing the switch into the PROGRAMMING position allows a locally attached programming device to make changes and to upload and download from the PLC. In this mode, none of the outputs will be changed. PLCs that are stand-alone systems—meaning they are not connected to a communications network—may be placed in RUN mode and have the key removed for safety reasons and to prevent tampering.

Some processors have a battery that is accessed from the exterior, while other PLC processor modules must be removed to replace the battery, which is mounted on the CPU card. Follow manufacturer recommendations on how often to replace the battery. In addition to the battery, a capacitor internally mounted on the circuit board also holds a charge. Some control circuits also have multiple power sources, to provide power to the PLC when the main breaker is open. All of these power sources are used to ensure that the program remains in the processor memory.

While there are a number of different kinds of memory, the most typical is *EEPROM (electrically erasable programmable read-only memory)*. This is a memory chip that has a copy of the original program and is added to the processor. When the PLC is powered, the EEPROM loads the copy of the program into working memory, and then runs the system from the working-memory program. If future changes are made to the program, they are only made to the working-memory program. In order to make changes to the original program located on the EEPROM, the engineering firm that programmed the machine is typically contacted to make a new EEPROM with the added changes, although other contractors can also do this. With EEPROM, if changes are made to the program—even timing changes—and the PLC power is turned off and the battery discharged, the changes to the program will be lost. The PLC will reload the original program from the EEPROM once it is powered, but all changes are lost.

PROCEDURE	Changing PLC Processor Batteries

Because changes to a program can be lost when using EEPROM if the power goes out, always perform the following steps when changing processor batteries in a PLC:

1. Download and save the working-memory program.
2. Remove power.
3. Take precautions against static discharge.
4. Check to ensure that there are no other power sources to the PLC.
5. Remove the CPU module.
6. Replace the battery.
7. Carefully install the CPU module.
8. Apply power.
9. Check for proper operation, or go online with the PLC and check the program.
10. If needed, reload the saved program into memory.
11. Dispose of the old battery properly.

11qq22/Shutterstock.com

Figure 32-4. A processor module in an expandable PLC is pictured here. Processors vary slightly by manufacturer.

PLC processor modules have LED indicators on the front, including a fault indicator. A fault indication on the processor module may be a fault condition that stops the processor. The processor may also have indication for a minor fault, which will allow the processor to continue to run. A major fault may be caused by improper programming, a program getting hung up on a repeated cycle in a subroutine, a problem with the processor memory, or a program taking too long to cycle. While some faults may be cleared by cycling power and therefore reloading the program, others must be cleared by establishing communications with the PLC and resetting the fault through software. Some PLCs have the option of resetting by external means, such as through a specific input programmed for that function, and this should be discussed with engineering.

A *force instruction* is an action performed by a programmer on a PLC to turn inputs on or off for the purposes of testing or troubleshooting. A force instruction will cause the PLC to turn on an input or an output internally, which results in an output action by the output module. Most PLCs require a two-step process to perform a force function. First, a force condition is enabled by use of software. Second, the force instruction is placed. This will result in both LEDs being on, if both are present. Only when the force instruction and condition are cleared will the indicating LEDs turn off.

SAFETY NOTE
Safety with Force Instructions

Technicians should be aware of any force instructions placed on a machine, or it may not act as expected. The programmer must have complete knowledge of how the system operates before applying force instructions. Force conditions could easily cause damage to personnel or machines if proper care is not taken. All technicians in the area should be notified of any applied force instructions.

32.1.2 Input and Output Modules

Input and output, or I/O, modules are another key component of a PLC system. **Figure 32-5** shows examples of input and output modules wired in a PLC system. PLC input modules can be digital or analog. Digital modules have two states, on and off, whereas analog modules have varying states to measure changing variables, such as temperature or pressure. Specialty input modules are available for systems requiring control of timing, temperature, pressure, and position.

Xmentoys/Shutterstock.com (top); EZAutomation (bottom)

Figure 32-5. Examples of PLC I/O modules are pictured here.

Figure 32-6 shows the circuit details of a simplified isolated input module. The use of a semiconductor device called an *optocoupler*—also called an *optoisolator*—to connect the signal to the CPU creates isolation, which prevents external higher voltage from damaging internal components. The higher incoming voltage is rectified (in the case of an AC input) and then drives the LED side of the optocoupler. The LED and phototransistor are housed in one small unit. As the LED turns on, the phototransistor conducts and provides the lower-level voltage input. While a particular input point could still be damaged, the damage will not extend beyond that input module. Each input point, or terminal, would have this circuit.

Output modules come in just as wide a variety as input modules. The most common modules have specific wiring requirements and methods. Output modules may be isolated just like input modules. Nonisolated output modules must always have fuse protection installed on the load side. In addition, output interposing relays may be used (shown in **Figure 32-3**), with integral fusing and both normally open (NO) and normally closed (NC) contacts.

TECH TIP
Input Voltage

Applying the wrong voltage level to the input terminal can burn out that input point. If the input module is not an isolated module, further damage can result. Always make sure of proper voltage levels and polarity before applying power to a terminal.

DC and AC Digital Input Modules

DC input modules are coordinated with the input sending the signal. Only the properly matched and wired input can be wired to the input module directly. Digital modules are available in sourcing and sinking options. With a *sourcing module*, current flows out of the PLC port. With a *sinking module*, current flows into the PLC port. **Figure 32-7** summarizes the ideas of sinking and sourcing input modules. In the United States, sinking input modules are typically used, but this may depend on where a machine was designed and wired.

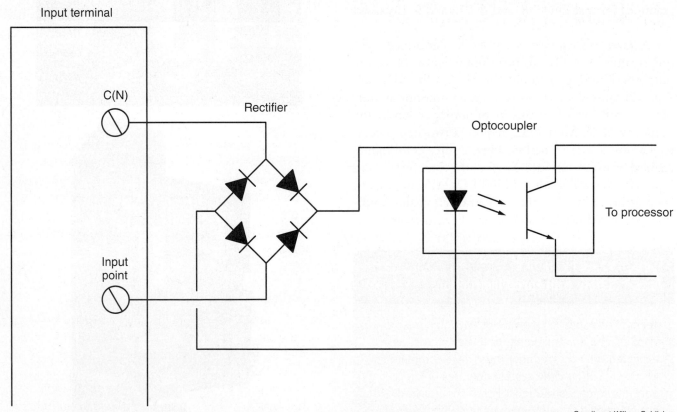

Goodheart-Willcox Publisher

Figure 32-6. Isolated input cards are slightly more expensive, but they protect the processor.

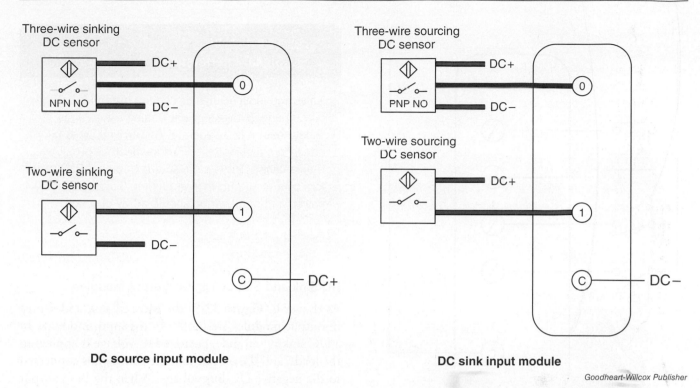

DC source input module

DC sink input module

Figure 32-7. An understanding of sinking and sourcing is required to be able to troubleshoot a PLC-controlled system.

For a three-wire NPN sensor, the sensor is wired to the sourcing input module. When the sensor detects, the current flows from the PLC DC+ out to the sensor output and then through the sensor's blue wire to the DC– side of the DC power supply. The indicating light on the PLC for this input will turn on only when the sensor is activated.

For a two-wire sinking sensor, when the sensor is activated, the current flows from the DC+ on the input module, through the sensor, to the blue wire to the DC– side of the power supply. Again, the PLC input indicating light will be on only when the sensor is activated. Remember that with two-wire sensors, there will be leakage current in order for the sensor to operate.

For a three-wire PNP sensor, the sensor is wired to the sinking input module. When the sensor detects, the current flows from the DC+ side of the sensor, through the sensor output (black wire), to the input terminal of the PLC, and then out the DC– (common) point of the input module. The indicating light on the input module will turn on when the sensor is activated.

For a two-wire sourcing DC sensor, when the sensor detects, current will flow from the DC+ (brown wire), through the sensor, out the blue wire to the input terminal, and then to the DC– terminal (common). Again, there will be leakage current, and the indicating light will only light up when the sensor is activated.

TECH TIP
Normally Closed (NC)

Remember that some sensors may be normally closed. This would mean the input indicating light is on until the sensor detects.

AC inputs switch on the "hot" side. Considering this, the AC neutral is attached to the PLC input module common. **Figure 32-8** shows the simplified wiring for an AC input module. As each input device is switched on, voltage is applied to the input terminal, and the indicating light on the module turns on. Again, no logic (or very limited logic) is wired externally to the PLC.

Analog Input Modules

Analog input modules take an incoming analog signal and convert it to a binary number for the PLC to use. Analog inputs can be either voltage or current. While some modules may be limited to a specific type of input, others are selectable. The *resolution* of a module refers to the number of bits available in the binary number to convert the analog signal. A *bit* is a unit of information that is the result between two alternatives, such as on or

L1   L2 (n)

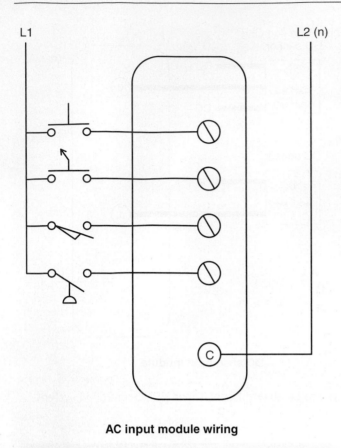

AC input module wiring

Figure 32-8. The wiring of an AC module is simple and straightforward.

off. The larger the binary number is, or the greater the number of bits, the higher the resolution will be. So a 16-bit binary number would translate to higher resolution than a 4-bit binary number. A 4-bit binary number can only represent 16 distinct numbers (0 through 15). If a 4–20 mA signal was converted into a 4-bit binary number, each binary number would represent a change of just over 1 milliamp from the transmitter. This may be an unacceptable resolution, depending on the purpose of the system. Typical input modules range from 12- to 16-bit resolution.

For two-wire transmitters, the negative (–) is connected to the module common, and the positive signal output is connected to an input terminal. Shielded twisted pair wire should be used to avoid signal interference. *Shielded twisted pair wire* or cable consists of insulated conductor wires arranged in pairs and twisted together, with the addition of a foil shield between the outer insulating jacket and the group of twisted conductor wire pairs. While the transmitter may have its own power supply, the negative lead should be brought back to the input module to the common point.

TECH TIP
Signal Interference

Noise and induced signals can interfere with both current and voltage signals. Analog signal wires should be run in a separate conduit and be separated from high-voltage lines as much as possible. If an analog input is not resulting in proper PLC system control output, then grounding, noise isolation, power, and input wiring should all be checked.

DC Sink and Source Digital Output Modules

As shown in **Figure 32-9**, the ideas of sink and source in output modules are similar to the input modules. In a DC sinking module, positive DC voltage is applied to the loads, and the output module common is connected to the negative DC bus voltage. When the PLC output point turns on, current flows from the DC+, through the PLC, to the DC–. This is similar in idea to switching an AC load on the neutral side. With the DC sourcing output, the DC+ voltage is connected to the common on the output module. When the output point turns on, the current flows from the DC+, through the PLC, to the load, and then to DC–. In the United States, the DC sourcing outputs are more popular. With either module, the indicating light on the PLC will be on only when the load is on.

Relay Output Modules

With *relay output modules*, miniature relays are mounted on the circuit board. These relay coils are energized when an output is given, which shuts a related normally open contact. Each relay contact may be completely separate or have a common terminal associated with several outputs. The concept is highlighted in **Figure 32-10**. These relays do have "hard" contacts, but they are limited in amperage rating. In addition, the entire module will be limited in amperage rating, based on the heat dissipation of the module and its connection to the backplane. The entire module will typically have an amperage rating that is less than the sum of the amperage of the separate outputs, and all outputs cannot exceed the rating for the entire module. For instance, if an output module has eight outputs that each have a maximum rating of 0.5 A, the rating for the module may be 3 A total. All outputs may be on at the same time, but they still cannot exceed the total rating. This arrangement of *grouped commons*, where a common terminal is associated with several outputs, allows the use of different voltages on the same output module. In **Figure 32-10**, for the grouped contacts, 120 VAC could be used for outputs 0 and 1, while 24 VDC could be used for outputs 2 and 3. This method of grouped commons is especially prevalent on smaller brick PLCs. PLC output indicating lights will be on when the miniature relay coil is energized—an externally applied voltage to the common of that group is not needed for indication.

TECH TIP
High-Resistance Relay

It is entirely possible to have a miniature relay coil energized (the indicating light on) and have a high-resistance relay contact, which will result in voltage drop across the contact and low voltage output.

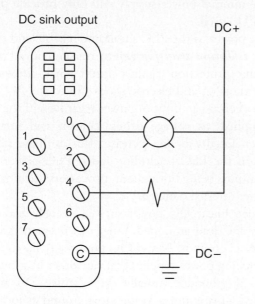

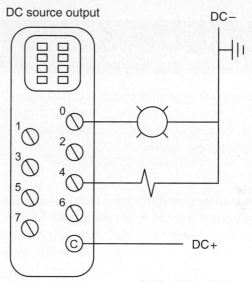

Goodheart-Willcox Publisher

Figure 32-9. Sinking and sourcing output modules are wired with reverse polarity.

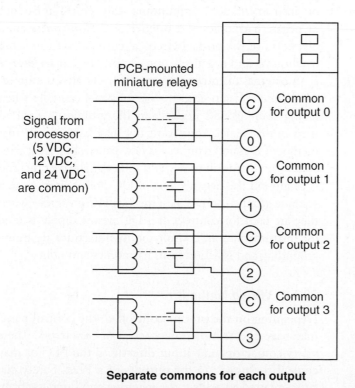

Separate commons for each output

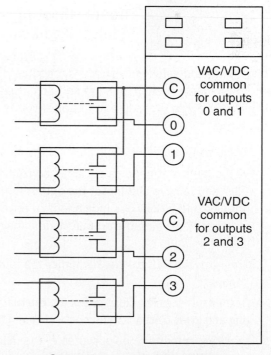

Commons grouped for outputs

Goodheart-Willcox Publisher

Figure 32-10. Relay outputs are normally grouped with common voltage input.

Analog Output Modules

Analog output modules are the reverse of the analog input module. These modules convert a binary number from the processor into a voltage or current output that is proportional to that number. Analog modules may be a combination of inputs and outputs, or strictly all inputs or all outputs. Output modules range from controlling 2 to 16 analog outputs. Output signals vary and may be in VDC or mA. Wiring should follow the same recommendations as the input analog devices.

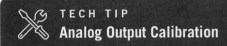

TECH TIP
Analog Output Calibration

Calibration of an analog output module is an involved process. The manufacturer's instructions should be closely followed by someone with a process control background.

Some analog modules have fault indicators. Faults may be caused by an open protective fuse, internal memory fault, or internal fault on the printed circuit board (PCB) because of environmental conditions, such as high heat, excessive current draw, or improper grounding.

PROCEDURE | Troubleshooting Analog Modules

If a fault is indicated with an analog module and a protective fuse is found to still be closed, use the following procedure to troubleshoot the issue:

1. Cycle power to the PLC to reinitialize the module.

2. If the fault is still indicated, de-energize the PLC and remove and reseat the output module, looking for obvious indications of physical damage. Reseating the output module may clear the fault.

3. Energize the PLC to check whether the fault clears.

4. If the fault does not clear, check the external wiring and loads closely for grounds and continuity.

5. If wiring is found to be acceptable, the module should be replaced.

32.1.3 Power Supply

Some manufacturers offer different power supplies categorized by maximum power output, while other manufacturers expect the system inputs and outputs to be separately powered. Typically, a PLC with a limited number of 24 VDC I/O modules can be powered by its own internal power supply. Larger circuits should have dedicated 24 VDC power supplies. PLCs with AC inputs and relay outputs will be supplied by line voltage, and the internal power supply will only provide power to the PLC itself.

The power to the PLC should be reliable and clean. Use an *isolation transformer* to transfer power while providing protection against interference, unwanted current transfer, and electric shock. Since the PLC will require a certain amount of power regardless of the voltage supplied, as voltage drops, current will increase. Heat generated will also increase. This can have serious effects on the PLC, including failed PLC power supplies, outputs being burned out, processors burning out, and other internal damage.

Older, linear DC power supplies usually had their negative DC lead grounded, usually to the same ground as local AC grounds. Newer DC power supplies are usually switching power supplies that can either be grounded or not. If grounded, usually the negative DC bus is grounded—many times to the same ground as local AC grounds. Depending on the system purpose (controls or instrumentation), grounding can either introduce or negate harmonics and interference. *Harmonics* are a distortion of the normal electrical current that can cause overheating and negatively impact efficiency. *Interference* is an external disturbance that negatively affects an electrical circuit. Examine each control panel carefully when taking readings and understand the grounding method used in that panel. Your company may have a particular method it uses and requires for new panels, but older panels may not conform to newer standards. The best method is to ground the negative bus of DC power supplies to a separate ground—other than the local AC ground—when they are used for controls. If a DC power supply is used for instrumentation, consult the manufacturer regarding grounding and isolation from harmonics and noise.

32.1.4 Wiring Methods

Depending on the process controlled, the control panel may have one or more safety relays or contactors. These safety contactors may input directly to the PLC or may control the power to the outputs of a PLC. Safety circuits are used to protect the machine and operator from damage and should never be overridden or bypassed.

As shown in **Figure 32-11**, when used as an emergency stop, the safety relay has several circuits that pass through any E-stops (emergency stops), exclusion zone limit switches, or other safety features, and it controls the power to the PLC outputs. When used as an input to the PLC, the safety feature is part of the programming and uses a specific command to not allow program operation. This is usually a master control relay (MCR)

command. A *master control relay (MCR)* is a type of external safety relay used to shut down a section of an electrical system. The discrete relay and programming function should not be confused.

Relay Logic Diagrams

Relay logic diagrams show the logical relationship between devices wired in the PLC system. Relay logic diagrams look like a ladder, with two vertical sides, called rails, with horizontal rungs in between. **Figure 32-12** shows an example of a relay logic diagram. The following rules guide the drawing of such diagrams:

1. All input and output devices are placed on the horizontal rungs.
2. Input devices are placed near the left rail on the rung.
3. Input devices can be connected in series, parallel, or a combination of both.
4. There can be only one output device on a rung, and it is placed near the right. Each output device is represented only once in the diagram.
5. Output devices cannot be connected in series.
6. Current should flow from left to right in the diagram.
7. Rung numbers are placed on the left side of each rung.
8. Rung comments are placed on the right side of each rung. These comments tell what is happening on the rung and help to identify the locations of contacts.

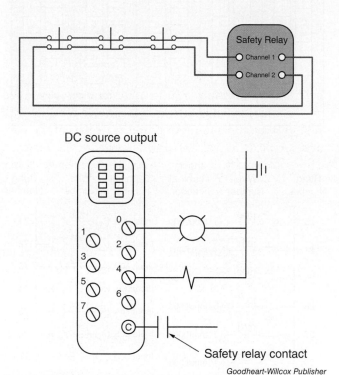

Goodheart-Willcox Publisher

Figure 32-11. Safety relays allow PLC inputs while preventing PLC outputs from energizing.

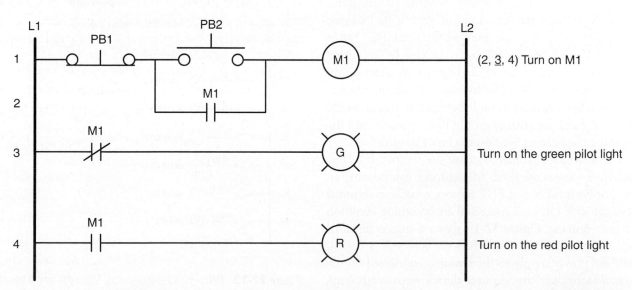

Goodheart-Willcox Publisher

Figure 32-12. This is an example of a simple relay logic diagram with comments. A bar underneath a number indicates that contact is normally closed.

A ladder logic diagram is similar to a relay logic diagram but is the actual program loaded into the PLC. Properly organized wiring schematics, wire numbers, and labeling can be a definite advantage in a PLC-controlled system. **Figure 32-13** shows one example of how these systems can relate. The following should be noted:

- The ladder logic diagram shows no external logic—all logic is obtained at the program level.

- Rung numbers start at 23—the entire system is not shown here.

- Wire numbers for the input module start with 1000. Everything above wire number 1000 is used for PLC inputs and outputs on this diagram. Those numbers below 1000 are used for other purposes, such as power or a safety circuit.

- The wire number has 4 digits. Take wire number 1001, for example: The "1" digit in the thousands place signifies this is a PLC input or output. The "0" in the hundreds place means this wire is attached to input card 0. The last two digits, "01," indicate that this wire is attached to terminal 01 on that particular input module.

- The outputs follow a similar pattern. The output from terminal 309 has a wire number of 1309.

- Wire numbers should be labeled at the PLC terminal and any terminal strip or other termination.

This is just an example, and panels from different manufacturers may use different systems. Each panel and its schematics must be carefully examined to determine the particular numbering system. In addition, not all PLCs label the first module with a 0. In some instances, the "0" may be reserved for the CPU. Input and output modules can appear in the rack in any order.

By understanding the numbering system, a technician can easily find a suspected faulty input, check the discrete input itself by actuating it, find the input terminal on the PLC, and check for voltage on the PLC terminal and the appropriate indicating light on the input module.

The majority of control systems use sensors. If a large number of sensors are used, wiring space can become an issue. For both NPN and PNP sensors, multilevel terminal blocks can save DIN rail space and are becoming common on newer systems. **Figure 32-14** shows a trilevel terminal block with LED indication when the sensor is actuated. While the power supply to the sensor is connected on the lower two terminals, the output of the sensor passes through the top level, and then to the input terminal on the PLC.

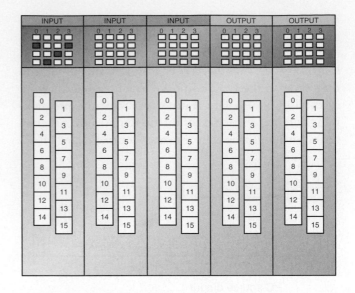

Rung Number	Wire Number	Input Terminal Number	Output Terminal Number	Wire Number	Rung Number
23	1000	Input 000	Input 300	1300	71
24	1001	Input 001	Input 301	1301	72
25	1002	Input 002	Input 302	1302	73
26	1003	Input 003	Input 303	1303	74
27	1004	Input 004	Input 304	1304	75
28	1005	Input 005	Input 305	1305	76
29	1006	Input 006	Input 306	1306	77
30	1007	Input 007	Input 307	1307	78
31	1008	Input 008	Input 308	1308	79
32	1009	Input 009	Input 309	1309	80
33	1010	Input 010	Input 310	1310	81
34	1011	Input 011	Input 311	1311	82
35	1012	Input 012	Input 312	1312	83
36	1013	Input 013	Input 313	1313	84
37	1014	Input 014	Input 314	1314	85
38	1015	Input 015	Input 315	1315	86

Figure 32-13. Proper I/O numbering systems are essential and simplify troubleshooting efforts.

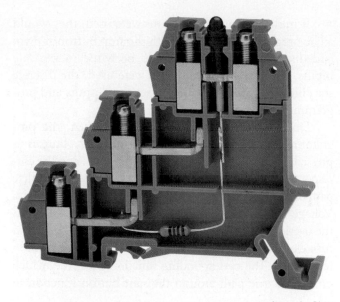

Automation Direct

Figure 32-14. Trilevel terminal blocks greatly reduce the DIN rail space needed.

Basic PLC Commands

Instruction	Description	Abbreviation
Examine if closed	Examines a normally open bit for an ON condition	XIC
Examine if open	Examines a normally closed bit for an OFF condition	XIO
Output (or output energize)	Turns ON a bit or an output port	OTE
Output latch	Latches a bit	OTL
Output unlatch	Unlatches a bit	OTU
Bit output (or one-shot rising)	Turns ON a bit for one scan only	OSR

Goodheart-Willcox Publisher

Figure 32-15. This table describes some basic PLC commands for controlling inputs and outputs.

32.2 PROGRAMMING

Many PLCs use very similar ladder logic structures for programming, but they may use a Windows-based drag-and-drop logic editor or function keys to build the ladder logic. You may never know the details of all the different types of programming software, but knowing the basics will allow you to adapt to each new PLC. This section covers some of the most basic programming concepts. Entire books are written on programming, and taking at least one specific programming class is a good investment.

32.2.1 Basic Commands

Basic commands are comprised of functions such as examine if closed, examine if open, output, output latch, output unlatch, and bit output. These commands are described in the table in **Figure 32-15**. Specific symbols and commands may differ slightly depending on the PLC manufacturer. **Figure 32-16** summarizes how commands operate. In addition to the actual command, an address is used to tell the processor where to look. The address can be an input address, output address, internal bit relay address, or any other memory location, such as timers or counters, for example. Similar to hard-wired logic, when a complete true path exists to an output, the output will be "energized," or turned on. Any address can be examined multiple times throughout the program, but an output command to an address—whether internal or external—can only name that address once in the program.

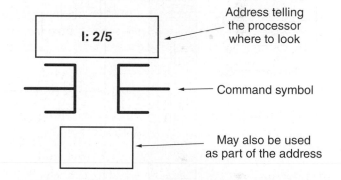

PLC Programming Command Operation

Command	Examined address ON	Examined address OFF
⊣ ⊢	When the examined address is ON, this command will result in a true condition.	When the examined address is OFF, this command will result in a false condition.
⊣/⊢	When the examined address is ON, this command will result in a false condition.	When the examined address is OFF, this command will result in a true condition.

Goodheart-Willcox Publisher

Figure 32-16. In the address, "I" stands for input, the "2" tells the processor that this is the second slot, and the "5" tells the processor that the input connected to terminal 5 on that slot should be examined. Commands result in a true or false state depending on the condition of the address they examine.

Examine Program 1 in **Figure 32-17** closely. Consider the following:

- The STOP pushbutton is a normally closed (NC) pushbutton, and input terminal 1.0 is ON.

- The M1 auxiliary contact is normally open (NO) and being used as an input for feedback.

- The discrete M1 coil is wired in series with the overload contact.

- The logic looks very much like that in a similar hard-wired system.

- The green box represents an ON or true state.

- Remember that when a path of true states exists, an output will happen on that rung.

If multiple stop pushbuttons were used, they would all be programmed in series with the stop button shown, and discrete stop buttons would be wired to separate inputs. If multiple start buttons were used, the discrete start buttons would be wired to separate inputs and programmed in parallel.

Since the stop button is normally closed, the programmed function is true. When the start button is pressed, voltage is applied to input terminal 1, and the corresponding function goes true in the program. A complete path of trues exist; therefore, the output 2.0 is turned ON. Voltage is applied to the output point 0, and as long as the overloads are closed, the M1 coil is energized. This in turn closes the M1 auxiliary contact and provides input to terminal 2. The corresponding function goes true, which provides a true path around the start button function in

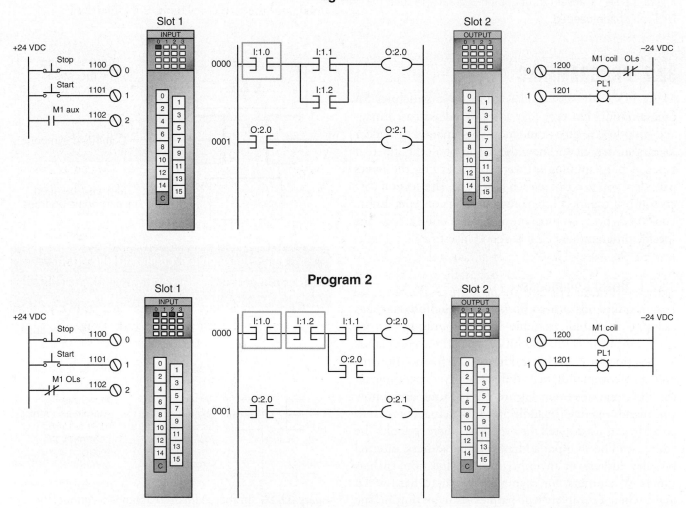

Goodheart-Willcox Publisher

Figure 32-17. A simple start/stop program is considered here, along with associated input and output modules. An output command can only occur once in a program, but the address of that output can be examined many times, as demonstrated in rung 0001.

the program. When the start button is released, the input turns off, and the programmed function goes false. The output is maintained because of the M1 auxiliary contact input. The second rung of programming examines (internally) output 2.0. If that output is ON, the programmed function on rung 0001 is true, and therefore turns the output on that rung ON, resulting in the pilot light (PL1) coming on. This takes place in milliseconds.

If the overloads trip at any point, the M1 coil drops out, causing the M1 auxiliary contact to open, which interrupts the true path and turns the output off. If the stop button is pressed at any time, this interrupts the true path, and the output turns off. If the M1 auxiliary contact were not used as a feedback device, and the overload contact opened, the M1 coil would drop out but would remain energized (the PLC output is still on). If someone were to reset the outputs, M1 would immediately pull in.

Examine Program 2 in **Figure 32-17** closely. Notice the following:

- The M1 overloads are now used as an input. They are normally closed and are programmed so that a true condition exists when the overloads are reached.

- The M1 coil is no longer wired in series with the overloads.

- The auxiliary contact is no longer used as a feedback device.

- The path around the programmed start button is now internal to the program and examines the output address internally.

The sequence of events is exactly similar to Program 1, but the memory function is performed internally within the program. When M1 is pulled in, if the overloads open or the stop button is pressed, the path of trues is interrupted and the output will turn off. Either method is acceptable, but your workplace may have a specific preference.

Latch and unlatch commands are programmed as output commands, OTL or OTU. Paired latch and unlatch commands are both labeled with the same output address. **Figure 32-18** shows the use of latch and unlatch commands. On a momentary true path, a latch command will energize the output and keep it on even when the true path goes false. In order for the output to turn off, the unlatch command must see a true path, momentarily. Even if power is lost to the PLC while the output is latched, the output will relatch when power is restored. Some brands of PLC have an internal bit that can be set on power-up to prevent the inadvertent relatching of previously latched outputs. Latch and unlatch commands can also be used to latch internal relay bits. In **Figure 32-18**, the hydraulic valve solenoid to extend is energized with a momentary push of the extend pushbutton. The extend solenoid remains energized until the retract pushbutton is momentarily activated, and the retract solenoid is then latched.

SAFETY NOTE
Use Proper Logic

The use of latch and unlatch commands is uncommon—they are rarely truly needed. Many beginning programmers use these commands instead of proper logic, and this can become a concern if proper safety measures are not taken.

All PLCs also offer a section of memory for the use of internal relay bits, also called binary bits. This section of memory can be used as relays to accomplish the following:

- Coordinate inputs and outputs in a more organized manner in a specific section of the program.

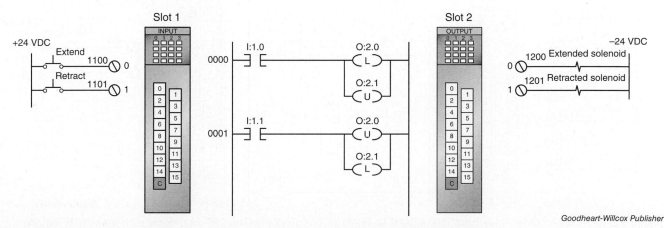

Goodheart-Willcox Publisher

Figure 32-18. This simple program extends and retracts a maintained two-position cylinder.

- Split up lengthy logic trees into several rungs so that it is easier to program and understand.

- Reference an address multiple times, on separate rungs, that affects many outputs or internal programming functions.

Notice that in **Figure 32-19**, the address of the bit function is different than inputs and outputs. The bit function specifies a word and a bit in that word of memory. The length of a word of memory depends on the processor.

32.2.2 Timers and Counters

Timers and counters appear as output functions. The timer function can be programmed as an on-delay timer (TON), off-delay timer (TOF), or retentive timer (RTO). An *on-delay timer (TON)* is used to delay the start of a process for a set period of time. An *off-delay timer (TOF)* is an instruction to delay the shutdown of a process for a set time. A *retentive timer (RTO)* retains, or holds, its accumulated value and can be used as an instruction to track the time a machine has been operating or to shut down a process after an accumulated time. Timers have several specifications that need to be set, including:

- The timer base time, possibly specified to thousandths of a second.

- The set point or preset, which is the amount of time the timer will time to. This number is multiplied by the base time, so a base time of 0.1 second and preset of 100 would equal 10 seconds ($0.1 \times 100 = 10$).

- A specific memory location of that timer, which will depend on the brand. (For example, this is typically file 4 (T4) in Allen-Bradley processors and T0 in Siemens processors.)

Timers have several status bits associated with them that can be used in other parts of the program, including the following:

- A timing bit, which is a bit that comes on when the timer is timing.

- A done bit, which is a bit that comes on when the timer has reached the set point.

- An enable bit, which is a bit that comes on when the logic upstream of the timer is true.

The on-delay timer (TON) starts timing when the upstream logic changes from false to true. **Figure 32-20** shows an example program using an on-delay timer. When the logic upstream turns true, the timer enable bit and timer timing bit will go true. If the logic stays true until the accumulated time equals the preset time, the timer timing bit will go false (0) and the timer done bit will go true (1). The enable bit will stay true as long as the logic upstream is true, whether the timer is timing or not. If the upstream logic goes false before the timer reaches the set point, the timer resets. In the situation depicted in **Figure 32-20**, when the start button is pressed, the timer is enabled. The enable bit and timing bit will both go true. The enable bit bypasses the start button and will keep the timer timing unless the logic upstream goes false. For 10 seconds, the timing bit will cause PL2 to be lit. Once the timer reaches 10 seconds, the timing bit will turn off, and the timer done bit will go true (ON). The timer done bit will turn output 0 and 1 on. When the stop button is pressed (or the overloads trip), the logic goes false upstream of the timer. This resets the timer, and the timer done bit goes false (OFF). Output 0 and 1 will de-energize.

A retentive timer (RTO) acts similarly to an on-delay timer, but maintains its accumulated value unless specifically

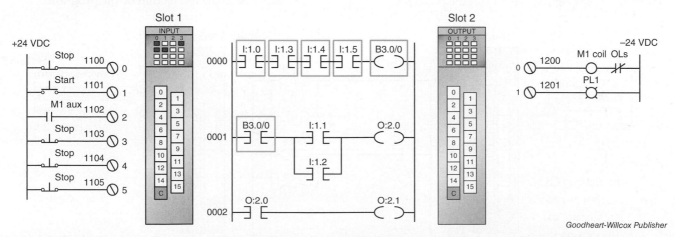

Goodheart-Willcox Publisher

Figure 32-19. This internal relay function is used to organize the program. The "B" in a command stands for binary, signifying an internal bit.

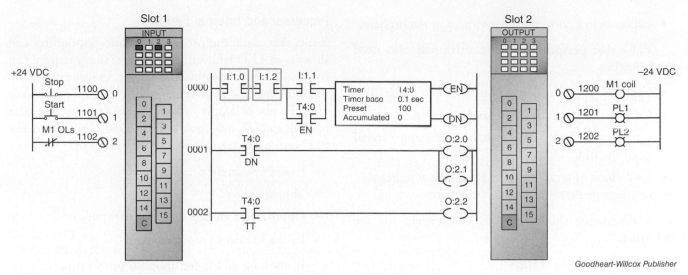

Figure 32-20. An on-delay timer delays this motor from starting for 10 seconds.

reset by a command. The accumulated value will be retained in the case of power loss, change of processor mode, or rung state transition (from true to false, for example). Retentive timers can be useful for predicting when maintenance might be needed on a motor, for example, since they can keep track of the total, accumulated running time. The timer will start when the motor runs, and it will retain that run time while the motor is not running, starting again where it left off when the motor again turns on.

The off-delay timer (TOF) begins timing when the logic upstream goes false. While the timer is timing, the timing bit is true. The enable bit follows the same sequence as the on-delay timer. The timer done bit is true until the timer accumulated value equals the preset value, and then turns false. When the upstream logic goes true again, the done bit turns on.

Counters appear similar to timers and also have status bits that are used for reference in the program. Counters may be programmed to count up (CTU) or down (CTD), starting at 0 or at a number loaded into that word of memory. Counters have a preset and accumulated value and take a reset instruction to return the counter to zero. A counter will count each time the logic upstream changes from false to true. As with timers, the memory location for counters will depend on the specific brand of PLC. Common status bits include the following:

- **Count up (CU).** This bit goes true when the logic upstream turns true, and it stays true until the logic goes false.

- **Count down (CD).** This bit goes true when the logic upstream turns true, and it goes false when the logic upstream turns false.

- **Done (DN).** This bit turns true when the counter accumulated value is equal to or greater than the preset.

- **Overflow (OV).** This bit goes true when the up-counter has exceeded the upper limit imposed by the size of the word the processor uses. The accumulated value will "wrap around" to a negative number, and the OV bit will stay on unless reset.

- **Underflow (UN).** This bit will turn on when a down-counter has exceeded the lower limit (a negative number). The accumulated value will "wrap around" to a positive number, and the UN bit will stay on unless reset.

32.3 MAINTENANCE

The majority of failures in a PLC-controlled system happen outside of the PLC itself. When properly protected, output points rarely fail. Processors also rarely fail, unless subjected to extremes in the surrounding environment. In this section, preventive maintenance, best practices, and troubleshooting a PLC-controlled system will be discussed.

32.3.1 Preventive Maintenance

Preventive maintenance for a PLC should consist of a review that ensures the PLC is operating within tolerances, and that the PLC is not exposed to excessive environmental conditions. Some of the environmental conditions that will affect a PLC include the following:

- Excessive heat (above 130°F).

- Excessive vibration.

- Excessive humidity (greater than 90% relative humidity, or any condensation).

- Exposure to any moisture or water within the control cabinet.

- Exposure to a corrosive environment or atmosphere.

PLCs that perform process control may also need the following:

- Calibration of all analog sensors that input to the PLC.
- Calibration of analog input and output modules.
- Calibration of all analog actuators receiving a signal from the PLC.
- Calibration of temperature sensors, such as resistance temperature detectors (RTDs) and thermocouples.

Maintenance should also include checks for the following:

- Proper incoming voltage.
- Proper voltage for DC power supplies.
- Proper grounding.
- Proper terminal connections.
- Proper torque of all terminals.
- Possible battery replacement.
- A backup of the program.
- Condition of the control cabinet and internals, including temperature, health of gaskets, cleanliness, and available diagrams.

32.3.2 Troubleshooting

The approach to troubleshooting a PLC-controlled system greatly depends on the failure being dealt with. In the great majority of cases, the failure will be external to the PLC. If the PLC has been running and controlling the system properly for some time, the failure is not with the programming logic. In order to properly troubleshoot a PLC-controlled system, the technician should have an in-depth understanding of how the system operates. This includes an understanding of the internal logic created with the PLC program.

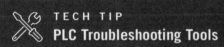

TECH TIP
PLC Troubleshooting Tools

To properly troubleshoot a PLC-controlled system, you will likely need the following items: Internet connection, appropriate software, diagrams, the current running program, a laptop, proper connecting cables, communication drivers, necessary passwords, and a copy of the original program with rung comments.

Processor and Internal Faults

Faults that cause the processor to stop responding and show as an LED fault indication most likely require you to go online with the PLC and reset it. In some instances, the fault may be cleared by cycling power. A problem with the power supply of the PLC itself is another type of fault, which may not be indicated. Failures of the PLC internal power supply can be caused by:

- Excessive current draw.
- Voltage spikes.
- Unprotected voltage loss or brownout.
- Excessive heat or moisture.

In the case of a failed internal power supply, if the PLC is a single, brick-type unit, the entire PLC may need to be replaced. If the PLC is modular, you may be able to simply replace the power supply and then transfer the processor and modules to the new rack. Be sure to check any internal power supply fuses.

If the PLC cannot be reset through the use of software, and the fault continues to reoccur after resetting, the processor will need to be replaced. This is a very rare failure. If an EEPROM program is installed, this should be transferred to the new processor.

I/O Failures

Single input and output points can fail occasionally. These may fail because of an excessive current draw—for instance, if an output is on when a power outage occurs. More common failures include an external failure with a discrete item such as a pushbutton, sensor, or limit switch, for example, or a problem with the wiring to or from the discrete item. **Figure 32-21** details the use of a digital multimeter (DMM) to check an input module.

PROCEDURE   Checking an Input

Use the following procedure to check to see if an input point has failed:

1. Place the PLC in programming mode or de-energize the safety relay/contactor by activating an emergency stop (E-stop), or do both. This should allow inputs to be made and LED indication to be given on the input module without any outputs energizing.

(continued)

2. Check to ensure that voltage is being applied to inputs (120 VAC, 24 VDC) and that a power supply fuse is not open. If voltage is not available, correct the failure.

3. Referring to the appropriate ladder logic, energize each input at the source and check that indication is shown on the PLC module for that input. This may take two people in constant communication. If a discrete input is activated but not registering on the PLC, use a DMM to check if voltage or current is being applied to that input point (this depends on the type of input, DC sink or source or 120 VAC).

4. If voltage and/or current is not being applied to that input point, the failure is either in the wiring to the discrete item or in the item itself. This is the most likely scenario.

5. If voltage and/or current is being applied to the input and the LED indication is not on, that input point, including the terminal and related internal circuit, has failed. This is a much less likely occurrence.

TECH TIP
Checking Sourcing DC Input Modules

For sourcing DC modules, using a DMM set to VDC to check terminal 0 in relation to C tells you nothing. The DC+ is connected internally to each output point, and that output terminal is at the same potential as C whether the sensor is ON or OFF. Checking from DC− to terminal 0 also tells you nothing because it will always show 24 VDC, whether the sensor is ON or OFF.

⚠ CAUTION

For sourcing DC modules, using a DMM set to mA to check from DC− to terminal 0 will result in a short. Either a fuse will open, or something will burn out, such as the power supply, module, meter, or sensor.

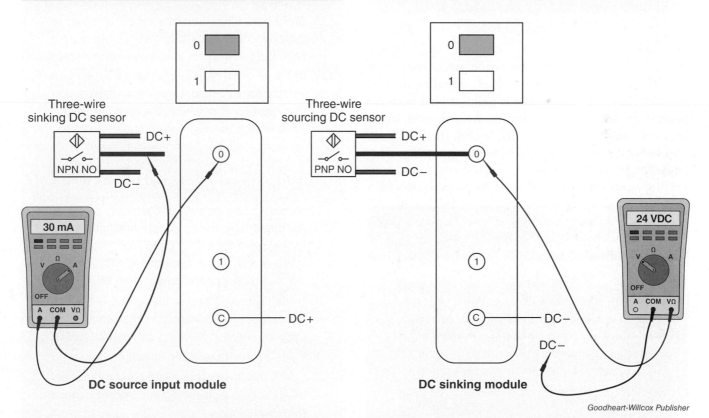

Goodheart-Willcox Publisher

Figure 32-21. With a sourcing DC module, you will only get an indication (a difference) between the OFF and ON states of the NPN sensor by either using a clamp-on meter that can read milliamps or by breaking the circuit and making current flow through the meter. When the sensor is ON, current will flow through the meter, and the indication light will be lit. With the DC sinking module, you will see voltage applied at the PLC terminal when the sensor is ON, but no voltage when it is OFF.

The method for checking a discrete element depends on the individual element. While a sensor can usually be easily checked using its indicating lights for power and activation, the wiring going back to the PLC may be more difficult to check. A continuity check will ensure that the wire is continuous. Use a megohmmeter on disconnected wires to check for possible shorts. If in doubt about the sensor's return signal to the PLC, run a separate wire directly from the sensor to the PLC input, activate the sensor, and check for a corresponding indicating LED on the PLC. Failed discrete devices need to be replaced. Failed wiring needs to be replaced. If an input point on the input module itself has failed, that input module needs to be replaced.

TECH TIP
Forcing an Input

Forcing an input by using software does nothing to the input terminal. Forcing an input is a software function and will not affect the actual input terminal or input LED. Forcing is used for bypassing an input's OFF state and for internal logic testing.

Checking a suspected failed output module requires the use of software. You can check the part of the system from the output terminal to the load, including the wiring and the load itself, without going online with the processor. While internal output circuitry does sometimes fail, the failure is more likely with the load itself. **Figure 32-22** details a check of output wiring using a DMM.

SAFETY NOTE
Energizing an Output

Energizing an output load manually may place the system in an unexpected condition. All people in the area need to be warned. Turning on an indicating light is one thing, turning on an output that energizes a motor starter is something else. A technician must know exactly what will happen when an output is energized, and how to de-energize it. It is entirely possible that an internally programmed interlock is preventing an output from energizing. If this interlock is bypassed, damage can occur.

PROCEDURE   Checking Output and Load Wiring

To check output and load wiring, use the following procedure:

1. First, recognize which type of output you have, such as DC sink or source, AC, or relay, for example.

2. On a source output, there should not be voltage present on the output terminal unless the output is energized and the indicating LED light is on. The same applies to relay and AC outputs. However, there will be voltage on the common terminal(s).

3. On a DC sink output, voltage will be applied to the output terminals of the PLC, and the indicating LEDs will be off.

4. Place the system in a safe mode by de-energizing safety relays/contactors. The PLC may also be keyed and should be placed in program mode.

5. De-energize the DC bus or AC bus, as applicable.

6. Disconnect the wire at the suspected faulty output terminal.

7. Check resistance from the output terminal to the opposite polarity bus. This should show a typical resistance for that load when you refer to the ladder diagram.

8. If resistance is much higher than usual, more investigation is needed to determine if the issue is with the wiring or load.

If using software:

1. Establish communications with the processor and go online.

2. Before forcing an output, check the programmed logic to see if there is a false logic that is preventing the output from energizing. If this is so, troubleshoot the applicable input that is preventing the output from energizing.

3. If the logic upstream of the output is true and the output should be on (in other words, the program shows it as being on), enable forces.

4. Place a force to turn the output on.

5. Check that the output came on and that the indicating LED on the output module came on.

6. Remove the force.

7. Check that the output turned off and that the indicating LED on the output module turned off.

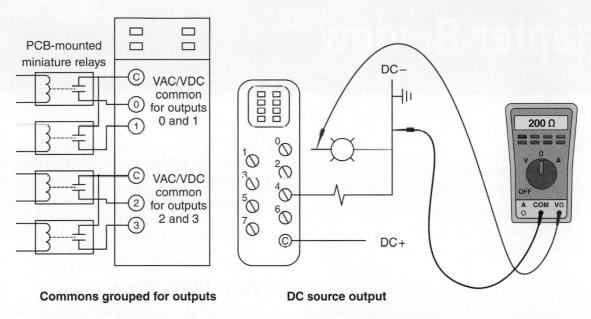

PCB-mounted miniature relays

VAC/VDC common for outputs 0 and 1

VAC/VDC common for outputs 2 and 3

DC−

200 Ω

DC+

Commons grouped for outputs

DC source output

Figure 32-22. Output modules can only be fully checked from within the processor, using software. To check a DC source output with a DMM, with the output wire removed, check resistance across the discrete output. Then, compare this resistance to a similar item—such as a solenoid or indicating light—that you know works.

If, when forced on, the output module did not respond, that output point has failed and the output module must be replaced. If the output module indicating light came on when forced on but the actual discrete output did not come on, there is an external failure that must be investigated further. The likely and common cause is in the wiring to the discrete output or in the output element itself. If output interposing relays are used, these must also be checked.

CHAPTER WRAP-UP

PLCs are common and will continue to advance. Every technician needs basic PLC skills to troubleshoot the increasingly complex control systems found throughout industry. Technology and software are constantly changing, and future operations will be even more reliant on microprocessors. A willingness to learn new concepts and become a lifelong learner will help you advance as a valued technician.

PROCEDURE   Replacing I/O Modules

To replace either an input or output module, use the following procedure:

1. Wear proper PPE and use protection from static discharge, such as a wrist grounding strap.

2. Make sure you have a backup copy of the program.

3. De-energize the PLC and remove other external power sources.

4. Use a DMM to make sure the panel and PLC are de-energized and no voltage is present.

5. Lockout and tagout the power supplies.

6. Remove the wiring harness or terminal strip, if available.

7. Remove the faulted input or output module.

8. Carefully insert the new module, making sure that it seats properly.

9. Replace the wiring harness, terminal strip, or wiring.

10. Remove lockout and tagout.

11. Energize the panel and PLC.

12. Reload the program if necessary.

13. Check for proper system operation and response.

Chapter Review

SUMMARY

- Most PLCs follow a specific scan cycle.
- A breakout board or interposing relays reduce the needed output current from a PLC.
- Very limited external logic is wired in a PLC-controlled system.
- A panel containing a PLC may have multiple power sources.
- A hard copy of the original program is included on an EEPROM.
- Forces, or force instructions, can turn on internal input bits and turn on outputs.
- Inputs and outputs should be isolated.
- It is important to understand the concepts of sink and source to properly troubleshoot a PLC system.
- Analog modules have specific wiring requirements to prevent unwanted noise and interference.
- Relay output modules use miniature relays for each output point. Some of these may have grouped commons.
- The power to a PLC should be through an isolation transformer that protects against interference, unwanted current transfer, and electric shock.
- Safety relays and contactors are common on production equipment.
- Wire numbers and terminal numbers are coordinated in a PLC-controlled system.
- Latched outputs remain latched until unlatched.
- Timers have status bits, such as timing, enable, and done.
- Counters will count on a false-to-true transition of the upstream logic.
- Most failures occur with discrete devices or the wiring.
- Some failures occur with input or output points and modules.
- The act of forcing inputs or outputs needs to be preceded by serious consideration for safety and equipment.
- You must be sure of how to take a voltage or current reading on a sink or source module in order to troubleshoot a PLC-controlled system.

REVIEW QUESTIONS

Answer the following questions using the information provided in this chapter.

1. Describe how input signals are converted into binary representation.
2. What is the role of the processor in a PLC?
3. *True or False?* The power supply converts incoming line voltage to low-level AC voltage to run the PLC.
4. List and describe the modes of a PLC.
5. _____ is the most typical kind of memory in PLCs, has a copy of the original program, and is added to the processor.
6. Describe a method to clear a fault indication on a module.
7. What is force instruction on a PLC?
8. Describe the function of an input module.
9. Explain the two options available for digital DC modules.
10. Discuss the wiring and current flow of a three-wire NPN sensor and a three-wire PNP sensor.
11. *True or False?* The larger the binary number is, or the greater the number of bits, the lower the resolution will be.
12. When is a twisted pair shielded wire used? Why?
13. Relay output modules are energized when an output is given, which shuts a related _____ contact.
14. *True or False?* In relay output modules, the entire module will have an amperage rating that is less than the sum of the amperage of the separate outputs.

15. When troubleshooting an analog module, what should you do if a fault is still indicated after power is cycled to reinitialize the module?

16. What is the purpose of an isolation transformer?

17. Describe two negative impacts that grounding can either introduce or negate in a power supply.

18. A(n) _____ is a type of external safety relay used to shut down a section of an electrical system.

19. In the following logic diagram, which devices cannot be connected in series?

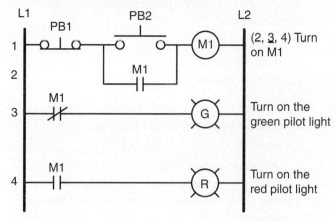

20. What does a wire number 1002 indicate in a PLC system?

21. With the command pictured here, what is the resulting condition when the examined address is ON?

22. In the program below, what happens when the start button is pressed?

23. *True or False?* When a complete true path exists to an output, the output will be de-energized, or turned off.

24. How are latch and unlatch commands used to turn on and off an output on a true path?

25. List and explain the three timer output functions.

26. What status bits are associated with timers and can be used in other parts of a program?

27. If a timer base was set to 0.01 second and the preset was set to 150, when would the done bit come on?

28. How are counters similar to timers? How do they differ?

29. *True or False?* In the majority of cases, a failure in a PLC-controlled system will be external to the PLC.

30. What are common causes of I/O failures?

31. On a DC sink output, voltage will be applied to the _____ terminals of the PLC, and the indicating lights will be off.

32. When should an output module be replaced?

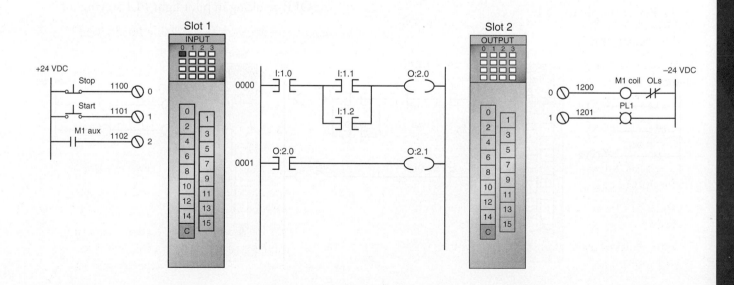

NIMS CREDENTIALING PREPARATION QUESTIONS

The following questions will help you prepare for the NIMS Industrial Technology Maintenance Level 1 Electronic Control Systems credentialing exam.

1. Which of the following is *not* a function of a PLC-controlled system?

   A. Current is converted to mechanical energy to operate pushbuttons.
   B. Input signals are converted into a binary representation.
   C. Output conditions are changed by programmed logic.
   D. Timing, counting, and math functions are performed.

2. A processor performs its regular duties with no limitations, but cannot be remotely programmed unless the key is switched, when it is in _____ mode.

   A. PROGRAMMING
   B. NORMAL
   C. REMOTE
   D. RUN

3. Which of the following is *not* a correct wiring connection in a PLC network?

   A. CPU to input module
   B. Input module to output module
   C. Output module to interposing relays
   D. Power supply to CPU

4. Which of the following is *not* true regarding relay logic diagrams?

   A. Input devices can be connected in series, parallel, or a combination of both.
   B. Output devices cannot be connected in series.
   C. Each output device can be represented several times in the diagram.
   D. Rung comments are placed on the right side of each rung.

5. _____ modules take an incoming signal and convert it to a binary number with a specific number of bits.

   A. AC source input
   B. Analog input
   C. DC sink output
   D. Relay output

6. In the following PLC ladder logic program, which is true about output 2.0 if it is ON?

   A. The programmed function on rung 0001 is false, and therefore turns the output on that rung OFF, resulting in pilot light PL1 staying off.
   B. The programmed function on rung 0001 is false, and therefore turns the output on that rung ON, resulting in pilot light PL1 coming on.
   C. The programmed function on rung 0001 is true, and therefore turns the output on that rung ON, resulting in pilot light PL1 coming on.
   D. The programmed function on rung 0001 is true, and therefore turns the output on that rung OFF, resulting in pilot light PL1 staying off.

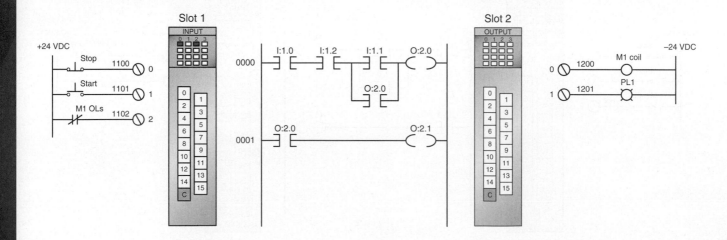

7. How do timers and counters differ?

   A. Timer functions can be programmed, and counters cannot.
   B. Timers appear as input functions, and counters are output functions.
   C. Both A and B.
   D. Neither A nor B.

8. In the following on-delay timer operation, what will occur once the timer reaches 10 seconds?

   A. The timing bit will go true, and the timer done bit will turn off.
   B. The timing bit will go true, and the timer done bit will go true.
   C. The timing bit will turn off, and the timer done bit will go true.
   D. The timing bit will turn off, and the timer done bit will turn off.

9. Which of the following would you use in a PLC-controlled system to learn the total accumulated time a motor has been running?

   A. Force condition
   B. Counter to count up
   C. On-delay timer
   D. Retentive timer

10. Which of the following is *not* a cause of PLC internal power supply failure?

   A. Excessive heat or moisture
   B. Not enough current draw
   C. Unprotected voltage loss or brownout
   D. Voltage spikes

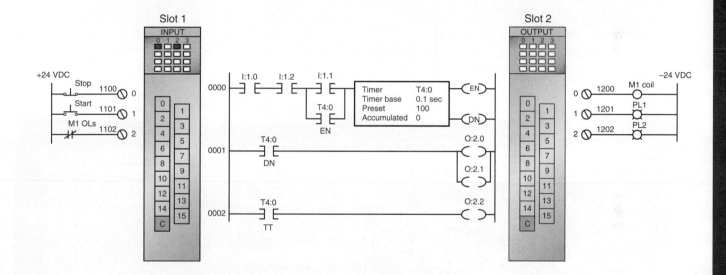

33 Human-Machine Interfaces

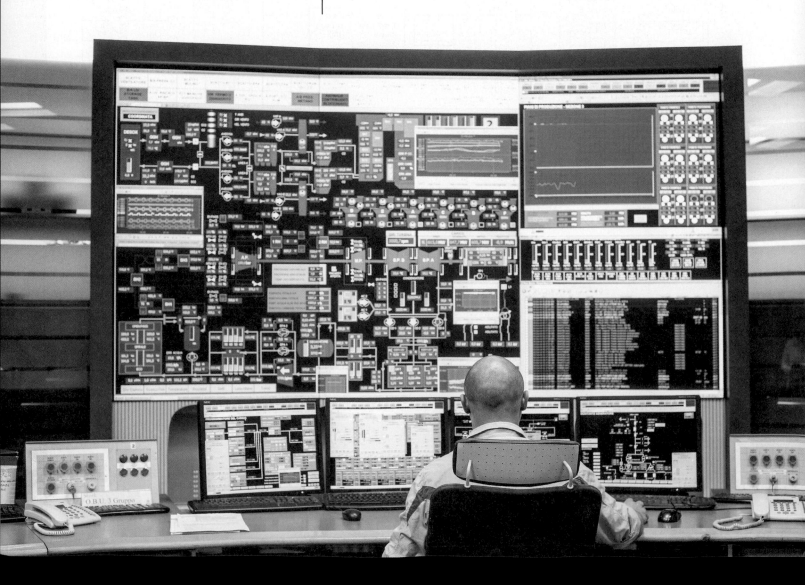

LEARNING OBJECTIVES

After completing this chapter, you will be able to:

☐ Describe the categories and topologies of communication networks.

☐ Discuss basic wiring considerations with TIA/EIA 568A and 568B cable wiring.

☐ Explain the differences between common industrial communication protocols.

☐ Understand the functions and performance of an HMI.

☐ Describe how to properly install an HMI and its operating principles.

☐ Discuss a symbol library and how it is used to design an HMI screen for indication or operator control.

☐ Describe how an operator can control system operation through an HMI.

☐ Discuss how to troubleshoot an HMI/PLC controlled system.

TECHNICAL TERMS

application
ControlNet
crosstalk
Data Highway
Data Highway Plus
DeviceNet
enclosure ground
EtherNet/IP
fail-safe
firmware
hardware
local area network (LAN)

network topology
radio local area network (RLAN)
supervisory control and data acquistion (SCADA)
serial communication
state
symbol library
Universal Service Ordering Code (USOC)
watchdog timer circuit
wide area network (WAN)

While a system or machine that is installed on the factory floor is likely to have a preprogrammed human-machine interface (HMI), it is important to know the basic operating principles of the system. An HMI and PLC pair could be a stand-alone system or part of a much larger system control network. HMIs range from the simplest information screen to multiple computer screens that provide information and system control for a city-wide area. **Figure 33-1** highlights a large HMI-controlled system called a SCADA system. *Supervisory control and data acquisition (SCADA)* is a system that uses a network of computers, HMIs, and PLCs to monitor and control all of the networked industrial operations and systems from a central control room, either on-site or remotely.

Alessia Pierdomenico/Shutterstock.com

Figure 33-1. SCADA (supervisory control and data acquisition) is a type of HMI that controls large systems, often from a central control room, as seen here.

33.1 COMMUNICATION NETWORKS

An industrial communication network is used to organize and centrally control the complex machines and operations of an industrial business. Industrial networking can involve numerous different computers, controllers, drives, HMI panels, and other field devices. In general, hardwired networks use cables to send and receive information between the various elements of the network, but technology continues its trend toward wireless communication. In this section, we will examine the basic principles, conductors, industrial protocols, and troubleshooting methods that apply to communication networks.

33.1.1 Basic Principles

The purpose of any type of network is the exchange of information. While some devices on a network produce and consume data, other devices only repeat or transfer the data to a new network or part of a network. Multiple remote I/O (input/output) nodes transferring data to and from a PLC that then transfers that data through an Ethernet connection is one example of a network. A node is simply a connection point for data transmission on a network. Data sent over the network is sent in discrete packets, unlike the data received when streaming a movie, which involves a continuous data transfer. Each attached device "sees" all of the data transmitted but only acts on the data addressed to it specifically. Therefore, each connected node must have its own identifying address. With many remote I/O nodes, a PLC processor must know which node the data was sent from in order to respond properly. The data packet has all of this information contained within the message.

Categories

The category of network used depends mainly on the size of the geographic area and the specific requirements. A *local area network (LAN)* is normally used within a closely situated area, such as one factory or area of a factory connected for the purpose of data transfer. Most industrial networks are LANs, but they may also connect to larger networks. **Figure 33-2** is a diagram of a typical industrial LAN. Many industrial LANs have

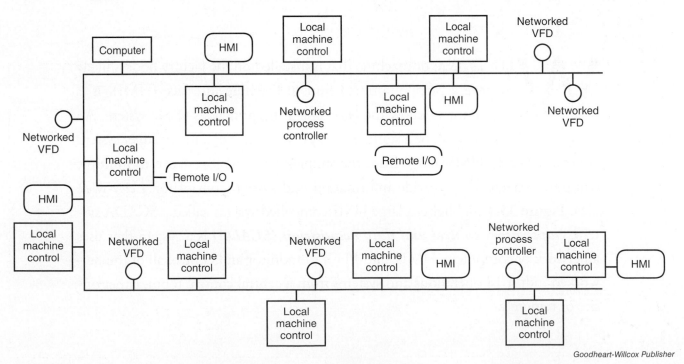

Goodheart-Willcox Publisher

Figure 33-2. A typical LAN for a manufacturing company is diagrammed here.

specific cabling requirements, use specific protocols, and may or may not be compatible with network devices from multiple manufacturers.

A *wide area network (WAN)* covers a larger area than a LAN and uses transmission conduits that are already in place. For example, a local municipality with a WAN may use dedicated phone lines to transfer data between remote stations and the controlling facility. A *radio local area network (RLAN)*—also called a *radio area network (RAN)*—is a slightly different stand-alone network that uses radio transmission to communicate. Depending on the geographic limitations and size of the network, using radio transmitters may be a more reliable way to transfer data. SCADA systems commonly use RLANs to control remote devices. **Figure 33-3** shows the organization of this type of system. System controls are typically a combination of computer-based programming logic and human controls.

Network Topology

Network topology is the way in which a network is organized and physically wired. The network topology will depend on the needs of the business. Network topologies, such as star, ring, and bus, all handle communication and data transfer differently. A star network relies on a central hub or server. A ring network is connected in series, with each node passing information to the next. A bus network connects all nodes to the bus, which is a single cable, and does not pass information through a central server. The bus network is a common network topology in an industrial setting.

Local communication and data transmission between stand-alone devices or used in the field when programming and troubleshooting is typically done through the use of serial communication via a USB (universal serial bus) connection and cable or an RS-232 pin cable connector. *Serial communication* is a method of communication in which data is sent and received via cable one bit at a time. The speed of data transfer through an RS-232 pin serial port connection can range from 200 to 460 kilobits per second (kbps), whereas USB 2.0 can handle up to 480 Mbps (megabits per second), and USB 3.0 can handle up to 640 Mbps. (A kilobit is 1000 bits, and a megabit is 1000 kilobits.) Since the RS-232 serial port connection method transfers data differently from a USB connection, a converter patch cord will not provide faster communication. Additional

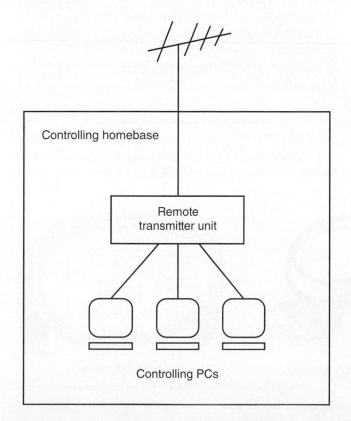

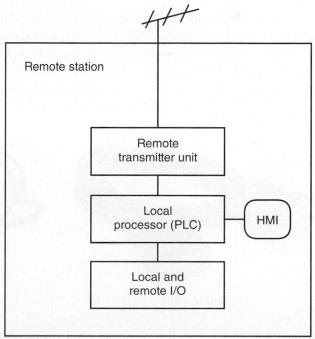

Figure 33-3. A SCADA system controlling remote elements through the radio transmission of data.

software drivers supplied by manufacturers must be used. **Figure 33-4** shows a nine-pin RS-232 serial cable and two USB cables.

Conductors

Shielded or unshielded twisted pair cable is the most common conductor used. A network installation needs to follow the *National Electrical Code (NEC)*, and the best installation results come from careful planning. Consider the following when installing a network:

- Cable lengths are limited depending on the size (category) of the cable.

- Do not use elements (plugs, jacks, patch cords) that do not meet the proper specifications.

- Remember that twisted pair cable is made up of small conductors—treat it accordingly. Overfilling the conduit will damage the cable. Using more than 20 lb of pulling tension will stress the cable beyond its strength limit.

- Run network cabling in its own conduit, not in a conduit that contains power circuits.

- Power lines, motors, high-intensity discharge (HID) lamps and fluorescent lighting, and variable speed drives can all cause electromagnetic interference (EMI).

- If pulling cable, never pull into or from a live panel. This is a safety consideration to minimize hazards and reduce the possibility of electric shock.

Categories of cable are specified in TIA/EIA 568, which is a set of commercial telecommunication cabling standards from the Telecommunications Industry Association (TIA). The standards categorize cables, with specifications for bandwidth, loss, and interference. Common cable categories include the following:

- **Category 3.** Older category of cable, for telephone communication and data speeds up to 16 MHz.

- **Category 4.** For data speeds up to 20 MHz.

- **Category 5.** For data speeds up to 100 MHz.

- **Category 5e.** Similar to Category 5, but supports higher speeds and cuts down on **crosstalk**, which is interference that can occur between the wires inside the cable.

- **Category 6.** For data speeds up to 250 MHz.

- **Category 6A.** For data speeds up to 500 MHz.

- **Category 8.** For data speeds up to 2000 MHz (2 GHz).

Category is often abbreviated as "Cat." Connections and jacks must meet the same standard as the cable. Using a Cat 3 jack connection with a Cat 5 cable renders the Cat 5 cable only as good as the lower-rated Cat 3 jack.

SAFETY NOTE
Safe Wiring

Always wear appropriate clothing when installing wiring—all conductive items should be removed or covered. No live work is permitted. Never break the plane of a control panel that is live without proper PPE, and then only to take troubleshooting readings. Never pull cable into or from a live panel or junction box. Know the cable run before you start to pull wires.

When making connections, a standard 8-pin jack called RJ45 is used, and the connections follow the pattern of TIA/EIA 568A, TIA/EIA 568B, or USOC. *Universal Service Ordering Code (USOC)* is a wiring system that

RS-232 cable

USB A-A cable

USB A-B cable

EZAutomation (left); 3d-eye/Shutterstock.com (middle); Automation Direct (right)

Figure 33-4. USB connections are common in digital communications. RS-232 serial communication is slower than USB. Serial communication is typically 9600 bits per second but can be slightly faster.

was invented to connect customer telecommunications equipment to public network lines. The wiring patterns of TIA/EIA 568A and TIA/EIA 568B are standards set by the Telecommunications Industry Association, an offshoot of the Electronic Industries Alliance, and are only differentiated by switching two pairs of wires. Older installations and their connections may use 568B. Newer installations typically use 568A.

When connecting wires to connectors, insulation must be included inside the body of the jack. No more than 0.5″ of insulation may be removed. With the insulation and because each pair of wires has different twist rates, crosstalk is reduced. The wires are placed into the jack and are carefully fed into each cavity before being crimped (**Figure 33-5**). Crimping compresses conductive members of the jack, which "bite" into the

individual wires through the insulation. This helps to prevent a degradation of the connection and increased resistance. **Figure 33-6** shows the tools commonly used in this process, and **Figure 33-7** summarizes these connections. Each wire or jumper made should be tested to ensure good connections. Testing instruments can check the wire mapping—testing the cable for proper connections—and test for proper speed, **Figure 33-8**. Testing tools range from under $100 to $500. The less expensive tools will test for proper connections and operation for purposes of troubleshooting communication issues, but they are not meant for certification.

When wiring communication networks, everything must be up to standard. Just one mistake can prevent the whole system from functioning properly. Troubleshooting at a later date can prove costly and take far too much time. (Imagine having to check every termination and jack in a factory-wide installation with hundreds of nodes.) Problems such as shorts, opens, and reversed or crossed pairs of wires can be difficult to detect. Technicians making jack connections should have training and experience with the wiring.

Aliaksandr Bukatsich/Shutterstock.com

Figure 33-5. Wires are placed into the RJ45 jack with insulation on them and are carefully fed into each cavity before being crimped. The setup pictured here is 568B.

> ### ⚒ TECH TIP
> ### Grouping Wires
>
> When routing wires, do not place excessive stress on them by making tight bundles. Zip ties should group the wires, but they should not be so tight that there is no give or flex in the group.

Stripping tool   Die insert   Crimping tool

Automation Direct

Figure 33-6. Common tools for installing network cabling include a stripping tool for the outer sheath of the networking cable, a die insert for the crimping tool for RJ45 jacks, and the crimping tool itself, which can accommodate a variety of die inserts.

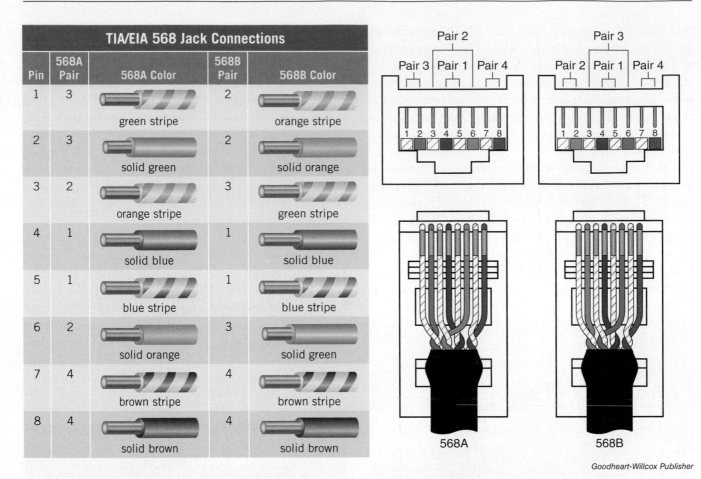

Pin	568A Pair	568A Color	568B Pair	568B Color
1	3	green stripe	2	orange stripe
2	3	solid green	2	solid orange
3	2	orange stripe	3	green stripe
4	1	solid blue	1	solid blue
5	1	blue stripe	1	blue stripe
6	2	solid orange	3	solid green
7	4	brown stripe	4	brown stripe
8	4	solid brown	4	solid brown

Goodheart-Willcox Publisher

Figure 33-7. Two pairs of wires are switched between the 568A and 568B setups.

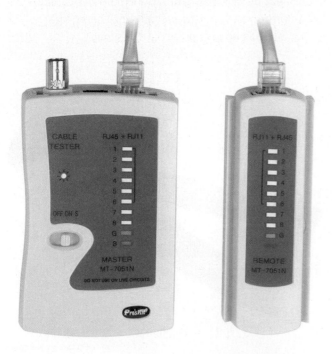

Eclipse Tools, CAT 5 Tester 8 LED with Remote #400-026

Figure 33-8. Testing instruments such as this can test the cable for proper connections and proper speed.

33.1.2 Industrial Protocols

Several different network communication protocols are used in digital networks. A communication protocol is a system of rules that define the format of communication in a network. *DeviceNet* was developed by Allen-Bradley to connect smart field devices to processors and data management PCs. DeviceNet allows different vendors' elements to be connected, as long as they meet the standard. Network size is limited to 64 nodes, and distance is limited depending on the amount of data transmitted and the conductor size. Nodes may be added or removed under power, which limits the impact on operating systems.

Also originally developed by Allen-Bradley, *ControlNet* is a bus topology with taps for nodes. Coaxial cable (RG-6 quad shield) makes up the bus, with a maximum length of 1000 meters and a maximum number of nodes of 99. As more nodes are added, the maximum length is reduced. Repeaters can be used to extend the bus but are limited to five.

Data Highway and *Data Highway Plus* are local area networks (LANs) developed to connect PLCs, HMIs, and PCs for communication and data transfer.

Data Highway is an older system, but it is still in use. DH-485 supports up to 32 devices, while DH+ can support 64 devices.

EtherNet/IP™ is quickly becoming the most popular protocol for data transmission in industrial applications. *EtherNet/IP*—or Ethernet Industrial Protocol—is a network communication standard capable of handling large amounts of data at high speeds. The availability of Ethernet hubs, switches, routers, and repeaters due to the expansion of online personal computing make this method appealing. While most controllers or processors from different manufacturers can be networked in this manner, this does not mean that one manufacturer's elements can talk to a different manufacturer's elements. Software on the receiving end must use the same protocols as the transmitting element in order to understand the information transmitted. Cat 5e or Cat 6 cable is typically used for Ethernet connections.

This was merely a brief summary of a few common protocols and methods. There is such a wide range of both open and proprietary protocols that becoming an automation technician requires specific training in this area.

33.2 HUMAN-MACHINE INTERFACES

Human-machine interfaces range in functionality and purpose depending on the system requirements. While some HMIs may be connected directly to the processor (PLC) through serial communication and comprise a stand-alone unit, others may be networked through EtherNet/IP for the purposes of information sharing and troubleshooting. All manufacturers have specific software for their HMIs. While some software packages may cover several models, having multiple types of HMI (brands and models) requires multiple software packages. This discussion focuses on troubleshooting and basic principles rather than strict programming.

33.2.1 Types and Options

The functions and performance of HMIs can include some or all of the following, depending on the system design:

- Binary (on-off) indications of system response and operation.
- Indication of system variables, such as temperature, pressure, times, counts.
- Overall system control (on-off).
- Control of process variable set points, such as for temperature or pressure.
- Troubleshooting aids.
- Data logging and graphing.

Depending on the model and manufacturer, HMI units come with a variety of communication ports, such as USB, micro SD card, Ethernet, RS-232, RS-422, and RS-485. RS-232 can be used to hook up one transmitter and one receiver—typically a stand-alone HMI/PLC pair. RS-422 can hook up one transmitter to up to 10 receivers, and RS-485 can connect up to 32 transmitter or receiver devices.

Most HMI panels are powered by 24 VDC. Some HMIs are available for an AC power supply module for powering with 110 VAC. A real-time clock's memory is backed up with a battery. The battery does not back up stored user programs. **Figure 33-9** shows a variety of HMI terminals.

33.2.2 Installation Considerations

Installation considerations include the environment surrounding the HMI and the electrical environment the HMI operates in. The HMI's physical environment must not exceed specific temperature guidelines from the manufacturer. Consider the typical surrounding temperature, but also the other components within the panel and the heat they give off. Follow the manufacturer's specified clearances around the panel. These clearances dissipate heat, and they also allow enough space to manipulate memory cards and required wiring connections. Most panels have a gasket placed between the panel and the cutout when the HMI panel is installed in an enclosure, such as a control station or cabinet. In order for this gasket to perform its intended function, the enclosure must be rigid, flat, and free of contamination or corrosion, and the edges must be deburred. Screws that mount the HMI panel to the enclosure should not be overtorqued, as this may result in damage to the panel. A panel may be mounted at an angle other than vertical, but this will affect heat dissipation, and manufacturer recommendations should be followed. A panel should never be mounted in a position that will expose it to direct sunlight. The ultraviolet rays will eventually damage the screen.

Connections to terminals should be done correctly, with proper torque. Carefully check the polarity of the DC power supply and terminals. Grounding terminals on the HMI panel are meant to be taken back to an enclosure ground, not the DC– bus. An *enclosure ground* is the ground bar or point located on the back of the control cabinet. Use a DC power supply dedicated to powering just the panel. Proper grounding is required to reduce noise and EMI (electromagnetic interference). Power and communication lines should not be tied together in a bundle. These lines should be run separately with as much distance as possible between them, especially with AC power lines. If the lines must cross, do so at 90° angles to minimize EMI.

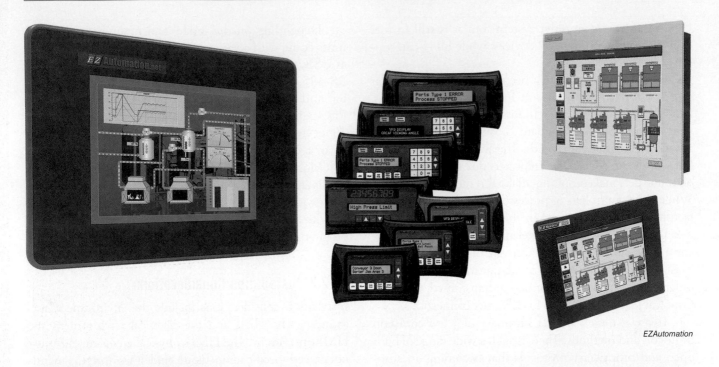

EZAutomation

Sporlan Division, Parker Hannifin Corporation

Used with permission of Rockwell Automation, Inc.

Figure 33-9. A variety of HMI panels is pictured here. While the component graphic terminals are being phased out, many are still in use.

33.2.3 Operating Principles

An HMI panel can be used to control a single machine or system or multiple systems, depending on the design. The most common arrangement is for an HMI panel to provide indication and operator control for a single system or a related group of systems. The control of the system is accomplished through operator input using the HMI and through PLC or computer control. **Figure 33-10** summarizes the organization of a simple HMI system.

An HMI can provide operators a convenient way to control a system, but it cannot be relied on for safety measures. HMIs can fail. Emergency stops and other critical safety functions must reside outside the HMI. An HMI's communication cannot solely be relied on, and systems must be designed to be fail-safe. *Fail-safe* means that during the failure of an element, the machine or system places itself in a safe shutdown condition, or the operator is able to place the system in a safe condition without the use of an HMI.

SAFETY NOTE
E-Stop

While an HMI may include graphical stop pushbuttons, a physical E-stop is used for machine safety.

System design directly influences what will be shown on the HMI. For instance, if a system's temperature is controlled by a PLC with analog input and output modules, this information may be shared with the HMI for indication purposes only. The set point of the temperature may be controlled by the PLC, with no opportunity for the operator to influence that system variable. In a different system design,

the operator may be required to input a system variable set point, such as weight, which is then transferred to the PLC. The PLC controls the system operation until that set point is reached—for example, filling a container until a weight is reached. As the system reacts, the data is transferred back to the HMI for indication purposes.

The back-and-forth communication between the HMI and PLC or controller may be transmitted as single occurrences or by block transfers. A more formalized approach of block transfers of memory and the organization of the PLC memory, or tags, increases the efficiency of the communication. HMI panels typically have a *watchdog timer circuit* set at 50–200 milliseconds. If communication between the PLC and HMI does not occur within this time frame, a communications error will occur.

Applications are uploaded and downloaded to and from the HMI using a memory card (SD or USB), locally from a connection to a laptop, or remotely from a PC. *Applications* are the user programs that contain screen layouts and tags, or memory addresses, that are used for indication and control purposes.

On initial start-up, HMIs without an application loaded will start in a configuration mode that allows the setup of communications and loading of applications from a memory device. The exact setup of the communication between the HMI and PLC depends on the protocol being used. Follow the manufacturer's instructions for that particular HMI panel. If using EtherNet/IP communications and DHCP (Dynamic Host Configuration Protocol) is on, the IP address and subnet will automatically be assigned and cannot be changed. If DHCP is off, addresses can be manually entered. The IP address is a unique number that is assigned to the HMI panel for communication purposes. It is a set of four numbers with delimiters in

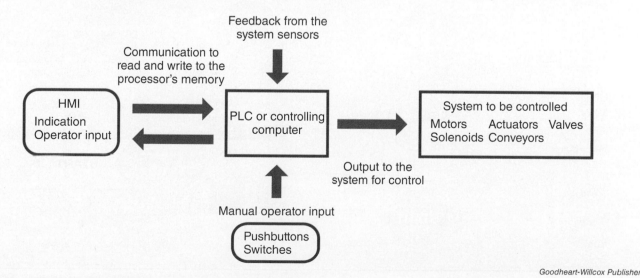

Goodheart-Willcox Publisher

Figure 33-10. HMI does not control the system directly, but only exchanges information with the processor.

between (XXX.XXX.XXX.XXX). Each number can be from 0 to 255. The subnet number is used when multiple networks are used. It follows the same pattern as the IP address.

33.2.4 Basic Programming

Many panels come with an extensive array of predesigned screens and a symbol library. A *symbol library* is a group of graphical symbols used to design a screen for indication or operator control. Most panels also allow programmer-defined and drawn symbols. The following basic design concepts should be considered when setting up an HMI:

- The more information placed on a screen, the harder it is for an operator to grasp.
- The smaller the font used, the harder it is for an operator to easily and quickly read the text.
- One main screen should be designed for indication and general control, such as main components being turned on and off. Linked or subscreens can be designed for specific information, recipes, and controlling physical subsystems.
- While there may be many screens, each should link back to the main screen.
- Standard symbols and color schemes should be used.

A screen can hold only so much information. Use a larger screen if your design requires many information points or numbers on the screen in addition to graphical symbols and graphics that simulate the controlled system. Make sure text and graphic elements are clear and legible. Using a very small font size could make it difficult for some operators to clearly see and understand the information.

The experience of the operator plays a role in understanding the screen that is designed. An operator with little experience may rely more on shape or color, while an operator with years of experience may rely more on the text on the screen. Using standard symbols and colors helps to prevent mistakes in operation. Consider the pushbutton symbols in **Figure 33-11**. Some symbols seem less logical and are harder to read than others. The colors and font sizes used clearly make a difference, with certain colors standing out and larger font sizes more quickly legible. Beveled symbols, which have a border around them, are often more recognizable as operator inputs. Nonbeveled symbols are more often recognized as indicating lights. Notice the square start pushbutton that is red. Some European manufacturers have pop-up flag indicators on breakers that become red when the breaker is closed and on—giving a warning that power is on. Some machines may have indications that they are running with a red indicating light being on. Using standard symbols and colors

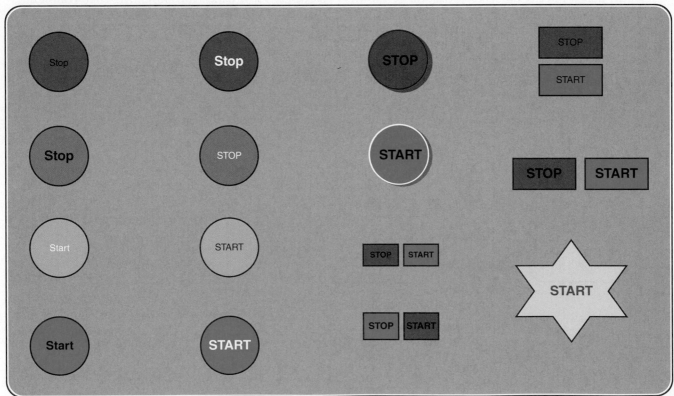

Goodheart-Willcox Publisher

Figure 33-11. The more easily a screen is understood, the less likely an operator error will occur.

that are easy to read and interpret is what is needed, not bright or wild colors. Animations and highlighted colors should show the operator that an alarm state is occurring, or that some system variable is out of the normal range.

Three basic types of symbols are operator inputs, indications, and system representation. Operator inputs may be simple pushbuttons, switches with multiple positions, thumbwheels, or a value entry. These types of symbols are shown in **Figure 33-12**. The following elements of each symbol can be controlled: size, position, color, associated tag (memory location), states, and font size. A *state* is a condition of each input or element. The "Hand-Off-Auto" switch in **Figure 33-12** has three states, which match each position. Each state may be linked to a PLC memory location, or tag. Some momentary pushbuttons, like the start button, may have three states. The normally open position (without operator intervention) is one state, and the pressed position is another state. A third state may be used that changes the color of the pushbutton to indicate when the machine is running, providing feedback just like a lighted pushbutton would. In this way, some operator inputs also act as indications. The tags, or memory locations, in a PLC are typically binary bits from internal PLC memory relays, which are included in the program to affect an output or program variable. Numbers that represent system variables, such as temperature, weight, level, and time, are input using a thumbwheel or on-screen keypad. This number is then transmitted to the PLC into an integer file or directly into the memory location that it controls, such as a timer preset value, for instance.

Indications on an HMI screen can take the form of indicating lights (such as for on-off states), gauges (which require a proportional signal to the PLC), numerical information (which requires an analog signal to the PLC), or simply highlighting a part of the system when it is in use (changing the color of a motor to indicate it is running, for example). Just as for operator inputs, indications rely on specific tags, or memory locations, in the PLC to show state. **Figure 33-13** shows several different types of indications. Gauges and bar graphs are specified with the range needed depending on the system and typical operating pressures or values. For instance, if a water cooling system typically operates at a pressure of 25 psig, you would not want to set the gauge to read from 0 to 25 psig. Common sense and industrial design considerations dictate that the normal operating point should be approximately midscale on the gauge. Gauges showing a position within a full range may be easier to read and immediately interpret than digital readouts with numerical digits.

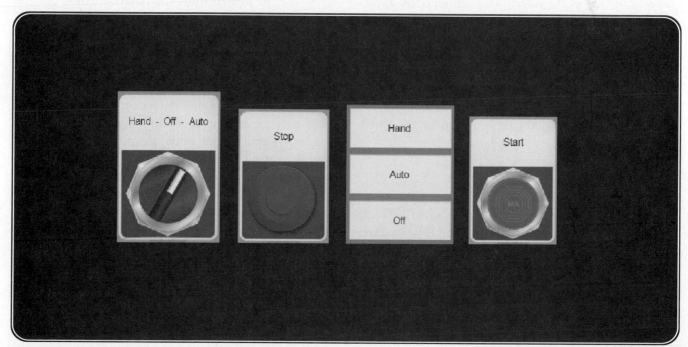

AdvancedHMI

Figure 33-12. Common operator inputs are shown here as they might appear on an HMI screen.

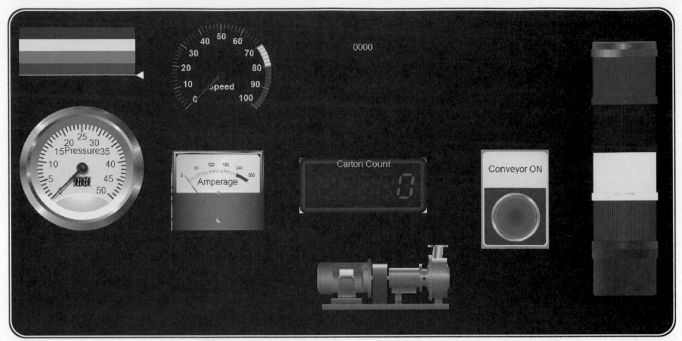

Figure 33-13. This HMI screen shows a sample of various indications.

System representation can be made using standard symbol libraries, but keep in mind that the complete system does not need to be shown. It may be confusing to an operator to see numerous pressures and temperatures on one screen, and this may not be needed for the operation or control of the system. Simple is sometimes better. **Figure 33-14** shows symbols for a simple tank system. Looking at the screen, it is immediately apparent what is happening without including every pipe or pressure gauge. The HMI does not perform logic functions. If the valve operator switch in **Figure 33-14** was in either position but the main control switch was in the Auto position, the HMI would not make the decisions. The programmed PLC makes the decision on system operation based on the position of the three-position switch.

33.2.5 Troubleshooting

Many HMI panels provide error codes when a failure occurs. These should be researched using the manufacturer's documentation. Errors generally fall into one of the two following categories:

- Errors caused when initially programming and placing the system online.

- Errors that appear after the system has been properly running for an extended amount of time.

Errors that appear when first programming and placing the system online are commonly caused by either the terminal not being able to read a memory location (tag) in the controller or PLC, or the terminal not being able to write to a memory location in the controller. This may be caused by the controller memory address not being properly configured, or a wrong tag/address being read or written to by the HMI. In either case, all external tags/memory addresses that the HMI is reading or writing to should be checked. Another common communication issue happens if the HMI and controller (PLC) are not properly set up to communicate, with each knowing the other's address.

Errors that appear after the system has been online for some time are commonly communication errors. Check that the communication cables are properly attached, and reset the panel by cycling power. If the panel hangs during reboot or start-up, a problem with the panel's memory, touch screen, or keypad points to a hardware problem. *Hardware* refers to the panel's physical elements. A hardware failure requires replacing the panel. *Firmware* refers to the panel's operating system (not including user programs). You may be able to recover from a firmware error by reloading or updating the firmware.

Common operating errors include damaging the screen or trying to activate several inputs at once with multiple touches. Touching two areas at one time may result in unpredictable action. Analog screens will take

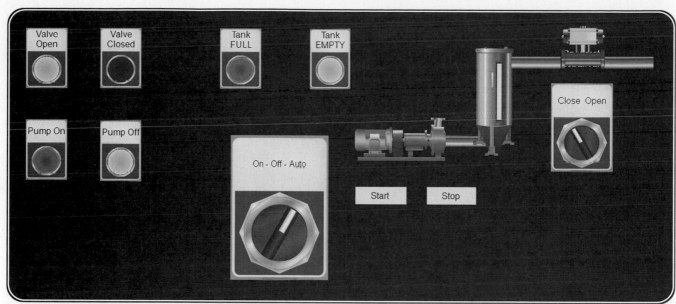

Figure 33-14. A simplified system diagram is easy to understand.

both inputs and average them to a point in between, resulting in action if that point has an address attached to it. Make inputs using only one finger at a time—the screen is not a smartphone. If a screen is damaged, it must be replaced.

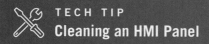

TECH TIP
Cleaning an HMI Panel

Do not use harsh chemicals or solvents to clean an HMI display. Use only a mild detergent or soap with little moisture (do not soak the display) to clean, and immediately dry the screen with a clean cloth. Power should be off and preferably disconnected from the panel. Never use a high-pressure stream of water to clean a panel.

HMI panels offer several methods for resetting the memory to factory conditions. This may include various degrees of resetting or restoring, including clearing part of the memory and completely clearing the memory. This typically does not clear the firmware from the panel. This resetting can be accomplished through software or by use of a physical button on the back of the panel.

With the proper operating knowledge, you can troubleshoot the failure of a system with the aid of an HMI. Referring to **Figure 33-15**, suppose a valve fails to open when an operator inputs to the HMI. For this example, assume the PLC is programmed to allow manual control

of the fill valve and that each time the tank reaches the full level, the pump automatically starts and drains the tank. Currently the tank is empty, the pump is off, and the valve is closed. When the operator changes the valve switch to open, nothing happens. The following should be considered when troubleshooting this problem:

- This is not a communication problem between the HMI and PLC, otherwise an error would be showing on the screen in the form of a timeout or communications error.

- You can further assume this is not a communications problem because the HMI is showing that the valve is in the closed position and the pump is off, which requires communication between the PLC and HMI.

PROCEDURE   HMI Troubleshooting

Considering **Figure 33-15**, once you have eliminated the possibility of a communications problem, use the following procedure to determine why nothing happens when the valve is switched to open:

1. First, make sure you have all appropriate prints of the wiring and PLC program. If a hard-copy print of the program is not available, you need to be online with the PLC.

(continued)

2. Check any conditions that might prevent operation of the system, such as E-stops, incoming power to the control panel, or in-line output fuses to the valve.

3. This valve is a motor-operated ball valve with limit switches for the open and closed position—power to the limit switches may be different from power to the valve motor.

4. While observing the output indicating light on the PLC, try to open the valve using the HMI switch. If the output indicating light comes on, there is a problem external to the PLC in the output circuit, such as an interposing relay, an output fuse, the valve motor itself, or wiring.

5. If the output light does not indicate an output to the valve, there may be a problem with that output point on the output module.

6. At this point, you will have to go online with the PLC in order to check the logic and determine if something is preventing the output from turning on. If the logic is true, and the program shows that the output should be on, the output module must be replaced.

7. If the logic does not show a true path for an output, the interruption must be investigated. You may find fault with an input that is required, such as from an E-stop or safety relay.

SAFETY NOTE
Remote Troubleshooting

A person remotely troubleshooting a system must know how the system is designed to work and how it is and is not performing the intended functions. Performing force functions on a PLC-controlled system while not actually in front of the PLC is potentially dangerous and requires an in-depth and unparalleled knowledge and understanding of the system.

Remember that the HMI is acting as a provider of inputs to the PLC. The only additional requirement is communication. If a system is not responding to an HMI, examine it in the same way you would a PLC-controlled system. Going online with the HMI is of little value as long as communication is occurring between the PLC and HMI.

CHAPTER WRAP-UP

While HMI used to be limited to small screens with only several lines of text, HMI systems have expanded to allow multiple screens, Ethernet communication, and advanced graphic interfaces. A facility may have numerous HMIs of various ages and from different manufacturers. As a technician, it is wise to review each HMI's documentation in regard to troubleshooting and the installation requirements to ensure the longevity of the panel.

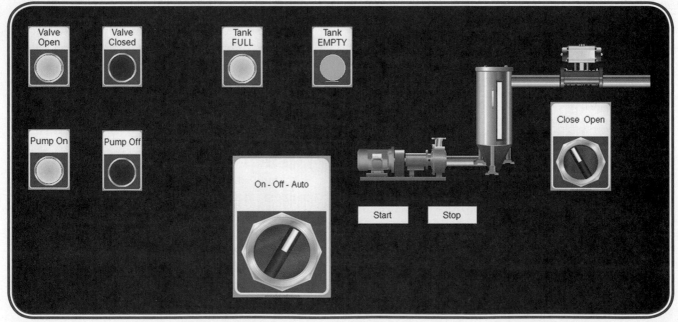

AdvancedHMI

Figure 33-15. In this troubleshooting example using this HMI screen, the valve is not opening when it should.

Chapter Review

SUMMARY

- A SCADA (supervisory control and data acquisition) system uses a network of computers, HMIs, and PLCs to monitor and control all of the networked industrial operations and systems from a central control room, either on-site or remotely.

- The purpose of a network is the exchange of information. LANs, WANs, and RLANs are the most commonly used networks, while LANs are the most popular in industrial settings.

- Network topology is the way in which a network is organized and physically wired. Topologies include star, ring, and bus.

- Serial communication is a method of communication in which data is sent and received via cable one bit at a time.

- When installing a network, cabling should be run in its own conduit or tray, not intermingled with power lines. Cable connections must be properly terminated, or they will reduce the efficiency and operation of networks.

- Several network communication protocols are used. DeviceNet has a maximum of 64 nodes. ControlNet has a maximum of 99 nodes. Data Highway has a maximum of 32 devices, and DH+ can have up to 64.

- An HMI does not perform logic functions, but transfers information to and from a PLC for purposes of indication and operator input. An HMI can include functions such as overall system control, binary indications of system response, and control of process variable set points.

- When installing an HMI, the surrounding environment and supplied power will both affect it. Connections to terminals should be done correctly, with proper torque.

- An HMI's communication cannot be solely relied on, and systems must be designed to be fail-safe.

- Three basic types of HMI symbols are operator inputs, indications, and system representation.

- A state is a condition of each input or element.

- Many HMIs provide error codes. These errors generally occur from initial programming and placing a system online or after the system has been properly running for an extended period of time.

- Hardware is the physical circuitry of the HMI. Firmware is the operating system of the HMI. Errors with these components generally occur after a long period of use.

- HMI failures are commonly linked to wiring, communication, and power issues.

- Remotely troubleshooting a system controlled by a PLC needs to be carefully considered and monitored.

REVIEW QUESTIONS

Answer the following questions using the information provided in this chapter.

1. What is a SCADA system?

2. What is the purpose of an industrial communication network?

3. *True or False?* The purpose of any type of network is to produce and consume data.

4. Explain the common categories of a network. What determines the category of network used?

5. What is network topology? Describe the bus network topology.

6. _____ is a method of communication in which data is sent and received via cable one bit at a time.

7. List five considerations when installing a network.

8. What is TIA/EIA 568? What standards does it include?

9. *True or False?* Cat 6A provides standards for data speeds up to 600 MHz.

10. How specifically are 568A and 568B different?

11. What is crosstalk, and how is it limited?

12. Discuss the different network communications protocols used in digital networks.

13. Describe the different purposes of an HMI.

14. A(n) _____ is the ground bar or point located on the back of the control cabinet.

15. *True or False?* While an HMI can provide operators a convenient way of controlling a system, it cannot be relied on for safety measures.

16. Discuss how an HMI and PLC communicate in a system design where an operator is required to input a system variable set point.

17. What are applications? What are they used for?

18. Describe a symbol library, including the basic types of symbols and elements of each symbol that are controlled.

19. A(n) _____ is a condition of each input or element that may be linked to a PLC memory location, or tag.

20. What forms can indications take? What do each of these forms require?

21. What two categories do HMI errors fall into when a failure occurs?

22. What is the difference between hardware and firmware?

23. Suppose a valve fails to open when an operator inputs to the HMI. The PLC is programmed to allow manual control of the fill valve, and each time the tank reaches the full level, the pump automatically starts and drains the tank. What should be considered when troubleshooting?

24. *True or False?* When troubleshooting, if the output light does not indicate an output to the valve, there may be a problem with the input point on the output module.

NIMS CREDENTIALING PREPARATION QUESTIONS

The following questions will help you prepare for the NIMS Industrial Technology Maintenance Level 1 Electronic Control Systems credentialing exam.

1. What type of network is most commonly used in industrial settings?
   A. Local area network
   B. Ring network
   C. Star network
   D. Wide area network

2. _____ is a method of communication in which data is sent and received via cable one bit at a time.
   A. Crosstalk
   B. EtherNet/IP
   C. Industrial networking
   D. Serial communication

3. HMIs *cannot* perform which of the following functions?
   A. Binary indications of system response
   B. Data logging and graphing
   C. Electromagnetic interference control
   D. Indication of system variables

4. Which of the following should *not* be done when installing an HMI?
   A. Mount an HMI panel at a vertical angle.
   B. Run power and communication lines together.
   C. Take grounding terminals on the HMI panel back to the DC– bus.
   D. Use a DC power supply dedicated to powering just the panel.

5. How do PLCs and HMIs interact with each other?
   A. PLCs control a system operation until a set point is reached and then transfer the data back to the HMI for indication purposes.
   B. HMIs control a system operation until a set point is reached and then transfer the data to the PLC for indication purposes.
   C. PLCs upload applications to and from the HMI using a local or remote connection from a PC.
   D. HMIs upload applications to and from the PLC using a local or remote connection from a PC.

6. A pushbutton is an example of a(n) _____ symbol that appears on an HMI screen.
   A. indication
   B. operator input
   C. state
   D. system representation

7. What is the most likely cause of an error when first programming and placing a system online?
   A. A terminal is not able to read a memory location in the PLC.
   B. The operator is trying to activate several inputs at once.
   C. There is a firmware problem with the network.
   D. There is a hardware problem with the HMI panel.

8. When troubleshooting an HMI, if the output indicating light comes on while trying to open the valve using the HMI switch, there may be a problem that is _____.
   A. external to the PLC in the output circuit
   B. internal to the PLC in the input circuit
   C. within the output point on the output module
   D. within the input point on the output module

Process Control Systems

The Process Control Systems section will introduce you to control systems, the operation of heating and cooling systems, and related maintenance and troubleshooting tasks. It is important to learn how to read and understand piping and instrumentation diagrams (P&IDs) and the standard symbols they contain, as these diagrams are a technician's road map in troubleshooting and maintaining industrial systems.

The content in this section will help prepare you to earn the NIMS Industrial Technology Maintenance Level 1 Process Control Systems credentials. By earning credentials, you are able to show employers and potential employers proof of your knowledge, skills, and abilities.

34 | Industrial Process Control

LEARNING OBJECTIVES

After completing this chapter, you will be able to:

□ Understand components of a process control system.

□ Explain the importance of process control.

□ Describe the function of a control loop in process control.

□ Discuss the differences between a closed loop and an open loop.

□ Interpret common piping and instruction diagrams (P&ID) and symbols.

□ Explain the function of common instrumentation and elements in a control loop.

□ Describe the differences in controller modes.

□ Describe the characteristics and applications of various types of control loops.

□ Discuss calibrating an instrument and troubleshooting a closed loop control system.

TECHNICAL TERMS

capacitance

capacitive level sensors

cascade control

closed loop

contact flow meters

control loop

deadband

differential pressure (DP) flow meter

disturbance

error

Faraday's law

feedback

gain

I/P transducer

instrument identification number

instrument index

magnetic inductive flow meters

magnitude

noncontact flow meters

offset

open loop

orifice

piping and instrumentation diagrams (P&ID)

process

process control system

process variable

proportional band

proportional-integral-derivative mode (PID)

proportional mode (P)

proportional plus integral mode (PI)

reset control

resistance temperature detector (RTD)

scaling

set point

signal conditioning

signal converters

system capacitance

tag number system

temperature switch

temperature transmitter

thermocouple

thermowell

transducer

tuning

ultrasonic flow meters

ultrasonic level sensors

venturi

P rocess control takes many forms. The most basic form relies on operator intervention to control the process. The operator must constantly supervise the machine or process operation and watch for system variables that diverge from specified standards. Imagine standing in front of a stove all day watching pasta boil, and making minute changes to the heat input—too much heat and it overflows, too little heat and the water does not boil.

Forms of process control vary within industrial areas and processes.

- *On-off process control* is used with controllers that have only two positions, usually ON and OFF. It is common for the majority of manufacturing machines or processes. When a temperature gets too high, for example, a cooling fan turns on. When the temperature is reduced to a set value, the cooling fan turns off.

- *Sequential process control* is used when a series of events or operations must be performed to produce a manufactured item. For example, an item may be clamped, stamped into a shape, drilled, and then reamed.

- *Continuous* or *batch process control* is used when a system variable must be maintained, either manually or automatically. An example is the automatic adjustment of air pressure in an air supply duct based on the output speed of a fan.

Automation has been integrated into many aspects of process control. The pneumatic signals that once controlled a diaphragm-operated valve in order to maintain level have been replaced by capacitive or sonic level detectors with a networked controller that outputs to a motor-operated valve. In this chapter, we will examine the basic concepts of process control, instrumentation, control loops, and troubleshooting.

34.1 BASICS OF CONTROL SYSTEMS

In a larger view of industry, control systems tend to be classified based on their purpose. While the specific applications may be different, the control is very similar. Many times the words *servo* and *servo system* are used interchangeably. In the area of process control, *servo system* refers to a specific type of process. A *servo* controls motion. This motion may be used to exactly place an item (controlling position), control the amount of fluid pumped, or control the amount of torque applied. The most well-known use of servos is controlling the motion of robots or CNC motion, as shown in **Figure 34-1**.

A ***process control system*** uses various devices and methods to monitor and control a process or environment in order to maintain a desired output. Process control systems use the same basic idea and concepts as servo systems, but with distinct differences. In a servo system, the output and feedback happen very quickly. In a process control system, the system response takes place over an extended amount of time. This response time is determined by a number of system variables that we will examine.

Gabriel Georgescu/Shutterstock.com

Figure 34-1. This CNC machine has numerous motion axes. Each is servo controlled.

SAFETY NOTE
Process Control Signals

Process control signals may be "low voltage"; however, this does not mean that the signals are incapable of causing severe shock and injury. In addition, the system component that receives the control signal, such as a motorized valve, is likely supplied with a higher voltage for its operation. Always follow proper safety procedures to prevent injury and equipment damage.

34.1.1 System Concepts

A ***process*** is an action performed on a material to achieve a desired outcome. Processes include heating and cooling, pumping, filling, draining, mixing, settling, and many others. A desired outcome might be to make a liquid more viscous, less viscous, evenly distributed, or evenly mixed. Many processes are controlled to maintain a variable in order to achieve a desired result. A ***process variable*** is a characteristic or factor that is measured within a process. For example, process variables in fluid systems include flow, pressure, and temperature.

Without process control, the resulting product may have such wide variances that it is unusable. Controlling process variables increases the output of a process and reduces costs. In some instances, not controlling a variable in the process results in the process not working at all. In wastewater treatment, for example, biological organisms break down waste products and produce methane, which can be captured and reused. If the biological organisms are not kept within a narrow band of temperature, they will not work effectively in the wastewater treatment process.

34.1.2 Control Loop

A *control loop* is a component of a control system that monitors a process variable. All control loops have the following basic elements:

- Process controller.
- Process or an actuator.
- Sensed variable.
- Sensor.
- Feedback signal or loop.

The example shown in **Figure 34-2** is a simplified on-off system of a home furnace. The process is heating and the medium is air. As the air is heated, it is distributed into the space of the home. The thermostat measures the current state of the variable (temperature). In this case, the

thermostat also compares the sensed variable with the set point. The *set point*, or *command point*, is an established system value or condition that the process works to maintain. A system's set point may be configured manually or programmed into the controller. When the sensed variable (temperature) exceeds the set point, the thermostat opens. This removes the signal to the furnace, and the furnace shuts down. This is an example of the three distinct functions of a loop:

- *Measure* the variable.
- *Compare* the measurement with the set point.
- *Provide output* to change the variable.

These are the three basic functions of all control loops. There may be more elements or complex functions included in a control loop, but these three are always present. Keep this basic loop in mind as we examine how changes affect the control of the system.

In the home furnace example, the process variable is the indoor air temperature. The process variable is what is

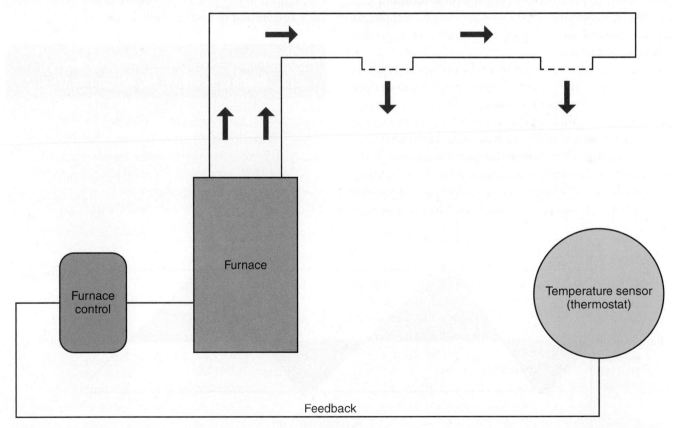

Goodheart-Willcox Publisher

Figure 34-2. A home heating system is a common example of how the basic components of a control loop operate. The furnace control is the process controller and the furnace is the process. The air temperature is the sensed variable, which is monitored by the thermostat (sensor). The thermostat generates and sends a feedback signal to the furnace control to complete the control loop.

being measured. The control loop has an effect on this process variable by controlling the process (the heating action of the furnace). The thermostat senses the current value of the process variable and compares it to a set point. In this way, the thermostat is sensing the error between the set point and process variable. *Error* is the difference between the current value of the process variable, or "where we are," and the set point of the system, or "where we want to be." In an industrial process control loop, an error signal is generated internally by the controller.

If we were to examine the amount of error in the home heating system with regard to time, we might see a graph similar to **Figure 34-3**. Notice that the rise in temperature when the furnace is on is similar in slope during each occurrence. Also note, when the furnace turns off, the drop in temperature has a similar slope during each occurrence. The upper and lower ends of the deadband impact how much error is measured. The *deadband*, or *control band*, is a range of values for a process within which no action is taken. If the deadband were larger, the process variable would have more variability. If the deadband were much smaller, a closer tolerance could be maintained, but the furnace would turn on and off constantly. In the home furnace example, we trade variability for reducing the number of starts and stops of the furnace.

If there were a variation in the home heating system, it would impact the process. Leaving the front door open during winter, for example, places additional load on the process to maintain the process variable. A variation that affects the system's ability to maintain the process variable is called a *disturbance*, or *load disturbance*. If the home furnace is large enough, it may be able to keep up with the load disturbance and keep the process variable within range. If so, the process is said to be *in-control*.

If the furnace cannot keep up with the disturbance and additional load, the process variable may go outside of the upper and lower ends of the deadband. In this case, the process is said to be *out-of-control*. Even though a process is responding properly, the load placed on it may be too great. For example, the furnace runs continuously to maintain the temperature set point when the front door is left open. This is not a process that needs troubleshooting, but a load that needs to be corrected.

In the home furnace example used in this section, there was feedback in the control system from the sensor. A control system that includes feedback is called a *closed loop*, or *closed control loop*. Outdoor lighting controlled by a light sensor is another example of a closed loop. When the sensor detects a preset value for the level of darkness, it sends a signal to turn on the outdoor lights. A typical industrial closed control loop is shown in **Figure 34-4**. A control system that does not incorporate feedback is an *open loop*. An open loop may be a simple ON/OFF system. For example, if a light in the room of a home is turned on, it may be left on all night if someone forgets to turn the light off. There is no mechanism in an open loop to evaluate system feedback, such as duration of time or light levels.

⚒ **TECH TIP**
Process Issue or Too Much Load?

An out-of-control process is different from a process experiencing too much load. An important step in troubleshooting is understanding the difference between these situations and recognizing when the process is working properly.

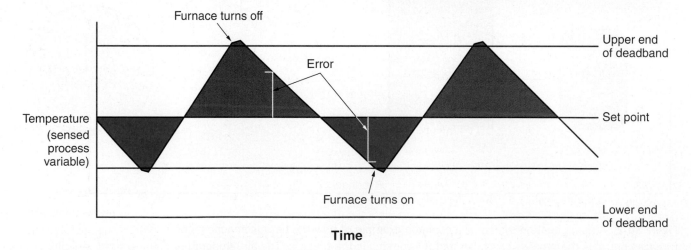

Goodheart-Willcox Publisher

Figure 34-3. Error over time can be predicted in the home heating system example.

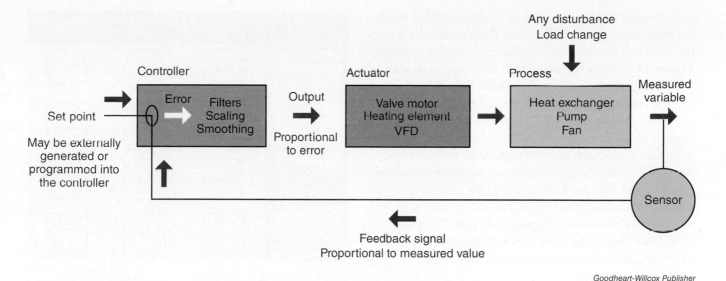

Figure 34-4. Block diagram of a closed control loop.

The **feedback** in a closed loop communicates the measurement of a process variable. The feedback signal is proportional to the measured variable and may be transmitted in current or voltage (4–20 mA or 1–5 V are common). The feedback signal returns to the controller and is compared against the set point. This comparison results in an error signal that is proportional to the difference between the feedback and set point. The controller may apply filters, or **scaling**, to the signal before it is output to the actuator. Scaling may be performed because the actuator responds to a different type of signal, or the signal needs to be changed to reflect the system process. The actuator responds to the controller output signal by directly affecting the process. This is accomplished differently depending on the process—a valve may open, a pump may ramp-up its speed, etc. The action of the actuator changes the process and drives the measured variable closer to the set point. Any changes in load (disturbance) are registered by the sensor, and the feedback cycle continues.

34.1.3 Piping and Instrumentation Diagrams

Piping and instrumentation diagrams (P&ID) are drawings of the interconnections of physical system components, including representations of instrumentation and control loops. ANSI/ISA-5.1 covers numbering systems and drawings for P&ID. In larger systems or facilities, there may be hundreds of control loops. Being able to troubleshoot or calibrate them depends on recognizing and using P&IDs.

Common representations of instrument (or sensor) signal lines are shown in **Figure 34-5**. These lines form the control loop and follow the signal through the process—from the instrument (sensor), to the controller, to

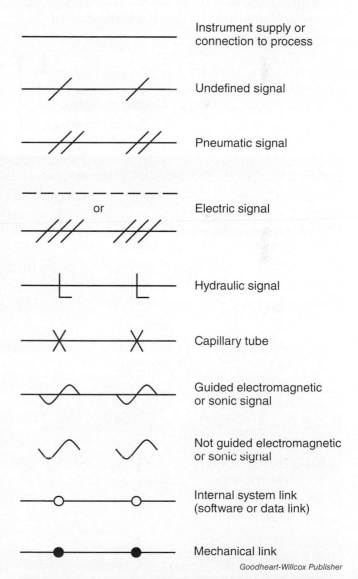

Figure 34-5. Using common symbols ensures that P&ID drawings are consistent across the field.

the actuator. Other lines may be used depending on the engineer and specific facility, but these should be noted or called out on the drawing. Instrumentation symbols, **Figure 34-6**, show manually read local gauges, as well as inaccessible instruments or functions. Functions that are PLC or computer controlled can also be shown.

Each instrument or function has an associated tag number in the drawing that indicates function and control loop. The *tag number system* is used to identify all components of a control loop and differentiate them from other control loops. Tag numbering follows the form of a two or three (or more) letter identifier followed by a loop or system number, as shown in **Figure 34-7**, with "XXX ###" or "XXX-###" being the most common forms. The position of letters and numbers provide specific information:

- First letter—Indicates the quantity being measured, such as level, flow, temperature, current, etc. In the example shown, level is the quantity measured.

	Primary Location	Field-Mounted	Auxiliary Location
Discrete Instruments	⊖	◯	⊖
Shared Display, Shared Control			
Computer Function			
Programmable Logic Control			

Goodheart-Willcox Publisher

Figure 34-6. Every item or element in the control loop is indicated with a symbol, including actuators and instrument line valves.

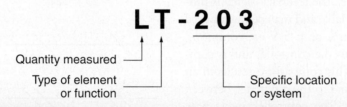

Quantity measured
Type of element or function
Specific location or system

Common Identification Letters					
	Meaning				
Letter	As first letter	As modifier (after first letter)	After first letter	As an output	As a modifier
A	Analysis		Alarm		
B	Burner				
C	not defined		Control	Control	
E	Voltage		Sensor		
F	Flow	Ratio			
G	not defined		Gauge glass		
H	Hand				High
I	Current		Indicator		
J	Power				
K	Time	Rate of change		Control station	
L	Level		Light		Low
P	Pressure		Test point		
S	Speed/Hz	Safety		Switch	
T	Temperature			Transmit	
V	Vibration			Valve (or mechanically actuated device)	
W	Weight		Well		

Goodheart-Willcox Publisher

Figure 34-7. Understanding tag numbers and how they are used at your facility will increase your efficiency as a technician.

- Second and third letters—Define the type of element or function, such as sensor, transmitter, valve, etc. In the example shown, the element is a transmitter.

- Group of numbers—Refers to a location or system, such as a specific building, room, part of a piping system, loop number, or other criteria defined by the process engineer.

The *instrument identification number*, or *instrument ID number*, is specific to an individual instrument. Many times the tag number/instrument ID number is present somewhere on the actual system element to aid in identification.

An *instrument index* lists all instruments in the facility, including the tag number/instrument ID number, purpose, and associated system of each. This list may also include items such as operational notes, material history, and calibration records. Depending on the facility, the instrument index may be kept as a physical printed document, such as an Excel spreadsheet, or part of a computerized maintenance management system (CMMS).

A control loop with tag numbers is shown in **Figure 34-8**. In this example, a temperature transmitter is connected by a well (TTW) to the system and outputs an electrical signal. The tag number 5-3 indicates that this is system 5 and temperature transmitter #3. The temperature controller with indication (TIC) in loop #102 receives the signal from TTW and outputs an electrical signal. The temperature relay/transducer (TY) receives a proportional signal and outputs a proportional pneumatic signal to control the valve.

Most P&IDs do not show the complete system, nor all mechanical elements. Only those mechanical elements of the system that are required to show function of the control loop are usually included. Additional piping diagrams of the system and other electrical diagrams would be needed to have a complete picture of the system.

34.2 INSTRUMENTATION

In this section, we will examine common instrumentation, controllers, and customary elements of a control loop. While the most common elements are the primary sensor (instrument), controller, and actuator, other elements may include a transducer, relay, indicators, and data recorders.

34.2.1 Temperature Sensors

Many processes control temperature through heating or cooling. Basic types of temperature sensors commonly seen in industrial settings are the temperature switch, thermocouple, and resistance temperature detector (RTD). A *temperature switch* senses temperature and provides an output once the set point is reached.

A *thermocouple* is a connected pair of wires made of dissimilar metals that produce a voltage when heated. Output from a thermocouple is in the millivolt range, so it must be connected to an external temperature transmitter (shown in **Figure 34-9**) to provide a standardized

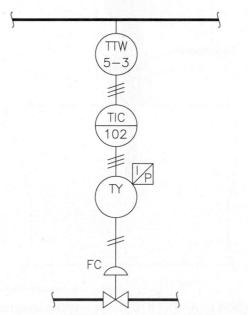

Goodheart-Willcox Publisher

Figure 34-8. A control loop example for a pneumatically operated valve with common symbols.

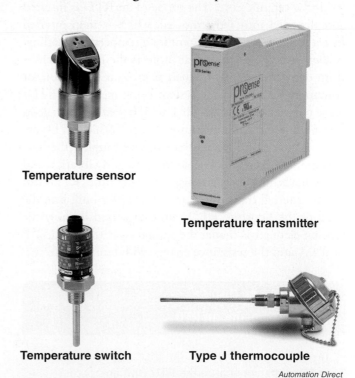

Temperature sensor

Temperature transmitter

Temperature switch

Type J thermocouple

Automation Direct

Figure 34-9. Various temperature switches, sensors, and thermocouples.

output or reading. Thermocouples are classified by type, each with different temperature ranges. Types J, K, and T are most commonly found in industry, **Figure 34-10**.

When replacing a thermocouple, an exact replacement is required. Consider the thermocouple type, whether it is grounded or not, sheath material (stainless steel or Inconel), and type of wire. While thermocouples are typically less expensive than resistance temperature detectors, they also have a nonlinear response to temperature change. This nonlinear response is small and does need to be compensated for using a temperature transmitter to achieve acceptable precision.

Common Thermocouple Types

Type	Range	Error
J	32° to 1380° F	.75%
T	–328° to 660° F	.75%
E	–328° to 1650° F	.5%
K	–328° to 2280° F	.75%

Goodheart-Willcox Publisher

Figure 34-10. Common types of thermocouples and corresponding temperature ranges.

A *resistance temperature detector (RTD)* is a sensor that reacts to temperature changes by producing changes in electrical resistance, **Figure 34-11**. This sensor is made of a thin wire, typically platinum, copper, or nickel, coiled around a ceramic core. The probe of an RTD is inserted and threaded into a thermowell, which is then inserted into a fitting within the system for a temperature reading. A *thermowell* is a sleeve that protects the inserted sensor from damaging elements, such as pressure, corrosion, or contaminants, within the system being measured. RTDs sense temperatures from –60°F to 570°F and have a good linear response to temperature change. While RTDs are more accurate and stable than thermocouples, they are also more expensive, less able to withstand vibration and shock loads, and have a longer response time. RTDs can use an external transmitter or a transmitter built into the head of the housing. RTDs are categorized by how the sensing element is made, the resistance at freezing (32°F or 0°C), and the resistance change with temperature.

TECH TIP
Removing an RTD

If removing an RTD from the thermowell, use an appropriate wrench on the RTD with an appropriate backup wrench on the thermowell. This will prevent you from unscrewing the thermowell when removing the RTD. If the thermowell is removed and the system is under pressure, you could be seriously injured or damage the RTD.

A *temperature transmitter* receives input from either a thermocouple or RTD and outputs a proportional signal to a controller or programmable logic controller (PLC). Newer models may be set up through software and direct communication from a PC. Older models are usually designated for a specific temperature range and are calibrated by adjusting potentiometers according to the manufacturer's documentation. Note that the selected transmitter range may affect the overall accuracy of the system. For example, if the temperature reading for a process normally lies within a band of 400°F–500°F, a transmitter with a range of 0°F–1000°F will not be as accurate as a transmitter with a range of 0°F–500°F. The 0°F–1000°F transmitter will scale 4–20 milliamps linearly between 0°F–1000°F (4 mA at 0°F, 20 mA at 1000°F). See **Figure 34-12**. Due to the system temperature range in this scenario, the output of the transmitter will never reach 20 mA.

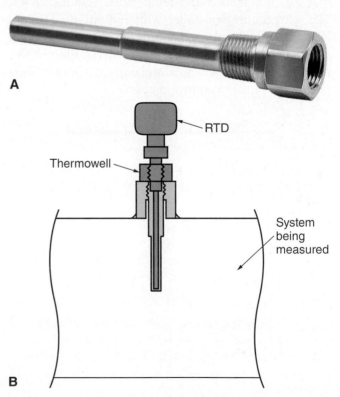

Automation Direct/Goodheart-Willcox Publisher

Figure 34-11. A—The RTD probe length is inserted and threaded into a thermowell. B—The RTD probe, protected by the thermowell, extends into the system to sense system temperature.

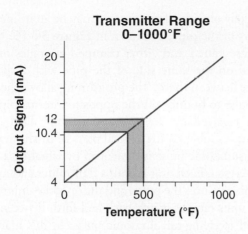

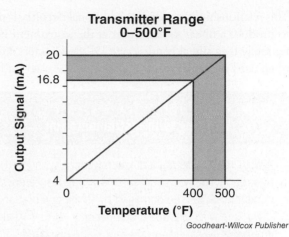

Figure 34-12. Graphs representing a Type J thermocouple reading a typical system process temperature of 400°–500° F. A—The typical output of a transmitter with a range of 0° F–1000° F will only be a 1.6 mA difference between 400° F and 500° F. B—A transmitter with a range of 0°F–500° F will typically produce a 3.2 mA difference between 400° F and 500° F.

THINKING GREEN
Range Adjustment

Adjusting a system's range can be an effective method of reducing energy consumption. Increasing both the deadband and overall temperature in a cooling system, for example, decreases the compressor motor current used and running time, which saves energy.

TECH TIP
No Voltage Thermocouple

When troubleshooting a thermocouple that is not producing a voltage (millivolt range), it likely needs to be replaced.

34.2.2 Flow Meters

Flow meters are categorized by the method they employ to measure flow. **Contact flow meters** are physically placed in the line of flow and are better suited for clean fluids. **Noncontact flow meters** are not in the line of flow and can be used with slurries or even gases. The most common types of flow meters used in industrial processes include the following:

- Differential pressure flow meters.
- Magnetic inductive flow meters.
- Ultrasonic flow meters.

A **differential pressure (DP) flow meter** senses pressure on each side of a fluid orifice or venturi. An **orifice** is usually a sharp-edged restriction in the fluid flow path, while a **venturi** is a narrow, nozzle-shaped area in the fluid flow path. Each of these shapes produces a pressure drop, or energy loss, which is compared by reading the pressure in two areas. See the illustration in **Figure 34-13**. As fluid is forced into the orifice or venturi, fluid velocity increases. As velocity increases, pressure drops at the throat of the restriction. The difference between the incoming pressure and pressure at the throat of the restriction is proportional to the velocity of the fluid. As the fluid flow rate increases, the pressure differential also increases.

Flow rate is proportional to the square root of the differential pressure drop across the orifice or venturi:

$$Q \approx \sqrt{\Delta P}$$

(34-1)

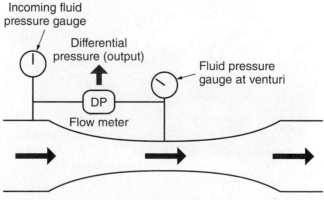

Figure 34-13. A differential pressure flow meter develops a signal based on the differential pressure across a restriction to flow.

This relationship is not linear. Additional circuitry is needed to produce a linear signal, either at the instrument itself or remotely by a signal conditioner. DP flow meters are more often used for gas or steam, but also can be used for fluids.

TECH TIP
Working with Instrumentation

Always review the manual when working with instrumentation. Many instruments have fragile internal components that can be damaged by shock or vibration. If working with electronic components, always wear a grounding wrist strap to prevent static discharge.

Magnetic inductive flow meters develop an output signal based on the voltage created by the conductive fluid passing through the meter. *Faraday's law* states that a changing magnetic flux will induce an electromotive force (voltage) on a closed circuit. Nearly all fluids are conductive and produce a voltage when moving through the meter's applied magnetic field. Exceptions to this include reverse osmosis and distilled water. **Figure 34-14** shows a smaller (2″ NPT) magnetic flow meter that is placed inline with the fluid flow. The output signal developed is proportional to the flow velocity, measured in gallons per minute (gpm).

Ultrasonic flow meters use an ultrasonic pulse to develop a signal that is proportional to velocity, or flow rate, of a fluid. These meters provide accurate measurement that is not affected by fluid characteristics. Ultrasonic flow meters can be a clamp-on style that attach

completely outside of the pipe or may be an inline, integral part of the pipe, as shown in **Figure 34-15**. Instruments are paired and either clamped at angles or both clamped on the same side of the pipe with a specified distance between them. The positioning allows the ultrasonic pulse to bounce off the opposite internal pipe wall and be received by the appropriate unit. Demanding applications may require several pairs of instruments.

An ultrasonic pulse is developed by vibrating a piezoelectric crystal. Each unit is both a transmitter and receiver. One unit produces the pulse, and the opposite unit receives it. The units rapidly switch back and forth between emitting and receiving an ultrasonic pulse. In this manner, a pulse is passed to the downstream receiver and then passed back. By measuring the difference in the time of travel, the

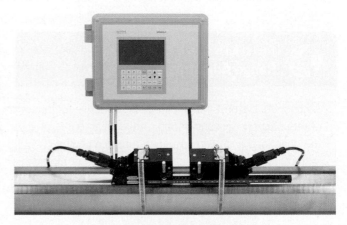

Clamp-on ultrasonic flow meter

Inline ultrasonic flow meter

Siemens AG

Figure 34-15. Clamp-on ultrasonic meters provide no interruption to flow while taking a measurement. Inline ultrasonic meters have greater accuracy, but require prior planning.

Pushbuttons for setup
Digital display

Automation Direct

Figure 34-14. On this magnetic flow meter, local digital reading is configured using the pushbuttons. Larger-sized magnetic flow meters are available for use by municipalities.

instrument develops a signal proportional to the velocity of the fluid. No difference in time means that there is zero velocity, or zero flow. The faster a fluid (or gas) moves, the greater the difference between the time a pulse is sent and received. Since the fluid does not change in pressure, temperature, or viscosity between pulses (milliseconds), the condition of the fluid does not impact the accuracy of the measurement. By using the time difference of the pulses and diameter of the pipe, an output signal proportional to flow rate is developed.

TECH TIP
Ultrasonic Flow Meters

Ultrasonic flow meters must be properly aimed and spaced in order to measure flow. Review the manufacturer's recommendations prior to installing the instruments.

34.2.3 Level Sensors

Level of a fluid can be accurately measured by weight of the fluid or by direct level measurement to develop a proportional signal. A proportional signal is developed by measuring the weight of a fluid and scaling the weight from a zero point to maximum level. Applications that measure the weight of a fluid may include batch operations and mixing operations usually associated with chemical processes or food processing. For direct level measurement, capacitive and ultrasonic instruments are the most commonly used for industrial processes. Applications that use direct level control range from steam systems in power generation facilities to the cooling systems used in plastic injection molding processes.

Capacitance is the ability to store an electrical charge. *Capacitive level sensors* use a change in capacitance to produce a signal that indicates the level of the liquid. This method can be used for many different fluids and with both conductive and nonconductive tanks. The dielectric medium between the plates of a capacitor determines the amount of capacitance. By increasing the dielectric strength between the plates, the capacitance increases.

An insulated probe is used for conductive liquids (water-based). As the liquid level rises, the capacitance increases because of the insulated probe. In nonconductive liquids, the liquid itself increases the capacitance. As liquid level rises, the capacitance increases. For nonconductive liquids, a grounding probe (noninsulated) is used. By calibrating the empty and full levels of the tank, a signal is developed that is proportional to the height of the

fluid. If the tank is of nonuniform size (sloped, cone, etc.), further scaling must be performed to accurately reflect the amount of fluid with changes in height.

Ultrasonic level sensors are noncontact sensors that use ultrasonic waves to measure the level of fluid and solid material. A proportional output is developed depending on the length of time it takes for an ultrasonic pulse to emit and return to the emitter. The sensors are calibrated to zero and maximum, which corresponds to the 4–20 mA (or voltage) output. Ultrasonic level sensors are used for level measurement in a wide range of applications:

- Tank liquid levels.
- Open channel flow.
- Nonliquid heights of piles (gravel pile).
- Nonliquid heights of cones or cylinders (grain or salt).

Ultrasonic level sensors typically include an integral controller. Integral controllers can be set up using onboard key functions or by connecting to a PC with the proper configuration software. Controllers may offer direct digital reading, multiple relay outputs, and proportional voltage or current outputs. A controller's digital reading may be offered in a number of different engineering units (gallons, pounds, cubic feet) and must be programmed to the specific application.

TECH TIP
Sensor Mounting Connection

Tightening a sensor's mounting connection too tight may produce a "ring" and interfere with the instrument's ability to properly measure level.

34.2.4 Signal Conditioning

Signal conditioning is the processing or manipulation of an output signal to make it usable by the next step in the system. The purpose of signal conditioning depends on the control system and may be needed for various reasons including the following:

- To interact with other systems of different voltages.
- To compensate for loop loading, such as voltage drop due to excessive wire distances or multiple outputs.
- To provide a response to a low, high, or other specific signal level in order to affect system control, such as an operator alarm.

Depending on the design of the control system, there may be signal conditioning components in the control loop. A *relay output* triggers an isolated relay contact output, normally open (NO) or normally closed (NC), at a set input current level. An analog to relay limit alarm is shown in **Figure 34-16**. While many controllers offer programmable relay outputs, a stand-alone adjustable relay output may be needed to interact with another control system.

Because of system loading or distance between the sensor output and controller, an *isolated signal conditioner* may be used. This element receives a 4–20 mA signal from the sensor and outputs a separate isolated 4–20 mA signal, which can be adjusted for load. **Signal converters** allow a wide range of proportional inputs to be rescaled to a selectable output range. Ranges include 0–5 V, 0–10 V, 0–20 mA, and 4–20 mA.

A **transducer** converts one signal type to another. An **I/P transducer**, **Figure 34-17**, receives an electrical signal (current or voltage) and converts it to a proportional pneumatic output (pressure). For example, an I/P transducer receives a 4–20 mA signal and outputs a proportional 3–15 psig pneumatic signal. An E/P switch is commonly used to convert an electrical signal to a proportional

pneumatic output signal. This is discussed in greater detail in a later chapter. Various output ranges are available, but 3–15 psig is the range commonly used for industrial and HVAC applications. These types of transducers come with calibration adjustments to match the output signal to the input signal level. A transducer can convert an electrical signal into a mechanical signal, or a mechanical signal into an electrical signal.

Figure 34-17. A typical use for this I/P transducer is to control a diaphragm-operated valve to control fluid flow.

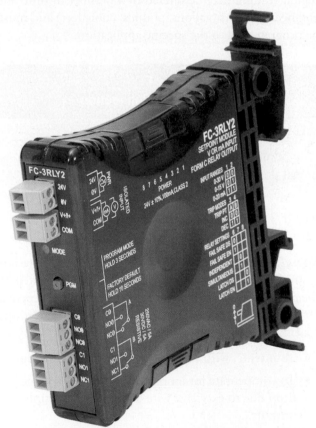

Figure 34-16. This relay module can be configured for a variety of alarm and control applications.

PROCEDURE	Connect and Test an I/P Transducer

To test an I/P transducer, assemble a test bench that includes regulated air pressure, indication gages, a signal generator (voltage and amperage), and digital multimeter. To check an I/P transducer with a properly equipped test bench, perform the following:

1. Supply a current or voltage signal to the unit.
2. Adjust the current or voltage input to the zero point.
3. With proper air pressure supplied to the I/P transducer, check that output air pressure is at the low end of the band specified for that I/P transducer.
4. Slowly ramp input current or voltage up to the top of the range (20 mA or 5 V), while checking for proper pressure response (15 psig, depending on range).

Controllers may be local stand-alone controllers, part of a PLC, or part of a computer as a *distributed control system (DCS)*. A local controller is shown in **Figure 34-18**. Controllers may be configurable or programmable by both keypad and PC software. Controllers typically include displays for input and output levels, and may show these in engineering units. The configuration of the controller greatly depends on the system being controlled, which will be covered in later sections of this chapter.

SAFETY NOTE
No Live Work

Remember that no live work is allowed by the code of federal regulations. Working inside of a control panel is only allowed to take troubleshooting readings. Proper care must be taken to prevent shock and equipment damage.

Automation Direct

Figure 34-18. Controllers may control several loops and may have numerous programmable relay outputs.

34.3 CONTROLLERS

The type of control loop implemented is determined by the required control, variables controlled, expected system response, and process type (such as batch, continuous, or cascade). This section will introduce the three main modes of operation for a controller, as well as different categories of control loops.

34.3.1 P, PI, and PID Modes

Instruments develop a proportional signal in response to a measurement. A controller operating in *proportional mode (P)* develops an output signal based on the error

between the set point and current controlled variable. In addition, the controller may have other filters in place, such as averaging, smoothing, or output delay. The set point may also be reset (or changed) by additional inputs. For example, a heat exchanger transferring energy from steam to water controls the flow of steam into the unit by measuring the temperature of the exiting water. If the temperature of the water drops below the set point, the controller responds and sends a higher signal to the steam valve. The normally closed valve is opened. More heat transfers into the water, and the temperature of the outgoing water rises. This type of negative feedback loop is common in control systems. **Figure 34-19** shows a system response after just starting up a similar system.

Proportional Control

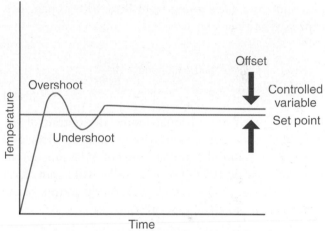

Goodheart-Willcox Publisher

Figure 34-19. This graph shows proportional system response to a large error with gain properly tuned.

A controller operating in proportional mode only is characterized by overshoot and an eventual *offset* (or error) between the controlled variable and set point. In some cases, offset may be minimized by manual adjustment. The amount of offset is referred to as *magnitude* or *magnitude of error* and used when comparing different values of error over time.

One important adjustment for controllers is gain. *Gain*, also called *amplification*, is a comparison or ratio of the output change compared to the input change. When a change of 20% to the input corresponds to a change of 60% to the output, the gain would be 3.

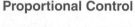

$$\text{gain} = \frac{\text{output \% change}}{\text{input \% change}}$$

(34-2)

On older controllers, gain may be adjusted by adjusting a potentiometer. On newer controllers, the gain is adjusted by making changes to parameters. **Figure 34-20** shows how adjusting gain can affect controller output. Note that this response can be seen with either a set point change or system load change, which increases error. When gain is adjusted to a greater ratio, the variable has more overshoots and undershoots, but a smaller eventual offset. When gain is decreased, there is little if any overshoot, but steady-state error is increased. Increasing gain improves system response, but only to a point. Increasing gain too high can result in an unstable system that fluctuates back and forth constantly, being too sensitive to inputs.

Another way to define gain is *proportional band*. **Proportional band** is the amount of change in the controller input that will cause a 100% change in the output, such as a valve going from 0 to 100% open. It is expressed as a percent.

$$\text{proportional band} = \frac{100}{\text{gain}}$$

(34-3)

In this type of proportional mode, the output of the controller is determined by the product of error and gain. With a gain of 1 and an error of 10%, the output would change by 10%. If the controller had a gain of 1.5 and the change in the input was 20%, the output of the controller would change by 30%.

$$\text{controller output } \Delta\% = \text{error } \Delta\% \times \text{gain}$$

(34-4)

Effect of Adjusting Gain

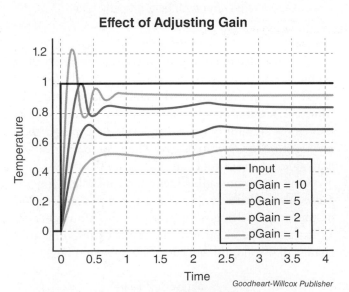

Figure 34-20. Gain adjustment has an effect on error, but also system response (controller output).

When operating in *proportional plus integral mode (PI)*, error over time is considered. The offset (error over time) between the set point and process variable changes the output of the controller through feedback, until the process variable equals the set point. An integral (in a mathematical sense) is an area under a curve. In **Figure 34-21**, the error over time (area under the curve) adds an additional element to the output of the proportional controller, and eventually eliminates the offset.

The *proportional-integral-derivative mode (PID)* considers the slope of the process variable over time. With increased error, PID mode allows a fast response. System response is shown in **Figure 34-22**. Depending on gain and system response, an overshoot may occur. With a larger magnitude of error, controller output is greatly increased until the process variable begins to reach the set point. Since the PID mode depends on the slope (or rate of change), when error is large and changing quickly, the derivative portion of the output is large. When the rate of change for error is small, the derivative portion of the output is small.

The mode used by the controller depends on a number of system factors:

- System response time.
- Allowed offset or error.

Proportional Plus Integral Mode (PI)

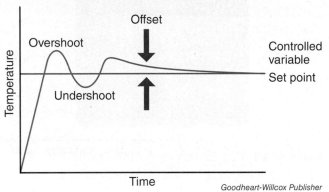

Figure 34-21. PI mode reduces offset over time.

Proportional-Integral-Derivative Mode (PID)

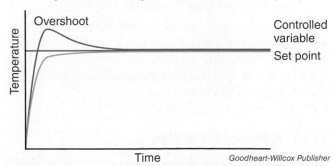

Figure 34-22. PID mode reduces overshoot and time to bring the control variable to the set point.

- Possibility of load disturbances.
- How often the set point is changed.
- System capacitance (storing of energy).

Proportional mode is the easiest to tune and has the most stability, as long as an offset is acceptable. In some cases, the offset may be reduced by adding bias. Proportional plus integral mode is commonly used for controllers. This mode reduces offset to nearly zero, depending on instrument accuracy, but can be more difficult to tune. Proportional-integral-derivative mode offers even better results than PI, but a noisy input signal leads to controller output instability. Note that both the integral and derivative portions are associated with time. These timing settings must be set appropriately to achieve the best system response. Refer to the documentation for the particular controllers (or PLCs) for appropriate settings.

TECH TIP

Changing Controller Settings

If a controller and system have been working for some time in an acceptable manner, you do *not* need to go into the controller's programming or settings to determine why the system has suddenly yielded strange results. Changing the settings may result in massive system instability or sluggish response.

34.3.2 Types of Control Loops

The types of control loops and systems implemented are based on the process they control. A simple and straightforward control loop is often preferred. Simple control loops are more easily tuned and have fewer possible failures. As control loops become more complex, more information, time, and effort are needed to properly understand, control, calibrate, and troubleshoot the system.

Single Control Loop

Most control loops fall into the category of a single variable loop. This type of control loop senses one process variable and maintains it by controlling an actuator. This loop is widely used in industrial process control and HVAC to maintain level, flow, temperature, or pressure. HVAC applications also commonly use a reset. When referring to a control loop for HVAC, a **reset control** directly resets the set point of the controlled variable.

This type of feedback loop is normally the easiest to maintain and tune, but may not be able to control larger systems or systems with a large capacitance. **Figure 34-23** shows a typical temperature control loop that falls into this category. In this example, steam heats the shell of the heat exchanger, which heats the water flowing through the tubes inside. The RTD monitors the temperature of water exiting the heat exchanger and develops a control signal proportional to the temperature of the water. The controller compares the signal from the RTD to an internal set point and develops an output signal that is proportional to error

Heat Exchanger Control Loop

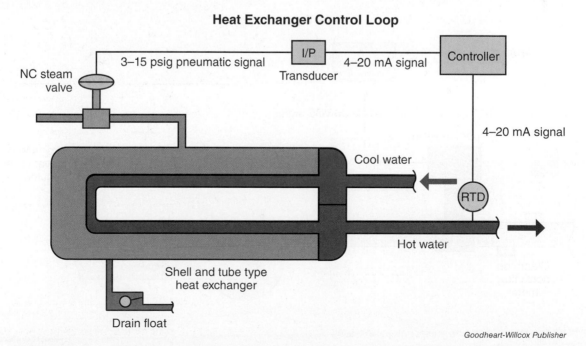

Figure 34-23. A simple control loop for heating water through a heat exchanger.

between the input signal and set point. The I/P transducer receives the signal and converts it to a pneumatic signal, which is applied to the normally closed steam valve to control its position. In this way, if the temperature of the water exiting the system falls, the feedback loop opens the steam valve to raise the temperature inside the heat exchanger. If the temperature of the water exiting the system is too high, the system responds by closing the steam valve slightly to lower the temperature inside the heat exchanger.

Flow Control Loop

Many times, flow is controlled in order to control temperature of a process fluid. Flow can be controlled by:

- Controlling the speed of a pump by controlling a variable speed drive.
- Controlling the resistance to flow on the discharge side with a control valve, as in the case of a centrifugal pump.

Controlling flow results in a quick feedback signal because fluids are incompressible. Overall response of the control loop is typically limited by the response time of the actuator. **Figure 34-24** shows a flow control loop using a centrifugal pump and discharge-side control valve. In this example, a sonic flow meter develops a signal proportional to the discharge flow rate and sends a signal to the controller. The controller compares this signal to an internal set point and develops an output signal proportional to the error between the input signal and set point. The signal from the controller is sent to a motorized proportional valve that opens or closes depending on the signal received. A manual valve allows a minimum amount of recirculation to prevent the pump from overheating on a low flow condition. The pump runs at a constant speed. If downstream pressure drops, more flow takes place across the control valve. The sonic meter detects this increase in flow and the controller responds by closing the valve slightly to reduce flow back to the normal set point.

Level Control Loop

Level control can be accomplished either by controlling the speed at which a vessel fills or the speed at which a vessel is emptied. The method depends on the purpose of the system. In water reclamation, for example, the incoming flow from households cannot be controlled—it increases as more people discharge to the drainage system. Instead, the output flow from the vessel (such as a tank or open sump) is controlled in order to ensure the vessel neither overflows nor empties. As flow into the tank increases, pump speed on the discharge side increases. The tank acts as a buffer. In this case, the system response time is dependent on the size of the storage tank, the flow coming into the tank, the flow going out of the tank, and the comparison between the two flows.

The response time of the control loop needs to be fast enough to prevent the tank from overflowing. **Figure 34-25** illustrates this type of loop. In this system, the fluid level in the tank level is maintained by varying the speed of a pump.

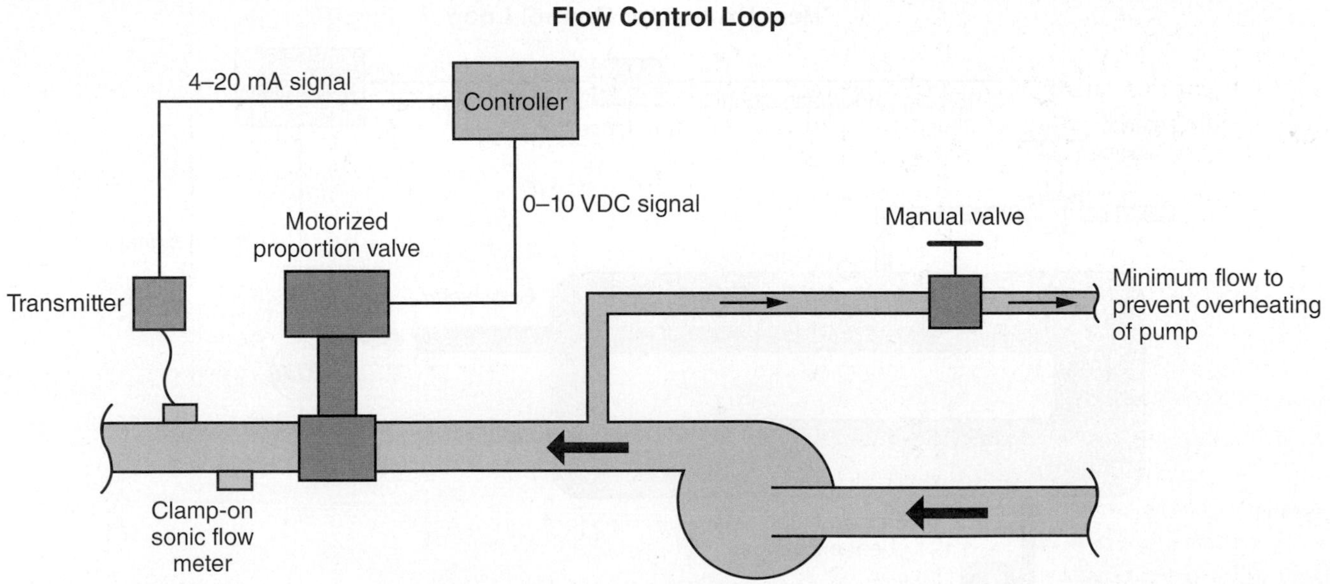

Flow Control Loop

Figure 34-24. The motorized valve is controlled by a 0–10 V signal, but actuated by either a 24 VDC or 120 VAC motor.

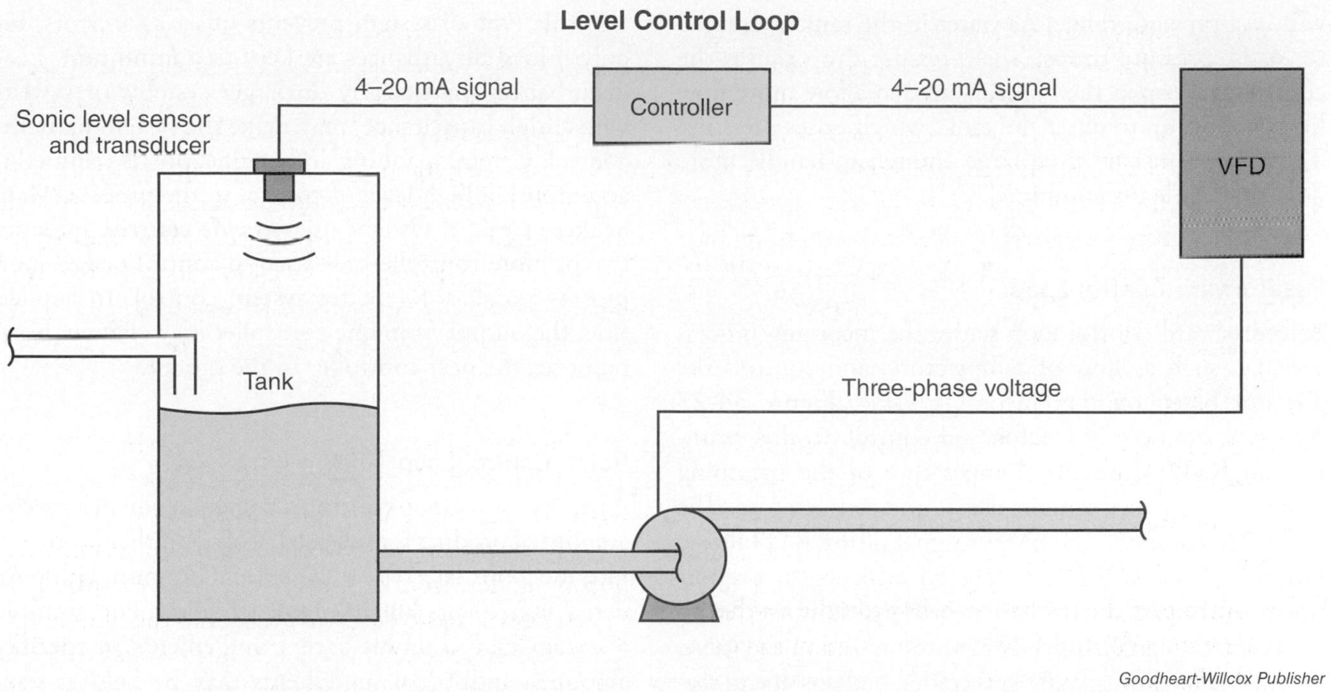

Level Control Loop

4–20 mA signal   Controller   4–20 mA signal

Sonic level sensor and transducer

VFD

Tank

Three-phase voltage

Figure 34-25. In a level control loop, the maximum output flow of the pump must exceed the maximum flow coming into the tank.

The input flow into the tank is variable and cannot be controlled. The fluid level inside the tank is controlled in order to prevent the pump from cavitating (damaging itself) and to prevent the tank from overflowing. As the fluid level in the tank increases, the sonic level sensor increases its signal output to the controller. The controller compares this signal to an internal set point and develops an error signal proportional to the difference between the incoming signal and set point. The output signal is sent to a variable frequency drive that responds by speeding up the pump to maintain the desired level in the tank. If the fluid level were to decrease, the control loop would slow the pump to maintain the set level.

Pressure Control Loop

A pressure control loop senses pressure and acts on an actuator to control that pressure. Systems may respond quickly or slowly, depending on the system design. Systems that involve a compressible gas, steam, or air tend to respond slower than fluid systems that are not compressible (hydraulics, water only).

Pressure may be controlled to ensure proper flow. As pressure varies across a valve, flow will also vary. **Figure 34-26** shows an example of a system that combines water with pressurized air. In this example, water flows to loads (other pump mechanical seals) and is kept

Pressure Control Loop

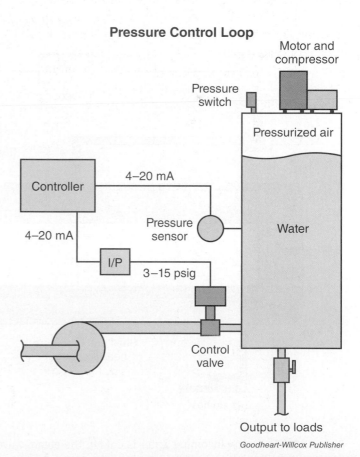

Motor and compressor

Pressure switch

Pressurized air

Controller   4–20 mA

Water

4–20 mA

Pressure sensor

I/P   3–15 psig

Control valve

Output to loads

Figure 34-26. This air over water system maintains pressurized water for pump mechanical seals.

within a pressure range. As water in the tank is drained to loads, pressure drops. This pressure drop causes the controller to open the control valve to allow more flow from the pump to enter the tank, which raises pressure. The pump must be sized large enough to handle more flow than all loads combined.

Feedforward Control Loop

A feedforward control loop senses the incoming process variable, such as flow or temperature, and controls the actuator based on that incoming value. **Figure 34-27** shows an example of feedforward control. In this example, an RTD senses the temperature of the incoming water in order to maintain the output temperature. The controller compares the output signal of the RTD to the internal set point and develops an error signal proportional to the difference. This error signal is converted to a pneumatic signal, which is used to reposition the valve. As outside incoming water gets colder, perhaps due to seasonal weather changes, the steam valve is opened to provide more heat to maintain the output water temperature.

This type of system prevents offset and errors, but only if load disturbances are kept to a minimum. Load disturbances, particularly on larger systems or systems with a high capacitance, may make the system harder to control. Careful planning and engineering is required to accomplish a high level of control in the process. Many of these types of systems use *cascade control*, in which two or more controllers are used to control one or more process variables for better system control. In application, the output from one controller may change the set point for the next controller in the process.

Batch Control Loop

A batch control loop controls a system in which a specific amount of product is processed before another is brought into the loop. This type of system is common with powdered ingredients and processed liquids. For example, a system may combine several ingredients in specified amounts. Individual ingredients may be held at separate temperatures using their own control loops. Various stages of manufacturing may be performed before a final

Feedforward Control Loop

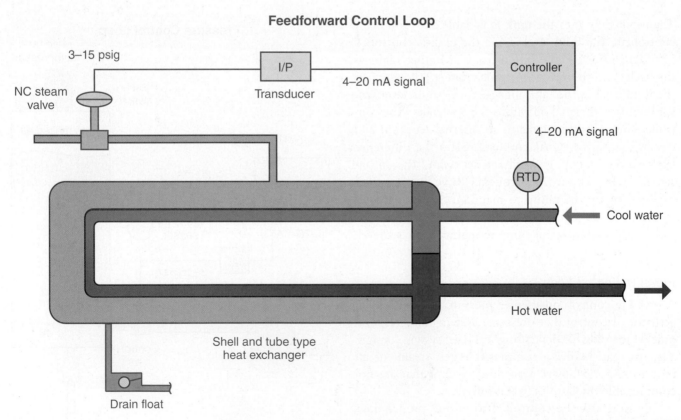

Figure 34-27. As the incoming water is colder, the steam valve opens.

product is packaged. Stages might include heating, mixing, cooling, resting, and curing. **Figure 34-28** shows a simple example of this type of process.

Some processes may need to be shut down and cleaned between batches, especially with items meant for consumption. This may entail a complete tear down and cleaning, or simply a flushing to rid the system of previous ingredients. Parts of the process may be mobile to help with cleaning efforts. Processes may last from minutes to months, depending on the product. While mixing and packaging dry ingredients may happen quickly, curing meats happens under strict conditions over time.

Because a batch process limits output, multiple parallel processes may run at the same time. Processes may be staged or staggered so that overall output is relatively constant. Since processes are started from a condition of maximum error (offset), PID modes are typically used.

34.4 CALIBRATION AND TROUBLESHOOTING

Control loop *tuning* is the process of adjusting parameters of a controller to achieve the desired performance of the controlled variable, which includes accuracy, precision, response, and stability. In this section, we will

Batch Process

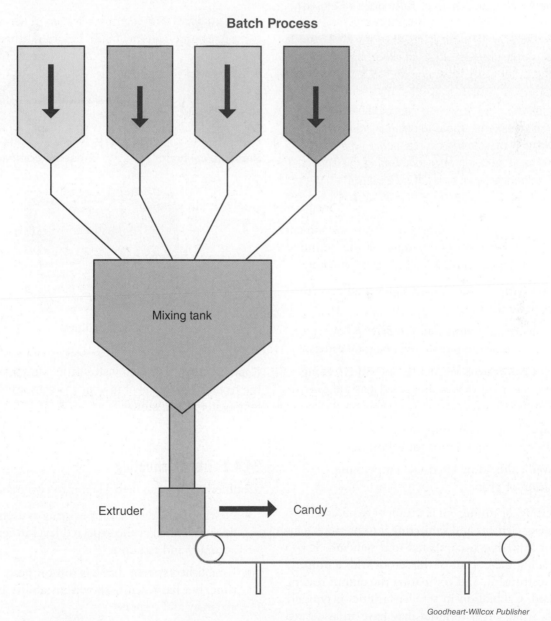

Goodheart-Willcox Publisher

Figure 34-28. Batch processes are used to combine and process ingredients.

examine tuning the process and troubleshooting the process. While tuning is performed initially, troubleshooting is performed after the system has shown acceptable response and control over a period of time.

34.4.1 Calibration

Checking for proper calibration and controller response can be accomplished using a signal generator. The actuator may also be checked at the same time. From the instrument or sensor end, sending a 4 mA or 20 mA signal to the controller should result in the proper output from the controller (high or low) and the proper response of the actuator (such as fully open or closed).

SAFETY NOTE
Stroking an Actuator

Stroking an actuator, or cycling the actuator from one position to an extreme opposite position, will have an effect that will cause a system reaction. Proper safety measures must be taken prior to initiating this action, such as isolating a valve or isolating a pump.

Checking for proper calibration of an instrument itself requires further tooling. Some of the common tools, shown in **Figure 34-29**, include the following:

- **Process clamp meter.** Allows signal reading without breaking the loop.

- **Temperature calibrator.** Can simulate and measure RTDs and thermocouples to calibrate transmitters.

- **Pressure calibrator.** Verifies the pressure reading of a measurement device; may read static and/or differential pressure.

- **Pressure comparator.** Supplies pressure to test device and reference gauge for calibration.

- **Dry-well calibrator.** Used for temperature calibration of probes.

Calibrating, or tuning, an instrument requires removing the sensing element and subjecting it to pressure, temperature, or flow that is compared to a standard. It also requires reading the output of the sensor with a properly calibrated instrument, and comparing the output reading to a standard. Calibration in smaller facilities is typically outsourced, while larger facilities may have an associated or on-site calibration shop just for this purpose. In a large facility, calibration can be a large portion of the work performed by a process technician.

Process clamp meter   **Temperature calibrator**

Pressure calibrator   **Pressure comparator**

Dry-well calibrator

Reproduced with Permission, Fluke Corporation

Figure 34-29. Calibration instruments are specific to the type of measurement read by the sensor (pressure, temperature, flow, etc.).

34.4.2 Troubleshooting

Troubleshooting can be broken into two main areas:

- Newly installed system or existing system that has never operated at the expected level in terms of accuracy and response.

- Established system that has run properly for some time, but has recently shown instability in terms of accuracy and response.

Understanding and being able to use P&ID schematics is an important part of troubleshooting a complicated process control system.

New System

A new or existing system that has never operated appropriately in terms of accuracy and response may have one or more of several possible problems:

- Wrong type of controller mode (P, PI, PID) for the process.
- Excessive noise (input side of controller) due to interference.
- Incorrect sizing of a final actuator; too small or too large to properly control the system.
- Wiring or grounding problems.
- Inaccurate tuning.
- Incorrect planning or design processes. For instance, a system is expected to maintain temperature within +/− 2°F, but the system was designed to maintain temperature within a band of +/− 10°F.

The mode of control is an important factor. With most systems, PI control is acceptable. In some systems, such as temperature control, PID is preferred if system capacitance is large. *System capacitance* is the ability of a system to store energy, or resist a change in total energy. Think of trying to heat a 15,000 gallon tank with a small heating element—the system capacitance is very large. In most processes, the response of the sensor is fast when compared to the response of the system to a control action. In other words, a flow, level, pressure, or temperature sensor will respond much faster to a change in the measured variable than the system whole will respond. Just because PID is a more complicated control mode does not mean that it is the correct mode to use for a process.

When using PID mode, noise on the incoming signal to the controller can have drastic results. Noise typically includes spikes in the incoming signal from the sensors or transmitters. The controller sees the spikes as part of the input signal, and they appear as if a massive fast-acting error has occurred. This directly affects the derivative portion of the output, and can result in poor response and control. To help alleviate the noise problem, ensure that wires are correctly shielded, shields are grounded at one end, and wires are kept away from other high voltage sources. For example, running signal wires in their own raceway or conduit can help reduce noise.

Actuators that are too small to control the system load will show through sluggish system response. For instance, a small heating or cooling valve in the full open position may take a very long time to produce a noticeable system response. An actuator that is too large may

result in faster system response, but show a poor ability to throttle the flow at low levels. Imagine cooling a small bedroom with a 2-ton cooling unit.

Inaccurate tuning (gain too high or too low) or incorrect time settings can drastically affect controller response. Too much gain makes a controller overly responsive, while too little gain produces sluggish response with more offset.

Established System

A system that has been running properly for some time, but suddenly does not respond appropriately points to specific causes:

- Intermittent or ongoing outside interference (RFI, etc.).
- Load disturbance that is beyond the ability of the system to correct.
- Occasional system disturbance that places the system in an uncontrolled region.
- Inaccurate calibration of input settings or parameters, or failure of an input impacted the control system.
- Corrosion, interruption, or severing of wires.
- Sudden failure of sensors, transmitters, or controllers.

The first step in troubleshooting this system is to visually inspect all elements of the system for damage. Correct, repair, or replace any damaged elements in the system before continuing to troubleshoot. Next, generate a signal at the instrument or transmitter and verify that the controller is receiving the signal. Last, use the controller or signal generator to generate a signal that makes the actuator respond from 0–100%. Visually verify that the actuator responds. This ensures the wiring is intact. If the actuator responded to the first input from the transmitter or sensor, you know that both the controller and actuator are working. Check any mechanical linkages for proper response and integrity. If an actuator is not responding, it may be stuck due to friction or some internal cause. If a sensor or transmitter does not respond, it must be investigated further. A sensor may fail due to internal or external physical damage, exposure to an environment beyond its designed use, vibration, or electrical circuitry failure in either the sensor or transmitter.

A system that suddenly stops responding appropriately may be the result of a load disturbance that is beyond the ability of the system to correct. In this case, the load disturbance must be removed. The system may, however, react perfectly normal to this disturbance. A controller that is not responding properly (or responding as it used to) needs to be checked. It is beneficial to speak with the person who previously adjusted the controller settings. Ask the following questions:

1. When did you last work on the controller?
2. Did you change any settings? Which settings were changed?
3. What caused you to change these settings?

An operator may have changed settings inadvertently, or something in the process may have recently changed and settings were adjusted to better control the process. Confer with someone very familiar with the system and its normal response to learn exactly how the system has changed recently.

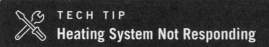

TECH TIP
Heating System Not Responding

If a plant-wide heating system is taxed to the point of not responding or maintaining temperatures, it may simply be a proper system response. Perhaps someone has left a bay door open in the middle of winter. No heating system can counteract this loss of heat and maintain the building temperature.

A system that has been running properly for some time, but is slowly showing signs of not responding properly points to either instrument calibration or a slow, natural process building up to prevent system control. In a hydraulic system that is cooled by an air cooler or heat exchanger, for example, the system will start to run at a higher equilibrium temperature as a layer of dust, dirt, or other contamination builds up on the heat transfer surface over time. If left unchecked, this loss of heat transfer will eventually impact the cooling system. The system will run excessively hot and may result in the failure of hoses. A controller will still respond when a sensor or transmitter is slowly drifting out of calibration; however, there will be more error or offset from the set point in the response.

CHAPTER WRAP-UP

Process control can involve different elements depending on the facility. In one instance, process control may take place with stand-alone controllers, while another process may use a programmable logic controller (PLC) or distributed control system (DCS). Process control may take place through purely mechanical feedback mechanisms, or can be complicated multi-variable inputs to a PLC in which the controller response is programmed. This is such a wide area of expertise that many technicians make this their life's work. From equipment calibration to controller tuning and process troubleshooting, the technology of control systems is always advancing. Technicians will always be challenged to learn the concepts, programming, and troubleshooting methods for new technology.

Chapter Review

SUMMARY

- A process control system uses various devices and methods to monitor and control a process to maintain a desired output.

- A control loop monitors the process variable within a process. Three basic functions of all control loops are to measure a variable, compare the measurement with the set point, and provide an output to change the variable.

- Error is the difference between the set point of a process variable and the current value.

- Deadband is the range of values for a process within which no action is taken. Larger deadband means a variable would have more variability. A variation that affects the system's ability to maintain the process variable is called *disturbance*.

- Feedback communicates the measurement of a process variable. Control systems that include feedback are closed loop, and those that do not incorporate feedback are open loop.

- P&ID shows the interconnections of physical system components, including control loops and instrument ID numbers. Troubleshooting and calibrating control systems depend on knowing how to use P&IDs.

- Common instrumentation and elements of a control loop include a primary sensor, controller, actuator, and various other elements. Depending on the process, the sensor may be a temperature sensor, flow meter, or level sensor, and may or may not involve signal conditioning.

- I/P transducers receive an electrical signal (voltage or current) and converts it to a proportional pneumatic output, such as pressure. These transducers come with calibration adjustments to match the output signal to the input signal.

- Gain is a ratio of the output change compared to the input change. Once gain is determined, proportional band can be calculated.

- The three main modes of operation for a controller are proportional mode (P), proportional plus integral mode (PI), and proportional-integral-derivative mode (PID).

- Various types of control loops are used for industrial processes, including single, flow, level, pressure, feedforward, and batch control loops. Most control loops fall into the category of a single variable loop.

- Tuning a process is adjusting the controller to achieve the desired performance.

- Troubleshooting is performed after the system has shown acceptable response and control over a period of time.

- If a process changes drastically, controllers may need to be retuned. If a process response slowly drifts out of normal, calibration of elements needs to be performed.

REVIEW QUESTIONS

Answer the following questions using the information provided in this chapter.

1. Describe three different forms of process control that occur in industry.

2. What is a process control system? Discuss how this system varies from a servo system.

3. Describe a process variable. What is an example of a process variable in a fluid system?

4. Explain the importance of process control.

5. A _____ is a component of a control system that monitors a process variable.

6. What are the three basic functions of all control loops?

7. *True or False?* When a thermostat senses the current value of indoor air temperature and compares it to a set point, it is sensing the error between the set point and process variable.

8. How does a disturbance affect a system? What happens if a system cannot keep up with a disturbance?

9. *True or False?* A control system that includes feedback is called an open loop.

10. Discuss feedback in a closed loop system. What is a feedback signal used for?

11. Any difference between the feedback signal and set point generates a(n) _____.

12. Scaling may be performed because the _____ responds to a different type of signal, or the signal needs to be changed to reflect the system.

13. What are piping and instrumentation diagrams (P&ID)?

14. *True or False?* The instrument identification number is used to identify all components of a control loop and differentiate them from other control loops.

15. What are common elements of instrumentation?

16. What are the similarities and differences between a thermocouple and a resistance temperature detector (RTD)?

17. Describe the three common types of flow meters used in industrial processes.

18. Nearly all fluids are conductive and produce a _____ when moving through a meter's applied magnetic field.

19. Explain how to receive a direct measurement of a fluid level.

20. What is the purpose of signal conditioning?

21. *True or False?* An I/P transducer receives an electrical signal and converts it to a proportional pneumatic output.

22. What is proportional mode? Discuss how it is characterized.

23. In a proportional controller, when the input changes by 80% and the output changes by 40%, what is the gain? What is the proportional band?

24. If gain is 1.5 and the percent of error is 5%, what will the controller output be?

25. How does PID control differ from PI mode?

26. What factors determine the mode used by a controller?

27. Describe how flow is controlled in order to control temperature of a process fluid.

28. Many feedforward control loop systems use a(n) _____ in which two or more controllers are used to control one or more process variables for better system control.

29. Give an example of a batch process controlled system.

30. How is an instrument calibrated?

31. Describe the two main areas of troubleshooting for systems.

32. *True or False?* PI is the preferred mode of control for new systems if the system capacitance is large.

33. List three potential causes of an established system suddenly not responding.

NIMS CREDENTIALING PREPARATION QUESTIONS

The following questions will help you prepare for the NIMS Industrial Technology Maintenance Level 1 Process Control Systems credentialing exam.

1. Which is an example of a process variable in a control system?

   A. A thermostat comparing temperature to its set point.
   B. A furnace running continuously with the front door open.
   C. Cooling a home using an air conditioner.
   D. The flow measured in a fluid system.

2. A feedback signal is compared against a set point, resulting in a(n) _____ that is proportional to the difference between the feedback and set point.

   A. deadband
   B. error signal
   C. load disturbance
   D. process variable

3. Based on the P&ID diagram, the following control loop indicates that _____.

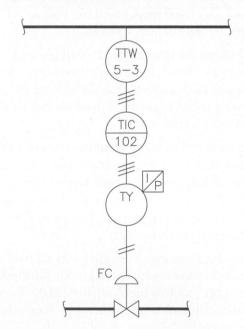

   A. a temperature transmitter outputs a pneumatic signal
   B. a temperature sensor outputs an electrical signal
   C. a temperature controller outputs a hydraulic signal
   D. a temperature transducer outputs a pneumatic signal

4. Which of the following statements is *not* true about sensors?

   A. They cannot be noncontact.
   B. An output is produced when a set point is met.
   C. They respond only to temperature changes.
   D. A proportional signal can be developed by measuring a variable in its system.

5. An I/P transducer comes with calibration adjustments to _____.

   A. make the mechanical signal proportional to the electrical signal
   B. match the mechanical signal to the electrical signal
   C. make the output signal proportional to the input signal
   D. match the output signal to the input signal

6. What mode considers the slope of the process variable over time?

   A. Proportional mode (P).
   B. Proportional-integral-derivative mode (PID).
   C. Proportional plus integral mode (PI).
   D. Proportional plus derivative mode (PD).

7. Which of the following is *not* a component of a single control loop?

   A. An actuator.
   B. A converter.
   C. A controller.
   D. A transducer.

8. How can an instrument be calibrated?

   A. By removing the sensing element and subjecting it to pressure, temperature, or flow that is compared to the standard.
   B. By reading the input of the sensor with a properly calibrated instrument and comparing it to the standard.
   C. By sending a 4 mA or 20mA signal to the actuator to check the output from the actuator and response of the controller.
   D. By troubleshooting the instrument before determining proper calibration.

35

Heating and Refrigeration

LEARNING OBJECTIVES

After completing this chapter, you will be able to:

☐ Describe the three different methods of heat transfer.

☐ Identify the major components of a steam heating system.

☐ Discuss the forms of heat that impact both heating and cooling processes.

☐ Discuss the processes involved in the refrigeration cycle.

☐ Explain the function of a hydronic heating system and steam boilers.

☐ Discuss common maintenance items for steam heating systems.

☐ Identify common refrigerants and their uses.

☐ Explain the function of compressors in cooling systems.

☐ Identify the pressures that act on a thermostatic expansion valve.

☐ Explain general maintenance tasks and troubleshooting for a refrigeration system.

TECHNICAL TERMS

azeotrope	latent heat of fusion
blowdown	latent heat of vaporization
capillary tube	make-up water
cavitation	positive displacement compressor
centrifugal compressor	
combustion	pressurized head tank
condensate	radiation
condenser coil	reciprocating compressor
conduction	rotary screw compressor
convection	rotary vane compressor
endothermic reaction	scroll compressor
exothermic reaction	sensible heat
feed water	slugging
fire tube boiler	superheat
heat exchanger	thermostatic expansion valve (TXV)
hydronic system	
insulator	water tube boiler
latent heat	zeotrope

Heating and refrigeration processes take many forms. From a residential furnace to a facility- or plant-wide heating and cooling system, HVAC (heating, ventilation, and air-conditioning) systems are required not only for comfort, but also to ensure proper operation of sensitive electronic components. Heating takes place using high volumes of heated forced air, heated water circulation (hydronic heating), steam, or a combination of these types. Likewise, cooling takes place either through cooled forced air or plant-wide chilled water. In this chapter, the basic concepts and components of heating and cooling will be examined, along with system operation and maintenance.

35.1 HEATING AND COOLING CONCEPTS

HVAC equipment and systems are continually integrating new technology to use energy more efficiently. HVAC control systems also continue to evolve by integrating new technologies, including automated digital control and wireless connectivity. However, the core concepts of heating and cooling have remained unchanged. This section examines the main concepts of heating and cooling systems.

35.1.1 Heat Transfer Methods

The main function of a heating or cooling system is to transfer heat from one place or system to another. Transferring heat takes place through three methods (**Figure 35-1**):

- Conduction.
- Convection.
- Radiation.

All of these methods are used in heating and cooling systems. Both conduction and convection are used in residential furnaces, floor heating or slab heating, and large industrial heat exchangers. Radiation is used with infrared lamps and infrared heating panels to heat localized areas for comfort.

Conduction

The most commonly used method of heat transfer is conduction. **Conduction** transfers heat from one molecule to another through physical contact. On a molecular scale, heat is transferred from the warmer material to the cooler material through collisions of atoms. The rate of heat transfer depends on the properties of the materials themselves, the size of the areas in contact, and the difference in temperature between the objects. For example, a section of copper pipe conducts heat more readily than a comparably sized cut of wood. Materials that are poor conductors of heat are called **insulators**. Common examples of insulator materials include wood, rubber, and fiberglass.

Convection

Convection transfers heat from one object or area to another through liquid or gas. In some instances, the flow of liquid or gas may take place naturally due to differences in density. In this case, the lower density of the hotter fluid causes movement or flow to produce convection. In other systems, fluid flow is forced using a

Conduction

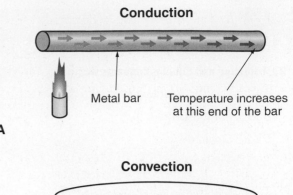

Metal bar   Temperature increases at this end of the bar

A

Convection

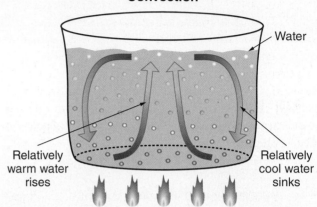

Water

Relatively warm water rises

Relatively cool water sinks

B

Radiation

Electromagnetic waves

Electromagnetic waves

Earth

Space

Sun

C

Figure 35-1. Heat is transferred via three methods. A—Molecules in the metal bar transfer heat from the end heated by the flame to other end of the bar. B—Water at the bottom of container becomes warmer due to heat from the flame (conduction), but heat is transferred in the water by convection as warmer water rises and cooler water sinks. C—The heat received by the Earth from the sun is transferred through radiation. Heat is released when the sun's rays strike an object.

pump. Both forced and natural convection transfer heat in the same manner. Much like conduction, the amount of heat transferred depends on the surface area in contact, the ability of the material to transfer heat, and the temperature difference between the materials.

Radiation

Radiation transfers energy through electromagnetic waves. All objects have internal movement at the atomic level. This molecular motion releases electromagnetic radiation that impacts other objects. Thermal radiation can take place over great distances, such as between the sun and Earth, or short distances, such as between a space heater and a person. Thermal radiation can be absorbed or reflected. While often within the infrared spectrum, this radiation uses a range of frequencies of emitted photons.

35.1.2 Combustion

Combustion is a chemical reaction between oxygen and fuel that produces energy in the form of heat. The process of combustion requires oxygen, fuel, and an ignition source to start the reaction. When a *complete* combustion process occurs, no fuel remains. This assumes the proper amount of oxygen is readily available to the reaction. Without the proper amount of oxygen, *incomplete* combustion occurs. Carbon monoxide (CO) and other contaminants are given off when incomplete combustion occurs. In industrial applications such as the use of boilers, flue gases are regularly checked to ensure maximum efficiency of the boiler and combustion process. Due to health concerns, carbon monoxide detectors are also used to detect a buildup of carbon monoxide. **Figure 35-2** shows the chemical equation for the combustion of propane.

Some reactions are exothermic and others are endothermic. An ***exothermic reaction*** is a chemical reaction that releases energy in the form of heat or light. Even though an exothermic reaction produces heat, it typically requires an input of energy to start. It may maintain itself thereafter with the heat produced by the reaction. A common exothermic reaction is the burning of fuel oils, natural gas, or wood to supply heating to a home. An ***endothermic reaction*** is a chemical reaction that absorbs energy, usually heat, from the surroundings. Cooking an egg is an example of an endothermic reaction. Heat energy from the stove burner is absorbed by the pan, which is applied to the egg inside the pan.

SAFETY NOTE
Exothermic Reactions

Some exothermic reactions are difficult to stop once started. Burning magnesium is one example of an exothermic reaction that takes energy to start, but will continue until the magnesium is completely burned with nearly no way of stopping the reaction.

35.1.3 Steam Cycle

At standard atmospheric pressure (14.7 psia), water boils at 212°F. At lower atmospheric pressures, water boils at lower temperatures. At higher atmospheric pressures, the temperature to boil is higher than 212°F. This relationship between pressure and temperature and its effect on vaporization is important in understanding both the steam and refrigeration cycles.

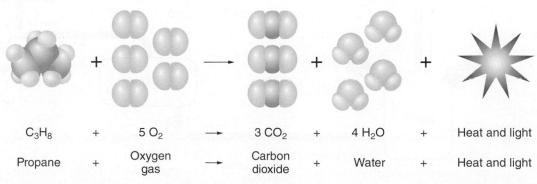

C_3H_8 + 5 O_2 → 3 CO_2 + 4 H_2O + Heat and light

Propane + Oxygen gas → Carbon dioxide + Water + Heat and light

Figure 35-2. This balanced equation shows how propane combusts and results in CO^2, water, heat, and light.

As pressure is increased, the temperature of boiling is increased. At 16 psig (30.7 psia), water will boil at approximately 250°F. At higher pressures, much higher temperatures are required to reach the boiling point. At nearly 600 psig, boiling temperature is almost 500°F.

Figure 35-3 illustrates a typical steam cycle used for heating purposes. This type of system typically runs with lower pressure steam than that used in power generation systems. The boiler adds heat into the system, which vaporizes water into steam. The steam travels to loads, such as radiators, convectors, or other heat exchangers, and transfers heat to the surrounding air or fluid being heated through a heat exchanger. As energy is transferred out of the steam, it condenses into water and returns to the boiler to continue the cycle. While the boiler adds the additional heat needed to cause vaporization, the loads remove this same heat. These cycles occur in many industrial heating and cooling processes.

Two forms of heat impact heating and cooling processes: sensible heat and latent heat. **Sensible heat** is heat that causes a change in temperature in a material and is measurable. For example, the heat applied to a pan of water on the stove raises the temperature of the water. The increase in temperature can be measured using a thermometer. When the heat is removed, the temperature of the water gradually drops. Sensible heat can be felt, sensed, and measured. **Latent heat** is heat that causes a change in state without changing the temperature of a substance. For example, a block of ice absorbs heat and starts to melt. It will not entirely melt all at once, therefore the temperature of the ice remains 32°F. The additional heat applied to keep the block of ice melting is latent heat. The temperature of the water rises only after all the ice is melted.

Latent heat of fusion is the heat required to cause a change in state from solid to liquid or liquid to solid. In the block of ice example, latent heat of fusion is the heat that must be added to turn the ice to liquid form.

As heat is added to the now melted ice, the water temperature rises. Once the temperature reaches 212°F, the water requires more heat in order to vaporize. The water does not instantly and completely flash to steam. This extra energy needed to vaporize a liquid into gas is called the **latent heat of vaporization**. Both latent heat of fusion and latent heat of vaporization are measures of how much energy is added to a substance without a change in temperature. Substances that are regularly gases at room temperature and have a very low melting point tend to require lower latent heat of fusion and vaporization to cause a change of state. **Figure 35-4** shows how water changes state based on heat input.

35.1.4 Refrigeration Cycle

Since refrigerants are in a gaseous state at standard atmospheric pressures and temperatures, the refrigeration cycle uses higher pressures. Just as a higher pressure requires a higher temperature for water to boil, a higher pressure in a refrigeration system allows the refrigerant to be in a liquid state.

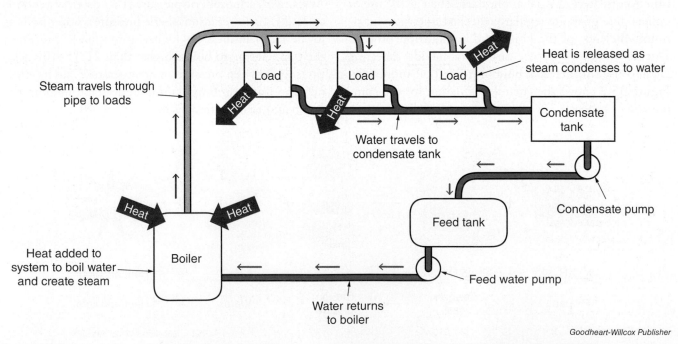

Figure 35-3. In a basic steam cycle, the boiler adds heat to cause vaporization and the loads, such as radiators or heat exchangers, remove the additional heat.

Temperature—Heat Diagram for Water

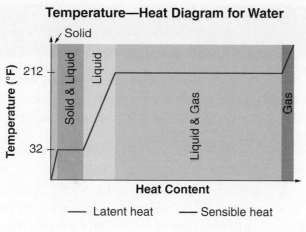

— Latent heat   — Sensible heat

Goodheart-Willcox Publisher

Figure 35-4. The additional heat needed to cause a change in state from solid to liquid is the latent heat of fusion. The additional heat needed to change a liquid to a gas is the latent heat of vaporization.

Figure 35-5 outlines a common refrigeration cycle. In this cycle, a compressor increases the pressure of the refrigerant vapor. This high-pressure vapor travels to the condenser, which transfers the heat of vaporization to outside air or another cooling fluid through a heat exchanger. The refrigerant is condensed into liquid at a high pressure. The high-pressure liquid is allowed to expand through a metering device into a low-pressure liquid, which then travels to the evaporator. In the evaporator, heat is transferred into the liquid through a heat exchanger. As heat is removed from the air or water being cooled, it adds the latent heat of vaporization back into the refrigerant. The refrigerant boils and becomes a low-pressure vapor. The low-pressure vapor is then drawn into the compressor to begin the cycle again.

Refrigeration Cycle

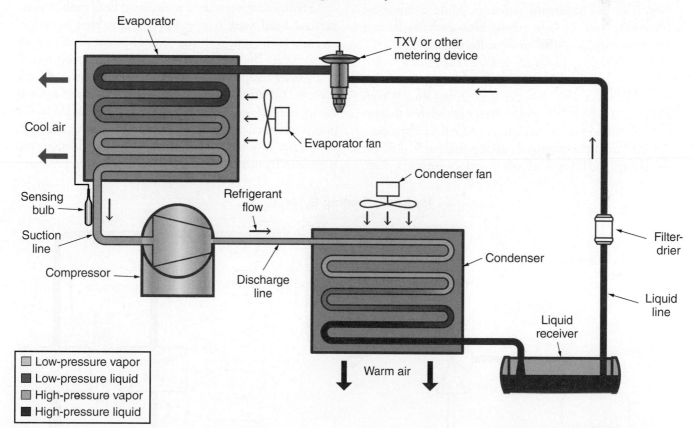

Low-pressure vapor
Low-pressure liquid
High-pressure vapor
High-pressure liquid

Goodheart-Willcox Publisher

Figure 35-5. The basic components and operation of the refrigeration cycle. The compressor changes refrigerant from a low-pressure vapor to a high-pressure vapor. The high-pressure vapor condenses to liquid and releases heat in the condenser. The high-pressure liquid moves through the liquid receiver and filter-drier to the TXV. The TXV restricts the flow of the liquid. The liquid refrigerant passes from the high-pressure liquid line though the TXV into the low pressure evaporator. As it passes through the evaporator, the liquid refrigerant absorbs heat and changes back to a vapor. This low-pressure vapor refrigerant returns to the compressor, and the cycle begins again.

35.2 HEATING

In commercial and industrial systems, the main modes of heating are hydronic boilers and steam boilers. Depending on the environment, large forced air furnaces or electric heating units may also be used. In this section, we will discuss both hydronic and steam systems.

35.2.1 Hydronic Boilers

A *hydronic system* uses water to transfer heat for both heating and cooling applications. Hydronic heating systems range from small residential units to plant-wide systems, and may include solar heating. **Figure 35-6** summarizes a hydronic heating system for a plant-wide application.

In a hydronic heating system, the pressure is kept high enough and the temperature kept low enough so that the water does not vaporize into steam. System pressures typically run at 15–60 psig. Loads include building heating, but also may include heating for other industrial processes specific to the site. While a typical system may use baseboard radiators, some commercial applications may include reheat heat exchangers in a larger air handling system or floor heating.

Loads within a system are grouped into zones, which offers better control. Zoning happens during the design phase of a project with the goal of grouping similar loads together. In a residential system, zones may be split to separate different floors of the residence or even different rooms. In an industrial environment, zoning maintains different temperatures and humidity levels in various areas of a plant.

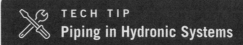

TECH TIP
Piping in Hydronic Systems

Newer residential installations use a flexible plastic piping called *PEX*, while commercial and industrial installations still use metal piping, typically black iron or copper.

A major safety consideration for high-temperature hydronic commercial or industrial installations is the possibility of rapid expansion of the heating water into steam. Consider a hydronic system pressurized to 40 psig running with a hot loop temperature of 230°F. If a system leak or rupture develops that reduces the system pressure to atmospheric pressure, the hot water will instantly flash to steam, as its temperature is above 212°F. Since steam takes up a much larger amount of volume than water, this rapid expansion can cause extreme damage.

Hydronic systems use a pressurized head tank. A *pressurized head tank* is a pressurized tank that maintains pressure in the system. It contains a bladder filled with air or nitrogen. The pressure allows the heating loop to run at a higher temperature and ensures that pumps do not cavitate. *Cavitation* occurs when vapor bubbles form at the suction side of a circulation pump, are drawn in by the pump, and then collapse as the fluid is pressurized through the pump. This can lead to internal system damage, but is avoided by maintaining proper pump suction pressure.

Hydronic Heating System

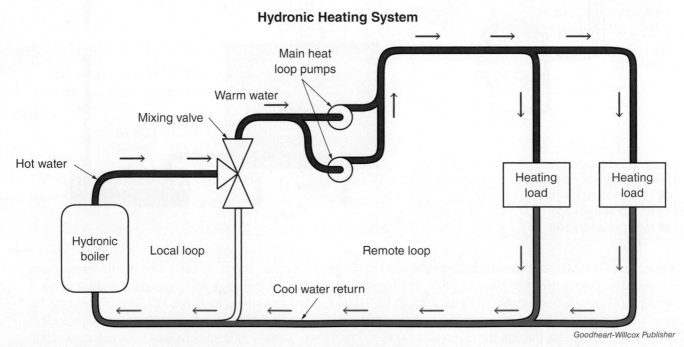

Goodheart-Willcox Publisher

Figure 35-6. The basic components of a plant-wide hydronic heating system.

THINKING GREEN
Instantaneous Boilers

Instantaneous boilers are similar to instantaneous, tankless water heaters. These boilers have no tank and heat water only when heat is called for. Because there are no standby heat losses, these boilers are much more energy efficient than conventional boilers and may be suitable for certain industrial applications.

35.2.2 Steam Boilers

Two common types of industrial boilers are the fire tube boiler and water tube boiler. A *fire tube boiler*, shown in **Figure 35-7**, is a type of boiler that heats a tank of water through conduction using hot gases in sealed tubes. With this type of boiler, combustion is supplied with a powered and modulated burner. The combustion hot air makes several passes through fire tubes, which transfer heat into the surrounding boiler water. Combustion gases leave the boiler through the flue and out the stack. As water vaporizes, it exits through the top of the boiler to system loads.

A *water tube boiler*, shown in **Figure 35-8**, is a type of boiler in which boiler water travels through the boiler in tubes and is heated by combustion taking place within the boiler. Combustion heats the water as it makes several passes through and into a steam drum at the top of the boiler. Water level in the steam drum is maintained automatically, and steam exits the top of the steam drum. Steam pressures range between 15–250 psig for commercial boilers. Power generation boilers operate at much higher pressures.

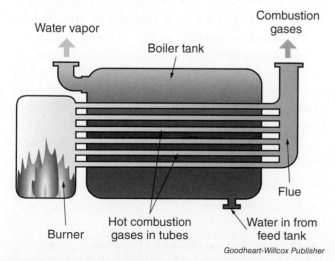

Goodheart-Willcox Publisher

Figure 35-7. A fire tube boiler heats a tank of water using hot gases in sealed tubes.

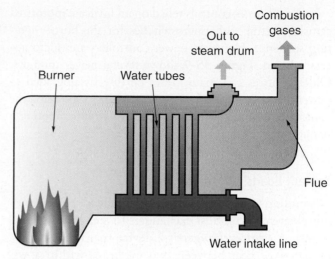

Goodheart-Willcox Publisher

Figure 35-8. In a water tube boiler, water is heated as it travels through the boiler in tubes.

Burner Assembly

The function of the burner assembly is to properly mix fuel and air, light the pilot and main flames, and provide combustion evaluation and control. Air in the burner assembly of a boiler is supplied by a forced draft blower. Older systems may be limited to only a low or high fire, while newer systems are modulated. Modulation and proper mixing is accomplished through the use of any or all of the following:

- Variable speed blower.
- Modulated inlet blower valve.
- Modulated inlet gas valve.
- Controller and HMI (human-machine interface).

Burner ignition occurs in a controlled sequence:

1. Blower starts, proven by the pressure switch.
2. Purge period to remove any gases present; timed by controller.
3. Pilot gas solenoid energizes.
4. Spark ignition for pilot.
5. Pilot ignites; proven by thermocouple or UV sensor.
6. Main flame starts at low fire with main gas valve and blower inlet valve opening to low fire position; proven by microswitches on actuators.
7. Main flame proven by UV sensor.
8. Main flame ramps up to higher setting.

If the sequence is interrupted or a step is not proven, such as limit switch failure, pressure switch failure or flame failure, the burner goes into a shutdown mode. All gas valves are shut, the combustion area is purged with air for a predetermined time (30–60 seconds), and then the draft blower is shut off.

Older burner controls relied on an intricate motorized camshaft setup with microswitches for the burner ignition sequence. However, newer controllers are microprocessor based. **Figure 35-9** shows several newer modules. Older controls can many times be directly upgraded to newer controllers without extensive wiring. Observing the controller during start-up will help in troubleshooting a flame-out.

System Loads and Condensate

Depending on the application, system loads can be various elements from heat exchangers to area heating units (coils). A *heat exchanger* is a device that facilitates the exchange of heat between two mediums within a system. In HVAC systems, the mediums used are typically liquid or gas. These heat exchangers may incorporate one pass straight through or multiple passes to increase the heat transfer rate. Steam typically flows through the shell, while the heated medium flows through the tubes. See **Figure 35-10**. As energy is transferred from the steam, steam condenses and drains to a condensate tank. *Condensate* is the liquid that forms as a result of the process of condensation. The condensate tank collects the condensed steam and intermittently pumps the condensate to the boiler feed system. Pumps are usually controlled by float switches inside the tank.

Heat exchanger

U-shaped tubes

Anton Moskvitin/Shutterstock.com

Figure 35-10. Heat exchangers are available in a variety of materials, including copper, stainless steel, carbon steel, and bronze. Those with tubes that make multiple passes in the heat exchanger, like a U-shaped configuration, increase the heat transfer rate.

Feed System

Feed water is water that is treated and supplied to a boiler system to produce steam. A boiler feed system collects condensate from several sources, conditions the water, and feeds it back into the boiler system. A large boiler feed tank and pumps are shown in **Figure 35-11**. *Make-up water* is water added to the boiler feed system to replace water depletion due to system losses.

Depending on the size and specifications of the system, a boiler feed system may also perform the following functions:

- Provide for a buffer volume for changing system loads.
- Preheat feed water to release oxygen and carbon dioxide.
- Boost pressure above boiler pressure.
- Add chemicals to minimize corrosion throughout the system.

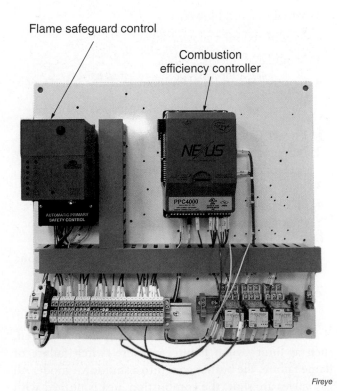

Flame safeguard control

Combustion efficiency controller

Fireye

Figure 35-9. Newer burner controls are microprocessor based and may include wireless communications.

wattana/Shutterstock.com

Figure 35-11. Feed tanks have safety relief valves, sightglasses, and feed pumps to increase pressure for boiler injection.

Chemical additions to the feed water may take different forms. A *chemical injection system* may be installed for the boiler itself to maintain a basic pH in the boiler, which prevents corrosion. Make-up water may be treated before being added to the system, as shown in **Figure 35-12**. This may involve the use of water softeners, dealkalizers, resin bed ion exchangers, or demineralizers.

Blowdown System

Another common system found on larger installations is a blowdown system. A **blowdown** is a periodic removal of water from the boiler through a drain. This helps to remove sludge and scale particles in the boiler. These particles are concentrated in the boiler as water is boiled because sediments do not travel with the steam. The blowdown system simply empties a predetermined amount of water on a regular schedule. A blowdown recovery system provides for heat transfer from the hot blowdown water to condensate that is fed to the feedwater system.

Alhim/Shutterstock.com

Figure 35-12. Feedwater may be treated by different systems.

TECH TIP
Blowdown Valves

When performing a blowdown, open the blowdown valve fully and quickly. Blowdown valves should not be partially opened, as this contributes to the wear of the valve disc and seat. The purpose is to produce a large flow quickly to get as much sludge or sediment out of the boiler as possible. Be aware of the boiler water level during this process—you do not want to drain the boiler completely.

35.2.3 Common Maintenance

Overall boiler efficiency can be estimated by observing flue temperature. In a fire tube boiler, for example, the more heat that is transferred into the water results in a lower flue temperature. As fire tubes become scaled on the inside, the amount of heat transferred into the water is reduced and the combustion gases leaving through the flue become hotter. A regular maintenance task for this system is to shut down the boiler and punch the tubes to remove scale. Tools to remove scale may be completely manual or air-driven, **Figure 35-13**. Tool heads must be sized correctly for the application. A wet-dry vac also aids in collecting removed scale.

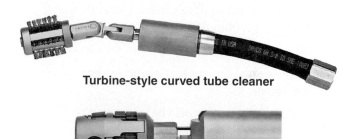

Turbine-style curved tube cleaner

Straight turbine-style tube cleaner

Portable cleaner with hydro-powered brush actuator

Elliott Tool Technologies

Figure 35-13. Proper cleaning tools are essential for tube cleaning.

Scale can also build up at the end portion of the burner assembly on the burner heads. The burner head divides the incoming natural gas into many different injectors. Each injector nozzle has an orifice opening that can become clogged over time. If efficiency is not corrected by cleaning the boiler tubes, the burner head may need to be cleaned also. The size of each orifice (nozzle) is important and the proper tooling must be used. Using an oversized drill bit to remove scale on the nozzles will increase the size of the orifice and alter the geometry of the nozzle.

SAFETY NOTE
Valve Adjustments

Do not adjust air or gas valves without the proper training and instruments. An improper adjustment can result in injury. Only a trained and certified technician should adjust the air/gas mixture of a boiler.

Common failures on boiler start-up include:

- Blower motor failure.
- Draft switch failure.
- UV sensor fails to detect flame (occasional cleaning is required).
- Pilot solenoids failing.
- Diaphragm valves stick or cracks in the diaphragm.
- Valve position microswitches failing.

Any failure must be carefully examined and remedied. Safeties and interlocks are never bypassed.

Repairing and rebuilding centrifugal pumps falls into the category of regular maintenance. The manufacturer's recommendations and procedures should be followed to rebuild a pump. **Figure 35-14** shows the internals of a common centrifugal pump. Internal wear rings that seal the pump's discharge side from the suction side eventually wear. This wear causes the pump to have reduced output flow and pressure, which eventually leads to no pumping at all—it just recirculates water internally. If a centrifugal pump is not developing proper discharge pressure or flow, temporarily shutting the discharge valve slightly should increase pressure on the discharge side. If the discharge pressure does not increase, the pump needs to be removed for service.

Wear rings

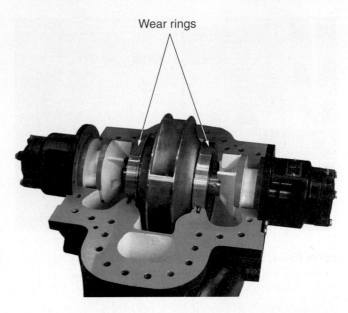

sspopov/Shutterstock.com

Figure 35-14. This double suction centrifugal pump has a steel impeller and two bronze wear rings, one at each end of the impeller. The wear rings seal the center (discharge) side of the pump from each suction side. Replacing wear rings is a common maintenance task on long-life centrifugal pumps.

TECH TIP
Wear Rings

Bronze is a common material used for wear rings. They may be installed with a locking compound that needs to be heated to remove.

Cavitation is more prevalent on condensate pumps and can cause damage to the pump impeller, which will also affect pump flow and pressure in the worst cases. Cavitation occurs when excessive gases form bubbles, or vapor bubbles form in the eye of the impeller. The vapor bubbles impact the impeller and collapse as the pressure is increased along the edge of the impeller.

An additional maintenance item is adjusting packing to prevent excessive leakage. Packing is material that forms a seal between internal components and the outside environment. Packing adjustment is covered in a later chapter.

With numerous valves in the system, the technician should be familiar with all of the common types. Valves may be motor operated or diaphragm operated as control valves. The manufacturer's documentation should be carefully reviewed before maintenance is planned.

35.3 COOLING

Cooling systems serve both refrigeration and climate control functions in industrial environments. The scale and specific application of a system determines the arrangement and design of its components.

35.3.1 Refrigerants

Air-conditioning systems and refrigeration systems use many different types of refrigerant. Each system is designed to use a specific type of refrigerant. Refrigerants are generally not interchangeable. For example, if an air-conditioning system uses R-410A refrigerant, you cannot replace the R-410A refrigerant with a R-134a refrigerant.

Refrigerants are named as "R-" followed by a number and sometimes a letter or letters. Refrigerants are classified based on their chemical composition. Common refrigerant groups include the following:

- **CFCs.** (Chlorofluorocarbons) These refrigerants were used for many years, but have been phased-out by an international treaty because they cause damage to the ozone layer. Molecules in these refrigerants contain chlorine, fluorine, and carbon.

- **HCFCs.** (Hydrochlorofluorocarbons) These refrigerants were developed to replace CFC refrigerants. While HCFC refrigerants cause less damage to the ozone layer than CFC refrigerants, they are still damaging. In addition, HCFC refrigerants have the potential to contribute to global warming when released into the atmosphere. HCFC refrigerants are currently in the process of being phased-out and eliminated. Molecules in these refrigerants contain hydrogen, chlorine, fluorine, and carbon.

- **HFCs.** (Hydrofluorocarbons) HFC refrigerants do no damage to the ozone layer and generally have far less potential to contribute to global warming than HCFC refrigerants. However, some HFC refrigerants have relatively high global warming potential. Molecules in these refrigerants contain hydrogen, fluorine, and carbon.

- **HFOs.** (Hydrofluoro-olefins) Like HFC refrigerants, HFO refrigerants contain hydrogen, fluorine, and carbon. However, HFOs have very low potential to contribute to global warming due to the molecular bonds in their molecules.

- **HCs.** (Hydrocarbons) Sometimes referred to as *natural refrigerants*, HC refrigerants are beginning to be used in some applications, such as commercial refrigeration systems in grocery stores. Hydrocarbons are naturally occurring substances that do not damage the ozone layer and have low global warming potential. HC refrigerants are flammable, so extra precautions are required when working on these systems.

Figure 35-15 summarizes the cylinder color codes used for common refrigerants.

Refrigerant Cylinder Color Code		
Refrigerant Number	Cylinder Color	Type
R-11	Orange	CFC
R-12	White	CFC
R-22	Light green	HCFC
R-23	Light blue-gray	HFC
R-113	Dark purple (violet)	CFC
R-123	Light blue-gray	HCFC
R-125	Medium brown (tan)	HFC
R-134a	Light blue (sky)	HFC
R-401A	Pinkish-red (coral)	HCFC
R-401B	Yellow-brown (mustard)	HCFC
R-401C	Blue-green (aqua)	HCFC
R-402A	Light brown (sand)	HCFC
R-402B	Green-brown (olive)	HCFC
R-404A	Orange	HFC
R-407A	Lime green	HFC
R-407B	Cream	HFC
R-407C	Medium brown	HFC
R-410A	Rose	HFC
R-500	Yellow	CFC
R-502	Light purple (lavender)	CFC
R-503	Blue-green (aqua)	CFC
R-507A	Blue-green (teal)	HFC
R-508B	Dark blue	HFC
R-717	Silver	Inorganic compound

Goodheart-Willcox Publisher

Figure 35-15. Some common refrigerants used in industrial and residential systems.

Additional types of refrigerants are created by mixing two more refrigerants together in specific proportion. These mixtures are called *refrigerant blends*. Refrigerant blends come in two forms: azeotropic mixtures and zeotropic mixtures. *Azeotropes* are refrigerant blends that have fixed boiling and condensing points. When boiled into a vapor, all of the refrigerants in the blend boil at the same temperature. Azeotropic mixtures make up the R-500 refrigerants. *Zeotropes* are refrigerant blends that contain refrigerants with different boiling and condensing points. Zeotropic refrigerant blends vaporize at a range of temperatures, with one of its constituent refrigerants boiling before another. Zeotropic mixtures are designated with an R number in the R-400 range.

SAFETY NOTE
Releasing Refrigerant into the Atmosphere

Never knowingly release any refrigerant into the atmosphere, nor charge a refrigerant system that you know to be leaking.

35.3.2 Ammonia Refrigeration Systems

Ammonia used in industrial refrigeration systems is becoming more popular as CFCs and other refrigerants are phased out or banned. The ammonia used in refrigeration systems is different from the ammonia used in household applications. As a refrigerant, ammonia is nearly pure, containing no water, and is called *anhydrous ammonia*. The ammonia used for household purposes is mixed with a large percentage of water, 90–95%. The main advantages of using ammonia instead of CFCs or HCFCs are its cost and zero damage to the ozone layer. The only drawback to using ammonia as a cooling medium is its toxicity in high concentrations.

The purity of refrigerant ammonia is very high. Water above 30 ppm (parts per million) and oil above 2 ppm are not allowed. Oxygen is also kept to less than just a few parts per million to prevent corrosion in steels. Corrosion can lead to leaks, vessel ruptures, and large releases of ammonia. Proper maintenance procedures in both the initial charging and ongoing repair work need to be followed to prevent contaminants from entering the system and to prolong the life of the system. The low boiling point of ammonia (–28°F) makes ammonia an excellent heat transfer medium. Ammonia refrigeration systems operate using the refrigeration cycle described earlier in the chapter.

An ammonia piping must be labeled as "ammonia" and properly marked to indicate the following information:

- Piping system or area of the system.
- State of fluid contained, such as vapor or liquid.
- Pressure level high or low with color indication.
- Flow direction arrow.

Symbols and abbreviations used to label ammonia systems are summarized in **Figure 35-16**.

SAFETY NOTE
Ammonia Exposure

The ammonia exposure limit is 50 ppm in the air. People can detect ammonia in the range of 10–50 ppm through smell. Therefore, if you can smell it, something needs to be done to fix it. Respiratory protection should be used when investigating a leak.

The main effects of ammonia exposure are to the eyes, respiratory functions, and skin. Depending on the concentrations, exposure can have a variety of effects, from mild eye irritation to coughing, and even suffocation. Skin exposed to liquid ammonia will absorb the ammonia, causing serious burns. Exposure to ammonia requires a complete wash with water.

35.3.3 Compressors

The function of a compressor in a cooling system is to increase the pressure of the refrigerant. By increasing the pressure of the gaseous refrigerant, it can then be condensed. Most refrigerant compressors are positive displacement compressors. *Positive displacement compressors* draw in a measured amount of refrigerant and reduce the volume of the chamber to compress the refrigerant. These compressors can achieve high discharge pressures required for most refrigeration applications. A *centrifugal compressor* uses an impeller to create centrifugal force to compress refrigerant. This type of compressor is used with lower compression ratios (lower pressures) in larger cooling systems.

Positive displacement compressors fall into four general categories:

- Reciprocating (piston).
- Rotary vane.
- Rotary screw.
- Scroll.

Abbreviations for Ammonia Piping	
Abbreviation	**Meaning**
BD	Booster discharge
CD	Condenser drain
DC	Defrost condensate
EQ	Equalizer
ES	Economizer suction
FS	Flooded supply
FR	Flooded return
HG	Hot gas
HGD	Hot gas defrost
HPL	High pressure liquid
HSD	High stage discharge
HSS	High stage suction
HTL	High temperature liquid
HTS	High temperature suction
HTRL	High temperature recirculated liquid
HTRS	High temperature recirculated suction
LTRL	Low temperature recirculated liquid
LTRS	Low temperature recirculated suction
LIC	Liquid injection cooling
LSD	Low stage discharge
LSS	Low stage suction
LTL	Low temperature liquid
LTS	Low temperature suction
OC	Oil charge line
OCWR	Oil cooling water return
OCWS	Oil cooling water supply
OD	Oil drain
PRG	Purge
RV	Relief vent
TSR	Thermo-syphon return
TSS	Thermo-syphon supply
LIQ	Refrigerant in liquid state
VAP	Refrigerant in vapor state
HIGH	High pressure level within piping
LOW	Low pressure level within piping

Goodheart-Willcox Publisher

Figure 35-16. Common abbreviations used to label ammonia piping.

Reciprocating Compressors

Reciprocating compressors use the linear movement of pistons to compress refrigerant. They are used for both smaller household applications and large industrial applications. These compressors operate in a similar fashion to an engine, without the internal combustion. Reciprocating compressors can operate efficiently for many years with

proper maintenance. **Figure 35-17** shows the operation of a small, single-piston reciprocating compressor.

Larger reciprocating compressors may have two or more pistons, which may be in an in-line configuration or rotary configuration. Larger compressors typically come with unloading mechanisms and safety systems that automatically shut down the compressor due to lack of suction, high discharge pressure, lack of oil, or overload.

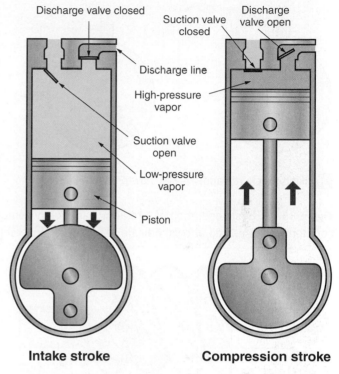

Goodheart-Willcox Publisher

Figure 35-17. As the piston in a reciprocating compressor moves back and forth, it draws in low-pressure refrigerant and compresses it into high-pressure refrigerant.

Rotary Vane Compressors

A *rotary vane compressor* compresses refrigerant using an off-center rotor that turns blades and creates pockets of refrigerant against the casing of the compressor. A simplified diagram of a cycle is shown in **Figure 35-18**. As the rotor spins, vanes seal against the inside casing and compress gas into a smaller volume. Since motion is rotary, the vibration and noise associated with reciprocating compressors is reduced. Units with multiple counter-rotating rotors are regularly used for larger capacity systems.

Rotary Screw Compressors

Rotary screw compressors compress refrigerant through the meshing action of two or more helical shaped rotors. **Figure 35-19** illustrates the compression of refrigerant in a rotary screw compressor. One screw is driven and the other is only powered because it is meshing with the driven screw. The screws rotate in opposite directions to force gases to the discharge outlet. This two screw arrangement can lead to axial and radial thrust on the rotors. Other models may also have opposing vertical screws. Rotary screw compressors used for refrigeration are usually seen only in larger or plant-wide systems. These types of compressors are used in pneumatic systems, refrigerant systems, as well as high-performance automobiles as a blower.

A three-screw compressor reduces or eliminates thrust on the bearing. In a three-screw compressor, the drive screw sits between two nonpowered screws. This effectively cancels thrust and extends bearing life. Larger installations are built from packaged units. Cooling capacities can range from 15 tons to several thousand tons.

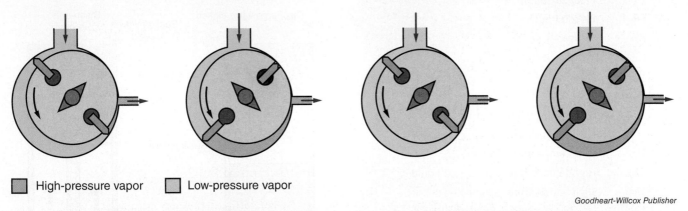

■ High-pressure vapor   ■ Low-pressure vapor

Figure 35-18. Basic operation of a rotating-vane rotary compressor. In this example, the rotor is rotating counterclockwise. Red arrows indicate the flow of refrigerant.

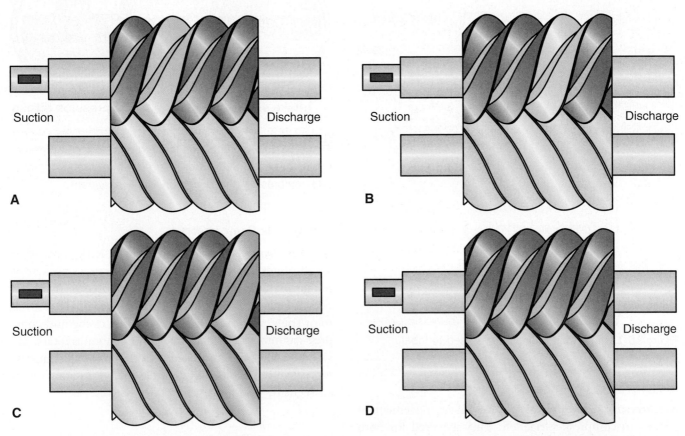

Figure 35-19. In a rotary screw compressor, the revolving rotors compress the refrigerant. A—Spaces between screw threads being filled with refrigerant. B—Beginning of compression. C—Full compression of trapped refrigerant. D—Compressed refrigerant discharged.

Scroll Compressors

A *scroll compressor* compresses refrigerant through the motion of two intertwined scrolls. One scroll is affixed to the casing and remains stationary. The other scroll is powered by a motor and orbits the stationary scroll in an eccentric path creating a radial displacement motion. This motion causes the two scrolls to compress gas from the outer to inner area of the sealing scrolls as the volume is decreased. **Figure 35-20** shows the motion of a scroll compressor. This type of compressor is commonly used for residential and small commercial applications.

The scroll compressor shown in **Figure 35-21** is a hermetically sealed motor and compressor package. A lubricant sump at the bottom of the unit is drawn up through an internal cavity in the shaft to bearings. Compressed refrigerant is discharged through the top of the unit. Since this unit is hermetically sealed, little maintenance can be performed.

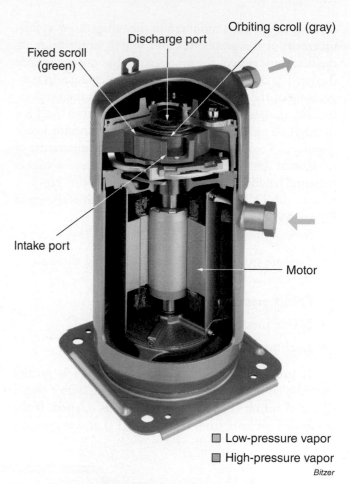

□ Low-pressure vapor
■ High-pressure vapor

Bitzer

Figure 35-21. Cutaway view of a scroll compressor.

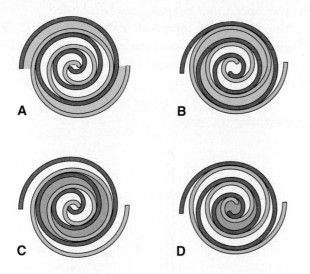

Goodheart-Willcox Publisher

Figure 35-20. Compression of refrigerant using a scroll compressor. A—Refrigerant enters pockets between the scrolls. B—As the gray scroll orbits, it seals the pockets. C—The pockets get progressively smaller and pressure increases. D—Pockets of compressed refrigerant are forced to the center of the scrolls, where the refrigerant escapes through the discharge port.

35.3.4 Thermal Expansion Valves

The *thermostatic expansion valve (TXV)* regulates the flow of liquid refrigerant into the evaporator based on the signal received from a sensing bulb. A cutaway of a TXV is shown in **Figure 35-22**. High-pressure refrigerant enters from the condenser, moves through the TXV, and exits as low-pressure atomized liquid and vapor.

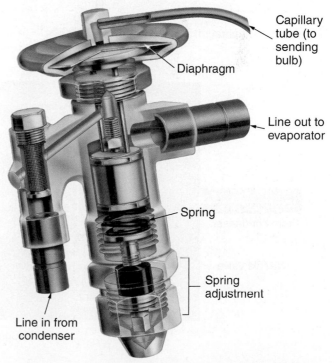

Sporlan Division, Parker Hannifin Corporation

Figure 35-22. The various components of a thermostatic expansion valve (TXV).

A *sensing bulb* is mounted in the refrigeration system upstream of the evaporator and senses temperature on the discharge side of the evaporator. The pressure inside the sensing bulb changes in response to changes in temperature of the refrigerant discharged from the evaporator. The sensing bulb sends a pressure signal to the TXV through a length of seamless tubing with a specific inside diameter called a **capillary tube**. As the temperature of refrigerant leaving the evaporator increases, a higher pressure feedback signal travels through the capillary tube to the TXV. An overview of the circuit is shown in **Figure 35-23**.

Several pressures are at play in the TXV:

- Pressure from the sensing bulb received through the capillary tube (P_1).
- Outlet pressure to the evaporator (P_2).
- Spring pressure (P_3).
- Inlet pressure from condenser (P_4).

If pressure from the sensing bulb (P_1) is greater than the combined pressure from the evaporator outlet (P_2) and spring (P_3), the needle valve is opened. If the combined pressure from the spring (P_3) and evaporator outlet (P_2) are greater than pressure from the sensing bulb pressure (P_1), the needle valve closes. Pressure from the incoming condenser line (P_4) also acts downward against the needle valve. As system load increases, the temperature of refrigerant leaving the evaporator will go up. As a result, a higher feedback pressure is sent to the TXV, which opens the needle valve and allows more flow to the evaporator.

By limiting the flow into the evaporator, the TXV ensures that the refrigerant entering the compressor is fully vaporized with no liquid. Sending liquid refrigerant to the compressor would damage the compressor, as liquid is not compressible, and is called **slugging**. Maintaining a superheat in the exiting refrigerant ensures it completely vaporized. **Superheat** is a measure of the energy added above the required boiling point. In relation to a TXV, superheat is maintained by both internal spring pressure and pressure from the capillary tube. While the spring itself cannot be changed once the system is in operation, the spring can be adjusted.

The amount of superheat maintained at the discharge side of the evaporator is a combination of the initial spring opposition force and the force needed to open the needle valve. The superheat needed to overcome the

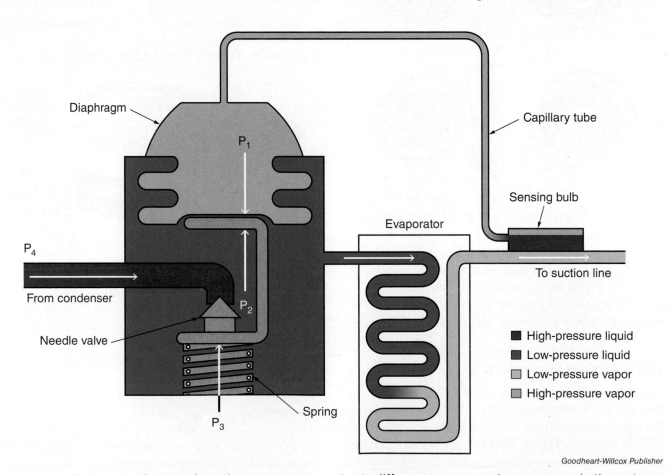

Figure 35-23. A thermostatic expansion valve operates by reacting to differences among various pressures in the system.

spring force in the needle valve is called *static superheat*. The static superheat is kept at 6–8°F. *Opening superheat* is the amount of superheat required to start to open the needle valve from the seat. By adjusting the spring force, a technician can adjust static superheat.

35.3.5 Coils

Condenser coils remove the heat of vaporization from high-pressure refrigerant vapor and condense the refrigerant back into liquid. These coils may be water-cooled, like the heat exchangers of an industrial chilled water system, or air-cooled, as in residential cooling systems. A typical residential unit has condenser coils exposed to outside air.

When refrigerant is pressurized by the compressor, it condenses at a higher temperature. Refrigerant leaving the condenser is slightly below the temperature required to condense into liquid, which means it is slightly *subcooled*. Subcooling ensures that only liquid is sent to the TXV and evaporator coil.

An evaporator coil multistage system is shown in **Figure 35-24**. As high-pressure liquid refrigerant passes through the needle valve in the TXV, it enters the low-pressure evaporator coil. Some of the liquid flashes to vapor immediately. As the refrigerant travels through the evaporator coils, it continues to absorb heat until all of the liquid refrigerant has vaporized. The refrigerant exits the evaporator as a superheated vapor.

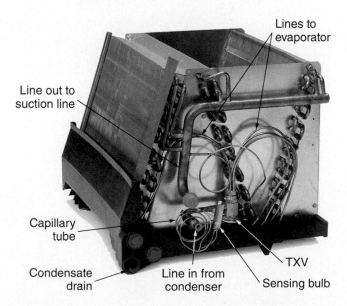

Rheem Manufacturing Company

Figure 35-24. Common components of a multistage evaporator coil. In this example, the pressure feedback line to the TXV is external.

One of the largest maintenance concerns with coils is cleanliness. Heat transfer depends on flow of both mediums and the rate of transfer through the material. When coils are dirty, the efficiency of heat transfer suffers. Coils should be cleaned on a regular basis, at least yearly, to maintain proper efficiencies. Even a thin layer of dust and dirt can drastically affect system temperatures and efficiency.

35.3.6 Troubleshooting and Common Maintenance

General maintenance for the proper long-term operation of a refrigerant system includes proper evacuation and cleanliness prior to charging, adding a proper charge, checking for and correcting leaks, general cleanliness of coils, occasional checking for proper temperatures, and other cleanliness items. As previously mentioned, cleanliness of coils is important for proper heat transfer. To this end, coil cleaning and air filter replacements are important tasks that directly affect the operation of the system. For ammonia systems, proper valve maintenance and packing adjustment is also needed to prevent leaks.

Before charging a system, certain preparation steps must be completed:

- All coils must be clean.
- Check for proper airflow.
- Purge charging lines (hoses, gauges) with refrigerant to avoid adding outside air to the system.
- Determine if the system has a fixed orifice or a TXV; the process of charging these systems is different.

When evacuating a system in preparation for maintenance, proper cleanliness must be followed. Every time a system is opened for maintenance, it must be evacuated prior to charging. In addition, air-driers and filters should be replaced every time a system is taken down and evacuated for maintenance.

There are several charging methods. The *weigh-in method* is the most accurate and is not dependent on outside temperature. This method uses an accurate scale and exact weight of the refrigerant to be charged into the system. The exact length of lines must be known to evaluate the volume of the system and manufacturer's recommendations must be followed. Other charging methods require strict outside temperature conditions.

An improper charge, either too little or too much refrigerant, can affect system performance. An undercharged system produces low compressor suction and discharge pressures, low current draw on the compressor, a low amount of subcooling, and a high amount of superheat. An overcharged system results in high compressor suction and discharge pressures, a high amount of subcooling, high compressor current draw, and low superheat. Other items, such as dirty coils, can mimic some of these symptoms.

Symptoms of compressor failure can be caused by improper charging. However, regular current draw tests can also point to other problems. If the system charge is correct and condenser coils are clean, a compressor may draw higher than normal current due to a higher than normal mechanical load caused by a soon-to-fail bearing. Over time and without proper lubrication, a bearing will fail. Scroll compressors rely on a lobed shaft or guide to provide the motion between scrolls. This guide can fail due to lack of lubrication. This failure causes higher than normal current draw, which results in higher than normal temperatures of the motor coils. Insulation will eventually break down and result in a short. Discharge or suction valves on larger compressors may need occasional rebuilding.

If TXV problems are suspected, check the sensing bulb for proper contact to the evaporator discharge line. Bulbs should be tightly clamped to a horizontal line off-center of the bottom of the line. If the TXV is not responding to a change in load, it is possible that the valve is jammed internally. In this case, the system should be completely evacuated in order to replace or repair the valve.

CHAPTER WRAP-UP

Study in the area of heating and cooling requires a basic understanding of the concepts and components used in these systems. While many of the problems encountered are electrical in nature, other problems, particularly in older systems, are mechanical in nature. Many manufacturers offer both online and classroom training on their systems. If you are involved in the maintenance of heating and cooling systems, take every opportunity for further training with manufacturing companies.

Chapter Review

SUMMARY

- Heating and cooling systems transfer heat from one place to another through conduction, convection, or radiation.

- Combustion produces energy in the form of heat. A complete combustion process occurs when the proper amount of oxygen is available to the reaction.

- At lower atmospheric pressures, water boils at lower temperatures. At higher pressures, much higher temperatures are required to reach the boiling point.

- Sensible heat and latent heat are two forms of heat that impact heating and cooling processes.

- The latent heat of vaporization is the amount of energy needed to vaporize a liquid without changing the temperature. In a steam system, the boiler adds the latent heat of vaporization while loads remove it.

- Refrigerants are in a gaseous state at standard atmospheric pressures and temperatures. Therefore, the higher pressure in a refrigeration system allows the refrigerant to be in a liquid state.

- In a hydronic heating system, the pressure is kept high enough and the temperature kept low enough so that the water does not vaporize into steam.

- Zoning groups similar loads together and offers better control.

- Two common types of industrial boilers are the fire tube boiler and water tube boiler. Both transfer the heat of combustion to water in the system.

- A boiler fed system collects condensate from several sources, conditions the water, and feeds it back into the boiler system. Both feed water and make-up water are needed for the system.

- To maintain the efficiency of heat transfer in boiler systems, removing scale from the inside of fire tubes and the end portion of the burner assembly on the burner heads should be a regular maintenance task.

- The most commonly known refrigerants are the halocarbon group, which consists of chlorofluorocarbons, hydrochlorofluorocarbons, and hydrofluorocarbons.

- Azeotropic refrigerant blends have fixed boiling and condensing points. Zeotropic refrigerant blends contain liquids that have different boiling and condensing points.

- The main advantages of using ammonia refrigeration systems instead of CFCs or HCFCs are its cost and zero damage to the ozone layer. However, ammonia is toxic in high concentrations.

- Positive displacement compressors draw in a measured amount of refrigeration and reduce the volume of the chamber to compress the refrigerant. The four general categories are reciprocating, rotary vane, rotary screw, and scroll.

- The thermostatic expansion valve ensures that the refrigerant entering the compressor is fully vaporized with no liquid. Maintaining a superheat in the exiting refrigerant ensures it completely vaporized and avoids slugging.

- Cleanliness of coils in a cooling system is important to heat transfer. Coil cleaning and air filter replacements are important tasks that directly affect the operation of the system.

- Improperly charging a refrigeration system, either too little or too much refrigerant, can affect system performance.

REVIEW QUESTIONS

Answer the following questions using the information provided in this chapter.

1. Identify and define each method of heat transfer.

2. What is the difference between exothermic and endothermic reactions? Give an example of each.

3. How does pressure affect the boiling point of fluids?

4. Briefly describe the steam cycle used for heating purposes.

5. Identify and describe the two forms of heat impact heating and cooling processes.

6. _____ is heat required to cause a change in state from solid to liquid or liquid to solid.

7. *True or False?* A higher pressure in a refrigeration system allows the refrigerant to be in a liquid state.

8. What occurs in the evaporator during a refrigeration cycle?

9. What is a hydronic heating system? How does this system ensure that water does not vaporize?

10. Describe safety considerations of a hydronic system.

11. Explain the function of the two types of steam boilers.

12. *True or False?* Burner ignition occurs in a controlled sequence and can go into shutdown mode if the sequence is interrupted or a step not proven.

13. Describe the difference between feed water and make-up water in a boiler feed system.

14. What is blowdown? What is the purpose of this process?

15. List some common failures that occur on boiler start-up.

16. If a centrifugal pump is not developing proper discharge pressure or flow, temporarily shutting the _____ slightly should increase pressure on the discharge side.

17. What are five common refrigerant groups?

18. Describe the difference between a zeotropic and azeotropic refrigerant mixture.

19. Discuss why ammonia is used in industrial refrigeration systems and explain how it differs from ammonia used in household applications.

20. Identify the four general categories of positive-displacement compressors and describe the typical application of each.

21. *True or False?* A thermostatic expansion valve regulates the flow of liquid refrigerant into the condenser based on the signal received from a sensing bulb.

22. Identify the different pressures that act on a TXV.

23. What is the purpose of superheat in a refrigeration system?

24. What is the function of condenser coils in a refrigeration system?

25. List four common maintenance items to perform or check on refrigerant systems.

26. What are the results of improperly charging a refrigerant system?

NIMS CREDENTIALING PREPARATION QUESTIONS

The following questions will help you prepare for the NIMS Industrial Technology Maintenance Level 1 Process Control Systems credentialing exam.

1. Which is an example of sensible heat?

   A. A block of ice absorbs heat and starts to melt.
   B. Heat applied to a pan of water on the stove.
   C. Steam released during a hot shower.
   D. Thermal radiation causing skin to burn.

2. When no fuel remains after the process of combustion, this is considered a(n) _____.

   A. complete combustion
   B. endothermic reaction
   C. exothermic reaction
   D. incomplete combustion

3. In a steam cycle, at what point is heat released as steam condenses to water?

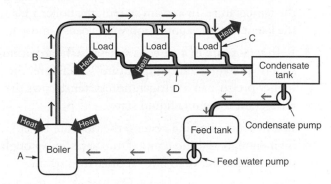

   A. A
   B. B
   C. C
   D. D

4. How is refrigerant condensed during the refrigeration cycle?

   A. Using high pressure.
   B. Using low pressure.
   C. Using high heat of vaporization.
   D. Using low heat of vaporization.

5. What is the role of a boiler feed system?

   A. Collect and condition condensate.
   B. Feed water back into the system.
   C. Both A and B.
   D. Neither A nor B.

6. What type of refrigerant may be used to have a fixed boiling and condensing point?

   A. Azeotropes
   B. CFCs
   C. Zeotropes
   D. None of the above.

7. What statement is true about positive-displacement compressors?

   A. They are found in systems only used in small household applications.
   B. They are used in boiler systems.
   C. They draw in a measured amount of refrigerant and help compress the refrigerant.
   D. The four categories are reciprocating, rotary vane, scroll, and centrifugal.

8. Based on the following illustration of a thermostatic expansion valve, if point A is greater than the combined pressure from B and C, the needle valve is _____.

   A. closed
   B. faulty
   C. open
   D. shut down

9. Which of the following is *not* a maintenance operation for a refrigerant system?

   A. Checking the sensing bulb for proper connections if TXV problems are evident.
   B. Cleaning condenser and evaporator coils to ensure proper heat transfer.
   C. Performing the weigh-in method to charge the system.
   D. Properly cleaning the system after evacuation is performed.

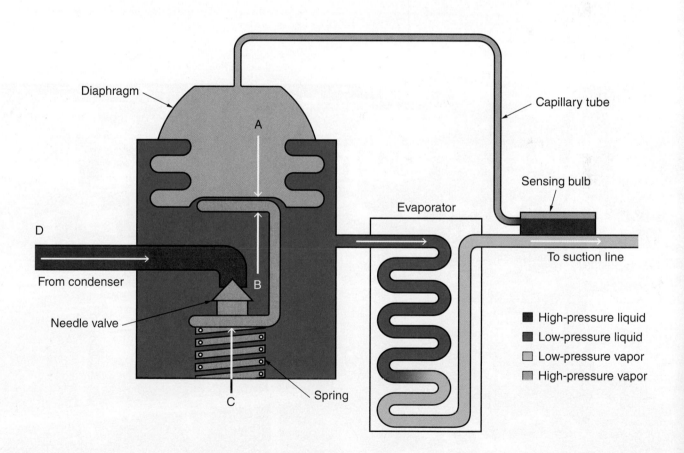

36

HVAC System Control and Design

LEARNING OBJECTIVES

After completing this chapter, you will be able to:

☐ Describe how pneumatics is used in HVAC control.

☐ Identify the components of pneumatic systems and the function of each within the system.

☐ Discuss the major functions and benefits of a direct digital control (DDC) system.

☐ Describe the function of electronic control components in an HVAC system.

☐ Explain the difference between air handling units and rooftop units.

☐ Discuss the controls in an air handling system.

☐ Discuss the control elements used in rooftop units.

☐ Explain common maintenance tasks and troubleshooting for an HVAC system.

TECHNICAL TERMS

air handling unit

bimetallic strip

damper

direct digital control (DDC)

enthalpy

filter-regulator

Java application control engine (JACE)

metal oxide varistor (MOV)

pneumatic transmitters

positive positioner

receiver controller

rooftop unit

In the area of commercial and industrial HVAC, large building-wide or plant-wide heating and cooling systems need almost constant monitoring and regular maintenance to ensure maximum efficiency. Every large warehouse, manufacturing site, high-rise office, logistics center, and commercial or industrial site has accompanying large HVAC systems. If you are planning to start a career as an industrial technician, expanding your knowledge into the area of HVAC gives you an advantage in the workplace. With multiple credentials, such as NIMS and Universal EPA refrigeration certification, potential employers will view you as a serious and versatile candidate.

36.1 HVAC CONTROL SYSTEMS

HVAC control systems follow many of the industrial control schemes presented elsewhere in this text. While engineering has advanced into programmable and computer controls for large HVAC systems, it is still common to see pneumatic systems in operation. In this section, we will examine the operation of pneumatic and electronic control systems, including maintenance and troubleshooting. These control systems may control the temperature of a localized area, as in *zone control*, or may be used to control a large building-wide heating or cooling system. As discussed in a previous chapter, a *zone* is a defined area of a building, which may contain several rooms with similar loads, that is supplied with heating or cooling.

36.1.1 Pneumatic Systems

While many pneumatic systems have been replaced with newer electronic controls, there are pneumatic systems currently in operation. Therefore, a basic understanding of pneumatic components is essential. Some control systems are a combination of electronic signals with pneumatic positioners. Examining your particular systems is the best way to gain knowledge of how they work.

Air Supply

Air supply for a pneumatic system is provided by a compressor, refrigerated air dryer, and filter-regulator, as shown in **Figure 36-1**. Air compressors are typically the piston reciprocating type and may be single or two stage configuration. The air compressor creates the high pressure in the lines that is needed for a pneumatic system to function. As free air is compressed, it is heated and directed into the receiver. In the receiver, air cools and moisture condenses. An automatic drain valve controlled by a timer intermittently drains condensate from the tank.

TECH TIP
Automatic Drain Malfunction

If the automatic drain is not operating, likely due to a solenoid failure, the receiver will slowly fill with water from the compression cycle. Eventually, this will cause the compressor to run more often. If half the receiver is filled with water, it is not filled with air, which results in the compressor starting twice as often.

Several maintenance items are associated with this part of the system:

- Routine run-time check.
- Compressor oil level check.
- Check belts for wear.
- Manually bypass the automatic drain to ensure proper operation.

Compressor run time is calculated by the following equation:

$$\% \text{ run time} = \frac{\text{ON time of compressor}}{\text{OFF time} + \text{ON time of compressor}} \times 100\%$$

(36-1)

If a compressor has an ON time of 3 minutes followed by an OFF time of 5 minutes, the percent run time would be 37.5%. Compressor oil should be changed as per manufacturer's recommendations. Compressor run times should not exceed 40%, and should be consistent from week to week. A large increase in run time indicates system leaks. A high run time (greater than 40%) will reduce compressor life and encourage oil carry-over into the system.

The refrigerated air dryer removes moisture from compressed air before the air enters the system. Most dryers have

System Air Supply

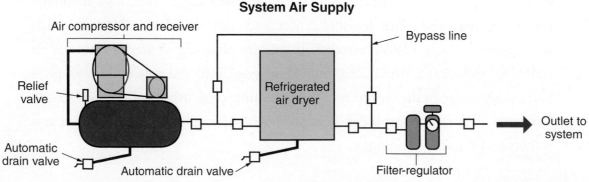

Goodheart-Willcox Publisher

Figure 36-1. A basic commercial compressed air supply system, including an air compressor and receiver, refrigerated air dryer, and filter-regulator.

a discharge temperature near 40°F. Condensate should be draining—this may also have a timed automatic drain that should be checked regularly. Just as with any HVAC unit, the coils should be cleaned on a regular basis.

The *filter-regulator* is a unit with two components. The filter component removes particles from the air exiting the dryer. Removing these particles from the air improves the operation of the systems and extends the life of the components. The regulator component of the filter-regulator controls the pressure in the portion of the system beyond the regulator. Maintaining the correct pressure is critical for proper system performance.

The filter-regulator should be checked on a regular basis, and the filter replaced on a seasonal schedule. Downstream pressures are normally set from 20–25 psig. Dual pressure regulators are used with pneumatic systems that use a day-night schedule or summer-winter schedule. Main pressure is adjusted depending on the application and controls manufacturer. Other remote filters may be present in pneumatic control panels and should be changed when a color change is noted. In-line desiccant dryers usually turn a pink or orange color depending on the type of desiccant. Check them at least on a quarterly basis.

System Components

Pneumatic systems are inherently proportional because of the action of the thermostat. Pneumatic thermostats can be reverse or direct acting (or both), and have day-night settings. On a direct acting thermostat, the output of the sensor increases as the measured variable increases. The action of a reverse acting thermostat is the opposite—the output of the sensor decreases as the measured variable increases. **Figure 36-2** shows a typical pneumatic thermostat. This two-pipe thermostat is supplied with low air pressure, typically between 20–25 psig, and controls the output air pressure using a bimetallic strip and spring pressure. A *bimetallic strip* has two layers made of different metals that react to temperature changes at different rates. A change in temperature causes a physical change in the bimetallic strip. A calibration screw alters the pressure on the bimetallic strip, which rests on an air

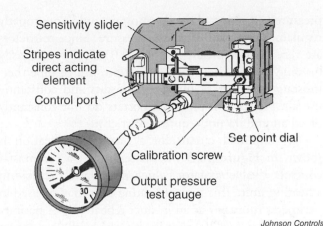

Johnson Controls

Figure 36-2. A common pneumatic thermostat.

bleed port. As the strip reacts to room temperature, the control port pressure is changed.

Pneumatic thermostats are not interchangeable between manufacturers due to different pressure settings. Thermostats should be calibrated on a regular schedule. The process to calibrate a thermostat depends on the specific model and manufacturer.

PROCEDURE Calibrating a Thermostat

To calibrate the model shown in **Figure 36-2**, an output pressure gauge is attached and the local room space temperature checked with a different instrument.

1. Set the set point of the thermostat to the actual temperature of the space.

2. The test gauge should read 8 psig.

3. If the test gauge does not read 8 psig, adjust with the calibration screw.

4. With hands off the equipment, observe the test gauge pressure to be 8 psig. Even the slight pressure of resting a miniature screwdriver on the adjustment screw can alter calibration.

5. While observing actuators, dampers, or valves, adjust the room set point 2° above and below the room temperature.

6. Output pressure should cycle between 3–13 psig on the test gauge.

It should be noted that just upstream of a single-pipe thermostat is a small restrictor that limits flow to the thermostat and controlled elements. Without this restrictor, all controlled elements would be at supply line pressure and action of the thermostat would not properly regulate thermostat discharge pressure. Since restrictors are very small, typically less than 0.010″, they can easily become plugged with contamination. Single-pipe thermostats are constant bleed thermostats and constantly use air. Newer two-pipe thermostats do not constantly bleed air and do not require a restrictor.

A simplified circuit for a heating application is shown in **Figure 36-3**. In this example, a thermostat controls a valve actuator, which modulates the flow to a heating unit. This type of heating unit may be used in perimeter rooms or as an in-duct reheat coil. A perimeter room has a wall(s) that lies on one of the outermost walls of the building. As temperature rises beyond the set point of the thermostat, the output pressure, also called *branch line pressure*, is increased. The normally open valve is modulated toward the shut position to reduce the flow through the heating unit. As temperature drops, the thermostat lowers the branch line pressure sent to the heating valve. This lower pressure allows the valve to modulate open with spring pressure, and more heat is supplied to the room. As long as the heat applied is greater than the heat being lost, the room will maintain temperature. Just as in other industrial proportional systems, there is some offset or error between the set point and actual room temperature.

Pneumatic transmitters, **Figure 36-4**, sense a controlled variable, such as temperature, humidity, or pressure, and develop a proportional air pressure output signal. The output is linear and is used as an input to a controller or a direct input to act on an actuator. Transmitters are generally interchangeable among different manufacturers. Temperature transmitters are available in various temperature ranges, from −40°F to 270°F. Different ranges are used depending on the application. A transmitter to sense outside temperature, for example, would have a different

Johnson Controls

Figure 36-4. A pneumatic temperature transmitter used to sense outside air temperature.

range from a transmitter measuring a zone. Other available transmitters include those that measure pressure, differential pressure, and humidity.

Receiver controllers, **Figure 36-5**, control actuators and can scale their output across a portion of the transmitter input. The set point, controlled by the output of the receiver controller, may be reset or rescaled by a separate input. Receiver controllers have adjustable gain and ratio, and can be direct or reverse acting.

A single input receiver controller may measure heat loop temperature and control a large mixing valve to mix boiler water with the main loop. Controllers that use a reset schedule are commonly called *dual input*. A dual input receiver controller performs the same functions as a single input receiver, but also measures outside air temperature. As the outside air temperature drops, the set point of the main heat loop is increased.

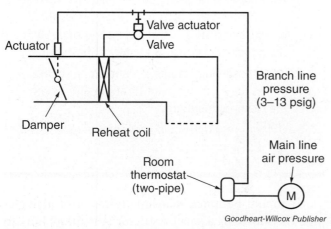

Goodheart-Willcox Publisher

Figure 36-3. A simple circuit for a pneumatic thermostat application with reheat coil.

The receiver controller is usually within a control panel, and once properly set it does not need calibration or adjustment. Ratio, gain, and set point should not be field adjusted, except by experienced personnel. Just as changing gain in an industrial process controller will make the controller more sensitive, too much gain in a receiver controller can result in wide fluctuations of the controlled variable.

Actuators include a wide range of pneumatic operators that may control dampers or valves. A typical pneumatically operated valve is shown in **Figure 36-6**. This valve is operated by pneumatic pressure against a diaphragm and spring. Heating valves are commonly open with 0 psig (fail-open) input, while cooling valves are closed with 0 psig input.

Valves rarely have maintenance issues with the exception of possible leakage at the packing or a cracked diaphragm. Springs for most actuators are color-coded to indicate their spring tension range. Different spring ranges are used depending on the purpose of the valve.

Figure 36-7 shows a constant airflow heating and cooling application for one zone. In this example, the heating valve uses a spring with a range of 2–7 psig and the cooling valve controlled by the same controller uses a spring range of 8–12 psig. If room temperature rises above the thermostat set point, branch line pressure increases above 8 psig and opens that valve actuator for the cooling coil. If room temperature drops below the thermostat set point, branch line pressure drops and the cooling valve shuts. If branch line pressure drops below 7 psig, the heating valve actuator opens.

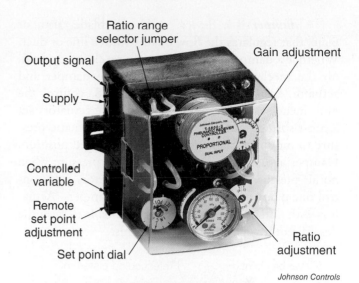

Johnson Controls

Figure 36-5. Pneumatic proportional receiver-controller.

Johnson Controls

Figure 36-6. A typical diaphragm-operated pneumatic valve actuator.

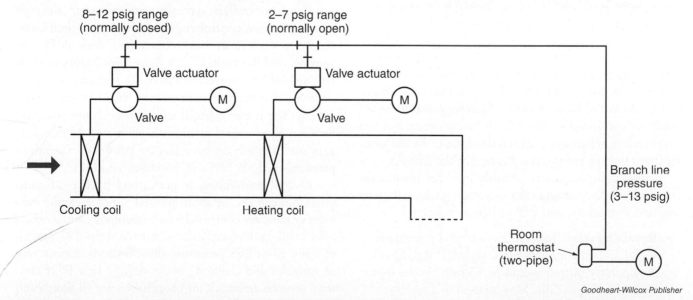

Goodheart-Willcox Publisher

Figure 36-7. A one-duct system with heating and cooling coils in each zone. In this example, the heating valve will be fully closed at 7 psig before the cooling valve starts to open at 8 psig.

A *damper* is a device, such as a blade, arm, or plate, that regulates the flow of air within a line or duct. Whether a damper is normally open or closed depends on the mechanical linkage between the damper and actuator. *Damper actuators*, **Figure 36-8**, control the movement of dampers in a system. These actuators act against spring return pressure and use pneumatic pressure to modulate the damper to the extended position. Damper actuators retract due to spring pressure when no air pressure is applied. Damper actuators may control one damper or multiple opposing dampers. Actuators with different spring ranges are available to meet the design requirements of various systems.

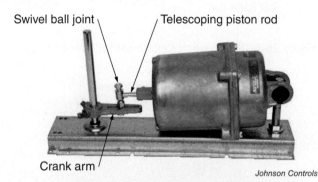

Swivel ball joint   Telescoping piston rod
Crank arm
Johnson Controls

Figure 36-8. A high torque damper actuator with a range of 8–13 psig.

TECH TIP
Spring Range

If replacing a damper actuator, check the spring range of the actuator. Be sure the replacement has the same spring range before installing it.

Positive positioners may be used for larger loads. A larger load may be a larger damper with multiple vanes. In this case, a normal-sized actuator may not have the torque to reposition the larger damper. *Positive positioners* are actuators controlled by the branch line pressure, but use main air line pressure to position the damper. In addition, feedback linkage ensures the position of the damper.

Other components commonly used for pneumatic circuits include reversing relays, averaging relays, high or low selection relays, and E/P or P/E switches.

- **Reversing relay.** Takes in a controlled pneumatic signal and outputs a reversed signal. If the input is 3 psig, relay output would be 13 psig. As the input rises, the output falls. May be used in a system where maximum airflow is desired for both heating and cooling.

- **Averaging relay.** Takes several input pneumatic signals and outputs an averaged signal. Typically used to control a larger zone with several thermostats in smaller rooms.

- **P/E switch.** Converts a pneumatic signal to an electrical output signal.

- **E/P switch.** Converts an electrical signal to a pneumatic output signal.

E/P and P/E switches may be proportional or snap acting. For a snap acting relay, electrical contacts will change state at a set air pressure. The contacts may be single or dual pole, single or dual throw, including both normally open and closed contacts. Proportional signal transducers take 0–10 VDC or 4–20 mA and convert it to a proportional pneumatic output signal—the reverse is also available.

36.1.2 Electronic Systems

Direct digital control (DDC) is a control system that uses various digital and analog inputs and outputs connected to a central computer that operates the HVAC system or automates other building systems. DDC improves system performance and allows systems to operate at a higher efficiency. These systems work on several levels, as shown in **Figure 36-9**. At the local level, sensors, actuators, and local controllers communicate and control local zones. These controllers typically control local *variable air volume (VAV)* dampers and controls. Local controllers are similar to an industrial PLC with inputs and outputs, and may have both digital (on-off) and analog (4–20 mA or voltage) signals for inputs and outputs.

Local controllers typically communicate through Java application control engine units, which then communicate with supervisory controllers, such as PCs or laptops, and the main PC. A *Java application control engine (JACE)* connects to a network and acts as a translator for the various components within a networked system. The use of a JACE unit allows communication between components from multiple manufacturers that may use different communication protocols. Common protocols include BACnet, Modbus, and OPC.

Overall monitoring is performed by a local computer, but can also be accomplished remotely. Some software systems are completely programmable, while others offer configurable platforms with fixed drop-down menu options. Most user programs allow both the monitoring of variables and changing of set points. In a DDC system, sensors, actuators, and controllers are all linked and controlled digitally. A digital touch-screen thermostat, shown in **Figure 36-10**, is a commonly used commercial

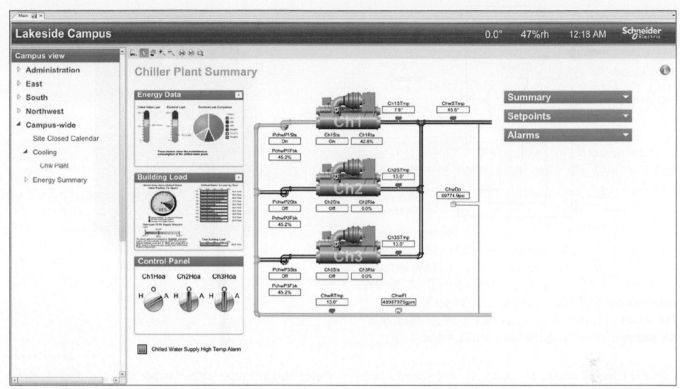

Figure 36-9. Central control of an HVAC system can be applied to systems that are small and very simple to complex plant-wide systems.

Figure 36-10. The interface on a digital touch-screen thermostat is user-friendly and can usually be customized for the unique characteristics of a system or location.

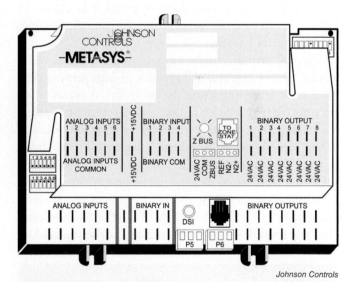

Figure 36-11. Local VAV controller with numerous inputs and outputs.

thermostat. Many models offer advanced features including password protection, a programmable occupancy schedule, multiple languages on the interface, and a memory card port for importing and exporting data.

At the local level, a VAV controller, shown in **Figure 36-11**, receives all local inputs and executes a program to control all zone elements. Inputs are typically from a thermostat, but may also include feedback signals from output elements. Output elements may be damper actuators and valve actuators.

An example application of a VAV controller circuit is shown in **Figure 36-12**. In this example, a hot deck (duct supplying heated air) and cold deck (duct supplying cooled air) supply heating and cooling throughout a building.

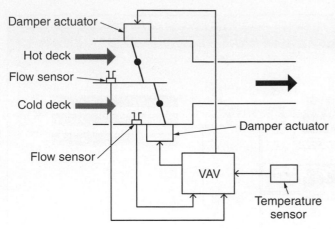

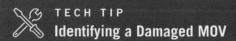

Goodheart-Willcox Publisher

Figure 36-12. The VAV controller receives input regarding the temperature in the conditioned space and the airflow in the hot and cold decks. Based on this information, the VAV controller can send signals to the two damper actuators. The actuators then move the dampers to adjust the mix of hot and cold air entering the conditioned space.

Johnson Controls

Figure 36-13. Proportional damper actuator rated at –4°F to 125°F ambient temperature.

Separate damper actuators control the two streams independently, depending on zone needs. Flow sensors send feedback to the controller. With this setup, full flow heat, full flow cooling, or a mixture can be provided to the zone. A common damper actuator is shown in **Figure 36-13** and can be controlled in on/off, floating, or proportional modes.

Several manufacturers suggest using a metal oxide varistor on the controls transformer to help prevent electrical noise from impacting the controller. A *metal oxide varistor (MOV)* is a device that diverts transient or excessive voltage away from circuits or system components. When a surge or spike in voltage is detected, the resistance of an MOV is reduced. The voltage travels through the MOV and is diverted to a ground or neutral line. A spike in voltage may occur due to local switching, power interruption, or a lightning strike. MOVs tend to short after diverting several voltage spikes and need to be replaced.

The indication and control of a DDC system simplifies troubleshooting. Since all indications are localized, a technician can easily see current conditions and system response. System-wide malfunctions will show as affecting all zones of the building. Malfunctions in specific zones will show as affecting all rooms in a particular zone—one VAV may supply several rooms grouped into a zone. If a zone controller is suspect, set points can be altered remotely in order to gauge system response. If the system responds, it is neither a communication or controller issue. Further troubleshooting at specific elements, such as actuators, is needed. Actuators can be checked by verifying that the signal is arriving at the actuator. If the signal is arriving at the actuator and no response is noted, the actuator needs to be replaced.

TECH TIP
Identifying a Damaged MOV

If an MOV is damaged, it may appear as a ground or an open. This may trip the circuit breakers supplying the 120 VAC to VAV controllers. Each controller should be checked for the damaged MOV. A shorted device may appear burned. If all MOVs have a normal physical appearance, the technician would need to test them with a multimeter.

TECH TIP
Troubleshooting Newly Installed Systems

Troubleshooting a system that has been properly running for an extended amount of time is a completely different undertaking than troubleshooting a system that is newly installed. Newly installed systems have to be properly commissioned and checked before being turned over to building occupants.

36.2 OVERALL SYSTEM DESIGN

There are many different system designs that largely depend on location needs and the engineering design team. Some general system types include the following:

- A *central heating and cooling plant*, **Figure 36-14**, distributes hot water for heating and chilled water for cooling to remote air handling units.

- A *local chiller and boiler* perform the same functions as a central heating and cooling plant, but there may be several buildings on the site, each with separate systems.

- A completely air-driven system that uses the *same rooftop units for both heating and cooling*, **Figure 36-15**. The rooftop units can provide heat through electric, gas-fired, or hot water coil.

- A completely air-driven system with *separate rooftop units for cooling and heating*, **Figure 36-16**. A two-duct system offers better control of each zone, but comes with a higher cost of installation and controls.

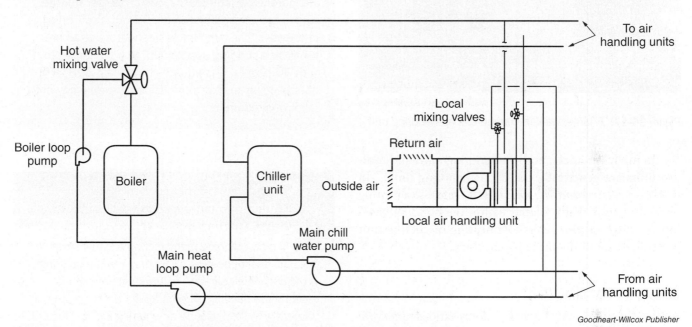

Goodheart-Willcox Publisher

Figure 36-14. A central plant with separate heating and cooling (four-pipe system) supplies each remote air handling unit.

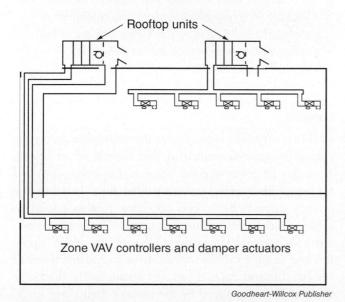

Goodheart-Willcox Publisher

Figure 36-15. Rooftop units may be designated to certain floors of the building or may serve multiple floors.

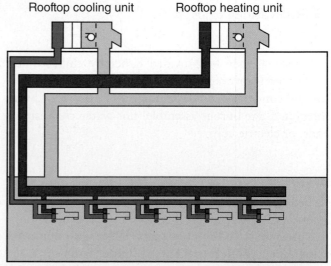

Goodheart-Willcox Publisher

Figure 36-16. A two-duct system with separate rooftop heating and cooling units.

- An *air-driven system for cooling loads* with heating provided by a boiler and reheat coils or radiators in each zone. An example of an air handler unit used in this type of system is shown in **Figure 36-17**.

Figure 36-17. A large, walk-in commercial air handling unit.

In many instances, both air handling units and rooftop units are generically called "air handling units." In this text, an ***air handling unit (AHU)*** refers to a heating or cooling unit that is placed within the building envelope. A ***rooftop unit*** refers to a heating or cooling unit placed outside of the building envelope.

36.2.1 Air Handling Units

Air handling units (AHU) may perform various functions:

- Air filtration.
- Heating, cooling, or a combination.
- Recirculation of air.
- Fresh air intake.

Cooling may be provided either by chilled water cooling coils or evaporator coils with outside compressor and condenser coils. Heating may be provided by an integrated gas burner assembly, hot water coils, steam coils, or electric coils.

Operation

Depending on the design, some units may have both return and supply air fans. Supply air fans provide positive pressure to conditioned spaces. Return air fans draw a vacuum to both recirculate air and exhaust air to the outside. In the mixing box area, return air combines with outside air when the dampers are properly positioned. Outside air, or fresh air, dampers may be controlled or

may be opened by vacuum only. As a general rule, if the mixing box is at a positive pressure, no outside air is drawn in. Outside air is important to reduce carbon dioxide levels within the conditioned space and to maintain or improve the quality of air.

SAFETY NOTE
ASHRAE Fresh Air Standard

The current American Society of Heating, Refrigerating, and Air-Conditioning Engineers (ASHRAE) standard 62.1 requires 17 cubic feet per minute (CFM) per person of fresh outside air at all times of the year. This amount of fresh air helps to prevent "sick building syndrome." In a typical room with 30 people, this equates to at least 510 CFM of outside air to that room. Older systems may not meet this requirement.

TECH TIP
Outside Air

Installing and running many rooftop exhaust fans does not ensure fresh air is coming into the building. If the outside air damper is closed, no fresh air is entering the building. The building may be at a slight vacuum with little outside air entering *except* through leakage. If opening an outside door is difficult due to the pressure difference across it, the building is at a vacuum. The only way to ensure that outside air is entering the building is by measuring it at the intake damper with an appropriate instrument.

Controls

AHU controls are typically locally controlled, with the option of remote monitoring and control of set points. Fans may be across-the-line starters (full voltage applied for maximum speed) or controlled by a variable frequency drive (VFD). Two common methods used to achieve variable air volume controls (VAV) of the system are throttling down the discharge side of the supply fan and using a variable speed drive to control speed of the fan. In most cases, a pressure sensor in the discharge duct controls the speed of the variable frequency drive. As dampers throttle shut, duct pressure increases and the VFD slows to maintain static pressure. As dampers throttle open, duct pressure decreases and the VFD speeds up in order to maintain static pressure. **Figure 36-18** shows a typical control loop to control flow.

Other common control loops within the AHU, depending on design and application, may include controlling the temperature of discharge air and maximizing outside air intake. Controlling the temperature of discharge air can be accomplished by controlling mixing valves, using multiple stages of cooling (multiple compressors or coils), and controlling electric or gas heat. Maximizing outside air intake considers the outside air temperature and humidity in addition to inside air requirements, and is accomplished by controlling damper positions.

These control loops may be controlled by a local microprocessor that is monitored remotely with a direct digital control (DDC) system. Common signals include 0–20 mA, 4–20 mA, 0–10 VDC, digital signals, and older applications may include pneumatic signals in the range of 0–20 psig.

Control schemes that include *enthalpy control* seek to compare the cost of outside air with return air. **Enthalpy** in HVAC is the amount of heat in the air measured in BTUs. Enthalpy control measures outside air temperature and humidity and determines if outside air can meet cooling requirements. Newer microprocessor-controlled enthalpy control units are a requirement for cost savings.

Common Maintenance

Scheduled maintenance items for an AHU include the following:

- Coil cleaning with 15 psig air, light detergent water, or specific coil cleaning solution from manufacturer.
- Checking belt drives for wear (both belt and sheave).
- Replacing filters.
- Inspecting valve and dampers for proper operation.

- Inspecting the local recirculation pump (if included), bearing lubrication, and coupling check.
- Greasing all bearings (if not sealed) approximately every season.
- Checking drains for proper drainage and proper seal.
- Cleaning damper blades or seals.
- If an extended period of shutdown is expected, rotate shafts several turns every month.

On a less frequent schedule, perhaps on a yearly basis, check the spring-loaded antivibration mounts for proper operation, and inspect, clean, and lubricate all linkages and their joints.

SAFETY NOTE
Lock Out and Tag Out

Always make sure to lock out and tag out the AHU before performing mechanical work, electrical work, and even cleaning.

TECH TIP
Damper Blade Bearings

Damper blade bearings, especially on an outside air damper, may build up dirt and particulate matter if lubricated with a liquid or grease. This can lead to dampers that require excessive torque from the actuator to operate—which can lead to actuator failure. A dry graphite lubricant is a better choice to help prevent this.

Supply Fan Control

Figure 36-18. Control loop for building supply air duct air pressure by varying fan speed.

> ### THINKING GREEN
> ### Dirty Coils and Fans
>
> Dirt on evaporator and condenser coils decreases the coils' ability to transfer heat and may, in extreme cases, increase resistance to airflow across the coil. To keep the system operating at maximum efficiency, the coils should be cleaned regularly.
>
> Fans and fan motors should also be kept clean to maximize efficiency. Dirt buildup on a fan can reduce its ability to move air efficiently. Motors with dirt buildup operate at higher temperatures, which can lead to motor failure.

Troubleshooting Scenarios

To troubleshoot a system, a technician must know what the system was designed to do, examine how the system is operating, and pinpoint which function the system is not performing. Asking several questions can help:

1. Is this a local problem, in one zone or room?
2. Does this problem extend to the entire area that this AHU serves?
3. Is this a plant-wide issue or an issue in multiple buildings?

In order to understand what is happening with the system, the technician must understand how the system is designed. This section presents common HVAC system scenarios and examples of troubleshooting steps related to the system issues.

The location for these troubleshooting scenarios is a two-story professional building. Multiple rooms on each level are split into control zones. There is a mechanical room and an AHU on each floor to serve that level of the building. The HVAC system is a one-duct system and each zone has a reheat coil. This type of system can supply only cooling air or heating air to all of the supplied zones, which may be locally reheated for a particular zone, as needed. While the AHU has both heating and cooling coils, these coils are modulated to maintain a discharge duct temperature from the AHU of 68°F.

For these troubleshooting scenarios, assume the following conditions:

- The building is located in the central United States. It is currently the fall season with outdoor temperatures ranging from 40°F at night to 75°F during the typically sunny afternoons. Consider how these factors will affect the system operation.

- Each floor of the building has numerous perimeter rooms and central rooms. Think about how this affects system control.

- The building is a standard rectangular shape, with the long side exposed east and west. Evaluate how this affects the operation of HVAC units.

In order to better understand system operations, review how the system reacts throughout the cycle of a normal day.

- In the cold morning, just before occupants arrive, the night setback schedule returns to day or occupied mode. The average building temperature may be on the cold side, around 50°F, with perimeter rooms even cooler. The AHU changes the set point of discharge air temperature from the night setting to day setting at 68°F. The heating valve cycles fully open.

- Air returning to the AHU from control zones is still cool and all zone dampers are fully open. Reheat coil valves at each zone are also likely fully open.

- The AHU brings in the minimum required outside air.

- As zones warm, the AHU return duct temperature goes up. The reheat valves close and the modulated dampers close to a minimum position in order to supply the minimum airflow to each zone.

- As the AHU return temperature increases, the main heating coil valve modulates toward the closed position until it reaches a steady position that matches load and return air temperature and maintains a discharge temperature at 68°F.

- Outside air temperature eventually rises to 75°F. As this happens, outside air dampers initially open and mix outside air with return air. The central zones of the building may become warmer than expected. All reheat valves in all control zones are now fully closed.

- When outside air temperature rises above 68°F, the AHU outside air damper returns to a minimum setting. The main heating coil valve is fully closed and the cooling valve starts to open in order to maintain the AHU discharge temperature of 68°F. Zone dampers modulate open to cool each control zone.

- In the late afternoon, zones on the east side of the building may start to become cooler. Zones on the west side of the building may be warm due to the thermal heating provided by the afternoon sun. The AHU is still cooling. Dampers in the west zones are fully open to cool those zones. Dampers in the east zones may open along with reheat valves to heat cooler zones.

- As the outside temperature drops, the temperature of the fresh air also drops, which helps the cooling loads. The cooling water valve modulates toward the closed position.

- Eventually, cooling is no longer needed and the cooling valve is fully closed. The main heating coil valve starts to modulate open as the building cools during the evening while the building is still occupied.

- Room dampers modulate the airflow to maintain heating in the control zones. Some of the colder east zones may be opening their reheat valves fully.

- At a predetermined point after occupants leave the building, the AHU switches to night setback (50°F) and heating valves modulate toward the closed position to maintain the set temperature.

TECH TIP
System Operation Based on Design

Understanding how an HVAC system operates and reacts during the day is an important part of troubleshooting. Each system operates slightly differently depending on design. Not every system can exactly maintain every zone temperature. There is a tradeoff between exact temperature in all control zones/spaces and operating expense.

Scenario 1. Just after arriving in the morning, an occupant in a west control zone complains that the area is cold and the system is not responding. Closer examination reveals that this particular zone is 10°F cooler than other zones. This is a zone problem. Something with this particular zone is not responding. This could point to a zone damper problem, a zone reheat valve problem, or a zone thermostat problem.

To troubleshoot this situation, observe the damper and reheat valve while changing the zone thermostat setting. Perform the following:

1. If the damper does not respond to a change in the zone thermostat setting, disconnect linkage and cycle the damper manually. With the linkage disconnected, observe the damper actuator while cycling the thermostat. If the damper actuator does not respond, check to make sure it is receiving a signal. If the damper is receiving a signal but not responding, the damper actuator should be replaced.

2. If neither the damper nor the reheat valve respond to the thermostat, this may be a thermostat

problem. If the thermostat is networked, check remotely for any indication that the setting is being changed. Also, try to remotely change the setting and observe damper and reheat valve actuation. Check the local output of the thermostat (current, voltage, or air pressure) to see if the output responds to a change in setting. Pneumatic thermostats need to be calibrated, which may be the cause of the problem. Otherwise, replace the thermostat. Check operation and response after replacement.

3. If the reheat valve does not respond to a change in thermostat setting, check the actuator of the valve. The actuator may be a simple solenoid that is burned out or a pneumatic valve that is stuck. If the entire valve needs to be replaced, properly isolate the valve before replacing it. Check for proper operation after replacement.

TECH TIP
Position of Dampers

Always examine the exact position of dampers before disconnecting linkage. If needed, mark the position with a permanent marker so that the exact linkage setup can be configured when reconnecting.

Scenario 2. Several building occupants are complaining that it is too hot in their areas. An initial examination of the situation shows that the occupants are from several zones across the building. This may be either an AHU or plant-wide problem. Checking several zones reveals that all temperatures are 10°F higher than the thermostat setting. After checking the operation of zone dampers in response to thermostat setting, the issue is narrowed down to at least an AHU problem. Perform the following to troubleshoot this situation:

1. Check that the AHU is powered and all fans are running.

2. Check the discharge air temperature and chill water supply valve.

3. If the discharge temperature is high and the chill water supply valve is fully open, this points to a plant-wide problem.

4. If a plant-wide problem is suspected, check both incoming and returning chill water supply temperatures to the AHU. The chill water supply temperature should be within the normal operating band. If it is not, the main cooling plant should be checked. This could point to a unit tripping offline due to protective functions or a number of other issues.

5. If the discharge temperature is high and the chill water supply valve is not responding, this points to an AHU problem.

6. To evaluate an AHU problem, change the set point of the discharge air temperature and observe the chill water response. If no response is observed from the valve, check the incoming signal to the valve actuator. The actuator may be receiving a signal, but the valve is stuck in position. The valve may be fine, but the actuator is not responding. Pneumatically actuated valves can develop a leak in a diaphragm. If this is suspected, disconnect the incoming air signal and check for valve operation. Cold water valves normally fail in the closed position, whereas hot water heating valves normally fail in the open position. If the valve is motorized or electrically operated, disconnect the linkage and check to see if the actuator responds to a change in input signal. If the actuator does not respond to an input signal and the actuator has power, the actuator must be replaced. If the valve is diaphragm operated, review the manufacturer's documentation before working on the valve. Most diaphragms can be replaced in place.

36.2.2 Rooftop Units

Rooftop units, **Figure 36-19**, are typically mounted outside the building or mounted in a mechanical room that is indirectly exposed to outside atmospheric conditions. These types of units may include the following:

- Several stages of cooling using multiple compressors and evaporator coils.
- Heating by hydronic, gas, or electric.
- Air filtration.
- An economizer section and enthalpy control.
- Control of return air, outside air supply, and building pressure.
- Specific modes of operation during a fire for smoke control.
- Humidity control.
- Stand-alone microprocessor controls with remote communications and supervisory control.

Operation

Operation for cooling follows a staged method. The first stage of cooling on units with an economizer is using outside air for cooling. The microprocessor senses outside air with the enthalpy switch. If the outside air meets the requirements for cooling, dampers modulate return air and outside air to provide for cooling to maintain the discharge air temperature. If this is not enough to maintain the discharge temperature, a remote two-stage thermostat will call for additional cooling. The second stage of cooling is mechanical cooling using a compressor and associated evaporator coil. If outside air temperature rises significantly, the enthalpy switch will lock out the first stage of cooling (economizer operation). Outside air dampers will close to a minimum position, and cooling will be provided strictly by mechanical means. Some systems may include multiple evaporators. In these systems, one evaporator is activated as the second stage of cooling. If more cooling is needed, a second evaporator is activated as a third stage of cooling.

TECH TIP
Outside Air Temperature

Units will not provide cooling if outside air temperature is below the protection setting. A protection setting prevents operation of the unit in cooling mode if the outside air temperature is so cold that operation of the unit might cause damage.

The system follows the typical refrigeration cycle, but it may have other controls:

- Electronically adjustable TXVs.
- Compressor head pressure controls.
- Hot gas bypass to reduce compressor output for low loading times.
- Multiple condenser coil fans or variable speed drive condenser fans.

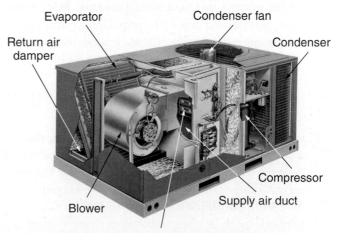

Rheem Manufacturing Company

Figure 36-19. This large, commercial rooftop unit is equipped with electric resistance heating elements.

Heating is commonly gas forced air, but can also be electric coils in geographic areas that have mild winters. Gas heating may be one stage or multiple stages. Heating controls follow a typical start-up procedure, controlled by microprocessors:

1. The call for heat is received; main fan is running.
2. Draft fans start, proven draft by pressure switch.
3. Gas valve opens and igniter lights flame (may also include a pilot sequence).
4. Flame is proven by thermocouple or UV sensor; flame may be modulated or have a low-high fire setup.

Other stages start in the same fashion if discharge air temperature is not met. Any interruption causes a shutdown. A unit may allow several tries, and then will lock out.

Nearly all units are microprocessor controlled, and may have multiple circuit boards to control different functions of the unit. These types of control circuits involve little outside wiring logic—inputs and outputs are directly wired to the processor or circuit logic boards. Outputs for larger loads, such as compressors and fans, will energize an interposing relay, which then energizes a main contactor or sends a signal to a VFD. In larger units, the controller cannot handle the current required to directly energize the main contactor's coil, so an interposing relay is used.

External Control

External controls to a rooftop unit are control elements that reside outside of the unit but are wired to the unit. Some of the more common control elements used include the following:

- **Space temperature averaging sensors.** Average the temperature of different areas in larger buildings; may be used to reset the control point.

- **Mode switch.** Remotely wired switch to control the unit from within the building.

- **Discharge duct pressure control sensor.** Controls the supply air pressure duct to the building, which allows consistent flow across dampers to zones.

- **Building air pressure control.** Controlling building pressure ensures fresh air is being brought in.

- **Economizer control.** Allows override controls to either limit fresh-air intake due to atmospheric conditions or maximize intake because of preferable conditions.

Larger buildings may require finer-tuned controls and multiple rooftop units.

Maintenance

Maintenance items for rooftop units include the following:

- Regularly change air filters.
- Routinely clean all coils in the unit.
- Check belt drives, belts, and pulleys for wear and damage.
- Check condensate drain for obstructions.
- Lubricate all fan shaft bearings and motor bearings as recommended by the manufacturer.
- Check operation of all panel hinges, lubricate if needed, and clean door gaskets.
- Check for leaks in the refrigerant system.
- Check for moisture in the refrigerant system (liquid line indicator).

An often overlooked maintenance item is coil cleaning. This directly affects efficiency, and therefore the electric bill. Dirty coils can also cause numerous problems within the unit itself, such as high compressor discharge pressures. If consistent preventive maintenance is not performed, failures will eventually occur.

Troubleshooting

Most units provide operator indications:

- Power indication LEDs on each board.
- Digital numerical readout that indicates current state and stage of cooling or heating; also indicates any error code received.
- Ability to cycle through all temperature and pressure indications, either on the main processor, remotely, or manually read gauges.
- Ability to set a minimum position for the outside air damper, either manually with a potentiometer or through the processor

The first examination of any problem should include the following:

1. Examine the processor for any error codes.
2. Visually examine all wiring, looking for problems.
3. Determine the function the unit is not performing. Is it cooling, but not cooling enough? Is it being asked for heat, but not firing up?
4. Examine the current conditions.

Error codes can most often be found in the manuals provided by manufacturers, many of which are online. Once the error code is known, the problem can be examined more fully. For example, a visual examination of the wiring may indicate something is burned or that wires are loose. Loose connections can happen simply due to the environment in which the unit operates. Changing temperatures and vibration can eventually cause loose wires. These should be checked occasionally with the unit locked out.

SAFETY NOTE
Checking Loose Wires

Do not start manually pulling on wires to check if they are firmly connected while the unit is powered. A loose wire could easily result in either electrocution or arc-flash that could blind you. Lockout/tagout procedures should be performed before checking the tightness of conductors.

The unit may not be performing because of a power failure (all phases or single phase loss), tripped breakers, a low outside temperature lockout, a mode switch out of position, or because it has met cooling or heating requirements.

Mode switch. A mode switch changes the operation mode of the unit, such as "auto, off, manual," "heating, cooling, off," or "day, night." A mode switch may be a simple, manually-controlled switch or a contact controlled by a programmable electronic timer. Most electronic programmable timers have a battery backup. Unless the timer calls for "occupied mode" (the contact shut), the rooftop unit will not respond. Timers can fail in a number of ways:

- Control transformer powering the timer has failed, or fuse blown.

- Timer has failed internally; miniature relay coil failed.

- Timer lost its program due to extended power outage and battery backup has failed.

With any failure, incoming power to the timer must be checked and the internal battery must be replaced, if available. If the timer calls for "occupied" mode but the contact is not shut due to an internal relay failure, the contact could be jumped out to check the response of rooftop units. Jumping out the contact involves placing a jumper across the open contact to simulate a closed contact. Check the manufacturer's documentation on the timer, as it may be possible to manually place it in "occupied" mode.

Economizer issues. The economizer is the first stage of cooling. When cooling is called for, the economizer compares outside conditions, using an enthalpy sensor and circuit board, with building return conditions. If outside conditions are favorable, outside air is brought into the unit. However, this may not happen due to failed enthalpy sensors or circuit board, or failed actuator(s).

The actuator(s) for return air and outside air should be able to be cycled using the main control board. This can be accomplished either through a minimum position rheostat or program set point. A minimum position rheostat is an adjustable resistor that can change the minimum position of the outside air damper. If the set point is changed and the actuator does not respond, the actuator must be checked. If the actuator responds, the problem is with either the enthalpy control circuit board or enthalpy sensors. If the enthalpy sensor is suspected, refer to manufacturer's documentation to test the sensor.

TECH TIP
Damper Connections

When changing a damper actuator, carefully mark the exact locations of all mechanical connections. Replacement may require recalibrating the actuator through the microprocessor for zero and maximum positions. Using a shim, brick, or other object is not an acceptable way of ensuring minimum outside airflow through the damper. This will only ensure that the damper actuators will fail because they will be continually trying to close against a hard stop.

Breakers tripped. Breakers trip because a motor is drawing too much current. If a system has been running properly for some time, through several seasons and environmental changes, then the problem is not with the load on the system. The following are some important items to check before resetting tripped breakers:

- Proper voltage is coming to the unit on each phase.

- Condition of all fuses.

- All three phase connections incoming and outgoing to all contactors.

- Check for grounds on all affected motors using a megohmmeter.

Breakers can trip for various reasons. For example, the unit loses one phase while a compressor is running during a storm or drastic change in seasonal temperatures.

Without phase protection, the compressor draws higher amperage long enough that the windings overheat, which breaks down insulation. This may result in an internal ground in the motor of the compressor. In this case, the fuses open and a breaker trips. A complete loss of power may also cause breakers to trip, but may result in no actual damage to components of the unit.

Find the cause of a blown fuse or circuit breaker trip *before* placing the unit back online. This may include checking associated contactors and wiring that supply voltage to that particular motor. Contacts may be damaged and may need to be replaced. Do not assume that resetting a tripped circuit breaker will solve the problem—it tripped for a reason.

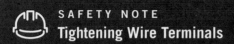

SAFETY NOTE
Tightening Wire Terminals

Never try to tighten terminals of wires while the unit is on. Lockout and tagout the unit. Failing to do this can result in accidentally stripping the threads on a terminal lug, which could easily result in a massive arc-flash and severe injury.

Heating issues. Heating failures usually show as failures of ignition. Common causes of ignition failure include:

- Failure of a draft switch.
- Failure of an ignitor.
- Failure of a control board.
- Failure of a high-temperature limit switch.
- Failure of a flame sensor.
- Failure of a gas control valve.

Draft switches are connected to the discharge side of the draft fan through a silicone or rubber tube. This tube can become blocked, broken, or split. Replace the connection tubing, if needed. Disconnecting the sensing line and checking pressure at the discharge side of the draft fan can help to determine the cause of the failure. Draft switches also have an internal diaphragm that can eventually crack and prevent contact actuation. If the draft fan is not running on start-up, it may be the draft motor itself or the relay energizing the motor. If the draft motor is running and proper pressure can be manually felt at the draft switch input, the draft switch itself should be checked. If the contact is not operating correctly, replace the draft switch. If the contact is operating correctly and the signal is being sent to the circuit board, the control circuit board must be replaced.

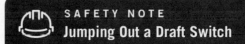

SAFETY NOTE
Jumping Out a Draft Switch

Never install a jumper wire around a draft switch—this is a safety interlock! Jumping out a draft switch could result in a fire in the unit. Instead of jumping out a safety interlock, test it with your multimeter. If you are unsure how to do this, check with another technician and stop work until you are absolutely sure of what you are doing. No amount of training allows anyone to jump out a safety interlock.

Failure of the ignition system shows as an error code. Observing a start-up will allow the technician to evaluate the sequence of events. If the burner never ignites and shows no evidence of sparking, the igniter should be replaced. This will typically solve the issue. If the igniter still does not spark after being replaced, the control board may need to be replaced. If the igniter works and a flame is present, but the unit still shuts down, this points to a problem with a flame sensor. The flame sensor should be cleaned or replaced, if needed.

High-temperature limit switches may open and prevent the unit from running. Do not jump out the switch. Check the switch with a multimeter and replace, if needed.

Gas valves occasionally fail. If a valve is being sent voltage to open and it is not opening, the valve must be replaced.

CHAPTER WRAP-UP

Commercial and industrial HVAC systems have a wide range of designs and controls. This offers technicians both a challenging and lifelong pursuit of knowledge. Choosing this area of controls troubleshooting will ensure that you are never bored. Manufacturers of controls, both software and hardware, offer extensive training at their sites and online. If an employer wants you to become knowledgeable about these systems and offers to send you to training, take advantage of the opportunity. The training will only increase your worth in the marketplace.

Chapter Review

SUMMARY

- Pneumatic and electronic control systems may be used for zone control or to control the temperature within a large building-wide heating or cooling system.

- While many pneumatic systems have been replaced with newer electronic controls, there are pneumatic systems currently in operation. Some control systems are a combination of electronic signals with pneumatic positioners.

- Air supply for pneumatic systems is provided by a compressor, refrigerated air dryer, and filter-regulator.

- Pneumatic systems are inherently proportional because of the action of the thermostat. Just as in other industrial proportional systems, there is some offset or error between the set point and actual room temperature.

- Pneumatic transmitters sense a controlled variable and develop a proportional air pressure output signal. The output is used as an input to a controller or a direct input to an actuator. Actuators include a wide range of pneumatic operators that may control dampers or valves.

- Direct digital control (DDC) systems operate the HVAC system or automate other building systems and improve system performance.

- At the local level, sensors, actuators, and local controllers communicate and control local zones. A VAV controller receives all local inputs and executes a program to control all zone elements.

- With a DDC system, all indications are localized so a technician can easily see current conditions and system response.

- Overall system design determines how a system maintains temperatures and reacts to changes in the environment.

- Air handling units and rooftop units perform the same major functions. Both systems have multiple control loops to maintain building supply temperature, humidity, intake of fresh outside air, cooling and heating stages, and building pressure.

- Air handling is typically locally controlled, with the option of remote monitoring and control of set points.

- External controls to a rooftop unit reside outside of the unit but are wired to the unit and can include space temperature averaging sensors, mode switch, discharge duct pressure control sensor, building air pressure control, and economizer control.

- Regular maintenance and cleaning of components in an HVAC system is necessary for optimal and efficient system operation.

- To troubleshoot a system, a technician must know what the system was designed to do, examine how the system is operating, and pinpoint which function the system is not performing.

REVIEW QUESTIONS

Answer the following questions using the information provided in this chapter.

1. Discuss the air supply process for a pneumatic system.

2. If a compressor has an ON time of 4 minutes and an OFF time of 3 minutes, what is the % run time?

3. What is the purpose of a filter-regulator?

4. What is the difference between a single-pipe and two-pipe thermostat?

5. Describe how to calibrate a pneumatic thermostat.

6. *True or False?* As temperature rises beyond the set point of a thermostat, the output pressure, or branch line pressure, is decreased.

7. What is a pneumatic transmitter?

8. A dual input receiver controller performs the same functions as a single input receiver, but also measures _____.

9. How are heating and cooling valves controlled? How do they differ?

10. What is a damper actuator? Discuss what type of actuator may be used for larger loads to control a damper and why.

11. What is the function of a direct digital control? What are its benefits?

12. Describe a common application of a VAV controller.

13. *True or False?* Since all indications of a DDC are localized, a malfunction in a specific zone will affect all rooms in a particular zone.

14. What is an air handling unit? Explain its functions.

15. What are two common methods used to achieve variable air volume controls of a fan system?

16. _____ controls measure outside air temperature and humidity and determine if outside air can meet cooling requirements.

17. A zone is not responding to a cold room. What questions would you ask? What must be understood first in order to troubleshoot this problem?

18. *True or False?* To better understand an AHU system operation, a technician must consider how the system reacts throughout the cycle of a normal day.

19. One area in the west control zone is 10°F cooler than other zones, which indicates there is a zone problem. What type of zone problem could it be?

20. Describe the stages of cooling that apply to rooftop units.

21. List the typical start-up procedure for heating controls.

22. How would you initially examine a problem with a rooftop unit to know how to troubleshoot the issue?

23. What are common external controls used with rooftop units?

24. What are common causes of ignition, or heating, failure?

NIMS CREDENTIALING PREPARATION QUESTIONS

The following questions will help you prepare for the NIMS Industrial Technology Maintenance Level 1 Process Control Systems credentialing exam.

1. Which component of a pneumatic system makes the system inherently proportional?

   A. Damper
   B. Pneumatic transmitter
   C. Receiver controller
   D. Thermostat

2. What is the function of a pneumatic transmitter?

   A. It develops a proportional air pressure input signal.
   B. It is used to control an actuator.
   C. It regulates the flow of air within a line or duct.
   D. It senses a controlled variable, such as temperature or pressure.

3. In a constant airflow heating and cooling application for one zone, if the room temperature rises above the thermostat set point, what then occurs?

   A. Branch line pressure increases and closes the valve actuator for the heating coil.
   B. Branch line pressure decreases and closes the valve actuator for the heating coil.
   C. Branch line pressure increases and opens the valve actuator for the cooling coil.
   D. Branch line pressure decreases and opens the valve actuator for the cooling coil.

4. Which of the following is *not* a task of local controllers in direct digital control?

   A. Signal for industrial inputs and outputs.
   B. Communicate through Java application control engine units.
   C. Communicate to local zones.
   D. Control local variable air volume (VAV) dampers.

5. A(n) _____ refers to a heating or cooling unit that is placed within a building envelope.

   A. AC control loop
   B. air handling unit
   C. enthalpy control
   D. rooftop unit

6. A particular zone in a building is 15°F cooler than other zones, which indicates that this variation is a _____ problem.

   A. zone damper
   B. zone thermostat
   C. Either A or B
   D. Neither A nor B

7. What occurs in the first stage of cooling by a rooftop unit?

   A. A compressor is used for mechanical cooling.
   B. An economizer uses outside air for cooling.
   C. A micropressor senses inside air with an enthalpy switch.
   D. An outside air damper closes to a minimum position.

8. Which of the following is *not* a maintenance consideration for rooftop units?

   A. Changing air filters.
   B. Checking condensate drain for obstructions.
   C. Checking for leaks in the refrigeration system.
   D. Checking vaporization from the heating loop.

Maintenance Welding

The ability to repair and fabricate metal equipment and surfaces is a necessary skill for a successful career in maintenance welding. This requires technicians to have a thorough understanding of proper safety measures and the welding and cutting processes used in their field. This section will explore the fundamentals of oxyfuel and arc welding and cutting.

The content in this section will help prepare you to earn the NIMS Industrial Technology Maintenance Level 1 Maintenance Welding credential. By earning credentials, you are able to show employers and potential employers proof of your knowledge, skills, and abilities.

37

Welding Safety, Equipment, and Processes

LEARNING OBJECTIVES

After completing this chapter, you will be able to:

☐ Discuss three areas of welding safety procedures.

☐ Understand the proper use of PPE when welding.

☐ Identify the two weld types and five basic weld joints.

☐ Identify the various welding positions used for operation.

☐ Understand the AWS welding symbol and basic weld symbols.

☐ Identify various types of welding power sources and their uses.

☐ Identify the equipment used with a welding power source.

☐ Explain the AWS Electrode Identification System.

☐ Explain the proper use and care of electrodes and filler wire.

☐ Describe the five welding processes and equipment regularly used in the maintenance field.

☐ Discuss the advantages and disadvantages of welding processes.

TECHNICAL TERMS

arc

arc length

arc welding

base metal

butt joint

constant current (CC)

constant voltage (CV)

corner joint

Dewar flask

direct current electrode negative (DCEN)

direct current electrode positive (DCEP)

duty cycle

edge joint

electrode

filler metal

fillet weld

flux

flux cored arc welding (FCAW)

gas metal arc welding (GMAW)

gas tungsten arc welding (GTAW)

groove weld

lap joint

oxyacetylene welding (OAW)

oxyfuel gas welding

polarity

shielded metal arc welding (SMAW)

shielding gas

slag

T-joint

weld

weld symbols

welding power source

welding symbol

wire feed speed

Maintenance technicians that are able to safely repair metal parts are a valuable part of every manufacturing operation. This chapter will outline the welding safety requirements, equipment, and processes technicians must know to start their career in maintenance welding. The material learned will provide a foundation for what will be examined in later chapters of this section.

37.1 WELDING SAFETY

Welding is a skill that can be performed safely with minimum risk if a technician uses common sense and follows appropriate safety rules. Establish proper safety habits, such as checking equipment regularly and ensuring your environment is always safe, as you work with these processes.

The American Welding Society (AWS) publishes the standard ANSI Z49.1 *Safety in Welding, Cutting, and Allied Processes*. This standard provides a detailed outline of the welding safety required for the protection of personnel and property. This chapter will cover three major areas addressed in ANSI Z49.1—personal safety, welding process safety, and welding environment safety.

37.1.1 Personal Safety

You are responsible for protecting yourself from conditions created during the welding process. The proper use of personal protective equipment (PPE) is necessary whenever an individual is in or around the welding environment. This equipment and protective clothing is used to minimize exposure to a variety of hazards, such as sparks, heat, and arcs. An *arc* is the flow of electricity across an air gap during welding. Contact with extreme heat released by an arc can often cause serious injury or death.

SAFETY NOTE
Ultraviolet and Infrared Rays

Always wear proper PPE when working on any arc welding process. This includes covering all bare skin and wearing proper eye protection. An electric arc creates ultraviolet and infrared rays. These light rays are harmful to the eyes and can cause severe burns on bare skin.

Proper welding PPE must be worn from head to feet. See **Figure 37-1**. Consider the following PPE checks when starting a welding procedure:

- **Head.** Welding cap, hard hat, welding hood, safety glasses or goggles, hearing protection, face shield, respiratory protective equipment.
- **Body.** Cotton, wool, or leather welding shirt, sleeves, or jacket; leather gloves; heavy cotton pants such as jeans; leather apron.
- **Feet.** Leather boots with nonslip soles and steel toes.

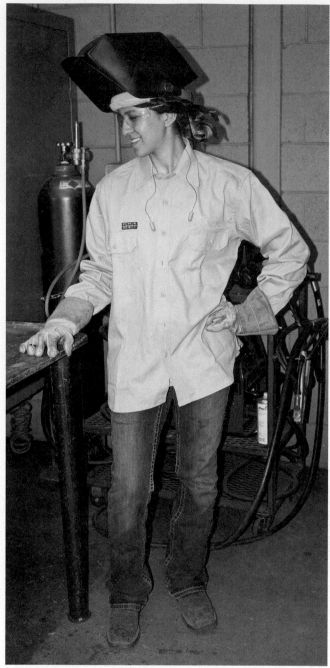

James Mosman

Figure 37-1. This welder is wearing proper PPE, including a welding hood, cap, earplugs, steel-toe boots, and hair pulled back.

Eye protection should be worn at all times in a manufacturing or construction environment. Any welding hood, cutting goggles, or face shield includes a shaded lens, which protects your eyes from the ultraviolet (UV) and infrared (IR) rays created during welding and cutting processes. As the shade number increases, the lens gets progressively darker. The higher the amperage required for the welding process, the higher the required shade number. Clear safety glasses (Z87 rated to eliminate ultraviolet radiation) or shields are to be used for eye protection during grinding, chipping, and brushing activities.

Hearing protection, such as earplugs or muffs, is recommended to help reduce noise levels often found in manufacturing operations. The ear protection will also keep sparks and welding debris out of your ears.

All welding operations create fumes and smoke that should be avoided. Fume extraction systems and increased airflow are highly recommended during welding and grinding operations. Some welding procedures produce toxic fumes. In these cases, respiratory protective equipment must be worn. This equipment may also be required when working in a confined space where oxygen may be reduced. The type of work being done determines the specific type of respiratory protection required.

Protective clothing should be of flame-resistant material such as cotton, wool, or leather. There should be no rips, tears, or frayed material. Welding gloves should fit comfortably and have proper insulation for the desired welding process. Processes that normally weld at higher amperages require a heavier glove to resist the additional heat created. See **Figure 37-2**.

37.1.2 Welding Process Safety

Welding equipment used in most welding processes consists of an electrical power source and compressed gas cylinders. Both of these components require proper care and handling for safe operation.

Electrical Safety

The primary voltage of an electrically-powered welding machine is usually 115 VAC to 440 VAC. This amount of voltage may cause extreme shock to the body and possibly death. For this reason, apply the following safety rules:

- Never install fuses of higher amperage than specified on the data label or the operation manual.
- Install electrical components in compliance with all electrical codes, rules, and regulations.
- Ensure all electrical connections are tight.
- Never open a welding machine cabinet while a machine is operating.
- Always lock primary voltage switches open and remove fuses when working on electrical components inside the welding machine.

James Mosman

Figure 37-2. Various welding gloves and thicknesses are needed depending on the welding process being used. For instance, GMAW and GTAW processes require heavier gloves to resist additional heat that is created.

Electricity supplied by constant current (CC) and constant voltage (CV) power sources has an open circuit voltage range of 40–80 V. This voltage can still produce an electric shock that may cause health problems or even death. To avoid electric shock, do the following:

- Keep the welding power supply, power cable, work cable, and welding gun dry.
- Make sure the work clamp is securely attached to the power supply and workpiece.

Shielding Gas Safety

A **shielding gas** is a gas that shields an electrode and protects the molten metal from contamination. Gases used in most welding processes are produced, stored, and distributed in either liquid or gaseous form. All storage vessels used for these gases are stamped and approved by the Department of Transportation (DOT).

The majority of gases used in welding are odorless and colorless. Thus, a dangerous leak can be difficult to detect. Therefore, special precautions must be taken when using them. Never enter any tank, pit, or vessel where gases may be present until the area is purged (cleaned) with air and checked for oxygen content. These gases can cause asphyxiation, or suffocation, in a confined area without sufficient ventilation. An atmosphere must contain at least 18% oxygen to prevent dizziness, unconsciousness, or even death.

High-pressure gas cylinders contain gases under very high pressure (approximately 2000–4000 psi) and must be handled with extreme care. Cylinders should always be stored in the vertical position with protective caps on and secured with safety chains, cables, or straps. See **Figure 37-3**.

PROCEDURE	Proper Cylinder Setup

Prior to using the contents of a cylinder for welding, perform the following steps to ensure safe and proper setup.

1. Confirm the contents of the cylinder by checking its label.

2. Ensure the protective cap of a cylinder is in place.

3. Using a cylinder cart with safety chains properly installed, secure and move the cylinder to the proper location.

4. Check the outlet threads and clean the valve opening by cracking (slightly opening and closing) the cylinder valves before attaching the regulator.

5. Ensure the proper equipment for the type of gas is being used.

6. Attach the regulator.

7. Securely mount the regulator or flow meter in the vertical position, **Figure 37-4**.

8. Stand aside and slowly open the cylinder valve. Never stand in front of the gauges when opening a cylinder.

9. Ensure that the regulator adjusting screw is free of pressure on the regulator diaphragm.

10. The cylinder is properly set up and ready for use in welding.

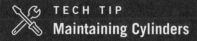

TECH TIP
Maintaining Cylinders

Cylinders are important pieces of welding equipment that must be maintained properly. When a cylinder is not in use, a cylinder valve should be closed and the diaphragm screw fully loosened. If a cylinder is defective, return it to the supplier. When a cylinder is empty, be sure to close the valve, replace the safety cap, and mark the cylinder as "Empty Cylinder" or "MT."

James Mosman

Figure 37-3. These dual oxyfuel cylinders are properly secured to the cylinder cart using safety straps. Protective caps are also fastened on the cylinders, ready for transport.

Liquefied gas cylinders, commonly called *Dewar flasks*, are similar to vacuum bottles. See **Figure 37-5**. Dewar flasks contain liquid oxygen originally reduced from gas to make it easier to store and ship from a supply plant. The liquid is later converted to a gas for welding by heat exchangers within the cylinder. Liquefied gases are extremely cold and can cause severe frostbite if they come in contact with the eyes or skin. Be sure to always wear proper PPE such as gloves and safety glasses when handling.

Like other cylinders, Dewar flasks must always be kept in the vertical position and secured on cylinder carts. Proper equipment for installing and connecting cylinders must always be used, and equipment components should never be interchanged.

37.1.3 Welding Environment Safety

Be aware of the environmental hazards encountered on the factory floor. Always be aware of what is happening in your welding area, and inform other technicians when you begin any potentially dangerous activity.

Welding areas must always remain clean and free of combustibles. Adequate ventilation must always be provided when performing any type of welding. When welding on cadmium-coated steels, copper, or beryllium copper, use equipment to remove fumes and vapors from your work area. **Figure 37-6** shows a fume exhaust system near the weld area to remove welding smoke, fumes, and vapors. Portable smoke removal systems can also be used for this purpose.

SAFETY NOTE
Exhaust Fumes

Exhaust fumes from an arc welding area can be toxic. A flexible exhaust pickup tube should be used to remove these fumes. Fumes should always be picked up and exhausted before they cross a welder's face.

Technicians must also ensure that welding equipment is safe and proper PPE is used before welding. Observe the following rules to help you and your fellow technicians maintain a safe welding environment:

- Repair or replace worn or frayed ground or power cables.
- Make sure the part to be welded is securely grounded.
- Be aware of moving equipment, such as forklifts and vehicles.
- Apply safe space requirements when around robotic equipment.

James Mosman

Figure 37-4. The flow meter is properly connected for use in the vertical position.

Goodheart-Willcox Publisher

Figure 37-5. A liquefied oxygen Dewar flask is shown hooked to a manifold system.

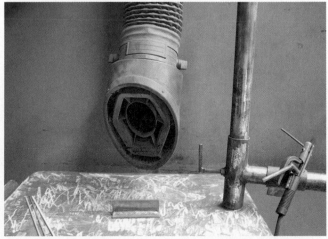

James Mosman

Figure 37-6. A fume ventilation system is used approximately 8″ to 10″ from the weld zone. Proper ventilation is needed to remove dangerous fumes and prevent suffocation.

- Ensure welding helmets have no light leaks.

- Use a proper shade number lens to protect the eyes from arc radiation. (Refer to the *Guide for Shade Numbers* chart in the *Appendix* to determine the proper lens shade number for a given application.)

- Always wear safety glasses or safety shields when using power brushes. This type of equipment is very dangerous since the brushes eject broken pieces of wire.

- Wear tinted safety glasses when others are tack welding or welding near you.

- Use safety screens or shields to protect your work area.

Fires may start in a number of ways, including by the ignition of combustible materials, misuse of fuel gases, shorting of electrical circuits, or improper ground connections. If a fire starts in your welding area, ensure that it is completely out before the area is left unattended. If the fire cannot be contained, contact the proper authorities and leave the area. Always be aware where your shop's fire extinguishers are located and how to properly use them.

SAFETY NOTE
Oxygen

Never use oxygen in place of compressed air. Oxygen supports combustion and makes a fire burn violently.

Other severe health hazards can arise during welding, making it imperative to take the proper safety measures to protect yourself and others around the welding area. Consider the following safety rules to prevent shop injury or death:

- Do not weld near trichloroethylene vapor degreasers. A welding arc produced changes this vapor to poisonous phosgene gas. A sweet taste in your mouth indicates that this gas is being formed.

- Never weld on a container that has previously held a fuel until you are sure that it has been purged with an inert gas and tested for fume content.

- Never enter a vessel or confined space that has been purged with an inert gas until the space is checked with an oxygen analyzer to determine that sufficient oxygen is present.

- Be alert to clamping operations when working with mechanical, hydraulic, or air clamps on tools, jigs, and fixtures. Serious injury may result if parts of your body are caught in the clamp.

37.2 BASIC WELDING

Welding is a process of making a weld on a joint. A ***weld*** is the blending of two or more metals by applying heat until they are molten and flow together. The metal to be welded is called the ***base metal***. Base metals are placed together and then heated by an arc or flame. Parts of the metals then melt and form a liquid pool or puddle of metal. Once the molten metal cools, a solid weld is created.

37.2.1 Basic Weld Types and Joints

There are two main weld types, or profiles, used in welding applications—the fillet weld and the groove weld. A ***fillet weld*** is a triangular cross section that connects perpendicular or uneven joint surfaces at right angles to each other. A ***groove weld*** connects two joint surfaces or edges against each other. See **Figure 37-7**.

A weld joint is the point where two metals are fit together during welding. As shown in **Figure 37-8**, the following are the five basic types of weld joints:

- ***Butt joint.*** Two pieces of metal welded together at their edges. See **Figure 37-9**.

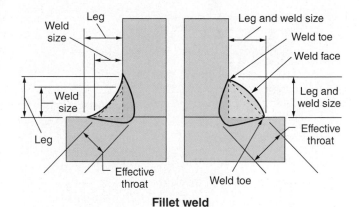

Fillet weld

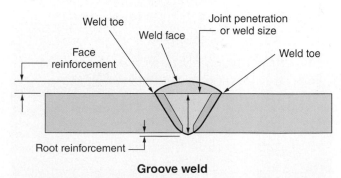

Groove weld

Goodheart-Willcox Publisher

Figure 37-7. Fillet welds and groove welds are two common weld types. Terminology for both of these weld types is shown here.

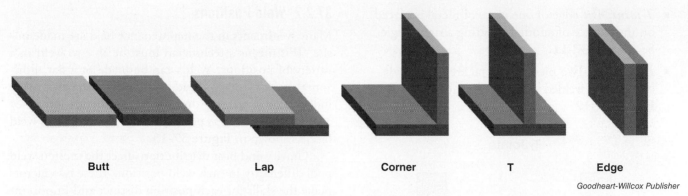

Butt   **Lap**   **Corner**   **T**   **Edge**

Goodheart-Willcox Publisher

Figure 37-8. There are five basic weld joints.

Butt Joints

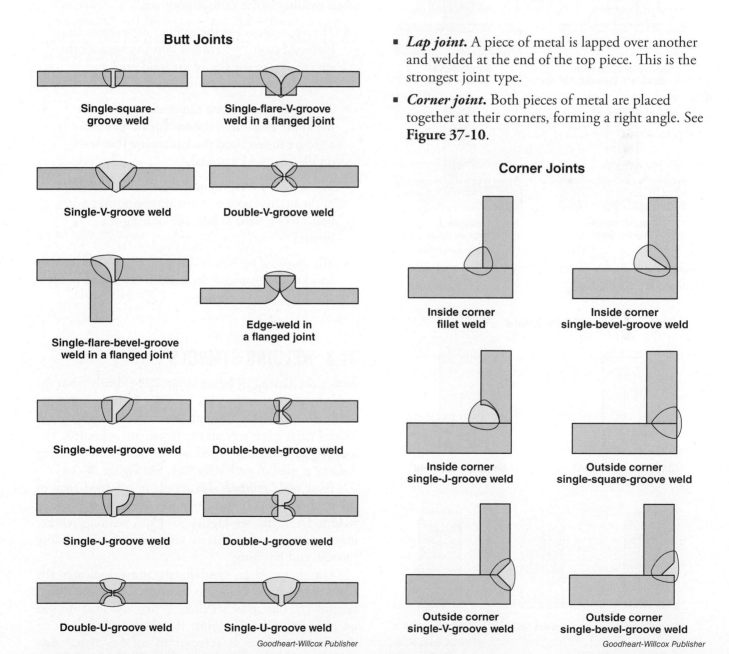

Single-square-
groove weld

Single-flare-V-groove
weld in a flanged joint

Single-V-groove weld

Double-V-groove weld

Single-flare-bevel-groove
weld in a flanged joint

Edge-weld in
a flanged joint

Single-bevel-groove weld

Double-bevel-groove weld

Single-J-groove weld

Double-J-groove weld

Double-U-groove weld

Single-U-groove weld

Goodheart-Willcox Publisher

Figure 37-9. Several of the many combinations of edge preparations and welds for butt joints.

- *Lap joint.* A piece of metal is lapped over another and welded at the end of the top piece. This is the strongest joint type.

- *Corner joint.* Both pieces of metal are placed together at their corners, forming a right angle. See **Figure 37-10**.

Corner Joints

Inside corner
fillet weld

Inside corner
single-bevel-groove weld

Inside corner
single-J-groove weld

Outside corner
single-square-groove weld

Outside corner
single-V-groove weld

Outside corner
single-bevel-groove weld

Goodheart-Willcox Publisher

Figure 37-10. A sample of corner joint edge preparation and welds are shown here.

- *T-joint.* The edge of one piece of metal is placed on the surface of another, creating a right angle. See **Figure 37-11.**

- *Edge joint.* Two pieces of metal are placed side by side and welded along one or more edge. See **Figure 37-12**.

T-Joints

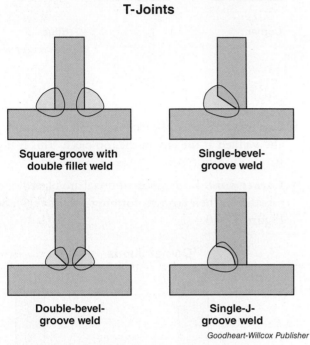

Square-groove with double fillet weld

Single-bevel-groove weld

Double-bevel-groove weld

Single-J-groove weld

Goodheart-Willcox Publisher

Figure 37-11. Some T-joint edge preparation and weld varieties are shown here.

Edge Joints

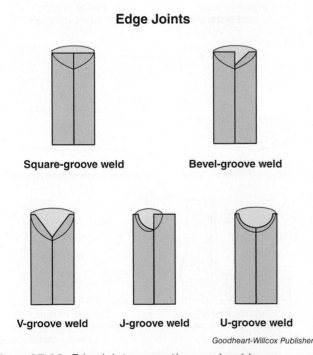

Square-groove weld

Bevel-groove weld

V-groove weld

J-groove weld

U-groove weld

Goodheart-Willcox Publisher

Figure 37-12. Edge joint preparations and welds.

37.2.2 Weld Positions

Many weldments in the maintenance field are made on-site. This means a technician must be able to weld in a variety of positions. Welds can be made in a flat, horizontal, vertical, or overhead position. These positions are indicated on a welding drawing as an abbreviation. They are determined by the position of the weld axis and weld face, as shown in **Figure 37-13.**

Gravity and heat distribution affect the molten weld pool differently in each weld position. These two factors make the skills for each position distinct and important to practice for welding. There are several considerations when welding in the various positions:

- The *flat position* has the greatest deposition rate. Flat welds usually have less porosity because the gas can rise to the top of the weld pool and escape before the metal solidifies.

- The *horizontal position* can cause undercut at the upper portion of the weld pool. Undercut is a groove melted into the base metal that is left unfilled by weld material.

- The *vertical position* requires close observation of the molten pool to prevent sagging and over-heating of a weld as heat rises during the process.

- The *overhead position* is the most tiring since it has the slowest metal deposition. Overhead welds are also prone to porosity.

37.3 WELDING SYMBOLS

Each weld joint must be communicated clearly from the weld designer to technician in order to make a proper weld. This is done through mechanical drawings of welded parts to convey all the information needed. The *welding symbol*, developed by the American Welding Society, is used in such drawings. See **Figure 37-14**.

Basic *weld symbols* also appear in specified areas of the welding symbol. These symbols represent the type of weld to be made. See **Figure 37-15**. A welding symbol may include specifications for joint preparation, welding process, and finishing.

On the welding symbol, an arrow indicates the point at which the weld is to be made. The arrow always touches the line to be welded. When a weld is needed on only one side of a joint, the side of the joint that the arrow touches is referred to as the *arrow side*. Whenever the basic weld symbol is placed below the

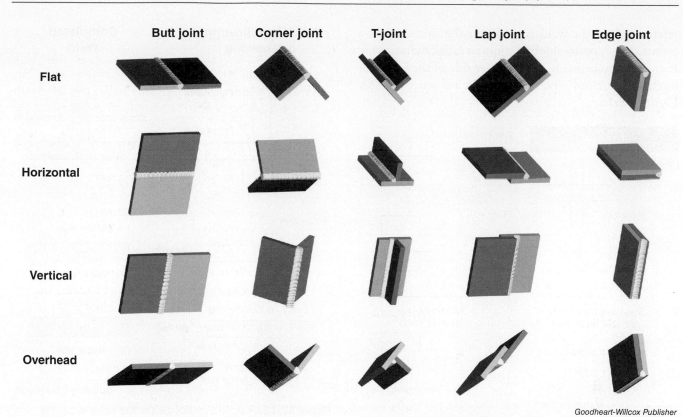

Goodheart-Willcox Publisher

Figure 37-13. Each of the five weld joint types can be completed in all four weld positions. The weld axis is shown as the blue line running lengthwise through the weld center. The weld face is the exposed surface of the finished bead.

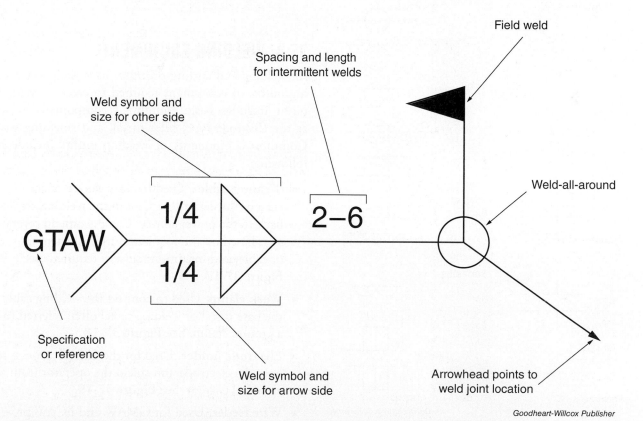

Goodheart-Willcox Publisher

Figure 37-14. The welding symbol conveys specific and complete information to the welder.

reference line, the weld is made on the side the arrow points. The opposite side of the joint is called the *other side*. If a weld is to be made on the other side of the joint, the basic weld symbol is placed above the reference line. See **Figure 37-16**.

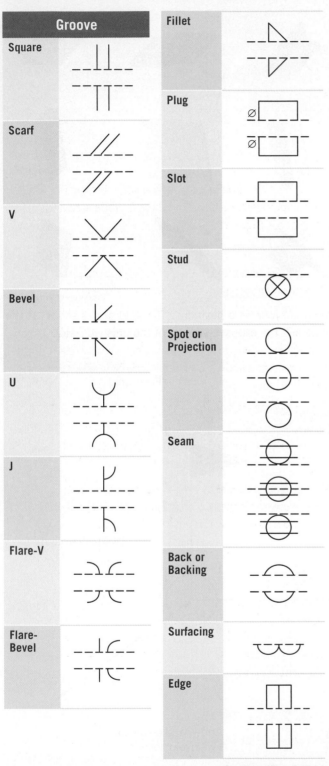

Figure 37-15. Basic weld symbols. The weld symbol is a part of the welding symbol.

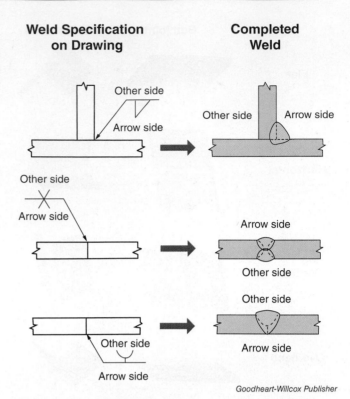

Figure 37-16. A weld symbol below the reference line indicates the weld is to be made on the arrow side. A weld symbol above the reference line indicates the weld is to be made on the side opposite of the arrow.

37.4 WELDING EQUIPMENT

All welding and cutting processes have their own welding outfit, or equipment required to create a weld. An outfit includes basic equipment components as well as the electrode type, filler metal, and shielding gases. Common components of welding outfits include the following:

- **Welding cables**. Used to carry the electrical current through the system to the welding arc and back to the power source. Cables come in several sizes based on the required duty cycle, length of cable, and maximum current required. See **Figure 37-17**.

- **Work clamp**. Used to connect the welding cable to the base metal or welding table; often referred to as a ground clamp. See **Figure 37-18**.

- **Electrode holder**. Used for the SMAW process to hold the electrode and shield the operator from the electrical current. See **Figure 37-19**.

- **Wire feeder**. Used for GMAW and FCAW processes. The wire feeder holds the spool of electrode wire and has controls for wire feed speed.

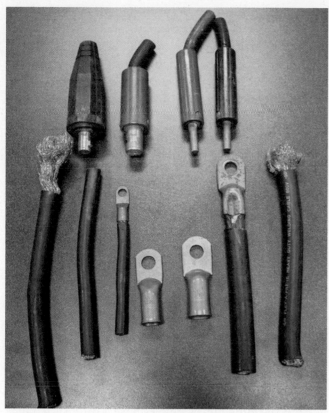

Figure 37-17. Proper welding cables and connections are determined by the machine duty cycle.

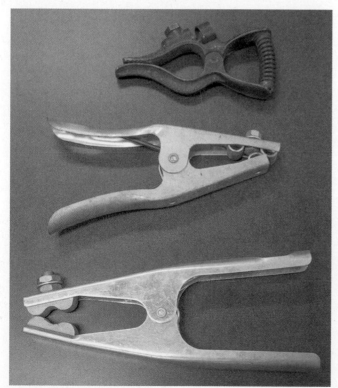

Figure 37-18. Work clamps hold a base metal in place for welding.

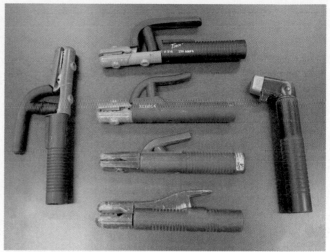

Figure 37-19. SMAW electrode holders are available in various designs based on the electrode size and amperage required.

- **Flow meter.** Used to control and regulate the volume of shield gas delivered to the weld zone.

- **Hand and power tools.** Additional tools used in the welding and material preparation processes that include grinders, chipping hammers, wire brushes, and files.

- **GMAW (MIG) gun.** Used for GMAW/FCAW electrode wire delivery. The gun has a trigger that controls the shielding gas solenoid, wire feeder motor, and electrical current.

- **GTAW torch.** Used for the GTAW process to hold the tungsten electrode and shield the operator from the electrical current.

37.4.1 Arc Welding Power Sources

A **welding power source**, or *arc welding machine*, supplies the electric current used to make the weld. Arc welding machines produce either direct current (DC) or alternating current (AC) and provide either constant current (CC) or constant voltage (CV).

37.4.2 Current Flow and Polarity

Alternating current (AC) reverses direction of electron flow at set intervals. An AC power source supplies high voltage and low current and outputs low voltage and high current that is required for welding. These machines are easy to use and generally lower in cost than DC welding machines. See **Figure 37-20**. They are used primarily for the SMAW process.

Direct current (DC) flows in only one direction. Current flows from one terminal—either the base metal or electrode—of a DC welding machine and then to the second terminal. Electric current can be manually reversed, thus changing the direction of current flow. This change in current is referred to as *polarity*. When electrons flow from the electrode to the base metal, the electrode has negative polarity, and the base metal has positive polarity. This is called *direct current electrode negative (DCEN)*, or *direct current straight polarity (DCSP)*. When current flows in the opposite direction, from the base metal to electrode, the base metal has negative polarity, and the electrode has positive polarity. This is called *direct current electrode positive (DCEP)*, or *direct current reverse polarity (DCRP)*. Since DC sources can change polarity, this allows for more versatility. They are used for GMAW and GTAW when set up in the DCEN direction.

Many power sources have the capability to use both AC and DC current, referred to as AC/DC combination machines. These multiprocess machines are ideal for the maintenance welder. These machines can be used for all arc welding processes and plasma arc cutting.

37.4.3 Constant Current and Constant Voltage

Constant current (CC) power sources are used for manual welding processes, such as SMAW and GTAW. CC power sources allow a welder to maintain a constant

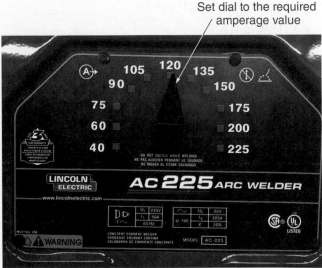

Set dial to the required amperage value

James Mosman

Figure 37-20. The front panel of this AC power source allows a technician to set and maintain current during welding.

current for welding as voltage is changed. When voltage changes, so does the arc gap distance. Manual processes require all welding variables to be controlled by the technician's hand, including the *arc length* (distance between the base metal and electrode) and the *wire feed speed* (rate at which an electrode is fed into a joint).

Power sources that have *constant voltage (CV)* maintain a set voltage and a constant arc length while large changes can be made to current. With a constant voltage machine, the current is established by the wire speed. Increasing or decreasing the wire speed results in an increase or decrease in current. CV machines are required for semiautomatic processes, such as in GMAW and FCAW, where arc length and wire feed speed are set and controlled by the machine.

37.4.4 Duty Cycle

Another important consideration when selecting a welding power source is machine duty cycle. *Duty cycle* is the percentage of time in a ten-minute period that a machine can be operated at a rated output without overheating. Welding power sources have a duty cycle range between 20% and 100%. Each percent equals one minute of use in a ten-minute period. Thus, a machine with a 60% duty cycle at 200 amps would be able to operate at that amperage for 6 minutes and then require 4 minutes to run without a load to cool down. If a machine is used above the duty cycle rating, it will overheat and shut down or cause possible damage to the electrical components.

37.4.5 Filler Metal and Electrodes

Most welding processes require the use of some type of filler metal to connect the base metals. *Filler metal* is metal that is added to replace material removed when the base metal is cut or machined to form a groove during welding. A welding rod is a common filler metal. Butt joints, T-joints, and some corner joints require filler metal in order to complete a weld.

Welding electrodes that melt and enter the weld serve as filler metal. This occurs in processes where the electrode is a consumable wire. An *electrode* is the point to which electricity is brought to produce the arc required for arc welding. An electrode is selected for a welding process based on the base metal used, design requirements, and process. The welding electrode and the power source must be compatible.

The proper care and storage of welding electrodes is essential for quality welding. Careful handling is required when moving or stacking electrode containers to avoid damaging the flux coating on the electrodes. Always keep electrodes dry at varying temperatures depending on the type of electrode. Any low-hydrogen electrode must be stored in electrode ovens at 250°F to 300°F. See **Figure 37-21**. Depending on the specific electrode designation, the exposure time outside this temperature may be from 4 to 10 hours. Any electrodes that have been exposed to humid conditions or moisture must be properly dried according to manufacturer's recommendations prior to use.

The American Welding Society created an electrode classification system for electrodes, electrode wire, and filler material. In this system, each number and letter is used to describe the electrode characteristics and suggested usage. See **Figure 37-22**.

James Mosman

Figure 37-21. Low-hydrogen electrodes must be stored in an electrode oven at a temperature between 250°F and 300°F. Outside exposure time varies with electrode designation.

37.4.6 Shielding Gases

Shielding gases, or secondary gases, cover and protect the weld area, which includes the base metals, arc, and electrode, and prevents contamination from the atmosphere.

SMAW Electrodes

E 70 1 8 - 1
E XX X X - X
① ② ③ ④ ⑤

GMAW Electrodes and GTAW Welding Rods

E 70 S - 6
ER XX X - X
① ② ⑥ ⑦

FCAW Electrodes

E 7 1 T - 1 C
E X X X - X X
① ② ③ ⑧ ⑨ ⑩

Key

● Designates an electrode (E) or rod (ER).

● Minimum tensile strength in ksi.

● Welding position.

● Type of flux coating and polarity.

● Additional alloy elements in the wire core.

● Filler metal solid (S) or (C) composite.

● Chemical composition.

● Tubular or cored.

● Usability of the electrode.

● Type of shielding gas.

James Mosman

Figure 37-22. The AWS classification systems for electrodes and welding rods.

These gases are used for GMAW, FCAW, and GTAW processes, but vary based on the specific process and type of metal to be welded. Shielding gases have several additional purposes:

- Transfer heat from the electrode to the metal.
- Stabilize the arc.
- Aid in controlling weld penetration.

- Assist in metal transfer from the electrode.
- Assist in cleaning the joint and providing a wetting action.

Shielding gases are usually inert gases—gases that are chemically inactive and do not combine with any product of the weld area. Argon and helium are two common inert gases. These gases are pure, colorless, and tasteless, and may be used as a single gas or as part of a mixed gas combination.

37.5 MAINTENANCE WELDING PROCESSES

There are two common groups of welding processes—oxyfuel gas welding and arc welding. **_Oxyfuel gas welding_** processes use heat produced by a gas flame to complete a weld or cut. **_Arc welding_** processes use heat produced from an electric arc to melt and weld two metals. Although there are many specialized welding processes documented by the American Welding Society, only a handful may be useful to a maintenance welder.

The following are the five common welding processes used in maintenance applications:

- Oxyacetylene welding (OAW).
- Shielded metal arc welding (SMAW).
- Gas metal arc welding (GMAW).
- Flux cored arc welding (FCAW).
- Gas tungsten arc welding (GTAW).

This section will provide a brief description of each of the five welding processes. Chapter 38, *Maintenance Welding Procedures and Inspection*, will provide a more thorough explanation of these welding procedures.

37.5.1 Oxyacetylene Welding and Brazing

Oxyacetylene welding (OAW) is a form of oxyfuel gas welding that is used to weld most low-grade steels. OAW melts and fuses metals using extreme heat created by a combination of acetylene gas and oxygen. When ignited, oxyacetylene produces a flame that can reach temperatures as high as 6000°F. See **Figure 37-23**.

Oxyacetylene welding uses dual oxygen and acetylene cylinders that store gases under pressure and are controlled by gas pressure regulators. These regulators deliver the gas mixture through a hose to the welding torch, which then produces a flame used to melt the base metal. See **Figure 37-24**. As the base metal is melted, a welding rod is added to increase the thickness of the weld.

Uniweld Products Inc.

Figure 37-23. Oxyacetylene welding produces a flame that has the highest temperature of any other fuels used to weld materials.

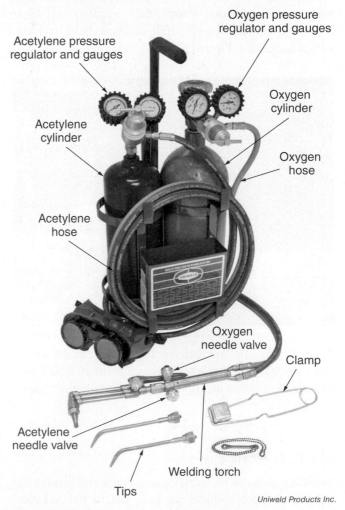

Uniweld Products Inc.

Figure 37-24. Basic welding outfit for oxyacetylene welding.

Brazing applications also use oxyacetylene gas for cast iron or dissimilar metal repairs. Because brazing requires less extreme temperatures than welding, the base metal is heated from 840°F to just below the melting point. A brass filler rod is melted to attach the two pieces.

The OAW process has several advantages and disadvantages when compared to other welding processes. Some advantages include the following:

- Equipment is relatively inexpensive.

- Equipment is more portable than most other welding types.

- OAW process does not require electricity or a power source unlike arc welding processes.

- OAW can be used to cut larger pieces of material.

Some disadvantages of OAW include the following:

- Acetylene gas is very flammable and can cause severe injury if not handled properly.

- When at high pressures, oxygen gas can be highly reactive.

- Due to high temperatures, weld lines and cuts may not be as neat as those produced by other processes.

37.5.2 Shielded Metal Arc Welding

Shielded metal arc welding (SMAW), often referred to as *stick welding*, is the most commonly used for maintenance applications. Welding technicians use this process to complete quick repairs while producing a solid and secure joint. In SMAW, metals are welded using the heat of an electric arc struck between an electrode and base metal. See **Figure 37-25**. Electrodes used in SMAW are covered by a material called *flux* that helps remove oxides and other unwanted substances from a metal surface. As flux melts on the electrode, it provides a shielding gas around the weld being formed. When it later solidifies, it forms a hard, metal layer over the weld bead called *slag*.

Manual welding processes like SMAW use CC (constant current) welding machines to alter arc length and set a constant amperage. Additional SMAW equipment includes a welding machine, electrode holder, workplace leads, and electrode.

Some advantages of SMAW include the following:

- SMAW electrodes can be coated with a variety of shielding fluxes to support a wide range of industry applications.

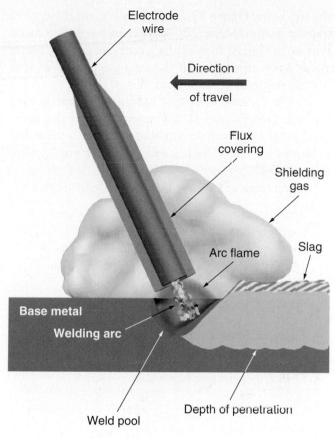

Figure 37-25. This drawing demonstrates the shielded metal arc welding process.

- The process is generally less expensive than oxyfuel gas welding types, such as OAW.

- No separate shielding gas is required.

Several disadvantages of SMAW include the following:

- SMAW requires frequent stops to replace electrodes, which makes it not as continuous or productive as processes like SMAW and FCAW.

- Welding speeds are generally slower than OAW, SMAW, and FCAW.

- Welds may contain slag inclusions.

37.5.3 Gas Metal Arc Welding

Gas metal arc welding (GMAW), often referred as *metal inert gas (MIG) welding*, produces an electrical arc to heat metals. GMAW electrodes are metal-cored, consumable wires that melt and are continuously fed into the weld. See **Figure 37-26.** Depending on the type of transfer used, GMAW can be used on nearly any metal type or thickness. A shielding gas or gas mixture serves as the main protection for the weld since GMAW does not use electrodes that produce flux or slag.

A CV (constant voltage) welding machine is used for both GMAW and FCAW semiautomatic processes. Additional equipment components include a cable and work clamp, a wire feeder, and a welding gun system, **Figure 37-27.**

Advantages of GMAW include the following:

- A continuous electrode is used, which allows a weld to be made without stopping to change electrodes. This eliminates the starts and stops that occur with SMAW and causes weld defects.

- Welding speeds are faster than SMAW.

- Electrode waste is limited—nearly 100% of the electrode becomes part of the weld.

- Deeper penetration is possible.

Miller Electric Mfg Co.

Figure 37-27. Equipment for gas metal arc welding and flux cored arc welding are identical except for the combination cable and gun used.

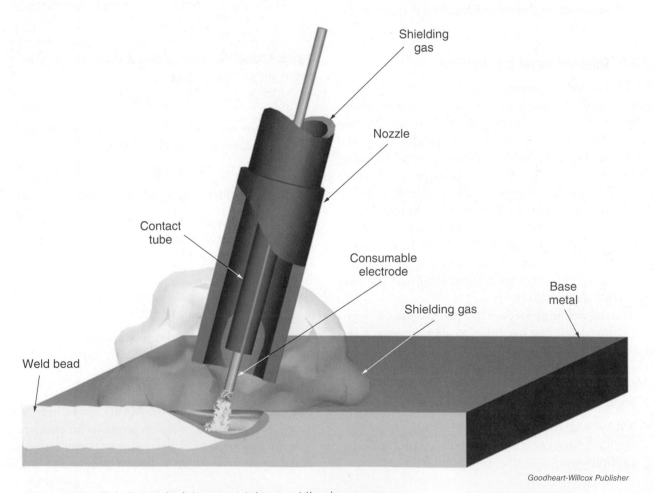

Goodheart-Willcox Publisher

Figure 37-26. This figure depicts gas metal arc welding in process.

Some disadvantages of GMAW include the following:

- Cost of equipment is higher than SMAW equipment, mainly due to cost of additional shielding gas.

- Some welds are hard to reach with a welding gun.

- Rapid air movement in different locations may decrease the impact of shielding gas. This may result in weld defects.

37.5.4 Flux Cored Arc Welding

Similar to GMAW, *flux cored arc welding (FCAW)*, or *flux core welding*, uses an electrical arc produced between a continuous consumable electrode and base metal. However, the electrodes used in this process are tubular flux-filled wires. As the electrode melts, flux creates a protective slag deposit on top of the weld. The process then continues similarly to SMAW. See **Figure 37-28.** Some flux-cored electrodes require the use of additional

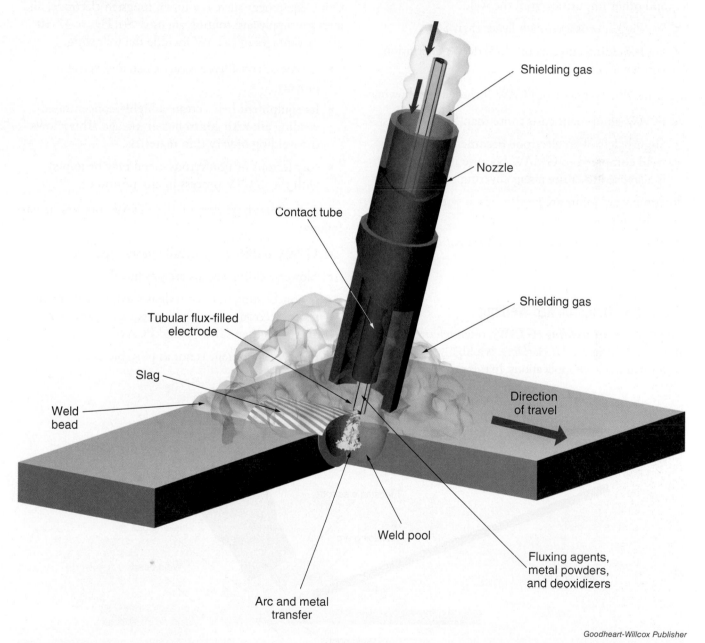

Shielding gas

Nozzle

Contact tube

Shielding gas

Tubular flux-filled electrode

Slag

Direction of travel

Weld bead

Weld pool

Fluxing agents, metal powders, and deoxidizers

Arc and metal transfer

Goodheart-Willcox Publisher

Figure 37-28. Flux cored arc welding in process. Unlike GMAW, FCAW uses an electrode that produces flux and slag layers to protect the weld formed.

shielding gas while other electrodes need only what is produced from their own physical makeup. Flux cored welding is used only on ferrous and nickel-based metals.

FCAW equipment, like GMAW equipment, includes a CV welding machine, cable and work clamp, a wire feeder, and a welding gun system.

The FCAW process provides several advantages:

- The use of a continuous and consumable electrode eliminates the starts and stops of SMAW.

- Flux cored electrodes can remove oxygen, nitrogen, and other impurities from the weld.

- Welding speeds are much faster than SMAW.

- High welding currents produce deep penetration and few defects.

Some disadvantages of FCAW include the following:

- FCAW equipment costs more than SMAW.

- Slightly less of an electrode becomes part of the weld compared to GMAW because its material is also used to create a slag covering.

- Some weld joints are hard to reach with a welding gun.

- Slag removal requires extra time spent on the welding process.

37.5.5 Gas Tungsten Arc Welding

Gas tungsten arc welding (GTAW), often referred to by its outdated designation *TIG welding*, is a highly versatile process for maintenance applications. In this process, a tungsten alloy electrode is used as the terminal point for an electrical arc. Like SMAW, the electrode is nonconsumable—it does not melt or join the weld. Because of this, filler metal may be needed to complete the weld. The filler material used are straight-wire rods that vary in thickness and are usually 36″ in length. Inert gases are used to shield the arc area from the atmosphere to prevent electrode and base metal oxidation. See **Figure 37-29**. Argon and helium are common gases used in this process.

Like SMAW, GTAW uses a CC welding machine. The current is often regulated with a foot pedal or thumb switch. Other equipment such as a torch, tungsten electrode, and inert gas supply and controls are used. See **Figure 37-30**.

Advantages of GTAW include the following:

- GTAW is considered a very clean and versatile process.

- Its equipment helps create a highly concentrated welding arc with controlled amperage. This allows the welding of very thin materials.

- Any ferrous or nonferrous metal may be joined with the GTAW process in any position.

A few disadvantages of the GTAW process are as follows:

- GTAW outfits are generally more expensive.

- Slower welding speeds are produced.

- A nonconsumable electrode is used, which creates more electrode waste. It also requires more starts and stops than GMAW and FCAW.

- GTAW equipment is not as portable as other processes.

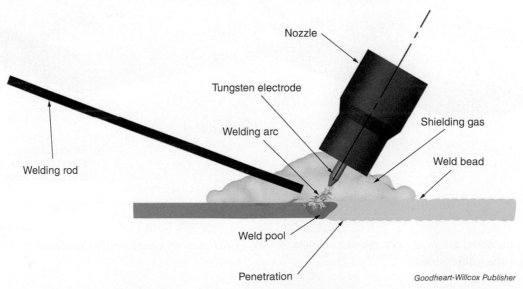

Nozzle

Tungsten electrode

Welding arc

Welding rod

Shielding gas

Weld bead

Weld pool

Penetration

Goodheart-Willcox Publisher

Figure 37-29. This drawing shows the gas tungsten arc welding process.

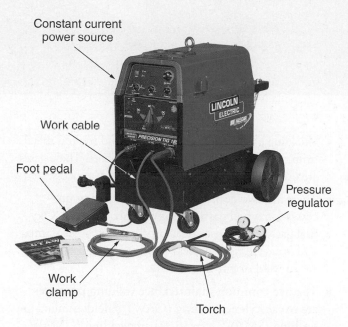

Constant current
power source

Work cable

Foot pedal

Work
clamp

Torch

Pressure
regulator

The Lincoln Electric Co.

Figure 37-30. Standard equipment used for gas tungsten arc welding.

CHAPTER WRAP-UP

This chapter discussed the fundamentals of welding safety, equipment, and processes that are necessary to help a technician conduct work in a successful manner. Thorough understanding and application of these concepts will allow you to be instrumental in developing proper techniques and practices in a shop setting. Continue to build upon the knowledge acquired in this chapter as you begin to look at maintenance welding processes more in-depth.

Chapter Review

SUMMARY

- The American Welding Society (AWS) publishes ANSI Z49.1, a national standard for *Safety in Welding, Cutting, and Allied Processes* that outlines welding safety for personnel and property protection.

- The three safety areas addressed in the national standard are personal safety, welding process safety, and welding environment safety. Each area has specific standards that must be followed.

- A weld joint is the point where two metals are fit together during welding. The five basic weld joints are butt, T, edge, corner, and lap joints.

- The welding symbol developed by the American Welding Society shows the proper weld configuration on drawings. Basic weld symbols are also used to indicate the type of joint, placement, and the type of weld to be made.

- Welding and cutting outfits include basic equipment and additional components such as electrodes, filler metal, and shielding gases.

- A welding power source, or arc welding machine, supplies electric current used to make a weld. These machines produce either direct current (DC) or alternating current (AC) and provide either constant current (CC) or constant voltage (CV).

- DC current can be manually reversed, thus changing polarity. When electrons flow from the electrode to base metal, it is called direct current electrode negative (DCEN). When electrons flow from the base metal to the electrode, it is called direct current electrode positive (DCEP).

- Constant current machines are used for manual welding processes, in which all welding variables are controlled by hand. Constant voltage machines are used for semiautomatic welding processes, in which welding variables such as arc length and wire feed speed are set by the machine.

- Filler metal is metal added to replace material removed during welding. An electrode can serve as a type of filler metal in certain welding processes. Both filler metal and electrodes are identified using the AWS electrode classification system.

- Shielding gases protect an electrode and molten metal from contamination from the atmosphere. These gases are used in GMAW, FCAW, and GTAW processes.

- There are two common groups of welding processes—oxyfuel gas welding and arc welding. Oxyfuel gas welding uses heat produced by a gas flame, and arc welding uses heat produced by an electric arc to weld or cut metals.

- The five common maintenance welding processes are oxyacetylene welding (OAW), shielded metal arc welding (SMAW), gas metal arc welding (GMAW), flux cored arc welding (FCAW), and gas tungsten arc welding (GTAW).

REVIEW QUESTIONS

Answer the following questions using the information provided in this chapter.

1. What is the purpose of ANSI Z49.1?

2. Explain the three areas of welding safety.

3. A(n) _____ is the flow of electricity across an air gap during welding.

4. List at least five items of PPE used for cutting and welding.

5. What can cause electric shock during welding?

6. A(n) _____ is a cylinder used to contain liquefied gas.

7. *True or False?* Fumes produced during welding are toxic.

8. What is the difference between a weld and a joint?

9. Describe the difference between a fillet weld and a groove weld.

10. List the five types of weld joints.

11. *True or False?* A butt joint is produced when two pieces of metal are placed side by side and welded along one or more edges.

12. Name the four welding positions, and discuss two factors that may impact a weld in each position.

13. Describe the purpose of the AWS welding symbol.

14. Identify five components of a welding outfit.

15. *True or False?* An electrode can serve as a type of filler metal.

16. Explain what each of the numbers and letters means for E7018-1 when using the AWS electrode classification system.

17. What is the purpose of a shielding gas?

18. Discuss the difference between oxyfuel gas welding and arc welding.

19. Discuss the process of oxyacetylene welding (OAW).

20. Shielded metal arc welding uses an electrode covered with _____ to provide a shielding gas around the weld being formed that later hardens.

21. Compare and contrast gas metal arc welding (GMAW) to flux cored metal arc welding (FCAW).

22. What welding processes use a constant current welding machine? Constant voltage machines?

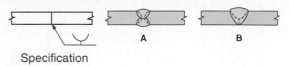 **NIMS CREDENTIALING PREPARATION QUESTIONS**

The following questions will help you prepare for the NIMS Industrial Technology Maintenance Level 1 Maintenance Welding credentialing exam.

1. Which of the following is *not* true about proper cylinder storage and transportation?

   A. A cylinder should be transported with safety chains, cables, or straps.
   B. An empty cylinder must have the safety cap replaced and valve closed.
   C. A high-pressure cylinder should be stored in the horizontal position.
   D. When attached to a cylinder, the regulator or flow meter must be in the vertical position.

2. Which action is a safe method for removing exhaust fumes or smoke during welding or grinding?

   A. Decreasing airflow in the welding area
   B. Increasing oxygen in the welding area
   C. Using a flexible exhaust pickup tube
   D. Wearing proper PPE

3. The following illustration depicts what type of weld and joint?

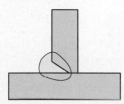

   A. Inside corner fillet weld
   B. Single-bevel groove weld
   C. Square-groove with single-fillet weld
   D. V-groove weld

4. Based on the specification provided, which of the following would be the completed weld?

   A. A
   B. B
   C. Both A and B
   D. Neither A nor B

5. In semiautomatic arc welding processes, what two variables can be set by the welding machine?

   A. AC and DC
   B. Amperage and voltage
   C. Arc length and wire feed speed
   D. Duty cycle and polarity

6. What is the purpose of an electrode?

   A. It increases the duty cycle of a welding power source.
   B. It determines the type of polarity used in AC welding machines.
   C. It determines the rate at which a filler metal is fed into a joint.
   D. It provides a terminal point where an electrical arc is struck for welding.

7. _____ produces the highest-temperature flame to melt and cut metal.

   A. Flux cored arc welding
   B. Gas tungsten arc welding
   C. Oxyacetylene welding
   D. Shielded metal arc welding

8. Which statement is true about GMAW?

   A. GMAW electrodes become filler metal in a weld.
   B. GMAW uses the same outfit as FCAW except it uses a CC welding machine.
   C. GMAW does not require separate shielding gas.
   D. GMAW is a less efficient process than SMAW since there are many starts and stops throughout the process.

38

Maintenance Welding Procedures and Inspection

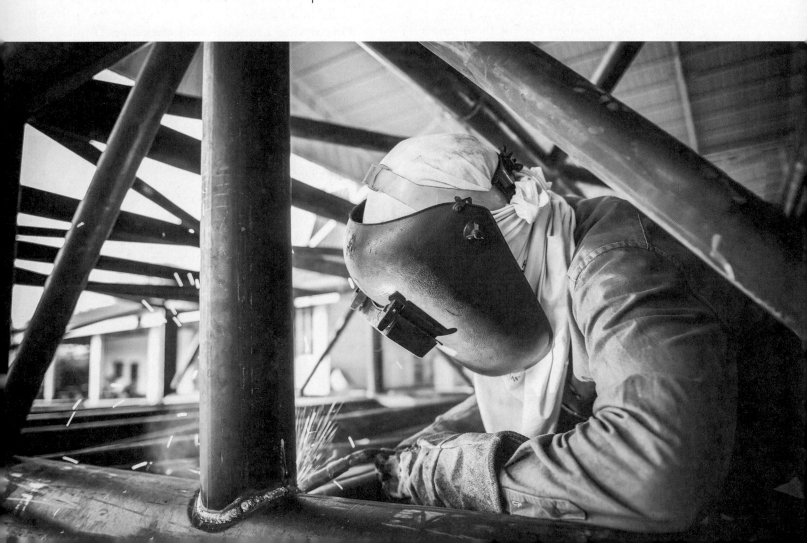

LEARNING OBJECTIVES

After completing this chapter, you will be able to:

☐ Discuss the purpose of a welding code and a welding procedure specification.

☐ Describe welding variables and how they impact the properties of a weld.

☐ Explain proper material preparation for welding.

☐ Define travel angle and work angle as it pertains to welding torch position.

☐ Demonstrate the proper procedure for setting up an SMAW power source.

☐ Perform the proper SMAW procedure for making a weld bead in the flat position.

☐ Compare GMAW with FCAW-G and FCAW-S processes.

☐ Demonstrate the proper procedure for setting up a GMAW power source.

☐ Perform the proper GMAW procedure for making a weld bead in the flat position.

☐ Describe metal transfer and the four modes of deposition.

☐ Perform the proper GTAW procedure for making a weld bead in the flat position.

☐ Identify the basic principles of weld inspection to identify common weld defects.

☐ Understand the various weld defects, causes, and corrective actions.

TECHNICAL TERMS

backhand welding

destructive testing

drag angle

electrode extension

forehand welding

gas-shielded flux cored arc welding (FCAW-G)

globular deposition

nondestructive testing

pulse spray-arc deposition

push angle

self-shielded flux cored arc welding (FCAW-S)

short-circuiting arc deposition

spray-arc deposition

stringer bead

travel angle

visual inspection

weave bead

weld bead

weld defect

welding code

welding procedure specification (WPS)

work angle

LEARNING OBJECTIVES

After completing this chapter, you will be able to:

☐ Discuss the purpose of a welding code and a welding procedure specification.

☐ Describe welding variables and how they impact the properties of a weld.

A maintenance welder must understand how to develop and perform a qualified welding procedure based on official codes and specifications. This includes consideration of various welding variables that must be set before a technician can begin any official welding operation and produce a final, quality weld. This section will provide a basic overview on qualified welding procedures, welding variables, performing common maintenance welding processes, and inspecting welds.

38.1 WELDING CODES AND SPECIFICATIONS

A *welding code* is a national standard that provides guidelines, requirements, and standards for welding. Welding codes may be used as reference by a manufacturer to ensure quality welding. In other cases, a customer may require that welding be performed in accordance with a specific code. Three common codes that address welding are the following:

- *American Society of Mechanical Engineers (ASME) Boiler Code.* Pertains to design, manufacturing, inspection, and installation requirements of boilers.

- *American Welding Society (AWS) Structural Welding Code.* Developed for structural applications. The codes are specified for structural steel, stainless steel, aluminum, sheet metal, bridges, and similar applications.

- *American Petroleum Institute (API) Pipeline Welding Code.* Developed for pipelines and related facilities, welding and construction of low-pressure storage tanks, and weld inspection and metallurgy.

A *welding procedure specification (WPS)* is established based on information from a welding code. A WPS is a document that records required welding variables for a specific weld application. This ensures that a trained welder can properly complete the weld. The document must be approved or qualified for a welder to use the WPS to perform a test weld. This test weld should have the same joint design, metal, and material thickness as the required weld. It must also be able to perform its intended function. When the test weld is complete, the weld must be inspected for quality.

TECH TIP
Testing Welds

Anytime you set up a machine, change an electrode, install a gas supply, or modify a parameter, run a test weld on a piece of scrap material to ensure that the setup is correct. It is far better to ruin a piece of scrap metal than to ruin a production part with a poor quality weld.

A WPS must be followed precisely. If a change to a welding variable is made after qualification, then the WPS is no longer being followed. When this occurs, weld testing must be repeated to requalify the new procedure.

38.2 WELDING VARIABLES

Welding variables are set when a welding procedure is established and impact the mechanical properties of a weld. Many variables are determined once the metal type, metal thickness, and welding process are selected. These may include the type of power source, electrode material, filler wire, shielding gas, welding position, and metal transfer used for welding.

Additional variables that impact the quality of a weld include the following:

- Electrode extension.
- Travel speed.
- Torch angle.
- Weld bead patterns.

38.2.1 Electrode Extension

Different welding processes require a certain amount of electrode extension to complete a successful weld. *Electrode extension* is the distance from the contact tip to the end of the unmelted electrode. See **Figure 38-1**. The electrode must extend beyond the contact tube to preheat the electrode and initiate the welding process. A technician

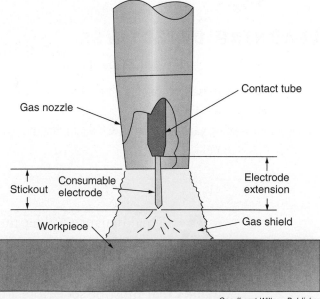

Goodheart-Willcox Publisher

Figure 38-1. This illustration shows the electrode extension distance from the contact tip to the end of the electrode.

must measure the actual distance from the contact tip to the end of the electrode—not only the amount of electrode extending from the end of the gas nozzle.

38.2.2 Travel Speed

The travel speed of the welding gun must be regulated to produce a consistent weld. In semiautomatic welding, the speed is set by the machine and measured in inches per minute (ipm). However, in manual welding, the technician must set and maintain a desired speed by hand. This requires the ability to adapt to conditions that may occur as the weld continues, such as changing shapes, improper fit-ups, or gaps. In manual welding, the skill of the welder plays an integral role in the quality of the weld.

38.2.3 Torch Angle

Two separate angles describe the position of a welding torch or gun in a welding process. These two angles are called the travel angle and the work angle. The *travel angle* is a longitudinal angle that describes the position of the torch tilted *along* the weld axis plane. The *work angle* is the transverse angle that describes the position of the torch tilted *across* the weld axis. The travel and work angles are shown in **Figure 38-2**.

A torch angle affects weld penetration, bead form, and final weld bead appearance. The angles may change as welding progresses in order to maintain good weld pool control and weld bead shape. **Figure 38-3** shows some typical work angles and bead placements.

The travel angle can be further broken down based on the angle related to the direction the gun travels. ***Backhand welding*** is completed with the welding gun pointed back at the weld as the weld progresses in the opposite

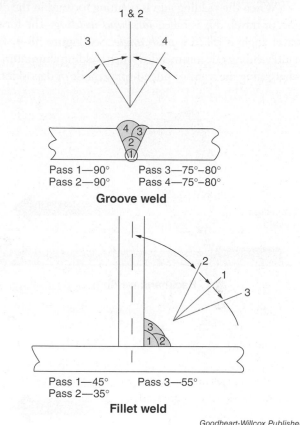

Pass 1—90°   Pass 3—75°–80°
Pass 2—90°   Pass 4—75°–80°

Groove weld

Pass 1—45°   Pass 3—55°
Pass 2—35°

Fillet weld

Goodheart-Willcox Publisher

Figure 38-3. Work angles and bead placement for multipass groove weld and fillet weld.

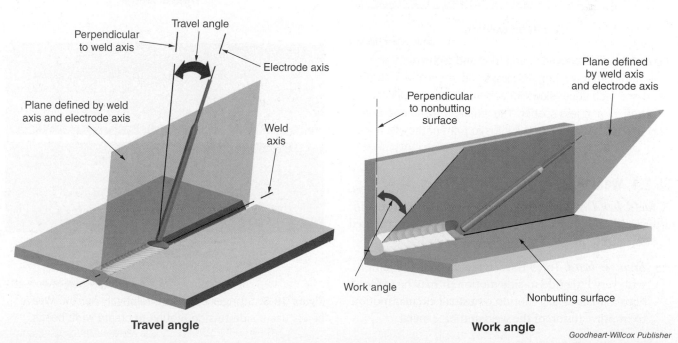

Travel angle

Work angle

Goodheart-Willcox Publisher

Figure 38-2. Gun angles are defined as either travel angle or work angle. Both are important for a proper weld bead profile.

direction. The travel angle for backhand welding is called the *drag angle*. Backhand welding produces a more stable arc, results in less spatter, and provides deeper penetration when compared with forehand welding.

When the welding gun is pointing forward in the direction of travel, this is called *forehand welding*. The forward travel angle is called a *push angle*. See **Figure 38-4**. Forehand welding is recommended when welding aluminum and when using the spray or pulsed-spray mode of deposition.

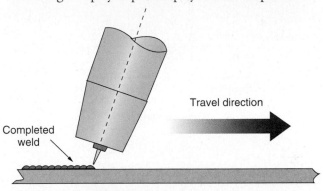

Backhand welding

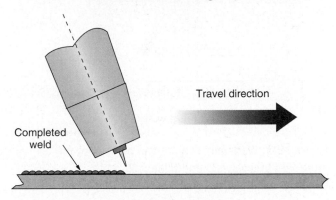

Forehand welding

Goodheart-Willcox Publisher

Figure 38-4. The backhand and forehand techniques are used in different welding processes and applications. The backhand technique allows for a more stable arc, deeper penetration, and less spatter. The forehand technique is an easy method since the weld joint is in front of the electrode.

38.2.4 Weld Bead Patterns

A *weld bead* is the shape of the finished welded joint after welding is complete. Two types of patterns are used for depositing metal, **Figure 38-5**:

- *Stringer bead.* Torch travel is made along the joint with very little side-to-side motion. It may be made with a small zig-zag motion or a small circular motion to improve fusion of the weld and base metal.

- *Weave bead.* Uses a side-to-side or similar motion to produce beads wider than a stringer bead. Requires a dwell (pause) of the gun at the end of each weave to ensure a strong weld. Also called a *wash bead* pattern.

Stringer bead

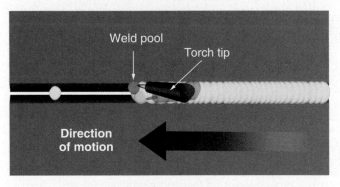

Weave bead

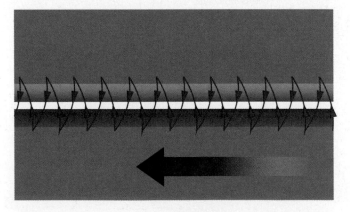

Weave bead motion

Goodheart-Willcox Publisher

Figure 38-5. Stringer beads are relatively narrow. Weave beads use a side-to-side motion to create wide beads.

38.3 MATERIAL PREPARATION

All metal to be welded must be prepared or cleaned. Ensure all dirt, oil, and grease is removed with a chemical such as acetone or alcohol. The presence of such materials could cause a weak or defective weld. If chemical cleaning is not used or the base metal has been coated or painted, it should be cleaned mechanically by grinding or using a wire brush. See **Figure 38-6**. A right-angle grinder with a grinding wheel, sanding disk, or wire brush can remove all rough edges and oxide film as well as help prepare a straight, flat surface for fit-up and welding.

If a torch specific to thermal cutting is used, an oxide film can form on the cut surface. This must be removed prior to welding to prevent pores and inconsistencies in the weld.

TECH TIP
Preheating Metals

It is necessary to preheat certain metals prior to welding. Preheating causes less local expansion of the part and slows down the cooling rate, thus producing less stress. Some materials like aluminum and high carbon steel require preheating with a torch. Specific welding procedures must be established and followed when working with such metals.

James Mosman

Figure 38-6. The material has been sanded to a "bright metal" finish to prepare the part for welding. This type of sander is often called a "PG" wheel.

38.4 PERFORMING WELDING PROCESSES

The ability to make a good, strong weld using any welding process comes after extensive practice and precision. Although welding procedures vary based on the needs of the application or industry, common maintenance welding processes are used to complete most tasks. A technician must know how to perform all common maintenance welding processes and understand each step in detail.

Chapter 37, *Welding Safety, Equipment, and Processes*, introduced four arc welding processes—SMAW, GMAW, FCAW, and GTAW. The following sections will help you to build on the knowledge you gained previously and learn how to perform each welding process.

38.5 SHIELDED METAL ARC WELDING (SMAW)

Shielded metal arc welding (SMAW), or stick welding, is the most basic electric arc welding process. See **Figure 38-7**. In this process, an arc is struck between the base metal and electrode by scratching the electrode on the metal and withdrawing it or using a straight up-and-down motion.

designbydx/Shutterstock.com

Figure 38-7. SMAW process being used by a technician.

SMAW uses a constant current power source and can be done in AC, DCEP, or DCEN. Since a CC power source is used for this process, the arc length is controlled by the technician's hand. The arc length should be approximately equal to the diameter of the electrode wire to be used. Electrode wires used in this process are covered with flux to create a protective shield gas around the weld area.

The type of metal and metal thickness are important factors when selecting which electrode to use. The strength of the electrode metal must be as strong as that of the base metal, and its diameter should be 2–3 times smaller than the weld bead. Additional factors that determine electrode selection include joint alignment, welding current, technician skill, welding position, and depth of penetration.

Two of the most commonly used electrodes for SMAW are the E6010 and the E7018. The E6010 electrode is a deep penetrating electrode that should be used with DCEP. The E7018 is a low-hydrogen electrode that has mild penetration and produces code-quality welds.

38.5.1 Assembling an SMAW Outfit

An SMAW outfit must be set up and inspected before any welding occurs. All connections must be tight, insulation intact, and equipment ready and safe to use. When inspection is performed, ensure that the welding machine is turned off to prevent any electrical hazards.

Like any arc welding process, SMAW produces an electrical arc, which can cause severe burn. Pockets on a technician's fire-resistant clothing should be covered to prevent sparks from being caught within them. A cap, helmet, leather gloves, flash safety glasses, and welding hood with a #10–#13 lens must always be worn during welding operation.

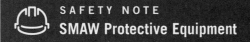

PROCEDURE   SMAW Setup

Follow the steps listed below to perform basic SMAW setup:

1. Connect the welding machine to an AC or DC source of electricity.

2. Ensure that the electrode and workpiece leads are the same diameter.

3. Connect and secure the electrode and workpiece leads to the terminals of the welding machine.

4. Connect the electrode lead to the electrode holder. Be sure that the electrode holder is not in contact with the worktable or workpiece.

5. Attach the workpiece lead to the worktable or workpiece.

6. Set the power source to DCEP and connect the work clamp to the welding table.

7. Select an amperage between 85 and 100 amps.

8. Select a 1/8″ E6010 electrode and clamp it in the electrode holder. Make sure it is not in contact with the worktable.

9. Select some 1/4″ thick mild steel 3″ × 6″ for practice welds.

38.5.2 SMAW Welding Procedure

SMAW can begin once setup is complete. Before turning on the power source, it may be useful to practice the motion of welding across the weld plate while standing in a comfortable position. This allows a technician to become familiar with the welding process.

Any arc welding process introduces hazards into the work area. This includes electrical shock, fumes and gases, hot metals, and fires. Refer to Chapter 37, *Welding Safety, Equipment, and Processes*, for a detailed review on proper welding safety.

PROCEDURE   SMAW Operation

Carefully follow these steps to produce a successful weld using shielded metal arc welding:

1. Turn the power source on while holding the electrode away from the welding area.

2. Bring the tip of the electrode about 1/8″ from the weld metal.

3. Create the welding arc using a scratching or tapping motion.

4. Once the arc is established and a weld puddle begins to form, slowly manipulate the electrode in

(continued)

a backhand direction to create a weld bead that is approximately 2–3 times the size of the electrode. This should be about 1/4″ to 3/8″ wide and about 1/8″ high above the base metal. Refer to **Figure 38-8** for the proper angle of the electrode.

5. Slowly feed the electrode toward the weld puddle as the electrode melts away. See **Figure 38-9** for a reference to a proper weld bead profile.

6. Stop the weld by quickly pulling the electrode away from the plate when it is about 1/4″ from the end, or when the electrode is about 2″ in length.

7. Practice making the weld beads and overlapping each bead to create a full weld pad. See **Figure 38-10**. Create several weld pads until the weld beads are consistent in size and shape.

After becoming comfortable with making weld beads using various electrodes, the next step is to begin attaching two pieces of plate together to form a 90° corner or a "T" by using a fillet weld. To practice this weld, deposit three weld beads, as shown in **Figure 38-11**. Pay attention to the variations in the electrode angles with each deposited weld bead.

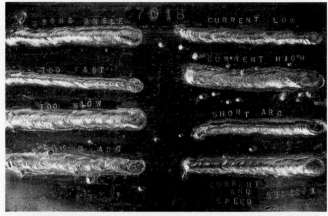

James Mosman

Figure 38-9. The proper amperage, arc length, and travel speed are important to obtain the correct bead profile and penetration.

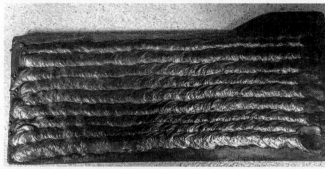

James Mosman

Figure 38-10. A buildup pad is good practice for overlapping weld beads that are a consistent size and shape.

James Mosman

Figure 38-8. The SMAW electrode should be held at the proper travel angle of 10° to 15° and work angle of 0°.

James Mosman

Figure 38-11. The fillet weld is practiced with three or more beads to obtain a proper weld profile.

38.6 GAS METAL ARC WELDING (GMAW) AND FLUX CORED ARC WELDING (FCAW)

GMAW and FCAW both use the same constant voltage power source and DCEP polarity, equipment, and consumable electrode. CV machines maintain a constant voltage but can cause large changes in current. Current is altered based on the setting of the wire feed speed on the machine. Increasing wire feed speed increases welding current, while decreasing wire feed speed decreases welding current.

Electrodes used in GMAW always require gas shielding. In FCAW, the electrodes may or may not need shielding gas. *Self-shielded flux cored arc welding (FCAW-S)*, also known as the *open arc process*, does not require additional shielding gas, since all of the ingredients required for the shielding are included in the core material. *Gas-shielded flux cored arc welding (FCAW-G)*, also known as the *dual-shield process*, uses carbon dioxide alone or in combination with argon to shield the weld pool and arc stream from the outside atmosphere. A technician must determine which process is best suited for a particular application and set up the appropriate equipment.

TECH TIP
Electrode Selection

Choosing the proper electrode wire for GMAW and FCAW is based on several factors. These include the base metal composition, base metal properties, cleanliness of a base metal, shielding gas used, metal transfer method, and welding position. See **Figure 38-12** for a list of common GMAW and FCAW electrodes.

TECH TIP
Shielding Gases

Shielding gases used in GMAW and FCAW are selected based on the type of base metal being welded and the metal transfer method to be used. Different shielding gases may have varying effects on the weld bead shape and penetration. See **Figure 38-13.**

38.6.1 Assembling a GMAW and FCAW Outfit

Proper setup for GMAW and FCAW ensures that welding work can be done safely without equipment malfunctions. Always refer to the equipment manufacturer recommendations and reference material before setting up and operating any welding equipment.

Recommended Electrodes for Common Base Metals	
Base Metal	**Recommended Electrodes**
GMAW	
Aluminum	ER4043, ER5356
Copper and copper alloys	ERCu, ERCuSi-A, ERCuAl-A1
Carbon steel	ER70S-2, ER70S-6
Low-alloy steel	ER80S-B2, ER80S-D2
Stainless steel	ER308, ER308L, ER316, ER347
Nickel and nickel alloys	ERNi-1, ERNiCr-3, ERNiCrMo-3
Magnesium	ERAZ61A, ERAZ92A
Titanium	ERTi-1
FCAW	
Carbon steel	E70T-1, E71T-1, E70T-2
Low-alloy steel	E80T1-B2, E80T1-Ni2
Stainless steel	E308T-3, E308LT-3, E316LT-3, E347T-3

Goodheart-Willcox Publisher

Figure 38-12. Common metal types and suggested electrodes to use for GMAW and FCAW processes.

PROCEDURE — GMAW and FCAW Setup

A technician must select the proper welding parameters based on the type of welding performed. These parameters include the type of metal and metal thickness to be welded, metal transfer to be used, and power source. Once these are considered, a technician can begin setup using the following steps:

1. Select the proper electrode recommended for the base metal.

2. Install the electrode wire on the hub of the wire feeder.

3. Select and install the drive rolls with the appropriate groove diameter. Adjust pressure setting to feed the electrode wire.

4. Install the contact tube into the welding gun.

(continued)

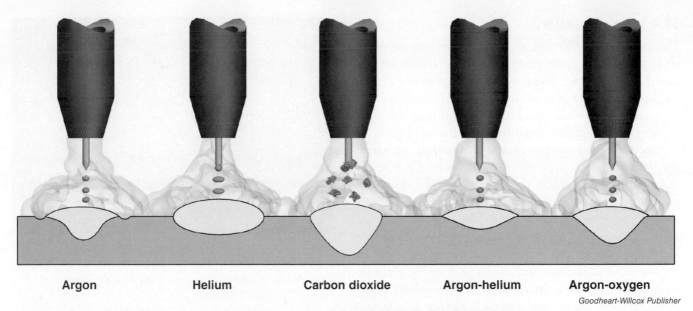

| Argon | Helium | Carbon dioxide | Argon-helium | Argon-oxygen |

Goodheart-Willcox Publisher

Figure 38-13. Penetration may vary based on the type of shielding gas used for GMAW operation.

5. Attach the gas nozzle to the end of the gun.

6. Attach the welding gun to the wire feeder.

7. Select the shielding gas and flow rate recommended for the base metal to be used.

8. Select the type of metal transfer that will be used for welding.

9. Set the flow rate on the flowmeter.

10. Set the voltage and wire feed speed as recommended for the metal transfer type selected.

38.6.2 GMAW and FCAW Welding Procedure

Gas metal arc welding is performed using a backhand or forehand welding method. This is different from FCAW, which primarily uses the backhand method. A technician should be familiar with each of these methods and be able to use them for any given welding process. In this following GMAW procedure, the forehand welding method is used.

SAFETY NOTE
GMAW and FCAW Protective Equipment

Head-to-feet PPE, including fire-resistant or leather clothing, helmet, and safety goggles with a #11–#14 lens, must always be worn. Because both SMAW and FCAW create fumes in operation, a fume exhaust should be positioned at the rear of the welding area.

PROCEDURE   Performing GMAW

Follow these instructions carefully to produce a successful weld using shielded metal arc welding.

1. Position the welding gun over the area where you plan to weld.

2. Tilt the gun to the desired travel angle.

3. Move the electrode wire to the base metal, so the two are slightly touching.

4. Raise the gun 1/16″ above the weld area.

5. Press the switch on the gun. The welding arc will form and almost immediately begin to melt the base metal.

6. Move the welding gun in the direction of travel.

7. Form the weld bead by moving the gun side to side.

8. Continue to move the arc and weld pool until reaching the end of the weld.

9. Move the gun backward over the weld pool 1/4″–1/2″ to fill the weld pool.

10. Release the switch.

GMAW and FCAW are more technical processes than SMAW. Quality welds are created only after a complete understanding of all of the variables and hours of practice and testing.

38.6.3 Metal Transfer

During the GMAW process, metal from the electrode melts off and becomes part of the weld. This process is called metal transfer. Metal transfer occurs in one of four methods, called modes of deposition. These modes are separated by how the arc is created and how the electrode wire melts and is deposited into the weld zone.

In the **short-circuiting arc deposition**, or *short-arc deposition*, the electrode wire short-circuits when it is fed into and contacts the workpiece. The short circuit causes the electrode to melt off and be deposited as molten metal into the weld joint. This deposition mode uses relatively low voltages and amperages. **Globular deposition** occurs when an electrode burns off above or in contact with the workpiece in a very erratic globular pattern. This mode is the result of using a voltage range between short-circuit and spray-arc ranges. It is generally undesirable and not widely used in today's industry.

Spray-arc deposition uses a higher voltage range than is used for short-circuit and globular deposition. In this deposition mode, the electrode is melted off above the workpiece, forming a fine spray of molten metal that is deposited into the weld zone. The **pulse spray-arc deposition**, or *spray-arc pulsed deposition*, is a method of melting off or pulsing a drop of molten filler metal from the electrode wire at a controlled rate and time in the weld cycle.

38.7 GAS TUNGSTEN ARC WELDING (GTAW)

Gas tungsten arc welding (GTAW) uses DCEN current for steel and stainless steel welding applications. The major difference between GTAW and the other welding processes is the use of a nonconsumable electrode. These electrodes have a tungsten base with small percentages of rare earth elements added for use with various materials.

The tungsten is held inside the GTAW torch by the collet, collet body, and backing cap. The main components of a GTAW torch are shown in **Figure 38-14**. The tungsten electrode should be sharpened to a selected angle that will help determine the weld bead profile. Always sharpen the tungsten perpendicular to the length, never to the side.

38.7.1 GTAW Welding Procedure

Before welding with the GTAW process, prepare your base metal by cleaning and removing any mill scale or other impurities that can cause defects. This procedure will help you begin welding on 1/4″ mild steel plate using a remote foot pedal and high frequency arc start.

James Mosman

Figure 38-14. The GTAW torch has several components that can be interchanged for various welding situations.

Gas tungsten arc welding can be successfully completed by following these steps carefully:

1. Set the constant current power source to DCEN.
2. Set the amperage to 110 amps.
3. Set the power source to remote when using a foot pedal amperage control.
4. Using argon gas for shielding, set the flowmeter between 25 and 35 cfh.
5. Set the preflow and postflow gas time to 5 seconds.
6. Select a properly prepared 3/32″ E-3 or similar tungsten electrode and install it in the torch so it extends approximately 1/8″.
7. Set the power source to high frequency arc start mode.
8. Holding the torch in one hand, with the tip of the tungsten electrode 1/8″ above the base metal for high frequency start, slowly press down on the foot pedal.
9. Wait until the arc heats the base metal and forms a molten weld puddle.
10. Once the puddle is formed, dip the filler material (that is held in the opposite hand) slowly into the front of the weld puddle.

(continued)

11. Continue to use a forehand direction, adding the filler metal and then moving forward slightly each time until the weld bead is finished.

12. At the end of the bead, add some filler metal and then slowly let off the foot pedal until the arc shuts off. Leave the torch over the weld puddle until the postflow gas stops.

Continue to practice this technique until consistent weld beads are produced. Practice overlapping the beads and also fillet welds before using this process for service and repair work.

James Mosman

Figure 38-15. A variety of tools for measuring, observing, and documenting welds are used during the visual inspection operations.

Weld defects common for GTAW include porosity, undercutting, overlap, and lack of penetration. With appropriate practice and technique using this process, you will learn to reduce these defects and produce quality welds.

38.8 WELD INSPECTION

Although welding procedures are used to ensure a weld quality meets specification, all welds have flaws. Welds may also have defects. A ***weld defect*** is an extensive flaw that may cause the weld to fail. A weld defect may involve a problem with size, shape, or makeup. A technician must be able to identify weld defects and know how to prevent them in the future.

Two types of testing are used to verify a welding procedure and welding performance. ***Destructive testing*** analyzes the physical properties of a weld by applying stress to a weld until it fails. ***Nondestructive testing*** verifies the quality of the weld without causing damage to the physical weld. This testing is usually completed in the form of visual inspection.

38.8.1 Visual Inspection

Visual inspection involves viewing a weld on the front or top surfaces (and penetration side). It can also involve using measuring tools, scales, squares, mirrors, flashlight, and other weld measurement instruments to determine the condition of a weld. See **Figure 38-15**.

Several common defects can be identified by nondestructive visual examination. They include the following:

- **Underfill.** Molten metal sagging downward and does not fill the top of the welding joint.

- **Undercut.** A groove formed at the top of the weld bead in horizontal fillet welds due to the base metal melting and not being filled.

- **Overlap.** A weld face that is larger than and protrudes over the weld toe.

- **Porosity.** Small internal voids, or pores, in the weld caused by gas bubbles that did not have enough time to rise to the surface of the molten weld pool before it froze.

- **Lack of penetration.** Insufficient weld metal penetration into the joint intersection.

- **Burn through.** A hole created when the molten pool melts through the base material.

Additional defects may include the following:

- Warpage.

- Excessive or inadequate crown height.

- Incorrect crown profile.

- Surface cracks.

- Crater cracks.

- Joint mismatch.

- Variation from dimensional tolerances.

- Inadequate or excessive root side penetration.

- Incorrect root side profile.

Fillet Weld Inspection

Inspecting horizontal leg size   **Inspecting vertical leg size**   **Inspecting weld throat**

James Mosman

Figure 38-16. Several checks are required to determine the proper size of a fillet leg.

Fillet welds require the fillet leg be a specific size. Fillet welds may also require inspection of the throat dimension. See **Figure 38-16**.

The ripples in the weld should be even, without high and low areas or undercut at the weld toe. A special inspection tool can be used to check undercut depth, crown height on a butt weld, and drop-through on the penetration side of the butt weld. **Figure 38-17** shows an indication that the weld face reinforcement is within tolerance. **Figure 38-18** indicates that there is excessive reinforcement on the weld root.

James Mosman

Figure 38-17. This bridge gauge measures the reinforcement on the groove weld face. This weld face is within tolerance.

Goodheart-Willcox Publisher

Figure 38-18. This illustration shows a welded metal with a lack of penetration.

38.9 WELD DEFECTS AND CORRECTIVE ACTION

A technician must be aware of defects common to particular weld types and know how to prevent them. Corrective actions to fix welding defects may be very similar for all processes, and more than one action may need to be taken.

Common defects that affect fillet welds are listed in **Figure 38-19**, along with suggestions for correction.

Fillet Weld Defects and Corrective Actions

Weld Defect	Corrective Actions
Lack of penetration	■ Decrease the travel angle. ■ Increase the amperage. ■ Decrease the voltage. ■ Decrease the size of the weld deposit. ■ Use stringer beads. ■ Do not make weave beads on root passes.
Lack of fusion	■ Remove oxides and scale from the previous weld passes. ■ Increase the amperage. ■ Decrease the voltage. ■ Decrease the travel speed. ■ Change the electrode, work angle, or travel angle. ■ Decrease the electrode stickout. ■ Keep the arc on the leading edge of molten pool.
Overlap	■ Reduce the size of the weld pass. ■ Reduce the amperage. ■ Change the electrode, gun or torch angle. ■ Increase the travel speed.
Undercut	■ Make a smaller weld. ■ Make a multiple-pass weld. ■ Change the electrode, gun or torch angle. ■ Use a smaller diameter electrode. ■ Decrease the amperage. ■ Decrease the voltage.
Convexity	■ Reduce the amperage. ■ Decrease the electrode stickout. ■ Decrease the voltage. ■ Decrease the gun angle.
Craters	■ Do not stop welding at the end of the joint (use run-off tabs). ■ Using the same electrode, gun or torch travel angle, move the weld puddle backward a quarter inch, over the completed weld, to backfill the crater before stopping the weld.

James Mosman

Figure 38-19. Common defects and corrective actions for fillet welds. *(continued)*

Fillet Weld Defects and Corrective Actions *(continued)*	
Cracks 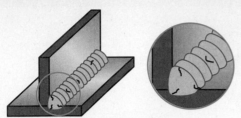	▪ Use the suggestions given for groove weld cracks.
Burn through 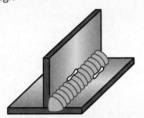	▪ Decrease the amperage. ▪ Increase the travel speed. ▪ Change the electrode, gun or torch angle.
Porosity 	▪ Remove all heavy rust, paint, oil, or scale on the joint before welding. ▪ Remove any oxide film (GMAW) or slag (SMAW & FCAW) from the previous passes or layers of the weld. ▪ Increase the electrode stickout distance for FCAW. ▪ Check the gas flow. ▪ Protect the welding area from wind. ▪ Remove spatter from the interior of the gas nozzle. ▪ Check the gas hoses for leaks. ▪ Check the gas supply for contamination.
Excessive spatter 	▪ Decrease the voltage. ▪ Decrease the drag angle. ▪ Decrease the travel speed. ▪ Increase the electrode stickout. ▪ Decrease the electrode wire feed speed. ▪ Use an antispatter spray. ▪ Do not use CO_2 for GMAW on steel. ▪ Use the pull technique for GMAW.

James Mosman

Figure 38-19.

CHAPTER WRAP-UP

There are several steps involved in producing a quality weld. In this chapter, you were able to build on your basic welding knowledge by examining a maintenance welding process at large. A welding technician must be able to establish a qualified welding procedure, consider welding variables that affect a weld, perform welding operations, and inspect completed welds. Continuing to learn and gain more experience in these areas will allow you to become a successful technician in the future.

Thermadyne

Gas metal arc welding (GMAW) process performed in the horizontal position.

Chapter Review

SUMMARY

- A welding code is a national standard that provides guidelines, requirements, and standards for welding.

- A welding procedure specification (WPS) records all required variables for a specific weld application and must be qualified and tested.

- Welding variables impact the mechanical properties of a weld. Some of these variables include electrode extension, travel speed, torch angle, and weld bead patterns.

- Two separate angles describe the position of a welding torch or gun during welding—travel angle and work angle.

- All metal must be properly prepared before welding can be completed. Any oil, grease, dirt, paint, or coating must be removed. All thermally cut surfaces should be ground to a clean, flat surface.

- Shielded metal arc welding (SMAW) uses electrode wires covered with flux. The type of metal and metal thickness are important factors when selecting which electrode to use.

- Gas shielded metal arc welding (GMAW) and gas-shielded flux cored arc welding (FCAW-G) require gas shielding, while self-shielded flux cored arc welding (FCAW-S) does not require additional shielding.

- Metal transfer is the process in which metal from the electrode melts off and becomes part of the weld. This occurs in one of four methods called modes of deposition.

- Gas tungsten arc welding (GTAW) electrodes should be sharpened to a selected angle to determine the weld bead profile.

- All welds have flaws. A weld defect is an extensive flaw that may cause the weld to fail and must be identified when a weld is completed.

- Both destructive and nondestructive inspection methods are used to determine the quality and performance of the completed weld and skill of the welding technician.

- Visual inspection involves viewing a weld on the front or top surfaces. This inspection should be performed before any other inspection or testing process.

- Common weld flaws can be corrected by using the proper corrective actions.

REVIEW QUESTIONS

Answer the following questions using the information provided in this chapter.

1. What is a welding code and what is its purpose?

2. *True or False?* A welding procedure specification must be qualified for a welder to perform a test weld using the procedure.

3. What are welding variables? List five welding variables.

4. What is the difference between a travel angle and a work angle?

5. In forehand welding, the forward angle is called the _____.

6. What is a weld bead?

7. Give two examples of how metal can be prepared and cleaned for welding.

8. What are the strength and size requirements when selecting an electrode for SMAW?

9. What PPE should be worn during SMAW?

10. *True or False?* During SMAW procedure, the electrode is slowly fed toward the weld puddle and melts into the weld.

11. Explain the two major differences between FCAW-G and FCAW-S processes.

12. How is a shielding gas selected for GMAW and FCAW?

13. When setting up GMAW and FCAW, _____ and _____ must be set by the machine.

14. Describe how a weld bead is formed during GMAW.

15. What is metal transfer? List the four modes of deposition.

16. How is an electrode prepared prior to GTAW operation?

17. Describe how filler material is added to a weld in GTAW.

18. *True or False?* All weld flaws are considered defects and must be removed.

19. Describe the two types of testing used to verify weld quality.

20. Name four welding defects that can be determined with visual inspection methods.

21. Describe measures that can be used to correct undercut.

NIMS NIMS CREDENTIALING PREPARATION QUESTIONS

The following questions will help you prepare for the NIMS Industrial Technology Maintenance Level 1 Maintenance Welding credentialing exam.

1. What is a welding procedure specification?
   A. A document that lists welding variables to follow for a welding application.
   B. A guideline on proper welding training.
   C. A national standard for welding.
   D. A standardized procedure to be followed for any welding process.

2. What welding variable determines the shape of the finished weld joint?
   A. Electrode extension
   B. Travel speed
   C. Torch angle
   D. Weld bead

3. Which is a proper preparation method of a base metal?
   A. Painting over any pores or inconsistencies.
   B. Preheating all ferrous metals to reduce local expansion.
   C. Removing oil or grease with acetone.
   D. Using a wire brush to remove coating.

4. Which is *not* a factor considered when selecting an SMAW electrode?
   A. Depth of penetration
   B. Electrode diameter
   C. Metal thickness
   D. Type of shielding gas

5. Which of the following is *not* a step used in SMAW?
   A. Clamp the electrode to the electrode holder without contact to the worktable.
   B. Create a weld bead that is approximately 2–3 times the size of the electrode.
   C. Set the amperage on the arc welding machine.
   D. Feed the electrode into the weld.

6. GMAW and FCAW-G processes require _____.
   A. additional shielding gases
   B. different electrodes
   C. Both A and B
   D. Neither A nor B

7. What step occurs first in GMAW?
   A. Bring the electrode wire near the base metal.
   B. Form the weld bead by moving the gun side to side.
   C. Move the gun backward over the weld pool.
   D. Move the weld gun in the direction of travel.

8. _____ is a weld defect in which small internal voids caused by gas bubbles are found on a weld.
   A. Overlap
   B. Porosity
   C. Undercut
   D. Underfill

9. Which of the following is a corrective method for this type of weld defect?

   A. Change the gun angle.
   B. Increase voltage.
   C. Increase amperage.
   D. Make a larger weld.

39 | Cutting Procedures

LEARNING OBJECTIVES

After completing this chapter, you will be able to:

☐ Understand thermal cutting safety procedures.

☐ Understand the proper use of PPE when thermal cutting.

☐ Demonstrate safe procedures for oxyfuel and plasma arc cutting processes.

☐ Describe the advantages and disadvantages of oxyfuel cutting.

☐ Describe the advantages and disadvantages of plasma arc cutting.

☐ Explain the proper methods of material preparation for cutting.

TECHNICAL TERMS

acetylene gas	neutral flame
carburizing flame	oxidation
constricting nozzle	oxidizing flame
cutting tip	oxyacetylene
cutting torch	oxyfuel gas cutting (OFC)
flashback	oxygen
gas pressure regulator	plasma
kerf	plasma arc cutting (PAC)

Oxyfuel and plasma arc cutting processes are the two most widely used methods used for cutting metal. This chapter will examine these two processes in-depth, focusing on required cutting equipment and operation. You will also learn how to follow the proper procedures to perform an oxyfuel gas cut and plasma arc cut safely and successfully.

39.1 OXYFUEL GAS CUTTING

Oxyfuel gas cutting (OFC) is a process in which metals are cut by rapid *oxidation*, or producing extreme heat by using a gas flame and pressurized oxygen. Although oxygen is not flammable, it becomes highly reactive when combined with other gases. The combination can produce a flame hot enough to cut any ferrous metal—including steel. See **Figure 39-1**. Oxyfuel gas cutting is also referred as *burning* or *flame cutting*.

There are many gases used as fuel for the OFC process. These gases include propylene, natural gas, propane, MAPP® gas, and methylacetylene-propadiene (MPS). *Oxyacetylene*, a combination of oxygen and acetylene gas, is the most common gas used in oxyfuel cutting. In this section, we will focus primarily on the oxyacetylene gas cutting process.

Oxyfuel gas cutting has several advantages:

- Oxyfuel torches can cut materials as thick as 6″–12″, while plasma arc cutting can cut only up to 1″–2″.

- Oxyfuel gas cutting equipment is portable and can be used indoors or outdoors.

- Equipment is relatively low cost and produces quality cuts, even on steel.

- Oxyacetylene may also be used for brazing, welding, and heating purposes.

This process also has a few disadvantages:

- Oxyfuel cutting is not effective on stainless steel and nonferrous materials, unlike plasma arc cutting.

- The heat-affected zone is larger than that produced by plasma arc cutting, which can introduce undesirable changes to the metal.

- Acetylene fuel gas and high-pressure oxygen cylinders can be dangerous and cause serious injury if not handled properly.

39.1.1 Oxyfuel Gas Cutting Equipment

Basic equipment for oxyfuel gas cutting includes dual oxygen and fuel gas cylinders, regulators, hoses, and a cutting torch. A typical cutting system is shown in **Figure 39-2**. The equipment required to perform oxyacetylene gas cutting is nearly identical to oxyacetylene gas welding equipment, except for differences in their pressure regulators and torch. Cutting processes may require pressure regulators with a larger volume capacity since a larger volume of oxygen at a higher pressure is needed to oxidize and cut metal. Cutting and welding processes always use a torch specific to its intended application. This will be examined later in the chapter.

Penka Todorova Vitkova/Shutterstock.com

Figure 39-1. Oxyfuel gas cutting is used to cut various ferrous metals and thicknesses up to 12″.

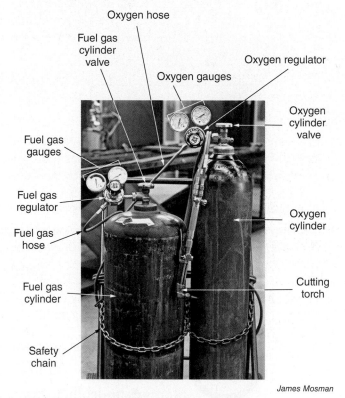

James Mosman

Figure 39-2. An oxyacetylene cutting outfit includes both oxygen and acetylene gas components as shown.

Oxygen and Acetylene

Oxygen is a nonflammable gas that is colorless and odorless, which makes it difficult to detect. Pure oxygen only burns when it is combined with a fuel source. Oxygen is supplied either in liquefied form in a Dewar flask container or in gaseous form in high-pressure cylinders. Dewar flasks, as shown in **Figure 39-3**, are hollow cylinders that have a valve on top that contains a safety pressure release. Gaseous oxygen stored in high-pressure

James Mosman

Figure 39-3. A Dewar flask containing liquefied oxygen is shown.

cylinders must not exceed a pressure of 2200 psig and should always be moved with caution. Always transport oxygen cylinders with a designed handcart or by tilting and rolling the cylinder along the bottom edge.

Acetylene gas is a compound of carbon and hydrogen formed by mixing calcium carbide in water. It has a strong garlic-like odor that is easily detected if the cylinder is left open or there is a leak in the system. When mixed with the oxygen, the produced flame exceeds 5500°F, which makes it useful for cutting, welding, and heat-treating steel.

SAFETY NOTE
Acetylene

Acetylene is a common fuel gas used for oxyfuel cutting and must be handled carefully since it is extremely flammable and explosive at high pressures. Pure acetylene should never be used at a gauge pressure that exceeds 15 psig.

Because of the explosive nature of acetylene, this gas cannot be stored in a hollow cylinder. Instead, a special cylinder filled with a porous concrete-like material is used to stabilize the liquid acetylene at higher pressures. See **Figure 39-4**. Acetylene cylinders must always be stored and used in an upright position or the acetylene may draw

out through the regulators. Acetylene cylinders have safety fuse plugs, usually located on the bottom, which melt in the event of a fire and allow the gas to escape slowly. These cylinders are also available in many sizes.

Pressure Regulators

Gas pressure regulators are designed to reduce the pressure in which oxygen and acetylene are delivered from the cylinders to the hoses. Most regulators for oxyacetylene systems have two gauges—one to indicate the remaining higher pressure in the cylinder and the other to indicate the lower pressure of the gas sent to the cutting torch. An adjusting screw or handle is used to control the amount of pressure needed for the torch to produce a flame. Refer to the torch manufacturer's chart for proper gauge pressure settings.

Oxygen regulators have green markings and higher-pressure scales than acetylene regulators, as shown in **Figure 39-5**. All connections have a smooth nut and are

James Mosman

Figure 39-4. Acetylene cylinders are attached to a manifold delivery system. These cylinders must be handled and stored carefully.

James Mosman

Figure 39-5. Oxygen regulators are equipped to handle higher pressures than acetylene regulators as indicated on the pressure gauges.

made with a right-hand thread. Acetylene and other fuel gas regulators have red markings. All connections are made with left-hand threads, and connection nuts also have a groove to distinguish from oxygen connections. Acetylene gauges always have a red line safety indication at the 15 psig pressure setting, **Figure 39-6**.

James Mosman

Figure 39-6. Acetylene gauges have a lower pressure gauge that includes red warning at 15 psig. Acetylene gas cannot exceed a gauge pressure of 15 psig.

SAFETY NOTE
Pressure Regulator Connectors

For safe operation, never switch connector nuts or gauges. Do not use a regulator not designed for its specific gas. Always remove gas regulators when cylinders are not going to be used for any extended length of time.

Hoses and Cutting Torches

A hose carries gas from the pressure regulator to the torch. Like pressure regulators, hoses used for the oxyacetylene outfit are normally green for oxygen and red for the acetylene. Various diameters are available and must be correctly selected for the application. For instance, gases that require more volume to complete a cut need a hose with a larger diameter to maintain appropriate pressure.

A *cutting torch* controls the amount of acetylene and oxygen supplied for the cut. The torch is also responsible for combining the two and directing the produced gas flame to the cutting area. These torches include a tip, tip nut, torch head, torch valves, and flashback arrestors.

Torches are available in two general styles. The standard, one-piece cutting torch is designed only for cutting and is mainly used for high-volume, continuous cutting operations. See **Figure 39-7**. The cutting torch attachment, or two-piece combination torch, is more common, **Figure 39-8**. This torch can be quickly converted and used for welding, brazing, or heating applications by switching out the cutting attachment. Attachments should only be tightened by hand. Never use a wrench to tighten any attachment to the torch body.

Cutting tips are attached to the end of the torch and are the exit point for the mixed fuel and oxygen. They are always separate from the torch body and are available in numerous styles and vary based on diameter of gas orifice, as shown in **Figure 39-9**. Cutting tip sizes are identified by numbers ranging from 00 to 8. Refer to **Figure 39-10** for suggested tip sizes to use with various metal thicknesses. The cutting tip nut should always be tightened with a wrench to ensure a good tip to torch seal.

SAFETY NOTE
Flashbacks

A *flashback* occurs when a flame produced in cutting moves into the mixing chamber of the torch, causing the flame to burn back and damage the torch component. Flashbacks can result in a violent explosion, which can ruin the regulator or cylinder or cause serious personal injury. To prevent this, flashback arrestors are installed between the torch and hoses for safety precaution, as shown in **Figure 39-11**. Always check valves and flashback arrestors during operation for loose or leaky connections.

39.1.2 Assembling Oxyfuel Cutting Equipment

Technicians must use the proper procedure and correct equipment to make clean, precise cuts. Before cutting, metals should be checked and cleaned of any areas with oil, dirt, or paint. This can be done with alcohol or acetone cleaners or removed by grinding or sanding.

Proper safety precautions should always be followed before and during cutting. Check your cutting station for any hazardous or flammable material. Refer to Chapter 37, *Welding Safety, Equipment, and Processes*, to ensure that you understand and apply the proper safety measures while cutting.

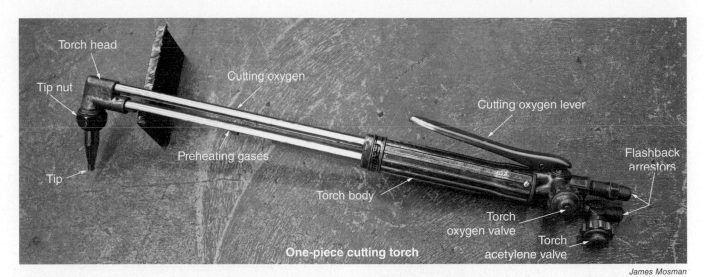

One-piece cutting torch

James Mosman

Figure 39-7. A one-piece oxyfuel cutting torch may be used to perform functions such as gouging or cutting sheet metal.

Two-piece combination torch

James Mosman

Figure 39-8. Similar to one-piece torches, a two-piece combination torch includes the torch body and the cutting attachment. The cutting attachment can be switched out to complete a welding or brazing application.

James Mosman

Figure 39-9. Cutting tips are available in several sizes based on the size of the gas orifices. This image shows the size differences between a 000, 00, 0, and 1 tip. The larger the cutting tip, the thicker the metal that can be cut or pierced.

A **B**

Thermadyne; Steve Hamann/Shutterstock.com

Figure 39-11. Flashback arrestors prevent flashbacks from occurring during cutting operation. A—A basic flashback arrestor. B—A flashback arrestor secured into an oxyacetylene outfit.

SAFETY NOTE
Oxyfuel Cutting PPE

Before performing any oxyfuel cut, be sure that you are wearing the proper PPE. Proper safety glasses or shield with a shade #3 to #5 lens should be worn. Additional required PPE include cotton, wool, or leather clothes, steel-toe boots, and leather gloves. Remove all lighters and matches from your pockets and place in a secured area.

Tip Cutting Sizes Based on Metal Thickness												
Material Thickness (in Inches)	1/8	1/4	1/2	3/4	1	1 1/2	2	4	5	6	8	10	12
Recommended Tip Number	00	0	1	1	2	2	3	3	4	6	6	7	8

Goodheart-Willcox Publisher

Figure 39-10. Recommended tip sizes for cutting various metal thicknesses from 1/8″ to 12″.

PROCEDURE

OFC Equipment Assembly

Follow these steps to set up oxyacetylene gas cutting equipment:

1. Fasten the oxygen and acetylene cylinders to a wall or hand truck in the vertical position. Chains or steel straps may be used to fasten the cylinders in their proper location. Remove safety caps once the cylinders are safely secured.

2. Clean the oxygen and acetylene cylinder outlets by quickly opening and closing the cylinder valves.

3. Attach the regulators to the cylinder outlets. Thread regulators onto the cylinder nozzles by hand. Ensure regulator nuts are tightened with a regulator wrench so they are leakproof.

4. Attach the acetylene hose to the acetylene regulator. Attach the oxygen hose to the oxygen regulator.

5. Attach flashback arrestors or check valves between each hose and regulator.

6. Attach the oxygen hose to the torch oxygen inlet fitting and the acetylene hose to the torch acetylene inlet fitting.

7. Select the appropriate size torch tip and place it into the torch or torch tube. Make sure the tip is tight.

39.1.3 Performing Oxyfuel Gas Cutting

Once the oxyacetylene outfit is assembled, complete a final visual inspection on the equipment. Check the torch, valves, hoses, fittings, regulators, and cylinders for any damage. Ensure that the oxygen and acetylene valves and torches are closed.

PROCEDURE

Oxyfuel Cutting Operation

By following these steps carefully, you will be able to safely light a torch, adjust the flame, and make a clean cut.

1. Loosen the pressure adjusting screws on the regulators by turning them counterclockwise.

2. Turn the acetylene cylinder valve 1/4 to 1/2 counterclockwise until the pressure registers on the regulator gauge.

3. Carefully open the oxygen cylinder valve by turning counterclockwise until fully open. The pressure gauge should provide a reading on both regulators.

4. Turn the acetylene torch valve 1/2 clockwise.

5. Set the correct working pressure by turning the adjusting screw on the acetylene regulator gauge clockwise until the desired pressure is indicated on the low-pressure gauge.

6. Close the acetylene torch valve.

7. Open the oxygen torch valve by turning counterclockwise.

8. Turn the oxygen regulator adjusting screw clockwise, approximately 1/4 of a turn, until the desired working pressure is found.

9. Set the working pressure for the acetylene torch valve.

10. Close the oxygen torch valve. Release the oxygen cutting lever.

Once the correct working pressures have been set, the cutting torch is ready. Follow these steps to light the torch and make a clean cut with oxyacetylene gas.

1. Use a spark lighter and, holding the tip downward and 1″ away, light the torch. **Figure 39-12** show various spark lighters.

2. Open the acetylene torch valve approximately 1/16 of a turn.

3. Light the preheating flames with a spark lighter. If the torch is clean and working properly, all flames should light at the same time.

4. Continue to increase the flow of acetylene through the torch valve until the black smoke is gone.

(continued)

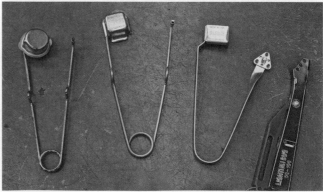

James Mosman

Figure 39-12. Various oxyfuel spark lighters.

5. Slowly turn on the oxygen torch valve on the cutting attachment. Increase the oxygen volume until a proper neutral flame is reached.

6. Press down on the oxygen cutting lever and readjust the flame to neutral if necessary.

7. In a safe position, hold the flame about 1/4″ above the metal until it turns red hot.

8. Slowly press down on the cutting lever and begin your cut. A slot in the base metal called a **kerf** is made as you continue in the direction of torch travel. See **Figure 39-13**.

9. Maintain a steady forward motion to complete the cut.

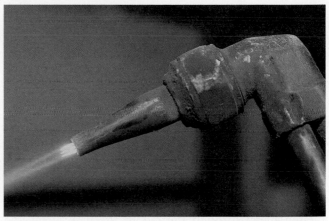

Figure 39-14. A neutral flame is desirable in oxyfuel gas cutting operation to produce a clean cut.

39.1.4 Shutting Down Cutting Equipment

Oxyacetylene equipment shutdown is a necessary step in the cutting process. Never leave the cutting station while the equipment is running. When shutting down, make sure that each step is followed in the correct order. If shutdown is completely successful, the pressure gauges will read zero.

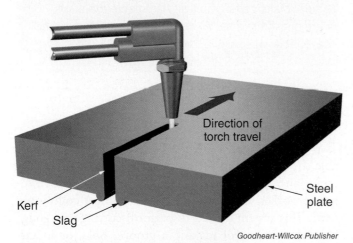

Goodheart-Willcox Publisher

Figure 39-13. A proper cut is being made in this illustration, producing both a kerf and slag.

PROCEDURE	OFC Equipment Shutdown

Use the following instructions to ensure the outfit is shut down correctly.

1. Release the cutting lever.

2. Turn off the flame by closing the acetylene torch valve first and then the oxygen torch valve. This must be done whenever the torch is not in hand.

3. Completely close the acetylene and oxygen valves on their cylinders.

4. Slightly release the pressure from the oxygen and acetylene hoses by reopening the valves on the torch and then closing them. This removes the gases in the system.

5. Check that both cylinder pressure gauges are decreased to zero.

6. Turn the adjusting screws on the regulators counterclockwise until loose.

7. Close the torch valves.

8. Wrap up the torch and hoses on the cart. Ensure the cylinders are secured.

TECH TIP
Flames

A **neutral flame**, as shown in **Figure 39-14**, results from combining oxygen and a fuel gas in proper proportion. When oxygen and a fuel gas are not in proportion, two other flames result. An **oxidizing flame** occurs when too much oxygen is present. This type of flame can oxidize or burn the surface of the base metal. A **carburizing flame** is produced when too much carbon or too little oxygen is produced. Slightly carburizing flames may be used for brazing, soldering, and welding certain metal alloys.

With practice and following each procedure carefully, these steps will become easier, and you will begin to have clean and consistent cuts.

39.2 PLASMA ARC CUTTING

Plasma arc cutting (PAC) cuts metals by using an arc and plasma gas forced through a constricting nozzle in the torch. Similar to arc welding processes, the plasma arc is produced between an electrode and the base metal. The arc is then used to heat and melt the metal. As the torch is moved along the cut path, plasma gas blows molten metal through the cut in the base metal and removes metal. See **Figure 39-15**.

Plasma is an ionized gas—a gas with free electrons that conducts electricity—and is often referred as the fourth state of matter. Plasma can be found in fluorescent and neon lightbulbs. It is also what makes up lightning. Although plasma is present in all welding processes, only processes that use a constricting nozzle to concentrate and direct plasma for cutting are considered a plasma arc process.

There are several advantages to plasma arc cutting when compared to oxyfuel gas cutting:

- Plasma cutting can cut both ferrous and nonferrous metals due to the high temperature produced in the process.
- No preheating is required.
- No dangerous or explosive gases, such as acetylene, are used.
- The cutting speed is much higher, creating minimal heat affected zones.
- PAC produces better cut quality with minimal preparation and postwork needed.

Thermadyne

Figure 39-15. The plasma arc cutting process can cut any ferrous and nonferrous metals up to 2″. A technician is cutting a 1″ metal block using the proper procedure.

There are only a few disadvantages of the plasma arc cutting process:

- The cost of equipment is generally higher than other cutting processes. Shielding gases are an additional expense.
- A source of electricity is needed.
- The process produces an electrical arc and metal fumes that can cause safety hazards.
- The process is not as portable as oxyfuel cutting.

> **SAFETY NOTE**
> **Shop Fire**
>
> Sparks and molten metal created during plasma arc welding can be fire hazards within the shop area. Make certain that all flammable materials are out of the area. Always have a fire extinguisher available for use.

39.2.1 Plasma Arc Cutting Equipment

Two general types of plasma cutting systems are available. These systems are mechanized, CNC controlled tables and handheld, portable systems. For a maintenance worker, the handheld system is most commonly used. This system is used to perform cutting and gouging—the process of making a groove, hole, or indentation—for most common steel, stainless steel, and aluminum. An average handheld system is capable of cutting a metal thickness of about 2″. This system consists of a constant current (CC) power source, gas or gases supply and regulators, work clamp and cable, and a plasma torch. Equipment for this process must be able to create plasma and properly direct the matter. See **Figure 39-16**.

One of the most important components in plasma arc cutting is the torch. The torch is responsible for providing a good electrical connection to the electrode as well as a path for the plasma gas. A *constricting nozzle* is attached to the torch body and has a small hole that forces plasma gas to pass directly through the path of the arc produced. This allows the arc to superheat the gas, ionize it, and push the plasma out at a very high speed and temperature. A second nozzle is located on the torch, which is used to direct the shielding gas, during operation. Additional parts of the torch include the shield, retaining cap, electrode, and swirl ring. See **Figure 39-17**.

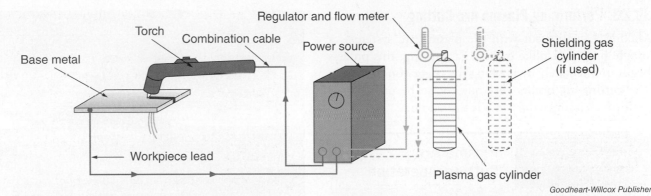

Figure 39-16. A portable handheld plasma arc cutting system and its components connected to a plasma gas cylinder and shielding gas cylinder. Air serves as a secondary gas when a shielding gas and cylinder is not used.

Goodheart-Willcox Publisher

James Mosman

Figure 39-17. The plasma torch contains the consumable parts, the shield, retaining cap, nozzle, electrode, and swirl ring.

39.2.2 Assembling Plasma Arc Equipment

There are only a few steps for equipment assembly before a technician is ready to begin cutting. It is still important to follow these steps carefully and set up the plasma arc equipment properly.

> **SAFETY NOTE**
> **Plasma Cutting PPE**
>
> Working with live electrical components and an electrical arc can cause severe injuries, such as electrical shock or burns. Always wear safety goggles with a lens shade of #5 to #7, insulated gloves and boots, and dry clothing. Keep your fingers covered and clear of the arc at all times. **Figure 39-18.**

shinobi/Shutterstock.com

Figure 39-18. This technician is wearing proper PPE while performing a plasma arc cut in a safe position.

> **PROCEDURE** **PAC Equipment Assembly**
>
> Set up a plasma arc welding outfit by following these instructions:
>
> 1. Plug the power supply into an electrical outlet or wire into an electrical source. Systems that use 110 V, 220 V, and 460 V power are available.
>
> 2. If a gas cylinder is used for cutting, attach a regulator to the cylinder. Connect the hose from the regulator to the power supply.
>
> 3. Connect the cutting torch to the power supply. Most PAC equipment have a combination cable that must be plugged and secured into the power supply.
>
> 4. Assemble the parts of the cutting torch following manufacturer directions. Secure each piece without overtightening.
>
> 5. Connect the work (ground) clamp to the metal.
>
> 6. Prepare the work area and remove any flammable items.

39.2.3 Performing Plasma Arc Cutting

Making a quality cut with the plasma arc system is a simple process. Unlike oxyfuel gas cutting, this process begins immediately by pressing the trigger on the torch. The cutting arc occurs very quickly, so be prepared to perform a cut once the torch is turned on.

PROCEDURE	Plasma Arc Cutting Operation

By using these steps, you will be able to turn on the outfit, perform a cut, and shut down the equipment successfully.

1. Turn the power on the machine.
2. Adjust the air or gas pressure to the torch. Common pressures for PAC are 65–70 psig.
3. Adjust the amperage required based on the thickness and type of material to be cut.
4. Hold the torch at approximately 1/16″–1/8″ standoff at the edge of the metal. Never let the nozzle come in contact with the work.
5. Press the trigger of the torch. The plasma arc will start between the electrode and constricting nozzle before the plasma gas is ionized and moves to the base metal.
6. Move the torch toward you to complete your cut, as shown in **Figure 39-19**. It may be helpful to use a guide or straightedge during cutting.
7. Point the nozzle in the direction of travel as you near the end of your cut. This is especially important when cutting thick metal.
8. Release the trigger at the end of the completed cut.
9. Check for cut quality.
10. Turn off the machine and the airflow. Roll up the torch and the work lead.

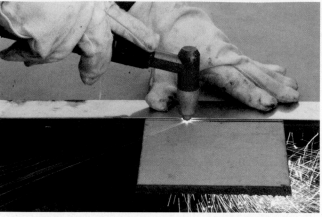

chuyuss/Shutterstock.com

Figure 39-19. When cutting, move the torch in a straight line toward your body to complete a plasma arc cut.

Plasma arc cutting is very versatile. Its equipment can create countless arc starts and cuts before cut quality decreases. When cut quality decreases, the electrode and nozzle must be checked. Both of these components are consumable parts. They deteriorate and wear out with extended use and when an arc is triggered without cutting. **Figure 39-20** shows a worn out electrode, nozzle, and drag shield.

James Mosman

Figure 39-20. This plasma electrode, nozzle, and drag shield are worn and should be replaced.

CHAPTER WRAP-UP

In order to repair or fabricate metal materials, a maintenance technician must be able to use proper equipment and processes to cut metals. This chapter examined the two metal cutting processes common to maintenance technicians—oxyfuel gas cutting and plasma arc cutting. Both processes have distinct equipment and operation procedures. With precise practice with these two cutting processes, you will be able to develop as a skilled technician.

Chapter Review

SUMMARY

- Oxyfuel gas cutting (OFC) is a process in which metals are cut by rapid oxidation, or producing extreme heat with a gas flame and pressurized oxygen.

- Oxyacetylene, a combination of oxygen and acetylene gas, is the most common gas used in oxyfuel cutting.

- The oxyfuel system can be used for brazing, welding, and heating applications and can cut materials as thick as 6″–12″.

- An oxyfuel gas cutting outfit includes oxygen and acetylene gas, pressure regulators, hoses, a cutting torch, and cutting tip.

- A flashback occurs when a flame produced in cutting moves into the mixing chamber of the torch, causing the flame to burn back and damage the torch component.

- Proper procedures and safety precautions must be followed when assembling, performing, and shutting down an oxyfuel cutting outfit.

- The three types of oxyfuel flames are carburizing, neutral, and oxidizing. The neutral flame is recommended for most cutting applications.

- Plasma arc cutting (PAC) cuts metals by using an arc generated and plasma gas forced through a constricting nozzle in the torch.

- Maintenance welders commonly use a handheld PAC system that can cut metals with a thickness of 2″.

- A PAC outfit includes a CC power source, gas supply and regulators, work clamp and cable, and a plasma torch with a constricting nozzle.

- When cut quality starts to decrease during plasma arc cutting, check to see if the electrode and constricting nozzle need to be replaced.

REVIEW QUESTIONS

Answer the following questions using the information provided in this chapter.

1. Describe the process of oxyfuel gas cutting.

2. _____ is a combination of oxygen and acetylene gas and is the most common gas used in oxyfuel gas cutting.

3. What thickness of material can be cut using oxyfuel gas cutting compared to plasma arc cutting?

4. *True or False?* Acetylene should *not* be used at a gauge pressure that exceeds 15 psig due to its high flammability.

5. Discuss the purpose of a cutting torch.

6. How are cutting tips identified?

7. What is flashback?

8. What PPE should be worn during oxyfuel gas cutting?

9. *True or False?* The acetylene hose should be attached to the oxygen regulator during oxyfuel equipment assembly.

10. Explain the process of setting the correct working pressure on the acetylene cylinder.

11. What is a kerf?

12. Discuss the three types of flames produced by the oxyfuel system. Which flame type is recommended for most procedures?

13. If OFC equipment shutdown is successful, the pressure gauges will read _____.

14. What is plasma? How is it used in PAC?

15. What plasma arc system is commonly used by maintenance technicians? What equipment components are included in the system?

16. What PPE is commonly worn for plasma arc cutting?

17. *True or False?* Plasma arc equipment assembly requires more steps and preparation to start a cut than oxyfuel gas cutting.

18. Describe how a cut is made in PAC.

19. What two PAC equipment components must be checked when cut quality starts to decrease? Why?

NIMS CREDENTIALING PREPARATION QUESTIONS

The following questions will help you prepare for the NIMS Industrial Technology Maintenance Level 1 Maintenance Welding credentialing exam.

1. Which statement is *not* true about oxyfuel gas cutting?

   A. Pure, pressurized oxygen exits the nozzle to fuel the cutting flame.
   B. It produces a temperature hot enough to cut through any ferrous metal.
   C. Its equipment includes dual cylinders, regulators, and hoses.
   D. Kerf is created as a cut is made in the direction of torch travel.

2. The working pressure needed for oxyacetylene cutting is set by the _____.

   A. acetylene fuse plug
   B. adjusting screw on a regulator gauge
   C. cylinder valves
   D. oxygen valve on the cutting torch

3. A cutting tip is selected based on the _____.

   A. cutting torch
   B. ferrous metal type
   C. pressure needed for cutting
   D. metal thickness

4. How is a neutral flame produced for oxyacetylene cutting?

   A. By supplying the same amount of oxygen and acetylene gas to the torch.
   B. By supplying more oxygen than acetylene gas to the torch.
   C. By supplying less oxygen than acetylene gas to the torch.
   D. By cutting oxygen supply to the torch once a flame is produced.

5. What can be done to prevent a flashback from occurring?

   A. Decrease the working pressure of the acetylene regulator.
   B. Ensure the proper torch is used during operation.
   C. Install flashback arrestors between the torch and hoses.
   D. Loosen the valve and flashback arrestor connections.

6. Which of the following is *not* true on how plasma gas is used to cut a metal?

   A. It blows molten metal through the cut in the base metal.
   B. It is pushed out of a constricting nozzle at a very high speed.
   C. It is superheated by an arc produced.
   D. It is used to melt the electrode and base metal.

7. What equipment component is connected to the plasma torch during assembly?

   A. CC power source
   B. CV power source
   C. Gas regulator
   D. Work clamp

8. How is a cut made during PAC?

   A. By adjusting the speed the torch is moving across the weld axis.
   B. By moving the torch away from your body.
   C. By moving the torch toward your body.
   D. By varying the distance of the torch to the base metal.

Maintenance Piping

The Maintenance Piping section will introduce you to the basics of piping systems and the tools and operations needed to maintain these systems. With basic knowledge of the engineering principles that apply to piping systems, you can be more effective when troubleshooting and repairing these systems. The content in this section will help prepare you to achieve the NIMS Industrial Technology Maintenance Level 1 Maintenance Piping credential.

40 Piping Systems, Components, and Materials

LEARNING OBJECTIVES

After completing this chapter, you will be able to:

☐ Discuss hazards associated with maintenance and repairs on fluid systems.

☐ Identify the differences between new construction safety and repair safety.

☐ Explain the lockout/tagout procedure for a piping system.

☐ Discuss the types of fluid flow inside piping.

☐ Describe how the design of a fluid system affects head loss.

☐ Explain the operation of a positive-displacement pump.

☐ Explain the operation of a nonpositive-displacement pump.

☐ Identify the types of schematics created for piping systems.

☐ Discuss the purpose of the common components in piping systems.

☐ Identify common sizing and marking practices used for piping.

☐ Discuss the common forms and related uses of steel pipe.

☐ Discuss common thermoplastics and how they are used in fluid systems.

☐ Discuss the common types of tubing used in fluid systems.

TECHNICAL TERMS

accumulation	lantern ring
blowdown	nominal pipe size (NPS)
cavitation	nonpositive-displacement
chattering	pump
durometer	positive-displacement pump
flaring	schedule
flow diagram	single-line diagram
head loss	Stellite
isometric drawing	turbulent flow
laminar flow	valve slamming

In most cases, you cannot see fluid flowing through a piping system. Instrumentation allows us to evaluate information about a fluid system in order to better understand and visualize what is happening within the system. In this chapter, you will examine how safety relates to fluid systems, the basic concepts of fluid dynamics, and components of fluid systems. Understanding basic engineering principles allows you to apply these core principles across a wide range of systems for troubleshooting and repair effectiveness.

40.1 GENERAL SAFETY

The steps involved in planning repair work are just as important as the processes used to perform the work. Planning can take various forms, from a manager coordinating work with outside contractors to a frontline technician evaluating the job at hand before starting it. Regardless of who is involved in the planning, properly considering safety factors is required before starting a job. This includes use of proper personal protective equipment (PPE), knowledge of the piping system, and safety-minded job preparation. A technician should not start a job without first examining aspects of the job, possible safety issues, and strategies to reduce potential hazards.

40.1.1 New Construction Safety

In the most general terms, controlling hazards can be accomplished by identifying the hazard, evaluating the hazard, and placing controls to limit the hazard. **Figure 40-1** presents some common hazards that should be evaluated, particularly with new construction piping work. New construction safety hazards may involve confined spaces, welding hazards, atmospheric hazards, and trench work.

Common Hazards for New Construction Piping

Type of Piping Work	Planning and Evaluation of Hazards
Confined spaces	■ Assess depth of trench ■ Perform atmospheric testing ■ Plan for fresh air supply ■ Employer confined space entry program ■ PPE ■ Provide proper training
Welding	■ Assess fire hazard ■ Identify electrical or gas hazard ■ Perform atmospheric testing ■ Plan for fire extinguishers ■ PPE ■ Person on fire watch
Trench work	■ Identify cave-in hazard ■ Perform soil evaluation ■ Plan for trench wall support/bracing ■ Perform atmospheric testing ■ Plan for fresh air supply ■ Provide proper training
Heavy lifting	■ Identify piping and heavy equipment hazards ■ Schedule inspection of lifting rigs ■ PPE ■ Provide proper training

Goodheart-Willcox Publisher

Figure 40-1. A summary of some common piping tasks in new construction and potential hazards to evaluate.

Confined space entry is common on new construction jobs. Procedures may differ slightly from one jobsite to the next, but all related safety programs include the following elements:

- Tracking of personnel.
- Tracking of time spent in the confined space.
- Atmospheric testing of area before, during, and after work.
- Supply of fresh air to the space.
- PPE, such as harnesses, hearing protection, and possibly a personal atmospheric tester.
- Supervision and monitoring of activities.

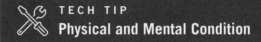

TECH TIP
Physical and Mental Condition

Simply working in a confined space can pose severe hazards to workers. You should not perform any task in a confined space when you are not at your best physical and mental condition. If you believe that you are taking a risk, evaluate your surroundings and methods. Many workers have met an untimely end because they ignored their best judgment. Carefully consider your surroundings, tools, and always inspect your safety equipment before entering a confined space.

PPE is an integral line of defense between you and a hazard. Always check the condition of your PPE before using it. Consider the job to be performed and how the PPE compares to the protection you will need. For instance, welding gloves come in several styles and uses. Gloves used for GTAW welding may not be appropriate for GMAW welding. Additionally, shades of welding goggles may need to be changed depending on the application and arc current.

Evaluate each job to be performed for the following:

- Potential hazards in the environment.
- Possible hazards from tooling.
- Options to remove or reduce the hazards.
- Your own mental and physical condition.

40.1.2 Repair Safety

Repairing existing systems typically involves the fluids contained within the piping system. This is the main difference between new construction safety and repair safety.

New construction of piping systems do not involve this added concern. **Figure 40-2** examines some common fluid systems and implications on safety and hazard control. Because of fluids in the system, lockout/tagout (LOTO) procedures are important in safely securing mechanical systems for repair.

Considering the fluid contained in the piping is an important aspect of LOTO. System pressures and temperatures should also be evaluated before performing LOTO and repair work. Venting a high-pressure fluid or gas is much different from draining a water line under normal pressure. Opening a valve to vent a high-pressure system can result in unexpected results such as the following:

- Heated water can instantly flash to steam and pose electrical and scalding hazards.

- High-pressure air may freeze a valve in the open position due to temperature/pressure drop.

- High-pressure air can create impact and hearing hazards.

- High-pressure gases can quickly displace oxygen and create an asphyxiation hazard.

In piping systems, LOTO prevents the release of energy that could cause injury. Performing a lockout/tagout involves the following:

- Communicating with others.

- Reviewing blueprints.

- Reviewing methods and procedures.

- Locking out the equipment.

- Safely releasing energy.

- Communicating with others after the work is complete.

- Properly restoring the system.

- Communicating that the system is back to normal.

The exact procedure implemented depends on the type of fluid in the system, maintenance to be performed, and possible hazards posed by the fluid. For example, one system may need to be completely purged of contaminants, while another system may only need to be drained and dried.

40.2 FLUID DYNAMICS

Fluid systems transfer energy in the form of heating or cooling, transfer fluid from one system to another, and transfer power to accomplish work. In the case of heating and cooling systems, fluid is the medium used to transfer heat energy. In water-cooled engines, for example, heat created by internal combustion is transferred into the fluid from the surrounding metal of the engine. The hot fluid is pumped to the radiator, where the heat in the fluid is transferred to the surrounding air.

Hazards for Various Piping Systems		
Fluid	**Potential Hazards**	**Considerations**
Low pressure water (<100 psig)	▪ Temperature ▪ Contaminants ▪ Drainage route	Where will the water be drained? Will this pose an additional hazard?
High pressure water (>100 psig)	▪ Safe venting and draining ▪ Temperature ▪ Skin hazards	Is the water heated? Could the water flash into steam?
Low pressure steam (<25 psig)	▪ Safely venting ▪ Temperature ▪ Skin hazards	Are there chemicals in the system?
High pressure steam (>25 psig)	▪ Safely venting ▪ Temperature ▪ Skin hazard	Are exhaust fans needed? What PPE is needed?
Infinite water system (water tower, large tank)	▪ Drowning ▪ Equipment damage	Can the system be properly isolated?
Gases	▪ Asphyxiation ▪ Inhalation hazard	What is the pressure of the gas? Can fresh air be supplied while venting? Will this be vented into the atmosphere?
Other fluids (hydraulic fluid, chemicals, petroleum products)	▪ Review SDS for possible hazards ▪ Potential for multiple hazards	What PPE is needed? How will drainage be collected?

Goodheart-Willcox Publisher

Figure 40-2. A comparison of common fluids, possible hazards, and safety considerations.

Some systems exist only to transfer fluid from one system to another. Large municipal water systems do exactly this. Water is treated and delivered for home use. Other transfer purposes may include chemical treatment, removal of solids, or mixing of separate solutions.

When a fluid system is used to transfer power to accomplish work, as in a hydraulic system, a hydraulic pump transfers energy by increasing the pressure of the fluid. The high-pressure fluid then causes motion at an actuator, which converts the energy (pressure) into mechanical motion.

40.2.1 Fluid Flow

Fluid flows from a high-energy state to a lower energy state. In most instances, fluid flow is caused by a difference in pressure, much as voltage causes electrical current to flow. Increasing pressure can be accomplished through raising the height of the fluid or through the use of a pump. While gases are quite compressible, fluid is nearly incompressible. When trying to compress fluid, pressure increases without a decrease in volume. When compressing gases, pressure increases and volume correspondingly decreases.

There are two main types of fluid flow inside piping: laminar flow and turbulent flow. See **Figure 40-3**. *Laminar flow* is a streamlined type of fluid flow in which all layers flow in parallel paths. Layers nearest the sidewall of a pipe have a lower velocity while layers in the center of the pipe have a higher velocity. There is no random or perpendicular flow, as all flow is in the same direction of fluid travel. *Turbulent flow* is a type of fluid flow in which the direction and velocity of the fluid fluctuate. Turbulent flow is characterized by mixed flows that generally move in the same direction. Laminar flow is associated with lower flow rates, while turbulent flow tends to take place with higher velocities. Transitional flow is characterized by turbulent flow in the center with laminar flow near the edges of the flow. Characteristics including fluid velocity, density, and pipe

roughness determine the type of fluid flow. For instance, PVC piping has very smooth walls compared to rough concrete piping. This will impact how the fluid flows.

40.2.2 Head Loss

Just as an electrical circuit has resistance to current flow, a piping system has resistance to fluid flow called *head loss*. *Head loss* is a measurement of the resistance to fluid flow in piping systems. The amount of head loss in a system is determined by elements of the system's design, including the length of piping, the type and number of valves and fittings, and the type of piping used. For example, a system with many fittings and smaller diameter piping will have substantially more head loss than a system with fewer fittings and larger diameter piping for the same fluid flow rate.

The design of fluid systems is extremely important when dealing with energy and power transmission. A system that has too much head loss results in inefficiency and wasted power. A system with too little head loss can produce negative effects on the pump. Compare the operation of the two systems illustrated in **Figure 40-4**. In this example, the larger piping system has less head loss for the

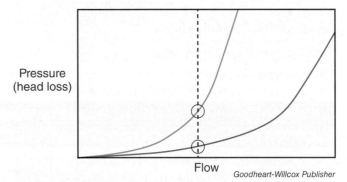

Goodheart-Willcox Publisher

Figure 40-4. This diagram compares head loss versus flow in two systems of different sizes. The larger system is represented by the blue line and the smaller system is represented by the green line.

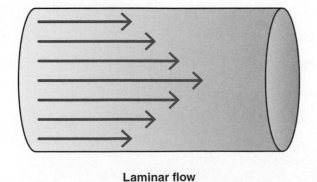

Laminar flow

Turbulent flow

Goodheart-Willcox Publisher

Figure 40-3. Fluid flow in a piping system can occur as laminar or turbulent flow.

same amount of flow. Increasing the flow in either system also increases head loss. Systems designed for large amounts of fluid flow are designed with very low head loss curves. Systems designed for power transmission where fluid flow is lower tend to have steeper head loss curves. Matching the pump to the system is an important part of system design that contributes to achieving maximum efficiency.

40.2.3 Pressure and Flow

Fluid pressure and flow in a system are a function of both pump design and the head loss curve for the system. Pumps can be classified into two broad categories: positive-displacement and nonpositive-displacement.

A *positive-displacement pump* moves a fluid by trapping a certain amount in a sealed area and then forcing that trapped volume into a discharge pipe. Examples of positive-displacement pumps include piston pumps, vane pumps, gear pumps, and progressive cavity pumps. Similar to an air compressor with a piston and cylinder, all of these pumps have a tight seal between the moving member and the casing. See **Figure 40-5**.

Positive-displacement pumps have a vertical pump curve. In other words, the output flow does not depend on the backpressure against the discharge head. The output flow is linear and depends only on the speed of the pump. The vertical pump curve shown in **Figure 40-6** represents the constant flow of a positive-displacement pump at a constant speed. A piston pump with a constant rpm discharges the same flow regardless if it is pumping against a head loss of 10 psig or 1000 psig. Because of this, a relief valve on the discharge side (output) of the pump ensures excess pressure is relieved back to the suction side or reservoir of the pump. If the discharge side valve is shut while the pump is running and there is no relief valve, the pump

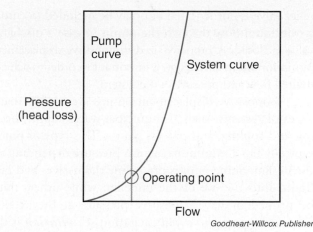

Figure 40-6. This diagram depicts a vertical pump curve intersecting the system curve.

would either damage itself or an electrical protective function would trip to stop the pump. Positive-displacement pumps are typically used in power transfer systems, such as hydraulics, or in a system where definite flow is required, such as in an oil system.

A *nonpositive-displacement pump* creates a continuous flow of fluid, but output varies because there is no seal between the rotating member and the pump casing. In this type of pump, there is usually a small gap between the impeller and case of the pump. Centrifugal pumps are a type of nonpositive-displacement pump commonly found in industry. Energy is transferred into the fluid by transferring kinetic energy (speed) into the fluid through the design of the impeller. The impeller flings the fluid outward into the casing and volute. Discharge pressure of the pump depends on system design and head loss factors. In the pump curve shown in **Figure 40-7**, the operating point is the point where the system will operate with regard to flow and pressure. A discharge side

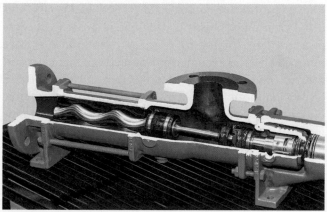

Surasak_Photo/Shutterstock.com

Figure 40-5. This positive-displacement pump develops pressure by a reducing volume and tight seal between the rotor and casing.

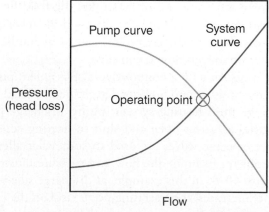

Goodheart-Willcox Publisher

Figure 40-7. In this diagram, the downward pump curve intersects the upward system curve at the point where the system will operate with regard to flow and pressure.

relief valve is not required but may be included to ensure a minimum flow through the pump in case a discharge valve is closed. A properly sized nonpositive-displacement pump for the system design is important in order to achieve desired flow and pressure in the system.

Nonpositive-displacement pumps are used in many types of systems, such as municipal water supply, heating and cooling, and process plants. This type of pump typically has a minimum suction pressure depending on the design. Some pumps can prime themselves and pull liquid into the eye of the impeller, while others must be primed or have a positive pressure at the suction side of the pump to avoid cavitation. *Cavitation* is the formation and collapse of air bubbles in the pumped fluid. Cavitation starts at the eye of the impeller in the suction area and ends when the pressure is increased as fluid is thrown out along the impeller. The collapse of air bubbles is noisy, typically sounding like marbles inside of the pump, and is damaging to internal components. Cavitation can be prevented by properly venting the pump prior to starting and ensuring there is enough suction pressure to the pump.

40.2.4 System Operating Characteristics

System operating pressures depend on the system design. Positive-displacement and nonpositive-displacement pumps operate differently within a piping system.

A positive-displacement pump produces a constant flow at a constant speed. System pressure, however, is determined by system design and may be controlled by a downstream pressure regulation valve or relief valve. In a hydraulic system, pressure at the discharge side of the pump is normally constant and some fluid is relieved back to the reservoir. Flow to the system changes depending on load. As load decreases, such as when an actuator reaches the end of its stroke, the fluid flow to that load decreases, causing flow through the relief valve to increase back to the reservoir. In most hydraulic systems, pump output flow is constant.

In a system with a nonpositive-displacement pump, discharge pressure and flow are determined by head loss. In a large fluid heating system, pump discharge pressure varies as valves open and shut to heating coils. In practice, specific valves are used to somewhat alleviate this problem. Examine the flow and pressure illustrated in **Figure 40-8**. In this example, as discharge side back-pressure increases, the operating point rises on the curve and a new operating point with higher pressure and less flow is achieved. As discharge valves open and pressure drops, a lower operating point on the curve is attained. Having a variable speed pump changes the pump curve

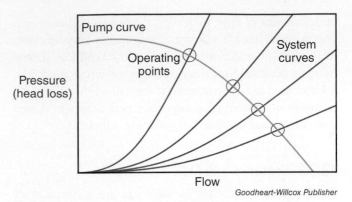

Figure 40-8. This diagram shows the effects of system demand or load on the operating point.

and the operating point of the system. In all cases, the operating point of the system (pressure and flow) is determined by the intersection of the two curves.

40.3 PIPING SCHEMATICS

This section presents a general guide for reading piping schematics. Many engineering companies may use some or all of the types and symbols discussed in this section. However, some may change symbols slightly.

Line use on a piping schematic is similar to line use on any blueprint and may include object lines, break lines, section lines, hidden lines, and dimension lines. The purpose of each line is determined by how it is used on each schematic. On an electrical ladder logic diagram, for example, a solid line represents the logical flow of current from one item to the next. On a piping diagram, a solid line indicates a pipe carrying fluid from one component to the next.

40.3.1 Line Identification Numbering System

Line identification numbers are used to specify and detail each pipe. This is particularly important in complicated, large piping or process systems. Just as wire numbers help a technician troubleshoot an electrical control system, piping identification systems help a technician locate, trace, and troubleshoot fluid systems. Numbering systems are not always provided on the piping, but may be listed on the piping diagrams. These numbering systems are not standardized, but instead rely on the engineer who designs the system and what the company standard requires. As in schematic symbols, a legend included with the blueprints should detail the identification method used. **Figure 40-9** shows several examples of the methods used. All methods usually show a reference number, piping material and size, and references to an area in the plant or system.

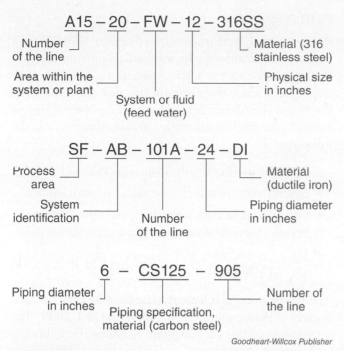

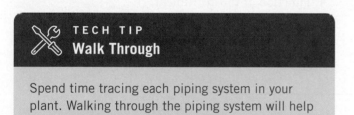

Goodheart-Willcox Publisher

Figure 40-9. Several examples of line identification systems for piping.

TECH TIP
Walk Through

Spend time tracing each piping system in your plant. Walking through the piping system will help you visualize flow through the system and aid in your understanding of how the system works.

40.3.2 Types of Schematics

The type of schematics created for a piping system depends on the system design, process, and complexity. Common types of schematics created for piping systems include the following:

- Single-line diagrams.
- Flow diagrams.
- Piping and instrumentation diagrams (P&ID).
- Isometric drawings.

Symbols for components of piping systems are covered by ASME Y32.2.3: 1949 (R1999), which is a standard for diagrams. In addition, ANSI Z32.23 is an older standard that is still used. In a group or system of piping blueprints, a detailed legend is provided that includes symbols used on those blueprints.

Single-Line Diagrams

Single-line diagrams show all components in a piping system, which may include fittings, through the use of standard symbols. See **Figure 40-10**. Typically, these diagrams do not show flow and are not drawn to scale. The purpose of a single-line diagram is to show system components and connections, and allow the technician to gain a logical understanding of system operation.

This type of diagram can show information such as the following:

- Piping sizes and material.
- Valve sizes, types, and actuators.

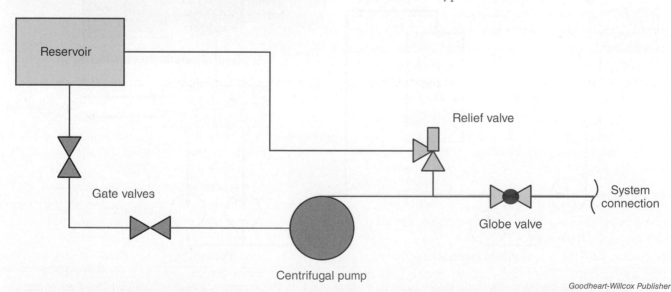

Goodheart-Willcox Publisher

Figure 40-10. A single-line diagram of a simplified piping system. A centrifugal pump takes its suction from a reservoir and discharges to another system. This diagram shows that the reservoir has an isolation valve, the pump has both a suction and discharge isolation valve, and the pump has a relief valve that discharges back to the reservoir.

- Fittings and fitting types (welded, flanged, screwed).
- Pump size, type, and horsepower (hp), or other prime mover type.
- Physical size and volume of reservoir.

Flow Diagrams

Flow diagrams show more complicated process plant piping systems and include the following:

- Major system components.
- Main piping lines in single line format.
- Flow arrows.
- Other system components, such as pumps, vessels, heat exchangers, and storage reservoirs.
- System connections.

These diagrams are not usually drawn to scale. Flow diagrams are organized to aid in the understanding of logical flow paths instead of the physical placement of items. In **Figure 40-11**, examine the fluid flow paths for a basic power plant steam system. Note that all piping is not shown, only major lines and structures.

P&ID Diagrams

A piping and instrumentation diagram (P&ID) is a diagram that combines the mechanical portion of a system (piping) and the instrumentation, including control signals. These diagrams are also not drawn to scale, but instead are drawn for understanding control system interactions. P&IDs normally contain items such as the following:

- Major structures (equipment) and valves.
- Specific control valves and related control signals.
- Gauges, transmitters, controllers, and signal lines between these elements, as well as coding information.
- Piping codes, sizes, and changes in piping sizes may also be included.

Symbols on P&IDs are standardized by International Society of Automation (ISA), ANSI/ISA-5.1-2009. The instrument coding on a diagram contains the following information:

- Location of the instrument.
- Function of the instrument.
- Identifying loop number or gauge number.

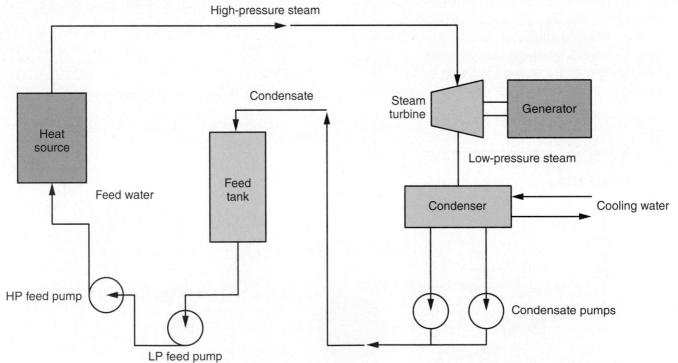

Goodheart-Willcox Publisher

Figure 40-11. A flow diagram of a basic steam cycle in a power plant. In this example, a heat source boils feed water into steam. High-pressure steam leaves the boiler and is transferred to the steam turbine, causing it to rotate. The turbine is the prime mover for the generator, which generates electricity. The low-pressure steam is condensed back into water (condensate) in the condenser, which transfers heat from the low-pressure steam to the cooling water (a heat exchanger).

Before looking at a P&ID, examine a simple loop and how it works. A simple control loop is shown in **Figure 40-12**. Each portion of the control loop performs a different function—measurement, control, and actuation. The purpose of the entire control loop is to control a variable (such as temperature, level, or pressure) within a set range.

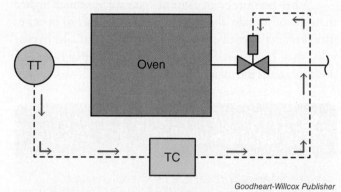

Goodheart-Willcox Publisher

Figure 40-12. This diagram shows a simplified P&ID with loop control. In this control loop example, temperature is sensed by an instrument (TT), a signal is developed, and sent on to the temperature controller (TC). This controller checks the current temperature against the set point temperature in memory and scales an output signal to control the valve position.

Notice that the piping lines are solid, while the signal lines are dashed. The control signals used can be of several mediums, including voltage, current, or air pressure. The type of control signal used largely depends on when the system was designed and choice of the process engineer. Many industrial applications use proportional signals of 4–20 mA.

TECH TIP
Maintain P&IDs

A P&ID is one of the most important schematics for troubleshooting and understanding the control of the process. A P&ID is the only type of drawing on which a technician can quickly see all system control signals, inputs, and outputs. It is important to keep this drawing up to date when piping changes take place.

Isometric Drawings

Isometric drawings show three-dimensional spatial relationships between components, valve positions, multiple piping runs, and dimensions of piping. These drawings are not normally scaled, so dimensional measurements

cannot be taken from isometric drawings unless dimensions are specifically listed. Vertical piping is oriented vertically in these drawings, while horizontal piping is shown at a 30° angle to horizontal. This type of drawing is normally used for construction and pipe fitting and typically shows smaller sections or single systems.

A variety of information can be shown on an isometric drawing including the following:

- Valve type, pressure rating, and size.
- Pipe size, lengths, and spatial position.
- Types of joints (screwed, flanged, welded).
- Flow direction.
- Pipe supports.

The information included on an isometric drawing depends on the engineering company and applicable piping standards. If used for construction, full dimension data is provided. In **Figure 40-13**, dimension data is provided from center-to-center of welded or screwed fittings. In other flanged piping, dimensions may be provided from flange face-to-flange face or may include the gasket, which would be noted.

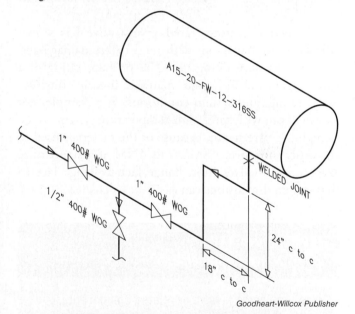

Goodheart-Willcox Publisher

Figure 40-13. An isometric drawing of a drain system.

40.4 PIPING SYSTEM COMPONENTS

Depending on the purpose of a piping system and the fluid within it, a wide variety of components may be needed to ensure proper operation. However, every system has the following general components:

- Valves.
- Pumps.

- Pressure vessels.
- Connections.
- Instrumentation.

40.4.1 Valves

Valves control fluid flow. Every type of valve has a specific purpose and certain restrictions to their use. All valves are made of the following common elements:

- **Body.** Directs flow through the valve.
- **Stem.** Connects the operator to the sealing mechanism.
- **Operator.** Allows the operation of the valve, such as a handwheel, solenoid, or lever.
- **Sealing mechanism.** Seats on the sealing surface to open, shut, or throttle flow.
- **Sealing surface.** Surface, such as a seat ring, on which the sealing mechanism seats.
- **Outer seal.** Barrier between system pressure and outside atmosphere, such as packing or Teflon™ rings.

Valves are pressure rated. Every valve has a specific pressure that it can withstand before damage and a typical pressure rating use. The pressure rating of a valve depends of its design, materials used in construction, sealing, and system conditions. For example, not all valves can withstand high-temperature systems, such as high-pressure steam, because of the materials used to construct the valve. ASME and ANSI also standardize exact sizes of valves from flange face-to-flange face so that system dimensions can easily be calculated.

TECH TIP
Valve Replacement

When replacing a valve or related components, ensure that the replacement meets or exceeds the original specification. Components made of different materials may not be compatible with the fluid in every system. Some materials may melt when exposed to incompatible fluids. Compatibility also applies to system pressures and temperatures. For example, brass and Grade 8 steel have vastly different characteristics. Brass cannot withstand the same temperature as Grade 8 steel.

Gate Valves

The purpose of a gate valve is to isolate one section of a system or piece of equipment from another. By the design of the valve body and gate, **Figure 40-14**, flow quickly increases as the valve is opened. Gate valves are not used for throttling flow, as throttling would damage the gate and seats of the valve.

When pressure on one side of a gate valve is much higher than the other side, the valve becomes difficult to operate. This is often the case with larger gate valves. A smaller bypass valve may be installed in parallel to allow equalization of pressure across the main valve before opening.

TECH TIP
Isolating a Valve

When repairing a valve, it is not acceptable to use that valve for isolation. You cannot open a gate valve and repack it without it being isolated. The valve must be isolated both upstream and downstream in order to work on it and avoid system pressure acting on the packing or outer seal during the repair.

Globe Valves

A globe valve may be used for isolation or throttling flow. There are several different globe valve body types, such as Y, angle, and straight, with specific applications. Straight globe valves are the most common. While the globe valve has a higher head loss than a gate valve, it has better throttling flow characteristics.

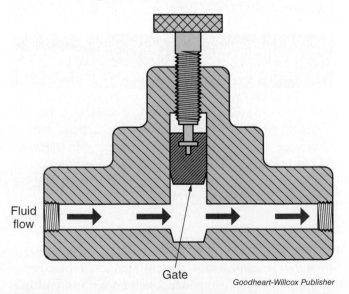

Fluid flow

Gate

Goodheart-Willcox Publisher

Figure 40-14. The flow path through a basic gate valve. Fluid flow quickly increases when the gate is opened, even partially.

Figure 40-15 illustrates the correct path of flow through a globe valve. Installing the valve so that flow enters above the disk can result in chattering. *Chattering* is noise caused by vibration of the disk due to flow disturbances inside the body of a valve. By installing the valve so that flow comes up through the disk and seat, the disk sustains a fairly constant pressure and will not chatter. Chattering eventually results in damage to the valve.

There are several different disk and seat arrangements for globe valves. One of the most common arrangements is a plug disk and a threaded seat ring. The seat ring is threaded into the cast body of the valve. The plug disk can rotate on the stem, which prevents creating an indentation of foreign material into the disk or seat ring.

The needle valve is a type of globe valve used to throttle fluid or gas flow, particularly in smaller lines. Depending on the size, the disk (or needle) may simply be the machined end of the valve stem. Use caution when closing a needle valve, as excessive force can damage the needle and seat.

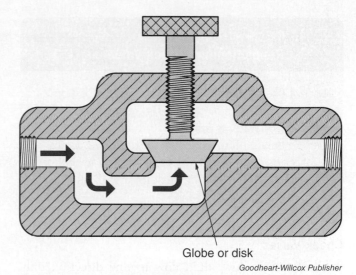

Goodheart-Willcox Publisher

Figure 40-15. The correct path of fluid flow through a globe valve is one in which flow enters below the disk.

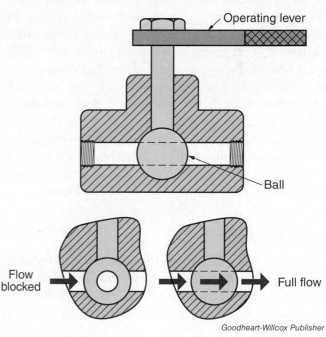

Goodheart-Willcox Publisher

Figure 40-16. The operation of a ball valve in controlling fluid flow.

TECH TIP
Checking Valve Position

If no visual position indicator exists when checking the position of a valve, check the valve in the shut direction by rotating the handwheel a small amount clockwise. If the valve is shut, placing more shutting force on the valve will not change system operation. If the valve was open and rotating the handwheel shut it slightly, flow was not appreciably changed and the valve can be returned to full-open.

Check for possible limit switches that can indicate position prior to rotating the handwheel. Also carefully consider valve position prior to checking. A valve that is mounted upside down and below you needs to be turned in the counterclockwise direction in order to close. Valves mounted in the ceiling with chain operators should be carefully evaluated before checking.

Ball Valves

A ball valve is used for isolation and limited throttling. As shown in **Figure 40-16**, the ball valve is closed when the operating lever is at a right angle to the piping. When the valve is opened, the handle is parallel with the direction of fluid flow.

Ball valves are classified by ball type and body type. Ball types include the following:

- **Full port.** Refers to the diameter of the hole through the spherical ball. The flow path through the ball remains the same diameter as the inside diameter of the pipe size.

- **Regular port.** Substantially smaller than full port. Results in pressure loss in high-flow systems.

- **Venturi.** Limits flow and equalizes pressure drops across loads in a large system to provide consistent flows. Also called a *flow balancer*.

Check Valves

A check valve allows fluid flow in one direction only and automatically prevents backflow. Check valves are available in different body styles including horizontal swing, horizontal lift, vertical lift, and vertical swing. The style used depends on the engineering application and specific system requirements. Swing check valves, **Figure 40-17**, generally experience more valve slamming than other types. *Valve slamming* occurs when the disk rapidly closes against the seat due to a drop off in flow through the system or if flow attempts to reverse. Excessive slamming can damage the disk, seat, and hinge. Check valves are usually maintenance-free for many years unless placed in a harsh fluid environment, such as high sediment or acid, which can block the action of the valves or corrode disk and seat.

Butterfly Valves

A butterfly valve, **Figure 40-18,** is mainly used for isolation, but can be used for throttling flow. Two main types of butterfly valves are the single-disk and split-disk. The single-disk butterfly valve has a flat, circular disk that

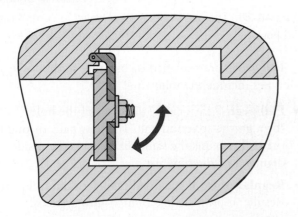

Goodheart-Willcox Publisher

Figure 40-17. A swing check valve pivots on a hinge to interrupt flow.

Yuthtana artkla/Shutterstock.com

Figure 40-18. This butterfly valve is a single-disk type with a hand lever operator.

rotates to be in line with flow through a 90° turn of the hand lever. The split-disk butterfly valve is made of two half disks that flap to an open position. The main advantages of this type of valve is its low head loss profile and simple design.

Relief Valves

The purpose of a relief valve is to prevent a system over-pressure condition. Relief valves are classified based on how they open to relieve pressure, which depends on the type of system the valve is protecting. *Accumulation* is a type of valve operation in which the relief valve starts to open at its set point and becomes fully open at a higher pressure. This is typically used when the fluid or gas discharged is hazardous and the engineer wants to minimize discharge. An accumulation type relief valve gradually opens to relieve pressure. *Blowdown* is a type of valve operation in which an open relief valve does not close until the pressure falls a specified amount below its set point. For example, this type of relief valve could be fully open at 150 psig to relieve pressure until the system pressure falls below 140 psig. Operating in this manner prevents rapid cycling/slamming of the valve, which can damage its disk and seat.

SAFETY NOTE
Relief Valve Safety

Under no circumstances should the pressure set-ting of a relief valve be changed. This setting was performed under testing conditions at the factory. If the relief valve is safety-wired, the safety wire should not be cut. The only way to test a relief valve setting adjustment is in a controlled system or bench test. There is virtually no way to measure the pressure adjustment made using the adjusting screw.

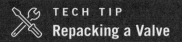

TECH TIP
Repacking a Valve

When repacking a valve and adding packing rings, there are several important things to keep in mind:

- Use packing that is the appropriate size and mate-rial and is compatible with the fluid in the system.
- Use packing extractors (pullers) that are the cor-rect size. Using a packing puller that is too large can scratch the stem of the valve, which causes the valve to leak.
- Cycle the valve open and shut several times between packing adjustment. Adjust the pressure on the packing slowly by tightening the packing gland nuts a small amount each time between cycling the valve.

40.4.2 Pumps

The majority of pumps used in industry are nonpositive-displacement pumps, such as centrifugal pumps. All nonpositive-displacement pumps have several common elements, including the following:

- A rotating impeller.
- A seal between the suction and discharge side of the impeller (wear ring).
- Some type of sealing mechanism on the pump shaft (packing or mechanical seal).

Pump materials vary depending on the application and can range from plastics, to high-end stainless steels. The two most common maintenance tasks associated with pumps are adjusting or replacing the packing or seal, and completely rebuilding the pump.

Packing and Seals

Many older pumps, and some newer pumps, use a packing gland and packing as the seal, as shown in **Figure 40-19**. Adjusting packing to minimize leaks, known as *tightening* the packing, should be performed slowly with the pump running and 1/2 to 1 flat turn on the packing gland nuts per adjustment. After an adjustment, observe the packing for leakage for a period of time before adjusting again. Some leakage may be required, which will be detailed in the manufacturer's specifications.

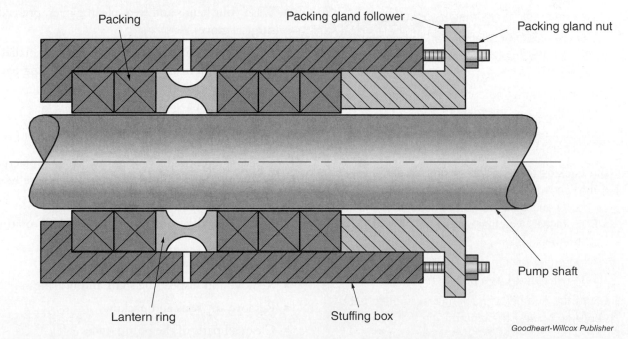

Goodheart-Willcox Publisher

Figure 40-19. A cross-sectional view of a packing gland with a lantern ring.

TECH TIP
Compressing Packing

The more force placed on the packing by tightening the packing gland nuts, the more the packing is compressed. This compression also presses against the pump shaft. Too much pressure on the pump shaft can eventually wear away the shaft and create an hour-glass shape, which will require extensive rework. It is better to have some leakage rather than zero leakage. If zero leakage is required, a mechanical seal should be used instead of packing.

Some pumps use a lantern ring to help keep the packing inside a pump cool. A *lantern ring* filters water between layers of packing along the pump shaft to cool the packing material. The lantern ring is supplied with pressurized water, which then is pushed out of the top and bottom of the lantern ring into the packing and along the shaft.

PROCEDURE
Repacking a Pump

If complete repacking is required, keep in mind the following general rules:

1. Use appropriately sized packing pullers.
2. Have a manufacturer's manual on hand.
3. Verify the amount of packing and placement of the lantern ring in the manufacturer's manual.
4. Use only packing material that is recommended by the manufacturer.
5. Soak bulk packing overnight to allow it to swell slightly.
6. Cut packing using a pipe or unused shaft. Pump packing is often cut at a butt end (90° cut) and placed into the packing gland at a 90° stagger, which allows some leakage to cool the packing.
7. When pulling packing, be careful not to scratch or damage the pump shaft.
8. Count the number of packing rings pulled both above and below the lantern ring.
9. Replace packing rings and lantern rings exactly as they were removed in order to line up the lantern ring with the incoming seal water.
10. Slowly adjust packing pressure over several hours while the pump is running until the desired leakage rate is achieved.

There are several different types of mechanical seals commonly used with pumps. Mechanical seals provide sealing using O-ring seals between the elements and the casing or packing gland, and seal between two ring-like components that mate against each other. There are a wide variety of materials used for mechanical seals. One common material is *Stellite*, which is a cobalt-chromium alloy with high wear resistance. The Stellite ring remains stationary, while the rotating member pressing against it is either graphite or ceramic. The two rings are highly polished, extremely flat, and mate exactly so that they prevent leakage.

Cleanliness is a high priority when replacing a mechanical seal. Consider the following:

- The gland of the pump must be expertly cleaned.
- The area in which the work will be performed must be very clean.
- Do not remove the new seal from its packaging until ready to install it.
- Use only lint-free cloths during installation.
- Wash your hands and wear clean gloves, preferably surgical gloves.
- Do not touch mating surfaces. Even the smallest amount of oil from your skin on the sealing parts can cause the seal to leak once installed.

Rebuilding Pumps

If rebuilding a pump is necessary, a manufacturer's manual should be readily available. A complete rebuild involves the following:

- Completely disassemble the shaft and all rotating elements.
- Remove the wear ring(s).
- Remove and inspect the shaft and bearings.
- Remove any seals and packing.
- Clean all parts of the pump once disassembled.

The task of rebuilding a large pump can be time-consuming and should, therefore, be done correctly the first time. Carefully consider the use of lubricants for assembly, locking fluids, and anti seize compounds. Not all of these fluids are compatible with all systems and should be closely examined before use. The work area needs to be clean during disassembly and reassembly. All mating surfaces should be perfectly clean. Check new bearings to ensure they match the manufacturer's recommendations and are intended for use with the specified shaft measurements. A new shaft may be needed depending on inspection results. Leave bearings in their original packing, as much as possible, until ready to be measured and installed. New gaskets should also be installed during a rebuild.

40.4.3 Pressure Vessels

Pressure vessels and tanks are found in nearly every fluid system. They serve a variety of purposes depending on the application, and can range in size from a few cubic feet for chemical mixing or be as large as a water tower. The following are some common applications of pressure vessels:

- A storage tank to ensure positive pressure for a system.
- A tank used to mix different fluids and chemicals.
- A tank used to hold pressurized gases.
- A tank used to store fluid.
- A vessel used for heat exchange.

Pressure vessels, including those at a vacuum with external pressure, are regulated by ASME Section VIII codes. Materials used, thickness, welding, flange requirements, and testing are just some of the regulated specifications. Vessels and tanks should be inspected on a regular basis for cracks, damage, and corrosion. Some vessels may have internal components that are not readily identifiable from the outside. Before inspection, review a technical manual from the manufacturer on that particular tank.

TECH TIP
Repair and Maintenance on Vessels and Tanks

Think safety first! Properly isolate and drain a pressure vessel. Even when completely drained, a vessel may contain vapors that are dangerous to your health. Entering the inside of a pressure vessel or tank is confined space entry and requires safety controls. Flammable fluids may be present. Do not perform hot work without completely flushing the tank or vessel.

40.4.4 Connections

There are a wide array of connection types used for tubing and piping, including compression fittings, push-on fittings, welding, and gluing. The most common connection methods are summarized in **Figure 40-20**. Certain steps apply to any method of connection:

1. Make sure the right tools are available.
2. Properly prepare and clean the sealing surfaces.
3. Remove any burrs.
4. Carefully inspect materials before use to ensure compatibility and condition.
5. Use proper lubricants or sealants on O-rings, seals, flange gaskets, and threaded fasteners.

TECH TIP
Pipe Joint Compounds and Pipe Thread Tape

Do not overuse pipe joint compounds and pipe thread tape. Using too much can result in contaminating the system. Particles can find their way to components with small clearances and cause problems.

Connection Methods and Applications	
Type of Connection	**Typical Use**
Compression fittings	Metal tubing and some plastic Hydraulic systems Instrumentation Control systems
Quick disconnect push-on fittings	Softer and smaller plastic tubing, such as beverage, low-pressure air, controls
Threaded	Metal and plastic pipe Piping systems between 1/8″ and 6″
Welded socket or butt welding	Metal pipe Large or high-pressure piping
Ultrasonic welding	Plastic pipe
Permanent glue	Plastic pipe
Flanged	Larger metal and plastic pipe
Grooved and clamped fittings	Metal pipe Thin-wall pipe Fire sprinkler systems Cast-iron pipe

Goodheart-Willcox Publisher

Figure 40-20. A chart summarizing common piping and tubing attachment methods.

When assembling piping, chains/straps and lifting rigs should not be used to pull pipe together. Lifting rigs are used only for lifting. Using chains, straps, or lifting rigs to pull together flanges while rapidly assembling flange bolts will result in problems later. If excessive force is needed, the piping needs to be corrected and redesigned. If not, either the piping or a component connected to the piping, such as the pump, will be strained.

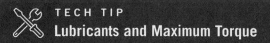

TECH TIP
Lubricants and Maximum Torque

When using lubricants on threaded fasteners, remember that the lubricant allows the fastener to move more easily. This reduces the maximum torque that can be applied to a fastener before exceeding its strength and possibly damaging the fastener.

40.4.5 Instrumentation

All piping systems have various forms of instrumentation. Some older systems may be limited to manually read gauges, while the instrumentation on newer systems can include remote locations and computer-generated graphics. In order to effectively troubleshoot these systems, you must have a general understanding of the instruments.

Basic manually read gauges are common in most piping systems and are used to measure pressure, differential pressure, temperature, level, weight, flow, and various other system conditions. Basic mechanical instruments have no external power supply and rely solely on mechanical devices—such as gears, springs, diaphragms, and Bourdon tubes—to transfer system conditions into a form that can be easily read. All types of mechanical instrumentation are subject to similar operational and maintenance issues. Eventually, the mechanical components wear or stick due to a buildup of contaminants, which causes the gauge to stop responding evenly. Shock, vibration, and impact also cause gauges to become less accurate. Because of this, larger industrial applications usually require instrumentation to be calibrated. A calibration program may involve routine calibration on all significant process instruments annually, semiannually, or even monthly.

The next level in sophistication for instrumentation involves electrically generated signals that are proportional to the item being measured. *Proportional* means that as one thing goes up, the other things follow. For example, as you turn the temperature control knob on your oven to a higher temperature setting, the oven gets hotter. This relationship is proportional because the response (temperature going up) to your action (turning the knob) is in the same direction and ratio.

Another level of innovation with instrumentation is advanced electronics, which provide heightened sensitivity and accuracy. For example, ultrasonic flow meters (**Figure 40-21**) use sound to measure open channel flow, depth, or volume of a tank. The controller can be programmed for a wide variety of measurements, which are all accomplished by an ultrasonic sound wave that is emitted, timed, and recorded. This type of instrument outputs a proportional signal of 4–20 mA that can then be sent to other recorders, industrial computers, remote loggers, or chart recorders.

Varying levels of system and process control take place depending on the technology being used. With local control, there may be observable local indications provided with simple electronically driven gauges, charts, or human-machine interface (HMI) screens. **Figure 40-22** shows an example of the architecture used in a local control scheme. Troubleshooting networked systems requires specific training in hardware and software.

40.5 PIPING AND TUBING MATERIALS

Piping is manufactured in a variety of ways and to a wide variety of standards. Pipe standards specify designations, use, material, material properties, and strength. While several types of piping may meet the needs of a

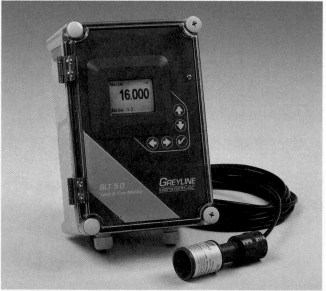

Greyline Instruments

Figure 40-21. This ultrasonic level and flow monitor uses a noncontact sensor to measure, display, and transmit liquid levels in tanks, and can monitor flow within a channel.

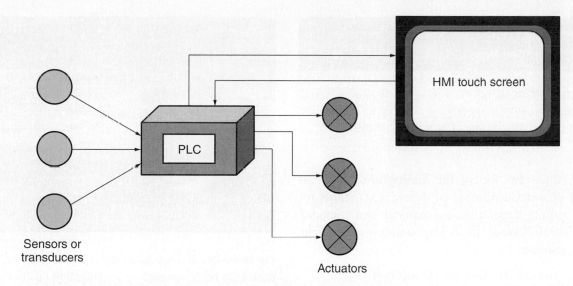

Goodheart-Willcox Publisher

Figure 40-22. This diagram illustrates the local architecture for a simplified control system and HMI. In this control system, variables are sensed by the sensors and transducers. A proportional signal is produced and sent to the programmable logic controller (PLC). The PLC examines these inputs (normally 4–20 milliamp), performs controller functions, develops output signals, and sends the signals to the actuators. The PLC also communicates with the HMI touch screen and displays all system information. The HMI touch screen allows operators to control the system through on-screen options and other manual controls.

specific installation, systems are designed to use a specific type of piping. The piping material should not be changed to another type without specific engineering approval.

Piping is sized in both thicknesses and schedule. The **schedule** of a pipe refers to the thickness of the pipe wall. Standard piping is available in a wide range of schedules from 10 (thin walls) to 160 (thick walls). However, the most common are 10, 40, and 80 for both steel and plastic piping. Outside diameter of piping remains the same while schedule changes. For example, a 2″ pipe has the same outside diameter regardless of the schedule. This is referred to as nominal pipe size. **Nominal pipe size (NPS)**

is a rounded value that is close to the actual inside diameter of pipe. NPS is commonly used to specify piping in the trades. Other terms used to describe piping include the following:

- *Standard* refers to schedule 40 piping.
- *Extra-heavy* refers to schedule 80 piping.
- *Extra double-heavy* refers to slightly above schedule 160 piping.

Figure 40-23 shows how NPS, outside diameter, and wall thickness compare among several common sizes of threaded piping.

Common Pipe Dimensions				
Nominal Pipe Size (NPS)	Outside Diameter (in inches)	Wall Thickness (in inches)		
		Standard (Schedule 40)	Extra Heavy (Schedule 80)	Extra Double-Heavy
1/4	0.540	0.088	0.119	NA
1/2	0.840	0.109	0.147	0.294
3/4	1.050	0.113	0.154	0.308
1	1.315	0.133	0.179	0.358
1 1/2	1.900	0.145	0.200	0.400
2	2.375	0.154	0.218	0.436

Goodheart-Willcox Publisher

Figure 40-23. A comparison of common threaded piping sizes and dimensions.

The American Society for Testing and Materials (ASTM) provides a number of different standards for marking piping depending on material and intended service. The following are common items among the different standards:

- Name of manufacturer (or ID number).
- ASTM specification.
- Material or method of manufacturing.
- Schedule.
- Length.
- Size.

40.5.1 Metal Piping

The two most common forms of steel pipe are black iron pipe and galvanized pipe. Black iron pipe is steel pipe without any corrosion inhibitors or coatings, with the exception of a coating that helps prevent surface rust during shipping and storage. Black iron pipe is inexpensive piping that is used for specific instances where possible contamination by zinc is undesirable. Systems in which black iron pipe is used include the following:

- Natural gas and propane systems.
- Fire sprinkler systems.
- Water piping systems.
- Heating and cooling systems.

Galvanized steel pipe provides some protection from corrosion and is used in some older water systems and fluid systems that are exposed to environmental moisture. Many of these systems have been replaced with either black iron or copper materials. Zinc plating coats both the outside and inside of the piping and can eventually flake off or react with minerals in the water and increase pressure loss. There are a wide variety of coatings and treatments for steel piping, **Figure 40-24**.

Stainless steel piping is used where corrosion resistance is of utmost importance. Corrosion resistance comes from chromium and nickel. Some stainless steel piping also

American Cast Iron Pipe Company

Figure 40-24. A large spiral welded steel pipe with epoxy coating is being lowered into installation position.

contains molybdenum, aluminum, titanium, and carbon. Stainless steels come in a number of series including 200, 300, 400, and 500. The series number indicates the alloying materials used, corrosion resistance, strength, and other characteristics of the metal. The two most common types of stainless steel piping are in the 300 series: 304 and 316. Both have a very high corrosion resistance, but 316 has a slightly higher nickel content and better resistance to salt for marine applications. An "L" designation, such as 304L and 316L, for stainless steel pipe refers to a low carbon content (less than 0.03%). This type is used in welded pipe applications to help prevent corrosion of the welds and surrounding metal. Common schedule sizes of stainless steel include Schedule 5, 10, 40, and 80. Common stainless steel piping sizes range from 1/8″ to 36″ NPS.

Copper and brass offer excellent corrosion resistance and are widely used in potable water and boiler systems. Seamless brass and copper pipe is readily available and come in standard NPS sizes up to 6″. Outside dimensions are the same as steel pipe, while the wall thickness and internal diameter differs. The piping may be threaded or not and is available in two strengths: regular and extra strong. The wall thickness of extra strong pipe is typically 140–170% greater than regular piping and the pressure rating is approximately double.

Ductile iron pipe is used by municipalities in water distribution and waste collection systems. This piping has excellent corrosion resistance and strength, with a variety of connection methods. Classes of pipe are based on pressure and include 150, 200, 250, 300, and 350. Ductile iron pipe comes in standard sizes from 3″ to 64″. Coatings such as zinc, polyethylene, and ceramic epoxy lining improve corrosion resistance when installed underground. A variety of fittings for ductile iron pipe are available, from welded outlets and flanges to couplings and saddles.

Cast-iron pipe, also called gray cast iron, was used for many years in municipal water, sewer, and drainage systems. While some cast-iron pipe is still produced, it is mainly used to repair old installations that cannot be replaced with better piping materials. Sizes of this piping range from 3″ to 48″, with smaller sizes being more common. **Figure 40-25** shows some of the most common fittings available.

40.5.2 Plastic Piping

The most common type of plastic piping used is thermoplastic. Thermoplastics are moldable at higher temperatures and turn to a liquid with increased temperature. After solidifying, they regain their strength and hardness, but become brittle. The thermoplastic piping often found in industry is schedule 40 and 80, but schedule 120 is also available. PVC and CPVC are commonly used thermoplastics in the piping

industry and come in standard sizes from 1/2″ to 24″. The following are some common forms of thermoplastics and their abbreviations:

- Polyvinyl chloride (PVC) is used in the majority of fluid and gas systems.
- Chlorinated polyvinyl chloride (CPVC) is resistant to most acids, bases, halogens, and alcohols.
- Acrylonitrile butadiene styrene (ABS) is used for drain-waste-vent systems.
- Polyethylene (PE) is used in many water, waste, and gas systems.

Schedule 40 piping is typically not threaded, although schedule 80 can be. The best pressure ratings occur with socket end joints through priming and gluing. As piping size increases, the pressure rating drops. For example, 1/2″ schedule 40 pipe is rated at 600 psi, while 10″ schedule 40 pipe is rated at 140 psi. Schedule 80 piping offers a higher pressure rating, but it is not double the rating of schedule 40. Schedule 80 PVC and CPVC are commonly used for strength and impact resistance, not just a higher pressure rating. **Figure 40-26** shows some common sizes of fittings and piping for schedule 40 PVC and ABS.

Charlotte Pipe and Foundry Company

Figure 40-26. A variety of PVC and ABS piping and fittings.

PVC, CPVC, and ABS are most commonly joined using a two-step process that involves solvent and gluing. Outside temperature affects the curing time of solvent joints. Most manufacturers recommend that a system not be pressurized for at least 24 hours after gluing. Larger piping diameter, outside temperature, and high humidity can drastically increase curing times.

40.5.3 Tubing

Tubing is usually a thin-wall material that is thinner than piping of the same diameter. There are a wide range of uses for tubing, including structural, assembly, controls, hydraulics, rollers, and mechanical frames. Although the uses of tubing are extensive, we will focus on fluid system applications.

Charlotte Pipe and Foundry Company

Figure 40-25. A variety of gray cast-iron piping and fittings.

Low Carbon Steel Tubing

The Society of Automotive Engineers (SAE) has a wide range of standards that cover seamless and welded tubing for hydraulic systems. Low carbon steel tubing for hydraulic systems must be able to be flared and bent without kinking. *Flaring* is a process where tubing is drawn over a mandrel to expand the end of the tube to a specific angle in order to make a seal. It is a common practice when making connections. Bends in hydraulic tubing have a minimum radius depending on the diameter of the tubing. Common tubing sizes range from 1/8″ to 2″ outside diameter, with varying wall thicknesses available depending on the pressure application. As an example, 1/2″ welded hydraulic tubing with a wall thickness of .035″ has a burst pressure of 6300 psi. With a wall thickness of .065″, burst pressure increases to 11,700 psi. The same seamless 1/2″ tubing with .035″ wall thickness has a bursting pressure of 6900 psi, while a wall thickness of .065″ has a bursting pressure of approximately 13,000 psi.

Stainless Steel Tubing

Stainless steel pressure tubing is commonly available in 304 and 316 series in the same sizes as low carbon steel tubing. Stainless steels have the added benefit of corrosion resistance, but require more costly fittings. This tubing is normally supplied in random 10–12′ straight lengths depending on the manufacturer. Applications range from medical to chemical, and environments where corrosion is a factor.

Aluminum Tubing

Aluminum tubing provides moderate corrosion resistance, lighter weight, and acceptable pressure ratings. It is typically used in aeronautics where weight is a consideration and lower pressure ratings are acceptable. A number of aluminum alloys are available commercially. Alloys used for aluminum pressure tubing adhere to ASTM B210. Aluminum tubing is easily bent and flared and is readily available from 1/16″ to 3″ OD in both welded and seamless types. Annealed coils of aluminum tubing are also available in 25′, 50′, and 100′ lengths. Both steel and anodized aluminum fittings are used for aluminum tubing applications. Anodized aluminum fittings are often colored, but the specific uses of a color code greatly depends on the engineer of the system.

TECH TIP
Tube Fitting Replacement

Never replace a tube fitting with a lower class fitting, such as replacing a stainless steel fitting with an aluminum fitting. Pressure rating, resistance to vibration, and corrosion resistance should be considered when replacing a fitting.

Copper Tubing

Copper tubing is used in water and refrigeration systems, as well as some hydraulic systems. **Figure 40-27** lists the most common types of copper tubing, their sizes, and color coding.

Air conditioning and refrigeration (ACR) tubing is purged and capped to retain cleanliness. ACR copper tubing is measured by its outside diameter instead of nominal size. Drain, waste, and vent (DWV) copper tubing is the thinnest tubing and usually only used for nonpressure application of drainage.

Hard (tempered) copper is normally connected by soldering or brazing. Soft (annealed) copper is usually

Types of Copper Tubing		
Type	**Available Sizes**	**Coding**
M	1/4″–12″ OD in 20′ tempered lengths	Red
L	1/4″–10″ OD in 20′ tempered lengths 1/4″–2″ OD in 40′–100′ annealed lengths (coil)	Blue
K	1/4″–12″ OD ranging from 20′–12′ tempered lengths 1/4″–2″ OD in 40′–100′ annealed lengths (coil)	Green
ACR	3/8″–4″ OD in 20′ tempered lengths 1/8″–1 5/8″ OD in 50′ annealed lengths (coil)	Blue
DWV	1 1/4″–8″ OD in 12′ or 20′ tempered lengths	Yellow

Goodheart-Willcox Publisher

Figure 40-27. A summary of common copper tubing used in fluid and gas systems.

connected using flared fittings, but may also be soldered or brazed. There are specific ferrule fittings that allow connections without sweating or flaring. The outside diameter of copper tubing is 1/8″ greater than its size, and fittings reflect this. The exception is ACR tubing, which is actual size.

TECH TIP
Creep in Joints

Vibration and heat can often cause creep in sweated copper joints. Creep occurs when a sweated joint pulls apart over a period of time. If temperature and vibration is a concern, brazing should be done with different materials.

Pressure ratings for tempered copper tubing range from approximately 400 psi to 1800 psi, depending on the tubing type and outside diameter. Annealed copper tubing has lower pressure ratings that range from 230 psi to 1000 psi, depending on type and diameter. When tubing is exposed to higher temperatures, pressure ratings decrease slightly.

Plastic Tubing

Some of the common materials used are rubber, PVC, Teflon, silicone, polyurethane, nylon, polyethylene, polypropylene, fluoroelastomers, and fluorosilicone. Tubing comes in various colors and levels of clarity, from crystal clear to totally opaque, in nearly all materials. While some tubing materials are very resistant to some chemicals, they may not be resistant to others. Reinforced tubing can be used for applications up to 350 psi. When replacing tubing or specifying tubing, it is important to ensure that the system fluid is compatible with the tubing material.

The hardness of tubing is measured using a durometer and Rockwell R scales. A **durometer** is a measuring instrument for soft materials with several scales. The scales used are called *shores* and range from Shore 0, 00, M, A, D, and Rockwell R from softest to hardest. Since scales overlap, compare similar materials to choose the best fit.

Tubing size is specified by outside diameter and is stated in common imperial sizes (1/8″–2″) and metric sizes. Because plastic tubing has specific temperature limitations (both high and low), both fluid system and environmental temperatures must be considered when specifying material.

The types of fittings used depend on the hardness of the plastic tubing. For instance, brass barbed fittings cannot be used on very hard tubing because the barbs will not penetrate the material in order to maintain grip and pressure rating. Fittings are available in a wide array of materials from brass and stainless steel to PVC, polypropylene, and polyethylene. Fittings may be barbed, quick-disconnect, compression, and push-to-connect types. **Figure 40-28** shows an array of fittings used for plastic tubing.

CHAPTER WRAP-UP

Fluid systems transfer energy in the form of heating or cooling, transfer fluid from one system to another, and transfer power to accomplish work. Understanding the principles that apply to fluid piping systems will allow you to be more effective when installing, troubleshooting, and repairing these systems. It is important to remember that safety is part of the planning and performance of every job or repair and includes use of proper personal protective equipment (PPE), knowledge of the piping system, and safety-minded job preparation. A technician should not start any job, whether it is new construction or a repair, without evaluating the potential hazards involved.

Figure 40-28. Several plastic and metal fittings used for plastic tubing.

Chapter Review

SUMMARY

- Safety is part of the planning and performance of every job or repair. Controlling hazards can be accomplished by identifying the hazard, evaluating the hazard, and placing controls to limit the hazard.

- Lockout/tagout procedures apply to all sources of energy, mechanical as well as electrical.

- Fluid systems transfer energy in the form of heating or cooling, transfer fluid from one system to another, and transfer power to accomplish work.

- Fluid flows from a high-energy state to a lower energy state. There are two main types of fluid flow inside piping: laminar flow and turbulent flow.

- Head loss is the resistance to fluid flow in piping and is affected by the design of the fluid system. Smaller diameter piping, more fittings, or more valves in series results in higher head loss and lower flow rates.

- Pumps can be classified into two broad categories: positive-displacement and nonpositive-displacement. A positive-displacement pump produces a constant flow at a constant speed. The discharge pressure and flow of a nonpositive-displacement pump are determined by head loss.

- Knowing how fluid flows is essential to understanding how a piping system schematic represents the operation of a fluid system. Common types of schematics created for piping systems include single-line diagrams, flow diagrams, piping and instrumentation diagrams, and isometric drawings.

- Every piping system has the same general components: valves to control fluid flow, pumps that act as the prime mover, pressure vessels, connections, and instrumentation to measure system conditions.

- Piping is manufactured to meet various standards that specify designations, use, material, material properties, and strength. Piping is sized in both thicknesses and schedule.

- Nominal pipe size (NPS) is a rounded value that is close to the actual inside diameter of pipe and is commonly used to specify piping in the trades.

- The two most common forms of steel pipe are black iron pipe and galvanized piping. Stainless steel piping is used where corrosion resistance is of utmost importance. Copper and brass offer excellent corrosion resistance and are widely used in potable water and boiler systems.

- The most common type of plastic piping used is thermoplastic. PVC and CPVC are commonly used thermoplastics in the piping industry.

- Tubing is usually a thin-wall material that is thinner than piping of the same diameter. Tubing used for fluid system applications include low carbon steel, stainless steel, aluminum, copper, and plastic. When replacing tubing or specifying tubing, it is important to ensure that the system fluid is compatible with the tubing material.

REVIEW QUESTIONS

Answer the following questions using the information provided in this chapter.

1. What procedures are included in new construction safety?

2. List the types of evaluation that should be performed for any repair job.

3. *True or False?* Repairing existing systems involves the fluids contained within the piping system.

4. A tank that involves _____-pressure gas is always more hazardous.

5. What is the purpose of LOTO in piping systems? When performing this, what is involved?

6. What can be the result of venting a large high-pressure fluid system in preparation for repair?

7. List three purposes of fluid systems.

8. Describe the two types of fluid flow inside piping. How does each relate to velocity?

9. What is head loss? What determines the amount of head loss in a system?

10. Describe the two broad categories of pumps and give an example of each.

11. What will happen if a positive-displacement pump is started with its discharge valve shut and there is no relief valve?

12. What is cavitation? How is it caused?

13. Discuss how system pressure and flow are determined for the two categories of pumps.

14. *True or False?* A solid line on a piping diagram indicates a pipe carrying fluid from one component to the next.

15. Identify the four types of schematics used for piping systems. Describe the purpose and components of each type.

16. List the general components of every piping system.

17. *True or False?* Gate valves, globe valves, and check valves can be used for both isolation and throttling flow purposes.

18. _____ occurs when the disk rapidly closes against the set due to a drop off in flow through the system or if flow attempts to reverse.

19. What is the purpose of a relief valve? Describe two valve operations of this valve.

20. What type of pump is commonly used in industry? What elements does this type of pump include?

21. Why are mechanical seals used with pumps?

22. List common applications of pressure vessels.

23. *True or False?* Using lubricant on threaded fasteners increases the maximum torque that can be applied to a fastener before exceeding its strength.

24. Describe the differences in the types of instrumentation used in older and newer piping systems.

25. How is piping sized?

26. What does NPS mean? What does it specify?

27. What are the most common thermoplastics used in the piping industry? Discuss their purposes.

28. _____ is usually a thin-wall material that is thinner than piping of the same diameter.

29. What is flaring?

30. Identify the considerations that should be made when replacing a tube fitting.

31. *True or False?* When tubing is exposed to higher temperatures, pressure ratings increase slightly.

32. What is a durometer? Why is it important to use when sizing plastic tubing?

NIMS CREDENTIALING PREPARATION QUESTIONS

The following questions will help you prepare for the NIMS Industrial Technology Maintenance Level 1 Maintenance Piping credentialing exam.

1. Which of the following procedures is most hazardous and must be handled carefully?

   A. Draining cool water from a reserve tank.
   B. Draining a water line under normal pressure.
   C. Venting a high-pressure gas from a system.
   D. Venting a low-pressure fluid from a system.

2. Which piping system will have the least amount of head loss?

   A. A system with fewer fittings and larger diameter piping.
   B. A system with fewer fittings and smaller diameter piping.
   C. A system with many fittings and larger diameter piping.
   D. A system with many fittings and smaller diameter piping.

3. The following schematic depicts system components using a _____ diagram.

   A. flow
   B. isometric
   C. P&ID
   D. single-line

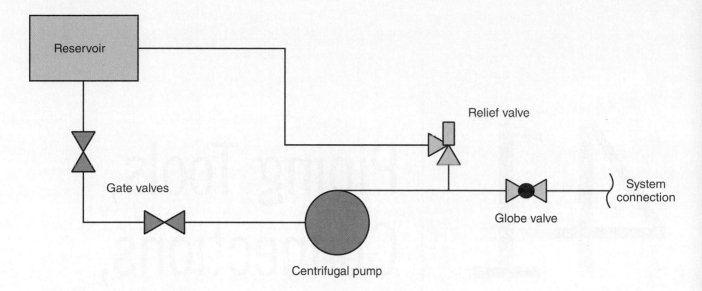

4. Based on the following schematic, which of the following is true?

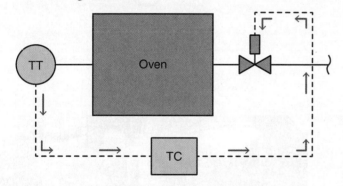

   A. A signal is developed at TC.
   B. A signal is sensed by instrument TT.
   C. Each portion of the control loop is performing the same function.
   D. Signal lines are solid, and piping lines are dashed.

5. When repairing a valve, which of the following is true?

   A. A valve can be opened and repacked before being isolated.
   B. A valve can be repaired without being isolated.
   C. A valve must be isolated both upstream and downstream in order to work on it.
   D. A valve must be isolated downstream to avoid system pressure acting on the outer seal.

6. What connection is needed if metal pipe or high-pressure piping is used?

   A. Compression fittings
   B. Flanged
   C. Threaded
   D. Welded socket

7. What is the difference between piping and tubing?

   A. Piping is a thicker-wall material than tubing of the same diameter.
   B. Piping material can be changed to another type while tubing cannot.
   C. Both A and B.
   D. Neither A nor B.

8. Which of the following is *not* a consideration when sizing piping?

   A. Hardness
   B. Nominal pipe size (NPS)
   C. Outside diameter
   D. Schedule

9. What type of fitting can be used with aluminum piping?

   A. Stainless steel fitting
   B. Steel fitting
   C. Brass fitting
   D. Brass barbed fitting

41 Piping Tools, Connections, and Fittings

LEARNING OBJECTIVES

After completing this chapter, you will be able to:

☐ Describe the most commonly used hand tools and their proper use.

☐ Explain how to cut and thread pipe with power tools and a threading machine.

☐ Discuss common threads used in piping systems and their differences.

☐ Identify the connection methods for piping systems.

☐ Describe specifications for flange face inspection.

☐ Discuss how welding pipe joining is used for fitment and repair purposes.

☐ Explain the similarities and differences between soldering and brazing.

☐ Explain the eight steps for adhesion joining of PVC.

☐ Describe the function of pipe fittings and the standard types used in piping systems.

☐ Discuss the procedure for tubing fitting assembly.

TECHNICAL TERMS

annealing	liquidus
compression	mechanical joining
die	pipe nipple
ferrule	reaming
flange	solidus
flaring	take-up
flux	tube gain
gasket	

In many industrial areas, pipefitting is a full-time union position. However, this is not always the case. Pipefitters normally perform the original construction, while company maintenance technicians perform repairs. To execute repairs, technicians must be familiar with the array of piping and tubing connection methods, tools, and materials. Some methods and tools are used specifically for a particular type of piping material, while others are used across the board. Hand tools are commonly used for assembly, but even these may be specific to the piping material. In this chapter, we will examine tools, connection methods, fittings, and some specific recommendations in their use.

41.1 PIPE WORK SAFETY

In the context of pipefitting, safety deals with personal safety in regard to the use of tools. While the need for some personal protective equipment (PPE) should be obvious—such as proper gloves and steel-toed boots—some may not be so obvious.

Figure 41-1 shows some common job performance categories and possible safety concerns. Consider the needs and concerns of each job before beginning. For example, pipe threading can be done by hand with manual pipe threaders or powered threaders, or it can be done with power threading machines. Each of these choices has its own safety concerns that must be taken seriously and carefully planned for prior to starting the job. Consider that a power threading machine, **Figure 41-2**, operates with a great deal of power and torque to remove material from a steel pipe to create threads. This amount of torque and power can easily cause serious harm to anything that gets in the way. Work must be performed with the utmost concentration and attention to detail.

Courtesy of RIDGID® (RIDGID® is the registered trademark of RIDGID, Inc.)

Figure 41-2. A RIDGID Model 535A automatic threading machine is pictured here.

Pipe Work Safety Concerns		
Job or Action	**Possible Safety Concern**	**Possible Corrective Action**
Pipe threading	Pinching, blunt force, exposure to lubricants	Proper gloves, proper shoes, clean work area with enough space, communication with others nearby, tool knowledge
Pipe weld repair	Arc burns, vision damage, atmospheric contaminants	Proper eye protection, removal of energy, purging the system, proper ventilation
Repair of system piping	Pressurized fluid or gases	Proper lockout, removal of energy, purging the system, proper ventilation

Goodheart-Willcox Publisher

Figure 41-1. Some common categories of piping work and possible corresponding safety concerns are listed in this table.

41.2 PIPE WORK TOOLS

Work with piping requires specific tools and knowledge of how to use them. Hand tools, such as specific wrenches, are necessities. Your workplace may also have power tools that make working with pipes easier. The use of these power tools requires specific knowledge and careful safety precautions.

41.2.1 Hand Tools

The most common hand tool used in the piping trades is the pipe wrench, which is used to grip and turn a pipe or other cylindrical object. Other hand tools used with piping and tubing include cutting, bending, and threading tools.

Wrenches

Pipe wrenches range from 6″ to 60″ in length and can be used with piping up to 8″ in diameter. Standard pipe wrenches are made from either ductile iron or cast iron. Higher-quality pipe wrenches have lower jaws that are replaceable. Larger pipe wrenches can weigh upward of 50 pounds. If performing a great amount of work on larger pipes, consider using lightweight wrenches. Pipe wrenches made of aluminum have exceptional strength, are offered in lengths of 10″ to 48″, commonly weigh approximately half as much as a standard pipe wrench, and use the same jaws that a standard pipe wrench uses. **Figure 41-3** shows a standard and a lightweight pipe wrench. Pipe wrenches also come in a range of offsets, from a slight offset to a nearly 90° offset. Offset pipe wrenches are typically used in cramped locations where a standard pipe wrench cannot be positioned correctly.

Chain wrenches and strap wrenches, **Figure 41-4**, are offered in smaller sizes than traditional pipe wrenches. Both chain and strap wrenches can be used at odd angles and in restricted locations, where a standard pipe wrench might not fit. The strap wrench is also used to grip a pipe or fitting without marring the surface of the piping. Chain wrenches allow for the rotation of a pipe without having to remove the wrench from the pipe to reposition it, which is ideal for trench work with threaded piping.

Standard pipe wrench

Lightweight aluminum pipe wrench

Courtesy of RIDGID® (RIDGID® is the registered trademark of RIDGID, Inc.)

Figure 41-3. A standard pipe wrench and aluminum pipe wrench look very similar but differ in weight.

Chain wrench

Strap wrench

Courtesy of RIDGID® (RIDGID® is the registered trademark of RIDGID, Inc.)

Figure 41-4. A chain wrench and a strap wrench are pictured here.

⚠ CAUTION

While a pipe wrench will work when used with PVC pipe, heavy use will damage the pipe when the wrench's jaws grab into the surface. Strap wrenches will not damage the surface finish of PVC pipe and will still allow proper torque to be applied.

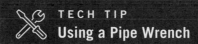

TECH TIP
Using a Pipe Wrench

When positioning a pipe wrench, leave a gap between the shank and the pipe itself. The shank includes the shank jaw (top jaw) and is the adjustable part of the wrench. Without a gap between the shank and pipe, the shank jaw will not be flexible enough to move and develop the gripping force needed, and it could slip.

⚠ CAUTION

Do not use extensions (cheater bars) on a pipe wrench to increase torque. Excessive torque can damage the tool. If greater force is needed, use a larger pipe wrench.

Smooth-jaw pipe wrenches, sometimes known as Ford wrenches or hex wrenches, are often used for non-marring applications, such as with chrome and stainless-steel unions, square stock, and square nuts.

Slip-joint pliers come in two general categories: straight jaw and curved jaw (see **Figure 41-5**). Straight-jaw adjustable pliers have limited use for piping work, with the exception of union assembly and disassembly. Curved-jaw adjustable pliers can be used on smaller piping (1/8″–3/4″ diameter) and fittings, but they typically

Straight-jaw adjustable pliers

Curved-jaw adjustable pliers

GearWrench (Apex Tool Group)

Figure 41-5. Straight-jaw adjustable pliers typically have more use in electrical trades or residential plumbing work. Curved-jaw adjustable pliers can develop more torque with less effort.

have an aggressive jaw that will leave marks. Straight-jaw adjustable pliers develop their gripping power and torque from the grip force applied by the operator, while adjustable pliers with curved or angled jaws require less operator gripping power to develop torque.

Standard open-end wrenches and adjustable wrenches have their place in piping work. However, if performing instrumentation and tubing work, flare-nut wrenches are also a valuable addition. **Figure 41-6** shows a flare-nut wrench used for hexagonal fittings typically found with instrumentation.

GearWrench (Apex Tool Group)

Figure 41-6. Flare-nut wrenches such as this provide for better-distributed torque on hexagonal fittings.

Cutting Tools

Cutting tools for pipe and tubing come in a wide array of sizes and types and are chosen depending on the piping material, **Figure 41-7**. A typical pipe cutter works well with steel piping but can deform plastic (PVC) piping. Smaller tubing is easily cut using cutters specifically designed for the particular material. A mini tubing cutter is normally used on softer materials, such as brass, copper, and aluminum. A harder cutting wheel and additional bearings and rollers are needed for use on stainless-steel tubing.

Once a cut is made, the pipe must be reamed where it was cut. *Reaming* is the act of smoothing a pipe where it was cut, in order to remove any remaining burrs. This is done using a device called a reamer.

Using a Pipe Cutter

All types of pipe cutter work in a similar manner:

1. Place the cutter on the material and slightly tighten it, until the roller wheel (or wheels) and the cutting wheel contact the surface.
2. Rotate the tool around the material.
3. During the rotation, the knob of the cutter gradually tightens until the cut is made.
4. After one complete rotation, tighten the cutter's adjustment screw and make another rotation, if this is necessary to cut through thicker material. Continue to tighten and rotate until the pipe is cut.
5. Loosen and disengage the pipe cutter. Some cutters have a quick release, which makes repeated cuts more efficient.

Cutters intended for plastic piping are normally used on smaller-diameter pipes (less than 2″ NPS) and work using a ratcheting method to develop cutting force, which cleanly cuts through the pipe in one stroke. Larger pipe sizes are normally cut using a variety of saws. Taller saw blades work best and allow a cleaner, square cut.

Soil pipe, also called cast-iron pipe, is best cut using either an abrasive cutoff saw or a soil pipe cutter. **Figure 41-8** shows a compact soil pipe cutter. The chain of the cutter is wrapped around the pipe and tightened. The surface of the

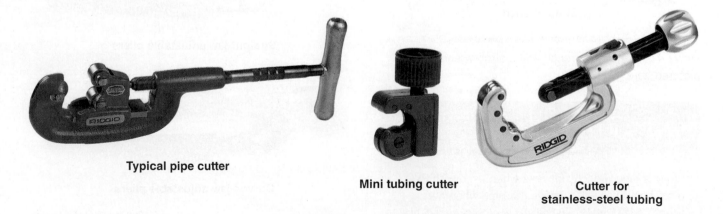

Typical pipe cutter

Mini tubing cutter

Cutter for stainless-steel tubing

Courtesy of RIDGID® (RIDGID® is the registered trademark of RIDGID, Inc.) (left and right)/GearWrench (Apex Tool Group)(center)

Figure 41-7. A typical heavy-duty pipe cutter, mini tubing cutter, and cutter for stainless-steel tubing are pictured here.

Courtesy of RIDGID® (RIDGID® is the registered trademark of RIDGID, Inc.)

Figure 41-8. This soil pipe cutter is used for in-place piping.

pipe is then scored by the cutting heads of the chain, and the chain is tightened until the pipe snaps at the score line. Some of these types of cutters do not require scoring of the pipe, but they do require careful adjustment of the chain tension prior to snapping.

TECH TIP
Making a Square Cut

Cutting a larger-diameter pipe squarely takes some practice. A shorter blade, such as a reciprocating saw blade or hacksaw blade, can cause the cut to angle. A taller, longer blade allows for visual inspection while the cut is taking place, as well as better control over the angle. Square cuts in PVC piping are important in order to achieve the maximum pressure rating with proper socket insertion.

Bending Tools

Tube benders are designed based on the material to be bent and can be used with copper, stainless steel, and aluminum. **Figure 41-9** shows a spring tube bender and a bending tool meant for use with pipes of several

different diameters made from soft material. While spring benders can be used to bend tubing, bends are less than ideal and can result in kinked lines. Spring tube benders can accommodate tubes with diameters up to 7/8″ in soft material. Spring benders are acceptable only for occasional work.

Two things must be considered in order to accurately bend tubing: take-up and tube gain. *Take-up* is the amount the bending tool must be moved back from the measured bend point to allow for the material taken up by the bend and still reach the desired 90° meeting point. The take-up amount is approximately the same as the radius of the bend being made. Back-to-back bends do not require take-up. *Tube gain*, or *tube stretch*, is the small amount of apparent lengthening of a tube when it is bent on a curve compared to a sharp 90° angle. The amount of gain is approximately 30% more than the diameter of the tubing (up to 1″ diameter) for each 90° bend. Bends of 45° have very little effect on apparent length.

TECH TIP
Bending Practice

Before starting a larger job, take the time to plan for the materials, space, layout, and fittings needed. Include extra material so that one or two practice bends can be made and checked, so that the finished project looks professional. Accurate tube bending takes practice and attention to detail.

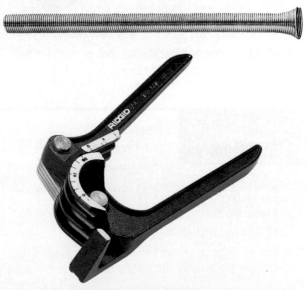

Courtesy of RIDGID® (RIDGID® is the registered trademark of RIDGID, Inc.)

Figure 41-9. A hand spring tube bender and multidiameter tube bending tool are pictured here.

Joining Tools

Tubing typically is joined by one of two methods: flaring or compression. Different fittings are used for each method, and these are discussed later in this chapter. *Flaring* is the widening of the opening at the end of a pipe in order to make a secure connection while using a fitting. The tools used for flaring fall into two categories: hammer driven and mechanical. Hammer-driven tools are acceptable for occasional work and are typically used on soft copper or thin-walled harder materials. **Figure 41-10** shows a mechanical flaring tool that produces excellent repeatable flares with up to 3/4″ tubing. Flares come in different angles depending on their use. Flares of 45° are normally used for potable water systems, while 37° flares are used for other fluids and gases.

Courtesy of RIDGID® (RIDGID® is the registered trademark of RIDGID, Inc.)

Figure 41-10. A mechanical flaring tool is used for soft tubing and thin-walled harder tubing.

PROCEDURE	Performing a Flare

To perform a flare, use the following procedure:

1. Place the fitting on the tubing prior to flaring, otherwise it will not fit over the flare.
2. Split the base of the tool in half, position it around the tubing, and reassemble it.
3. Lock the tool into the base and press it into the tubing, expanding it into a flare.

Compression is the joining of pipes using compression fittings, which make a watertight seal with the tightening of a threaded compression nut. Once a pipe is properly cut and prepared, compression joining requires few tools. After tightening the compression nut by hand, use a wrench to further tighten it and ensure a watertight seal. Compression joining is often used in plumbing applications.

Threading Tools

Threading piping is accomplished through two methods: hand threading or powered threading. Hand threading is best used for smaller-diameter piping (less than 1″ NPS) and as repair work. Hand threading requires a large amount of force, and therefore strong pipe vises should be used. **Figure 41-11** shows the tools used for hand threading a pipe. Different pipe materials require different dies. The *die* is the device that cuts the threads into the pipe. Dies are available for threading steel, PVC, and stainless-steel pipe. Lubricant can be used to keep the dies and pipe cool and smooth the process. Hand threaders are also used in conjunction with power heads to minimize die adjustments and improve productivity when threading multiple sizes of pipe. The table in **Figure 41-12** shows that common sizes of threads, in threads per inch, cover several sizes of pipe.

Courtesy of RIDGID® (RIDGID® is the registered trademark of RIDGID, Inc.)

Figure 41-11. This pipe hand threading set includes the ratchet head and all necessary dies.

Threading Pipe	
Pipe Size	Threads per Inch
1/4" and 3/8"	18
1/2" and 3/4"	14
1"–2"	11 1/2

Goodheart-Willcox Publisher

Figure 41-12. Dies for pipe threading are used for multiple pipe sizes. While threads-per-inch stays constant, the diameter of the pipe changes.

PROCEDURE **Threading Pipe**

Use the following general procedure for pipe threading:

1. Check measurements.
2. Cut the pipe to required lengths.
3. Ream the pipe on the inside and outside to remove all burrs caused by cutting.
4. Thread the pipe, ensuring the proper cut depth.
5. Ream the inside of the pipe if burrs are present, and check the outside threads.
6. Clean the threads and inside of the pipe, removing any excess lubricant used for threading.

⚠ **CAUTION**

Using the incorrect die for a particular pipe material can result in damage to the die and poorly made threads on that pipe. Special attention must be given to the length of the threaded portion of pipe. If the threaded portion of a pipe section is overthreaded, the pipe could bottom out in a fitting thread and not properly seal. If underthreaded, there may not be enough thread engagement with a fitting, and the joint may be mechanically weak. When in doubt, use a standard pipe nipple for comparison. *Pipe nipples* are short pieces of pipe threaded on both ends and used to join two fittings that are close together.

41.2.2 Power Tools

Power cutting and threading equipment is more efficient than manual tools when tackling a large project. Both handheld and stand-alone power equipment are available. **Figure 41-13** shows a powered threading machine that uses the same dies as a manual threader. While this tool can be used for production work, it is mainly used for field work because of its adaptability, relatively light weight, and changeable dies. Because of the tool's power, a strong piping vise is still needed to hold the pipe securely while threading.

A full-size threading machine is typically used for production and larger work. **Figure 41-14** shows a typical example with an integral pipe reamer, cutter, threading

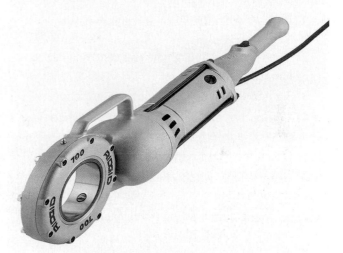

Courtesy of RIDGID® (RIDGID® is the registered trademark of RIDGID, Inc.)

Figure 41-13. This is a handheld powered threading machine for use with 1/8" to 2" pipe.

Courtesy of RIDGID® (RIDGID® is the registered trademark of RIDGID, Inc.)

Figure 41-14. A staple of any piping shop, this threading machine can thread up to 4" NPS piping.

head, vise, and power head. This threading machine has an adjustable automatic feature that opens the die head at the end of the cut. Review the machine operation manual before using it, and closely inspect the die head. As with a manual tool, the proper die is needed prior to threading, and the die head must be set appropriately.

PROCEDURE

Pipe Cutting and Threading with a Threading Machine

Use the following procedure when cutting and threading pipe with a power tool:

1. Measure the pipe for length and mark the measurements.
2. Insert the pipe through the rear centering device and the jaw chuck to the proper position.
3. Tighten the centering device and jaw chuck.
4. Ensure the reversing switch is in the proper position.
5. Check the rpm for the proper application.
6. Use the foot switch while cutting the pipe to the desired length.
7. Move the cutter, swing the reamer down into position, and ream the pipe.
8. Move the reamer and swing the die head down into position.
9. Thread the pipe.
10. Remove the pipe, check the threads, and clean off any excess lubricant.

TECH TIP
Threading Machine Know-How

Prior to starting a piping job, always check that the machine is in good repair, enough cutting fluid is available, and used fluid readily drains back to the supply tank. With extensive use, die shavings can easily clog fluid drains. Without proper lubrication, dies can quickly become worn or damaged. Only an experienced operator should adjust the die head and replace dies. When working with longer lengths of pipe, use an additional piping brace or stand to offset the weight on the chuck and keep the pipe level.

Pipe roll grooving is a connection method that can be used with PVC, steel, copper, and other piping materials. It is especially used in fire sprinkler systems. Grooving essentially compresses a small section of the pipe until a groove is formed. Smaller grooving machines use manual adjustments to develop the needed pressure, while larger units use hydraulic pressure. **Figure 41-15** shows a common grooving machine, which is attached to a power head. The power head provides the power to rotate the pipe, while hydraulic pressure compresses the pipe to form the groove. The pipe to be grooved is inserted into the attachment, held between rollers, and then rolled while pressure is applied through the grooving tool until a stop limit is reached.

Typical grooving machines can groove pipe sizes of schedule 10, schedule 40, and up to large-diameter piping. Careful setup and adjustment are required to properly groove piping. Fully review the manual for the particular machine prior to operating it. Check finished grooves to ensure proper depth. The stop limit is adjustable and should be checked for proper depth with each change in piping material or schedule.

Crimping is an increasingly popular method for connecting piping systems that use copper, stainless steel, and PEX tubing, which is a type of plastic tubing made from polyethylene. The crimping method uses specific fittings and is best performed with a portable powered crimper, **Figure 41-16**. The powered crimping tool applies clamp force until the fitting is mechanically reduced over the tubing to prevent leakage. For smaller-diameter tubing, the tool is placed directly on

Courtesy of RIDGID® (RIDGID® is the registered trademark of RIDGID, Inc.)

Figure 41-15. With pipe roll grooving, the power head provides the power to rotate the pipe, while hydraulic pressure compresses the pipe to form the groove.

Courtesy of RIDGID® (RIDGID® is the registered trademark of RIDGID, Inc.)

Figure 41-16. This battery-powered crimping tool can be used with tubing up to 4″ in diameter, depending on the tubing material.

the fitting. For larger-diameter tubing, an intermediary press ring is placed around the fitting, and the crimping tool then applies force to this press ring. The crimping method requires no heat and ensures a consistent joint when performed properly.

41.3 CONNECTION METHODS

The connection method used for piping depends on the type and use of the piping. The system temperature, pressure, and fluid type are all considered to choose the best and safest possible connection method. Broad categories of connection methods include mechanical joining, welding, and adhesion.

41.3.1 Mechanical Joining

Mechanical joining is connecting or assembling components without heat or adhesives. Materials may be joined using force, fasteners, or means that lock components together. This section covers the following mechanical joining methods:

- Threading.
- Flanges.
- Flaring and swaging.

	Thread Length and Engagement	
Pipe Size	Length of Thread (in inches)	Typical Thread Engagement (in inches)
1/4	5/8	5/16
1/2	13/16	1/2
3/4	13/16	9/16
1	1	11/16
1 1/2	1	11/16
2	1 1/16	3/4

Goodheart-Willcox Publisher

Figure 41-17. This table compares piping size and thread length.

Threading

One of the most common joining methods for smaller-diameter piping and fittings is threading. Pipe threads are standardized in length and taper so they properly fit with national pipe thread (NPT) fittings. **Figure 41-17** lists some common pipe sizes and the typical length of threads for each. While NPT tapered pipe thread is the most common type used, there are others used depending on the type of system. These include the following:

- **National pipe taper fuel (NPTF) dryseal.** Used in hydraulics, fuel, and some pneumatic systems. NPTF has an interference fit that causes thread crush and a mechanical seal.
- **ISO 7-1 tapered pipe threads.** Developed from British standards. Found on fluid systems outside the United States or on imported machinery and systems.
- **ISO 228-1 parallel pipe threads.** Developed from British standards. Similar to ISO 7/1, but without taper.

Even though they are tapered, NPT threads do not produce a mechanical seal and therefore require some type of sealing compound. Sealing compounds are available in hardening and nonhardening forms, and have specific properties and uses including compatible liquids or gases, temperature range, sealing component, and color. Pipe thread tape is also used for sealing and comes in a wide variety of widths, thicknesses, and colors. The following are some general rules that apply to the use of any sealant:

- Do not overuse tape or sealing compound, as either can contaminate a fluid or gas system.
- Ensure compatibility with the fluid or gas in the system.
- Do not contaminate the sealing compound by adding thinners. Thinners can contaminate the system or degrade the piping.

- Apply the tape or sealing compound only when ready to assemble in order to prevent contaminants from adhering to the threads.
- If using pipe thread tape, wrap clockwise so that making the joint does not unwrap the tape.
- With pipe thread tape, use one or two wraps around the thread sealing area. Do not wrap tape at the beginning of the threads.

Flanges

A *flange* is a rim or collar used to make connections in piping systems. Flanges are specified by ANSI and follow guidelines depending on pressure rating (ANSI B16.5). Pressure class is divided into 150, 300, 450, 600, 900, 1500, and 2500. These pressure classes are not strict pressure ratings, as rating also depends on temperature. Flanges are marked with a manufacturer code, pressure class, pipe size, and material.

Flanges are categorized by type, attachment, seal, and face finish. Flange types include blind flanges, reducing flanges, regular, and spectacle. The attachment of a flange to piping can be accomplished through welding, a screwed attachment, or slip joint (also known as lap-joint). Sealing type refers to the pressure barrier type used and may include paper gaskets, O-rings (tongue and groove), gaskets, and ring-gaskets. Face finish refers to the geometry of the flange face, or sealing area, and can be tongue and groove (ring), flat, lap-joint, or raised face. The most common types of flanges are regular, welded (or screwed), gasket, and raised-face flanges. See **Figure 41-18**.

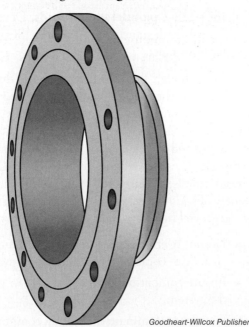
Goodheart-Willcox Publisher

Figure 41-18. A welded-neck, raised-face flange is one of the most common types.

A *gasket* is a piece made of rubber or a similar material that is used to create a tight seal between two parts. The gaskets used with flanges are classified by type and material. Gasket types include full-face, ring, O-ring, metallic ring, and ring-backed. Materials range from a simple paper to spiral wound metallic materials, depending on application, fluid, and temperature.

TECH TIP
Gasket Types

Know the piping systems and common gasket types used in your facility. Gaskets should be preordered and kept on hand. When a fluid system starts to leak, a lack of planning will result in a critical process being stopped for several days while waiting for a gasket delivery.

A common method of making a gasket uses a soft hammer, such as a dead-blow mallet, to gently hammer the outline of the flange with the gasket material placed over the flange. The inside rim can be accomplished in the same fashion. A smaller hammer or gasket punch, **Figure 41-19**, is used for the bolt holes. This method can easily be used on paper gaskets for nearly any application, including pump casings, flanges, and valve gaskets.

PROCEDURE   Assembling a Flange

When preparing to assemble a flange:

1. Ensure both mating surfaces are clean and free of extensive wear or scratches.
2. Use a soft scraper or wire brush to clean the mating surfaces.
3. Consider using some type of anti-seize compound on the gasket that is compatible with the fluid system to help future disassembly.
4. The mating surfaces should be aligned without the use of excessive force.
5. Use lubricant or anti-seize compound on the threaded fasteners.
6. Assemble and tighten bolts in a crisscross pattern to the proper torque.

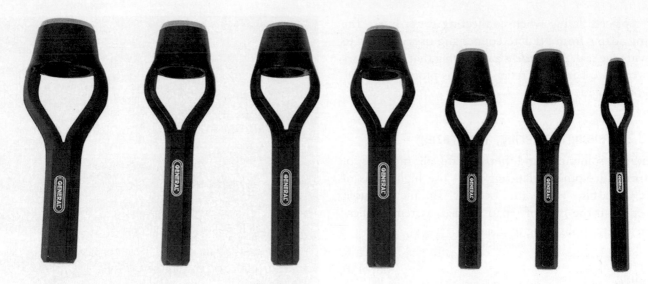

Figure 41-19. Gasket punches should be kept clean with the cutting edge oiled to prevent rust during storage.

Be aware of the condition of the mating surfaces. Damaged surfaces, such as low spots, degraded face serrations, or deep scratches that cross the mating surface, may prevent the flange from sealing properly and should be repaired prior to assembly.

When disassembling flanges, consider that the system may not be fully drained and pressure may still be present inside the piping system. Loosen fasteners, break the seal to relieve pressure or let fluid drain, and then remove fasteners from the flange.

SAFETY NOTE
Piping Support

Keep in mind that the piping may not be properly supported and may move when fasteners are loosened or removed. Supporting the piping prior to starting work makes disassembly easier and safer.

Fasteners used for flanges vary depending on system pressures, mechanical strength, temperatures, environment, corrosion, and other requirements. The fasteners are typically standard nut and bolt, stud bolts and nuts, or T-bolts and nuts. Each of these types is available in strengths specified by ASTM and are categorized as low strength, intermediate strength, high strength, and special alloy. Lubricant allows a fastener to turn with less friction. If using lubricants on threaded fasteners, the maximum torque applied should be reduced. Larger flanges with a greater number of fasteners need to be tightened in several steps up to the final torque.

TECH TIP
Replacing Flange Bolting

If replacing flange bolting, be certain that the replacement meets or exceeds the original design requirement. For instance, you would never replace a high-strength fastener with a low-strength fastener.

Flaring

Flaring is another method of mechanically joining tubing. Flaring is typically used in higher-pressure hydraulic lines and automotive applications. For the most part, the use of specific compression fittings, **Figure 41-20,**

Richard z/Shutterstock.com

Figure 41-20. Pictured here is a compression fitting with a set screw used with copper tubing.

has replaced flaring when connecting water lines. The fitting adapts from NPT to copper and uses a ferrule to provide the seal. A *ferrule* is a small ring that compresses between two harder materials, such as piping and a fitting, to provide a seal.

41.3.2 Welding, Soldering, and Brazing

Welding, soldering, and brazing are all methods of connecting piping and fittings that use high levels of heat and a filler compound or alloy. The method used depends on the type of repair needed, system application, and type of piping in place.

Welding

Pipe welding for fitment is performed by qualified and experienced tradesmen. Extensive practice in both basic welding techniques and pipefitting is required to achieve acceptable results. Pipe welding covers a wide range of materials and applications in nearly all types of systems and piping. This section focuses on pipe welding for repair work. While all welding methods (SMAW, GMAW, and GTAW) can be used for pipe fitting, SMAW is the most common for in-the-field applications.

When constructing a piping system in the field, the piping materials are typically prepared in the shop beforehand. Piping and fitting ends are often beveled and require proper alignment, spacing, and positioning before welding. Since the piping is welded into a specific position, all-position welding is required after several tack welds. This includes flat, vertical, and overhead welding positions. Some welds may be performed flat, if the sections can be premade. However, most of the welding is performed in the final position of the piping and fittings.

The process of welding follows a procedure, which can include preparation, heating, spacing, specific welding electrodes, multiple passes, and finish. **Figure 41-21** shows a typical SMAW multipass pipe weld for a butt-weld application. Pipe welding is usually in the form of socket weld fittings or butt-weld fittings. Socket weld fittings allow the insertion of the pipe into a fitting and require a fillet joint to be welded around the circumference of the pipe. Butt-weld fittings are made end-to-end and must be properly spaced and aligned when performing the weld.

Preparation is one of the most important aspects of ensuring a good joint or weld. Initial cleaning and proper preparation are particularly important when performing GTAW on high-pressure tubing. Cleaning the weld between passes to ensure that inclusions are not present in the final weld is also important in ensuring a good joint or weld.

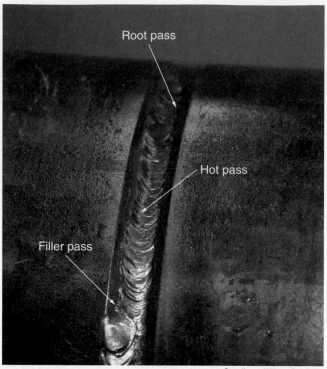

Goodheart-Willcox Publisher

Figure 41-21. A multipass weld on schedule 80 steel pipe is pictured, showing the root, hot, and filler passes.

Repairing piping by welding can be challenging, particularly when the piping system is several years old with questionable integrity. Not all piping failures can be repaired by welding, but using welding as a repair method may be appropriate for the following conditions:

- Corrosion to the point of leakage or failure.
- Mechanical impact resulting in deformation and leakage.
- Failure due to exposure to the environment, such as freezing and subsequent ruptures.
- Redesign to include other fittings or allowances.

Many of the preparation steps for repair welding are the same as construction welding. The piping system needs to be isolated with appropriate lockout/tagout. The pressure must be relieved and the system should be drained and flushed of flammable contents or vapors. The weld area needs to be cleaned by a mechanical method (grinding) to remove any foreign material or rust that may interfere with the weld. General maintenance welding is covered more thoroughly in the *Maintenance Welding* section.

If the welded repair will not meet original standards for the piping, it may be better to replace the piping material or use repair clamps, **Figure 41-22**. A failure in a cast-iron pipe, for instance, would be better addressed by replacement or a clamp.

Figure 41-22. Piping repair clamps are available in a variety of materials and sizes.

Soldering

Soldering is a common method of joining copper tubing that involves filling the gap between the fitting and pipe with a melted metallic compound to create a leak-free joint. Several compounds are used, each melt at different temperatures and have specific benefits. Soldering is performed at a temperature below 840°F, while brazing requires temperatures above 840°F. *Liquidus* is the temperature at which a metal melts and becomes liquid. *Solidus* is the temperature at which a metal *begins* to melt. In soldering, the metals joined together stay below the solidus, as only the solder melts. Solder is meant only to fill small clearances between parts and does not combine with the joined metals, except at the surface.

Several gases are used for soldering depending on the specific use, **Figure 41-23**. Propane is commonly used, but it is limited to smaller copper joints due to its temperature. MAPP gas is used for larger-diameter copper tubing and solders with higher melting temperatures. Joints that will be buried or subject to vibration should be brazed instead of soldered. Soldered joints exposed to heat and vibration can creep, or separate. Creep occurs with lower-strength soldered joints over time when exposed to high vibration and temperature.

Solders containing lead are still used in industrial applications, but they are not allowed in potable water systems. Solders for industrial applications come in a mixture of tin and lead. Each mixture melts at a different temperature—a higher percentage of lead yields a higher melting temperature. Common solder ratios of tin to lead are 63-37 (361°F melting point), 60-40 (374°F melting point), and 50-50 (419°F melting point). Potable water systems and applications that require stronger

Soldering and Brazing Gases		
Gas	**Use**	**Temperature**
Propane	Low-temperature soldering	2700°F (hottest point)
MAPP (propyne and propadiene)	Higher-temperature soldering	3670°F
MAPP and oxygen	Highest-temperature soldering and small welding	5300°F
Acetylene and oxygen	Brazing	5720°F

Figure 41-23. This table lists various gases used for soldering and brazing with related flame temperatures.

joints require lead-free solder. Using silver-cadmium, zinc-aluminum, and cadmium-zinc solders require a hotter flame and MAPP gas is normally used. **Figure 41-24** shows many common solders used with corresponding solidus and liquidus temperatures.

Flux cleans the joint, prevents oxidation of the base metal when at high temperature, and helps the solder flow through the joint with better capillary action. It comes in liquid and paste forms for piping applications. When selecting a flux for soldering, consider that you may be soldering not only tubing, but fittings and valves as well. The flux used should be compatible with copper and brass. When soldering tubing-valve joints, more heat is needed due to the mass of the brass or bronze valve.

Typically, the best cleaning fluxes are also the most corrosive, which requires the surface be cleaned after soldering is complete. Inorganic fluxes are corrosive and are not to be used on electrical or electronic materials due to the difficulty in cleaning the residual after soldering. Rosin-based fluxes are noncorrosive and made of a solid form of resin. This type of flux is best for electronic and electrical soldering at a low temperature. Organic fluxes have little or no corrosive effects and are often water-soluble. This type of flux can be used for copper tubing applications, but it may not perform well with higher heat.

SAFETY NOTE
Soldering PPE

When soldering, pay attention to the proper PPE required depending on the flux used. Some flux contains hydrochloric acid and should be handled carefully.

Solder Alloy Compositions and Melting Temperatures

Alloy	Composition (% by Weight)								Solidus Temperature		Liquidus Temperature	
	Tin	Lead	Silver	Antimony	Cadmium	Zinc	Aluminum	Indium	°F	°C	°F	°C
Tin-Antimony	95			5					450	232	464	240
Lead-Tin-Silver	96		4						430	221	430	221
	62	36	2						354	180	372	190
	5	94.5	0.5						561	294	574	301
	2.5	97	0.5						577	303	590	310
	1.0	97.5	1.5						588	309	588	309
Tin-Zinc	91					9			390	199	390	199
	80					20			390	199	518	269
	70					30			390	199	592	311
	60					40			390	199	645	340
	30					70			390	199	708	375
Silver-Cadmium			5		95				640	338	740	393
Cadmium-Zinc					82.5	17.5			509	265	509	265
					40	60			509	265	635	335
					10	90			509	265	750	399
Zinc-Aluminum						95	5		720	382	720	382
Tin-Lead-Indium	50							50	243	117	257	125
	37.5	37.5						25	230	138	230	138
		50						50	356	180	408	209

Goodheart-Willcox Publisher

Figure 41-24. Soldering alloys melt at different temperatures depending on their composition. Notice that as the percentage of silver increases, so does the solidus temperature of the solder.

PROCEDURE   General Soldering Technique

The following are general steps for soldering copper tubing:

1. Cut tubing to the desired length. If a tubing cutter is used, it will leave a small internal lip on the inside diameter of the tubing. Ream the inside diameter to remove the lip.

2. Use crocus cloth or emery paper to remove any outside burrs, coatings, or oxidation from the tubing and internal mating surface of the fitting.

3. Apply flux to both the tubing and fitting.

4. Assemble the joint.

5. Light the torch and adjust, if needed.

6. Heat the joint while applying the solder near the joint.

7. When the joint is hot enough, the solder will melt quickly and wick into the joint. This action is caused by the flux.

8. Be certain enough solder is applied and the joint is not overheated.

9. Remove the heat and let the joint cool.

10. Clean the outside of the joint and piping to prevent oxidation, depending on solder and flux used.

Eliminate any remaining or leaking fluid from the pipe or tubing before soldering. Design the joint so that if a leak is present, fluid flows away from the joint. Another method is to use plumbers putty to temporarily stop the leak by blocking it. It may be possible to change the tubing configuration by adding a union or coupling so that only one joint needs to be soldered with leakage present.

Brazing

Brazing requires temperatures above 840°F, but the base metals are not melted. The temperature must be high enough to melt the filler alloy (above the liquidus), but below the solidus of the base metal. The bond that is created between the base metals is made through the filler material, which makes a strong surface bond with the base material. Brazing is preferred in higher-temperature applications, buried piping, piping that will be subjected to vibration, or where a mechanically stronger joint is needed.

As with soldering, the base material must be cleaned by mechanical or chemical cleaning, or both. With brazing, a brass alloy or silver alloy rod is used instead of solder. Rods are commercially available in diameters of 1/16″ to 1/4″, and are supplied in both bare and flux-coated varieties. If a bare rod is used, another type of flux must be applied. The fluxes used for brazing melt at a higher temperature than those used for soldering. For brazing at higher temperatures, a separate flux is needed.

SAFETY NOTE
Brazing Safety

The standard safety precautions used with oxyfuel welding equipment must be followed for brazing. While the base material will not be melted, the materials will be very hot. Take proper precautions for hot work and PPE.

The technique used for brazing is essentially the same as for soldering. With brazing, watch that the joint is not overheated. Use only enough heat to allow the brazing rod to flow into the joint. The size of the tip used should be larger than the tip used for oxyfuel welding. By selecting a larger tip, heat is more evenly spread across a large area. Once completed, allow the joint to cool on its own—do not use water or soaked rags. Since brazing heats tubing to a higher temperature than soldering, some annealing will occur. *Annealing* is the process of heating a metal and allowing it to cool slowly in order to reduce the metal's hardness. This may reduce the maximum working pressure of the joint, but it also makes the joint less brittle.

SAFETY NOTE
Heat from Brazing and Soldering

Remember that both brazing and soldering use heat. Piping, tubing, and fittings will stay hot for a long time after the operation is completed. Use the appropriate gloves to prevent injury.

41.3.3 Adhesion Joining

Adhesion joining is typically used for plastic piping. However, PVC is also joined using flanges, threading, and slip-fit grooved/O-ring fitment. Threaded PVC is used sparingly and normally on smaller-diameter piping and fittings. For most adhesion joints, both primer and cement is used. Primer softens the PVC piping surface to ensure a good bond with the cement. Preparation is important to achieve maximum working pressures.

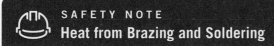

PROCEDURE   Adhesion Joining Plastic Pipe

1. Cut pipe ends square. If not square, the pipe will not seat properly in the socket.

2. Ream ends, both inside and on the outside diameter. A slight bevel is acceptable to help insertion into the joint. No burrs or foreign material should be left. Internal or external ridges can prevent proper insertion into the socket or fluid flow.

3. Clean both the inside and outside of the piping and fitting in the area of the joint. No scraps from cutting should be left.

4. With a clean applicator, apply primer to both the piping and fitting where the joint is to be made. Some primers require two coats. Let the primer fully dry before proceeding.

5. Apply cement to both the fitting and pipe.

6. Insert the piping into the joint until it is bottomed out in the joint. Turn the piping slightly in the joint to ensure consistent joint coverage.

7. Apply pressure to the joint to ensure that the piping does not back out. If a joint backs out, it loses some mechanical strength.

8. Allow the joint to properly cure, or set, before any system pressure is introduced. Curing time can vary widely depending on environmental conditions (such as humidity and temperature) and piping diameter. In colder temperatures, higher humidity, and with larger pipe diameters, longer curing times are required. Consult the manufacturer of the cement to determine proper curing times.

Flanged connections are also used with PVC piping. Flanges are typically glued to piping and must be allowed to fully cure before use.

O-ring and slip-fit joints are commonly used with larger PVC lines, **Figure 41-25**, found in water and sewer systems and are not prevalent in industrial fluid applications. The outside ends of this piping are normally beveled to allow easier insertion into the joint. If the piping is cut to length, ensure the cut ends are square and beveled. Neither primer nor cement is used in making these joints, but a lubricant is applied to both the O-ring and pipe end prior to fitment. Joints are lined up and pressed into place until the pipe bottoms out in the joint.

41.4 FITTINGS

Fittings perform a variety of functions in a piping system, such as:

- Connecting piping to valves, equipment, and other systems.
- Providing for multiple flow paths.
- Providing for increases or reductions in diameter.
- Changing the layout of the piping system, including direction, flow, angle, and elevation.

Fittings are specified using the type of connection (such as threaded, welded, or slip), the type of fitting (such as tee, wye, or cross), the type of material (such as PVC, stainless steel, or cast iron), the piping schedule (such as 10, 40, or 80), and the pressure rating of the fitting.

41.4.1 Standard Pipe Fittings

Tees, or T-shaped fittings, are a common pipe fitting for all pipe materials, schedules, and types of connection. Tees come as either straight fitting, with no reduction, or as reducing tees. A straight-fitting tee is named for the NPS size—a 2″ tee, for example, which means all of its connections are 2″. Reducing tees, **Figure 41-26**, follow a naming pattern in which the 90° takeoff is the last named in the sequence of three. For example, a 2 × 2 × 1 tee has a straight run of 2″ NPS, with a takeoff (the 90° connection) of 1″ NPS.

Several subcategories of tees accomplish specific purposes related to the type of system. Cleanout tees are used in sanitary or drain systems and have the takeoff connection at a sweeping angle to allow for the insertion of rodding or cleaning equipment. A cross tee has four openings, which may be of the same diameter or include a reduction in size. Cross tees are specified by a series of four numbers representing the NPS size in inches of each connection,

Figure 41-25. This larger-diameter O-ring and slip-joint piping is stacked and ready for a municipal construction project.

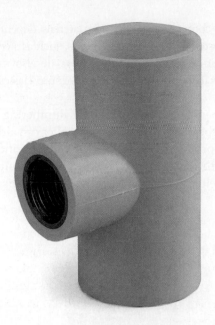

Figure 41-26. This reducing tee fitting is for PVC piping.

Figure 41-27. This is a standard elbow fitting with a short radius.

such as $4 \times 4 \times 4 \times 2$. A test pipe tee, also called a service tee, has a threaded opening that is plugged or capped and is used for testing, maintenance, or cleanout. A bullhead tee is a tee with the largest connection on the takeoff side and is used in heating and cooling systems.

Wyes have a naming convention similar to tees, in that the takeoff portion of the Y shape is the last to be named. A 2″ wye would have all connections at 2″ NPS. A $2 \times 2 \times 1$ wye would have a takeoff of 1″ NPS.

Elbows are typically two-opening fittings that change the direction of piping layout. Elbows are categorized by size, material, connection method, radius, and the angle at which they change the piping direction. Elbows can be standard with the same dimension at both ends, **Figure 41-27**, or be a reducing elbow. For example, a 2×1 elbow reduces from 2″ NPS to 1″ NPS. Standard elbows are available in 90°, 60°, 45°, 22 1/2°, and 11 1/4°.

Unions are generally used for smaller threaded pipe diameters and allow ease of disassembly for equipment maintenance. **Figure 41-28** shows a steel union made of three distinct pieces:

- Female threaded end with metal seats.
- Male end with metal seats and an integral retaining lip.
- Hex nut that compresses the male end into the female end by transferring force to the retaining lip.

Figure 41-28. Markings on a union may include pressure class and a manufacturer's mark.

Unions are available in various pressure classes, such as 150, 250, and 300, depending on material. The actual pressure rating depends not only on the class of a fitting, but the temperature of the system as well. Pressure ratings fall as the temperature of fluid increases. This is an example of an inversely proportional relationship.

Some unions may include a gasket between the female and male pieces. The gasket may be made of Teflon, paper, or rubber. One application for this type is the dielectric fitting used in a residential water heater. The gasket and other plastic liners prevent metal-to-metal contact to inhibit corrosion.

Couplings connect piping where there is little chance for disassembly of the piping. They come in threaded and welded configurations for all pressure classes and materials. Couplings are also manufactured in categories of reducing, or concentric reducing, and eccentric reducing. A standard reducing coupling maintains the centerline between the reductions in pipe size. An eccentric reducer maintains the outside diameter straight line on one side of the coupling. Smaller fittings, such as plugs, caps, and hexagonal bushings, are routinely used to provide access for maintenance, drainage, testing, and instrumentation ports.

41.4.2 Tube Fittings

Tube fittings come in as wide a range as standard pipe fittings. In addition, tube fittings have numerous styles of adapter fittings. Adapter fittings adapt from one size of NPT to a tube diameter. The tube end of an adapter fitting has different threads than the NPT end. Tube fittings are made up of the following:

- A machined body.

- One or more tubing nuts.

- One or more ferrules.

The machined body may take the form of a straight compression coupling, a tee, a 90° elbow, or a small valve. The ends of the machined body may be male or female NPT, compression tubing, a cross, or a bulkhead fitting. Tube fittings are widely used for instrumentation connections to piping systems. **Figure 41-29** shows examples of fittings.

Fittings come in a variety of materials depending on their fluid system use and requirements, such as steel, stainless steel, aluminum, and brass. Fittings are also manufactured to specifications for different pressure classes, up to 60,000 psi.

Each manufacturer has its own numbering system to specify tube fittings. Your local supplier can aid you in finding the proper manufacturer for your specific needs. Reviewing the instrumentation fittings at your workplace will help in determining your future needs.

Cegli/Shutterstock.com

Sanit Ratsameephot/Shutterstock.com

Figure 41-29. Tube fittings.

PROCEDURE

PROCEDURE | Assembling Fittings

Fittings are easily assembled and require no special tooling. Use the following procedure to assemble ferruled fittings:

1. Ensure external burrs and internal lips have been removed from the cut tubing.
2. Ensure tubing is free of dirt, moisture, and contaminants.
3. Place the ferrules and nut on the tubing.
4. Insert the tubing into the body of the fitting until it bottoms out.
5. Holding the tubing in its bottomed-out position, hand tighten the tubing nut.
6. Use a wrench to tighten the nut one to one-and-a-quarter more turns beyond the tightening by hand.
7. As a further check, disassemble the fitting and ensure the ferrules are seated properly on the tubing. Use a thread-locking sealant where applicable.

CHAPTER WRAP-UP

As a technician, you may have to repair piping and tubing systems where you work. With a knowledge of the proper tools, methods for bending pipe, and how to make proper piping and tubing system connections, you will be ready for the challenge.

Chapter Review

SUMMARY

- The most common hand tool used in the piping trades is the pipe wrench. When using a pipe wrench, a gap should exist between the material and the top jaw, otherwise slippage could result. Pipe wrenches can be used with piping up to 8″ in diameter.

- Slip-joint pliers come in two general categories: straight jaw and curved jaw.

- Once a cut is made, the pipe must be reamed—smoothed to remove any remaining burrs—where it was cut.

- Take-up and tube gain must be considered in order to accurately bend tubing. Take-up describes how far a tubing bender must be moved to compensate for a bend. Tube gain is the apparent lengthening of tubing because of the difference in distance of a square 90° turn and an arc.

- Threading different piping materials requires different threading dies. A die is the device that cuts the threads into the pipe.

- Power cutting and threading equipment is more efficient than manual tools for a bigger project. A full-size threading machine is typically used for production and larger work.

- The connection method used for piping depends on the type and use of the piping. Broad categories of connection methods include mechanical joining, welding, and adhesion.

- Methods for mechanical joining include threading, flanges, and flaring.

- Threads are standardized. Taper on NPT pipe threads and length of threads should be consistent.

- NPTF threads seal through interference fit on the threads.

- Flanges are categorized by type, attachment, seal, and face finish. Sealing type refers to the type of pressure barrier used and may include paper gaskets, O-rings, gaskets, and ring gaskets.

- Using lubricant on threaded fasteners requires applying less torque.

- Welding, soldering, and brazing are all methods of connecting piping and fittings that use high levels of heat and a filler compound or alloy.

- Pipe welding is usually in the form of socket-weld fittings or butt-weld fittings. Using welding as a repair method may be appropriate.

- Soldering is a common method of joining copper tubing.

- Soldering is performed at a temperature below 840°F, while brazing requires temperatures above 840°F.

- Solidus is the temperature at which a metal starts to melt. Liquidus is the temperature at which a metal is fully melted.

- Annealing is the process of heating a metal and allowing it to cool slowly in order to reduce the metal's hardness.

- Adhesion joining is typically used for plastic piping. However, PVC is also joined using flanges, threading, and slip-fit grooved/O-ring fitment.

- Fittings are specified using the pipe size(s), the type of connection, the type of fitting, the type of material, the piping schedule, and the pressure rating of the fitting.

- Tees come as either straight (with no reduction) or as reducing tees. Elbows are fittings that change the direction of the piping layout.

REVIEW QUESTIONS

Answer the following questions using the information provided in this chapter.

1. Give an example of a safety concern that must be considered when working with a power threading machine.

2. Discuss when offset wrenches, chain wrenches, and strap wrenches are used instead of the standard pipe wrench and why.

3. *True or False?* When positioning a pipe wrench, a gap should be left between the shank and the pipe itself due to the inflexibility of the shank.

4. Describe the two general categories of slip-joint pliers. How do they each develop their gripping power and torque?

5. List the types of cutters used for pipe and tubing and what material each cutter is used for.

6. Once a cut is made, the pipe must be _____ to smooth the pipe and remove any remaining burrs.

7. What is the process for using a pipe cutter?

8. What are tube benders? What are they used for?

9. Explain the two things that must be considered in order to accurately bend tubing.

10. *True or False?* The amount of gain is approximately 40% more than the diameter of the tubing (up to 1″ diameter) for each 90° bend.

11. Flares of 45° are normally used for _____, while 37° flares are used for other _____.

12. When is hand threading used? How are dies used in hand threading?

13. How does "underthreading" and "overthreading" affect a pipe joint?

14. _____ are short pieces of pipe threaded on both ends and used to join two fittings that are close together.

15. *True or False?* Power cutting and threading equipment is more efficient than manual tools when tackling a larger project.

16. What pipe sizes and materials can roll-grooving machines work with?

17. Describe the process of a common grooving machine.

18. List three considerations needed to choose the best and safest possible connection method.

19. What is mechanical joining? What are the three connection methods it uses?

20. *True or False?* NPT threads do not produce a mechanical seal and therefore require some type of sealing compound.

21. What are flanges? Describe how they are categorized and their types.

22. _____ are used with flanges, and their types include full-face, ring, O-ring, metallic ring, and ring-backed.

23. *True or False?* If using lubricants on threaded fasteners, the maximum torque applied should be increased.

24. What is a ferrule? When is it used?

25. The pipe welding method _____ is the most common for in-the-field applications.

26. What is the difference between butt-weld fittings and socket-weld fittings?

27. Give four examples where welding is appropriate to be used as a repair method.

28. What is soldering? How does it differ from brazing?

29. What is flux? Describe the three types of flux used after soldering is complete.

30. Discuss how a bond is created during brazing. What applications use brazing?

31. _____ is the process of heating a metal and allowing it to cool slowly in order to reduce the metal's hardness.

32. *True or False?* Adhesion joining is typically used for stainless-steel piping.

33. How can PVC piping be joined?

34. What functions do fittings perform?

35. A 3″ tee fitting means that all of its connections are _____ ″.

36. What are elbow fittings? Describe how they are categorized.

37. Explain when union fittings are used. What are the three distinct pieces of this fitting?

38. Tube fittings are made up of a _____, one or more _____, and one or more _____.

NIMS CREDENTIALING PREPARATION QUESTIONS

The following questions will help you prepare for the NIMS Industrial Technology Maintenance Level 1 Maintenance Piping credentialing exam.

1. Which of the following is *not* a consideration when using a tube bender?

   A. The amount of lengthening of a tube when it is bent on a curve compared to a sharp 90° angle

   B. The amount the tool must be moved back from the measured bend point

   C. The type of joining method

   D. The type of material to be bent

2. Which step occurs first when cutting and threading a pipe with a threading machine?

   A. Move the cutter, and swing the reamer down into position.
   B. Move the reamer, and swing the die head down into position.
   C. Ream the pipe.
   D. Thread the pipe.

3. Which of the following is *not* a connection method used for piping?

   A. Adhesion
   B. Bending
   C. Mechanical joining
   D. Welding

4. What does *not* need to be completed when assembling a flange?

   A. Assemble and tighten in a crisscross pattern to proper torque.
   B. Break the seal to relieve pressure.
   C. Use lubricant or anti-seize on the threaded fasteners.
   D. Use a soft scraper or wire brush to clean the mating surfaces.

5. How are gaskets classified?

   A. By attachment and face finish
   B. By seal and flange type
   C. By system temperature and pressure
   D. By type and material

6. When should soldering be used instead of welding?

   A. To join copper tubing with a melted metallic compound
   B. To join metal tubing for a butt-weld application
   C. To redesign a system to include other fittings and allowances
   D. To repair a pipe with corrosion to the point of leakage or failure

7. Which of the following is *not* a specification type for fittings?

   A. Pressure rating of the system
   B. Piping schedule
   C. Type of connection
   D. Type of fitting

8. _____ fittings are two-opening fittings that change the direction of piping layout.

   A. Coupling   C. Tee
   B. Elbow   D. Union

Appendix A

Math Review

WHY MATH

All technical trades involve using math for numerous tasks. Industrial maintenance technicians use math to make precise measurements and convert units. They must be able to read and interpret prints. Perhaps most importantly, technicians must be able to determine if equipment is maintained and operated within specific tolerances. Some maintenance tasks require other specialized or more advanced math, but all technicians require an understanding of what is presented here.

CALCULATORS

Calculators can be a great time-saver, but you should not rely on a calculator to replace your knowledge of basic math. There will be times in the field when a calculator is not available. However, a calculator can be a handy tool when you have one. Knowing some basic math and using common sense observation can help prevent big errors when using a calculator.

It is important to do operations in the correct order when using a calculator. For example, the formula for the area of a circle requires multiplying π (pronounced "pi" and roughly equal to 3.1416) by the radius of the circle squared (multiplied by itself). Say the radius is 5″. The area is found by multiplying 5″ × 5″ (5″ squared), then multiplying the result by 3.1416. If you know 5 × 5 (written 5^2) is 25, and 3 times 25 is 75, then you know the answer is slightly more than 75. Now you can check to see if the answer you find on your calculator is close. The correct answer is 78.54 in^2 as expected. As stated earlier, it is important to know which order to multiply the numbers. Using this example, if you multiply 3.1416 × 5 and then multiply that by itself, you get 246.7413—not even close! Knowing the correct order to do the operations on paper will help you enter numbers in the correct order on your calculator.

There are several types of calculators commonly used in technical trades. The most familiar type is a *general calculator*, which allows the user to perform basic math functions. *Basic calculators* have a memory function that allows the user to store the result of a calculation so that number can be recalled and used in a subsequent calculation. See **Figure A-1**. Most also have a % key. When a number is shown in the display, pressing the % key allows the user to then press a number key to calculate that percent of the displayed number. The √ key is for finding the square root of a number. (Squares, roots, and exponents are covered later in this supplement.) The ± key changes the value of a displayed number from positive to negative or negative to positive.

Scientific calculators are more advanced, having the ability to do many trigonometric functions and other advanced operations, **Figure A-2**. Scientific calculators can be very useful when working on electrical systems or other systems that include angles in calculations, such as load or welding angles. (Review sections on working with angles and vectors later in this supplement.) Depending on your particular responsibilities, a scientific calculator may not be necessary.

Another type of calculator that can be useful for trades workers is the *construction calculator*. Construction calculators can be used to convert inches to feet and

Shawn C. Ballard

Figure A-1.

Shawn C. Ballard

Figure A-2.

feet to inches. Most allow the user to convert between metric and US Customary measurements. Construction calculators vary from one manufacturer and model to the next, so it is not possible to give a detailed description or instructions here. All calculators come with instructions that explain their functions.

WHOLE NUMBERS

Whole numbers are simply numbers without fractions or decimal points, numbers such as 1, 2, 3, 4, etc. Adding, subtracting, multiplying, and dividing whole numbers primarily requires memorizing a few math facts.

Adding and Subtracting Whole Numbers

For example, adding this column of whole numbers requires memorizing the *sum* of 3 + 5 and the sum of 8 + 2.

$$
\begin{array}{r}
3 \\
5 \\
+\ 2 \\
\hline
10
\end{array}
$$

The same type of memorization of math facts is required to subtract whole numbers. We know that the result of subtracting 12 from 37 is 25 because we know that 2 from 7 is 5 and 1 from 3 is 2.

$$
\begin{array}{r}
37 \\
-\ 12 \\
\hline
25
\end{array}
$$

The key to both addition and subtraction is to line up the columns of digits correctly. Whole numbers should be aligned on the right.

In subtraction, if the number being subtracted (the number on the bottom) is larger than the number it is being subtracted from (the number on the top), borrow 10 from the next digit to the left and add it to the one on the right. Write small numerals above the column to help you keep track. Consider this example:

$$
\begin{array}{r}
{}^{2\ 16} \\
3\!\!\!/6 \\
-\ 19 \\
\hline
17
\end{array}
$$

Multiplying Whole Numbers

Multiplication of whole numbers requires memorization of a multiplication table. The only way to get $6 \times 5 = 30$ is to know that multiplication fact or to add 6 + 6 + 6 + 6 + 6. Longhand addition quickly becomes tedious for bigger multiplication problems. To multiply numbers whose values are 10 or more (those with more than 1 digit), align the digits representing 0 through 9 (the 1s digit) in the right-hand column. Then multiply the top row by the 1s digit in the second row:

$$
\begin{array}{r}
31 \\
\times\ 15 \\
\hline
155
\end{array}
$$

Next, multiply the top row by the 10s digit in the second row. Because you multiplied by the 10s digit, the *product* (the result of multiplication) is written with its right-most digit in the 10s column:

$$
\begin{array}{r}
31 \\
\times\ 15 \\
\hline
155 \\
31
\end{array}
$$

If the problem has more digits in the second row, repeat the above steps for each digit and write the products in rows beneath one another. Be sure to record the right-most digit in each row in the column for the place it represents: 100s, 1000s, etc.

When all the multiplication is complete, add the products just as you would for a simple addition problem. The result of this addition is the product (answer) of the multiplication problem.

$$
\begin{array}{r}
31 \\
\times\ 15 \\
\hline
155 \\
31 \\
\hline
465
\end{array}
$$

Dividing Whole Numbers

Division of whole numbers is simply the reverse of multiplication, but the problem must be set up differently. The *dividend* (the number being divided) is written inside the division symbol. The *divisor* (the number the dividend will be divided by) is written to the left of the symbol:

divisor ⟶ 7 ⟌ 28 ⟵ dividend

From the multiplication table, we know that $7 \times 4 = 28$. So, if 28 is divided into 7 parts, each part will have 4, or $28 \div 7 = 4$. 4 is the *quotient* (the answer to a division

problem). It is written above the division symbol and above the 1s place of the 28:

$$
\begin{array}{r}
4 \leftarrow \text{quotient} \\
7\,\overline{)\,28}
\end{array}
$$

When the dividend is large compared to the divisor, the process is divided into steps as follows:

$$
4\,\overline{)\,320}
$$

4 goes into 32 8 times. Write the 8 above the 2 (the right column of the 32). Now multiply 4×8, which is 32. Write the 32 beneath the 32 in the division symbol.

$$
\begin{array}{r}
8 \\
4\,\overline{)\,320} \\
32
\end{array}
$$

Subtract the product of your multiplication (32) from the number above it (32). Because 4 goes into 32 exactly 8 times, the numbers are the same, so the result of your subtraction is 0.

$$
\begin{array}{r}
8 \\
4\,\overline{)\,320} \\
\underline{32} \\
0
\end{array}
$$

Drop the next digit to the right in the dividend (0 in this case) down beside the result of your subtraction. That makes the number at the bottom 00. 4 will not go into 0 (or 00), so the quotient of that step is 0. The quotient (answer) of the division problem is 80.

$$
\begin{array}{r}
80 \\
4\,\overline{)\,320} \\
\underline{32} \\
00
\end{array}
$$

320 can be divided by 4 80 times. If there are more places under the division symbol, just keep doing the same division, multiplication, subtraction, and drop down for each digit moving to the right. Consider the following example:

$$
\begin{array}{r}
102 \\
6\,\overline{)\,616} \\
\underline{6} \\
01 \\
\underline{0} \\
016 \\
\underline{12} \\
4 \leftarrow \text{remainder}
\end{array}
$$

If the last number produced by the drop-down cannot be divided evenly by the divisor, that number is called the *remainder*. In the example above, the quotient is 102 with a remainder of 4.

Practice A-1

Test your skills with the following problems. Do *not* use a calculator.

A.
$$342 \atop + \ 16$$

B.
$$79 \atop + 29$$

C.
$$68 \atop - 13$$

D.
$$124 \atop - \ 35$$

E.
$$18 \atop \times \ 4$$

F.
$$213 \atop \times 24$$

G. $3\overline{)36}$

H. $7\overline{)214}$

FRACTIONS

A fraction is a part of something larger. If there are three machine bolts in a pound, each machine bolt weighs one-third of one pound. One-third can be written as follows:

$$\frac{1}{3} \ \leftarrow \text{numerator} \atop \leftarrow \text{denominator}$$

The number above the fraction bar is called the *numerator*. It indicates how many parts are in the fraction, in this case 1 machine bolt. The number below the fraction bar is the *denominator*. The denominator indicates how many parts are in the whole, in this case 3 machine bolts in the whole pound. If there are 50 machine bolts in a carton and you take 7 of them out, you have taken

$$\frac{7}{50}$$

of the machine bolts.

Equivalent Fractions

If two fractions represent the same value, they are said to be *equivalent fractions*. For example, 1/3 and 2/6 are equivalent fractions because they both represent one-third of the whole. If both the numerator and denominator of a fraction are multiplied by the same amount, the result is an equivalent fraction.

$$\frac{1}{3} \frac{\times 2 =}{\times 2 =} \frac{2}{6}$$

Adding Fractions

Fractions must have a *common denominator* in order to be added. If the denominator of one of the fractions is 8, the other fraction must be written as an equivalent fraction with a denominator of 8. For example, to add 3/4 and 1/8, write 3/4 as an equivalent fraction with a denominator of 8.

$$\frac{3}{4} \times \frac{2}{2} = \frac{6}{8}$$

When both fractions have the same denominator, add the numerators.

$$\frac{6}{8} + \frac{1}{8} = \frac{7}{8}$$

6 eighths plus 1 eighth is a total of 7 eighths.

A common denominator can be found by multiplying all the denominators in a problem. Consider the following example:

$$\frac{1}{3} + \frac{2}{5} + \frac{9}{14}$$

$$3 \times 5 \times 14 = 210$$

$$\frac{70}{210} + \frac{84}{210} + \frac{135}{210} = \frac{289}{210}$$

In this example, the numerator is larger than the denominator. This is because the value of the fraction is greater than 1. 289 is 79 parts larger than the whole of 210 parts. The same number could be written as follows:

$$1 \frac{79}{210}$$

This is called a *mixed number* because it is made up of a whole number plus a fraction.

Subtracting Fractions

Subtracting fractions is similar to adding them. Both fractions must have a common denominator, then the numerators are subtracted just like whole numbers.

$$\frac{2}{3} - \frac{1}{4} = ?$$

Find common denominators and subtract the numerators.

$$\frac{8}{12} - \frac{3}{12} = \frac{5}{12}$$

Multiplying Fractions

To multiply fractions, multiply the numerators. Then multiply the denominators.

$$\frac{3}{4} \times \frac{2}{5} = \frac{3 \times 2}{4 \times 5} = \frac{6}{20}$$

To make the result easier to work with, always reduce it to its lowest terms. That is, write it as an equivalent fraction with the lowest possible denominator. For example, both 6 and 20 can be divided by 2 to make an equivalent fraction with a smaller denominator.

$$\frac{6 \div 2}{20 \div 2} = \frac{3}{10}$$

To multiply a fraction by a mixed number, first change the mixed number to a fraction. Then multiply as common fractions. Reduce the product to its lowest terms.

$$\frac{2}{3} \times 4\frac{1}{2} = \frac{2}{3} \times \frac{9}{2}$$

$$\frac{2}{3} \times \frac{9}{2} = \frac{18}{6}$$

$$\frac{18}{6} = \frac{3}{1} = 3$$

Dividing Fractions

To divide fractions, invert the divisor (swap the numerator and denominator) then multiply as common fractions. It may help you to remember "keep it, change it, flip it"—keep the first fraction as it is, change the division to multiplication, and flip the second fraction.

divisor

$$\frac{3}{4} \div \frac{2}{3} = \frac{3}{4} \times \frac{3}{2}$$

$$\frac{3 \times 3}{4 \times 2} = \frac{9}{8}$$

In this example, the quotient is a fraction with a larger numerator than its denominator. This indicates that the value is greater than 1. To express this in its simplest form, convert it to a mixed number.

$$\frac{9}{8} = \frac{8+1}{8} = 1\frac{1}{8}$$

Sometimes it is necessary to combine operations in a single problem. Some such problems also involve more than one type of unit, such as inches and feet. The first step in solving more complex problems is to decide which operation should be done first. It often helps to write the problem in such a way that it states the order of operations. Next, convert everything to the same unit(s). Where the operations will include adding or subtracting fractions, convert them to their least common denominators. Now solve the problem, doing all operations in the planned order. Finally, convert the units to whatever makes sense for the problem. For example, you would not write fine measurements in square feet.

Work through the following example:

How long is the shaded space in this drawing?

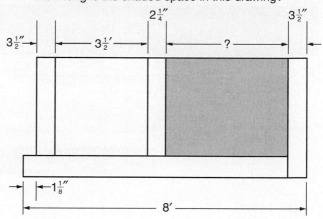

1. Convert the overall length to inches. 96″
2. Subtract $1\frac{1}{8}″$. $95\frac{7}{8}″$
3. Convert $3\frac{1}{2}'$ to inches. 42″
4. Add dimensions at top. $3\frac{1}{2}″ + 42″ + 2\frac{1}{4}″ + 3\frac{1}{2}″ = 51\frac{1}{4}″$
5. Subtract $51\frac{1}{4}″$ from $95\frac{7}{8}″$. $95\frac{7}{8}″ - 51\frac{2}{8}″ = 44\frac{5}{8}″$

Practice A-2

A. $\frac{1}{4} + \frac{5}{8}$ B. $\frac{3}{4} - \frac{1}{3}$

C. $\frac{1}{2} \times \frac{3}{4}$ D. $\frac{1}{3} \div \frac{1}{2}$

E. $\frac{2}{3} + \frac{7}{12}$ F. $\frac{13}{16} - \frac{2}{5}$

G. $\frac{2}{3} \times 2\frac{1}{2}$ H. $1\frac{1}{4} \div \frac{2}{5}$

READING A RULER

Measuring devices, such as rulers and tape measures, may be marked for measuring inches and fractions of an inch; meters, centimeters, and millimeters; feet, inches and tenths of an inch; or by any other system. The measuring system used to divide the spaces on a measuring device is called the *scale*. The most common linear (in a line) scale in construction uses yards, feet, inches, and fractions of an inch. There are 3 feet in a yard and 12 inches in a foot. Inches are most often divided into halves, fourths, eighths, and sixteenths. See **Figure A-3**.

On finer measuring devices, the longest marks on the scale may indicate inches, **Figure A-4**. The inches on a measuring device may be divided into eighths, sixteenths, or even thirty-seconds. The second longest marks on the scale represent halves, the next longest

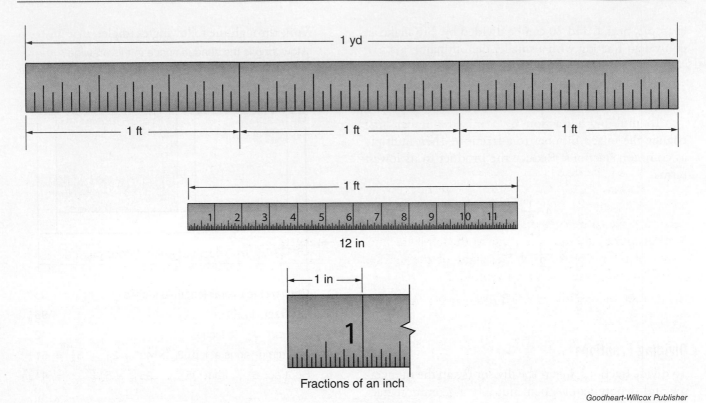

Figure A-3.

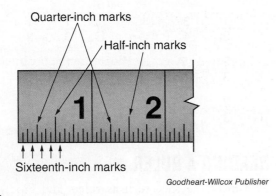

Figure A-4.

represent fourths, and so on. The first step in reading the scale is determining what the smallest marks on the scale represent. Count down from the whole inch to the halves, then the quarters, eighths, sixteenths, and thirty-seconds, if they are used. Then count the number of marks from the last inch mark to the mark you are reading.

Practice A-3

What measurements are represented by the letters on **Figure A-5**?

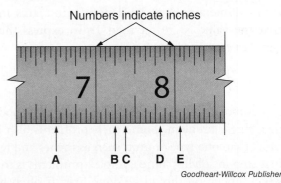

Figure A-5.

DECIMAL FRACTIONS

Decimal fractions are commonly called *decimals*. Decimal fractions are fractions whose denominators are multiples of 10. If the denominator is 10, the fraction is tenths. If the denominator is 100, the fraction is so many hundredths.

Decimal fractions are often written on a single line with a dot separating digits. The dot between the whole number and the decimal fraction is the *decimal point*. Every place to the left of the decimal point increases the value of the digit in that place tenfold. That is why the second place to the left of the decimal point is called the tens place, and the third place to the left is the hundreds place, etc. Moving to the right of the decimal point, the place values decrease tenfold. A decimal fraction of $\frac{5}{10}$ can be written as 0.5. A decimal fraction of $\frac{12}{1000}$ can be written as 0.012.

```
        1 2 3 . 4 5 6
hundreds ─┘ │ │   │ │ └─ thousandths
    tens ───┘ │   │ └─── hundredths
    ones ─────┘   └───── tenths
              └───────── decimal point
```

The value of the number in the example above is one hundred twenty-three and four hundred fifty-six thousandths.

Adding and Subtracting Decimals

To add decimals, line up the decimal points in a column, add the numbers, and put the decimal point in the result in the decimal point column.

$$
\begin{array}{r}
1.4 \\
19.2 \\
+\ 31.7 \\
\hline
52.3
\end{array}
$$

Subtracting decimals is very similar. Line up the decimal points in the problem and the answer, and subtract as usual.

$$
\begin{array}{r}
27.74 \\
-\ 2.23 \\
\hline
25.51
\end{array}
$$

If there are more decimal places in the number being subtracted than there are in the number it is being subtracted from, zeros can be added to the right without affecting the value of the number.

$$
\begin{array}{r}
5.70 \leftarrow \text{added zero} \\
-\ 2.02 \\
\hline
3.68
\end{array}
$$

Multiplying Decimals

Decimals are multiplied the same as whole numbers, except for the placement of the decimal point in the product (answer). Add the number of decimal places to the right of the decimal point in both the number being multiplied and the number it is being multiplied by. The decimal point should be placed that many places to the left in the product.

```
   12.25   ← two decimal places to the right
 ×  3.75   ← two decimal places to the right
   6125       (total of four decimal places)
  8575
 3675
 45.9375  ← decimal point is four places
             to the left
```

Dividing Decimals

Dividing decimals is also much like dividing whole numbers, except for keeping track of the placement of the decimal point. As a reminder, the number being divided is the dividend, the number it is divided by is the divisor, and the answer is the quotient. To start the division problem, move the decimal point in the divisor all the way to the right. Move the decimal point in the dividend the same number of places to the right. Add zeros to the right of the dividend, if necessary. Divide as you would for whole numbers.

```
divisor → .4 )‾20    ← dividend
          4. )‾200.    move decimal points

             50
          4 )‾200
             20
             ‾‾‾
             00
```

Practice A-4

Test your skills with the following problems.

$$
\text{A.}\quad
\begin{array}{r}
2.12 \\
17.01 \\
+\ 9.05 \\
\hline
\end{array}
\qquad
\text{B.}\quad
\begin{array}{r}
34.09 \\
12.125 \\
+\ 2.899 \\
\hline
\end{array}
$$

$$
\text{C.}\quad
\begin{array}{r}
18.48 \\
-\ 12.25 \\
\hline
\end{array}
\qquad
\text{D.}\quad
\begin{array}{r}
134.02 \\
-\ 8.14 \\
\hline
\end{array}
$$

$$
\text{E.}\quad
\begin{array}{r}
5.25 \\
\times\ 5 \\
\hline
\end{array}
\qquad
\text{F.}\quad
\begin{array}{r}
15.34 \\
\times\ 6.25 \\
\hline
\end{array}
$$

G. $35 \div .07$ \qquad H. $2.25 \div .25$

CONVERTING COMMON FRACTIONS TO DECIMAL FRACTIONS AND ROUNDING OFF

To change a common fraction to a decimal fraction, divide the numerator by the denominator. For example, change 1/4 to a decimal fraction.

$$\begin{array}{r} .25 \\ 4\,\overline{)\,1.00} \\ \underline{8} \\ 20 \end{array}$$

$$\frac{1}{4} = .25$$

Sometimes the division yields a number with a repeating decimal, as in converting 2/3 to a decimal.

$$\begin{array}{r} .666 \\ 3\,\overline{)\,2.000} \\ \underline{1\,8} \\ 20 \\ \underline{18} \\ 20 \end{array}$$

These numbers should be rounded off to the desired number of places. When rounding off, the last digit should be increased by 1 if the next digit is 5 or more. If the next digit is less than 5, the last digit used stays the same. In the above example, round the answer to two places. The second digit is rounded up to 7 because .666 is closer to .67 than it is to .66.

To convert a mixed number as a decimal, keep the whole number as is and convert the fractional part as above.

Express $12\,\dfrac{3}{4}$ as a decimal.

$$12 + 4\,\overline{)\,\begin{array}{r} .75 \\ 3.00 \\ \underline{2\,8} \\ 20 \end{array}} = 12.75$$

Practice A-5

Convert these fractions to decimals and round the answers to three places.

A. $\dfrac{1}{3}$ B. $\dfrac{22}{7}$ C. $\dfrac{10}{15}$

CONVERTING DECIMALS TO COMMON FRACTIONS

To convert a decimal to a fraction, drop the decimal point and write the given number as the numerator. The denominator will be 10, 100, 1000, or 1 with as many 0s as there were places in the decimal number.

$$.42 = \frac{42}{100} \text{ or } .125 = \frac{125}{1000}$$

READING A VERNIER SCALE

The vernier scale is named after French mathematician Pierre Vernier. It uses two scales that slide past one another to allow accurate measurements beyond what would be possible with a single scale. The vernier scale principle can be used with US Customary units or the metric system.

Reading an Inch-Based Vernier Scale

The main scale is divided into inches, tenths of an inch, and 25 thousandths of an inch. The vernier scale is divided into slightly smaller increments, so that only one mark on the vernier scale can be aligned with a mark on the main scale. See **Figure A-6**.

To read the vernier scale, start with the location of the 0 on the scale. See **Figure A-7**. Find the first whole inch mark to the left of the vernier scale 0 mark. In **Figure A-7**, that mark is 2″. We will be adding the numbers found in each step, so write down "2." Next, find the largest 0.1″ mark to the left of the 0 on the vernier scale. In **Figure A-7**, that is the 0.2″ mark. Write that under "2," with their decimal points aligned for addition later.

2.0
0.2

Add the decimal point after the top 2 and additional 0s as necessary to help keep things aligned for later addition. The third step is to find the largest 0.025″ mark on the main scale to the left of the 0 on the vernier scale. In this case, that is the second one after the 0.2″ mark, so it represents 0.025″ + 0.025″ or 0.050″. Write that number in your addition column, adding 0s as necessary.

2.000
0.200
0.050

Main scale

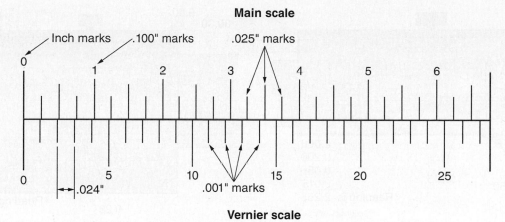

Vernier scale

Goodheart-Willcox Publisher

Figure A-6.

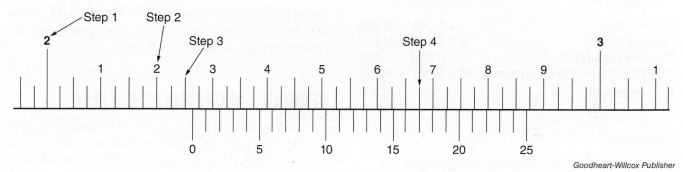

Figure A-7.

Goodheart-Willcox Publisher

The final reading is on the vernier scale. Find a line on the vernier scale that lines up with any mark on the main scale. In **Figure A-7**, that is the mark representing 17, or 0.017″. Write that in your addition column, adding any necessary 0s, and do the addition.

$$
\begin{array}{r}
2.000 \\
0.200 \\
0.050 \\
\underline{0.017} \\
2.267
\end{array}
$$

The vernier scale in **Figure A-7** is reading 2.267″.

Review two additional examples of reading a US Customary vernier scale in **Figure A-8** and **Figure A-9**. In **Figure A-8**, the "0" line on the vernier plate is

Past the 2:	2×1	= 2.000
Past the 3:	3×0.100	= 0.300
Plus 2 graduations:	2×0.025	= 0.050
Plus 18 vernier scale graduations:	18×0.001	= 0.018
Total reading		= 2.368″

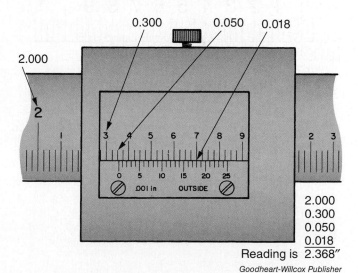

2.000
0.300
0.050
0.018
Reading is 2.368″

Goodheart-Willcox Publisher

Figure A-8.

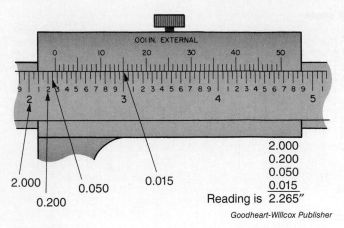

Figure A-9.

Goodheart-Willcox Publisher

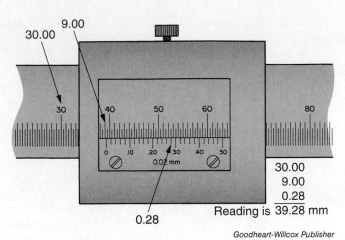

Figure A-10.

Goodheart-Willcox Publisher

In **Figure A-9**, the "0" line on the vernier plate is

Past the 2:	2 × 1.000	= 2.000
Past the 2:	2 × 0.100	= 0.200
Plus one graduation:	1 × 0.050	= 0.050
Plus 15 vernier scale graduations:	15 × 0.001	= 0.015
Total reading		= 2.265″

Practice A-6

Take readings from the following vernier scales.

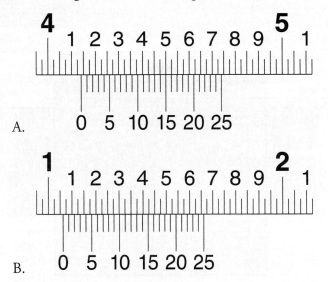

A.

B.

Reading a Metric Vernier Scale

The principles used in reading metric vernier measuring tools are the same as those used for US Customary measure. However, the readings on the metric vernier scale have 0.02 mm precision. A 25-division metric vernier scale is illustrated in **Figure A-10**, and a 50-division metric vernier scale is shown in **Figure A-11**. Note that each division on both scales corresponds to two-hundredths of a millimeter (0.02 mm).

READING A MICROMETER CALIPER

The micrometer caliper is a precision measuring tool capable of measuring to 0.001″ or 0.01 mm. When fitted with a vernier scale, it will read to 0.0001″ or 0.002 mm.

Reading an Inch-Based Micrometer

The micrometer is read by recording the highest number on the sleeve (1 = 0.100, 2 = 0.200, etc.). To this number, add the number of vertical lines visible between the number on the sleeve and the thimble edge (1 = 0.025,

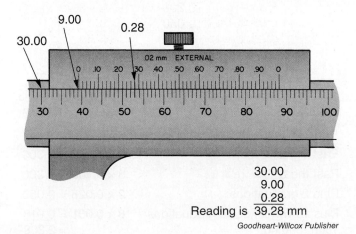

Figure A-11.

Goodheart-Willcox Publisher

2 = 0.050, etc.). To this total, add the number of thousandths of an inch indicated by the line that corresponds with the horizontal sleeve line. To practice, add the readings from the sleeve and thimble in **Figure A-12**:

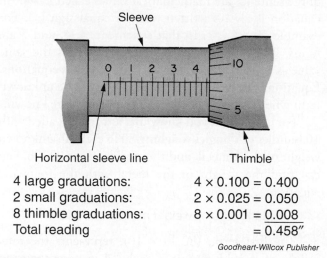

4 large graduations: 4 × 0.100 = 0.400
2 small graduations: 2 × 0.025 = 0.050
8 thimble graduations: 8 × 0.001 = 0.008
Total reading   = 0.458″

Goodheart-Willcox Publisher

Figure A-12.

Another example is presented in **Figure A-13**. As before, add the readings from the sleeve and thimble to determine the total reading:

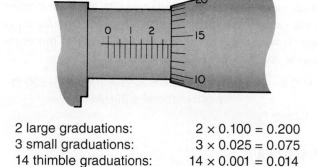

2 large graduations: 2 × 0.100 = 0.200
3 small graduations: 3 × 0.025 = 0.075
14 thimble graduations: 14 × 0.001 = 0.014
Total reading   = 0.289″

Goodheart-Willcox Publisher

Figure A-13.

Vernier micrometers have an additional set of lines on the sleeve, **Figure A-14**, to allow even more precision.

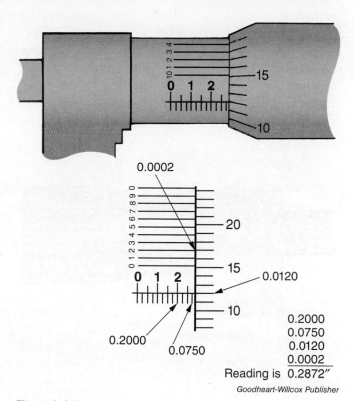

0.2000
0.0750
0.0120
0.0002
Reading is 0.2872″

Goodheart-Willcox Publisher

Figure A-14.

As before, add the readings from the sleeve and thimble, then add the 1/10,000″ (0.0001″) reading to find the total.

Practice A-7

Take the readings from the micrometers shown below.

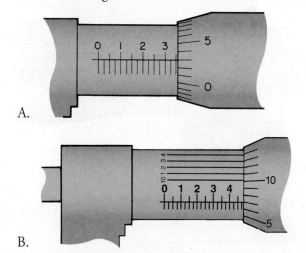

A.

B.

Reading a Metric Micrometer

Metric micrometers, **Figure A-15**, and vernier micrometers, **Figure A-16**, are read much like inch-based micrometers. The only difference is that you will read tenths, hundredths, thousandths, and two-thousandths of a millimeter, instead of fractions of an inch.

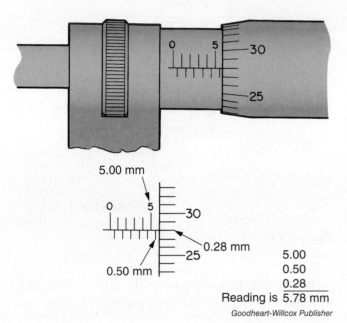

5.00
0.50
0.28
Reading is 5.78 mm

Goodheart-Willcox Publisher

Figure A-15.

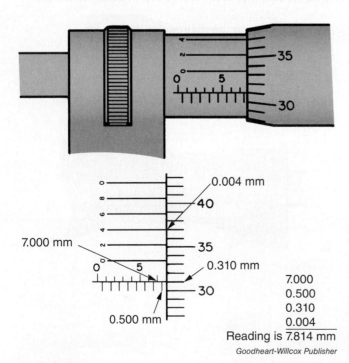

7.000
0.500
0.310
0.004
Reading is 7.814 mm

Goodheart-Willcox Publisher

Figure A-16.

EQUATIONS

An *equation* is a mathematical statement that two things have the same or equal value. An equation can be thought of as a mathematical sentence. The words of the sentence are mathematical values called *terms*. An equation is always written with an equal sign (=). For example, 3 + 4 = 7. In that statement, 3, 4, and 7 are terms. The statement says that 3 plus 4 has the same value as 7. Many useful formulas are stated as equations. Equations can be used to find the value of one unknown term when the other values in the equation are known.

For example, if you know that a truck is loaded with 10 bundles of shingles weighing 80 lb each, an unknown weight of sheet metal, and the total load is 1000 lb, you can find the weight of the metal with the following equation:

$$(80 \text{ lb} \times 10) + \text{weight of metal} = 1000 \text{ lb}$$

The first term, (80 lb × 10), represents the total weight of the shingles. It is enclosed in parentheses to indicate that it is a single term that should be computed before the rest of the equation. Whenever a mathematical term is enclosed in parentheses, that computation should be done first. Now, write the equation with the shingle weight computed:

$$800 \text{ lb} + \text{weight of metal} = 1000 \text{ lb}$$

When a mathematical operation is done on one side of an equation, the equation remains a true statement if the same thing is done on the other side of the equal sign. If we subtract 800 lb from both sides of our equation, it is still a true equation.

$$800 \text{ lb} + \text{weight of metal} - 800 \text{ lb} = 1000 \text{ lb} - 800 \text{ lb}$$
$$\text{weight of metal} = 200 \text{ lb}$$

Practice A-8

Find the unknown value in each equation.

A. cost = ($.60 − $.04) × 5

B. $X = \dfrac{3}{4} + 20$

C. 240 lb = 2 × weight of crate

D. $\dfrac{1}{4} \div \dfrac{2}{3} = Y$

AREA MEASURE

The area of a surface is always measured in units of square inches, square meters, square feet, etc. When a number is squared, that means it is multiplied by itself.

For example, 3 squared is 9. Square units are written with a superscript 2, indicating that it is units × units.

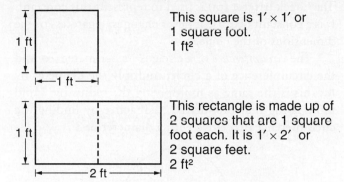

This square is 1′ × 1′ or 1 square foot. 1 ft²

This rectangle is made up of 2 squares that are 1 square foot each. It is 1′ × 2′ or 2 square feet. 2 ft²

Finding the Area of Squares and Rectangles

The area of a square or rectangle is the number of units it is wide multiplied by the number of units it is long.

16′–0″

12′–0″

12 ft × 16 ft = 192 ft²

The width and length must be expressed in the same units. For example, to find the area of this rectangle, convert all feet to inches, then multiply.

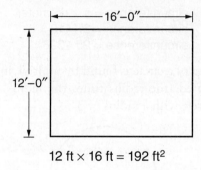

4′–6″

2′–4″

2′–4″ = 24″ + 4″ = 28″
4′–6″ = 48″ + 6″ = 54″
28 in × 54 in = 1512 in²

A square foot is 12″ × 12″, or 144 in². If an area is given as a large number of square inches (in²), it can be converted to square feet (ft²) by dividing by 144. Consider the example above where the area of the rectangle is 1512 in². Divide by 144 to find the area in ft².

$$\frac{1512 \text{ in}^2}{144} = 10.5 \text{ ft}^2$$

Finding the Area of Triangles

To find the area of any triangle, multiply the height times 1/2 the base.

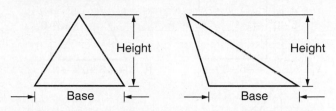

Height   Base   Height   Base

Find the area of this triangle.

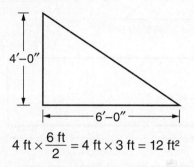

4′–0″

6′–0″

$$4 \text{ ft} \times \frac{6 \text{ ft}}{2} = 4 \text{ ft} \times 3 \text{ ft} = 12 \text{ ft}^2$$

Another way to achieve the same result is to multiply the base times the height, then divide by 2.

$$4 \text{ ft} \times 6 \text{ ft} = 24 \text{ ft}^2$$

$$\frac{24 \text{ ft}^2}{2} = 12 \text{ ft}^2$$

Some figures may be made up of squares, rectangles, and triangles of varying sizes. To find the area of such a figure, break it into its various parts and find the area of each part, then add those areas.

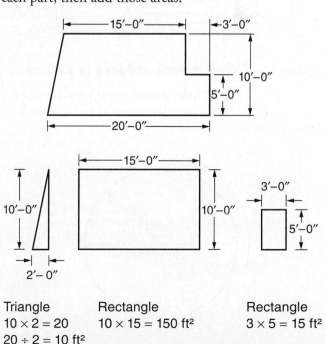

15′–0″   3′–0″

10′–0″

5′–0″

20′–0″

15′–0″

10′–0″   10′–0″   3′–0″

5′–0″

2′–0″

Triangle	Rectangle	Rectangle
10 × 2 = 20	10 × 15 = 150 ft²	3 × 5 = 15 ft²
20 ÷ 2 = 10 ft²		

10 ft² + 150 ft² + 15 ft² = 175 ft²

Practice A-9

Find the areas of the figures below.

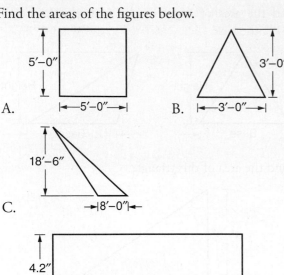

A.

B.

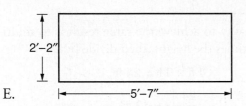

C.

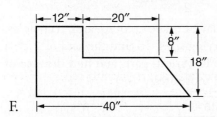

D.

E.

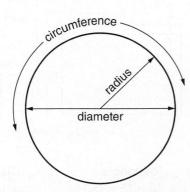

F.

Finding the Circumference and Area of a Circle

The distance from a circle's center point to its outer edge is its *radius*. The total distance across a circle through its center point is its *diameter*.

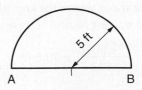

Many calculations involving circles or parts of circles use a constant of approximately 22/7, or 3.1416. The Greek letter π (pi) is used to represent this constant. It is a constant because it never changes, regardless of the dimensions of the circle.

The *circumference* of a circle is its perimeter. To find the circumference of a circle, multiply the diameter by π. This is the same as multiplying the radius by 2 and multiplying that product by π. Practice by finding the circumference of a circle with a diameter of 8'.

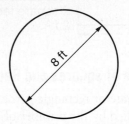

circumference = π × diameter
circumference = 3.1416 × 8'
circumference = 25.1328'

The area of a circle is found by multiplying π by the radius squared (the radius times the radius). Find the area of a circle with a radius of 3".

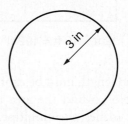

area = π × radius²
area = 3.1416 × (3")²
area = 28.2744 in²

Notice that in both examples using π, the answer is rounded to four decimal places. That is because π was rounded to four places. The answer cannot be accurate to more decimal places than the values used to calculate it.

Sometimes shapes encountered in the workplace are semicircles or quarter-circles. The areas and perimeters of these shapes can be found using the formulas for circles and dividing the result by 2 for a semicircle or by 4 for a quarter-circle. For example, find the perimeter of the semicircular shape below.

$$\text{diameter} = 2 \times 5 \text{ ft} \qquad = 10 \text{ ft}$$
$$\text{circumference} = \pi \times 10 \text{ ft} \qquad = 31.416 \text{ ft}$$

$$\frac{\text{circumference of}}{\text{semicircular portion}} = \frac{\text{circumference}}{2} = 15.708 \text{ ft}$$

$$\text{line AB} = 10 \text{ ft}$$

$$\frac{\text{perimeter of}}{\text{semicircle}} = 10 \text{ ft} + 15.708 \text{ ft} = 25.708 \text{ ft}$$

Practice A-10

Find the perimeter and the area of each of these figures.

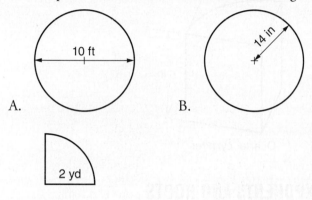

A.   B.

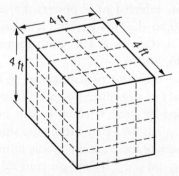

C.   Quarter circle

VOLUME MEASURE

The volume of a solid is always measured in units of cubic inches, cubic meters, cubic feet, etc. When a number is cubed, that means it is multiplied by itself, then by itself again. For example, 3 cubed is 27 ($3 \times 3 \times 3$). Cubic units are written with a superscript 3, indicating units × units × units.

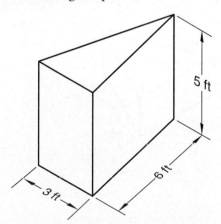

$4 \text{ ft} \times 4 \text{ ft} \times 4 \text{ ft} = 64$ cubic ft or 64 ft³

This cube is made up of 64 individual cubes, each measuring 1 foot by 1 foot by 1 foot.

As long as a solid (a three-dimensional shape) has the same size and cross-sectional shape throughout its depth, its volume can be found by multiplying the area of one

surface by the depth from that surface. To find the volume of a *cube* (all edges are the same size) or a rectangular solid (a rectangle with a third dimension), multiply the length, width, and height. For a 3-inch cube,

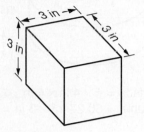

$$\text{volume} = 3 \text{ in} \times 3 \text{ in} \times 3 \text{ in} = 27 \text{ in}^3$$

For a rectangular solid,

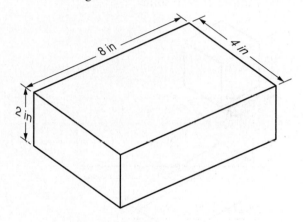

$$\text{volume} = 4 \text{ in} \times 8 \text{ in} \times 2 \text{ in} = 64 \text{ in}^3$$

A solid with two opposite triangular faces is called a *triangular prism*. A solid with two circular faces is a *cylinder*. The volume of a triangular prism or a cylinder is found by multiplying the area of its face by its height. Consider a triangular prism:

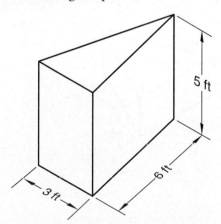

$$\text{area of face} = 1/2 \times 3 \text{ ft} \times 6 \text{ ft} = 9 \text{ ft}^2$$
$$\text{volume} = 9 \text{ ft}^2 \times 5 \text{ ft} \qquad = 45 \text{ ft}^3$$

Consider a cylinder:

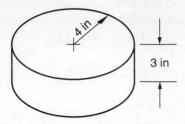

area of face = π × 4 in × 4 in = 50.27 in²
volume = 50.27 in² × 3 in = 150.81 in³

Practice A-11

Find the volume of each solid.

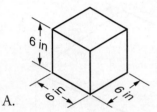

A.

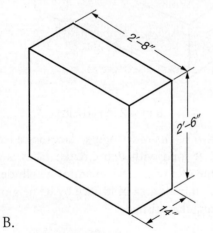

B.

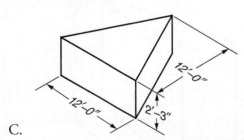

C.

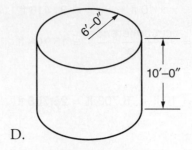

D.

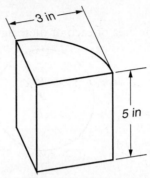

E.   Quarter cylinder

EXPONENTS AND ROOTS

When a number is squared or cubed, the little superscript number written to the right is called an *exponent*. For example, in the number 10^2, the exponent is 2, indicating that the number is 10 multiplied by itself. Another way of saying this is "10 to the second power." If the number were to be $10 \times 10 \times 10$, it could be written as 10^3 and it could be called "10 cubed" or "10 to the third power." Exponents of 2 and 3 have names—squared and cubed, respectively—because they are the exponents used with area and volume measure. Higher exponents are only referred to as powers. For example, 10^5 is read as "10 to the 5th power." It is easier than saying "$10 \times 10 \times 10 \times 10 \times 10$." Both forms of that number equal 100,000.

Roots are the opposite of exponents. When you take the square root of a number, you are asking, "What number multiplied by itself is equal to this number?" For example, 4^2 equals 16. That means the square root of 16 is 4. Most square roots are not whole numbers. To find a square root, use the $\sqrt{}$ button on your calculator.

Practice A-12

A. What is 12 squared?

B. What is 8.5 cubed?

C. Calculate 4 to the 6th power.

D. What are two other ways to write $3 \times 3 \times 3 \times 3$?

E. What is the square root of 25?

F. Calculate $\sqrt{215}$.

WORKING WITH RIGHT ANGLES

A *right angle* measures 90°. A triangle having a right angle is called a *right triangle*. It will be helpful to know a few terms associated with right triangles, **Figure A-17**.

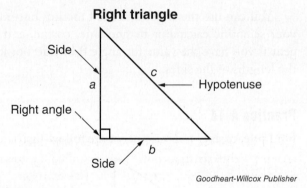

Right triangle

Side

a

c

Hypotenuse

Right angle

Side

b

Goodheart-Willcox Publisher

Figure A-17.

A triangle can only have one 90° corner. The sum total of all three angles in a triangle is always 180°, so if one angle is 90°, the other two must add up to 90° together. *Hypotenuse* is a special term for the longest side of a triangle. The hypotenuse will always be across from the largest angle in a triangle.

The *Pythagorean theorem* is a principle that makes right triangles convenient to work with. Named after the Greek mathematician Pythagoras, the Pythagorean theorem states that the sum of the squares of the sides of a right triangle is equal to the square of the hypotenuse. To help keep track of the terms used in the Pythagorean theorem, it is common to label the two sides a and b and the hypotenuse c. Then the theorem can be stated as an equation:

$$a^2 + b^2 = c^2$$

If the lengths of the two sides of a right triangle are known, the Pythagorean theorem can be used to find the length of the hypotenuse. Consider a right triangle with sides equal to 3 ft and 6 ft:

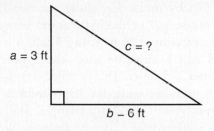

$a = 3$ ft

$c = ?$

$b - 6$ ft

$$c^2 = a^2 + b^2$$
$$c^2 = 3^2 + 6^2$$
$$c^2 = 9 \text{ ft}^2 + 36 \text{ ft}^2$$
$$c^2 = 45 \text{ ft}^2$$
$$c = \sqrt{45 \text{ ft}^2}$$
$$c = 6.7 \text{ ft}$$

The Pythagorean theorem can be used to find the length of any side of a right triangle if the other two are known. For example, if side b and the hypotenuse, c, are known, $a^2 + b^2 = c^2$ can be rearranged to $a^2 = c^2 - b^2$.

Explanation:

- $a^2 + b^2 = c^2$

- The equation stays in balance if you do the same thing on both sides of the equal sign.

- Subtract b^2 from both sides of the equation.

- $a^2 + b^2 - b^2 = c^2 - b^2$

- $a^2 = c^2 - b^2$

Practice with the following example:

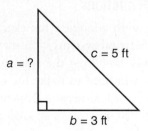

$a = ?$

$c = 5$ ft

$b = 3$ ft

$$a^2 = c^2 - b^2$$
$$a^2 = 5^2 - 3^2$$
$$a^2 = 25 \text{ ft}^2 - 9 \text{ ft}^2$$
$$a^2 = 16 \text{ ft}^2$$
$$a = \sqrt{16 \text{ ft}^2}$$
$$a = 4 \text{ ft}$$

The same can be done to find side b when side a and the hypotenuse are known.

Special Right Triangles

The Pythagorean Theorem can be used to verify that a corner is 90° by measuring along the two sides, then checking the length of the hypotenuse between them. This is usually simplified by using 3 and 4 units as the sides. If 3 and 4 units are used, the hypotenuse of a square corner is 5 units. This is called a 3-4-5 triangle. If 6 and 8 units are used, the hypotenuse is 10 units. This is a multiple of the 3-4-5 triangle. These right triangles are preferred for easy calculations. Note that all sides are whole numbers in a 3-4-5 triangle or any of its multiples.

Practice A-13

Find the length of the unknown side to the nearest 1/10 of an inch.

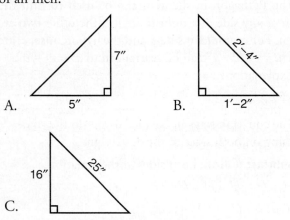

Trigonometric Functions

When working with triangles, for example, as part of your work on electrical systems, you may need to know more than simply how long the sides of a triangle are. Sometimes, you will need to make use of basic trigonometric functions: sine, cosine, and tangent. Trigonometric functions are defined in terms of an angle, θ, in a right triangle, **Figure A-18**.

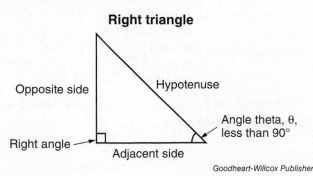

Right triangle

Goodheart-Willcox Publisher

Figure A-18.

Use the mnemonic device SOH CAH TOA to remember each trig function:

- **SOH.** Sine (S) equals opposite (O) over hypotenuse (H).

$$\sin(\theta) = \frac{\text{opposite}}{\text{hypotenuse}}$$

- **CAH.** Cosine (C) equals adjacent (A) over hypotenuse (H).

$$\cos(\theta) = \frac{\text{adjacent}}{\text{hypotenuse}}$$

- **TOA.** Tangent (T) equals opposite (O) over adjacent (A).

$$\tan(\theta) = \frac{\text{opposite}}{\text{adjacent}}$$

You can use the trigonometric function buttons on your scientific calculator to find sine, cosine, and tangent if you have the value for angle θ but do not know the lengths of the sides.

Practice A-14

Find sine, cosine, and tangent for the following triangles.

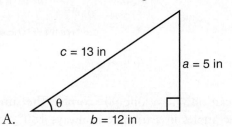

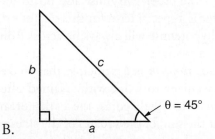

VECTOR ADDITION

Vectors are unlike other values you have calculated so far in that they have both magnitude and direction. For example, traveling at 15 mph north is not the same as traveling 15 mph south. Though they have the same magnitude, these speeds are in opposite directions. Vectors are useful when such distinctions are necessary.

As an industrial maintenance technician, you may need to use vectors to describe electrical systems, such as when resistive and reactive components are out of phase with each other. (See Chapter 22, *Alternating Current*.) A common way to add vectors is by the tip-to-tail method, **Figure A-19**. Simply place the start of one vector (tip) at the end of the other (tail) while preserving the direction of the vectors. Draw the resultant vector from the tail of the first vector to the tip of the second vector. Then, use the Pythagorean theorem to find the resultant vector. For the example given in **Figure A-19**, the resultant vector can be calculated as follows:

$$c^2 = a^2 + b^2$$
$$V^2 = R^2 + X^2$$
$$V^2 = 10^2 + 12^2$$
$$V^2 = 100\ \Omega^2 + 144\ \Omega^2$$
$$V = \sqrt{244\ \Omega^2}$$
$$V = 15.6\ \Omega$$

Tip-to-Tail Method of Vector Addition

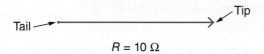

$R = 10\ \Omega$

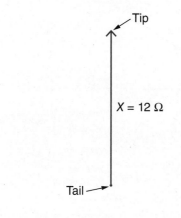

$X = 12\ \Omega$

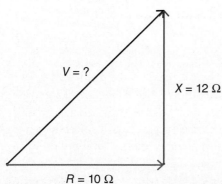

Goodheart-Willcox Publisher

Figure A-19.

Practice A-15

Find the magnitude of the resultant vector shown below. Confirm that your answer is less than the sum of the magnitudes of the two given vectors.

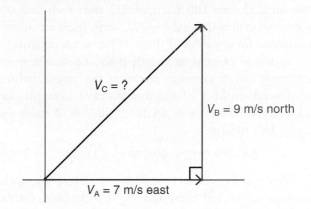

$V_C = ?$

$V_B = 9$ m/s north

$V_A = 7$ m/s east

COMPUTING AVERAGES

An average is a typical value of one unit in a group of units. For example, if 4 windows have areas of 12.0 ft², 11.2 ft², 11.2 ft², and 14.5 ft²; the average size of one of those windows is 12.2 ft². The average is computed by adding all of the units and dividing that sum by the number of units in the group.

```
 12.0 ft²
 11.2 ft²
 11.2 ft²
 14.5 ft²
 48.9 ft²

       12.22
 4 )48.90    The quotient should be rounded off to the
    4        same number of decimal places as is used
    08       in the problem.
     8
    09
     8
    10
```

Average window size is 12.2 ft²

Practice A-16

Compute the averages of these groups.

A. 14, 14.4, 14.5, 15

B. 80 lb, 83 lb, 88 lb, 79.5 lb, 81.6 lb, 84 lb

C. 11 cubic yards, 13 cubic yards, 11.5 cubic yards, 12 cubic yards, 12.8 cubic yards

PERCENT AND PERCENTAGE

A *percent* is one part in a hundred. One penny is 1% of a dollar. Twenty-five cents is 25% of a dollar. On the other hand, 25 cents is 50% of a half-dollar. If the half-dollar were divided into 100 parts of 1/2 cent each and the quarter were also divided into 1/2-cent increments, the quarter would equal 50 of those 1/2-cent increments.

Think of percent as hundredths. To find a given percentage of an amount, multiply the amount times the desired number of hundredths. For example, say you want to find 12% of $4.40. 12% is 0.12 times the whole. This means,

$$\$4.40 \times 0.12 = \$0.528 \text{ or } 53 \text{ cents}$$

Percent is sometimes interchanged with *percentage* in common usage, but there is a slight difference. *Percent* should be used when a specific number is used, such as 15% of the labor force. *Percentage* should be used when no specific number is used with the term, such as a large percentage of our homes are green.

Using your knowledge from the earlier section on equations and solving for an unknown, you can calculate the whole if you know the percentage. For instance, what was the total spent on tools if $22.00 was spent on sales tax and the tax rate is 8%?

1. Write an equation with the facts you know.

$$\$22.00 = 8\% \times \text{total}$$

2. Write percent as hundredths.

$$\$22.00 = 0.08 \times \text{total}$$

3. Divide each side of the equation by 0.08.

$$\text{total} = \$275$$

Practice A-17

A. What is 10 percent of 225?

B. What is 65 percent of $350.50?

C. If merchandise cost $22.25 and the bill comes to $23.14, what percentage was added for sales tax?

D. If a 500-gallon tank is 60 percent full of water, how much water does it contain?

Appendix B

Reference Section

The following pages contain tables, charts, and other materials that will be useful as reference in a variety of technical maintenance areas. To make locating information easier, the material in this section is listed below, along with the page number.

Conversion Table: US Customary to SI Metric

When You Know:	Multiply By:		To Find:
	Very Accurate	Approximate	
Length			
inches	*25.4		millimeters
inches	* 2.54		centimeters
feet	* 0.3048		meters
feet	*30.48		centimeters
yards	* 0.9144	0.9	meters
miles	* 1.609344	1.6	kilometers
Weight			
grains	15.43236	15.4	grams
ounces	*28.349523125	28.0	grams
ounces	* 0.028349523125	0.028	kilograms
pounds	* 0.45359237	0.45	kilograms
short ton	* 0.90718474	0.9	tonnes
Volume			
teaspoons		5.0	milliliters
tablespoons		15.0	milliliters
fluid ounces	29.57353	30.0	milliliters
cups		0.24	liters
pints	* 0.473176473	0.47	liters
quarts	* 0.946352946	0.95	liters
gallons	* 3.785411784	3.8	liters
cubic inches	* 0.016387064	0.02	liters
cubic feet	* 0.028316846592	0.03	cubic meters
cubic yards	* 0.764554857984	0.76	cubic meters
Area			
square inches	* 6.4516	6.5	square centimeters
square feet	* 0.09290304	0.09	square meters
square yards	* 0.83612736	0.8	square meters
square miles	* 2.58998811	2.6	square kilometers
acres	* 0.40468564224	0.4	hectares
Temperature			
Fahrenheit	*5/9 (after subtracting 32)		Celsius
Density			
pounds per cubic feet	1.602×10	16	kilograms per cubic meter
Force			
ounces (F)	2.780×10^{-1}		newtons
pounds (F)	4.448×10^{-3}		kilonewtons
kips	4.448		meganewtons
Stress			
pounds/square inch (psi)	6.895×10^{-3}		megapascals
kips/square inch (ksi)	6.895		megapascals
Torque			
ounce-inches	7.062×10^{3}		newton-meters
pound-inches	1.130×10^{-1}		newton-meters
pound-feet	1.356		newton-meters

* = Exact

Conversion Table: SI Metric to US Customary

When You Know:	Multiply By:		To Find:
	Very Accurate	Approximate	
Length			
millimeters	0.0393701	0.04	inches
centimeters	0.3937008	0.4	inches
meters	3.280840	3.3	feet
meters	1.093613	1.1	yards
kilometers	0.621371	0.6	miles
Weight			
grains	0.00228571	0.0023	ounces
grams	0.03527396	0.035	ounces
kilograms	2.204623	2.2	pounds
tonnes	1.1023113	1.1	short tons
Volume			
milliliters		0.2	teaspoons
milliliters	0.06667	0.067	tablespoons
milliliters	0.03381402	0.03	fluid ounces
liters	61.02374	61.024	cubic inches
liters	2.113376	2.1	pints
liters	1.056688	1.06	quarts
liters	0.26417205	0.26	gallons
liters	0.03531467	0.035	cubic feet
cubic meters	61023.74	61023.7	cubic inches
cubic meters	35.31467	35.0	cubic feet
cubic meters	1.3079506	1.3	cubic yards
cubic meters	264.17205	264.0	gallons
Area			
square centimeters	0.1550003	0.16	square inches
square centimeters	0.00107639	0.001	square feet
square meters	10.76391	10.8	square feet
square meters	1.195990	1.2	square yards
square kilometers	0.38610216	0.4	square miles
hectares	2.471054	2.5	acres
Temperature			
Celsius	*9/5 (then add 32)		Fahrenheit

* = Exact

Goodheart-Willcox Publisher

Decimal Conversion Chart

Fraction		Inches	mm
	1/64	.01563	.397
1/32		.03125	.794
	3/64	.04688	1.191
1/16		.0625	1.588
	5/64	.07813	1.984
3/32		.09375	2.381
	7/64	.10938	2.778
1/8		.12500	3.175
	9/64	.14063	3.572
5/32		.15625	3.969
	11/64	.17188	4.366
3/16		.18750	4.763
	13/64	.20313	5.159
7/32		.21875	5.556
	15/64	.23438	5.953
1/4		.25000	6.350
	17/64	.26563	6.747
9/32		.28125	7.144
	19/64	.29688	7.541
5/16		.31250	7.938
	21/64	.32813	8.334
11/32		.34375	8.731
	23/64	.35938	9.128
3/8		.37500	9.525
	25/64	.39063	9.922
13/32		.40625	10.319
	27/64	.42188	10.716
7/16		.43750	11.113
	29/64	.45313	11.509
15/32		.46875	11.906
	31/64	.48438	12.303
1/2		.50000	12.700

Fraction		Inches	mm
	33/64	.51563	13.097
17/32		.53125	13.494
	35/64	.54688	13.891
9/16		.56250	14.288
	37/64	.57813	14.684
19/32		.59375	15.081
	39/64	.60938	15.478
5/8		.62500	15.875
	41/64	.64063	16.272
21/32		.65625	16.669
	43/64	.67188	17.066
11/16		.68750	17.463
	45/64	.70313	17.859
23/32		.71875	18.256
	47/64	.73438	18.653
3/4		.75000	19.050
	49/64	.76563	19.447
25/32		.78125	19.844
	51/64	.79688	20.241
13/16		.81250	20.638
	53/64	.82813	21.034
27/32		.84375	21.431
	55/64	.85938	21.828
7/8		.87500	22.225
	57/64	.89063	22.622
29/32		.90625	23.019
	59/64	.92188	23.416
15/16		.93750	23.813
	61/64	.95313	24.209
31/32		.96875	24.606
	63/64	.98438	25.003
1		1.00000	25.400

Goodheart-Willcox Publisher

Key and Keyseat Dimensions for Various Shaft Diameters*

Shaft Diameter		Key Size			Keyseat Depth	
				Height		
From	To (Incl)	Width	Square	Rectangular	Square	Rectangular
5/16	7/16	3/32	3/32	NA	3/64	NA
7/16	9/16	1/8	1/8	3/32	1/16	3/64
9/16	7/8	3/16	3/16	1/8	3/32	1/16
7/8	1 1/4	1/4	1/4	3/16	1/8	3/32
1 1/4	1 3/8	5/16	5/16	1/4	5/32	1/8
1 3/8	1 3/4	3/8	3/8	1/4	3/16	1/8
1 3/4	2 1/4	1/2	1/2	3/8	1/4	3/16
2 1/4	2 3/4	5/8	5/8	7/16	5/16	7/32
2 3/4	3 1/4	3/4	3/4	1/2	3/8	1/4
3 1/4	3 3/4	7/8	7/8	5/8	7/16	5/16
3 3/4	4 1/2	1	1	3/4	1/2	3/8
4 1/2	5 1/2	1 1/4	1 1/4	7/8	5/8	7/16
5 1/2	6 1/2	1 1/2	1 1/2	1	3/4	1/2

Goodheart-Willcox Publisher

*All dimensions given in inches.

ISO 4406 Hydraulic Fluid Cleanliness Recommendations

System Components	System Pressure	Cleanliness Recommendation
Servo valves	any	15/13/10
Slip-in cartridge valves	>2000 psig	19/17/14
Screw-in cartridge valves	>2000 psig	17/15/12
Directional valves	>2000 psig	19/17/14
Variable piston pump	>2000 psig	16/14/12
Cylinders	<2175 psig	20/18/15

Goodheart-Willcox Publisher

ISO 4406 Cleanliness Standard	
ISO Range Code	Range of Particles in 1 ml of Fluid
30	5,000,000–10,000,000
29	2,500,000–5,000,000
28	1,300,000–2,500,000
27	640,000–1,300,000
26	320,000–640,000
25	160,000–320,000
24	80,000–160,000
23	40,000–80,000
22	20,000–40,000
21	10,000–20,000
20	5000–10,000
19	2500–5000
18	1300–2500
17	640–1300
16	320–640
15	160–320
14	80–160
13	40–80
12	20–40
11	10–20
10	5–10
9	2.5–5
8	1.3–2.5
7	0.64–1.3
6	0.32–0.64
5	0.16–0.32
4	0.08–0.16
3	0.04–0.08
2	0.02–0.04
1	0.01–0.02

Goodheart-Willcox Publisher

		Steel Pipe Dimensions							
		Schedule 40			Schedule 80				
Nominal Pipe Size (inches)	Outside Diameter (inches)	Wall Thickness (inches)	Inside Diameter (inches)	Weight (lb/ft)	Wall Thickness (inches)	Inside Diameter (inches)	Weight (lb/ft)	Threads per Inch
1/8	0.405	0.07	0.27	0.24	0.10	0.22	0.31	27
1/4	0.540	0.09	0.36	0.42	0.12	0.30	0.54	18
3/8	0.675	0.09	0.49	0.57	0.13	0.42	0.74	18
1/2	0.840	0.11	0.62	0.85	0.15	0.55	1.00	14
3/4	1.050	0.11	0.82	1.13	0.15	0.74	1.47	14
1	1.315	0.13	1.05	1.68	0.18	0.96	2.17	11 1/2
1 1/4	1.660	0.14	1.38	2.27	0.19	1.28	3.00	11 1/2
1 1/2	1.900	0.15	1.61	2.72	0.20	1.50	3.65	11 1/2
2	2.375	0.15	2.07	3.65	0.22	1.94	5.02	11 1/2
2 1/2	2.875	0.20	2.47	5.79	0.28	2.32	7.66	8
3	3.500	0.22	3.07	7.58	0.30	2.90	10.30	8
3 1/2	4.000	0.23	3.55	9.11	0.32	3.36	12.50	8
4	4.500	0.24	4.03	10.79	0.34	3.83	14.90	8
5	5.563	0.26	5.05	14.61	0.38	4.81	20.80	8
6	6.625	0.28	6.07	18.97	0.43	5.76	28.60	8
8	8.625	0.32	7.98	28.55	0.50	7.63	43.40	8
10	10.750	0.37	10.02	40.48	0.59	9.56	64.40	8
12	12.750	0.41	11.94	53.60	0.69	11.38	88.60	8
14	14.000	0.44	13.13	63.00	0.75	12.50	107.00	8
16	16.000	0.50	15.00	78.00	0.84	14.31	137.00	8
18	18.000	0.56	16.88	105.00	0.94	16.13	171.00	8
20	20.000	0.59	18.81	123.00	1.03	17.94	209.00	8
24	24.000	0.69	22.63	171.00	1.22	21.56	297.00	8

Goodheart-Willcox Publisher

PVC Pipe Dimensions

Nominal Pipe Size (inches)	Outside Diameter (inches)	Schedule 40		Schedule 80	
		Min. Wall Thickness (inches)	Inside Diameter (inches)	Min. Wall Thickness (inches)	Inside Diameter (inches)
1/2	0.840	0.109	0.622	0.147	0.546
3/4	1.050	0.113	0.824	0.154	0.742
1	1.315	0.133	1.049	0.179	0.957
1 1/4	1.660	0.140	1.380	0.191	1.278
1 1/2	1.900	0.145	1.610	0.200	1.500
2	2.375	0.154	2.067	0.218	1.939
2 1/2	2.875	0.203	2.469	0.276	2.323
3	3.500	0.216	3.068	0.300	2.900
4	4.500	0.237	4.026	0.337	3.826
5	5.563	0.258	5.047	0.375	4.813
6	6.625	0.280	6.065	0.432	5.761
8	8.625	0.322	7.981	0.500	7.625
10	10.750	0.365	10.020	0.593	9.564
12	12.750	0.406	11.938	0.687	11.376
14	14.000	0.437	13.124	0.750	12.500
16	16.000	0.500	15.000	0.843	14.314

Goodheart-Willcox Publisher

Temperature Ratings of Common Conductor Insulation Types

Insulation Type	140°F (60°C)	167°F (75°C)	194°F (90°C)
THHN			✓
THHW		✓ (wet)	✓ (dry)
THW		✓	
THW-2			✓
THWN		✓	
THWN-2			✓
TW	✓		
XHH			✓
XHHN			✓
XHHW		✓ (wet)	✓ (dry)
XHHW-2			✓
XHWN		✓	
XHWN-2			✓
UF	✓		
USE		✓	
USE-2			✓

Goodheart-Willcox Publisher

Allowable Ampacity for Insulated Conductors*							
		Copper			Aluminum		
		Temperature Rating of Insulation			Temperature Rating of Insulation		
Size (AWG or kcmil)	Area (circular mils)	140°F (60°C)	167°F (75°C)	194°F (90°C)	140°F (60°C)	167°F (75°C)	194°F (90°C)
18	1620	—		14	—	—	—
16	2580	—		18	—	—	—
14	4110	15	20	25	—	—	—
12	6530	20	25	30	15	20	25
10	10,380	30	35	40	25	30	35
8	16,510	40	50	55	35	40	45
6	26,240	55	65	75	40	50	55
4	41,740	70	85	95	55	65	75
3	52,620	85	100	115	65	75	85
2	66,360	95	115	130	75	90	100
1	83,690	110	130	145	85	100	115
1/0	105,600	125	150	170	100	120	135
2/0	133,100	145	175	195	115	135	150
3/0	167,800	165	200	225	130	155	175
4/0	211,600	195	230	260	150	180	205
250	250,000	215	255	290	170	205	230
300	300,000	240	285	320	195	230	260
350	350,000	260	310	350	210	250	280
400	400,000	280	335	380	225	270	305

*Allowable ampacity for no more than three insulated conductors in conduit, cable, or underground at 86°F (30°C). Provided for reference use only. Refer to the complete and appropriate code requirements for actual design information and conductor sizing.

Goodheart-Willcox Publisher

Ampacity Reduction for More Than Three Current-Carrying Conductors	
Number of Current-Carrying Conductors	Ampacity Factor (%)
4–6	80
7–9	70
10–20	50
21–30	45
31–40	40
41+	35

Goodheart-Willcox Publisher

Ambient Temperature Correction Factors				
Ambient Temperature		**Conductor Temperature Rating**		
		140°F	167°F	194°F
°F	°C	(60°C)	(75°C)	(90°C)
50	10 or less	1.29	1.20	1.15
51–59	11–15	1.22	1.15	1.12
60–68	16–20	1.15	1.11	1.08
69–77	21–25	1.08	1.05	1.04
78–86	26–30	1.00	1.00	1.04
87–95	31–35	0.91	0.94	0.96
96–104	36–40	0.82	0.88	0.91
105–113	41–45	0.71	0.82	0.87
114–122	46–50	0.58	0.75	0.82
123–131	51–55	0.41	0.67	0.76
132–140	56–60	—	0.58	0.71
141–149	61–65	—	0.47	0.65
150–158	66–70	—	0.33	0.58
159–167	71–75	—	—	0.50
168–176	76–80	—	—	0.41
177–185	81–85	—	—	0.29

Goodheart-Willcox Publisher

Maximum Overcurrent Protection for Small Conductors	
Conductor Size	**Maximum Overcurrent Protective Device Rating**
Copper	
18 AWG	7 A
16 AWG	10 A
14 AWG	15 A
12 AWG	20 A
10 AWG	30 A
Aluminum or Copper-Clad Aluminum	
12 AWG	15 A
10 AWG	25 A

Goodheart-Willcox Publisher

Guide for Shade Numbers

Operation	Electrode Size 1/32 in. (mm)	Arc Current (A)	Minimum Protective Shade	Suggested[1] Shade No. (Comfort)
Shielded metal arc welding	Less than 3 (2.5)	Less than 60	7	—
	3–5 (2.5–4)	60–160	8	10
	5–8 (4–6.4)	160–250	10	12
	More than 8 (6.4)	250–550	11	14
Gas metal arc welding and flux cored arc welding		Less than 60	7	—
		60–160	10	11
		160–250	10	12
		250–500	10	14
Gas tungsten arc welding		Less than 50	8	10
		50–150	8	12
		150–500	10	14
Air carbon arc cutting	(Light)	Less than 500	10	12
	(Heavy)	500–1000	11	14
Plasma arc welding		Less than 20	6	6 to 8
		20–100	8	10
		100–400	10	12
		400–800	11	14
Plasma arc cutting	(Light)(2)	Less than 300	8	9
	(Medium)(2)	300–400	9	12
	(Heavy)(2)	400–800	10	14
Torch brazing	—	—	—	3 or 4
Torch soldering	—	—	—	2
Carbon arc welding	—	—	—	14

	Plate Thickness			
	in.	mm		
Gas welding				
Light	Under 1/8	Under 3.2		4 or 5
Medium	1/8 to 1/2	3.2 to 12.7		5 or 6
Heavy	Over 1/2	Over 12.7		6 or 8
Oxygen cutting				
Light	Under 1	Under 25		3 or 4
Medium	1 to 6	25 to 150		4 or 5
Heavy	Over 6	Over 150		5 or 6

(1) As a rule of thumb, start with a shade that is too dark to see the weld zone. Then go to a lighter shade that gives sufficient view of the weld zone without going below the minimum. In oxyfuel gas welding or cutting where the torch produces a high yellow light, it is desirable to use a filter lens that absorbs the yellow or sodium line in the visible light of the (spectrum) operation.

(2) These values apply where the actual arc is clearly seen. Experience has shown that lighter filters may be used when the arc is hidden by the workpiece.

Goodheart-Willcox Publisher

Tap Drill Sizes

Probable Percentage of Full Thread Produced in Tapped Hole Using Stock Sizes of Drill

Tap	Tap Drill	Decimal Equivalent of Tap Drill	Theoretical % of Thread	Probable Oversize (Mean)	Probable Hole Size	Percentage of Thread
0–80	56	.0465	83	.0015	.0480	74
	3/64	.0469	81	.0015	.0484	71
1–64	54	.0550	89	.0015	.0565	81
	53	.0595	67	.0015	.0610	59
1–72	53	.0595	75	.0015	.0610	67
	1/16	.0625	58	.0015	.0640	50
2–56	51	.0670	82	.0017	.0687	74
	50	.0700	69	.0017	.0717	62
	49	.0730	56	.0017	.0747	49
2–64	50	.0700	79	.0017	.0717	70
	49	.0730	64	.0017	.0747	56
3–48	48	.0760	85	.0019	.0779	78
	5/64	.0781	77	.0019	.0800	70
	47	.0785	76	.0019	.0804	69
	46	.0810	67	.0019	.0829	60
	45	.0820	63	.0019	.0839	56
3–56	46	.0810	78	.0019	.0829	69
	45	.0820	73	.0019	.0839	65
	44	.0860	56	.0019	.0879	48
4–40	44	.0860	80	.0020	.0880	74
	43	.0890	71	.0020	.0910	65
	42	.0935	57	.0020	.0955	51
	3/32	.0938	56	.0020	.0958	50
4–48	42	.0935	68	.0020	.0955	61
	3/32	.0938	68	.0020	.0958	60
	41	.0960	59	.0020	.0980	52
5–40	40	.0980	83	.0023	.1003	76
	39	.0995	79	.0023	.1018	71
	38	.1015	72	.0023	.1038	65
	37	.1040	65	.0023	.1063	58
5–44	38	.1015	79	.0023	.1038	72
	37	.1040	71	.0023	.1063	63
	36	.1065	63	.0023	.1088	55
6–32	37	.1040	84	.0023	.1063	78
	36	.1065	78	.0026	.1091	71
	7/64	.1094	70	.0026	.1120	64
	35	.1100	69	.0026	.1126	63
	34	.1110	67	.0026	.1136	60
	33	.1130	62	.0026	.1156	55
6–40	34	.1110	83	.0026	.1136	75
	33	.1130	77	.0026	.1156	69
	32	.1160	68	.0026	.1186	60

Tap	Tap Drill	Decimal Equivalent of Tap Drill	Theoretical % of Thread	Probable Oversize (Mean)	Probable Hole Size	Percentage of Thread
8–32	29	.1360	69	.0029	.1389	62
	28	.1405	58	.0029	.1434	51
8–36	29	.1360	78	.0029	.1389	70
	28	.1405	68	.0029	.1434	57
	9/64	.1406	68	.0029	.1435	57
10–24	27	.1440	85	.0032	.1472	79
	26	.1470	79	.0032	.1502	74
	25	.1495	75	.0032	.1527	69
	24	.1520	70	.0032	.1552	64
	23	.1540	67	.0032	.1572	61
	5/32	.1563	62	.0032	.1595	56
	22	.1570	61	.0032	.1602	55
10–32	5/32	.1563	83	.0032	.1595	75
	22	.1570	81	.0032	.1602	73
	21	.1590	76	.0032	.1622	68
	20	.1610	71	.0032	.1642	64
	19	.1660	59	.0032	.1692	51
12–24	11/64	.1719	82	.0035	.1754	75
	17	.1730	79	.0035	.1765	73
	16	.1770	72	.0035	.1805	66
	15	.1800	67	.0035	.1835	60
	14	.1820	63	.0035	.1855	56
12–28	16	.1770	84	.0035	.1805	77
	15	.1800	78	.0035	.1835	70
	14	.1820	73	.0035	.1855	66
	13	.1850	67	.0035	.1885	59
	3/16	.1875	61	.0035	.1910	54
1/4–20	9	.1960	83	.0038	.1998	77
	8	.1990	79	.0038	.2028	73
	7	.2010	75	.0038	.2048	70
	13/64	.2031	72	.0038	.2069	66
	6	.2040	71	.0038	.2078	65
	5	.2055	69	.0038	.2093	63
	4	.2090	63	.0038	.2128	57
1/4–28	3	.2130	80	.0038	.2168	72
	7/32	.2188	67	.0038	.2226	59
	2	.2210	63	.0038	.2248	55
5/16–18	F	.2570	77	.0038	.2608	72
	G	.2610	71	.0041	.2651	66
	17/64	.2656	65	.0041	.2697	59
	H	.2660	64	.0041	.2701	59

(continued)

Tap Drill Sizes (continued)

Probable Percentage of Full Thread Produced in Tapped Hole Using Stock Sizes of Drill

Tap	Tap Drill	Decimal Equivalent of Tap Drill	Theoretical % of Thread	Probable Oversize (Mean)	Probable Hole Size	Percentage of Thread
5/16–24	H	.2660	86	.0041	.2701	78
	I	.2720	75	.0041	.2761	67
	J	.2770	66	.0041	.2811	58
3/8–16	5/16	.3125	77	.0044	.3169	72
	O	.3160	73	.0044	.3204	68
	P	.3230	64	.0044	.3274	59
3/8–24	21/64	.3281	87	.0044	.3325	79
	Q	.3320	79	.0044	.3364	71
	R	.3390	67	.0044	.3434	58
7/16–14	T	.3580	86	.0046	.3626	81
	23/64	.3594	84	.0046	.3640	79
	U	.3680	75	.0046	.3726	70
	3/8	.3750	67	.0046	.3796	62
	V	.3770	65	.0046	.3816	60
7/16–20	W	.3860	79	.0046	.3906	72
	25/64	.3906	72	.0046	.3952	65
	X	.3970	62	.0046	.4016	55
1/2–13	27/64	.4219	78	.0047	.4266	73
	7/16	.4375	63	.0047	.4422	58
1/2–20	29/64	.4531	72	.0047	.4578	65
9/16–12	15/32	.4688	87	.0048	.4736	82
	31/64	.4844	72	.0048	.4892	68
9/16–18	1/2	.5000	87	.0048	.5048	80
	33/64	.5156	65	.0048	.5204	58
5/8–11	17/32	.5313	79	.0049	.5362	75
	35/64	.5469	66	.0049	.5518	62
5/8–18	9/16	.5625	87	.0049	.5674	80
	37/64	.5781	65	.0049	.5831	58
3/4–10	41/64	.6406	84	.0050	.6456	80
	21/32	.6563	72	.0050	.6613	68
3/4–16	11/16	.6875	77	.0050	.6925	71
7/8–9	49/64	.7656	76	.0052	.7708	72
	25/32	.7812	65	.0052	.7864	61
7/8–14	51/64	.7969	84	.0052	.8021	79
	13/16	.8125	67	.0052	.8177	62
1–8	55/64	.8594	87	.0059	.8653	83
	7/8	.8750	77	.0059	.8809	73
	57/64	.8906	67	.0059	.8965	64
	29/32	.9063	58	.0059	.9122	54
1–12	29/32	.9063	87	.0060	.9123	81
	59/64	.9219	72	.0060	.9279	67
	15/16	.9375	58	.0060	.9435	52

Tap	Tap Drill	Decimal Equivalent of Tap Drill	Theoretical % of Thread	Probable Oversize (Mean)	Probable Hole Size	Percentage of Thread
1–14	59/64	.9219	84	.0060	.9279	78
	15/16	.9375	67	.0060	.9435	61
1 1/8–7	31/32	.9688	84	.0062	.9750	81
	63/64	.9844	76	.0067	.9911	72
	1	1.0000	67	.0070	1.0070	64
	1 1/64	1.0156	59	.0070	1.0226	55
1 1/8–12	1 1/32	1.0313	87	.0071	1.0384	80
	1 3/64	1.0469	72	.0072	1.0541	66
1 1/4–7	1 3/32	1.0938	84			
	1 7/64	1.1094	76			
	1 1/8	1.1250	67			
1 1/4–12	1 5/32	1.1563	87			
	1 11/64	1.1719	72			
1 3/8–6	1 3/16	1.1875	87			
	1 13/64	1.2031	79	No test results available		
	1 7/32	1.2188	72			
	1 15/64	1.2344	65			
1 3/8–12	1 9/32	1.2813	87	Reaming recommended		
	1 19/64	1.2969	72			
1 1/2–6	1 5/16	1.3125	87			
	1 21/64	1.3281	79			
	1 11/32	1.3438	72			
	1 23/64	1.3594	65			
1 1/2–12	1 13/32	1.4063	87			
	1 27/64	1.4219	72			

Taper Pipe		Straight Pipe	
Thread	Drill	Thread	Drill
1/8–27	R	1/8–27	S
1/4–18	7/16	1/4–18	29/64
3/8–18	37/64	3/8–18	19/32
1/2–14	23/32	1/2–14	47/64
3/4–14	59/64	3/4–14	15/16
1–11 1/2	1 5/32	1–11 1/2	1 3/16
1 1/4–11 1/2	1 1/2	1 1/4–11 1/2	1 33/64
1 1/2–11 1/2	1 47/64	1 1/2–11 1/2	1 3/4
2–11 1/2	2 7/32	2–11 1/2	2 7/32
2 1/2–8	2 5/8	2 1/2–8	2 21/32
3–8	3 1/4	3–8	3 9/32
3 1/2–8	3 3/4	3 1/2–8	3 25/32
4–8	4 1/4	4–8	4 9/32

Goodheart-Willcox Publisher

RESISTOR COLOR CODES

Resistors are commonly used to control voltage and current levels in a circuit. The value of a resistor is typically indicated by colored bands where each color indicates a specific value. The number of colored bands and where they are placed on a resistor is significant to resistor specification. Refer to the following chart for guidance on how to decode a resistor and determine its value.

Resistors may have as few as three bands or as many as six, which denote their value. If a resistor has only three bands, it is decoded the same as a four-band resistor, except the tolerance is 20%. Six-band resistors are read the same as five-band resistors, except the last band indicates the temperature coefficient. Refer to the following examples of how to decode resistors:

Resistor Color Codes				
Color	Value	Multiplier	Tolerance	Temperature Coefficient
Black	0	1		
Brown	1	10	1%	100 ppm
Red	2	100	2%	50 ppm
Orange	3	1k		15 ppm
Yellow	4	10k		25 ppm
Green	5	100k	0.5%	
Blue	6	1M	0.25%	
Violet	7	10M	0.1%	
Gray	8	100M		
White	9	1G		
Gold		0.1	5%	
Silver		0.01	10%	

Goodheart-Willcox Publisher

Four-Band Resistor, 20 kΩ ± 5%

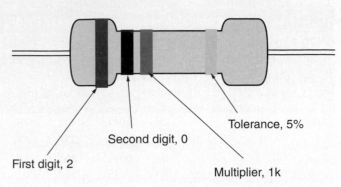

First digit, 2
Second digit, 0
Tolerance, 5%
Multiplier, 1k

Goodheart-Willcox Publisher

Five-Band Resistor, 460 kΩ ± 1%

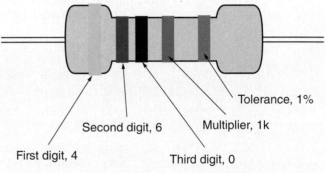

First digit, 4
Second digit, 6
Third digit, 0
Multiplier, 1k
Tolerance, 1%

Goodheart-Willcox Publisher

SCIENTIFIC NOTATION AND SI PREFIXES

In some cases, you may have to deal with very large or very small numbers. It is inconvenient to write all the decimal places for extremely small amounts or all the zeros for extremely large amounts. Scientists have a system of using powers of 10 written as exponents to represent the number of decimal places or the number of zeros in a number. This is called *scientific notation*. This notation is frequently used in the equations that describe electrical systems, so it is important to understand how the system of scientific notation works and how to convert between units.

For example, 5×10^3 is the same as writing 5000, and 5×10^9 is equivalent to 5,000,000,000. Observe that scientific notation is simpler and clearer, and it leaves less room for error. The previous examples use a positive exponent, which tells you how many zeros to add to a given value. Positive exponents correspond to larger numbers. If written with a negative exponent, scientific notation indicates the number of decimal places. For example, $2 \times 10^{-3} = 0.002$ and $27 \times 10^{-6} = 0.000027$.

Negative exponents indicate smaller numbers, usually values less than zero.

You probably noticed that all exponents in the examples provided above are multiples of 3. This practice is not an absolute rule, but it is a convention in scientific and engineering fields. The value 100,000 written as 1×10^5 is perfectly acceptable in a mathematical context. In *engineering notation*, a variation on scientific notation, 1×10^5 is not inherently incorrect, but a more accepted notation would be 100×10^3. Using multiples of 3 facilitates another kind of shorthand, SI prefixes.

Now that you have an understanding of scientific notation, the next step is learning the SI prefixes. SI stands for *Systeme International*, the French name for the internationally accepted system of conventions and units of measurement. SI units (such as the ampere, volt, coulomb, and ohm) and prefixes are used throughout this text and in most scientific and technical fields. SI prefixes provide a method of mathematical shorthand that makes it easier to write large or small numbers. Refer to the chart below to see how scientific notation, longhand notation, and symbolic representation relate to each other.

SI Prefixes				
Prefix	Symbol	Value	Exponent	Name
Tera	T	1,000,000,000,000	10^{12}	Trillion
Giga	G	1,000,000,000	10^9	Billion
Mega	M	1,000,000	10^6	Million
Kilo	k	1,000	10^3	Thousand
				(Base unit)
Milli	m	0.001	10^{-3}	One thousandth
Micro	μ	0.000001	10^{-6}	One millionth
Nano	n	0.000000001	10^{-9}	One billionth
Pico	p	0.000000000001	10^{-12}	One trillionth

Goodheart-Willcox Publisher

Glossary

A

abrasive cutoff saw. A power tool similar to an angle or die grinder, but fitted with a thin abrasive wheel used for cutting metal. (4)

abrasive wear. A type of mechanical wear that takes place when particles contained in lubricant scrape against the mating surfaces and remove material. (14)

accelerometer. A device that measures the changes in motion of an object, including vibrations. (7)

accumulation. A type of valve operation in which the relief valve starts to open at its set point and becomes fully open at a higher temperature. (40)

accumulator. A subsystem on a conveyor line that provides an overflow or staging area for products on the line. (13)

accuracy. A term that expresses how close a measurement is to a target or standard value. (14)

acetylene gas. A compound of carbon and hydrogen formed by mixing calcium carbide in water. (39)

actuator. A cylinder or motor in a fluid power system that uses the fluid power to create mechanical motion, converting the fluid power into actual work. (15)

addendum. The top half of a gear tooth. (12)

adhesive wear. A type of mechanical wear that occurs when sliding surfaces contact each other with enough force to remove material from one or both of the surfaces. (14)

adjustable wrench. A wrench with one fixed jaw and one movable jaw, with a screw-like adjustment located on the portion of the wrench between the jaws and the handle. Also called *crescent wrench*. (4)

adsorption. The process by which moisture is attached to the surface area of a media. (17)

AED (automatic electrical defibrillator). A device that may be used to restart the heart and save an individual's life. The AED shocks the victim's heart back into a normal rhythm, allowing it to once again begin pumping blood. (2)

airend. The screw element of a rotary compressor. (17)

air handling unit (AHU). A heating or cooling unit that is placed within the building envelope. (36)

air impact wrench. A wrench that delivers short, concentrated bursts of torque and is capable of quickly tightening or loosening nuts or bolts that require extreme torque to install or remove. (4)

air ratchet. A ratchet powered by an air mechanism. (4)

alphabet of lines. A system of line conventions that define how lines appear on drawings. (6)

alternating current (AC). An electric current that changes direction at regular time intervals (periodically). (19, 22)

alternator. A device that turns rotational mechanical energy into electricity and produces alternating current. (25)

American Wire Gage (AWG). A standard unit classification for wire diameter. As gage increases, the diameter of the wire decreases. (23)

ampacity. The amperage, or amount of current, a conductor is capable of safely carrying. (23, 27)

ampere. The SI unit for electric current or the rate of electron flow, abbreviated as A. One ampere equals one coulomb per second. (19)

amplifier. An electronic device that can increase the power of a signal. Also called *amp*. (30)

anaerobic adhesive. A sealing agent that hardens to form a tight seal under conditions without the presence of oxygen. (17)

analog sensor. A sensor that provides a signal that is proportional to what is being measured, allowing knowledge of the exact position of a cylinder. (17)

angle grinder. A handheld power tool that has an abrasive, motor-driven grinding wheel mounted at a 90° angle to the shaft. Most commonly used for welding and cutting. (4)

anion. A negatively charged ion. (19)

annealing. The process of heating a metal and allowing it to cool slowly in order to reduce the metal's hardness. (41)

antioxidant. An oil additive that minimizes thickening of lubricants when oxidation takes place. (10)

anti-wear additive. An oil additive that reduces metal-to-metal contact with high load. (10)

apparent power. The power consumed by a circuit. Measured in volt-amperes (VA) and designated by the variable S. (22)

applications. The user programs that contain screen layouts and tags, or memory addresses, that are used for indication and control purposes. (33)

arbitrary function generator (AFG). A programmable device that can generate almost any analog or digital signal. (21)

Note: The number in parentheses following each definition indicates the chapter in which the term can be found.

arc. The flow of electricity across an air gap during welding. (37)

arc-fault circuit interrupter (AFCI). A circuit breaker that senses arcing and interrupts the circuit. (23)

arc welding. A welding process that uses heat produced from an electric arc to melt and weld two metals. (37)

asperity. A small imperfection left on the surface of a material after machining. (10)

assembly drawing. A drawing that shows the individual parts of an assembly and indicates how they are put together. (6)

atom. The smallest part of an element. (19)

attitude. A person's outlook on life. (1)

autonomous maintenance. A part of a total productive maintenance or predictive approach involving the machine operator and the maintenance technician working as a team, with the operator trained to perform small routine maintenance tasks and to detect abnormalities before they become failures in the machine's operation. (3)

auto-ranging. The automatic calculation of the correct range on a digital multimeter without operator intervention. (21)

autotransformer. A type of transformer with only one winding that shares one connection between the primary and secondary. It may have one or more taps and may be used as a step-up or step-down transformer. (24)

auxiliary contact. An additional contact often used in motor control circuits with relay logic. Auxiliary contacts may be directly operated by the coil or mechanically connected to the main action. (26)

axial load. A load that exerts force parallel to the shaft. Also called *thrust*. (10)

axial movement. A movement parallel to a shaft. Also called *axial runout*. (10)

azeotrope. A refrigerant blend that has a fixed boiling and condensing point. (35)

B

back face. The portion of a seal that provides transverse strength and normally faces out. (10)

backhand welding. A type of welding completed with the welding gun pointing back at the weld as the weld progresses in the opposite direction. (38)

backlash. The side-to-side clearance between mating gear teeth. (12)

backpressure. The amount of force applied between products on a conveyor line. Also known as *line pressure*. (13)

balanced dial indicator. A dial indicator that reads from 0 to 100 thousandths of an inch and can travel several revolutions. (11)

ball bearing. A sphere-shaped rolling element bearing that reduces friction and supports moderate loads. (10)

ball-end hex key. A type of hex key that allows a screw to be driven at an angle. (4)

ball-peen hammer. A hammer on which one side of the head is spherical in shape, used when working with metal or heavy-duty mechanical assemblies. (4)

band-pass filter. A circuit that allows the passing of only a specific group of frequencies near the resonant frequency and rejects all others. Also called *acceptor circuit*. (22)

band-reject filter. A circuit that rejects the passing of frequencies at or near the resonant frequency and passes frequencies further away from resonance. Also called *reject circuit*. (22)

bandwidth. The range of frequencies near resonant frequency, f_0, that produce resonance effects in a circuit. (22)

bar sag. The amount that an indicator's mounting and supporting hardware bends as a result of gravity. (11)

base metal. The metal to be welded. (37)

basic dimension. A theoretically exact dimension. (6)

bathtub curve. A graphical representation of the failure rate of products divided into three regions: early-life failure, useful life, and wearout. (14)

battery. A simple source of electricity with two terminals, a positive terminal (cathode) and a negative terminal (anode). The negative terminal has a surplus of electrons and the positive terminal has a deficit of electrons. (19)

bearing. A mechanical component that supports a moving part, generally a rotating shaft. (10)

belt. A continuous band used to transfer power from one pulley wheel to another. (12)

belt tension gauge. A measurement tool that reads the amount of deflection of a belt, when used with a straight-edge, and determines the tension of the belt. Often used in combination with a steel bench rule. (7)

bench rule. A tool capable of moderately accurate measurements for objects larger than 6″ but not large enough to require a steel tape. (4)

beta ratio. A filter rating representing a comparison of the number of particles of a particular size upstream of the filter to the number downstream, thus indicating how many have been removed from the fluid that flows through the filter. (15)

bimetallic overload. A type of overload that relies on current sensing through heat generation. A bimetallic overload contains a bimetallic strip located near a heating element that warps as temperature rises. (26)

bimetallic strip. A type of contact that consists of two layers made of different metals that react to temperature changes at different rates. (36)

binary search. A troubleshooting technique carried out by dividing a system in half and making a measurement to methodically determine where the problem is. Also called *divide-and-conquer approach* and *half-split method*. (7, 28)

bipolar junction transistor (BJT). A semiconductor device with three terminals, called the emitter, collector, and base, with one of the three terminals shared in common between the input and output circuits. The two types of bipolar junction transistors are referred to as the NPN and the PNP type. (30)

bit. A unit of information that is the result between two alternatives, such as on or off. (32)

blanking. The disabling of a section of a laser curtain's sensing field. (31)

bleed-off resistor. A resistor that drains the capacitor of a power supply circuit within a few minutes of it being de-energized, for safety purposes. Also called *bleeder resistor*. (31)

block diagram. A diagram that shows the basic operations of a machine and steps related to machine function. (28)

blocking. The use of physical barriers to prevent a machine or part from moving unexpectedly. (2)

blowdown. A periodic removal of water from the boiler through a drain to help remove sludge and scale particles in the boiler. The term can also refer to a type of valve operation in which an open relief valve does not close until the pressure falls a specified amount below its set point. (35, 40)

body language. A part of nonverbal communication that includes facial expressions and body posture. (1)

bolt. A fastener with a head at one end and external threads at the other to accept a nut. (5)

boundary film lubrication. A separation between opposing surfaces created by a thin film of lubricant on asperities serving as sole separation between surfaces. Also called *thin film lubrication*. (10)

bowline knot. A knot made on the end of a rope to secure a load. When tied properly, the bowline will not loosen or slip. (8)

bowline on a bight knot. A loop or eye knot in the middle of a rope that will not slip and can be easily adjusted up or down the length of the rope. (8)

Boyle's law. A statement that pressure and volume are inversely proportional in a gas at fixed temperature, expressed as $p \times V = k$. (15)

branched system. A distribution system in which the main line branches off into other lines. Larger loads must be supplied from the main header, or unacceptable pressure drops can result. (17)

breakdown maintenance. The maintenance performed after apparent problems arise. Also called *reactive maintenance*. (3)

breaker bar. A long bar that provides more torque when loosening especially difficult-to-remove fasteners. (4)

break line. A line used to show an imaginary break in an object where a portion is omitted. (6)

brittleness. A material's tendency to fracture under load with little or no deformation before fracturing. (14)

busbar. A flat, heavy copper conductor capable of carrying high currents. (27)

busway. A rectangular metal duct that encloses busbars. Also called *bus duct*. (27)

butt joint. Two pieces of metal welded together at their edges. (37)

C

cable. An assembly of multiple conductors within a sheath or outside covering. (23)

cable tie. A thin, flexible nylon strap with teeth that engage with the open of the tie, preventing it from loosening. (5)

caliper. A tool capable of very accurately measuring external, internal, and depth dimensions, especially of small parts. (4)

capacitance. The opposition to any change in voltage by a capacitor. The term also refers to the ability of a material to store an electrical charge. Designated by the variable C. (22, 34)

capacitive level sensor. A level sensor that uses a change in capacitance to produce a signal that indicates the level of the liquid. (34)

capacitive reactance. A description of the reactance of a capacitor. Designated by the variable X_C and measured in ohms, Ω. (22)

capacitive sensor. A proximity sensor that works by detecting a change in capacitance created by the proximity of a detected object. Sensor ranges depend largely on the diameter of the sensor, but are rarely greater than 10 mm. (31)

capacitor. A device formed when two conductors are in close proximity to each other and separated by an insulator or dielectric. (23)

capacitor-run motor. A two-phase motor created from a single-phase induction motor. A second set of windings is set 90° apart from the primary set and wired in series with a capacitor. Also called a *permanent-split capacitor motor*. (25)

capacitor-start, capacitor-run motor. A variation of the single-phase motor. It starts with both start and run capacitors in circuit, and the start winding is not switched out in run mode. (25)

capacitor-start motor. A variation of the capacitor-run motor. It has a larger value capacitor, heavier gage wire, and more turns in the start winding than in the primary winding, and provides more starting torque for heavier loads. (25)

capillary tube. A length of seamless tubing with a specific inside diameter that carries a pressure signal to the thermostatic expansive valve (TXV) in a refrigeration system. (35)

cap screw. A fastener that is similar to a bolt, but extends through a clearance hole in one part and screws into a second part without using a nut. (5)

carburizing flame. A flame produced when too much carbon or too little oxygen is produced. (39)

carcass. The center layer of a conveyor belt that is composed of various reinforcing materials. (13)

carryway. The low-friction top surface on which the belt of a plastic belt conveyor rides. (13)

cartridge valve. A valve with a compact design that can be used in manifold systems. (18)

cascade control. A control loop in which two or more controllers are used to control one or more process variables for better system control. (34)

catenary sag. The amount of belt that hangs on the underside of a conveyor between the supporting rollers or pulleys. (13)

cation. A positively charged ion. (19)

cat's-paw knot. A hitch knot allowing the use of the middle of a rope to attach to a hook or clevis. (8)

cavitation. A condition that occurs when vapor bubbles form at the suction side of a circulation pump, are drawn in by the pump, and then collapse as the hydraulic fluid is pressurized through the pump, creating noise and potentially damaging internal components. (15, 35, 40)

centerlines. The lines that represent axes of symmetrical objects. (6)

center punch. A tool used in combination with the prick punch to mark and dimple a metal object for drilling, specifically to create a larger dimple at the location where the prick punch has marked. (4)

centrifugal compressor. A type of compressor that uses an impeller to create centrifugal force to compress refrigerant. (35)

centrifugal switch. A mechanical device that opens once the shaft rotation reaches a specified rate of speed. (25)

ceramic capacitor. A capacitor formed when a disk of ceramic material has silver plating on each side. Often used when a high working voltage is required. (23)

chain pitch. The distance between the centers of the pins on each link of a roller chain. (12)

Charles's law. The statement that temperature and volume are directly related when pressure is held constant, expressed as $V / T = k$. (15)

chattering. The noise caused by the vibration of a disk in a fluid system due to flow disturbances inside the body of a valve. (40)

chisel. A tool used for cutting or chipping, especially if a job does not require a precise or clean cut. Also called *cold chisel*. (4)

chordal action. An action that occurs when a straight chain enters and revolves around a sprocket. (12)

circuit board. A board built of layered fiberglass mats and epoxy resin and used to connect electrical components. (23)

circuit breaker. A resettable overcurrent protection device. (23)

circular pitch. The distance between a point on the face of one gear tooth to the same point on the next tooth, measured along the pitch circle. (12)

circular saw. A power cutting tool that has a round blade. Often used to cut wood, though it is capable of cutting other materials as well. (4)

clamp-on ammeter. A device used to measure electric current through a noncontact method. It is capable of measuring higher currents than a digital multimeter, and it also includes a voltmeter, ohmmeter, diode tester, and capacitor tester. (7, 21)

claw hammer. A hammer used for driving and removing nails and prying pieces of wood apart when they have been nailed or stapled together. (4)

clearance. The amount of space between the top of the tooth on one gear and the bottom of the mating space on the opposite gear. (12)

clearance fit. A fit in which clear space exists between two parts when assembled. (6)

clevis. A U-shaped hook that is attached at the end of a chain by use of a pin. (8)

clevis fastener. An assembly consisting of a clevis, a clevis pin, and a cotter pin. (5)

clip. An external ring fastener installed into shaft grooves to hold mechanical parts in place. (5)

closed loop. A control system that includes feedback. Also called *closed control loop*. (34)

closed-loop system. A type of VFD control system that provides for direct feedback to the drive concerning the motor speed. (31)

combination wrench. A common type of wrench with one open end and one box or socket end. Also called *box-end/open-end wrench*. (4)

combined gas law. A law that compares a system's pressure, volume, and temperature before and after an event, expressed as $(p_1 \times V_1) / T_1 = (p_2 \times V_2) / T_2$. (15)

combustion. A chemical reaction between oxygen and fuel that produces energy in the form of heat. (35)

come-along. A manually powered hoist that uses a lever and gear reducers to lift and suspend a heavy load from a chain. (8)

command signal. An input signal that determines where the spool of the pilot valve will position. (18)

common-base circuit. A transistor circuit configuration in which the emitter input and collector output circuits share the base as a common terminal. (30)

common-collector circuit. A transistor circuit configuration in which the input is applied to the base, the output is taken from the emitter, and the common terminal for both circuits is the collector. Also called *emitter follower*. (30)

common-emitter circuit. A transistor circuit configuration in which an emitter/base current or voltage controls the emitter-to-collector current. (30)

commutator. A device that controls the speed and direction of a motor's rotation by periodically switching the direction of current, which reverses the magnetic field in the rotor. (25)

comparator. A decision-making operational amplifier (op amp) circuit that compares one analog voltage with another analog voltage. Often, one of the two voltages is a reference voltage, while the other is an actual input signal. (30)

complete circuit. The closed path of electricity from the negative terminal of a voltage source to a load and then from the load to the positive terminal of the voltage source. (19)

complex circuit. A circuit with some components connected in series and other components connected in parallel. (19, 20)

component terminal number. A number that helps clarify ladder diagrams when multiple terminals are used on a single component. (27)

compression. The joining of pipes using compression fittings, which make a watertight seal with the tightening of a threaded compression nut. (41)

compression spring. A spring that resists axial shortening as a load is applied. (9)

compressor. A component that produces compressed air for a pneumatic system by increasing the pressure of air by reducing its volume through one or several stages. (17)

computer-aided drafting (CAD). A process in which a computer is used with CAD software to produce drawings for manufacturing or construction. (6)

computerized maintenance management system (CMMS). An electronic computer-based system where all documented maintenance records can be stored. (3)

condensate. The liquid that forms as a result of the process of condensation. (35)

condenser coil. A component of a cooling system that removes the heat of vaporization from high-pressure refrigerant vapor and condenses the refrigerant back into the liquid. (35)

condition monitoring. The process of observing a machine's operation using tools to examine such items as vibration, temperature, and voltage/current draw that help identify any symptoms of malfunctions. (3)

conduction. A method of heat transfer in which heat is transferred from one molecule to another through physical contact. (35)

conductor. A low-resistance material that allows electricity to flow easily. (19)

conflict. A hostile situation resulting from opposing views. (1)

conflict management. An effort to prevent conflict from becoming a destructive force. (1)

constant current (CC). A current that is maintained at a constant amount for welding as voltage is changed. (37)

constant voltage (CV). A voltage that is maintained at a constant amount for welding as current is changed. (37)

constricting nozzle. An attachment to a plasma arc cutting torch body that has a small hole to force plasma gas to pass directly through the path of the arc produced. (39)

contact-controlled timer. A timer that is controlled by an off-delay as well as more programmable functions. (26)

contact flow meter. A flow meter that is physically placed in the line of flow. (34)

contactor. A device that provides resistance to a motor at startup and removes resistance when the motor reaches running speed. (26)

contact seal. A seal that comes into contact with rolling or sliding surfaces. (10)

continuing education. The training that goes beyond regular classes. This can include coursework for workforce training or personal enrichment. (1)

continuity tester. A test lamp with an added power source used to indicate whether a circuit is complete. (21)

continuous dial indicator. A dial indicator that reads from 0 at the twelve o'clock position to 50 at the six o'clock position in both the positive and negative direction. (11)

control loop. A component of a control system that monitors a process variable. (34)

ControlNet. An industrial network protocol developed by Allen-Bradley that is a bus topology with taps for nodes. Coaxial cable (RG-6 quad shield) makes up the bus, with a maximum length of 1000 meters and a maximum number of nodes of 99. As more nodes are added, the maximum length is reduced. Repeaters can be used to extend the bus but are limited to five. (33)

convection. A method of heat transfer in which heat is transferred from one object or area to another through liquid or gas. (35)

conveyor. A mechanical system that moves materials. (13)

copper losses. The losses resulting from the resistance of the transformer windings. (24)

corner joint. A joint made of two metal pieces placed together at their corners, forming a right angle. (37)

corrosion. A chemical reaction that causes the deterioration of metal components. (14)

corrosion inhibitor. An oil additive that prevents corrosion to metal surfaces when water or high humidity is present. (10)

corrosive wear. A type of mechanical wear caused by a chemical reaction between the material of a mechanical component and a corrosive agent. (14)

coulomb. The fundamental unit of electrical charge, equal to the charge of 6.241×10^{18} electrons, and abbreviated as C. (19)

coupling. A device used to connect two shafts and allow the transfer of power from one shaft to another. (11)

cover letter. A formal letter sent to a potential employer when applying for a job, the contents of which include the title of the job sought, where the candidate heard about the job, the candidate's skills and qualities that might apply to the job, reasons the candidate should be considered for the job, and a request for an interview. Also called *letter of application*. (1)

credentialing. The act of establishing and documenting a specific set of qualifications, competencies, or skill standards. (1)

creep. The permanent separation of manila rope fibers over time. In a fluid power or piping system, the term refers to the gradual shift of position or pulling apart of piping joints. (8, 15)

crest. The top surface of a screw thread. (5)

crimper. A pliers-like tool used to compress (crimp) a connector onto the stripped end of a wire to make a tight connection. (4)

crimping. The termination of wire by bending a metal crimp terminal securely around the wire. (23)

critical speed. The speed at which a rotating machine element reaches resonance. (14)

critical-thinking skills. The higher-level skills that enable a person to think beyond the obvious in order to make decisions and solve problems. (1)

crosstalk. The interference that can occur between the wires inside a cable. (33)

cubicle. A small box or cabinet that provides for the organized connection of drop cords or conduits containing conductors for connecting machinery. (27)

current ratio. The ratio of secondary current to primary current, or the inverse of the voltage ratio or the turns ratio of a transformer. (24)

current source inverter (CSI). A type of AC variable frequency drive (VFD) in which the current output is controlled. Current source inverters control the rectifier section by using silicon-controlled rectifiers (SCRs) or other high-speed switching devices, and inductors in the DC bus circuit smooth the current. (31)

current transformer. A single winding on a toroidal core that becomes a transformer when a current-carrying conductor is passed through the "doughnut hole" of the toroid. (24)

cursor. A pair of lines, either horizontal or vertical, whose position may be changed to take measurements on an oscilloscope. (21)

cutting-plane line. A line that indicates where an imaginary cut has been made through an object in order to show interior features. (6)

cutting tip. An attachment for a cutting torch, a cutting tip serves as the exit point for the mixed fuel and oxygen needed to perform a cut. (39)

cutting torch. The equipment that controls the amount of gas supplied for a cut. (39)

C_v. An indication of how much flow would occur through a valve with a 1 psi pressure difference across the valve. Also called *C sub v* or *flow coefficient*. (15)

cycle. The description of one compete rise and one compete fall of a wave. (22)

cylindrical bearing. A cylinder-shaped rolling element bearing that increases contact area to distribute load. (10)

D

damper. A device that regulates the flow of air within a line or duct. (36)

damping. The various methods used to reduce the transmission of vibration through machinery. (14)

Darlington pair. A structure made up of two bipolar transistors that act as one transistor to provide a higher current gain, with the gain of the first transistor in the pair multiplied by the gain in the second transistor. (30)

Data Highway. A local area network (LAN) developed to connect PLCs, HMIs, and PCs for communication and data transfer. DH-485 supports up to 32 devices. (33)

Data Highway Plus. A local area network (LAN) developed to connect PLCs, HMIs, and PCs for communication and data transfer. Data Highway Plus can support 64 nodes. (33)

datum. An exact point, axis, or plane from which locations of other features and geometric controls are established. (6)

DC power supply. A device that uses AC from a higher-voltage source and converts it to DC voltage appropriate to the circuitry requiring DC power. (29)

deactivator. An oil additive that protects nonferrous metals from oxidation. (10)

deadband. The range of change of the input signal that does not show any change in the output signal—the range of values within which no action is taken. Also called *control band*. In motor controls, the range of values between the make and break points for a switch or contact. (18, 26, 34)

dead-blow hammer. A hammer filled with lead shot to prevent bounce when striking an object. (4)

dedendum. The length of a gear tooth below the pitch circle. (12)

delta (Δ) connection. A three-phase connection configuration used in generators, transformers, and motors, in which the phase windings are connected in the shape of a delta (Δ), where the start of one winding is connected to the end of the next winding. (22)

demulsifier. An oil additive used to separate emulsions. (10)

depth micrometer. A measuring tool used when trying to measure the depth of a feature that would be impossible to measure with a conventional micrometer. (7)

destructive testing. A type of weld inspection that analyzes the physical properties of a weld by applying stress to the weld until it fails. (38)

detent. A spring-loaded ball that keeps the spool in position by providing pressure on one of several grooves. (16)

detergent. An oil additive that prevents contaminants from accumulating on surfaces and forming varnishes. (10)

DeviceNet. An industrial network protocol developed by Allen-Bradley to connect smart field devices to processors and data management PCs. DeviceNet allows different vendors' elements to be connected, as long as they meet the standard. Network size is limited to 64 nodes, and distance is limited depending on the amount of data transmitted and the conductor size. Nodes may be added or removed under power, which limits the impact on operating systems. (33)

Dewar flask. A cylinder containing liquid oxygen reduced from gas to make easier storage and shipment. (37)

diagonal cutting pliers. Pliers with insulated handles and hardened jaws ground to a V shape, used for cutting copper wire and component leads. (4)

dial indicator. A gauge used to measure small linear distances for precision shaft alignment and calibration. (11)

diametral pitch (DP). The number of teeth per inch of length on the pitch circle of a gear. (12)

die. The device in pipefitting that cuts the threads into the pipe. (5, 41)

die grinder. A power grinding tool used for removing small amounts of material by hand with as much precision as possible. (4)

dielectric. The insulator between a capacitor's conducting plates. (23)

dielectric constant. The description of an insulator's ability to support the electric field between two oppositely charged conducting plates. Denoted by the variable *K*. Also called *permittivity*. (23)

differential amplifier. An operational amplifier that has been configured to amplify the difference between two input voltages while suppressing any voltage common to both inputs. (30)

differential pressure (DP). The loss of pressure through a component. Also called *head loss*. (15)

differential pressure (DP) flow meter. A flow meter that senses pressure on each side of the fluid orifice or venturi. (34)

diffuse mode. A sensing mode in which the light transmitted from the sensor shines on the object and is scattered, or diffused, in different directions, with a small portion reflected back to the sensor receiver. (31)

digital integrated circuit. A circuit device that only has two states on the output, which are on and off (1 and 0). (30)

digital multimeter (DMM). A device capable of taking several electrical measurements, including voltage, current, resistance, capacitance, and diode function, and displaying the values. (7, 21)

digital sensor. A sensor that provides feedback through an ON or OFF signal to sense the position of a cylinder. (17)

dimension. The precise measurement of a feature. (6)

dimensioning. The process in which the designer adds dimensions to a part drawing to communicate the size and location of each feature. (6)

dimension line. A line used to indicate the extent and direction of a dimension. (6)

DIN rail. A metal rail that devices can be mounted to and follows the standards set by the DIN. (18)

diode. A two-terminal (cathode and anode) semiconductor device that conducts when current flows in one direction and does not conduct when current attempts to flow in the opposite direction—a one-way "valve" for direct current. (23, 29)

direct current (DC). The flow of electrons in only one direction. (19)

direct current electrode negative (DCEN). A situation in which direct current flows from the electrode with negative polarity to the base metal with positive polarity. Also called *direct current straight polarity (DCSP)*. (37)

direct current electrode positive (DCEP). A situation in which direct current flows from the base metal with negative polarity to the electrode with positive polarity. Also called *direct current reverse polarity (DCRP)*. (37)

direct digital control (DDC). A control system that uses various digital and analog inputs and outputs connected to a central computer that operates the HVAC system or automates other building systems. (36)

directional control valve. A valve that controls the flow of fluid from one or more sources into various paths of a hydraulic or pneumatic system. (16)

directional flow regulator. A valve that allows free, unrestricted flow in one direction while regulating flow in the opposite direction with the use of a needle valve. (17)

discharge pressure. The pressure of the fluid as it leaves the pump. (16)

disconnecting means. A safety feature by which connected devices can be immediately disconnected from their source of power. (27)

discrete device. A separate, external device. Also called *field device*. (32)

dispersant. An oil additive that prevents sludge from accumulating and clumping together. (10)

displacement. The movement of something from one place to another. (16)

disturbance. A variation that affects a system's ability to maintain the process variable. Also called *load disturbance*. (34)

dither. An AC signal placed on top of the DC control signal that causes the valve to constantly micromove, which helps to reduce friction between the spool and body of the valve. (18)

doping. A process in which materials called impurities are added to silicon crystal to create positive (P-type) and negative (N-type) electrical conducting characteristics. (30)

double bowline knot. A knot tied on the end of a rope to secure a load, post, pipe, or other object. Stronger than the simple bowline knot. (8)

drag angle. The travel angle used for backhand welding. (38)

drop length. The measurement from the centerline of the stem to the sealing face or seat for a hydraulic hose with a 90° fitting. (15)

dual-inline package (DIP) switch. A manual electric switch packaged in groups. These switches are used to set various operating options or modes for equipment. (28)

dual-pole, dual-throw (DPDT). A contact with two arms or poles that are each capable of making two connections or throws. (26)

dual relay. A device that contains two relays in one with two coils and two sets of contacts. (23)

dual-voltage coil. A type of coil that develops a magnetic field when energized, which shuts the main contacts against spring pressure and indirectly changes the state of any auxiliary contacts. (26)

ductility. A measure of a material's ability to bend and stretch rather than break. (8, 14)

durable good. A manufactured product that lasts over time and wears slowly, such as a car or building materials. (1)

durometer. An instrument used to measure soft materials with several scales called shores. (40)

duty cycle. The amount of time a signal is on as a percentage of the total time of one complete cycle. With welding equipment, the term refers to the percentage of time in a 10-minute period that a machine can be operated at a rated output without overheating. (31, 37)

dynamic balance. The ability of an object to stay in balance while rotating. (12)

dynamic seal. A seal that is used when there is motion between surfaces. Dynamic seals are usually made of more durable material than static seals and may require more lubrication to avoid tearing and wear from the movement they experience. (17)

E

eccentric locking collar. A mounting device used with bearings that have a machined, off-center shoulder to create a squeeze fit on the shaft. (11)

eddy current losses. The losses resulting from a transformer's primary magnetic field linking to other parts of the transformer, such as the core material, rather than the secondary winding. (24)

edge joint. A joint made from two pieces of metal placed side by side and welded along one or more edge. (37)

EEPROM. A memory chip, used in a PLC, that has a copy of the original program and is added to the processor. When the PLC is powered, the EEPROM loads the copy of the program into working memory, and then runs the system from the working-memory program. If future changes are made to the program, they are only made to the working-memory program. (32)

elastic deformation. The temporary deformation of a material caused by stretching beyond its limit. (8)

elastic limit. The point beyond which an object will permanently deform when stretched or compressed. (9)

elastohydrodynamic lubrication. The separation between opposing surfaces created by increased lubricant pressure and viscosity in a localized area due to opposing surfaces under very high loads temporarily elastically deforming. (10)

electric current. The flow of electrons through a circuit. Unit of measurement is the ampere (A). (19)

electrode. The point to which electricity is brought to produce an arc required for welding. (37)

electrode extension. The distance from the contact tip to the end of the unmelted electrode. (38)

electrolyte. A nonmetallic conductor of electricity. (23)

electrolytic capacitor. A capacitor that uses electrolytes to move charged ions. (23)

electromagnet. A temporary magnet created by a current flowing in a conducting coil. (19)

electromagnetic induction. An electric current produced by the presence of a magnet or magnetic field. May also refer to a magnetic field produced by the presence of an electric current. (19)

electromagnetic interference (EMI). A signal generated by an external source that can be conducted through the air or any wiring circuit and that can impact low-voltage controls and communication equipment. Also called *radio frequency interference (RFI)*. (31)

electron. A negatively charged particle that orbits around the nucleus. (19)

electrostatic field. The field surrounding a charged object. (19)

elongation. The tendency of a rope to stretch when a load is applied, specified by rope manufacturers as a percentage. (8)

emulsifier. An oil additive that encourages the mixing and suspension of one liquid in another, such as water and oil. (10)

enclosure ground. The ground bar or point located on the back of a control cabinet. (33)

end-float. The axial displacement, or lengthwise movement, of shafts. (11)

endothermic reaction. A chemical reaction in which energy, usually in the form of heat, is absorbed in order to produce the reaction. (8, 35)

energy. The capacity to do work. (9)

enthalpy. The amount of heat in the air measured in British thermal units (BTUs). (36)

envelope. A rectangular box that includes all of the diagramed positions of a valve. (16)

Environmental Protection Agency (EPA). A government organization focused on the protection of public health and the environment. (2)

error. The difference between the current value of a process variable and the set point of a system. (34)

error signal. A signal indicating the required amount of change to get the valve spool to the desired position. (18)

EtherNet/IP. A popular industrial network communication standard capable of handling large amounts of data at high speeds. (33)

ethical behavior. A behavior that conforms to accepted standards of fairness and good conduct. (1)

exothermic reaction. A chemical reaction that releases energy in the form of heat or light. (8, 35)

extension line. A line used to indicate where the ends of a dimension line terminate. (6)

extension spring. A spring that resists axial lengthening as a pulling load is applied. (9)

extreme pressure additive. An oil additive that forms a protective thin boundary film for high-load equipment with low speeds. (10)

eyebolt. A bolt with a loop at the end that is securely attached to an object so that ropes can be tied to it. (8)

F

face velocity. The speed of the air passing across a filter. (15)

fail-safe. A situation in which, during the failure of an element, the machine or system places itself in a safe, shutdown condition, or the operator is able to manually place the system in a safe condition without the use of an HMI. (33)

farad. The SI unit for capacitance, F. (22)

Faraday's law. A physics principle that states a changing magnetic flux will induce an electromotive force on a closed circuit. (34)

fastener. A device used to hold, or fasten, parts together. (5)

feature control frame. In GD&T, a rectangular box divided into compartments containing the geometric characteristic symbol, tolerance specification, and datum feature reference(s). (6)

feedback. A signal that communicates the measurement of a process variable in a closed loop. (34)

feedback signal. A signal based on output that indicates the actual position of the valve spool. (18)

feed water. The water that is treated and supplied to a boiler system to produce steam. (35)

ferroresonant transformer. A transformer with an extra winding connected to a capacitor, forming a resonant circuit. Also called *constant voltage transformer* or *voltage regulating transformer*. (24)

ferrule. A type of crimp terminal that, once crimped, resembles a pin and is an alternative for tinning stranded wire if it is to be used with a mechanical connection. The term also refers to a small ring that compresses between two harder materials, such as piping and a fitting, to provide a seal. (23, 41)

fiber core. A rope core made of either natural or synthetic fibers, which offers more flexibility than other cores but less strength. (8)

field-effect transistor (FET). A three-terminal, voltage-controlled device that uses an input voltage applied to the gate terminal to control output current in a source-to-drain circuit. (30)

file. A tool used to remove burrs and small amounts of excess material when working with metal. (4)

file card. A flat device with extremely short, straight bristles used to remove any material that has loaded a file. (4)

filler metal. The metal that is added to replace material removed when the base metal is cut or machined to form a groove during welding. (37)

fillet weld. A weld with a triangular cross section that connects perpendicular or uneven joint surfaces at right angles to each other. (37)

filter. The part of an electronic assembly that allows certain frequencies to pass while blocking others. (22)

filter capacitor. A capacitor added to provide some filtering or smoothing, bringing pulsating DC closer to steady-state DC. (29)

filter-regulator. A unit within a pneumatic system that houses two components. The filter removes particles from the air exiting the dryer. The regulator controls the pressure in the portion of the system beyond the regulator. (36)

fire tube boiler. A type of boiler that heats a tank of water through conduction using hot gases in sealed tubes. (35)

firmware. The operating system of an HMI panel or other computer device. (33)

fishbone diagram. A visual tool for cause-and-effect analysis to determine a root cause. Also called *Ishikawa diagram.* (14)

fit. The tightness or looseness between mating parts. (6)

five-why method. A method of root cause analysis (RCA) that involves asking why each successive event or failure happened. After five whys, a technician should arrive at the root cause. (28)

flange. A rim or collar used to make connections in piping systems. (41)

flange bearing. A type of bearing that mounts outside of the machine housing to provide shaft support. (11)

flaring. A process where tubing is drawn over a mandrel (metal bar) to expand the end of the tube to a specific angle in order to make a seal using a fitting. (40, 41)

flashback. A situation in which a flame produced in cutting moves into the mixing chamber of the torch, causing the flame to burn back and damage the torch component. (39)

float switch. A switch that floats on liquid and activates when the liquid reaches a certain level. (26)

flow. In a fluid power system, the amount of movement of fluid in a system in a certain time. (15)

flow control valve. A valve that controls the amount of flow of the fluid. (16)

flow diagram. A type of schematic for piping systems that includes major system components, main piping lines, fluid flow, and system connections. (40)

flow meter. A device that measures the rate of flow of a fluid through an area, such as a pipe. (7)

flux. A liquid or paste material used on electrodes that helps remove oxides and other unwanted substances from a metal surface. It is also used to prepare a joint for soldering, cleaning it, preventing oxidation of the base metal when at high temperature, and helping solder flow through the joint with better capillary action. (23, 37, 41)

flux cored arc welding (FCAW). A form of arc welding that uses heat from an electric arc produced between a continuous consumable electrode and base metal. Also called *flux core welding.* (37)

foaming. The production of a mass of small bubbles in a froth, caused by gas coming out of solution in the reservoir of a hydraulic system. (15)

foam inhibitor. An oil additive that reduces foaming in the lubricant sump. It also helps trapped air escape from oil and settle out. (10)

force. A pushing or pulling effort that changes or tries to change an object's motion. (9)

force instruction. An action performed by a programmer on a PLC to turn inputs on or off for the purposes of testing or troubleshooting. (32)

forcing. The ability to force the state of an input or output in a PLC program. (28)

forehand welding. A type of welding completed with the welding gun pointing forward in the direction of travel. (38)

forward biased. The description of a diode that is conducting, with electrons flowing from the cathode to the anode. (29)

frequency. The periodic rate at which AC current alternates. Designated by the variable f and measured in cycles per second, or hertz in SI units. (22)

fretting corrosion. A form of corrosion that takes place at the asperities of close mating components and results in surface grooves and a fine, dry, red dust. (14)

friction. A resistive force that acts counter to sliding motion. (9)

FRL unit. A unit in a pneumatic system made up of a filter, regulator, and lubricator that prepares compressed air for end use. (15, 17)

fulcrum. The axis about which a lever rotates to apply force and movement. (9)

full-wave bridge rectifier. A rectifier circuit that uses four diodes to accomplish full-wave rectification, rather than two diodes and a center-tapped transformer. (29)

full-wave rectification. A circuit situation in which both halves of the AC cycle are converted to provide pulsating DC with more pulses and less time below peak than half-wave rectification. (29)

fuse. A single-use overcurrent protection device. (23)

G

gain. A measure of sensitivity to an input signal. The higher the gain, the more sensitive the system is to a changing input signal. Also called *amplification*. In electronic control systems, the term refers to the amount of amplification provided by a transistor amplifier, measured as the ratio of signal output over signal input. (18, 30, 34)

galling. A situation in which two surfaces bond together due to extreme friction. (14)

galvanic corrosion. A type of corrosion that takes place when dissimilar metals are in contact with each other in the presence of an electrolyte. (14)

garter spring. A component of a seal that provides additional pressure to keep the seal against the shaft. (10)

gasket. A piece made of rubber or a similar material that is used to create a tight seal between two parts. (41)

gas metal arc welding (GMAW). A form of arc welding that uses heat from an electric arc struck between a metal-cored consumable wire and base metal. Also called *metal inert gas (MIG) welding*. (37)

gas pressure regulator. A piece of equipment for oxyfuel gas cutting or welding designed to reduce the pressure in which a gas is delivered to a hose. (39)

gas-shielded flux cored arc welding (FCAW-G). A type of flux cored arc welding that requires the use of carbon dioxide or a carbon dioxide and argon combination to shield the weld pool and arc stream from the outside atmosphere. Also called *duel-shield process*. (38)

gas tungsten arc welding (GTAW). A form of arc welding that uses heat from an electric arc produced between a tungsten alloy electrode and base metal. Also called *TIG welding*. (37)

Gay-Lussac's law. The pressure of an enclosed gas is directly proportional to its temperature, expressed as $p / T = k$. (15)

gearbox. A set or system of gears that may or may not be contained within a casing. (12)

gear pump. A pump that uses the rotary motion of meshing gears to directly move fluid. (16)

generator. A device that turns rotational mechanical energy into electricity and produces direct current. (25)

geometric dimensioning and tolerancing (GD&T). A dimensioning system used to control interpretation of tolerances defining geometric relationships of features. (6)

GFCI circuit breaker. A circuit breaker that combines the protection of a standard circuit breaker and senses ground-fault current. GFCI circuit breakers trip out whenever current to ground is detected. (23)

globular deposition. A method of metal transfer in which an electrode burns off above or in contact with the workpiece in a very erratic globular pattern. (38)

Grafcet. A graphic model, developed in the mid-1970s by a group of industrial and process engineers, that is a representation of a sequential process including each step of the process, conditions to prove a step has taken place, and logic that proves all previous steps have taken place in order for the next step to happen. (18)

gravity conveyor. A conveyor system that has unpowered rollers and is pitched to use gravity to move material along the conveyor. Also called *static conveyor*. (13)

gravity flow rack. An angled product storage system that allows product to roll or slide toward the front of the rack. Also called *first-in-first-out rack*. (13)

grease. A soft or semisolid lubricant that has a larger amount of soap than oils have. (10)

groove weld. A weld that connects two joint surfaces or edges against each other. (37)

ground-fault circuit interrupter (GFCI). A circuit protection device that measures the difference between the current flowing into a circuit and the current flowing out through the neutral and trips when a predetermined current difference is detected, disconnecting the source of current. (2, 23)

grounding. A process that uses a neutral, or grounding, wire to connect to earth and allow any extra current to flow back to the source rather than become a hazard and flow through anyone who might touch the device or appliance. (2)

ground ring. A device used to conduct shaft currents to ground before they reach motor bearings and cause damage. (25)

grouped commons. With a PLC output module, a situation in which a common terminal is associated with several outputs. (32)

grout. A fluid form of concrete used to fill gaps as an adhesive element between materials. (8)

H

hacksaw. A hand tool used to cut metal. The two main parts are the blade and handle. (4)

half-wave rectification. A circuit situation in which the positive half cycle of the input signal is converted into pulsating DC output signal. (29)

Hall effect sensor. A noncontact sensor that senses a voltage difference when in close proximity to a magnetic field. (31)

hammer drill. A power tool used for drilling holes in concrete and masonry. (4)

hand drill. A power tool that uses a spinning bit to make a hole. (4)

hard contact. A contact made of copper, sometimes with a silver coating, that has very little resistance when closed and nearly infinite resistance when open. (26)

hardness. A measure of a material's resistance to penetration, such as surface scratching, abrasion, or denting. (14)

hardware. The physical elements of an HMI panel or other device. (33)

harmonic distortion. A signal superimposed over the regular AC sine wave that distorts the sine wave. This distortion can be caused by external electrical components (usually nonlinear loads) or by the drive itself. (31)

harmonics. A distortion of the normal electrical current that can cause overheating and negatively impact efficiency. (32)

head loss. The loss of pressure through a component in a fluid power system. Also called *differential pressure (DP)*. In piping systems, the term also refers to a measurement of the resistance to fluid flow. (15, 40)

head pulley. The pulley located at the discharge end of a conveyor that is responsible for driving a conveyor belt. Also known as a *drive pulley*. (13)

heat exchanger. A device that facilitates the exchange of heat between two mediums within a system. (16, 35)

heat sink. A device that accepts heat and transfers it to the surrounding environment. (16)

henry. The SI unit for inductance, H. (22)

hertz. The SI unit for frequency, Hz, measured in cycles per second. 1 cycle per second equals 1 Hz. (22)

hex key. A typically L-shaped tool commonly used with machine bolts and screws. Also called *Allen wrench*. (4)

hidden lines. The lines that represent object edges and contours that are located behind other features and not visible in a given view. (6)

highway diagram. A diagram used by a panel builder to wire a control panel and make it ready for field connections. (27)

hitch. The manner in which a sling is configured for lifting. Three standard types of hitch are a vertical straight hitch, a choker hitch, and a basket hitch. (8)

hitch knot. A knot used to temporarily secure a line. (8)

hoist. A device that provides a mechanical advantage when lifting heavy loads. (8)

Hooke's law. A principle of spring force stating that the restoring force of a spring is proportional to the distance the spring is compressed or stretched. (9)

horsepower (hp). A measure of the mechanical power of an engine or motor, or the rate at which work is accomplished. One horsepower is equal to 550 ft-lb/s or 33,000 ft-lb/min. (9, 16)

human-machine interface (HMI). A device and software that allows technicians to interact with a machine, usually taking the form of a control panel and digital touch screen that tells technicians about the status of alarms, sensors, or other pertinent troubleshooting data. (3, 28)

hunt. A situation in which a system seesaws back and forth dramatically from fully open or on, to fully closed or off, due to gain being set too high. (18)

hydraulic cylinder. A type of actuator that transfers the energy from the fluid into linear motion and mechanical force. Also called *linear actuator*. (16)

hydraulic pump. A mechanical device that converts mechanical energy into hydraulic fluid energy by generating flow. (16)

hydraulic reservoir. A device in a hydraulic system that stores hydraulic fluid in a tank and performs several other functions in the system, including the removal of heat, settled particulates, air, and moisture. (16)

hydrodynamic lubrication. A separation between opposing surfaces created by a lubricant film that is maintained by the motion of the surfaces themselves. The film develops as speed increases between surfaces; no film is present at low speeds or at rest. (10)

hydronic system. A system that uses water to transfer heat for both heating and cooling applications. (35)

hydrostatic lubrication. A separation between opposing surfaces created by pressurized lubricant that is injected between the two surfaces to prevent contact. (10)

hysteresis losses. The losses resulting from the orientation and reorientation of the grains in the ferromagnetic material of a transformer core. (24)

I

idler pulley. The pulley in a conveyor system that is responsible for redirecting the belt and providing belt tension in longer conveyor units. (13)

impedance. The total resistance to the flow of alternating current. Designated by the variable Z and measured in ohms, Ω. (22)

inclined plane. A simple machine composed of a sloped surface used to move an object in a vertical direction by applying force along a horizontal direction. (9)

independent wire rope core (IWRC). A single wire-wound rope core, which is very strong but less flexible than other cores. (8)

inductance. The opposition to any change in current by an inductor. Designated by the variable L. (22)

induction motor. An AC motor in which current is induced by a rotating magnetic field which causes the motor to run. (25)

inductive reactance. A description of the reactance of an inductor. Designated by the variable X_L and measured in ohms, Ω. (22)

inductive sensor. A proximity sensor that is comprised of a coil that develops a field about it when voltage is applied and that can detect both ferrous and nonferrous metals, if designed for that purpose. Inductive sensors are best used to detect repeated movements of ferrous objects at close to very close distances, usually around 5–30 mm or less. (31)

inductor. A conductor, often a coil of wire, used as a circuit component to produce voltage in the presence of a magnetic field or electric current. (19)

infrared radiation (IR). A part of the electromagnetic spectrum that is not visible to the human eye but is experienced as heat. (7)

infrared (IR) thermometer. A noncontact electronic device that measures infrared radiation emitted from a surface or object to determine its temperature. (7, 21)

initiative. The act of starting a task or activity without being told. (1)

input image table. A record of the binary representation that reflects all of the input conditions of a PLC and is held in a specific section of the processor's memory. (32)

input/output (I/O) modules. The components connected to PLCs to provide analog or digital electronic connections between the PLC and external actions, such as the push of a button (analog) or an electronic signal. (18)

instantaneous voltage. The voltage at any instant during a cycle. (22)

instrument identification number. A unique number used to identify an individual instrument within a system. Also called *instrument ID number*. (34)

instrument index. A document that lists all instruments in a facility, including the tag number/instruction ID number, purpose, and associated system of each. (34)

insulated-gate bipolar transistor (IGBT). A transistor capable of high efficiency and fast switching that is actually a cross between a bipolar junction transistor (BJT) and a field-effect transistor (FET) and is used primarily as an electronic switch. (30)

insulator. A high-resistance material that does not permit the flow of electrons. Also called *nonconductor*. The term can also refer to a material that is a poor conductor of heat. (19, 35)

integrity. The characteristic of firmly following one's moral beliefs. (1)

intensifier. A type of cylinder that may use compressed air, hydraulic fluid, or both to increase pressure and add efficiency to a pneumatic system. (17)

intercooler. A component located between the two stages of a two-stage pneumatic compressor, which removes heat from the compressed air, thus increasing the efficiency of compression. (17)

interference. An external disturbance that negatively affects an electrical circuit. (32)

interference fit. A fit in which the external dimension of one part is slightly larger than the internal dimension of the mating part. Also called *press fit*. (6)

interlock. A device used to prevent a possible short on a reversing motor starter by preventing the coil in the opposite phase from shutting its contacts when the main coil is energized and actuated. (26)

International Electrotechnical Commission (IEC). A nonprofit, nongovernmental international standards organization that prepares and publishes standards for electrical and electronic systems and related technologies. (21, 26)

interposing relay. A relay used to separate two different devices or circuits, such as in cases where each connected device requires operation at a different voltage. (32)

interrupt rating. The maximum current at which a fuse is guaranteed to clear a fault. (26)

inverter. The simplest logic gate, in which the output is the opposite of the state of that on the input. Also called *NOT gate*. (30)

I/P transducer. A type of transducer that receives an electrical signal and converts the signal to a proportional pneumatic output. (34)

ISO 11727. A standard from the International Organization for Standardization that establishes port numbering for fluid power systems. (17)

ISO 1219-1. A standard from the International Organization for Standardization that establishes valve symbols for fluid power systems. (17)

ISO 15552. A standard from the International Organization for Standardization that establishes specific guidelines for metric series pneumatic cylinders for interchangeability purposes. (17)

isolation transformer. A transformer with a turns ratio of 1:1 that transfers power from a source to a device while isolating the device from the power source, thus protecting against interference, unwanted current transfer, and electric shock. (24, 32)

isometric drawing. A type of schematic that shows three-dimensional spatial relationships between components, valve positions, multiple piping runs, and dimensions of piping. (40)

J

Java application control engine (JACE). A unit that connects to a network and acts as a translator for the various components within a networked system. (36)

job safety analysis (JSA). A procedure that promotes safety by identifying hazards before they occur. Also called *job hazard analysis (JHA)*. (2)

joule. The SI unit for energy or work, abbreviated as J. (20)

junction field-effect transistor (JFET). A three-terminal field-effect transistor that only operates in depletion mode. JFETs are the simplest type of FET. (30)

K

kelvin. The unit of temperature measurement often used in physical sciences, abbreviated as K. (15)

kerf. The portion of metal removed during the sawing process. In welding, the term refers to a slot in the base metal created as a cut is made in the direction of torch travel. (4, 39)

key. A metal fastener used to prevent shaft rotation and transfer torque between parts. (5)

keyseat. In a key assembly, the slot where the key fits on the shaft. (5)

keyway. In a key assembly, the slot where the key fits on the hub of the mating part. (5)

kinetic energy. The energy of motion of an object. (9)

kinetic friction. The friction between two objects that are moving in relation to each other. (9)

Kirchhoff's current law (KCL). The sum of the currents entering any point in a circuit must equal the sum of the currents leaving the same point. (20)

Kirchhoff's voltage law (KVL). The sum of the voltage sources and voltage drops around any closed circuit is zero. (20)

L

ladder diagram. A type of schematic diagram that shows the function of an electrical circuit and resembles a ladder. It is used to develop an understanding of a control circuit's operation and troubleshoot a circuit that is not operating properly. (6, 27)

laitance. The particles that accumulate on the surface of a foundation. (8)

laminar flow. A streamlined type of fluid flow in which all layers flow in parallel paths and in the same direction of travel. (40)

land. A sealing area as it pertains to a hydraulic or pneumatic system. (16)

lang lay. A type of rope lay in which the lay of wires in each individual strand is the same direction as the lay of the strands that form the rope. (8)

lantern ring. A component in some types of pumps that filters water between layers of packing along the pump shaft to cool the packing material. (40)

lap joint. A metal piece lapped over another and welded at the end of the top piece. This is the strongest type of weld joint. (37)

lapped/shear. A type of spool that has no seals. (17)

laser curtain. A presence-sensing device that uses lasers to detect and keep workers safe around machines and potentially dangerous areas. (31)

latency. The amount of delay between the activation of an input or output and when the status change shows up on the computer screen when monitoring a PLC program. (28)

latent heat. The heat that causes a change in state without changing the temperature of the substance. (35)

latent heat of fusion. The heat required to cause a change in state from solid to liquid or liquid to solid. (35)

latent heat of vaporization. The extra energy needed to vaporize a liquid into a gas. (35)

lay. The direction and method of how strands of wire rope are weaved around the core. Types include right-hand, left-hand, alternating, and nonrotating. (8)

LCR meter. A device that measures inductance (L), capacitance (C), and resistance (R). (21)

lead. The axial distance a screw thread advances in one complete revolution. (5, 9)

leadership. The ability to guide and motivate others to complete tasks or achieve goals. (1)

least material condition (LMC). The size condition of a feature containing the least amount of material within the tolerance limits. (6)

lever. A simple machine composed of a rigid bar and a fulcrum. (9)

lift capacity. The size of the load that a sling or hitch can actually lift in its particular configuration. Also called *load capacity*. (8)

lifting chain. A chain used for lifting and rigging, in which the links are welded closed, heat-treated, and tested. Lifting chain typically comes in surface treatments of bare, galvanized, lacquered, and powder coated, and is rated by a grade based on its capacity. (8)

light-emitting diode (LED). A small electronic device that lights up when electric current passes through it. (28)

limit switch. A switch that physically contacts an element and converts mechanical movement into an electrical output. (26)

linear integrated circuit. An analog circuit device that works with varying voltage levels. (30)

linear variable differential transformer (LVDT). A method of detecting spool position in which an output signal develops based on the position of the core that travels through the transformer as the main spool moves. (18)

lineman's pliers. Heavy duty, versatile pliers with wide, flat jaws and considerable weight. (4)

line reactor. A wired device in a variable frequency drive (VFD) that provides protection from intermittent or transient voltage spikes, current spikes, and harmonic distortion. (31)

line splitter. An adapter that separates hot and neutral conductors so that each may be individually placed into a clamp-on ammeter. (28)

lip. The portion of a seal that holds the sealing edge against the shaft and is made of an elastomeric or other type of material. (10)

liquidus. The temperature at which a metal melts and becomes liquid. (41)

"live-dead-live" (LDL) test. A test used to verify that a meter is measuring correctly and a circuit is dead before proceeding with work. First, a voltmeter is used to measure a known live circuit, then to measure the circuit to be verified as dead, and then to measure the known live circuit once again. (2)

local area network (LAN). A digital communication network normally used within a closely situated area, such as one factory or area of a factory connected for the purpose of data transfer. Most industrial networks are LANs, but they may also connect to larger networks. (33)

locking pliers. Pliers that can be locked into position, often used for clamping and for removing damaged nuts and rounded-off bolt heads. Also called *clamping pliers*. (4)

lockout/tagout (LOTO). A process used to isolate sources of energy from a piece of equipment and notify others while maintenance and repair operations are being performed. (2)

logic function. A simple processing function in a circuit that leads to the performance of a logical operation through the use of binary inputs and outputs. Basic logic functions provide system control and include AND, OR, and NOR functions. (18)

logic gate. A switching circuit that governs whether inputs pass to outputs, with input and output values being one of two states, on (1) or off (0). (30)

long-nose pliers. Pliers with long, narrow, pointed jaws used for gripping and bending and that are useful in tight places. Also called *needle-nose pliers*. (4)

loop system. A distribution system in which the main line forms a continuous loop, which provides maximum flow with a minimal pressure drop between workstations. Isolation valves allow larger sections of the loop to be isolated for maintenance while maintaining supply to the rest of the system. (17)

lubrication plan. A record of the lubrication requirements for all machines and equipment. Also called *lubrication schedule*. (10)

M

machinist's rule. A steel rule with markings for 1/32 and 1/64 of an inch used for measuring small parts. (4)

magnet. A material able to attract ferromagnetic material and retain a magnetic field. (19)

magnetic inductive flow meter. A flow meter that develops an output signal based on the voltage created by the conductive fluid passing through the meter. (34)

magnetic overload. A device that works similarly to a clamp-on ammeter. An electronic control senses the voltage generated by a motor transformer and opens a contact when it exceeds the set point. (26)

magnetic position sensor. A digital sensor that uses a magnetic strip on a piston to sense its position. An electrical contact is closed when a switch senses the magnetic field as the piston moves. (17)

magnetic reed contact. A sealed contact that is actuated by an internal magnet. (26)

magnetic reed sensor. A sensor with an electrical reed switch that operates through the use of a magnetic field. (17)

magnetic sensor. A proximity sensor that uses a magnetic field to sense objects at a distance of usually less than 3″. The magnetic field can penetrate nonmagnetic or nonmetallic materials, so the sensor can detect a magnet on the other side. (31)

magnitude. The amount of offset in a system or process. Also called *magnitude of error*. (34)

maintained contacts. A contact that remains active until it is deactivated. (26)

major diameter. The largest diameter of a screw thread, measured from crest to crest for external thread or from root to root for internal thread. (5)

make-up water. The water added to a boiler feed system to replace water depletion due to system loss. (35)

manifold. A component with numerous ports, or ways, to which valves and stacks of valves can be attached to form circuits and functions. Also called *manifold block*. (18)

manifold stack. A manifold block with several stacked valves attached. Also called *stack*. (18)

master control relay (MCR). A type of external safety relay used to shut down a section of an electrical system. (32)

maximum material condition (MMC). The size condition of a feature containing the greatest amount of material within the tolerance limits. (6)

measurement by comparison (MBC). A troubleshooting method in which a known good circuit is measured against a potentially faulty one to determine whether a circuit is operating properly. (28)

mechanical advantage. The proportional increase in output force relative to the input force when moving a load. (9)

mechanical joining. The connecting or assembling of components without heat or adhesives. (41)

megohmmeter. An ohmmeter that uses high voltage to make resistance measurements. Also called *megger* or *insulation tester*. (7, 21)

melting alloy overload. A type of overload that relies on current sensing through heat generation. A melting alloy overload holds a contact closed against spring pressure until the alloy it contains melts at a predetermined temperature and opens the contact. (26)

metal-oxide semiconductor field-effect transistor (MOSFET). A four-terminal field-effect transistor that operates in both depletion and enhancement mode. Generally, MOSFETs are more expensive and are used in very large-scale integration (VLSI) circuits, such as those found in computers. Also called *insulated-gate field-effect transistor (IGFET)*. (30)

metal oxide varistor (MOV). A small, usually circular device that protects against voltage surges by diverting transient or excessive voltage away from circuits or system components. (31, 36)

micrometer. A tool used when the required measurement accuracy is within 1/10,000 of an inch. (7)

microstepping. A process in which small steps are taken between the fixed mechanical steps a motor is normally able to take. (25)

minimum breaking strength. The amount of force required to break a rope. Also called *breaking strength*. (8)

minor diameter. The smallest diameter of a screw thread, measured from root to root for external thread or from crest to crest for internal thread. (5)

mixed film lubrication. A separation between opposing surfaces created by lubrication using characteristics between elastohydrodynamic and boundary film lubrication. Distances between surfaces are smaller and some contact may occur. (10)

modulated light. A light wave that is varied in its amplitude or frequency in order to transmit information. (31)

molded-case circuit breaker (MCCB). The most common type of circuit breaker in a three-phase motor supply circuit. These circuit breakers are available at various sizes and voltages and offer additional protections for motors beyond protecting against overloads and short circuits. (26)

momentary contact. A contact that returns quickly to its normal state when pressure is released. Also called *snap-acting contact*. (26)

monoblock. A manifold that is a cast one-piece body that houses all of the valve functions in one integrated unit. (18)

motor nameplate. A tag attached to a motor that contains information about the motor's construction and operation. (25)

motor starter. A device that provides resistance to a motor at startup and removes resistance when the motor reaches running speed. A motor starter includes an overload to protect the motor from current spikes. (26)

multiview drawing. A representation that shows the different views of a part on one drawing. (6)

mutual inductance. A process whereby two coils are brought close to each other and the varying magnetic field produced by one coil induces a voltage in the other. (24)

N

National Electrical Code (NEC). A standard for the safe installation of electrical wiring and equipment. This standard is published by the National Fire Protection Association (NFPA) and is revised every three years. (27)

National Electrical Manufacturers Association (NEMA). A US-based association of electrical equipment manufacturers that standardizes the sizes and ratings of starters. (26)

National Fire Protection Association (NFPA). A nonprofit organization that focuses on preventing fires due to electrical hazards. (27)

National Institute for Metalworking Skills (NIMS). An organization that sets industry skill standards, certifies individual skills against those standards, and accredits training programs that meet NIMS quality requirements. (1)

National Institute for Occupational Safety and Health (NIOSH). The research section of OSHA, which develops recommended occupational safety and health standards. (2)

natural frequency. A measure of how much an object vibrates without damping. (14)

needle bearing. A cylindrical-shaped rolling element bearing that is long and narrow. (10)

negative slope. The configuration of an analog-output ultrasonic sensor that results in the output being highest (20 mA) when the detected object is closest to the sensor and to fall as the object moves away. (31)

negotiation. The process of reaching an agreement that requires all parties to give and take. (1)

network topology. The way in which a network is organized and physically wired. (33)

neutral flame. A flame that results from combining oxygen and a fuel gas in proper proportion. (39)

neutron. An electrically neutral particle located in the nucleus of an atom. (19)

newton. A unit of measure equal to the force needed to accelerate a mass of one kilogram at the rate of one meter per second squared. (9)

noncontact flow meter. A flow meter that is not placed in the line of flow. (34)

non-contact seal. A seal that creates gaps or chambers between rotating and stationary components. (10)

noncontact voltage tester. A device used to verify the presence of voltage in a conductor even through conductor insulation. (21)

nondestructive testing. A type of weld inspection that verifies the quality of the weld without causing damage to the physical weld. (38)

nondurable good. A manufactured product that is consumed quickly, such as bread or beer. (1)

nonpositive-displacement pump. A pump that creates a continuous flow of fluid, but output varies because there is no seal between the rotating member and the pump casing. (40)

nonverbal communication. A form of communication that deals with the sending and receiving of messages without the use of words. (1)

normally closed (NC). A contact that is closed when in a de-energized or "normal" state. (26)

normally closed, timed-open (NCTO) contact. A timed contact that is closed when de-energized. (26)

normally open (NO). A contact that is open when in a de-energized or "normal" state. (26)

normally open, timed-closed (NOTC) contact. A timed contact that is open when de-energized. (26)

normal pipe size (NPS). A rounded value that is close to the actual inside diameter of a pipe used to specify piping. (40)

nut. An internally threaded fastener screwed to a bolt to hold parts together. (5)

nut driver. A tool similar to a screwdriver used to loosen or tighten small nuts and bolts. Also called *nut runner*. (4)

O

object lines. The lines that represent the visible edges and contours of an object. (6)

observation. The most important troubleshooting tool, it helps a technician discover obvious defects and provides knowledge about how a machine operates. (28)

Occupational Safety and Health Administration (OSHA). The federal agency tasked with ensuring that employers provide a safe and healthful workplace. (1, 2)

off-delay timer. A timer device that de-energizes after a set time has elapsed. Off-delay timers come in two basic configurations: two- or three-pole and contact-controlled. With a programmable logic controller (PLC), the term refers to a programming instruction (TOF) that accomplishes the same thing as the off-delay timer device. (26, 32)

offset. A valve adjustment that allows position change for the correction of a constant error. In process control systems, the term refers to the amount of error between the controlled variable and the set point. (18, 34)

Ohm's law. An equation that describes the relationship between voltage, current, and resistance: $E = I \times R$. (20)

on-delay timer. A timer device that energizes an output after a preset time elapses. With a programmable logic controller (PLC), the term refers to a programming instruction (TON) that accomplishes the same thing as the on-delay timer device. (26, 32)

open loop. A control system that does not incorporate feedback. (34)

open-loop system. A type of variable frequency drive (VFD) control system that does not provide feedback to the drive from the system. (31)

operational amplifier (op amp). A common, basic amplifier linear integrated circuit, capable of mathematical operations, that uses a small input signal to control a larger output signal. (30)

optocoupler. A semiconductor device used to connect the signal of an I/O module to the CPU in a PLC while isolating the signal, which prevents external higher voltage from damaging internal components. Also called *optoisolator*. (32)

orifice. A sharp-edged restriction in a fluid flow path. (34)

O-ring seal. An O-shaped or doughnut-shaped seal made of elastomer, or rubber. (17)

oscilloscope. An electrical troubleshooting device that shows the change in electrical signal or voltage over time as a continuous graph, usually a wave. Also called *scope*. (7, 21)

outside diameter. The portion of a seal in contact with and that seals against that housing and provides rigidity to the seal. (10)

overhung load. A load perpendicular to the shaft that is applied beyond the outermost bearing of the shaft. (12)

overload. The part of a motor starter that provides protection to the motor by sensing current and opening a normally closed contact when current exceeds a predetermined value. (26)

overvoltage. An undesirable voltage spike in excess of intended voltage. Also called *transient*. (21)

oxidation. A process in which pressurized oxygen combines with a gas flame to produce extreme heat used for cutting. (39)

oxidizing flame. A flame produced when too much oxygen is present. (39)

oxyacetylene. A combination of oxygen and acetylene gas. Oxyacetylene is the most common gas used in oxyfuel gas cutting. (39)

oxyacetylene welding (OAW). A form of oxyfuel gas welding that uses a combination of oxygen and acetylene gas to produce a flame to melt a base metal. (37)

oxyfuel gas cutting (OFC). A process in which metal is cut by producing extreme heat by using a gas flame and pressurized oxygen. Also called *burning* or *flame cutting*. (39)

oxyfuel gas welding. A welding process that uses heat produced by a gas flame to complete a weld or cut. (37)

oxygen. A nonflammable gas that is colorless and odorless. (39)

P

panelboard. A part of an electrical distribution system that is wall mounted or otherwise supported by mounting to some sort of structure. (27)

panel layout diagram. A diagram that shows the positions of all panel components without showing wiring. (27)

parallel circuit. A circuit containing more than one path for current to flow. (19)

parallel system. A distribution system in which one large header supplies all loads in parallel down a single line. If the header is not sized properly, loads at the far end of the line will experience pressure drop. (17)

Pascal's law. A statement that when there is an increase in pressure at one point, this pressure is transmitted through the fluid to all other points in the system. (15)

pass transistor capacitance multiplier. A circuit that multiplies the capacitance by the gain (β) of the transistor, creating greater capacitance when use of a larger capacitor is not practical. (29)

peak inverse voltage (PIV). A parameter that indicates the highest voltage that a diode can hold off in the reverse direction. Also called *peak reverse voltage (PRV)*. (29)

period. The amount of time it takes to complete one cycle. (22)

permanent-magnet motor. A motor that has permanent magnets in the stator instead of windings. (25)

personal protective equipment (PPE). The safety equipment that protects an individual in the workplace. Examples are safety glasses, hard hats, aprons, and gloves. (2)

P/E switch. A device that converts a pneumatic signal to an electrical signal. (18)

phantom line. A line used to show an imaginary break in an object where a portion is omitted. (6)

phantom voltage. A reading on a digital multimeter of electrical potential, or voltage, between conductors that have no voltage. Also called *ghost voltage*. (21)

phase. The position of a specific point in time on a sine wave. Represented by lowercase Greek letter theta, θ. (22)

phase angle. The measurement of the difference in phase between two waves. Represented by the lowercase Greek letter theta, θ. (22)

phase control. A method of DC motor voltage control that uses silicon-controlled rectifiers (SCRs) to control the applied voltage. (31)

phase sequence tester. A device used to detect the presence of voltage and the direction of rotation in three-phase conductors. (21)

phasing. The direction transformer windings are connected, either in phase or out of phase. (24)

phasing dot. A dot printed next to one of the leads on each winding of a transformer to indicate the phase relationships between the windings. (24)

Phillips screwdriver. A screwdriver that fits Phillips-head screws, which have a cross-like slot in the top. (4)

photo sensor. A noncontact sensor that can measure an object's reflective ability, lack of reflection, presence of light, or lack of light. Also called *photoelectric sensor* or *photo eye*. (31)

pick-and-pack operation. A process that gathers and packages various components for distribution. (13)

piezoelectric effect. The ability of a material, such as a quartz crystal, to generate electricity in response to applied pressure. (31)

pillow block. A housing or bracket that contains a bearing and is used to support a rotating shaft. (11)

pilot valve. A smaller, easily operated valve used to control operations that would require a great deal of force to otherwise operate directly. (16)

pin. A cylindrical fastening device used to align and secure parts. (5)

pinch point. A point where two mechanical parts come together, creating a danger of catching a body part or loose article of clothing. (2, 12)

pin punch. A tool used to remove roll or dowel pins. Also called *drift punch*. (4)

pipe nipple. A short piece of pipe threaded on both ends and used to join two fittings that are close together. (41)

piping and instrumentation diagram (P&ID). A drawing of the interconnections of physical system components, including representations of instrumentation and control loops. (34)

pi (π) section filter. A filter circuit shaped like the Greek letter π that has capacitors along the vertical sides of the π and inductors along the top and that filters out unwanted frequencies. (29)

piston. A cylindrical piece moved back and forth within a cylindrical chamber by means of fluid pressure. (16)

pitch. The distance from one point on a screw thread to the corresponding point on the next thread. On a timing belt, the term refers to the distance between the same location on adjacent teeth on the belt. (5, 9, 12)

pitch circle. The length of a belt at the location of the tension members. (12)

pitch diameter. The diameter corresponding to a theoretical cylinder passing through the points on the screw thread profile where the width of the thread ridge is equal to the width of the groove. (5)

pitting. The formation of cavities or holes in the surface of a material caused by corrosive wear. (14)

pitting corrosion. A localized form of corrosion that produces cavities or holes on the surface of a metal. (14)

plain bearing. A type of bearing that is a bushing that supports a shaft on a nonmoving, interior cylindrical surface. Also called *sleeve bearing*. (10)

planned maintenance. A type of maintenance that is best scheduled ahead of time, often to coincide with a planned equipment shutdown. (3)

plasma. An ionized gas, consisting of free electrons that conduct electricity. Plasma is often referred as the fourth state of matter. (39)

plasma arc cutting (PAC). A process in which metals are cut by using an arc and plasma gas forced through a constricting nozzle in the torch. (39)

plastic deformation. The permanent deformation of a material caused by stretching beyond its limit. (8)

plates. The parallel conductors that form a capacitor. (23)

pneumatic transmitter. A type of transmitter that senses a controlled variable, such as temperature, humidity, or pressure, and develops a proportional air pressure output signal. (36)

point of operation. The point where a cutting tool contacts a workpiece in a machining process. (2)

polarity. The direction of instantaneous current flow in a transformer winding. Also called *phase*. With a DC welding machine, the term refers to the direction of current flow between terminals, which can be manually reversed. (24, 37)

polarized. A laser position sensor in which polarized filters are used so that only the light reflected from the object is sensed and no other reflected light. (31)

port. The open end of a passage that fluid passes through. Also called *way*. (16)

portable band saw. A power tool consisting of a toothed belt, or band, stretched between wheels and powered for continuous cutting. (4)

portable filtration cart. A movable device with series filters and a separate motor that can help to remove unacceptable particle or moisture contamination from a hydraulic system and can also be used as added filtration in a separate loop to help clean up systems that are not meeting cleanliness requirements. (15)

positive displacement compressor. A type of compressor that draws in a measured amount of refrigerant and reduces the volume of the chamber to compress the refrigerant. (35)

positive-displacement pump. A pump that moves a fluid by trapping a certain amount in a sealed area and then forcing the trapped volume into a discharge pipe. (16, 40)

positive drive system. A system in which no slip occurs between the belt and the sprocket. (12)

positive positioner. An actuator controlled by branch line pressure, but uses main air line pressure to position the damper. (36)

positive slope. The configuration of an analog-output ultrasonic sensor that causes the output to be lowest (4 mA) when the detected object is closest to the sensor and to increase (up to 20 mA) as the object moves away from the sensor. (31)

positive suction head. A situation provided when the fluid height is above the suction height of the pump and there is minimal restriction to the pump. (16)

potential energy. The stored energy of an object resulting from its position or internal stresses. (9)

potentiometer. A three-terminal, adjustable resistor device that measures voltage difference and is used to control current by varying resistance. Also called *pot* and *variable resistor*. (23, 31)

power. A measure of work that takes place within a specific period of time. (9, 20)

power equation. An equation used to calculate electrical power, $P = I \times E$. (20)

power factor. The ratio of true power to apparent power in an AC circuit, which is equal to the cosine of the phase angle, $\cos(\theta)$. Abbreviated *PF*. (22)

power factor correction (PFC). The stage that adjusts the rectified pulsed DC from the rectifier to a standard voltage regardless of the input voltage, allowing for a wide input range while still maintaining the same output voltage. (29)

power rating. The maximum amount of current a resistor can handle without overheating. Also called *wattage rating*. (23)

predictive maintenance. An approach to proactive maintenance that consists of testing and monitoring machines to help predict breakdowns. (3)

preloading. A practice in which the bearing has a load placed on it by outside means in order to reduce excessive radial internal clearance. (10)

pressure. A measure of force applied to a unit area. (9)

pressure angle. The angle at which gear teeth surfaces mesh together. (12)

pressure compensation. The varying of an orifice, or opening, in order to maintain fluid flow regardless of differences in pressure between areas of a hydraulic or pneumatic system. (16)

pressure control valve. A device used to keep pressure below a set limit in a hydraulic system and maintain a set pressure in the circuit. (16)

pressure gauge. A device used to determine the pressure behind hydraulic fluid or compressed air. (7)

pressure-reducing valve. A type of pressure-regulating valve that helps to reduce upstream pressure to a lower, acceptable level for downstream branches in the system. (16)

pressure switch. A switch that uses a bellows or diaphragm to sense pressure and convert it to mechanical movement against a spring. (26)

pressurized head tank. A pressurized tank containing a bladder filled with air or nitrogen that maintains pressure in a hydronic system. (35)

preventive maintenance. A type of proactive maintenance intended to reduce unexpected downtime and machine failures. (3)

prick punch. A tool used in combination with the center punch to mark and dimple a metal object for drilling, specifically to accurately locate the spot for drilling. (4)

primary air treatment. The initial treatment or conditioning of air before it is used in a pneumatic system, in which particles, moisture, and heat are removed from the system air. (17)

proactive maintenance. A type of maintenance performed before problems arise. (3)

process. An action performed on a material to achieve a desired outcome. (34)

process control system. A system that uses various devices and methods to monitor and control a process or environment in order to maintain a desired output. (34)

processor. The part of a PLC that runs the operating system, performs programmed functions, and communicates with the human-machine interface (HMI) and remote input/output (I/O). (32)

process variable. A characteristic or factor that is measured within a process. (34)

programmable air controller (PAC). A pneumatic sequencer that is powered by air, in which all sensors, logic, and sequencing functions are mechanical. (18)

programmable logic controller (PLC). A digital microprocessing device used to control the inputs and outputs of industrial systems. (18, 28, 32)

proportional band. The amount of change in the controller input that will cause a 100% change in the output. (34)

proportional-integral-derivative mode (PID). A controller operating mode that considers the slope of the process variable over time. (34)

proportional mode (P). A controller operating mode in which an output signal is developed based on the error between the set point and current controlled variable. (34)

proportional plus integral mode (PI). A controller operating mode in which error over time is considered. (34)

proportional valve. A valve that allows for the control of the position of the spool by using an electric input signal, which allows for control of direction and speed. (18)

proton. A positively charged particle located in the nucleus of an atom. (19)

proximity sensor. A noncontact sensor that senses a material based on its metallic or capacitive qualities or by using a generated sound wave and measuring the returned signal. (31)

psia (pounds per square inch absolute). A measurement of atmospheric pressure or absolute pressure. (15)

psig (pounds per square inch gauge). A measure of pressure relative to ambient air pressure, with ambient air pressure always measured as 0 psig. (15)

pull-down resistor. A resistor used to connect input pins to ground to make sure the input is low and recognized as a 0 by a logic gate. (30)

puller. A device designed to remove gears, pulleys, bearings, and wheels without damaging machinery. (4)

pulley. A simple machine consisting of a wheel-and-axle assembly used to transfer power through a belt or cable that sits along the edge of the wheel. (9, 12)

pull-up resistor. A resistor used to connect input pins to the DC supply voltage to make sure the input is high enough to be recognized as a 1 by a logic gate. (30)

pulse spray-arc deposition. A method of metal transfer in which the melting off or pulsing a drop of molten filler metal from the electrode is at a controlled rate and time in the weld cycle. Also called *spray-arc pulsed deposition*. (38)

pulse width modulation (PWM). A method of generating an analog signal from a digital source, thus producing an action, such as controlling a motor or dimming a light. (29)

punctual. The act of being prompt and on time. (1)

push angle. The forward travel angle used for forehand welding. (38)

push-to-test. A type of indicator in which a contact is incorporated to allow a user to check that the indicating light is working. (26)

Q

quality factor. A unitless ratio of inductive reactance to resistance of a circuit. Abbreviated *QF*. (22)

quick exhaust valve. A valve that is mounted close to or directly on an actuator for quick response in one direction. (17)

R

raceway. An enclosed channel designed to help route and organize electrical infrastructure. (27)

radial load. A load that exerts force at a right angle to the shaft. (10)

radial movement. A shaft deflection or runout that is perpendicular to a shaft and housing. Also called *radial runout*. (10)

radiation. A method of heat transfer in which energy is transferred through electromagnetic waves as motion releases electromagnetic radiation that impacts other objects. (35)

radio local area network (RLAN). A digital communication network that uses radio transmission to communicate and transfer data. Also called *radio area network (RAN)*. (33)

rail. Another name used for the power supply in an op amp circuit. (30)

ramp. A valve adjustment that determines how quickly a signal is allowed to increase or decrease. (18)

ratchet. A handle and a wheel that allows motion in only one direction at a time, to which sockets of various sizes attach for loosening and tightening bolts. (4)

RC circuit. A circuit that contains resistive and capacitive loads. (22)

reactance. The opposition to change in electrical current resulting from inductance or capacitance in an AC circuit. Measured in ohms, Ω and designated by the variable, X. (22)

reactive maintenance. A type of maintenance performed after apparent problems arise. Also called *breakdown maintenance*. (3)

reactive power. The power from reactive components in a circuit. Measured in volt-amperes-reactive (VAR) and designated by the variable *VAR*. (22)

reaming. The act of smoothing a pipe where it was cut, in order to remove any remaining burrs. (41)

receiver. A sealed, fixed-volume chamber that holds compressed air until needed by the pneumatic system and that also assists in cooling the air and removing water vapor. (17)

receiver controller. A component of a pneumatic system that controls an actuator and can scale its output across a portion of the transmitter input. (36)

receptacle tester. A device with three indicator lights used to verify that an electrical outlet is wired correctly. (21)

reciprocating compressor. A type of compressor that uses the linear movement of pistons to compress refrigerant. (35)

reciprocating machine. A machine that involves repetitive up-and-down or back-and-forth motion, such as through pistons. (8)

reciprocating saw. A power tool with a relatively short blade that moves in a reciprocating motion (in and out). Also sometimes referred to as a saber saw or by the trademarked name Sawzall™. (4)

rectifier. A circuit that turns alternating current into pulsating direct current. (29)

reference. An individual who will provide important information about someone seeking employment to a prospective employer. Examples are a teacher, school official, or a previous supervisor or coworker. (1)

refrigerated air dryer. A pneumatic system component that removes moisture from the air by cooling the air and condensing moisture. (17)

relay. A switch controlled by electromagnets. Also called *motor starter, solenoid*, and *contactor*. (23)

relay output module. A PLC module that includes miniature relays mounted on a circuit board. These relay coils are energized when an output is given, which shuts a related normally open contact. (32)

reliability-centered maintenance (RCM). A type of maintenance that prioritizes equipment by its importance to the operation, as well as considering downtime cost, safety concerns, quality implications, and the cost of repair should a failure happen. (3)

relief valve. A valve that protects a hydraulic system from overly high pressure and allows excess flow from the pump to flow back to the reservoir. (16)

repeatability. A description of how well a system reproduces an established outcome under uniform conditions. In fluid power systems, the term refers to the ability to produce the same output signal for the same position over many cycles. (14, 17)

reset control. A control that directly resets the set point of the controlled variable in a control loop for HVAC. (34)

resistance. The opposition to the flow of current through a circuit. Unit of measurement is the ohm (Ω). (19)

resistance temperature detector (RTD). A sensor that measures temperature by relying on predictable changes in the electrical resistance of metals that are proportional to changes in temperature. (31, 34)

resistor. A common electrical component used to create voltage drops or limit current. (23)

resolution. The smallest increment a system can measure or recognize. In an analog PLC input module, the term refers to the number of bits available in the binary number to convert the analog signal. The larger the binary number is, or the greater the number of bits, the higher the resolution will be. (14, 32)

resonance. The natural ringing frequency. With basic mechanical systems, the term refers to vibration that occurs when an input force is equal to or close to the natural frequency of an object. With electrical systems, the term refers to a circuit with no net reactance due to X_L and X_C being equal. (8, 14, 22)

résumé. A brief outline of a person's education, work experience, and other qualifications for work. (1)

retaining ring. A fastener installed into machined grooves in a hole or shaft, used to accurately position and hold mechanical parts together. (5)

retentive timer (RTO). A timer programming instruction that retains, or holds, its accumulated value and can be used as an instruction to track the time a machine has been operating or to shut down a process after an accumulated time. (32)

retroreflective. A type of laser position sensor that contains the light source and receiver in one housing. (31)

returnway rollers. The rollers in a plastic belt conveyor that support the returning belt from the drive end of the conveyor and ensure the belts wraps 180° around the sprockets. (13)

reversing drum switch. A device used to reverse the rotation of DC, single-phase AC, and three-phase AC motors by switching the line voltage to the motor. (26)

reversing motor starter. A device used to reverse the rotation of DC, single-phase AC, and three-phase AC motors by shutting main contacts when one starter energizes and swapping specific motor connections. (26)

revolutions per minute (rpm). A measure of the frequency of rotation, or the speed, of a rotating mechanism. (16)

rheostat. A type of adjustable variable resistor that uses only two terminals and is used to control current. (23, 31)

rivet. A headed cylindrical shaft used to fasten parts together in permanent installations. (5)

rivet nut. An internally threaded tubular rivet installed from one side of a joint. Also called *blind thread insert*. (5)

RLC circuit. A circuit that contains a resistor (*R*), a capacitor (*C*), and an inductor (*L*) in which reactive elements from the inductance and capacitance oppose each other. (22)

RL circuit. A circuit that contains inductive and resistive loads. (22)

rolling-element bearing. A type of bearing composed of rollers arranged between two races. (10)

rooftop unit. A heating or cooling unit placed outside of the building envelope. (36)

root. The bottom surface of a screw thread between two sides, or flanks, of the thread. (5)

root cause. The underlying source of a problem. (7)

root cause analysis (RCA). A process performed after repairs to determine the source of the problem. (3, 14, 28)

root-mean-square (RMS). The square root of the mean over time of the square of the peak value of a waveform. Also called *effective value*. (22)

rope torque. The tendency of rope to turn when a load is applied. Also called *rope twist*. (8)

rotary screw compressor. A type of compressor that compresses refrigerant through the meshing action of two or more helical shaped rotors. (35)

rotary vane compressor. A type of compressor that compresses refrigerant using an off-center rotor that turns blades and creates pockets of refrigerant against the casing of the compressor. (35)

rotating machine. A machine that has one or more components that turn around an axis or point. (8)

rotor. A rotating part located inside the stator of a motor. An electromagnetic field from the stator induces a voltage in the rotor to cause current to flow. (25)

run time. A comparison of the time the compressor runs to the total cycle time. Also called *duty time*. (17)

S

sacrificial anode. A highly active metal that will corrode first, sacrificing itself in place of the more valuable metal it is protecting. (16)

safety data sheet (SDS). An OSHA-required document provided by chemical manufacturers, distributors, and importers that communicates potential dangers and necessary precautions to individuals using a hazardous chemical. (2)

safety factor. A rating arrived at through a comparison of minimum breaking strength and allowable working load. Also called *design factor*. (8)

saturation. A situation wherein the magnetic field in a

transformer core can go no higher regardless of how high the input voltage to the primary becomes. (24)

scale. The proportional relationship between the actual, real-life size of an item and the size it is shown in the drawing (6)

scaling. The application of filters to a signal by the controller before the signal is output to the actuator. (34)

schedule. A designation for the thickness of the wall of a pipe. (40)

scheduled maintenance. A type of maintenance that uses either time or the amount of use to determine when maintenance tasks should be performed. Examples are checking the air pressure in tires every 500 miles and checking the oil every fifth fill-up. (3)

schematic diagram. An electrical diagram that shows the components, electrical connections, and operation of a circuit. Also called *line diagram*. (6, 28)

screw. A cylindrical body about which screw threads are formed. (9)

screwdriver. A tool used to fasten screws. There are a variety of types for fastening screws of different styles and sizes. (4)

screwdriver-adjustable potentiometer. A variable resistor used to adjust circuit current by varying resistance. These devices require special tools to adjust and so are difficult to change. (28)

screwdriver voltage tester. A device for testing the "hot" slot of a receptacle to determine whether the receptacle is wired properly. (21)

screw-in valve. A type of cartridge valve that screws into a threaded cavity of a manifold. (18)

scroll compressor. A type of compressor that uses the motion of two intertwined scrolls to compress refrigerant. (35)

seal. A mechanical device that joins parts of a system together tightly to prevent leakage, contain pressure, and reduce contamination. (16)

seal-in current. A motor's normal running current that is several times lower than the initial current draw. (26)

sealing edge. The portion of a seal that contacts the shaft and provides a seal. (10)

secondary air treatment. The treatment or conditioning of air that takes place at the point of use in a pneumatic system, which may be specific to a machine, area, or control panel. (17)

section lines. The lines used in section views to show features of a part that have been cut by a cutting plane. (6)

self-excited generator. A type of generator where no voltage is externally applied to the stator windings. Instead, the DC voltage needed to create the magnetic field in the stator comes from the rotor. (25)

self-motivation. An inner urge to perform well. (1)

self-shielded flux cored arc welding (FCAW-S). A type of flux cored arc welding that does not require additional shielding gas since all the ingredients require for the shielding are included in the core material. Also called *open arc process*. (38)

semiconductor. A term that indicates that under certain

circumstances the device will conduct while in other circumstances the device is nonconducting. (29)

sensible heat. The heat that causes a change in temperature in a material and is measurable. (35)

sequence valve. A pressure control valve that senses upstream pressure to control a sequence of operations with more than one actuator. Once the first actuator has completed a cycle, the spool of the sequence valve allows flow to the next actuator in order for it to move. (16)

sequential function chart (SFC). A graphical model that displays process flow as a diagram, allowing control of the sequential processes by describing the conditions and actions for each step. (18)

serial communication. A method of communication in a digital network in which data is sent and received via cable one bit at a time. (33)

series circuit. A circuit containing only one path for current to flow. (19)

series resonant circuit. A circuit where the resistance, inductance, and capacitance are in series. (22)

series-wound DC motor. A motor with rotor windings that are connected through a commutator in series with its stator windings. It should never be operated without a load. Also called a *series motor*. (25)

service factor. A multiplier on a motor nameplate that indicates how much load beyond the rated load can be safely placed on the motor. (25)

servo amplifier. A device that reads a servo motor's feedback device and set point, and then drives the motor until it reaches the desired position. (25)

servo motor. A DC motor used to control movement in various industrial applications. A feedback device is connected to the motor shaft to detect position, direction of travel, or distance of travel. (25)

set point. With industrial process control, an established system value or condition that the process works to maintain. Also called *command point*. With motor controls, the term refers to the adjustable point at which certain switches can be programmed to open or close. (26, 34)

shading ring. A component that offsets the main motor coil and assists in maintaining the magnetic field of the coil. Also called *shading coil*. (26)

shaft current. An undesirable current that can occur and cause negative effects when a motor shaft is in the center of a rotating magnetic field and voltage is induced. (25)

shaft runout. The amount a shaft deviates from centerline rotation. (11)

sheave. The wheel component of a pulley that contains a groove or channel into which a belt or cable sits. (12)

shielded metal arc welding (SMAW). A form of arc welding that uses heat from an electric arc struck between an electrode and base metal. Also called *stick welding*. (37)

shielded twisted pair wire. A wire or cable that consists of insulated conductor wires arranged in pairs and twisted together, with the addition of a foil shield between the outer insulating jacket and the group of twisted conductor wire pairs. (32)

shielding gas. A gas that shields an electrode and protects the molten metal from contamination. (37)

short circuit. An unintended path in a circuit allowing electricity to return to the source without first going through the load. (19)

short-circuiting arc deposition. A method of metal transfer in which the electrode wire short-circuits when it is fed into the contacts and the workpiece. Also called *short-arc deposition*. (38)

shotgun approach. A troubleshooting approach that involves replacing random parts until a system functions again. This random method requires a considerable amount of time and resources and is not preferable for troubleshooting. (28)

shunt-wound DC motor. A motor with its stator windings and commutator connected in parallel. Also called *constant-speed motor* or *shunt motor*. (25)

shuttle valve. A valve used as a logic element, which allows an output if either input is received. (17)

signal conditioning. The processing or manipulation of an analog signal so that it will be usable for the next stage in the system or process. (31, 34)

signal converter. A signal conditioning component that allows a wide range of proportional inputs to be rescaled to a selectable output range. (34)

silencer. A part attached to an exhaust valve that reduces the noise from air escaping through the valve and helps protect the system from debris. Silencers can be made from a metallic mesh, sintered bronze, a fabric mesh, or other materials. (17)

silting. The accumulation of fine particles within the body of the valve. (16)

simple machine. A mechanical device that takes an input force, changes it in one or more ways, and outputs a force to a load. (9)

sine. A trigonometric function of an angle of a triangle that is determined by the ratio of the length of the side that is opposite the angle to the length of the longest side of the triangle (the hypotenuse). (8)

sine wave. A curve representing alternating current values over a time period. (22)

single-line diagram. With industrial wiring, a diagram that represents a three-line diagram with only one line, offering a simplified depiction of three-phase distribution systems. With piping systems, this is a type of schematic that shows all components in the system but does not show fluid flow and is not drawn to scale. (27, 40)

single-pole, dual-throw (SPDT). A contact that has a single arm with two connection positions. (26)

single-pole, single-throw (SPST). A contact with a single arm or pole that can make a single connection or throw. (26)

sinking module. A PLC module in which current flows into the PLC port. (32)

6S program. A program implemented by companies to

increase efficiency, reduce costs, improve quality, and promote safety. The six *S*'s stand for "Sort, Sweep, Sanitize, Set to order, Sustain, and Safety," although specific programs vary in terminology. (2)

slag. A hard, metal layer that is formed over a weld bead. (37)

sling angle. The angle formed between the horizontal plane of the load and the lifting sling. (8)

slip. The amount that the rotor falls behind the rotation of the stator field in a motor. With AC motors, this is expressed as a percentage and represents the slip speed divided by the synchronous speed. (25, 31)

slip-in valve. A type of cartridge valve that attaches to a manifold and is typically used for high flow rates, greater than 40 gpm. (18)

slip-joint pliers. Pliers in which the joint where the two arms are connected can be adjusted to change the gripping range of the jaws. Also called *gas pliers*. (4)

slip ring. A device that allows constant contact with each end of a rotor winding in an alternator. (25)

slip speed. The difference between a rotor's actual speed and the synchronous speed. (25)

slugging. A condition that occurs when liquid refrigerant is sent to the compressor and damaging the compressor. (35)

snub pulley. The pulley in a conveyor system that is responsible for adjusting the amount that the belt wraps around the head pulley. (13)

socket. A part that fits over a bolt or nut and is used to tighten or loosen it. (4)

socket adapter. An attachment that allows for ratchets of different drive sizes to be used with sockets having a different drive recess. (4)

socket extension. An attachment that can be installed between a socket and a ratchet, allowing the ratchet to be positioned farther away from the fastener. (4)

soft foot. A condition where the feet of a machine do not make perfect and even contact with the baseplate or foundation. (11)

solder. A nonferrous metallic alloy with a low melting point used in soldering. (23)

soldering. A connecting method that joins metals by the fusion of alloys known as solders. (23)

solenoid. A cylindrical coil of wire that converts current to mechanical movement by producing a magnetic field. Solenoids can be used to turn something on or off or to open or close something. (15, 26)

solidus. The temperature at which a metal begins to melt. (41)

sorter. A subsystem on a conveyor line that separates items on a conveyor line according to predetermined factors. (13)

sourcing module. A PLC module in which current flows out of the PLC port. (32)

speed wrench. A tool that allows the user to rapidly install or remove a fastener with low torque. (4)

spherical bearing. A barrel-shaped rolling-element bearing designed to carry heavy loads at low speeds. (10)

spider. The intermediary element between the hubs of a jaw coupling that helps the hubs mesh together. (11)

splicing. The act of connecting the end of a wire rope back upon itself to form an eye. (8)

spool. A cylindrical internal valve component with seals along its surface. The spool shifts within the valve, directing fluid flow to the various valve ports. (16)

spray-arc deposition. A method of metal transfer in which the electrode is melted off above the workpiece, forming a fine spray of molten metal that is deposited into the weld zone. (38)

spring. A device that stores energy when compressed or extended by a force and exerts an equivalent amount of energy when released. (9)

spring constant. A measure of the resisting force of the spring. (9)

sprocket. A gear or wheel with teeth that mesh with a chain in a chain drive system. (12)

squirrel-cage rotor. A rotor commonly used in induction motors. It is made of laminated electric steel and has aluminum or copper bars instead of windings. (25)

stability. A description of a hydraulic fluid's ability to keep performing its functions over time. (15)

stack valve. A valve that is vertically stacked with other valves that together fit onto a single spot on a manifold. (18)

standard operating procedure (SOP). An established and accepted method for performing a task. (3)

state. The condition of each input or element as shown through symbols on an HMI panel. (33)

static balance. The ability of an object to maintain balance while stationary on its axis. (12)

static electricity. A charge at rest generated when materials are rubbed together Also called *triboelectric charging*. (19)

static friction. The friction between two objects that are at rest. (9)

stator. A part that is one of two main parts of a motor. It is made of stationary windings that produce electromagnetic fields when power is applied. (25)

steel tape. A long, thin, flexible metal measuring tape that comes in lengths of 10′ or greater and is used for common measurements not requiring extreme accuracy. (4)

Stellite. A cobalt-chromium alloy with high wear resistance used for mechanical seals. (40)

step-by-step troubleshooting. A method for determining exactly where a fault lies. By following a circuit step by step and taking measurements, a fault can be found between the first point a technician fails to find a measurement and the previous correct measurement. (28)

step-down transformer. A transformer with a greater number of turns in the primary than in the secondary and a lower output voltage in the secondary than the primary. (24)

stepper motor. A motor used to convert electrical impulses into discrete mechanical rotational movements. (25)

step-up transformer. A transformer in which the number of turns in the secondary is greater than the number of turns in the primary. This causes the output voltage in the secondary winding to be higher than the input voltage in the primary. (24)

strength. A measure of how much stress can be applied to a material before it deforms. (14)

stress-strain curve. A graphical representation of the relationship between the stress applied to a material and the strain produced in the material. (14)

stringer bead. A weld bead that is produced as a welding torch travels along the joint with very little side-to-side motion. (38)

superheat. A measure of the energy added above the required boiling point to ensure refrigerant is completely vaporized. (35)

supervisory control and data acquisition (SCADA). A system that uses a network of computers, HMIs, and PLCs to monitor and control all of the networked industrial operations and systems from a central control room, either on-site or remotely. (33)

surface-mount resistor. A resistor placed on the surface of a circuit board and held in place by soldering. (23)

swaging. The act of joining a fitting to the hose through compression. (15)

switch. A device used to interrupt or complete the flow of electricity. (19)

switchboard. A free-standing, three-phase unit that serves as a main disconnecting means for electrical service entering a building from the power company transformer. (27)

switching power supplies. The DC power supplies that employ switching technology, with several stages and a feedback path to regulate the output voltage. Switching power supplies are capable of much higher current ranges, have a much smaller footprint, and—without a transformer—are much lighter than linear power supplies. Also called *switching-mode power supplies, switch-mode power supplies, SMPSs, switched power supplies,* or *switchers.* (29)

symbol library. A group of graphical symbols that comes with an HMI and is used to design a screen for indication or operator control. (33)

synchronous speed. The speed at which the magnetic field rotates about the stator in a motor. (25, 31)

system capacitance. The ability of a system to store energy, or resist a change in total energy. (34)

T

tag number system. A numbering system used to identify all the components of a control loop and differentiate them from other control loops. (34)

tail pulley. The pulley located at the input end of a conveyor that is responsible for directing the belt back toward the head pulley. (13)

take-up. In tube and pipe bending, the amount the bending tool must be moved back from the measured bend point to allow for the material taken up by the bend and still reach the desired 90° meeting point. The take-up amount is approximately the same as the radius of the bend being made. (41)

take-up pulley. A type of idler pulley that provides tension and can remove slack as the conveyor belt wears. (13)

tap. A tool used to cut internal threads. In electrical systems,

the term refers to a connection at an intermediate point within a winding of a transformer. (5, 24)

tapered adapter sleeve. A mounting device used in conjunction with a locknut and washer to lock a bearing with tapered bore onto the shaft. (11)

tapered bearing. A conical-shape rolling element bearing in which both races and rolling elements are tapered. (10)

teaching input. An input with laser sensors used to "teach" the receiver to recognize a new sensing range. (31)

team. A small group of people working together for a common purpose. (1)

temperature compensation. The use of a valve to compensate for changes in fluid temperature in order to manage flow rate. (16)

temperature switch. In motor controls, a switch that opens or closes once a certain temperature is reached. These switches are typically bimetallic or gas/vapor filled. In industrial process control systems, this is a type of sensor that senses temperature and provides an output once the set point is reached. (26, 34)

temperature transmitter. A device that receives input from either a thermocouple or RTD and outputs a proportional signal to a controller or programmable logic controller (PLC). (34)

termination. The connection of a wire to a device or another wire. (23)

test light. A lamp with two wires used to test for the presence of electricity. (21)

thermal expansion. The growth that occurs as a machine and its foundation warm to operating temperature. (8)

thermal imager. A device that provides a picture of various temperatures and their locations, showing a range of temperatures at the locations within the captured image. (7)

thermistor. With motor controls, a motor protection device used to sense winding temperatures and change resistance as temperature changes. With electronic control systems, this refers to a temperature sensor that is a resistor that detects temperature changes through corresponding changes in electrical resistance. (26, 31)

thermocouple. A sensor that detects temperature based on the voltage produced when two joined wires made of dissimilar metals are heated. (31, 34)

thermostatic expansion valve (TXV). A type of valve that regulates the flow of liquid refrigerant into the evaporator based on the signal received from a sensing bulb. (35)

thermowell. A sleeve or device that isolates and protects the inserted sensor from damaging elements within the system being measured. (26, 34)

thread class. A screw thread designation that specifies the amount of tolerance permitted in engagement between mating threads. (5)

thread form. The standard profile of a screw thread. (5)

thread series. A classification of threads in a thread system. (5)

three-line diagram. A diagram that represents a three-phase system with all phases distinct. (27)

three-phase alternating current. An electrical current

comprising three separate phases, each out of phase with the others by 120°. (22)

three-phase rectification. An important part of the variable-frequency drive process in which three-phase AC is converted into DC output. (29)

three-terminal voltage regulator integrated circuit (IC). A single chip containing an entire voltage regulator circuit, with the exception of a couple of capacitors. Three pin-out terminals represent input, output, and ground, respectively. (29)

through-hole resistor. A resistor with wire leads that are inserted through the holes in a circuit board and soldered in place. (23)

thru-beam mode. The sensing mode in which the light transmitted from the sensor travels to a separate receiver and an object is detected when it interrupts that beam. (31)

thrust bearing. A type of rolling-element bearing designed to support high axial loading. (10)

tinning. The process of coating leads with solder to protect them and increase wetting during the soldering process. (23)

title block. A bordered area that contains information about a print, usually located in the lower-right corner of the print. (6)

T-joint. An edge of one piece of metal placed on the surface of another, creating a right angle. (37)

tolerance. The total amount a part dimension can vary. In electrical systems, the term refers to a specification indicating how far from the published resistance value an actual resistor can be. (6, 23)

tongue-and-groove pliers. Pliers with one fixed jaw and one movable jaw often used as a pipe wrench. Also called *channel-lock pliers*. (4)

torque. A turning force applied to an object that is on a fixed axis. With motors, the term refers to a force created by the effect of a stator's magnetic field on the conductors in a rotor. Also called *rotational force*. (9, 25)

torque arm. The distance from the axis of rotation to the point of force where torque is applied. (9)

torque wrench. A wrench that allows the user to set the amount of tightening torque. (4)

torsion spring. A spring that resists twisting as torque is applied about the central axis. (9)

total harmonic distortion (THD). The total, cumulative amount of distortion caused by harmonics in the electrical current. (31)

total indicator reading (TIR). The difference between the minimum and maximum readings taken by an indicator. (11)

total productive maintenance (TPM). A maintenance method developed from preventive and productive maintenance concepts that involves a company-wide system for maintaining manufacturing equipment and facilities. (3)

transducer. A signal conditioning component that converts one signal type to another. (34)

transformer. A device consisting of two coils of wire wound on a common core used to transfer energy from the primary winding to the secondary winding through mutual inductance. (24)

transistor. A solid-state, three-terminal semiconductor device that is used as an amplifier or a switch. (30)

transition fit. A fit in which a clearance or interference fit results when two parts are assembled. (6)

travel angle. A longitudinal angle that describes the position of the torch tilted along the weld axis plane. (38)

triggering. The point at which an oscilloscope begins sweeping the displayed waveform from left to right. (21)

true power. The power equal to the product of voltage and current in a pure resistive circuit. Measured in watts (W) and designated by the variable *P*. Also called *real power*. (22)

trunnion. The machined ends on the journal cross of a universal joint. (11)

tube gain. The small amount of apparent lengthening of a tube when it is bent on a curve compared to a sharp 90° angle. Also called *tube stretch*. (41)

tuned circuit. A series or parallel RLC circuit at resonance. (22)

tuning. The process of adjusting parameters of a controller to achieve the desired performance of the control variable, which includes accuracy, precision, response, and stability. (34)

turbulent flow. A type of fluid flow in which the direction and the velocity of the fluid fluctuate. (40)

turns ratio. The ratio of the number of turns in the primary winding to the number of turns in the secondary winding of a transformer. (24)

U

ultrasonic flow meter. A flow meter that uses an ultrasonic pulse to develop a signal that is proportional to velocity, or flow rate, of a fluid. (34)

ultrasonic level sensor. A noncontact sensor that uses ultrasonic waves to measure the level of fluid and solid material. (34)

ultrasonic sensor. A proximity sensor that emits a high-frequency sound and measures the time it takes for a return echo from the detected object. These sensors may be used to detect solid objects or to sense liquids and liquid levels within a tank. (31)

undercutting. The wear on the lower part of the dedendum. (12)

Underwriters Laboratories (UL). An independent testing organization that approves electrical devices related to safety. (27)

universal joint. A connection used to transmit mechanical force between two shafts while allowing them to move out of alignment. Also called a *U-joint* or *Cardan joint*. (11)

universal joint socket adapter. An attachment that allows for a ratchet to be used at an angle. (4)

universal motor. A motor that can operate from either AC or DC power and includes a wound rotor, commutator, and wound stator. It is most often used in single-phase AC appliances. (25)

Universal Service Ordering Code (USOC). A wiring system

that was invented to connect customer telecommunications equipment to public network lines. (33)

unloading. The act of removing the load from the mechanical side of the compressor when starting up or, in the case of a screw compressor, when reaching its maximum pressure. (17)

unmodulated light. A light wave that is not modified. (31)

V

valence orbital. The outermost shell of electrons surrounding an atom. (19)

valve actuator. A manually operated or power operated mechanism used to control valves by opening or closing them. (16)

valve slamming. A situation that occurs when a disk in a fluid system rapidly closes against a seat due to a drop off in flow through the system or if flow attempts to reverse. (40)

vane pump. A pump that uses the rotary motion of vanes to produce a pumping action. (16)

variable voltage inverter (VVI). A type of AC variable frequency drive (VFD) in which the voltage and frequency are controlled. Variable voltage inverters typically use diodes as a bridge rectifier. Both inductors and capacitors are used in the DC bus circuit to smooth the DC voltage applied to the inverter section. Smaller drives tend to use transistors in the inverter section. (31)

variable voltage transformer. A variation of the autotransformer wound on a toroidal core. The winding has a movable tap that may be repositioned to create a variable turns ratio. (24)

varnish. A hard coating in hydraulic components resulting from a breakdown of hydraulic fluid over time due to the introduction of oxygen and moisture. (16)

vector force. A force that has magnitude and direction, causing horizontal movement during a lift. (8)

venturi. A short tube with a narrowing point of constriction that creates an increase in fluid velocity and decrease in fluid pressure. (16, 34)

verbal communication. A form of communication that involves speaking, listening, and writing. (1)

vernier scale. A scale of measurement that allows a user to measure more precisely than could be done when reading a normal uniformly divided straight scale. (7)

vibration meter. A device that measures and analyzes the vibrations emanating from a machine to help determine the possible cause or causes and the severity of the vibration. (7)

viscosity. The measurement of a fluid's resistance to flow. (10, 15)

viscosity index (VI). A measure of the amount that a fluid's viscosity is affected by temperature change. (15)

visual inspection. A weld inspection that involves viewing a weld on the front or top surface (and penetration side). (38)

volt. The SI unit for voltage, abbreviated as V. (19)

voltage. The force or difference in electric potential that causes current to flow. Also called *electromotive force (EMF)* or *potential difference*. Unit of measurement is the volt (V). (19)

voltage divider. A series circuit that acts as a series of resistances that divides the source voltage into smaller increments. (20)

voltage drop. The voltage lost across each individual load in a circuit. (20)

voltage ratio. The ratio of primary voltage to secondary voltage in a transformer. (24)

W

washer. A device used with threaded fasteners to increase their contact area with the material being fastened. (5)

watchdog timer circuit. An HMI timer circuit set at a particular time range, usually in milliseconds. If communication between the PLC and HMI does not occur within this time frame, a communications error will occur. (33)

water tube boiler. A type of boiler in which boiler water travels through the boiler in tubes and is heated by combustion taking place within the boiler. (35)

watt. The SI unit for power abbreviated as W and defined as the amount of power required to move one volt of electrical potential one coulomb in one second. (20)

weave bead. A weld bead that is produced using a side-to-side or similar motion to create a wide weld. Also called *wash bead*. (38)

wedge. A simple machine composed of two inclined planes joined back-to-back along a common base. (9)

weld. The blending of two or more metals by applying heat until they are molten and flow together. (37)

weld bead. The shape of the finished welded joint after welding is complete. (38)

weld defect. An extensive flaw that may cause the weld to fail. (38)

welding code. A national standard that provides guidelines, requirements, and standards for welding. (38)

welding power source. A machine that supplies electric current used to make a weld in arc welding. Also called *arc welding machine*. (37)

welding procedure specification (WPS). A document that records required welding variables for a specific weld application to ensure a trained welder can properly complete the weld. (38)

welding symbol. A symbol developed by the American Welding Society used in mechanical drawings of welded parts to provide a complete specification to make a weld. (6, 37)

weld symbol. A basic symbol that appears in specified areas of the welding symbol to represent the type of weld to be made. (6, 37)

Wheatstone bridge. An electrical bridge circuit used to measure an unknown electrical resistance on one of its sides. (31)

wheel and axle. A simple machine composed of a wheel attached to an axle. (9)

wide area network (WAN). A digital communication

network that covers a larger area than a LAN and uses transmission conduits that are already in place. (33)

wiggler bar. A tool support device used to take readings on the face of a coupling when there is limited space. (11)

wiper. A movable contact located on the third terminal of a potentiometer used to adjust resistance. (23)

wire. A standard electrical conductor available in stranded and solid varieties. (23)

wire connector. A device used to splice stranded wire to stranded wire. Also called *wire nut*, *cone connector*, or *twist-on wire connector*. (23)

wire rope. A steel-braided rope with a core surrounded by wire strands made up of smaller steel wires. (8)

wire run. The space inside a panel where the wire is actually placed, usually by use of adhesive pads and wire ties or plastic wire channels. (27)

wire strand core (WSC). A wire rope core of multiple wire strands, which offers more strength than a fiber core but less flexibility. (8)

wire strippers. A tool used to remove a small amount of insulation from the ends of wires intended for termination. (4)

wiring diagram. An electrical diagram that shows the general arrangement and location of components in a circuit, including exact terminal-to-terminal wiring, the placement of each wire termination, and the number of wires per terminal. (6, 27)

work. The application of force through a distance to move an object. (9)

work angle. The transverse angle that describes the position of the torch tiled across the weld axis. (38)

working depth. The entire length of a gear tooth, minus the clearance. (12)

working load limit (WLL). The maximum load that can be safely lifted. Also called *capacity*. (8)

working voltage. The maximum voltage that is safe to place across a capacitor without the dielectric material breaking down. (23)

work order. A basic document or form used for planning and controlling maintenance activities and tasks. (3)

worm. A section of threaded rod that is the driving element of a worm drive. Also called *worm screw*. (12)

worm drive. A gearbox used to transmit power between two shafts and change the direction of rotation by 90°. (12)

worm gear. A wheel-shaped gear with meshing teeth that is the driven element of a worm drive. Also called *worm wheel*. (12)

wound rotor. A rotor commonly used in induction motors. It is made of laminated electric steel and features windings on a rotating cylinder. (25)

wrench. A tool used for tightening and loosening nuts and bolts. (4)

wye (Y) connection. A three-phase connection configuration used in generators, transformers, and motors, in which the phase windings are connected in a Y shape that allows for two separate voltages. (22)

Z

zener diode. A voltage regulator that allows a current to flow in the forward direction much like a conventional rectifier diode, but also allows current to flow in its reverse direction once the zener voltage level has been reached. (29)

zeotrope. A refrigerant blend that contains refrigerants with different boiling and condensing points. (35)

zero energy state. The condition when a system is safe from the possibility of becoming reenergized or experiencing a release of contained internal energy from any source. (2)

zero point. A valve adjustment that aligns the zero position of the signal with valve position. (18)

Index